Optical Mineralogy

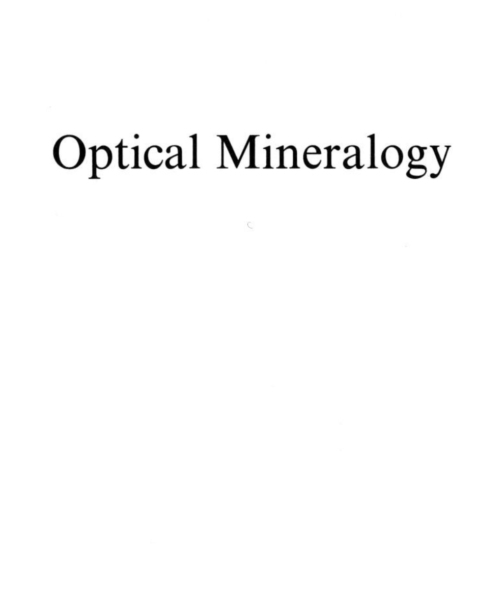

A Series of Books in Geology
EDITOR: *A. O. Woodford*

OPTICAL MINERALOGY

THE NONOPAQUE MINERALS

Wm. Revell Phillips
BRIGHAM YOUNG UNIVERSITY

Dana T. Griffen
BRIGHAM YOUNG UNIVERSITY

W. H. FREEMAN AND COMPANY
San Francisco

Sponsoring Editor: Gunder Hefta
Project Editors: Nancy Flight and Judith Wilson
Copy Editor: Linda Purrington
Designer: Robert Ishi
Production Coordinator: William Murdock
Compositor: Syntax International
Printer and Binder: Kingsport Press

Library of Congress Cataloging in Publication Data

Phillips, William Revell, 1929–
Optical mineralogy.

(A Series of books in geology)
Includes bibliographies and index.
1. Optical mineralogy. I. Griffen, Dana T., 1943– joint author. II. Title.
QE369.06P44 549′.125 80-12435
ISBN 0-7167-1129-X

Printed in the United States of America

9 8 7 6 5 4 3 2 1

Contents

CHAPTER 3 The Nesosilicates 101

CHAPTER 4 Sorosilicates and Cyclosilicates 145

Preface

This book, *Optical Minerology: The Nonopaque Minerals*, was originally intended as a companion volume to *Mineral Optics: Principles and Techniques* to supply the tables of optical constants and mineral descriptions so obviously missing in the early volume. These data were always intended to be more than the few pages appended to most textbooks of optical mineralogy; however, as the work progressed, the need for a truly comprehensive summary of the optical properties of minerals seemed obvious. It is hoped that this volume will be accepted as the logical successor to Winchell, Larsen and Berman, and other classics that are still standard references, although long out of date and out of print (A. N. Winchell, 1967, *Elements of Optical Mineralogy*, II: *Descriptions of Minerals*, New York: Wiley; and E. S. Larsen and H. Berman, 1934, *The Microscopic Determination of the Nonopaque Minerals*, U.S. Geological Survey Bulletin 848).

This work has two parts: Part I, a detailed description of the more common mineral varieties of the common rock-forming mineral groups, and Part II, an abbreviated summary of the essential optical constants and occurrence of "all" nonopaque mineral varieties. In view of the questionable status of many mineral varieties, the lack of published optical data on others, and the rapid discovery of new minerals, no single published work may ever contain truly all critical data on all nonopaque minerals; however, we have done our best to gather and filter as much data as possible and organize them into a usable form for beginning students and seasoned researchers.

After experimenting with grouping the rock-forming minerals by their most common occurrences, we yielded to tradition, and the minerals described herein are arranged in structural groups as pyroxenes, feldspars, micas, and so forth.

March 1980

Wm. Revell Phillips
Dana T. Griffen

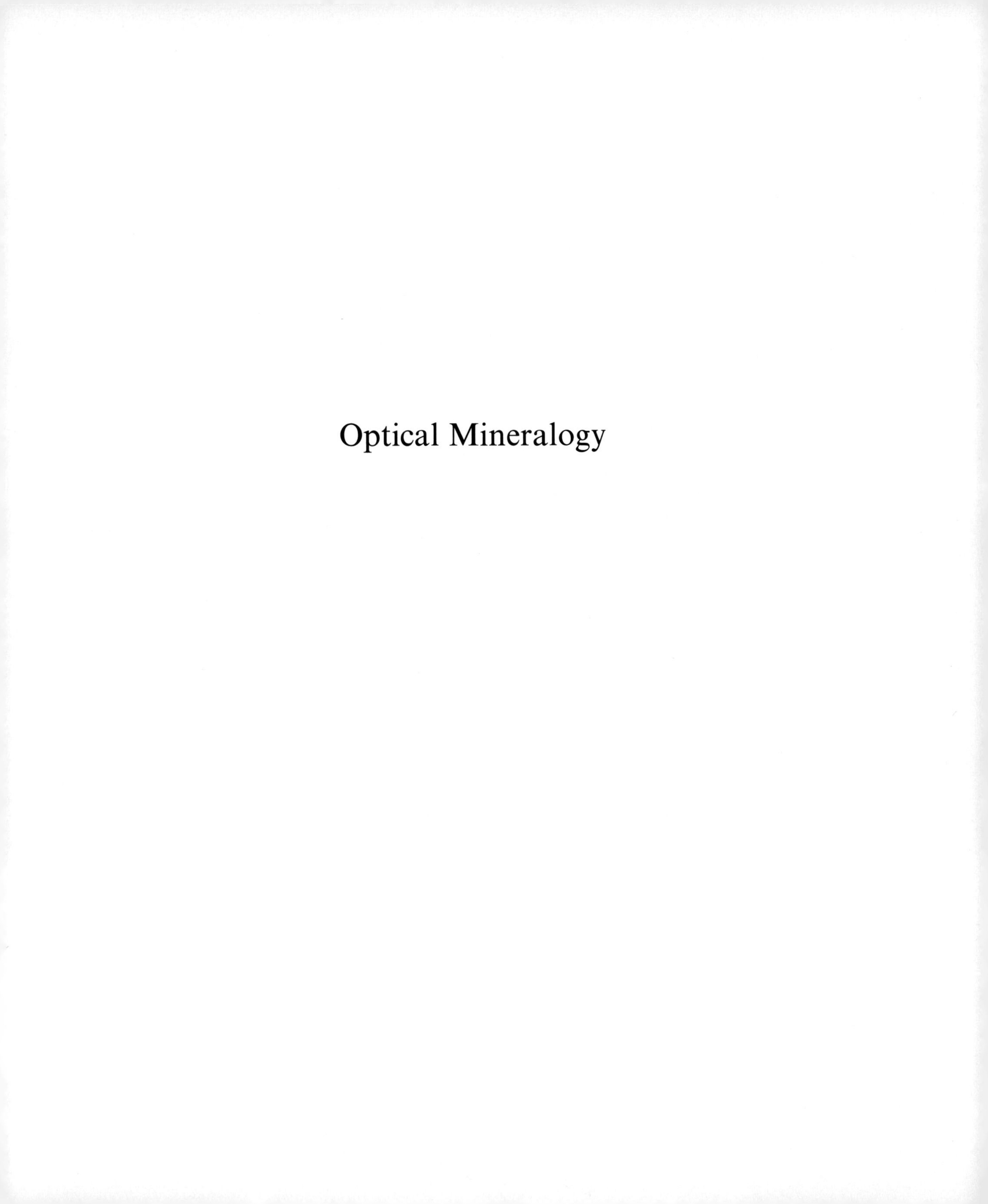

Optical Mineralogy

PART I

Detailed Description of the Common Rock-Forming Minerals

CHAPTER 1

Minerals Without Complex Anions

Native elements, halides, oxides and hydroxides, and sulfides are grouped together here because, in general, their crystal structures tend to be fairly simple, often based on packing schemes such as cubic closest-packing, hexagonal closest-packing, or the body-centered cubic arrangement. Beyond this and the absence of complex anions (SO_4^{6-}, CO_3^{2-}, and so on), these groups are unrelated in physical and chemical properties, including optical characteristics.

Only about 20 elements occur alone in nature as minerals, but several (such as Au, Ag, Cu, and Pt) are of considerable economic importance. Except for sulfur and diamond, they are opaque, due to metallic or semimetallic bonding, and hence not amenable to identification by transmitted light microscopy. In diamond, the chemical bonding is entirely covalent, and the refractive index and luster are therefore high. Sulfur contains both van der Waals and covalent bonds, and because of the latter its refractive indices are also high. Halides, which are common only in evaporite and hydrothermal deposits, have generally low refraction indices and vitreous lusters, because of their high proportion of ionic character. Oxides and hydroxides exist in nearly all geologic environments and exhibit a wide range of properties, depending largely on the metal atoms involved. Most have densities between 3.0 and 5.0, although a few (such as brucite) are considerably lighter, and some (such as cassiterite) are heavier. Indices of refraction are moderate to very high, and some are opaque and metallic. Most sulfides appear metallic, but a few (notably sulfides of Group II-A elements and of arsenic) are transparent, with high refractive indices. Many sulfides are very important ore minerals.

NATIVE ELEMENTS

Diamond — C — ISOMETRIC

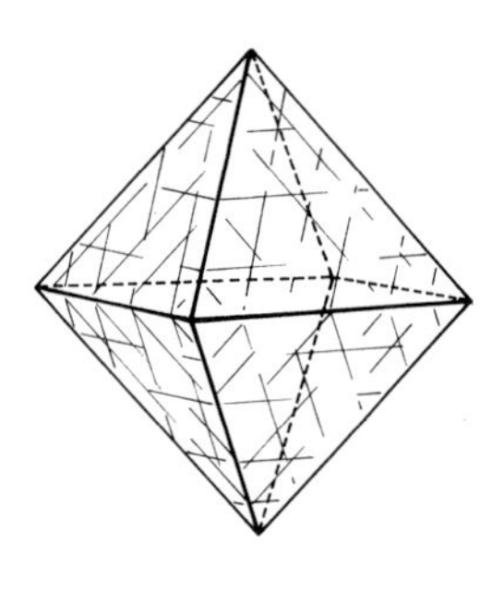

$n = 2.4193$

Colorless in thin sections or fragments

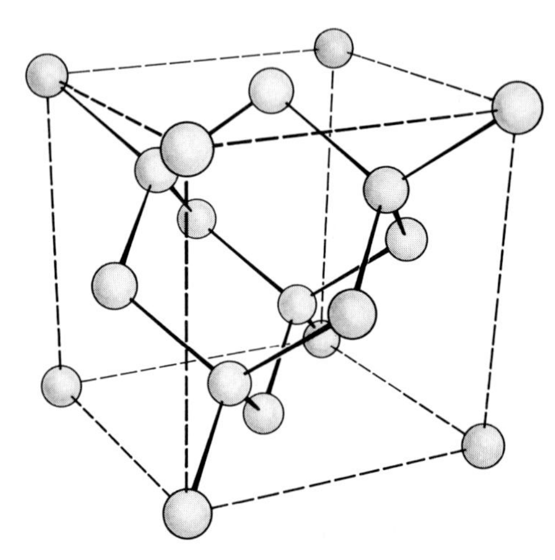

FIGURE 1-1. The diamond structure. Each carbon atom is in fourfold (tetrahedral) coordination bonded to four other carbon atoms by covalent sp^3 bonds.

COMPOSITION AND STRUCTURE. Diamond is the only completely covalent mineral. Its structure (Fig. 1-1) is isometric, with tetrahedral symmetry in which each carbon atom is surrounded by four other carbon atoms in tetrahedral distribution.

Diamond is ideally pure carbon. It may contain tiny inclusions of graphite, magnetite, garnet, chrome diopside, chlorite, olivine, phlogopite, and so on. Chemical substitutions include Al and N for C.

PHYSICAL PROPERTIES. H = 10. Sp. Gr. = 3.511. In hand sample, diamond is colorless or pale to deep yellow or brown, rarely orange, pink, blue, red, green, violet, or black. Color and crystal form appear to be related. Luster is distinctly adamantine.

COLOR AND PLEOCHROISM. Diamond is rarely found in rock thin sections, because diamond sections cannot be prepared with standard abrasives. Fragments are colorless or, rarely, weakly colored, and without pleochroism.

FORM. Diamond usually occurs as small euhedral to subhedral crystals, normally octahedral, sometimes dodecahedral, cubic, or tetrahedral. Crystal faces are commonly curved, striated, or otherwise deformed. Granular to cryptocrystalline forms are called *bort*, and black compact forms are called *carbonado*.

CLEAVAGE. Perfect octahedral cleavages {111} may not be obvious in fragments.

BIREFRINGENCE. Diamond is isotropic. It commonly shows weak birefringence near inclusions and fractures, or as strain patterns due to dislocations or plastic deformation.

TWINNING. Simple contact twinning on {111} is very common, forming spinel-type twins. Twins may also be lamellar, penetration, or even cyclic groups. Tetrahedral forms may show twinning on {001}.

DISTINGUISHING FEATURES. Very limited occurrence, extreme hardness, crystal form, and extreme refractive index are distinctive. Garnet and spinel have much lower indices of refraction and are usually highly colored.

ALTERATION. Diamond is highly resistant to weathering processes and appears without alteration in placer deposits.

OCCURRENCE. As a primary mineral, diamond occurs only in olivine-rich peridotites, notably kimberlite, associated with olivine, pyroxene, garnet, magnetite, and phlogopite. It persists as a heavy mineral in the placer gravels of streams and beaches, both modern and ancient (consolidated conglomerates).

REFERENCES: DIAMOND

Dawson, B. 1967. The covalent bond in diamond. *Proc. Roy. Soc. London* Series A, *298*, 264–288.

Eremenko, G. K., and Yu. A. Polkanov. 1969. Color and luminescence of diamonds of various habit. *Zap. Vses. Mineral. Obshchest.*, *98*, 334–338.

Lang, A. R. 1967. Causes of birefringence in diamond. *Nature*, *213*, 248–251.

Lubinsky, A. R. 1974. Optical and x-ray properties of diamond and silicon carbide. Unpublished doctoral dissertation, Northwestern University, Evanston, Illinois.

Sobolev, E. V., V. E. Il'in, E. N. Gil'bert, S. V. Lenskaya, V. A. Pronin, L. Pelekis, I. Mednis, and V. M. Moralev. 1970. Role of impurity aluminum in the optical properties of diamond. *Zh. Strukt. Khim*, *11*, 1048–1052.

Sulfur — S — ORTHORHOMBIC

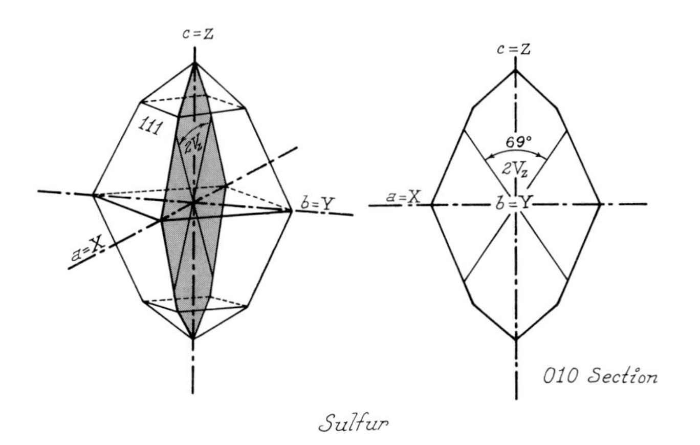

Sulfur

$n_\alpha = 1.958$

$n_\beta = 2.038$

$n_\gamma = 2.245$

$n_\gamma - n_\alpha \simeq 0.287$

Biaxial positive, $2V_z \simeq 69°$

$a = X, b = Y, c = Z$

$v > r$ weak

Pale yellow and weakly pleochroic in thin section

COMPOSITION AND STRUCTURE. Sulfur contains puckered, covalent rings of eight sulfur atoms. The S_8 rings are stacked parallel to *c* and held by residual bonds, accounting for great softness, low melting point, and other physical properties.

Sulfur is a native element. Crystals may contain very minor Se or Te and are often contaminated by clays or bituminous impurities.

PHYSICAL PROPERTIES. H = $1\frac{1}{2}$–$2\frac{1}{2}$. Sp. Gr. = 2.07. In hand sample, sulfur is sulfur-yellow, sometimes brownish, reddish, or greenish. Luster is distinctly resinous (adamantine).

COLOR AND PLEOCHROISM. Sulfur is pale yellow and weakly pleochroic in section or as fragments.

FORM. Sulfur crystals are common, are usually dipyramidal or thick tabular (001), and are commonly skeletal or hopper types. Colloform and massive forms are also common.

CLEAVAGE. Cleavages on {001}, {110}, and {111} are all rather poor. Parting may be prominent on {111}

BIREFRINGENCE. Maximum birefringence, seen in {010} sections is very extreme ($n_\gamma - n_\alpha = 0.287$), interference colors are strictly high-order white.

TWINNING. Twinning is uncommon and usually simple on {101}, {011}, or {110}.

INTERFERENCE FIGURE. Basal sections yield centered acute bisectrix figures showing countless isochromes and weak optic axis dispersion $v > r$. Fragments assume essentially random orientation.

OPTICAL ORIENTATION. The optic plane is {010}. Extinction is symmetrical to pyramidal parting and parallel to poor cleavages. Cleavages do not favor elongated fragments.

DISTINGUISHING FEATURES. Limited occurrence, extreme relief and birefringence, and low hardness are distinctive.

ALTERATION. Sulfur oxidizes to sulfurous and sulfuric acids, which, in turn, may form sulfate minerals.

OCCURRENCE. Sulfur commonly appears in bedded evaporite deposits associated with gypsum, calcite, aragonite, halite, and celestite. It possibly forms by the reduction of sulfates (gypsum or anhydrite to sulfur) through the action of reducing bacteria.

Sulfur is intimately associated with volcanic gases and is deposited in fumeroles and volcanic vents by direct pneumatolitic precipitation or by the interaction of sulfur gases. It is also precipitated in many thermal springs by the decomposition of H_2S.

REFERENCE: SULFUR

Cooper, A. S., W. L. Bond, and S. C. Abrahams. 1961. The lattice and molecular constants in orthorhombic sulfur. *Acta Crystallogr.*, *14*, 1008.

SULFIDES

Sphalerite (Zincblende) — ZnS — ISOMETRIC

Sphalerite

$n = 2.37$–2.50*

Isotropic

Colorless to pale yellow-brown in section

COMPOSITION AND STRUCTURE. Sphalerite is a strongly covalent structure, with both Zn and S in regular tetrahedral coordination held by sp^3-type bonds. A high-temperature polymorph, wurtzite,† is stable above 1020°C. Sphalerite is isometric with zinc atoms, or sulfur, in open cubic closest packing (Fig. 1-2A); wurtzite is hexagonal with zinc atoms, or sulfur, in open hexagonal closest packing (Fig. 1-2B). Packing arrangements that are intermediate between cubic and hexagonal close packing are not uncommon.

Most natural sphalerite (ZnS), contains significant iron, and iron increases with formation temperature to 40 molecular percent (marmetite). Mn and Cd are common impurities, especially in the wurtzite polymorph, and Ga, Ge, In, Co, and Hg are known to replace Zn in very limited amounts.

PHYSICAL PROPERTIES. H = $3\frac{1}{2}$–4. Sp. Gr. = 3.9–4.1. Pure ZnS is nearly colorless, but natural sphalerite is honey-yellow, brown to black, in hand sample. Luster is resinous or adamantine.

* Refractive index increases with iron content (Fig. 1-3).

† Wurtzite is uniaxial positive. $n_\omega = 2.356$, $n_\varepsilon = 2.378$.

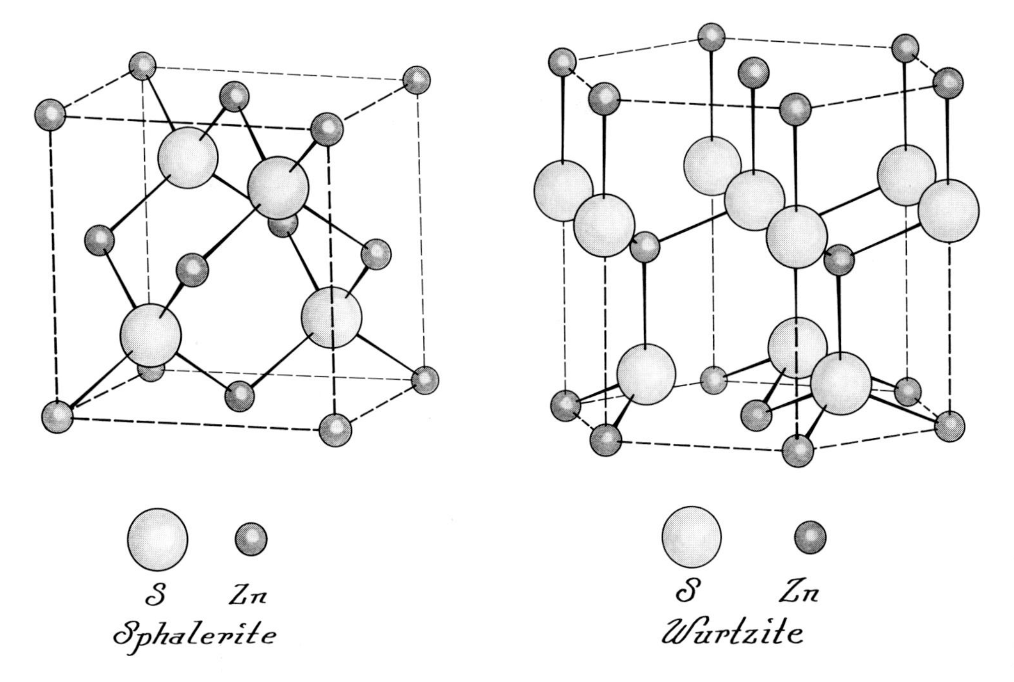

FIGURE 1-2. (A) The structure of sphalerite (ZnS). (B) The structure of wurtzite (ZnS). In both ZnS polymorphs, both zinc and sulfur are bonded to four of the opposite-type atoms in tetrahedral co-ordination. The sphalerite structure is isometric and analogous to that of diamond, in which one-half of the carbon atoms are replaced by sulfur atoms and one-half by zinc. In wurtzite, the tetrahedral units are arranged for hexagonal symmetry. Zn—S bonds are highly covalent and are responsible for the large indices of refraction and strong resinous-adamantine luster of the ZnS polymorphs.

COLOR AND PLEOCHROISM. Sphalerite is colorless to pale yellow or pale brown without pleochroism in standard section or as fragments.

FORM. Crystals of sphalerite show tetrahedral or dodecahedral habit, frequently with curved faces, and are coarse to fine granular; rarely, fibrous or banded concretionary.

CLEAVAGE. Perfect {110} cleavages (six directions) control fragment orientation.

BIREFRINGENCE. Sphalerite may show weak strain birefringence. In addition, "sphalerites" that contain significant proportions of hexagonal closest-packed sequences are birefringent, since they are not truly isometric (Fig. 1-4).

TWINNING. Simple or multiple twinning on {111} (spinel law) is common but may not be visible in section.

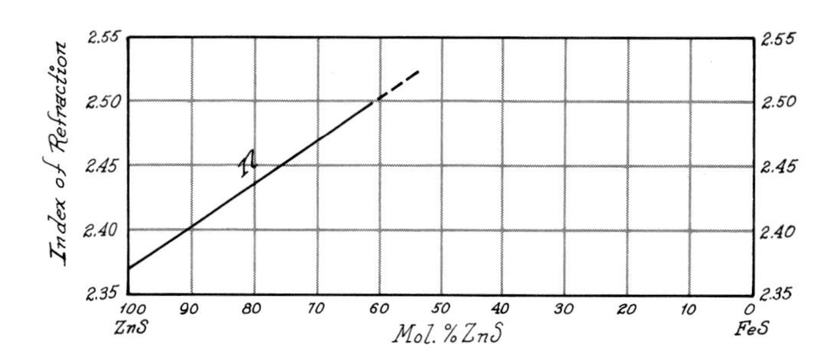

FIGURE 1-3. Variation in the refractive index of sphalerite with FeS content.

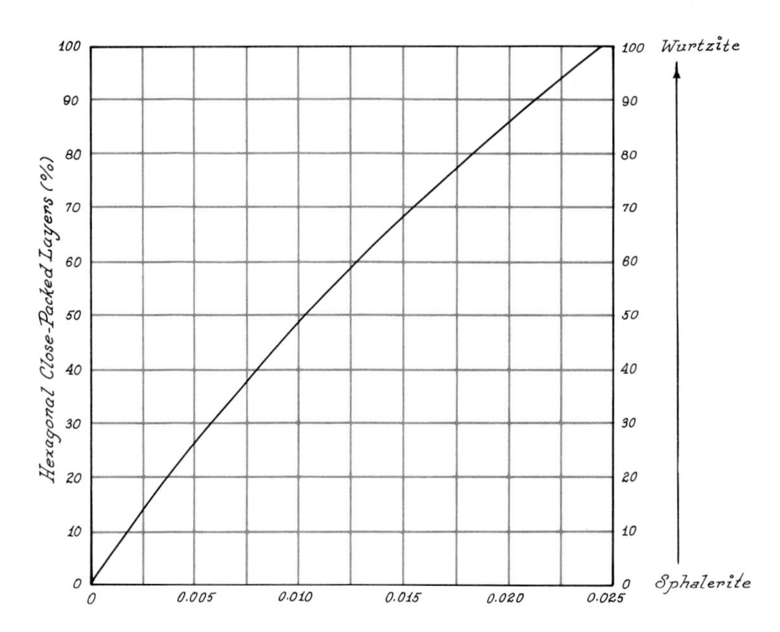

FIGURE 1-4. Increasing birefringence with structural state variation between sphalerite (cubic close-packed layers) and wurtzite (hexagonal close-packed layers). (After Fleet, 1977)

ZONING. Color banding may appear in concretionary forms.

DISTINGUISHING FEATURES. Extreme relief, isotropism, cleavage, and occurrence are distinctive.

ALTERATION. Sphalerite may alter to smithsonite, hemimorphite, goslarite or limonite. It may replace galena, tetrahedrite, barite, or calcite.

OCCURRENCE. Sphalerite is nearly always intimately associated with galena and commonly with chalcopyrite. It is a rather common primary mineral in hydrothermal sulfide veins associated with magnetite, garnet, rhodonite, gahnite, fluorite, and apatite, in deep-seated high-temperature veins, and with dolomite, barite, chalcedony, and fluorite, in shallow low-temperature veins. Sphalerite, with galena, is common in contact metamorphic and replacement deposits in carbonate sediments near siliceous intrusions. Oriented intergrowths of sphalerite with chalcopyrite or wurtzite are characteristic of high- and low-temperature deposits respectively. Sphalerite is a rare accessory mineral in granite or granite pegmatite.

In sedimentary rocks, sphalerite is a rare syngenetic mineral in low-grade coals and may occur in concretionary masses with galena and marcasite.

REFERENCES: SPHALERITE

Brafman, O., and I. T. Steinberger. 1966. Optical band gap and birefringence of ZnS polytypes. *Phys. Rev.*, *143*, 501–505.

Chernyshev, L. V., and V. N. Anfilogov. 1967. Experimental data on the composition of sphalerite in association with pyrrhotite and pyrite at 350°–500°. *Dokl. Akad. Nauk SSSR*, *176*, 925–928.

Dunham, K. C., ed. 1950. The geology, paragenesis, and reserves of the ores of lead and zinc. *Rept. of 18th Intern. Geol. Cong. 1948*. VII. London: International Geological Congress.

Fleet, M. E. 1977. The birefringence-structural state relation in natural zinc sulfides and its application to a schalenblende from Pribram. *Can. Mineral.*, *15*, 303–308.

Roedder, E., and E. J. Dwornik. 1968. Sphalerite color banding: Lack of correlation with iron content, Pine Point, Northwest Territories, Canada. *Amer. Mineral.*, *53*, 1523–1529.

Skinner, B. J. 1961. Unit-cell edges of natural and synthetic sphalerties. *Amer. Mineral.*, *46*, 1399–1411.

Pyrite FeS_2 ISOMETRIC

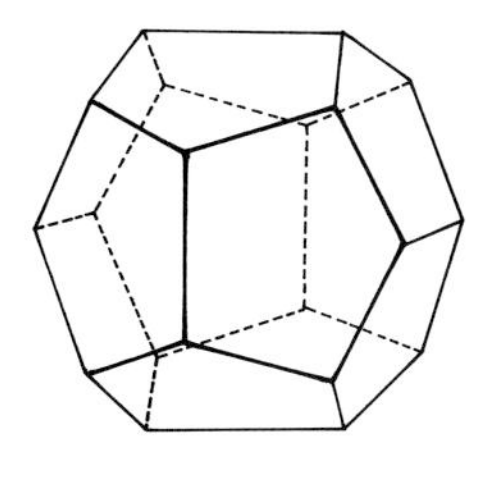

Completely opaque to transmitted light
Pale brass-yellow in reflected light

COMPOSITION AND STRUCTURE. Pyrite (FeS_2) is the most common of the small group of iron sulfide minerals, which includes the pyrite dimorph marcasite (FeS_2). Many metals and semimetals are reported in pyrite analyses; however, all are minor, and many are probably present as tiny inclusions (for example, Au, Cu, and Pb). Pyrite may form significant, but incomplete solid solution with its Ni-isomorph, vaesite (NiS_2), and its Co-isomorph, cattierite (CoS_2), and pyrite containing a few percent Co or Ni is rather common. Pyrite with major Co or Ni is rare and has been named *bravoite.*

Hauverite MnS_2, cobaltite (Co,Fe)AsS, and gersdorffite (Ni,Fe,Co)AsS also possess the pyrite structure but form very little solid solution with it.

The pyrite structure is analogous with the simple NaCl structure (Fig. 1-5) where covalent S_2 "dumb-bell" pairs form the anion alternating with Fe in three mutually perpendicular directions to yield a cubic cell. S—S dumb-bells are elongated parallel to cube cell diagonals to reduce $\frac{4}{m}\bar{3}\frac{2}{m}$ symmetry to $\frac{2}{m}\bar{3}$.

PHYSICAL PROPERTIES. H = $6–6\frac{1}{2}$. Sp. Gr. = 4.80–5.03. (Low hardness and density values derive from high Co or Ni.) Pale brass-yellow and splendent metallic luster in hand sample. Surfaces may show iridescent tarnish. Streak is black to greenish black or brownish black.

COLOR AND PLEOCHROISM. Even in the thinnest slivers, pyrite is opaque to transmitted light. A light directed on the surface of fragments or thin section reveals a shiny to granular surface with pale brass-yellow reflections.

FORM. Euhedral crystals of pyrite are very common. Cube {100} or pyritohedron {120} habits are most common; however, octadedrons {111} are often seen, as are combinations of these basic forms. Oscillatory combination of cube and pyritohedron forms produces striations on cube faces to betray $\frac{2}{m}\bar{3}$ symmetry, and may result in curved or rounded crystal faces.

Pyrite is very often massive and may appear as radiating, subfibrous aggregates in colloform, globular, or stalactitic masses. In thin section section, opaque squares, triangles, and other sections of simple isometric forms are often seen.

CLEAVAGE. Pyrite cleavage is usually very poor; however, some specimens may show rather well-defined cleavage or parting on {100}, {311}, {110}, or {111}.

TWINNING. Highly symmetrical penetration twins, called *iron cross twins*, are common (twin plane [011], twin axis [011]).

DISTINGUISHING FEATURES. Pyrite is a widely distributed, isometric, opaque mineral. Yellowish surface reflection may separate it from magnetite, ilmenite, hematite, and other grayish opaque minerals. Marcasite, arsenopyrite, chalcopyrite, pyrrhotite, and other brassy opaque minerals are usually much softer than pyrite and may show tabular or other nonisometric forms.

Polished sections of pyrite may show weak anisotropism between the crossed polars of a reflecting microscope; however, it does not show the strong anisotropism and polarization colors of marcasite or arsenopyrite. Chalcopyrite has a distinctly deeper yellow color than pyrite, and pyrrhotite shows a pale bronze color and is distinctly magnetic.

ALTERATION. Pyrite normally oxidizes to hematite, limonite, and other iron hydroxides, which often form pseudomorphs after the common pyrite forms. Other minerals (such as chalcocite and graphite) may also form pseudomorphs after pyrite, and pyrite itself may replace other crystals (such as pyrrhotite, marcasite, and hematite, or even fossils, assuming strange forms.

OCCURRENCE. Pyrite is a "persistent" mineral formed under a very large range of geologic conditions. It is the most common sulfide mineral and one of the most common opaque minerals in a wide variety of igneous, sedimentary, and metamorphic rocks.

In plutonic igneous rocks, pyrite tends to be associated with other sulfides in the full composition range from felsic (granites) to ultramafic. Pyrite is introduced by hydrothermal solutions and tends to occur on fracture surfaces in granitic intrusions and is ubiquitous in all forms of hydrothermal metalliferous veins. Unusually large, massive bodies of pyrite are attributed to moderate- to high-temperature hydrothermal deposition. The alteration of near-surface pyrite to iron oxides (gossan) and sulfate solutions is very important in the secondary enrichment processes of hydrothermal ore deposits.

Pyrite is rare in pegmatite veins. Gabbros, norites, and similar mafic rocks often contain pyrite, pyrrhotite, chalcopyrite, and other sulfides as orthomagmatic minerals formed by the crystallization of an immiscible sulfide magma fraction in the mafic sulfide melt.

Volcanic rocks may contain pyrite crystals as a product of sublimation.

In sedimentary rocks, pyrite tends to form in a reducing environment in association with the carbonaceous matter or glauconite of black or green shales, limestones, coals, and so on. It may be deposited in shallow marine or lacustrine muds by the action of microorganisms. Pyrite may replace wood or shells or appear as radiated crystals in colloform masses.

Marcasite occurs in similar environments and in low-temperature hydrothermal veins.

Metamorphic rocks contain pyrite crystals in black slates, chlorite schists, and so on and as great masses in contact sulfide deposits.

REFERENCES: PYRITE

Brostigen, G., and A. Kjekshus. 1969. Redetermined crystal structure of FeS_2 (pyrite). *Acta Chem. Scand.*, *23*, 2186–2188.

Frenzel, G., and F. D. Bloss. 1967. Cleavage in pyrite. *Amer. Mineral.*, *52*, 994–1002.

HALIDES

Halite — NaCl — ISOMETRIC

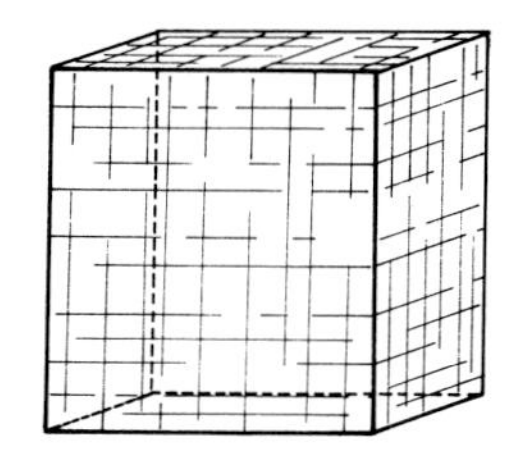

$n = 1.544$*

Colorless or very pale tints in thin section

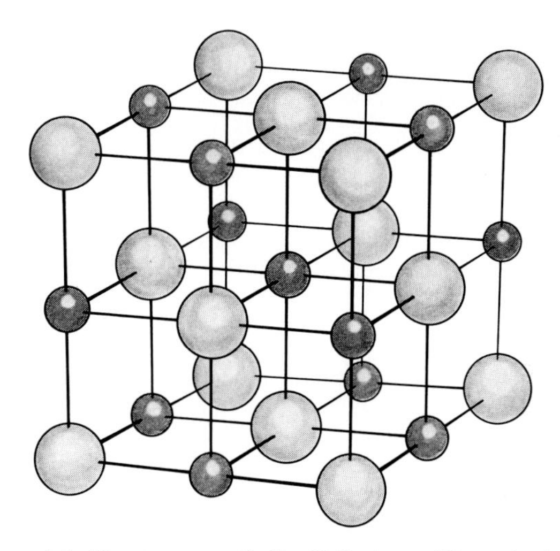

FIGURE 1-5. The structure of halite. Halite is an *AX*-type ionic compound having a medium-size cation appropriate to sixfold (octahedral) coordination. The structure is essentially a cubic close-packed configuration of Cl^- ions on {111}, with Na^+ ions filling all octahedral sites between close-packed layers. Light nontransition elements and weak ionic bonding favor a colorless mineral with a small index of refraction.

COMPOSITION AND STRUCTURE. The structure of halite (Fig. 1-5) was the first to be determined by x-ray methods (Bragg, 1914). It is a very simple structure of alternating Na^+ and Cl^- ions in three mutually perpendicular directions; each ion is coordinated by six (octahedral coordination) ions of opposite charge.

Halite is nearly pure NaCl; massive rock salt, however, commonly contains admixed inclusions of clay, iron oxides, gypsum, sylvite, and other evaporite minerals. Brine inclusions are common. Minor Br^- or I^- may replace Cl^-. Very little K^+ replaces Na^+; however, halite NaCl and sylvite KCl commonly crystallize together and may form submicroscopic mixtures with intermediate properties.

PHYSICAL PROPERTIES. H = 2–$2\frac{1}{2}$. Sp. Gr. = 2.165. In hand sample, halite is usually white or colorless. It is often gray, yellow, brown, or reddish, due to admixed clays or iron oxides. Blue or violet coloration is commonly due to lattice disturbances by radiation and disappears with heating; it may also be due to free colloidal sodium. Luster is vitreous. Halite is water soluble, with a saline taste.

COLOR AND PLEOCHROISM. Halite is usually colorless in standard section or as fragments. Deep-colored crystals may show pale color usually in zones or color centers. Crystals under pressure may show weak pleochroism.

FORM. Halite crystals are cubic, rarely octahedral, and may show hopperlike forms. Halite is often coarse to fine granular as anhedral grains and is sometimes stalactitic.

CLEAVAGE. Perfect cubic cleavages {100} (three mutually perpendicular directions) form square to rectangular fragments.

TWINNING. Synthetic crystals are known to twin on {111} but twinning is not apparent in section or fragments.

---

* Halite may form submicroscopic mixtures with sylvite (KCl) and the refractive index of the mixture may range from that of halite ($n = 1.544$) to that of sylvite ($n = 1.490$).

ZONING. Color often occurs in zones parallel to cube faces.

DISTINGUISHING FEATURES. Halite is isotropic, with perfect cubic cleavage and very low relief in balsam. Saline taste is also distinctive. Sylvite has moderate negative relief in balsam ($n = 1.490$), fluorite shows high negative relief and octahedral cleavage, cryolite shows high negative relief and weak birefringence, and villiaumite (NaF) has high negative relief.

ALTERATION. Halite is very soluble, and casts of halite crystals are common. Dolomite, gypsum, anhydrite, and soluble salts commonly form pseudomorphs after halite.

OCCURRENCE. Halite occurs mainly as sedimentary beds of all geologic ages resulting from the evaporation of isolated bodies of marine water or saline lakes. It is commonly associated with dolomite, anhydrite, gypsum, sylvite, polyhalite, carnallite, and other evaporite salts.

Halite precipitates from marine water when it has evaporated to about 10 percent of its volume. It follows the deposition of carbonates and gypsum-anhydrite and precedes the soluble sulfates and chlorides of K and Mg. Deeply buried beds may rise as diapiric plugs or salt domes. Halite is a common precipitate in modern saline lakes and appears as surface efflorescence in playas and soils.

REFERENCES: HALITE

Barbieri, M., and A. Penta. 1968. Geochemical observations of San Cataldo (Caltanissetta) Miocene evaporites. *Period. Mineral.*, *37*, 777–807.

Bragg, W. L. 1914. The structure of some crystals as indicated by their diffraction of X-rays. *Proc. Roy. Soc. London* (*Series A*), *89*, 248–277.

Sylvite — KCl — ISOMETRIC

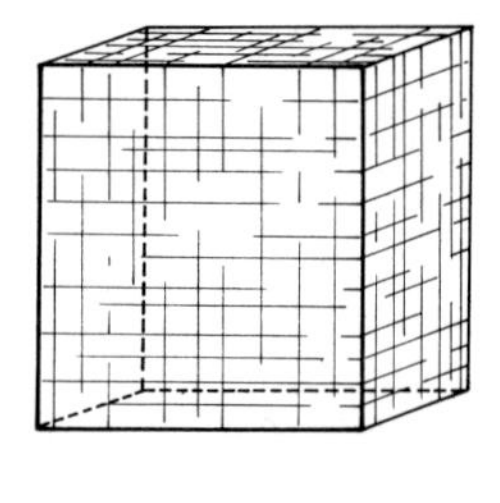

Sylvite

$n = 1.4903$*

Colorless in thin section

COMPOSITION AND STRUCTURE. The structure of sylvite is analogous to that of halite (Fig. 1-5) and consists of alternating K^+ and Cl^- in three mutually perpendicular directions; each ion in sixfold (octahedral) coordination.

* Sylvite may form submicroscopic mixtures with halite, and the refractive index of the mixture may range from that of sylvite ($n = 1.490$) to that of halite (1.544).

Sylvite is essentially pure KCl, but is usually intimately intergrown or mixed with crystals of halite. Very little Na^+ replaces K^+, although submicroscopic intergrowths show intermediate properties. Minor Br^- may replace Cl^-, and hematite inclusions are common.

PHYSICAL PROPERTIES. H = 2. Sp. Gr. = 1.993. In hand sample, sylvite is colorless, gray, or bluish. Oriented hematite inclusions may cause reddish or orange color. Luster is vitreous. Sylvite is highly water soluble, and its taste is saline but more bitter than halite.

COLOR AND PLEOCHROISM. Sylvite is colorless in section or as fragments.

FORM. Crystal habit is cubic, rarely octahedral. Cleavable masses and granular forms are common.

CLEAVAGE. Perfect cubic cleavages {100} yield cubic fragments.

BIREFRINGENCE. Sylvite is isotropic, but minor strain may show as weak birefringence.

TWINNING. Simple twinning on {111} should be possible.

ZONING. Inclusions of hematite or halite are aligned on cubic planes {100}, rarely on {111} or {110}.

DISTINGUISHING FEATURES. Sylvite has very limited occurrence; it is isotropic, with perfect cubic cleavages and moderate negative relief in balsam. Bitter saline taste is distinctive.

Halite has greater indices of refraction ($n = 1.544$) and simple saline taste; fluorite has octahedral cleavage and often uneven color; villiaumite has extreme negative relief and very limited occurrence.

ALTERATION. Sylvite is highly water soluble.

OCCURRENCE. Sylvite occurs in bedded evaporite deposits intimately associated, intergrown, or intermixed with halite. It represents one of the final salts deposited in saline lakes or isolated marine basins and is much less common than halite. Other associates are dolomite, clays, gypsum, anhydrite, carnallite, polyhalite, kainite, and kieserite.

REFERENCE: SYLVITE

Borshcherskii, Yu. A. 1964. Nature of red color in potassium salts. *Geokhimiya*, no. 2, pp. 289–290. Translated 1964, in *Geochemistry International*, no. 2, pp. 289–290.

Fluorite (Fluorspar) CaF_2 ISOMETRIC

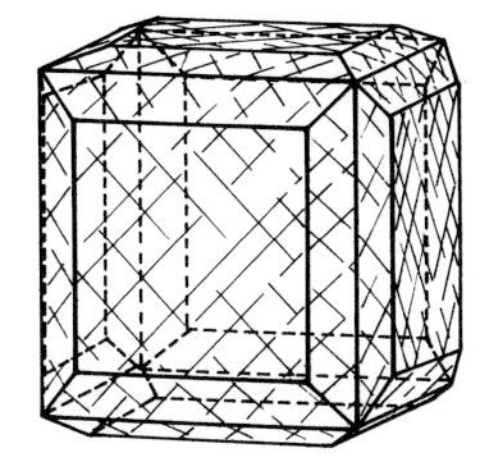
Fluorite

$n = 1.433$–1.435*

Colorless, pale violet, pale green

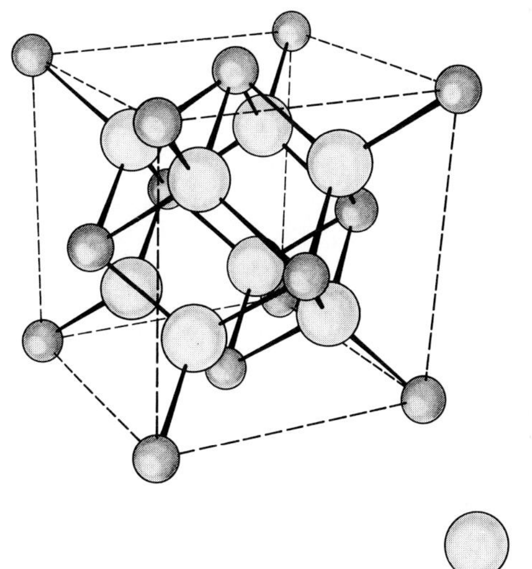

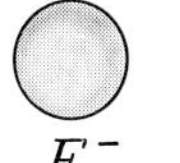

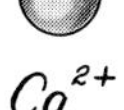

FIGURE 1-6. The structure of fluorite. Fluorite is an AX_2-type ionic compound with a large cation appropriate for eightfold (cubic) coordination. The large F^- anions are coordinated to four Ca^{2+} cations. The structure yields a face-centered, hexoctahedral isometric unit cell, and light nontransition elements and weak ionic bonding favor a colorless mineral with small index of refraction.

COMPOSITION AND STRUCTURE. The fluorite structure is characteristic of AX_2 compounds where the radius ratio $R_a:R_c > 0.732$. Ca^{2+} cations, in eightfold coordination, lie on the points of a cubic face-centered lattice; F^- anions, in fourfold (tetrahedral) coordination, lie at the center of each octant of the cubic cell (Fig. 1-6).

Fluorite is nearly pure CaF_2; however, minor rare earths, particularly Y^{3+} or Ce^{3+}, may replace Ca^{2+} toward the

* Index of refraction increases slightly with substitution of Y for Ca. Brown and reddish fluorites tend to have slightly greater index of refraction than green and blue varieties.

artificial compound YF_3 (yttrofluorite), which requires excess, interstitial fluorine. Free fluorine is common in some fluorites and is released by grinding. Minor Sr^{2+} may also replace Ca^{2+}.

PHYSICAL PROPERTIES. H = 4. Sp. Gr. = 3.180. Color, in hand sample, is extremely variable. Pure CaF_2 is colorless; however, fluorite is usually colored, most commonly pale green, pale yellow, or purple from pale purple to almost black; less commonly pale blue, pink or brown. Color is often visibly zoned or uneven. The intensity of purple coloration is often directly related to radiation received from radioactive minerals; it is apparently the result of disrupted crystal structure and disappears with heating. Luster is vitreous.

COLOR AND PLEOCHROISM. Fluorite is usually colorless in standard section or as fragments; however, deeply colored varieties may show pale green or pale to deep purple, usually in zoned or irregular distribution. Pleochroism is, of course, absent.

FORM. Fluorite usually appears as euhedral crystals, usually cubes, rarely octahedrons or even dodecahedrons. It is sometimes coarse to fine granular or earthy; rarely columnar to fibrous as globular, banded masses.

CLEAVAGE. Perfect octahedral {111} cleavages (four directions) are distinctive and commonly produce fragments as trigonal pyramids or triangular shapes (Fig. 1-7).

BIREFRINGENCE. Fluorite is usually isotropic but may sometimes show weak anomalous birefringence.

TWINNING. Twins on {111} are common, usually as penetrating cubes, sometimes as spinel twins. Twins are not obvious in thin section or fragments, as fluorite is isotropic.

ZONING. Color is often in distinct zones parallel to cube faces.

DISTINGUISHING FEATURES. Fluorite has unusually low index of refraction ($n \simeq 1.434$); it is isotropic and shows perfect octahedral cleavage {111} and often uneven color, usually purple. Cryolite has even lower index of refraction ($n \simeq 1.338$) and shows weak birefringence and pseudocubic cleavage; halite ($n = 1.544$) has perfect cubic cleavage.

ALTERATION. Fluorite is relatively soluble in carbonated waters, and casts of fluorite crystals are common. Fluorite crystals are commonly replaced by chalcedony or quartz. Numerous other minerals form pseudomorphs after fluorite, especially Fe- or Mn-oxides, calcite, dolomite, siderite, sphalerite, pyrite, marcasite, cerussite, smithsonite, hemimor-

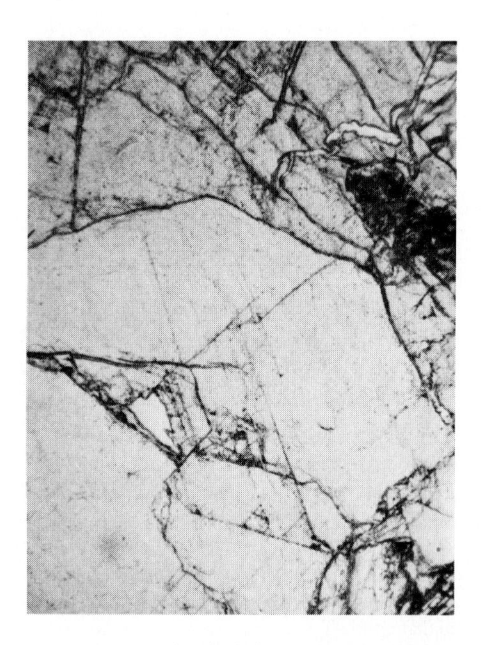

FIGURE 1-7. Fluorite is isotropic and commonly shows irregular color distribution and octahedral cleavages, which may appear as patterns of equilateral triangles.

phite, chlorite, and talc. Fluorite may form pseudomorphs after calcite, barite, or galena.

OCCURRENCE. Fluorite is a very common vein mineral and may constitute nearly the entire vein. It appears in hydrothermal veins or carbonate replacement deposits with quartz, calcite, dolomite, barite, sphalerite, galena, and U-minerals; in pneumatolitic veins and greisens with topaz, tourmaline, lepidolite, zinnwaldite, apatite, and cassiterite; and in pegmatite veins (rare earth fluorite) with topaz, apatite, Li-minerals, and rare earth minerals. Veins of calcite and fluorite commonly dissect syenites and feldspathoidal rocks.

Fluorite may appear in drusy cavities and on joint surfaces in granites and similar rocks. It is also a constituent of some carbonatites and is common in veins and cavities in carbonate sediments in association with celestite, dolomite, gypsum, anhydrite, and sulfur. Earthy varieties may be deposited by hot springs. It is relatively common as a detrital mineral and rarely cements detrital fragments.

REFERENCES: FLUORITE

Calas, G. 1972. Blue color of some natural fluorites. *Bull. Soc. Fr. Mineral. Cristallogr.*, *95*, 470–474.

Gorski, W. 1961. Hardness values of colored fluorite crystals. *Z. Kristallogr.*, *115*, 468–469.

Il'inskii, G. A. 1962. Effect of radioactive radiation on the color of fluorite. *Zap. Vses. Mineral Obshchest.*, *91*, 613–615.

Mueller, G., W. Recke, and R. Vera Mege. 1961. Interrelation between coloration, absorption spectra, and radioactivities of fluorites from diverse localities. *Bol. Soc. Chilena Quim.*, *11*, 8–12.

Nozhkin, A. D. 1971. Rare-earth thorium fluorite. *Zap. Vses. Mineral. Obshchest.*, *100*, 334–337.

Recker, K., A. Neuhaus, and R. Leekebusch. 1968. Comparative studies on color and luminescence properties of natural and grown, doped fluorites. *Int. Mineral. Assoc., Proc. 5th Gen. Mtg.*, *1966*, 145–152.

Carnallite

$KMgCl_3 \cdot 6H_2O$ ORTHORHOMBIC

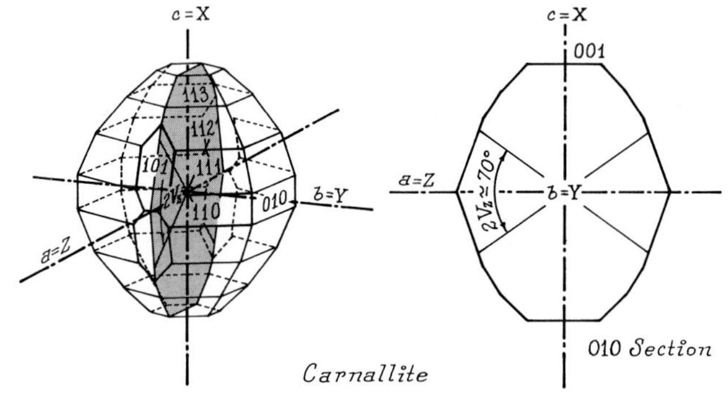

$n_\alpha = 1.465–1.467$
$n_\beta = 1.472–1.475$
$n_\gamma = 1.494–1.497$
$n_\gamma - n_\alpha \simeq 0.030$
Biaxial positive, $2V_z \simeq 70°$
$a = Z, b = Y, c = X$
Weak $v > r$
Colorless in thin section

COMPOSITION. Carnallite $KMgCl_3 \cdot 6H_2O$ may contain minor Rb, Cs, NH_4, or Tl in substitution for K; minor Fe^{2+} for Mg; and minor Br for Cl. Scales of hematite are common inclusions.

PHYSICAL PROPERTIES. H = $2\frac{1}{2}$. Sp. Gr. = 1.60. In hand sample, carnallite is colorless to white and sometimes reddish due to hematite inclusions. Luster is greasy. Carnallite is highly water soluble, with a bitter taste.

COLOR AND PLEOCHROISM. Carnallite is colorless in section or as fragments.

FORM. Carnallite is usually coarse to fine granular. Crystals are rare and pseudohexagonal as barrellike bipyramids or thick basal plates.

CLEAVAGE. Carnallite shows no distinct cleavage.

BIREFRINGENCE. Maximum birefringence, seen in {010} sections, is moderate ($n_\gamma - n_\alpha \simeq 0.030$). Interference colors are brilliant, ranging to middle second order.

TWINNING. Twinning lamellae may form on prismatic planes by stress pressure.

INTERFERENCE FIGURE. Sections parallel to {100} show acute bisectrix figures, with numerous isochromes and weak optic axis dispersion $v > r$.

OPTICAL ORIENTATION. Carnallite is orthorhombic $a = Z$, $b = Y$, and $c = X$, and the optic plane is {010}. Extinction is parallel to possible crystal elongation.

DISTINGUISHING FEATURES. Carnallite occurs intermixed with highly soluble salts. It shows no cleavage, negative relief, and moderate birefringence. Kieserite has greater indices of refraction ($n_\gamma = 1.584$) and greater birefringence; polyhalite has indices of refraction above balsam and is biaxial negative.

ALTERATION. Carnallite is highly water soluble.

OCCURRENCE. Carnallite occurs in bedded evaporite deposits, associated with kieserite, polyhalite, sylvite, halite, and other late-depositing, highly soluble salts.

REFERENCES: CARNALLITE

Baimuradov, R., E. E. Plyushchev, T. A. Slivko, and A. Ashirov. 1973. Physical and chemical properties of rubidium carnallite. *Izv. Akad. Nauk Turkm. SSR, Ser. Fiz.-Tekh., Khim. Geol. Nauk, 1973*, 91–93.

Borshchevskii, Yu. A. 1964. Nature of the red color of potassium salts. *Geokhimiya, 1964*, 289–290.

Choudhari, B. P., and D. S. Datar. 1967. Constitution of mixed salt. I. Microscopic examination of mixed salts and related substances. *Salt Res. Ind., 4*, 113–125.

Fischer, W. 1973. Crystal structure of carnallite, $KMgCl_3 \cdot 6H_2O$. *Neues Jahrb. Mineral., Monatsh., 1973*, 100–109.

Cryolite

Na_3AlF_6

MONOCLINIC
$\angle\beta = 90°11'$

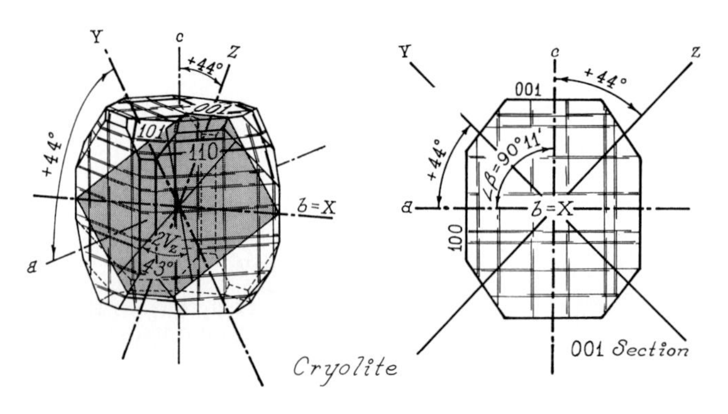

$n_\alpha = 1.338$
$n_\beta = 1.338$
$n_\gamma = 1.339$
$n_\gamma - n_\alpha \simeq 0.001$
Biaxial positive, $2V_z \simeq 43°$*
$b = X, c \wedge Z = +44°, a \wedge Y = +44°$*
Weak $v > r$ with horizontal bisectrix dispersion
Colorless in standard section

* Note that, because $n_\alpha \simeq n_\beta \simeq n_\gamma$, very minor changes in refractive indices may cause major changes in the magnitude of $2V_z$, and little distinction exists between X and Y.

COMPOSITION AND STRUCTURE. Independent, octahedral structural units (AlF_6^{3-}) are linked by Na^+ ions to yield an essentially isometric structure similar to garnet (a high-temperature polymorph is strictly isometric above 560°C).

Cryolite is very near its ideal composition Na_3AlF_6; however, minor Fe^{3+}, Mn^{3+}, and Ca^{2+} are sometimes reported in analyses, and carbonaceous inclusions may cause dark coloration.

PHYSICAL PROPERTIES. H = $2\frac{1}{2}$. Sp. Gr. = 2.97. In hand sample, cryolite is colorless or white, rarely red-brown, brown, or black. Luster is vitreous to greasy.

COLOR AND PLEOCHROISM. Cryolite is colorless in section or as fragments.

FORM. Cryolite is usually fine to coarse granular. Crystals are pseudocubic with {001} and {110} forms.

CLEAVAGE. Pseudocubic parting on {001} and {110} yield cubic fragments; also on {$\bar{1}$01}.

BIREFRINGENCE. Maximum birefringence is very weak ($n_\gamma - n_\alpha = 0.001$) and a $\frac{1}{4}\lambda$ accessory plate may be necessary to confirm birefringence. Interference colors, in standard thin section, do not exceed dark first-order gray.

TWINNING. Twinning is very common, often as several complex, polysynthetic twin laws occurring simultaneously. Lamellar twinning is reported on {001}, {110}, {$\bar{1}$01}, {100}, {112}, and other planes. Some laws may be secondary glide twinning.

INTERFERENCE FIGURE. Cryolite is biaxial positive, with $2V_z$ reported near 43°. Interference figures on standard sections or small fragments should be extremely dark and confused by polysynthetic twin laws. Thick cleavage plates near {$\bar{1}$01} may yield near-centered acute bisectrix figures showing broad isogyres on a gray field, dispersion $v > r$, and horizontal bisectrix dispersion.

OPTICAL ORIENTATION. Cryolite is monoclinic, with $b = X$, $c \wedge Z = +44°$, $a \wedge Y = +44°$. Maximum extinction, when perceptible, is seen in {010} sections and is essentially 45° to pseudocubic cleavages. Cleavages do not favor elongation.

DISTINGUISHING FEATURES. Cryolite is a rare mineral of limited occurrence; it has very low indices of refraction, shows complex multiple twinning, and is sensibly isotropic. Fluorite is isotropic, with a greater refractive index ($n = 1.434$) and octahedral cleavages. Cryolite is most likely confused with other aluminum-fluorides, which probably represent alteration products of cryolite.

ALTERATION. Numerous aluminum-fluorides occur with cryolite and are presumed to result from its alteration (for example, elpasolite, pachnolite, thomsenolite, jarlite, gearksutite, prosopite, chiolite, ralstonite, and weberite). All of these fluorides show low indices of refraction and low to negligible birefringence.

OCCURRENCE. Cryolite occurs in pegmatitelike veins associated with granitic intrusions. Associated minerals are quartz, fluorite, microcline, muscovite, topaz, zircon, siderite, cassiterite, wolframite, columbite, sulfide ores, and aluminum-fluorides.

REFERENCES: CRYOLITE

Bøggild, O. B. 1953. The mineralogy of Greenland. *Medd. om Grønland, 149*, 1–442.

Hawthorne, F. C., and R. B. Ferguson. 1975. Refinement of the crystal structure of cryolite. *Can. Mineral., 13*, 377–382.

Marchenko, E. Ya. 1972. Fluorine-rich metasomatites of sub-alkaline granitic rocks. *Dokl. Akad. Nauk SSSR, 203*, 1170–1172.

Naray-Szabó, I., and K. Sasvári. 1938. The structure of cryolite, Na_3AlF_6. *Z. Kristallogr., 99*, 27–31.

Pabst, A. 1950. A structural classification of fluoaluminates. *Amer. Mineral., 35*, 149–165.

OXIDES

Periclase — MgO — ISOMETRIC

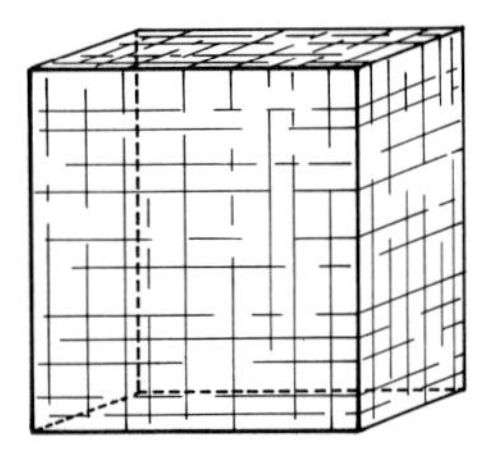

Periclase

$n = 1.736–1.745$*
Colorless in thin section

COMPOSITION AND STRUCTURE. Natural periclase is nearly pure MgO, with a NaCl-type structure (Fig. 1-5). It may form complete solid solution with a ferrous analog FeO, at least in synthetic compositions, and extensive solid solution may also exist with MnO and NiO end members; however, natural compositions seldom contain more than 10 percent FeO (ferropericlase) and traces of Mn^{2+} or Ni^{2+}.

PHYSICAL PROPERTIES. $H = 5\frac{1}{2}–6$. Sp. Gr. = 3.58–3.68. Hand sample color is white to pale gray, yellow, or brown.

COLOR AND PLEOCHROISM. Periclase is colorless in thin section or as fragments. Iron-rich varieties may show very weak coloration in yellow or brown, without pleochroism.

FORM. Periclase appears as cubic fragments or isometric crystals, often surrounded by fibrous aggregates of brucite.

CLEAVAGE. Perfect cubic cleavage {100} is a consequence of NaCl-like structure.

* Index of refraction increases with increasing iron to about 2.32 for the ferrous end member FeO.

BIREFRINGENCE. Periclase is isotropic and shows zero birefringence in all orientations.

TWINNING. Spinel-type twins on {111} are known but are not optically apparent.

DISTINGUISHING FEATURES. Periclase is isotropic, with high relief and cubic cleavages. Its occurrence is severely restricted, and it is usually more or less altered to brucite. Garnet and spinel tend to be colored and show little or no cleavage.

ALTERATION. Periclase consistently alters to brucite, $Mg(OH)_2$. Serpentine and hydromagnesite are rarely reported as alteration products, and iron oxides may separate from altered ferropericlase.

OCCURRENCE. Periclase occurs only in high-temperature contact metamorphic rocks, due to the dissociation of dolomite or magnesite. Unaltered periclase cores are commonly surrounded by fibrous brucite and appear in association with calcite, forsterite, serpentine, and Ca-Mg silicates. Periclase forms at high temperatures above the wollastonite range and below monticellite.

REFERENCES: PERICLASE

Hazen, R. M. 1976. Effects of temperature and pressure on the cell dimensions and x-ray temperature factors of periclase. *Amer. Mineral.*, *61*, 266–271.

Kennedy, G. C. 1956. The brucite-periclase equilibrium, *Amer. Jour. Sci.*, *254*, 567–573.

Piriou, B. 1964. Optical constants of corundum and magnesium oxide. *C. R. Acad. Sci. Paris*, *259*, 1052–1055.

Tresvyatskii, S. G., Z. A. Yaremenko, and L. M. Lopato. 1966. Crystallooptical properties of synthetic periclase single crystals. *Kristallografiya*, *11*, 459–463.

Rutile

TiO_2

TETRAGONAL

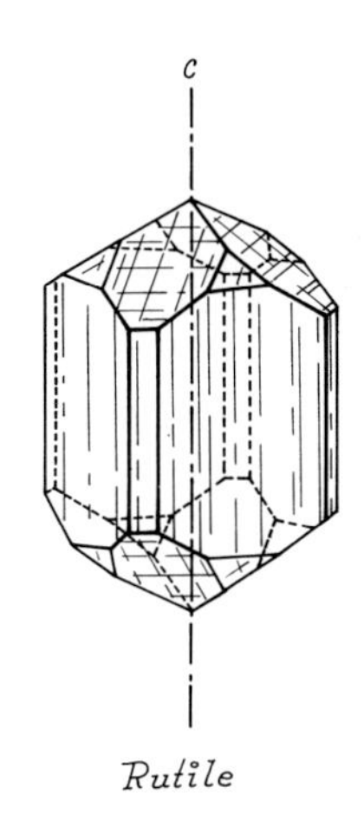

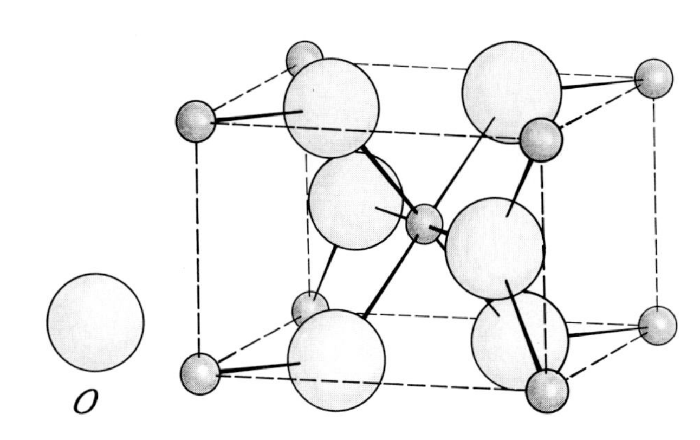

FIGURE 1-8. The structure of rutile. Rutile is an AX_2-type compound, with a medium-size cation appropriate to sixfold (octahedral) coordination. Large oxygen anions are bonded to three cations. The structure yields a tetrahedral cell, only part of which is shown in the figure.

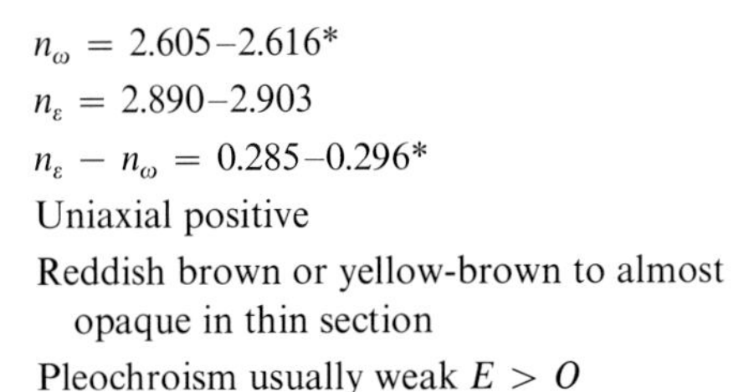

$n_\omega = 2.605–2.616$*
$n_\varepsilon = 2.890–2.903$
$n_\varepsilon - n_\omega = 0.285–0.296$*
Uniaxial positive
Reddish brown or yellow-brown to almost opaque in thin section
Pleochroism usually weak $E > O$

COMPOSITION AND STRUCTURE. The rutile structure is the usual atomic arrangement for AX_2 compounds with moderate-size cations. Each cation (Ti^{4+}) is surrounded by six O^{2-} ions in slightly deformed octahedral coordination and each O^{2-} is bonded to three cations in triangular coordination (Fig. 1-8).

Rutile is basically TiO_2; it forms some degree of solid solution with tapiolite $Fe(Nb,Ta)_2O_6$; and Fe^{2+}, Fe^{3+}, Nb^{5+}, and Ta^{5+} may be present in rutile as major constituents. Columbian rutile (ilmenorutile) tends to contain about half as many Fe^{2+} ions as $Nb^{5+} + Ta^{5+}$ ions, as might be expected from charge balance considerations. Lesser amounts of Cr^{3+}, V^{5+}, or Sn^{4+} may appear in rutile.

PHYSICAL PROPERTIES. $H = 6–6\frac{1}{2}$. Sp. Gr. 4.2–5.6. Hand sample color is brick-red or brown to black, rarely yellowish, violet, or green. Varieties rich in Fe, Nb, or Ta tend to be black, and pure synthetic rutile is colorless to pale yellow-green. Luster is distinctly adamantine to submetallic.

COLOR AND PLEOCHROISM. Color in sections or fragments is usually deep red-brown to yellow-brown, and impure varieties may be almost opaque. Pleochroism is usually weak $E > O$ but may be moderately strong.

FORM. Rutile crystals are usually more or less euhedral tetragonal crystals yielding square or octagonal basal sections. Equidimensional anhedral grains are less common. It also occurs in long, acicular needles as hairlike inclusions in quartz, phlogopite, or corundum. Knee-shaped twins, cyclic twins, and netlike parallel crystal growths are common.

CLEAVAGE. Prismatic cleavages, distinct {110} and fair {100}, effect minor control on fragment orientation. Pyramidal cleavage {111} is poor, but parting on {902} or {101} may be prominent as a result of multiple twinning.

* Indices of refraction and birefringence show little obvious variation with chemical composition.

BIREFRINGENCE. Maximum birefringence is very extreme (0.285–0.296), and even basal sections show high-order interference colors in the slightly convergent light of orthoscopic illumination. Interference colors appear essentially independent of thickness or crystal orientation and form a very high-order uniform white, usually masked by inherent mineral color.

TWINNING. Twinning is most common on {011}, yielding highly varied forms that may be simple contact (knee-shaped twins), cyclic (sixlings or eightlings), or, rarely, polysynthetic. Glide twinning is also known on {092}, and simple contact twins form, rarely, on {031}.

ZONING. Color zoning due to composition change may be anticipated, but birefringence is uniformly extreme. Rutile itself may be zonally included in ilmenite, biotite, and quartz and is also the cause of the misty "blue quartz" of granulities.

INTERFERENCE FIGURE. Interference figures commonly show a uniaxial cross on a field of many isochromes, and the entire field is often deep to pale red-brown, due to inherent mineral coloration.

OPTICAL ORIENTATION. Elongated crystals show parallel extinction and positive elongation (length-slow), which may be difficult to detect due to extreme birefringence.

DISTINGUISHING FEATURES. Red-brown color, extreme relief, and extreme birefringence are diagnostic. Cassiterite and baddeleyite (ZrO_2) have much lower birefringence. Anatase and hematite are optically negative, brookite is biaxial, and limonite is isotropic.

ALTERATION. Rutile is a very stable mineral and is itself a common alteration product of other Ti-bearing minerals. Rutile rarely alters to sphene, leucoxene, or anatase.

OCCURRENCE. Rutile is more common than the other TiO_2 polymorphs, anatase and brookite; it is the dense, high-temperature, high-pressure polymorph that occurs in both igneous and metamorphic plutonic rocks.

In igneous rocks, rutile appears as small accessory crystals or grains. It is common in hornblende-rich rocks, anorthosite, eclogite, pegmatites (granite and basic), and in quartz veins.

Rutile appears in many metamorphic rocks especially schist, gneiss, amphibolite, and marble.

Sedimentary rocks may contain rutile as tiny needles in recrystallized shales or as common detrital grains. Rutile is also a common alteration product of Ti-rich minerals (for example, ilmenite, sphene, and perovskite).

Leucoxene is largely fine-grained rutile.

REFERENCES: RUTILE

Abrahams, S. C., and J. L. Bernstein. 1971. Rutile. Normal probability plot analysis and accurate measurement of crystal structure. *Jour. Chem. Phys.*, *55*, 3206–3211.

Cardona, M., and G. Harbeke. 1965. Optical properties and band structure of wurtzite-type crystals and rutile. *Phys. Rev.*, *137*, 1467–1476.

Cromer, D. T., and K. Harrington. 1955. The structures of anatase and rutile. *Jour. Amer. Chem. Soc.*, *77*, 4708–4709.

Czanderna, A. W., C. N. Ramachandra Rao, and J. M. Honig. 1958. The anatase-rutile transition. I. Kinetics of the transformation of pure anatase. *Trans. Faraday Soc.*, *54*, 1069–1073.

Davis, T. A. 1968. Pressure dependence of the refractive indices of tetragonal crystals: Ammonium dihydrogen phosphate, potassium dihydrogen phosphate, calcium molybdate (VI), calcium tungstate (VI), and rutile. *Jour. Opt. Soc. Amer.*, *58*, 1446–1451.

MacChesney, J., and A. Muan. 1959. Studies in the system iron oxide-titanium oxide. *Amer. Mineral.*, *44*, 926–945.

Nodop, G. 1956. Anatase, rutile and cassiterite. *Hamburger Beitr. Angew. Mineral. u. Kristallphys.*, *1*, 239–284.

Anatase (Octahedrite) — TiO_2 — TETRAGONAL

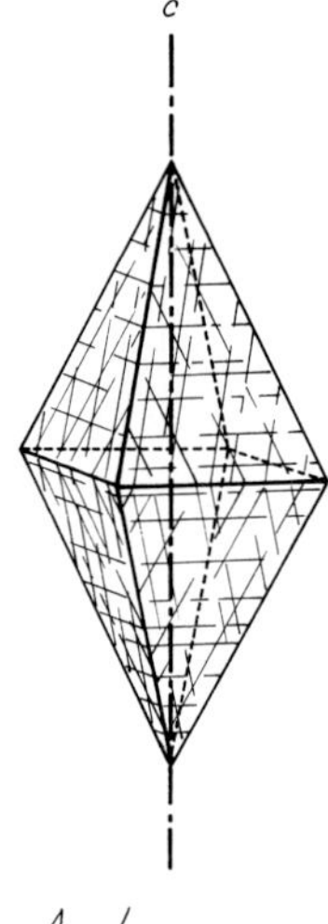

$n_\omega = 2.561$

$n_\varepsilon = 2.488$

$n_\omega - n_\varepsilon = 0.073$

Uniaxial negative

Color yellow-brown to dark brown, green or deep blue

Weakly pleochroic $E > O$, rarely $O > E$

COMPOSITION AND STRUCTURE. Anatase, like rutile and brookite, has a structure of TiO_2 octahedra, each sharing two octahedral edges with adjacent octahedra. Different edges are shared in each polymorph. Anatase crystals are commonly stubby dipyramids resembling octahedrons, inspiring the alternate name *octahedrite*.

Anatase, the low-temperature polymorph, is essentially TiO_2. Minor Fe^{3+} and Sn are reported, and, rarely, Nb and Ta are important constituents.

PHYSICAL PROPERTIES. $H = 5\frac{1}{2}$–6. Sp. Gr. = 3.90. In hand sample, anatase is yellow-brown or red-brown ranging to black or deep blue. Luster is adamantine to submetallic.

COLOR AND PLEOCHROISM. In thin section, anatase is pale yellow-brown, red-brown to deep brown, green, or dark blue. Pleochroism is usually weak $E > O$, sometimes $O > E$.

FORM. Anatase appears as anhedral grains to euhedral triangular or rectangular sections.

CLEAVAGE. Perfect basal {001} and pyramidal {111} cleavages yield triangular fragments.

BIREFRINGENCE. Sections parallel to *c* show strong maximum birefringence (0.073) and fourth-order interference colors often masked by inherent mineral color.

TWINNING. Rare on {112}.

ZONING. Crystals may show growth zoning or rutile overgrowths.

INTERFERENCE FIGURE. Uniaxial negative interference figures show numerous isochromes often obscured by deep mineral color. Darker-colored varieties may show small $2V$.

OPTICAL ORIENTATION. The fast wave vibrates parallel to the *c*-crystallographic axis, so that crystals elongated on *c* are length-fast. Extinction is parallel, inclined, or symmetrical to cleavage traces.

DISTINGUISHING FEATURES. Anatase is highly colored and uniaxial negative (rarely showing small $2V$). Rutile is uniaxial positive, and brookite is biaxial positive.

ALTERATION. Anatase is a common alteration product of sphene and, less commonly, ilmenite and other Ti-minerals. It may alter or invert to rutile.

OCCURRENCE. Anatase is the low-temperature TiO_2 polymorph, occurring largely in open veins and cavities in granite, pegmatites, volcanics, schists, and gneisses as a product of hydrothermal introduction. It is a common alteration product of sphene, ilmenite, and other Ti-rich minerals.

REFERENCES: ANATASE

Horn, M., C. F. Schwerdtfeger, and P. E. Meagher. 1972. Refinement of the structure of anatase at several temperatures. Z. *Kristallogr.*, *136*, 273–281.

Thienchi, N. 1946. Transformation de lánatase en rutile. *C. R. Acad. Sci. Paris*, *222*, 1178–1179.

Weiser, H. B., W. O. Milligan, and E. L. Cook. 1941. X-ray studies on the hydrous oxides. X. Anatase and rutile modifications of titania. *Jour. Physical Chem.*, *45*, 1227–1234.

Brookite TiO_2 ORTHORHOMBIC

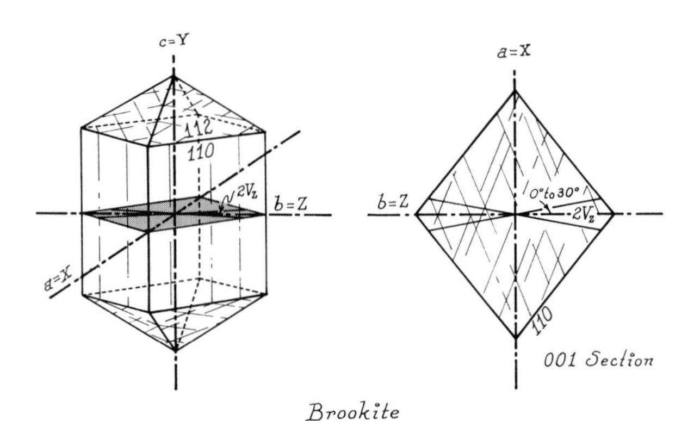

Brookite

$n_\alpha = 2.583$
$n_\beta = 2.584$
$n_\gamma = 2.700$
$n_\gamma - n_\alpha = 0.117$
Biaxial positive, $2V_z = 0°–30°$
$a = X, b = Z, c = Y$ (red, yellow light); $a = Y$, $b = Z, c = X$ (green, blue light)
Very strong crossed axial plane dispersion (see text)
Yellow-brown or red-brown to deep brown in thin section
Pleochroism very weak $Z > Y > X$

COMPOSITION AND STRUCTURE. Brookite structure is similar to that of rutile or anatase; however, TiO_6^{8-} octahedra share edges to form zigzag chains of octahedra. Brookite is essentially TiO_2, commonly with minor Fe^{3+} and even less Nb, Sn, or Pb.

PHYSICAL PROPERTIES. H = $5\frac{1}{2}$–6. Sp. Gr. = 4.12. Color in hand sample is yellow-brown or red-brown to black. Luster is adamantine to submetallic.

COLOR AND PLEOCHROISM. Brookite is yellow-brown or red-brown to deep brown in thin section with very weak pleochroism in shades of yellow, orange, and brown $Z > Y > X$.

FORM. Brookite occurs only as crystals, usually tabular on {010} or elongated on *c*.

CLEAVAGE. Cleavage is indistinct on {120} and very poor on {001}.

BIREFRINGENCE. Maximum birefringence is seen in {001} or {100} sections and is extreme (0.117 or greater). Interference colors are high order in most sections and are commonly obscured by mineral color and commonly abnormal, due to strong crossed axial plane dispersion. Sections or fragments may show dispersion colors and fail to extinguish completely in any position of stage rotation.

TWINNING. Twinning is very rare but has been reported on {120}.

ZONING. Growth color zoning is known.

INTERFERENCE FIGURE. Brookite is normally biaxial positive with small $2V_z = 0°–30°$. Crossed axial plane dispersion produces anomalies in the interference figure. Isogyres are not black, and dispersion fringes are dominant, being symmetrical to optic plane and optic normal.

OPTICAL ORIENTATION. All optical constants vary seriously with both temperature and light wavelength. For long wavelengths (red to yellow-green 5555 Å), the optic plane is {001}, and $2V_z$ decreases from 27° to 0° at 5555 Å, where brookite is uniaxial positive. With decreasing wavelength (green to violet) the $2V_z$ opens in the {100} plane from 0° at 5555 Å to 24° at 4800 Å (violet) and {100} is the optic plane for short wavelengths. For long wavelengths, $a = X$, $b = Z$, and $c = Y$; for short wavelength, $a = Y, b = Z$, and $c = X$.

Extinction is parallel, inclined, or symmetrical to poor cleavages and crystal faces; crystals elongated on *c* may be either length-fast or length-slow.

DISTINGUISHING FEATURES. Brookite shows strong color, extreme relief and birefringence, and anomalous dispersion. Rutile is uniaxial positive, anatase is uniaxial negative, and both show good cleavages.

ALTERATION. Brookite may invert to rutile.

OCCURRENCE. Brookite usually occurs as small accessory crystals in schists and gneisses or in hydrothermal veins and pegmatites. It is commonly associated with both rutile and anatase. Quartz, adularia, sphene, albite, and calcite are also associates of brookite. Brookite is common as detrital grains in coarse sediments and is found in many gold or diamond placer deposits.

REFERENCES: BROOKITE

Baur, W. H. 1961. Atomic distances and bond angles in brookite, TiO_2. *Acta Crystallogr.*, *14*, 214–216.
Ikornikova-Laemmlein, N. G. 1946. On the optical anomalies of brookite. *Dokl. Acad. Sci. USSR*, *53*, 251–254.
Pauling, L., and J. H. Sturdivant. 1928. The crystal structure of brookite. *Z. Kristallogr.*, *68*, 239–256.
Phillips, F. C. 1932. Crystals of brookite tabular parallel to the basal plane. *Mineral. Mag.*, *23*, 126–129.
Rastall, R. H. 1938. On brookite crystals in the Dogger. *Geol. Mag.*, *75*, 433–440.

Cassiterite

SnO_2

TETRAGONAL

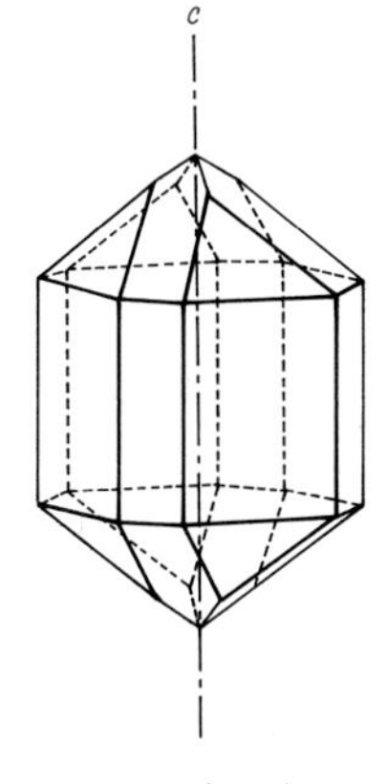

$n_\omega = 1.990–2.010$*
$n_\varepsilon = 2.091–2.100$
$n_\varepsilon - n_\omega = 0.10–0.09$
Uniaxial positive
Yellow, brown, red or greenish, rarely nearly colorless, in thin section
Pleochroism ranges from weak to strong $E > O$

* Refractive indices increase slightly with increasing iron content.

COMPOSITION AND STRUCTURE. Cassiterite is tin oxide (SnO_2) with a rutile-type crystal structure. Iron, usually as Fe^{3+}, is the most common impurity; however, minor Ti^{4+}, Nb^{5+}, and Ta^{5+} may also substitute for Sn^{4+}.

PHYSICAL PROPERTIES. H = 6–7. Sp. Gr. = 6.9–7.1. Hand sample color is dark brown to black, rarely red or yellow. Splendent adamantine luster is a consequence of extreme refractive indices.

COLOR AND PLEOCHROISM. In thin section, cassiterite is normally highly colored in shades of yellow, brown, reddish, or green, rarely almost colorless. Color is commonly uneven, being blotchy or zoned, and is often caused by microinclusions of wolframite or other minerals. Fragments and detrital grains may be almost opaque. Pleochroism may be very weak to very strong $E > O$.

FORM. Crystals of cassiterite are tetragonal, as stubby bipyramids or short prisms, euhedral to anhedral, rarely acicular needles. Vein alignment of anhedral grains is common. Wood tin shows colloform structures of radiated acicular crystal layers with intergrown needles of quartz, hematite, or topaz.

CLEAVAGE. Cleavages are poor on {100}, {110}, and {111}.

BIREFRINGENCE. Maximum birefringence, seen in prismatic sections, is very strong ($n_\varepsilon - n_\omega = 0.10–0.09$). Interference colors in standard thin section are pale fifth-order colors and are often obscured by intense mineral coloration.

TWINNING. Normal twinning on {101} is common. Twinning may be simple or repeated as polysynthetic {101} lamellae or as cyclic twins.

ZONING. Composition zoning of impurities (Fe, Nb, Ta, Ti, and so on) often appears as distinct color zones.

INTERFERENCE FIGURE. Cassiterite is uniaxial positive, and basal sections show distinct isogyres fanning broadly toward the edge of the visible field. Numerous isochromatic color rings are concentric about the melatope. Rarely, isogyres separate with stage rotation to yield a small $2V_z$.

OPTICAL ORIENTATION. Extinction is parallel to prismatic cleavages or crystal faces, and elongated crystals show positive elongation (length-slow).

DISTINGUISHING FEATURES. Extreme relief and birefringence and deep coloration are characteristic of cassiterite. It strongly resembles rutile, which usually shows even stronger relief, birefringence, and color. Brookite is biaxial, anatase is uniaxial negative, allanite and melanite garnet show moderate to weak birefringence, and sphalerite is isotropic.

ALTERATION. Cassiterite is a very stable mineral not very subject to near-surface weathering alteration. Cassiterite itself is a common alteration product of tin-bearing sulfides (for example, stannite, herzinbergite, and teallite).

OCCURRENCE. Cassiterite is most commonly found in highly siliceous, plutonic igneous rocks. It commonly appears in granites and similar rocks either as a primary mineral or by hydrothermal introduction. It appears in granite pegmatites and hypothermal veins with quartz, topaz, tourmaline, muscovite, and minerals of tungsten and molybdenum.

Cassiterite occurs in greisens and skarn rocks of metasomatic contact zones. It is formed as fine-grained aggregates by the alteration of Sn-bearing sulfide minerals, and it is a common detrital mineral in alluvial tin deposits.

Wood tin is formed by low-temperature colloidal precipitation and may appear in rhyolite cavities with topaz.

REFERENCES: CASSITERITE

Ecklebe, F. 1932. Optische Untersuchungen am Zinnerz im Temperaturbereich von 16°–1100°. *Neues Jahrb. Min., Abt. A*, *66*, 47–88.

Evzikova, N. Z. 1972. Practical aspects of cassiterite crystallomorphology. *Zap. Vses. Mineral. Obshchest.*, *101*, 237–249.

Frankel, J. J. 1947. Crystals of artificial cassiterite. *Mineral. Mag.*, *28*, 111–117.

Gotman, Ya. D. 1938. On the properties of cassiterite in connection with conditions of its formation. *Bull. Soc. Nat. Moscou*, *46*, 130–157.

Grubb, P. L. C., and P. Hannaford. 1966. Ferromagnetism and color zoning in some Malayan cassiterite. *Nature*, *209*, 677–678.

Corundum Al_2O_3 TRIGONAL

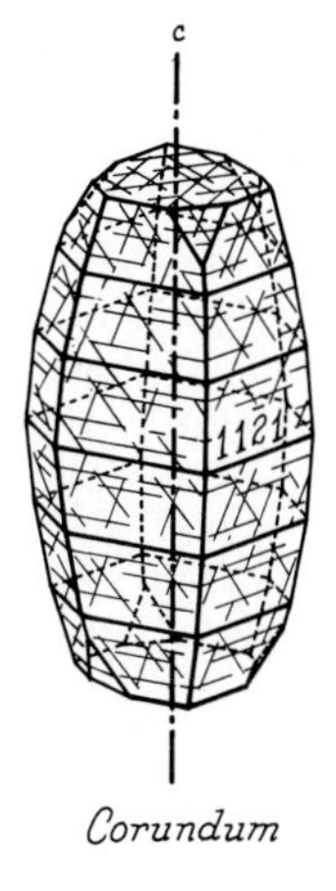

Corundum

$n_\omega = 1.767–1.772$*
$n_\varepsilon = 1.759–1.763$
$n_\omega - n_\varepsilon = 0.007–0.010$
Uniaxial negative (commonly anomalous biaxial)
Colorless to pale blue or red in thin section

COMPOSITION AND STRUCTURE. Corundum is essentially pure alumina (Al_2O_3). Colored varieties owe their colors to minor substitution of small metal ions for Al^{3+}. Red corundum

* Minor impurities, notably Cr^{3+} and Fe^{3+} cause maximum values for indices of refraction and birefringence.

(ruby) contains significant Cr^{3+}; blue corundum (sapphire) is colored by Fe^{2+} and Ti^{4+}; green corundum shows significant Fe^{3+} and Fe^{2+} with possible V^{5+}, Co^{2+}, Ni^{2+}; and yellow corundum may contain Fe^{2+}, Fe^{3+}, and Ni^{2+} or Mn^{3+} in minor amounts.

The corundum structure is a dense arrangement of oxygen ions in essentially hexagonal closest packing with Al^{3+} ions in two-thirds of the available sites of sixfold coordination between close-pack oxygen layers.

PHYSICAL PROPERTIES. H = 9. Sp. Gr. = 3.98–4.10. Hand sample color is highly variable, usually gray-blue or red, less commonly brown, yellow, green, colorless, and so on. Luster is vitreous to adamantine, sometimes pearly on basal planes.

COLOR AND PLEOCHROISM. Thin sections and fragments are commonly colorless, however, pale colors may be present, usually pale blue or red. Color is often nonuniform as irregular patches or hexagonal zones.

Pleochroism is usually weak $O > E$. Sapphire may show O = deep to pale blue, E = blue-green to yellow-green, and ruby is usually pleochroic from O = red-purple to E = pale yellow.

FORM. Well-formed crystals are most common, often rather large, showing hexagonal and rectangular sections. Crystals are commonly bounded by hexagonal prisms or steep hexagonal dipyramids yielding barrel-shaped forms. Tabular $\{0001\}$ and rhombohedral forms are less common.

CLEAVAGE. Corundum shows no cleavage; however, parting is usually well developed on basal $\{0001\}$ planes and commonly on rhombohedral $\{10\bar{1}1\}$ planes (Fig. 1-9). Parting may well control fragment orientation.

BIREFRINGENCE. Maximum birefringence, seen in prismatic sections, is weak (0.007 to 0.010), and interference colors are first-order gray and white to pale yellow, in standard thin sections. Corundum sections may be thicker than sections of associated minerals due to the extreme hardness of corundum.

FIGURE 1-9. Corundum crystals, shown here with spinel in an emery rock, show prismatic sections of high relief and rhombohedral partings.

TWINNING. Multiple lamellar twinning on rhombohedral planes $\{10\bar{1}1\}$ is very common. Simple contact twins on $\{10\bar{1}1\}$ or $\{0001\}$ are less common. Glide twinning supplies basal and rhombohedral parting planes.

ZONING. Color zoning is common on hexagonal prism or pyramidal planes or as irregular patches. Zones of tiny mineral inclusions (for example, hematite, rutile, zircon, garnet, and spinel) are common, and needle crystals of rutile or elongated liquid inclusions may be oriented parallel to hexagon diagonals in basal sections to yield the asterism of star sapphire and star ruby gems.

INTERFERENCE FIGURE. Corundum is uniaxial negative, and basal sections yield a broad isogyre cross on a white field. Anomolous biaxial figures with $2V_x$ as large as 58° are reported and are possibly the result of twinning.

OPTICAL ORIENTATION. Since the optic axis parallels *c* and the extraordinary ray is fast, elongated crystals show parallel extinction and negative elongation (length-fast). Extinction is also parallel to basal parting and symmetrical to rhombohedral parting planes. Basal parting may produce lineation parallel to the slow wave.

DISTINGUISHING FEATURES. Extreme relief and weak birefringence with possible parting and lamellar rhombohedral twinning are the most diagnostic features. Irregular coloration, extreme hardness, and high specific gravity are also distinctive. Sapphirine is biaxial and untwinned, and apatite has lower indices of refraction and lower birefringence.

ALTERATION. Corundum alters readily to fine-grained, Al-rich minerals, especially margarite or muscovite; less commonly, to diaspore, gibbsite, zoisite, andalusite, kyanite, sillimanite, spinel, or chloritoid.

OCCURRENCE. Corundum appears in a remarkably wide variety of rock types. It is characteristic only of Al-rich, Si-poor environments, however, and appears in association with other aluminous minerals and silica-deficient minerals.

Plutonic igneous rocks that are light colored (Fe-Mg-poor) and without quartz (have no free silica) may contain corundum. Syenites, nepheline syenites, and monzonite may bear corundum in association with andesine-oligoclase, orthoclase, nepheline, and scapolite. Quartz-free pegmatites, perhaps the result of desilication of normal acidic magmas invading ultramatic igneous rocks or carbonate sediments, commonly contain corundum in association with orthoclase, muscovite, Na-plagioclase, andalusite, tourmaline, and so on.

Volcanic igneous rocks rarely contain corundum; however, it has been reported in a few basalts. Andesite dikes that cut crystalline limestones are also known to yield corundum.

Metamorphic rocks, both regional and contact, may yield corundum, especially those derived from aluminous or carbonate sediments. In mica or chlorite schists, corundum may appear with kyanite, sillimanite, dumortierite, and other aluminous minerals. Corundum is also known in hornfels and shale xenoliths resulting from contact metasomatism. Emery deposits are commonly formed in contact zones between igneous rocks and limestone and contain some combination of magnetite, spinel, garnet, hoegbomite, and hematite in addition to fine-grained corundum. Much gem corundum occurs as individual isolated crystals in crystalline limestone as the result of metasomatism. Spinel, rutile, chondrodite, hornblende, or phlogopite may form with corundum in crystalline limestone.

Corundum grains are locally abundant in the heavy fraction of detrital deposits.

REFERENCES: CORUNDUM

Brandt, J. W. 1946. Corundum "indicator" basic rocks and associated pegmatites in northern Transvaal. *Trans. Geol. Soc. South Africa*, *49*, 51–102.

Carlson, H. D. 1957. Origin of the corundum deposits of Renfrew County, Ontario, Canada. *Bull. Geol. Soc. Amer.*, *68*, 1605–1636.

Harder, H. 1968. Color-imparting trace elements in natural corundums. *Neues Jahrb. Mineral. Abh.*, *110*, 128–141.

Jeppesen, M. A. 1958. Some optical, thermo-optical and piezo-optical properties of synthetic sapphire. *Jour. Opt. Soc. Amer.*, *48*, 629–632.

Neuberger, M. 1965. Optical properties and thermal conductivity of aluminum oxide. NASA Accession No. N65–34038, Rept. No. AD 464823. NASA Contract AF 33 (615)-2460. Hughes Aircraft Co., Culver City, Calif.

Pauling, L., and S. B. Hendricks 1925. The crystal structures of hematite and corundum. *Jour. Amer. Chem. Soc.*, *47*, 781–790.

Piriou, B. 1964. Optical constants of corundum and magnesium oxide. *C. R. Acad. Sci. Paris*, *259*, 1052–1055.

Tait, A. S. 1955. Asterism in corundum. *Jour. Gemmology*, *5*, 65.

Webster, R. 1957. Ruby and sapphire. *Jour. Gemmology*, *6*, 101.

Wells, A. J. 1956. Corundum from Ceylon. *Geol. Mag.*, *93*, 25–31.

Hematite

Fe_2O_3

TRIGONAL

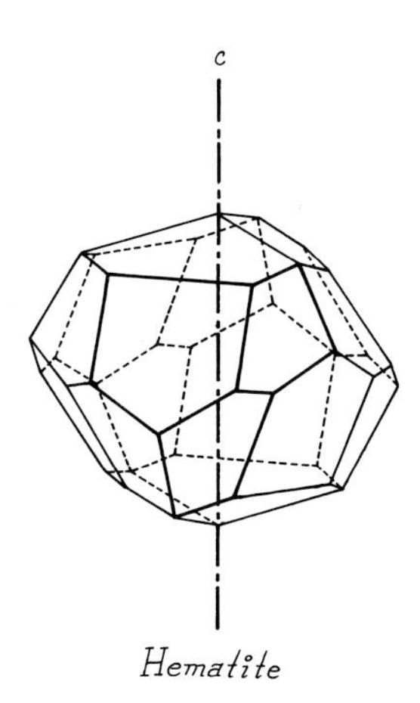

$n_\omega = 3.15–3.22$
$n_\varepsilon = 2.87–2.94$
$n_\omega - n_\varepsilon \simeq 0.28$
Uniaxial negative
Extreme dispersion*
Opaque to deep blood-red in section and pleochroic $O > E$.

COMPOSITION AND STRUCTURE. Hematite consists of oxygen ions in essentially hexagonal closest packing, with Fe^{3+} ions in two-thirds of the interlayer, octahedral sites. It is analogous to the structure of corundum (see page 25).

Most hematite is relatively pure Fe_2O_3 with only minor Fe^{2+}, Ti, Al, and Mn. Solid solution is very limited between hematite and magnetite (Fe_3O_4), ilmenite ($FeTiO_3$), corundum (Al_2O_3) or bixbyite [$(Mn,Fe)_2O_3$].

PHYSICAL PROPERTIES. H = 5–6. Sp. Gr. = 5.26. Color, in hand sample, is steel-gray to bright red or red-brown. Luster is bright metallic (specular hematite) or submetallic to dull. Hematite gives a red-brown to cherry-red streak.

COLOR AND PLEOCHROISM. Hematite is usually opaque except near thin edges, where it is deep red-brown. Thin plates of the specularite variety are deep blood-red and pleochroic. O = deep red-brown, E = yellow-brown.

FORM. Hematite shows extreme variety of form. Well-crystallized forms, called *specularite*, tend to form flat trigonal crystals, and basal parting may produce micaceous specularite as small scales or flakes. Reniform (kidney ore) or botryoidal forms are common as dehydrated goethite and break down to fibers or splinters of hematite. Oolitic and fossiliferous hematite is probably due to carbonate replacement. Hematite may also be massive and structureless or soft and earthy (red ochre).

CLEAVAGE. Hematite shows no cleavage, but parting may be very pronounced on $\{0001\}$ or $\{10\bar{1}1\}$.

BIREFRINGENCE. Maximum birefringence is very extreme (0.28) and very weak, high-order interference colors are completely masked by red-brown mineral color.

TWINNING. Lamellar twinning on $\{0001\}$ and $\{10\bar{1}1\}$ is common and results in parting.

INTERFERENCE FIGURE. Hematite is uniaxial negative, but worthwhile interference figures are rarely seen, due to deep color and extreme dispersion and birefringence.

OPTICAL ORIENTATION. Since hematite is uniaxial negative, the fast wave vibrates parallel to c. Basal parting should show parallel extinction and produce positive (length-slow) elongation. Hematite may show no extinction with stage rotation, due to extreme dispersion, and elongation is difficult to determine, due to extreme birefringence.

DISTINGUISHING FEATURES. Hematite is distinguished from most opaque minerals by red internal reflections in reflected light and blood-red color as thin plates or edges in transmitted light. Hydrated iron oxides tend to be yellow-brown in transmitted light.

ALTERATION. Hematite itself is the alteration product of many Fe-bearing minerals, especially, magnetite, siderite, and pyrite.

---

* As hematite is uniaxial, dispersion fringes do not appear in interference figures; however, due to extreme dispersion, hematite grains may show no extinction with stage rotation.

It may even replace noniron-bearing minerals (such as galena, cuprite, barite, and aragonite). Hematite is sometimes replaced by, or altered to, magnetite, limonite, siderite, or pyrite.

OCCURRENCE. Sedimentary deposits of hematite may be huge and are the major source of iron. Hematite is derived largely from the weathering of Fe-bearing minerals and is precipitated in seas and lakes by chemical or organic processes. It is commonly associated with chert or quartz as alternating bands and intermixed layers of chert and hematite, especially in Precambrian sediments that have been mildly metamorphosed. Hematite is exceedingly common as surface stains and intergranular cement in detrital sediments and other rocks, accounting for red coloration, and as the dehydration product of geothite.

Igneous rocks, especially granite, rhyolite, and similar rocks, may contain primary, accessory grains of hematite, suggesting oxygen-rich magmas. Specularite is commonly deposited as euhedral crystals or plates with quartz, in veins and cavities by volcanic gases, and in quartz veins by high-temperature hydrothermal solutions. Lattice-type exsolution intergrowths of ilmenite and hematite are very common, and flakes of hematite are common inclusions in quartz, feldspar, pyroxene, and many other mineral varieties.

Metamorphic rocks contain hematite in metamorphosed ferruginous sediments and in metasomatic contact deposits with magnetite.

REFERENCES: HEMATITE

Blake, R. L., R. E. Hessevick, T. Zoltai, and L. W. Finger. 1966. Refinement of the hematite structure. *Amer. Mineral.*, *51*, 123–129.

Bose, M. K. 1958. Goethite-hematite relation—an ore microscope observation. *Amer. Mineral.*, *43*, 989.

Edwards, A. B. 1949. Natural ex-solution intergrowths of magnetite and hematite. *Amer. Mineral.*, *34*, 759.

Saito, T. 1965. The anomalous thermal expansion of hematite at high temperature. *Bull. Chem. Soc. Japan*, *38*, 2008–2009.

Von Gehlen, K., and H. Piller. 1965. Optical properties of hematite and ilmentite. *Neues Jahrb. Mineral., Monatsh.*, *1965*, 97–108.

Willis, B. T. M., and H. P. Rooksby. 1952. Crystal structure and antiferromagnetism of haematite. *Proc. Physical. Soc.* London, Sect. B., *65*, 950–954.

Ilmenite $FeTiO_3$ TRIGONAL

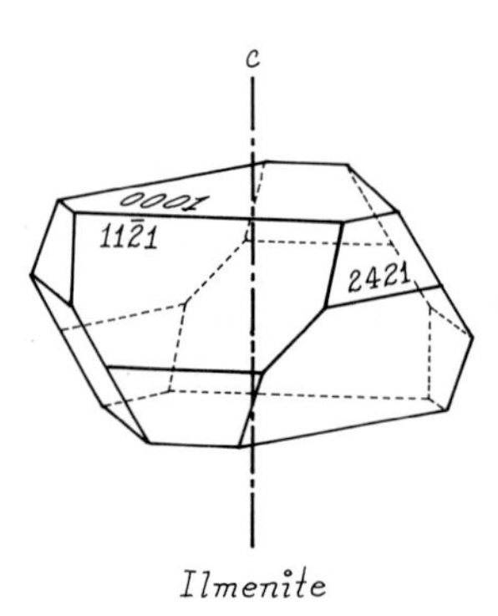

Ilmenite

	n_ω	n_ε
Ilmenite	$\simeq$ 2.7	(essentially opaque)
Geikielite	2.31	1.95
Pyrophanite	2.48	2.21

$n_\omega - n_\varepsilon$ = very large
Uniaxial negative
Opaque in standard section

COMPOSITION AND STRUCTURE. Ilmenite ($FeTiO_3$) has the same structure as corundum, Al_2O_3, or hematite, Fe_2O_3; that is, hexagonal-close-packed oxygen ions with metal cations in 2/3 of the available interlayer sites. One Fe^{2+} and one Ti^{4+}

replace two Al^{3+} or Fe^{3+} to form ilmenite, and symmetry drops to $\bar{3}$ from $\bar{3}$ $2/m$. A magnesium isomorph, geikielite, $MgTiO_3$, and a manganese isomorph, pyrophanite, $MnTiO_3$, are known. Ilmenite, $FeTiO_3$, commonly contains significant MgO and/or MnO and probably forms complete solid solution with both magnesium and manganese end members. Ilmenite usually contains a few weight percent Fe_2O_3 in partial solid solution above 1050°C. At lower temperatures, exsolution occurs, and ilmenite commonly shows lattice-type intergrowth, with hematite on the basal planes {0001} and with magnetite on the {111} magnetite planes

PHYSICAL PROPERTIES. H = 5–6. Sp. Gr. = 4.70–4.79 (ilmenite = 4.79, geikielite = 4.05, and pyrophanite = 4.54). Color in hand sample is iron-black. Streak is black or reddish, if rich in Mg-Mn.

COLOR AND PLEOCHROISM. In standard thin section, ilmenite is totally opaque. Very thin fragments may show deep red on thin edges. Geikielite, $MgTiO_3$, and pyrophanite, $MnTiO_3$, are somewhat more transparent and show reddish-violet and red-orange respectively on thin edges. Ilmenite shows whitish-gray in reflected light.

FORM. Ilmenite commonly occurs as thick tabular crystals, as anhedral masses imbedded in igneous or metamorphic rocks, or as opaque sand grains. In thin section, ilmenite grains are anhedral or lamellar plates with long rectangular outline. Ilmenite also commonly shows skeletal forms in thin section.

CLEAVAGE. None; however, ilmenite may show distinct parting on either {0001} or $\{10\bar{1}1\}$ due to twinning. Geikielite shows good rhombohedral cleavage $\{10\bar{1}1\}$ and pyrophanite shows cleavage perfect on $\{02\bar{2}1\}$ and good on $\{10\bar{1}2\}$.

BIREFRINGENCE. Only very thin edges transmit any light. Only red light is transmitted with very large birefringence.

TWINNING. Ilmenite may show simple twinning on or repeated lamellar twinning on $\{10\bar{1}1\}$.

DISTINGUISHING FEATURES. Ilmenite is an opaque mineral that is white to gray in reflected light. Reflected light may also reveal partial alteration to a whitish aggregate called *leucoxene*, which is distinctive. Ilmenite is most likely to be confused with magnetite, which is strongly magnetic and isometric, commonly showing squares, triangles, diamonds, and other octahedron sections, in contrast to the long rectangular or skeletal sections of ilmenite. Graphite is an opaque mineral that is soft and of limited occurrence.

ALTERATION. The alteration product of ilmenite is called *leucoxene* and is a fine-grained aggregate of one or more titanium oxides. Rutile is most common, but leucoxene may contain only anatase or brookite or may represent a mixture with rutile. In thin section, leucoxene appears opaque, or nearly so, but in reflected light, it is white to grayish, yellowish, or red-brown, due to the presence of various ferric oxides.

OCCURRENCE. Although some anorthosite rocks may contain very large amounts of ilmenite, it is most common as an opaque accessory in a wide range of igneous and metamorphic rocks, especially gabbroic varieties, where it is associated more with pyroxenes than olivines.

Ilmenite is quite common in ore veins, where it appears most commonly with chalcopyrite and pyrrhotite. It is also a constituent of some pegmatites. Sandy sediments often contain weather-resistant ilmenite grains, and some black beach sands are largely ilmenite. Ilmenite is often intimately associated with hematite or magnetite as lattice-type intergrowths.

REFERENCES: ILMENITE

Gruner, J. W. 1959. The decomposition of ilmenite. *Econ. Geol.*, *54*, 1315.

Karkhanavala, M. D., and A. C. Momin. 1959. The alteration of ilmenite. *Econ. Geol.*, *54*, 1095.

Shvetsova, I. V. 1970. Mixed rutite-anatase leucoxene. *Dokl. Acad. Sci. USSR, Earth Sci. Sect.*, *194*, 130–133.

Wallace, R. M. 1953. A proposed petrographic method for the rapid determination of ilmenite. *Amer. Mineral.*, *38*, 729–730.

Perovskite

$CaTiO_3$

PSEUDOISOMETRIC (MONOCLINIC OR ORTHORHOMBIC)

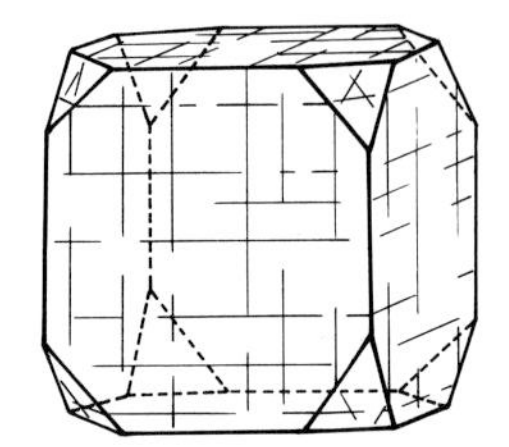

$n = 2.30–2.38$
Very weak birefringence
Colorless, pale yellow-brown to deep brown
Weak pleochroism

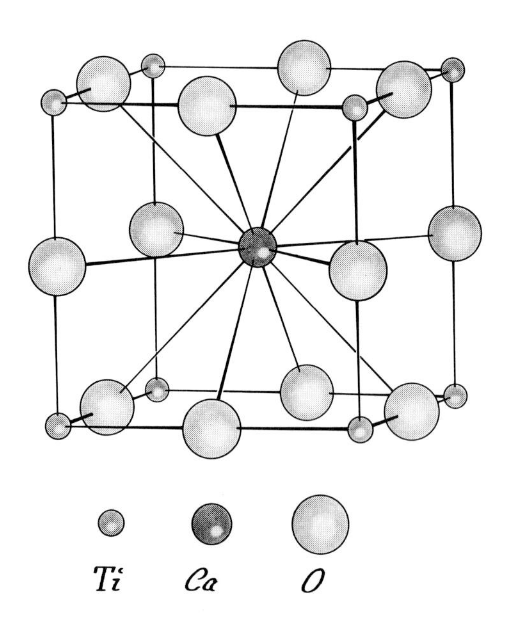

FIGURE 1-10. The structure of perovskite. The idealized structure is essentially a cubic-close-packed configuration of Ca^{2+} and O^{2-} ions with Ti^{4+} in some octahedral sites between close-packed layers.

COMPOSITION AND STRUCTURE. The idealized structure of perovskite is simple isometric with small cations in sixfold, octahedral coordination and with large cations in twelvefold, close pack, coordination (Fig. 1-10). As Ca^{2+} is small for twelvefold coordination with oxygen, the structure is deformed slightly to probably monoclinic or orthorhombic symmetry.

Common perovskite is essentially $CaTiO_3$; however, numerous single and coupled substitutions are possible, and varieties rich in cerium earths (knopite), niobium (dysanalyte), and niobium and rare earths (loparite) have been described. These varieties tend to be alkali-rich through coupled substitution of Na^+ and Ce^{5+} or Nb^{5+} for Ca^{2+} and Ti^{4+}, and their structure varies even further from ideal isometric. Numerous other elements may be present in significant amounts, including Th, La-earths, U, Cr, Fe, Zr, and Al.

PHYSICAL PROPERTIES. $H = 5\frac{1}{2}$. Sp. Gr. = 3.98–4.26. Color, in hand sample, is yellow, red-brown to dark brown, or black. Luster is adamantine to metallic.

COLOR AND PLEOCHROISM. Perovskite is colorless, pale yellow, or pale red-brown to dark brown as fragments or in thin section. Pleochroism is weak $Z > X$.

FORM. Perovskite occurs most frequently as tiny crystals of either cubic or octahedral habit (Fig. 1-11).

CLEAVAGE. Poor cubic cleavage {100} is visible only in large crystals.

BIREFRINGENCE. Small crystals appear isotropic; large crystals and thick sections of especially rare-earth varieties may show very weak birefringence (dark first-order gray), which is obvious only because of ubiquitous polysynthetic twinning.

TWINNING. Birefrigent forms invariably show complex lamellar twinning on {111}.

ZONING. Color zoning of reddish core and yellowish periphery is reported.

INTERFERENCE FIGURE. As perovskite is essentially isotropic, with very weak birefringence, complex lamellar twinning, extreme relief, and, commonly, deep color, interference figures

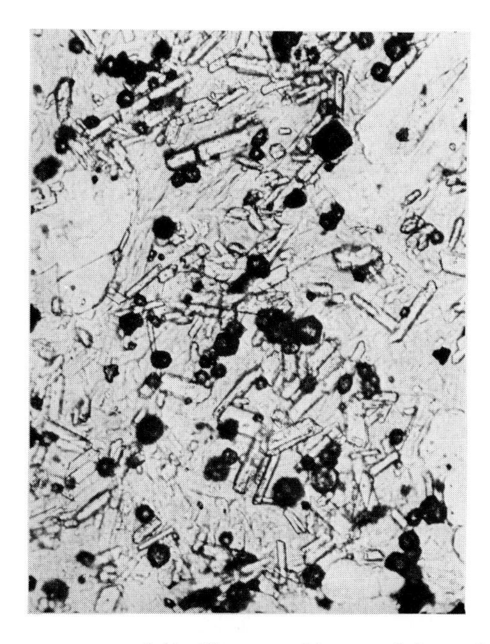

FIGURE 1-11. Tiny perovskite crystals in an altered peridotite show isometric forms and very high relief; they are pale yellow and isotropic.

are largely impractical. Perovskite is reported, however, to be biaxial positive, with $2V_z$ ranging from small to essentially 90° and optic axis dispersion $r > v$.

OPTICAL ORIENTATION. If perovskite is considered to be orthorhombic, the optic plane is said to be {001} and $a = X, b = Z, c = Y$; if monoclinic, the optic plane is {010} and $b = Y$, $c \wedge Z = 45°$.

DISTINGUISHING FEATURES. Small, euhedral crystals (cubes or octahedrons) with extreme relief, yellow or brownish color, and possibly very weak birefringence and lamellar twinning are diagnostic. Rutile and its polymorphs show extreme birefringence, and brown spinels and garnet have lower relief and different crystal habit.

ALTERATION. Perovskite itself is a possible alteration product of sphene or ilmenite. It may alter to leucoxene.

OCCURRENCE. Perovskite is most common as tiny accessory crystals of either magmatic or deuteric origin in silica-deficient igneous rocks. In highly alkaline rocks (such as nepheline syenites or ijolites), perovskite forms in association with nepheline or melilite. Leucite-bearing volcanic rocks may also contain perovskite. Tiny crystals of perovskite are common in peridotites, olivinites, and pyroxenites and are reported in basic pegmatites and carbonatites.

Metamorphism of impure carbonate sediments, especially in contact with alkaline intrusions, may yield perovskite crystals in contact marbles with melilite, spinel, larnite, and wollastonite. Rarely in talc or chlorite schists.

REFERENCES: PEROVSKITE

Bonshtedt-Kupletskaya, E. M. 1946. Certain data on minerals of the perovskite group. *Problems in Mineralogy, Geochemistry and Petrography.* Fersman memorial volume. Moscow and Leningrad (Acad. Sci. USSR), *43*, 53–59.

Megaw, H. D. 1946. Crystal structure of double oxides of the perovskite type. *Proc. Physical Soc. London*, *58*, 133–152.

Murdock, J. 1951. Perovskite. *Amer. Mineral.*, *36*, 573–580.

Smith, A. L. 1970. Sphene, perovskite, and coexisting Fe-Ti oxide minerals. *Amer. Mineral.*, *55*, 264–269.

Turi, B. 1968. Chemical analysis of perovskite. *Period. Mineral.*, *37*, 567–575.

THE SPINEL GROUP

The spinel group is represented by at least a dozen minerals and many more artificial compounds, all conforming to the generalized formula $A^{2+}B_2{}^{3+}O_4$ and the structural pattern shown in Fig. 1-12. Three series are defined by the dominant trivalent ion:

Spinel series: Al^{3+} is the dominant trivalent ion.

Magnetite series: Fe^{3+} is the dominant trivalent ion.

Chromite series: Cr^{3+} is the dominant trivalent ion.

Compositional end members within each series are further defined by the dominant bivalent ion:

Spinel series	
Common spinel	$MgAl_2O_4$
Hercynite	$FeAl_2O_4$
Gahnite	$ZnAl_2O_4$
Galaxite	$MnAl_2O_4$
Magnetite series	
Magnetite	$Fe^{2+}Fe_2{}^{3+}O_4$
Magnesioferrite	$MgFe_2{}^{3+}O_4$
Franklinite	$ZnFe_2{}^{3+}O_4$
Jacobsite	$MnFe_2{}^{3+}O_4$
Trevorite	$NiFe_2{}^{3+}O_4$
Chromite series	
Chromite	$Fe^{2+}Cr_2O_4$
Magnesiochromite	$MgCr_2O_4$

Ideal end-member compositions are rare in nature, and almost complete solid miscibility is possible within each series; solid solution between series, however, is considerably more restricted.

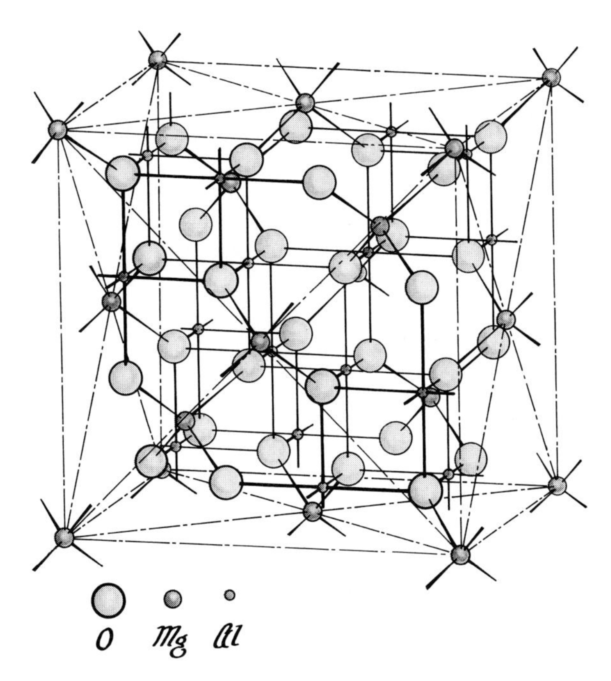

FIGURE 1-12. The structure of spinel. The spinel structure is essentially a cubic close-packed oxygen arrangement with two different cation types in sixfold (octahedral) and fourfold (tetrahedral) coordination between the {111} close-packed layers. Twice as many octahedral sites are occupied as tetrahedral sites, and in true spinels small trivalent cations (such as Al^{3+}) occupy large octahedral sites while large bivalent cations (such as Mg^{2+}) occupy small tetrahedral sites. In antispinel, large cations occupy large octahedral sites, and in some spinels (such as magnetite) one cation type (such as Fe^{3+}) may be distributed between octahedral and tetrahedral sites.

Two structural modifications are recognized in the spinel group. In the normal structure (Fig. 1-12), all bivalent ions (A^{2+}) appear in tetrahedral (fourfold) coordination with oxygen and all trivalent ions (B^{3+}) in octahedral (sixfold) coordination, even though the bivalent ions are larger. All members of the spinel and chromite series are normal. The inverse structure is characterized by all bivalent ions (A^{2+}) in octahedral coordination and the trivalent ions (B^{3+}) distributed evenly between tetrahedral and remaining octahedral sites. Magnetite has the following inverse spinel structure: One-half the Fe^{3+} ions are in tetrahedral sites and one-half in octahedral sites alternating with an equal number of Fe^{2+} ions. As the cations in tetrahedral coordination and those in octahedral coordination have antiparallel magnetic moments, magnetite displays ferrimagnetic properties.

The Spinel Series — $(Mg,Fe^{2+},Zn,Mn)Al_2O_4$ — ISOMETRIC

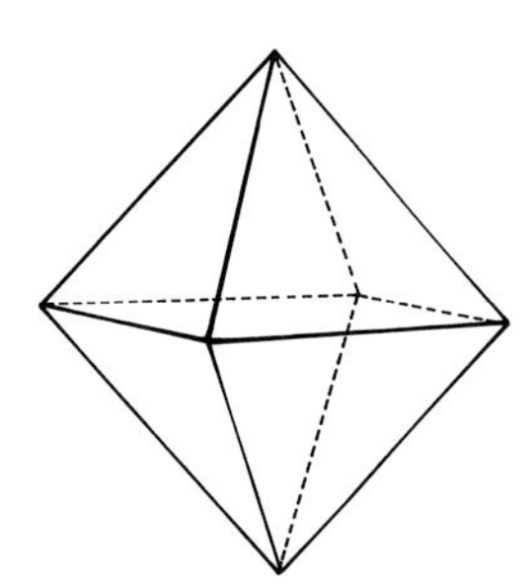

Spinel

		n*	
Spinel	$MgAl_2O_4$	1.719	red, blue, green, etc.
Hercynite	$FeAl_2O_4$	1.835	green
Gahnite	$ZnAl_2O_4$	1.805	blue-green, yellow
Galaxite	$MnAl_2O_4$	1.920	red-brown

COMPOSITION. Common spinel, $MgAl_2O_4$, forms complete solid solution with hercynite, $FeAl_2O_4$; gahnite, $ZnAl_2O_4$; magnesiochromite, $MgCr_2O_4$; and probably galaxite, $MnAl_2O_4$. Spinel containing appreciable Fe^{2+} (Mg:Fe = 1–3) is called *pleonast* or *ceylonite*, and spinel containing both Fe^{2+} and Cr^{3+} is called *mitchellite*. Hercynite, $FeAl_2O_4$, forms complete solid solution with spinel, $MgAl_2O_4$, and chromite, $FeCr_2O_4$, but only limited solid solution with magnetite, $Fe^{2+}Fe_2{}^{3+}O_4$ (exsolution intergrowths of hercynite and magnetite are known). *Picotite* is Cr-rich hercynite. Gahnite, $ZnAl_2O_4$, forms complete solid solution with spinel, $MgAl_2O_4$, and possibly galaxite, $MnAl_2O_4$, with *dysluite* (Mn-rich gahnite) as an intermediate member. Galaxite is rare but may contain appreciable Fe^{2+} and Fe^{3+}. Ti^{4+} is moderately common in many spinels.

PHYSICAL PROPERTIES. H = $7\frac{1}{2}$–8. Sp. Gr. = 3.55–4.04. Color in hand sample is highly variable: spinel is colorless, red, blue, yellow, green, and so on; hercynite is dark green to black; gahnite is dark blue-green, brown or yellow; and galaxite is deep red-brown to black. Luster is vitreous to submetallic.

COLOR AND PLEOCHROISM. In transmitted light, spinel is colorless or pale to deep colors (red, blue, green, yellow, and so on), hercynite is deep green, gahnite is pale to deep blue-green or yellow-brown, and galaxite is deep red-brown. Although spinels are isometric, anomalous pleochroism is sometimes observed.

FORM. Spinels appear largely as independent grains and granular groups, anhedral to euhedral. Euhedral crystals are common as octahedrons, yielding square to triangular sections (Fig. 1-13).

CLEAVAGE. Although spinels have no cleavage, parting is sometimes prominent on {111}.

TWINNING. Twinning is common on {111} (spinel law) usually simple, sometimes multiple.

ZONING. Composition zoning is rarely apparent.

DISTINGUISHING FEATURES. The spinel series is isotropic, showing extreme relief and commonly deep color and euhedral outline. Other series in the spinel group are essentially opaque. Pink and brown spinels are easily mistaken for garnet, but garnets do not show octahedral sections.

ALTERATION. Members of the spinel group are relatively stable under conditions of ordinary weathering. Spinel is known, however, to alter to talc, mica, serpentine, chlorite (amesite), corundum, hydrotalcite, and diaspore. Gahnite may alter to sphalerite, chlorite, or muscovite. Spinel itself may form by the alteration of sillimanite, corundum, forsterite, and chlorite.

OCCURRENCE. The minerals of the spinel series are high-temperature minerals characteristic of high-grade metamorphism of aluminous rocks.

Common spinel occurs in schists and gneisses with sillimanite, talc, garnet, chondrodite, phlogopite, corundum, and so on; in marble and crystalline limestone, contact or regional, with chondrodite, cordierite, forsterite, and enstatite;

* Indices of refraction vary in almost straight-line relationship between any two ideal end members.

FIGURE 1-13. Spinel crystals are isotropic, with high relief, and may show isometric sections. Note the octahedral parting.

and, as pleonast, in emery deposits with magnetite, corundum, and hematite. Spinel is incompatible with quartz and occurs with it only when surrounded by a reaction rim.

Spinel occurs as accessory grains in mafic and subsilicic igneous rocks and is developed in aluminous xenoliths. It is a common detrital mineral where spinel-bearing rocks are weathered, and it may appear in gem gravels with corundum, zircon, diamond, and garnet.

Hercynite occurs largely in metamorphic rocks derived from iron-rich, argillaceous sediments (for example, laterites), in granitic granulites (hercynite can form in the presence of free silica), and, rarely, in ultramafic igneous rocks intergrown with magnetite.

Gahnite is found chiefly in granite pegmatites; less commonly, in contact carbonate zones and metasomatic zinc ores.

Galaxite occurs in Mn-vein deposits with rhodonite, spessartite, tephroite, and so on.

REFERENCES: THE SPINEL SERIES

Hoekstra, K. E. 1969. Crystal chemical considerations of the spinel structural group. Unpublished doctoral dissertation. Ohio State University, Columbus, Ohio.

Okamoto, S. 1971. Crystal chemistry of spinel compounds. *Kagaku Kogyo*, *22*, 814–824.

Ponomareva, M. N., N. V. Pavlov, and I. I. Chuprynina. 1964. Determination of the composition of some mineral species of chrome-spinelides by their refractive indexes. *Geol. Rudn. Mestorozhd.*, *6*, 103–106.

Satoh, T., R. Tsushima, and K. Kudo. 1974. Classification of normal spinel-type compounds by ionic packing factor. *Mater. Res. Bull.*, *9*, 1297–1300.

Zhelyazkova-Panaiotova, M. 1971. Rational classification of the spinel group from ultrabasic rocks. *Int. Mineral. Assoc.*, *Proc. 7th Gen. Mtg.*, *1970*, 174–179.

The Chromite Series $(Fe^{2+},Mg)Cr_2O_4$ ISOMETRIC

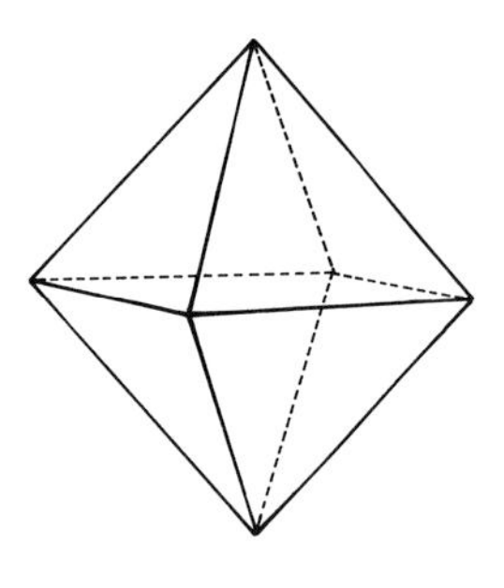

Chromite

Chromite	$FeCr_2O_4$
Magnesiochromite	$MgCr_2O_4$

Essentially opaque in sections or as fragments

Light gray (slightly brownish) in reflected light
Thin edges of fragments are dark brown in transmitted light ($n \simeq 2.1$)

COMPOSITION AND STRUCTURE. Spinel minerals in which Cr^{3+} is the dominant trivalent ion constitute the chromite series. Bivalent ions are largely Fe^{2+} or Mg^{2+}, and complete solid solution exists between $FeCr_2O_4$ and $MgCr_2O_4$. Ferrous members are called *chromite* (that is, $Fe^{2+} > Mg$), and magnesian members are *magnesiochromite.* Natural minerals always contain appreciable Mg, and magnesiochromite may well be more common than chromite. Minor Zn^{2+} or Mn^{2+} may replace Mg-Fe. Significant Al^{3+} or Fe^{3+} may replace Cr^{3+}, and at high temperatures (>510°C), complete solution may exist to spinel-hercynite, $(Mg,Fe)Al_2O_4$.

PHYSICAL PROPERTIES. $H = 5\frac{1}{2}$ (chromite) to 7. Sp. Gr. = 5.1 (chromite), 4.2 (magnesiochromite). In hand sample, chromite and magnesiochromite are iron-black and yield a brown streak. Luster is submetallic to metallic, and magnetism is weak.

COLOR AND PLEOCHROISM. Minerals of the chromite series are essentially opaque in section or as fragments but may be yellow-brown or deep brown and nonpleochroic on thin edges. They are pale gray or brownish gray in reflected light.

FORM. Chromite and magnesiochromite are usually massive or coarse to fine granular. Euhedral crystals are rare; octahedrons sometimes modified by cube faces.

CLEAVAGE. Cleavage is not present; however, parting on {111} should be possible.

TWINNING AND ZONING. Twinning on {111} (spinel law) and exsolution lamellae of magnetite or hematite on {111} planes are possible.

DISTINGUISHING FEATURES. Chromite and magnesiochromite are opaque minerals that show gray in reflected light and exhibit very restricted occurrence. Thin fragment edges are brown in transmitted light. Magnetite, pyrite, and marcasite commonly show euhedral outline, and the latter two are pale brass-yellow in reflected light. Ilmenite may show whitish leucoxene alteration in reflected light.

ALTERATION. Minerals of the chromite series alter to limonite; stichtite $Mg_6Cr_2(OH)_{16}CO_3 \cdot 4H_2O$; maghemite, γ-Fe_2O_3; and chlorite (amesite or kämmererite).

OCCURRENCE. Chromite-magnesiochromite is essentially confined to ultramafic plutonic igneous rocks and serpentine rocks derived from them. Most peridotites and dunites contain some chromite as an accessory mineral, often in layers or bands as the result of crystal fractionation. The most common composition is probably Fe-rich magnesiochromite.

Chromite is a common heavy detrital mineral in sediments and is well known in meteorites.

REFERENCES: THE CHROMITE SERIES

Den Tex, E. 1955. Secondary alteration of chromite. *Amer. Mineral., 40,* 353–355.

DeWaal, S. A., and I. Copelowitz. 1972. Interdependence of the physical properties and chemical compositions of chrome spinels from the Bushveld igneous complex. *Int. Geol. Congr., 24th Rep. Sess., 14,* 171–179.

Fisher, L. W. 1929. Chromite: Its mineral and chemical composition. *Amer. Mineral., 14,* 341–357.

Ponomareva, M. N., N. V. Pavlov, and I. I. Chuprynina. 1964. Determination of the composition of some mineral species of chrome-spinelides by their refractive indexes. *Geol. Rudn. Mestorozhd.*, *6*, 103–106.
Satch, T., R. Tsushima, and K. Kudo. 1974. Classification of normal spinel-type compounds by ionic packing factor. *Mater. Res. Bull.*, *9*, 1297–1300.
Stevens, R. E. 1944. Composition of some chromites of the Western Hemisphere. *Amer. Mineral.*, *29*, 1–34.
Winchell, A. N. 1941. The spinel group. *Amer. Mineral.*, *26*, 422.
Zhelyazkova-Panaiotova, M. 1971. Rational classification of the spinel group from ultrabasic rocks. *Int. Mineral. Assoc., Proc. 7th Gen. Mtg., 1970*, 174–179.

The Magnetite Series

$(Fe^{2+},Mg,Zn,Mn,Ni)Fe_2{}^{3+}O_4$ ISOMETRIC

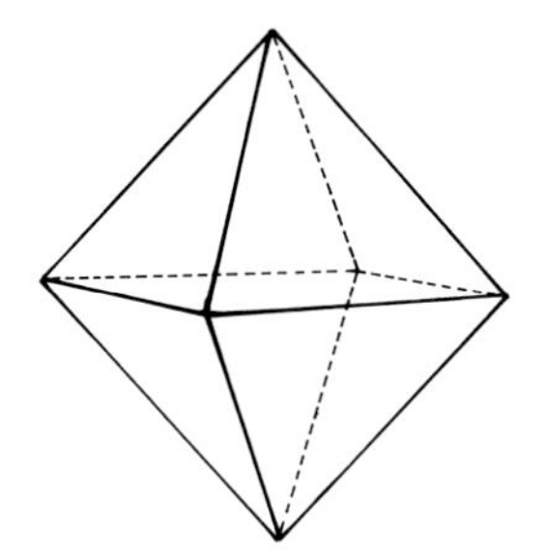

Magnetite

Magnetite	$Fe^{2+}Fe_2{}^{3+}O_4$
Magnesioferrite	$MgFe_2{}^{3+}O_4$
Franklinite	$ZnFe_2{}^{3+}O_4$
Jacobsite	$MnFe_2{}^{3+}O_4$
Trevorite	$NiFe_2{}^{3+}O_4$

Opaque and gray in reflected light

COMPOSITION AND STRUCTURE. Members of the magnetite series have an inverted spinel structure (Fig. 1-12) in which the bivalent cations are some combination of Fe^{2+}, Mg, Zn, Mn, and Ni. Magnetite is by far the most common member of the series. Magnetite, $Fe^{2+}Fe_2{}^{3+}O_4$, forms continuous solid solution with magnesioferrite, $MgFe_2{}^{3+}O_4$, but most natural magnetite contains little Mg. Minor Mn or Ca and lesser Ni, Co, or Zn may replace Fe^{2+} in magnetite, and minor Al, V, or Cr may replace Fe^{3+}. Magnetite may contain major amounts of Ti^{4+} and may form complete solid solution with ulvöspinel, $Fe_2{}^{2+}TiO_4$; it may also contain excess Fe^{2+} and form partial solid solution with maghemite, γ-Fe_2O_3. Maghemite is unstable and inverts to hematite, α-Fe_2O_3, which may appear as exsolution intergrowth lamellae on the {111} planes in magnetite. Ulvöspinel may also form exsolution intergrowths with magnetite.

Franklinite shows highly variable composition and normally shows considerable substitution of Fe^{2+} and Mn^{2+} for Zn^{2+} and Mn^{3+} for Fe^{3+}.

Jacobsite and trevorite are rare minerals and may show considerable substitution of Mg^{2+} or Fe^{2+} for Mn^{2+} and Mg^{2+} or Fe^{2+} for Ni^{2+}.

PHYSICAL PROPERTIES. $H = 5$–$6\frac{1}{2}$. Sp. Gr. $= 5.2$ (magnetite) to 4.6 (magnesioferrite). In hand sample, the magnetite minerals are iron-black to brownish black. Luster is metallic to submetallic. Magnetite, magnesioferrite, and trevorite are strongly magnetic; franklinite and jacobsite less so.

COLOR AND PLEOCHROISM. All members of the magnetite series are opaque as fragments or in section; they are gray in reflected light.

FORM. Magnetite and other members of the magnetite series commonly occur as euhedral, octahedral crystals (rarely dodecahedral) that yield triangular, square, and trapezoid sections; may also be anhedral granular.

CLEAVAGE. Octahedral parting {111} is rather common, especially in magnetite.

TWINNING. Twinning on {111} is common as simple spinel twins and may be repeated as lamellar twins.

DISTINGUISHING FEATURES. Minerals of the magnetite series are opaque and strongly magnetic, showing gray in reflected light and commonly showing sections of octahedrons. Pyrite shows cube sections and yellowish reflection, hematite is reddish in reflected light, ilmenite commonly shows whitish leucoxene alteration in reflected light, and chromite may transmit deep brown color in thin splinters.

ALTERATION. Magnetite normally alters to hematite, goethite, or limonite but may be replaced as pseudomorphs by pyrite, siderite, chlorite, chalcopyrite, or serpentine; titanian magnetite may also yield sphene, rutile, or anatase. Pseudomorphs of hematite after magnetite are called *martite*. Magnetite, itself, is an alteration product of many iron-bearing minerals; it appears as a finely dispersed by-product in the alteration of ferromagnesian silicates (as when olivine alters to serpentine or talc) or as pseudomorphs after hematite, serpentine, siderite, pyrite, pyrrhotite, and so on.

Franklinite may alter to chalcophanite (Zn,Mn,Fe) $Mn_2O_5 \cdot 2H_2O$ or to a complex aggregate of impure Mn- and Fe-oxides (wad).

OCCURRENCE. Magnetite is one of the most ubiquitous of all minerals, appearing in a vast variety of igneous, metamorphic, and even sedimentary rocks, usually as small disseminated crystals or grains making up less than 1 percent of the rock in which it appears.

Nearly all varieties of plutonic igneous rocks commonly contain magnetite as highly disseminated, small accessory crystals. Magnetite grains may also be concentrated by crystal fractionation to form the major or only constituent of large rock masses. In mafic rock types, magnetite is commonly titaniferous and closely associated with pyroxenes, olivine, Ca-plagioclase, and apatite.

Metamorphic rocks derived from ferruginous sediments in regional or contact environments commonly yield magnetite by the reduction of hematite and ferric-hydroxides. Quartz-magnetite or hematite-magnetite are common associates, and magnetite may also form with chlorite, grunerite, hedenbergite, andradite, or corundum-spinel (emery). Magnetite may form by metasomatic replacement of limestone in association with andradite, hematite, epidote, apatite, diopside, olivine, pyrite, and chalcopyrite. It is sometimes deposited by fumerolic gases and may appear in high-temperature, hydrothermal sulfide veins.

Detrital grains of magnetite commonly are the major constituents of black sands and the heavy fraction of sands and sandstones.

Magnesioferrite is formed by fumerolic action on magnesian rocks; franklinite appears with metasomatic zinc ores (for example, zincite and willemite) at Franklin, New Jersey; jacobsite is known to occur in limestone skarns and trevorite in talc schist or phyllite.

REFERENCES: THE MAGNETITE SERIES

Lepp, H. 1957. Stages in the oxidation of magnetite. *Amer. Mineral.*, *42*, 679–681.

Nemec, D. 1973. Occurrence of zinc spinels in the Bohemian massif. *Tschermaks Mineral. Petrogr. Mitt.*, *19*, 95–109.

Petzold, D. R. 1971. Temperature dependence of cation distribution in inverse spinels. *Krist. Tech.*, *6*, K43–K45.

Shchepetkin, A. A., V. I. Dvinin, and G. I. Chufarov. 1971. X-ray diffraction study of the crystal structure of spinel oxides in the magnesium-iron-titanium-oxygen system. *Kristallografiya*, *16*, 527–531.

Chrysoberyl

$BeAl_2O_4$

ORTHORHOMBIC

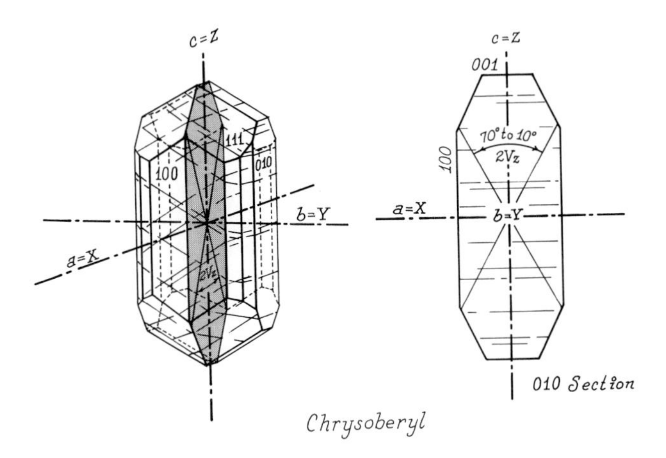

Chrysoberyl

$n_\alpha = 1.732–1.747$*
$n_\beta = 1.734–1.749$
$n_\gamma = 1.741–1.758$
$n_\gamma - n_\alpha = 0.008–0.011$
Biaxial positive, $2V_z = 70°$ to $\sim 10°$*
$a = X, b = Y, c = Z$†
$r > v$ weak
Colorless in standard section

COMPOSITION AND STRUCTURE. The chrysoberyl structure is isomorphous with that of olivine. Chrysoberyl has a rather constant composition ($BeAl_2O_4$); however, minor Fe^{3+} commonly replaces Al^{3+}, accounting for the common greenish color. The highly valued gem variety, alexandrite, results when some Cr^{3+} substitutes for Al^{3+}. Very minor Mg, Ti, and Ca are reported in some analyses.

PHYSICAL PROPERTIES. H = $8\frac{1}{2}$. Sp. Gr. = 3.68–3.75. Color in hand sample is usually yellow to yellow-green, rarely colorless. Alexandrite is yellowish or bluish green in sunlight and red or red-violet in incandescent indoor light. Cat's eye, another gem variety of chrysoberyl, displays chatoyancy.

COLOR AND PLEOCHROISM. Most chrysoberyl is colorless in thin section or as small fragments. Large fragments may be pleochroic: X = slightly rose, Y = greenish yellow, and Z = pale green. Alexandrite shows strong pleochroism: X = red-purple, Y = orange, and Z = green.

FORM. Crystals are short, prismatic, elongated on c, and tabular on {100}. Vertical striations are usually present on {100} and rarely on {010}. Twinned crystals are very common as triangular or hexagonal shapes. Chrysoberyl may also be massive or granular.

CLEAVAGE. Cleavages are rather distinct on {011}, indistinct on {010}, and poor on {100}.

TWINNING. Twinning on {031} is very common. Simple contact twins yield triangular, heart-shaped, or kite-shaped forms. Repeated cyclic twinning on {031} produces pseudohexagonal trillings.

BIREFRINGENCE. Weak birefringence (0.008–0.011) yields first-order yellow, white, or gray interference colors in standard sections.

INTERFERENCE FIGURE. Chrysoberyl yields good biaxial positive figures with large (70°) to small ($\sim$10°) $2V_z$, few isochromes, and weak $r > v$ optic axis dispersion. At high temperatures, $2V_z$ gradually decreases to 0°. The acute bisectrix lies normal to no cleavage; however, most fragments lie on fracture surfaces.

OPTICAL ORIENTATION. Chrysoberyl is described herein as orthorhombic $a = X$, $b = Y$, and $c = Z$. {100} sections show symmetrical extinction to {011} cleavages. Prismatic crystals have parallel extinction and positive elongation ($c = Z$); however, cleavage fragments {011} show parallel extinction and negative elongation ($a = X$).

A second crystal setting is often used where $a = Z$, $b = Y$, and $c = X$, resulting in tabular crystals {001} and prismatic cleavages {110}.

* Large indices of refraction result from Fe^{3+} or Cr^{3+} substitution for Al^{3+}. Indices of refraction and $2V_z$ commonly vary somewhat even within a single crystal.

† A second crystal setting is commonly used: $a = Z$, $b = Y$, *and* $c = X$.

With increasing temperature, $2V_z$ decreases, closing on $Z = c$ in the {010} optic plane, to 0° and opening again in the {100} plane to yield a new optical orientation $a = Y$, $b = X$, and $c = Z$ and a new optic plane {100}.

DISTINGUISHING FEATURES. Chrysoberyl is characterized by very high relief, low birefringence, and great hardness and specific gravity. Beryl and tourmaline are uniaxial negative with rather low relief, phenacite is uniaxial with low relief, euclase has greater birefringence and smaller indices of refraction than chrysoberyl, and corundum is uniaxial negative.

ALTERATION. Chrysoberyl is highly resistant to normal weathering processes and commonly appears in gem placers with other resistant minerals. Carbonate solutions may act on chrysoberyl to produce dolomite and mica. Kaolin clays are also reported to replace chrysoberyl, and chlorite is often found in fractures.

OCCURRENCE. Chrysoberyl is a rather common mineral in complex granite pegmatites, where it occurs with tourmaline, topaz, columbite, apatite, spinel and other Be-minerals (such as beryl, phenacite, and euclase). In contact metamorphic deposits, chrysoberyl appears in dolomitic marble with corundum and in fluorine-skarns with fluorite, garnet, idocrase, and magnetite. Mica schists occasionally contain chrysoberyl, and it has been reported in aluminous rocks with garnet, kyanite, staurolite, sillimanite, and muscovite. Much chrysoberyl is recovered from placer deposits with columbite, cassiterite, ilmenite, rutile, monazite, and many other gem minerals: beryl, corundum, diamond, tourmaline, topaz, spinel, and garnet.

REFERENCES: CHRYSOBERYL

Dostal, J. 1969. Some new data for chrysoberyl from Maršikov, northern Moravia. *Acta Univ. Carolinae, Geologica*, 261–270.

Ferrell, E. F., J. H. Fang, and R. E. Newnham. 1963. Refinement of the chrysoberyl structure. *Amer. Mineral.*, *48*, 804–810.

Hudson, D. R. 1971. Gemstones in the system BeO-MgO-Al_2O_3. *Australian Gemmologist*, *11*(2), 5–9,36.

Okrusch, M. 1971. Zur Genese von Chrysoberyll—und Alexandrit—Lagerstätten. *Zeits. d. Deutschen Gemmologischen Gesell.*, *20*, 114–124.

Tröger, W. E. 1959. *Optische Bestimmung der gesteinsbildenden Minerale.* Stuttgart: E. Schweizerbart'sche Verlagsbuchhandlung.

Vlasov, K. A., ed. 1966. *Geochemistry and Mineralogy of Rare Elements and Genetic Types of their Deposits.* Vol. 2: *Mineralogy of Rare Elements.* Jerusalem: Israel Program for Scientific Translations.

HYDROXIDES

Brucite

$Mg(OH)_2$ TRIGONAL

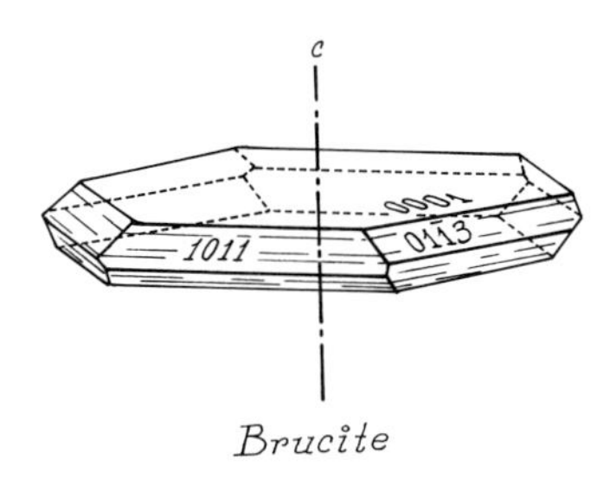

Brucite

$n_\omega = 1.559–1.590$
$n_\varepsilon = 1.580–1.600$
$n_\varepsilon - n_\omega = 0.02–0.01$
Uniaxial positive
Colorless in standard section

COMPOSITION AND STRUCTURE. Each brucite layer consists of two closest-packed layers of $(OH)^-$ ions with Mg^{2+} ions sandwiched between, filling all sites of octahedral coordination. The $(OH)^-$ ions are polarized, and brucite layers are held together by a combination hydrogen and Van der Waal's bonding.

Brucite is essentially $Mg(OH)_2$, with limited replacement of Mg^{2+} by Fe^{2+} (ferrobrucite), Mn^{2+} (manganobrucite), and possibly Zn^{2+}. Pyrochroite, $Mn(OH)_2$, and portlandite, $Ca(OH)_2$, are analogous isomorphs.

PHYSICAL PROPERTIES. H = $2\frac{1}{2}$. Sp. Gr. = 2.40. Color in hand sample is white to pale green, blue, or brownish. Luster is pearly to vitreous or waxy.

COLOR AND PLEOCHROISM. Colorless as fragments or sections.

FORM. Brucite most commonly appears as fine aggregates of folia, often concentric whorls, or sometimes fibers elongated on *c*. Small crystals are tabular on {0001}. Fibrous brucite is often called *nemalite*.

CLEAVAGE. Cleavage is perfect {0001}.

BIREFRINGENCE. Maximum birefringence is mild (0.01–0.02) seen parallel to foliation. Interference colors are first-order, usually highly anomalous olive or red-brown.

TWINNING AND ZONING. None reported.

INTERFERENCE FIGURE. Brucite is rarely biaxial with small $2V_z$, especially fibrous varieties. Fine grain size may make interference figures impractical. Cleavage plates yield centered optic axis figures.

OPTICAL ORIENTATION. The single optic axis lies normal to basal cleavage or parallel to fiber elongation. Extinction is parallel to cleavages or fibers. Elongation parallel to cleavages is negative (length-fast); fibers are uncommon and are length-slow.

DISTINGUISHING FEATURES. Anomalous birefringence, foliated whorls, and limited occurrence are characteristic of brucite. Talc and micas have higher birefringence and are optically negative, as is gypsum. Chlorite and serpentine minerals may show low anomolous birefringence; however, they are commonly pale green and usually show positive elongation (length-slow) parallel to cleavages.

ALTERATION. Brucite itself is formed by the hydration of periclase. It alters readily to hydromagnesite, $3MgCO_3 \cdot Mg(OH)_2 \cdot 3H_2O$ and, less commonly, to serpentine, deweylite, or brugnatellite. Through dehydration, brucite may return to a fine aggregate of periclase crystals.

OCCURRENCE. In association with calcite, brucite forms "brucite marble," resulting from the alteration of periclase in contact-metamorphosed limestones or by the direct conversion of dolomite.

Brucite is also found in low-temperature hydrothermal veins in serpentinite or chlorite schists with talc, aragonite, hydromagnesite, magnesite, and so on.

REFERENCES: BRUCITE

Gordon, R. S., and W. D. Kingery. 1966. Thermal decomposition of brucite. I. Electron and optical microscope studies. *Jour. Amer. Ceram. Soc.*, *49*, 654–660.

Haines, M. 1968. Two staining tests for brucite in marble. *Mineral. Mag.*, *36*, 886–888.

Liebling, R. S., and A. M. Langer. 1972. Optical properties of fibrous brucite from Asbestos, Quebec. *Amer. Mineral.*, *57*, 857–864.

Zigan, F., and R. Rothbauer. 1967. Neutron diffraction measurements on brucite. *Neues Jahrb. Mineral. Monatsh.*, *1967*, 137–143.

Gibbsite (Hydrargillite)

$Al(OH)_3$

MONOCLINIC
$\angle\beta = 94°34'$

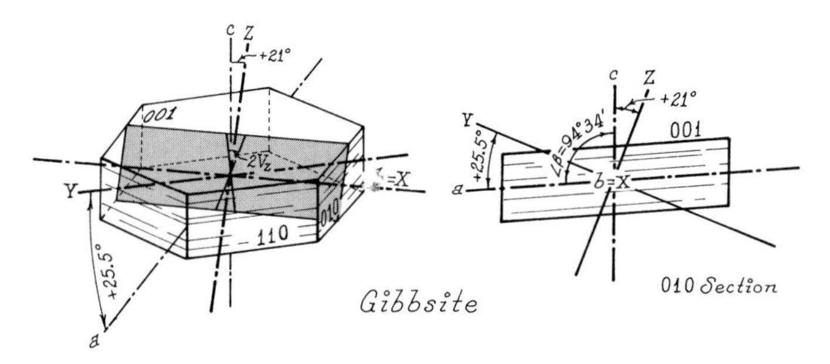

$n_\alpha = 1.568–1.580$
$n_\beta = 1.568–1.580$
$n_\gamma = 1.587–1.600$
$n_\gamma - n_\alpha \simeq 0.019$
Biaxial positive; $2V_z = 0°–40°$
$b = X, c \wedge Z = +21°, a \wedge Y = +25\frac{1}{2}°$*
Strong optic axis dispersion $r > v$, sometimes $v > r$; possible horizontal bisectrix dispersion
Colorless

COMPOSITION AND STRUCTURE. The structure of gibbsite is analogous to that of brucite: Cations in sixfold (octahedral) coordination are sandwiched between two closest-pack $(OH)^-$ layers to form a single layer of gibbsite. Because the aluminum ion is trivalent, only two-thirds of the available octahedral sites are occupied by cations. In brucite and probably the gibbsite variety (polymorph) *bayerite*, successive layers are stacked such that $(OH)^-$ ions are in hexagonal close-pack arrangement (ABABAB . . .). Successive layers of normal gibbsite are stacked with the hydroxyl ions in one gibbsite layer (a double layer of hydroxyl ions) essentially in line with hydroxyl ions in the gibbsite layers above and below (ABBAABBAA . . .). Layers are slightly offset, to produce monoclinic symmetry. Minor $Al^{3+} \rightleftharpoons Fe^{3+}$ substitution is common in gibbsite.

* On gentle heating, $2V_z$ closes to 0°, and above $56\frac{1}{2}$°C it opens in the {010} plane such that $b = Y$, $c \wedge Z = +45°$, and $a \wedge X = +49\frac{1}{2}°$.

PHYSICAL PROPERTIES. H = $2\frac{1}{2}$–$3\frac{1}{2}$. Sp. Gr. = 2.38–2.42. Color, in hand sample, is white to pale gray, brown, green, or pink. Luster is pearly to vitreous; likely dull.

COLOR AND PLEOCHROISM. Gibbsite is colorless to pale brown as fragments or in thin section. Pleochroism is very weak.

FORM. Gibbsite commonly appears as tiny tabular crystals (tabular on {001}) showing pseudohexagonal outline bounded by {100} and {110}. Compact lamellar aggregates, or rarely fibers, may occur as whorls or as botryoidal or stalactitic forms.

CLEAVAGE. Perfect basal {001} cleavage may control fragment orientation.

BIREFRINGENCE. Maximum birefringence, seen at a high angle to cleavage surfaces, is mild to moderate ($n_\gamma - n_\alpha \simeq 0.020$). Interference colors are first-order, usually gray, white, or yellow.

TWINNING. Normal twinning is common on {001} and less common on {100} or {110}. Parallel twinning about {130} is also common.

INTERFERENCE FIGURES. Gibbsite is biaxial positive, with small $2V_z$ ($<40°$), which decreases to 0° on heating to 56.5°C. Cleavage plates and tabular crystals lie with the acute bisectrix (Z) 21° from vertical to yield off-center acute bisectrix figures. Moderate isogyre separation occurs in a field showing a few isochromes. Optic axis dispersion is strong usually $r > v$, sometimes $v > r$ with possible horizontal bisectrix dispersion.

OPTICAL ORIENTATION. At normal temperatures $b = X$, $c \wedge Z = +21°$, $a \wedge Y = +25\frac{1}{2}°$. Maximum extinction is 25.5° to cleavage traces on {010} sections, and cleavage elongation is negative (length-fast).

Above $56\frac{1}{2}$°C, $b = Y$, $c \wedge Z = +45°$, and $a \wedge X = +49\frac{1}{2}°$, and the optic plane is {010}.

DISTINGUISHING FEATURES. Gibbsite is a foliated, fine-grained, secondary mineral of tropical and subtropical occurrence and is most likely to be confused with other claylike minerals. The micas and talc are optically negative and show higher birefringence than gibbsite. Kaolin and montmorillonite clays are optically negative, kaolin shows weak birefringence, and montmorillonite shows low relief. Boehmite and diaspore have higher relief, larger $2V_z$ and parallel extinction, and brucite shows parallel extinction and anomalous interference colors.

ALTERATION. Gibbsite is an alteration product of feldspar, nepheline, corundum, and other aluminum-rich minerals. It often results from the hydration of boehmite γ-AlO(OH) in laterites and may be converted to boehmite by partial dehydration. Gibbsite is very stable under surface conditions but may be silicified to become kaolin.

OCCURRENCE. Gibbsite is a product of intense chemical weathering of aluminous minerals. As such, it appears in lateritic soils and clays, especially bauxite, where it may be the dominant mineral, in association with diaspore, boehmite, and ferric oxides. Gibbsite is less commonly a low-temperature hydrothermal mineral appearing in veins and voids in aluminum-rich igneous rocks.

REFERENCES: GIBBSITE

Saalfeld, H., and M. Wedde. 1974. Refinement of the crystal structure of gibbsite. *Z. Kristallogr.*, *139*, 103–115.

Wefers, K. 1962. The structure of aluminum trihydroxides. *Naturwissenschaften*, *49*, 204–205.

The Diaspore-Boehmite Series

Diaspore	α-AlO(OH)	ORTHORHOMBIC
Boehmite	γ-AlO(OH)	ORTHORHOMBIC

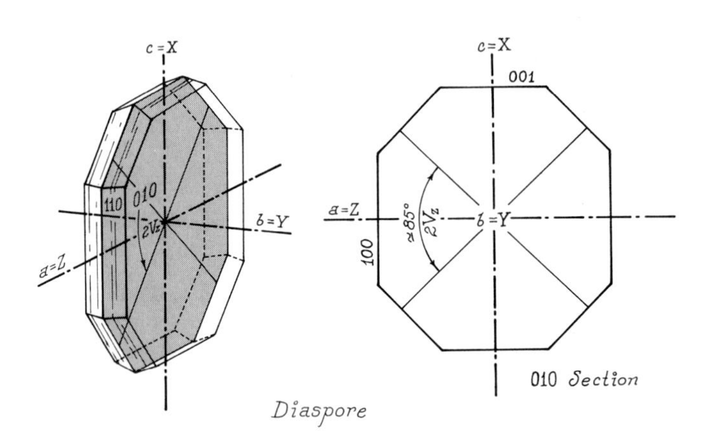

Diaspore

Diaspore	Boehmite*
$n_\alpha = 1.682\text{–}1.706$	$n_\alpha = 1.64\text{–}1.65$
$n_\beta = 1.705\text{–}1.725$	$n_\beta = 1.65\text{–}1.66$
$n_\gamma = 1.730\text{–}1.752$	$n_\gamma = 1.65\text{–}1.67$
$n_\gamma - n_\alpha = 0.052\text{–}0.046$	$n_\gamma - n_\alpha \simeq 0.015$
Biaxial positive, $2V_z = 84°\text{–}86°$	Biaxial positive, $2V_z \simeq 80°$
$a = Z, b = Y, c = X$	$a = Z, b = Y, c = X$
Weak dispersion $v > r$	Colorless in standard section
Colorless, rarely pale colors in standard section	

COMPOSITION AND STRUCTURE. Diaspore, α-AlO(OH), and boehmite, γ-AlO(OH), are polymorphs of hydrated alumina, AlO(OH), and are isomorphous equivalents of the hydrated ferric oxides goethite, α-FeO(OH), and lepidocrocite, γ-FeO(OH), respectively.

The structure of all these minerals is basically a framework of oxygen ions in closest-packing arrangement, with cations (Al^{3+} or Fe^{3+}) in octahedral coordination between the oxygen layers.

In diaspore and goethite, oxygen ions are in hexagonal close-packing arrangement, and the structures are analogous to corundum and hematite. With dehydration, diaspore, α-AlO(OH), and goethite, α-FeO(OH), become corundum, α-Al_2O_3, and hematite, α-Fe_2O_3, respectively. In boehmite and lepidocrocite, oxygen ions are cubic close packing, and the structures are analogous to spinel. With dehydration, boehmite, γ-AlO(OH), and lepidocrocite, γ-FeO(OH), become the isometric forms, γ-Al_2O_3, and γ-Fe_2O_3, respectively.

The anhydrated structures Al_2O_3 and Fe_2O_3 contain Al^{3+} or Fe^{3+} in two-thirds of the available octahedral sites; hydrated structures AlO(OH) and FeO(OH) contain fewer cat-

* Boehmite is usually so fine grained that optical properties are almost useless.

ions, due to the presence of H^+ ions. Hydroxyl ions $(OH)^-$ probably do not exist in the normal sense, in that hydrogen ions H^+ lie between pairs of oxygen ions.

PHYSICAL PROPERTIES. H = $6\frac{1}{2}$–7 (diaspore), H = $3\frac{1}{2}$–4 (boehmite). Sp. Gr. = 3.3–3.5 (diaspore), Sp. Gr. = 3.01–3.06 (boehmite). Both minerals are white when pure; Fe^{2+}, Mn^{2+}, Cr^{3+}, or Fe^{3+} impurities may cause the minerals to be brown, green, yellow, pink, violet, or deep red. Coarse folia may show pearly luster.

COLOR AND PLEOCHROISM. Both minerals are colorless in section. Thick fragments of diaspore may show pale color and absorption $Z > Y > X$.

FORM. Both minerals appear as fine aggregates of tiny scales or plates with foliation on {010}. Diaspore may be fibrous. The texture of boehmite aggregates is usually submicroscopic.

CLEAVAGE. Perfect to distinct cleavage on {010} is characteristic of both diaspore and boehmite. Diaspore may show imperfect prismatic {110} cleavages and poor cleavage on {100}. Boehmite may show good cleavage on {100}.

BIREFRINGENCE. The maximum birefringence of diaspore is seen on cleavage plates {010} and is strong (0.04–0.05). Maximum interference colors are vivid third-order in standard section. Boehmite birefringence is much less (0.015), yielding dull interference color of low first order. Grain size may be much less than section thickness.

TWINNING. Twinning is reported only in diaspore and is uncommon on {061} or {021}.

INTERFERENCE FIGURE. Both diaspore and boehmite are biaxial positive, with large $2V_z \simeq 85°$. Fine grain size frequently makes interference figures impractical or totally impossible. Diaspore may show weak optic axis dispersion $v > r$. Cleavage plates {010} should yield flash figures.

OPTICAL ORIENTATION. Extinction is parallel to cleavage traces and elongation is negative (length-fast) to diaspore cleavage and positive (length-slow) to boehmite cleavage traces.

DISTINGUISHING FEATURES. Diaspore and boehmite are commonly found together and in the presence of gibbsite. Gibbsite shows the lowest refractive indices (low to moderate relief), inclined extinction and negative (length-fast) elongation. Diaspore shows the greatest refractive indices, parallel extinction, and negative (length-fast) elongation. Boehmite is length-slow.

ALTERATION. Diaspore and boehmite are alteration products of aluminous minerals, especially nepheline, formed by intense chemical weathering. They may form corundum (α-Al_2O_3) or γ-Al_2O_3 by dehydration, or kaolin clays by silicification.

OCCURRENCE. Diaspore, α-AlO(OH); boehmite, γ-AlO(OH); and gibbsite, $Al(OH)_3$, are the basic components of the aluminum ore bauxite in association with analogous ferric hydroxides and minor quartz. They may occur in essentially equal proportions, or any one of the three minerals may dominate to the almost complete exclusion of the other two. The three Al-hydroxides appear in lateritic soils and aluminous clays and shales. Boehmite is largely confined to the occurrences just mentioned, and even there it is probably metastable. Diaspore, however, is an uncommon mineral in many environments.

In metamorposed aluminous sediments, diaspore is a rather common constituent of emery in association with corundum, magnetite, hematite, rutile, margarite, spinel, chloritoid, and so on; less commonly, in chlorite or quartz schists with chloritoid, corundum, kyanite, and other Al-rich minerals. Corundum-bearing crystalline limestone or marble may also contain diaspore.

Diaspore may form as the product of hydrothermal introduction or hydrothermal alteration of aluminous minerals in association with alunite, pyrophyllite, kaolin, or topaz. Rather large crystals may form as drusy crusts in fractures and cavities.

REFERENCES: THE DIASPORE-BOEHMITE SERIES

Biais, R., A. Bonnemayre, X. De Gramont, M. Michel, H. Gilbert, and C. Janot. 1972. Aluminum-iron substitutions in synthetic oxides and hydroxides. Preparation of iron diaspore. *Bull. Soc. fr. Mineral. Cristallogr.*, *95*, 308–321.

Ervin, G., Jr. 1952. Structural interpretation of the diaspore-corundum and boehmite-γ-Al_2O_3 transitions. *Acta Crystallogr.*, *5*, 103–108.

Kostov, I. 1969. Habit types and crystallography of diaspore and boehmite. *God. Sofii Univ., Geol.-Geogr. Fak.*, *61*, 167–176.

Sahama, Th. G., M. Lehtinen, and P. Rehtijarvi. 1973. Natural boehmite single crystals from Ceylon. *Contrib. Mineral. Petrol.*, *39*, 171–174.

Shelley, D., D. Smale, and A. J. Tullock. 1977. Boehmite in syenite from New Zealand. *Mineral. Mag.*, *41*, 398–400.

The Goethite-Lepidocrocite Series

Goethite (Xanthosiderite) α-FeO(OH) ORTHORHOMBIC

Lepidocrocite (Pyrosiderite) γ-FeO(OH) ORTHORHOMBIC

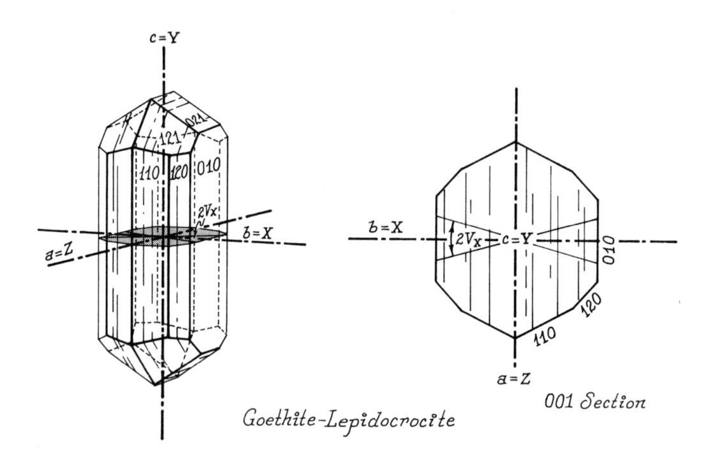

Goethite-Lepidocrocite

Goethite	**Lepidocrocite**
$n_\alpha = 2.260–2.275$	$n_\alpha = 1.94$
$n_\beta = 2.393–2.409$	$n_\beta = 2.20$
$n_\gamma = 2.398–2.515$	$n_\gamma = 2.51$
$n_\gamma - n_\alpha = 0.138–0.140$	$n_\gamma - n_\alpha \simeq 0.57$
Biaxial negative, $2V_x = 0°–27°$*	Biaxial negative, $2V_x \simeq 83°$
$a = Z, b = X, c = Y$*	$a = Z, b = X, c = Y$
Extreme dispersion $r > v$	Weak dispersion
Yellow to orange or brown in section and strongly pleochroic $Z > Y > X$	Yellow to red in section and strongly pleochroic $Z > Y > X$

COMPOSITION AND STRUCTURE. The structures of goethite, α-FeO(OH), and lepidocrocite, γ-FeO(OH), are analogous to those of diaspore, α-AlO(OH), and boehmite, γ-AlO(OH), respectively (see page 44). With dehydration, goethite becomes hematite, α-Fe_2O_3, which is paramagnetic, and lepidocrocite becomes maghemite, γ-Fe_2O_3, which is ferromagnetic. Both minerals are essentially FeO(OH), but minor $Fe^{3+} \rightleftharpoons Mn^{3+}$ substitution may occur, and admixed silica is common.

PHYSICAL PROPERTIES. $H = 5–5\frac{1}{2}$ (both minerals). Sp. Gr. $\simeq$ 4.3 (goethite), Sp. Gr. $\simeq$ 4.1 (lepidocrocite). Color, in hand sample, is yellow-brown or red-brown to brown-black. Luster is metallic adamantine to dull; sometimes silky (fibrous forms).

* The optical orientation of goethite varies widely with changing temperature or light wavelength. The orientation described is for medium (yellow) and short (green to violet) wavelengths. With increasing wavelength $2V_z$, closes to 0° in the {001} plane and opens again in the {100} plane so that, for long wavelengths (red) $a = Y$, $b = X$, and $c = Z$. Angle $2V_x$ also closes with increasing temperature, and at 59°C goethite is uniaxial negative (λ = 5,780 Å).

COLOR AND PLEOCHROISM. Both goethite and lepidocrocite are highly colored from yellow to brown and strongly pleochroic $Z > Y > X$. Fragments may be almost opaque. X = yellow to colorless; Y = yellow-brown, red-orange; and Z = yellow-orange, deep red-orange.

FORM. Goethite is usually fibrous or acicular on *c*, commonly grouped as radiating botryoidal forms; rarely as tabular scales on {010}. Lepidocrocite is usually tabular on {010} as scaley aggregates or single crystals; less commonly elongated on *a* as acicular or bladed crystals.

CLEAVAGE. Perfect {010} cleavages control fragment orientation for both minerals. Both minerals show imperfect cleavage on {100}, and lepidocrocite shows a third fair cleavage on {001}.

BIREFRINGENCE. Maximum birefringence is very extreme (lepidocrocite, 0.57) to extreme (goethite, 0.138) but appears only in rare basal sections. Maximum interference colors are weak (very high-order) and are largely obscured by inherent mineral color. Birefringence normal to the tabular habit is much less, however, especially for goethite, and bright low-order colors may be evident.

TWINNING. Rare cruciform twins are reported in goethite.

ZONING. Color banding is common normal to fiber elongation in radiating, colloform masses.

INTERFERENCE FIGURE. Both minerals are biaxial negative; however, $2V_x$ is markedly different for goethite ($2V_x$ is small, $<30°$), and lepidocrocite ($2V_x$ is large, $\simeq 83°$). Goethite is uniaxial negative at long wavelengths (near red) and at elevated temperatures (59°C). The acute bisectric (X) is normal to the platy habit and dominant cleavage of both minerals. Figures show numerous isochromes on an orange field.

OPTICAL ORIENTATION. For both minerals, the optic plane is {001} and $a = Z, b = X$, and $c = Y$. Extinction is parallel to all cleavages and elongation. Acicular crystals elongate on $c = Y$ may show positive or negative elongation. Fibers and cleavage elongation of lepidocrocite show positive (length-slow) elongation.

DISTINGUISHING FEATURES. Ferric oxides and hydroxides are often difficult to separate. Hematite is opaque or distinctly red, with reddish streak, and is uniaxial negative; goethite is yellowish, with small $2V_x$ and strong dispersion; lepidocrocite is usually red-brown, with large $2V_x$; and limonite is isotropic or cryptocrystalline.

OCCURRENCE. Paragenesis is essentially the same for both goethite and lepidocrocite, and the two minerals are commonly intimately associated; layers or crusts of lepidocrocite overlying those of goethite are typical. Both minerals are largely the result of surface weathering of iron-rich minerals, especially sulfides and oxides. Goethite is a very common mineral and is the stable iron hydroxide in a humid environment at normal temperatures and pressures.

In laterites, goethite and lepidocrocite are associated with hematite, limonite, the Al-hydroxides, and equivalent Mn-hydroxides. Both minerals are precipitated from marine and groundwaters in springs and bogs, and the bog iron ores are locally of major economic significance. Low-temperature hydrothermal deposits sometimes yield goethite as a primary mineral or as the hydrothermal alteration product of primary Fe-sulfides. Quartz (often amethyst), siderite, and calcite are common associates.

REFERENCES: THE GOETHITE-LEPIDOCROCITE SERIES

Fasiska, E. J. 1967. Structural aspects of the oxides and oxide hydroxides of iron. *Corros. Sci.*, *7*, 833–839.

Herzog, F. 1972. Goethite or nadeleisenerz. *Schweiz. Strahler*, *2*, 356–360.

Sampson, C. F. 1969. Lattice parameters of natural single crystal and synthetically produced goethite (α-FeOOH). *Acta Crystallogr.*, *B 25*, 1683–1685.

Schwertmann, U., and R. M. Taylor. 1972. Influence of silicate on the transformation of lepidocrocite to goethite. *Clays and Clay Minerals*, *20*, 159–164.

Schwertmann, U., and R. M. Taylor. 1972. Transformation of lepidocrocite to goethite. *Clays and Clay Minerals*, *20*, 151–158.

White, W. B., and R. Roy. 1965. Infrared spectra-crystal structure correlations. II. Comparison of simple polymorphic minerals. *Amer. Mineral.*, *49*, 1670–1687.

Limonite

$FeO(OH) \cdot nH_2O$

AMORPHOUS

$n = 2.0$–2.1*

Isoptropic or finely cryptocrystalline

Yellow to yellow-brown or red-brown in section

COMPOSITION AND STRUCTURE. Limonite is largely a hardened, amorphous gel of hydrated iron. It is a pseudomineral of highly variable water content, and much material formerly called *limonite* has been shown to be fine-crystalline goethite with adsorbed water. Common impurities are admixed colloids of silica, manganese, and phosphates, clays, aluminum hydroxides, and bitumins.

PHYSICAL PROPERTIES. $H = 4$–$5\frac{1}{2}$ (highly variable). Sp. Gr. = 2.7–4.3. Color, in hand sample, is yellow-brown to black-brown, and luster is dull and earthy to vitreous. Limonite is commonly tarnished to irridescent shades of pink and green.

COLOR AND PLEOCHROISM. Limonite in thin section or as fragments is yellow to yellow-brown or red-brown without pleochroism.

FORM. Being an amorphous, hardened colloid, limonite shows typical colloform structures (for example, botryoidal, mammalary, and stalactitic).

BIREFRINGENCE. Limonite (probably cryptocrystalline goethite or hematite) sometimes shows anomalous birefringence up to 0.04.

DISTINGUISHING FEATURES. Limonite may be difficult to separate from fine-grained varieties of other Fe-oxides and hydroxides, and no sharp boundary exists between limonite and cryptocrystalline goethite. Limonite should be isotropic or nearly so.

ALTERATION. Limonite is the alteration product of Fe-bearing minerals and may form as pseudomorphs after pyrite and numerous minerals, even noniron-bearing minerals. It may even replace organic materials.

OCCURRENCE. Limonite is always secondary resulting from surface weathering of Fe-bearing minerals and is always intimately associated with goethite and frequently with hematite and Mn-hydroxides. It is most common in swamps or spring deposits as the result of organic or inorganic colloidal deposition; it also exists as gossan caps over sulfide deposits.

REFERENCE: LIMONITE

Kanurkov, G. 1970. Classification of limonites. *Ann. Soc. Geol. Belg.*, *93*, 331–336.

* Limonite fragments may absorb index liquids, and its refractive index may increase on standing in the liquid.

CHAPTER 2

Minerals with Complex Anions (Excluding Silicates)

Complex anions are discrete entities within a crystal structure and consist of three or four oxygen atoms (anions) relatively tightly bound to a central, positively charged atom (cation). Anions other than oxygen (for example, halogens) are not common in complex ions of minerals, so they are often called *oxyanions*. They may be bound directly together by sharing of common oxygens, or indirectly by relatively weaker bonds to common metal atoms, or both. Such minerals are usually classified according to the type of complex anion present. Hence there are phosphates (containing PO_4^{3-} oxyanions), sulfates (SO_4^{2-}), carbonates (CO_3^{2-}), nitrates (NO_3^{-}), borates (BO_3^{3-} and BO_4^{5-}), arsenates (AsO_4^{3-}), vanadates (VO_4^{3-}), chromates (CrO_4^{2-}), tungstates (WO_4^{2-}), molybdates (MoO_4^{2-}), beryllates (BeO_4^{6-}), and silicates (SiO_4^{4-}). Some other oxyanions occur in nonmineral inorganic compounds. Silicates are of sufficient mineralogical importance that they are treated separately, and some of the other groups are represented only by minerals too uncommon for inclusion in this text.

Phosphates and sulfates are similar in several respects. Many are hydrous. Most consist of isolated TO_4 (T = P,S) tetrahedra connected by metal atoms, although a few sulfates and several phosphates contain polymerized oxyanions (T_2O_7 groups, for example). Boron occurs in both tetrahedral (BO_4^{5-}) and triangular (BO_3^{3-}) oxygen coordination, often with $(OH)^-$ substituting for O^{2-}, and borates are unique in that they contain complex insular groups made up of both triangles and tetrahedra. For example, the $[B_4O_5(OH)_4]^{2-}$ group of borax consists of two tetrahedra and two triangles, with five oxygen atoms shared among them.

Carbonates

Many carbonate minerals are hydrated or contain other complex anions (for example, SO_4^{2-}, PO_4^{3-}, and OH^-), but the simple carbonates are far more common in nature.

The carbonate ion CO_3^{2-} is the unique structural unit of the carbonate minerals; it is a strongly covalent, planar unit of three oxygen ions centered on the corners of an equilateral triangle about a tiny, central carbon ion. The carbonate ion carries a net double negative charge and commonly alternates with bivalent cations to yield trigonal symmetry when cations are small ($<$ about 1.0 Å radius) or orthorhombic symmetry when cations are large ($>$ about 1.0 Å radius).

Structure of the *trigonal carbonates* (Table 2-1), characterized by that of calcite, $CaCO_3$ (Fig. 2-1), is analogous to the simple structure of halite, NaCl, with alternating cations and anions in three dimensions. The cubic cell and cleavage of halite are distorted to rhombohedral symmetry by the planar carbonate ions, which lie in a single plane {0001} perpendicular to

TABLE 2-1. Trigonal Carbonates

		n_ω	n_ε
Calcite	$CaCO_3$	1.658	1.486
Magnesite	$MgCO_3$	1.700	1.509
Rhodochrosite	$MnCO_3$	1.816	1.597
Smithsonite	$ZnCO_3$	1.850	1.625
Cobaltocalcite	$CoCO_3$	1.855	1.600
Siderite	$FeCO_3$	1.875	1.633
Dolomite	$CaMg(CO_3)_2$	1.679	1.500
Ankerite	$Ca(Mg, Fe)(CO_3)_2$	1.690–1.750	1.510–1.548

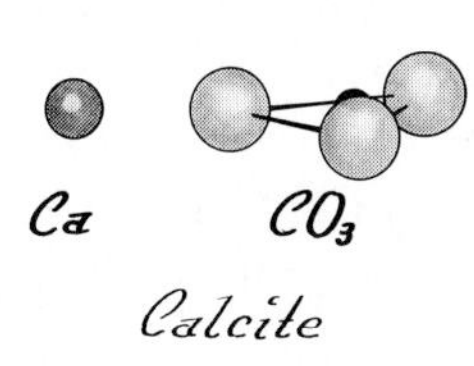

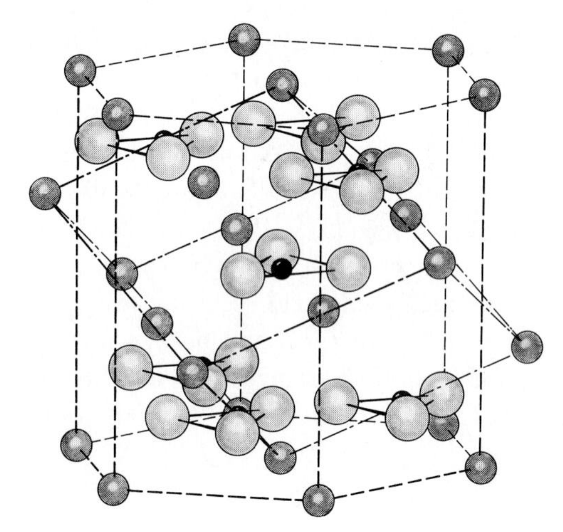

FIGURE 2-1. The structure of calcite. Rhombohedral calcite has a structure analogous to cubic halite, where complex CO_3^{2-} anions replace Cl^- anions. The flat, triangular carbonate anions lie in one (111) plane of the cubic cell, making it unique and normal to the *c*-crystallographic axis of the rhombohedral cell. Highly polarized, oriented, anisotropic carbonate ions are responsible for the optic sign (−) and for the large difference between n_ω and n_ε (large birefringence).

FIGURE 2-2. Rhombohedral cleavages $\{10\bar{1}1\}$ and strong birefringence characterize all trigonal carbonate minerals.

one cube diagonal (one isometric triad), which becomes the unique crystallographic axis (trigonal triad). Cations lie in an open, distorted cubic close-pack and are bivalent, with radius less than about 1.0 Å. Rhombohedral cleavages $\{10\bar{1}1\}$ of the trigonal carbonates (Fig. 2-2) are analogous to the cubic cleavages $\{100\}$ of halite.

Dolomite-type trigonal carbonates have a calcite structure, with two small (radius < 1.0Å) bivalent cations in equal numbers (such as Ca^{2+} and Mg^{2+}) substituting for Ca^{2+} ions in calcite. Ca^{2+} and Mg^{2+} ions alternate parallel to c making two nonequivalent structural sites and lowering symmetry from $\bar{3}\ 2/m$ (calcite) to $\bar{3}$ (dolomite). Magnesian calcite is essentially disordered dolomite.

Orthorhombic carbonates (Table 2-2), typified by aragonite, $CaCO_3$, contain large cations (radius > 1.0 Å) in open hexagonal close pack alternating with planar carbonate ions lying in

TABLE 2-2. Orthorhombic Carbonates

		n_α	n_β	n_γ	$2V_x$	Dispersion
Strontianite	$SrCO_3$	1.516	1.664	1.666	7°	$v > r$
Witherite	$BaCO_3$	1.526	1.676	1.677	16°	$v > r$
Aragonite	$CaCO_3$	1.530	1.680	1.685	18°	$v > r$
Cerussite	$PbCO_3$	1.803	2.074	2.076	$\simeq 9°$	$r > v$
Alstonite	$CaBa(CO_3)_2$	1.526	1.671	1.672	$\simeq 6°$	$r > v$

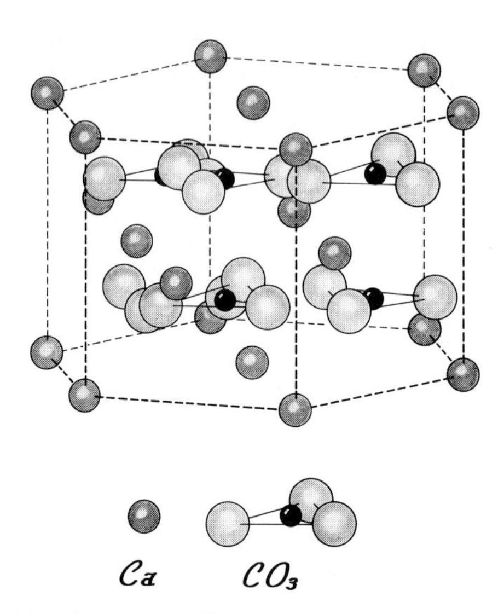

FIGURE 2-3. The structure of aragonite. The aragonite $CaCO_3$ polymorph is orthorhombic with the flat, triangular CO_3^{2-} ion oriented in the basal (001) plane. Highly polarized, oriented, anisotropic carbonate ions are responsible for the optic sign (−) and for the large difference between n_γ and n_α (large birefringence).

basal planes {0001} (Fig. 2-3). Crystals are commonly pseudohexagonal, although alternate reversal of carbonate ions in the basal plane lowers true symmetry to orthorhombic.

Alstonite, $CaBa(CO_3)_2$, has an aragonite structure with two large cations in equal numbers and ordered array analogous to the dolomite-calcite relationship.

Calcium carbonate is dimorphous, as the Ca^{2+} ion is very near the 1.0 Å radius that separates trigonal and orthorhombic structures. Calcite is the low-pressure, high-temperature form (Fig. 2-1) and is the only stable form under normal surface conditions. The crystallization of aragonite is favored by the presence of large cations Sr^{2+}, Ba^{2+}, or Pb^{2+} and may be stabilized by their presence as substituents for Ca^{2+}. At least three other polymorphs of $CaCO_3$ are known, although none are certainly known in natural environments.

Calcite $CaCO_3$ TRIGONAL

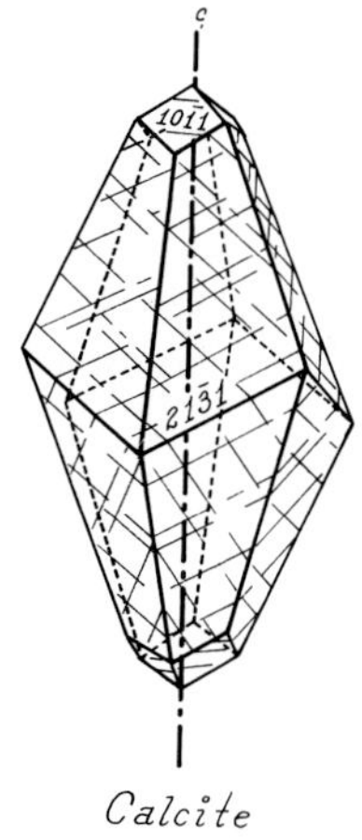

$n_\omega = 1.658$*
$n_\varepsilon = 1.486$†
$n_\omega - n_\varepsilon = 0.172$*
Uniaxial negative
Colorless in standard section

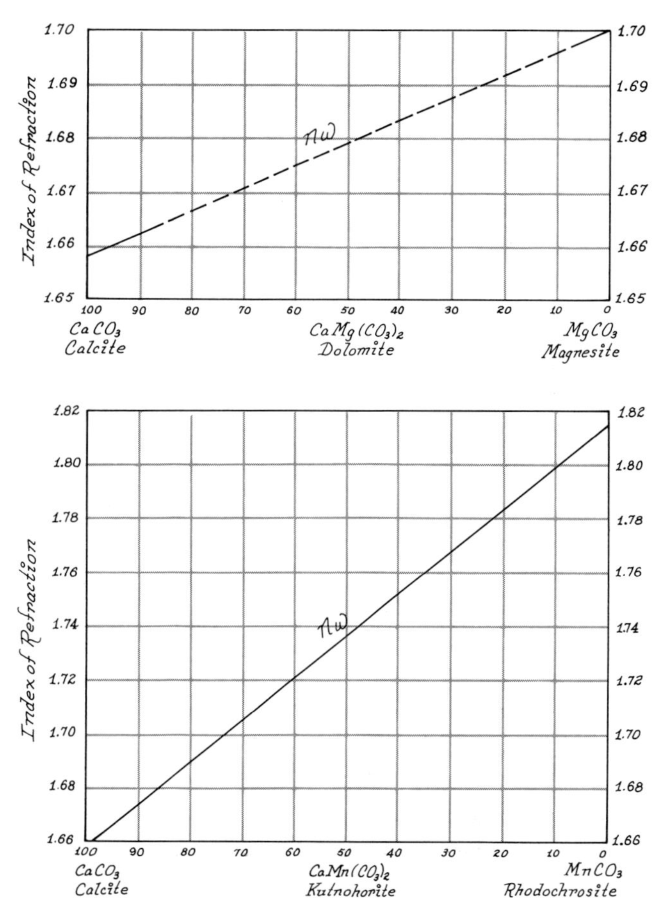

FIGURE 2-4. Variation of the n_ω refractive index in the calcite-rhodochrosite and calcite-magnesite series.

COMPOSITION AND STRUCTURE (see page 50). Calcite is the low-pressure $CaCO_3$ polymorph and is the only truly stable form under normal surface conditions and in most other geologic environments. Aragonite appears in special environments characterized by low temperature, high pressure, and recent age; other polymorphs (such as vaterite and elatolite) are known but are rarely, if ever, encountered in nature.

Calcite is usually almost pure $CaCO_3$, although several cations may substitute for calcium and admixed inclusions of other minerals are common. Calcite, $CaCO_3$, forms complete solid solution with rhodochrosite, $MnCO_3$, at least at high temperatures, and partial solid solution with siderite, $FeCO_3$ (Fe:Ca < 1:4); smithsonite, $ZnCO_3$; magnesite, $MgCO_3$; and cobaltocalcite, $CoCO_3$. Magnesian calcite is most common, although the Ca $\rightleftharpoons$ Mg substitution is very limited (Mg:Ca < 1:4). Manganoan calcite may range through kutnohorite $CaMn(CO_3)_2$ to rhodochrosite but is common only in association with ore minerals of manganese. Large cations Sr^{2+}, Ba^{2+}, and Pb^{2+} may replace Ca^{2+} in limited

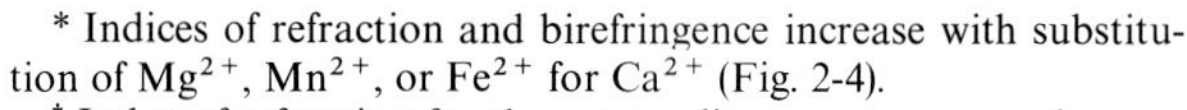

* Indices of refraction and birefringence increase with substitution of Mg^{2+}, Mn^{2+}, or Fe^{2+} for Ca^{2+} (Fig. 2-4).

† Index of refraction for the extraordinary ray seen on cleavage rhombs is $n_\varepsilon' = 1.566$.

amounts, although they are more common in the orthorhombic polymorph aragonite. Barian calcite (baricalcite) may be surprisingly Ba-rich (Ba:Ca = 1:1).

Microscopic inclusions of iron oxides, pyrite, chlorite, graphite, and other admixed minerals may impart color to calcite.

PHYSICAL PROPERTIES. H = 3. Sp. Gr. = 2.710. Color, in hand sample, is highly variable, usually white or colorless. Mn varieties tend to be pink; Fe varieties, yellow to brown; Co varieties, rose-red. Microscopic mineral inclusions may impart the color of the admixed mineral. Luster is vitreous to (sometimes) pearly on cleavage surfaces.

COLOR AND PLEOCHROISM. Calcite is colorless to neutral gray or turbid as fragments or in standard section.

FORM. The megascopic crystal habit of calcite is highly variable. Euhedral crystals are common, usually showing rhombohedron and/or scalenohedral forms with possible prisms or basal pinacoid. Thin sections of calcite usually show fine to coarse anhedral aggregates. Fibrous or columnar forms are rare.

CLEAVAGE. Perfect rhombohedral cleavages $\{10\bar{1}1\}$ intersect at about 75° and dominate fragment orientation. Parting may develop on negative rhombohedron twin planes $\{01\bar{1}2\}$.

BIREFRINGENCE. Maximum birefringence is extreme ($n_\omega - n_\varepsilon = 0.172$), and even when the optic axis is near vertical birefringence is large as light rays are convergent. Most rays are somewhat inclined to the optic axis at any orientation. Interference colors are high-order and show only uniform gray or pale shades of pink or green. Adjoining twin lamellae show reversed optical orientation; zones of overlap, due to inclined twin planes, may fail to extinguish with stage rotation; and twinning may produce an oblique plaid design of pale interference colors.

TWINNING. Lamellar twinning on the negative rhombohedron $\{01\bar{1}2\}$ is very common and is a characteristic feature of calcite (Fig. 2-5). Lamellae are parallel to rhomb edges or parallel to the long diagonal of the cleavage diamond (rhomb section), in contrast to dolomite, where twin traces usually parallel the short diagonal. Repeated twinning is commonly glide twinning, due to natural or artificial shear; it may even

FIGURE 2-5. Calcite crystals are commonly cut by multiple sets of polysynthetic twins.

be the result of grinding thin sections. Simple contact twinning on {0001} is common, and twins are reported on the positive rhombohedron $\{10\bar{1}1\}$ and on $\{0\bar{2}21\}$ but are rare.

ZONING. Microscopic mineral inclusions may produce color zones.

INTERFERENCE FIGURE. Calcite yields an optic axis figure showing thin, distinct isogyres that broaden rapidly toward the edge of the field. Myriads of thin isochrome bands encircle the melatope, each band representing an entire order of interference colors. Off-centered figures show very broad, diffuse isogyres, and cleavage fragments favor highly off-centered figures.

Calcite in metamorphic rocks is sometimes biaxial negative with $2V_x$ up to 15°.

OPTICAL ORIENTATION. Extinction is symmetrical or inclined to rhombohedral cleavage traces, and the slow ray is parallel to the shorter rhomb diagonal. The angle between the plane of the extraordinary wave and the trace of the twin lamellae is greater than 55°.

DISTINGUISHING FEATURES. Trigonal carbonates are characterized by perfect rhombohedral cleavages $\{10\bar{1}1\}$, extreme birefringence, variable relief; and effervescence with weak acids. Calcite very commonly shows lamellar glide twinning. Traces of the composition plane range from parallel to cleavage traces to parallel to the long diagonal of the cleavage diamond; the twin lamellae of dolomite and ankerite are parallel to the short diagonal. Calcite shows lower indices of refraction than other trigonal carbonates and n_ε is appreciably below the index of balsam. A universal stage method of distinguishing calcite, dolomite, and magnesite is described in Phillips (1971, p. 201), following Emmons' (1943) original description.

Orthorhombic carbonates show extreme birefringence but are biaxial (small $2V_x$) and do not show rhombohedral cleavage.

ALTERATION. Calcite is easily dissolved by weakly acidic waters and may be subsequently replaced by quartz, chalcedony, opal, hematite, goethite, pyrolusite, or other minerals to yield pseudomorphs after calcite.

OCCURRENCE. Calcite is one of the most common and most widespread minerals at or near the earth's surface, where it is the only stable form of $CaCO_3$.

In sedimentary rocks, calcite appears as a direct chemical precipitate or as shells and skeletons of marine organisms. It is the major and essential constituent of limestone, chalk, travertine, coquina, oolite, and similar sediments. Most direct chemical precipitation of $CaCO_3$ in marine waters or freshwater lakes and the shells of many marine organisms are aragonite, although aragonite thus formed inverts to the stable calcite polymorph. Calcite is a common cementing agent in detrital sediments and may be crystallographically (optically) continuous over several centimeters between sand grains (luster-mottled sandstone) or may form euhedral calcite crystals enclosing sand grains (sand crystals). Fine calcite and clay minerals are intimately associated in marl and calcareous shale. It is common in large cavities as geodes and cave deposits (stalactites, stalagmites, and so on) and appears as coarse calcite veins in any calcareous sediment. Thermal springs and streams may precipitate porous deposits of calcite as calc-sinter or tufa due to the surface loss of CO_2 and the subsequent conversion of soluble $CaH_2(CO_3)_2$ to insoluble $CaCO_3$.

In metamorphic rocks, coarse, recrystallized aggregates of calcite form from former calcareous sediments. Calcite may be the only major mineral in marble or may occur with Ca—Mg silicates (for example, wollastonite, diopside, idocrase, tremolite, epidote, and grossular) in skarn and calc schists and gneisses. Forsterite marble and ophicalcite contain calcite in associate with forsterite and serpentine.

In igneous rocks, primary magmatic calcite is largely confined to alkali-rich, silica-poor plutonic rocks in association with nepheline and other feldspathoids, orthoclase, and Nb-Ta minerals. Carbonate magmas are postulated as the origin of rare calcite intrusions (carbonatite).

Vesicles and amygdales in basalt or other mafic volcanics commonly contain calcite or aragonite in association with zeolites, chalcedony, or chlorite.

Calcite is a common mineral in the late stages of hydrothermal deposition. It is commonly found as euhedral crystals or crusts in vugs or fissures in association with fluorite, barite, dolomite, siderite, and quartz and with sphalerite, galena, and other late sulfides.

Calcite is also a common alteration product of plagioclase and many other Ca-rich minerals appearing as fine granular aggregates recognized by variable relief and extreme birefringence.

REFERENCES: CALCITE

Boettcher, A. L., and P. J. Wyllie. 1967. Biaxial calcite inverted from aragonite. *Amer. Mineral.*, *52*, 1527–1529.

Cowley, E. R. 1970. Refractive indices of some carbonate minerals in the point dipole approximation. *Can. Jour. Phys.*, *48*, 297–302.

DeWys, E. C. 1973. Optical surface of calcite containing various amounts of manganese. *Libyan. Jour. Sci.*, *3*, 81–83.

Emmons, R. C. 1943. The universal stage. *Geol. Soc. Am. Mem. 8*.

Friedman, G. M. 1959. Identification of carbonate minerals by staining methods. *Jour. Sed. Petrol.*, *29*, 87–97.

Graf, D. L., and J. E. Lamar. 1955. Properties of calcium and magnesium carbonates and their bearing on some uses of carbonate rocks. *Econ. Geol.*, *50th Anniv. Vol.*, 639–713.

Kennedy, G. C. 1947. Charts for correlation of optical properties with chemical composition of some common rock-forming minerals. *Amer. Mineral.*, *32*, 561–573.

Pecora, W. T. 1956. Carbonatites: A review. *Bull. Geol. Soc. Amer.*, *67*, 1537–1556.

Phillips, W. R. 1971. *Mineral Optics: Principles and Techniques.* San Francisco: W. H. Freeman and Company.

Rath, R., and D. Pohl. 1971. Biaxiality phenomenon in calcite and dolomite. *Contrib. Mineral. Petrol.*, *32*, 74–78.

Rath, R., and D. Pohl. 1971. Optical behavior of twin lamellae. *Contrib. Mineral. Petrol.*, *33*, 239–244.

Magnesite — $MgCO_3$ — TRIGONAL

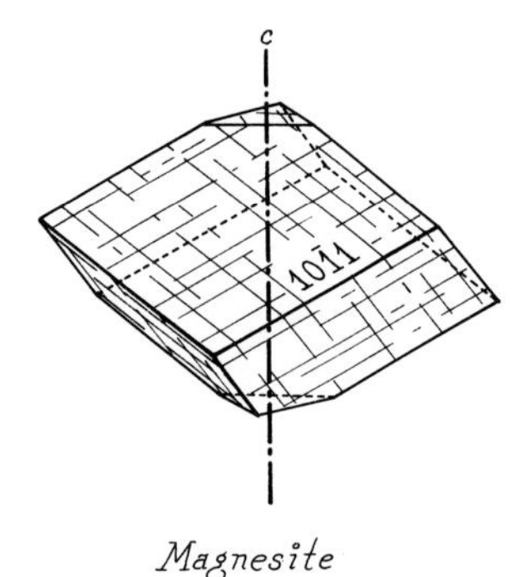

Magnesite

$n_\omega = 1.700$*

$n_\varepsilon = 1.509$†

$n_\omega - n_\varepsilon = 0.191$*

Uniaxial negative

Colorless in standard section

* Indices of refraction and birefringence increase with substitution of Fe^{2+} for Mg. When Mg:Fe $\simeq$ 1:1, $n_\omega = 1.788$, $n_\varepsilon = 1.570$, and $(n_\omega - n_\varepsilon) = 0.218$. Indices of refraction and birefringence also increase with substitution of Mn^{2+}, Co, or Zn for Mg and decrease with Ca for Mg (Fig. 2-6).

† Index of refraction for the extraordinary ray seen on cleavage rhombs is $n_\varepsilon' = 1.602$.

COMPOSITION. Pure magnesite is $MgCO_3$. It may, however, form complete solid solution with siderite, $FeCO_3$ (the composition range 5–50 molecular percent siderite is called *breunnerite*) and partial solid solution with calcite, $CaCO_3$, or rhodochrosite, $MnCO_3$, up to Mg:Ca or Mn $\simeq$ 9:1.Minor Co, Zn, or Ni may also replace Mg. Porcelainlike magnesite commonly contains much admixed opal. Hydromagnesite is one of several forms of hydrated magnesium carbonate ($Mg_4(CO_3)_3(OH)_2 \cdot 3H_2O$) and is structurally unrelated to magnesite.

PHYSICAL PROPERTIES. H = $3\frac{1}{2}$–$4\frac{1}{2}$. Sp. Gr. = 3.00–3.48. Color, in hand sample, is white or gray to yellow, brown, or pink. Luster is pearly and vitreous or dull and earthy.

COLOR AND PLEOCHROISM. Colorless or cloudy in section or as fragments.

FORM. Magnesite normally occurs as compact aggregates of coarse to fine anhedral grains; rarely, as euhedral crystals. It may also be very fine grained and earthy or porcelainlike, commonly mixed with opal.

CLEAVAGE. Perfect rhombohedral cleavages $\{10\bar{1}1\}$ are characteristic of trigonal carbonates and dominate fragment orientation.

BIREFRINGENCE. Maximum birefringence is extreme ($n_\omega - n_\varepsilon = 0.191$) and increases with substitution of Fe^{2+} for Mg.

Cleavage rhombs show lower, but still strong, birefringence ($n_\omega - n'_\varepsilon = 0.098$), and even orientations near the optic axis yield high-order interference colors, due to converging light rays. Interference colors are high-order gray or white.

TWINNING AND ZONING. Obvious twinning and zoning are rare in magnesite, although translational gliding may take place on {0001}.

INTERFERENCE FIGURE. Basal sections show a distinct uniaxial cross (isogyres broaden quickly toward outer edge of figure), with myriads of fine color rings. Cleavage fragments yield highly off-centered figures.

OPTICAL ORIENTATION. Extinction is symmetrical or inclined to rhombohedral cleavages.

DISTINGUISHING FEATURES. Magnesite shows higher refractive indices than either calcite or dolomite and lower than siderite or rhodochrosite. Grains are normally anhedral and lack the lamellar twinning of dolomite and calcite. Magnesite, dolomite, and calcite may be distinguished by universal stage procedures (Phillips, 1971, p. 201) or staining techniques.

ALTERATION. Magnesite seldom shows alteration, although Fe-rich varieties may yield iron hydroxides. It may be dissolved and replaced by other minerals as pseudomorphs (such as limonite and talc).

OCCURRENCE. Magnesite occurs most commonly as masses and veins in serpentine, where it is formed by mild metamorphism in the presence of CO_2. The serpentine, in turn, has resulted from the alteration of peridotite, dunite, or other Mg-rich igneous or metamorphic rocks.

In metamorphic rocks, magnesite appears as veins or disseminated grains in talc, chlorite, or mica schists or may form in metasomatic zones between dolomite or limestone sediments and granitic intrusions.

In igneous rocks, magnesite is a rare mineral of primary crystallization with other Mg-rich minerals (for example, talc and enstatite). It is an uncommon mineral in pegmatites or hydrothermal ore veins and may appear in amygdules in basic volcanics.

In sedimentary rocks, magnesite is most common as bedded deposits associated with evaporite salts. It may be a primary precipitate or a product of recrystallization of carbonate sediments in the presence of Mg-brines.

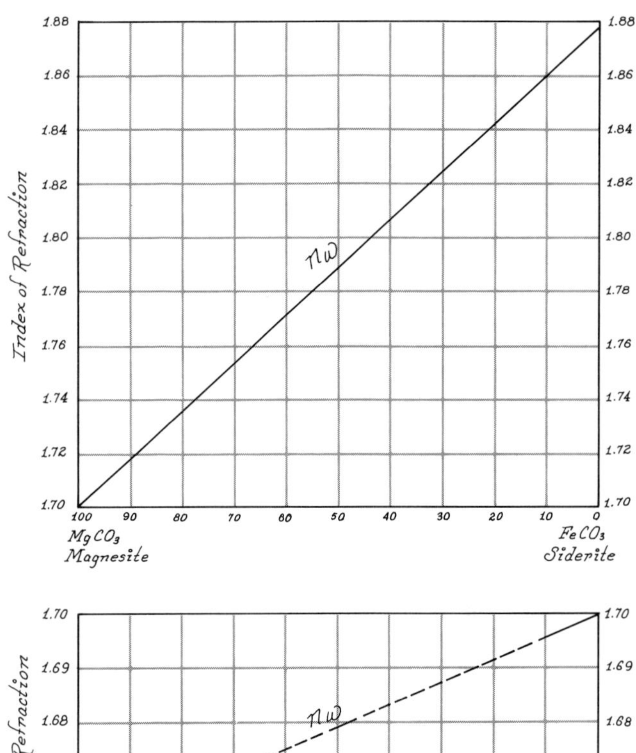

FIGURE 2-6. Variation of the n_ω refractive index in the magnesite-siderite and calcite-magnesite series.

REFERENCES: MAGNESITE

Graf, D. L., and J. E. Lamar. 1955. Properties of calcium and magnesium carbonates and their bearing on some uses of carbonate rocks. *Econ. Geol., 50th Anniv. Vol.*, 639–713.

Johannes, W. 1969. Formation of deposits of magnesite and siderite in the system Mg^{2+}-Fe^{2+}-$CO_3{}^{2+}$-$Cl_2{}^{2-}$-H_2O. *Beitr. Mineral. Petrogr., 21*, 311–318.

Oh, K. D., H. Moriakawa, S. Iwai, and H. Aoki. 1973. The crystal structure of magnesite, *Amer. Mineral. 58*, 1029–1033.

Phillips, W. R. 1971. *Mineral Optics: Principles and Techniques.* San Francisco: W. H. Freeman and Company.

Uzan, E., H. Damany, and V. Chandrasekharan. 1971. Optical properties of magnesite in the Schumann region. *Opt. Commun., 2*, 452–454.

Siderite

$FeCO_3$ TRIGONAL

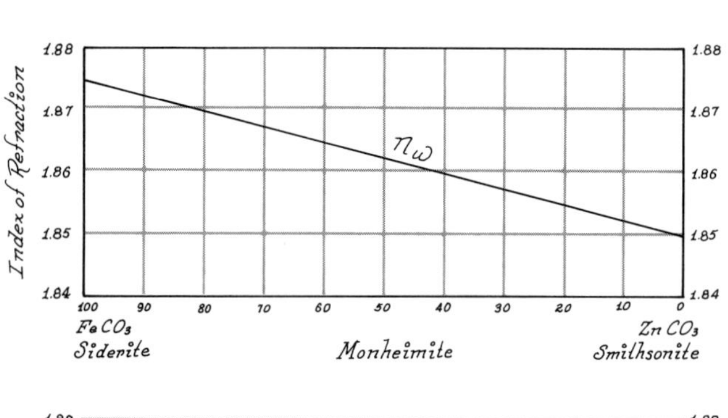

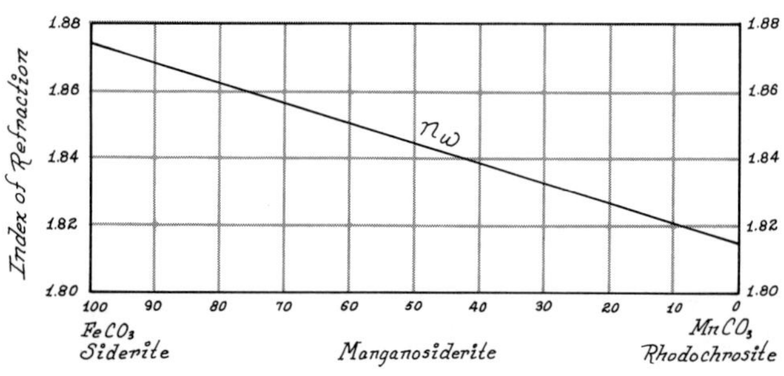

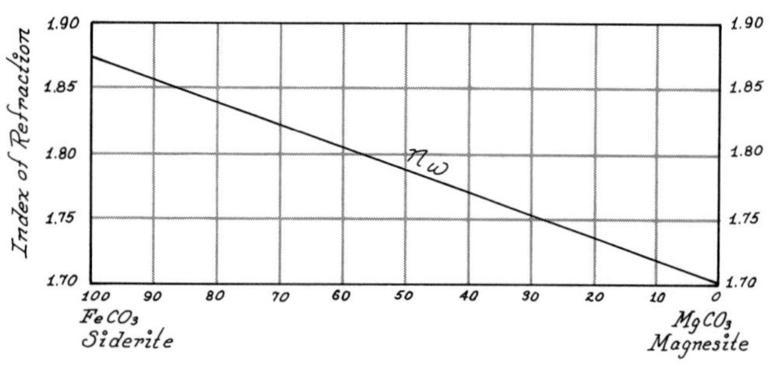

FIGURE 2-7. Variation of the n_ω refractive index in the siderite-smithsonite, siderite-rhodochrosite, and siderite-magnesite series.

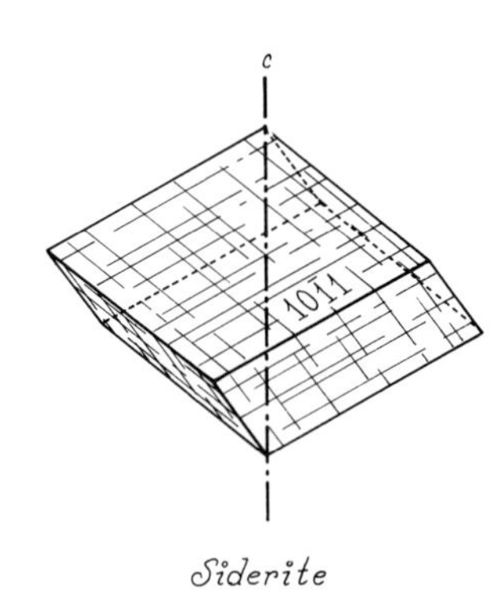

$n_\omega = 1.875$*
$n_\varepsilon = 1.633$†
$n_\omega - n_\varepsilon = 0.242$*
Uniaxial negative
Colorless to pale yellow or yellow-brown

COMPOSITION. Pure siderite, $FeCO_3$, is rare, as complete solid solution exists with magnesite, $MgCO_3$, and rhodochrosite, $MnCO_3$. Manganoan varieties have been called *oligonite* (5–40 percent $MnCO_3$) and magnesian varieties, *sideroplesite* (5–30 percent $MgCO_3$) and *pistomesite* (30–50 percent $MgCO_3$). Limited solid solution exists with $CaCO_3$, and minor Co and Zn may substitute for Fe^{2+}. Impure siderite is often admixed with clays or microcrystalline silica.

PHYSICAL PROPERTIES. H = 4–4½. Sp. Gr. = 3.96. Color, in hand sample, is yellow-brown or reddish-brown to gray or greenish-gray. Luster is vitreous to pearly.

---

* Indices of refraction and birefringence decrease with substitution of Mn^{2+}, Mg^{2+}, or Ca^{2+} for Fe^{2+} (Fig. 2-7).

† The extraordinary ray seen on cleavage rhombs shows $n'_\varepsilon = 1.748$.

COLOR AND PLEOCHROISM. Colorless to pale yellow or yellow-brown in section or as fragments.

FORM. Siderite normally occurs as fine to coarse-grained anhedral aggregates. Euhedral crystals are usually rhombohedrons $\{10\bar{1}1\}$, and faces are often curved. Nodular or botryoidal forms show coarse radiated fibers. Siderite may also be oolitic or earthy.

CLEAVAGE. Perfect rhombohedral cleavages $\{10\bar{1}1\}$ are characteristic of trigonal carbonates and dominate fragment orientation.

BIREFRINGENCE. Maximum birefringence is extreme $(n_\omega - n_\varepsilon) = 0.242$ decreasing with substitution of Mg, Mn, or Ca for Fe^{2+}. Interference colors are very high-order uniform white or gray even for orientations near the optic axis.

TWINNING. Lamellar twinning on the negative rhombohedron $\{01\bar{1}2\}$ is sometimes observed, and simple twinning on $\{0001\}$ is rare.

INTERFERENCE FIGURE. Siderite is uniaxial negative. Basal sections show a distinct uniaxial cross with myraids of fine isochromatic rings, and isogyres fan broadly toward the periphery of the figure. Cleavage fragments yield highly off-centered figures.

OPTICAL ORIENTATION. Extinction is symmetrical or inclined to cleavage traces.

DISTINGUISHING FEATURES. Perfect rhombohedral cleavages, extreme birefringence, and solution in weak acids are characteristic of the trigonal carbonates. Siderite has higher indices of refraction than other common carbonates, and iron oxide alteration is distinctive.

ALTERATION. Siderite alters to iron oxides—usually goethite or limonite; rarely, hematite or magnetite. Pseudomorphs after siderite are commonly formed by iron oxides, quartz, cacoxenite, chlorite, calcite, and barite.

OCCURRENCE. The most common occurrence of siderite is in bedded sedimentary deposits. Fine-grained siderite in association with clays, chert, chamosite, greenalite, and minnesotaite form clay ironstones, banded siderite cherts, and siderite mudstones and limestones; concretions of botryoidal or fibrous siderite are common in many rock types. Sedimentary siderite represents the precipitation of soluble ferrous bicarbonates in oxygen-poor environments, sometimes by biogenic processes, and is commonly interbedded with coal, underclays, and similar bog deposits.

Siderite, often Mn-rich, is a rather common gangue mineral in high-temperature hydrothermal veins where it occurs with calcite, ankerite, fluorite, barite, sphalerite, galena, cassiterite, tetrahedrite, and silver minerals. It is a rare mineral in pegmatites and occurs with cryolite at Ivigtut, Greenland, in a pegmatite-pneumatolitic deposit. Globules of fibrous siderite or euhedral rhombohedrons appear in fissures or cavities in basalt, andesite, and diabase. Siderite may be the mafic mineral in unusual carbonatites.

Metamorphic siderite results from the recrystallization of sideritic sediments or metasomatism of carbonate rocks. It is a rare constituent of mica schists.

REFERENCES: SIDERITE

Babcan, J. 1970. Low-temperature synthesis of siderite. *Geol. Zh.*, *21*, 89–97.

Bagin, V. I., and R. S. Rybak. 1970. Temperature conversion of siderite. *Izv. Akad. Nauk SSSR, Fiz. Zemli, 1970*, 101–106.

Johannes, W. 1969. Formation of deposits of magnesite and siderite in the system Mg^{2+}-Fe^{2+}-$CO_3{}^{2-}$-$Cl_2{}^{2-}$-H_2O. *Beitr. Mineral. Petrogr.*, *21*, 311–318.

Rhodochrosite

$MnCO_3$ TRIGONAL

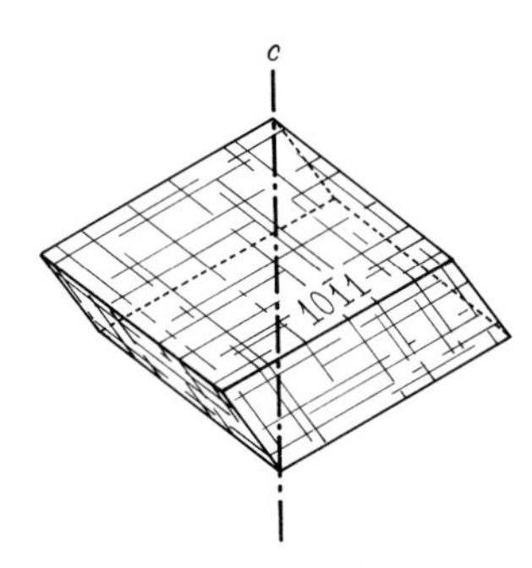

$n_\omega = 1.816$*
$n_\varepsilon = 1.597$†
$n_\omega - n_\varepsilon = 0.219$*
Uniaxial negative
Colorless or pale pink, absorption $O > E$

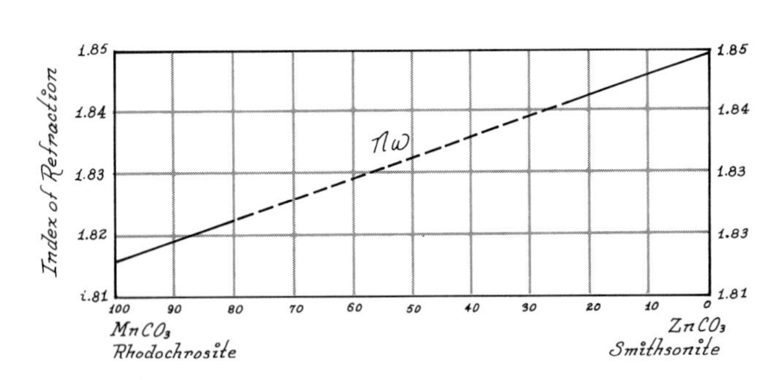

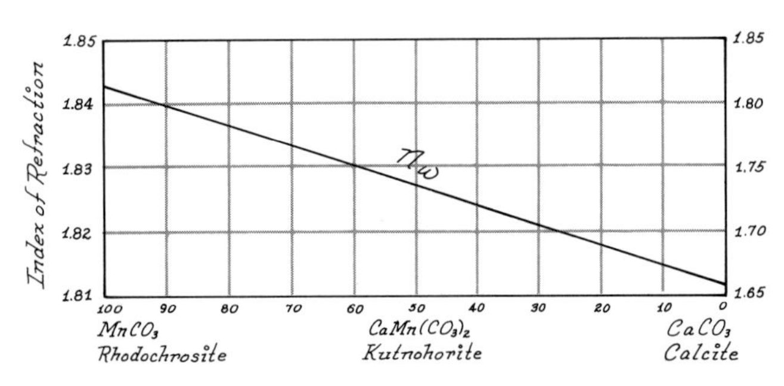

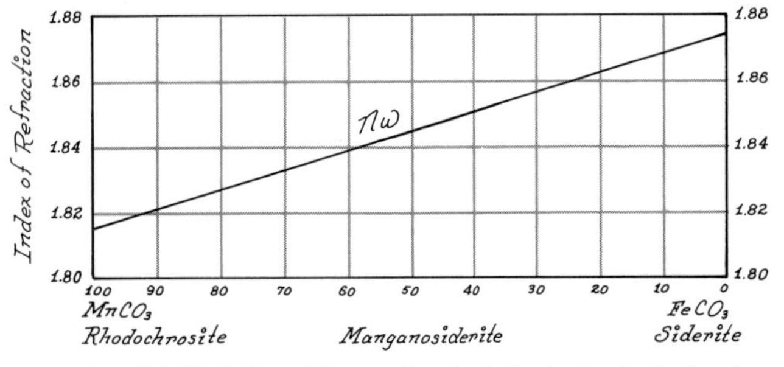

FIGURE 2-8. Variation of the n_ω refractive index in the rhodochrosite-smithsonite, rhodochrosite-calcite, and rhodochrosite-siderite series.

COMPOSITION. Pure rhodochrosite, $MnCO_3$, is rare, as complete solid solution probably exists with both calcite, $CaCO_3$, and siderite, $FeCO_3$. *Ponite* is Fe^{2+}-rich rhodochrosite that becomes *manganosiderite* when Mn:Fe $\simeq$ 1:1. *Kutnohorite* $CaMn(CO_3)_2$ is a distinct dolomite-type mineral between rhodochrosite and calcite. Partial solid solution exists with magnesite, $MgCO_3$; smithsonite, $ZnCO_3$, and cobaltocalcite, $CoCO_3$, and a Fe^{2+}-Zn variety is called *capillite.*

PHYSICAL PROPERTIES. H = $3\frac{1}{2}$–4. Sp. Gr. = 3.70. Color, in hand sample, is pink to rose-red, becoming yellow or brownish with Fe^{2+}; also gray. Luster is vitreous to pearly.

COLOR AND PLEOCHROISM. Colorless to pale pink in section or as fragments with pleochroism in pink $O > E$.

---

* Indices of refraction and birefringence decrease with increasing Ca or Mg and increase with Fe^{2+}, Co, or Zn (Fig. 2-8).

† The extraordinary ray seen on cleavage rhombs shows $n_\varepsilon' = 1.702$.

FORM. Rhodochrosite normally occurs as massive coarse to fine cleavage aggregates. Euhedral crystals are rare (rhombohedron or scalenohedron forms and commonly rounded). Banded botryoidal masses are common.

CLEAVAGE. Perfect rhombohedral cleavages $\{10\bar{1}1\}$ are characteristic of trigonal carbonates and dominate fragment orientation.

BIREFRINGENCE. Maximum birefringence is extreme ($n_\omega - n_\varepsilon = 0.219$) and increases with substitution of Fe^{2+}, Zn^{2+}, or Co^{2+} for Mn^{2+}. Birefringence is less on cleavage rhombs ($n_\omega - n_\varepsilon' = 0.114$) but still high. Interference colors are very high-order, uniform white or gray even for orientations near the optic axis.

TWINNING. Rare lamellar twinning is reported on $\{01\bar{1}2\}$, and translational gliding may be present on $\{0001\}$.

ZONING. Composition zoning may be evident by color or refractive index; deep colors accompany high index.

INTERFERENCE FIGURE. Rhodochrosite is uniaxial negative. Basal sections show a distinct uniaxial cross with myriads of fine isochromatic rings and isogyres that fan broadly toward the periphery. Cleavage fragments yield highly eccentric figures.

OPTICAL ORIENTATION. Extinction is symmetrical or inclined to rhombohedral cleavages.

DISTINGUISHING FEATURES. Perfect rhombohedral cleavages, extreme birefringence, and solution in weak acids are characteristic of the trigonal carbonates. Rhodochrosite is normally associated with other manganese minerals and alters to black Mn-oxides. Its indices of refraction are greater than those of calcite, magnesite, dolomite, or ankerite and less than siderite or smithsonite. Rhodochrosite is usually rose-red in hand sample.

ALTERATION. Rhodochrosite alters on exposed surfaces to a mixture of black or brown oxides of Mn and Fe. Complete alteration normally yields pyrolusite, MnO_2, or manganite, $MnO \cdot OH$. Rhodochrosite may be dissolved and replaced by other minerals (such as quartz) as pseudomorphs.

OCCURRENCE. Rhodochrosite is a common gangue mineral in moderate- to low-temperature hydrothermal veins and limestone replacement bodies with calcite, siderite, dolomite, fluorite, barite, quartz, and the ore minerals sphalerite, galena, alabandite, tetrahedrite, and Ag-minerals. It is a rare mineral in pegmatites with lithiophilite and amblygonite.

In metamorphic rocks, rhodochrosite is most common in metamorphic zones in association with other Mn-minerals (such as rhodonite, braunite, hausmannite, spessartine, friedelite, tephroite, and alabandite). It may also appear in schists derived from Mn-rich sediments.

In sedimentary rocks, rhodochrosite may be either syngenetic or diagenetic and tends to be associated with iron silicates and carbonates. It may also result from the alteration of other Mn-minerals.

REFERENCES: RHODOCHROSITE

Bank, H. 1975. Low refractive indices and birefringence of nearly transparent rhodochrosite from Argentina. *Dtsch. Gemmol. Ges. Z.*, *24*, 160–161.

Egan, W. G., and T. Hilgemann. 1971. Optical properties of naturally occurring rhodochrosite between 0.33 and 2.5 μ. *Appl. Opt.*, *10*, 2132–2136.

Smithsonite

$ZnCO_3$

TRIGONAL

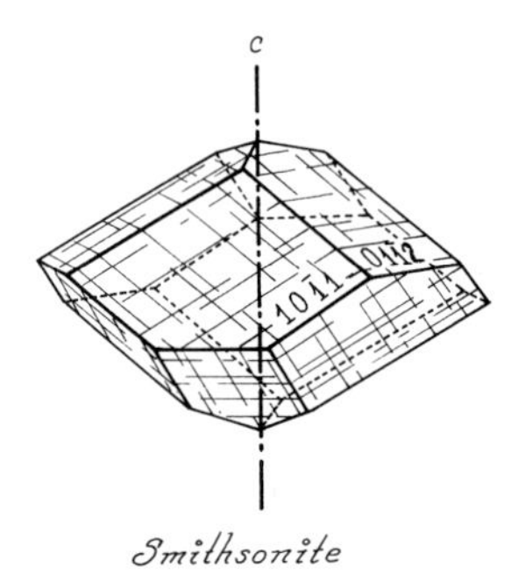

$n_\omega = 1.850$*
$n_\varepsilon = 1.625$†
$n_\omega - n_\varepsilon = 0.225$*
Uniaxial negative
Colorless in standard section

COMPOSITION. Pure smithsonite, $ZnCO_3$, is rare as Fe^{2+}, Mn, Mg, and Ca commonly substitute for some Zn; substitution of Cu, Co, Cd, and Pb is less common. Solid solution is probably nearly complete with siderite, $FeCO_3$, and rhodochrosite, $MnCO_3$, and the composition nearly intermediate between smithsonite and siderite has been called *monheimite*.

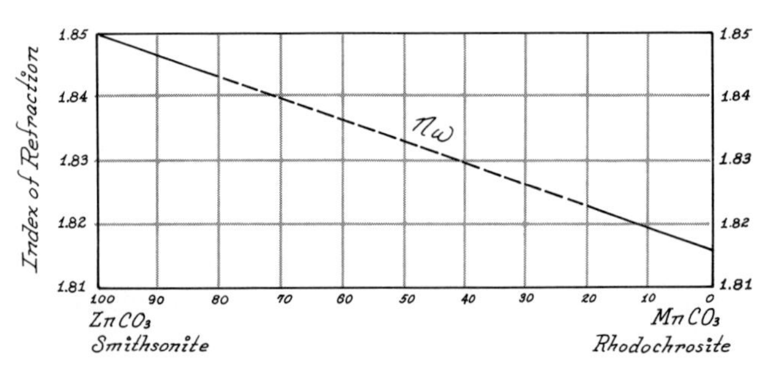

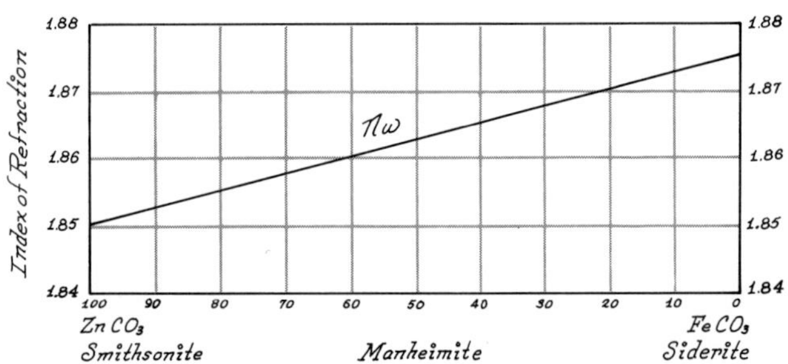

FIGURE 2-9. Variation of the n_ω refractive index in the smithsonite-rhodochrosite and the smithsonite-siderite series.

PHYSICAL PROPERTIES. H = 4–4½. Sp. Gr. = 4.43. Color, in hand sample, is white or gray to pale blue, green, or brown. Luster is vitreous to pearly.

COLOR AND PLEOCHROISM. Smithsonite is colorless in section or as fragments.

FORM. Smithsonite normally appears as botryoidal or stalactitic masses, often as secondary incrustations. Euhedral crystals (rhombohedrons) are rare, and faces are generally rough and curved.

CLEAVAGE. Rhombohedral cleavages $\{10\bar{1}1\}$ are somewhat less perfect than in other trigonal carbonates.

BIREFRINGENCE. Maximum birefringence is extreme ($n_\omega - n_\varepsilon = 0.225$), and interference colors are very high-order white even for orientations near the optic axis.

* Indices of refraction and birefringence increase slightly with Fe^{2+} and decrease with substitution of Mn, Mg, or Ca for Zn (Fig. 2-9).

† Index of refraction for the extraordinary ray seen on cleavage rhombs is $n_\varepsilon' = 1.723$.

TWINNING. None reported.

INTERFERENCE FIGURE. Smithsonite is uniaxial negative. Basal sections show a distinct uniaxial cross with innumerable isochromatic rings and isogyres that fan away from the melatope. Cleavage fragments yield highly off-centered figures.

OPTICAL ORIENTATION. Extinction is symmetrical or inclined to cleavage traces.

DISTINGUISHING FEATURES. Rhombohedral cleavages, extreme birefringence, and solution in weak acids are characteristic of the trigonal carbonates. Botryoidal form, lack of twinning, and limited occurrence are distinctive of smithsonite.

ALTERATION. Smithsonite may alter to hemimorphite and limonite. It may be dissolved and replaced as pseudomorphs by quartz, limonite, or pyrolusite, and, being a secondary mineral itself, smithsonite may form pseudomorphs after calcite, dolomite, or anglesite, and is a common alteration product of sphalerite.

OCCURRENCE. Smithsonite is a secondary mineral found in the oxide zones of ore deposits containing primary zinc minerals, especially sphalerite. It is associated with hemimorphite, willemite, cerussite, anglesite, malachite, azurite, pyromorphite, hydrozincite, and so on.

REFERENCE: SMITHSONITE

Chen, T. T., and L. L. Y. Chang. 1975. High-magnesium smithsonite from Broken Hill, New South Wales, Australia. *Mineral. Mag.*, *40*, 307–308.

The Dolomite-Ankerite Series

$CaMg(CO_3)_2$–$Ca(Mg,Fe)(CO_3)_2$ TRIGONAL

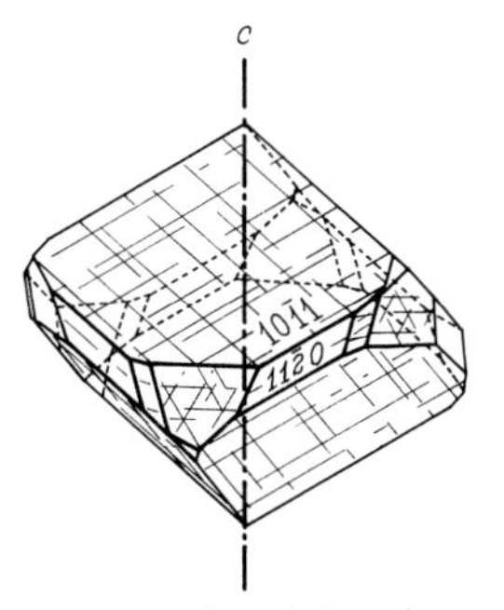

Dolomite-Ankerite

Dolomite	Ankerite
$n_\omega = 1.679$*	$n_\omega = 1.690$–1.750*
$n_\varepsilon = 1.500$†	$n_\varepsilon = 1.510$–1.548
$n_\omega - n_\varepsilon = 0.179$*	$n_\omega - n_\varepsilon = 0.180$–$0.202$*
Uniaxial negative	Uniaxial negative
Colorless in standard section	Colorless in standard section

* Indices of refraction and birefringence increase with substitution of Fe^{2+} for Mg; less rapidly with Mn for Mg (Fig. 2-10).
† Index of refraction for the extraordinary ray seen on cleavage rhombs is $n_\varepsilon' = 1.588$.

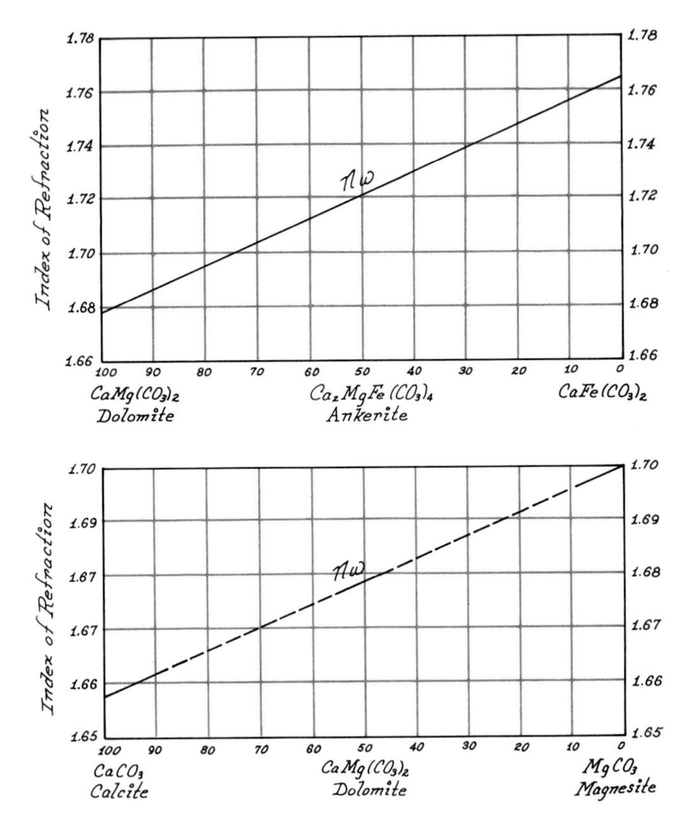

FIGURE 2-10. Variation of the n_ω refractive index in the dolomite-$CaFe(CO_3)_2$ and calcite-dolomite-magnesite series.

COMPOSITION. The structure of dolomite is similar to that of calcite with ordered distribution of equal numbers of Ca and Mg ions. Minor excesses of Mg or especially Ca are possible, but dolomite composition is usually near the ideal $CaMg(CO_3)_2$.

Continuous replacement of Mg by Fe leads to ankerite compositions $Ca(Mg,Fe^{2+})(CO_3)_2$, although the ferrous end member $CaFe^{2+}(CO_3)_2$ does not exist in nature, and any line of separation between dolomite and ankerite is necessarily arbitrary and has not been firmly established. Deer, Howie, and Zussman (1962) suggest Mg:Fe > 4:1 for dolomite. On the ankerite end, solid solution extends beyond "ideal" ankerite to at least $Ca(Mg_{0.3}Fe_{0.7})(CO_3)_2$.

Although most dolomite contains little Mn, a similar solid solution series exists between dolomite, $CaMg(CO_3)_2$, and kutnohorite, $CaMn(CO_3)_2$, with Mg:Mn < 1:1 for kutnohorite. Dolomite compositions containing both Fe^{2+} and Mn are common.

Minor Co and Zn may also replace Mg, and Ba and Pb may replace some Ca.

PHYSICAL PROPERTIES. H = $3\frac{1}{2}$–4. Sp. Gr. = 2.86–3.10, increasing with Fe^{2+}. Color, in hand sample, is white to yellow or brown. Varieties containing Mn, or Co, tend to be pinkish. Luster is vitreous to pearly.

COLOR AND PLEOCHROISM. Both dolomite and ankerite are colorless in section or as fragments.

FORM. Dolomite and ankerite appear most commonly as fine to coarse, anhedral cleavage aggregates; euhedral crystals, however, are rather common as rhombohedrons usually with curved faces. Dolomite is rarely columnar, fibrous, or pisolitic.

CLEAVAGE. Perfect rhombohedral cleavages $\{10\bar{1}1\}$ are characteristic of trigonal carbonates and dominate fragment orientation.

BIREFRINGENCE. Maximum birefringence is extreme: $n_\omega - n_\varepsilon = 0.179$, increasing to about 0.202 as Fe^{2+} replaces Ca. Interference colors are very high-order white even for orientations near the optic axis. Twin lamellae, which fail to extinguish with stage rotation, show somewhat lower birefringence.

TWINNING. Shear pressures produce glide twin lamellae on $\{02\bar{2}1\}$; not as ubiquitous as in calcite. Simple twinning is also common on $\{0001\}$, $\{10\bar{1}0\}$, and $\{11\bar{2}0\}$; rarely on $\{10\bar{1}1\}$.

ZONING. Single crystals may show composition zoning resulting from the replacement of Mg by Fe^{2+} or Mn, which is expressed by graduation in indices of refraction.

INTERFERENCE FIGURE. Both dolomite and ankerite are uniaxial negative; rarely biaxial negative with small $2V_x$ due to strain. Basal sections show isogyres, which broaden rapidly away from the melatope, and very numerous isochromatic rings. Cleavage rhombs yield highly eccentric figures.

OPTICAL ORIENTATION. Extinction is symmetrical or inclined to cleavage traces. The angle between the plane of the extraordinary wave and the trace of twin lamellae is usually less than 40°.

DISTINGUISHING FEATURES. Perfect rhombohedral cleavages $\{10\bar{1}0\}$, extreme birefringence, variable relief, and effervescence in weak acids are characteristic features of trigonal carbonates. Calcite grains are more commonly anhedral and more commonly twinned. Twin lamellae in calcite are parallel or oblique to the long diagonal of cleavage rhombs; in dolomite, lamellae may parallel the long or short diagonal. Dolomite (Sp. Gr. = 2.86) is more dense than calcite (2.71) and much less responsive to weak acids, and its refractive indices are greater ($n'_\varepsilon = 1.588$ *vs* 1.566). A universal stage procedure for distinguishing calcite, dolomite, and magnesite is outlined in Phillips (1971, p. 201).

Indices of refraction increase from dolomite to ankerite, passing through those of magnesite. Lamellar twinning is rare in magnesite.

Orthorhombic carbonates are biaxial (small $2V_x$) and do not show rhombohedral cleavage.

ALTERATION. Dolomite is commonly a secondary mineral. It replaces calcite as pseudomorphs or as entire sedimentary formations through the action of Mg-rich meteoric or hydrothermal solutions on limestone. Pseudomorphs of dolomite are also common after aragonite, barite, fluorite, and halite. Dolomite, in turn, may be replaced by pseudomorphs of calcite, siderite, smithsonite, quartz, limonite, hematite, pyrite, and so on.

With high-grade, thermal metamorphism, dolomite may dissociate to calcite and periclase or brucite.

OCCURRENCE. Most dolomite occurs as the major consituent of extensive, bedded sedimentary formations associated with more or less calcite. Most dolomite formations are considered to have been derived from limestone beds by the action of magnesian waters from marine or hydrothermal sources. Dolomitization may influence a single, widespread sedimentary horizon of marine limestone before consolidation or irregular zones of well-lithified limestone where solutions gain access through fracture zones.

Primary dolomite is formed by direct precipitation from abnormally saline, isolated marine waters or saline lakes and is associated with anhydrite, gypsum, halite, polyhalite, sylvite, and other evaporite minerals.

Ankerite is not a major mineral in carbonate sediments but may appear as veins or concretions in Fe-rich sediments with siderite, iron oxides, and carbonaceous matter.

Regional or contact metamorphism may cause the recrystallization of dolomite or magnesian limestone to form dolomitic marble. Contact thermal metamorphism may decompose dolomite to calcite and periclase or brucite, and silicate impurities commonly combine with free magnesia to yield magnesian silicates (for example, talc, forsterite, chlorite, tremolite, phlogopite, chondrodite, dravite, and diopside) in association with calcite. High-grade regional metamorphism may yield schists containing ankerite with cummingtonite and garnet.

Well-crystallized dolomite occurs in fissures and cavities in carbonate sediments or in hydrothermal veins in association with calcite; siderite, fluorite, barite, quartz, sphalerite, chalcopyrite, and galena. Ankerite is characteristic of ore zones containing fluorite, galena, and sulfosalt minerals and is usually present with siderite. Dolomite is a rare mineral in pegmatites, and ankerite may appear in hypothermal veins with gold, tourmaline (dravite), scheelite, pyrite, and rutile.

Primary igneous dolomite or ankerite appears in a few highly alkaline dike rocks and rare carbonatite bodies. Ultramafic igneous rocks altered by hydrothermal solutions may yield dolomite in association with magnesite, serpentine, and talc.

REFERENCES: THE DOLOMITE-ANKERITE SERIES

Althoff, P. L. 1977. Structural refinements of dolomite and a magnesian calcite and implications for dolomite formation in the marine environment. *Amer. Mineral., 62*, 772–783.

Barron, B. J. 1974. Use of coexisting calcite-ankerite solid solutions as a geothermometer. *Contrib. Mineral Petrol., 47*, 77–80.

Beran, A. 1975. Microprobe analyses of ankerites and siderites from the Styrean Erzberg. *Tschermaks Mineral. Petrogr. Mitt., 22*, 250–265.

Deer, W. A., R. A. Howie, and J. Zussman. 1962. *Rock-Forming Minerals.* Vol. 5: *Non-Silicates.* New York: Wiley.

Howell, J. E., and K. R. Dawson. 1958. Technique for optical determination of iron-bearing dolomites. *Can. Mineral., 6*, 292–294.

Land, L. S., E. Mutio, and W. F. Bradley. 1972. Crystallochemical and geochemical comparisons of recent with older dolomites. *Fla. Geol. Surv. Spec. Publ.* No. 17.

Nakissa, M., and P. Paulitsch. 1972. Biaxial dolomite lamellae. *Contrib. Mineral. Petrol., 34*, 224–228.

Phillips, W. R. 1971. *Mineral Optics: Principles and Techniques.* San Francisco: W. H. Freeman and Company.

Rath, R., and D. Pohl 1971. Biaxiality phenomenon in calcite and dolomite. *Contrib. Mineral. Petrol., 32*, 74–78.

Schneider, H. 1976. The progressive crystallization and ordering of low-temperature dolomites. *Mineral. Mag., 40*, 579–587.

Aragonite

$CaCO_3$

ORTHORHOMBIC

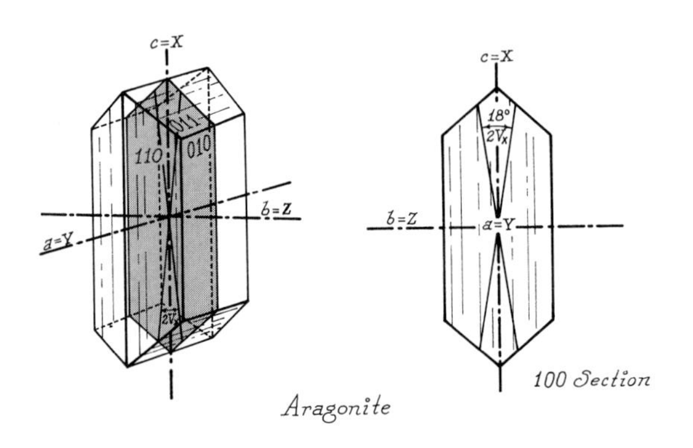

$n_\alpha = 1.530$*
$n_\beta = 1.680$
$n_\gamma = 1.685$
$n_\gamma - n_\alpha = 0.155$
Biaxial negative $2V_x = 18°$†
$a = Y, b = Z, c = X$
$v > r$ weak
Colorless in standard section

COMPOSITION. Aragonite composition commonly approaches the theoretical $CaCO_3$. Strontian and plumbian varieties, however, are common and are called *mossottite* and *tarnowitzite*, respectively. Solid solutions to about Sr:Ca = 1:25 and Pb:Ca = 1:12 are reported. Minor Zn or Ba may also replace Ca.

PHYSICAL PROPERTIES. H = $3\frac{1}{2}$–4. Sp. Gr. = 2.95, increasing rapidly with Pb. Color, in hand sample, is white, colorless, or gray ranging to yellow, blue, violet, green, or rose-red. Luster is vitreous.

COLOR AND PLEOCHROISM. Aragonite is colorless in section or as fragments.

FORM. Aragonite occurs mostly as parallel or radiated aggregates of stout columnar to acicular or fibrous crystals elongated on *c*. Simple crystals are much less common than pseudohexagonal columns formed by cyclic twinning. Stalactitic, colloform, and corallike (flos-ferri) forms are characteristic of hot spring and cave deposits.

CLEAVAGE. Cleavage on {010} is imperfect; very poor on {110} and {011}.

BIREFRINGENCE. Maximum birefringence, seen in {100} sections, is extreme ($n_\gamma - n_\alpha = 0.155$); cleavage fragments also show extreme birefringence ($n_\beta - n_\alpha = 0.150$). Interference colors are usually very high-order white; however, basal sections show low-order colors.

TWINNING. Extremely common cyclic twinning on {110} produces pseudohexagonal columns that show deep striations parallel to *c* and subtle radial striations on basal planes. Basal sections of cyclic twins show sixfold radiated sectors.

Polysynthetic twinning on {110} also yields striations parallel to *c* and parallel lamellae in thin section. Twinning requires only reorientation of CO_3^{2-} groups in {001} planes.

ZONING. Plumbian varieties may show composition zoning by gradation in refractive index.

INTERFERENCE FIGURE. Aragonite is biaxial negative, with small $2V_x$ near 18°, which may range to about 23° with increasing Pb. Basal sections show distinct, centered, acute bisectrix figures with very numerous, nearly circular isochromatic bands. Isogyres are very thin at melatopes but broaden rapidly toward the periphery of the figure and separate only slightly with stage rotation. Dispersion is optic axis type only, $v > r$ weak. Cleavage fragments yield Bxo figures indistinguishable from flash figures.

OPTICAL ORIENTATION. Aragonite is orthorhombic with $a = Y, b = Z$, and $c = X$. Extinction is parallel to cleavage traces and crystal elongation. Since $c = X$, crystals elongated on *c*

* Indices of refraction decrease as Sr replaces Ca to about $n_\alpha = 1.527$, $n_\beta = 1.670$, $n_\gamma = 1.676$, and increase with Pb to about $n_\alpha = 1.542$, $n_\beta = 1.695$, $n_\gamma = 1.699$.

† As Pb replaces Ca, $2V_x$ widens to a maximum of about 23°.

are length-fast (negative elongation). Cleavage traces always parallel the faster wave.

DISTINGUISHING FEATURES. Aragonite resembles calcite because of its extreme birefringence but is biaxial, lacks rhombohedral cleavages, and has somewhat higher indices of refraction. Refractive indices of aragonite are also slightly higher than those of strontianite ($SrCO_3$) or witherite ($BaCO_3$) but much lower than those of cerussite ($PbCO_3$). Each of these orthorhombic carbonates has much greater specific gravity than aragonite Fragments of all three sink in methylene iodide (the 1.74 index oil), and flame tests for Sr, Ba, and Pb are distinctive. Zeolites may resemble aragonite as colorless, columnar, or accicular crystals in vesicles, but zeolites show weak birefringence and negative relief.

ALTERATION. Aragonite inverts to its stable polymorph calcite in most environments, and pseudomorphs of calcite after aragonite are very common. High strontium content inhibits the inversion. Aragonite is known to replace gypsum, and pseudomorphs after aragonite are sometimes composed of dolomite, deweylite, native copper, and so on.

OCCURRENCE. Sedimentary aragonite is much less common than calcite and is formed only in low-temperature, near-surface deposits of rather recent geologic origin.

Aragonite is deposited as sinter in hot springs and as travertine dripstone in caves (often rich in Sr, Ba, and Pb). It is formed by direct precipitation in modern oceans as aragonite needles or ooliths in calcareous mud and may appear with gypsum, sulfur, or celestite in marl or shales of evaporite deposits. Shells of certain marine organisms are largely aragonite, with or without calcite, and pearls are usually aragonite.

Aragonite may occur in amygdules in basalt or andesite with zeolites and in veins in serpentine with dolomite, magnesite, or brucite. It is a rare mineral in the oxide zones of ore deposits, occurring with limonite, malachite, azurite, smithsonite, cerussite, and so on, and may occur with limonite and siderite in sedimentary iron ores.

Aragonite is a characteristic mineral of high-pressure, low-temperature (blueschist facies) metamorphism, where it occurs with glaucophane-crossite, lawsonite, and aegirine-albite rocks.

REFERENCES: ARAGONITE

Boettcher, A. L., and P. J. Wyllie. 1967. Biaxial calcite inverted from aragonite. *Amer. Mineral.*, *52*, 1527–1529.

Cowley, E. R. 1970. Refractive indices of some carbonate minerals in the point dipole approximation. *Can. Jour. Phys.*, *48*, 297–302.

De Villiers, J. P. R. 1971. Crystal structures of aragonite, strontianite, and witherite. *Amer. Mineral.*, *56*, 758–767.

Dickens, B. and J. S. Bowen. 1971. Refinement of the crystal structure of the aragonite phase of calcium carbonate. *Jour. Res. Nat. Bur. Stand., Sect. A*, *75*, 27–31.

Folk, R. L. 1974. Natural history of crystalline calcium carbonate. Effect of magnesium content and salinity. *Jour. Sed. Petrol.*, *44*, 40–53.

Hiragi, Y., S. Kachi, T. Takada, and N. Nakanishi. 1966. The superstructure in fine aragonite particles. *Bull. Chem. Soc. Japan*, *39*, 2361–2364.

Isherwood, B. J., and J. A. James. 1976. Structural dependence of the optical birefringence of crystals with calcite and aragonite type structures. *Acta Crystallogr. A32*, 340–341.

MacDonald, G. J. F. 1956. Experimental determination of calcite-aragonite equilibrium relations at elevated temperatures and pressures. *Amer. Mineral.*, *41*, 744–756.

Rath, R., and D. Pohl. 1971. Optical behavior of twin lamellae. *Contrib. Mineral. Petrol.*, *33*, 239–244.

Strontianite $SrCO_3$ ORTHORHOMBIC

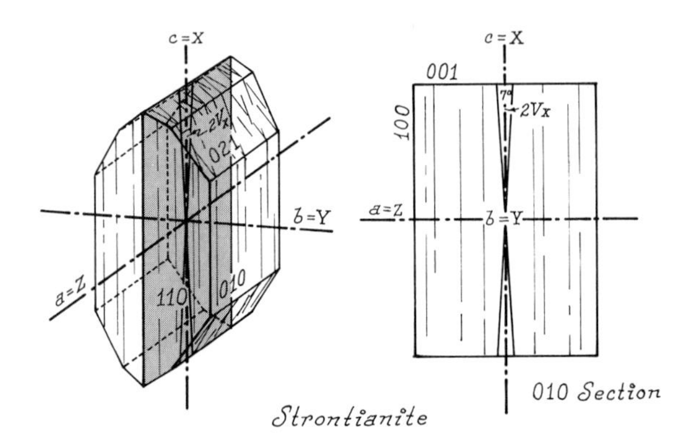

$n_\alpha = 1.516$*
$n_\beta = 1.664$
$n_\gamma = 1.666$
$n_\gamma - n_\alpha = 0.150$
Biaxial negative $2V_x = 7°$
$a = Z, b = Y, c = X$
$v > r$ weak
Colorless in standard section

COMPOSITION. Strontianite is a member of the aragonite group; it usually contains some Ca in substitution for Sr to about Ca:Sr ≃ 1:3. Complete solid solution is possible between strontianite $SrCO_3$ and witherite $BaCO_3$, although Ba substitution appears to be highly restricted in most natural occurrences.

PHYSICAL PROPERTIES. H = $3\frac{1}{2}$. Sp. Gr. ≃ 3.75. Color, in hand sample, is white or gray to pale yellow, brown, or green. Luster is vitreous.

COLOR AND PLEOCHROISM. Strontianite is colorless in section or as fragments.

FORM. Strontianite occurs mostly as crude, pseudohexagonal, prismatic to acicular crystals elongated on c; also fibrous and granular.

CLEAVAGE. Prismatic cleavages are distinct on {110}, and poor cleavages may be visible on {021} and {010}.

BIREFRINGENCE. Maximum birefringence, seen in {010} sections, is extreme ($n_\gamma - n_\alpha = 0.150$), and interference colors are very high-order whites and gray except in basal sections {001}, where birefringence is small. Cleavage fragments show near maximum birefringence.

TWINNING. Twinning is very common on {110} as simple contact, cyclic, or polysynthetic lamellae.

INTERFERENCE FIGURE. Strontianite is biaxial negative with small $2V_x \simeq 7°$ to 10°. Basal sections show centered acute bisectrix figures with very numerous isochromatic rings nearly circular about closely spaced melatopes. Fragments on prismatic cleavages yield eccentric flash figures.

OPTICAL ORIENTATION. Extinction is parallel to crystal elongation and prominent cleavages. Elongation of crystals and cleavage fragments is negative (length-fast).

DISTINGUISHING FEATURES. Extreme birefringence, parallel extinction and vigorous reaction to weak acids is characteristic of orthorhombic carbonates. Strontianite has somewhat lower indices of refraction than witherite and greater specific gravity than aragonite (fragments of strontianite sink in the 1.74 index oil—methylene iodide). The intense vermillion flame of Sr is distinctive.

ALTERATION. Strontianite may alter to celestite and is easily soluble in acid solutions.

OCCURRENCE. Strontianite is deposited in veins, cavities, and concretions in limestone and calcareous clays by low-temperature hydrothermal solutions, usually associated with calcite, celestite, barite, or gypsum. It may also appear as a gangue mineral in metalliferous veins with fluorite, galena, and silver minerals. Strontianite is a product of celestite alteration.

REFERENCES: STRONTIANITE

De Villiers, J. P. R. 1971. Crystal structures of aragonite, strontianite, and witherite. *Amer. Mineral.*, *56*, 758–767.

Speer, J. A., and M. L. Hensley-Dunn. 1976. Strontianite composition and physical properties. *Amer. Mineral.*, *61*, 1001–1004.

* Indices of refraction increase with Ca substitution to about $n_\alpha = 1.520$, $n_\beta = 1.667$, $n_\gamma = 1.668$.

Witherite $BaCO_3$ ORTHORHOMBIC

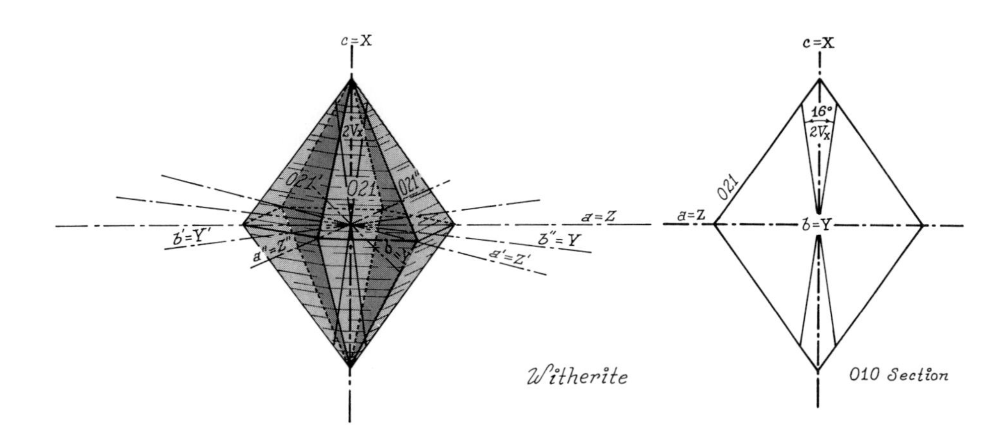

$n_\alpha = 1.529$
$n_\beta = 1.676$
$n_\gamma = 1.677$
$n_\gamma - n_\alpha = 0.148$
Biaxial negative $2V_x = 16°$
$a = Z, b = Y, c = X$
$v > r$ very weak
Colorless in standard section

COMPOSITION. Witherite is commonly nearly pure $BaCO_3$. Although complete solid solution is possible with strontianite, $SrCO_3$, and partial solution with aragonite, $CaCO_3$, substitution of either Sr^{2+} or Ca^{2+} for Ba^{2+} appears very limited in natural occurrences. Alstonite $CaBa(CO_3)_2$ is a distinct mineral variety with Ca and Ba ions in ordered arrangement.

PHYSICAL PROPERTIES. H = $3\frac{1}{2}$. Sp. Gr. = 4.29. Color, in hand sample, is white or gray to pale yellow, brown, or green. Luster is vitreous.

COLOR AND PLEOCHROISM. Witherite is colorless in section or as fragments.

FORM. Crystals of witherite are always stubby, pseudohexagonal dipyramids usually having very rough, horizontally striated faces. Witherite sometimes appears in botryoidal or globular forms; rarely coarse fibrous or granular.

CLEAVAGE. Cleavage is distinct on {010}; poor on {110} and {012}.

BIREFRINGENCE. Maximum birefringence, seen on {010} sections and cleavage fragments, is extreme ($n_\gamma - n_\alpha = 0.148$). Interference colors are very high-order white and gray except on basal {001} sections, where birefringence is weak.

TWINNING. Repeated twinning is always present on {110}, yielding pseudohexagonal crystal forms and radial sectors of different crystallographic orientation in basal sections.

INTERFERENCE FIGURE. Basal sections show centered acute bisectrix figures with very numerous isochromatic rings nearly circular around closely spaced melatopes. Fragments on dominant cleavages yield flash figures.

OPTICAL ORIENTATION. Witherite is orthorhombic with $a = Z$, $b = Y$, and $c = X$. Extinction is parallel to distinct cleavages. Elongation, defined by cleavage traces, is negative (length-fast).

DISTINGUISHING FEATURES. Witherite shows the extreme birefringence characteristic of carbonates but is distinguished from trigonal carbonates by the absence of rhombohedral cleavage. It effervesces in cold, weak acids, and the yellow-green flame of Ba is distinctive. Witherite shows higher indices of refraction than strontianite and much lower indices than cerussite. It has higher specific gravity than aragonite, and fragments sink in methylene iodide (1.74 index oil).

ALTERATION. Witherite may alter to barite or may be formed by the alteration of barite.

OCCURRENCE. Witherite occurs mostly in low-temperature hydrothermal veins and open cavities, usually in limestone or other calcareous sediments. It is commonly associated with galena and barite; less commonly, with anglesite and alstonite.

REFERENCE: WITHERITE

De Villiers, J. P. R. 1971. Crystal structures of aragonite, strontianite, and witherite. *Amer. Mineral.*, *56*, 758–767.

Cerussite — $PbCO_3$ — ORTHORHOMBIC

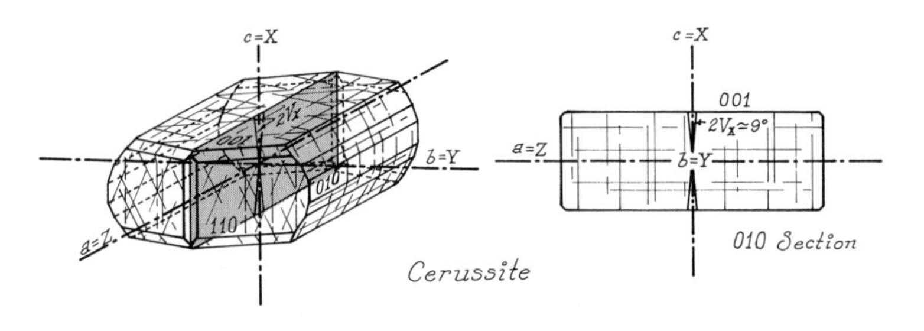

$n_\alpha = 1.803$
$n_\beta = 2.074$
$n_\gamma = 2.076$
$n_\gamma - n_\alpha = 0.273$
Biaxial negative $2V_x \simeq 9°$
$a = Z, b = Y, c = X$
$r > v$ strong
Colorless in standard section

COMPOSITION. Cerussite is usually nearly pure $PbCO_3$. Minor Sr or Zn may replace Pb and tiny inclusions of sulfide minerals or manganese oxides are common.

PHYSICAL PROPERTIES. H = $3–3\frac{1}{2}$. Sp. Gr. = 6.56. Color, in hand sample, is colorless or white; often gray to black due to sulfide inclusions; rarely green or blue due to copper impurities. Luster is distinctly adamantine.

COLOR AND PLEOCHROISM. Colorless or gray in sections or as fragments.

FORM. Cerussite usually occurs as distinct crystals or crystal aggregates, often with reticulate relationship, of extremely variable habit. Crystals are commonly tabular on {010} but may be columnar or even acicular on *c* or *a*.

CLEAVAGE. Cleavages are distinct in four directions on {110} and {021} and poor on {010} and {012}.

BIREFRINGENCE. Birefringence is very extreme ($n_\gamma - n_\alpha =$ 0.273), and interference colors are extremely high-order, showing uniform white.

TWINNING. Repeated twinning on {110} is very common, yielding pseudohexagonal cyclic forms or polysynthetic lamellae. Simple contact twinning on {130} yields heart-shaped twins.

INTERFERENCE FIGURE. Basal sections show centered acute bisectrix figures, with countless isochromatic rings nearly circular about closely spaced melatopes. Cleavage fragments yield highly eccentric figures.

OPTICAL ORIENTATION. Cerussite is orthorhombic, with $a = Z$, $b = Y$, and $c = X$. Extinction may be parallel, inclined, or symmetrical to prominent cleavages that do not favor distinct elongation.

DISTINGUISHING FEATURES. Very extreme birefringence and indices of refraction are distinctive. Anglesite is biaxial positive, with large $2V_z \simeq 75°$ and mild birefringence.

ALTERATION. Cerrusite itself is a common alteration product of galena, with anglesite representing a common intermediate alteration stage. Cerussite may replace anglesite and other secondary ore minerals as pseudomorphs and may be replaced by malachite, limonite, pyromorphite, crysocolla, calcite, and other secondary minerals.

OCCURRENCE. Cerrusite is a secondary mineral formed in the oxide zones of ore deposits in association with anglesite, limonite, pyromorphite, mimetite, smithsonite, caledonite, malachite, azurite, and other secondary ores of lead, zinc, and copper.

REFERENCE: CERUSSITE

Sahl, K. 1974. Refinement of the crystal structure of cerussite. *Z. Kristallogr.*, *139*, 215–222.

Malachite

$Cu_2(OH)_2CO_3$

MONOCLINIC
$\angle\beta = 98°44'$

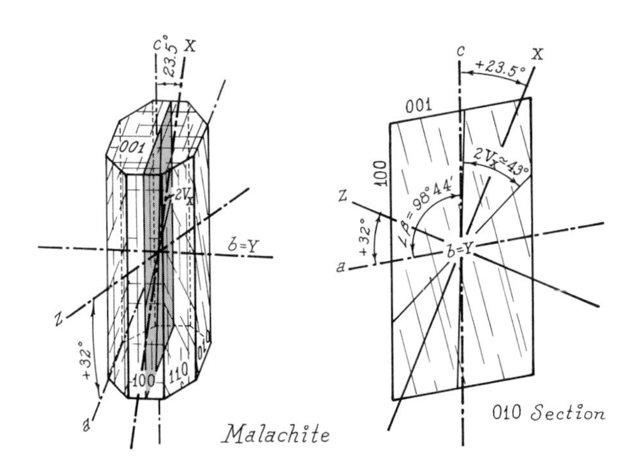

$n_\alpha = 1.655$*
$n_\beta = 1.875$
$n_\gamma = 1.909$
$n_\gamma - n_\alpha \simeq 0.244$
Biaxial negative, $2V_x \simeq 43°$*
$b = Y$, $c \wedge X = +23.5°$, $a \wedge Z \simeq +32°$
$v > r$ distinct, weak inclined bisectrix dispersion
Green and yellow-green in thin section, with distinct pleochroism $Z > Y > X$

* With Zn substitution for Cu, n_α appears to increase and n_β and n_γ decrease; birefringence decreases and $2V_x$ decreases to almost 0°.

COMPOSITION. Most malachite compositions are near the ideal $Cu_2(OH)_2CO_3$; however, appreciable Zn^{2+} may replace Cu^{2+} and zinc-rich varieties (Cu:Zn $\simeq$ 3:2) are called *rosasite*. A cobaltian malachite (14–18.5 percent Co) is known.

PHYSICAL PROPERTIES. H $= 3\frac{1}{2}$–4. Sp. Gr. $=$ 3.9–4.1. In hand sample, malachite is bright green to dark green. Luster is adamantine to vitreous, sometimes silky.

COLOR AND PLEOCHROISM. Malachite is distinctly colored and pleochroic in section or as fragments. X = colorless to pale green; Y = yellow-green; and Z = dark green.

FORM. Single crystals of malachite are rare, stubby to elongated prismatic, and nearly always twinned on {100}. Thin acicular crystals are common as divergent or parallel groups, usually as successive layers forming mammillary or botryoidal masses or incrustations.

CLEAVAGE. Perfect cleavage on $\{\bar{2}01\}$ and distinct cleavage on {010} intersect at 90° and largely control fragment orientation.

BIREFRINGENCE. Maximum birefringence, seen in {010} sections, is very extreme ($n_\gamma - n_\alpha \simeq 0.244$). Interference colors, in section or fragments, are high-order grays distinctly masked by strong green mineral color. Cleavage fragments on {010} show maximum birefringence ($n_\gamma - n_\alpha$), and fragments on $\{\bar{2}01\}$ yield ($n_\gamma - n_\beta$).

TWINNING. Simple twinning on {100} is almost universal, sometimes penetration or polysynthetic.

INTERFERENCE FIGURE. Fragments on perfect $\{\bar{2}01\}$ cleavages yield almost-centered acute bisectrix figures showing myriads of isochromes and rather strong optic axis dispersion $v > r$ and weak inclined bisectrix dispersion. Fragments on {010} yield flash figures.

OPTICAL ORIENTATION. Malachite is monoclinic with $b = Y$, $c \wedge X = +23.5°$, and $a \wedge Z = +32°$, and the optic plane is {010}. Maximum extinction angle, seen in {010} sections is 23.5° to acicular crystals elongated on c but is nearly parallel to cleavage traces. Elongation of acicular crystals lies nearest the fast wave (negative elongation). Strictly cleavage fragments should be elongated almost parallel to Z; they show near-parallel extinction and positive elongation (length-slow).

DISTINGUISHING FEATURES. Acicular green crystals showing cross cleavages, inclined extinction, and extreme birefringence are diagnostic. Other secondary copper minerals may be acicular and deep pleochroic green. Atacamite shows parallel extinction, lower birefringence ($n_\gamma - n_\alpha \simeq 0.05$), and large $2V_x$; brochantite shows near parallel extinction, and prismatic crystals are length-slow.

ALTERATION. Malachite commonly forms pseudomorphs after cuprite, azurite, chalcopyrite, tetrahedrite, pyrite, and many other minerals. Malachite rarely alters to azurite or cuprite.

OCCURRENCE. Malachite is a secondary mineral appearing in the oxide zones of copper deposits. It is very closely associated with azurite, although more widespread. It is also associated with chrysocolla, cuprite, tenorite, limonite, chalcedony, calcite, atacamite, brochantite, and many other secondary minerals of copper, lead, zinc, iron, and manganese.

REFERENCES: MALACHITE

Deliens, M., R. Oosterbosch, and T. Verbeek. 1973. Cobaltiferous malachites from southern Shaba (Zaire). *Bull. Soc. Fr. Mineral. Cristallogr.*, *96*, 371–377.

Suesse, P. 1967. Refinement of the crystal structure of malachite, $Cu_2(OH)_2CO_3$. *Acta Crystallogr.*, *22*, 146–151.

Azurite $Cu_3(OH)_2(CO_3)_2$ MONOCLINIC $\angle\beta = 92°25'$

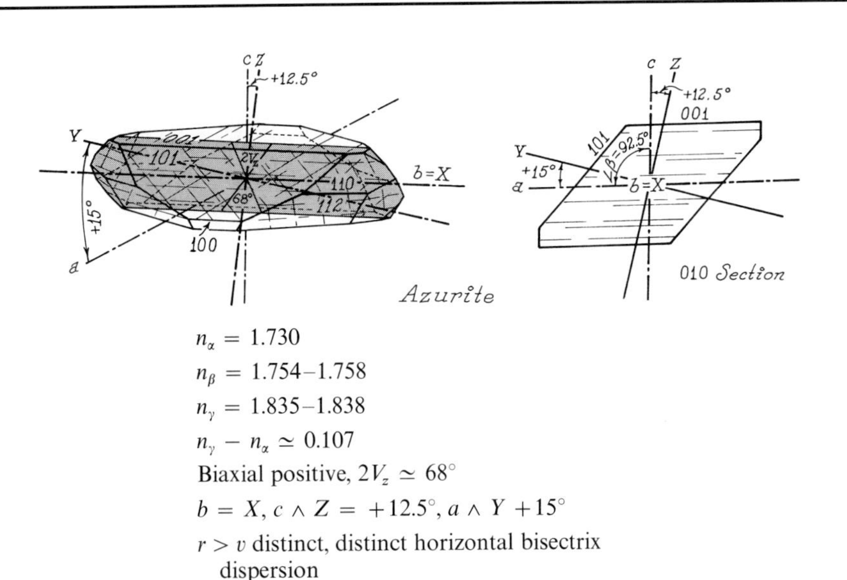

$n_\alpha = 1.730$
$n_\beta = 1.754–1.758$
$n_\gamma = 1.835–1.838$
$n_\gamma - n_\alpha \simeq 0.107$
Biaxial positive, $2V_z \simeq 68°$
$b = X, c \wedge Z = +12.5°, a \wedge Y +15°$
$r > v$ distinct, distinct horizontal bisectrix dispersion
Azure blue to blue-violet, in standard thin section, with weak pleochroism $Z > Y > X$

COMPOSITION. The composition of azurite appears to vary little from its ideal $Cu_3(OH)_2(CO_3)_2$.

PHYSICAL PROPERTIES. H = $3\frac{1}{2}$–4. Sp. Gr. = 3.77–3.89. In hand sample, azurite is azure to berlin-blue from light to very dark. Luster is vitreous to adamantine.

COLOR AND PLEOCHROISM. Azurite is distinctly blue, in section or as fragments, and weakly pleochroic with absorption $Z > Y > X$. X = clear blue; Y = azure blue; and Z = deep blue-violet.

FORM. Crystals of azurite are common, usually tabular on {001} or {101}, sometimes prismatic and slightly elongated on c or b, or pseudorhombohedral. Radiated aggregates of columnar crystals may assume globular or stalactitic forms.

CLEAVAGE. Cleavage is perfect on {011}, distinct on {100}, and poor on {110}.

BIREFRINGENCE. Maximum birefringence, seen in sections near {100} is extreme ($n_\gamma - n_\alpha \simeq 0.107$). Interference colors are high-order gray and are masked by deep blue mineral color.

TWINNING. Twins are uncommon but are reported on {$\bar{1}01$} also on {$\bar{1}02$} and {001}.

INTERFERENCE FIGURE. Azurite is biaxial positive, with rather large $2V_z \simeq 68°$. Basal sections yield off-centered acute bisectrix figures showing myriads of isochromes on a blue field and distinct optic axis dispersion $r > v$ and horizontal bisectrix dispersion. Cleavage fragments on {011} yield highly eccentric figures, and those on {100} show near-centered flash figures.

OPTICAL ORIENTATION. Azurite is monoclinic, with $b = X$, $c \wedge Z = +12\frac{1}{2}°$, $a \wedge Y = +15°$, and the optic plane is normal to {010} $12\frac{1}{2}°$ from c.

Maximum extinction angle seen in {010} sections is $12\frac{1}{2}°$ from {100} cleavage traces or 15° from {011} traces. Extinction is symmetrical to {011} cleavages in {100} sections.

Elongation defined by {100} traces is positive (length-slow). Cleavage fragments do not show consistent elongation.

DISTINGUISHING FEATURES. Deep blue color, extreme birefringence, and association with malachite are diagnostic.

ALTERATION. Azurite is, itself, an alteration product of primary copper minerals. It is commonly replaced by malachite; less commonly, by cuprite or native copper.

OCCURRENCE. Azurite is common in the oxide zone of copper deposits, where it forms by the interaction of carbonated solutions and copper minerals or soluble copper salts and carbonate rocks. It is nearly always closely associated with malachite and is less common than the latter mineral; it is also commonly associated with cuprite, tenorite, chrysocolla, limonite, native copper, and numerous secondary copper minerals.

REFERENCE: AZURITE

Zigan, F., and H. D. Schuster. 1972. Refinement of the structure of azurite, $Cu_3(OH)_2(CO_3)_2$, by neutron diffraction. *Z. Kristallogr.*, *135*, 416–436.

PHOSPHATES

Xenotime — YPO_4 — TETRAGONAL

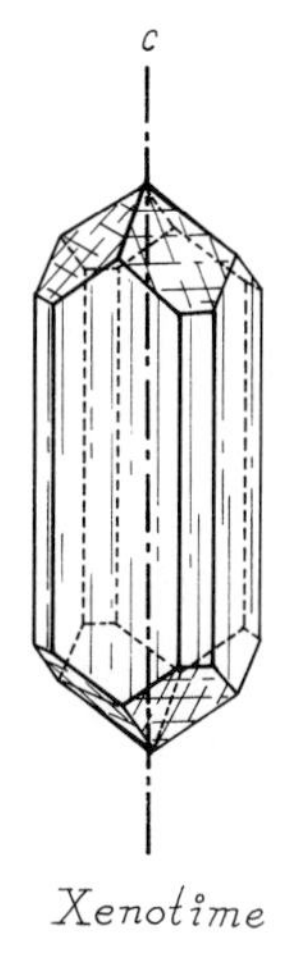

Xenotime

$n_\omega = 1.719–1.724$*

$n_\varepsilon = 1.816–1.827$

$n_\varepsilon - n_\omega = 0.095–0.107$

Uniaxial positive

Colorless to pale yellow-green or pale yellow-brown in thin section

Weak pleochroism

* Indices of refraction probably increase with Th or U.

COMPOSITION AND STRUCTURE. The structure of xenotime, YPO_4, is analogous to that of zircon, $ZrSiO_4$, with PO_4^{3-} tetrahedral units replacing SiO_4^{4-} tetrahedra (see page 131).

Other rare earth elements, especially those of the yttrium subgroup (and especially erbium), may partially replace Y. Zr, Th, U, or Ca may also replace some Y. The PO_4^{3-} tetrahedron may be partially replaced by SO_4^{2-} or especially SiO_4^{4-}. SiO_4^{4-}-PO_4^{3-} substitution is probably balanced by equal Zr^{4+}, Th^{4+} U^{4+} substitution for Y^{3+}, and Ca^{2+} for Y^{3+} may compensate SO_4^{2-} for PO_4^{3-}.

PHYSICAL PROPERTIES. H = 4–5. Sp. Gr. = 4.3–5.1. In hand sample, xenotime is yellow-brown to red-brown, sometimes yellow, salmon-pink, or gray. Luster is resinous to adamantine.

COLOR AND PLEOCHROISM. Xenotime may be colorless to pale yellow or pale brown in standard section with weak pleochroism. O = pale pink, pale yellow-brown, pale yellow; E = yellow, gray-brown, yellow-green.

FORM. Simple euhedral crystals are most common and are stubby bipyramidal or elongated prismatic on *c*. Crystal forms resemble those of zircon; rarely appears as coarse radiated aggregates. Rounded detrital grains of xenotime are common and often mistaken for zircon.

CLEAVAGE. Distinct prismatic cleavages {110} intersect at 90°

BIREFRINGENCE. Maximum birefringence, seen in prismatic sections, is very strong to extreme ($n_\varepsilon - n_\omega = 0.095–0.107$), and interference colors show many orders in section or fragments. High-order gray is unchanged by introduction of microscope accessories. Even basal sections show high-order colors in convergent light.

TWINNING. Simple contact twinning on {101} is known.

ZONING. Xenotime may form parallel intergrowths with zircon. Radioactive halos in colored host minerals commonly surround small xenotime crystals.

INTERFERENCE FIGURE. Basal sections yield centered optic axis figures showing distinct, flaring isogyres on a field of very numerous isochromatic rings. Fragments tend to yield flash figures but are not strongly oriented by prismatic cleavages.

OPTICAL ORIENTATION. Extinction is parallel to elongated crystals and cleavage traces, and elongation is positive (length-slow).

DISTINGUISHING FEATURES. Small, euhedral accessory crystals with extreme relief and birefringence are distinctive but may well be confused with zircon, monazite, sphene, cassiterite, bastnäsite, or rutile. Compared to xenotime, zircon has higher indices of refraction and less birefringence; monazite is biaxial and shows lower birefringence, but is often intergrown with xenotime; sphene is biaxial and shows greater indices of refraction and strong dispersion; cassiterite has higher indices of refraction and, commonly, multiple twinning; bastnäsite effervesces in warm acid and commonly shows good basal parting; and rutile shows much higher indices of refraction and deep red-brown color.

ALTERATION. Xenotime is quite resistant to surface weathering and is commonly found as fresh detrital grains.

OCCURRENCE. Xenotime is a rather widespread accessory mineral as small, euhedral crystals in granites, syenites and similar siliceous or alkaline igneous rocks; it is commonly overlooked as zircon. Large crystals may form in complex granite pegmatites with zircon, monazite, allanite, and other rare earth minerals.

Metamorphic rocks may contain xenotime as a constituent of quartz-mica gneisses; rarely in contact marbles.

Xenotime is a rather common heavy mineral in sands and detrital sediments; also in placers containing cassiterite, gold, or diamond.

REFERENCES: XENOTIME

Graeser, S., H. Schwander, and H. A. Stalder. 1973. Solid solution series between xenotime ($Y + PO_4$) and chernovite ($Y + AsO_4$). *Mineral. Mag., 39,* 145–151.

Krstanovic, I. 1965. Redetermination of oxygen parameters in xenotime, YPO_4. *Z. Kristallogr. 121,* 315–316.

Snetsinger, K. G. 1967. Nuclei of pleochroic haloes in biotites of some Sierra Nevada granitic rocks. *Amer. Mineral., 52,* 1901–1903.

Monazite

$(Ce,La,Th)PO_4$

MONOCLINIC
$\angle\beta \simeq 104°$

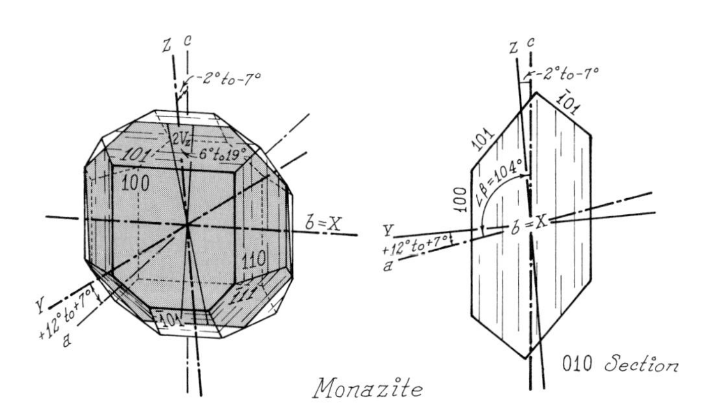

$n_\alpha = 1.770–1.800$*
$n_\beta = 1.777–1.801$
$n_\gamma = 1.828–1.851$
$n_\gamma - n_\alpha = 0.045–0.075$
Biaxial positive $2V_z = 6°–19°$
$b = X$, $c \wedge Z = -2°$ to $-7°$, $a \wedge Y = +12°$ to $+7°$
Usually $v > r$ weak, sometimes $r > v$; weak horizontal bisectrix dispersion
Colorless to pale yellow or yellow-brown in section
Weak pleochroism $Y = X \simeq Z$

* Large indices of refraction are shown by Th-rich varieties, especially those containing much Si and little Ca.

COMPOSITION AND STRUCTURE. Monazite, (Ce,La,Th)PO_4, is a rare earth phosphate of highly variable composition. Unlike xenotime, which has smaller metal atoms, monazite does not possess the zircon crystal structure. Rather, the Ce (La,Th) ions are surrounded by nine oxygens of the tetrahedral phosphate groups. All rare earth elements may be present in monazite, but Ce is usually the most abundant, followed by Nd, La, Sm, and Pr, not necessarily in that order. Significant coupled substitution of Th^{4+} + Ca^{2+} for two trivalent rare earth ions (for example, Ce^{3+}) yields the variety *cheralite*. Tetrahedral $SiO_4{}^{4-}$ or $SO_4{}^{2-}$ groups usually replace some tetrahedral $PO_4{}^{3-}$ groups, probably with coupled substitution of Th^{4+} for Ce^{3+} and major solid solution appears to exist with huttonite, a polymorph of $ThSiO_4$. Th-free varieties are rare. Numerous cations may replace minor amounts of rare earth elements (such as Mg^{2+}, Fe^{2+}, Fe^{3+}, Mn^{2+}, Be^{2+}, Sn^{4+}, and U^{4+}).

PHYSICAL PROPERTIES. H = 5–5½. Sp. Gr. = 4.6–5.4. In hand sample, monazite is yellow-brown to red-brown. Luster is resinous to vitreous or waxy.

COLOR AND PLEOCHROISM. Monazite is colorless or gray to yellow in standard section, with weak pleochroism in yellow $Y > X \simeq Z$.

FORM. Monazite appears most commonly as small, blocky, euhedral crystals, sometimes flattened on {100} or elongated on *b*. Crystals in pegmatites may be large, with rough faces.

CLEAVAGE. Cleavage is distinct on {100} and indistinct on {010}, intersecting at 90°. Several very poor cleavages are sometimes reported on {110}, {010}, and {011}. Distinct basal parting {001} may be present; rarely, parting occurs on {$\bar{1}$11}.

BIREFRINGENCE. Maximum birefringence, seen in sections near {100}, is strong ($n_\gamma - n_\alpha = 0.045$–$0.075$), and interference colors are bright, ranging to upper third or fourth order. Rounded detrital grains show many orders of distinct color rings but crystals in standard section show mostly brilliant second- or third-order colors. Fragments on basal parting show weak birefringence.

TWINNING. Simple contact twins are rather common on {100}; repeated lamellar twinning on {001} is rare.

INTERFERENCE FIGURE. Monazite is biaxial positive, with small $2V_z = 6°$ to 19°. Basal sections and fragments on {001} parting yield nearly centered acute bisectrix figures that show numerous isochromatic bands. Optic axis dispersion is usually weak $r > v$ and sometimes $v > r$, and bisectrix dispersion is horizontal. Fragments on dominant cleavage {100} yield near-flash figures.

OPTICAL ORIENTATION. Monazite is monoclinic with $b = X$, $c \wedge Z = -2°$ to $-7°$, and $a \wedge Y = +12°$ to $+7°$. Maximum extinction, seen in {010} sections is 2°–7° to {100} cleavage traces and 12°–7° to basal parting. Fragments on {100} cleavage or basal parting show extinction parallel to cleavage traces. Crystals or fragments elongated on *b* are length-fast (negative elongation).

DISTINGUISHING FEATURES. Extreme positive relief, strong birefringence, pale yellow color with weak pleochroism, and limited occurrence are distinctive of monazite. Sphene shows stronger birefringence and extreme dispersion $r > v$ and moderate $2V$; zircon is uniaxial and has very extreme relief; bastnäsite is uniaxial with lower relief than monazite; and epidote has lower relief, lower birefringence, and larger $2V$.

ALTERATION. Monazite commonly shows an opaque, yellowish-brown, earthy surface alteration of questionable composition. It is relatively resistant to weathering and occurs as detrital grains in sands and placers.

OCCURRENCE. Monazite, as tiny euhedral crystals, is an uncommon accessory mineral in granite, granodiorite, syenite, and similar plutonic igneous rocks. Large crystals occur in complex pegmatites, associated with zircon, xenotime, apatite, columbite, gadolinite, samarskite, fergusonite, and magnetite and, in Alpine-type veins and quartz veins, with anatase, rutile, sphene, albite, wolframite, and cassiterite.

Metamorphic rocks contain monazite as rare constituents of schists, gneisses, granulites, or dolomitic marble with biotite, muscovite, kyanite, staurolite, sillimanite, quartz, and carbonates.

Detrital grains of monazite appear in sands and some sandstones derived from the rock types just listed and in tin placers. They are common in the heavy fraction of sands with ilmenite, magnetite, zircon, rutile, garnet, and tourmaline.

REFERENCES: MONAZITE

Ghouse, K. M. 1965. The refinement of the crystal structure of an Indian monazite. *Naturwissenschaften*, *52*, 32–33.

Heinrich, E. W., R. A. Borup, and A. A. Levinson. 1960. Relationships between geology and composition of some pegmatitic monazites. *Geochim. Cosmochim. Acta*. *19*, 222–231.

Mateos, J. P., and Garcia, J. G. 1967. The morphology of detrital minerals in relation to their structural properties. II. Monazite. *An. Edafol. Agrobiol.*, *26*, 1227–1244.

Molloy, M. W. 1959. A comparative study of ten monazites. *Amer. Mineral.*, *44*, 510–532.

Serdyuchenko, D. P., A. M. Pap, V. M. Borkovskaya, and A. V. Bykova. 1967. A thorium-free monazite from Precambrian gneisses of Belorussia, and its genesis. *Dokl. Akad. Nauk SSSR*, *175*, 917–919.

Snefsinger, K. G. 1967. Nuclei of pleochroic haloes in biotites of some Sierra Nevada granitic rocks. *Amer. Mineral.*, *52*, 1901–1903.

Amblygonite

$LiAl(PO_4)F$

TRICLINIC
$\angle\alpha = 111°59'$
$\angle\beta = 97°46'$
$\angle\gamma = 68°16'$

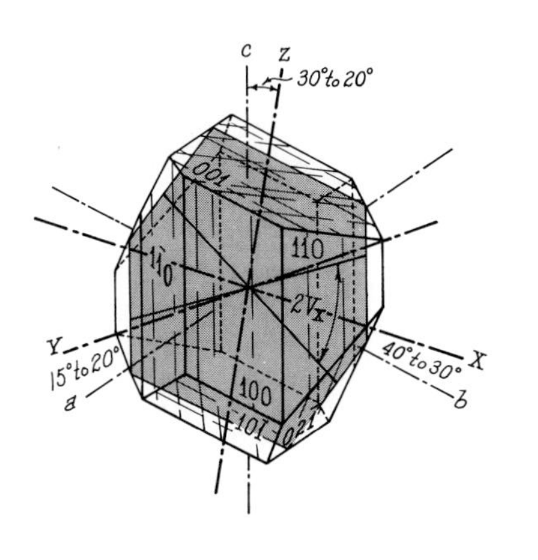

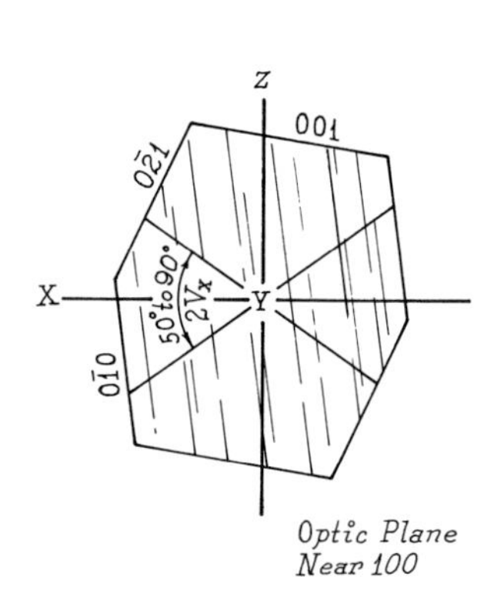

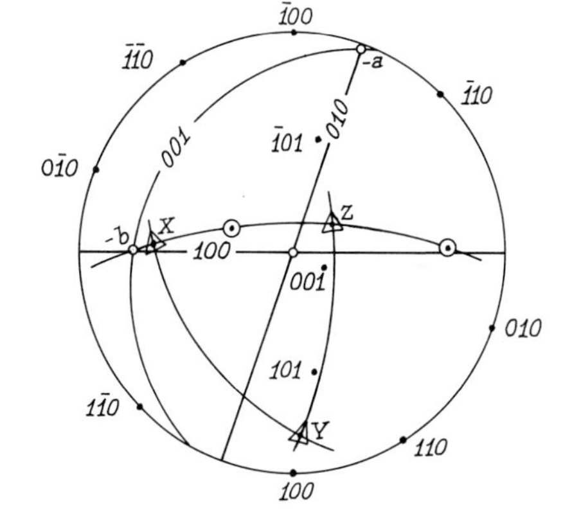

Amblygonite

$n_\alpha = 1.575–1.595$*
$n_\beta = 1.587–1.610$
$n_\gamma = 1.590–1.622$
$n_\gamma - n_\alpha = 0.014–0.027$*
Biaxial negative, $2V_x = 50°$ to $90°$*
$a \wedge Y \simeq 15°–20°$, $b \wedge X \simeq 40°–30°$,
$c \wedge Z \simeq 30°–20°$
$r > v$ weak, strong bisectrix dispersion is combined (crossed, inclined)
Colorless in standard section

* Indices of refraction, birefringence, and $2V_x$ increase with increasing OH and decrease with increasing Na (Fig. 2-11).

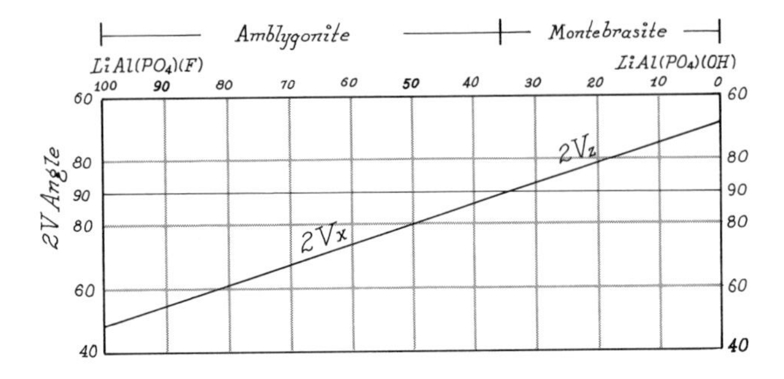

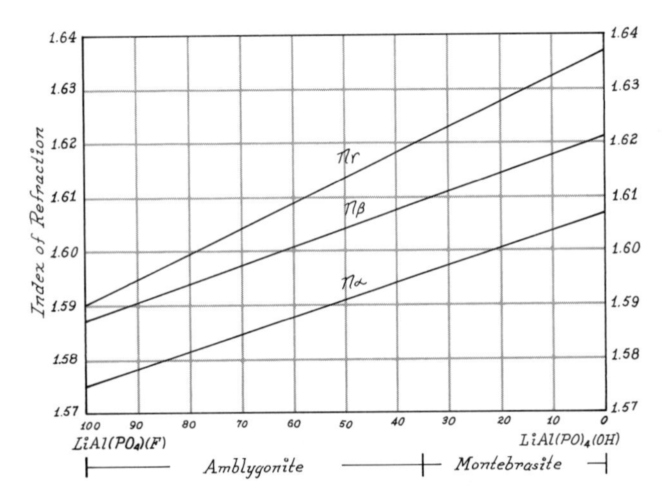

FIGURE 2-11. Variation of $2V$ angle and indices of refraction in the amblygonite-montebrasite series. (Adapted from Winchell and Winchell, 1951)

COMPOSITION. Amblygonite, $LiAl(PO_4)F$, forms a complete solid solution with montebrasite, $LiAl(PO_4)OH$, and a partial solid solution with fremontite, $NaAl(PO_4)(OH,F)$. Montebrasite is biaxial positive, $2V_z = 90°$ to $70°$, and $n_\alpha = 1.595$–1.605, $n_\beta = 1.610$–1.621, $n_\gamma = 1.622$–1.635.

PHYSICAL PROPERTIES. $H = 5\frac{1}{2}$–6. Sp. Gr. = 3.0–3.1. In hand sample, amblygonite is white to gray, pale violet, or pink; rarely, yellow, pale green, pale blue, or colorless. Luster is vitreous to pearly.

COLOR AND PLEOCHROISM. Amblygonite is colorless in thin section or as fragments.

FORM. Large crystals may be huge, weighing many tons. Small crystals are stubby prisms and often rough; however, anhedral cleavage masses are most common.

CLEAVAGE. Cleavages are perfect on $\{100\}$, near perfect on $\{110\}$, distinct on $\{0\bar{1}1\}$, and indistinct on $\{001\}$.

BIREFRINGENCE. Maximum birefringence is usually about 0.020 ($n_\gamma - n_\alpha = 0.014$–$0.027$). Interference colors may range to low second-order in standard thin section.

TWINNING. Polysynthetic twinning is common on $\{\bar{1}\bar{1}1\}$ and parallel to $\{110\}$, and twin lamellae are rare parallel to $\{111\}$.

INTERFERENCE FIGURE. Amblygonite, $LiAl(PO_4)F$, is biaxial negative, and montebrasite, $LiAl(PO_4)OH$, is biaxial positive. $2V_x$ increases from about 50°, for ideal amblygonite, to 90° at about 65 percent $LiAl(PO_4)OH$, and converges about Z to $2V_z \simeq 70°$ at ideal montebrasite. Amblygonite and montebrasite are logically divided by the sign change. Optic axis dispersion is weak $r > v$ about the Z bisectrix ($r > v$ for amblygonite, and $v > r$ for montebrasite). Bisectrix dispersion is combined inclined and crossed, often quite strong. All cleavage fragments yield eccentric figures.

OPTICAL ORIENTATION. Amblygonite is triclinic, with its optic plane near the perfect $\{100\}$ cleavage. X lies almost on the $\{100\}$ cleavage plane and Z in the $\{110\}$ plane. Cleavage fragments tend to be elongated on c. They show inclined extinction up to about 30° and are length-slow (positive elongation).

DISTINGUISHING FEATURES. Amblygonite shows very limited occurrence, and it may be associated with common pegmatite minerals. Topaz shows less birefringence than amblygonite and has a single cleavage. Spodumene has greater indices of refraction and dispersion $v > r$ without crossed bisectrix dispersion.

ALTERATION. Amblygonite alters to a dense mixture of kaolin and mica, often surrounding unaltered cores. Wavellite, wardite, morinite, and turquois are also reported as alteration products.

OCCURRENCE. Amblygonite is a mineral of granite pegmatites, where it occurs with other phosphates (such as apatite, lithophilite-triphylite, and monazite) and other lithium minerals (such as spodumene, lepidolite, petalite, and rubellite). It may also form in high-temperature hydrothermal veins and greisen with topaz and cassiterite.

REFERENCES: AMBLYGONITE

Černá, I., P. Černý, and R. B. Ferguson. 1973. Fluorine content and some physical properties of the amblygonite-montebrasite minerals. *Amer. Mineral.*, *58*, 291–300.

Dubois, J., J. Marchand, and P. Bourguignon. 1972. Mineralogical data on the amblygonite-montebrasite series. *Ann. Soc. Geol. Belg.*, *95*, 285–310.

Solomkin, S. G., and G. A. Sidorenko. 1962. Supplementary characteristics of minerals of the amblygonite group. *Mineral. Syr'e*, *1962*, 75–82.

Timchenko, T. I., L. P. Tsareva, and Yu. N. Yarmukhamedov. 1968. Extreme member of the montebrasite-amblygonite series. *Nestn. Mosk. Univ., Geol.*, *23*, 89–91.

Winchell, A. N., and H. Winchell. 1951. *Elements of Optical Mineralogy*. New York: Wiley.

The Lithiophilite-Triphylite Series

$Li(Mn,Fe)PO_4$ ORTHORHOMBIC

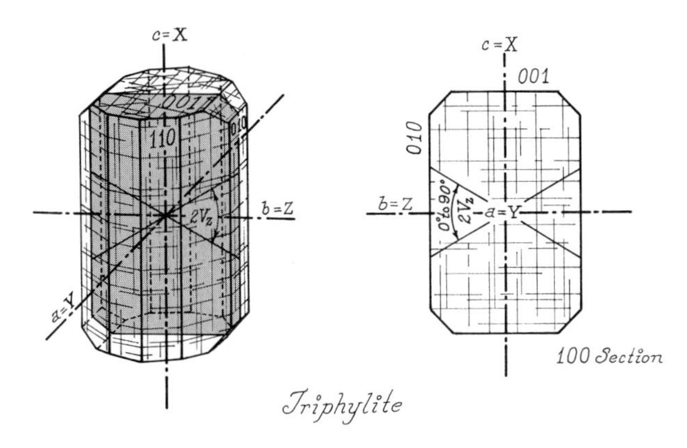

Triphylite

Lithiophilite

$n_\alpha = 1.670–1.689$*
$n_\beta = 1.677–1.689$
$n_\gamma = 1.684–1.696$
$n_\gamma - n_\alpha = 0.014–0.007$
Biaxial positive
$2V_z = 63°–0°$
$a = X, b = Z, c = Y$
$v > r$ strong
Colorless to pale pink in thin section

Triphylite

$n_\alpha = 1.689–1.705$*
$n_\beta = 1.689–1.710$
$n_\gamma = 1.696–1.720$
$n_\gamma - n_\alpha = 0.007–0.015$
Biaxial positive (<75 percent $LiFePO_4$)
$2V_z = 0°–90°, r > v$
$a = Y, b = Z, c = X$ (<85 percent $LiFePO_4$)
Biaxial negative (>75 percent $LiFePO_4$)
$2V_x = 90°–0°, v > r$
$a = Z, b = Y, c = X$ (>85 percent $LiFePO_4$)
Colorless in thin section

* Indices of refraction increase with increasing Fe^{2+} but are lowered appreciably by the presence of Mg^{2+} (Fig. 2-12).

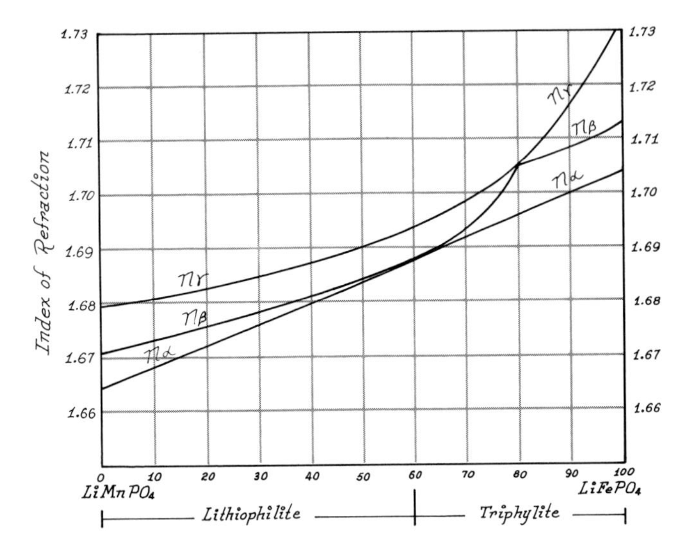

FIGURE 2-12. Variation of indices of refraction in the lithiophilite-triphylite series. Note that the series becomes uniaxial + ($n_\beta = n_\alpha$) at about 60 percent ("molecular percent") $LiFePO_4$ and uniaxial − ($n_\beta = n_\gamma$) at about 82 percent. (After Chapman, 1943)

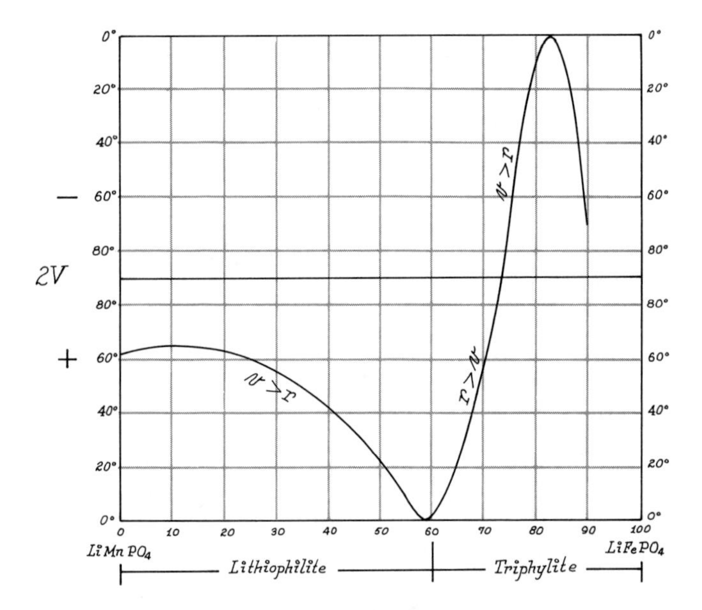

FIGURE 2-13. Variation of $2V$ angle in the lithiophilite-triphylite series. Note that the series becomes uniaxial + at about 60 percent, uniaxial − at about 82 percent, and changes from biaxial + to biaxial − at about 74 percent $LiFe(PO_4)$; ("molecular percent").

COMPOSITION AND STRUCTURE. The structure of lithiophilite and triphylite is similar to that of olivine (page 102), where PO_4^{3-} tetrahedral groups replace SiO_4^{4-} tetrahedral groups.

A complete isomorphous solid solution exists between lithiophilite, $LiMnPO_4$, and triphylite, $LiFePO_4$; the logical optical division appears where $2V_z = 0°$ and the optical orientation changes. Major amounts of Mg^{2+} and at least minor Ca^{2+} may replace (Mn^{2+},Fe^{2+}); however, only minor Na^+ replaces Li^+ toward the sodic analog natrophilite $NaMnPO_4$.

PHYSICAL PROPERTIES. H = $4\frac{1}{2}$–5. Sp. Gr. = 3.34 (Mn) to 3.58 (Fe). In hand sample, lithiophilite is salmon-pink to clove-brown and grades to blue-gray triphylite. Luster is vitreous to somewhat resinous.

COLOR AND PLEOCHROISM. Lithiophilite and triphylite are usually colorless in standard thin section, although thicker sections or fragments of lithiophilite may appear pale pink with weak pleochroism. X = deep pink; Y = pale greenish yellow; and Z = pale buff.

FORM. Crystals are commonly very large but are rarely euhedral; normally massive as cleavable anhedral masses.

CLEAVAGE. Cleavages are perfect on {001}, less perfect on {010}, and imperfect on prismatic planes {110}.

BIREFRINGENCE. Maximum birefringence decreases with increasing Fe^{2+} from about 0.014 (Tr_0) to about 0.007 (Tr_{80}) and then increases again. Interference colors are normally gray or white in thin section but may reach first-order orange. Cleavage fragments tend to show near-maximum birefringence.

TWINNING. Twinning is not reported.

ZONING. Concentric zones of alteration minerals may be observed.

INTERFERENCE FIGURE. Members of the lithiophilite-triphylite series may be biaxial positive or negative or uniaxial positive or negative, but are usually biaxial positive (Fig. 2-13). The $2V_z$ angle begins at about 60° at Tr_0, it closes to 0° on $b = Z$ in the basal {001} plane at about Tr_{60}; it opens again about $b = Z$ in the {100} plane passing through 90° to become negative at about Tr_{75} and closes to 0° on $c = X$ at about Tr_{85}. $2V$ apparently opens again, with increasing Fe, possibly in the {010} plane (see Fig. 2-13). Optic axis dispersion is strong and appears to be $v > r$ from Tr_0, to Tr_{60} and for compositions more iron-rich than Tr_{75}; for compositions Tr_{60-75}, probably $r > v$. Figures are usually distinct with few isochromes. Cleavage fragments tend to show either flash figures or centered bisectrix figures.

OPTICAL ORIENTATION. The lithiophilite-triphylite series is orthorhombic. In the range Tr_{0-60}, $a = X$, $b = Z$, and $c = Y$ and the optic plane is {001} $2V_z$ closes to 0° on $b = Z$ and opens again in {100}; between Tr_{60} and Tr_{85}, $a = Y$, $b = Z$, and $c = X$, and the optic plane is {100}. For compositions more Fe-rich than Tr_{85}, the optic plane is apparently {010}, and $a = Z$, $b = Y$, and $c = X$.

Extinction is parallel to all cleavages except in basal sections, where extinction is symmetrical to prismatic cleavage traces. Cleavages do not favor fragments with consistent elongation.

DISTINGUISHING FEATURES. The lithiophilite-triphylite minerals show widely varying optical properties and are distinguished largely by their limited occurrence (in granite pegmatites). They have much higher relief than quartz or feldspar. Apatite and beryl are uniaxial negative, amblygonite shows lower relief than lithiophilite and inclined extinction, and topaz has a single cleavage and a hardness of 8.

ALTERATION. Lithiophilite-triphylite alters readily to uncommon phosphate minerals. Hydrothermal alteration may yield some combination of the following minerals: triploidite-wolfeite, $(Mn,Fe)_2PO_4(OH)$; reddingite-phosphoferrite, $(Mn,Fe)_3(PO_4)_2 \cdot 3H_2O$; eosphorite, $(Mn,Fe)Al(PO_4)(OH)_2 \cdot H_2O$; fairfieldite, $Ca_2(Mn,Fe)(PO_4)_2 \cdot H_2O$; dickinsonite-fillowite, $H_2Na_6(Mn,Fe,Ca,Mg)_{14}(PO_4)_{12} \cdot H_2O$; rhodochrosite, $MnCO_3$; and siderite, $FeCO_3$. Near-surface weathering commonly alters lithiophilite-triphylite to siklerite, $(Li,Fe^{3+},Mn^{2+})PO_4$, or ultimately to heterosite, $(Fe^{3+}, Mn^{3+})PO_4$. Rims of successive alteration products may surround an unaltered core.

OCCURRENCE. Lithiophilite-triphylite minerals are essentially confined to granite pegmatites, where they are associated with Li-minerals (such as lepidolite, spodumene, amblygonite, and rubellite), other phosphate minerals (such as apatite, eosphorite, triploidite-wolfeite, and graftonite) and other minerals characteristic of complex pegmatites (for example, beryl, albite, and quartz).

REFERENCES: THE LITHIOPHILITE-TRIPHYLITE SERIES

Chapman, C. A. 1943. Magnesia-rich triphylite. *Amer. Mineral.*, *28*, 90–98.

Finger, L. W., and G. R. Rapp, Jr. 1970. Refinement of the crystal structure of triphyllite. *Carnegie. Inst. Wash. Year Book*, *1968–69*, 290–292.

Apatite $Ca_5(PO_4)_3(F,OH,Cl)$ HEXAGONAL

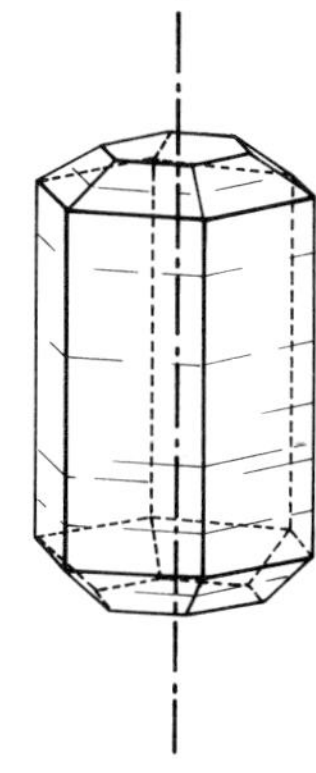

		n_ω*	n_ε	$n_\omega - n_\varepsilon$*
Fluorapatite	$Ca_5(PO_4)_3F$	1.633–1.650	1.629–1.646	0.003–0.005
Hydroxylapatite	$Ca_5(PO_4)_3OH$	1.643–1.658	1.637–1.654	0.007–0.004
Chlorapatite	$Ca_5(PO_4)_3Cl$	1.650–1.667	1.647–1.665	0.004–0.001
Dahllite-Francolite				
Dahllite	$Ca_5(PO_4,CO_3OH)_3(OH)$	1.603–1.628	1.598–1.619	0.007–0.017
Francolite	$Ca_5(PO_4,CO_3OH)_3(F)$			

Uniaxial negative

Colorless or very pale colors, with weak to moderate pleochroism $E > O$

* Fluorapatite is by far the most abundant apatite variety. Indices of refraction increase toward chlorapatite and less rapidly toward hydroxylapatite (Fig. 2-14). Varieties rich in Mn or rare earth elements tend toward high refractive indices, and carbonate varieties favor low indices of refraction. Birefringence decreases toward chlorapatite and increases toward hydroxylapatite and is especially large for carbonate-apatites.

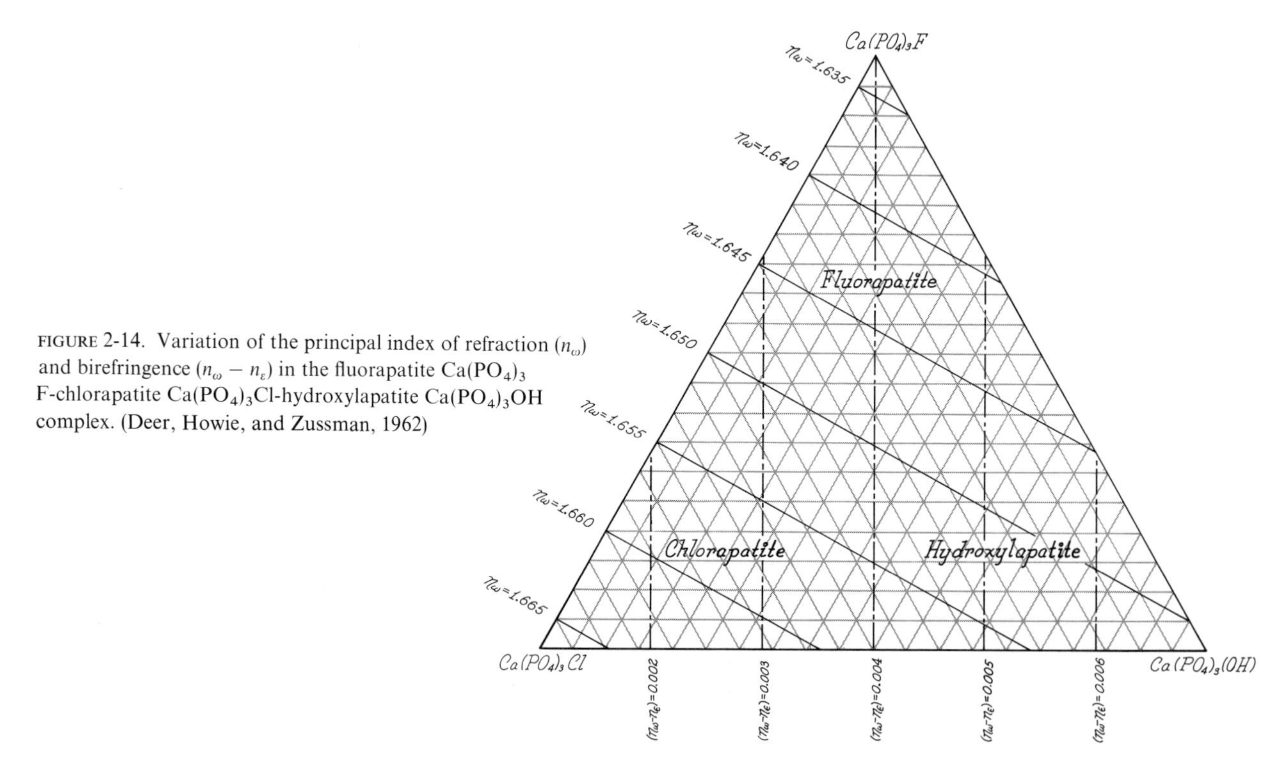

FIGURE 2-14. Variation of the principal index of refraction (n_ω) and birefringence ($n_\omega - n_\varepsilon$) in the fluorapatite $Ca(PO_4)_3$ F-chlorapatite $Ca(PO_4)_3Cl$-hydroxylapatite $Ca(PO_4)_3OH$ complex. (Deer, Howie, and Zussman, 1962)

COMPOSITION AND STRUCTURE. Phosphate oxyanions in apatite are linked by Ca^{2+} cations to form a hexagonal framework with F^- anions lying at the corners of the hexagonal cell and coordinated by three Ca^{3+} cations.

Composition of the apatite minerals is highly variable. Complete solid solution probably exists among the three end members fluorapatite, $Ca_5(PO_4)_3F$; chlorapatite, $Ca_5(PO_4)_3Cl$; and hydroxylapatite, $Ca_5(PO_4)_3OH$; the variety being determined by the dominant anion (F, Cl, or OH). Fluorapatite is by far the most common variety.

Carbonate-rich apatites probably result from the substitution of $(CO_3OH)^{3-}$ groups, of questionable configuration, for some tetrahedral $PO_4{}^{3-}$ groups. Carbonate-fluorapatite has been called *francolite*, and carbonate-hydroxylapatite, *dahllite*.

Tetrahedral phosphate groups ($PO_4{}^{3-}$) may be replaced by other tetrahedral anions in variable amounts ($SO_4{}^{2-}$ or $SiO_4{}^{4-}$ and to a lesser degree $MnO_4{}^{2-}$, $AsO_4{}^{3-}$, $CrO_4{}^{2-}$ and $VO_4{}^{3-}$). Coupled substitution $SiO_4{}^{4-} + SO_4{}^{2-}$ for $2PO_4{}^{3-}$ may produce rare end members, such as *ellestadite*, $Ca_5[(P,Si,S)O_4]_3(Cl,F,OH)$.

The Ca^{2+} cation is commonly partially replaced by Mn^{2+} to at least Mn:Ca $\simeq$ 1:6, especially in pegmatites, and Sr^{2+} may replace Ca^{2+} to Ca:Sr $\simeq$ 1:7. Rare earth elements, especially Ce, may replace some Ca in apatites of alkaline igneous rocks and many other cations are reported to substitute for Ca^{2+} in minor amounts or traces (for example, Mg^{2+}, Na^+, K^+, Cr^{3+}, U^{4+}, Th^{4+}). Alkali-rich apatites result from the coupled substitution $SiO_4{}^{4-} + (Na^+,K^+)$ for $PO_4{}^{3-} + Ca^{2+}$. Surprisingly little Pb^{2+} replaces Ca^{2+} in view of the existence of the lead analog pyromorphite, $Pb_5(PO_4)_3Cl$.

PHYSICAL PROPERTIES. H = 5. Sp. Gr. = 2.9–3.5. Apatite, in hand sample, shows an extremely wide range of colors. Yellow-green to blue-green is probably most common. Apatite, however, may be colorless, yellow, blue, violet, brown, orange, rose-red, and so on. Luster is vitreous to subresinous.

COLOR AND PLEOCHROISM. Apatite is normally colorless in standard section. Deep-colored varieties, however, may show pale color in section or fragments; pleochroism ranges from weak to moderate, with absorption $E > O$.

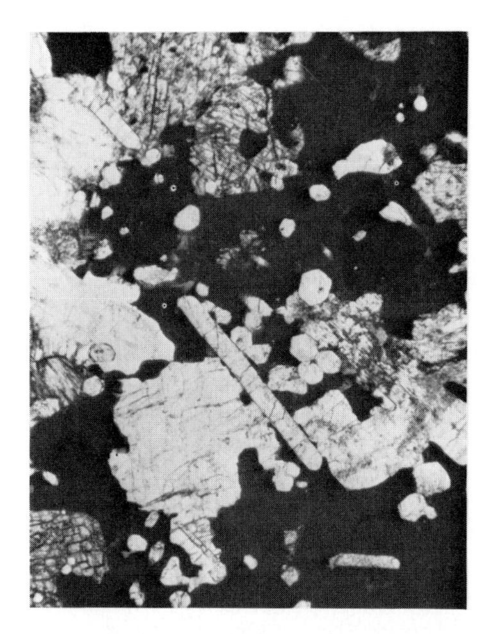

FIGURE 2-15. Apatite, in thin section, usually seen as tiny euhedral crystals imbedded in ferromagnesian minerals. Crystals show hexagonal cross-sections, which are dark between crossed nicols, and show elongated prismatic sections with white to gray interference colors.

FORM. Apatite normally appears as tiny euhedral prismatic crystals elongated on *c* showing hexagonal basal sections and rectangular to acicular prismatic sections; rarely tabular on {0001} (Fig. 2-15). Crystals often show corroded and rounded surfaces. It may appear in fibrous to columnar aggregates and is sometimes coarse granular. *Collophane* is the name for a secondary, cryptocrystalline mineralloid of essentially apatite composition that commonly shows oolitic, globular, botryoidal, or other colloform structures.

CLEAVAGE. Apatite cleavages vary somewhat in quality but are usually poor parallel to {0001} and very poor on $\{10\bar{1}0\}$.

BIREFRINGENCE. Maximum birefringence, seen in prismatic sections, is very weak to moderate ($n_\omega - n_\varepsilon = 0.001$ to about 0.015); only carbonate-apatites show birefringence above 0.007. Interference colors are usually first-order black to gray. Dahllite and francolite, however, may show colors to first-order red. Cryptocrystalline varieties (collophane) may appear essentially isotropic.

TWINNING. Apatite twins are rare. Simple contact twins are known on $\{11\bar{2}1\}$ also on $\{10\bar{1}3\}$; possibly on $\{10\bar{1}0\}$ and $\{11\bar{2}3\}$. Basal sections may show radiating twin sectors.

ZONING. Apatite in pegmatites often shows color zonation, and composition zoning may produce overgrowths of carbonate-apatite on normal compositions. Tiny rutile or monazite needles may appear as oriented inclusions parallel to the *c*-axis of apatite.

INTERFERENCE FIGURE. Apatite is uniaxial negative, but anomalous biaxial varieties are known, with $2V_x$ up to 20°. Basal sections of apatite yield a dark figure showing a broad, diffuse uniaxial cross on a gray background devoid of rings or color. Poor cleavages favor essentially random orientation of apatite fragments.

OPTICAL ORIENTATION. Prismatic crystal sections elongated on *c* show parallel extinction and negative elongation (length-fast).

DISTINGUISHING FEATURES. Moderately high relief and weak birefringence are distinctive; basal sections are hexagonal and almost black in crossed polarized light. In pegmatites and similar rocks, topaz may resemble apatite; topaz, however, is biaxial positive and cleavage traces parallel the fast wave. Quartz has less relief than apatite and is optically positive; melilite, idocrase, and zoisite usually show anomalous interference colors; andalusite and sillimanite are biaxial with distinct cleavages; beryl has much lower relief; and tourmaline is usually colored, with absorption $O > E$. Eudialyte closely resembles apatite but shows both uniaxial positive and negative varieties and distinct basal {0001} cleavage.

ALTERATION. Apatite is a stable mineral in most geological environments, and fresh grains are common in detrital sediments. Serpentine, kaolinite, wavellite, and turquois are known to form pseudomorphs after apatite, and apatite may replace pyromorphite or brushite. Apatite may result from the hydrothermal alteration of other phosphate minerals.

OCCURRENCE. Apatite is a very common, early-formed accessory mineral in nearly all types of igneous rocks. It usually appears as tiny euhedral crystals closely associated with ferromagnesian minerals. Fluorapatite is far more abundant than all other apatite varieties, although chlorapatite is common in mafic igneous rocks. Apatite is also concentrated by late magmatic segregation, and it appears in granite pegmatites, in association with spodumene, lepidolite, amblygonite, topaz, beryl, tourmaline, lithiophilite, triplite, and so on and, in high-temperature hydrothermal Alpine-type veins, with quartz and adularia. Mn-apatite is essentially confined to pegmatite occurrences, and Sr-apatite and rare earth varieties are characteristic of nepheline syenites and other feldspathoidal rocks. Carbonatites usually contain abundant apatite with calcite and uncommon rare earth minerals.

Metamorphic rocks, both contact or regional, may yield abundant apatite, especially where calcium is readily available. Crystalline limestones, marbles, and calc-silicate skarn rocks commonly contain apatite in association with diopside, augite, amphiboles, sphene, phlogopite, vesuvianite, and so on. Chlorapatite may appear with scapolite, due to Cl-metasomatism. Magnetite ores of contact replacement paragenesis often contain abundant pale yellow apatite. Talc-chlorite schists and some mica schists may contain apatite.

Sedimentary rocks contain apatite as relatively stable detrital grains or in secondary, cryptocrystalline calcium phosphates (collophane). Carbonate-apatites are the major minerals in phosphate beds and phosphatic limestones, shales, and oolitic ironstones and appear normally in oolite, ovulitic, or concretionary pellets of complex mineralogy. These ovalloid pellets often show concentric layers about a mineral grain core, usually quartz.

Bone fragments, fish scales, and teeth are highly phosphatic and may form the nucleus of phosphate nodules.

REFERENCES: APATITE

Deer, W. A., R. A. Howie, and J. Zussman. 1962. *Rock-Forming Minerals.* Vol. 5: *Non-Silicates.* New York: Wiley.

Frondel, C. 1943. Mineralogy of the calcium phosphates in insular phosphate rock. *Amer. Mineral., 28,* 215–232.

Larsen, E. S., Jr., M. H. Fletcher, and E. A. Cisney. 1952. Strontian apatite. *Amer. Mineral., 37,* 656–658.

McClellan, G. H., and J. R. Lehr. 1969. Crystal chemical investigation of natural apatites. *Amer. Mineral., 54,* 1372–1389.

McConnell, D. 1937. The substitution of SiO_4- and SO_4- groups for PO_4-groups in the apatite structure; ellestadite, the end-member. *Amer. Mineral., 22,* 977–986.

McConnell, D. 1938. A structural investigation of the isomorphism of the apatite group. *Amer. Mineral., 23,* 1–20.

McConnell, D. 1960. The crystal chemistry of dahllite. *Amer. Mineral., 45,* 209–216.

McConnell, D. 1973. *Apatite: Its Crystal Chemistry, Mineralogy, Utilization, and Geologic and Biologic Occurrences.* New York: Springer Verlag.

McConnell, D. 1974. Crystal chemistry of apatite. *Bull. Soc. Fr. Mineral. Cristallogr., 97,* 237–240.

Portnov, A. M., E. G. Litvintsev, V. N. Filippova, and L. T. Rakov. 1976. Pleochroism of apatite. *Dokl. Akad. Nauk SSSR, 228,* 952–953.

Sudarsanan, K., P. E. Mackie, and R. A. Young. 1972. Comparison of synthetic and mineral fluorapatite, $Ca_5(PO_4)_3F$, in crystollographic detail. *Mater. Res. Bull., 7,* 1331–1337.

Taborszky, F. K. 1972. Chemistry and optical properties of apatite. *Neues Jahrb. Mineral. Monatsh., 5,* 79–91.

Whippo, R. E., and B. L. Murowchick. 1967. The crystal chemistry of some sedimentary apatites. *Trans. Soc. Mining Eng. AIME, 238,* 257–263.

Young, R. A. 1967. Dependence of apatite properties on crystal structural detail. *Trans. N.Y. Acad. Sci., 29,* 949–959.

Wavellite

$Al_3(OH)_3(PO_4)_2 \cdot 5H_2O$

ORTHORHOMBIC

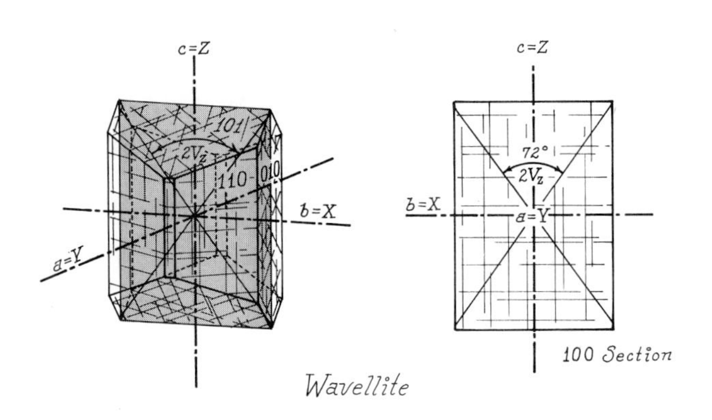

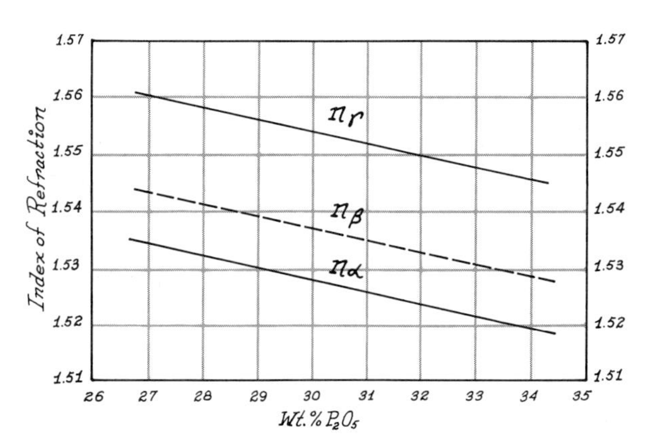

FIGURE 2-16. Variation of indices of refraction with P_2O_5 content of nonferroan wavellite. Most likely substitute for P^{5+} is V^{5+}.

$n_\alpha = 1.518–1.535$*
$n_\beta = 1.526–1.543$
$n_\gamma = 1.545–1.561$
$n_\gamma - n_\alpha = 0.025–0.027$
Biaxial positive, $2V_z \simeq 72°$
$a = Y, b = X, c = Z$
$r > v$ weak
Colorless or pleochroic in green, blue, yellow, or brown $X > Y > Z$

COMPOSITION AND STRUCTURE. Wavellite is a hydrated aluminum phosphate $Al_3(OH)_3(PO_4)_2 \cdot 5H_2O$. Significant F^- commonly replaces $(OH)^-$ and minor Fe^{3+}, or even Cr^{3+}, may replace Al^{3+}. Very minor Fe^{2+}, Ca^{2+}, or Mg^{2+} may appear in wavellite analyses. Significant V^{5+} may replace P^{5+}, imparting blue or green color. The atomic arrangement consists of corner-sharing chains of $AlX_6(X = O,OH,H_2O)$ octahedra parallel to the *c* crystallographic axis, cross-linked by PO_4 tetrahedra. Additional water molecules occupy cavities between the chains.

PHYSICAL PROPERTIES. H = $3\frac{1}{2}$–4. Sp. Gr. = 2.32–2.37. In hand sample, wavellite is white or gray to yellow, green, blue, brown, or black; vitreous to pearly luster.

* Indices of refraction should increase with substitution of Fe^{3+} for Al^{3+} and with substitution of VO_4 for PO_4 (Fig. 2-16).

COLOR AND PLEOCHROISM. Wavelite is usually colorless in standard thin section; however, it may be weakly to highly colored in green, yellow, blue, or brown. Pleochroism is weak to strong $X > Y > Z$. X = deep blue, green; Y = yellow-brown; and Z = pale brown, pale yellow, colorless.

FORM. Single crystals are rare and stout to thin acicular on *c*. Radiated aggregates of acicular crystals are most common, forming spherical or globular masses, usually in veins or cavities or as crusts.

CLEAVAGE. Cleavages are perfect on {110} (prismatic), less perfect on {101} and distinct on {010}.

BIREFRINGENCE. Maximum birefringence, seen in {100} sections, is moderate ($n_\gamma - n_\alpha = 0.025–0.027$). Interference colors are mostly bright first-order ranging to low second-order in standard section. Acicular crystals show maximum ($n_\gamma - n_\alpha$) to mild ($n_\gamma - n_\beta$) birefringence.

TWINNING AND ZONING. No twinning or zoning is reported.

INTERFERENCE FIGURE. Basal sections of wavellite crystals yield acute bisectrix figures that show a few isochromes and weak optic axis dispersion $r > v$. Cleavage fragments yield eccentric figures between obtuse or acute bisectrix and flash figures.

OPTICAL ORIENTATION. Wavellite is orthorhombic, with $a = Y$, $b = X$, and $c = Z$, and {100} is the optic plane. Acicular crystals, elongated on $c = Z$, and prismatic cleavage fragments, show parallel extinction and positive elongation (length-slow). Extinction is symmetrical to prismatic cleavages in basal sections and to {101} cleavages in {010} sections.

DISTINGUISHING FEATURES. Radiated acicular crystals with parallel extinction, positive elongation, and indices of refraction both above and below balsam are distinctive properties of wavellite. Zeolites show much lower birefringence. Acicular pectolite crystals show higher birefringence and high positive relief. Prehnite also has high positive relief in balsam.

ALTERATION. Wavellite is itself a secondary mineral and is relatively stable.

OCCURRENCE. Wavellite is a rather widespread secondary mineral deposited by hydrothermal solutions in open fractures and crevices in many rock types, especially in limonite ores, phosphate rocks, and low-grade aluminous metamorphic rocks. It may also appear in pegmatites or high-temperature hydrothermal veins with cassiterite, amblygonite, apatite, or vauxite.

REFERENCES: WAVELLITE

Araki, T., and T. Zoltai. 1968. Crystal structure of wavellite. *Z. Kristallogr.*, *127*, 21–33.

Bouska, V., and P. Pavondra. 1969. Colored wavellites from Czechoslovakia. *Cas. Mineral. Geol.*, *14*, 205–210.

Foster, M. D., and W. T. Schaller. 1966. Cause of colors in wavellite from Dug Hill, Arkansas. *Amer. Mineral.*, *51*, 422–428.

SULFATES

Anhydrite

$CaSO_4$

ORTHORHOMBIC

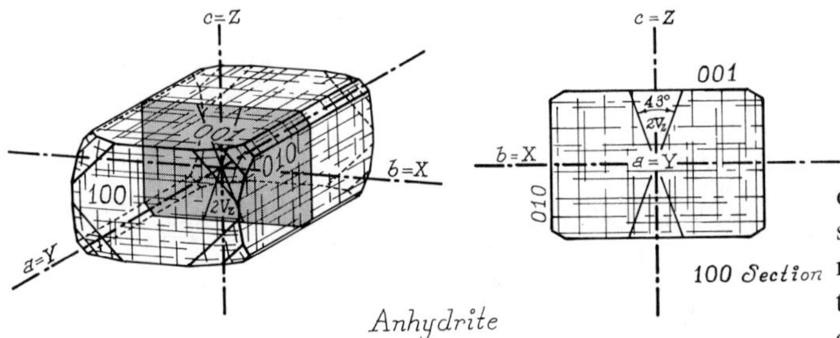

$n_\alpha = 1.570$
$n_\beta = 1.576$
$n_\gamma = 1.614$
$n_\gamma - n_\alpha \simeq 0.040$
Biaxial positive, $2V_z \simeq 43°$
$a = Y, b = X, c = Z$
$v > r$ strong
Colorless in standard section

COMPOSITION AND STRUCTURE. The structure of anhydrite is somewhat similar to that of zircon in that it contains alternating, edge-sharing SO_4 tetrahedra and CaO_8 polyhedra in three mutually perpendicular directions. Chainlike repetition of Ca-SO_4 is most prominent parallel to *c*, accounting for rare fibrous habit on *c*, while planar arrangement of polyhedra is best defined in {100} and {010}, explaining these perfect cleavages and imperfect cleavage on {001}.

The composition of anhydrite $CaSO_4$ shows little variation, and only very minor Sr^{2+} and Ba^{2+} is reported to substitute for Ca^{2+}. Minor water content suggests partial hydration to gypsum.

PHYSICAL PROPERTIES. $H = 3\frac{1}{2}$. Sp. Gr. = 3.0. In hand sample, anhydrite is usually colorless, white, or gray; however, blue to violet varieties are common; rarely red to brown. Luster is vitreous to somewhat pearly on {010}.

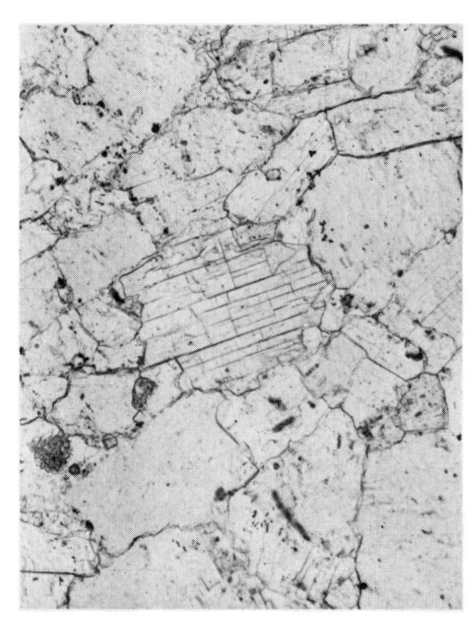

FIGURE 2-17. Anhydrite shows pseudocubic cleavages and brilliant interference colors.

COLOR AND PLEOCHROISM. Anhydrite is colorless in standard section or as fragments. Violet varieties may show weak color, in coarse fragments, and pleochroism $Z > Y > X$.

FORM. Anhydrite is most commonly fine to coarse granular aggregates; rarely fibrous as parallel or radiating aggregates. Single crystals are blocky or tabular, elongated on *a* or *c*.

CLEAVAGE. Pseudocubic cleavages are perfect on {010}, nearly perfect on {100}, and distinct on {001}—all mutually perpendicular (Fig. 2-17).

BIREFRINGENCE. Maximum birefringence, seen in {010} sections, is moderate to strong ($n_\gamma - n_\alpha \simeq 0.04$), and interference colors are mostly brilliant second order ranging to low third order in standard section.

TWINNING. Twinning is rather common on {011}; simple contact or polysynthetic lamellae. Simple contact twins on {120} are rare.

INTERFERENCE FIGURES. Anhydrite is biaxial positive, with moderate $2V_z = 42°–44°$. Isogyres, on the acute bisectrix, remain well within the visible field, which shows numerous isochromatic bands. Optic axis dispersion is $v > r$; no bisectrix dispersion. Cleavage fragments yield centered figures, usually obtuse bisectrix or flash figures; less commonly, acute bisectrix.

OPTICAL ORIENTATION. Extinction is parallel to cleavage traces and crystal elongation. Pseudocubic cleavages do not favor fragment elongation, but fibrous crystals elongated on *c* are length-slow (positive elongation).

DISTINGUISHING FEATURES. Pseudocubic cleavages with moderately high birefringence are distinctive of anhydrite. Gypsum shows negative relief in balsam, weak birefringence, and inclined extinction to most cleavage traces. Barite and celestite have higher refractive indices, weak birefringence, and high specific gravity. Fragments of barite and celestite sink in methylene iodide (1.74 refractive index oil).

ALTERATION. Anhydrite hydrates readily to gypsum, beginning on pseudocubic cleavages. Entire sedimentary beds may alter to gypsum, requiring a volume increase that may squeeze and deform gypsum beds. Pseudomorphs after anhydrite may be of gypsum, polyhalite, calcite, dolomite, siderite, or quartz; and anhydrite may form pseudomorphs after gypsum or calcite.

OCCURRENCE. Anhydrite occurs mostly as beds in sedimentary evaporite sequences in contact with beds of gypsum, dolomite, limestone, or halite; less commonly, with polyhalite or magnesite. Primary precipitation of anhydrite is favored by warm and highly saline waters, but most anhydrite beds probably form by dehydration of gypsum. The cap rocks of many salt domes contain anhydrite masses with gypsum and sulfur.

The oxidation of pyrite may produce sulfuric acid, which reacts with carbonates to yield anhydrite druses disseminated in carbonate sediments or in the oxide zones of ore deposits in association with smithsonite, anglesite, cerussite, malachite, and other secondary ore minerals of Pb, Zn, Ag, or Cu. Amygdales in mafic volcanics may contain anhydrite with zeolites, quartz, prehnite, or glauberite. Anhydrite may be formed by sulfur gases reacting with calcite in fumerolic or caliche deposits.

REFERENCES: ANHYDRITE

Cheng, G. C. H., and J. Zussman. 1963. The crystal structure of anhydrite ($CaSO_4$). *Acta Crystallogr.*, *16*, 767–770.

Hawthorne, F. C., and R. B. Ferguson. 1975. Anhydrous sulfates. II. Refinement of the crystal structure of anhydrite. *Can. Mineral.*, *13*, 289–292.

Hoehne, E. 1962. The crystal structure of anhydrite, $CaSO_4$. *Monatsber. Deut. Akad. Wiss., Berlin*, *4*, 72–74.

Gypsum

$CaSO_4 \cdot 2H_2O$

MONOCLINIC
$\angle\beta = 127°24'$

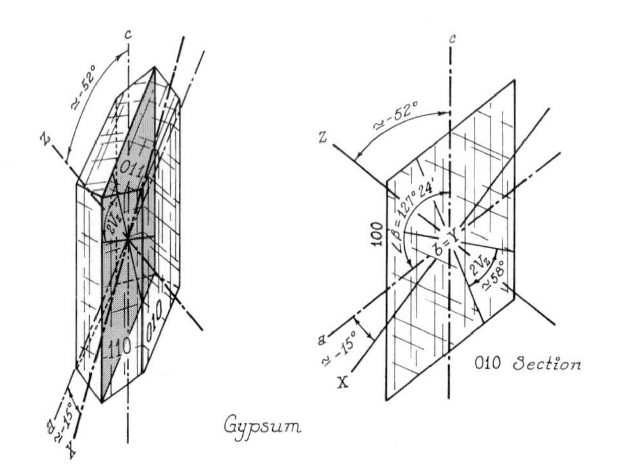

$n_\alpha = 1.519–1.521$*
$n_\beta = 1.522–1.526$
$n_\gamma = 1.529–1.531$
$n_\gamma - n_\alpha \simeq 0.010$
Biaxial positive $2V_z \simeq 58°$*
$b = Y, c \wedge Z \simeq -52°, a \wedge X \simeq -15°$
$r > v$ strong; inclined bisectrix dispersion
Colorless in standard section

* Indices of refraction decrease with increasing temperature to $n_\alpha = 1.518, n_\beta = 1.519$, and $n_\gamma = 1.527$ at 105°C. $2V_z$ also decreases, with increasing temperature, to $2V_z = 0°$ at 91°C.

COMPOSITION AND STRUCTURE. The gypsum structure is layered parallel to {010}. Tetrahedral SO_4^{2-} anion units are linked by Ca^{2+} cations as crude "chains" parallel to *c*, which form as double sheets interlayered, parallel to {010}, with sheets of water molecules.

Gypsum, $CaSO_4 \cdot 2H_2O$, is one of the few mineral varieties to show no significant chemical variation. Only trace amounts of Ba^{2+} or Sr^{2+} may replace Ca^{2+}.

Dehydration forms of gypsum probably range isostructurally from $CaSO_4 \cdot H_2O$ to $CaSO_4$; the hemihydrate $CaSO_4 \cdot \frac{1}{2}H_2O$, however, is a stable form known commercially as plaster of paris and in nature as *bassanite*.†

Gypsum commonly contains numerous inclusions of sand, clays, or iron oxides.

PHYSICAL PROPERTIES. H = 2. Sp. Gr. = 2.31. In hand sample, gypsum is colorless and transparent or white; sometimes gray, yellow, pink, brown, and so on. Luster is vitreous to pearly (on {010}).

COLOR AND PLEOCHROISM. Gypsum is colorless in section or as fragments.

FORM. Gypsum commonly occurs in single euhedral crystals with forms {010}, {110}, and {011}, as shown earlier. Prismatic crystals stubby to acicular on *c* are also common, frequently with curved faces or even warped or twisted into spiral forms. Lenticular crystal forms are often intergrown as rosettelike

† Bassanite is uniaxial positive; $n_\omega = 1.558$, $n_\varepsilon = 1.586$.

aggregates. Distinct crystal forms or broad, transparent plates are designated *selenite* gypsum. Fibrous forms, elongated on *c* and having splintery fracture and silky luster, are called *satin-spar* gypsum, and rare fibrous forms elongated on *a* are called *fascicular* gypsum. Massive and granular forms are *alabaster* gypsum, and *kopi* gypsum is an earthy variety.

CLEAVAGE. Perfect cleavage on {010} commonly controls fragment orientation. Cleavages are also distinct on {100} and {$\bar{1}$11}.

BIREFRINGENCE. Maximum birefringence is weak $(n_\gamma - n_\alpha) = 0.010$ and is seen in {010} sections and cleavage fragments on the prominent cleavage. Interference colors are first-order grays and white to pale yellow in standard thin section.

TWINNING. Simple contact twins on {100} are very common ("butterfly" twins); rarely multiple. Twinning on {$\bar{1}$01} is less common.

ZONING. Inclusions of clay may show zonal distribution.

INTERFERENCE FIGURE. With increasing temperature, $2V_z$ decreases to 0° at 91°C and opens above 91°C in a plane perpendicular to {010}. Isogyres are normally broad and diffuse on a nearly white field. Optic axis dispersion is strong $r > v$, and bisectrix dispersion is strongly inclined. Above 91°C, dispersion becomes $v > r$ in a plane normal to {010}, and bisectric dispersion becomes horizontal. Fragments on dominant cleavage {010} yield only flash figures.

OPTICAL ORIENTATION. The optic plane (*ac*) is parallel to dominant cleavage {010}. Maximum *extinction angle* is seen on {010} sections and is about 38° to {100} cleavage traces. Sections parallel to *b* show parallel extinction to {010} cleavage and either positive or negative elongation. On {010} cleavage surfaces, {100} cleavage traces lie nearer the fast wave.

With increasing temperature, $2V_z$ closes to 0° at 91°C and opens again in an optic plane perpendicular to {010}.

DISTINGUISHING FEATURES. Gypsum is characterized by low negative relief, low birefringence, and distinctive cleavages. Anhydrite shows much higher indices of refraction (moderate positive relief), higher birefringence, and pseudocubic cleavage. Bassanite has positive relief in balsam, moderate birefringence, and is uniaxial.

ALTERATION. Anhydrite readily hydrates to become gypsum at normal surface conditions. Gypsum may form pseudomorphs after anhydrite, calcite, or halite. It may be replaced by chalcedony, opal, calcite, aragonite, anhydrite, or celestite as pseudomorphs.

OCCURRENCE. Gypsum occurs most commonly as massive, stratified, sedimentary beds associated with beds of dolomite, limestone, red clays, anhydrite, and halite and other evaporite minerals. It is a chemical precipitate of isolated marine basins, both modern and ancient, where it is ideally preceded by clays and limestone or dolomite and followed by anhydrite, halite and salts of magnesium and potassium. Much gypsum doubtless results from the hydration of anhydrite, beginning on pseudocubic cleavage planes and possibly ending in the hydration of entire massive beds. Gypsum beds may, conversely, be dehydrated to form anhydrite, and which mineral was the primary evaporite is often obscure. Gypsum should precipitate first in the normal evaporite sequence, as anhydrite is favored by warm temperatures and high salinity. Gypsum appears in the cap rocks of salt domes with anhydrite and native sulfur, the latter possibly due to bacterial reduction of the sulfates.

Salt pans and dry lake beds commonly yield gypsum and ascending groundwater may deposit gypsum as single crystals, crystal aggregates ("desert roses"), or efflorescence in dry desert soils; rarely as colloform masses in caves.

Gypsum may form near fumeroles and volcanic vents by the action of sulfurous gases on carbonate rocks and may appear in association with sulfur, aragonite, and celestite.

In the gossan or oxide zones of metalliferous deposits, sulfuric acid may form from the oxidation of sulfide minerals and may react with carbonates to yield gypsum.

Bassanite forms in cavities in leucite volcanics.

REFERENCES: GYPSUM

Cole, W. F., and C. J. Lancucki. 1974. Refinement of the crystal structure of gypsum, $CaSO_42H_2O$. *Acta Crystallogr., B 30*, 921–929.

Edinger, S. E. 1973. Growth of gypsum: Factors which affect the size and growth rates of the habit faces of gypsum. *Jour. Cryst. Growth, 18*, 217–224.

Gordashevskii, P. F. 1966. Properties and crystal structure of gypsum. *Stroit. Materialy, 12*, 17–18.

Rohleder, J. W. 1972. Polarizability of the water molecule and optical anisotropy of natural gypsum, $CaSO_4 \cdot 2H_2O$. *Rocz. Chem., 46*, 2089–2097.

Strunz, H. 1942. Isotypie zwischen $YPO_4 \cdot 2H_2O$ and $CaSO_4 \cdot 2H_2O$. *Naturwissenschaften, 30*, 64.

Barite (Barytes) $BaSO_4$ ORTHORHOMBIC

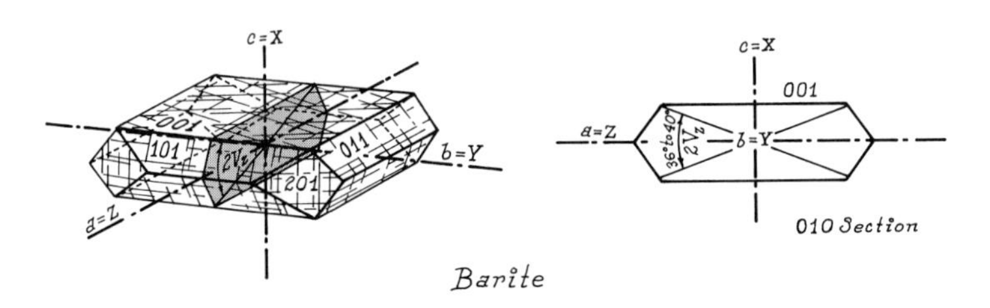

n_α = 1.634–1.637*
n_β = 1.636–1.639
n_γ = 1.646–1.649
$n_\gamma - n_\alpha$ = 0.010–0.013*
Biaxial positive, $2V_z$ = 36°–40°*
$a = Z, b = Y, c = X$
Weak dispersion $v > r$
Colorless to very pale yellow, brown, blue, and so on Weakly pleochroic, usually
$Z > Y > X$

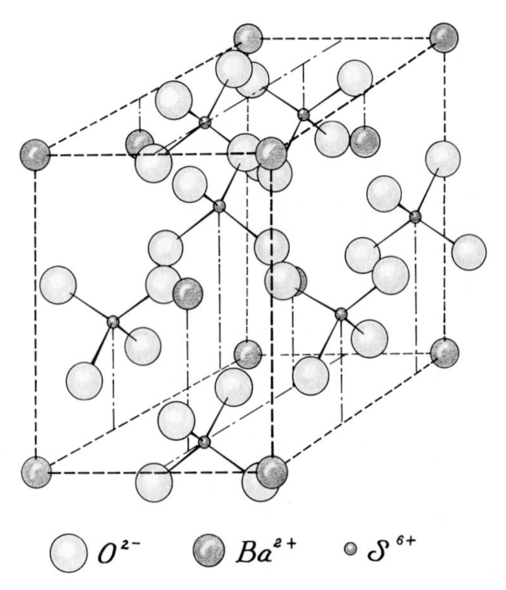

FIGURE 2-18. The structure of barite. Tetrahedral SO_4^{2-} units alternate with Ba^{2+} ions forming an orthorhombic lattice.

COMPOSITION AND STRUCTURE. The barite structure is characterized by sulfate ions (SO_4^{2-}) alternating with Ba^{2+} as shown in Fig. 2-18; each Ba is coordinated by 12 oxygen ions from seven SO_4 groups. Barite, $BaSO_4$, is isostructural with celestite, $SrSO_4$, and anglesite, $PbSO_4$, but not with anhydrite, $CaSO_4$; small cations (such as Ca^{2+}) require lower coordination.

Barite, $BaSO_4$, probably forms complete solid solution with celestite, $SrSO_4$, and possibly with anglesite, $PbSO_4$. Most natural barites, however, contain remarkably little Sr and are essentially without Pb. Varieties rich in Sr (*celestobarite*) or Pb (*hokutolite*) are rare but well documented. Solid solution with anhydrite, $CaSO_4$, is very limited, and Ca is a minor, but persistent impurity (up to 6 percent $CaSO_4$). Other very minor replacements are Hg^{2+}, Co^{2+}, and Ra^{2+} for Ba^{2+} and CrO_4^{2-}, SeO_4^{2-}, and MnO_4^{2-} groups for SO_4^{2-}.

* Refractive indices and birefringence decrease slightly with increasing Sr and increase with Pb. $2V_z$ increases with both Sr and Pb.

PHYSICAL PROPERTIES. H = 3–3½. Sp. Gr. = 4.50. Color, in hand sample is white ranging to yellow, brown, blue, reddish, and so on. Luster is vitreous.

COLOR AND PLEOCHROISM. In standard section or as fragments, barite is usually colorless; rarely shows pale tints of yellow, brown, blue, and so on. Pleochroism is highly variable with color and locality; absorption usually $Z > Y > X$. Blue color is probably due to radiation from Ra^{2+}.

FORM. Crystals of barite are usually tabular, thick to thin plates, on {001}, rarely prismatic parallel to *a* or *b*. Clusters or rosettes of random, platy crystals are common, as are also nodular concretions with concentric banding; barite is rarely fibrous or granular.

CLEAVAGE. Cleavage is perfect on {001} (basal), slightly less to slightly more perfect on {210}, (prismatic cleavages intersecting at 78°) and imperfect on {010}.

BIREFRINGENCE. Maximum birefringence, seen on {010} sections, is mild (0.010 to 0.013), and interference colors are low first-order (gray and white to yellow).

TWINNING. Secondary glide twinning is common in massive barite; otherwise, twinning is rare.

ZONING. Color zoning may reflect growth patterns.

INTERFERENCE FIGURE. Barite is biaxial positive, with moderate $2V_z \simeq 38°$. Sections parallel to {100} show an acute bisectrix figure with distinct isogyres on a white field; optic axis dispersion is weak $v > r$. The obtuse bisectrix $X = c$ lies normal to tabular habit and dominant cleavage {001}, making Bxo figures very common with fragments.

OPTICAL ORIENTATION. The optic plane of barite is {010}, and $a = Z$, $b = Y$, and $c = X$. Extinction is parallel to tabular habit {001} and pinacoidal cleavages {001} and {010} and symmetrical to prismatic cleavages {210}.

Sections normal to {001} show elongation due to platy habit and basal cleavage, which is positive (length-slow). Fibrous or columnar crystals elongated on *a* are also length-slow.

DISTINGUISHING FEATURES. Occurrence in veins and cavities is distinctive. A flame test may be necessary to separate barite and celestite. Celestite shows smaller refractive indices and birefringence and larger $2V_z$. Anhydrite has higher birefringence, gypsum shows inclined extinction, and anglesite shows higher refractive indices and birefringence.

ALTERATION. Barite may alter to witherite or may be replaced by any one of many minerals (for example, calcite, dolomite, siderite, rhodochrosite, magnesite, cerrusite, fluorite, quartz or chalcedony, pyrite, marcasite, chalcopyrite, and marcasite).

OCCURRENCE. The most characteristic occurrence of barite is as a gangue mineral in metalliferous hydrothermal veins in association with fluorite, sphalerite, galena, calcite, dolomite, chalcopyrite, stibnite, and so on. Barite may appear in vesicles and other fractures or cavities in rhyolite, granite, basalt, and other igneous rocks as a consequence of magmatic emanations and may also be deposited by thermal springs.

Veins and lenses of barite are quite common in limestone and dolomite and less common in other sedimentary rocks deposited by either magmatic or meteoric solutions. Barite is a rare cementing agent in detrital sediments and may be a constituent of residual clays formed by the weathering of baritic carbonate sediments.

REFERENCES: BARITE

Brower, E. 1973. Synthesis of barite, celestite, and barium-strontium solid solution crystals. *Geochim. Cosmochim. Acta*, *37*, 155–158.

Coleville, A. A., and K. Staudhammer. 1967. Refinement of the structure of barite. *Amer. Mineral.*, *52*, 1877–1880.

Hanor, J. S. 1968. Frequency distribution of compositions in the barite-celestite series. *Amer. Mineral.*, *53*, 1215–1222.

Hill, R. J. 1977. A further refinement of the barite structure. *Can. Mineral.*, *15*, 522–526.

Isetti G. 1967. Color and pleochroism of barite. *Period. Mineral.*, *36*, 25–41.

Narasimhamurty, T. S., and K. V. Rao, 1968. A new technique to study the dispersion of natural birefringence in uniaxial and biaxial crystals. *Indian Jour. Pure Appl. Phys.*, *6*, 367–370.

Patel, A. R., and J. Koshy, 1968. Cleavage and etching of barite. *Can. Mineral.*, *9*, 539–546.

Celestite (Celestine) $SrSO_4$ ORTHORHOMBIC

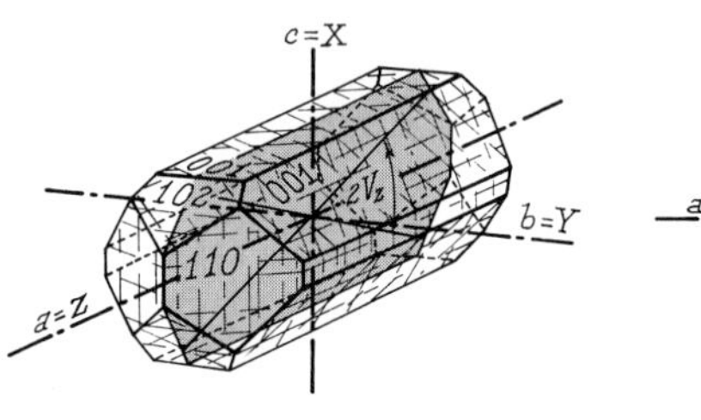

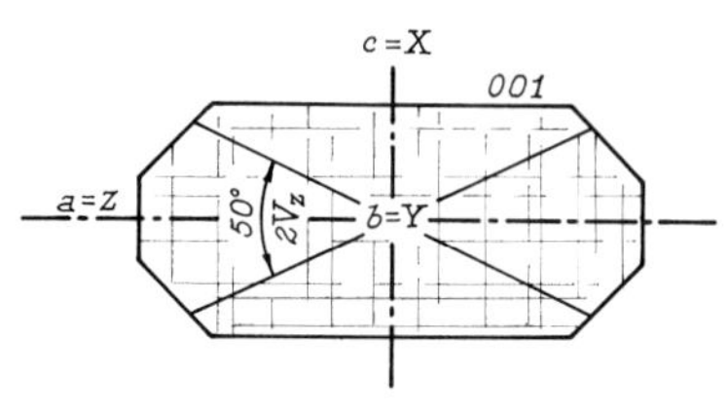

Celestite

$n_\alpha = 1.621–1.622$
$n_\beta = 1.623–1.624$
$n_\gamma = 1.630–1.632$
$n_\gamma - n_\alpha = 0.008–0.009$
Biaxial positive, $2V_z \simeq 50°$
$a = Z, b = Y, c = X$
Moderate dispersion $v > r$
Colorless; thick fragments may be weakly pleochroic in pale blue and violet
$Z > Y > X$

COMPOSITION AND STRUCTURE. Celestite, $SrSO_4$, is isomorphous with barite, $BaSO_4$, and forms complete solid solution with it. Natural celestite, however, seldom contains more than 2 or 3 percent Ba. Very limited solid solution exists with anhydrite, $CaSO_4$, and probably anglesite, $PbSO_4$, and both Ca and Pb may appear as minor impurities. Other ions (such as Hg) may appear in traces.

PHYSICAL PROPERTIES. $H = 3–3\frac{1}{2}$. Sp. Gr. $\simeq 3.95$. Color, in hand sample, is pale blue, white, red-brown, pale green, colorless.

COLOR AND PLEOCHROISM. Celestite is colorless in standard section; large fragments may be very pale blue and weakly pleochroic in blue, lavender, or blue-green, $Z > Y > X$.

FORM. Celestite crystals are commonly tabular, thick to thin, on {001}, or prismatic, elongated on *a*; less commonly, elongated on *b* or *c*. It may be fibrous, as parallel or radiating fibers, granular or earthy.

CLEAVAGE. Basal {001} cleavage is perfect, and prismatic cleavages {210} are distinct; cleavage on {010} is poor.

BIREFRINGENCE. Maximum birefringence, seen in {010} sections, is weak (0.008 to 0.009), and interference colors are first-order gray to white, in standard section.

TWINNING. Secondary glide twinning is reported in massive forms.

ZONING. Pale blue color in thick fragments may be uneven and arranged as growth zones. Growth zones may be turbid.

INTERFERENCE FIGURES. Celestite is biaxial positive, with moderate $2V_z \simeq 50°$. Sections parallel to {100} show an acute bisectrix figure, with distinct isogyres on a white field; optic axis dispersion is weak $v > r$. The obtuse bisectrix X lies normal to the major cleavage {001}, and tabular habit and fragments yield largely Bxo figures.

OPTICAL ORIENTATION. The optic plane of celestite is {010}, and $a = Z$, $b = Y$, and $c = X$. Extinction is parallel to pinacoidal cleavages {001} and {010} and tabular {001} or fibrous habit. It is symmetrical or inclined to prismatic {210} cleavages.

Sections normal to {001} show elongation due to tabular habit and basal cleavage, which is positive (length-slow). Fibrous or prismatic crystals elongated on a are also length-slow, those elongated on b may be either length-fast or length-slow, and those elongated on c are length-fast.

DISTINGUISHING FEATURES. Celestite has lower refractive indices, larger $2V$, and lower specific gravity than barite. Anhydrite has three mutually perpendicular cleavages and lower indices of refraction. Gypsum has inclined extinction.

ALTERATION. Celestite commonly alters to strontianite and may be replaced by calcite, quartz, chalcedony, barite, or sulfur.

OCCURRENCE. Celestite is most common in carbonate sedimentary rocks and in evaporite deposits. It fills fissures and cavities in limestone or dolomite, in association with strontianite, calcite, dolomite, gypsum, anhydrite, and fluorite. In beds of gypsum, anhydrite, or rock salt, it is often associated with crystals of native sulfur. Celestite may appear as disseminated grains or cement in shale, clay, sandstone, or marl, as a result of direct marine deposition or groundwater replacement.

Celestite is much less common than barite as a primary mineral in hydrothermal veins or as cavity filling in volcanic rocks.

REFERENCES: CELESTITE

Brower, E. 1973. Synthesis of barite, celestite, and barium-strontium sulfate solid solution crystals. *Geochim. Cosmochim. Acta*, *37*, 155–158.

Hanor, J. S. 1968. Frequency distribution of compositions in the barite-celestite series. *Amer. Mineral.*, *53*, 1215–1222.

Hawthorne, F. C., and R. B. Ferguson. 1975. Anhydrous sulfates. I. Refinement of the crystal structure of celestite with an appendix on the structure of thenardite. *Can. Mineral.*, *13*, 181–187.

Isetti, G. 1970. Coloration of celestite. *Doriana*, *4*, 1–7.

Roedder, E. 1969. Varvelike banding of possible annual origin in celestite crystals from Clay Center, Ohio, and in other minerals. *Amer. Mineral.*, *54*, 796–810.

Anglesite

$PbSO_4$

ORTHORHOMBIC

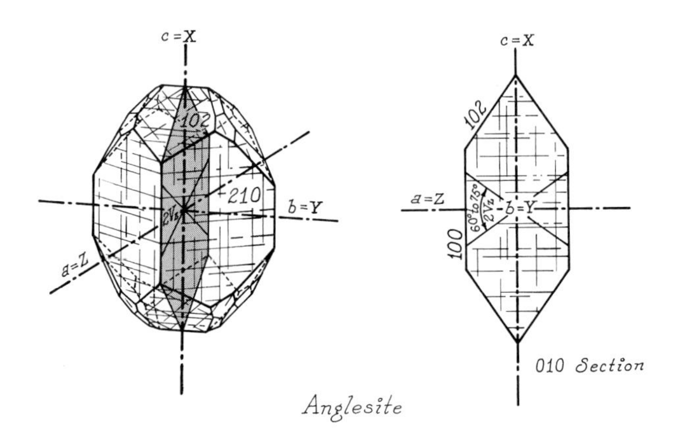

$n_\alpha = 1.878$
$n_\beta = 1.883$
$n_\gamma = 1.895$
$n_\gamma - n_\alpha = 0.017$
Biaxial positive, $2V_z = 60°–75°$
$a = Z$, $b = Y$, $c = X$
$v > r$ strong
Colorless, transparent to opaque in standard section

COMPOSITION AND STRUCTURE. The structure of anglesite should be analogous to that of celestite and barite, and some Sr^{+2} or Ba^{+2} may be expected to replace Pb^{+2}.

PHYSICAL PROPERTIES. H = $2\frac{1}{2}$–3. Sp. Gr. = 6.38. Color, in hand sample, is colorless or white to dark gray; rarely, yellow, green, or blue. Luster is adamantine.

COLOR AND PLEOCHROISM. Anglesite is colorless and transparent to nearly opaque in standard section or as fragments.

FORM. Anglesite shows a great variety of crystal habits, usually prismatic, tabular, or bipyramidal. It is commonly fine granular, massive, or nodular masses about a galena core.

CLEAVAGE. Cleavage is distinct on {001} and {210} and poor on {010}. Prismatic cleavages intersect at about 76°.

BIREFRINGENCE. Maximum birefringence, seen in {010} sections, is mild ($n_\gamma - n_\alpha = 0.017$). Interference colors range to first-order red in standard section.

TWINNING. Twins are rare; possible glide twinning.

ZONING. Megascopic banding is often visible as zoned replacement of galena on cubic cleavages.

INTERFERENCE FIGURE. Anglesite is biaxial positive with large $2V_z \simeq 60°–75°$. Sections parallel to {100} show acute bisectrix figures, with few isochromes and strong optic axis dispersion $v > r$.

OPTICAL ORIENTATION. Anglesite is orthorhombic, with $a = Z$, $b = Y$, and $c = X$. Extinction may be parallel inclined or symmetrical to dominant cleavages, and neither cleavages nor crystal habit yield consistent elongation.

DISTINCTIVE FEATURES. Extreme relief with mild birefringence, limited occurrence and close association with galena and cerussite are distinctive of anglesite. Cerussite shows very extreme birefringence and is biaxial negative, with very small $2V_x \simeq 9°$.

ALTERATION. Anglesite is a common oxidation product of galena and, in turn, commonly alters to cerussite. Unaltered cores of galena are commonly surrounded successively by zones of anglesite and cerussite. Anglesite may form pseudomorphs after galena, cerussite, or palmierite and may, itself, be replaced by pseudomorphs of cerussite, mimetite, or smithsonite.

OCCURRENCE. Anglesite is a secondary mineral resulting from the oxidation of galena and occurs in the oxide zones of ore deposits with cerussite, pyromorphite, mimetite, sulfur, smithsonite, and other secondary minerals of Pb, Zn, Cu, Ag, and so on.

REFERENCES: ANGLESITE

Quagliarella-Asciano, F. 1973. Green anglesites from Montevecchio (Sardinia). *Period. Mineral.*, *42*, 15–21.

Sahl, K. 1963. Refinement of the crystal structures of $PbCl_2$ (cotunnite), $BaCl_2$, $PbSO_4$ (anglesite) and $BaSO_4$ (barite) *Beitr. Mineral. Petrogr.*, *9*, 111–132.

Alunite — $KAl_3(SO_4)_2(OH)_6$ — TRIGONAL

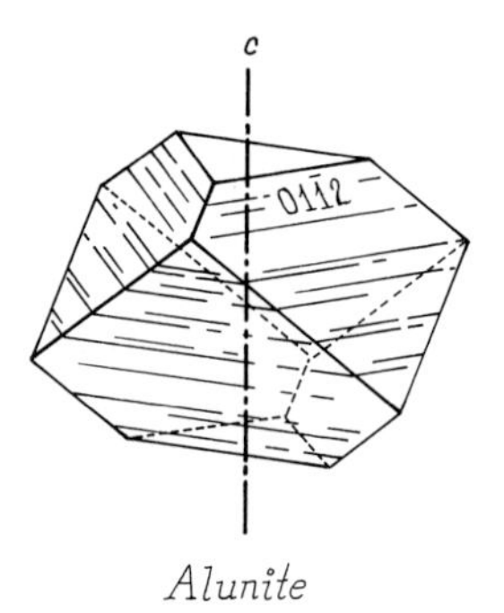

Alunite

$n_\omega = 1.568–1.585$
$n_\varepsilon = 1.590–1.601$
$n_\varepsilon - n_\omega = 0.010–0.023$
Uniaxial positive
Colorless in thin section

COMPOSITION. Extensive solid solution appears to exist between alunite, $KAl_3(SO_4)_2(OH)_6$, and natroalunite, $NaAl_3(SO_4)_2(OH)_6$, and, although natroalunite is a very rare mineral, most natural alunite contains significant Na. Only minor Fe^{3+} replaces Al^{3+}, and solid solution with jarosite, $KFe_3^{3+}(SO_4)_2(OH)_6$, is very limited in nature. Minor phosphorus generally replaces some sulfur.

PHYSICAL PROPERTIES. H = $3\frac{1}{2}$–4. Sp. Gr. = 2.6–2.9. In hand sample, alunite is white to gray; sometimes pale yellow, pink, or pale red-brown. Luster is vitreous to somewhat pearly.

COLOR AND PLEOCHROISM. Alunite is colorless in thin section or as fragments.

FORM. Anhedral to euhedral crystal aggregates may show cubelike rhombohedral habit or tabular crystals on {0001} with rectangular sections. Aggregates may also be fibrous to columnar or granular.

CLEAVAGE. Distinct basal cleavage {0001} may dominate fragment orientation; {01$\bar{1}$2} is very poor.

BIREFRINGENCE. Maximum birefringence, seen in prismatic sections is mild ($n_\varepsilon - n_\omega$ = 0.010–0.023), and interference colors, in standard section, are usually white or yellow ranging through first-order red. Basal cleavage fragments tend to show very low interference colors.

TWINNING. No twinning is reported.

INTERFERENCE FIGURES. Alunite is uniaxial positive, and basal sections and cleavage plates yield optic axis centered figures with broad isogyres and few isochromatic rings.

OPTICAL ORIENTATION. Extinction is parallel to basal cleavages or rectangular sections, and such elongation is negative (length-fast).

DISTINGUISHING FEATURES. Alunite is a secondary mineral showing mild birefringence and moderate positive relief. It is uniaxial positive and resembles brucite; the latter usually shows anomalous interference colors and different occurrence.

ALTERATION. Alunite is an alteration product formed by the action of sulfate waters on alkali feldspars.

OCCURRENCE. The most common occurrence of alunite is in felsic and intermediate volcanic flows and pyroclastics where sulfur gases and sulfate hydrothermal solutions alter alkali feldspars. It is usually associated with fine quartz and other products of feldspar alteration; kaolinite, allophane, and diaspore.

Alunite may form in the weathered zones of ore veins, where pyrite is oxidized to yield sulfuric acid.

REFERENCES: ALUNITE

Belov, N. V. 1967. Essays on structural mineralogy. XVIII. Silicates. *Mineral Sb.*, *21*, 231–245.

Botinelly, T. 1976. A review of the minerals of the alunite—jarosite, beudantite, and plumbogummite groups. *Jour. Res. U.S. Geol. Surv.*, *4*, 213–216.

Brophy, G. P., E. C. Scott, and R. A. Snellgrove. 1962. Sulfate studies. II. Solid solution between alunite and jarosite. *Amer. Mineral.*, *47*, 112–126.

Kashkai, M. A. 1969. Alunite group and its structural analogs. *Zap. Vses. Mineral. Obshchest.*, *98*, 150–165.

Parker, R. L. 1962. Isomorphous substitution in natural and synthetic alunite. *Amer. Mineral.*, *47*, 127–136.

Jarosite $KFe_3(SO_4)_2(OH)_6$ TRIGONAL

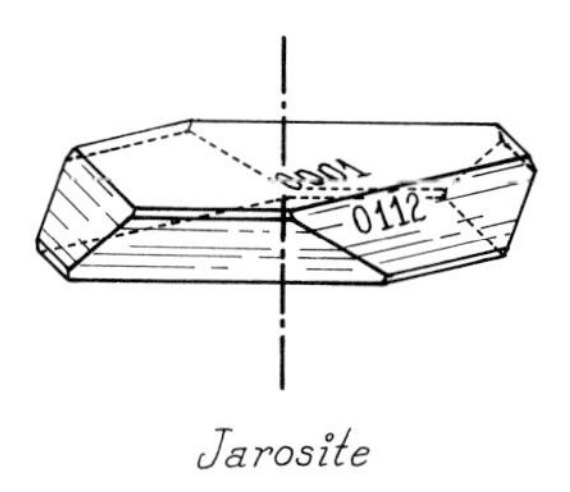

Jarosite

n_ω = 1.815–1.820
n_ε = 1.713–1.715
$n_\omega - n_\varepsilon$ = 0.101–0.105
Uniaxial negative
Golden yellow in thin section and pleochroic $O > E$

TABLE 2-3. Basic Sulfate Minerals Related to Jarosite

		n_ω	n_ε	Color
Natrojarosite[a]	$NaFe_3(SO_4)_2(OH)_6$	1.832	1.750	Pale yellow
Ammoniojarosite	$(NH_4)Fe_3(SO_4)_2(OH)_6$	1.800	1.750	Pale yellow
Argentojarosite	$AgFe_3(SO_4)_2(OH)_6$	1.882	1.785	Pale yellow
Carphosiderite	$HFe_3(SO_4)_2(OH)_6$	1.816	1.785	Deep yellow
Plumbojarosite	$Pb[Fe_3(SO_4)_2(OH)_6]_2$	1.875	1.786	Yellow-brown

[a] Often found with jarosite

COMPOSITION. Extensive solid solution exists between jarosite, $KFe_3(SO_4)_2(OH)_6$, and natrojarosite, $NaFe_3(SO_4)_2(OH)_6$. Solid solution with alunite, $KAl_3(SO_4)_2(OH)_2$, is very limited in nature, although intermediate compositions have been synthesized. Jarosite may contain minor Ag or Pb. Several basic sulfate minerals are structurally related to jarosite; see Table 2-3.

PHYSICAL PROPERTIES. H = 3. Sp. Gr. = 3.25. In hand sample, jarosite is golden yellow to dark brown, with resinous to adamantine luster.

COLOR AND PLEOCHROISM. Jarosite is strongly pleochroic in yellow and brown $O > E$. O = deep golden brown, or red-brown; E = pale yellow, pale yellow-green, or colorless.

FORM. Tiny crystals of jarosite are usually tabular plates on {0001} or rhombohedrons, sometimes rounded forms, and appear as encrusting coatings on other minerals. Crystal aggregates may be granular, fibrous, or scaley, and small colloform structures are common.

CLEAVAGE. Distinct basal cleavage {0001} and tabular crystal habit {0001} dominate fragment orientation.

BIREFRINGENCE. Maximum extinction, seen in prismatic sections, is extreme ($n_\omega - n_\varepsilon = 0.101–0.105$), and interference colors may range through fourth-order. Basal sections and cleavage plates yield low-order colors owing to their orientation.

TWINNING. Basal sections of biaxial varieties may show six radiating sector twins. The optic plane of each sector is parallel to its outer edge.

INTERFERENCE FIGURE. Jarosite and related sulfates are uniaxial negative; however, biaxial negative forms are very common, with $2V_x$ up to 10°. Basal sections and cleavage plates yield centered optic axis figures showing very numerous isochromes.

OPTICAL ORIENTATION. Extinction is parallel to rectangular sections and cleavage traces, and such elongation is positive (length-slow).

DISTINGUISHING FEATURES. Secondary occurrence, extreme relief and birefringence, and golden-yellow pleochroism are distinctive of jarosite. Limonite is isotropic, goethite is biaxial negative with extreme dispersion $r > v$ and small $2V_x$, and lepidocrocite is biaxial, with large $2V_x$. All three ferric hydroxides show indices of refraction of 2.0 or above. Jarosite may be optically indistinguishible from structurally related basic sulfates; however, argentojarosite and plumbojarosite have higher indices of refraction.

ALTERATION. Jarosite tarnishes quickly and weathers to limonite or goethite.

OCCURRENCE. Jarosite is a secondary mineral formed by the weathering of iron-bearing sulfides. It appears most commonly as scaley coatings or colloform crusts in the oxide zones of sulfide deposits and in cracks and cavities in iron ores and adjoining rocks. It is commonly associated with limonite, goethite, hematite, pyrite, kaolin clays, and secondary minerals of base metals. It is often overlooked when intermixed with iron oxides.

Jarosite may form in sandstone, shale, or clays by the alteration of glauconite, pyrite, or marcasite.

REFERENCES: JAROSITE

Belov, N. V. 1967. Essays on structural mineralogy, XVIII. *Mineral. Sb., 21*, 231–245.

Botinelly, T. 1976. A review of the minerals of the alunite-jarosite, beudantite, and plumbogummite groups. *Jour. Res. U.S. Geol. Surv., 4*, 213–216.

Brophy, G. P., E. C. Scott, and R. A. Snellgrove. 1962. Sulfate studies. II. Solid solution between alunite and jarosite. *Amer. Mineral., 47*, 112–126.

Kato, T., and E. W. Radoslovich. 1968. Crystal structures of soil phosphates. *Trans., 9th Int. Congr. Soil Sci., 2*, 725–731.

Zirkl, E. J. 1962. Jarosite and natrojarosite from the graphite workings at Weinberg near Trandorf, Lower Austria. *Neues Jahrb. Mineral. Monatsh., 1962*, 27–31.

BORATES

Borax

$Na_2B_4O_7 \cdot 10H_2O$

MONOCLINIC
$\angle\beta = 106°35'$

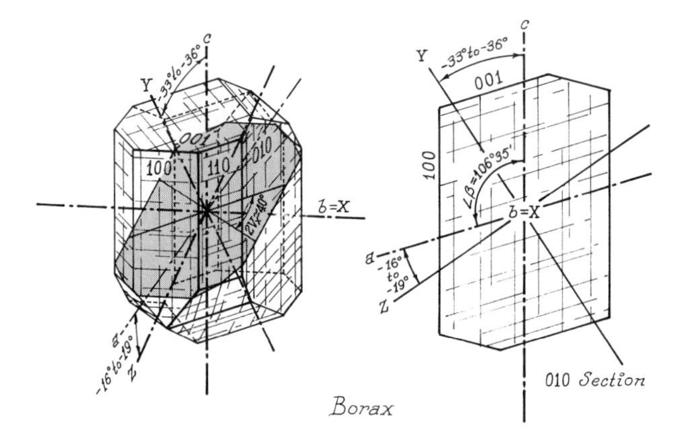

$n_\alpha = 1.447$
$n_\beta = 1.469$
$n_\gamma = 1.472$
$n_\gamma - n_\alpha = 0.025$
Biaxial negative, $2V_x \simeq 40°$
$b = X, c \wedge Y = -33°$ to $-36°, a \wedge Z = -16°$ to $-19°$
$r > v$ strong, crossed bisectrix dispersion
Colorless in thin section

COMPOSITION. Borax is usually very close to its ideal composition, $Na_2B_4O_7 \cdot 10H_2O$; however, it may lose water and crumble on exposure to air.

PHYSICAL PROPERTIES. H = $2–2\frac{1}{2}$. Sp. Gr. = 1.71. In hand sample, borax is white and sometimes gray, greenish, or bluish. Luster is vitreous or earthy. Borax is highly water soluble, with a weak alkaline taste.

COLOR AND PLEOCHROISM. Borax is colorless in standard section or as fragments.

FORM. Crystals are stubby prismatic with octagonal and rectangular sections resembling pyroxene. Sometimes exists as granular aggregates.

CLEAVAGE. Cleavage is perfect on {100} and distinct on {110} and resembles the parting and cleavages of pyroxene.

BIREFRINGENCE. Maximum birefringence, seen in {010} sections is moderate ($n_\gamma - n_\alpha = 0.025$). Interference colors reach low second-order in standard thin section; minimum birefringent sections show anomalous blue or olive-brown interference colors.

TWINNING. Twinning on {100} is uncommon.

INTERFERENCE FIGURE. Sections parallel to {010} show centered acute bisectrix figures with a few anomalous isochromes, strong optic axis dispersion $r > v$, and distinct crossed bisectrix dispersion. Cleavage fragments yield eccentric figures.

OPTICAL ORIENTATION. Borax is monoclinic, with $b = X$, $c \wedge Y = -33°$ to $-36°$, $a \wedge Z = -16°$ to $-19°$. The optic plane is normal to {010}, 16° to 19° from *a*. Maximum extinction angle, seen in {010} sections is 33° to 36° to cleavage traces, and the fast wave lies nearer to cleavage traces (length-fast). Sections parallel to {100} show parallel extinction and positive elongation (length-slow). Near-basal sections show symmetrical extinction to prismatic cleavages.

DISTINGUISHING FEATURES. Borax is highly water soluble, with restricted occurrence; it has very low indices of refraction, anomalous interference colors, strong optic axis dispersion $r > v$, and distinct crossed bisectrix dispersion.

ALTERATION. On exposure to the air, borax dehydrates to tincalconite, $Na_2B_4O_7{\cdot}5H_2O$, and crumbles to granules.

OCCURRENCE. Borax is a highly soluble evaporite mineral deposited in the clay and muds of saline lake beds and playas. It also appears as an efflorescence on desert soils and in thermal spring deposits. Borax is commonly associated with montmorillonite clays, calcite, gypsum, and numerous soluble salts of sodium (for example, halite, thenardite, glauberite, trona, ulexite, gay-lussite, hanksite, and soda-niter).

REFERENCES: BORAX

Bowser, C. J., and F. W. Dickson. 1966. Chemical zonation of the borates of Kramer, California. *Symp. Salt, 2nd, Cleveland, 1*, 122–132. Cleveland: Northern Ohio Geological Society.

Inan, K., A. C. Dunham, and J. Esson. 1973. Mineralogy, chemistry and origin of Kirka borate deposit, Eskishehir Province, Turkey. *Inst. Mining Met. Trans., 82*, B114–B123.

Colemanite

$Ca_2B_6O_{11}{\cdot}5H_2O$

MONOCLINIC
$\angle\beta = 110°07'$

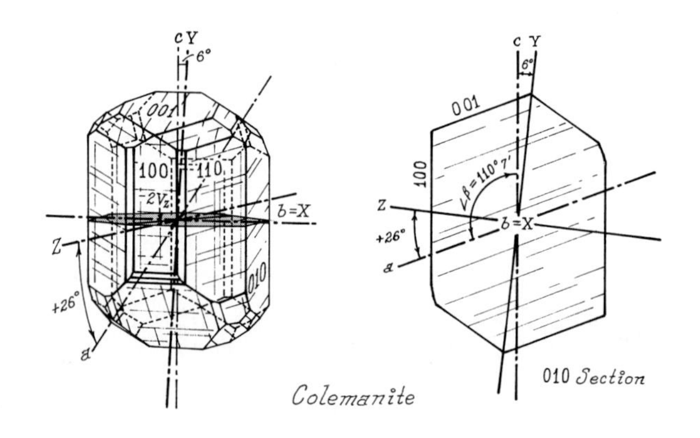

$n_\alpha = 1.586$
$n_\beta = 1.592$
$n_\gamma = 1.614$
$n_\gamma - n_\alpha = 0.028$
Biaxial positive, $2V_z \simeq 55°$
$b = X, c \wedge Y \simeq +6°, a \wedge Z \simeq +26°$
Weak $v > r$
Colorless in standard thin section

COMPOSITION. Colemanite, $Ca_2B_6O_{11} \cdot 5H_2O$, shows very little ionic substitution; very minor Mg^{2+} or alkalis may replace Ca^{2+}.

PHYSICAL PROPERTIES. $H = 4$–$4\frac{1}{2}$. Sp. Gr. = 2.42. In hand sample, colemanite is colorless to white or pale yellow. Luster is brilliant vitreous.

COLOR AND PLEOCHROISM. Colemanite is colorless in section or as fragments.

FORM. Crystals are common, usually stubby prismatic with numerous complex faces, sometimes pseudorhombohedral; also massive, as cleavable to granular masses.

CLEAVAGE. Perfect cleavage on {010} and distinct on {001} intersect at 90° and dominate fragment orientation.

BIREFRINGENCE. Maximum birefringence, seen in sections normal to {010} and 26° from {001}, is moderate ($n_\gamma - n_\alpha = 0.028$). Interference colors, in standard section, are mostly brilliant colors ranging to middle second-order.

TWINNING. No twinning is reported.

INTERFERENCE FIGURES. Sections near {100} yield acute bisectrix figures showing numerous isochromes and weak dispersion $v > r$ (possible horizontal bisectrix dispersion). Cleavage fragments on {010} and {001} yield obtuse bisectrix and near-flash figures respectively.

OPTICAL ORIENTATION. Fragments lying on {010} cleavage show extinction 26° to basal cleavages and positive elongation (length-slow); fragments on {001} cleavage show parallel extinction and positive elongation to {010} cleavage traces.

DISTINGUISHING FEATURES. Colemanite has very limited occurrence and appears only in association with other borate minerals in playa deposits. Kernite, borax, and ulexite show negative relief in balsam, and kernite, ulexite, hydroboracite, and howlite are fibrous, acicular, or splintery.

ALTERATION. Colemanite may result from the alteration of ulexite, hydroboracite, or priceite. Calcite may result from the alteration of colemanite.

OCCURRENCE. Colemanite is found in the evaporites of borate playas. It is probably largely the result of borax or ulexite alteration. Associated minerals are borax, ulexite, kernite, howlite, hydroboracite, and other borate minerals as well as gypsum, celestite, and calcite.

REFERENCES: COLEMANITE

Clark, J. R., D. E. Appleman, and C. L. Christ. 1964. Crystal chemistry and structure refinement of five hydrated calcium borates. *Jour. Inorg. Nucl. Chem.*, *26*, 93–95.

Hainsworth, F. N., and H. E. Petch. 1966. The structural basis of ferroelectricity in colemanite. *Can. Jour. Phys.*, *44*, 3083–3107.

Muehle, G. 1974. Colemanite pseudomorphs from the Corkscrew Mine, Death Valley, California. *Mineral. Rec.*, *5*, 174–177.

Schröder, A. 1957. The determination of the optical constants of colemanite. *Neues Jahrb. Mineral. Monatsh.*, *1957*, 135–140.

Schröder, A., and W. Hoffmann. 1956. Optics of colemanite. *Neues Jahrb. Mineral., Monatsh.*, *12*, 265–271.

The Ludwigite-Vonsenite Series

$(Mg,Fe^{2+})_2Fe^{3+}BO_3 \cdot O_2$ ORTHORHOMBIC

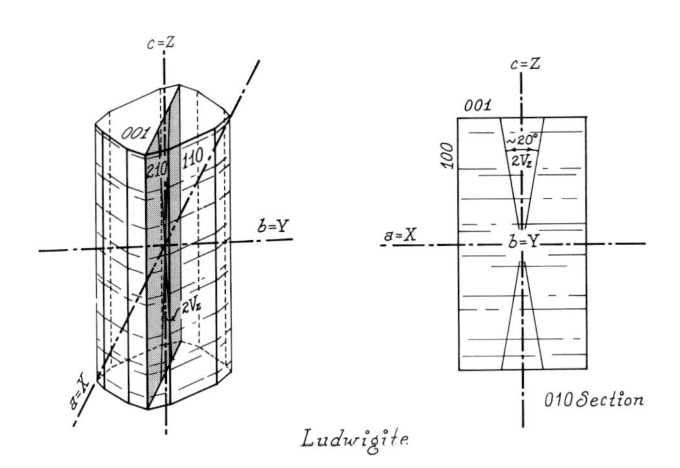

$n_\alpha = 1.83–1.85$*
$n_\beta = 1.83–1.85$
$n_\gamma = 1.97–2.02$
$n_\gamma - n_\alpha = 0.14–0.17$*
Biaxial positive, $2V_z \sim$ very small to 20°–45°
$a = X, b = Y, c = Z$
$r > v$ extreme
$X \simeq Y =$ deep olive-green, greenish black*
$Z =$ dark brown, dark red-brown

COMPOSITION AND STRUCTURE. The ludwigite structure contains independent BO_3 units with bivalent (Mg or Fe^{2+}) and trivalent (Fe^{3+} or Al) interunit cations. A complete isomorphous series almost certainly exists between Mg_2-

* Indices of refraction, birefringence, color intensity and density increase with increasing Fe^{2+} (Fig. 2-19). Al-rich varieties are reported, with smaller indices of refraction ($n_\alpha = 1.79$, $n_\gamma = 1.886$), birefringence (0.095), and color intensity.

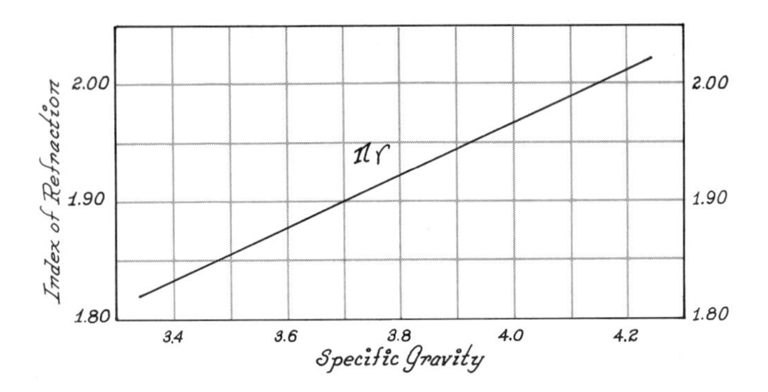

FIGURE 2-19. Variation of index of refraction (n_γ) and specific gravity for the ludwigite-vonsenite series. Substitution of Al^{3+} for Fe^{3+} as well as Fe^{2+} for Mg^{2+} precludes determination of composition by either refractive index or specific gravity, although the densest minerals of the series are certainly the most iron-rich. (Data from Panova, 1968)

$Fe^{3+}BO_3{\cdot}O_2$ (magnesioludwigite) and $Fe^{2+}Fe^{3+}BO_3{\cdot}O_2$ (ferrovonsenite), with ludwigite ($Mg > Fe^{2+}$) and vonsenite or paigeite ($Fe^{2+} > Mg$) as the common intermediate members.

Chemical analyses commonly report significant amounts of Al, Mn, Sn, or Ti. Aluminous ludwigite, $(Mg,Fe^{2+})_2$-$(Fe^{+3},Al)BO_3{\cdot}O_2$, is well known, as is the related Mn-mineral *pinakiolite*, $(Mg_3Mn^{2+})Mn_2{}^{3+}(BO_3{\cdot}O_2)_2$. The stannous analog *hulsite*, $(Fe^{2+},Mg)_2(Fe^{3+},Sn^{4+})BO_3{\cdot}O_2$, is monoclinic and structurally unique.

PHYSICAL PROPERTIES. H = 5. Sp. Gr. = 3.6 (Mg) to 4.7 (Fe). Color in handsample is dark green (ludwigite) to coal-black (vonsenite).

COLOR AND PLEOCHROISM. As fragments, the ludwigite-vonsenite minerals are almost opaque. Thin sections and splinters show deep colors in deep greens and browns with strong absorption and pleochroism $Z > Y \simeq Z$.

FORM. The ludwigite-vonsenite minerals usually appear as parallel, interwoven, or radiating aggregates of tiny acicular-fibrous crystals elongated on c; sometimes columnar, up to 2 cm long, or hairlike.

CLEAVAGE. Perfect basal {001} cleavage is reported.

BIREFRINGENCE. Very strong birefringence (0.14–0.17) yields high-order gray interference color, which is obscured by deep mineral color.

INTERFERENCE FIGURE. Fragments on basal cleavages yield centered acute bisectrix figures, with small $2V_z \simeq 0°$–$45°$ and extreme optic axis dispersion $r > v$. Deep mineral color absorption may darken out interference figures in any orientation, especially along the acute bisectrix.

OPTICAL ORIENTATION. Extinction is parallel to both cleavage and crystal elongation. Crystal elongation (c) is positive (length-slow).

DISTINGUISHING FEATURES. The ludwigite-vonsenite series resembles many almost opaque green-brown minerals. Amphiboles show inclined extinction and smaller indices of refraction. Dark tourmaline also has lower indices of refraction and is uniaxial.

ALTERATION. Limonite and other iron oxides are common alteration products.

OCCURRENCE. Ludwigite-vonsenite minerals are high-temperature phases of contact skarn rocks, often appearing as fibrous masses in crystalline limestone. They are often associated with forsterite, magnetite, clinohumite, diopside, garnet, and other borate minerals: tourmaline, axinite, szaibelyite, danburite, or datolite. Ludwigite-vonsenite minerals have been reported in cassiterite ores, magnetite deposits, and contact ores of iron and of copper.

REFERENCES: THE LUDWIGITE-VONSENITE SERIES

Butler, B. S., and W. T. Schaller. 1917. Magnesioludwigite, a new mineral. *Jour. Wash. Acad. Sci.*, *7*, 29–31.

Frederico, M. 1971. Oxidation products of vonsenite. *Period. Mineral.*, *40*, 1–6.

Giussani, A., and L. Vighi. 1964. Characteristics and genesis of boron minerals, ludwigite, ferroludwigite, szaibelyite, and camsellite, from the mines of Brosso, Ivrea. *Period. Mineral.*, *33*, 471–500.

Mokeeva, V. I., and S. M. Aleksandrov. 1969. Magnesium and iron distribution in the structure of borates of the ludwigite-vonsenite series. *Geokhimiya*, *1969*, 428–435.

Panova, M. A. 1968. Properties of minerals of ludwigite-vonsenite series. *Mineral. Syr'e*, *1968*, no. 18, 121–122.

Schaller, W. T., and A. C. Vlisidis. 1961. The composition of the aluminian ludwigite from Crestmore, California, *Amer. Mineral.*, *46*, 335–339.

Vlisidis, A. C., and W. T. Shaller. 1974. The identity of paigeite with vonsenite and chemical analyses of vonsenite, ludwigite and halsite. *Neues. Jahrb. Min.*, *Mh.*, 95–105.

Woodford, A. O. 1943. Crestmore minerals. *Calif. Jour. Mines & Geology*, *39*, 333–365.

CHAPTER 3

The Nesosilicates

The silicate minerals constitute by far the largest group of minerals in the crust of the earth and have consequently been more intensively studied than other groups. Their common feature is the presence of the silicate oxyanion, SiO_4^{4-}, but their structural diversity and abundance have led to several methods of further classification. The system in most common use among mineralogists, and the one used in this work, is based on the degree of polymerization of the tetrahedral SiO_4^{4-} groups. Minerals in which each silicate tetrahedron shares all four oxygens with other tetrahedra are called *tektosilicates*, or framework silicates. Sharing only three oxygens leads to the formation of sheets, yielding the *phyllosilicates*, or sheet silicates. The *inosilicates*, or chain silicates, contain tetrahedra that share two or two and a half (on the average) oxygen atoms with other tetrahedra. An alternate configuration of tetrahedra sharing two oxygen atoms yields *cyclosilicates*, or ring silicates. The *sorosilicates* contain $Si_2O_7^{6-}$ groups, wherein two tetrahedra share a common oxygen, and, finally, *nesosilicates* (or *orthosilicates*) contain no oxygen atoms shared by two SiO_4^{4-} tetrahedra. The term "bridging oxygen" is applied to oxygen atoms that are shared by (form a "bridge" between) two silicons. As the degree of polymerization of silicate oxyanions increases, their influence on the optical properties of the minerals becomes more pronounced.

Silicate minerals containing independent (insular) SiO_4^{4-} tetrahedra are known as *nesosilicates*, or *orthosilicates*. The presence of "$(SiO_4)_n$" in the chemical formula (where n is any positive integer) is a strong, but not a necessary, indication that the mineral is an orthosilicate; thus, $Ca_3Al_2(SiO_4)_3$, Mg_2SiO_4, and $Al_2SiO_4(OH)_2$ are all orthosilicates. Likewise, Al_2OSiO_4 is also an orthosilicate, but the formula is usually written Al_2SiO_5.

Because the only structural requirement for an orthosilicate is that the silicate tetrahedra be nonpolymerized, the variations in crystal structure and chemistry are many. Thus no consistency in crystal symmetry, morphology, or optical properties is to be expected among the minerals that make up this group. Some are isotropic, some uniaxial(+ and −), and others biaxial (+ and −). Birefringence may be large or small and depends on factors other than the presence of insular silicate tetrahedra.

Garnets, olivines, zircon, the aluminosilicate polymorphs (kyanite, andalusite, and sillimanite), staurolite, topaz, the humites, and chloritoid are among the nesosilicate minerals.

THE OLIVINE GROUP

Olivine minerals constitute an isomorphous group of orthosilicates. Independent silicate tetrahedra, in layers parallel to {100}, point alternately up and down along *a*, linking zigzag bands of edge-sharing MO_6 octrahedra (M = metal atom) that extend along *c*. The metal cations occupy two different lattice positions, which may accommodate different cations (for example, $CaMgSiO_4$). Extensive solid solution exists between end-member compositions of the olivine group when large cations are of similar size (for example, Mg $\rightleftharpoons$ Fe and Fe $\rightleftharpoons$ Mn). The following list shows end members of the olivine group:

Forsterite	Mg_2SiO_4
Fayalite	Fe_2SiO_4
Tephroite	Mn_2SiO_4
Monticellite	$CaMgSiO_4$
Kirschsteinite	$CaFeSiO_4$
Glaucochroite	$CaMnSiO_4$
Picrotephroite	$(Mg,Mn)_2SiO_4$

Complete solid solution exists between forsterite and fayalite to form the common olivines. Complete solid solution is also present between fayalite and tephroite, with *knebelite*, $(Mn,Fe)_2SiO_4$, as intermediate composition, and should be expected between monticellite and kirschsteinite, if kirschsteinite exists in nature. *Larnite*, Ca_2SiO_4, does not possess the

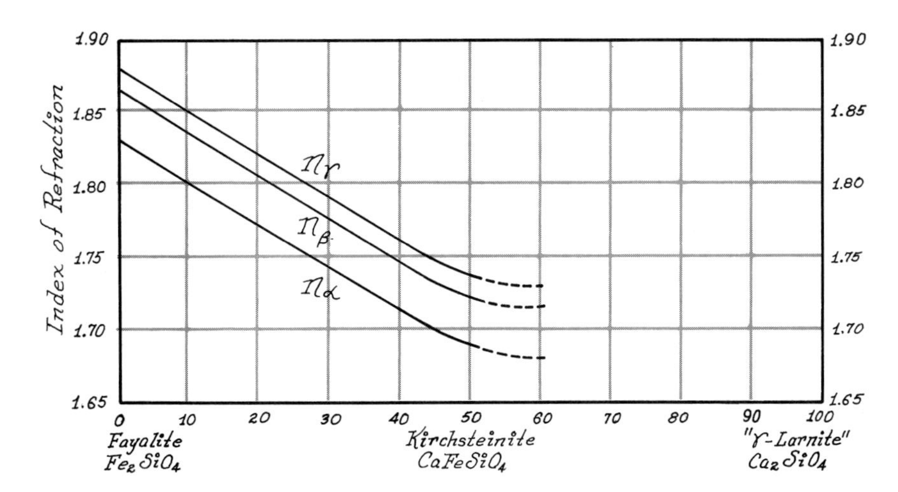

FIGURE 3-1. Variation of the indices of refraction in the fayalite–γ-larnite series. (Data from Wyderks and Mazanek, 1968)

olivine structure, but the synthetic mineral *γ-larnite* does. (See Fig. 3-1 for variation of refractive indices from fayalite toward γ-larnite.) A summary of the optical and physical properties of minerals of the olivine group is found in Table 3-1.

TABLE 3–1. Optical Properties of the Olivine Group

n_β	n_α	n_γ	$n_\gamma - n_\alpha$	Mineral	System	$2V$	Optic Sign	Cleavage	Optical Orientation	Color in Sections	Physical Properties
1.651	1.635	1.670	0.035	Forsterite $Mg_2SiO_4(Fo_{100})$	Orthorhombic	82° $v > r$	+	{010} and {110} poor	$a = Z, b = X, c = Y$	Colorless, pale green	H = 7. Sp. Gr. = 3.22. Pale olive-green
1.646–1.674	1.639–1.663	1.653–1.680	0.014–0.020	Monticellite $CaMgSiO_4$	Orthorhombic	90°–70° $r > v$	–	{010} poor	$a = Z, b = X, c = Y$	Colorless	H = 5½. Sp. Gr. = 3.1–3.3. Colorless to gray
1.693	1.672	1.712	0.040	Crysolite $(Fo_{80}–Fa_{20})$	Orthorhombic	87° $r > v$	–	{010} and {110} poor	$a = Z, b = X, c = Y$	Colorless, Pale green	H = 7. Sp. Gr. = 3.4. Pale olive-green
1.735	1.711	1.753	0.042	Hyalosiderite $(Fo_{60}–Fa_{40})$	Orthorhombic	78° $r > v$	–	{010} and {110} poor	$a = Z, b = X, c = Y$	Colorless, pale green	H = 7. Sp. Gr. = 3.6. Pale olive-green
1.777	1.750	1.794	0.044	Hortonolite $(Fo_{40}–Fa_{60})$	Orthorhombic	68° $r > v$	–	{010} and {110} poor	$a = Z, b = X, c = Y$	Colorless, pale yellow	H = 6½. Sp. Gr. = 3.9. Yellow-green
1.81	1.77	1.82	0.04	Tephroite $Mn_2SiO_4(Tp_{100})$	Orthorhombic	70° $r > v$	–	{010} imperfect {001} poor	$a = Z, b = X, c = Y$	Pale green, pale brown	H = 6. Sp. Gr. = 3.8. Olive-green
1.820	1.789	1.835	0.046	Ferrohortonolite $(Fo_{20}–Fa_{80})$	Orthorhombic	58° $r > v$	–	{010} indistinct {110} poor	$a = Z, b = X, c = Y$	Colorless, pale yellow	H = 6½. Sp. Gr. = 4.1. Yellow-green
1.82	1.78	1.83	0.05	Manganknebelite $(Tp_{80}–Fa_{20})$	Orthorhombic	57° $r > v$	–	{010} imperfect {001} poor	$a = Z, b = X, c = Y$	Colorless, pale green	H = 6. Sp. Gr. = 3.9. Deep olive-green
1.84	1.80	1.85	0.05	Knebelite $(Tp_{50}–Fa_{50})$	Orthorhombic	44° $r > v$	–	{010} imperfect {110} poor	$a = Z, b = X, c = Y$	Colorless, pale yellow	H = 6½. Sp. Gr. = 4.1. Brownish-black
1.86	1.82	1.87	0.05	Iron Knebelite $(Tp_{20}–Fa_{80})$	Orthorhombic	45° $r > v$	–	{010} imperfect {001} poor	$a = Z, b = X, c = Y$	Colorless, pale yellow	H = 6½. Sp. Gr. = 4.2. Brown
1.869	1.827	1.879	0.052	Fayalite $Fe_2SiO_4(Fa_{100})$	Orthorhombic	46° $r > v$	–	{010} distinct {001} poor	$a = Z, b = X, c = Y$	Colorless, pale yellow	H = 6½. Sp. Gr. = 4.39. Yellow-green

The Forsterite-Fayalite Series

$(Mg,Fe)_2SiO_4$ ORTHORHOMBIC

	n_α*	n_β	n_γ	$n_\gamma - n_\alpha$*	$2V$*	Dispersion
Forsterite (Fo_{100})	1.635	1.651	1.670	0.035	82°+	$v > r$ weak
Chrysolite (Fo_{80})	1.672	1.693	1.712	0.040	87°−	$r > v$ weak
Hyalosiderite (Fo_{60})	1.711	1.735	1.753	0.042	78°−	$r > v$ weak
Hortonolite (Fo_{40})	1.750	1777	1.794	0.044	68°−	$r > v$ weak
Ferrohortonolite (Fo_{20})	1.789	1.820	1.835	0.046	58°−	$r > v$ weak
Fayalite (Fo_0)	1.827	1.869	1.879	0.052	46°−	$r > v$ weak

$a = Z, b = X, c = Y$

Colorless to pale yellow or yellow-green in section, with weak pleochroism $Y > X = Z$

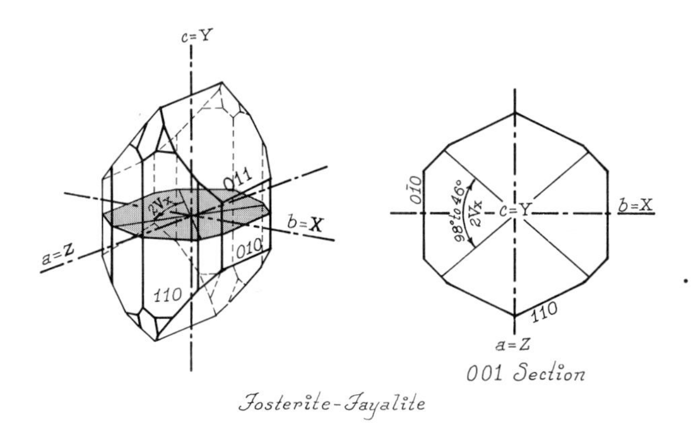

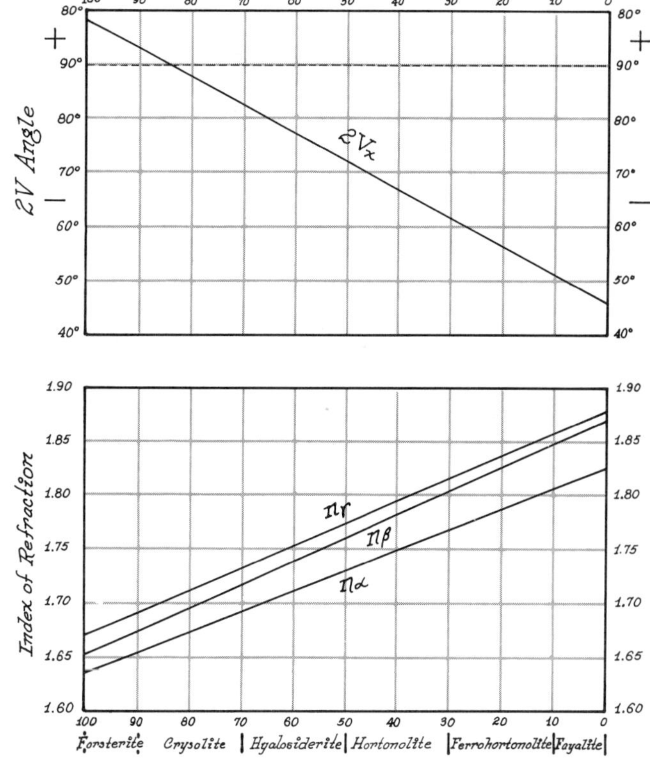

FIGURE 3-2. Variation of $2V$ angle and indices of refraction for the forsterite-fayalite (olivine) series. (After Bowen and Schairer, 1935)

COMPOSITION. The common olivines form a complete solution series between forsterite, Mg_2SiO_4, and fayalite, Fe_2SiO_4. Fayalite also forms a complete series with tephroite, Mn_2SiO_4, and partial series with *kirschsteinite*, $CaFeSiO_4$. Iron-rich members of the olivine series may contain appreciable Mn or Ca. Ferric iron (Fe^{3+}) is often reported in analyses but may be due largely to oxidation. Mg-rich compositions may contain minor Cr or Ni, and exsolution may yield tiny inclusions of magnetite or chromite.

PHYSICAL PROPERTIES. H = $6\frac{1}{2}$–7. Sp. Gr. = 3.22–4.39. Hardness decreases and specific gravity increases with increasing iron. Color, in hand sample, is commonly pale olive-green to yellow-green. With oxidation, olivine becomes reddish,

* (See Fig. 3-2 for continuous variation with composition).

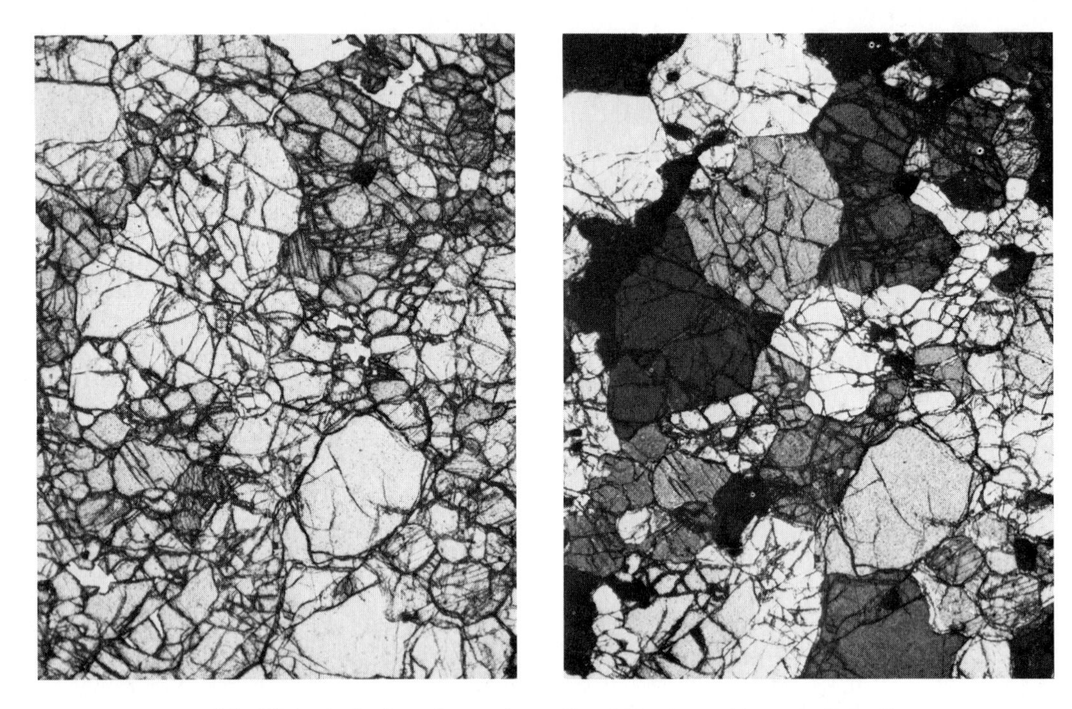

FIGURE 3-3. Olivine in dunite rocks may be unaltered. Fracture patterns are distinctive, and *Y*-shaped crystal junctions at 120° suggest simultaneous growth of individual crystals.

brown, or nearly black, with darker colors for Fe-rich compositions. Luster is vitreous.

COLOR AND PLEOCHROISM. Mg-olivine is colorless to pale green in section or as fragments, becoming lemon-yellow with increasing Fe. Colored varieties show moderate pleochroism $X = Z =$ pale yellow-green to yellow, $Y =$ yellow-orange. Oxidation may cause fayalite to be red-brown to nearly black in section.

FORM. In plutonic and metamorphic rocks, olivine crystals tend to be anhedral to subhedral as coarse, equidimensional grains (Fig. 3-3). Phenocrysts in volcanic rocks and crystals of differentiated intrusions may show euhedral outline of six or eight sides slightly elongated on *c*.

CLEAVAGE. Rather poor cleavages on {010} and {110} are less obvious than a coarse, irregular fracture pattern. The {010} cleavage improves with increasing iron to become distinct.

BIREFRINGENCE. Maximum birefringence, seen in basal {001} sections, increases with iron from moderate (0.035) to strong (0.052). Interference colors, in standard section, range to third-order and are characteristically bright colors noticeably above those of the pyroxenes.

TWINNING. Twinning in olivine is not common, but is most common and varied in Mg-compositions. Simple twinning on {100}, {011}, and {012} may be anticipated, and cyclic twinning on {031} reportedly produces trillings.

ZONING. Normal crystallization begins with Mg-rich compositions and proceeds with continuing Fe-enrichment; normal zoning shows Mg-rich cores and Fe-rich borders. Normal zoning is common in volcanic rocks and shallow intrusions and becomes obvious through birefringence change. In {010} sections, birefringence decreases toward the margin, as $(n_\gamma - n_\beta)$ decreases with Fe^{2+}, and, in {100} and {001} sections, birefringence increases toward the margin, as $(n_\beta - n_\alpha)$

and $(n_\gamma - n_\alpha)$ increase with Fe^{2+}. Composition zoning commonly ranges over one-third of the complete composition range (for example, Fa_{30}–Fa_{60}).

Reaction rims may appear on olivine crystals as corona structures or kelyphitic borders due to late magmatic reaction or metamorphic processes. That border zone in immediate contact with olivine is pyroxene (orthopyroxene or diopside) or amphibole, and outer zones of garnet or green spinel may be present.

Olivine may also appear in graphic intergrowth with orthopyroxene.

INTERFERENCE FIGURE. Forsterite is biaxial positive, with large $2V_z = 82°$, which increases with increasing iron. At Fa_{15}, $2V = 90°$, and the major portion of the series is biaxial negative as $2V_x$ converges on X to fayalite, where $2V_x = 46°$. Olivine in most rocks is near forsterite or chrysolite, and interference figures commonly show isogyres of a large $2V$ on a field of bright-colored isochromes. Poor cleavages favor bisectrix figures, but nearly random orientation is typical of fragments and sections. Fragments of compositions near fayalite tend to show acute bisectrix figures. Dispersion is weak through the series with $r > v$ about X, and bisectrix dispersion is absent.

OPTICAL ORIENTATION. The olivine series is orthorhombic $a = Z$, $b = X$, and $c = Y$. Both crystal sections and fragments tend to be elongated on c. Extinction is parallel to both cleavage and elongation in so far as cleavage or elongation is discernible. Crystal sections or fragments elongated on c show either positive (length-slow) or negative (length-fast) elongation, as $c = Y$.

DISTINGUISHING FEATURES. Olivine is most commonly confused with colorless pyroxenes, with which it is commonly associated. Pyroxenes show lower birefringence, good cleavages, and inclined extinction. Epidote shows low refractive index and large $2V_x$ compared to fayalite. The alteration of olivine is often distinctive.

ALTERATION. Olivine is highly subject to weathering, hydrothermal alteration, and metamorphism and usually shows some degree of alteration. Alteration products are usually complex mixtures of fine-grained minerals that are not often distinguished by optical means (Fig. 3-4). Three characteristic forms of olivine alteration show obvious differences and are given the generalized names *serpentine, iddingsite, and chlorophaeite.*

"Serpentine" alteration appears mostly in plutonic igneous and metamorphic rocks. It begins along fracture surfaces as the result of hydrothermal or deuteric processes and may completely convert an olivine rock to serpentine with a structure characteristic of the olivine fracture pattern (mesh structure). "Serpentine" alteration appears as an aggregate of pale green, low-birefringent fibers, actually consisting of serpentine minerals (antigorite and chrysotile), talc and magnesite, with tiny grains of by-product magnetite.

"Iddingsite" or "bowlingite" alteration is most characteristic of oxidation, deuteric alteration or weathering of volcanics and shallow intrusions. It appears as a reddish-brown replacement of olivine phenocrysts as distinct rims or complete replacement and may appear optically homogeneous, with high refractive index and extreme birefringence. Iddingsite and bowlingite are both fine-grained aggregates containing goethite and some combination of smectite (montmorillonite), chlorite, silica, calcite, talc, and periclase.

"Chlorophaeite" alteration is also characteristic of volcanics and shallow intrusions. This fine-grained aggregate tends to be isotropic, with low refractive index (1.6–1.5) and ranges from orange to green. It probably contains isotropic limonite and low-birefringence minerals such as chlorite or serpentine.

Metamorphism or late-stage magmatic processes commonly form pyroxene or amphibole reaction rims on olivine crystals.

OCCURRENCE. Olivine crystallizes very early in the crystallization sequence of a magma and is characteristic of mafic and ultramafic igneous rocks. It appears as a major constituent of stony meteorites and is, presumably, characteristic of subcrustal layers of the earth's interior. Early crystals may form layered intrusions by gravity settling, and early olivine, about Fo_{90}, may appear with magnetite or chromite or, essentially without associates, as dunite. Peridotites associate olivine, about Fo_{88}, with Mg-pyroxenes, and common mafic rocks (such as gabbro, norite, diabase, and basalt) show intermediate olivines, Fo_{80}–Fo_{50}, in association with augite, hornblende, biotite, and calcic plagioclase.

Fe-rich olivines are less common but appear in alkaline

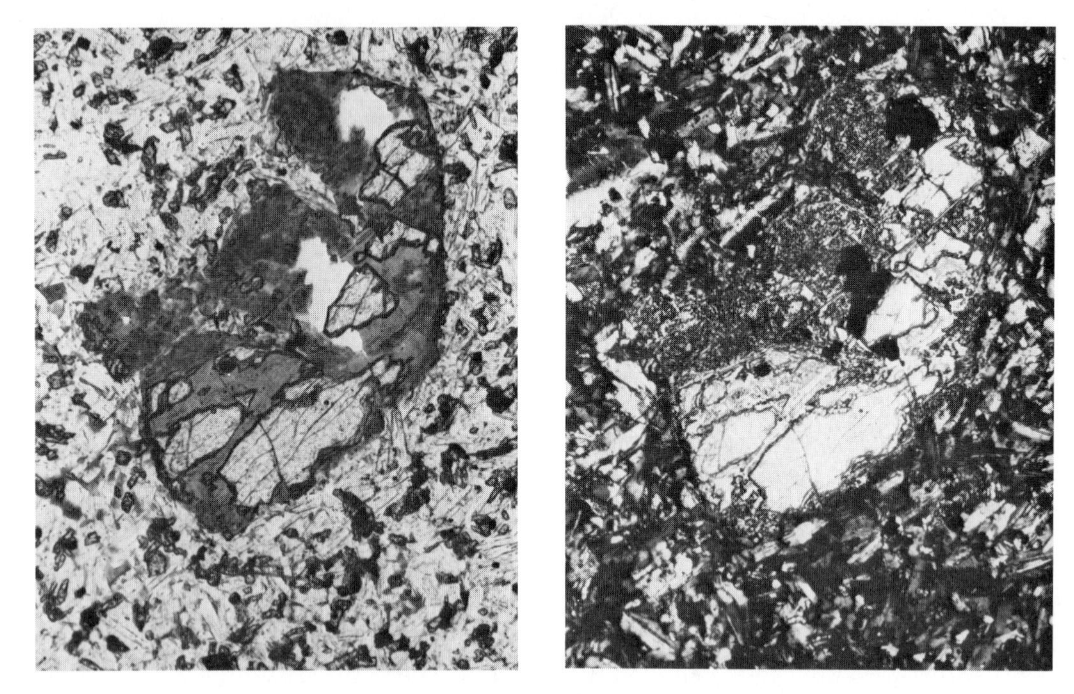

FIGURE 3-4. Olivine phenocrysts in basaltic rocks may be largely altered to iddingsite, serpentine, or chlorophaeite. Alteration begins in fracture surfaces and may leave only small islands of unaltered olivine.

rocks with nepheline and other feldspathoids and with Na- and Fe-pyroxenes and amphiboles and Fe-micas. Fayalite may appear with either quartz or nepheline and is occasionally found in salic igneous rocks, either intrusive (for example, alkali granite or pegmatite) or extrusive (such as rhyolite). Fayalite also appears in gas cavities in volcanic rocks and glasses with tridymite, cristobalite, sanidine, tourmaline, and other high-temperature minerals.

In metamorphic rocks, forsterite ($\simeq Fo_{100}$) is formed by thermal metamorphism of dolomites and dolomitic limestone. It appears in marbles, forsterite marbles, and ophicalcite. Pure forsterite is unknown in igneous rocks. Forsterite may also form by the regional metamorphism of dolomitic sediments, and fayalite may appear in metamorphosed Fe-rich sediments.

REFERENCES: THE FORSTERITE-FAYALITE SERIES

Birle, J. D., G. V. Gibbs, P. B. Moore, and J. V. Smith. 1968. Crystal structures of natural olivines. *Amer. Mineral.*, *53*, 807–824.

Bowen, N. L., and J. F. Schairer. 1935. The System MgO-FeO-SiO_2. *Amer. Jour. Sci.*, Series 5, *29*, 197.

Mossman, D. J., and D. J. Pawson. 1976. X-ray and optical characterization of the forsterite-fayalite-tephroite series, with comments on knebelite from Bluebell Mine, British Columbia. *Can. Mineral.*, *14*, 479–486.

Werk, H. R., and K. N. Raymond. 1973. Four new structure refinements of olivine. *Z. Kristallogr.*, *137*, 86–105.

Wyderks, M., and Mazanek, E. 1968. The mineralogical characteristics of calcium-rich olivines. *Mineral. Mag.*, *36*, 955–961.

The Tephroite-Fayalite Series

$(Mn,Fe)_2SiO_4$ ORTHORHOMBIC

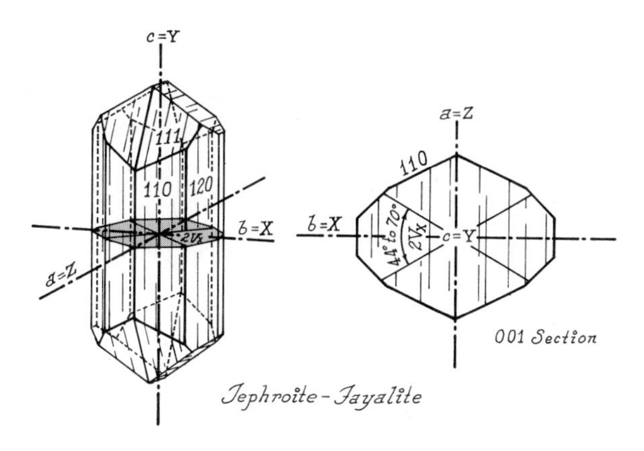

	n_α*	n_β	n_γ	$n_\gamma - n_\alpha$*	$2V_x$*	Dispersion
Tephroite (Tp_{100})	1.77	1.81	1.82	0.04	70°	$r > v$ weak
Manganknebelite (Tp_{80})	1.78	1.82	1.83	0.05	57°	$r > v$ weak
Knebelite (Tp_{50})	1.80	1.84	1.85	0.05	44°	$r > v$ weak
Iron knebelite (Tp_{20})	1.82	1.86	1.87	0.05	45°	$r > v$ weak
Fayalite (Tp_0)	1.827	1.869	1.879	0.052	46°	$r > v$ weak

$a = Z, b = X, c = Y$

Colorless to pale green or yellow in section, with weak pleochroism in brown, yellow, and green.

COMPOSITION. The continuous olivine series between tephroite, Mn_2SiO_4, and fayalite, Fe_2SiO_4, is divided as follows:

Tephroite	Tp_{100}–Tp_{90}
Manganknebelite	Tp_{90}–Tp_{70}
Knebelite	Tp_{70}–Tp_{30}
Iron knebelite	Tp_{30}–Tp_{10}
Fayalite	Tp_{10}–Tp_0

The series is commonly Mg-poor, and the name *picrotephroite* is reserved for compositions between forsterite and tephroite. Ferric iron is often present through oxidation, and Zn^{2+} may replace significant Mn^{2+} or Fe^{2+} to form *roepperite*.

PHYSICAL PROPERTIES. H = 6–6½. Sp. Gr. = 3.78–4.35. Color in hand sample is green or reddish to yellow, becoming almost black near the center of the series.

COLOR AND PLEOCHROISM. Tephroite is pale greenish or brown in standard section, showing weak pleochroism X = reddish brown, Y = pale red, and Z = blue-green; knebelite is colorless to pale yellow or green, with weak pleochroism X = pale yellow, Z = bluish. Fayalite is pale yellow, with pleochroism $X = Z$ = pale yellow, Y = yellow-orange.

* Refraction indices and birefringence decrease with increasing manganese, and $2V_x$ decreases slightly to 44° at Tp_{50} and then increases rapidly to 70° at Tp_{100}. The mild influence of manganese on optical constants is often overshadowed by the dominant control of magnesium.

FORM. In metamorphic rocks, crystals should be largely anhedral, becoming euhedral in vein occurrences.

CLEAVAGE. Mg-poor olivines show fairly distinct cleavage on {010}, and poor basal cleavage {001} may be noted.

BIREFRINGENCE. Maximum birefringence, seen in basal sections, is strong (0.04–0.05), yielding low third-order colors in standard section.

TWINNING. Twinning is uncommon.

INTERFERENCE FIGURE. All members of the series are biaxial negative, with moderate $2V_x$, which closes slightly, with increasing manganese, to 44° at Tp_{50} and opens again to 70° at tephroite. Cleavage sections yield acute bisectrix isogyres on a field brilliant with isochromatic colors and weak optic axis dispersion $r > v$.

OPTICAL ORIENTATION. Sections showing the {010} cleavage yield parallel extinction and negative (length-fast) elongation.

DISTINGUISHING FEATURES. The occurrence of Mn-olivines and their association with other manganese minerals is usually distinctive. Mg-olivines show larger $2V$ and lower indices of refraction.

ALTERATION. Mn-olivines may be expected to alter to Mn-rich chlorites and serpentines with exsolved magnetite.

OCCURRENCE. Minerals of the Mn-Fe olivine series are most commonly related to the regional or metasomatic metamorphism of Fe- and Mn-rich sediments or Mn ore deposits. They are consistently associated with other Mn-rich minerals (such as rhodonite and bustamite). Tephroite and fayalite may appear as primary vein minerals.

REFERENCES: THE TEPHROITE-FAYALITE SERIES

Henriques, A. 1956. Optical and physical properties of knebelite. *Arkiv. Min. Geol.*, *2*, 255.

Mossman, D. J., and D. J. Pawson. 1976. X-ray and optical characterization of the forsterite-fayalite-tephroite series with comments on knebelite from Bluebell Mines, British Columbia. *Can. Mineral.*, *14*, 479–486.

Russell, A. 1946. On rhodonite and tephroite from Treburland manganese mine Altarnum, Cornwall, and on rhodonite from other localities in Cornwall and Devonshire. *Mineral. Mag.*, *27*, 221.

Monticellite

$CaMgSiO_4$ ORTHORHOMBIC

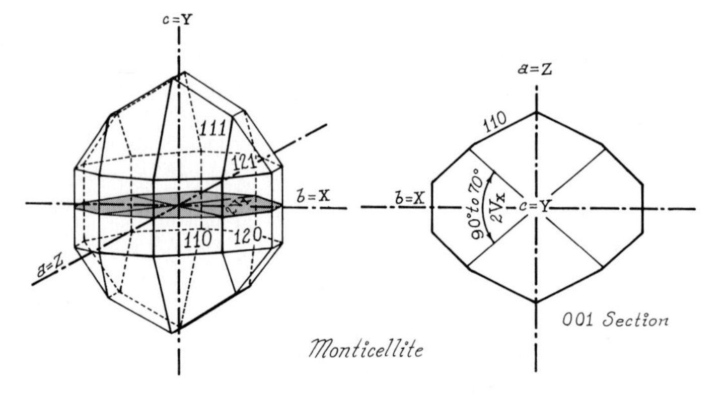

$n_\alpha = 1.639–1.663$*
$n_\beta = 1.646–1.674$
$n_\gamma = 1.653–1.680$
$n_\gamma - n_\alpha = 0.014–0.020$
Biaxial negative, $2V_x = 90°–70°$
$a = Z, b = X, c = Y$
$r > v$ distinct
Colorless in section

* Low refractive indices (synthetic $CaMgSiO_4$) and birefringence increase and $2V_x$ decreases with increasing Fe^{2+} of Mn. Optical constants for monticellite probably grade into those of kirschsteinite ($CaFeSiO_4$) $n_\alpha = 1.674$, $n_\beta = 1.694$, $n_\gamma = 1.735$, $2V_x = 50°$ and glaucochroite ($CaMnSiO_4$) $n_\alpha = 1.686$, $n_\beta = 1.723$, $n_\gamma = 1.736$, $2V_x = 61°$.

COMPOSITION. The composition of monticellite is usually near the ideal $CaMgSiO_4$. Calcium ions are too large to be replaced by Mg^{2+}, Fe^{2+}, or Mn^{2+}, so series with forsterite, fayalite, and tephroite are very restricted. Magnesium, however, is easily replaced by Fe^{2+} or Mn^{2+}, and a complete series with kirschsteinite, $CaFeSiO_4$, is known from synthetic mixtures. A series with glaucochroite has been proposed, but neither Mn^{2+} nor Fe^{2+} are commonly abundant in monticellite paragenesis.

PHYSICAL PROPERTIES. $H = 5\frac{1}{2}$. Sp. Gr. = 3.1–3.3. Monticellite is colorless to grayish in hand sample.

COLOR AND PLEOCHROISM. Monticellite is colorless in section or as fragments.

FORM. Monticellite crystals are usually anhedral, commonly granular. Euhedral crystals are slightly elongated on *c*.

CLEAVAGE. Poor cleavage on {010} is consistent with the olivine structure.

BIREFRINGENCE. Maximum birefringence, seen in basal sections, is mild (0.014–0.020) and much less than other olivines. In standard section, interference colors do not exceed first-order.

TWINNING. Cyclic twinning on {031} may yield stubby six-pointed stars.

ZONING. Granular monticellite may appear as reaction rims on crystals of forsterite or diopside.

INTERFERENCE FIGURE. Monticellite is biaxial negative, with large $2V_x = 90°$ to $70°$, which decreases with increasing Fe^{2+} or Mn^{2+}. Fragments show near-random orientation, and figures show few isochromes. Dispersion is distinct $r > v$.

OPTICAL ORIENTATION. Stubby crystals and poor cleavage on {010} make extinction angle and elongation almost meaningless, but extinction is parallel and elongation either negative or positive ($c = Y$) where it can be observed.

DISTINGUISHING FEATURES. Monticellite is to be expected in carbonate contact zones with other Ca or Ca-Mg silicates, most of which are monoclinic, showing inclined extinction to good cleavages. Forsterite has much higher briefringence, as do other olivines, and diopside is a clinopyroxene with good cleavage.

ALTERATION. Monticellite may alter to serpentine and a pyroxene near augite. It may replace or partially replace calc-silicates of lower temperature (such as tremolite, forsterite, diopside) and may be replaced by forms of higher temperature (such as merwinite or larnite).

OCCURRENCE. Monticellite is largely characteristic of metamorphic zones in dolomite containing silicate impurities and is especially characteristic of contact zones between granite and dolomitic limestone. It is a high-temperature calc-silicate forming at the expense of low-temperature calc-silicates through their reaction with remaining carbonates:

$$\underset{\text{Diopside}}{CaMg(SiO_3)_2} + \underset{\text{Forsterite}}{Mg_2SiO_4} + \underset{\text{Calcite}}{2CaCO_3} \longrightarrow \underset{\text{Monticellite}}{3CaMgSiO_4} + 2CO_2$$

At higher temperatures, monticellite is replaced by minerals of yet more complete reaction—for example, åkermanite, $Ca_2MgSi_2O_7$, and merwinite, $Ca_3Mg(SiO_4)_2$. Monticellite should be anticipated in association with carbonates and contact calc-silicates (merwinite, melilite, spurrite, diopside, forsterite, cuspidine).

In igneous rocks, monticellite occasionally appears in mafic and ultramafic plutonic rocks (such as peridotites, carbonatites, lamprophyres, and nepheline- or melilite-bearing rocks). It occurs in association with calcite, melilite, diopside, Mg-olivine, perovskite, nepheline, and so on.

REFERENCES: MONTICELLITE

Bowen, N. L. 1940. Progressive metamorphism of siliceous limestone and dolomite. *Jour. Geol., 48*, 225–274.

Ferguson, J. B., and H. E. Merwin. 1919. The ternary system CaO-MgO-SiO_2. *Amer. Jour. Sci., 48*, 81–123.

Onken, H. 1965 Refinement of the crystal structure of monticellite. *Mineral. Petrog. Mitt., 10*, 34–44.

Schaller, W. T. 1935. Monticellite from San Bernardino County, California and the monticellite series, *Amer. Mineral., 20*, 815–827.

Turner, F. J., and J. Verhoogen. 1960. *Igneous and Metamorphic Petrology*. New York: McGraw-Hill.

THE GARNET GROUP

The garnets are an isomorphous group of orthosilicates of highly variable composition. Their isometric structure comprises independent SiO_4^{4-} tetrahedra linked by trivalent cations in sixfold (octahedral) oxygen coordination and divalent cations in eightfold coordination to yield the basic garnet formula, $R_3^{2+}R_2^{3+}(SiO_4)_3$.

The chemistry of common garnets is based on six ideal end-member compositions.

Pyralspite Garnets		Ugrandite Garnets	
Pyrope	$Mg_3Al_2(SiO_4)_3$	Uvarovite	$Ca_3Cr_2(SiO_4)_3$
Almandine	$Fe_3Al_2(SiO_4)_3$	Grossular	$Ca_3Al_2(SiO_4)_3$
Spessartine	$Mn_3Al_2(SiO_4)_3$	Andradite	$Ca_3Fe_2(SiO_4)_3$

The compositions of natural garnets seldom approximate an ideal end-member as several end members are generally well represented, and the garnet name designates only the most abundant end member.

Calcium commonly occurs in eightfold coordination and the Ca-garnets form a stable composition group (ugrandite) showing extensive solid solution within the group but only minor solid solution with other garnets. Mg (pyrope), Fe^{2+} (almandine), and Mn (spessartine) are found in eightfold coordination and form a garnet group (pyralspite) showing only minor solid solution with the Ca-garnets.

A few rare end-member compositions are defined—eg., *calderite*, $Mn_3Fe_2^{3+}(SiO_4)_3$. An ill-defined composition range between pyrope and almandine is called "*rhodolite*"—$(Mg,Fe)_3Al_2(SiO_4)_3$. Minor impurities form other recognized varieties, such as *melanite*, $Ca_3(Fe^{3+}Ti)_2(SiO_4)_3$, and *schorlamite*, $Ca_3(Fe,Ti)_2[(Si,Ti)O_4]_3$. Fig. 3-5 shows the variation of refractive index with weight percent TiO_2.

A few SiO_4^{4-} tetrahedra may be replaced by PO_4^{3-} groups, and four $(OH)^-$ ions may form tetrahedral groups to replace SiO_4 units, forming the hydrogarnets. *Hydrogrossular*, $Ca_3Al_2(SiO_4)_2(OH)_4$, is commonly regarded as an intermediate range in a proposed series between grossular, $Ca_3Al_2(SiO_4)_3$, and *hibschite*, $Ca_3Al_2 \cdot 3(OH)_4$.

Because garnets exhibit a substantial compositional range, it is not possible to uniquely specify composition with measurements of refractive index alone. Nomograms (Fig. 3-6) have been developed, however, for obtaining composition in terms of major end members from measurements of the cell dimension, specific gravity, and refractive index. Even these three variables do not always yield a unique composition, but the color and mineralogical associations usually resolve the ambiguity. Physical and optical properties of the natural garnet end members are summarized in Table 3-2.

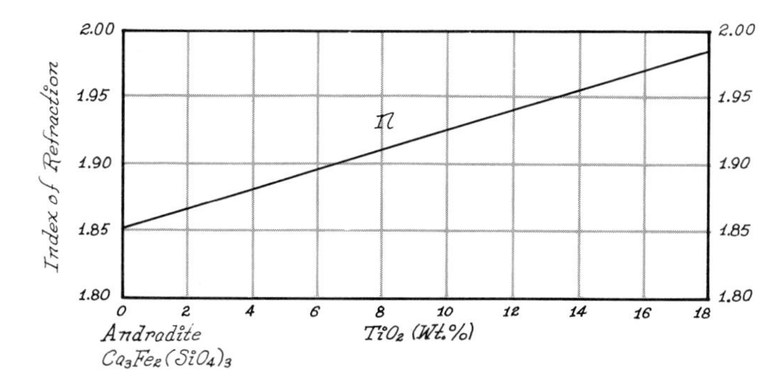

FIGURE 3-5. Variation in the refractive index of andradite with TiO_2 content. (Data from Howie and Wooley, 1968)

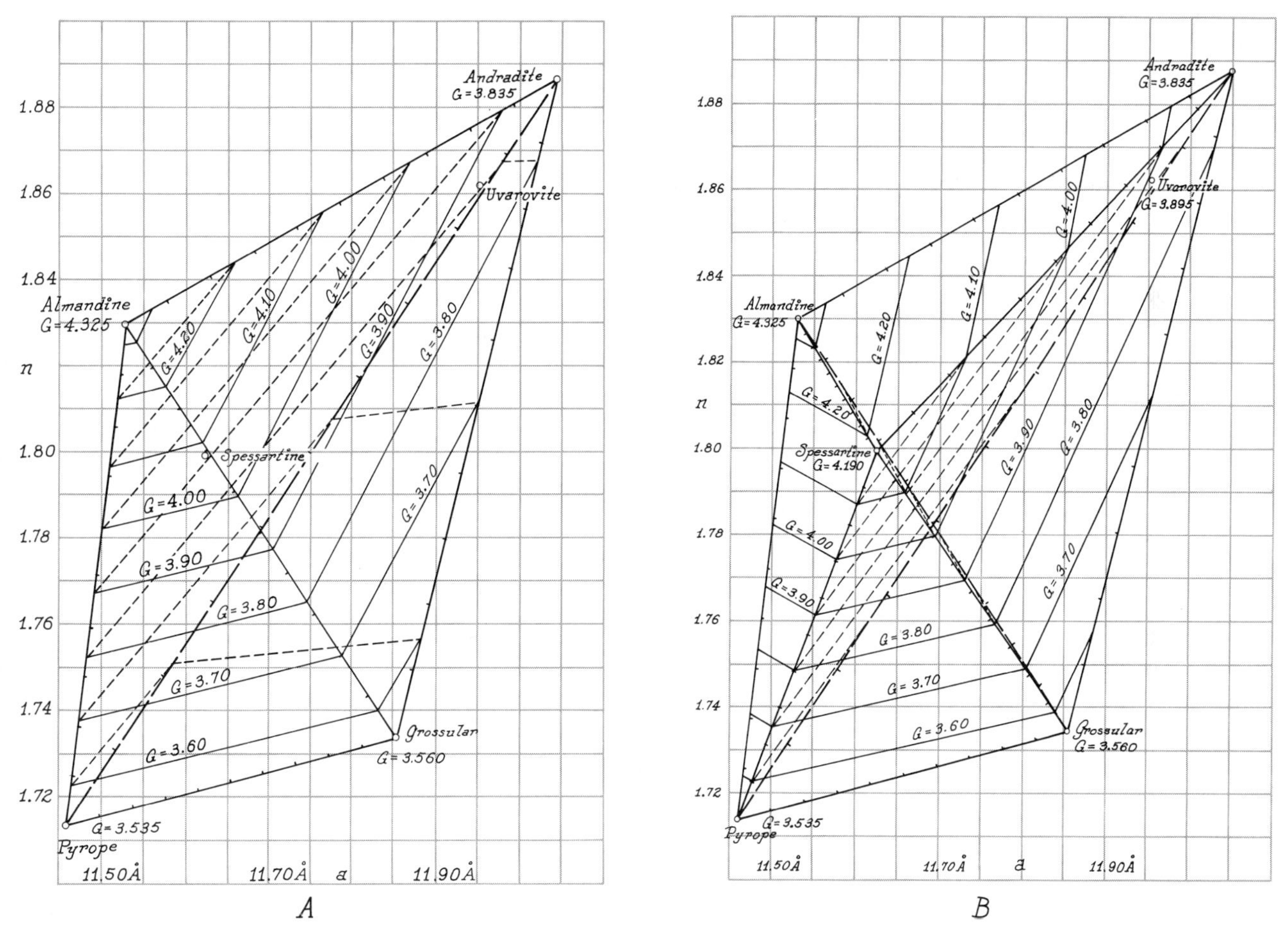

FIGURE 3-6. Garnet composition as a function of index of refraction (n), unit cell dimension (a) and specific gravity (G). (A) represents a skewed tetrahedron of four triangular faces, with the four most common garnet components at the corners. (B) includes the spessartine composition as an end member that requires six triangular fields. Composition may be read from any triangular field as from a ternary-phase diagram. Ticks on the triangle edges divide each binary series into 10 percent increments. (Winchell, The composition and physical properties of garnet, *Amer. Mineral.*, *43*, p. 597, Fig. 1, 1958, © by the Mineralogical Society of America)

TABLE 3–2. Optical Properties of the Garnet Group

n	$n_\gamma - n_\alpha$	Mineral	System	$2V$	Optic Sign	Cleavage	Color in Sections	Physical Properties
1.71	0.000	Pyrope $Mg_3Al_2(SiO_4)_3$	Isometric			Possible parting on {110}	Colorless, pale red	H = $7\frac{1}{2}$. Sp. Gr. = 3.6. pink, deep red, purplish
1.67–1.73	0.000	Hydrogrossular $Ca_3Al_2(SiO_4)_2(OH)_4$	Isometric			Possible parting on {110}	Colorless	H = $6\frac{1}{2}$. Sp. Gr. = 3.1–3.6. white, pale brown, pink, pale green, gray
1.73	0.000–0.005	Grossular $Ca_3Al_2(SiO_4)_3$	Isometric	0°–90° $v > r$ strong	–	Possible parting on {110}	Colorless	H = $6\frac{1}{2}$–7. Sp. Gr. = 3.6. colorless, yellow, cinnamon, red-brown, gray green
1.80	0.000–0.004	Spessartine $Mn_3Al_2(SiO_4)_3$	Isometric			Possible parting on {110}	Pink, pale brown	H = 7–$7\frac{1}{2}$. Sp. Gr. = 4.2. dark red, violet, black, brown-red
1.83	0.000	Almandine $Fe_3Al_2(SiO_4)_3$	Isometric			Possible parting on {110}	Colorless, pale pink-red, pale brown	H = 7–$7\frac{1}{2}$. Sp. Gr. = 4.3. dark red, brown-red, black
1.86	0.000–0.005	Uvarovite $Ca_3Cr_2(SiO_4)_3$	Isometric	0°–90° $v > r$ strong	–	Possible parting on {110}	Pale emerald green	H = $7\frac{1}{2}$. Sp. Gr. = 3.9. dark green, emerald green
1.89	0.000–0.005	Andradite $Ca_3Fe_2(SiO_4)_3$	Isometric	0°–90° $v > r$ strong	–	Possible parting on {110}	Yellow, pale brown, deep brown	H = $6\frac{1}{2}$–7. Sp. Gr. = 3.9. brown, yellow, green, red-brown, black

Pyralspite Garnets

$(Mg,Fe^{2+},Mn)_3Al_2(SiO_4)_3$ ISOMETRIC

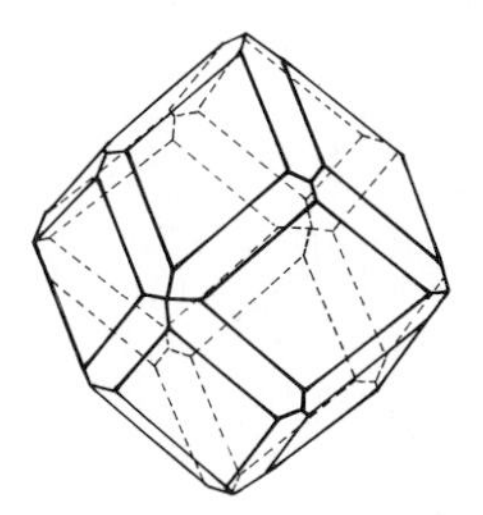

Pyralspite Garnets

		n
Almandine	$Fe_3Al_2(SiO_4)_3$	1.830*
Spessartine	$Mn_3Al_2(SiO_4)_3$	1.800
Pyrope	$Mg_3Al_2(SiO_4)_3$	1.714

Colorless to pink or brown in standard section

COMPOSITION. Pyralspite garnets show essentially complete solid solution among three end-member compositions: pyrope, $Mg_3Al_2(SiO_4)_3$; almandine, $Fe_3Al_2(SiO_4)_3$; and spessartine, $Mn_3Al_2(SiO_4)_3$; and minor solid solution with the ugrandite garnets. Minor yttrium appears in some spessartines.

PHYSICAL PROPERTIES. H = 6–$7\frac{1}{2}$. Sp. Gr. = 4.32–3.58. Color, in hand sample, is commonly deep wine-red but may range from black to deep pink or brown. Color tends to darken with increasing almandine (Fe) and may tend toward violet with Mn content.

COLOR AND PLEOCHROISM. Pyralspite garnets are colorless to pale red or brown in standard section and are without pleochroism. Fragments may show distinct red color.

FORM. The garnets are high in the crystalloblastic series and appear most commonly as euhedral or near-euhedral crystals of dodecahedral {110} or trapozohedral {112} habit or as some combination of the two forms.

CLEAVAGE. The garnets show no cleavage, although almandine may exhibit pronounced parting on {110}.

BIREFRINGENCE. The garnets are isometric, and pyrope and almandine are truly isotropic. Spessartine may show weak anisotropism.

TWINNING. Multiple wedge-shaped twins may radiate from the crystal center in birefringent crystals.

ZONING. Color zoning is fairly common in spessartine, and corona rims of alteration minerals may encircle individual crystals.

INTERFERENCE FIGURE. Most pyralspite garnets are isotropic. Birefringent forms may be biaxial.

DISTINGUISHING FEATURES. Pyralspite garnets commonly appear as nearly euhedral crystals that are isotropic or weakly birefringent, with extreme relief and possibly pale red color. The mode of occurrence is often distinctive.

Spinel is usually strongly colored in green or brown and may show octahedral {111} parting.

ALTERATION. Chlorite is the most common alteration product of the garnets; however, epidote, hornblende, and the iron oxides may appear as fibrous aggregates in corona rims. Black Mn-oxides may form on altered spessartine.

OCCURRENCE. Almandine is the garnet in schists and gneisses of regional metamorphism and is especially characteristic of mica schists, where it forms the index mineral for the almandine barrovian zone (Fig. 3-7). Increasing metamorphic grade in aluminous rocks (such as shales) produces almandine from chlorite or biotite, and staurolite, kyanite, or tourmaline may appear in association. Almandine forms in the amphibolite through granulite facies, and intermediate almandine-pyrope compositions form the garnet of the eclogites.

In contact metamorphic zones, almandine may appear in hornfels with cordierite or spinel. Spessartine forms, by metasomatism, in skarn rocks with other Mn-minerals (such as rhodonite and tephroite).

In plutonic igneous rocks, pyrope appears in ultramafic varieties (such as peridotites and associated serpentines) with

* Refractive indices are for ideal end-member compositions. Indices increase with Fe^{2+} and decrease with Mg, and natural garnets show intermediate values compatible with composition.

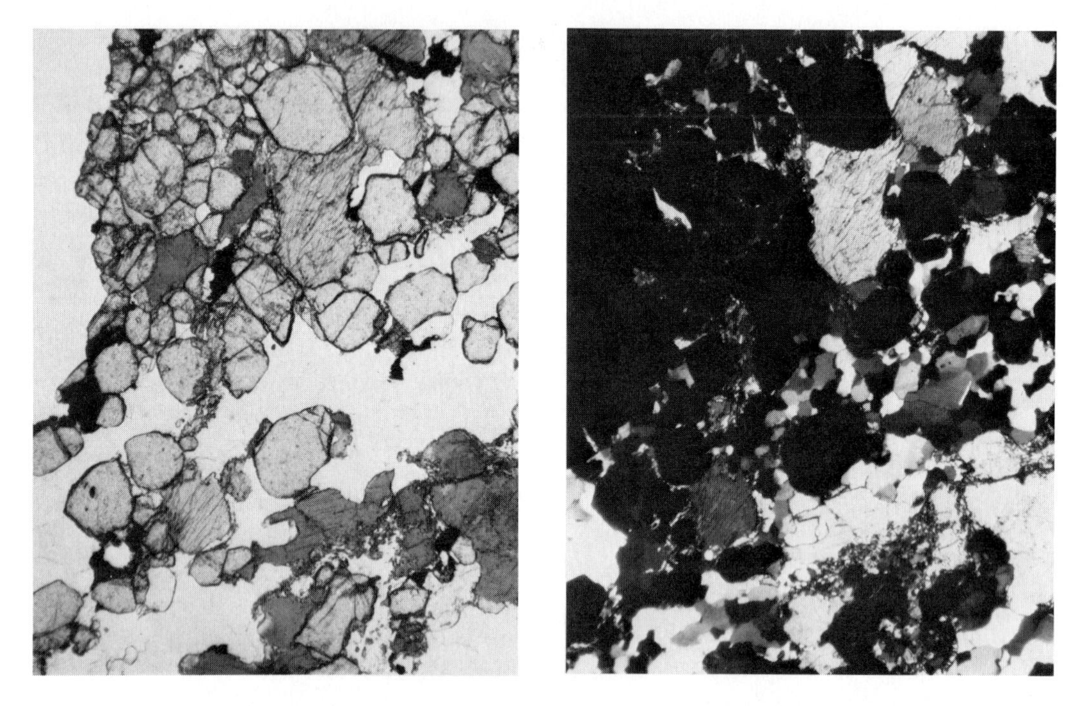

FIGURE 3-7. Pyralspite garnets commonly appear as more-or-less euhedral crystals in schists, gneisses, and certain igneous rocks. Note equant outline, high relief, and isotropism.

Mg-rich minerals, and almandine occurs in acidic types (such as granites) with Fe-rich minerals. Spessartine is a relatively rare garnet but may form in granite pegmatites. Volcanic rocks do not commonly contain garnets; however, almandine is reported in andesites, dacites, and other intermediate volcanics. Sedimentary rocks commonly contain detrital grains of garnet, usually almandine.

Ugrandite Garnets

$Ca_3(Al,Fe^{3+},Cr)_2(SiO_4)_3$ ISOMETRIC

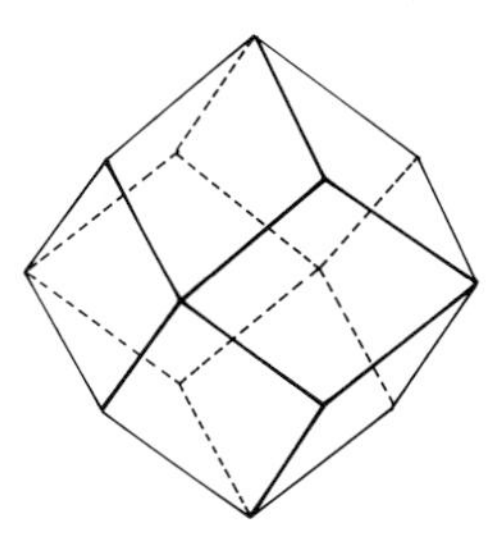

Ugrandite Garnets

		n
Grossular	$Ca_3Al_2(SiO_4)_3$	1.734*
Andradite	$Ca_3Fe_2^{3+}(SiO_4)_3$	1.887
Uvarovite	$Ca_3Cr_2(SiO_4)_3$	1.86

Grossular is colorless to pale yellow, andradite pale to deep brown, and uvarovite emerald-green in standard section

COMPOSITION. Extensive solid solution exists among the three end members of the ugrandite garnets: uvarovite, $Ca_3Cr_2(SiO_4)_3$; grossular, $Ca_3Al_2(SiO_4)_3$; and andradite, $Ca_3Fe_2^{3+}(SiO_4)_3$. Less solid solution exists between ugrandite and pyralspite garnets. Uvarovite is a rare composition, and most natural uvarovite contains much grossular and andradite.

Andradite compositions are often Ti-rich, with accompanying deep brown or black color. *Melanite* is a dark, Ti-rich andradite and *schorlomite* is very Ti-rich and resembles black tourmaline (schorl).

Hydrogrossular, $Ca_3Al_2[(SiO_4),(OH)_4]_3$, contains tetrahedral groups of four $(OH)^-$ in place of some SiO_4^{4-} units.

PHYSICAL PROPERTIES. $H = 6\frac{1}{2}$–7. Sp. Gr. = 3.4–4.1. Color is highly variable in hand sample. Garnets with appreciable Cr (uvarovite) are deep green to emerald-green; andradite is most commonly brown ranging from black (melanite to schorlomite) to reddish, green (demantoid), or yellow (topazolite); and grossular is commonly yellow or greenish ranging from colorless to cinnamon-brown (hessonite).

* Refractive indices given are for ideal end members. Indices of natural garnets are intermediate.

COLOR AND PLEOCHROISM. In standard section, grossular is usually colorless, uvarovite is pale emerald-green, and andradite is commonly yellow or pale brown; however, Ti-varieties may be deep brown. Color is not pleochroic but may be zoned.

FORM. Garnets show a strong tendency for euhedral crystals as dodecahedron {110}, trapezohedron {112}, or combined forms. Garnet in contact zones is often massive.

CLEAVAGE. Garnets show no cleavage, but distinct parting is possible on {110}.

BIREFRINGENCE. The ugrandite garnets often show weak birefringence, usually less than 0.005, rarely appreciably greater. True isotropism is not uncommon, however.

TWINNING. The birefringent garnets commonly show multiple twinning of 8, 12, 24, or 48 pie-shaped wedges radiating from the crystal center. Thin lamellar twinning on {111} is also reported. Some crystals show more than one type of twinning, and some areas may be truly isotropic.

ZONING. Color zoning due to slight composition changes is quite common and often very pronounced in the ugrandite garnets. Corona rims may also appear.

INTERFERENCE FIGURE. Ugrandite garnets may be isotropic but are often anisotropic, with birefringence. Interference figures are uniaxial or biaxial, usually negative, and $2V$ may range from 0° to 90°. Dispersion may be strong $v > r$.

OPTICAL ORIENTATION. The optical orientation of anisotropic garnets appears highly variable.

DISTINGUISHING FEATURES. The ugrandite garnets are common in limestone contact zones and commonly show weak birefringence, twinning, and zoning. Pyralspite garnets are rare in limestone contact zones and rarely show birefringence.

ALTERATION. The most common alteration product of the garnets is chlorite; however, the Ca-garnets may also yield calcite, epidote, and feldspar. Andradite may yield additional limonite or nontronite as products of weathering.

OCCURRENCE. The characteristic occurrence of Ca-garnets is due to contact metamorphism of impure limestones, where grossular and/or andradite appears with idocrase, wollastonite, diopside, and other calc-silicates. They may also appear in schists formed from impure limestones by regional metamorphism and in metasomatic skarns, where andradite is especially common. The light-colored andradites (demantoid and topazolite) appear mostly in chlorite schists and serpentinites.

Dark andradite varieties, such as melanite and schorlomite, are largely restricted to alkaline igneous rocks such as nepheline syenite, ijolite, and their extrusive equivalents. Grossular may also appear in alkaline igneous rocks and is reported in basalt vesicles, hydrothermal veins, and pegmatites.

Uvarovite is a very rare garnet found largely as seams and veins in peridotite or serpentinite associated with chromite ores.

REFERENCES: GARNET

Amthauer, G. 1976. Crystal chemistry and color of chromium bearing garnets. *Neues Jahrb. Mineral. Abh.*, *126*, 158–186.

Biswas, D. K. 1974. Quantitative relation between chemical composition and physical properties of natural garnets. *Indian Jour. Earth Sci.*, *1*, 141–147.

Biswas, D. K. 1975. Indexes of refraction (*n*) and reflectivity (*R*) of natural garnets. *Indian Mineral.*, *14*, 74–79.

Fleischer, M. 1937. The relation between chemical composition and physical properties in the garnet group. *Amer. Mineral.*, *22*, 751–759.

Geller, S. 1967. Crystal chemistry of the garnets. *Z. Kristallogr.*, *125*, 1–47.

Gentile, A. L., and R. Roy. 1960. Isomorphism and crystalline solubility in the garnet family. *Amer. Mineral.*, *45*, 701–711.

Howie, R. A., and A. R. Wooley. 1968. The role of titanium and the effect of TiO_2 on the cell size, refractive index and specific gravity in the andradite-melanite-schorlamite series. *Mineral. Mag.*, *36*, 775–790.

McConnell, D. 1964. Refringence of garnets and hydrogarnets. *Can. Mineral.*, *8*, 11–22.

Nandi, K. 1967. Garnets as indices of progressive regional metamorphism. *Mineral. Mag.*, *37*, 89–93.

Němec, D. 1967. The miscibility of the pyralspite and grandite molecules in garnets. *Mineral. Mag.*, *37*, 389–402.

Novak, G. A., and G. V. Gibbs. 1971. The crystal chemistry of the silicate garnets. *Amer. Mineral.*, *56*, 791–825.

Rickwood, P. C. 1968. Recasting analyses of garnet into end-member molecules. *Contrib. Mineral. Petrol.*, *18*, 175–198.

Skinner, B. J. 1956. Physical properties of end-members of the garnet group. *Amer. Mineral.*, *41*, 428–436.

Sriramadas, A. 1957. Diagrams for the correlation of unit cell edges and refractive indices with the chemical composition of garnets. *Amer. Mineral.*, *42*, 294–298.

Winchell, H. 1958. The composition and physical properties of garnet. *Amer. Mineral.*, *43*, 595–600.

Wright, W. I. 1938. The composition and occurrence of garnets. *Amer. Mineral.*, *23*, 436–449.

ALUMINOSILICATES

Andalusite

Al_2SiO_5

ORTHORHOMBIC

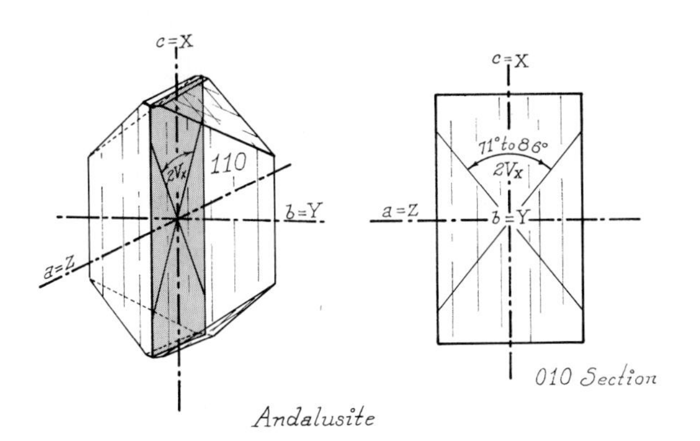

$n_\alpha = 1.629–1.640$
$n_\beta = 1.633–1.644$
$n_\gamma = 1.638–1.650$
$n_\gamma - n_\alpha = 0.009–0.011$*
Biaxial negative, $2V_x = 71° - 86°$
$a = Z, b = Y, c = X$
$v > r$ weak, rarely $r > v$*
Colorless in standard section, rarely pink or green, with distinct pleochroism in pink, green, and yellow

COMPOSITION AND STRUCTURE. Like sillimanite and kyanite, andalusite is made of chains of aluminum-containing octahedra sharing edges and elongated parallel to *c*. In andalusite, chains are linked laterally by Si^{4+} in tetrahedral coordination alternating with Al^{3+} in unusual fivefold coordination.

Andalusite is relatively pure Al_2SiO_5, although minor Fe^{3+}, Mn^{3+}, and Ti^{4+} may replace Al^{3+}. *Viridine* is a Mn-rich variety that may be a similar, but distinct, mineral species.

Andalusite formed at relatively low temperatures (*chiastolite*) commonly contains graphite inclusions in dark bands as diagonals of the prismatic cross section.

PHYSICAL PROPERTIES. $H = 6\frac{1}{2}–7\frac{1}{2}$. Sp. Gr. = 3.13–3.16. Color, in hand sample, is highly variable and often patchy. It is most commonly pink to maroon but may be white, gray, tan, green, yellow, or bluish. Luster is vitreous.

COLOR AND PLEOCHROISM. Andalusite is commonly colorless in standard section but may show rather strong color and distinct pleochroism: X = rose-pink or yellow; Y = colorless, pale yellow, or green; and Z = colorless, yellow, or olive-green.

FORM. Euhedral, prismatic crystals of nearly square cross section and parallel, columnar groups, elongate on *c*, are common. Fine, fibrous aggregates and radial groups are less common. Andalusite may form early as anhedral grains, which soon become euhedral by virtue of high position in the crystalloblastic series.

CLEAVAGE. Good prismatic cleavages {110} intersect at nearly 90°, parallel to prominent crystal faces, and tend to control fragment orientation.

BIREFRINGENCE. Maximum birefringence, seen in {010} sections, is weak (0.009–0.011). Interference colors are first-order grays and white.

TWINNING. Twinning is rare on {101}.

ZONING. Color zoning is sometimes distinct, and color is often irregular, indicating local concentrations of ferric iron or manganese. In the low-grade variety chiastolite, carbonaceous

* Refractive indices and birefringence increase with increasing Fe^{3+} or Mn^{3+}. The Mn^{3+}-rich variety *viridine* shows significantly higher refraction indices (about 1.66–1.69) and birefringence (about 0.025) and is optically positive, with large $2V_z$. Dispersion is strong $v > r$ in Fe^{3+}-rich compositions.

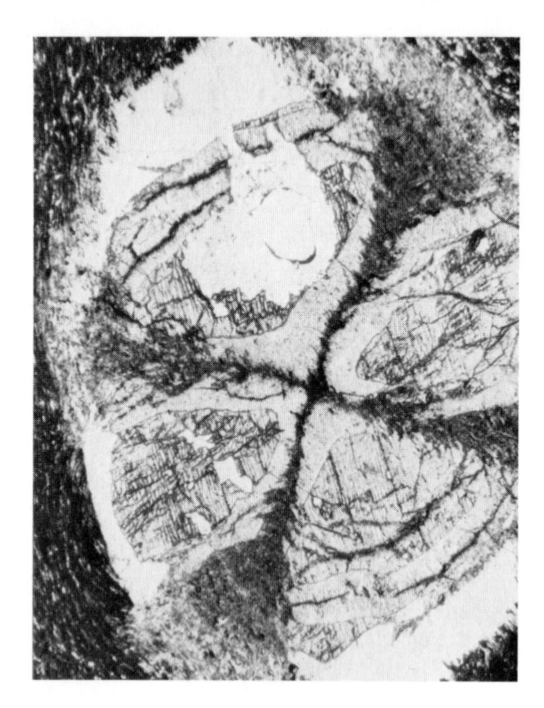

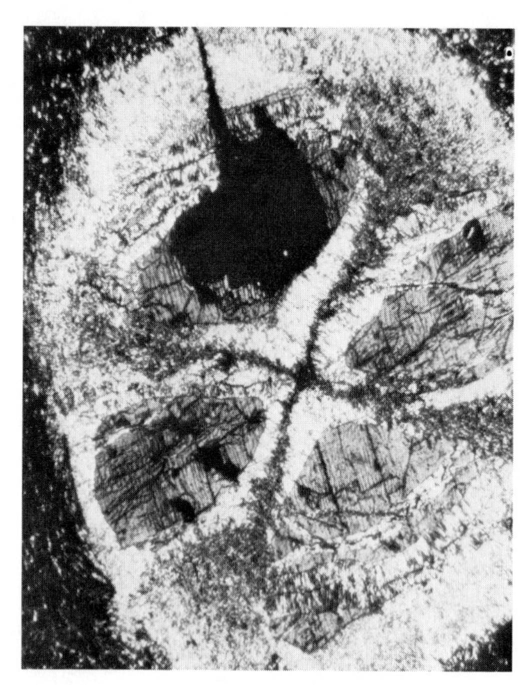

FIGURE 3-8. Andalusite crystals in carbonaceous hornfels often show crude diamond sections and inclusions aligned on the diamond diagonals (chiastolite).

inclusions form square zones and crosses parallel to front and side pinacoids (Fig. 3-8).

INTERFERENCE FIGURE. Andalusite is biaxial negative with large $2V_x = 71°$ to $86°$, or more, becoming positive for Mn-rich compositions. Basal sections yield acute bisectrix figures and cleavage fragments, elongated on $c = X$, yield off-center obtuse bisectrix or flash figures symmetrical on the optic normal. Optic axis dispersion is weak $v > r$, rarely $r > v$.

OPTICAL ORIENTATION. Good prismatic cleavages $\{110\}$ produce elongation on $c = X$. Extinction is parallel except in basal sections, where extinction is symmetrical to intersecting cleavages. Elongation on $c = X$ is negative (length-fast).

DISTINGUISHING FEATURES. Andalusite shows moderately high relief, weak birefringence, parallel extinction, and negative elongation. Euhedral rectangular to square cross sections are characteristic, as are carbonaceous inclusions.

Sillimanite is length-slow and biaxial positive, kyanite shows higher relief and birefringence and inclined extinction; topaz is biaxial positive with moderate $2V_z$; and the orthopyroxenes are length-slow and may be biaxial positive. Negative varieties of orthopyroxene show stronger birefringence.

ALTERATION. Andalusite alters readily to sericite and, less commonly, to fine-grained aggregates of spinel, corundum, and feldspar. Pyrophyllite and kaolinite are also possible alteration products, and increasing metamorphic intensity may cause andalusite to invert to sillimanite or kyanite.

OCCURRENCE. Andalusite is a metamorphic, "antistress" mineral appearing most commonly in hornfels at igneous-shale contacts in association with biotite and cordierite. Where stress is weak, it may appear in regional schists with sillimanite, kyanite, almandine, or staurolite. In igneous rocks, andalusite appears rarely in pegmatites, quartz veins, or granites, probably by argillaceous assimilation. In sedimentary rocks, andalusite is fairly common as detrital grains.

REFERENCES: ANDALUSITE

Albee, A. L., and A. A. Chodos. 1969. Minor element content of coexistent Al_2SiO_5 polymorphs. *Amer. Jour. Sci.*, *267*, 310–316.

Burnham, C. W., and M. J. Buerger. 1961. Refinement of the crystal structure of andalusite. *Z. Kristallogr.*, *115*, 269–290.

Chinner, G. A., J. V. Smith, and C. R. Knowles. 1969. Transition metal contents of Al_2SiO_5 polymorphs. *Amer. Jour. Sci.*, *267-A*, 96–113.

Kyanite

Al_2SiO_5

TRICLINIC
$\angle\alpha = 90°5'$
$\angle\beta = 101°2'$
$\angle\gamma = 105°44'$

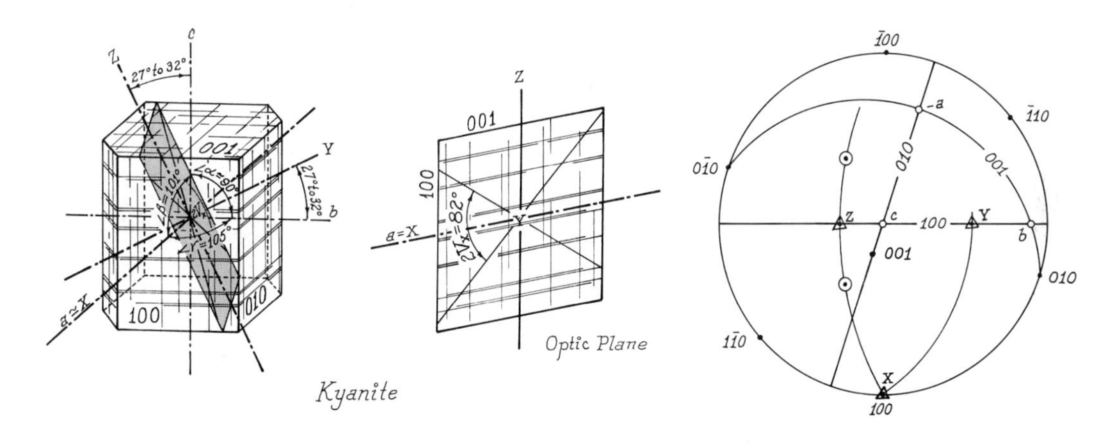

$n_\alpha = 1.712–1.718$
$n_\beta = 1.720–1.725$
$n_\gamma = 1.727–1.734$
$n_\gamma - n_\alpha = 0.012–0.016$
Biaxial negative, $2V_x = 82°$
$a \wedge X' = 0°$ to $3°$ (in $\{001\}$), $b \wedge Y' = 27°$ to $32°$ (in $\{100\}$), $c \wedge Z' = 27°$ to $32°$ (in $\{100\}$)
$r > v$ weak
Colorless to pale blue in standard section, with weak pleochroism $Z > Y > X$

COMPOSITION AND STRUCTURE. Chains of edge-sharing aluminum-containing octahedra are parallel to *c* and linked laterally by Si^{4+} in tetrahedral coordination and Al^{3+} in octahedral coordination (Fig. 3-9). The structure is similar to those of sillimanite and andalusite. Kyanite is favored by high pressure, and its structure is more close packed.

Kyanite closely approximates its ideal composition, with only very minor ferric iron, titanium, or chromium replacement of aluminum. Tiny inclusions of graphite or rutile are common.

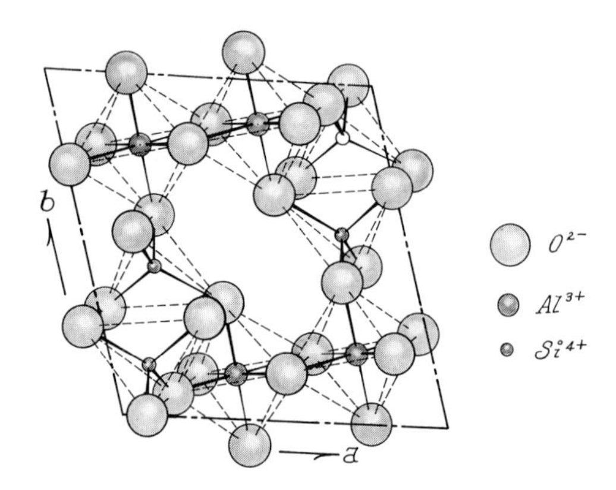

FIGURE 3-9. The structure of kyanite.

PHYSICAL PROPERTIES. H = 4 (parallel to *c* on {100}) to $7\frac{1}{2}$ (perpendicular to *c* on {1$\bar{1}$0}). Sp. Gr. ≃ 3.6. Hand sample color is blue to white or gray, with pearly luster; rarely, green or black.

COLOR AND PLEOCHROISM. Kyanite is most commonly colorless; in standard section, however, fragments and some sections are pale blue and pleochroic $Z > Y > X$. X = colorless, Y = violet-blue, and Z = deep cobalt blue.

FORM. Crystals are bladed to columnar, elongated on *c*, and often crumpled or wavy. Fragments are fibrous to splintery.

CLEAVAGE. Perfect cleavage on {100} and distinct on {010} produce fragments elongated on *c*. Parting on {001} commonly appears as cross fractures.

BIREFRINGENCE. Maximum birefringence is mild (0.012–0.016) and interference colors are white and yellow to first-order red, in standard section. Maximum birefringence is seen on *Y*, at 30° from {010}, and is unlikely to appear on fragments.

TWINNING. Multiple twinning is common on {100} as normal or parallel twins (twin axis perpendicular to {100} or parallel to *b* or *c*). Lamellar twins on {001} may be the result of shear pressure.

ZONING. Kyanite may form parallel intergrowth with staurolite, and kyanite color is often unevenly distributed.

INTERFERENCE FIGURE. The acute bisectrix *X* lies almost perpendicular to {100}, and cleavage fragments commonly yield good Bxa figures, nearly centered. Figures may show a few isochromatic color rings, and the dispersion is weak $r > v$.

OPTICAL ORIENTATION. Kyanite is triclinic, with the optic plane almost normal to {100} and inclined about 30° to both *c* and *b*. $a \wedge X' = 0°$ to 3°, in the {001} plane; $b \wedge Y' = 27°$ to 32°, in the {100} plane; and $c \wedge Z' = 27°$ to 32°, in the {100} plane.

Maximum extinction angle is near 30° for {100} sections and fragments lying on {100}. Fragments on {010} and sections normal to {100} show near-parallel extinction.

Elongation is positive (length-slow) for fragments and crystals.

DISTINGUISHING FEATURES. Kyanite shows high relief, weak birefringence, good cleavage, and inclined extinction. Sections showing maximum extinction yield Bxa figures, with large $2V$. Sillimanite and andalusite are orthorhombic, with parallel extinction and lower refractive indices. Andalusite is length-fast, and sillimanite is biaxial positive, with small $2V_z$.

ALTERATION. Kyanite alters quite readily to fine- or coarse-grained sericite. Pyrophyllite and chlorite are also reported as alteration products, and kyanite may invert to sillimanite with increasing temperature or to andalusite or mullite when reheated without shear pressure.

OCCURRENCE. Kyanite is a mineral of regional metamorphism forming in pelitic schists, gneisses, and granulites in association with quartz and muscovite. Almandine, staurolite, biotite, and sillimanite are also common associates. Kyanite is favored by stress and tends to form in the metamorphic zone between staurolite and sillimanite. In igneous rocks, kyanite is reported rarely in granite pegmatites, in quartz veins that cut schists, and in eclogites. In sedimentary rocks, kyanite is fairly common as detrital grains.

REFERENCES: KYANITE

Albee, A. L., and A. A. Chodos, 1969. Minor element content of coexistent Al_2SiO_5 polymorphs. *Amer. Jour. Sci.*, *267*, 310–316.

Burnham, C. W. 1963. Refinement of the crystal structure of kyanite. *Z. Kristallogr.*, *118*, 337–360.

Chinner, G. A., J. V. Smith, and C. R. Knowles. 1969. Transition metal contents of Al_2SiO_5 polymorphs. *Amer. Jour. Sci.*, *267-A*, 96–113.

Faye, G. H., and E. H. Nickel. 1969. Origin of color and pleochroism of kyanite. *Can. Mineral.*, *10*, 35–46.

Rost, F., and E. Simon. 1972. Geochemistry and color of kyanite. *Neues Jahrb. Mineral. Monatsh.*, *1972*, 383–395.

Sillimanite

Al_2SiO_5

ORTHORHOMBIC

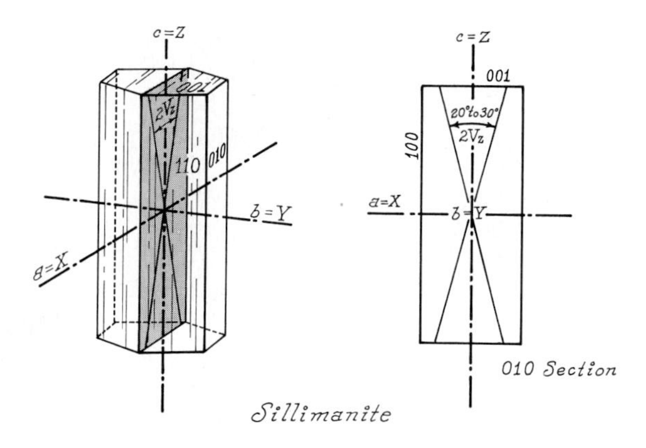

$n_\alpha = 1.653–1.661$*
$n_\beta = 1.654–1.670$
$n_\gamma = 1.669–1.684$
$n_\gamma - n_\alpha = 0.020–0.023$
Biaxial positive, $2V_z = 20°–30°$
$a = X, b = Y, c = Z$
$r > v$ strong
Colorless in standard section

COMPOSITION AND STRUCTURE. Aluminum-containing octahedra (AlO_6) share edges to form chains elongate on c. Parallel chains are linked by double chains of alternating Si^{4+} and Al^{3+} in tetrahedral coordination; thus, Al^{3+} occurs in two distinctly different sites (Fig. 3-10). The composition of sillimanite shows only minor variation from the ideal Al_2SiO_5. Minor Fe^{3+} may replace octahedral aluminum.

PHYSICAL PROPERTIES. H = $6\frac{1}{2} - 7\frac{1}{2}$. Sp. Gr. = 3.25. Color, in hand sample, is colorless or white to dark brown. Yellowish, pale green, and blue varieties are also reported.

COLOR AND PLEOCHROISM. Sillimanite is usually colorless in thin section or as fragments; some fragments, however, are pale brown, yellow, or blue, with weak pleochroism $Z > Y > X$. Z = pale brown or pale yellow, Y = brown or greenish, and X = dark brown or blue.

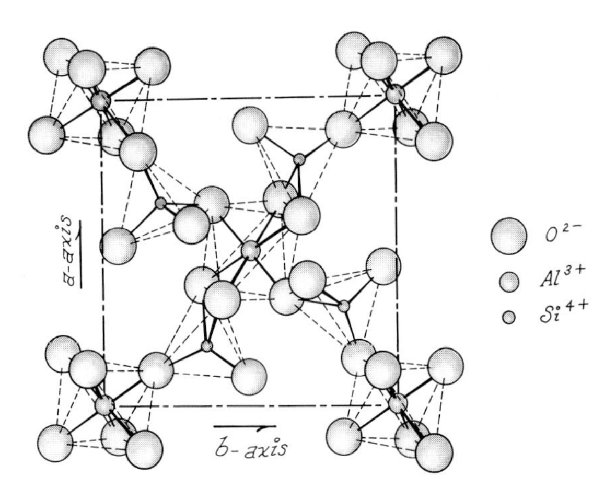

FIGURE 3-10. The structure of sillimanite.

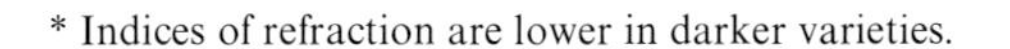

* Indices of refraction are lower in darker varieties.

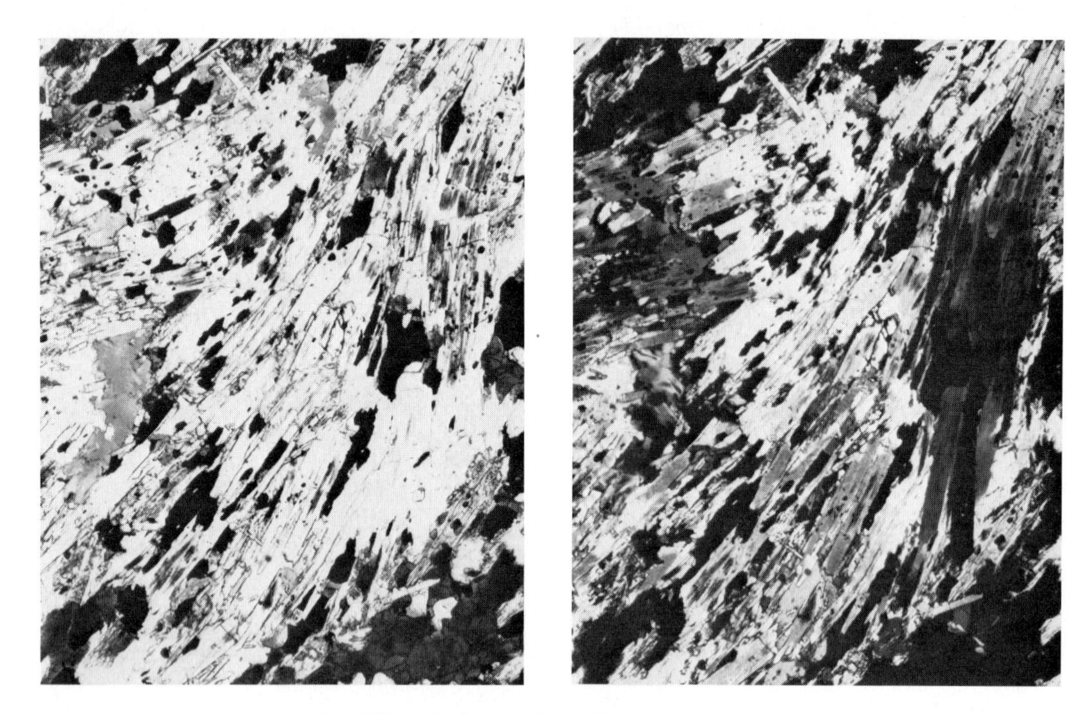

FIGURE 3-11. Sillimanite shows a dense, fibrous aggregate in thin section.

FORM. The characteristic form of sillimanite is prismatic, needlelike crystals highly elongated on *c* (Fig. 3-11) or as felted aggregates of very fine, matted fibers (*fibrolite*).

CLEAVAGE. Perfect {010} cleavage parallels crystal elongation and is not always visible in section. Cross fractures commonly divide elongated crystals into segments.

BIREFRINGENCE. Maximum birefringence, seen in {010} sections, is moderate (0.020–0.023), and interference colors, in standard section, range to the bright colors of low second order. Cleavage fragments tend to show maximum birefringence; however, elongated crystals and fibers commonly show colors in middle or low first order.

TWINNING. Twinning is not reported in sillimanite, nor is crystal zoning, although patchy color distribution is common.

INTERFERENCE FIGURE. Small, elongated sillimanite crystals rarely yield good interference figures. The acute bisectrix lies perpendicular to basal sections, which are nearly square and very small. Fragments on cleavage {010} yield flash figures. Sillimanite is biaxial positive, with moderately small $2V_z$ = 20° to 30° and strong optic axis dispersion $r > v$.

OPTICAL ORIENTATION. Crystals and fibers are elongated on $c = Z$ and, consequently, show parallel extinction and positive elongation (length-slow).

DISTINGUISHING FEATURES. Sillimanite shows high relief, moderate to low birefringence, parallel extinction, positive (length-slow) elongation, fibrous habit, and restricted occurrence. Andalusite is length-fast and biaxial negative, with large $2V_x$. Kyanite is biaxial negative, with large $2V_x$, has higher refractive indices than sillimanite, and may show inclined extinction and blue color. Zoisite and apatite are length-fast, with weak birefringence. Mullite is very rare in nature but cannot be distinguished from sillimanite by optical data alone.

ALTERATION. Sillimanite commonly alters to sericite. Pyrophyllite and kaolin and montmorillonite clays are other possible alteration products. Metamorphic stress conditions may cause sillimanite to invert to kyanite, and andalusite may form from sillimanite under static conditions of metasomatism.

OCCURRENCE. Sillimanite is normally the product of high-temperature metamorphism of aluminous rocks, either in regional or contact zones. Schist, gneiss, and granulite of high-grade regional metamorphism may contain sillimanite in association with biotite, quartz, garnet, cordierite, corundum, staurolite, and graphite. It may appear with either kyanite or andalusite. Thermal metamorphism in contact zones, especially near granite, may form sillimanite in hornfels or slates with micas and andalusite. In igneous rocks, sillimanite appears as a rare mineral in pegmatites and quartz veins. In sedimentary rocks, sillimanite is a locally important detrital mineral.

REFERENCES: SILLIMANITE

Albee, A. L., and A. A. Chodos. 1969. Minor element content of coexistent Al_2SiO_5 polymorphs. *Amer. Jour. Sci.*, *267*, 310–316.

Burnham, C. W. 1963. Refinement of the crystal structure of sillimanite. *Z. Kristallogr.*, *118*, 127–148.

Burnham, C. W. 1964. Composition limits of mullite, and the sillimanite-mullite solid solution problem. *Carnegie Inst. Wash. Year Book*, *63*, 227–228.

Cameron, W. E., and J. R. Ashworth. 1972. Fibrolite and its relation to sillimanite. *Nature, Phys. Sci.*, *235*, 134–136.

Chinner, G. A., J. V. Smith, and C. R. Knowles. 1969. Transition metal contents of Al_2SiO_5 polymorphs, *Amer. Jour. Sci.*, 267-A, 96–113.

Topaz

$Al_2(SiO_4)(F,OH)_2$ ORTHORHOMBIC

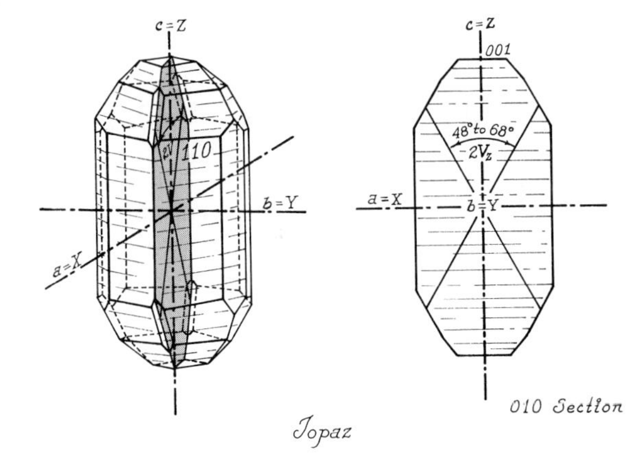

$n_\alpha = 1.606–1.630$*

$n_\beta = 1.609–1.631$

$n_\gamma = 1.616–1.638$

$n_\gamma - n_\alpha = 0.008–0.011$*

Biaxial positive, $2V_z = 48°–68°$*

$a = X, b = Y, c = Z$

$r > v$ moderate

Colorless in standard section; fragments may show weak color, with distinct pleochroism

* Refractive indices increase and $2V_z$ and birefringence become less as $(OH)^-$ replaces F^- (Fig. 3-12).

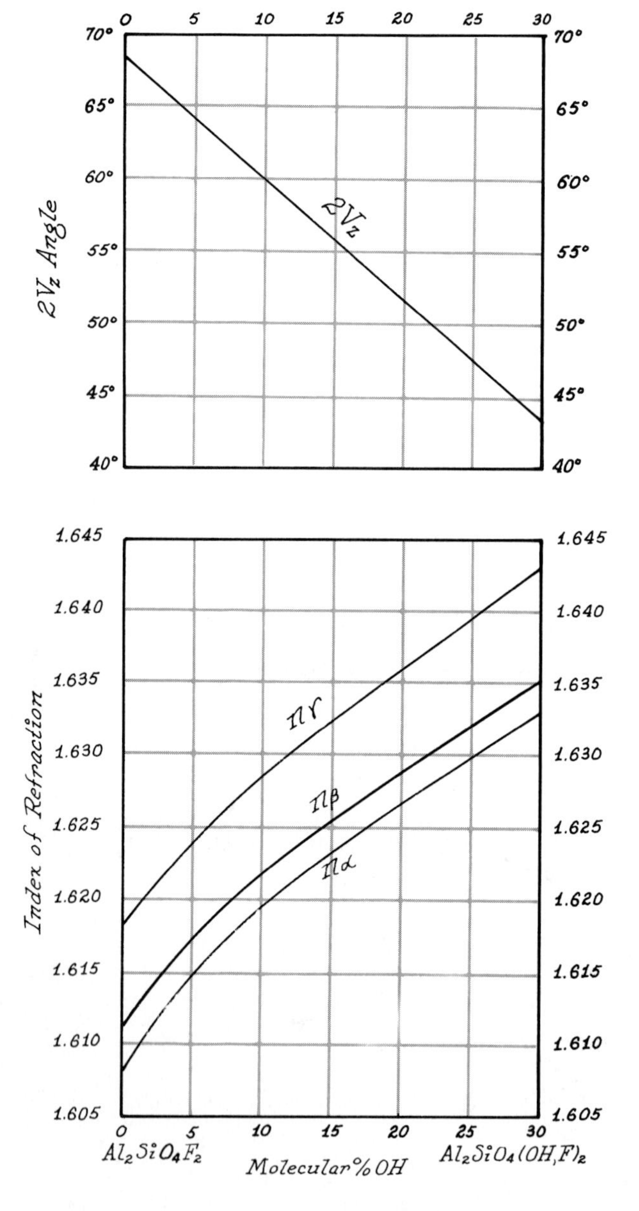

FIGURE 3-12. Variation in the indices of refraction and $2V$ angle with OH content in topaz. (After Ribbe and Rosenberg, 1971)

COMPOSITION AND STRUCTURE. Topaz is an orthosilicate containing Al^{3+} in octahedral (sixfold) coordination and Si^{4+} in tetrahedral coordination. The structure consists of "crankshaft" chains of edge-sharing AlO_4F_2 octahedra running parallel to *c*, with SiO_4 tetrahedra attached through corner-sharing oxygens.

The composition of topaz varies little from its ideal $Al_2SiO_4F_2$ except that $(OH)^-$ may replace F^- to at least 30 percent. Small amounts of Fe^{3+} and traces of Cr^{3+} and V^{3+} may replace Al^{3+} and may supply color.

PHYSICAL PROPERTIES. H = 8. Sp. Gr = 3.57–3.49. Color, in hand sample, varies widely, commonly colorless, yellow, to smoky yellow or wine-red; rarely pale blue or green. Luster is vitreous.

COLOR AND PLEOCHROISM. Topaz is most commonly colorless as fragments or in standard section. Fragments or thick sections may show pale colors, with distinct qualitative pleochroism. X = pale yellow-brown, Y = pale orange, and Z = pale pink.

FORM. Topaz is commonly euhedral as stubby crystals elongated slightly on *c*, with square to diamond-shaped or eight-sided cross sections. Crystals may be very large in pegmatites or may appear as tiny projections in open veins and cavities. Topaz may appear as coarse- to fine-grained masses in gneisses and hydrothermally decomposed granites. Parallel, columnar masses of topaz have been called *pyonite*.

Microscopic inclusions of foreign minerals (for example, specular hematite or magnetite) are common, as are tiny fluid inclusions of water or carbon dioxide.

CLEAVAGE. Very perfect basal {001} cleavage is distinctive and largely controls fragment orientation.

BIREFRINGENCE. Maximum birefringence, seen in {010} sections, is weak (0.008–0.011), and interference colors are first-order gray and white to pale yellow in standard sections. Fragments normally show less than maximum birefringence (0.003) and low-order colors.

TWINNING. Twinning is rare in topaz.

INTERFERENCE FIGURE. Topaz is biaxial positive, with moderately large $2V_z$ = 48°–68°, smaller values representing (OH)-rich compositions. Cleavage fragments and basal sections yield good, centered, acute bisectrix figures with relatively well-defined isogyres on a largely white field. Optic axis dispersion is easily noticeable $r > v$.

OPTICAL ORIENTATION. Topaz is orthohombic; $a = X$, $b = Y$, $c = Z$. Extinction is parallel to cleavage traces and prismatic sections and symmetrical to crystal cross sections Elongation is positive (length-slow) to cross section or crystal elongation and negative (length-fast) to cleavage traces.

DISTINGUISHING FEATURES. Topaz is characterized by moderately high relief, weak birefringence, moderate $2V_z$, and a single perfect cleavage favoring centered Bxa figures.

Quartz and feldspars show similar birefringence but lower relief. Apatite crystals are hexagonal and length-fast, andalusite is length-fast and biaxial negative (large $2V_x$), and idocrase is uniaxial, with very weak anomalous birefringence.

ALTERATION. Topaz is quite resistant to weathering but alters readily by hydrothermal solutions to sericite, illite, or kaolinite. Fluorite and margarite are also reported as alteration products.

OCCURRENCE. Topaz is formed largely in pneumatolitic or hypothermal stages of igneous activity. It appears in granites and granite pegmatites with quartz, microcline, muscovite, tourmaline, and Li-minerals and in hypothermal ore veins with cassiterite, wolframite, specular hematite, and native gold. Topaz forms in vugs, cavities, and fissures in acid volcanic rocks, notably rhyolite, and is often related with fluorite.

In metamorphic rocks, topaz forms by F-metasomatism in greisen and altered granites and quartzites with fluorite, zinnwaldite, corundum, and rutile and rarely in schists with quartz, muscovite, tourmaline, and kyanite. In sedimentary rocks, topaz is locally abundant as detrital grains.

REFERENCES: TOPAZ

Kôzu, S., and J. Ueda. 1929. Optical and thermal properties of topaz from Naegi, Japan. *Sci. Rep. Tohoku Univ.*, Series 3, *3*, 161–170.

Ribbe, P. H., and G. V. Gibbs. 1971. Crystal structure of topaz and its relation to physical properties. *Amer. Mineral.*, *56*, 24–30.

Ribbe, P. H., and P. E. Rosenberg. 1971. Optical and X-ray determinative methods for fluorine in topaz. *Amer. Mineral.*, *56*, 1812–1821.

Rosenberg, P. E. 1972. Compositional variations in synthetic topaz. *Amer. Mineral.*, *57*, 169–187.

Stuckey, J. L., and J. J. Amero. 1941. Physical properties of massive topaz. *Jour. Amer. Ceram. Soc.*, *24*, 89–92.

Mullite

$3Al_2O_3 \cdot 2SiO_2 - 2Al_2O_3 \cdot SiO_2$

ORTHORHOMBIC

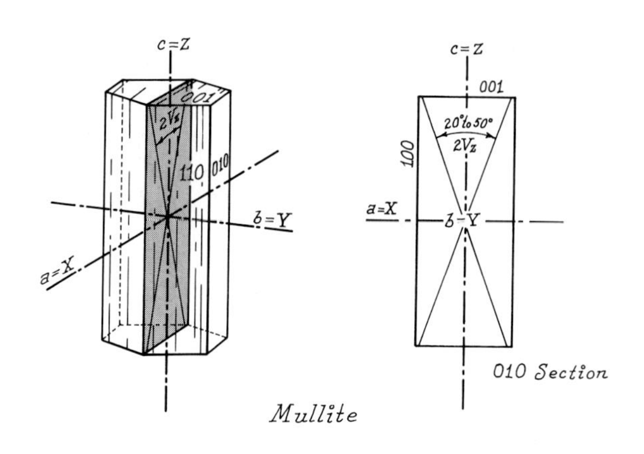

$n_\alpha = 1.634–1.666$*
$n_\beta = 1.635–1.670$
$n_\gamma = 1.644–1.690$
$n_\gamma - n_\alpha = 0.010–0.024$
Biaxial positive, $2V_z \simeq 20°–50°$
$a = X, b = Y, c = Z$
$r > v$ weak
Colorless in standard section

* The lower refraction indices are for pure, synthetic mullite. Natural mullites show larger refraction indices, which increase with Fe^{3+} and Ti^{4+} impurities. β-Mullite may show slightly higher indices than α-mullite.

COMPOSITION AND STRUCTURE. Mullite is the stable high-temperature polymorph of the aluminosilicates. Its structure is similar to that of sillimanite and is derived from it by the replacement of some tetrahedral Si^{4+} by Al^{3+} and by the loss of some oxygen ions to maintain charge balance. Some tetrahedral and oxygen sites are shifted to sites not occupied in sillimanite.

Mullite ranges in composition from at least $3Al_2O_3 \cdot 2SiO_2$ (α-mullite) to $2Al_2O_3 \cdot SiO_2$ (β-mullite). Ferric oxide, Fe_2O_3, and titania, TiO_2, are the major impurities and are commonly present in quantities of 1 or 2 weight percent each. The maximum substitution of ferric iron is probably about 11 percent Fe_2O_3. Complete solid solution between mullite and sillimanite has not been observed and may be crystallochemically inhibited.

PHYSICAL PROPERTIES. H = 6–7. Sp. Gr. = 3.12–3.26. Color, in hand sample, is white, light gray or brown, pale blue or green.

COLOR AND PLEOCHROISM. Mullite is colorless in thin section or as fragments.

FORM. Prismatic, needle like crystals are elongated on *c* and may form fibrous aggregates that resemble sillimanite.

CLEAVAGE. Distinct pinacoidal cleavage on {010}.

BIREFRINGENCE. Mild birefringence (0.010–0.024) yields first-order colors in standard section.

INTERFERENCE FIGURE. Mullite is biaxial, with moderate $2V$ angle of rather indefinite range. Cleavage fragments and acicular crystals yield flash and obtuse bisectrix figures. Some forms show weak orthorhombic dispersion $r > v$.

OPTICAL ORIENTATION. Cleavage fragments and elongated acicular crystals are length-slow (positive elongation) with parallel extinction.

DISTINGUISHING FEATURES. Mullite closely resembles sillimanite and andalusite. Andalusite is biaxial negative and length-fast; however, the optical properties of sillimanite are essentially identical to those of mullite, and the two cannot be distinguished practically with optical methods alone. Mullite does have much stronger dispersion (0.023) than sillimanite (0.012), and the association of mullite with very high-temperature minerals and glass may be distinctive.

ALTERATION. We might expect mullite to invert to lower-temperature forms of the aluminosilicates (sillimanite, andalusite, and kyanite) or to alter to the hydrated aluminum silicates.

OCCURRENCE. Mullite is a high-temperature, low-pressure metamorphic mineral of the sanidinite facies. It appears in buchites and procellainite hornfels where shales and other aluminous sediments are fused in xenoliths or contact zones with basic magmas. Mullite is commonly associated with fused glass, cordierite, corundum, spinel, hematite, pseudobrookite, and the high-temperature polymorphs of silica and feldspar.

Mullite is formed in slags, bricks, clayware, and other ceramics and refractories.

REFERENCES: MULLITE

Agrell, S. O., and J. V. Smith. 1960. Cell dimensions, solid solution, polymorphism and identification of mullite and sillimanite. *Jour. Amer. Ceram. Soc.*, *43*, 69–76.

Bowen, N. L., and J. W. Grieg. 1924. The system $Al_2O_3 \cdot SiO_2$. *Jour. Amer. Ceram. Soc.*, *7*, 238–254.

Burnham, C. W. 1964. Composition limits of mullite, and the sillimanite-mullite solid solution problem. *Carnegie Inst. Wash. Year Book*, *63*, 227–228.

Rooksby, H. P., and J. H. Partridge. 1939. X-ray study of natural and artificial mullites. *Jour. Soc. Glass Tech.*, *23*, 338–346.

Sadanaga, R., M. Tokonami, and Y. Takéuchi. 1962. The structure of mullite, $2Al_2O_3 \cdot SiO_2$, and relation with the structures of sillimanite and andalusite. *Acta Crystallogr.*, *15*, 65–68.

Varley, E. R. 1965. *Sillimanite*. Overseas Geological Survey, Mineral Resources Division. London: Her Majesty's Stationary Office.

Winchell, A. N. and H. Winchell. 1967. *Elements of Optical Mineralogy*. II. 4th ed. New York: Wiley.

OTHER NESOSILICATES

Dumortierite $(Al,Fe^{3+})_7O_3(BO_3)(SiO_4)_3$ ORTHORHOMBIC

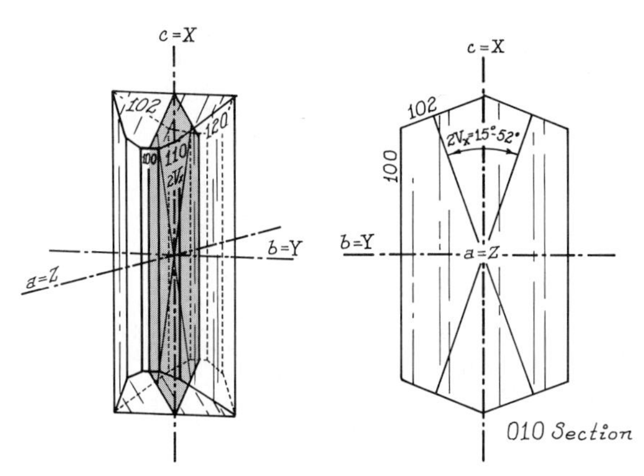

Dumortierite

$n_\alpha = 1.655–1.686$*
$n_\beta = 1.675–1.722$
$n_\gamma = 1.684–1.723$
$n_\gamma - n_\alpha = 0.011–0.027$*
Biaxial negative, $2V_x = 15°–52°$
$a = Z, b = Y, c = X$
Strong dispersion $v > r$ or rarely $r > v$
Highly colored and strongly pleochroic $X > Y > Z$
X = deep blue, blue-violet, green, brown;
Y = colorless, pale yellow-green, yellow, pale violet;
Z = colorless, pale yellow-green, pale blue

* Indices of refraction and birefringence increase with increasing Fe^{3+} or Ti^{4+}.

COMPOSITION AND STRUCTURE. The structure of dumortierite is unlike that of any other nesosilicate, as it contains triangular BO_3^{3-} units and AlO_6^{9-} octahedra that share faces to form chains parallel to c elongation. Major amounts of Fe^{3+} may replace Al^{3+} and minor Ti^{4+} is usually present.

PHYSICAL PROPERTIES. H = 7–8½. Sp. Gr. = 3.26–3.41. In hand sample, dumortierite is bright blue, violet, or greenish blue. Luster is vitreous.

COLOR AND PLEOCHROISM. Dumortierite is vividly colored and strongly pleochroic with absorption $X > Y > Z$. X = deep blue, blue-violet, green, brown; Y = colorless, greenish yellow, yellow, pale violet; and Z = colorless, greenish yellow, pale blue.

FORM. Dumortierite appears as aggregates of fibrous, bladed, or columnar crystals elongated on c (Fig. 3-13).

CLEAVAGE. Cleavage is distinct on {100} and poor on {110}.

BIREFRINGENCE. Maximum birefringence, seen in {010} sections, is mild to moderate ($n_\gamma - n_\alpha = 0.011–0.027$), and interference colors are mostly first-order, sometimes low second-order, in standard section.

TWINNING. Cyclic twinning on {110} may yield trillings and pseudohexagonal forms.

ZONING. Uneven color distribution is common.

INTERFERENCE FIGURE. Basal {001} sections yield centered acute bisectrix figures having few isochromes and strong optic axis dispersion $v > r$, rarely $r > v$. Cleavage fragments on {100} yield obtuse bisectrix figures.

OPTICAL ORIENTATION. Crystals and fragments are elongated on $c = X$; they show parallel extinction and negative elongation (length-fast).

DISTINGUISHING FEATURES. Deep blue or violet color and strong pleochroism is distinctive of dumortierite. The blue sodalite minerals (noselite, haüynite, or lazurite) are isotropic

FIGURE 3-13. Dumortierite crystals are accicular and columnar on *c* and show crude hexagonal sections and strong pleochroism in blue, violet, and yellow-green.

and show neither pleochroism nor elongation. Na-amphiboles (glaucophane, riebeckite, crossite) show amphibole cleavages {110} and inclined extinction. Tourmaline (schorl) and corundum are uniaxial. Piemontite is pleochroic from deep red-violet to yellow and is biaxial positive.

ALTERATION. Dumortierite alters readily to colorless mica. It inverts to mullite at high temperatures.

OCCURRENCE. Dumortierite appears in aplite, pegmatite, and quartz veins with cordierite, tourmaline, apatite, zircon, monazite, zenotime, garnet, rutile, anatase, and leucoxene. It may form in hydrothermally altered quartz-feldspar rocks. Dumortierite is also a constituent of granite gneiss, mica schists, and quartzites in association with cordierite, kyanite, andalusite, sillimanite, muscovite, and other aluminous minerals. It is a rare, but easily recognized, heavy mineral in detrital sediments.

REFERENCES: DUMORTIERITE

Anderson, B. W. 1968. Triple bill: Three items of interest to gemmologists. *Jour. Gemmology*, *11*, 1–6.

Golovastikov, N. I. 1965. The crystal structure of dumortierite $(Al,Fe)_7O_3[BO_3][SiO_4]_3$. *Soviet Physics—Doklady*, *10*, 493–495.

Van Dyck, D., P. Tambuyser, J. van Landuyt, and S. Amelinck. 1976. High-resolution electron microscopy of dumortierite. *Amer. Mineral.*, *61*, 1016–1019.

Zircon

$ZrSiO_4$

TETRAGONAL

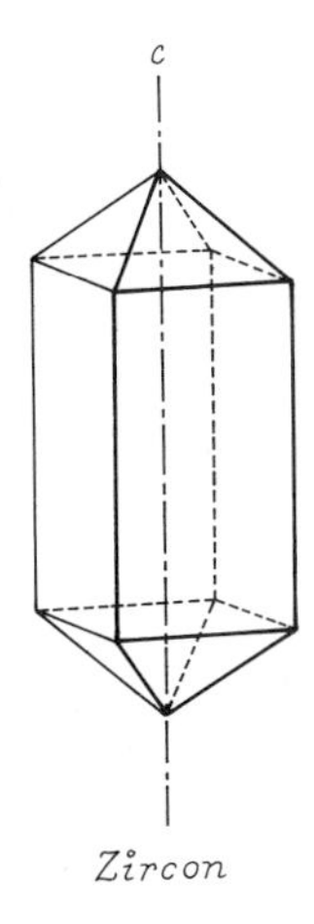

Zircon

$n_\omega = 1.920–1.960$*
$n_\varepsilon = 1.967–2.015$
$n_\varepsilon - n_\omega = 0.042–0.065$
Uniaxial positive (rarely biaxial positive $2V_z < 10°$)
Colorless to pale brown in standard section

COMPOSITION AND STRUCTURE. Zircon is a simple orthosilicate with Zr^{4+} ions lying between independent SiO_4^{4-} tetrahedra in complex coordination with four near and four distant O^{2-}. Owing to bombardment from radioactive impurities, the zircon structure commonly breaks down in a series of steps to a disordered, isotropic glass (metamict zircon or *malacon*).

Hafnium ions (Hf^{4+}) replace Zr^{4+} up to 6Hf:10Zr and Fe^{3+}, Y, and other rare earths and U and Th commonly replace some Zr. A few PO_4 tetrahedrons and groups of $(OH)_4$ may replace SiO_4 units.

* Refractive indices and birefringence decrease rapidly to an essentially isotropic form ($n_\varepsilon - n_\omega \simeq 0$ and $n \simeq 1.80$), as the zircon structure becomes metamict.

PHYSICAL PROPERTIES. H = 7.5. Sp. Gr. = 4.67 (Metamict forms to H = 6 and Sp. Gr. = 4.0.) Color, in hand sample, is most commonly reddish brown or yellow-brown, less commonly colorless, pink, lavender, or blue. Metamict forms tend to be greenish.

COLOR AND PLEOCHROISM. Zircon is normally colorless in standard section. Fragments are colorless or pale yellow, brown, gray, and so on, and color is often patchy. Pleochroism is usually weak $E > O$.

FORM. Zircon normally appears as tiny, euhedral, tetragonal crystals often enclosed in biotite or other early minerals. Zircon crystals are normally simple prismatic forms elongated on c; however, many intricate forms are known. Rounded and embayed crystals are formed by resorption in magmas, and rounded grains are common in detrital sediments.

CLEAVAGE. Poor prismatic cleavages {110} are seldom obvious in sections and fail to control fragment orientation.

BIREFRINGENCE. Maximum birefringence, seen in sections parallel to c, is strong, yielding bright interference colors of third and fourth-order in standard section. Detrital grains commonly show high-order white.

TWINNING. Zircon crystals are not commonly twinned; although knee-shaped (geniculate), simple twins on {111} are known.

ZONING. Color distribution is often patchy in zircon crystals, and distinct zoning commonly shows outer metamict zones on an unaltered core.

INTERFERENCE FIGURE. Fragments and sections show essentially random orientation. Near-basal sections of large crystals yield good, uniaxial positive, optic axis figures with many isochromatic rings. Metamict varieties may show biaxial positive, with small $2V_z < 10°$. Dispersion is very strong but does not influence uniaxial figures.

OPTICAL ORIENTATION. Small crystals, normally elongated on c, show parallel extinction and positive (length-slow) elongation.

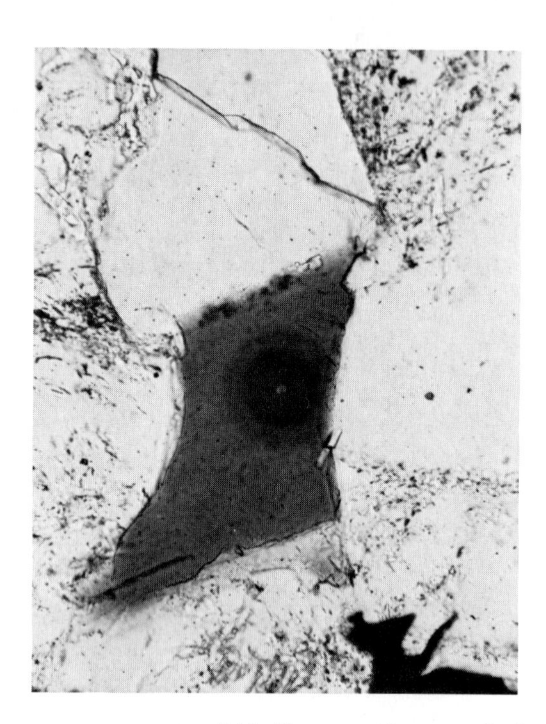

FIGURE 3-14. Zircon crystals are usually tiny, with high relief and bright interference colors, and are associated with ferromagnesian minerals in thin sections of granitic rocks. Note the radiation rings in this biotite grain produced by minor thorium in a tiny zircon crystal.

DISTINGUISHING FEATURES. Tiny, euhedral crystals with extreme relief and strong birefringence appearing in siliceous or alkaline plutonic rocks are commonly zircon. Crystals enclosed in biotite, hornblende, or other colored minerals are commonly surrounded by pleochroic halos of high color intensity (Fig. 3-14). Sphene is biaxial positive, with extreme birefringence, and tends to yield diamond-shaped cross sections; monazite is biaxial positive (small $2V_z$), with lower refraction indices than zircon, and xenotime shows less relief and extreme birefringence. Rutile and cassiterite commonly show color, extreme birefringence, and extreme relief.

ALTERATION. Zircon is relatively resistant to weathering and other forms of chemical attack, appearing as fresh grains in detrital sediments. Zircon is somewhat sensitive to alkaline solutions and may be hydrated when $(OH)_4$ groups replace SiO_4 units. Radioactive impurities favor metamict disorder with the formation of amorphous silica and isometric ZrO_2.

OCCURRENCE. Zircon is a common, minor accessory mineral in siliceous and alkaline, plutonic igneous rocks (for example, granite, diorite, syenite, and nepheline syenite). It appears most commonly as small, euhedral, tetragonal crystals associated with biotite, hornblende, and other early minerals and may form the nucleus of a deeply colored, pleochroic halo in the host mineral. Large zircon crystals are found in granite pegmatites and pegmatite-type veins in nepheline syenite and gabbro.

Zircon is much less common in volcanic rocks but may appear in most volcanics, especially trachyte. In metamorphic rocks, zircon is found in mica schists and granite gneisses, either as rounded relict grains from detrital sediments or as original euhedral crystals of high-grade metamorphism. Sedimentary rocks yield euhedral to rounded grains of zircon in the heavy fraction of detrital sediments.

REFERENCES: ZIRCON

Kresten, P. 1970. Metamictization of zircon. *Geol. Foeren. Stockholm Foerh.*, *92*, 110–113.

Makarov, E. S. 1970. Physicochemical factors in mineral metamictization—in particular, that of zircons. *Geokhimiya, 1970*, 54–58.

Robinson, K., G. V. Gibbs, and P. H. Ribbe. 1971. Structure of zircon: Comparison with garnet. *Amer. Mineral.*, *56*, 782–790.

Samaddar, M., and P. K. Ghosh. 1969. Origin of color in zircon crystals. *Technology*, *6*, 42–44.

Sphene (Titanite)

$CaTiSiO_5$

MONOCLINIC
$\angle\beta = 120°$

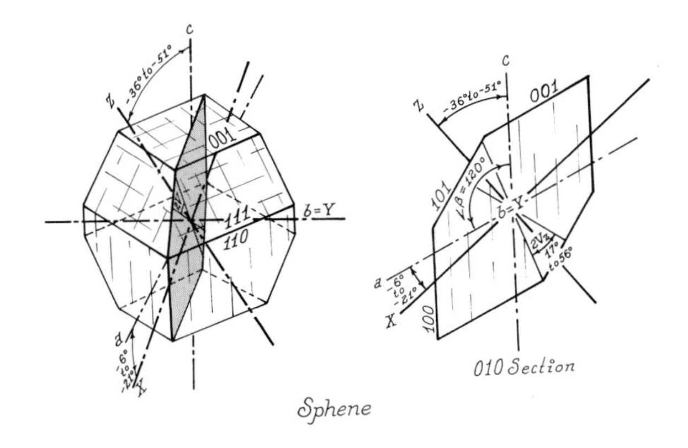

$n_\alpha = 1.840–1.950$*

$n_\beta = 1.870–2.034$

$n_\gamma = 1.943–2.110$

$n_\gamma - n_\alpha = 0.100–0.192$

Biaxial positive; $2V_z = 17°–56°$

$b = Y, c \wedge Z = -36°$ to $-51°$, $a \wedge X = -6°$ to $-21°$

$r > v$ strong, weak inclined dispersion

Colorless to pale gray-brown in section, with weak pleochroism $Z > Y > X$

Fragments may show deep coloration and distinct pleochroism

* Substitution of Al and Fe^{3+} for Ti lowers refractive indices and increases $2V_z$.

COMPOSITION AND STRUCTURE. Sphene is an orthosilicate with independent SiO_4 tetrahedra united by Ti^{4+} in octahedral coordination and Ca^{2+} in sevenfold coordination.

Al^{3+} or Fe^{3+} (rarely Cr^{3+}) may partially replace Ti^{4+} and are the most common impurities. One oxygen ion in the formula unit is not bound to any silicon and may be replaced by $(OH)^-$ or F^- to balance Al^{3+} for Ti^{4+}. Some rare earths—Nb^{5+}, Ta^{5+}, Sn^{5+}, or V^{5+}—may replace minor Ti^{4+} with equivalent Na^+ for Ca^{2+}.

PHYSICAL PROPERTIES. H = 5 – $5\frac{1}{2}$. Sp. Gr. = 3.45 – 3.56. Color, in hand sample, is commonly dark brown but may be yellow, greenish, or orange to black.

COLOR AND PLEOCHROISM. In standard section, sphene is colorless to pale gray-brown or yellow-brown, and colored varieties show weak pleochroism $Z > Y > X$. X = colorless to pale yellow, Y = pale yellow-brown to pale yellow-green or pink, and Z = orange-brown to green or pink.

FORM. Sphene most commonly appears as small to moderately large, euhedral, accessory crystals with nearly diamond-shaped or wedge-shaped sections (Fig. 3-15).

CLEAVAGE. Distinct prismatic {110} cleavages commonly show well in hand sample but are seldom obvious in section.

BIREFRINGENCE. Birefringence is strong in most sections and maximum birefringence, in {010} sections, is extreme (0.100–0.192) yielding very high-order interference colors of pearl-gray. Strong dispersion may prevent complete extinction in low-birefringent sections and may resemble high-order interference colors.

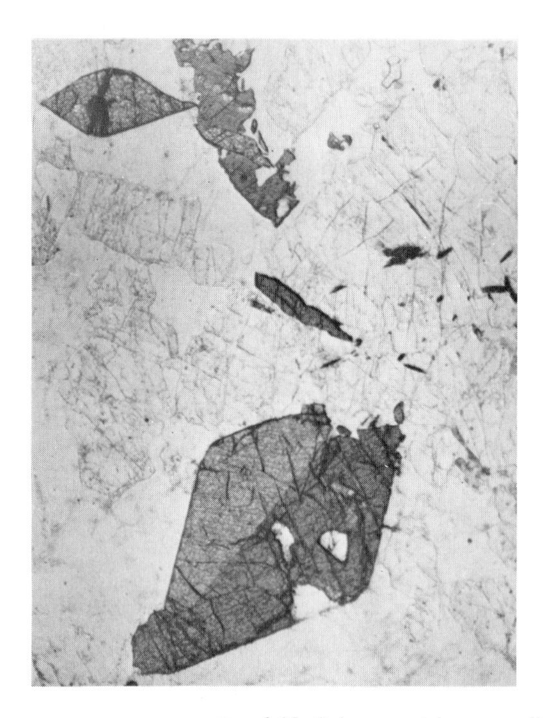

FIGURE 3-15. Sphene crystals are usually small, often showing diamond-shaped sections and sometimes twinned on the long diagonal.

TWINNING. Simple twinning on {100} is common, dividing diamond sections near the long diagonal. Multiple twinning is rare on {221}.

ZONING. Crystals may show color zoning with increased refractive index in outer zones.

INTERFERENCE FIGURE. Sphene is biaxial positive, with small to moderate $2V_z$ (17°–56°) and very strong dispersion $r > v$. Near-basal sections yield distinct acute bisectrix figures showing numerous isochromatic bands and isogyres that remain well within the visible field. Strong optic axis dispersion $r > v$ forms strong color fringes on isogyres, and weak inclined dispersion may be apparent. Fragments on cleavage surfaces yield off-center figures that are not symmetrical on either optic plane or optic normal.

OPTICAL ORIENTATION. Sphene is monclinic $b = Y$, $c \wedge Z = -35°$ to $-51°$, and $a \wedge X = -6°$ to $-21°$. The optic plane is {010}, and maximum extinction to cleavage traces is 36° to 45°. Extinction is often nearly symmetrical to diamond sections. The long diagonal of diamond sections lies near the faster wave.

DISTINGUISHING FEATURES. Diamond sections of extreme relief and very strong birefringence and dispersion are characteristic of sphene. Monazite shows lower refractive indices, lower birefringence, and weak dispersion; cassiterite, rutile, and xenotime are uniaxial, and baddeleyite is biaxial negative.

ALTERATION. Sphene may alter to a fine-grained, earthy, white or yellow aggregate that is nearly opaque in section. The aggregate is mostly Ti-oxides (anatase or rutile), with minor quartz, magnetite, or ilmenite, and is called *leucoxene*. Sphene may appear as a minor secondary product in the alteration of titanaugite or biotite.

OCCURRENCE. Sphene may appear in nearly any plutonic igneous rock but is most characteristic of undersaturated (for example, nepheline syenite) and intermediate types (for example, granodiorite, monzonite, syenite, and diorite). It is usually a minor accessory mineral as small euhedral crystals associated with ferromagnesian minerals and other early-formed crystals. Sphene, however, may be a major rock-forming mineral in a few rare rock types. Sphene may form in granite pegmatites and other coarse-grained, low-temperature dike rocks.

In volcanic rocks, sphene is much less common, but small crystals are reported in phonolites. Metamorphic rocks commonly contain sphene in schist, granite gneiss, amphibolite, and other rocks of regional metamorphism. Sphene is also common in marble, skarn, and blueschists. In detrital sediments, grains of sphene are quite durable, and sphene may even form in some sediments as an authigenic mineral.

REFERENCES: SPHENE

Higgins, J. B., and P. H. Ribbe. 1976. The crystal chemistry and space groups of natural and synthetic titanites. *Amer. Mineral.*, *61*, 878–888.

Sahama, T. G. 1946. On the chemistry of the mineral titanite. *Bull. Comm. Geol. Finlande*, *138*, 88–120.

Speer, J. A., and G. V. Gibbs. 1976. The crystal structure of synthetic titanite, $CaTiOSiO_4$, and the domain textures of natural titanites. *Amer. Mineral.*, *61*, 238–247.

Datolite

$CaB(SiO_4)OH$

MONOCLINIC
$\angle\beta = 90°09'$

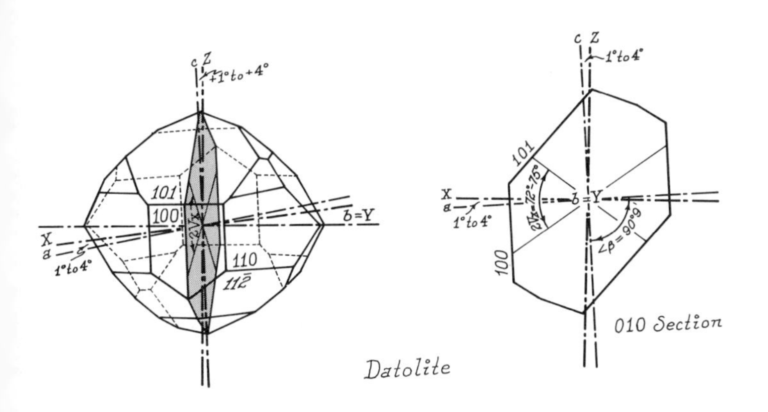

$n_\alpha = 1.622–1.626$
$n_\beta = 1.649–1.654$
$n_\gamma = 1.666–1.670$
$n_\gamma - n_\alpha = 0.044–0.046$
Biaxial negative, $2V_x = 72°–75°$
$b = Y$, $c \wedge Z = +1°$ to $+4°$, $a \wedge X = +1°$ to $+4°$
$r > v$ weak
Colorless in standard section

COMPOSITION AND STRUCTURE. Datolite is a nesosilicate consisting of sheets of 4- and 8-membered rings of alternating SiO_4 and $BO_3(OH)$ tetrahedra. Sheets of tetrahedra parallel to {100} are linked by Ca^{2+} ions coordinated by six O^{2-} and two $(OH)^-$.

Composition is usually close to the ideal $CaB(SiO_4)OH$, although, minor Al^{3+} or Fe^{3+} may replace Si^{4+} or B^{3+}, and minor Mn^{2+} or Mg^{2+} may replace Ca^{2+}.

PHYSICAL PROPERTIES. $H = 5–5\frac{1}{2}$. Sp. Gr. = 2.9–3.0. In hand sample, datolite is usually colorless to white or pale yellow; rarely green or pink. Luster is distinctly vitreous.

COLOR AND PLEOCHROISM. Colorless in thin section or as fragments.

FORM. Datolite crystals are commonly euhedral and blocky, with numerous and complex crystal forms.

CLEAVAGE. Datolite has no observable cleavage, and fragments lie in random positions on conchoidal fracture surfaces.

BIREFRINGENCE. Maximum birefringence, seen in {010} sections is strong ($n_\gamma - n_\alpha = 0.044–0.046$). Interference colors are mostly bright first- and second-order colors, ranging to middle third-order in standard section.

TWINNING. No twinning is reported.

INTERFERENCE FIGURE. Datolite is biaxial negative, with rather large $2V_x = 72°$ to $75°$. Fragments yield randomly oriented figures showing many isochromes and weak dispersion $r > v$.

OPTICAL ORIENTATION. Datolite is monoclinic, with $b = Y$, $c \wedge Z = +1°$ to $+4°$, and $a \wedge X = +1°$ to $+4°$; the optic plane is {010}. Neither fragments nor sections show consistent elongation.

DISTINGUISHING FEATURES. Datolite shows massive form without cleavage. It is colorless and has positive relief and much higher birefringence than zeolites, topaz, danburite, or pectolite. Prehnite has distinct cleavage.

ALTERATION. Datolite is, itself, a secondary mineral and is quite stable under surface conditions.

OCCURRENCE. The characteristic occurrence for datolite is in amygdales, cavities, and fissures in basalt, andesite, or diabase with zeolites, prehnite, calcite, or danburite. It may also occur in contact-metamorphic marbles or skarns, due to boron metasomatism of carbonate sediments, with calcite, grossular, prehnite, aegirine-augite, and fluorite. Datolite is a rare constituent of granite, serpentinite, basic schists, and hydrothermal ore veins.

REFERENCES: DATOLITE

Foit, F. F., M. W. Phillips, and G. V. Gibbs. 1973. Refinement of the crystal structure of datolite, $CaBSiO_4(OH)$. *Amer. Mineral.*, *58*, 909–914.

Ito, T., and H. Mori. 1953. The crystal structure of datolite. *Acta Crystallogr.*, *6*, 24–32.

Staurolite

$Fe_2^{2+}Al_{\sim 9}O_6(SiO_4)_4(OH)_{\sim 2}$

MONOCLINIC PSEUDOORTHORHOMBIC
$\angle\beta \simeq 90°$

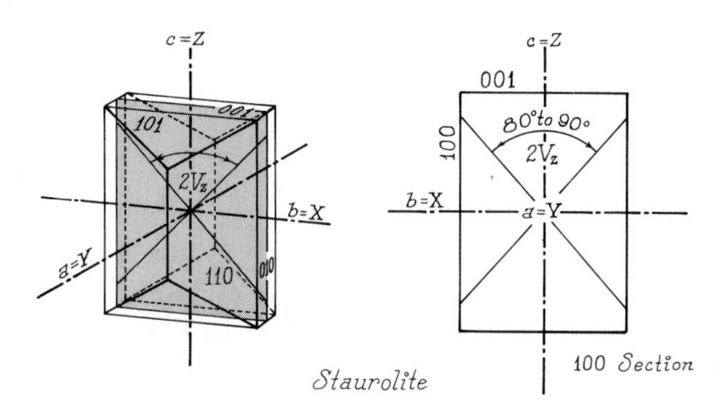

$n_\alpha = 1.736–1.747$*

$n_\beta = 1.740–1.754$

$n_\gamma = 1.745–1.762$

$n_\gamma - n_\alpha = 0.009–0.015$*

Biaxial positive, $2V_z = 80°–90°$

$a = Y, b = X, c = Z$

$r > v$ weak to moderate

Golden-yellow in standard section, with pronounced pleochroism $Z > Y > X$

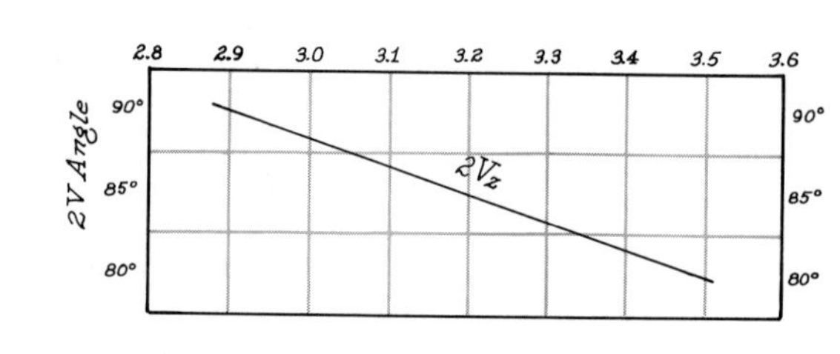

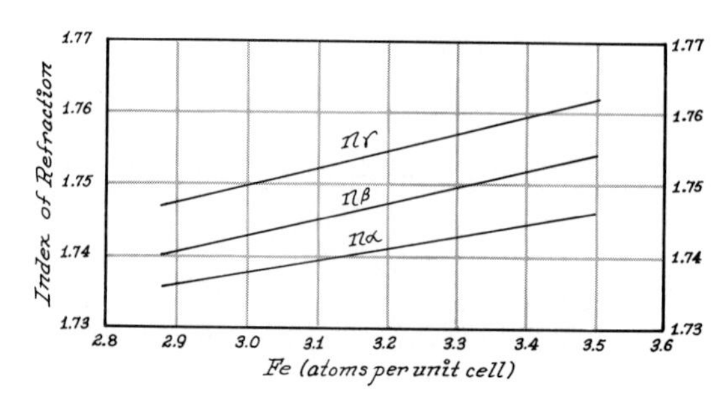

FIGURE 3-16. Variation in the indices of refraction and 2V angle with iron content in staurolite. (After Griffen and Ribbe, 1973)

* Refractive indices and birefringence increase with iron content, while $2V$ decreases (Fig. 3-16).

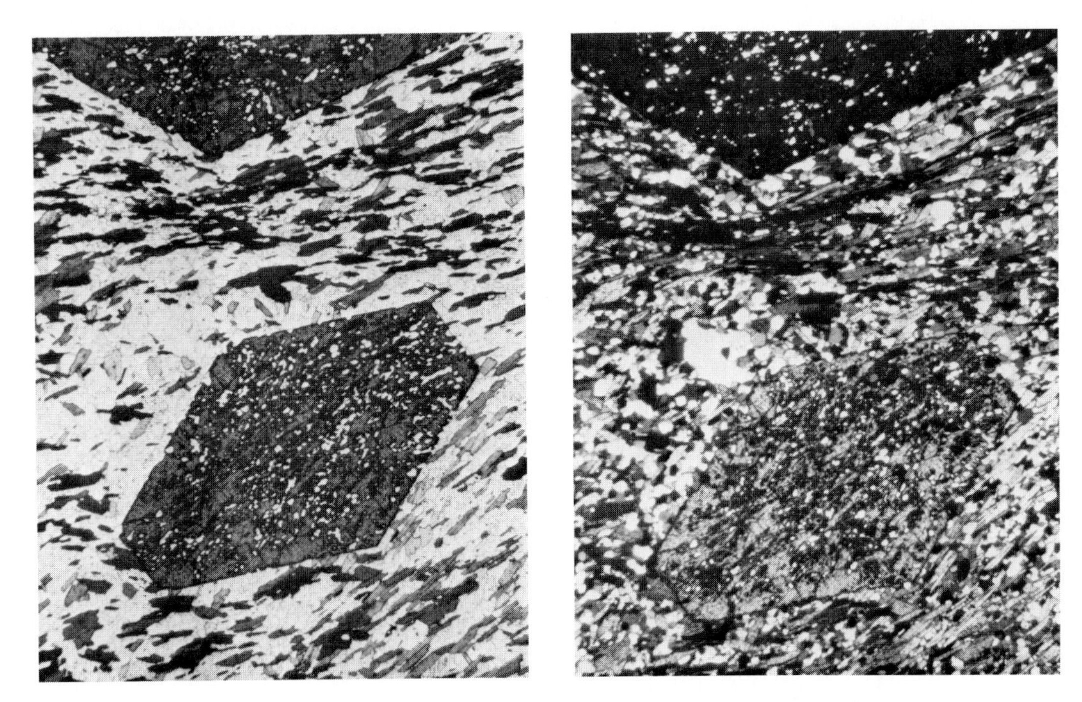

FIGURE 3-17. Staurolite crystals in mica schists may show diamond-shaped cross sections with flattened ends and are usually impregnated with tiny inclusions of other minerals (sieve structure).

COMPOSITION AND STRUCTURE. The structure of staurolite consists basically of layers of kyanite structure and composition (Al_2SiO_5) and layers of $Al_{0.7}Fe_2O_2(OH)_2$ alternating parallel to {010}, and kyanite and staurolite are often found in parallel intergrowth; namely, {100} plane of kyanite parallel to the {010} plane of staurolite. The ideal formula is approximated by $Fe_2{}^{2+}Al_{\sim 9}O_6(SiO_4)_4(OH)_{\sim 2}$, but the numerous substituents consistently include Mg^{2+}, Ti^{4+}, Zn^{2+}, Co^{2+}, Mn^{2+}, and V^{5+}, and often Cr^{3+}. The rare, blue, cobalt-rich staurolite is called *lusakite*.

PHYSICAL PROPERTIES. H = 7–$7\frac{1}{2}$. Sp. Gr. = 3.74–3.83. Color, in hand sample, is commonly dark brown ranging to reddish or yellow-brown.

COLOR AND PLEOCHROISM. In thin section, staurolite shows a characteristic golden-yellow, distinctly pleochroic from nearly colorless to deep yellow. $Z > Y > X$. X = colorless, pale yellow; Y = pale yellow, yellow-brown; and Z = golden-yellow, red-brown. The cobalt variety lusakite is pleochroic in blue and violet.

FORM. Staurolite occurs most commonly as large euhedral crystals elongated on c, showing rectangular prismatic sections and six-sided ({010} and {110}) cross sections (Fig. 3-17). Penetration twins at 90° or 60°31′ are very common but seldom obvious in section. Staurolite porphyroblasts commonly show ragged outline and numerous mineral inclusions (sieve structure) of quartz and other associated minerals.

CLEAVAGE. Rather poor pinacoidal {010} cleavage is seldom obvious in section and fails to control fragment orientation.

BIREFRINGENCE. Maximum birefringence, seen in {100} sections, is weak to mild (0.009–0.015). Interference colors, in standard section, are first-order gray, white, or yellow, usually showing as yellow, due to inherent mineral color.

TWINNING. Simple penetration twinning on {023} or {232} form cruciform twins at approximately 90° and 60° respectively, which are nearly as common as single crystals. This twinning is seldom obvious in thin section but contact twins on {031} are reported.

ZONING. Color zoning is sometimes seen in staurolite.

INTERFERENCE FIGURE. Staurolite is usually positive with $2V_z$ from 80° to 90° but may be negative, as $2V_z$ slightly exceeds 90°. The optic plane is {100}, with the acute bisectrix $Z = c$. Basal sections {001} and sections parallel to {010} yield bisectrix-centered figures of widely separating isogyres on a white field. Optic axis dispersion is weak to moderate $r > v$.

OPTICAL ORIENTATION. Staurolite is sensibly orthorhombic with $a = Y, b = X, c = Z$. Extinction is parallel to prismatic sections and cleavage traces and symmetrical to {110} faces in basal section. Elongation on $c = Z$ produces crystals of positive elongation (length-slow).

DISTINGUISHING FEATURES. Colorless to deep yellow pleochroism, high relief, weak birefringence, sieve structure, and limited paragenesis are usually sufficiently distinctive for staurolite. Melanite garnet is isotropic, brown tourmaline is uniaxial and length-fast with moderate birefringence, and idocrase is uniaxial with very weak birefringence and anomalous interference colors.

ALTERATION. Staurolite is quite resistant to weathering but may alter to sericite, chlorite, and limonite. With increasing metamorphic grade, staurolite becomes almandine and kyanite or sillimanite, and, by retrograde effects, staurolite inverts to sericite or chlorite, with possible magnetite, corundum, spinel, or quartz.

OCCURRENCE. Staurolite is largely restricted to medium-grade regional metamorphic rocks where it appears as porphyroblasts in pelitic schists in association with almandine, kyanite, muscovite, biotite, tourmaline, and quartz. In somewhat lower-grade schists and phyllites, staurolite sometimes appears with chloritoid and chlorite.

In sedimentary rocks, staurolite is locally abundant as detrital grains.

REFERENCES: STAUROLITE

Griffen, D. T., and P. H. Ribbe. 1973. The crystal chemistry of staurolite. *Amer. Jour. Sci.*, 273-A, 479–495.

Hollister, L. S., and A. E. Bence. 1969. Staurolite: Sectorial compositional variations. *Science*, *158*, 1053–1056.

Hurst, V. J., J. D. H. Donnay, and G. Donnay. 1956. Staurolite twinning. *Mineral. Mag.*, *31*, 145–163.

Juurinen, A. 1956. Composition and properties of staurolite. *Acad. Sci. Fennicae Ann.*, Series A, *47*, 1–53.

Richardson, S. W. 1966. Staurolite. *Carnegie Inst. Wash. Year Book*, *65*, 248–252.

Smith, J. V. 1968. The crystal structure of staurolite. *Amer. Mineral.*, *53*, 1139–1155.

Sapphirine

$(Mg,Fe^{2+})_2Al_4O_6(SiO_4)$

MONOCLINIC
$\angle\beta = 125°20'$
TRICLINIC

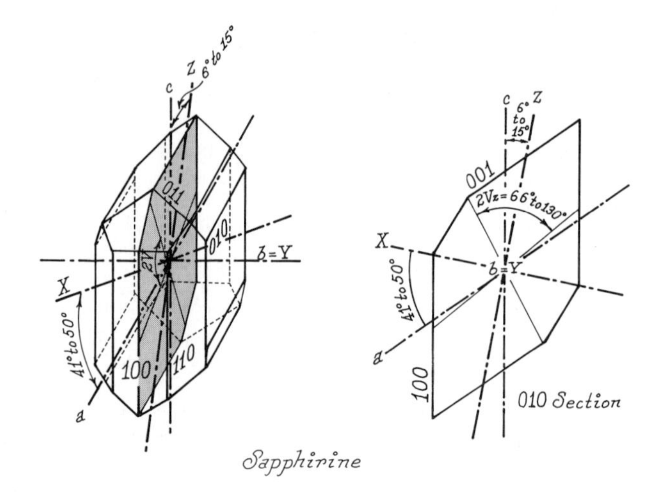

$n_\alpha = 1.701–1.729$*

$n_\beta = 1.703–1.732$

$n_\gamma = 1.705–1.734$

$n_\gamma - n_\alpha = 0.004–0.007$

Biaxial positive, $2V_z = 66°–90°$ *or*

Biaxial negative, $2V_x = 90°–50°$

$b = Y$, $c \wedge Z = +6°$ to $+15°$, $a \wedge X = +41°$ to $+50°$

$v > r$ strong, distinct inclined bisectrix dispersion

Colorless to pale blue, with pleochroism $Z > Y > X$

* Indices of refraction increase with increasing Fe^{2+}.

COMPOSITION AND STRUCTURE. Mg-Fe and Al, in six-coordination, form octahedral walls parallel to *c* and connected along *b* by an unusual $(Si_6O_{18})_\infty$ tetrahedral chain. Thus it is somewhat like a layer silicate, and both the common two-layer monoclinic polytype and an uncommon one-layer triclinic polytype are known. Sapphirine typically contains excess Al^{3+}, substituting in equal amounts for Si^{4+} and Mg^{2+}.

PHYSICAL PROPERTIES. $H = 7\frac{1}{2}$. Sp. Gr. = 3.40–3.58. In hand sample, sapphirine is pale blue, green, or gray.

COLOR AND PLEOCHROISM. Sapphirine is colorless to pale blue in standard thin section. Pleochroism is distinct: X = colorless, pale yellow, green-blue, buff; Y = pale blue, blue-green, pale blue-violet; and Z = deep blue, green.

FORM. Euhedral or subhedral crystals are somewhat elongated on *c*; granular aggregates are common.

CLEAVAGE. All cleavages, {010}, {001}, and {100}, are poor.

BIREFRINGENCE. Interference colors are shades of gray in standard section; some sections show somewhat anomalous indigo-gray color.

TWINNING. Broad lamellar twinning on either or both {010} and {100} is reported but is uncommon.

ZONING. Rims of sapphirine about feldspar, corundum, or spinel have been reported. Intergrowths of sapphirine and cordierite are known.

INTERFERENCE FIGURE. Acute bisectrix figures, seen in sections normal to {010}, show broad isogyres on a gray background, strong optic axis dispersion $v > r$, and distinct inclined bisectrix dispersion. Fragments yield essentially randomly oriented figures.

OPTICAL ORIENTATION. Neither crystals nor fragments show consistent elongation; crystal sections elongated on *c*, however, show extinction up to 6°–15° and are length-slow (positive elongation).

DISTINGUISHING FEATURES. Pale blue color, weak birefringence, and occurrence are distinctive. Kyanite has excellent cleavages and mild birefringence; corundum is uniaxial; cordierite has low relief, sometimes negative in balsam; Na-amphiboles and zoisite show good cleavages; lazulite shows moderate birefringence; and lazurite is isotropic.

ALTERATION. Sapphirine may alter to a mixture of corundum, biotite, and possibly talc.

OCCURRENCE. Sapphirine is a high-temperature mineral of regional or contact metamorphism. It is characteristic of the granulite-amphibolite facies and silica-poor, Mg-rich rocks. It appears in schists, gneisses, granulites, amphibolites, and hornfels with corundum, biotite, anorthite, almandine, anthophyllite, hornblende, cordierite, augite, hypersthene, spinel, diaspore, and sillimanite. Sapphirine may be a constituent of contact emery deposits with corundum, magnetite, and spinel. It is not compatible with quartz, olivine, periclase, or clinoenstatite.

REFERENCES: SAPPHIRINE

Kuzel, H. J. 1961. The formula and unit cell of sapphirine. *Neues Jahrb. Mineral. Monatsh.*, 1961, 68–71.

Merlino, S. 1973. Polymorphism in sapphirine. *Contrib. Mineral. Petrol.*, *41*, 23–29.

Moore, P. B. 1969. Crystal structure of sapphirine. *Amer. Mineral.*, *54*, 31–49.

		MONOCLINIC	TRICLINIC
Chloritoid	$(Fe^{2+},Mg,Mn)_2(Al,Fe^{3+})Al_3O_2(SiO_4)_2(OH)_4$	$\angle\beta = 101°39'$	$\angle\alpha \approx 90°$ $\angle\beta \approx 102°$ $\angle\gamma \approx 90°$

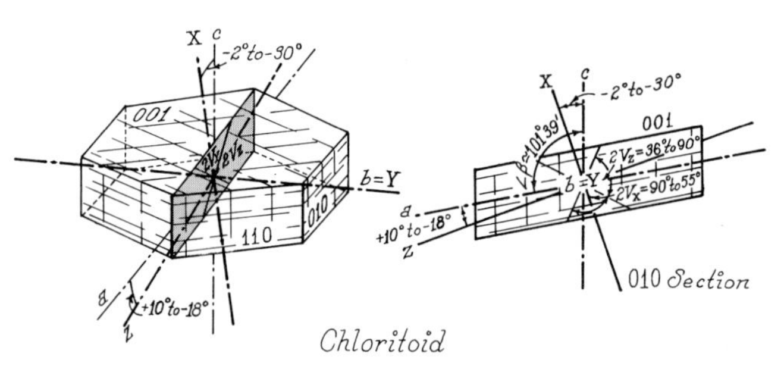

$n_\alpha = 1.713–1.730$*
$n_\beta = 1.719–1.734$
$n_\gamma = 1.723–1.740$
$n_\gamma - n_\alpha = 0.006–0.22$
Biaxial positive, $2V_z = 36°–90°$
Biaxial negative, $2V_x = 90°–55°$
$b = Y$ or X, $c \wedge Z = -2°$ to $-30°$ (monoclinic polymorph)
$r > v$ strong
Colorless to green-gray or blue-gray in standard section, with pleochroism $Y > X > Z$

COMPOSITION AND STRUCTURE. Chloritoid is an unusual sheet silicate (phyllosilicate) consisting of brucite-like layers (Fe^{2+}, Mg, and Mn in octahedral $(OH)^-$ coordination and layers of Al^{3+} in octahedral O^{2-} coordination, alternating normal to {001} and linked together by intervening layers of independent SiO_4 tetrahedra. The common monoclinic polymorph appears to be favored by high stress and temperature. A triclinic form is less common and appears independent of stress.

Mg^{2+} or Mn^{2+} may replace Fe^{2+} in significant amounts with minor Ca^{2+}, Ti^{4+}, and Fe^{3+}. The Mg-rich variety is called *sismondine* and the Mn-rich variety *ottrelite*. Minor Fe^{3+} and possibly Ti^{4+} may replace Al^{3+}.

* Refractive indices decrease slightly a^J $^\rho$eplaces Fe^{2+} and increase as Fe^{3+} replaces Al^{3+}.

PHYSICAL PROPERTIES. H = $6\frac{1}{2}$. Sp. Gr. = 3.26–3.80. Color, in hand sample, is dark gray to dark green, somewhat resembling chlorite, although chloritoid is brittle and much harder than chlorite.

COLOR AND PLEOCHROISM. In standard thin section, chloritoid is colorless to distinctly pleochroic in green, blue, or yellow $Y > X > Z$. X = pale gray-green, olive-green, green; Y = slate-blue, indigo-blue, pale green; and Z = colorless, pale yellow, pale green.

FORM. Small porphyroblasts of chloritoid are common, as are anhedral foliated aggregates (Fig. 3-18). Chloritoid is distinctly foliated or tabular on the basal plane {001}, and platy aggregates may show parallel to random arrangement. Chloritoid commonly contains inclusions of associated minerals (such as quartz, muscovite, rutile, tourmaline, magnetite, and garnet).

CLEAVAGE. Perfect basal cleavage {001} produces foliation. Cleavages are imperfect on {110}, and parting may appear on {010}.

BIREFRINGENCE. Maximum birefringence, seen in {010} or near {100} sections is weak to moderate, ranging from 0.006 to 0.022, and basal sections and cleavage plates show very weak birefringence. Interference colors are mostly first-order gray and white, sometimes anomalous by inherent mineral color; however, colors up to second-order are possible.

TWINNING. Twinning is very common on {001}, either simple or, more commonly, multiple.

ZONING. Color zoning and hourglass structure are common.

INTERFERENCE FIGURE. Chloritoid is either optically positive or negative, and $2V$ ranges widely from about 36° (+) through 90° to about 55° (−). It is usually positive, with only moderate $2V_z$. As Z lies within a few degrees of the normal to {001}, basal plates yield nearly centered Bxa figures. Optic axis dispersion is strong $r > v$, and bisectrix dispersion is reported as crossed, presumably in triclinic varieties.

OPTICAL ORIENTATION. Orientation of chloritoid is highly variable, probably due to several polymorphic forms. The monoclinic varieties show $b = Y$ or X, $c \wedge Z = -2°$ to $-30°$

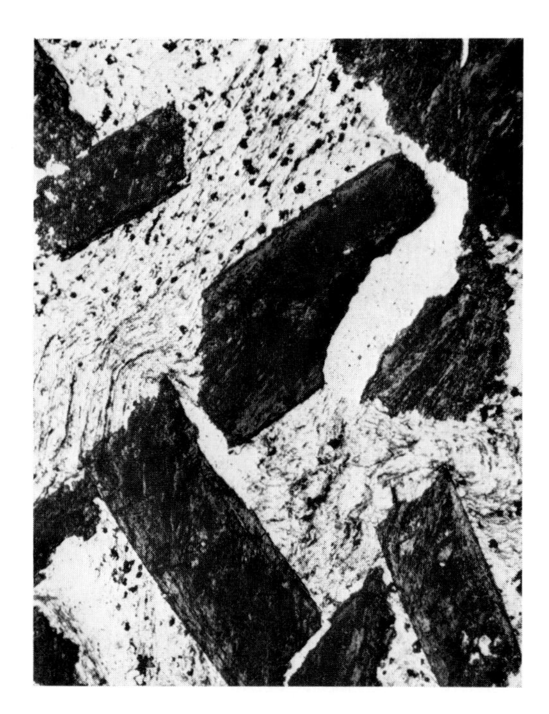

FIGURE 3-18. Ottrelite appears as tabular crystals with single cleavage and multiple twinning.

and $a \wedge X$ or $Y = +10^\circ$ to -18°. Maximum extinction to basal cleavage is 10° to 18° in {010} sections and symmetrical or inclined to prismatic cleavages {110} in basal plates. Elongation is negative (length-fast), as Z is almost normal to the major cleavage.

DISTINGUISHING FEATURES. Chloritoid may resemble chlorite, mica, and other foliated minerals, but its color, pleochroism, high relief, multiple cleavages, and twinning are characteristic. Chlorite has small $2V$ and lower refraction indices; biotite shows lower refractive indices, small $2V_x$, and strong birefringence; the brittle micas are biaxial negative (moderate $2V_x$) with lower refraction indices and dispersion $v > r$; and stilpnomelane is biaxial negative, with $2V_x$ near 0°.

ALTERATION. Chloritoid may alter to chlorite, sericite, or kaolinite, and with increasing metamorphic intensity chloritoid may invert to staurolite, quartz, and magnetite or hercynite, or to cordierite and hercynite.

OCCURRENCE. Chloritoid is commonly regarded as a metamorphic stress mineral characteristic of low- to medium-grade schists. It results largely from the regional metamorphism of sediments or igneous rocks rich in Fe and Al and deficient in Ca, Mg, Na, and K. The triclinic polymorph appears in low-grade schists with chlorite and albite, and the monoclinic form seems favored by increasing stress and is expected in association with muscovite, almandine, staurolite, and kyanite. Hydrothermal processes also form triclinic chloritoid in quartz-carbonate veins and hydrothermal cavity fillings.

REFERENCES: CHLORITOID

Brindley, G. W. and F. W. Harrison. 1952. The structure of chloritoid, *Acta Crystallogr.*, *5*, 698–699.

Foster, M. D. 1961. Analyses of chloritoid. *U.S. Geol. Surv., Prof. Paper. 424-C.*

Halferdahl, L. B. 1961. Chloritoid, its composition, x-ray and optical properties, stability, and occurrence. *Jour. Petrol.*, *2*, 49–135.

Hanscom, R. H. 1975. Refinement of the crystal structure of monoclinic chloritoid. *Acta Crystallogr.*, *B31*, 780–784.

The Humite Group

$nMg_2SiO_4 \cdot Mg_{1-x}Ti_x(OH,F)_{2-2x}O_{2x} (x < 1)$

ORTHORHOMBIC
MONOCLINIC

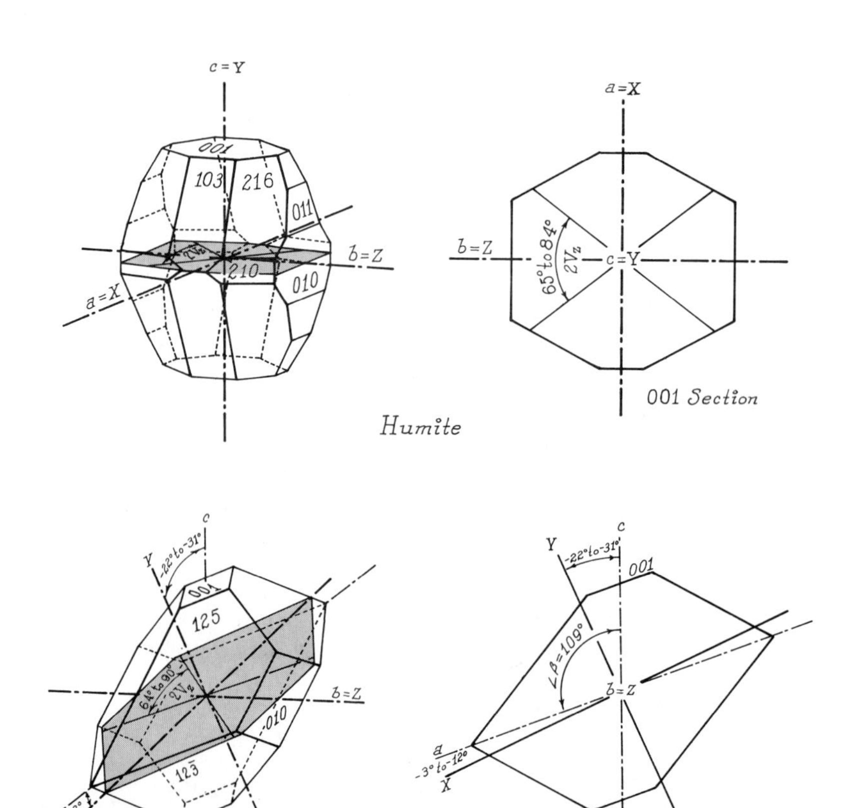

	n_α*	n_β	n_γ	$n_\gamma - n_\alpha$	$2V_z$*	$c \wedge Y$
Norbergite	1.561–1.567	1.566–1.579	1.587–1.593	0.026–0.027	44°–50° +	0°
Chondrodite	1.592–1.643	1.602–1.655	1.619–1.675	0.025–0.037	64°–90° +	−22° to −31°
Humite	1.607–1.643	1.619–1.653	1.639–1.675	0.028–0.036	65°–84° +	0°
Clinohumite	1.623–1.702	1.636–1.709	1.651–1.728	0.028–0.045	52°–90° +	−7° to −15°

* Refractive indices, birefringence, and $2V_z$ tend to increase from norbergite to clinohumite. These optical constants also increase with Fe^{2+} and Ti^{4+} and with $(OH)^-$ for F^-.

Biaxial positive, $2V_z = 33°–90°$

Orthorhombic forms (norbergite and humite): $b = Z, c = Y, a = X$

Monoclinic forms: $b = Z$

Chondrodite: $\angle\beta = 109°, c \wedge Y = -22°$ to $-31°, a \wedge X = -3°$ to $-12°$

Clinohumite: $\angle\beta = 101°, c \wedge Y = -7°$ to $-15°, a \wedge X = +4°$ to $-4°$

$r > v$ weak to strong

Colorless to pale yellow or yellow-brown and mildly pleochroic $X > Y = Z$ in standard section

COMPOSITION AND STRUCTURE. Like olivines, humites consist of a nearly hexagonal close-packed array of anions (O, OH, and F), with metal ions filling octahedral sites so as to result in zigzag chains of $M(O,OH,F)_6$ octahedra. These chains are staggered in a way that is unlike the olivine arrangement, so that none of the independent tetrahedral sites occupied by Si has OH or F corners. The following list shows simplified formulas with $x = 0$.

Norbergite	$Mg_2SiO_4 \cdot Mg(OH,F)_2$	Orthorhombic
Chondrodite	$2Mg_2SiO_4 \cdot Mg(OH,F)_2$	Monoclinic $\angle\beta = 109°$
Humite	$3Mg_2SiO_4 \cdot Mg(OH,F)_2$	Orthorhombic
Clinohumite	$4Mg_2SiO_4 \cdot Mg(OH,F)_2$	Monoclinic $\angle\beta = 101°$

The major replacement is OH F; fluorine is always present, usually exceeding OH. Mg^{2+} is replaced by Fe^{2+}, Mn^{2+}, Ca^{2+}, and Zn^{2+}, in that order of abundance, although magnesium always predominates.

PHYSICAL PROPERTIES. H = 6–6½. Sp. Gr. = 3.1–3.4, increasing toward clinohumite. In hand sample, the humite minerals range from yellow to brown; rarely, they are reddish or orange.

COLOR AND PLEOCHROISM. The humite minerals are colorless to yellow or pale brown, in section or as fragments, and show moderate pleochroism $X > Y = Z$. X = pale yellow, deep yellow, brownish-yellow; Y = colorless, pale yellow, golden yellow, pale yellow-green; and Z = colorless, pale yellow, pale brown, pale green.

FORM. Humite minerals commonly appear as anhedral masses or grains or as plates flattened on {010}, {001}, or {100}.

CLEAVAGE. Basal {001} cleavage is poor.

BIREFRINGENCE. Maximum birefringence, seen in near-basal sections, is moderate (0.025–0.045) yielding bright, upper second-order, interference colors in standard section. Birefringence increases somewhat from norbergite to clinohumite and with increasing Fe, Ti, and OH.

TWINNING. In monoclinic forms (chondrodite and clinohumite), simple or multiple twinning is common on {001}. Twinning is also reported on {105} and {305}.

ZONING. Zones of differing color intensity are common, and intergrown plates of chondrodite, humite, clinohumite, forsterite, and monticellite have been reported.

INTERFERENCE FIGURE. Fragment orientation is not seriously affected by poor basal cleavages, but basal sections should yield near-flash figures. Bright interference colors should be evident as isochromatic rings, and dispersion $r > v$ is commonly visible, especially in monoclinic forms, where crossed bisectrix dispersion is possible.

OPTICAL ORIENTATION. Orthorhombic forms (norbergite and humite) are oriented with $b = Z$, $c = Y$, and $a = X$, and the optic plane is {001}. Extinction is parallel to poor basal cleavage traces, and elongation is positive or negative.

Monoclinic forms (chondrodite and clinohumite) are oriented with $b = Z$, the optic plane is near {001}. Chondrodite $c \wedge Y = -22°$ to $-31°$, $a \wedge X = -3°$ to $-12°$; clinohumite $c \wedge Y = -7°$ to $-15°$, $a \wedge X = +4°$ to $-4°$. Extinction to basal cleavage traces is parallel to near parallel (3°–12°), and elongation is positive and negative, as Y is essentially perpendicular to cleavage.

DISTINGUISHING FEATURES. Olivine resembles colorless or weakly colored humite minerals, but olivines are mostly biaxial negative, except forsterite, which shows large $2V_z$ near 90°. Forsterite shows higher indices and birefringence than most humites and shows parallel extinction. Staurolite shows higher refractive indices, lower birefringence, and larger $2V_z$ and appears largely as euhedral crystals in schists.

Members of the humite group are very difficult to distinguish from each other by optical properties, but we may note that refractive index, birefringence, and $2V_z$ tend to increase from norbergite to clinohumite, and chondrodite and clinohumite are monoclinic, with inclined extinction to poor cleavage traces.

ALTERATION. The humite minerals commonly alter to serpentine or Mg-rich chlorites and dissolve by weathering, leaving iron oxide residues.

OCCURRENCE. The humite minerals are largely confined to metamorphic contact zones between carbonate rocks (limestone, dolomite, and marble) and acidic or alkaline intrusions where fluorine has been introduced by metasomatic processes. Chondrodite is the most common member of the humite group, and the orthorhombic members, especially norbergite, are really quite rare. Associated minerals are largely carbonates and the Ca- and Mg-silicates characteristic of carbonate contact zones (such as wollastonite, tremolite, grossular, diopside, monticellite, forsterite, and chlorite). Ti-rich clinohumite is reported in altered peridotites, and gabbros and chondrodite are reported in carbonatite rocks.

REFERENCES: THE HUMITE GROUP

Bragg, W. L., and G. F. Clairingbull. 1965. *Crystal Structures of Minerals*. Ithaca, N.Y.: Cornell University Press.

Gibbs, G. V., and P. H. Ribbe. 1969. Crystal structures of the humite minerals. I. Norbergite. *Amer. Mineral.*, *54*, 376–390.

Gibbs, G. V., P. H. Ribbe, and C. P. Anderson. 1970. Crystal structures of the humite minerals. II. Chondrodite. *Amer. Mineral.*, *55*, 1182–1194.

Jones, N. W., P. H. Ribbe, and G. V. Gibbs. 1969. Crystal chemistry of the humite minerals. *Amer. Mineral.*, *54*, 391–411.

Larsen, E. S. 1928. The optical properties of the humite group. *Amer. Mineral.*, *13*, 354–359.

Ribbe, P. H., and G. V. Gibbs. 1971. Crystal structures of the humite minerals. III. Magnesium/iron ordering in humite and its relation to other ferromagnesian silicates. *Amer. Mineral.*, *56*, 1153–1171.

Sahama, T. G., K. Hytönen, and H. B. Wiik. 1953. Mineralogy of the Humite Group. *Ann. Acad. Sci. Fennicae*, Series A, *31*, 1–50.

CHAPTER 4

Sorosilicates and Cyclosilicates

The simplest way to polymerize silicate tetrahedra is to link two SiO_4^{4-} groups so that they share one common oxygen, resulting in an $Si_2O_7^{6-}$ oxyanion. Minerals that contain $Si_2O_7^{6-}$ groups are known as *sorosilicates* and are relatively uncommon. The "type" sorosilicate is the rare mineral thortveitite, $Sc_2Si_2O_7$. The more widespread sorosilicates include the melilites, the epidotes (which also contain independent SiO_4^{4-} groups), lawsonite, and hemimorphite. As in the orthosilicates, the silicate tetrahedra of the sorosilicates are insufficiently polymerized for them to control the optical properties (although their presence must contribute to them). Thus the rest of the crystal structure largely determines the optical properties, and this is completely different for each of the sorosilicates just mentioned.

Cyclosilicates, as suggested by the name, are minerals whose common feature is a ring of silicate tetrahedra. The general formula for such a ring is $Si_nO_{3n}^{2n-}$, where n is any positive integer: $n = 6$ gives the most common kind of ring, $Si_6O_{18}^{12-}$, which is found in tourmaline, beryl, cordierite*, and dioptase. $Si_3O_9^{6-}$ rings and $Si_4O_{12}^{8-}$ rings occur in benitoite and axinite, respectively.

The Epidote Group

The epidote structure consists of octahedral AlO_6 groups sharing edges to form continuous chains parallel to b. In one-third of the octahedral groups, Al^{3+} is surrounded by four O^{2-}

* If the AlO_4^{5-} tetrahedra are included with the silicate tetrahedra in cordierite and beryl, then they can be classed as tektosilicates; some mineralogists prefer that classification.

and two $(OH)^-$. The octahedral chains are linked by silicon, in SiO_4 and Si_2O_7 groups, sharing oxygen ions with aluminum. Larger Ca^{2+} ions fill interchain voids to produce electrostatic neutrality. The resulting basic composition is $Ca_2Al_3O \cdot SiO_4 \cdot Si_2O_7 \cdot OH$, and the structure is dimorphous, as orthorhombic and monoclinic forms.

Significant Al^{3+} may be replaced by Fe^{3+}, Mn^{3+}, or even Cr^{3+}, and Ca^{2+} may be partly replaced by Ce^{3+}, La^{3+}, Y^{3+}, and other rare earth elements with equal replacement of Fe^{3+} and Al^{3+} by Fe^{2+} to maintain electrostatic balance. Only compositions near the ideal $Ca_2Al_3O \cdot SiO_4 \cdot Si_2O_7 \cdot OH$ are dimorphous; other epidote varieties are monoclinic.

Orthorhombic varieties

Zoisite	$Ca_2Al_3O \cdot SiO_4 \cdot Si_2O_7 \cdot OH$
Thulite	$Ca_2(Al,Fe^{3+},Mn^{3+})_3O \cdot SiO_4 \cdot Si_2O_7 \cdot OH$

Monoclinic varieties

Clinozoisite	$Ca_2Al_3O \cdot SiO_4 \cdot Si_2O_7 \cdot OH$
Epidote (Pistacite)	$Ca_2Fe^{3+}Al_2O \cdot SiO_4 \cdot Si_2O_7 \cdot OH$
Piemontite	$Ca_2(Al,Fe^{3+},Mn^{3+})_3O \cdot SiO_4 \cdot Si_2O_7 \cdot OH$
Tawmawite	$Ca_2(Al,Fe^{3+},Cr^{3+})_3O \cdot SiO_4 \cdot Si_2O_7 \cdot OH$
Hancockite	$(Ca,Pb,Sr)_2(Al,Fe^{3+})_3O \cdot SiO_4 \cdot Si_2O_7 \cdot OH$
Allanite	$(Ca,Ce,La)_2(Al,Fe^{3+},Fe^{2+})_3O \cdot SiO_4 \cdot Si_2O_7 \cdot OH$

Clinozoisite and epidote form a solid solution series from $Ca_2Al_3O \cdot SiO_4 \cdot Si_2O_7 \cdot OH$ to $Ca_2Fe^{3+}Al_2O \cdot SiO_4 \cdot Si_2O_7 \cdot OH$, the latter composition representing maximum $Fe^{3+} \rightleftharpoons Al$ substitution, and the separation of clinozoisite and epidote being defined by the change of optic sign. Substitutions of Mn^{3+} or Cr^{3+} for Al^{3+} are even more restricted; however, very minor amounts of Mn^{3+} produce the vivid color pleochroism characteristic of piemontite. The common epidote minerals are formed largely under conditions of low- to medium-grade metamorphism in association with Ca-poor plagioclase. At higher grades, epidote combines with albite to form a Ca-rich plagioclase. Epidote has been considered a metamorphic stress mineral, although its formation by hydrothermal processes is also well known. Directed pressure seems to favor the monoclinic polymorph and uniform pressure the orthorhombic form. Allanite is a minor accessory mineral in granites and certain pegmatites where it appears in association with other rare earth and radioactive minerals. The physical and optical properties of the end-member epidotes are summarized in Table 4-1.

TABLE 4-1. Optical Properties of the Epidote Group

n_β	n_α	n_γ	$n_\gamma - n_\alpha$	Mineral	System
1.688–1.711	1.685–1.707	1.697–1.725	0.005–0.020	Zoisite $Ca_2Al_3O \cdot SiO_4 \cdot Si_2O_7 \cdot OH$	Orthorhombic
1.707–1.725	1.703–1.715	1.709–1.734	0.004–0.012	Clinozoisite $Ca_2Al_3O \cdot SiO_4 \cdot Si_2O_7 \cdot OH$	Monoclinic $\angle\beta = 115°$
1.725–1.784	1.715–1.751	1.734–1.797	0.012–0.049	Epidote (Pistacite) $Ca_2FeAl_2O \cdot SiO_4 \cdot$ $Si_2O \cdot \cdot OH$	Monoclinic $\angle\beta = 115°$
1.730–1.807	1.725–1.794	1.750–1.832	0.025–0.082	Piemontite $Ca_2(Al,Fe,Mn)_3O \cdot SiO_4 \cdot$ $Si_2O_7 \cdot OH$	Monoclinic $\angle\beta = 115°$
1.700–1.815	1.690–1.791	1.706–1.828	0.013–0.036	Allanite $(Ca,Ce,La)_2(Al,Fe^{+3},Fe^{+2})_3O \cdot$ $SiO_4 \cdot Si_2O_7 \cdot OH$	Monoclinic $\angle\beta = 115°$

$2V$	Optic Sign	Cleavage	Optical Orientation	Color in Sections	Physical Properties
0°–60° $v \gtrless r$ strong	+	{010} perfect	$a = Z, b = Y, c = X$ or $a = Z, b = X, c = Y$	Colorless, pink	H = 6. Sp. Gr. = 3.15–3.36. Gray, green, brown, pink
14°–90° $v > r$ strong	+	{001} perfect	$b = Y$ $c \wedge X = -85°$ to $\theta°$ $a \wedge Z = -60°$ to $+25°$	Colorless	H = 6–7. Sp. Gr. = 3.3. Gray, yellow, green
90°–64° $r > v$ strong	−	{001} perfect	$b = Y$ $c \wedge X = 0°$ to $+15°$ $a \wedge Z = +25°$ to $+40°$	Colorless, pale yellow-green	H = 6–7. Sp. Gr. = 3.4. Olive green, yellow-green
50°–86° $r > v$ strong $v > r$ rare	+	{001} perfect	$b = Y$ $c \wedge X = +2°$ to $+9°$ $a \wedge Z = +27°$ to $+35°$	n_α = yellow n_β = red-violet n_γ = deep red	H = 6–6½. Sp. Gr. = 3.4–3.5. Deep maroon-black
40°–90° 90°–57° $r \gtrless v$ strong	− +	{001} imperfect, {100} poor, {110} poor	$b = Y$ $c \wedge X = +1°$ to $+47°$ $a \wedge Z = +26°$ to $+72°$	n_α = yellow n_β = pale brown n_γ = dark brown	H = 5–6½. Sp. Gr. = 3.4–4.2. Dark brown-black

The Clinozoisite-Epidote Series

$Ca_2(Al,Fe^{3+})_3O \cdot SiO_4 \cdot Si_2O_7 \cdot OH$

MONOCLINIC
$\angle\beta = 115°25'$

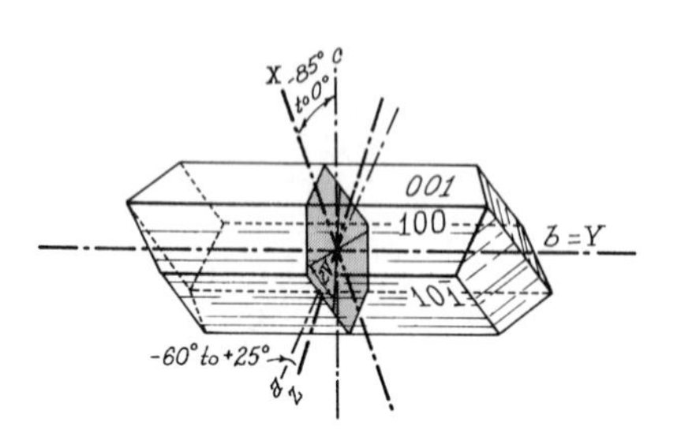

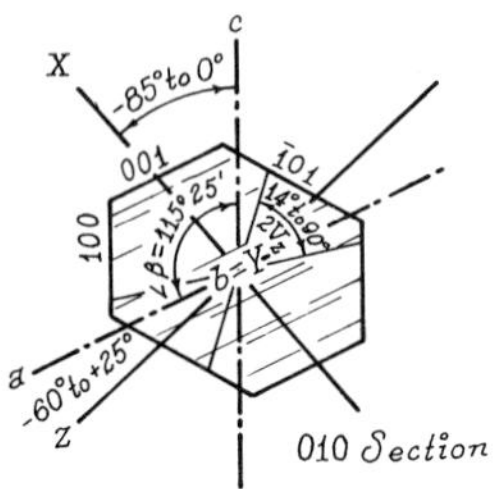

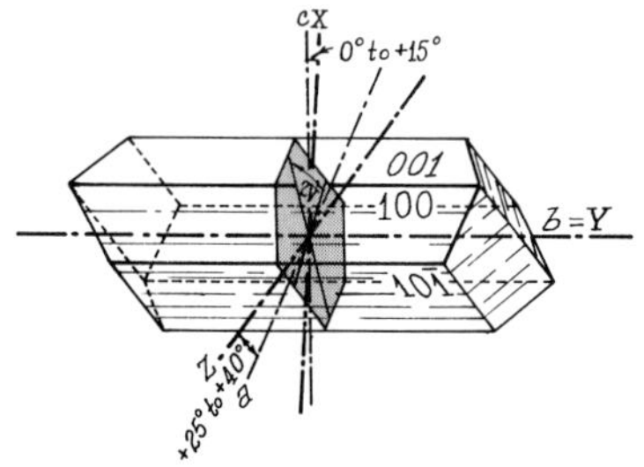

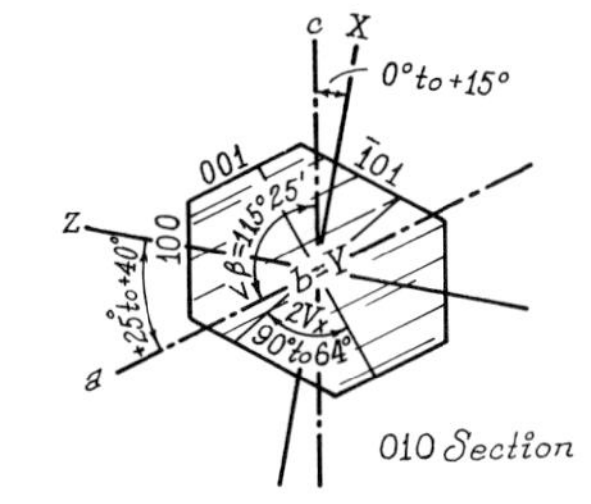

Clinozoisite

$n_\alpha = 1.703–1.715$*
$n_\beta = 1.707–1.725$
$n_\gamma = 1.709–1.734$
$n_\gamma - n_\alpha = 0.004–0.012$*
Biaxial positive
$2V_z = 14°–90°$
$b = Y, c \wedge X = -85°$ to $0°, a \wedge Z = -60°$ to $+25°$
$v > r$ strong
Colorless in standard section

Epidote

$n_\alpha = 1.715–1.751$*
$n_\beta = 1.725–1.784$
$n_\gamma = 1.734–1.797$
$n_\gamma - n_\alpha = 0.012–0.049$*
Biaxial negative
$2V_x = 90°–64°$
$b = Y, c \wedge X = 0°$ to $+15°, a \wedge Z = +25°$ to $+40°$
$r > v$ strong
Colorless to pale yellow-green in standard section, with weak pleochroism $Y > Z > X$

COMPOSITION. Continuous solid solution exists from the Fe-free end member $Ca_2Al_3O \cdot SiO_4 \cdot Si_2O_7 \cdot OH$ (clinozoisite) to about 35 percent of the nonexistent, Al-free end member $Ca_2Fe_3{}^{3+}O \cdot SiO_4 \cdot Si_2O_7 \cdot OH$; that is, to about $Ca_2Al_2Fe^{3+}O \cdot SiO_4 \cdot Si_2O_7 \cdot OH$ (epidote). The division between clinozoisite and epidote is arbitrarily made at the change in optic sign, which occurs at about 15 percent of the ferric end member (about 8 percent Fe_2O_3). The series is probably also continuous with piemontite, where Mn^{+3} replaces Fe^{3+} and Al^{3+} to about $Ca_2Mn_2{}^{3+}AlO \cdot SiO_4 \cdot Si_2O_7 \cdot OH$. Some Mn^{2+} may replace Ca, and minor Fe^{2+} is commonly reported. *Tawmawite* is a rare Cr-rich epidote.

* Refractive indices, birefringence, and $2V_x$ increase with substitution Fe^{3+} for Al^{3+} (Fig. 4-1).

PHYSICAL PROPERTIES. H = 7–6. Sp. Gr. = 3.21–3.49. Color, in hand sample, ranges from colorless to dark olive green in shades of yellow, green, or gray.

COLOR AND PLEOCHROISM. Clinozoisite is colorless in standard section. Increasing Fe^{3+} produces pale yellow-green in standard sections of epidote, which is pleochroic in yellow and green $Y > Z > X$. X = colorless, pale yellow, pale green; Y = yellow-green, brownish-green; and Z = colorless, pale yellow-green. The presence of even small amounts of Mn^{3+} results in reddish colors and strong pleochroism from deep pink or red to yellow, which is characteristic of piemontite.

FORM. Crystals elongate on b are common as columnar or acicular aggregates in parallel or radiated masses. Fine to coarse granular aggregates are also common.

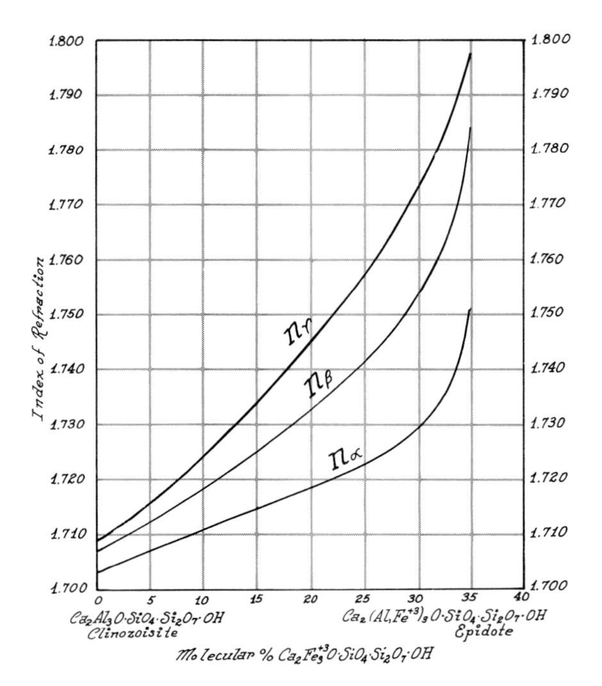

FIGURE 4-1. Variation of the indices of refraction in the clinozoisite-epidote series. (After Deer, Howie, Zussman, 1962)

CLEAVAGE. Perfect basal cleavage {001} largely controls fragment orientation. Cleavage on {100} is less perfect.

BIREFRINGENCE. Maximum birefringence, seen in {010} sections, increases with substitution Fe^{3+} for Al from 0.004 (clinozoisite) to 0.049 (epidote). Clinozoisite shows weak birefringence from 0.004 to about 0.012 and first-order interference colors of gray, white, and yellow; often anmalous blue or yellow-brown is seen, due to strong dispersion. Epidote shows mild to strong birefringence from about 0.012 to about 0.049 and brilliant interference colors, mostly of second order. Anomalous effects of strong dispersion are less obvious in epidote owing to strong interference colors.

TWINNING. Multiple normal twinning on {100} is moderately common.

ZONING. Composition zoning is very common and may be expressed as color zoning or drastic changes in optical variables (for example, $2V$, extinction angle, birefringence, index of refraction, and dispersion). The composition range may be considerable, and inner zones may be either more or less Fe-rich than outer zones.

INTERFERENCE FIGURE. By definition, clinozoisite is biaxial positive and epidote biaxial negative. Iron-free zoisite shows small $2V_z$ near 14°, which widens very rapidly on Z with increasing Fe^{3+}. $2V_z$ exceeds 90° to form epidote at a composition about 15 percent $Ca_2Fe_3{}^{3+}O \cdot SiO_4 \cdot Si_2O_7 \cdot OH$ and further converges on X to $2V_x$ = about 64° as the ferric end-member approaches 35 percent of the total composition.

Basal sections and cleavage plates yield figures symmetrical about the optic plane ranging from centered optic axis to obtuse bisectrix figures. Isogyres are distinct to diffuse, and isochromatic bands range from none to moderately abundant with increasing Fe^{3+}.

Optic axis dispersion is usually, but not always, strong, $v > r$ for clinozoisite and $r > v$ for epidote. Inclined bisectrix dispersion is often strong enough to reverse the optic axis dispersion fringes on the one isogyre that basal cleavage makes most accessible to view when viewing fragments.

OPTICAL ORIENTATION. Clinozoisite and epidote are monoclinic, with $b = Y$, and the optic plane is {010}. $c \wedge X$ varies widely, tending to be negative for clinozoisite (−85° to 0°) and positive for epidote (0° to +15°). For clinozoisite, $a \wedge Z = -60°$ to $+25°$; and for epidote, $a \wedge Z = +25°$ to $+40°$. Maximum extinction angle to cleavage in {010} sections ranges from 60° to 0°, usually less than 40°, and extinction to crystal elongation ($=b$) is parallel. Sign of elongation may be either positive or negative (length-slow or length-fast), for crystals elongate on $b = Y$. Cleavage traces usually show positive elongation (length-slow), as X is most nearly normal to cleavage planes.

DISTINGUISHING FEATURES. Clinozoisite (+) and epidote (−) are defined by optic sign; however, clinozoisite also shows lower birefringence and smaller refractive indices and lacks color. Zoisite is biaxial positive with extinction parallel to cleavage traces, and piemontite is highly colored and pleochroic in red and yellow.

Idocrase and melilite may show low-order, anomalous interference colors, but both are uniaxial, or nearly so, and idocrase and aluminous melilite are optically negative. Pyroxenes and amphiboles possess two directions of cleavage and lower (nonanomalous) birefringence than most epidote.

ALTERATION. Epidote and clinozoisite are relatively resistant to weathering and are themselves the products of hydrothermal alteration of Ca-feldspars. With high-grade metamorphism, epidote and clinozoisite disappear in favor of Ca-feldspars, pyroxene, and magnetite.

OCCURRENCE. Clinozoisite and epidote show similar paragenesis, although clinozoisite suggests a Fe-poor environment and is much less common. Epidote is commonly considered a metamorphic stress mineral and is very common in schists, slates, and phyllites of low- to medium-grade regional metamorphism. It appears with chlorite and actinolite in greenschists and is the associate of albite and hornblende in the albite-epidote-amphibolite facies. Epidote is to be expected in low- to medium-grade metamorphic assemblages deriving from limestone containing argillaceous impurities.

In metamorphic contact zones and zones of metasomatism in limestone hosts, epidote and clinozoisite appear in hornfels and similar rocks with grossularite, idocrase, anorthite, diopside, and calcite. Ca-metasomatism may introduce epidote in fissures and veins.

In igneous rocks, epidote may rarely crystallize in mafic rocks, or even granites, as a magmatic mineral, usually due to Ca-assimilation. It is more common as a product of hydrothermal introduction appearing in vugs and vesicles of basalts and other volcanics and in veins and fissures of granites, and it may even replace ferromagnesian minerals in granite.

As a product of hydrothermal alteration of Ca-feldspars and ferromagnesian silicates, granular epidote appears with zoisite or clinozoisite, albite, calcite, and possibly other fine-grained minerals as a possible constituent of "saussurite."

Epidote and clinozoisite are locally important as detrital grains in sediments.

REFERENCES: THE CLINOZOISITE-EPIDOTE SERIES

Deer, W. A., R. A. Howie, and J. Zussman. 1962. *Rock-Forming Minerals*. Vol. I: *Ortho- and Ring Silicates*. New York; Wiley.

Dollase, W. A. 1968. Refinement and comparison of the structures of zoisite and clinozoisite. *Amer. Mineral.*, *53*, 1882–1898.

Dollase, W. A. 1971. Refinement of the crystal structures of epidote, allanite, and hancockite. *Amer. Mineral.*, *56*, 467–474.

Kepezhinskas, K. B., and V. V. Khlestov. 1967. Correlation of compositions in minerals of the epidote group. *Dokl. Akad. Nauk SSSR*, *175*, 435–437.

Liou, J. G. 1973. Synthesis and stability relations of epidote, $Ca_2Al_2FeSi_3O_{12}(OH)$. *Jour. Petrol.*, *14*, 381–413.

Myer, G. H. 1965. X-ray determinative curve for epidote. *Amer. Jour. Sci.*, *263*, 78–86.

Raith, M. 1976. The aluminum-iron (III) epidote miscibility gap in a metamorphic profile through the penninic series of the Tavern Window, Austria. *Contrib. Mineral. Petrol.*, *57*, 99–117.

Strens, R. G. J. 1966. Properties of the Al-Fe-Mn epidotes. *Mineral. Mag.*, *35*, 928–944.

Zoisite

$Ca_2Al_3O{\cdot}SiO_4{\cdot}Si_2O_7{\cdot}OH$

ORTHORHOMBIC

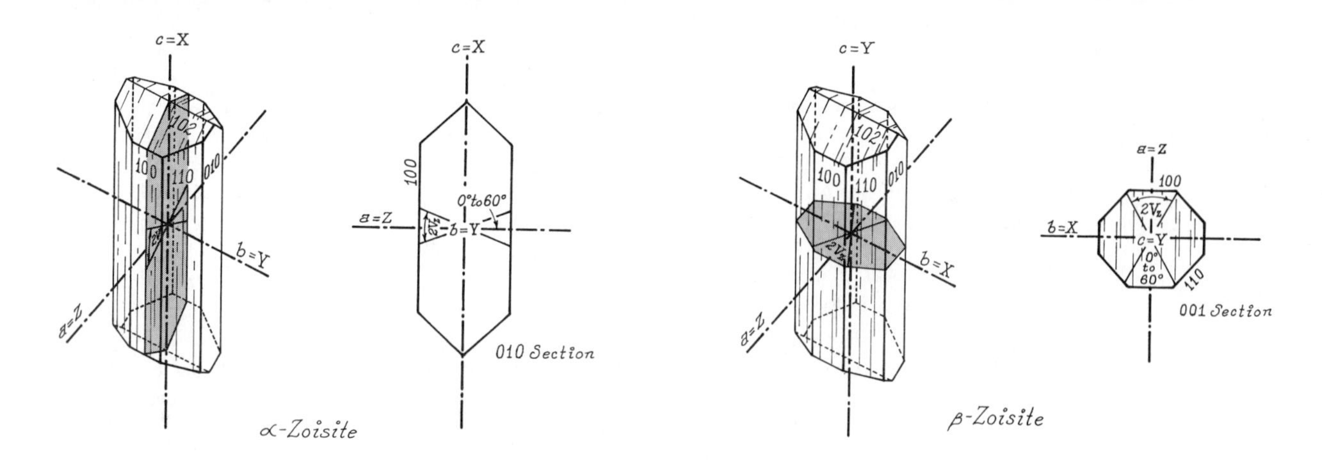

$n_\alpha = 1.685–1.707$*
$n_\beta = 1.688–1.711$
$n_\gamma = 1.697–1.725$
$n_\gamma - n_\alpha = 0.005–0.020$*
Biaxial positive, $2V_z = 0°–60°$*
$a = Z, b = Y, c = X$ (α-zoisite)
$a = Z, b = X, c = Y$ (β-zoisite)
$v > r$ strong (α-zoisite), $r > v$ strong (β-zoisite)
Colorless to pleochroic (thulite) in pink and yellow in standard section

COMPOSITION. Ionic substitution in the ideal zoisite composition $Ca_2Al_3O \cdot SiO_4 \cdot Si_2O_7 \cdot (OH)$ is very limited, as major substitution causes the monoclinic polymorph to form. Up to one-tenth of the available octahedral (Al^{3+}) sites may be occupied by Fe^{3+} and lesser amounts of Mn^{3+} or Cr^{3+} may be present. Iron-poor (Fe^{3+}) zoisites are designated α-zoisite, which shows different optical orientation from β-zoisite or Fe^{3+}-rich zoisites. Even very minor Mn^{3+} produces intense pink color and pleochroism, forming the poorly defined variety *thulite*. Minor Sr^{2+} and even less Ba^{2+} may replace Ca^{2+}.

PHYSICAL PROPERTIES. H = 6. Sp. Gr. = 3.15–3.36. Color, in hand sample, is usually gray, green, or brown.

COLOR AND PLEOCHROISM. Zoisite is most commonly colorless in standard section, although thulite is pleochroic $X > Y > Z$. X = deep pink, Y = pale pink, and Z = pale yellow.

FORM. Crystals, elongated on c, are columnar, bladed, or fibrous as parallel to radial aggregates. Granular masses are also common.

CLEAVAGE. Perfect {010} cleavage tends to orient fragments.

BIREFRINGENCE. Maximum birefringence, seen in {010} sections and on cleavage fragments (α-zoisite) or in basal {001} sections (β-zoisite), is weak to mild (0.005–0.020) showing interference colors to upper first order in standard section. Birefringence is greatest for Fe^{3+}-rich compositions and some thulite. Most zoisite shows weak birefringence and interference colors of first-order gray and white, often anomalous blue or yellow, owing to strong dispersion.

* Refraction indices and birefringence increase with Fe^{3+}, and $2V_z$ decreases to 0° and opens again with increasing Fe^{3+}.

TWINNING. Twinning is not common in zoisite, although it has been proposed that the orthorhombic polymorph results from the monoclinic structure by submicroscopic twinning on {100}.

ZONING. Composition zoning is common in zoisite, expressed as color zoning in thulite and as concentric birefringence zones and $2V_z$ variation in common colorless zoisite. α-zoisite may form a crystal core with β-zoisite forming the outer rim and $2V_z$ passing through 0° somewhere in the interior. Zoisite may be intergrown with monoclinic epidote.

INTERFERENCE FIGURE. Zoisite is biaxial positive; otherwise, the interference figure is highly variable with regard to orientation, $2V_z$, and dispersion. Fe^{3+}-free zoisite (α-zoisite) shows $2V_z$ near 30°, which closes to 0° and opens again to near 60° in the most Fe^{3+}-rich compositions (β-zoisite), and α and β forms are separated where $2V_z = 0°$ (uniaxial +). Cleavage fragments of α-zoisite yield flash figures and those of β-zoisite show the obtuse bisectrix. The acute bisectrix lies normal to {100} for both orientations. Dispersion is strong for all compositions, $v > r$ (α-zoisite) and $r > v$ (β-zoisite).

OPTICAL ORIENTATION. For iron-poor varieties (α-zoisite), the optic plane lies parallel to {010} with $a = Z$, $b = Y$, and $c = X$. With increasing Fe^{3+}, $2V$ closes to 0° on Z and opens again in the plane parallel to {001}, making this the optic plane for β-zoisite, with $a = Z$, $b = X$, and $c = Y$.

Extinction is parallel to cleavage, and cleavage elongation may be either positive or negative (α-zoisite) or positive, length-slow (β-zoisite).

DISTINGUISHING FEATURES. Elongated sections of high relief, parallel extinction, weak birefringence, often anomalous, and strong dispersion are characteristic of zoisite. Clinozoisite, epidote, and piemontite are monoclinic, showing inclined extinction in some sections. Piemontite is highly colored and pleochroic, as is thulite, and epidote is commonly pale yellow-green and shows higher birefringence than zoisite. Idocrase is normally uniaxial negative; however, it may show small $2V_x$. Melilite and apatite are uniaxial negative.

ALTERATION. Zoisite is quite resistant to weathering and is itself an alteration product of some feldspars. Zoisite is stable only in the lower grades of metamorphism and is incorporated into the Ca-feldspars at high temperatures.

OCCURRENCE. The most characteristic occurrence of zoisite is in schists and granulites of medium-grade regional metamorphism, where it appears with albite, calcite, quartz, garnet, hornblende, and biotite. Zoisite, in contrast to the monoclinic epidotes, is favored by nondirected pressure and may appear in contact zones with calcite, grossular, and idocrase. It is rarely found in high-pressure rocks such as glaucophane schist or eclogite. Zoisite tends to form from Mg-poor rocks containing silica, clay, and calcite.

In igneous rocks, zoisite appears rarely as a primary mineral in ultramafic rocks and veins, and thulite forms in pegmatites and hydrothermal veins.

Zoisite also forms as a hydrothermal alteration product of Ca-rich plagioclase where it appears in a fine-grained mass, called *saussurite*, with albite and other minerals (such as prehnite, orthoclase, chlorite, sericite, and zeolites).

REFERENCE: ZOISITE

Dollase, W. A. 1968. Refinement and comparison of the structures of zoisite and clinozoisite. *Amer. Mineral.*, *53*, 1882–1898.

Piemontite

$Ca_2(Al,Fe^{3+},Mn^{3+})_3O \cdot SiO_4 \cdot Si_2O_7 \cdot OH$

MONOCLINIC

$\angle\beta = 115°42'$

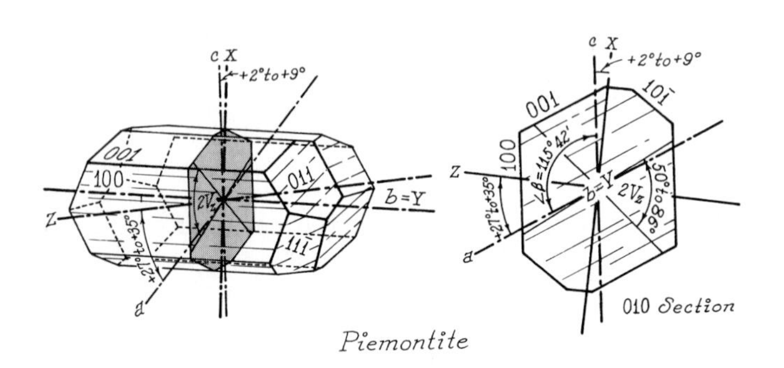

$n_\alpha = 1.725–1.794$*

$n_\beta = 1.730–1.807$

$n_\gamma = 1.750–1.832$

$n_\gamma - n_\alpha = 0.025–0.082$*

Biaxial positive, $2V_z = 50°–86°$

$b = Y$, $c \wedge X = +2°$ to $+9°$, $a \wedge Z = +27°$ to $+35°$

$r > v$ strong; rarely, $v > r$

Pink or red, in standard section, with strong pleochroism from deep red or violet to yellow, usually $Z > Y > X$ or $Z > X > Y$

* Refractive indices and birefringence increase rapidly with increasing Fe^{3+} for Al and less rapidly with Mn^{3+} for Al.

COMPOSITION. Piemontite probably forms complete solid solution with all compositions of the clinozoisite-epidote series and differs from it by the presence of manganese. Less than 1 percent Mn_2O_3 is sufficient to produce the pink-red color and strong pleochroism that are probably the best defining characteristics of piemontite. Mn^{3+} may replace two-thirds of the Al^{3+}, which is much more than the allowable replacement of Al^{3+} by Fe^{3+}. Minor Mn^{2+} may replace Ca^{2+}.

PHYSICAL PROPERTIES. H = $6–6\frac{1}{2}$. Sp. Gr. = 3.40–3.52. Color, in hand sample, is maroon or reddish brown to black.

COLOR AND PLEOCHROISM. Piemontite is very conspicuous in thin section, owing to deep color and strong pleochroism from red to yellow. X = pale yellow, pale orange, pink; Y = pale violet, red-violet; and Z = deep red, brownish red.

FORM. Columnar, bladed, or acicular crystals are elongated on b and form as fine to coarse, parallel to radiated aggregates. Crystal sections are roughly long rectangular, hexagonal, or octagonal.

CLEAVAGE. Perfect basal cleavage {001} commonly controls fragment orientation.

BIREFRINGENCE. Maximum birefringence, seen in {010} sections, ranges from moderate to strong (0.025–0.082). Interference colors are mostly brilliant colors of second and third order, often masked by intense mineral color.

TWINNING. Lamellar twinning on {100} is uncommon.

ZONING. Composition zoning is very common, expressed as obvious color changes and widely varying optical constants, which may reflect only minor composition changes.

INTERFERENCE FIGURE. Piemontite is biaxial positive, with moderate to large $2V_z$: 50° to 86°. Negative varieties are rare and are usually called *manganepidote*. Cleavage fragments and basal sections yield off-center optic axis figures that are symmetrical about the optic plane. Isochromatic rings are numerous but often obscured by inherent mineral color and strong dispersion $r > v$ (rarely $v > r$). Inclined bisectrix dispersion may be detectable.

OPTICAL ORIENTATION. Maximum extinction angle to cleavage traces is 27° to 35°, seen in {010} sections, but extinction to crystal elongation is parallel. Sign of elongation is positive or negative for crystals elongate on $b = Y$; however, cleavage traces are length-slow, as X lies nearly normal to cleavage.

DISTINGUISHING FEATURES. Strong pleochroism from deep red to yellow is usually sufficient to distinguish piemontite. Thulite shows parallel extinction, smaller $2V_z$, and lower refractive indices, and manganepidote is biaxial negative. Dumortierite crystals may be strongly colored and pleochroic in violet or red-violet, but they show parallel extinction and are biaxial negative with small $2V_x$.

ALTERATION. Piemontite is relatively resistant to weathering.

OCCURRENCE. The most common occurrence of piemontite is in schists of low-grade regional metamorphism in association with sericite, chlorite, quartz, and, rarely, with glaucophane. Piemontite may appear in manganese deposits of metasomatic or hydrothermal origin with quartz, calcite, and Mn-bearing minerals. In volcanic rocks, piemontite is found in vesicles or fissures and as spherulites or radiating aggregates in rhyolites, felsites, and andesites.

REFERENCES: PIEMONTITE

Dollase, W. A. 1969. Crystal structure and cation ordering of piemontite. *Amer. Mineral.*, *54*, 710–717.

Strens, R. G. J. 1966. Properties of the Al-Fe-Mn epidotes. *Mineral. Mag.*, *35*, 928–944.

Allanite

$(Ca,Ce,La)_2(Al,Fe^{3+},Fe^{2+})_3O \cdot SiO_4 \cdot Si_2O_7 \cdot OH$

MONOCLINIC
$\angle\beta = 115°$

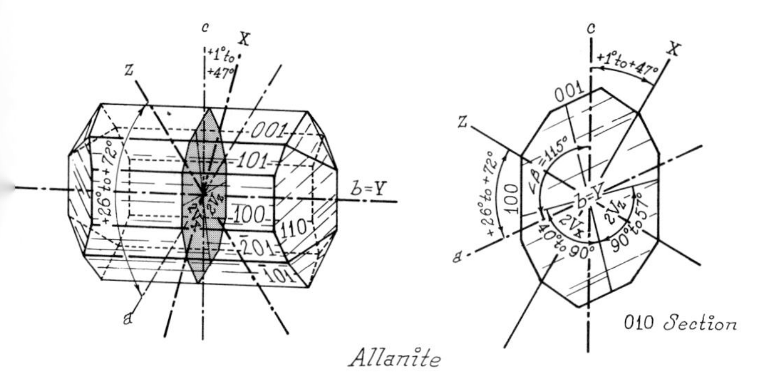

$n_\alpha = 1.690–1.791$*
$n_\beta = 1.700–1.815$
$n_\gamma = 1.706–1.828$
$n_\gamma - n_\alpha = 0.013–0.036$*
Biaxial negative, $2V_x = 40°–90°$
Biaxial positive, $2V_z = 90°–57°$
$b = Y, c \wedge X = +1°$ to $+47°, a \wedge Z = +26°$ to $+72°$*
$r > v$ strong, less common $v > r$
Pale to deep brown, yellow-brown, or green in standard section, with distinct pleochroism, usually $Z > Y > X$

* The optical properties of allanite are highly variable, and it is often metamict, appearing as a noncrystalline, isotropic "glass" with refractive index $n = 1.53–1.72$, lower values reflecting high water content. Refraction indices of crystalline allanite, as given, increase with the substitution of rare earths for Ca and Fe^{2+}, Fe^{3+}, Mn^{2+}, and Ti^{4+} for Al^{3+}. Two patterns of optical orientation are reported for allanite. The more common one, described here, places the optic plane parallel to {010}, although it has also been reported normal to {010}.

COMPOSITION. Allanite is epidote in which appreciable Ca^{2+} has been replaced by rare earth ions (such as Ce^{3+},La^{3+} and Y^{3+}) and an equal number of (Fe^{3+},Al^{3+}) ions have been replaced by bivalent ions (Fe^{2+}, Mn^{2+}) to maintain electrical balance. Minor amounts of many other elements may appear: Na^{+} and Mn^{2+} may replace some Ca^{2+}; Ti^{4+}, Sn^{4+}, V^{3+}, and Zn^{2+} may proxy for Al^{3+}; $PO_4{}^{3-}$ groups may replace some $SiO_4{}^{4-}$ units; F^{-} and O^{2-} may replace some $(OH)^{-}$; and a Be-rich allanite is known. There is usually sufficient U and Th to make allanite noticeably radioactive, to form pleochroic halos about tiny allanite crystals in biotite and other colored minerals, and to produce some degree of disarrangement in the allanite structure. Crystalline structure can often be partially or completely restored by carefully heating the metamict mineral.

PHYSICAL PROPERTIES. $H = 5$–$6\frac{1}{2}$. Sp. Gr. = 3.4–4.2 (to as low as 2.7 in metamict samples). Color, in hand sample, is dark brown to pitchy black.

COLOR AND PLEOCHROISM. In standard section, allanite is light to dark brown, rarely reddish brown, yellow-brown, or green and is distinctly pleochroic, usually $Z > Y > X$. X = colorless, light brown, yellow, or red-brown; Y = pale brown, yellow-brown, red-brown, pale green; and Z = dark brown, yellow-brown, red-brown, green.

FORM. Crystals are commonly euhedral as tablets on {100} or as acicular groups, elongate on *b*. Granular anhedra are also common. Allanite crystals are commonly overgrown by epidote or some other member of the epidote group.

CLEAVAGE. Cleavage is imperfect on {001} and poor on {100} and {110}.

BIREFRINGENCE. Maximum birefringence, seen in {010} sections, is mild to moderate (0.013–0.036). Interference colors, in standard section, range through brilliant second-order colors but are dimmed or obscured by deep mineral color.

TWINNING. Twinning sometimes forms on {100} but is not common.

ZONING. Composition zoning is commonly expressed as color zoning, usually dark core and lighter margins. Irregular, patchy color zones are also common. Marginal zones may be epidote, clinozoisite, or zoisite.

INTERFERENCE FIGURE. Allanite is usually biaxial negative with $2V_x = 40$ to $90°$; however, positive varieties are known, with $2V_z = 90°$ to $57°$. Dispersion is also highly variable, usually strong $r > v$, rarely $v > r$, with possible inclined bisectrix dispersion. Cleavage causes little preferred fragment orientation, and figures may be largely obscured by inherent color. Isotropic, and even uniaxial, areas are not uncommon in a given crystal.

OPTICAL ORIENTATION. Allanite is monoclinic, with the optic plane usually parallel {010}, and $b = Y$, $c \wedge X = +1°$ to $+47°$, $a \wedge Z = +26°$ to $+72°$. A second possible orientation is reported, in which the optic plane lies normal to {010}, and $b = Z$. Crystals elongate on $b = Y$ show parallel extinction and either positive or negative elongation.

DISTINGUISHING FEATURES. High relief, poor cleavage, uneven color, and pleochroism are characteristic of unaltered allanite. Brown amphibole or pyroxene show perfect prismatic cleavages and inclined extinction. Metamict allanite is isotropic and may not be separable optically from gadolinite, samarskite, pyrochlore, euxenite, and other dark brown isotropic minerals.

ALTERATION. Allanite is less stable than other members of the epidote group and weathers rather easily to a mixture of limonite, silica, and alumina. Radioactive constituents often cause allanite to become metamict as an isotropic, hydrated "glass."

OCCURRENCE. Allanite is a rather uncommon accessory mineral in granites, granodiorites, syenites, nepheline syenites, and similar light-colored, plutonic igneous rocks, where it is usually associated with epidote and iron-rich ferromagnesian silicates. Complex pegmatites commonly contain allanite in association with monazite, fergusonite, zircon, samarskite, and other rare earth or radioactive minerals; the association of allanite and cassiterite is worthy of note. In metamorphic rocks, allanite is a rare accessory of some schists and gneisses.

REFERENCES: ALLANITE

Dollase, W. A. 1971. Refinement of the crystal structures of epidote, allanite, and hancockite. *Amer. Mineral.*, *56*, 467–474.

Holmqvist, A. 1975. Low-2V allanite and magnesium-bearing allanite from the Kallmorberg Mine, Norberg, Sweden. *Geol. Foeren. Stockholm Foerh.*, *97*, 162–166.

Kepezhinskas, K. B., and V. V. Khlestov. 1967. Correlation of composition in minerals of the epidote group. *Dokl. Akad. Nauk SSSR*, *175*, 435–437.

Pumpellyite

$Ca_2Al_2(Mg,Fe^{2+},Fe^{3+},Al)(SiO_4)(Si_2O_7)(OH)_2(H_2O,OH)$

MONOCLINIC
$\angle\beta = 97°36'$

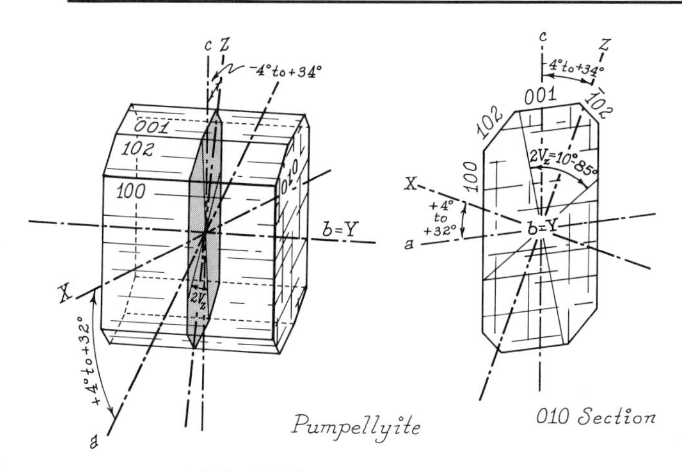

$n_\alpha = 1.674–1.748$*

$n_\beta = 1.675–1.754$

$n_\gamma = 1.688–1.764$

$n_\gamma - n_\alpha = 0.002–0.022$*

Biaxial positive, $2V_z = 10°–85°$ (rarely negative)

$b = Y, a \wedge X = +4°$ to $+32°$, $c \wedge Z = -4°$ to $+34°$

rarely: $b = Z, a \wedge X \simeq 40°, c \wedge Y \simeq 50°$

Strong $v > r$; rarely, $r > v$

Highly colored and strongly pleochroic in green and yellow, with absorption $Y > Z \geq X$

PHYSICAL PROPERTIES. H = 6. Sp. Gr. = 3.2–3.3. In hand sample, pumpellyite is green to brownish; luster is vitreous.

COLOR AND PLEOCHROISM. In standard section, pumpellyite is highly pleochroic in green and yellow $Y > Z \geq X$. X = colorless, pale greenish-yellow, pale blue-green; Y = blue-green, deep green; and Z = colorless, yellow, yellow-brown, red-brown. Optically negative varieties are usually brown.

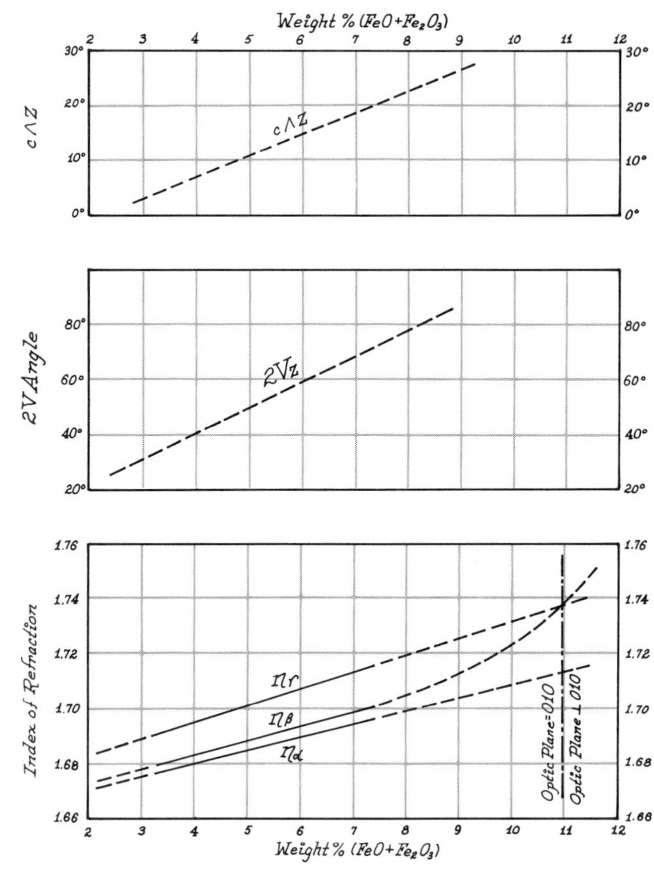

FIGURE 4-2. Variation of the indices of refraction. $2V$ angle and $c \wedge Z$ with FeO + Fe_2O_3 in pumpellyite. (After Deer, Howie, Zussman, 1962)

COMPOSITION AND STRUCTURE. Pumpellyite is a neso-sorosilicate containing both independent SiO_4^{4-} tetrahedra and $Si_2O_7^{6-}$ units cross-linking octahedral chains occupied by metals.

It is a complex, hydrated silicate with essential Ca and Al and major substitutions of Fe^{2+} for Mg^{2+} and Fe^{3+} for Al^{3+}. Minor Ti^{4+}, Mn^{2+}, or Na^+ may be present. Pumpellyite is related structurally and chemically to the epidote minerals.

* The optical constants of pumpellyite are highly variable. Indices of refraction and birefringence increase with increasing iron (Fe^{2+} or Fe^{3+})(see Fig. 4-2).

FORM. Acicular, fibrous, or bladed crystals of pumpellyite may show as radiated, subparallel, or random aggregates. Rosettes of needlelike crystals elongated on $b = Y$ are most common.

CLEAVAGE. Cleavage is distinct on {001}; less so on {100}. Crystals and fragments are usually elongated on $b = Y$.

BIREFRINGENCE. Maximum birefringence, usually seen in {010} sections, is highly variable, from as weak as 0.002 to about 0.022. Interference colors, in thin section, are usually first-order grays and white but may range to low second order in sections normal to crystal length. Colors are often anomalous blue-gray or yellow-brown due to strong dispersion.

TWINNING. Twins are common on {001} and {100}, yielding fourfold sectors with irregular composition planes.

ZONING. Outer crystal zones are commonly Fe-rich, showing deeper color, greater birefringence, and higher indices of refraction.

INTERFERENCE FIGURE. Pumpellyite is biaxial positive (negative varieties have been reported) with highly variable $2V_z$; values from near 0° to near 90° are reported. Sections parallel to {001} and fragments or crystals on the major cleavage {001} yield near-centered acute bisectrix figures symmetrical about the optic plane. Such figures show few, if any, isochromes and usually strong optic axis dispersion $v > r$. Rarely, when $b = Z$, dispersion is $r > v$. Weak inclined bisectrix dispersion is possible; crossed when $b = Z$.

OPTICAL ORIENTATION. Pumpellyite is monoclinic with highly variable, and somewhat uncertain, relationships between optical and crystallographic directions. Usually $b = Y$, $a \wedge X = +4°$ to $+32°$, and $c \wedge Z = -4°$ to $+34°$. Rarely, $b = Z$, $a \wedge X \simeq 40°$, and $c \wedge Y \simeq 50°$. Crystals and fragments elongated on b show parallel extinction and elongation, either positive or negative.

DISTINGUISHING FEATURES. Pumpellyite is most likely to be confused with members of the epidote group, although it normally shows deeper color. Epidote is optically negative (negative varieties of pumpellyite are brown); clinozoisite and zoisite are nearly colorless, and zoisite shows strictly parallel extinction in all orientations. Lawsonite is colorless has slightly lower indices of refraction, dispersion $r > v$, and strictly parallel extinction.

ALTERATION. Pumpellyite is stable at low temperatures and may, itself, be a product of deuteric-type alteration. It may be replaced by epidote and actinolite. At high temperatures, pumpellyite reacts to form anorthite, diopside, or hydrogrossular.

OCCURRENCE. Pumpellyite is a rather common mineral in rocks of low-grade regional metamorphism but is commonly overlooked in favor of the epidote minerals. It is a characteristic mineral of the glaucophane schist and greenschist facies and appears in schistose rocks with glaucophane, lawsonite, epidote, chlorite, albite, prehnite, axinite, zoisite, and other low-grade minerals.

Pumpellyite may form by hydrothermal action in low-temperature veins, amygdales, or altered mafic rocks. Amygdales in basalt, andesite, or spilitic basalt may contain pumpellyite in association with epidote, chlorite, analcite, prehnite, babbingtonite, albite, or quartz. It appears in quartz veins that cut schists, basic volcanics, or intrusives. Pumpellyite forms by deuteric or hydrothermal action in highly altered mafic rocks (such as diabase, spilites, greenstones, and hornfelses) at the expense of plagioclase or biotite. It may be quite common in the heavy mineral fraction of detrital sediments but is usually dismissed as epidote.

REFERENCES: PUMPELLYITE

Bloxam, T. W. 1958. Pumpellyite from South Ayrshire. *Mineral. Mag.*, *31*, 811–813.

Coombs, D. S. 1953. The pumpellyite mineral series. *Mineral. Mag.*, *30*, 113–135.

Deer, W. A., R. A. Howie, and J. Zussman. 1962. *Rock-Forming Minerals*. Vol. 1: *Ortho- and Ring Silicates*. New York: Wiley.

De Villiers, J. E. 1941. Optical properties and crystallography of zoned pumpellyite from the Witwatersrand. *Amer. Mineral.*, *26*, 237–246.

Galli, E. 1969. Crystal structure of pumpellyite. *Acta Crystallogr.*, *B25*, 2276–2281.

Gottardi, G. 1965. Die Kristallstruktur von Pumpellyite. *Tschermaks Mineral. Petrogr. Mitt., Series 3*, *10* (Machalschki Vol.), 115–119.

Lupanova, N. P., and V. T. Kurdyavtsev. 1961. The pumpellyite formula. *Dokl. Akad. Nauk SSSR*, *141*, 1457–1460.

Passaglia, E., and G. Gottardi. 1973. Crystal chemistry and nomenclature of pumpellyites and julgoldites. *Can. Mineral.*, *12*, 219–223.

Lawsonite $CaAl_2Si_2O_7(OH)_2{\cdot}H_2O$ ORTHORHOMBIC

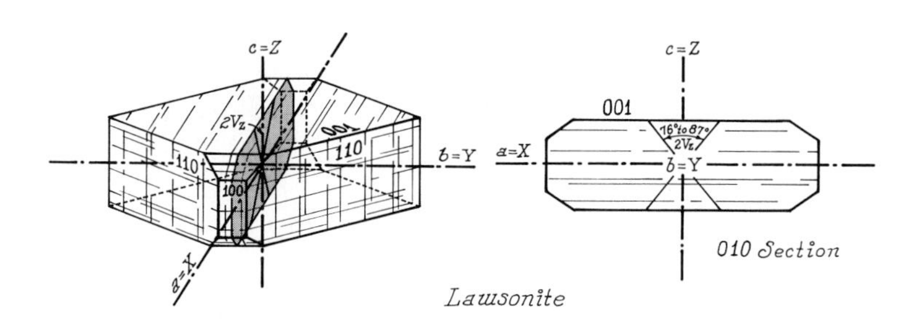

Lawsonite

$n_\alpha = 1.665$
$n_\beta = 1.672–1.676$
$n_\gamma = 1.684–1.686$
$n_\gamma - n_\alpha = 0.019–0.021$
Biaxial positive, $2V_z = 76°–87°$
$a = X, b = Y, c = Z$
Strong $r > v$
Usually colorless; rarely, very pale bluish green in standard section, with weak pleochroism in blue and yellow $X > Y > Z$

COMPOSITION AND STRUCTURE. The structure of lawsonite contains chains of edge-sharing $Al(O,OH)_6$ octahedra parallel to c. The chains are united laterally by Si_2O_7 double tetrahedron units that share corners with chain octahedra. Calcium ions and H_2O molecules fill independent voids in the structure and satisfy electrostatic neutrality.

The composition of lawsonite is near the ideal composition $CaAl_2Si_2O_7(OH)_2{\cdot}H_2O$. Minor foreign elements (such as Fe^{3+}, Ti^{4+}, Mg^{2+}, and Na^+) may replace Al^{3+} and Ca^{2+}.

PHYSICAL PROPERTIES. H = 7–8.Sp.Gr. $\simeq$ 3.1. Hand samples are colorless to bluish green or blue-gray.

COLOR AND PLEOCHROISM. Lawsonite is normally colorless in standard section; however, fragments and thick sections may show pale color and pleochroism from blue to yellow $X > Y > Z$. X = pale blue, pale yellow-brown; Y = pale yellow, bluish-green; and Z = colorless, pale yellow.

FORM. Crystals are commonly euhedral and usually tabular on {001}; rarely, prismatic with rhombic cross section and elongated on c.

CLEAVAGE. Perfect cleavages on {001} and {010} intersect at right angles, and imperfect prismatic cleavages {110} intersect at about 67°.

BIREFRINGENCE. Maximum birefringence, seen in {010} sections, is mild (0.019–0.021). Interference colors, in standard section, are mostly first-order white and yellow ranging through first-order red.

TWINNING. Simple or multiple twinning on {110} is common parallel to one or both prismatic directions.

ZONING. Zoning is not reported in lawsonite.

INTERFERENCE FIGURE. Lawsonite is biaxial positive, with large $2V_z = 76°$ to $87°$ and strong dispersion $r > v$. The optic plane is {010}, and the acute bisectrix *Z* lies normal to basal cleavage. Cleavage plates on {010} show flash figures.

OPTICAL ORIENTATION. Most cleavage fragments are elongated on *a* ({001} and {010} cleavages) to show parallel extinction and negative elongation (length-fast). Crystals elongate on $c = Z$ are length-slow, with parallel extinction.

DISTINCTIVE FEATURES. Lawsonite is most likely to be mistaken for some member of the epidote group. Zoisite shows smaller birefringence and $2V_z$; clinozoisite also shows low birefringence, inclined extinction to cleavages, and dispersion $v > r$; and epidote is biaxial negative with inclined extinction. Lawsonite does not exhibit anomalous birefringence, which is common in the epidote minerals. Andalusite is biaxial negative, with weak birefringence; scapolite is uniaxial negative, with low relief; and prehnite shows moderately high birefringence.

ALTERATION. Lawsonite is relatively stable at low temperatures and is itself an alteration product of plagioclase. It is commonly replaced by pumpellyite and returns to plagioclase at high temperatures.

OCCURRENCE. Lawsonite is derived from plagioclase by low-temperature metamorphism or hydrothermal alteration. It is especially characteristic of glaucophane schists and is very commonly associated with pumpellyite. As lawsonite is derived from the Ca-feldspar composition, it is commonly associated with Na-rich minerals formed from residual Na-feldspar (for example, albite, glaucophane, aegirine, and jadite). Lawsonite is formed by the alteration of gabbro, diorite, and similar rocks, along with the epidote minerals, calcite and albite.

REFERENCES: LAWSONITE

Deer, W. A., R. A. Howie, and J. Zussman. 1962. *Rock-Forming Minerals*. Vol. 1: *Ortho- and Ring-Silicates*. New York: Wiley.

Fry, N. 1973. Lawsonite pseudomorphed in Tauern greenschist. *Mineral. Mag.*, *39*, 121–122.

Romanova, I. M., and N. V. Belov. 1960. False symmetry in lawsonite structure. *Kristallografiya*, *5*, 215–217.

THE MELILITE GROUP

The Gehlenite-Åkermanite Series (Melilites)

$(Ca,Na)_2(Mg,Al)(Si,Al)_2O_7$ TETRAGONAL

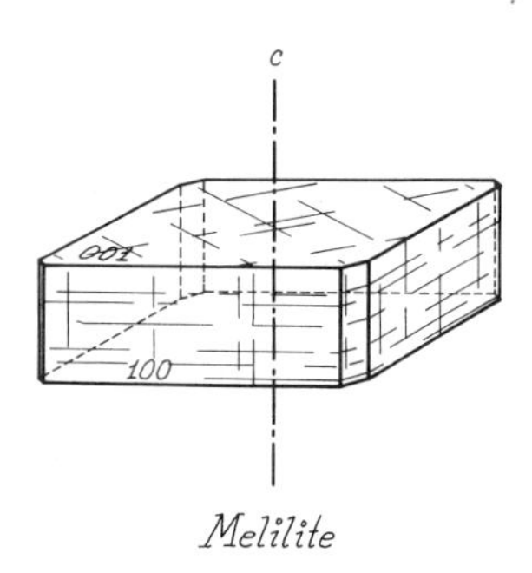

Melilite

Gehlenite	**Åkermanite**
$Ca_2MgSi_2O_7$	$Ca_2Al(SiAl)O_7$
$n_\omega = 1.670–1.651$*	$n_\omega = 1.651–1.632$*
$n_\varepsilon = 1.658–1.651$	$n_\varepsilon = 1.651–1.640$
$n_\omega - n_\varepsilon = 0.012–0.000$*	$n_\varepsilon - n_\omega = 0.000–0.008$*
Uniaxial negative	Uniaxial positive

Colorless to pale yellow in thin section. Weak pleochroism $O > E$

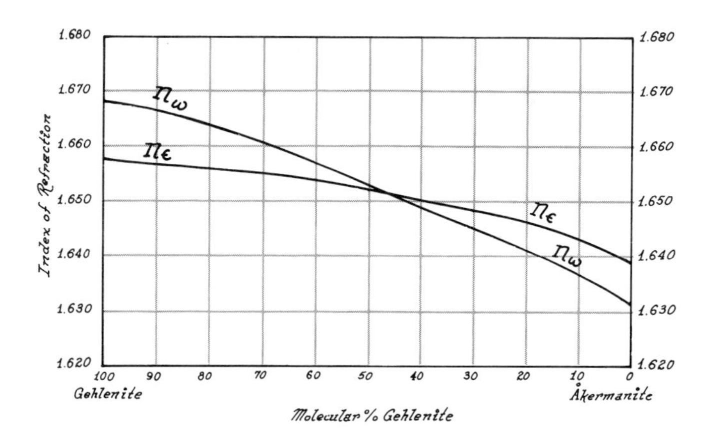

FIGURE 4-3. Variation of the indices of refraction in the gehlenite-åkermanite (melilite) series. (Nurse and Midgley, Studies in the melitite solid solutions, *Jour. Iron and Steel Inst.*, *174*, p. 121, 1953)

* Both principal indices of refraction decrease from pure gehlenite to pure åkermanite; n_ω decreases at a faster rate (Fig. 4-3). For gehlenite, $n_\omega > n_\varepsilon$, and gehlenite is optically negative; at about 54% Ak, $n_\omega = n_\varepsilon$, and the mineral is isotropic; and for åkermanite $n_\varepsilon > n_\omega$, and åkermanite is optically positive. Refractive indices decrease with increasing Na and increase with Fe. Birefringence is negative and almost 0.012, for pure gehlenite; it decreases to zero at about 54% Ak and becomes positive, increasing to 0.008 at pure åkermanite.

		n_ω	n_ε
Gehlenite	$Ca_2Al(SiAl)O_7$	1.669	1.659
Åkermanite	$Ca_2MgSi_2O_7$	1.632	1.639
Iron gehlenite	$Ca_2Fe^{3+}(SiAl)O_7$	1.726	1.723
Iron åkermanite	$Ca_2Fe^{2+}Si_2O_7$	1.690	1.673
Sodium melilite	$NaCaAlSi_2O_7$	1.580	1.575

COMPOSITION AND STRUCTURE. The melilites are sorosilicates containing Si_2O_7 groups of twin SiO_4 tetrahedra united by a single bridging oxygen. Both Al^{3+} and Mg^{2+} ions, however, are in tetrahedral coordination, and all tetrahedra are bound in sheets to yield the properties of a sheet silicate (phyllosilicate).

The melilite solid solution series is complete from pure gehlenite $Ca_2Al(SiAl)O_7$ to pure åkermanite $Ca_2MgSi_2O_7$, although pure end members are very rare. Much sodium and minor potassium commonly replaces calcium as $Na^+ + Si^{4+} \rightleftharpoons Ca^{2+} + Al^{3+}$ or $Na^+ + Al^{3+} \rightleftharpoons Ca^{2+} + Mg^{2+}$ and synthetic sodium melilites are known. Most melilites contain some Fe; $Fe^{2+} \rightleftharpoons Mg^{2+}$ or $Fe^{3+} \rightleftharpoons Al^{3+}$. A few rare melilites contain Zn (*hardystonite*), and some show a significant water content, probably as adsorbed H_2O.

PHYSICAL PROPERTIES. H = 5–6. Sp. Gr. = 3.04–2.94. Hand sample color is colorless or honey yellow to brown or gray-green.

COLOR AND PLEOCHROISM. Melilites are usually colorless as fragments or in thin section, although thick fragments or sections may show pleochroism in yellow or brown, $O > E$. O = deep golden yellow; E = colorless to pale yellow.

FORM. Melilite commonly appears as subhedral to euhedral crystals tabular on {001}. Sections are roughly rectangular, with the short dimension parallel to the *c*-axis.

CLEAVAGE. Basal cleavage {001} is indistinct, and very poor cleavage is reported parallel to {110}.

BIREFRINGENCE. Birefringence is usually weak (0.000–0.005) and interference colors are commonly anomalous, showing Berlin blue in place of first-order gray. Maximum birefringence is seen in prismatic sections; it is greatest for end-member compositions (gehlenite, 0.012, and åkermanite, 0.008) and decreases to zero (isotropic) at about 54 percent Ak. Sodium substitution has little effect on birefringence, but iron substitution raises the birefringence.

TWINNING. Twinning is not reported.

ZONING. Oscillatory composition zoning may show in color or birefringence, and adjacent zones may even reverse optic sign. Melilite crystals commonly contain rodlike inclusions aligned parallel to the *c*-axis (peg structure).

INTERFERENCE FIGURE. Weak birefringence leads to poor interference figures showing broad, diffuse isogyres on a gray field.

OPTICAL ORIENTATION. Extinction is parallel to cleavage or rectangular sections, and the ordinary wave commonly parallels cleavage and crystal elongation. The ordinary wave may be either slow (length-slow or positive elongation), for gehlenite compositions, or fast (length-fast or negative elongation), for åkermanite compositions.

DISTINCTIVE FEATURES. Weak, anomalous birefringence and limited occurrence are distinctive. Zoisite is biaxial, and idocrase has higher indices of refraction than melilite. Apatite may show hexagonal outline.

ALTERATION. Melilite commonly alters to brown, fibrous masses of cebollite $Ca_5Al_2Si_3O_{14}(OH)_2$, juanite, or prehnite. Calcite, zeolites, melanite garnet, perovskite, idocrase and diopside are all reported as alteration products of melilite. High-grade thermal metamorphism of carbonate mineral suites may form merwinite at the expense of melilite.

OCCURRENCE. Melilite is formed in high-temperature, low-pressure metamorphic zones where subsilicic magmas contact carbonate sediments. It is an index mineral of the sanidinite facies and appears in association with diopside, calcite, wollastonite, merwinite, larnite, spurrite, and other calc-silicates.

In igneous rocks, melilite implies the assimilation of carbonate sediments by silica-poor magmas. Melilite is most common in volcanics such as melilite basalt, leucitite, and melilite tuff. It appears largely as grains in a fine groundmass, but it may form euhedral phenocrysts in association with olivine, augite, nepheline, perovskite, leucite, and analcime. A few rare intrusive rocks (such as uncompahgrite and okaite) may contain 50 percent or more melilite with less olivine, monticellite, labradorite, aegirine, titanaugite, nepheline, biotite, haüynite, and perovskite.

REFERENCES: THE GEHLENITE-ÅKERMANITE SERIES

Berman, H. 1929. Composition of the melilite group. *Amer. Mineral.*, *14*, 389–407.

Buddington, A. F. 1922. On some natural and synthetic metilite. *Amer. Jour. Sci.*, Series 5, *3*, 35–87.

Deer, W. A., R. A. Howie, and J. Zussman. 1962. *Rock-Forming Minerals.* Vol. 1:*Ortho- and Ring Silicates.* New York: Wiley.

Ferguson, J. B., and A. F. Buddington. 1920. The binary system åkermanite-gehlenite, *Amer. Jour. Sci.*, Series 4, *50*, 131–140.

Goldsmith, J. R. 1948. Some melilite solid solutions. *Jour. Geol.*, *56*, 437–447.

Louisnathan, S. J. 1969. Refinement of the crystal structure of hardystonite. *Z. Kristallogr.*, *130*, 427–437.
Louisnathan, S. J. 1970. Crystal structure of synthetic soda melilite, $CaNaAlSi_2O_7$. *Z. Kristallogr.*, *131*, 314–321.
Louisnathan, S. J. 1971. Refinement of the crystal structure of a natural gehlenite, $Ca_2Al(Al,Si)_2O_7$. *Can. Mineral.*, *10*, 822–837.
Nurse, R. W., and H. G. Midgley. 1953. Studies in the melilite solid solutions. *Jour. Iron and Steel Inst.*, *174*, 121–131.
Plache, C. 1937. The minerals of Franklin and Sterling Hill, Susez County, New Jersey. *U.S. Geol. Surv., Prof. Paper*, *180*.
Sahama, T. G. 1967. Iron content of melilite. *Bull. Comm. Geol. Finlande*, no. 229, 17–28.

OTHER SOROSILICATES

Idocrase (Vesuviante)

$Ca_{10}(Mg,Fe^{2+})_2Al_4(Si_2O_7)_2(SiO_4)_5(OH,F)_4$ TETRAGONAL

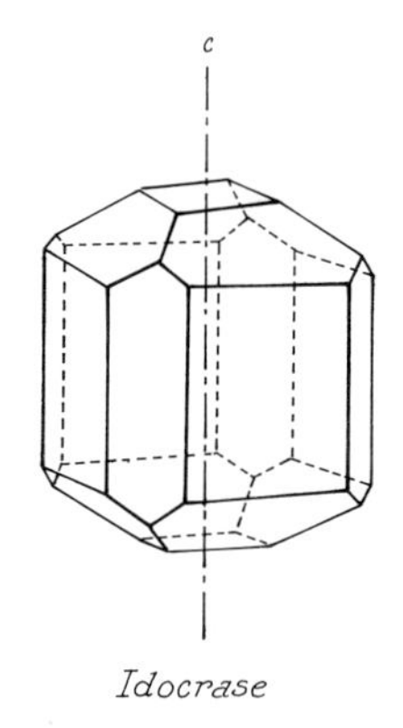

Idocrase

$n_\omega = 1.702–1.752$*
$n_\varepsilon = 1.698–1.746$
$n_\omega - n_\varepsilon = 0.001–0.012$*
Uniaxial negative
Colorless to pale green or brown in section
Pleochroism is very weak $O > E$

COMPOSITION AND STRUCTURE. The crystal structure of idocrase closely resembles that of garnet. It contains both independent silicate tetrahedra (SiO_4), like garnet, and sorosilicate units of two tetrahedra (Si_2O_7). These are united by Al^{3+} ions in one site of sixfold (octahedral) coordination, Mg^{2+} or Fe^{2+} ions in a second octahedral site, and Ca^{2+} ions in sites of eightfold coordination.

The composition of idocrase is highly variable. Calcium may be partially replaced by Na^+ or K^+, (Mg^{2+},Fe^{2+}) by Mn^{2+} or Zn^{2+}, and Al^{3+} by Fe^{3+} or Ti^{4+}. A Be-rich variety (*viluite*) is commonly biaxial, and Cu (*cyprine*) and Cr varieties are known.

PHYSICAL PROPERTIES. H = 6–7. Sp. Gr. = 3.33–3.45. Color, in hand sample, is commonly yellow-green to green-brown or brown; rarely, yellow, red, or blue.

COLOR AND PLEOCHROISM. Idocrase is commonly colorless in standard section but may be pale green or brown, with very weak pleochroism. Fragments may be pale green, brown, red, or blue with noticeable pleochroism $O > E$. O = pale brown, yellow, green, peach-pink, colorless; E = pale yellow-brown, colorless, yellow-green, deep blue.

FORM. Idocrase commonly forms euhedral tetragonal crystals. Prisms and pyramids of the same order commonly appear, and either form may predominate. It is also commonly massive as granular or fibrous aggregates.

CLEAVAGE. Poor cleavages {110} or {001} are seldom apparent.

BIREFRINGENCE. Maximum birefringence, in prismatic sections, is weak, usually about 0.005. Interference colors are no more than dark first-order gray and are commonly anomalous as deep indigo (Berlin) blue or dull olive-yellow.

* Refractive indices increase with increasing Ti^{4+} and Fe^{2+}. Birefringence decreases with increasing $(OH)^-$.

TWINNING. Twinning is not common, but basal sections of biaxial varieties are often divided as four diagonal sectors.

ZONING. Color zoning and uneven color distribution are very common.

INTERFERENCE FIGURE. Commonly uniaxial negative. Isogyres may separate with stage rotation, indicating a biaxial negative nature, and $2V_x$ may open to 65°. Some idocrase is optically positive, uniaxial or biaxial, and other varieties are sensibly isotropic.

DISTINGUISHING FEATURES. Idocrase is a high-relief, uniaxial negative mineral with very weak birefringence, often showing anomalous Berlin blue. Tetragonal outline and color zoning may be distinctive. Zoisite and clinozoisite may show similar relief and weak anomalous interference colors, but they show cleavage and are biaxial positive. Andalusite is biaxial negative, with large $2V_x$. It has lower refraction indices and higher birefringence than most idocrase. Melilite may be uniaxial negative, with weak anomalous birefringence. It has lower refractive indices and different paragenesis than vesuvianite. Grossular and hydrogrossular may closely resemble idocrase and may require more than optical means for distinction.

ALTERATION. Idocrase is stable over a wide range of conditions, and its alteration products are not defined. Chlorites, serpentine, epidote, and calcite are logical alteration products.

OCCURRENCE. The characteristic occurrence of idocrase is in limestone contact zones, where it appears in marble, tactite, skarn, and calc-silicate schist in association with calcite, Ca-garnets, wollastonite, diopside, epidote, and other calc-silicate minerals. In igneous rocks, idocrase may form by Ca-metasomatism in veins or cavities in gabbro and ultramafic rocks, in nepheline syenite, and in granite pegmatite.

REFERENCES: IDOCRASE

Isetti, G. 1961. Relations between birefringence, dispersion, and pleochroism of some minerals. *Period. Mineral.*, *30*, 139–163.

Ito, J., and J. Arem 1970. Idocrase: synthesis, phase relations, and crystal chemistry. *Amer. Mineral.*, *55*, 880–912.

Manning, P. G. 1975. Charge-transfer processes and the origin of color and pleochroism in some titanium-rich vesuvianites. *Can. Mineral.*, *13*, 110–116.

Rucklidge, J. C., V. Kocman, S. H. Whitlow, and E. J. Gabe. 1975. Crystal structures of three Canadian vesuvianites. *Can. Mineral.*, *13*, 15–21.

Hemimorphite (Calamine)

$Zn_4Si_2O_7(OH)_2 \cdot H_2O$ ORTHORHOMBIC

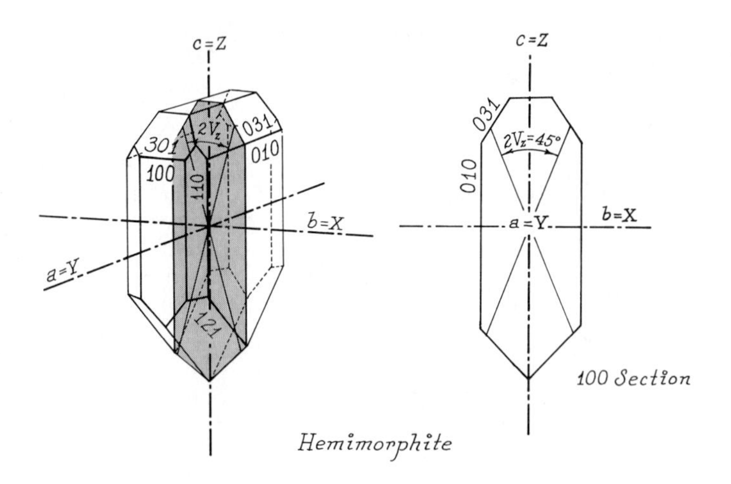

$n_\alpha = 1.611–1.617$
$n_\beta = 1.614–1.620$
$n_\gamma = 1.632–1.639$
$n_\gamma - n_\alpha = 0.022$
Biaxial positive, $2V_z = 44°–47°$
$a = Y, b = X, c = Z$
$r > v$ strong
Colorless in standard section

COMPOSITION AND STRUCTURE. Most analyses of hemimorphite report it to be a rather pure hydrated zinc silicate, $Zn_4Si_2O_7 \cdot (OH)_2 \cdot H_2O$. Trace amounts of ferric iron, copper, and lead are often reported but quite possibly represent surface stains or minor inclusions.

Hemimorphite is one of the rare true sorosilicates containing Si_2O_7 groups. Its structure is best considered, however, as (010) sheets of Zn—O tetrahedra and Si—O tetrahedra; the sheets are linked by Si—O—Si bridging oxygens.

PHYSICAL PROPERTIES. $H = 4\frac{1}{2}$–5. Sp. Gr. = 3.40–3.50. Color, in hand sample, is usually transparent colorless or white; less commonly, yellow, brown (iron-stained), gray, pale green, or blue.

COLOR AND PLEOCHROISM. Colorless in thin section or as fragments.

FORM. Hemimorphite usually appears as fanlike or sheaflike aggregates of thin, tabular, vertically striated crystals. Other aggregates may form stalactitic, mammillary, granular, or fibrous masses.

CLEAVAGE. {110} perfect, {101} poor, and {001} in traces.

BIREFRINGENCE. Moderate birefringence (0.022) yields bright first-order and low second-order interference colors. These may be somewhat anomalous, showing deficiency of yellow and orange colors.

TWINNING. Twinning on {001} is reported as uncommon.

INTERFERENCE FIGURE. Strong optic axis dispersion $r > v$ is obvious in bisectrix figures. Isochromatic bands may show an anomalous color progression lacking yellow, orange, and perhaps green. Cleavage fragments {110} yield off-center flash or obtuse bisectrix figures.

OPTICAL ORIENTATION. Cleavage fragments, elongated on $c = Z$, are length-slow (positive elongation) and show parallel extinction. Basal sections {001} show symmetrical extinction to {110} cleavages, and random sections may show inclined extinction to cleavage traces.

DISTINGUISHING FEATURES. Unique occurrence, strong optic axis dispersion ($r > v$), and anomalous interference colors may be distinctive. Both smithsonite and willemite are uniaxial; smithsonite is a carbonate with extreme birefringence, and willemite has much higher indices of refraction than hemimorphite, as do the secondary lead minerals.

ALTERATION. Hemimorphite itself is a secondary mineral formed by the alteration of primary zinc minerals (sphalerite). Under rare conditions, it may decompose to willemite plus water.

OCCURRENCE. Hemimorphite is a secondary ore mineral found in the oxide zones of ore deposits containing zinc and associated with stratified carbonate rocks. Hemimorphite is usually associated with remnants of the primary sulfides of zinc, lead, copper, and iron and with their secondary derivatives (smithsonite, cerussite, anglesite, aurichalcite, malachite, azurite, cuprite, limonite, hematite, and so on). Hemimorphite is a very rare mineral in granite pegmatites.

REFERENCES: HEMIMORPHITE

Dolter, C. 1914. *Handbuch der Mineralchemie.* 1 Dresden and Leipzig: Verlag von Theodor Steinkopff, p. 787.

Kostov, I. 1968. *Mineralogy.* London: Oliver and Boyd, pp. 324–325.

McDonald, W. S., and D. W. J. Cruickshank. 1967. Refinement of the structure of hemimorphite. *Z. Kristallogr., 124,* 180–191.

CYCLOSILICATES

Beryl

$Be_3Al_2(SiO_3)_6$ HEXAGONAL

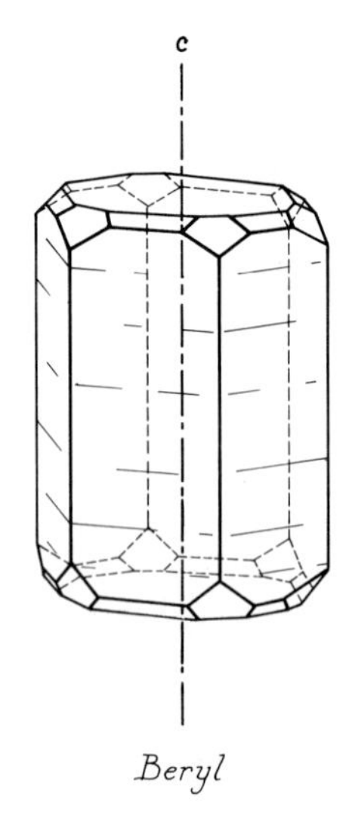

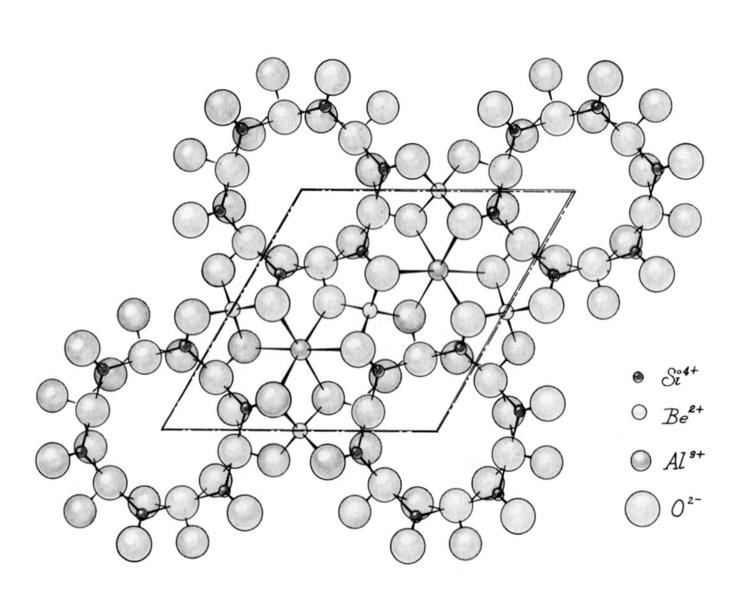

FIGURE 4-4. The structure of beryl. Hexagonal silicate rings stacked parallel to the *c*-axis normal to the page are held together by Al^{3+} and the Be^{2+} in octahedral and tetrahedral coordination respectively.

$n_\omega = 1.568–1.602$*
$n_\varepsilon = 1.563–1.594$
$n_\omega - n_\varepsilon = 0.004–0.008$*
Uniaxial negative
Colorless in thin section

COMPOSITION AND STRUCTURE. The beryl structure contains hexagonal rings of six SiO_4 tetrahedra, with the rings stacked vertically such that the hexad symmetry axis becomes the *c* crystallographic axis (Fig. 4-4). Adjacent stacks and successive rings in each stack are joined by Be^{2+} and Al^{3+} cations in four- and sixfold coordination between the stacks of hexagonal rings. Two of four O^{2-} ions in each silicate tetrahedron are bridging oxygens shared by two Si^{4+} cations, yielding the Si:O ratio of 1:3 characteristic of cyclosilicates. If, however, Be^{2+} in tetrahedral coordination is considered a substitute for Si^{4+}, then all oxygen ions are bridging, and beryl is a framework silicate (tektosilicate) $Al_2(Si_6Be_3)O_{18}$. Beryl shares many of the properties of quartz and similar tektosilicates.

The most common impurities in beryl are alkali ions (Li^+, Na^+, K^+, and Cs^+), which probably balance a charge deficiency created when extra Be^{2+} substitutes for Al^{3+} in sixfold coordination. Some small alkali ions (Li^+ or Na^+) may occupy octahedral Al^{3+} sites; most alkali ions, however, especially large ones, must lie in wide tunnels through the centers of hexagonal rings. These open channels may also

* Indices of refraction and birefringence increase with increasing alkali impurities.

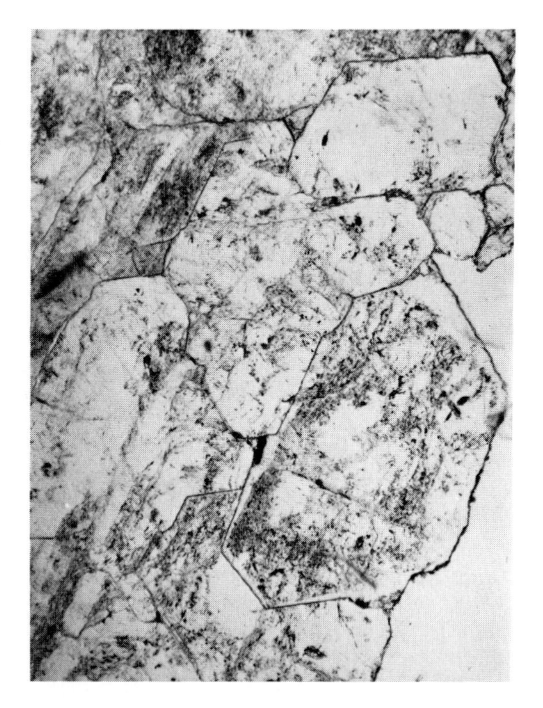

FIGURE 4-5. Beryl crystals in a granitic igneous rock may show crude hexagonal sections.

contain water molecules or helium atoms, the latter possibly formed by the splitting of unstable beryllium atoms by radiation from neighboring radioactive minerals.

Minor substitution of Fe^{3+} or Cr^{3+} for Al^{3+} is reportedly responsible for blue and green colors of aquamarine and emerald respectively. Pink beryl (*morganite*) and yellow beryl are possibly due to traces of lithium and uranium respectively.

PHYSICAL PROPERTIES. $H = 7\frac{1}{2}$–8, Sp. Gr. = 2.68–2.80. Hand sample color is highly variable, usually white, pale blue or green, yellow, rose, or emerald-green. Luster is vitreous.

COLOR AND PLEOCHROISM. Beryl is usually colorless as fragments or sections; however, thick sections of highly colored varieties may show pleochroism in pale colors, usually $E > O$. O = yellow-green, pale blue; E = green, blue.

FORM. Beryl crystals tend to be euhedral, with hexagonal basal sections and rectangular prismatic sections, and range from microscopic needles to huge crystals weighing tons (Fig. 4-5). More rarely, beryl appears as radiating masses in granites or as interstitial anhedral grains.

CLEAVAGE. Poor basal cleavage $\{0001\}$ is seldom visible.

BIREFRINGENCE. Weak maximum birefringence (0.004–0.008) is seen in prismatic sections. Interference colors in standard section never exceed first-order white.

TWINNING. Twinning is very rare and simple on pyramidal planes $\{31\bar{4}1\}$ or $\{40\bar{4}1\}$.

ZONING. Compositional zoning (alkali content) may produce outer zones of higher birefringence and refractive index. Liquid inclusions or specks of quartz, alkali feldspar, or mica may cloud the crystal core.

INTERFERENCE FIGURE. The interference figure for beryl is usually broad, diffuse isogyres on a plain white or gray field. The figure is normally uniaxial negative, but some figures show biaxial isogyre separation, with $2V_x$ up to 17°. Fragments tend to exhibit random orientation.

OPTICAL ORIENTATION. Crystals are normally elongated on c to yield prismatic sections that are length-fast (negative elongation) and have parallel extinction.

DISTINGUISHING FEATURES. Quartz, apatite, and nepheline resemble beryl in thin section, and each may show hexagonal cross sections. Quartz is optically positive, and crystals are length-slow; apatite has higher refractive indices and nepheline lower refractive indices than does beryl.

ALTERATION. Beryl most commonly alters to soft, kaolin-type clays, with lesser sericite or illite. Bavenite, $Ca_4(Be,Al)_4$-$Si_9(O,OH)_{28}$ is a less common alteration product, which may alter, in turn, to bertrandite, $Be_4Si_2O_7(OH)_2$.

OCCURRENCE. Beryl is a characteristic mineral of silica-rich granites and granite pegmatites, in association with quartz, alkali feldspars, and muscovite. Common pegmatite accessory minerals such as spodumene, topaz, lepidolite, tourmaline, amblygonite, Nb—Ta minerals, and rare earth minerals are likewise common associates of beryl.

Crystals of beryl may be huge and commonly appear in central pegmatite zones, often in open cavities. Beryl may also appear in high-temperature hydrothermal veins with cassiterite and tungsten minerals and, rarely, in nepheline syenite.

In metamorphic contact zones, beryl may appear in mica schists, gneisses, skarns, and marbles, as the result of metasomatism. Emerald is largely confined to this latter occurrence.

REFERENCES: BERYL

De Almeida Sampiao Filho, H., G. P. Sighinolfi, and E. Galli. 1973. Crystal chemistry of beryl. *Contrib. Mineral. Petrol.*, *38*, 279–290.

Gibbs, G. V., D. W. Breck, and E. P. Meagher. 1968. Structural refinement of hydrous and anhydrous synthetic beryl, $Al_2(Be_3Si_6)O_{18}$, and emerald, $Al_{1.9}Cr_{0.1}(Be_3Al_6)O_{18}$. *Lithos*, *1*, 275–286.

Isetti, G. 1961. Relations between birefringence, dispersion, and pleochroism of some minerals. *Period. Mineral.*, *30*, 139–163.

Samoilovich, M. I., L. I Tsinober, and R. L. Dunin-Barkovskii. 1971. Nature of the color of beryl containing an iron impurity. *Kristallografiya*, *16*, 186–189.

Vedam, K., and T. A. Davis. 1969. Pressure dependence of the refractive indices of the hexagonal crystals beryl, α-cadmium sulfide, α-zinc sulfide, and zinc oxide. *Phys. Rev.*, *181*, 1196–1201.

Cordierite

$Mg_2Al_3(Si_5Al)O_{18}$

ORTHORHOMBIC (PSEUDOHEXAGONAL)

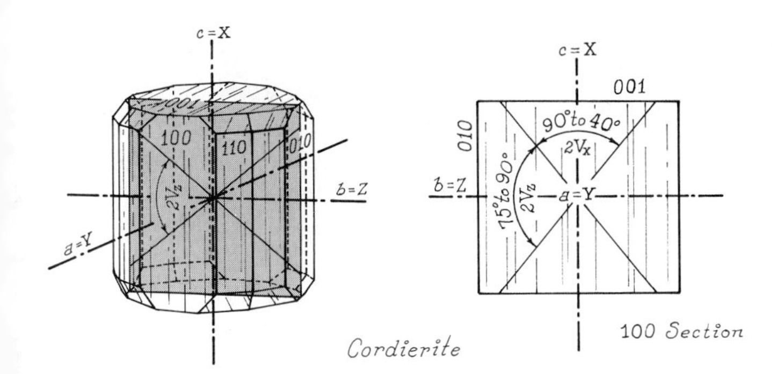

n_α = 1.522–1.560*
n_β = 1.524–1.574
n_γ = 1.527–1.578
$n_\gamma - n_\alpha$ = 0.005–0.018*
Usually biaxial negative, $2V_x = 40°–90°$*
Less commonly biaxial positive, $2V_z = 90°–75°$
Rarely uniaxial negative (indialite)
$a = Y, b = Z, c = X$
Dispersion $v > r$ weak, rarely $r > v$
Usually colorless in thin section, rarely pale blue, and pleochroic $Z > Y > X$

COMPOSITION AND STRUCTURE. Cordierite, $Al_3Mg_2(Si_5AlO_{18})$, is isostructural with beryl, $Be_3Al_2(Si_6O_{18})$. Like beryl, cordierite contains hexagonal rings of six silicate tetrahedra $(Si_6O_{18})^{12-}$, where one-half of the oxygen ions of each SiO_4 tetrahedron are bridging oxygens shared with adjoining tetrahedra in the ring. The rings are stacked vertically (parallel to *c*) and adjoining rings and stacks are united by $3Al^{3+}$ and $2Mg^{2+}$ in tetrahedral and octahedral coordiation respectively, replacing $3Be^{2+}$ and $2Al^{3+}$ in beryl. The single excess positive charge is balanced by substitution of one Al^{+3} for one Si^{+4} in the hexagonal rings.

Indialite is a high-temperature polymorph of cordierite and is structurally equivalent to beryl. It is hexagonal, indicating a strictly random ordering of Al^{3+} in ring tetrahedra. At lower temperatures, Al^{3+} ions seek specific ring sites, and the hexagonal rings are progressively distorted as Al-Si ordering becomes more complete. Ring distortion reduces hexagonal symmetry to orthorhombic (pseudohexagonal), and nearly all natural cordierites are orthorhombic with more or less Al-Si ordering.

The normal cordierite structural series, defined by Al-Si order (low cordierite), is further complicated by a second orthorhombic, higher-temperature, polymorphous series (high cordierite) independent of Al-Si order and possibly defined by Mg^{2+}-Fe^{2+}-Al^{3+} ordering in inter-ring sites.

The major chemical substitution in cordierite is Fe^{2+} for Mg^{2+} and a complete solid solution is possible to a ferric end member, $Al_3Fe_2{}^{2+}(Si_5Al)O_{18}$. Most natural cordierites, however, are very Mg-rich. Very minor Mn^{2+} or Ti^{4+} may also replace Mg^{2+}, and minor Fe^{3+} may replace Al^{3+}. A few alkali ions (Na^+, K^+) usually lie in broad tunnels through the hexagonal ring centers to balance isolated charge deficiencies; and these channels, in low-temperature cordierites, are commonly crowded with nonessential water molecules, which influence refractive indices.

Osumilite $(K,Na,Ca)(Mg,Fe^{+2})_2(Al,Fe^{+3})(Si,Al)_{12}O_{30} \cdot H_2O$ is a cordieritelike mineral with double hexagonal rings in which two of four O^{2-} are shared by adjoining tetrahedra within a single ring, as in cordierite, and a third O^{2-} is shared with a tetrahedron in the underlying ring to form the double hexagonal ring. One H_2O molecule occupies the center of each double ring, and an alkali ion, or Ca^{2+}, lies between stacked rings. Other cations bond adjacent rings and adjoining ring stacks. Osumilite is hexagonal, and n_ω = 1.545–1.547, n_ε = 1.549–1.551.

* Optical constants vary with structural distortion, composition and polymorphic forms. Low distortion favors high indices of refraction and small $2V_x$. Zero distortion is represented by the hexagonal form *indialite*, in which $2V_X = 0°$ and where refractive indices are maximum. Indices of refraction and birefringence also increase with the substitution of Fe^{2+} for Mg^{2+} and with increasing water content. A rare, high-temperature polymorphous series shows unusually low indices of refraction.

FIGURE 4-6. Cordierite may show polysynthetic twinning and birefringence resembling that of plagioclase.

Cordierite commonly contains numerous tiny inclusions of quartz and magnetite aligned with lineation in the host rock.

PHYSICAL PROPERTIES. H = 7–7½. Sp. Gr. = 2.53–2.78. Hand sample color is shades of blue or blue-violet from pale smoky gray-blue to deep indigo. Luster is vitreous.

COLOR AND PLEOCHROISM. Cordierite is normally colorless in standard section; Fe-rich varieties, however, may show pale color and pleochroism; X = colorless, Z = pale violet. Fragments or thick sections show strong pleochroism, $Z > Y > X$. X = colorless, pale yellow; Y = pale blue; and Z = blue, indigo. Small, enclosed zircon crystals are commonly surrounded by yellow, pleochroic halos.

FORM. Cordierite usually appears in poorly lineated metamorphic rocks as irregular porphyroblastic clots or subhedral to euhedral, pseudohexagonal crystals. Crystals tend to be more euhedral in rocks of higher metamorphic grade.

CLEAVAGE. Cleavage is moderately distinct on {010} and poor {100} and {001}. Parting may be pronounced on {001}, especially in altered crystals.

BIREFRINGENCE. Maximum birefringence is weak to mild (0.005–0.018) and is seen in {100} sections. Interference colors never exceed first-order red in standard thin section and are usually gray and white. Birefringence is greatest for cordierites rich in Fe^{2+}.

TWINNING. Cordierite crystals, especially anhedral clots, may be untwinned, or twinning may be simple, on {110} or {130}. However, most crystals are complexly twinned by polysynthetic or cyclic forms, all with {110} and/or {130} composition planes. Polysynthetic twinning may resemble that in plagioclase (Fig. 4-6), although twin lamellae, in basal sections, commonly radiate from a common center to yield cyclic star patterns of three, six, or twelve rays, with composition planes 60° or 30° to positions of extinction. Twin lamellae may also parallel pseudohexagonal prisms, forming concentric and discontinuous hexagons about a central core. Those cordierite crystals formed at high temperatures tend to be most complexly twinned.

ZONING. Cordierite has many reasons to show zoning, owing to different structural states, ionic substitutions, and common inclusions, and optical constants commonly vary widely in a single crystal. Zoning is not often conspicuous by color or birefringence; small $2V_x$ may suggest high-temperature crystal cores. Cordierite may appear as reaction rims over kyanite, and quartz rims over cordierite are known.

INTERFERENCE FIGURE. Most natural cordierites are biaxial negative with $2V_x$ between 65° and 85°. Figures are normally broad, dark isogyres on a white field showing weak optic axis dispersion, usually $v > r$.

Cordierites have been reported positive and negative, uniaxial and biaxial, and any size $2V$ appears possible from $2V_x = 0°$ through $2V = 90°$ to $2V_z = 0°$. High-temperature (low-distortion) normal cordierites (low cordierite) tend to show small $2V$ and are negative; indialite (no distortion) is uniaxial negative; $2V_x$ tends to increase with structural distortion. The rare, high-temperature polymorph (high cordierite) found in volcanic rocks tends to show large $2V$ and is commonly positive.

Fragments lying on basal parting {001} or dominant cleavage planes {010} yield centered bisectrix figures. Interference figures are commonly confused, however, by complex, multiple twinning.

OPTICAL ORIENTATION. Cordierite is orthorhombic, rarely hexagonal (indialite), with $a = Y$, $b = Z$, and $c = X$. Extinction is parallel to all cleavages and parting surfaces, and extinction directions lie 30° or 60° to the composition planes of all common twin laws.

Elongation is obvious in neither fragments nor sections, although the fast wave vibrates normal to basal parting (lineation of basal parting is length-slow—positive elongation), and the slow wave vibrates normal to the major {010} cleavage (lineation of {010} cleavage is length-fast—negative elongation).

DISTINGUISHING FEATURES. Untwinned cordierite most closely resembles quartz, K-feldspar, or nepheline, and twinned varieties are commonly mistaken for plagioclase. Quartz is uniaxial positive; K-feldspar shows well-developed cleavage and tends to have lower indices of refraction than cordierite; nepheline is uniaxial negative and does not appear in metamorphic rocks. Plagioclase has perfect cleavages and shows highly variable extinction angles to composition planes of common twin laws. Multiple twinning, dustlike opaque inclusions, blue-violet coloration, and yellow pleochroic halos about tiny zircon inclusions are inessential but characteristic features of cordierite. Indialite is uniaxial negative, and osumilite is uniaxial positive.

ALTERATION. Cordierite alters rather quickly, beginning on basal parting planes and causing the entire cordierite crystal to assume a yellowish cast. The alteration product is usually a fine-grained, greenish aggregate of largely chlorite and muscovite or biotite, and is popularly called *pinite*. Pinite may also contain minor, zoisite, serpentine, tourmaline, or garnet.

OCCURRENCE. Cordierite is most characteristic of metamorphic rocks formed by high temperature and low stress (confining pressure may range widely) acting on aluminous, often silica-deficient, sediments.

In thermal contact zones adjoining igneous intrusions, cordierite porphyroblasts form in argillaceous sediments in the early stages of heating at the expense of chlorite and is characteristic of the resulting hornfels rocks. It is associated with micas, plagioclase, orthoclase, quartz, andalusite, orthopyroxenes, and anthophyllite. With increasing thermal grade, cordierite may well persist as an important mineral in various mineral associations until sediments are literally fused to glass (buchite rocks). In high-grade hornfelses and buchites, cordierite appears with spinel, sillimanite, anorthite, corundum, and magnetite. Indialite appears in fused sediments in India, where it is the result of natural burning of coal.

In rocks of regional metamorphism, cordierite is formed only in antistress environments where pressures are high and uniform. It is common in high-grade gneisses, granulites and migmatites in association with plutonic minerals such as microcline, sillimanite, kyanite, andalusite, muscovite, and garnet. Kyanite and cordierite are not always mutually stable, and reaction rims of cordierite over kyanite cores are noted. Cordierite is very commonly associated with anthophyllite in both contact hornfelses and regional amphibolites and gneisses.

Cordierite in plutonic igneous rocks is largely the result of magmatic assimilation of aluminous sediments (such as shale). It appears in a few granites also containing muscovite, biotite, and, possibly, garnet. A few granite pegmatites contain large, euhedral crystals of cordierite, rarely gem quality, usually highly altered by later hydrothermal stages. In pegmatites or quartz veins, cordierite may be associated with other aluminum-rich minerals such as andalusite, topaz, or garnet and may form eutectic intergrowth with quartz. In gabbroic rocks, especially norites, cordierite may be present with orthopyroxene, biotite, plagioclase, quartz, and spinel.

Cordierite is rare in volcanic rocks but is reported in andesite and other flow rocks of intermediate composition. Volcanic cordierite is usually the rare high-temperature polymorph, sometimes indialite, and, possibly, much is the cordieritelike mineral osumilite.

REFERENCES: CORDIERITE

Cohen, J. P., F. K. Ross, and G. V. Gibbs. 1977. An x-ray and neutron diffraction study of hydrous low cordierite. *Amer. Mineral.*, *62*, 67–78.

Gibbs, G. V. 1966. The polymorphism of cordierite: I. The crystal structure of low cordierite. *Amer. Mineral.*, *51*, 1068–1087.

Lee, J. D., and J. L. Pentecost. 1976. Properties of flux-grown cordierite single crystals. *Jour. Amer. Ceram. Soc.*, *59*, 183.

Srivastava, G. S., and S. V. L. N. Rao. 1967. Correlation of optical properties and chemical composition of cordierites. *Proc. Indian Acad. Sci.*, Sect. B., *65*, 167–173.

Venkatesh, V. 1952. Development and growth of cordierite in paralavas. *Amer. Mineral.*, *37*, 831–848.

Venkatesh, V. 1954. Twinning in cordierite. *Amer. Mineral.*, *39*, 636–646.

Zeck, H. P. 1973. Symmetry, crystal structure, polymorphism, crystallographic orientation, and axial ratio of cordierite literature review. *Bull. Geol. Soc. Den.*, *22*, 39–49.

Tourmaline

$Na(Mg,Fe,Li,Al)_3Al_6(Si_6O_{18})(BO_3)_3(OH,F)_4$

TRIGONAL

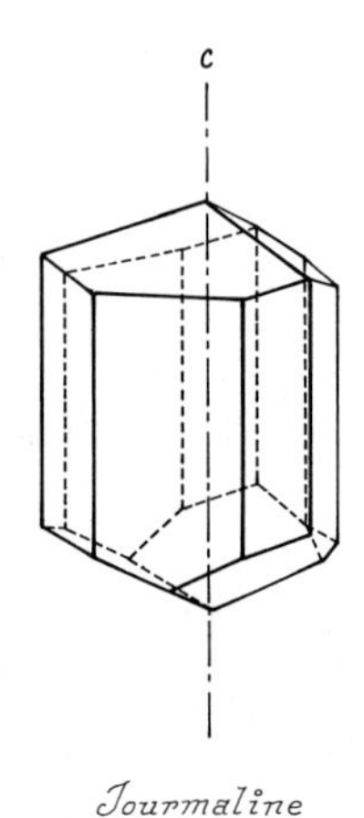

Tourmaline

Elbaite-Schorl Series

$n_\omega = 1.635–1.698$*

$n_\varepsilon = 1.615–1.675$

$(n_\omega - n_\varepsilon) = 0.019–0.035$*

Uniaxial negative

Colorless to black, also shades of gray, blue, green, or pink

Highly pleochroic $O > E$

Dravite-Schorl Series

$n_\omega = 1.631–1.698$*

$n_\varepsilon = 1.610–1.675$

$(n_\omega - n_\varepsilon) = 0.015–0.035$*

Uniaxial negative

Colorless to black, also shades of yellow and brown

Highly pleochroic $O > E$

* Indices of refraction and birefringence increase rapidly with increasing iron (toward schorl)—see Fig. 4-7.

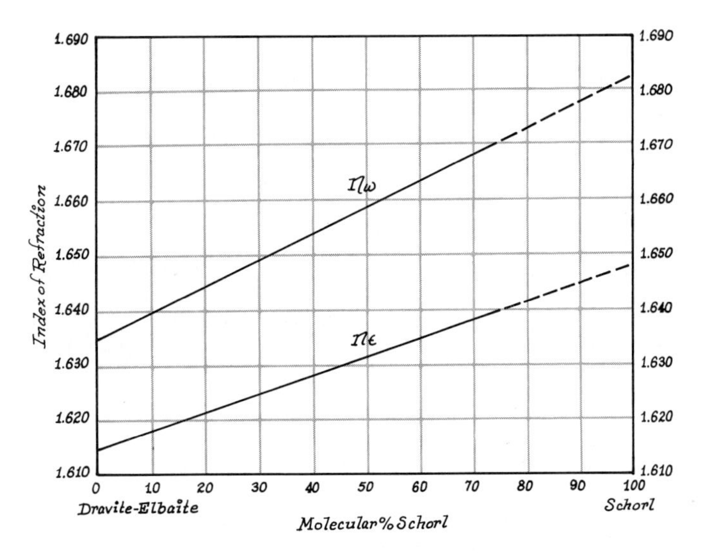

FIGURE 4-7. Variation of the indices of refraction in the tourmaline series.

COMPOSITION AND STRUCTURE. The crystal structure of tourmaline is based on six-unit rings of SiO_4 tetrahedra, each tetrahedron sharing two of the four O^{2-} with adjoining tetrahedra in the ring (Si_6O_{18}). The rings are distorted to trigonal symmetry and stacked "vertically" along the *c*-axis. Three planar BO_3^{3-} units lie in trigonal distribution between adjacent Si_6O_{18} rings and Mg^{2+} (also Fe^{2+} or Li^+ and Al^{3+}) lie with trigonal distribution within the ring in octahedral coordination with O^{2-} and OH^- around the ring and $(OH)^-$ at the ring center. Ions of Na^+ also lie along the ring axis, and Al^{3+} ions in distorted octahedral coordination connect adjacent ring stacks.

Two major solid solution series, elbaite-schorl and dravite-schorl are defined by end members:

Schorl	$NaFe_3{}^{2+}Al_6(Si_6O_{18})(BO_3)_3(OH)_4$
Dravite	$NaMg_3Al_6(Si_6O_{18})(BO_3)_3(OH)_4$
Elbaite	$Na(Li,Al)_3Al_6(Si_6O_{18})(BO_3)_3(OH)_4$

The dravite-schorl series is surely a complete solid solution ($Mg^{2+} \rightleftharpoons Fe^{2+}$) and at least schorl is commonly rich in Mn^{2+}, forming solid solution with another end member, *tsilaisite*, $NaMn_3Al_6(Si_6O_{18})(BO_3)_3(OH)_4$.

The elbaite-schorl series is probably complete ($2Fe^{2+} \rightleftharpoons Li^+ + Al^{3+}$), but near-end member compositions are most common. There appears to be very little solid solution between elbaite and dravite.

Complete substitution of Na^+ by Ca^{2+} or K^+ may be possible, and calcium end members have been proposed. Partial substitutions of Fe^{3+}, Cr^{3+}, Ti^{4+}, or Cu^{2+} for Al^{3+} or Mn^{2+} and F^- for $(OH)^-$ are noted, especially in elbaite compositions.

PHYSICAL PROPERTIES. $H = 7\text{–}7\frac{1}{2}$. Sp. Gr. = 3.0–3.25. Hand sample color is extremely variable. Iron-rich tourmaline (schorl) is black and is most common. Toward dravite, color lightens from black (schorl) through shades of brown and yellow to colorless (*achroite*). Elbaite is often brilliant shades of red (*rubellite*), green (*verdelite*), and blue (*indicolith*), which depend on minor impurities and become dark toward the schorl composition. Nonuniform coloration is very common, and color zoning commonly produces crystals with red cores and green outer zones or distinct color changes along the crystal length. Luster is vitreous and fracture is subconchoidal.

COLOR AND PLEOCHROISM. In fragments or sections tourmaline color is highly variable and often irregular.

The schorl-dravite series ranges from black to colorless through shades of brown and yellow with normally strong pleochroism $O \gg E$. Basal sections are strongly colored, and prismatic sections are dark when elongated perpendicular to

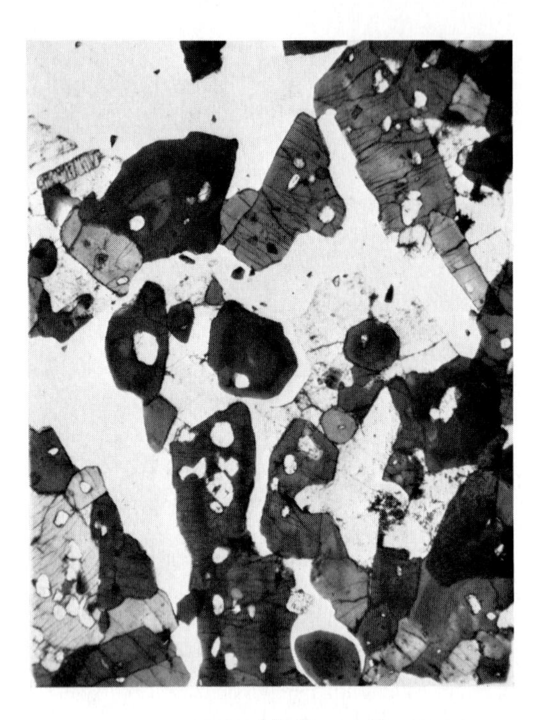

FIGURE 4-8. Tourmaline crystals commonly show color zoning and strong pleochroism. Note also the crude round-triangular cross sections and prismatic sections with crude cross fractures.

the vibration plane of the polarizer. O = dark brown, pale yellow-brown; E = pale yellow, colorless.

The schorl-elbaite series ranges from black to colorless through shades of blue, green, or pink. Elbaite is nearly colorless in thin section and only weakly colored as fragments, with little pleochroism $O > E$. Compositions near schorl are highly colored and pleochroic $O \gg E$ in shades of gray, green, or blue. Basal sections are dark as are prismatic sections elongated normal to the vibration direction of the lower nicol. O = black, deep blue-black, deep green, gray; E = gray, pale blue, pale green, buff, colorless.

FORM. Tourmaline crystals are usually euhedral, ranging from single, stubby, columnar crystals to acicular aggregates of parallel or radiated needles. Crystal cross sections may be six-sided but are most commonly barrellike with three curved sides (Fig. 4-8). Prismatic sections are crudely rectangular.

CLEAVAGE. Prismatic $\{11\bar{2}0\}$ and pyramidal $\{10\bar{1}1\}$ cleavages are both very poor; however, fracture cracks roughly perpendicular to c ($\{0001\}$ planes) are common and cause prismatic sections to appear segmented.

BIREFRINGENCE. Iron-tourmaline shows the strongest birefringence with interference colors up to low third-order in standard section. Lithium-tourmaline shows minimum birefringence and interference colors below second order.

TWINNING. Twinning in tourmaline is rare, as in most uniaxial minerals; however, it is sometimes observed as simple twinning on $\{10\bar{1}1\}$ or on other odd pyramidal planes.

ZONING. Concentric color zoning is very common in almost any compositional range, especially in basal sections but also in prismatic sections. Outer zones tend to be darker (iron-rich). Elbaite crystals are often complexly zoned, with obvious color changes both concentric on c and progressive along c. Elbaite colors tend to be weak in standard section, however, and are much more obvious in hand sample.

Alteration effects are often zoned around c, and small inclusions of quartz, micas, and feldspar are often concentrated in core zones. Core zones may also be intricate intergrowths of quartz and albite.

INTERFERENCE FIGURE. Tourmaline is uniaxial negative, and basal sections yield optic axis figures of broad isogyres on a rather dark field, commonly showing a few dark isochromes of anomalous colors. Strong absorption along *c* tends to make figures dark and obscure. Rarely, tourmaline shows small isogyre separation, with $2V_x$ up to 10°.

OPTICAL ORIENTATION. Crystals elongated on *c* and prismatic sections show parallel extinction and are length-fast (negative elongation). Prismatic sections are often segmented by "basal" cracks.

DISTINGUISHING FEATURES. Tourmaline is normally characterized by strong pleochroism, moderate birefringence, parallel extinction, and "rounded" basal sections, often showing color zoning. Biotite and hornblende may show color and pleochroism similar to that of tourmaline, but display perfect cleavage parallel to probable elongation and show greatest absorption of light vibrating parallel to elongation. Elbaite is uniaxial negative, with moderate birefringence in contrast to topaz, which is biaxial, and to apatite, which has low birefringence.

ALTERATION. Tourmaline alters to fine-grained secondary micas (illite, sericite, biotite) or chlorites. Lithium-tourmalines may yield secondary lepidolite or talclike cookeite $LiAl_4 \cdot (Si,Al)_4O_{10}(OH)_8$. Tourmaline may itself be an alteration mineral formed by late pneumalolitic alteration of biotite, plagioclase, and so on, through the introduction of boron (tourmalinization).

OCCURRENCE. Tourmaline may form in late-stage granites, pegmatites, or hydrothermal pneumatolitic veins by pneumatolitic introduction of boron and in contact or regional metamorphic rocks by boron metasomatism.

In granites, tourmaline is usually iron-rich schorl-dravite and is black. It may be formed by late-stage alteration of common granitic minerals by pneumatolitic fluids containing boron, and extreme tourmalinization may yield a rock of only quartz and tourmaline.

In pegmatites and hydrothermal pneumatolitic veins, tourmaline is schorl-elbaite. Black schorl is characteristic of quartz veins and outer pegmatite zones, while colored elbaite varities appear in pegmatite core zones with quartz, albite, muscovite, topaz, Li-minerals, and other common pegmatite minerals.

In contact metamorphic rocks, tourmaline is schorl-dravite formed by boron metasomatism of dolomites and magnesian limestones. Mica schists may contain brown dravite tourmaline, perhaps, resulting from evaporite borates or adsorbed boron in argillaceous sediments.

Tourmaline grains are locally important as heavy accessory grains in detrital sediments.

REFERENCES: TOURMALINE

Donnay, G., and R. Barton, Jr. 1972. Refinement of the crystal structure of elbaite and the mechanism of tourmaline solid solution. *Tschermaks Mineral. Petrogr. Mitt.*, *18*, 273–286.

Donnay, G., and Buerger, M. J. 1950. The determination of the crystal structure of tourmaline, *Acta Crystallogr.*, *3*, 379–388.

El-Hinnawi, E. E. 1966. Optical and chemical investigation of nine tourmalines. *Neues Jahrb. Mineral.*, *Monatsh.*, *1966*, 80–88.

Isetti, G. 1961. Relations between birefringence, dispersion, and pleochroism of some minerals. *Period. Mineral.*, *30*, 139–163.

Ito, T., and R. Sadanaga. 1951. A Fourier analysis of the structure of tourmaline. *Acta Crystallogr.*, *4*, 385–390.

McCurry, P. 1971. Relation between optical properties and occurrence of some black tourmalines from northern Nigeria. *Mineral. Mag.*, *38*, 369–373.

Townsend, M. G. 1970. Dichroism of tourmaline. *Jour. Phys. Chem. Solids*, *31*, 2481–2488.

Ward, G. W. 1931. Chemical and optical study of the black tourmalines. *Amer. Mineral.*, *16*, 145–190.

TRICLINIC

Axinite

$(Ca,Fe^{2+},Mn)_3Al_2BO_3(SiO_3)_4OH$

$\angle\alpha = 91°48'$
$\angle\beta = 98°10'$
$\angle\gamma = 77°18'$

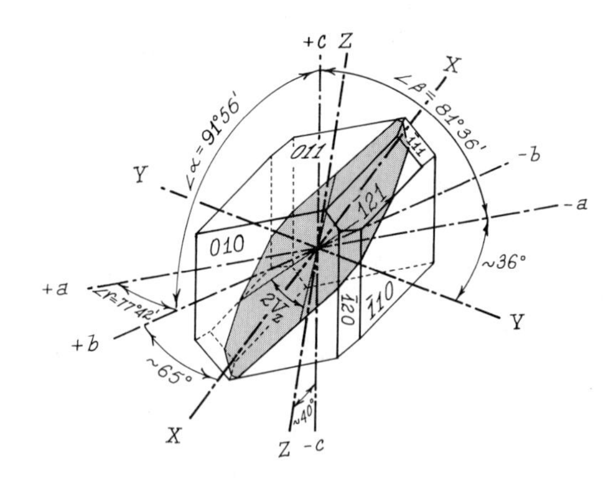

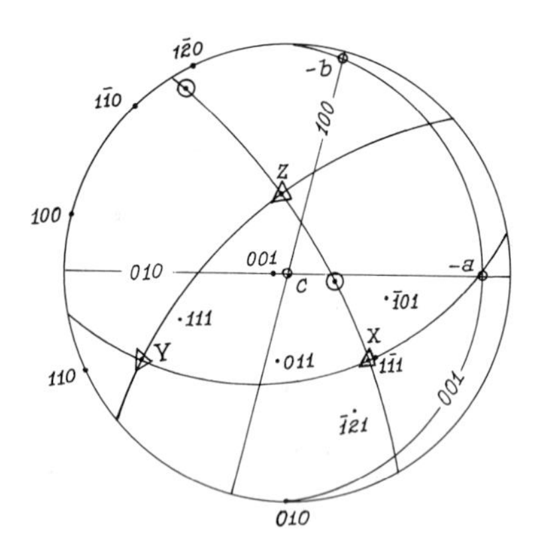

$n_\alpha = 1.672\text{–}1.693$*
$n_\beta = 1.677\text{–}1.701$
$n_\gamma = 1.681\text{–}1.704$
$n_\gamma - n_\alpha = 0.009\text{–}0.013$
Biaxial negative, $2V_x = 63°\text{–}90°$
Optical orientation varies somewhat with composition
$X \simeq \perp(\bar{1}11)$
Strong dispersion $v > r$
Colorless to pale purple, brown, or yellow in section
Weak pleochroism; $Y > X > Z$

COMPOSITION AND STRUCTURE. Axinite is a cyclosilicate with fourfold rings of silicate tetrahedra, $(SiO_3)_4$ joined to form $B_2Si_8O_{30}$ groups. These connect sheets consisting of Al^{3+} and Fe^{2+} (or Mn^{2+}) in octahedral coordination and Ca^{2+} in irregular tenfold coordination.

Normal compositions are Fe^{2+}-rich near $Ca_2Fe^{2+}Al_2BO_3(SiO_3)_4OH$, but a Mn^{2+}-rich variety (*tinzenite*) $Ca_2MnAl_2BO_3(SiO_3)_4OH$ is known. Appreciable Fe^{2+} and Mn^{2+} may replace Ca^{2+} with minor Na^+, Mg^{2+} may replace some Fe^{2+}, and minor Fe^{3+} and Ti^{4+} may replace Al^{3+}.

* Indices of refraction decrease as Mn^{2+} replaces Fe^{2+}.

PHYSICAL PROPERTIES. H = $6\frac{1}{2}$–7. Sp. Gr. = 3.26–3.36. Color, in hand sample, is light to dark brown, with a distinctly purple cast. Tinzenite is yellow to red. Luster is vitreous.

COLOR AND PLEOCHROISM. Axinite is normally colorless to pale brown or violet in standard section, with weak absorption $Y > X > Z$. X = pale brown, yellow, pale green; Y = violet, deep yellow, blue; and Z = pale violet, yellow, pale green.

FORM. Axinite crystals are commonly euhedral or subhedral, with wedge-shaped sections; less commonly, granular or radiated.

CLEAVAGE. Cleavage is distinct on {100} and poor on {001}, {110}, and {011}.

BIREFRINGENCE. Maximum birefringence is weak (0.009–0.011), and interference colors are grays and white ranging to possibly first-order yellow.

TWINNING. Most axinite is untwinned; however, repeated normal twinning is reported on {110}, and parallel twinning on {100} is known.

ZONING. Color zoning appears in some crystals, and $2V_x$ may vary in successive zones.

INTERFERENCE FIGURE. Fragments should show off-center orientation with broad, nearly straight isogyres sweeping across a white field. Dispersion is strong $v > r$, with possible bisectrix dispersion.

OPTICAL ORIENTATION. Optical directions X, Y, Z show somewhat variable relationships to crystallographic directions. X is reported almost normal to $(\bar{1}11)$ or (011) and is near b ($b \wedge X \simeq 15°$). As axinite is triclinic, no optical direction parallels a crystallographic direction, and extinction is inclined to all cleavages and crystal faces.

DISTINGUISHING FEATURES. Wedge-shaped sections, pale violet-brown color, high relief, low birefringence, inclined extinction, and occurrence are characteristic of axinite. Datolite has much higher birefringence, serendibite shows lamellar twinning, and K-feldspar shows perfect cleavage and low negative relief.

ALTERATION. Axinite may alter to chlorite and calcite.

OCCURRENCE. The characteristic occurrence of axinite is in metamorphic contact aureoles between granitic intrusives and carbonate sediments in association with calc-silicate minerals (epidote, zoisite, diopside, grossular, andradite, idocrase, datolite, tourmaline, calcite). It is less common in veins, cavities, and open fissures in granites, mafic intrusives, and basalt, with epidote, hornblende, prehnite, chlorite, datolite, danburite, pectolite, tourmaline, and calcite.

Axinite is reported in some biotite gneisses and amphibolites and is a rare accessory mineral in granite pegmatites and hydrothermal veins.

REFERENCES: AXINITE

Ito, T., and Y. Takéuchi. 1952. The crystal structure of axinite. *Acta Crystallogr.*, *5*, 202–208.

Takéuchi, Y., T. Ozawa, T. Ito, T. Arakai, T. Zoltai, and J. J. Finney. 1974. Boron silicon oxide ($B_2Si_8O_{30}$) groups of tetrahedra in axinite and comments on the deformation of silicon tetrahedra in silicates. *Z. Kristallogr.*, *140*, 289–312.

CHAPTER 5

Inosilicates

Two of the very large and important families of rock-forming minerals—the pyroxenes and amphiboles—constitute the bulk of the inosilicate group. Members of this group are often called *chain silicates* because the silicate tetrahedra are linked so as to form one-dimensionally "infinite" chains; that is, the chains extend along one crystallographic direction without interruption throughout an entire crystal. The morphological result is that crystals of inosilicates tend to be elongated along the direction of the chains, which in all of them has been designated the *c* axis.

Although a large number of distinct kinds of chains can be imagined, only two are found. These are single chains and double chains, as shown in Figs. 5-1 and 5-24. Although there may be substitution for silicon atoms (principally by aluminum), the "ideal" chemical formula for the single chain is $(SiO_3)_{\infty}{}^{2-}$, and for the double chain it is $(Si_4O_{11})_{\infty}{}^{6-}$. Thus the chains act as very large complex anions and are bound to one another by metal cations in six, seven-, and/or eightfold coordination.

All of the double-chain silicates belong to the amphibole group. The pyroxenes consist of those single-chain silicates for which the geometrical repeat along the chain (ignoring possible chemical differences) is two tetrahedra; this is the case shown in Fig. 5-1. If the tetrahedra are periodically tilted or rotated so that the repeat is larger—say, three, five, or seven tetrahedra—then we have a related but structurally somewhat different group called the *pyroxenoids.*

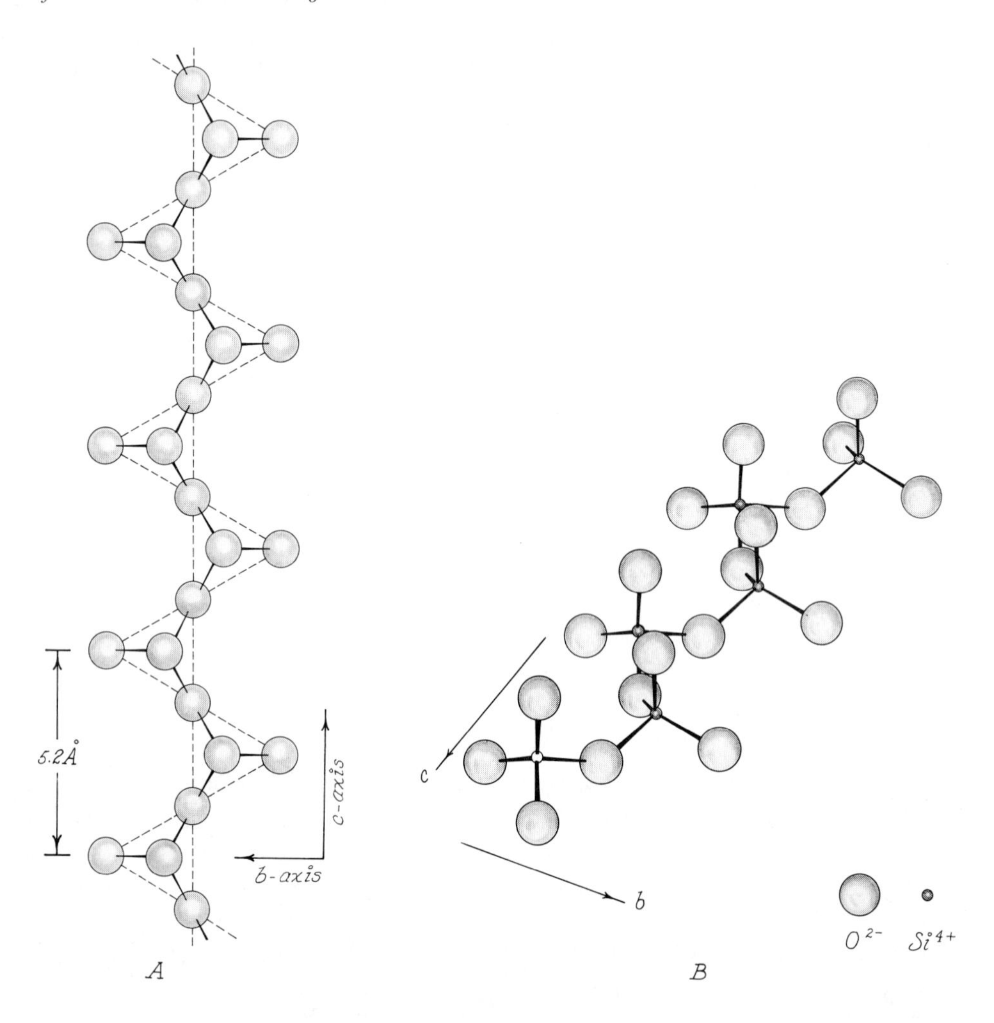

FIGURE 5-1. The idealized, single-pyroxene chain. Repeat unit along the chain (*c*) is two tetrahedra (5.2 Å). (A) View normal to plane of the chain (*ab* plane). (B) Inclined view.

Because optical properties are controlled by the crystal structure and chemistry of minerals, optical anisotropy is to be expected in the inosilicates. Although a few have low maximum birefringence, that of most of them is sufficient to give second- and third-order colors in standard section. The great diversity in chemistry of inosilicates makes it impossible to predict optical orientation from structure alone—a reasonably successful exercise for the phyllosilicates, to be considered later.

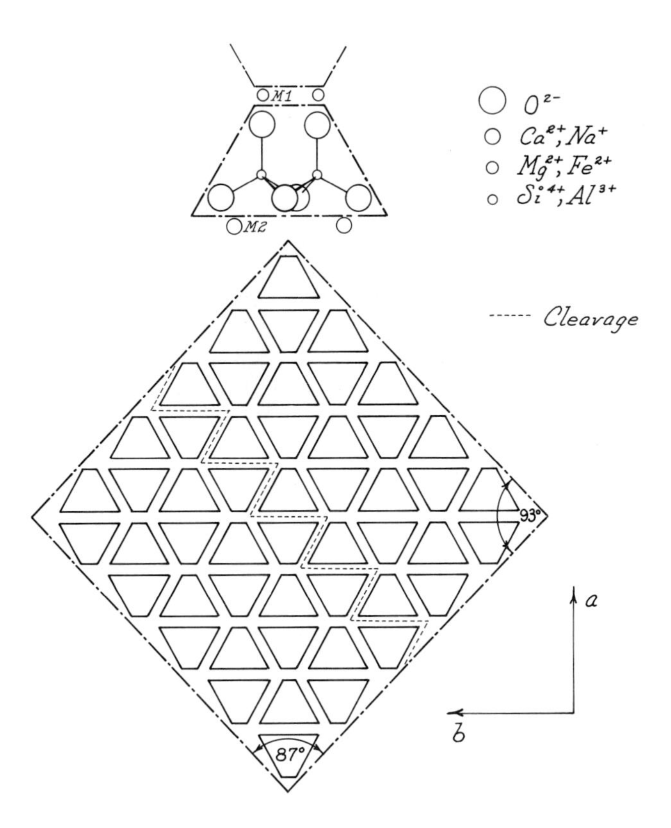

FIGURE 5-2. Arrangement of pyroxene chains as seen along *c*. Prismatic cleavages on (110) and (1$\bar{1}$0) intersect at about 87° and 93° and are characteristic of all pyroxene varieties.

The Pyroxene Group

Pyroxenes are single-chain inosilicates, with the "one-dimensionally infinite" $(SiO_3)_\infty$ chains (Fig. 5-1) extended along the crystallographic *c*-axis. Each silicate tetrahedron shares two oxygens with other tetrahedra to form the chains, with two left to participate strongly in bonds with other atoms. (Actually, the two oxygens involved in Si—O—Si linkages also participate, although weakly, in bonds to other atoms.) Fig. 5-2 shows the general pyroxene structure as viewed along the *c*-axis. Note that the $(SiO_3)_\infty$ chains are stacked so as to "point" in alternate

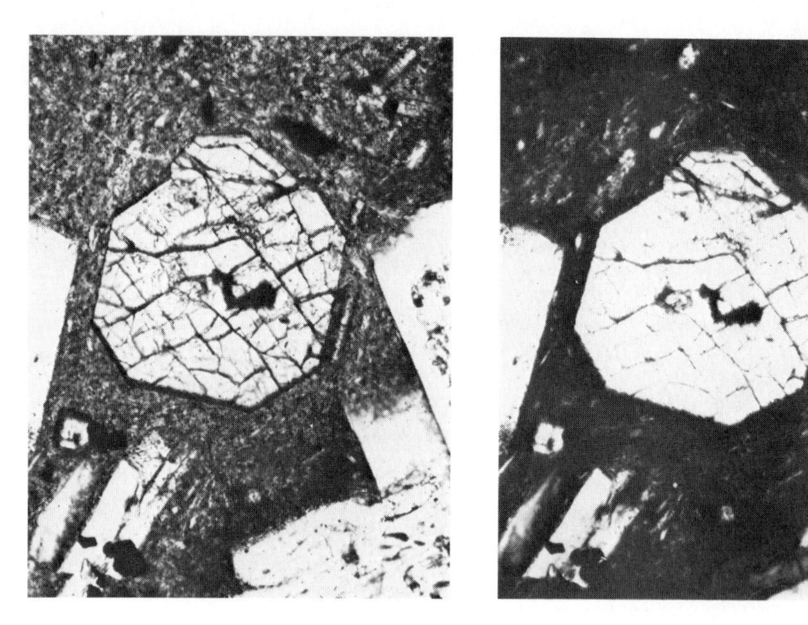

FIGURE 5-3. Pyroxene cross sections are commonly square to octagonal and show distinct prismatic {110} cleavages intersecting at almost 90°.

directions along the *a* and *b* directions, thus producing the diagnostic {110} pyroxene cleavages at 87°–93° (Fig. 5-3). The smaller of the two metal sites is known as *M*1, and is a six-coordinated position in all pyroxenes. *M*2, the larger site, is six-, seven-, or eight-coordinated, depending on the chemistry of the pyroxene. Those in which *M*2 is occupied by Fe^{2+} or Mg^{2+} are generally orthorhombic, forming a continuous solid solution from orthoenstatite, $Mg_2(SiO_3)_2$, to orthoferrosilite, $Fe_2(SiO_3)_2$. Compositions near the latter are unknown in nature but have been synthesized. There is also a rarely occurring monoclinic series of identical chemistry (clinoenstatite-clinoferrosilite). The substitution of atoms larger than Fe^{2+} or Mg^{2+} into the *M*2 site causes shifts in atomic positions that preclude the possibility of orthorhombic symmetry, and so all pyroxenes except for the orthenstatite-orthoferrosilite series are monoclinic. The general terms for the two symmetrically distinct groups are *orthopyroxenes* and *clinopyroxenes*.

Clinopyroxenes that contain from 0.05 to about 0.15 Ca:(Ca + Mg + Fe) are called *pigeonites*, and they constitute essentially the central part of a Ca-rich clinoenstatite-clinoferrosilite series. With increased concentration of calcium in the mid-Mg:Fe range (see Fig. 5-4) a miscibility gap is encountered from approximately 0.15 to 0.25 Ca:(Ca + Mg + Fe). Single-phase pyroxenes whose compositions lie within this gap are called *subcalcic augites*, and are found only in rapidly quenched volcanic rocks. More slowly cooled clinopyroxenes

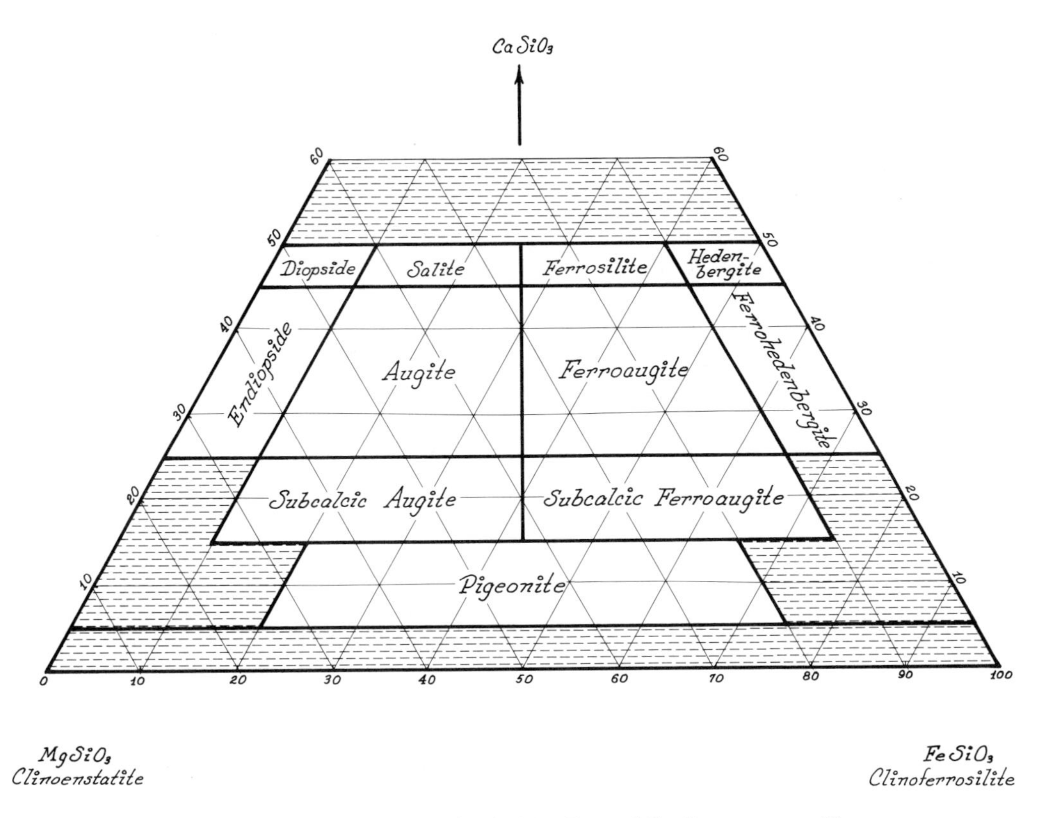

FIGURE 5-4. Composition fields for the Mg-, Fe-, and Ca-clinopyroxenes. The composition of few natural pyroxenes fall in the gray areas. Subcalcic augites and ferroaugites are found only in quickly cooled volcanic rocks. (Poldervaart and Hess, Pyroxenes in the crystallization of basaltic magmas, *Jour. Geol.*, 59, © by the University of Chicago Press)

with compositions inside the miscibility gap exsolve on cooling to orthopyroxene, or pigeonite, and augite. Augites have Ca:(Ca + Mg + Fe) ratios of 0.25 to 0.45, and may exhibit Mg:(Mg + Fe) ratios of nearly 0 to 1.0. Most natural augites, however, are close to 0.40 Ca:(Ca + Mg + Fe), with 0.10 < Mg:(Mg + Fe) < 0.90. The diopside-hedenbergite solid solution series has Ca: total metal ratios of 0.45 to 0.50, the ideal compositions of the end members being $CaMg(SiO_3)_2$ and $CaFe(SiO_3)_2$, respectively. The addition of more calcium causes a symmetry change to the triclinic system, accompanied by alteration in the geometry of the $(SiO_3)_\infty$ chain; the result is the pyroxenoid group, which is discussed separately (p. 207).

The pyroxenes discussed thus far have been represented on the ideal "pyroxene quadrilateral" (Fig. 5-4) and their approximate optical constants are summarized in Figs. 5-5 and

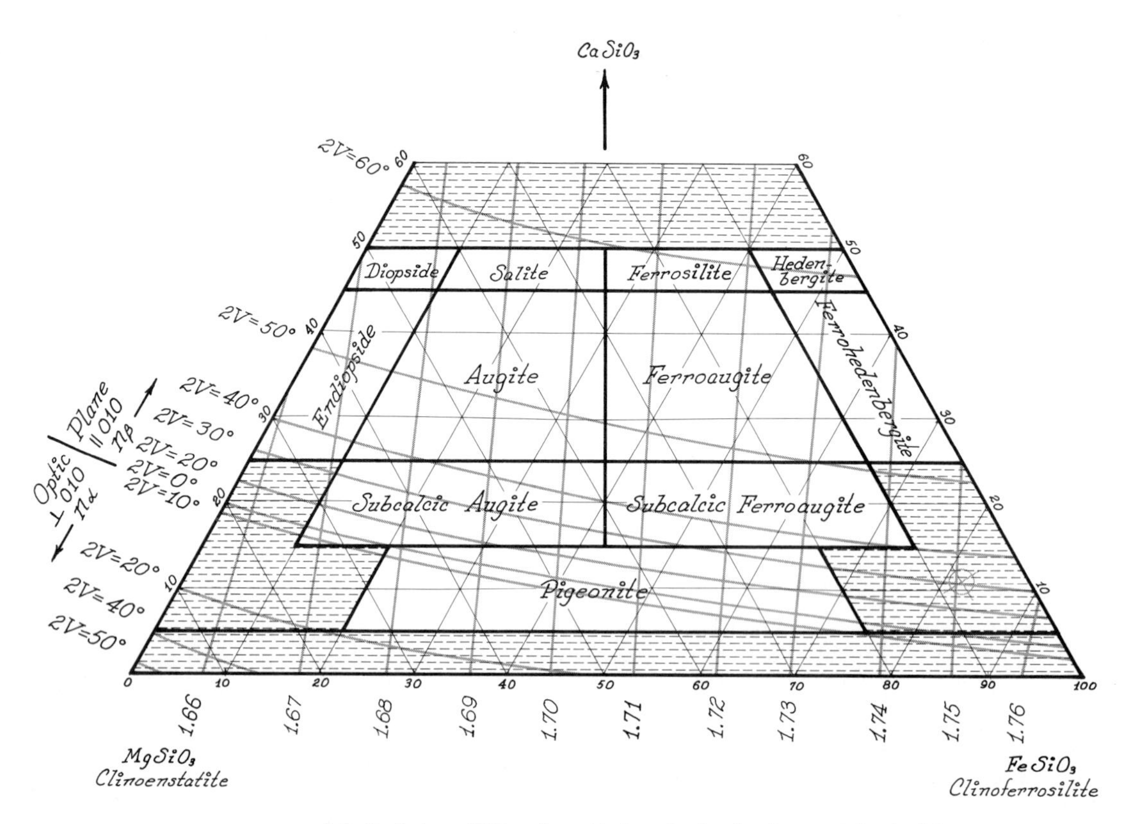

FIGURE 5-5. Variation of $2V$ angle and index of refraction (n_β or n_α) for the Mg-, Fe-, and Ca-clinopyroxenes. (Modified from Hess, 1949)

5-6. In nature these minerals often contain significant Al^{3+}, Fe^{3+}, Cr^{3+}, Ti^{4+}, and Mn^{2+} as substituents in the $M1$ and $M2$ sites, with the concomitant substitution of Al^{3+} for Si^{4+} where necessary to maintain electrostatic neutrality. Johannsenite, $CaMn(SiO_3)_2$, forms a complete solid solution series with hedenbergite, and at least a partial one with diopside. Aegirine (also called *acmite*), $NaFe^{3+}(SiO_3)_2$, and jadeite, $NaAl(SiO_3)_2$, are sodic pyroxenes that form at least a partial solid solution series in nature. Compositions intermediate between aegirine and augite are known as *aegirine-augite*, while *omphacites* lie compositionally between aegirine-jadeite and diopside-hedenbergite, with $0.25 < Ca:(Ca + Na) < 0.75$. Spodumene is the lithium clinopyroxene, $LiAl(SiO_3)_2$. The following list summarizes the ideal end-member pyroxene compositions, as well as the general chemistry of the some of the common intermediate pyroxenes.

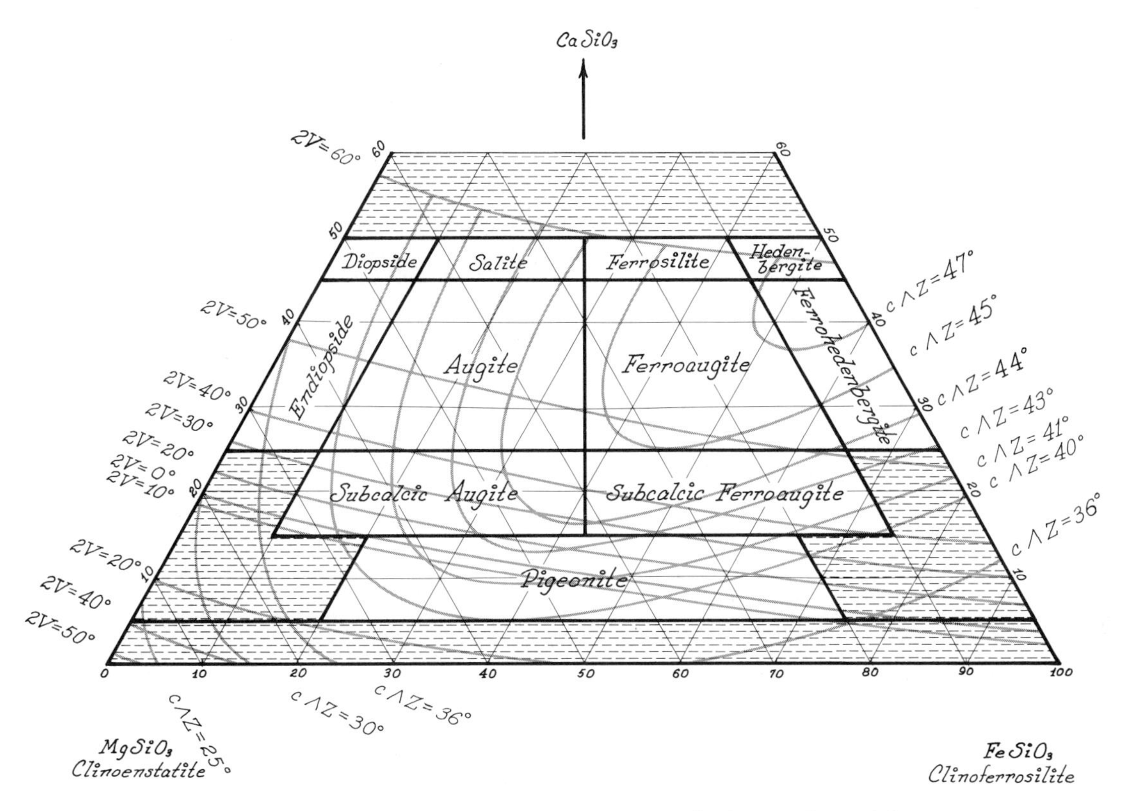

FIGURE 5-6. Variation of $2V$ angle and $c \wedge Z$ for the Mg-, Fe-, and Ca-clinopyroxenes. (Modified from Deer and Wager, 1938)

Orthopyroxenes

Enstatite	$Mg_2(SiO_3)_2$
Hypersthene	$(Mg,Fe^{2+})_2(SiO_3)_2$
Orthoferrosilite	$Fe_2{}^{2+}(SiO_3)_2$

Clinopyroxenes

Clinoenstatite	$Mg_2(SiO_3)_2$
Clinoferrosilite	$Fe_2{}^{2+}(SiO_3)_2$
Diopside	$CaMg(SiO_3)_2$
Hedenbergite	$CaFe^{2+}(SiO_3)_2$
Johannsenite	$CaMn(SiO_3)_2$
Fassaite	$Ca(Mg,Fe^{2+},Fe^{3+},Al)[(Si,Al)O_3]_2$
Augite	$(Ca,Mg,Fe^{2+})(Mg,Fe^{2+},Al)[(Si,Al)O_3]_2$
Pigeonite	$(Mg,Fe^{2+},Ca)(Mg,Fe^{2+})(SiO_3)_2$
Aegirine	$NaFe^{3+}(SiO_3)_2$
Jadite	$NaAl(SiO_3)_2$
Omphacite	$(Ca,Na)(Mg,Fe^{2+},Fe^{3+},Al)(SiO_3)_2$
Spodumene	$LiAl(SiO_3)_2$

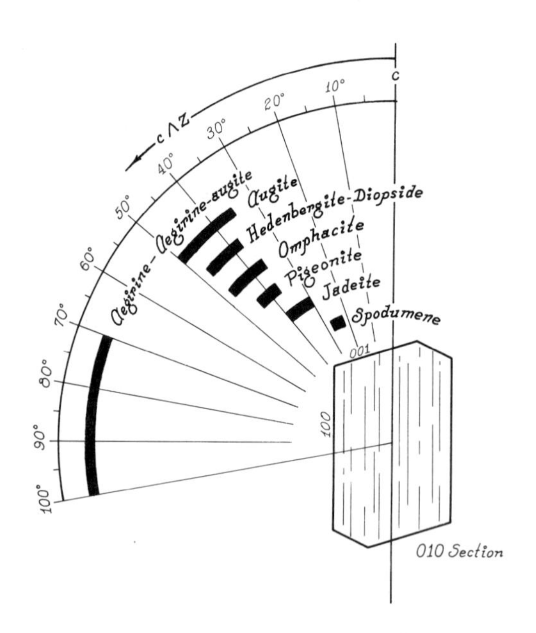

FIGURE 5-7. Ranges of $c \wedge Z$ (maximum extinction angles) for several common pyroxene series.

Fig. 5-7 compares the optical orientation (that is, $c \wedge Z$ ranges) for some of the major pyroxene series. Physical and optical properties of the various pyroxenes are summarized in Table 5-1.

REFERENCES: THE PYROXENE GROUP

Hess, H. H. 1941. Pyroxenes of common mafic magmas. II. *Amer. Mineral.*, *26*, 573–594.

Hess, H. H. 1949. Chemical composition and optical properties of common clinopyroxenes. *Amer. Mineral.*, *34*, 621–626.

Deer, W. A., and L. R. Wager. 1938. Two new pyroxenes included in the system clinoenstatite, clinoferrosilite, diopside, and hedenbergite. *Mineral. Mag.*, *25*, 15–22.

TABLE 5-1. Optical Properties of the Pyroxenes

n_β	n_α	n_γ	$n_\gamma - n_\alpha$	Mineral	System	$2V$	Optic Sign	Cleavage	Optical Orientation	Color in Sections	Physical Properties
1.65–1.67	1.65–1.66	1.66–1.68	0.014–0.027	Spodumene $LiAl(SiO_3)_2$	Monoclinic $\angle\beta=110°20'$	54°–69° $v>r$ weak	+	{110} distinct; parting {010} or {100}	$b=Y$ $c\wedge Z=-22°$ to $-27°$ $a\wedge X=-2°$ to $-7°$	Colorless	H = $6\frac{1}{2}$–7. Sp. Gr. = 3.0–3.2 White, gray, pale green, lavender
1.65–1.67	1.65–1.67	1.67–1.68	0.008–0.010	Enstatite $Mg_2(SiO_3)_2$ En_{100-88}	Orthorhombic	35°–90° $r>v$ moderate	+	{110} distinct; parting {010} or {100} (rare)	$a=X$ $b=Y$ $c=Z$	Colorless	H = 5–6. Sp. Gr. = 3.2–3.3 Gray, yellow, tan, pale green
1.66–1.67	1.65–1.67	1.67–1.69	0.012–0.032	Jadeite $NaAl(SiO_3)_2$	Monoclinic $\angle\beta=107°26'$	70°–75° $v>r$ moderate	+	{110} distinct; parting {100}	$b=Y$ $c\wedge Z=-30°$ to $-40°$ $a\wedge X=-13°$ to -23	Colorless, pale green	H = 6. Sp. Gr. = 3.2–3.4 White to light green
1.67–1.68	1.66–1.67	1.69–1.70	0.030–0.029	Diopside $CaMg(SiO_3)_2$ Di_{100-90}	Monoclinic $\angle\beta\simeq106°$	58°–57°	+	{110} distinct; parting {001} or {100}	$b=Y$ $c\wedge Z=-38°$ to $-40°$ $a\wedge X=-22°$ to $-24°$	Colorless, pale green	H = $5\frac{1}{2}$–$6\frac{1}{2}$. Sp. Gr. = 3.2–3.4 White to pale gray-green
1.67–1.70	1.67–1.69	1.68–1.70	0.010–0.012	Bronzite $(Mg,Fe^{2+})_2(SiO_3)_2$ En_{88-70}	Orthorhombic	90°–57° $r>v$ moderate	−	{110} distinct; parting {010} or {100} (rare)	$a=X$ $b=Y$ $c=Z$	Colorless	H = 5–6. Sp. Gr. = 3.3–3.4 Pale greenish-brown (bronze)
1.67–1.70	1.66–1.69	1.69–1.72	0.018–0.027	Omphacite (Ca,Na) (Mg,Fe^{2+},Fe^{3+},Al) $(SiO_3)_2$	Monoclinic $\angle\beta=106°$	58°–83° $r>v$ moderate	+	{110} distinct;	$b=Y$ $c\wedge Z=-36°$ to $-48°$ $a\wedge X=-20°$ to $-32°$	X = colorless $Y=Z$ = very pale green	H = 5–6. Sp. Gr. = 3.3–3.4 Green, dark green
1.67–1.74	1.67–1.73	1.70–1.76	0.018–0.030	Augite (Ca,Mg,Fe^{2+},Na) $(Mg,Fe^{2+},Fe^{3+},Al,Ti,Cr)$ $(Si,Al)_2O_6$	Monoclinic $\angle\beta\simeq105°$	25°–60°	+	{110} distinct; parting {001} or {100}	$b=Y$ $c\wedge Z=-35°$ to $-50°$ $a\wedge X=-20°$ to $-35°$	Pale green or pale brown	H = 5–6. Sp. Gr. = 3.2–3.6 Gray-green, black, brown
1.68–1.71	1.67–1.70	1.70–1.73	0.029–0.028	Salite $Ca(Mg,Fe^{2+})$ $(SiO_3)_2$ Di_{90-50}	Monoclinic $\angle\beta\simeq105°$	57°–58° $r>v$ moderate	+	{110} distinct; parting {001} or {100}	$b=Y$ $c\wedge Z=-40°$ to $-44°$ $a\wedge X=-24°$ to $-29°$	Colorless, pale green	H = $5\frac{1}{2}$–$6\frac{1}{2}$. Sp. Gr. = 3.3–3.5 Gray-green, brownish green
1.68–1.72	1.67–1.71	1.70–1.74	0.018–0.028	Fassaite $Ca(Mg,Fe^{3+},Al)$ $[(Si,Al)O_3]_2$	Monoclinic $\angle\beta\simeq106°$	51°–62°	+	{110} distinct; parting {100}	$b=Y$ $c\wedge Z=-41°$ to $-47°$ $a\wedge X=-25°$ to $-31°$	X = pale green Y = pale yellow-green Z = pale green	H = 6. Sp. Gr. = 3.0–3.3 Green, dark green
1.68–1.72	1.68–1.72	1.71–1.75	0.023–0.029	Pigeonite (Mg,Fe^{2+},Ca) (Mg,Fe^{2+}) $(SiO_3)_2$	Monoclinic $\angle\beta=108°$	0°–32° $r\gtrless v$ weak	+	{110} distinct; parting {001}	$b=X$ or Y $c\wedge Z=-37°$ to $-44°$ $a\wedge Y$ or $X=-20°$ to 26°	Colorless, $X=Z$ = pale green Y = green, brown	H = 6. Sp. Gr. = 3.3–3.5 Green-brown, black

(continued)

TABLE 5-1. Optical Properties of the Pyroxenes (*continued*)

n_β	n_α	n_γ	$n_\gamma - n_\alpha$	Mineral	System	$2V$	Optic Sign	Cleavage	Optical Orientation	Color in Sections	Physical Properties
1.70–1.72	1.69–1.71	1.70–1.73	0.012–0.015	Hypersthene $(Mg,Fe^{2+})_2(SiO_3)_2$ En_{70-50}	Orthorhombic	57°–50° $r > v$ weak	–	{110} distinct; parting {010} or {100} (rare)	$a=X$ $b=Y$ $c=Z$	X = pale pink Y = yellow Z = pale green	H = 5–6. Sp. Gr. = 3.4–3.6 Dark greenish brown
1.71–1.73	1.70–1.73	1.73–1.75	0.028–0.026	Ferrosalite $Ca(Fe^{2+},Mg)_2(SiO_3)_2$ Di_{50-40}	Monoclinic $\angle\beta \simeq 105°$	58°–62° $r > v$ moderate	+	{110} distinct; parting {001} or {100}	$b=Y$ $c \wedge Z = -44°$ to $-46°$ $a \wedge X = -29°$ to $-32°$	X = pale green Y = blue-green Z = yellow-green	H = $5\frac{1}{2}$–$6\frac{1}{2}$. Sp. Gr. = 3.4–3.6 Dark green, green-brown
1.71–1.73	1.70–1.72	1.73–1.75	0.028–0.029	Johannsenite $Ca(Mn,Fe^{2+})(SiO_3)_2$	Monoclinic $\angle\beta = 105°$	68°–70° $r > v$ moderate	+	{110} distinct; parting {001}, {010} or {100}	$b=Y$ $c \wedge Z = -46°$ to $-48°$ $a \wedge X = -33°$ to $-31°$	Colorless	H = 6. Sp. Gr. = 3.4–3.6 Brown, gray, green
1.71–1.74	1.70–1.72	1.73–1.76	0.030–0.036	Aegirine-augite $(Na,Ca)(Fe^{2+},Fe^{3+},Mg,Al)(SiO_3)_2$	Monoclinic $\angle\beta = 106°$	90°–70° $r > v$ strong	+	{110} distinct; parting {001} or {100}	$b=Y$ $c \wedge X = +12°$ to $+20°$ $a \wedge Z = +6°$ to $-4°$	X = green Y = yellow-green Z = pale yellow	H = 6. Sp. Gr = 3.4–3.6 Dark green, black
1.72–1.75	1.71–1.73	1.73–1.75	0.015–0.018	Ferrohypersthene $(Fe^{2+},Mg)_2(SiO_3)_2$ En_{50-30}	Orthorhombic	50°–57° $v > r$ weak	–	{110} distinct; parting {010} or {100} (rare)	$a=X$ $b=Y$ $c=Z$	X = pale brown Y = yellow Z = pale green	H = 5–6. Sp. Gr. = 3.6–3.7 Dark green, black
1.73–1.74	1.72–1.73	1.75–1.76	0.026–0.025	Hedenbergite $CaFe^{2+}(SiO_3)_2$ Di_{10-0}	Monoclinic $\angle\beta \simeq 104°$	62°–63° $r > v$ strong	+	{110} distinct; parting {001} or {100}	$b=Y$ $c \wedge Z = -46°$ to $-48°$ $a \wedge X = -32°$ to $-34°$	X = green Y = blue-green Z = yellow-green	H = $5\frac{1}{2}$–$6\frac{1}{2}$. Sp. Gr. = 3.5–3.6 Dark green, brown, black
1.74–1.82	1.72–1.78	1.76–1.84	0.036–0.060	Aegirine $NaFe^{3+}(SiO_3)_2$	Monoclinic $\angle\beta = 106°$	58°–90° $r > v$ strong	–	{110} distinct; parting {001} or {100}	$b=Y$ $c \wedge X = -10°$ to $+12°$ $a \wedge Z = +28°$ to $+6°$	X = bright green Y = yellow-green Z = pale yellow-green	H = 6. Sp. Gr. = 3.4–3.6 Dark green, black, brown
1.75–1.76	1.73–1.75	1.75–1.78	0.018–0.021	Eulite $(Fe^{2+},Mg)_2(SiO_3)_2$ En_{30-12}	Orthorhombic	57°–90° $v > r$ moderate	–	{110} distinct; parting {010} or {100} (rare)	$a=X$ $b=Y$ $c=Z$	X = pale brown Y = yellow Z = green	H = 5.6. Sp. Gr. = 3.7–3.9 Dark green, brown, black
1.76–1.77	1.75–1.77	1.78–1.79	0.021–0.022	Orthoferrosilite $Fe_2^{2+}(SiO_3)_2$ En_{12-0}	Orthorhombic	90°–35° $r > v$ strong	+	{110} distinct; parting {010} or {100} (rare)	$a=X$ $b=Y$ $c=Z$	X = brown Y = yellow Z = green	H = 5.6. Sp. Gr. = 3.9–4.0 Dark green, brown, black

The Enstatite-Orthoferrosilite Series $Mg_2(SiO_3)_2$–$Fe_2(SiO_3)_2$ ORTHORHOMBIC

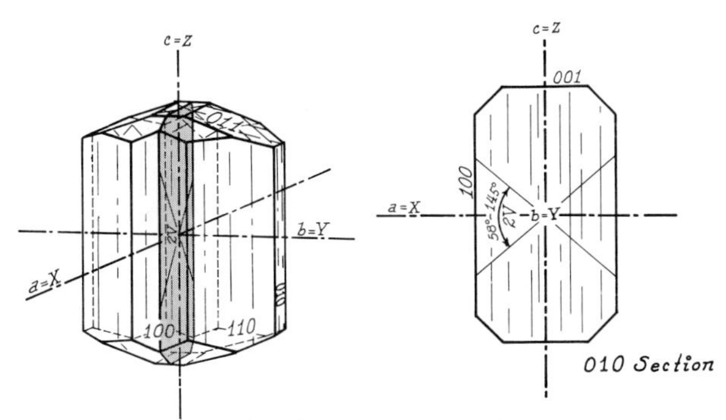

	n_α*	n_β	n_γ*	$n_\gamma - n_\alpha$*	2V*	Dispersion
Enstatite (En_{100})	1.654	1.655	1.665	0.008	35°+	$r > v$ weak
Bronzite (En_{80})	1.677	1.685	1.690	0.011	77°−	$r > v$ moderate
Hypersthene (En_{60})	1.699	1.711	1.714	0.014	53°−	$r > v$ weak
Ferrohypersthene (En_{40})	1.722	1.737	1.739	0.017	53°−	$v > r$ weak
Eulite (En_{20})	1.745	1.756	1.764	0.019	77°−	$v > r$ moderate
Orthoferrosilite (En_0)	1.768	1.771	1.788	0.022	35°+	$r > v$ strong

$a = X, b = Y, c = Z$

Colorless (enstatite) to weakly colored in section, with weak to strong pleochroism in pale red-brown to green

COMPOSITION. The orthopyroxene series is essentially a simple, solid solution series from enstatite (En), $Mg_2(SiO_3)_2$, to orthoferrosilite (Fs), $Fe_2(SiO_3)_2$. The series was divided by Poldervaart (1947) into six composition ranges, after the manner of the plagioclase series. It appears more appropriate, however, to displace two limiting compositions a few mole percent to make them conform to the changes in optic sign.

Enstatite	(En_{100}–En_{88})
Bronzite	(En_{88}–En_{70})
Hypersthene	(En_{70}–En_{50})
Ferrohypersthene	(En_{50}–En_{30})
Eulite	(En_{30}–En_{12})
Orthoferrosilite	(En_{12}–En_0)

* See Fig. 5-8 for continuous variation with composition.

Mg^{2+} and Fe^{2+} replace one another in interchain sites of octahedral coordination where several other ions appear in minor amounts. Volcanic rocks commonly yield orthopyroxenes with notable Ca or Fe^{3+} (quick cooling); Al-rich varieties are derived from metamorphic environments; Mn increases with ferrous iron in late-forming varieties of basic igneous rocks and Ti, Cr, Ni, and alkalis are reported in many analyses. Slow crystallization of orthopyroxene, under plutonic conditions, may cause a Ca-rich, monoclinic phase (pigeonite) to separate as microscopic lamellae on {010} planes. Platy or rodlike inclusions of titanium minerals (brookite, ilmenite) may appear on {010} or {001} planes to produce the schiller structure responsible for bronzelike luster in bronzite and hypersthene. These tiny crystals represent exsolved Ti, by slow cooling, and are less common in volcanic rocks and outer crystal zones.

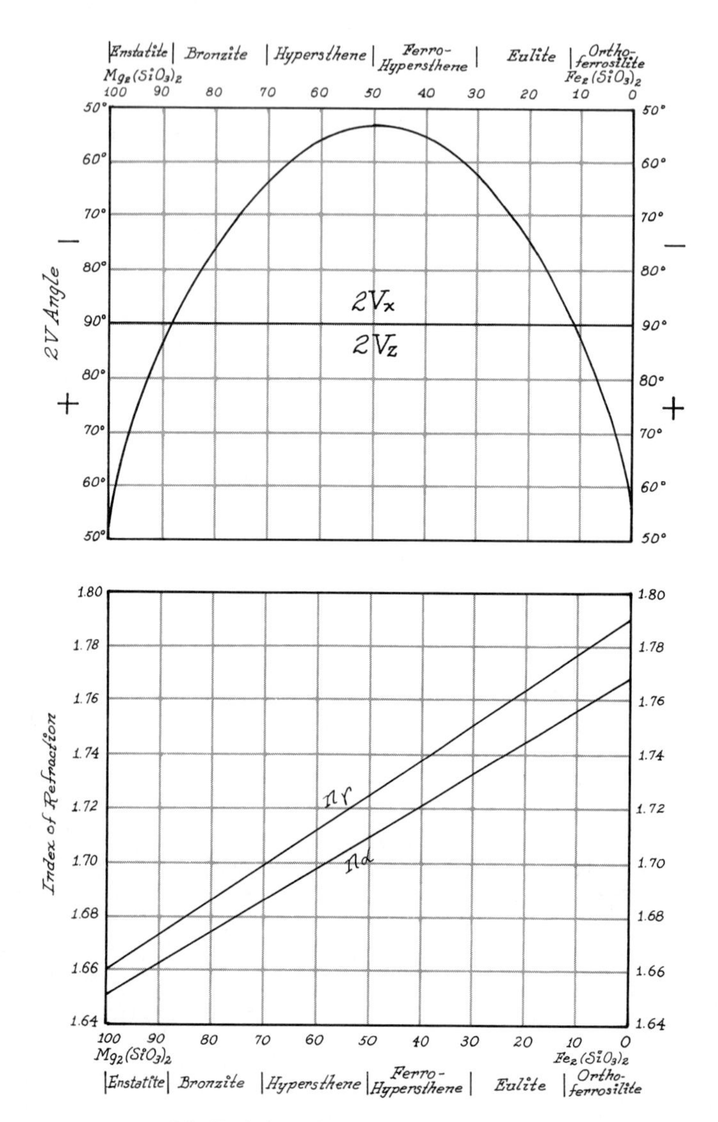

FIGURE 5-8. Variation of $2V$ angle and indices of refraction in the enstatite-orthoferrosilite (orthopyroxene) series. (After Leake, 1968)

The high-temperature clinoenstatite-clinohypersthene series appears unstable at surface conditions, and, excluding meteorites, only the Fe-rich member is reported (rarely) in nature (see Fig. 5-9). A third polymorph of $Mg_2(SiO_3)_2$, *protoenstatite*, is also a high-temperature form unknown in terrestrial rocks.

PHYSICAL PROPERTIES. H = 5–6. Sp. Gr. = 3.2–3.9, increasing with Fe. Color, in hand sample, ranges from pale gray, yellow, brown, or greenish (enstatite) through shades of green or brown to almost black. Luster is typically pearly to metalloid in the bronzite range.

COLOR AND PLEOCHROISM. In section or small fragments, enstatite is colorless. Color increases with Fe, and colored varieties are pleochroic, but pleochroism intensity bears no obvious relationship to composition. Hypersthene is usually lightly colored but highly pleochroic. X = salmon-pink, pale brown; Y = pale brown, yellow; and Z = pale green, bluish green.

FORM. Orthopyroxenes usually occur as short, prismatic crystals ranging from euhedral, in volcanics, to anhedral, in pyroxenites. Cross sections are square to octagonal and show characteristic prismatic cleavages.

CLEAVAGE. All pyroxenes display distinct prismatic cleavages parallel to (110) and $(1\bar{1}0)$ intersecting at 87° and 93°. Parting is relatively common on {010} and rare on {100}.

BIREFRINGENCE. Maximum birefringence, seen in {010} sections, increases uniformly, with ferrous iron, from 0.008 (enstatite) to 0.022 (orthoferrosilite). In standard section, interference colors are first-order grays and white, for enstatite, increasing to high first-order colors, near orthoferrosilite.

TWINNING. Rare normal twinning is reported on {101}, and some writers propose parallel twinning about c as the source of thin (0.001 mm ±), parallel lamellae on {010}. Multiple twinning is almost universal in the clinoenstatite-clinoferrosilite series.

ZONING. Mg-pyroxene crystallizes early and reacts with remaining magma to become richer in iron. Normal zoning

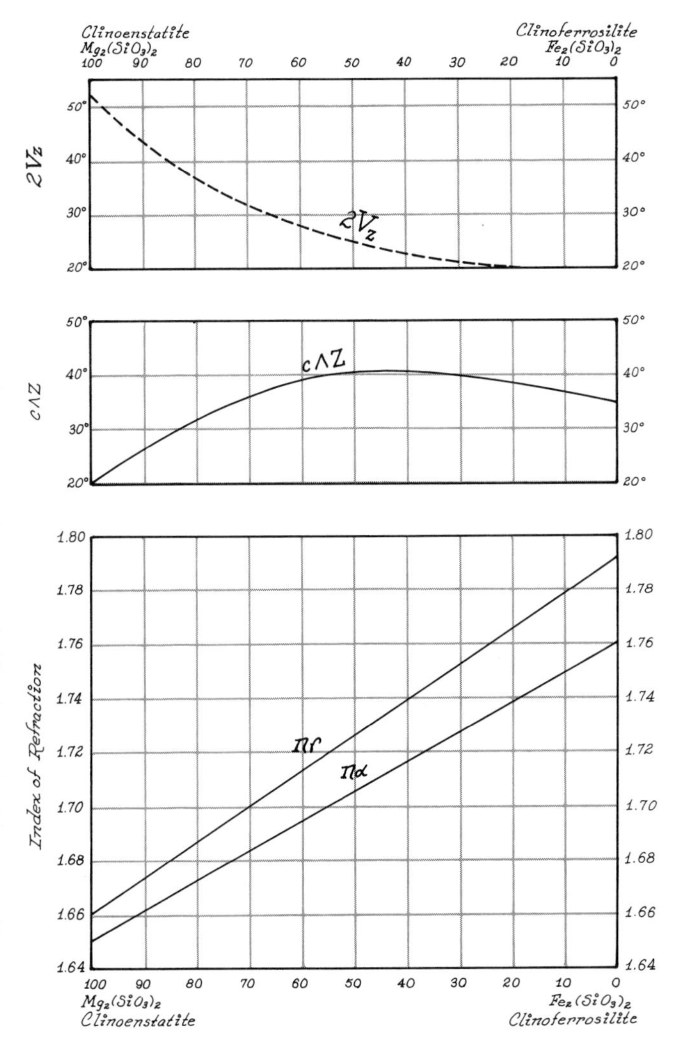

FIGURE 5-9. Variation of $2V$ angle, $c \wedge Z$, and indices of refraction in the clinoenstatite-clinoferrosilite series (After Bowen and Schairer, 1935). Minerals of this series are very rare in terrestrial nature.

is, therefore, represented by Fe-enrichment (and darker color) in outer zones and may range over 10 percent En. Reverse zoning is not uncommon, and a form of oscillatory zoning is reported. Slow crystallization in plutonic environments favors equilibrium, and plutonic igneous or metamorphic orthopyroxenes are largely unzoned. Volcanic crystals are normally zoned, however, and core zones may even show calcic exsolution lamellae.

INTERFERENCE FIGURE. Enstatite is biaxial positive and $2V_z$, increases rapidly with iron from 35° to 90°; it is usually large (see Fig. 5-8). As $c = Z$ is the acute bisectrix, Bxa figures are seen on basal sections showing both prismatic cleavages. Bronzite appears, by definition, when $2V_z$ exceeds 90° and compositions to orthoferrosilite are biaxial negative, with acute bisectrix figures forming about $a = X$ on {100} sections. $2V_x$ is large, for bronzite and eulite, reaching a minimum of about 50° between hypersthene and ferrohypersthene. Orthoferrosilite is positive, with large $2V_z$, decreasing with Fe^{+2}.

Dispersion is without bisectrix dispersion and ranges from imperceptible to moderately strong without obvious relation to composition. Measured on X, dispersion changes from $v > r$ to $r > v$ at En_{85} and back again to $v > r$ at En_{50}. Measured on the acute bisectrix, $r > v$ for enstatite, bronzite, hypersthene, and orthoferrosilite, and $v > r$ for ferrohypersthene and eulite.

All members of the clinoenstatite-clinoferrosilite series are biaxial positive with small $2V_z$ (20°), increasing with Mg to about 50°.

OPTICAL ORIENTATION. The optic plane is parallel to $c = Z$ and is usually designated as {010},* making $a = X$ and $b = Y$. $c = Z$ parallels {110} cleavage traces, giving fragments and primatic sections positive elongation (length-slow). Extinction is parallel for prismatic sections and fragments and symmetrical to {110} in basal sections.

DISTINGUISHING FEATURES. Orthopyroxenes are distinguished, generally, from clinopyroxenes by parallel extinction. Enstatite shows lower birefringence, and iron varieties show characteristic brown-to-green pleochroism. Orthopyroxenes

* Deer, Howie, and Zussman (1963) define the optic plane parallel to {100} and $a = Y$, $b = X$.

may occur with and be confused with anthophyllite, andalusite, or sillimanite. Anthophyllite is an amphibole pleochroic in browns with higher birefringence than enstatite and refractive indices lower than hypersthene. Andalusite is orthorhombic with prismatic cleavages at nearly 90°, but $c = X$, making fragments and elongated sections length-fast. Sillimanite has high birefringence and small $2V$ compared to enstatite or hypersthene.

ALTERATION. Serpentine resulting from the alteration of Mg-rich orthopyroxene is called *bastite* when it retains schiller inclusions and bronzelike luster of the original pyroxene. Secondary, fibrous, pale green amphibole (uralite) is a common alteration product of the orthopyroxenes, and talc and magnesite are less common. Early-forming orthopyroxene often reacts with magma or lava to yield reaction rims of clinopyroxene or olivine or inverts to garnet with retrograde metamorphism

OCCURRENCE. The orthopyroxenes are common constituents of mafic and ultramafic plutonic igneous rocks, volcanics, and high-grade metamorphic rocks, both regional and contact, and occur rarely as detrital grains in sediments.

Plutonic igneous rocks contain Mg-rich orthopyroxenes in ultramafic varieties and Fe-rich orthopyroxenes in mafic types. Enstatite, in association with olivine, Mg-rich clinopyroxenes and phlogopite, is characteristic of pyroxenites, peridotites, dunites (and serpentinites, which are derived from them), and is a common constituent of meteorites. Some pyroxenites are almost entirely enstatite. Hypersthene is the characteristic mineral of norites and some gabbros. Ferrohypersthene and eulite are occasional components of diorites, monzonites, granites, and similar intermediate and silicic rock types, where crystals of orthopyroxene are commonly surrounded by reaction rims of clinopyroxene.

Volcanic rocks tend to yield single-phase orthopyroxene, possibly rich in Ca, Fe^{+3}, or Ti. In volcanics, orthopyroxene is a rare associate of olivine and appears mostly in basalts or andesites having a silica content about 58 percent.

Metamorphic rocks of both regional and contact origin may contain orthopyroxenes; high-grade antistress conditions are implied. Orthopyroxene is most characteristic of the granulite facies, although it may appear in gneisses of the milder amphibolite facies, in association with labradorite and amphiboles, and is found in eclogites formed by extreme metamorphism. Orthopyroxene near bronzite is the characteristic mineral of the charnockites, where it appears with diopside and hornblende. Mafic metamorphic rocks tend to contain Mg-rich orthopyroxenes, and silicic rocks favor Fe-rich varieties. Ferrohypersthene and eulite may appear with Fe-olivine and Fe-amphiboles in metamorphosed iron sediments. Orthopyroxene is common, with cordierite and microcline, in high-grade hornfels.

REFERENCES: THE ENSTATITE-ORTHOFERROSILITE SERIES

Bowen, N. L., and J. F. Schairer. 1935. The system MgO-FeO-SiO_2. *Amer. Jour. Sci.*, Series 5, *229*, 151–217.

Bown, M. G., and P. Gay. 1959. The identification of oriented inclusions in pyroxene crystals. *Amer. Mineral.*, *44*, 592–602.

Deer, W. A., R. A. Howie, and J. Zussman. 1963. *Rock-Forming Minerals.* Vol. 2: *Chain Silicates.* London: Longmans Green.

Henriques, A. 1958. The influence of cations on the optical properties of orthopyroxenes. *Arkiv. Min. Geol.*, *2*. 385–390.

Henry, N. F. M. 1935. Some data on the iron-rich hypersthenes. *Mineral. Mag.*, *24*, 221–226.

Henry, N. F. M. 1938. A review of the data of Mg-Fe-clinopyroxenes. *Mineral. Mag.*, *25*, 23–29.

Henry, N. F. M. 1942. Lamellar structure in orthopyroxene. *Mineral. Mag.*, *26*, 179–189.

Hess, H. H. 1941. Pyroxenes of common mafic magmas. II. *Amer. Mineral.*, *26*, 573–594.

Hess, H. H., and A. H. Phillips. 1940. Optical properties and chemical composition of magnesian orthopyroxenes. *Amer. Mineral.*, *25*, 271–285.

Iijima, S., and P. R. Buseck. 1975. High-resolution electron microscopy of enstatite. I. Twinning, polymorphism, and polytypism. *Amer. Mineral.*, *60*, 758–770.

Kuno, H. 1954. Study of orthopyroxenes from volcanic rocks. *Amer. Mineral.*, *39*, 30–46.

Leake, B. E. 1968. Optical properties and composition in the orthopyroxene series. *Mineral. Mag.*, *36*, 745–747.

Norton, D. A., and W. S. Clavan. 1959. Optical mineralogy, chemistry and X-ray crystallography of ten pyroxenes. *Amer. Mineral.*, *44*, 844–874.

Poldervaart, A. 1947. The relation of orthopyroxene to pigeonite. *Mineral. Mag.*, *28*, 164–172.

Poldervaart, A. 1950. Correlation of physical properties and chemical composition in the plagioclase, olivine and orthopyroxene series. *Amer. Mineral.*, *35*, 1067–1079.

Poldervaart, A., and H. H. Hess. 1951. Pyroxenes in the crystallization of basaltic magma. *Jour. Geol.*, *59*, 472–489.

Smith, J. V. 1969. Crystal structure and stability of the $MgSiO_3$ polymorphs; physical properties and phase relations of Mg,Fe pyroxenes. *Mineral. Soc. Amer. Spec. Paper*, *2*, 3–29.

Walls, R. 1935. A critical review of the data for a revision of the enstatite-hypersthene series. *Mineral. Mag.*, *24*, 165–172.

The Diopside-Hedenbergite Series

$CaMg(SiO_3)_2$–$CaFe(SiO_3)_2$

MONOCLINIC

$\angle\beta = 106°–104°$

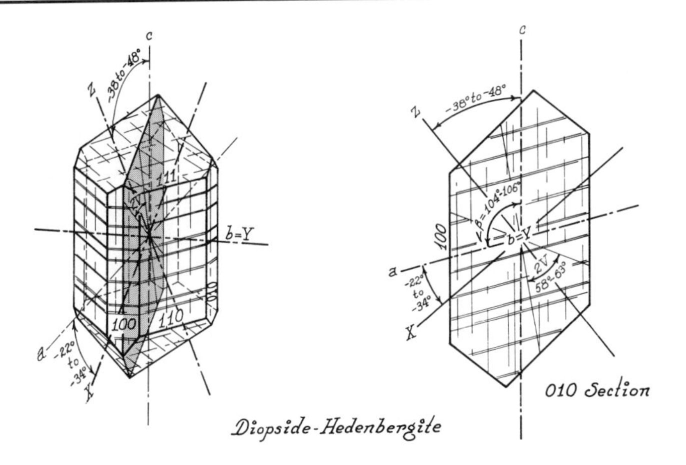

	n_α*	n_β*	n_γ*	$(n_\gamma - n_\alpha)$*	$c \wedge Z$*	$2V_z$*	Dispersion
Diopside (Di_{100})	1.664	1.672	1.694	0.030	−38°	58°+	$r > v$ weak
Salite (Di_{70})	1.685	1.692	1.714	0.029	−42°	57°+	$r > v$ moderate
Ferrosalite (Di_{30})	1.713	1.719	1.739	0.026	−45°	60°+	$r > v$ moderate
Hedenbergite (Di_0)	1.732	1.739	1.757	0.025	−48°	63°+	$r > v$ strong

$b = Y, c \wedge Z = -38°$ to $-48°$, $a \wedge X = -22°$ to $-34°$

Colorless (diopside) to weakly colored in section

Weak to strong pleochroism in blue-green and brown

COMPOSITION. Diopside (Di), $CaMg(SiO_3)_2$, and hedenbergite (Hd), $CaFe(SiO_3)_2$, form a relatively simple solid solution series that probably forms complete solid solution with a third end member, johannsenite (Jh), $CaMn(SiO_3)_2$—see Fig 5-11.

Diopside	(Di_{100}–Di_{90})
Salite	(Di_{90}–Di_{50})
Ferrosalite	(Di_{50}–Di_{10})
Hedenbergite	(Di_{10}–Di_0)

* See Fig. 5-10 for continuous variation of optical properties with composition. Indices of refraction increase with Fe^{3+} and Mn^{2+} and birefringence decreases with Al^{3+}.

Ca^{2+} ions and (Mg^{2+} + Fe^{2+}) occur in essentially equal numbers, and end members closely approach ideal composition. Minor Al^{3+} and Fe^{3+} may appear in intermediate compositions, and significant Cr^{3+} or Zn^{2+} may accompany Mn^{2+} in intermediate members.

PHYSICAL PROPERTIES. H = $5\frac{1}{2}$–$6\frac{1}{2}$. Sp. Gr. = 3.2–3.5, increasing with Fe. Color, in hand sample, is pale green to nearly white for diopside, becoming dark green or green-brown to nearly black for the ferrous end member.

COLOR AND PLEOCHROISM. Diopside is colorless to pale green, with little pleochroism. Color and pleochroism increase with iron in shades of bluish or brownish green. X = pale to strong green or blue-green; Y = brownish green or blue-green; and Z = yellowish green.

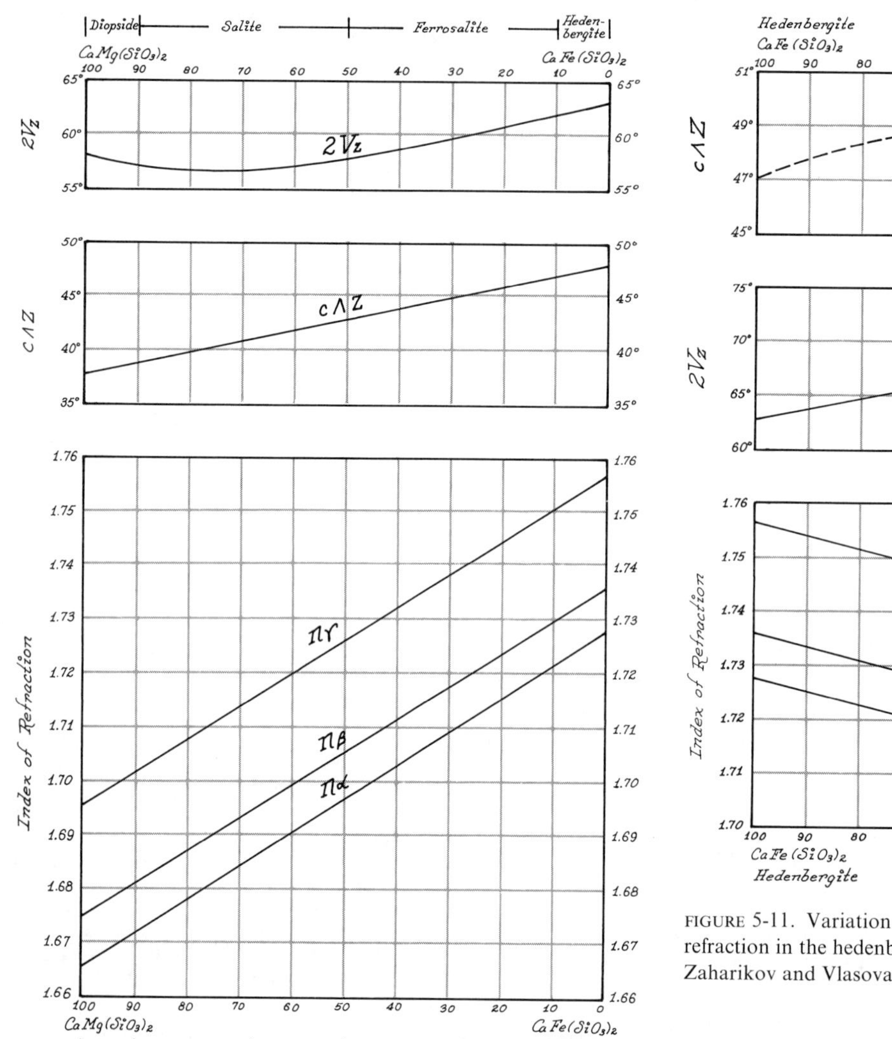

FIGURE 5-10. Variation of $2V$ angle, $c \wedge Z$, and indices of refraction in the diopside-hedenbergite series. (Data from Deer, Howie, and Zussman, 1962)

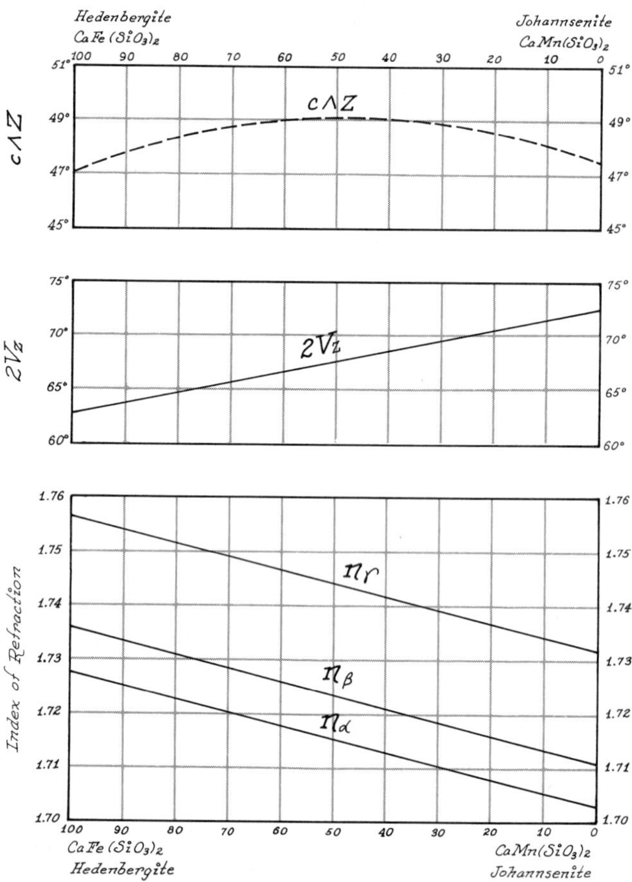

FIGURE 5-11. Variation of $2V$ angle, $c \wedge Z$, and indices of refraction in the hedenbergite-johannsenite series. (After Zaharikov and Vlasova, 1955)

FORM. Magnesium-rich varieties may appear as granular aggregates or stubby euhedral crystals, with rectangular sections and square or octagonal cross sections showing characteristic prismatic cleavages at 93°. Hedenbergite tends to occur as columnar or accicular aggregates in parallel to radial arrangement.

CLEAVAGE. Distinct prismatic {110} cleavages intersecting at 93° and 87° are characteristic of all pyroxenes. Parting on {001} and {100} are often well developed. Diopside or augite showing very prominent basal parting with magnetite or ilmenite dust on parting surfaces is often called *diallage.*

BIREFRINGENCE. Maximum birefringence, seen in {010} sections, ranges from 0.025 (hedenbergite) to 0.030 (diopside), decreasing somewhat with Al content. Interference colors, in standard 0.03 mm sections, are brilliant first order and low second order with low first-order colors (near white) in basal section.

TWINNING. Polysynthetic or simple twinning is relatively common on {100} or {001}.

INTERFERENCE FIGURE. All members of the series are optically positive, with $2V_z$ increasing slightly with Fe^{2+} from 58° (diopside) to 63° (hedenbergite) (see Fig. 5-10). $2V_z$ increases with Mn and rapidly with minor Na. {010} sections show flash figures, and basal sections {001} show off-center optic axis figures, with few isochromes, symmetrical about the optic plane. Cleavage fragments on {110} produce poor, eccentric figures.

OPTICAL ORIENTATION. The optic plane is {010} and $b = Y$. Within {010}, $c \wedge Z = -38°$ (diopside) increasing with Fe^{2+} to $-48°$ (hedenbergite) and $a \wedge X = -22°$ (diopside) to $-34°$ (hedenbergite).

Maximum extinction to cleavage traces ranges from 38° to 45° in {010} sections. Basal sections {001} show symmetrical extinction to prismatic cleavages, and {100} sections show parallel extinction.

Sign of elongation is difficult to determine, as Z and X lie almost 45° to elongation in {010} sections. For diopside, the slow wave (Z) is slightly nearer elongation, and for hedenbergite the fast wave (X) is slightly nearer. {100} sections are length-slow (positive elongation).

Optic axis dispersion is $r > v$ increasing with Fe^{2+} from weak to strong. Bisectrix dispersion is weakly inclined.

DISTINGUISHING FEATURES. Members of the diopside-hedenbergite series may be almost indistinguishable from augite-ferroaugite. Diopside-hedenbergite members usually show slightly larger $2V$ and occur mostly in metamorphic rocks and basic volcanics, while augite is most characteristic of mafic and ultramafic plutonic igneous rocks.

Compared to diopside-hedenbergite, pigeonite has low birefringence and small $2V$, orthopyroxenes show low birefringence and parallel extinction, and wollastonite is colorless, with low birefringence and low refractive index.

ALTERATION. Diopside-hedenbergite alters commonly to uralite (fibrous tremolite-actinolite) and rarely to talc, serpentine, or chlorite. Very high-temperature metamorphism causes diopside to become unstable with respect to other calc-silicate minerals (such as monticellite).

OCCURRENCE. At high temperatures, the ratio of Ca^{2+} to ($Mg^{2+} + Fe^{2+}$) tends to decrease, forming augite-ferroaugite in preference to diopside-hedenbergite in most igneous rocks and rocks of high-grade regional metamorphism.

Metamorphism forms diopside in Ca- and Mg-rich sediments by regional or contact metamorphism. It is especially characteristic of the epidote-amphibole or pyroxene-hornfels facies of carbonate contact zones, where it appears in magnesian marbles and skarns in association with calcite, forsterite, wollastonite, grossular, tremolite, idocrase, and other Ca-Mg silicates. Hedenbergite appears in contact zones in carbonate rocks and ferruginous silica sediments, often associated with grunerite.

Pyroxene granulites and gneisses commonly contain diopside with hypersthene, plagioclase, and almandine.

Igneous rocks contain less diopside-hedenbergite than do metamorphic rocks but diopside may be anticipated in very mafic volcanics (olivine basalt, picrite, and so on), especially alkali-rich varieties with olivine, nepheline, and alkali amphiboles. Diopside is occasionally found in peridotites or anorthosites, while hedenbergite tends to accompany quartz and fayalite in alkali granites. Diopside may appear in pegmatites through limestone assimilation.

REFERENCES: THE DIOPSIDE-HEDENBERGITE SERIES

Clark, J. R., D. E. Appleman, and J. J. Papike. 1969. Crystal chemical characterization of clinopyroxenes based on eight new structure refinements. *Mineral. Soc. Amer. Spec. Paper*, *2*, 31–50.

Deer, W. A., R. A. Howie, and J. Zussman. 1962. *Rock-Forming Minerals*. Vol. 2: *Chain Silicates*. New York: Wiley.

Hess, H. H. 1949. Chemical composition and optical properties of common clinopyroxenes. I. *Amer. Mineral.*, *34*, 621–666.

Myer, G. H., and D. H. Lindsley. 1969. Optical properties of synthetic clinopyroxenes on the join hedenbergite-clinoferrosilite. *Carnegie Inst. Wash. Year Book*, *67*, 92–94.

Norton, D. A., and W. S. Clavan. 1959. Optical mineralogy, chemistry and x-ray crystallography of ten pyroxenes. *Amer. Mineral.*, *44*, 844–874.

Zaharikov, V. A., and D. K. Vlasova. 1955. The diagram: Composition properties of the isomorphous series diopside-hedenbergite-johannsenite. *Dokl. Acad. Sci. USSR*, *105*, 814.

Pigeonite

$(Mg,Fe^{2+},Ca)(Mg,Fe^{2+})(SiO_3)_2$

MONOCLINIC
$\angle\beta = 108°$

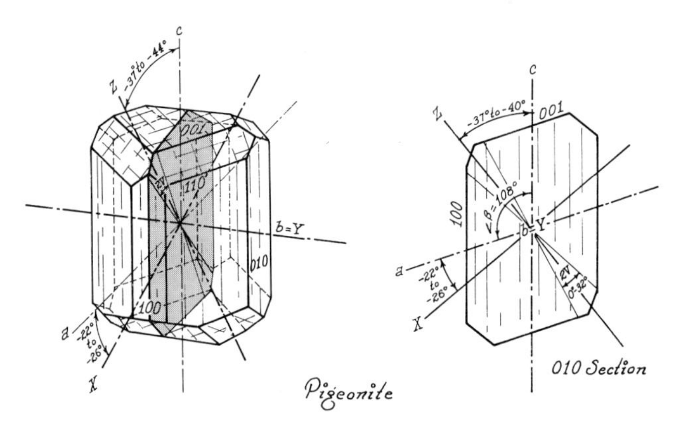

$n_\alpha = 1.682–1.722$*
$n_\beta = 1.684–1.722$
$n_\gamma = 1.705–1.751$
$n_\gamma - n_\alpha = 0.023–0.029$
Biaxial positive, $2V_z = 0°–32°$
Two orientation patterns:

1. Optic plane ⊥ (010) (most common)
   $b = X, c \wedge Z = -37°$ to $-44°$,
   $a \wedge Y = -19°$ to $-26°$
2. Optic plane|| (010) (less common)
   $b = Y, c \wedge Z = -40°$ to $-44°$,
   $a \wedge X = -22°$ to $-26°$

$r > v$ or $v > r$ weak to moderate
Colorless to pale green or brown in section, with faint to moderate pleochroism
$Y > X = Z$

COMPOSITION. Pigeonite is Ca-poor augite in the range $CaSiO_3$ = 5 to 15 percent (see Fig. 5-4). All Ca^{2+} ions occupy the interchain cation site of eightfold coordination but are insufficient to fill the site. Fe^{2+} ions fill out this site, which collapses to effectively sevenfold coordination. The repetition distance in adjacent chains changes from that of augite, and pigeonite assumes a different space group and is structurally distinguishable from augite.

Minor Al, Cr, and Na accompany Mg in early stages of crystallization, and Mn^{2+} replaces Fe^{2+} in late crystals. Fe^{3+} and Ti^{4+} also appear as minor constituents. Some degree of immiscibility appears to exist between augite and pigeonite, and with slow cooling, pigeonite exsolves Ca-rich augite as lamellae on {001}.

PHYSICAL PROPERTIES. H = 6. Sp. Gr. = 3.30–3.46. Color, in hand sample, is green-brown to black.

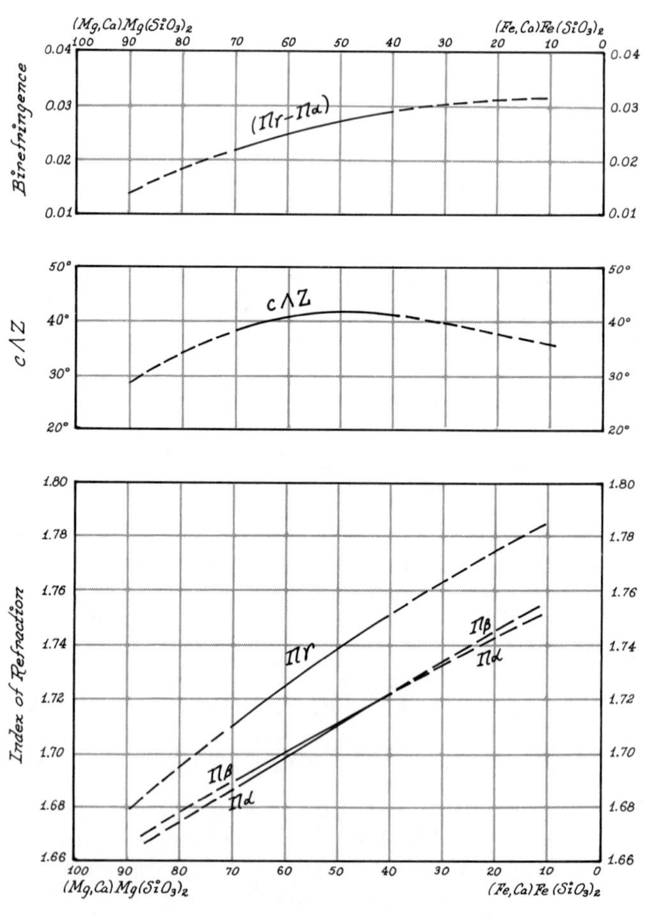

FIGURE 5-12. Variation of birefringence ($n_\gamma - n_\alpha$), $c \wedge Z$, and indices of refraction in the pigeonite pyroxenes. (After Hess, 1949)

* With decreasing Ca, the $2V_z$ closes, in the (010) plane, to 0° at about 10 percent $CaSiO_3$ and opens again in a plane normal to (010). Refractive indices and birefringence increase with Fe^{2+} (Fig. 5-12).

COLOR AND PLEOCHROISM. Pigeonite is colorless to pale green or brown, in section, with deeper color, as fragments. Pleochroism is absent to occasionally moderate in pale greens and browns, with maximum absorption on Y.

FORM. Pigeonite appears as small, euhedral crystals to anhedral grains in the groundmass of intermediate volcanic rocks and, less commonly, as euhedral phenocrysts. Crystals are prismatic, elongated on c, with typical octagonal pyroxene outline and prismatic cleavage.

CLEAVAGE. Pigeonite shows typical pyroxene prismatic $\{110\}$ cleavages intersecting at 87° and 93°. Parting on $\{001\}$ may be present and may reflect exsolution lamellae on $\{001\}$.

BIREFRINGENCE. Maximum birefringence increases with iron from 0.023 to 0.029 producing bright colors to middle second order, in standard section. Maximum birefringence appears in $\{010\}$ sections of Ca-poor pigeonite and in sections containing b and inclined about 42° to c of Ca-rich pigeonites, and all sections containing Z show near-maximum birefringence.

TWINNING. Both simple and multiple twinning on $\{100\}$ are commonly observed. Exsolution lamellae on $\{001\}$ may yield the appearance of polysynthetic twinning and, in combination with $\{100\}$ twinning, produce a herringbone pattern.

ZONING. Zoning is common, showing Ca-poor cores.

INTERFERENCE FIGURE. Pigeonite is optically positive and may appear essentially uniaxial. Acute bisectrix and optic axis figures are rare, show small $2V_z$ with isogyres that remain in the visible field and, indeed, may separate only slightly with stage rotation. Fragments on $\{110\}$ cleavages show greatly off-center figures without symmetry, and inherent color may obscure broad isochromes. Fragments on $\{001\}$ parting show near-optic axis figures or off-center Bxa symmetrical on Y.

Optic axis dispersion may be absent to moderate $r > v$ or $v > r$, and bisectrix dispersion may show as weakly inclined or horizontal.

OPTICAL ORIENTATION. Ca-poor pigeonite shows an optic plane perpendicular to $\{010\}$. With increasing Ca or, to a lesser degree, Fe^{2+}, $2V_z$ closes to zero (uniaxial) near the center of the pigeonite range and opens again about Z in the $\{010\}$ plane. Pigeonite, thus, shows two orientation patterns: When the optic plane is perpendicular to $\{010\}$ (Ca-poor), $b = X, c \wedge Z = -37°$ to $-44°$ and $a \wedge Y = -19°$ to $-26°$. When the optic plane is parallel to $\{010\}$ (Ca-rich), $b = Y$, $c \wedge Z = -40°$ to $-44°$, and $a \wedge X = -22°$ to $-26°$.

The maximum extinction angle of pigeonite is 44° to 37° to cleavage traces, seen in $\{010\}$ sections. Extinction is parallel in $\{100\}$ sections, symmetrical in basal sections $\{001\}$ and about 32° for fragments on $\{110\}$ cleavages.

Elongation sign is always positive (length-slow), as Z is always nearer c. In $\{010\}$ sections, however, Z may lie nearly 45° to cleavage traces.

DISTINGUISHING FEATURES. Pigeonite is distinguished from other pyroxenes by its small $2V_z < 32°$ and its restricted occurrence in volcanic rocks. Pigeonite is most commonly mistaken for augite, in which $2V_z > 32°$, or for orthopyroxene, which shows lower birefringence and parallel extinction. Members of the clinoenstatite-clinoferrosilite series may show $2V_z$ only slightly greater than 32°, but these minerals are essentially unknown in natural occurrences. Olivine is a common associate of pigeonite, showing higher birefringence and minor cleavage.

ALTERATION. Pigeonite may alter to a fibrous aggregate of amphibole needles or plates (uralite) beginning at the outer margin; less commonly to chlorite or serpentine.

OCCURRENCE. Pigeonite is limited to igneous rocks that have cooled quickly as volcanics or shallow intrusives. It is most characteristic of andesites and similar intermediate volcanics, where it occurs in the groundmass as small crystals or inter-feldspar grains and, less commonly, as distinct phenocrysts. Pigeonite may appear in association with augite or hypersthene and may form, with either, as corona rims or layered intergrowths. Although plutonic igneous rocks do not contain pigeonite, it apparently forms on chilled borders and subsequently inverts to augite, which retains intergrowth planes.

REFERENCES: PIGEONITE

Brown, G. E., C. T. Prewitt, J. J. Papike, and S. Sueno. 1972. Comparison of the structures of low and high pigeonite. *Jour. Geophys. Res.*, *77*, 5778–5789.

Brown, G. M. 1966. Experimental studies on inversion relations in natural pigeonitic pyroxenes. *Carnegie Inst. Wash. Year Book*, *66*, 347–353.

Hess, H. H. 1949. Chemical composition and optical properties of common clinopyroxenes. *Amer. Mineral.*, *34*, 621–666.

Schwab, R. G., and K. H. Jablonski. 1973. Polymorphism in pigeonites. *Fortschr. Mineral.*, *50*, 223–263.

Turner, F. J. 1940. Note on determination of optical axial angle and extinction angle in pigeonite. *Amer. Mineral.*, *25*, 821–823.

Augite

$(Ca,Mg,Fe^{2+},Na)(Mg,Fe^{2+},Fe^{3+},Al,Ti,Cr)(Si,Al)_2O_6$

MONOCLINIC
$\angle\beta = 105°$

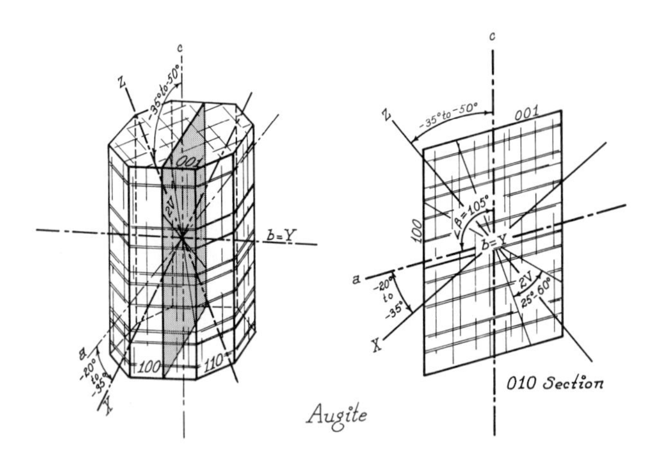

$n_\alpha = 1.671–1.735$*
$n_\beta = 1.672–1.741$
$n_\gamma = 1.703–1.761$
$n_\gamma - n_\alpha = 0.018–0.030$
Biaxial positive, $2V_z = 25°–60°$
$b = Y, c \wedge Z = -35°$ to $-50°$, $a \wedge X = -20°$ to $-35°$
$r > v$ weak to distinct, inclined bisectrix dispersion
Colorless to pale green or pale brown-green in section, with weak pleochroism
Titanaugite is moderately pleochroic in shades of brown and violet
$Y > Z > X$

* Refractive index, birefringence and $2V_z$ increase with Fe^{2+}. Refractive index increases with Fe^{3+} and Ti and decreases with Cr and Al, birefringence decreases with Al, and $2V_z$ is enlarged by Fe^{3+} and Ti.

COMPOSITION. Augite represents central, complex pyroxene compositions not easily defined in terms of simple end members (Fig. 5-4). It is a ferromagnesian clinopyroxene containing less Ca than diopside-hedenbergite, more Ca than pigeonite, and less Na than aegirine-augite. Ferroaugite contains more Fe^{2+} than Mg, and subcalcic augite and subcalcic ferroaugite contain less Ca (15–25 percent $CaSiO_3$) than normal augite. Substitution of Al^{3+} for Si^{4+} or Na^+ for Ca^{2+} allows substitution of trivalent ions (Al, Fe^{3+}, Cr, Ti, and so on) for bivalent ones (Mg, Fe^{2+}, Mn) and the augite complex results. *Fassaite* is an augite variety containing significant $(Al,Fe^{3+}) \rightleftharpoons (Mg,Fe^{2+})$ coupled with $Al \rightleftharpoons Si$ substitution (see Fig. 4-13).

Ca tends to increase with Mg† and the Ca-Mg-rich augites may contain significant Al, Ti, Fe^{3+}, and Cr with minor Ni, Co, V, Cu, and Zr. Titanaugites are usually Al-rich, and Fe^{3+} and Na replace Mg and Ca to form the series with aegirine augite. Aluminum appears largely as Si-replacement and, occasionally, minor Ti^{4+} or Fe^{3+} may prove necessary to fill tetrahedral sites. Ferroaugites may contain significant Mn and traces of Li, Sr, and rare earths.

Exsolution lamellae of two pyroxene phases commonly form as single crystals with the slow cooling of a subsiliceous magma. Early crystallization stages produce Mg-rich pyroxenes interlayered as Ca-rich (augite) and Ca-poor (enstatite or pigeonite) phases. Exsolution lamellae of orthopyroxene form on {100} planes of Mg-rich ($Mg:Fe^{2+} > 3:1$) augite host, and pigeonite layers form on {001} in augite containing more iron. Late crystallization of Fe^{2+}-rich magmas yields essentially a single pyroxene, ferroaugite, with perhaps a few very thin exsolution layers on {001}, appearing as schiller structure. The Ca-poor phase (Fe-rich pigeonite) may fail to exist in favor of fayalite.

PHYSICAL PROPERTIES. H = 5–6. Sp. Gr. = 3.2–3.6. Color, in hand sample, is usually gray-green to black. Titanaugite is brown or brown-violet.

COLOR AND PLEOCHROISM. Even varieties highly colored, in hand sample, are colorless to neutral gray or pale green in

† Subcalcic augite and Ca-rich ferroaugite are rare compositions formed only by rapid cooling.

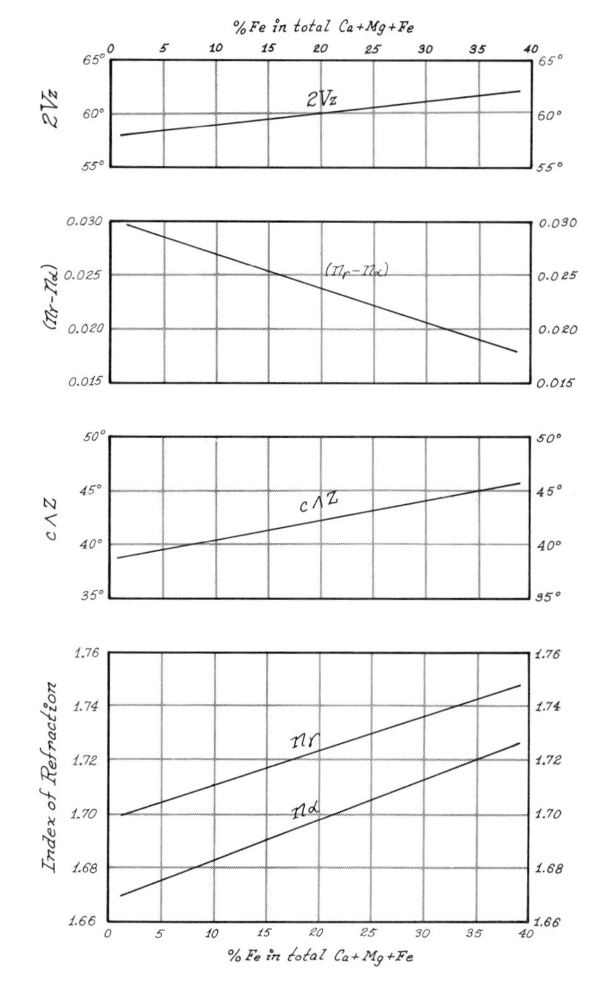

FIGURE 5-13. Variation of $2V$ angle, birefringence, $c \wedge Z$, and indices of refraction for the fassaite pyroxenes. (After Majmudar, 1970)

section. Fragments tend to be green, with no more than weak pleochroism in pale blue-green or pale yellow-green. Both color intensity and pleochroism increase with iron.

Titanaugite is distinctly colored in section and moderately pleochroic in brown and violet. X = pale violet-brown; Y = violet-pink; and Z = pale brown-violet.*

* Directions of maximum color absorption apparently do not parallel optical directions.

FORM. Augite in plutonic rocks and as volcanic phenocrysts is euhedral to subhedral, with stubby rectangular prismatic sections and square to octagonal basal sections. Ragged overgrowths of hornblende are common on augite crystals, and augite rims appear on clinopyroxenes. Exsolution lamellae of orthopyroxene, on {100} planes, or pigeonite, on {001} planes. are common in augite hosts, and grains of plagioclase and olivine appear in poikilitic crystals of augite. Small, anhedral grains of augite commonly appear in ophitic relationship with Ca-plagioclase in the groundmass of mafic volcanics.

CLEAVAGE. Distinct prismatic cleavage exists parallel to (110) and $(1\bar{1}0)$ intersecting at 87° and 93°. Closely spaced parting may exist on {100} or {001}. *Diallage* is the name applied to clinopyroxenes showing very highly developed parting on {100}, often with plates of ilmenite on parting surfaces. It is common in intrusive rocks. *Malacolite* is a less common name, restricted to diopside or augite showing similar distinct parting on {001}.

BIREFRINGENCE. Maximum birefringence is seen in {010} sections and is moderate, ranging from about 0.018 to 0.030, increasing with iron. Interference colors are bright colors of first order and low second order. Fragments commonly show similar bright colors, which may be dulled by inherent mineral color and which likely represent less than maximum birefringence (fragments usually lie on {110} cleavages).

TWINNING. Normal twinning on {100} is common and may be simple or repeated. Normal, polysynthetic twinning on {001} is responsible for basal parting and, in combination with the first law, produces herringbone structure in {010} sections. Penetration twinning on {010} and cyclic twinning on {122} are rare.

ZONING. Normal composition zoning is common in volcanic rocks and shows increasing Fe^{2+} and darker color in outer zones. Reverse zoning is rare, but titanaugite may well show oscillatory or "hourglass" zoning.

INTERFERENCE FIGURE. Cleavage fragments normally lie on {110} and yield off-center figures. Fragments lying on parting surfaces {100} or {001} show low birefringence and off-center optic axis figures symmetrical about the optic plane. Positive acute bisectrix figures show a moderate $2V_x$ increasing, with Fe^{2+}, from about 25° to 60°. Dispersion is normally weak $r > v$, but titanaugite shows strong dispersion $r > v$, with distinct inclined bisectrix dispersion.

OPTICAL ORIENTATION. $b = Y$, $c \wedge Z = -35°$ to $-50°$ and $a \wedge X = -20°$ to $-35°$. Maximum extinction angle to {110} cleavage is seen in {010} sections and ranges from 35° to 50°. Extinction is parallel in {100} sections and symmetrical to cleavage traces in basal sections.

Sign of elongation is essentially meaningless, since the extinction angle is near 45° and either the fast or slow wave may lie slightly nearer the elongation. Normally, the slow wave is slightly nearer elongation in {010} sections, and elongation is positive (length-slow) in {100} sections.

DISTINGUISHING FEATURES. Olivine is without good cleavage and shows higher birefringence than augite, and amphiboles show cleavage intersections of 56° and 124° and tend to be highly colored, with small extinction angles. Augite is distinguished from some other pyroxene varieties with great difficulty, owing to continuous chemical variation. High relief, moderate birefringence, and octagonal cross sections with prismatic cleavages at 87° are distinctive of pyroxenes as a group. Aegirine-augite is distinctly green in section, orthopyroxenes have parallel extinction and low birefringence, and pigeonite normally shows $2V$ less than 30°. It may prove impossible to distinguish augite-ferroaugite from diopside-hedenbergite by optical means. Augite is most characteristic of subsiliceous, plutonic igneous rocks, and diopside occurs mostly in metamorphic rocks and mafic volcanics. Violet-brown color and strong pleochroism is distinctive for titanaugite.

ALTERATION. Deuteric or hydrothermal alteration commonly rims augite crystals with a fibrous aggregate of secondary, light-colored amphibole called *uralite*. In intermediate, plutonic igneous rocks, augite is often surrounded by reaction rims of hornblende, indicating disequilibrium in late magmatic stages. Chlorite is a common alteration product of augite, and epidote, carbonates, serpentine, biotite, chalcedony, and talc are less common.

OCCURRENCE. Augite is the common pyroxene of common mafic igneous rocks and is especially characteristic of gabbros. It may occur in intrusive rocks, with either orthopyroxene or pigeonite as layered intergrowth or as reaction rims. Augite is fairly common in intermediate, or even silicic, igneous rocks, where it is usually rimmed by hornblende and may be abundant in certain ultramafics, usually as diallage. In basalts, diopside is probably more common than augite as phenocrysts, and pigeonite is most common in the groundmass. Volcanic augite is usually subcalcic and often zoned.

In mafic alkaline rocks, we may expect augite rich in Al, Fe^{3+}, Ti, Ca, and Na, without orthopyroxene or pigeonite phases. Iron enrichment in late stages favors aegirine-augite, and titanaugite is most commonly associated with feldspathoids in dark, alkali-rich plutonic rocks.

Pyroxenes, in general, are antistress minerals and not characteristic of low-grade metamorphism. Augite does appear in gneisses and granulites of high-grade regional metamorphism, but it tends to be Ca-rich, near diopside, and diopside is more common in metamorphic environments.

REFERENCES: AUGITE

Clark, J. R. D. E. Appleman, and J. J. Papike. 1969. Crystal chemical characterization of clinopyroxenes based on eight new structure refinements. *Mineral. Soc. Amer. Spec. Paper*, *2*, 31–50.

Hess, H. H. 1941. Pyroxenes of common mafic magmas. II. *Amer. Mineral.*, *26*, 573–594.

Hess, H. H. 1949. Chemical composition and optical properties of common clinopyroxenes. *Amer. Mineral.*, *34*, 621–666.

Kuno, H. 1955. Ion substitution in the diopside-ferropigeonite series of clinopyroxenes. *Amer. Mineral.*, *40*, 70–93.

Majmudar, H. H. 1970. Fassaite from Madagascar. *Southeastern Geology*, *11*, 269–278.

Muir, I. D. 1951. The clinopyroxenes of the Skaergaard intrusion, eastern Greenland. *Mineral. Mag.*, *29*, 690–714.

Norton, D. A., and W. S. Clavan. 1959. Optical mineralogy, chemistry and X-ray crystallography of ten pyroxenes. *Amer. Mineral.*, *44*, 844–874.

Poldervaart, A., and H. H. Hess. 1951. Pyroxenes in the crystallization of basaltic magma. *Jour. Geol.*, *59*, 472–489.

Ruegg, N. R. 1964. Use of the angle $A \wedge c$ in optical determination of the composition of augite. *Amer. Mineral.*, *49*, 599–606.

The Aegirine to Aegirine-Augite Series

$NaFe^{3+}(SiO_3)_2$–$(Na,Ca)(Fe^{3+},Fe^{2+},Mg,Al)(SiO_3)_2$

MONOCLINIC

$\angle\beta = 106°$

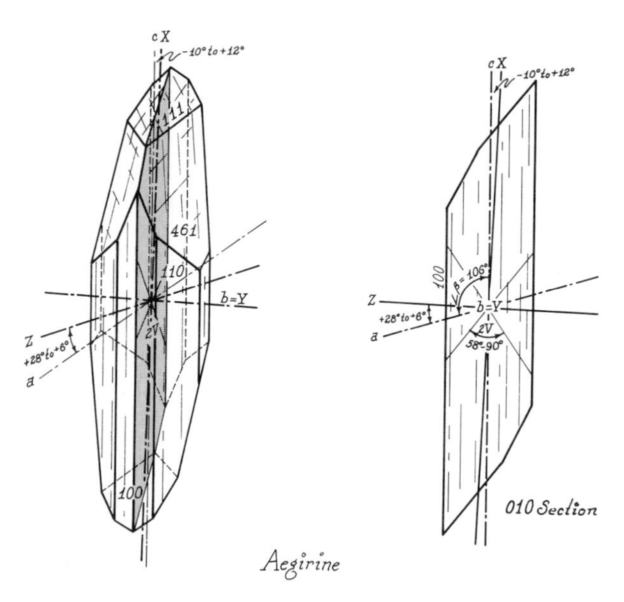

Aegirine (Acmite)

$n_\alpha = 1.776$–1.722*

$n_\beta = 1.820$–1.742

$n_\gamma = 1.836$–1.758

$n_\gamma - n_\alpha = 0.060$–$0.036$*

Biaxial negative

$2V_x = 58°$–$90°$*

$b = Y$, $c \wedge X = -10°$ to $+12°$, $a \wedge Z = +28°$ to $+6°$

$r > v$ moderate to strong

Bright green to yellow-green in section, with pronounced pleochroism in deep green and yellow-green

$X > Y > Z$

Aegirine-Augite

$n_\alpha = 1.722$–1.700

$n_\beta = 1.742$–1.710

$n_\gamma = 1.758$–1.730

$n_\gamma - n_\alpha = 0.036$–$0.030$

Biaxial positive

$2V_z = 90°$–$70°$

$b = Y$, $c \wedge X = +12°$ to $+20°$, $a \wedge Z = +6°$ to $-4°$

$r > v$ moderate to strong

Pale green to yellow-green in section, with weak pleochroism in pale yellow-green or brown-green

COMPOSITION. The ideal aegirine composition, $NaFe^{3+}(SiO_3)_2$, is analogous to diopside, $CaMg(SiO_3)_2$, with the replacement Na^+ and Fe^{3+} for Ca^{2+} and Mg^{2+}, and, indeed, aegirine-diopside represents a complete synthetic series (see Fig. 5-15). Natural compositions tend, however, to form essentially complete solid solution from $NaFe^{3+}(SiO_3)_2$ to augite compositions $(Ca,Na)(Mg,Fe^{2+},Fe^{3+},Al)[(Si,Al)O_3]_2$. Optical properties obviously become less definite toward the indefinite augite composition. Aegirine-augite is generally recognized as the intermediate member of the series, although its composition range is not generally agreed on. A change in

* See Fig. 5-14 for continuous optical variation with composition. Refractive indices, birefringence, and $2V_z$ increase toward aegirine.

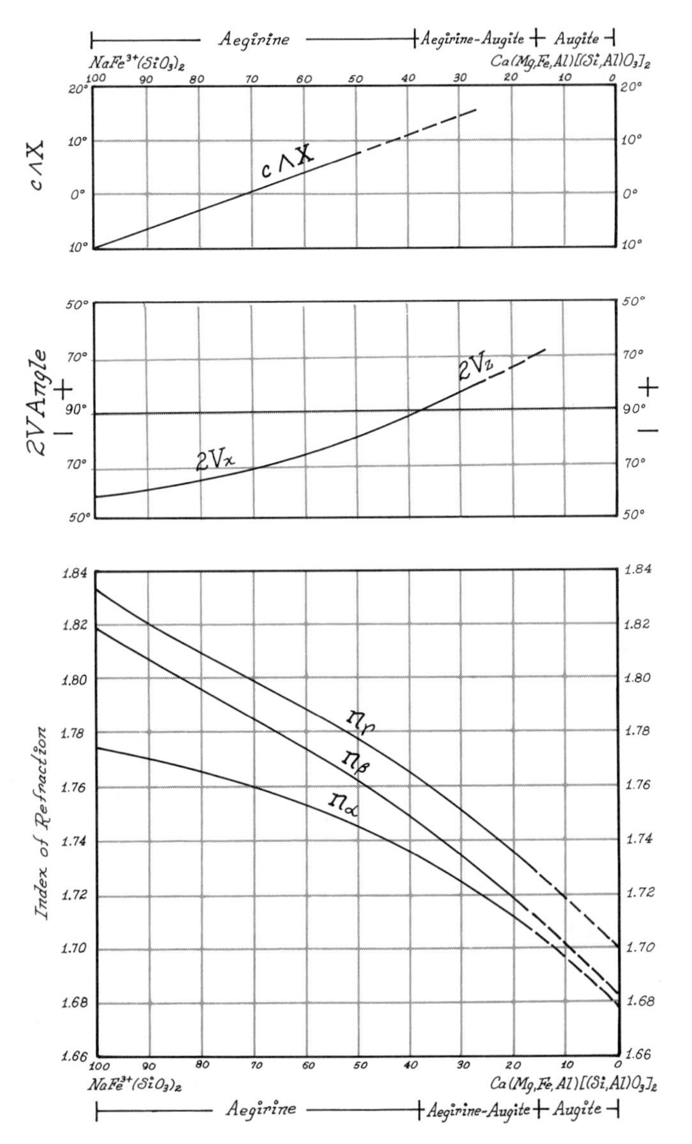

FIGURE 5-14. Variation of $c \wedge X$, $2V$ angle, and indices of refraction in the aegirine-augite series. (Data from Deer, Howie, and Zussman, 1962)

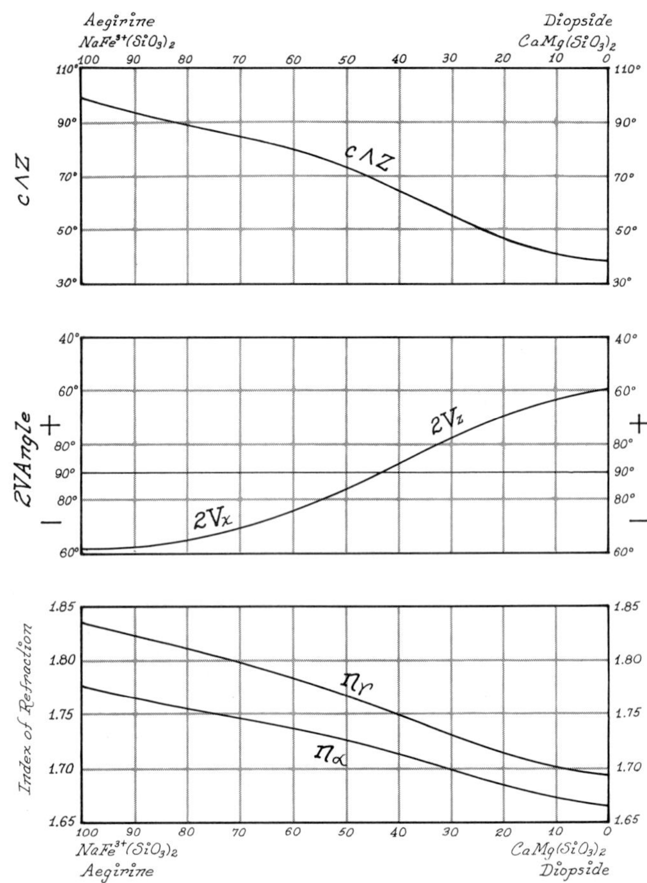

FIGURE 5-15. Variation of $c \wedge Z$, $2V$ angle, and refraction indices in the aegirine-diopside series.

optical sign at about Ae_{38} makes convenient optical division between aegirine and aegirine-augite and is adopted here. In natural compositions, the sign change appears over a narrow range due to indefinite composition and the augite to aegirine-augite boundary can only be arbitrarily defined in terms of the aegirine "molecule" at about Ae_{15}.

Aegirine	Ae_{100}–Ae_{38}
Aegirine-augite	Ae_{38}–Ae_{15}
Augite	Ae_{15}–Ae_{0}

FIGURE 5-16. Aegirine-augite crystals show the typical octagonal sections of pyroxenes and prismatic cleavages at $\sim 90°$. Color zoning suggests outer zones enriched in iron.

Brown aegirine is commonly called *acmite*. The brown color is probably due to significant Mn, although Zr and rare earths are also common impurities of acmite, and vanadium-rich varieties are known. Minor Ti, Cr, and Al may appear in octahedral sites, and minor K may replace Na. Composition zoning is very common throughout the series with outer zones enriched in aegirine.

PHYSICAL PROPERTIES. H = 6. Sp. Gr. = 3.4–3.6. Color, in hand sample, is dark green to black for aegirine and aegirine-augite, and acmite is commonly defined as a brown or reddish brown variety of aegirine.

COLOR AND PLEOCHROISM. Aegirine and aegirine-augite are the most highly colored of the common pyroxenes. Color, in standard section, is usually pleochroic in bright green and yellow-green, both color and pleochroism decrease toward augite compositions. X = emerald-green, bright green; Y = grass-green, yellow-green; and Z = pale yellow-green, yellow. The acmite variety of aegirine is weakly pleochroic in pale brown and yellow.

FORM. Aegirine and aegirine-augite appear most commonly as stubby euhedral crystals showing eight-sided cross sections, with prismatic pyroxene cleavages and rectangular prismatic sections (Fig. 5-16). Less commonly, aegirine occurs as acicular crystals elongated on c, in parallel, radiated, or felty aggregates.

CLEAVAGE. Characteristic prismatic $\{110\}$ pyroxene cleavages intersect at 87° and 93°, and parting may be well developed parallel to $\{001\}$ or $\{100\}$.

BIREFRINGENCE. Maximum birefringence appears in $\{010\}$ sections yielding second-order colors at standard section thickness. Birefringence is strong (0.060) at aegirine, decreasing toward the augite end of the series. In thin section, interference colors are commonly brilliant first and second order. In fragments, interference colors are often obscured by inherent mineral color.

TWINNING. Normal twinning may be expected on $\{100\}$.

ZONING. Composition zoning is very common and appears as distinct color zoning. Normal crystallization proceeds from augite to aegirine with increasing color intensity in Fe^{3+}-enriched outer zones (Fig. 5-16). The composition range of zoned crystals may be very appreciable, as shown by extinction angles.

INTERFERENCE FIGURE. Aegirine is here defined as optically negative with $2V_x$ increasing from about 58° (Ae_{100}) to 90° at about Ae_{38}. Aegirine-augite constitutes a small composition range of positive character, with $2V_z$ decreasing from 90° to about 70° toward augite. Basal {001} and {100} sections show off-center bisectrix figures symmetrical about the optic plane with large $2V$ and several isochromatic color bands. Since {010} is the optic plane, {010} sections yield flash figures. Cleavage fragments tend to lie on prismatic cleavage and yield poor figures due to inclined orientation and strong color absorption.

Optic axis dispersion is moderate to strong $r > v$, and bisectrix dispersion is weakly inclined. Sections showing weak birefringence may become alternately pale pink and bluish with stage rotation, due to strong dispersion.

OPTICAL ORIENTATION. The optic normal Y parallels b and the optic plane (XZ) is {010} with X near c and Z near a. $c = X$ at about Ae_{73}, and $c \wedge X$ becomes $-10°$ at pure aegirine and $+20°$ at Ae_{15} (see Fig. 5-14). $a \wedge Z$ ranges correspondingly from $+28°$ at aegirine to $-4°$ toward augite.

Fragments and rectangular sections tend to be elongated on c and are length-fast (negative-elongation), since X lies near c.

Maximum extinction angle, in {010} sections, ranges from about 10° at aegirine through 0° to 10°–20° for aegirine-augite, {100} sections show parallel extinction, and basal sections show extinction symmetrical to prismatic {110} cleavages.

DISTINGUISHING FEATURES. Aegirine is distinguished from aegirine-augite by optic sign and is easily separated from other proxenes by its strong color and pleochroism, high birefringence and refractive indices, negative sign, and small extinction angle. Aegirine-augite is optically positive, colored, and generally intermediate between aegirine and the more common pyroxenes. Amphiboles do not show pyroxene cleavage intersections and are usually length-slow.

ALTERATION. Aegirine and aegirine-augite may alter to chlorite or epidote with ferric oxides, or may be replaced by Na-uralite.

OCCURRENCE. Aegirine and aegirine-augite crystallize relatively late in Na-rich intrusives. They appear in association with quartz and Na-amphiboles (such as riebeckite or arfvedsonite) in alkali granites, syenites, and monzonites and are associated with nepheline and rare alkali-, Ti-, or Zr-silicates in nepheline syenite and unusual feldspathoid plutons and dike rocks. Aegirine and aegirine-augite enter into intergrowths with feldspar, appear as reaction rims on quartz or other pyroxenes, and may be partially replaced by rims of amphibole.

Volcanic rocks seldom contain aegirine or aegirine-augite, but microlites or needles of these minerals may appear in alkali extrusives.

Regional metamorphic rocks occasionally contain aegirine or aegirine-augite in Na-rich schists, gneisses, and granulites, due possibly to alkali metasomatism. The Na-pyroxenes are commonly associated with Na-amphiboles (such as glaucophane).

Acmite appears in metamorphosed, iron-rich silica sediments through the Na-metasomatism of hematite cherts.

REFERENCES: THE AEGIRINE TO AEGIRINE-AUGITE SERIES

Clark, J. R., D. E. Appleman, and J. J. Papike. 1969. Crystal chemical characterization of clinopyroxenes based on eight new structure refinements. *Mineral. Soc. Amer. Spec. Paper*, *2*, 31–50.

Deer, W. A., R. A. Howie, and J. Zussman. 1962. *Rock-Forming Minerals*. Vol. 2: Chain *Silicates*. New York: Wiley.

Dos Santos, A. R. 1973. Extinction curves method for optical study of the alkaline pyroxene series. *Bol. Soc. Geol. Port.*, *18*, 179–198.

Heinrich, E. W. 1965. *Microscopic Identification of Minerals*. New York: McGraw-Hill.

King, B. C. 1962. Optical determination of aegirine-augite with the universal stage. *Mineral. Mag.*, *33*, 132–137.

Sabine, P. A. 1950. The optical properties and composition of acmitic pyroxenes. *Mineral. Mag.*, *29*, 113–125.

Jadeite $NaAl(SiO_3)_2$ MONOCLINIC $\angle\beta = 107°26'$

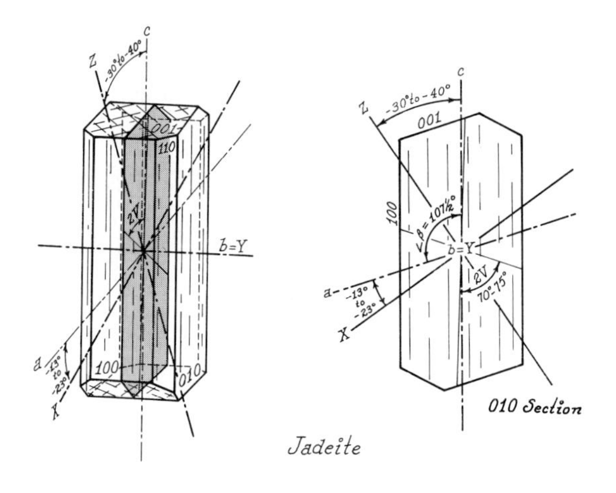

Jadeite

$n_\alpha = 1.654–1.665$*
$n_\beta = 1.659–1.674$
$n_\gamma = 1.667–1.688$
$n_\gamma - n_\alpha = 0.012–0.023$
Biaxial positive, $2V_z = 70°–75°$
$b = Y, c \wedge Z = -30°$ to $-40°$, $a \wedge X = -13°$ to $-23°$
$v > r$ moderate
Colorless in standard section

COMPOSITION. Jadeite composition commonly approaches the ideal $NaAl(SiO_3)_2$. It is the structural analog of diopside $CaMg(SiO_3)_2$, and an incomplete series probably exists where Na^+-Al^{3+} replaces Ca^{2+}-Mg^{2+}. The most significant impurity in jadeite seems to be Fe^{3+} as replacement for Al^{3+}, and a series may exist with aegirine $NaFe^{3+}(SiO_3)_2$. Omphacite $(Ca,Na)(Mg,Fe^{2+},Al,Fe^{3+})(SiO_3)_2$ appears as an intermediate member of the jadeite-diopside-aegirine complex.

Chemically, jadeite,$NaAl(SiO_3)_2$, is anhydrous analcite, $Na(AlSi_2O_6) \cdot H_2O$, and lies intermediate between albite, $Na(AlSi_3)O_2$, and nepheline, $Na(AlSi)O_4$. The latter three, however, are framework silicates.

* Ideal jadeite composition shows indices of refraction, birefringence, and $2V_z$ near the minimum values shown. These increase to somewhat indefinite upper limits with Fe^{3+}, Ca, Mg, and other impurities.

PHYSICAL PROPERTIES. H = 6. Sp. Gr. = 3.25–3.40. Color, in hand sample, is white to light green or blue-green.

COLOR AND PLEOCHROISM. Jadeite is colorless to pale green in section or as fragments. Color seems related to Fe^{3+} substitution and may show distinct pleochroism in pale green and yellow.

FORM. Jadeite tends to occur as fine, fibrous aggregates, although coarse aggregates of anhedral granules to stubby euhedral crystals are also common.

CLEAVAGE. Jadeite shows typical pyroxene cleavages, distinct prismatic on {110} intersecting at 87° and 93°. Parting on {100} is reported.

BIREFRINGENCE. Jadeite tends to show low birefringence and low first-order colors in standard thin section. Ideal jadeite compositions show maximum birefringence near 0.012, in {010} sections, yielding no more than first-order yellow in standard section. Birefringence increases to yield low second-order colors toward diopside and aegirine compositions.

TWINNING. Fine multiple twinning on {100} and rarely on {001} is reported.

INTERFERENCE FIGURE. Jadeite is biaxial positive, with large $2V_z$ near 70°. Fragments on cleavages yield off-center figures without symmetry. {010} sections show flash figures, and other pinacoidal sections show off-center optic axis figures symmetrical about the optic plane. Few isochromes surround broad isogyres. Dispersion is moderately strong $v > r$ with possible inclined bisectrix dispersion. With increasing Fe^{3+} substitution, dispersion apparently reverses to $r > v$.

OPTICAL ORIENTATION. The optic plane of jadeite is {010}, $b = Y$. Z and X lie in {010} at angles to crystallographic directions $c \wedge Z = -30°$ to $-40°$ and $a \wedge X = -13°$ to $-23°$. Maximum extinction angle, seen in {010} sections, ranges from 30° to 40°. Extinction is parallel in {100} sections, symmetrical in basal sections, and somewhat less than maximum on cleavage fragments. Sign of elongation is positive (length-slow), as Z lies nearer elongation.

DISTINGUISHING FEATURES. Jadeite is distinguished from the fibrous, amphibole form of jade (nephrite) by pyroxene cleavage, higher refraction index, and larger extinction angles. Jadeite shows lowest refractive indices among the pyroxenes, except spodumene, and lower birefringence than other clinopyroxenes. Spodumene shows higher birefringence and slightly lower extinction angles.

ALTERATION. At high temperatures, approaching melting, jadeite recrystallizes to albite and nepheline, and at low temperatures of weathering we may expect analcime to appear.

OCCURRENCE. Jadeite appears only in metamorphic rocks and is most common as a monomineralic rock. It is most characteristic of high uniform pressure and relatively low temperature but appears also in low-grade metamorphic rocks. Albite is a constant associate of jadeite, and other minerals may also appear in association with jadeite (nepheline, analcime, zoisite, muscovite, lawsonite, actinolite).

REFERENCES: JADEITE

Coleman, R. G. 1955. Optical and chemical study of jadeite from California, (abs.). *Amer. Mineral.*, *40*, *312*.

Prewitt, C. T., and C. W. Burnham. 1966. The crystal structure of jadeite, $NaAlSi_2O_6$. *Amer. Mineral.*, *51*, 956–975.

Omphacite

$(Ca,Na)(Mg,Fe^{2+},Fe^{3+},Al)(SiO_3)_2$

MONOCLINIC
$\angle\beta = 106°$

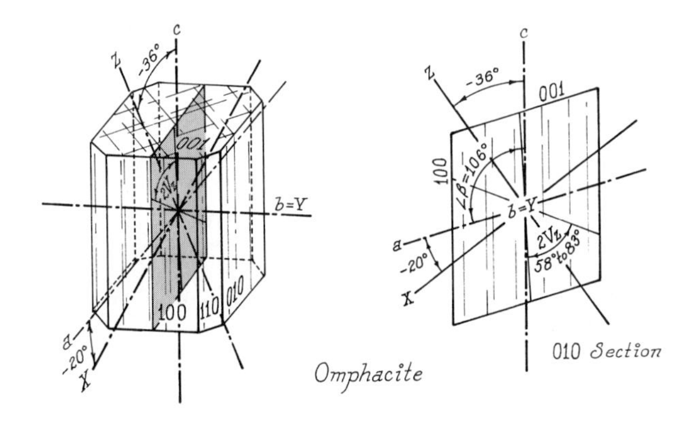

$n_\alpha = 1.662–1.691$*

$n_\beta = 1.670–1.700$

$n_\gamma = 1.688–1.718$

$n_\gamma - n_\alpha = 0.018–0.027$

Biaxial positive, $2V_z = 58°–83°$

$b = Y, c \wedge Z = -36°$ to $-48°$, $a \wedge X = -20°$ to $-32°$

$r > v$ moderate

Very pale green and very weak pleochroism
$Z = Y > X$

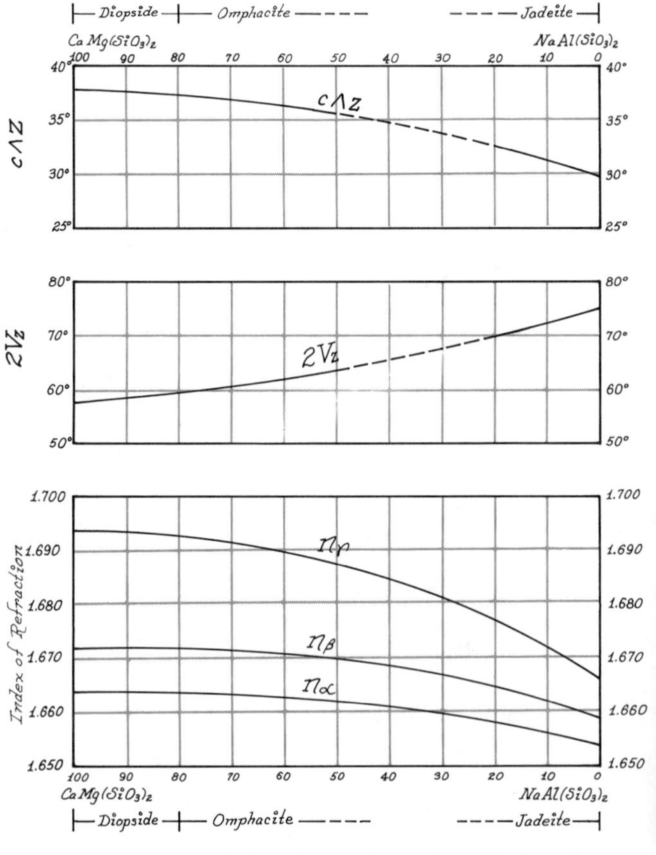

FIGURE 5-17. Variation of $c \wedge Z$, $2V$ angle, and indices of refraction in the diopside-omphacite-jadeite series.

* Omphacite shows a wide range of optical properties in response to its highly variable chemical composition. The presence of the jadeite "molecule" $NaAl(SiO_3)_2$, favors low refractive indices and low birefringence, while the aegirine "molecule", $NaFe^{3+}(SiO_3)_2$, raises refractive indices and birefringence (see Fig. 5-17).

COMPOSITION. Omphacite embraces an indefinite range in the compositional complex lying between three end-member compositions, diopside, $Ca(Mg,Fe^{2+})(SiO_3)_2$; jadeite, $(NaAl(SiO_3)_2$; and aegirine, $NaFe^{3+}(SiO_3)_2$. Omphacite composition normally lies near diopside, with significant Al^{3+} and lesser Fe^{2+} and Fe^{3+}.

PHYSICAL PROPERTIES. H = 5–6. Sp. Gr. = 3.29–3.37. Color, in hand sample, is medium to dark green.

COLOR AND PLEOCHROISM. Omphacite is colorless to very pale green in section, and shows minor pleochroism: X = nearly colorless, and $Y = Z$ = very pale green.

FORM. Omphacite appears most commonly as coarse, anhedral granules in association with garnet.

CLEAVAGE. Omphacite shows typical, distinct, prismatic {110} pyroxene cleavages intersecting at 87° and 93°.

BIREFRINGENCE. Maximum birefringence, seen in {010} sections, varies from about 0.018 to 0.027, yielding interference colors up to low second order, in standard section.

INTERFERENCE FIGURE. Omphacite is biaxial positive, showing a wide range for $2V_z = 58°–83°$. Fragments on cleavages show poor, off-center figures, {010} sections yield flash figures, and {100} and {001} may show nearly centered optic axis figures symmetrical about the optic plane. Dispersion is moderate $r > v$; weakly inclined bisectrix.

OPTICAL ORIENTATION. The optic plane parallels {010}, as with most pyroxenes. $b = Y$, $c \wedge Z = -36°$ to $-48°$, $a \wedge X = -20°$ to $-32°$. Z makes a large angle with c, which may exceed 45, and the maximum extinction angle may range from 36° to 45°. Elongated fragments and pinacoidal sections tend to be extinct near the 45° position, and the sign of elongation may be either slightly positive or negative. Usually the slow wave lies slightly nearer elongation.

DISTINGUISHING FEATURES. The most distinctive feature of omphacite is its limited occurrence in eclogites. It most nearly resembles diopside but usually shows significantly larger $2V_z$. It has higher refractive indices than jadeite and less color and lower refractive indices than aegirine.

ALTERATION. Omphacite is most commonly replaced by secondary, fibrous glaucophane or green hornblende (*smaragdite*). Alteration usually takes the form of retrograde metamorphic effects and begins at exposed boundaries and fractures.

OCCURRENCE. Omphacite appears almost exclusively in eclogite in association with garnet containing Mg and Fe^{2+}. Rutile, plagioclase, quartz, enstatite, ilmenite, and magnetite are common associated accessories. Omphacite, being formed by high-grade plutonic metamorphism, is occasionally found in granulites, migmatites, or gneisses with kyanite, zoisite, glaucophane, and so on.

REFERENCES: OMPHACITE:

Clark, J. R., D. E. Appleman, and J. J. Papike. 1969. Crystal chemical characterization of clinopyroxenes based on eight new structure refinements. *Mineral. Soc. Amer. Spec. Paper*, *2*, 31–50.

Edgar, A. D., A. Mottana, and N. D. Macrae. 1969. Chemistry and cell parameters of omphacites and related pyroxenes. *Mineral. Mag.*, *37*, 61–74.

Phakey, P. P., and S. Ghose. 1973. Direct observation of antiphase domain structure in omphacite. *Contrib. Mineral. Petrol.*, *39*, 239–245.

Spodumene

$LiAl(SiO_3)_2$

MONOCLINIC
$\angle\beta = 110°20'$

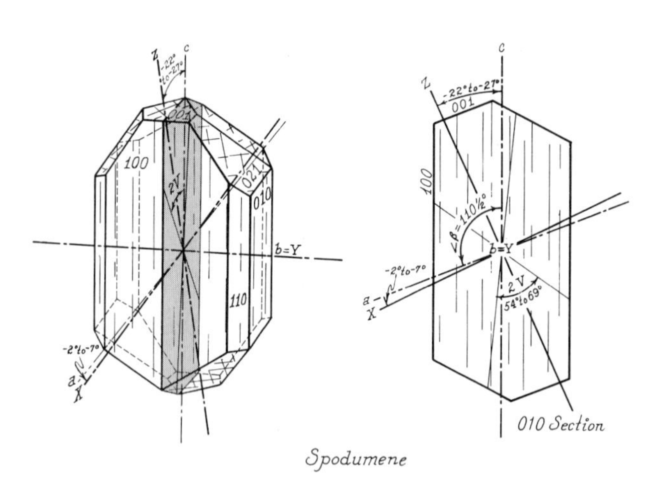

Spodumene

$n_\alpha = 1.648\text{–}1.663$*
$n_\beta = 1.655\text{–}1.670$
$n_\gamma = 1.662\text{–}1.679$
$n_\gamma - n_\alpha = 0.014\text{–}0.027$
Biaxial positive, $2V_z = 54°\text{–}69°$
$b = Y, c \wedge Z = -22°$ to $-27°, a \wedge X = -2°$ to $-7°$
$v > r$ weak
Colorless in standard section

COMPOSITION. Natural spodumene varies little from its ideal composition, $LiAl(SiO_3)_2$, and has the same structure as diopside, $CaMg(SiO_3)_2$, with logically smaller a and b cell dimensions. Na may replace Li in significant amounts, and minor Fe^{3+} or Cr^{3+} may appear in place of Al^{3+}. *Hiddenite* (emerald-green) and *kunzite* (lilac) are gem varieties colored by minor Cr and Mn, respectively.

PHYSICAL PROPERTIES. H = $6\frac{1}{2}$–7. Sp. Gr. = 3.0–3.2. Color, in hand sample, is white to pale gray, yellow, green, or lavender. Gem varieties hiddenite and kunzite are transparent emerald-green and lilac, respectively. Luster is pearly to vitreous.

COLOR AND PLEOCHROISM. Spodumene is normally colorless in section or as fragments; however, fragments of deep-colored gem varieties show color and distinct pleochroism in green or lilac, $X = Y > Z$.

FORM. Spodumene appears most commonly as euhedral or corroded, prismatic crystals elongated on c and sometimes flattened on {100}, giving octagonal cross sections. Crystals may be small and acicular to huge crystals weighing tons.

CLEAVAGE. Spodumene shows typical, prismatic, pyroxene cleavages {110} at 87° and 93°, with parting on {010} and (rarely) on {100}.

BIREFRINGENCE. Maximum birefringence is mild to moderate (0.014–0.027) showing upper first-order or low second-order colors in {010} sections of standard thickness. Second-order colors suggest significant Na or Fe^{3+} content.

TWINNING. Twinning on {100} has been noted.

INTERFERENCE FIGURE. Spodumene is biaxial positive with moderate $2V_z = 54°\text{–}69°$. In Bxa sections, broad isogyres, surrounded by a few isochromes, may separate about to field limits. Fragments on cleavages show off-center figures without symmetry, {010} sections show flash figures, and {100} sections show off-center Bxo symmetrical on the optic plane. Dispersion is weak $v > r$, possibly inclined.

OPTICAL ORIENTATION. The optic plane of spodumene is {010}, and $b = Y$. $c \wedge Z = -22°$ to $-27°$, and X is near a ($a \wedge X = -2°$ to $-7°$). Maximum extinction angle, in {010} sections, is low for a pyroxene at 22° to 27°. Other pinacoid sections show parallel {100} and symmetrical {001} extinction and fragments, on cleavages, show extinction angles less than maximum.

DISTINGUISHING FEATURES. Pegmatite occurrence is distinctive, and spodumene shows smaller extinction angles than other pyroxenes, except aegirine, which is colored.

ALTERATION. Li-mica and albite are probably the most common alteration products of spodumene with eucryptite ($LiAlSiO_4$) as an intermediate stage. Kaolinite is a common replacement product, and other clays, secondary micas, and cookeite may appear.

* Replacement of Li by Na lowers n_α, which increases birefringence ($n_\gamma - n_\alpha$), and replacement of Al by Fe^{3+} increases refractive indices and birefringence.

OCCURRENCE. Spodumene is characteristic of lithium-bearing, granite pegmatites where it appears as large crystals in intermediate zones between wall zone and core in association with quartz, microcline, albite, muscovite or lepidolite, beryl, and tourmaline. Apatite, amblygonite, topaz, and cassiterite are less prominent associates. Small spodumene crystals are reported in a few aplites and rarely in metamorphic gneisses.

REFERENCES: SPODUMENE

Clark, J. R., D. E. Appleman, and J. J. Papike. 1969. Crystal chemical characterization of clinopyroxenes based on eight new structure refinements. *Mineral. Soc. Amer. Spec. Paper*, *2*, 31–50.

Hess, F. L. 1946. The spodumene pegmatites of North Carolina. *Econ. Geol.*, *35*, 942–966.

THE PYROXENOID GROUP

Wollastonite $CaSiO_3$

TRICLINIC
$\angle\alpha = 90°02'$
$\angle\beta = 95°22'$
$\angle\gamma = 103°26'$

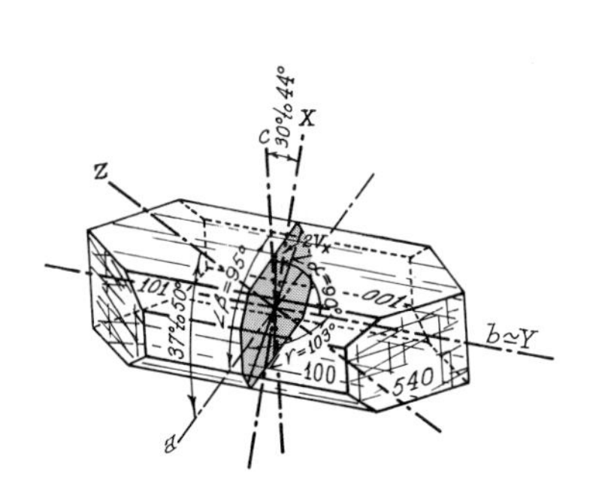
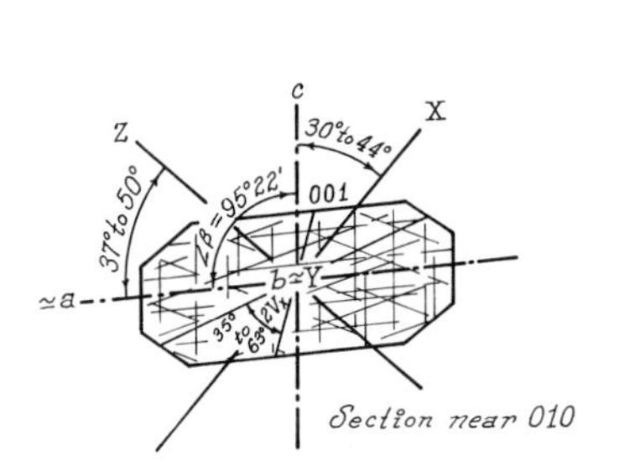

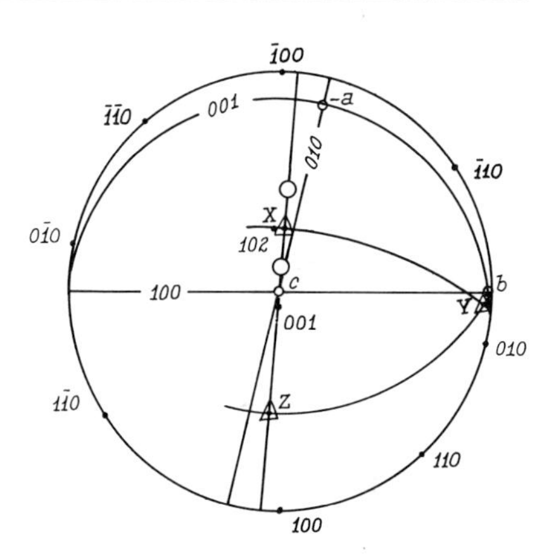

Wollastonite

$n_\alpha = 1.615–1.646$*

$n_\beta = 1.627–1.659$

$n_\gamma = 1.629–1.662$

$n_\gamma - n_\alpha = 0.013–0.017$*

Biaxial negative, $2V_x = 35°–63°$*

Optic plane nearly parallels {010}, $b \wedge Y = 0°$ to $5°$, $c \wedge X = 30°$ to $44°$, $a \wedge Z = 37°$ to $50°$

$r > v$ weak

Colorless in section

*Refraction indices, birefringence, and $2V_x$ for pure $CaSiO_3$ are near the minimum values given and increase uniformly with $FeSiO_3$ content (Fig. 5-18). Refractive indices also increase with $MnSiO_3$.

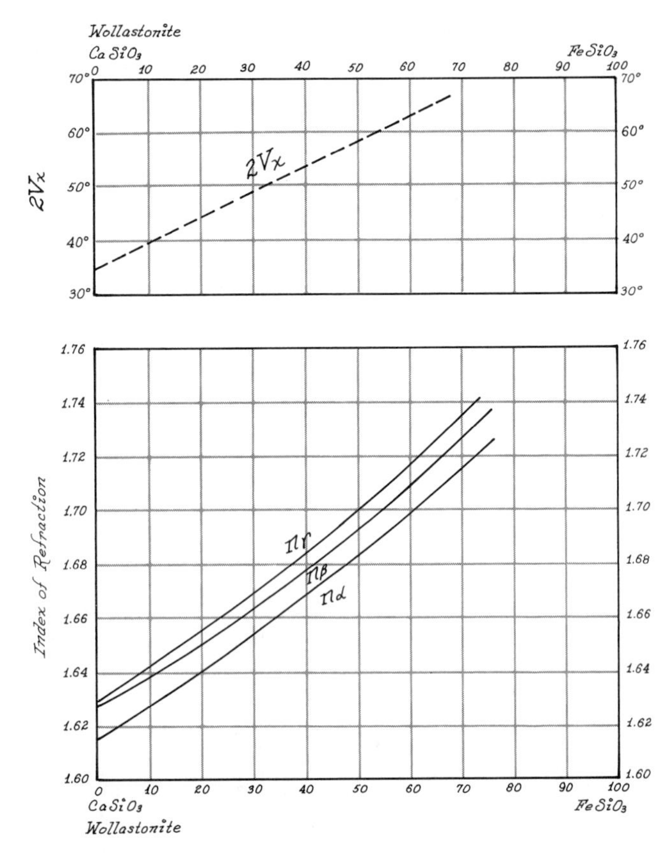

FIGURE 5-18. Variation of $2V$ angle and indices of refraction with Fe^{2+} in wollastonite. (After Bowen, Schairer, and Posjnak, 1933)

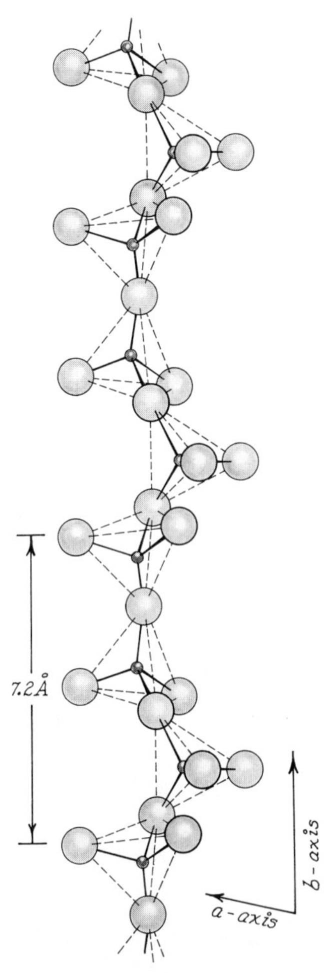

FIGURE 5-19. The idealized, single wollastonite chain: Repeat unit along the chain (b) is three tetrahedra (7.2 Å).

COMPOSITION AND STRUCTURE. Natural wollastonite is usually near ideal composition, $CaSiO_3$, although at high temperatures extensive solid solution exists with $FeSiO_3$ and $MnSiO_3$ (rhodonite), so that natural wollastonite may contain significant Fe^{2+} and Mn^{2+}. At low temperatures, hedenbergite, $CaFe(SiO_3)_2$, and *bustamite*, $(Mn,Ca,Fe)SiO_3$, form as intermediate members.

Although wollastonite is a single-chain structure, it is distinct from the pyroxenes in that silicate tetrahedra are arranged differently, with a three-tetrahedra repeat distance along chains running parallel to b (Fig. 5-19).

Three structural modifications of $CaSiO_3$ exist: wollastonite, *parawollastonite* (both are low-temperature forms of $CaSiO_3$), and *pseudowollastonite* (high-temperature $CaSiO_3$) (see Table 5-2). Pseudowollastonite (triclinic) is very rare as a natural mineral, and parawollastonite (monoclinic) is much less common than wollastonite, occurring in direct association with it.

PHYSICAL PROPERTIES. H = 4.5–5. Sp. Gr. = 2.9–3.1. Color, in hand sample, is white to gray, rarely very pale green, with pearly luster.

COLOR AND PLEOCHROISM. Colorless in section or as fragments.

TABLE 5-2. Three $CaSiO_3$ Structures

	n_α	n_β	n_γ	$n_\gamma - n_\alpha$	$2V$	$b \wedge Y$	$c \wedge X$
Wollastonite	1.618	1.630	1.632	0.014	39° −	4°	39°
Parawollastonite	1.618	1.630	1.632	0.014	44° −	0°	38°
Pseudowollastonite	1.610	1.611	1.654	0.044	Small +	—	9°

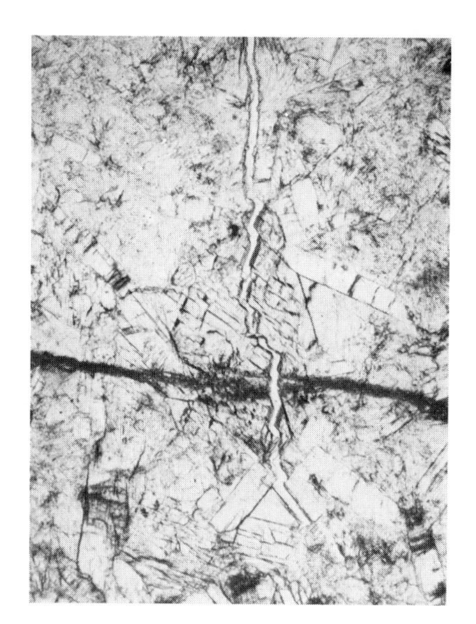

FIGURE 5-20. Wollastonite crystals show octagonal sections and prominent cleavages, intersecting at ~85°, which are suggestive of the pyroxenes. Wollastonite cleavages are {100} perfect and {001} good. Note multiple twinning on {100}.

FORM. Wollastonite is commonly columnar, bladed to fibrous. Crystals tend to be elongated on *b* and flattened on {100} or {001} (Fig. 5-20).

CLEAVAGE. Cleavage is perfect on {100} and good on {001} and {$\bar{1}$02}. {100} ∧ {001} = 85° and {100} ∧ {$\bar{1}$02} = 70° in {010} sections.

BIREFRINGENCE. Maximum birefringence, seen in essentially {010} sections, is near 0.014 for pure $CaSiO_3$, ranging to 0.017 with increasing iron. Interference colors, in standard section, range up to first-order red.

TWINNING. Multiple twinning on {100} in common and parawollastonite may prove to be only wollastonite with submicroscopic multiple twinning. Twinning is commonly parallel with twin axis parallel to *b*, although a twin axis parallel to *c* is also reported.

ZONING. Crystal borders may prove richer in iron, producing larger extinction angles and refractive indices for outer zones.

INTERFERENCE FIGURE. Wollastonite is biaxial negative, with $2V_x$ ranging from 35° to about 63°, with angles near 40° most common. Acute bisectrix isogyres will probably remain in the visible field, surrounded by only minor isochrome bands. Fragments on {$\bar{1}$02} cleavages show a nearly centered optic axis figure, but other cleavages {100} and {001} yield off-center bisectrix figures essentially symmetrical on the optic plane, and {010} sections produce flash figures. Dispersion is weak to moderate $r > v$, with distinct inclined bisectrix dispersion, at least for parawollastonite.

OPTICAL ORIENTATION. The optic plane of wollastonite is nearly parallel to {010} and Y lies within 4° or 5° of b. X lies essentially in {010}, 30° to 44° from c, in the acute $\angle \beta$, and Z falls about 37° to 50° from a, in obtuse $\angle \beta$.

All cleavage fragments are elongated on b and show nearly parallel extinction (to 5°) and either positive (length-slow) or negative (length-fast) elongation. Most fragments lie on {100} or {010} cleavages and are length-slow; those on {$\bar{1}$02} are length-fast.

DISTINGUISHING FEATURES. Tremolite and pectolite elongation is always positive (length-slow); wollastonite elongation may be either positive or, less commonly, negative, and wollastonite has lower birefringence. Wollastonite shows lower refractive indices and smaller $2V$ than zoisite, clinozoisite, or diopside, and diopside is optically positive. Parawollastonite differs from wollastonite only by strictly parallel extinction but pseudowollastonite is easily distinguished by strong birefringence and small $2V$ of positive sign.

ALTERATION. Both wollastonite and parawollastonite are replaced by calcite, rarely pectolite, at low temperatures. At high temperatures, wollastonite becomes unstable with respect to the formation of larnite, monticellite, and other high-grade calc-silicates.

OCCURRENCE. Wollastonite most commonly appears in contact metamorphic zones in limestones containing silica. At temperatures above 600°–700°C, wollastonite forms from calcite and quartz by the reaction $CaCO_3 + SiO_2 \rightarrow CaSiO_3 + CO_2$. Silica may be introduced by metasomatism. Wollastonite is commonly associated with calcite, tremolite, diopside, Ca-garnets, idocrase, epidote, anorthite, and a whole series of rare calc-magnesium silicates (larnite, monticellite, spurrite, merwinite, and so on). Wollastonite is less common in calcareous rocks of regional metamorphism and rarely appears in alkaline igneous rocks. Parawollastonite is rare, occurring with wollastonite. Pseudowollastonite appears in furnace slags and devitrified glass.

REFERENCES: WOLLASTONITE

Bowen, N. L., J. F. Schairer, and E. Posnjak. 1933. The system $CaO\text{-}FeO\text{-}SiO_2$. *Amer. Jour. Sci.*, *26*, 193–284.

Harker, R. I., and O. F. Tuttle. 1956. Experimental data on the P_{CO_2}–T curve for the reaction: calcite + quartz ⇄ wollastonite + carbon dioxide. *Amer. Jour. Sci.*, *254*, 239–256.

Mason, B. 1975. Compositional limits of wollastonite and bustamite. *Amer. Mineral.*, *60*, 209–212.

Prewitt, C. T., and M. J. Buerger. 1963. Comparison of the crystal structures of wollastonite and pectolite. *Mineral. Soc. Amer. Spec. Paper*, *1*, 293–302.

Sabine, P. A. 1975. Refringence of iron-rich wollastonite. *Bull. Geol. Survey Gr. Brit.*, *52*, 65–67.

Sathe, R. V., and P. D. Choudhary. 1967. Stellate wollastonite from calc-silicate scarns of Jothwad Hill, Panchmahal District, Gujrat, India. *Mineral. Mag.*, *36*, 616–618.

Tolliday, J. 1958. Crystal structure of β-wollastonite. *Nature*, *182*, 1012–1013.

Wenk, H. -R. 1969. Polymorphism of wollastonite. *Beitr. Mineral. Petrol.*, *22*, 238–247.

TRICLINIC

Rhodonite

$(Mn,Ca)SiO_3$

$\angle\alpha = 90°02'$
$\angle\beta = 95°22'$
$\angle\gamma = 103°26'$

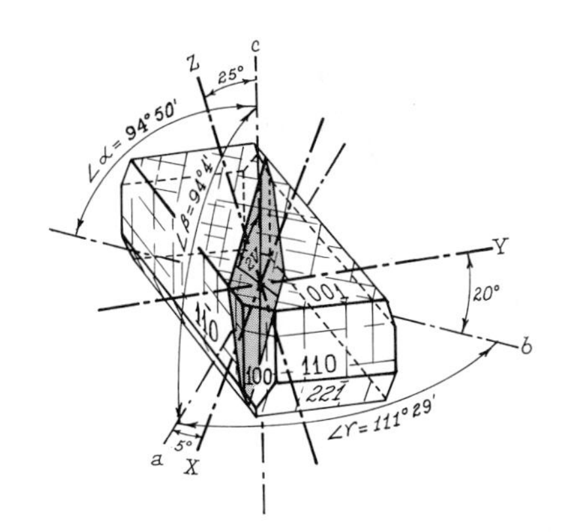

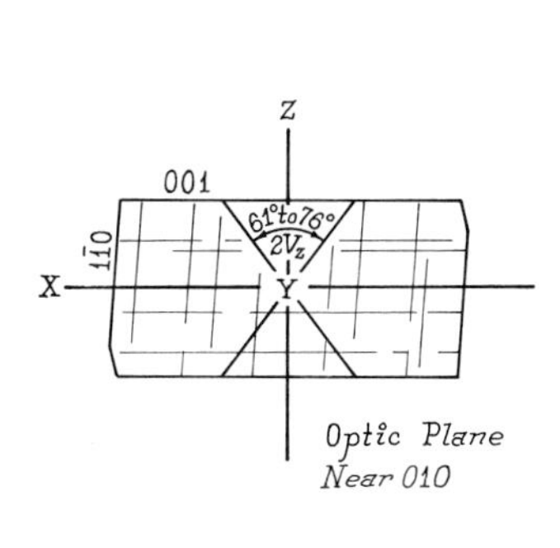

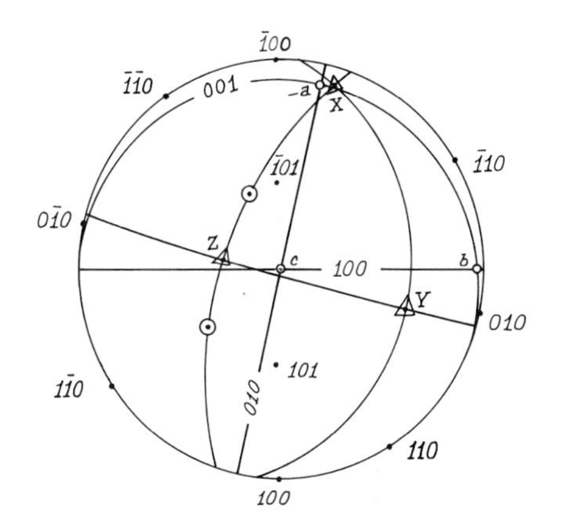

Rhodonite

$n_\alpha = 1.711–1.738$*
$n_\beta = 1.715–1.741$
$n_\gamma = 1.724–1.751$
$n_\gamma - n_\alpha = 0.011–0.014$*
Biaxial positive, $2V_z = 61°–76°$*
$a \wedge X \simeq 5°, b \wedge Y \simeq 20°, c \wedge Z \simeq 25°$
$v > r$ weak, crossed bisectrix dispersion
Colorless to pale pink, in thin section, with weak pleochroism

COMPOSITION AND STRUCTURE. Rhodonite chains, with five-tetrahedra repeat units (Fig. 5-22) are parallel to *b* and linked laterally by Mn^{2+} and Ca^{2+} ions in six- or seven-fold octahedral coordination. A high-temperature form of $MnSiO_3$ may form complete solid solution with wollastonite, $CaSiO_3$, but, although natural rhodonite always contains appreciable Ca, a distinct structural and optic discontinuity appears between "pure" rhodonite, $MnSiO_3$, and *bustamite*, $(Mn,Ca)\cdot SiO_3$. Bustamite is biaxial negative. Considerable Fe^{2+} and lesser Mg^{2+} may replace Ca^{2+}, and a Zn-rich variety is called *fowlerite*.

PHYSICAL PROPERTIES. H = $5\frac{1}{2}–6\frac{1}{2}$. Sp. Gr. = 3.57–3.76. Color, in hand sample, is pink or rose-red to red-brown.

COLOR AND PLEOCHROISM. Rhodonite is usually colorless in standard section, however, fragments may be pale pink

* Indices of refraction and birefringence decrease and $2V_z$ increases with increasing Ca or Mg (see Fig. 5-21).

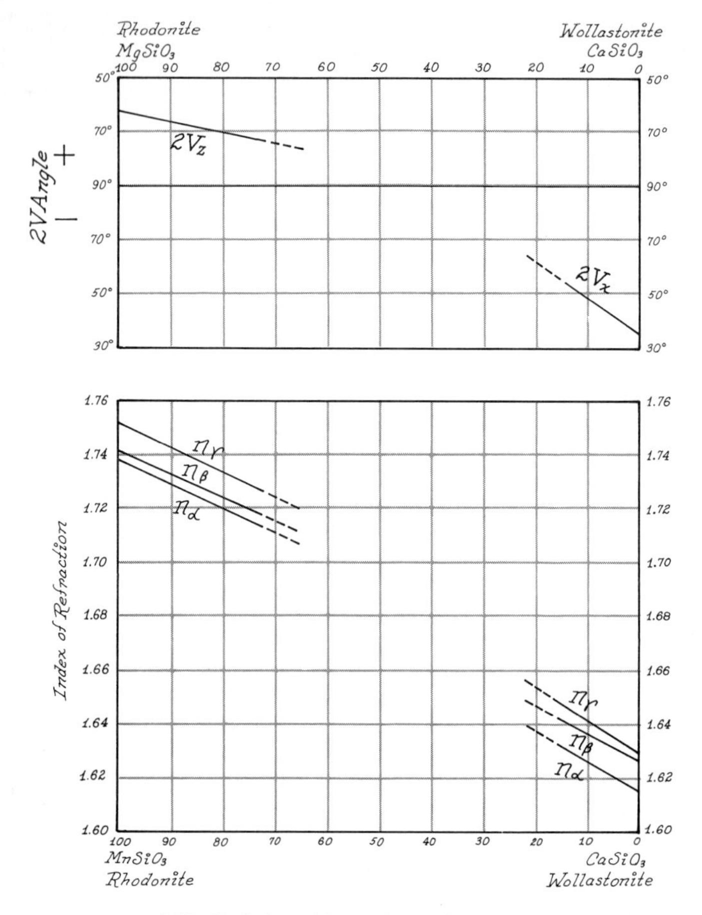

FIGURE 5-21. Variation of $2V$ angle and indices of refraction in the rhodonite-wollastonite partial series. (After Winchell and Winchell, 1951)

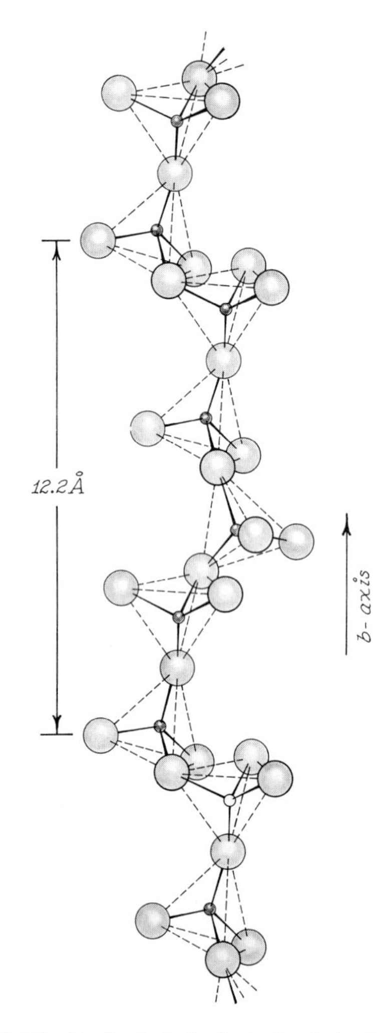

FIGURE 5-22. The idealized, single rhodonite chain: Repeat unit along the chain (b) is five tetrahedra (12.2 Å).

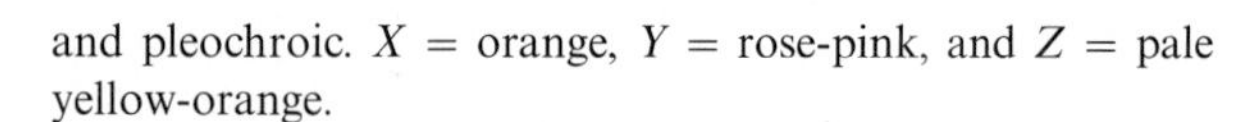

and pleochroic. X = orange, Y = rose-pink, and Z = pale yellow-orange.

FORM. Crystals are often large and anhedral to euhedral and somewhat tabular on {001}.

CLEAVAGE. Cleavages are {100} and {010} perfect and {001} distinct. (100) $\wedge$ (010) = $92\frac{1}{2}°$.

BIREFRINGENCE. Maximum birefringence, seen in sections near {010} is mild ($n_\gamma - n_\alpha$ = 0.011–0.014). Interference colors, in standard section, are gray and white, ranging to first-order yellow.

TWINNING. Lamellar twinning on {010} is uncommon.

ZONING. Color zoning usually grades from a darker core. Lattice-type exsolution lamellae of diopside may parallel {001}.

INTERFERENCE FIGURE. Rhodonite is biaxial positive, with rather large $2V_z = 61°–76°$. Centered figures are accidental, and isochromes are few. Optic axis dispersion is weak $v > r$; however, crossed bisectrix dispersion may be prominent. The obtuse bisectrix is almost normal to {001}, but figures on other cleavage fragments are eccentric.

OPTICAL ORIENTATION. Rhodonite is triclinic, and $a \wedge X \simeq 5°$, $b \wedge Y \simeq 20°$, and $c \wedge Z \simeq 25°$. Extinction is inclined to all cleavage directions, and fragments lack pronounced elongation.

DISTINGUISHING FEATURES. Restricted occurrence (association with other manganese minerals), black alteration, pink color, and consistently inclined extinction are distinctive of rhodonite. Bustamite is biaxial negative, pyroxmangite has smaller $2V_z$, and clinopyroxenes show larger birefringence.

ALTERATION. Rhodonite normally alters to rhodochrosite or black Mn-oxides (pyrolusite). It may also become hydrated to yield penwithite, neotocite, or Mn-serpentine.

OCCURRENCE. Rhodonite occurs largely in hydrothermal ore bodies of iron, zinc, lead, or copper. It is usually associated with rhodochrosite, Mn-oxides, and other manganese minerals.

It is also formed in contact metamorphic deposits and appears in marbles and skarns with bustamite, tephroite, pyroxmangite, spessartine, bemmentite, and grunerite. Rhodonite is very rare in pegmatites.

REFERENCES: RHODONITE

Dickson, B. L. 1975. Iron distribution in rhodonite. *Amer. Mineral.*, *60*, 98–104.

Hey, M. H. 1929. The variation of optical properties with chemical composition in the rhodonite-bustamite series. *Mineral. Mag.*, *22*, 193–205.

Peacor, D. R., and N. Niizeki. 1963. The redetermination and refinement of the crystal structure of rhodonite, $(Mn,Ca)SiO_3$. *Z. Kristallogr.*, *119*, 98–116.

Prewitt, C. T., and D. R. Peacor. 1964. Crystal chemistry of the pyroxenes and pyroxenoids. *Amer. Mineral.*, *49*, 1527–1542.

Sundius, N. 1931. On the triclinic manganiferous pyroxenes. *Amer. Mineral.*, *16*, 411–429.

Winchell, A. N., and H. Winchell. 1951. *Elements of Optical Mineralogy*. II. New York: Wiley.

TRICLINIC
$\angle\alpha = 90°31'$
$\angle\beta = 95°11'$
$\angle\gamma = 102°28'$

Pectolite

$Ca_2NaH(SiO_3)_3$

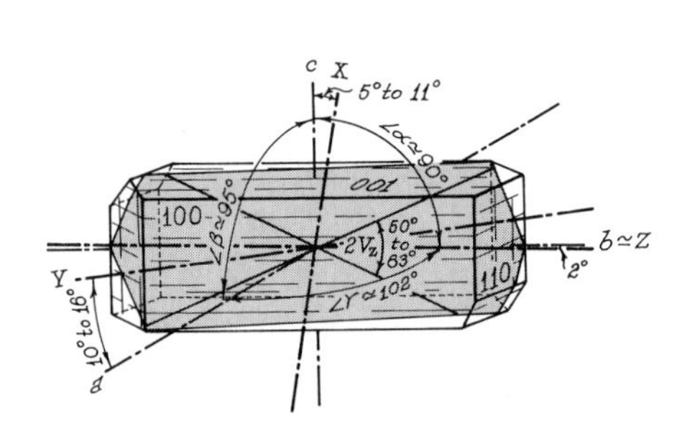

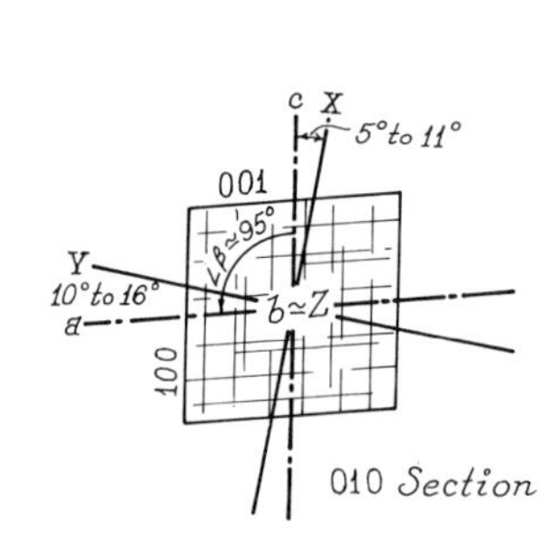

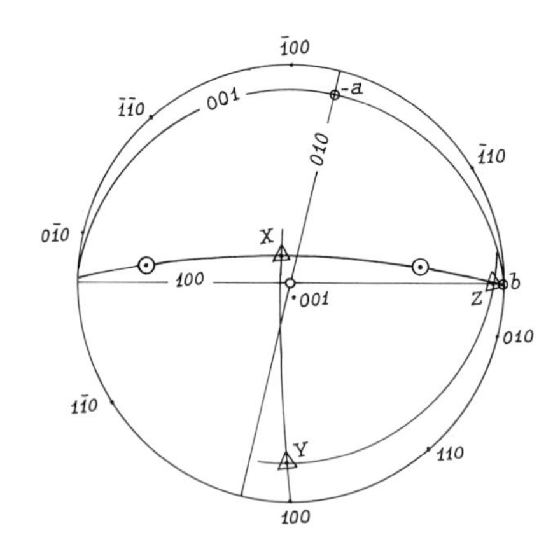

Pectolite

$n_\alpha = 1.595–1.610$*
$n_\beta = 1.604–1.615$
$n_\gamma = 1.632–1.645$
$n_\gamma - n_\alpha = 0.032–0.038$*
Biaxial positive, $2V_z = 50°–63°$*
$a \wedge Y \simeq 10°–16°, b \wedge Z \simeq 2°, c \wedge X \simeq 5°$ to $11°$
$r > v$ weak
Colorless in thin section

COMPOSITION AND STRUCTURE. Pectolite has nearly the same crystal structure as wollastonite. Chains of silicate tetrahedra, running parallel to *b*, are like those of wollastonite and are held together by Ca^{2+} ions in octahedral coordination and Na^+ in distorted octahedral coordination. Hydrogen bonds also link some oxygen ions in adjacent tetrahedra.

Although pectolite, $Ca_2NaH(SiO_3)_3$, is usually near its ideal composition, it apparently forms isomorphous solid solution with a manganese analog *serandite*, $Mn_2NaH \cdot (SiO_3)_3$, with *schizolite* as an intermediate member. Minor Mg and Fe^{2+} may also replace Ca, and magnesian pectolite is called *walkerite*.

PHYSICAL PROPERTIES. H = $4\frac{1}{2}$–5. Sp. Gr. = 2.75–2.90. Pectolite is white and acicular as radiating groups with vitreous to silky luster.

COLOR AND PLEOCHROISM. Pectolite is colorless in section or as fragments.

FORM. Pectolite occurs in radiating groups of rigid acicular crystals elongated on *b* (Fig. 5-23).

* Indices of refraction increase and birefringence and $2V_z$ decrease with increasing Mn or Fe^{2+}. Increasing Mg lowers refractive indices slightly.

FIGURE 5-23. Pectolite is characterized by radiating aggregates of acicular crystals formed in voids of basaltic rocks.

CLEAVAGE. Perfect cleavages on {100} and {001} intersect at 85° and 95° and resemble pyroxene cleavages in cross section.

BIREFRINGENCE. Maximum birefringence, seen in sections essentially normal to *b* (crystal length), is moderate ($n_\gamma - n_\alpha =$ 0.032–0.038). Interference colors are bright and vivid, ranging to high second order or low third order. Cleavage fragments may show near-maximum birefringence.

TWINNING. Parallel twins sometimes form on {100}, with twin axis parallel to *b*.

INTERFERENCE FIGURE. Pectolite is biaxial positive, with moderate $2V_z = 50°–63°$. Sections normal to crystal length (near {010}) show near-centered acute bisectrix figures. Distinct isogyres appear on a field of many isochromes and separate to near the edge of the visible field with state rotation. Dispersion is weak $r > v$. Cleavage fragments show somewhat eccentric obtuse bisectrix or flash figures.

OPTICAL ORIENTATION. Pectolite is triclinic, with *b* near *Z* ($b \wedge Z \simeq 2°$) and $a \wedge Y \simeq 10°–16°$, $c \wedge X \simeq 5°–11°$. Acicular crystals are elongated on *b*, which is very near to *Z*. Acicular crystals and cleavage fragments show near-parallel extinction (maximum extinction angle = 2°) and positive elongation (length-slow).

DISTINGUISHING FEATURES. Wollastonite shows less birefringence and is length-fast.

ALTERATION. Pectolite may be altered to stevensite, $Mg_3Si_4 \cdot O_{10}(OH)_2$, by waters containing magnesium bicarbonate.

OCCURRENCE. Pectolite is deposited by hydrothermal solutions in veins, cavities, and amygdales in basalt flows or diabase dikes, where it is associated with zeolites, calcite, prehnite, or datolite. It is a rare primary igneous mineral in feldspathoidal rocks (nepheline syenite, foyaite, phonolite) or peridotites. Pectolite may also appear in contact marbles and skarns.

REFERENCES: PECTOLITE

Buerger, M. J. 1956. The determination of the crystal structure of pectolite, $Ca_2NaHSi_3O_9$. *Z. Kristallogr.*, *108*, 248–261.

Prewitt, C. T., 1967. Refinement of the structure of pectolite, $Ca_2NaHSi_3O_9$. *Z. Kristallogr.*, *125*, 298–316.

Prewitt, C. T., and M. J. Buerger. 1963. Comparison of the crystal structures of wollastonite and pectolite. *Mineral. Soc. Amer. Spec. Paper*, *1*, 293–302.

Prewitt, C. T., and D. R. Peacor. 1964. Crystal chemistry of the pyroxenes and pyroxenoids. *Amer. Mineral.*, *49*, 1527–1542.

Schaller, W. T. 1955. The pectolite-schizolite-serandite series. *Amer. Mineral.*, *40*, 1022–1031.

Takeuchi, Y., Y. Kudoh, and T. Yamanaka. 1976. Crystal chemistry of the serandite-pectolite series and related minerals. *Amer. Mineral.*, *61*, 229–237.

The Amphibole Group

The amphiboles constitute the most chemically complex mineral group in nature, but they all possess the distinctive arrangement of atoms known as the *double silicate chain* (Fig. 5-24). Along the chain, successive tetrahedra share two and three oxygen ions, respectively, so that the "chemical formula" for the infinite chain is Si_8O_{22}. The chains are elongated along the *c* crystallographic direction, with the bases of the tetrahedra nearly in the *bc* plane; successive chains are inverted along the *b* direction and stacked alternately base to base and apex to apex along the *a* direction (Fig. 5-25). The array of double chains is bonded together by intervening cations: two *X* cations (per Si_8O_{22}) in sites coordinated by six to eight oxygen ions, five *Y* cations in sites of six-coordination, and sometimes one *W* cation in a large six- to ten-coordinated site that is often vacant. Two additional anion sites, generally occupied by hydroxyl groups, are located at the centers of the rings formed by the apical oxygen ions of the double chains.

The general formula for an amphibole is, then, $W_{0-1}X_2Y_5Z_8O_{22}(OH)_2$, where W represents Na^+ or K^+; X is Ca^{2+}, Na^+, K^+, Mg^{2+}, Fe^{2+}, or Mn^{2+}; Y is Mg^{2+}, Fe^{2+}, Fe^{3+}, Al^{3+}, Ti^{4+}, Mn^{2+}, Cr^{3+}, Li^+, or Zn^{2+}; Z is Si^{4+} with up to 25 percent substitution of Al^{3+}; and $(OH)^-$ may be replaced by F^- or O^{2-}.

The principal classification of amphiboles is based on the chemistry of the *X* cation; thus there are the Mg-Fe amphiboles, the calcium amphiboles, and the sodium amphiboles. Within these major chemical groups, there are solid solution series formed by $Mg^{2+} \leftrightarrows Fe^{2+}$ substitution in the *Y* sites, and there are subgroups formed by substitution of Al^{3+} for Si^{4+} in

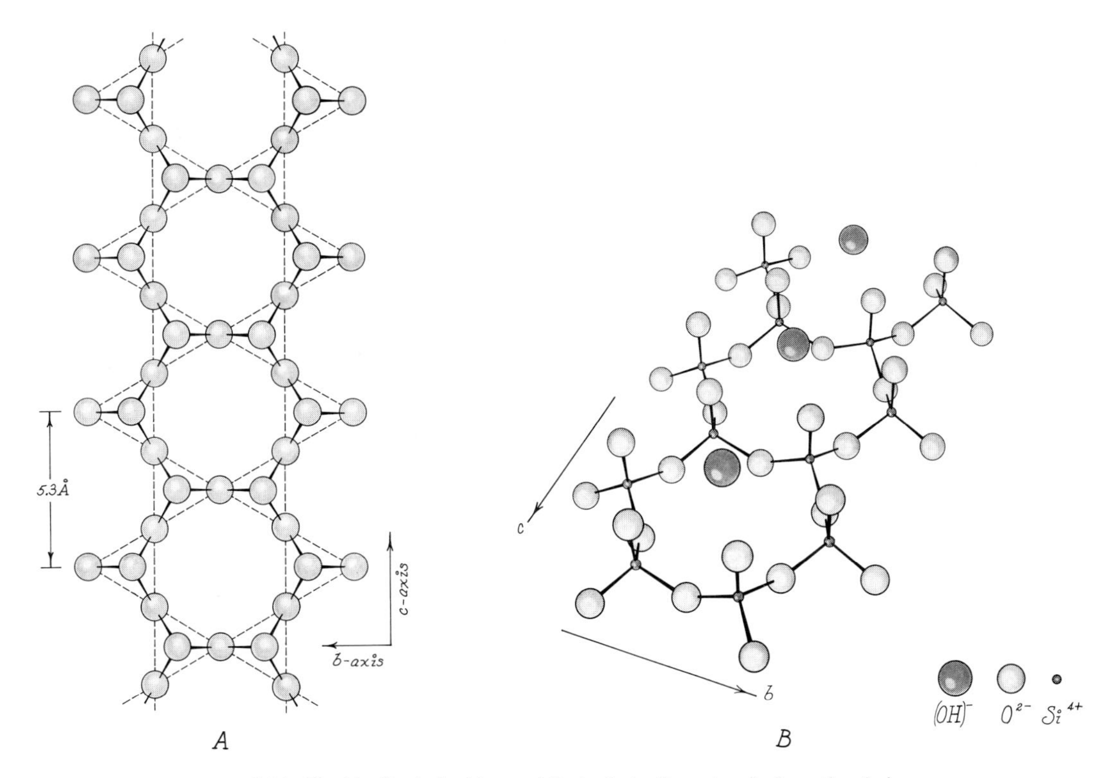

FIGURE 5-24. The idealized, double-amphibole chain. Repeat unit along the chain (c) is two tetrahedra (5.3 Å) (A) View normal to plane of the double chains (ab plane); (B) Inclined view.

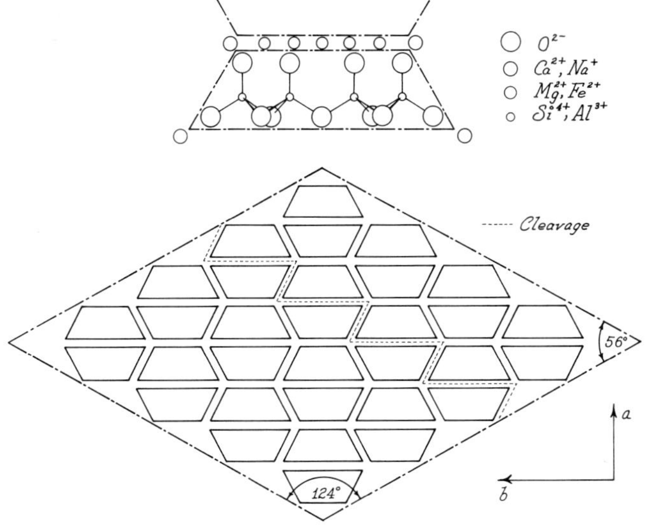

FIGURE 5-25. Arrangement of amphibole chains as seen along c. Prismatic cleavages on (110) and ($1\bar{1}0$) intersect at about 56° and 124° and are characteristic of all amphibole varieties.

the *Z* sites with concomitant replacement of divalent by trivalent atoms in the *Y* sites. Major groups are as follows (Fig. 5-26, Fig. 5-27).

1. Cummingtonite-grunerite series
   $(Fe^{2+},Mg)_2(Mg,Fe^{2+})_5Si_8O_{22}(OH)_2$
2. Calcium amphiboles (*X* sites occupied by Ca)
   a. Tremolite-actinolite series
      $Ca_2(Mg,Fe^{2+})_5Si_8O_{22}(OH)_2$
   b. Hornblende series
      $Ca_2(Mg,Fe^{2+})_4(Al,Fe^{3+})(Si_7Al)O_{22}(OH)_2$

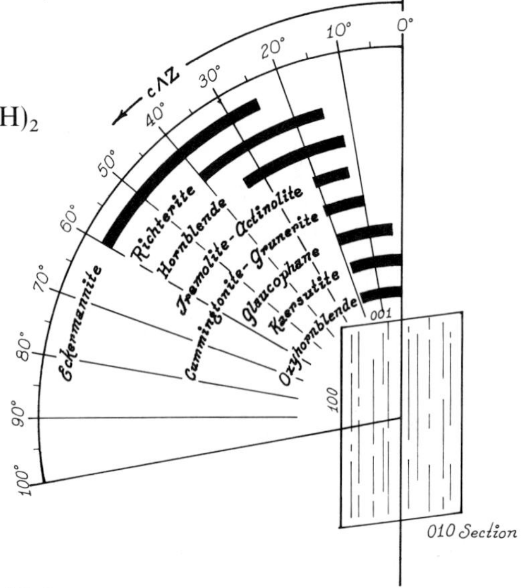

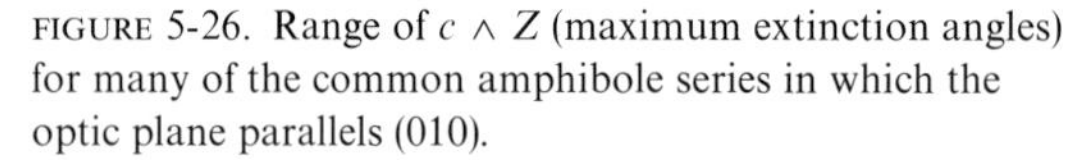
FIGURE 5-26. Range of $c \wedge Z$ (maximum extinction angles) for many of the common amphibole series in which the optic plane parallels (010).

3. Sodium Amphiboles (*X* sites occupied by Na)
   a. Glaucophane-riebeckite series
      $Na_2(Mg,Fe^{2+})_3(Al,Fe^{3+})_2Si_8O_{22}(OH)_2$
   b. Katophorite series
      $Na(NaCa)(Mg,Fe^{2+})_4Fe^{3+}(Si_7Al)O_{22}(OH)_2$
   c. Arfvedsonite series
      $Na(Na_{1.5}Ca_{0.5})(Mg,Fe^{2+})_4Fe^{3+}(Si_{7.5}Al_{0.5})O_{22}(OH)_2$

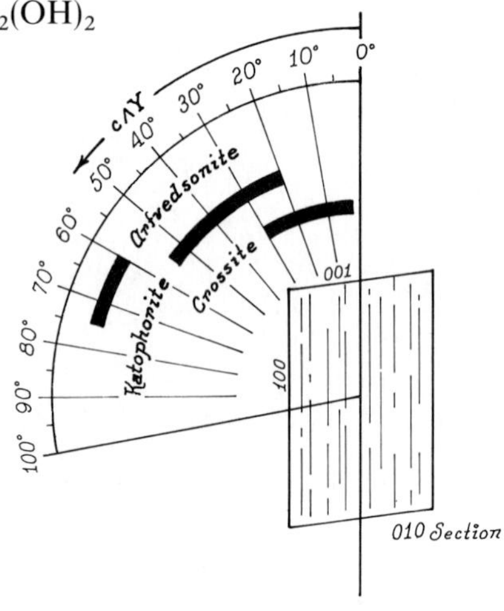

FIGURE 5-27. Range of $c \wedge Y$ (maximum extinction angles) for several amphibole series where the optic plane is perpendicular to (010).

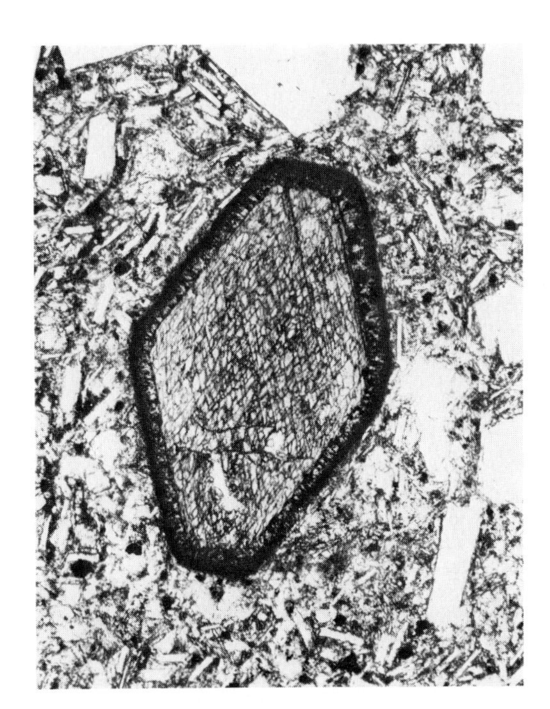

FIGURE 5-28. Amphibole crystals commonly yield diamond shaped cross sections, flattened on the ends as shown. Perfect prismatic cleavages {110} intersect at 56° and 124°. Note the fine pyroxene alteration rim on this hornblende, caused by rapid dehydration in the extruded lava.

Additional substitutions involving K^+, Mn^{2+}, Ti^{4+}, Cr^{3+}, Li^+, and Zn^{2+} in sites *W*, *X*, and Y and replacement of $(OH)^-$ by F^- or O^{2-} lead to a very complex network of isomorphous series, some of which are continuous and others only partial. Many variety names exist to define members of the above series and other less common series, and all possible varieties cannot be considered here nor, in fact, distinguished optically; thus the following optical properties are necessarily somewhat generalized. Optical properties of the major series, cited earlier, are rather well established, and the amphiboles will be treated by series according to their major occurrence. Optical and physical properties are summarized in Table 5-3.

All amphibole varieties have one physical property in common, which is a function of the structural scheme and commonly observable in thin section, and, which, in effect, defines the amphibole group: Perfect prismatic cleavages parallel to (110) and $(1\bar{1}0)$ intersect at 56° and 124° to betray the identity of the mineral group (see Fig. 5-28).

TABLE 5-3. Optical Properties of the Amphibole Group

n_β	n_α	n_γ	$n_\gamma - n_\alpha$	Mineral	System	$2V$	Optic Sign	Cleavage	Optical Orientation	Color in Sections	Physical Properties
1.60–1.65	1.59–1.64	1.62–1.66	0.024–0.016	Glaucophane $Na_2Mg_3Al_2(Si_4O_{11})_2(OH)_2$	Monoclinic $\angle\beta \simeq 104°$	50°–0° $v > r$ strong	−	{110} perfect	$b = Y$ $c \wedge Z = -6°$ to $-9°$ $a \wedge X = +8°$ to $+5°$	X = colorless Y = pale violet Z = deep blue	H = 6. Sp. Gr. = 3.1–3.3. Dark blue, blue-gray
1.61–1.63	1.60–1.62	1.63–1.65	0.027–0.022	Tremolite $Ca_2Mg_5(Si_4O_{11})_2(OH)_2$	Monoclinic $\angle\beta \simeq 105°$	88°–84° $v > r$ weak	−	{110} perfect parting {100}	$b = Y$ $c \wedge Z = -21°$ to $-7°$ $a \wedge X = -6°$ to $-2°$	Colorless	H = 5–6. Sp. Gr. = 3.0–3.2. White, gray, violet
1.61–1.71	1.60–1.69	1.62–1.72	0.028–0.013	Anthophyllite $(Mg,Fe^{2+})_7(Si_4O_{11})_2(OH)_2$	Orthorhombic	65°–90° 90°–59° $r \gtrless v$ weak	− +	{110} perfect {100}, {010} poor	$a = X$ $b = Y$ $c = Z$	X = pale yellow Y = yellow-brown Z = dark brown	H = 5–5½. Sp. Gr. = 2.9–3.6. Brown, green, gray
1.62–1.66	1.61–1.66	1.63–1.67	0.020	Pargasite $NaCa_2Mg_4(Al,Fe^{3+})[(Si_3Al)O_{11}]_2(OH)_2$	Monoclinic $\angle\beta \simeq 105°$	70°–90° $r > v$ weak	+	{110} perfect	$b = Y$ $c \wedge Z \simeq -26°$ $a \wedge X \simeq -11°$	X = colorless Y = blue-green Z = deep green-blue	H = 5–6. Sp. Gr. = 3.05. Light brown, brown
1.62–1.66	1.61–1.65	1.63–1.67	0.016–0.023	Edenite $NaCa_2Mg_5(Si_7Al)O_{22}(OH)_2$	Monoclinic $\angle\beta \simeq 105°$	50°–80° $r > v$ weak inclined	+	{110} perfect parting {100} or {001}	$b = Y$ $c \wedge Z \simeq -17°$ to $-27°$ $a \wedge X \simeq -2°$ to $-12°$	X = pale yellow Y = green Z = green-blue	H = 5–6. Sp. Gr. ≃ 3.1. Green
1.62–1.70	1.60–1.69	1.63–1.71	0.015–0.029	Richterite $Na_2Ca(Mg,Fe^{2+},Mn,Fe^{3+},Al)_5(Si_4O_{11})_2(OH)_2$	Monoclinic $\angle\beta \simeq 104°$	66°–87°	−	{110} perfect parting {100} or {001}	$b = Y$ $c \wedge Z = -15°$ to $-40°$ $a \wedge X = -1°$ to $-25°$	X = pale yellow Y = orange, green Z = pale green	H = 5–6. Sp. Gr. = 3.0–3.4. Yellow, brown, dark green
1.62–1.70	1.61–1.70	1.63–1.71	0.015–0.034	Common Hornblende $Ca_2(Mg,Fe^{2+})_4(Al,Fe^{3+})(Si_7Al)O_{22}(OH)_2$	Monoclinic $\angle\beta = 105°30'$	35°–90° $r \gtrless v$ moderate	−	{110} perfect parting {100} or {001}	$b = Y$ $c \wedge Z \simeq -12°$ to $-34°$ $a \wedge X \simeq +3°$ to $-19°$	X = pale green Y = green Z = blue-green	H = 5–6. Sp. Gr. = 3.0–3.5. Dark green, black
1.63–1.65	1.61–1.64	1.63–1.65	0.016–0.020	Eckermannite $Na_2CaMg_4Fe^{2+}(Si_7Al)O_{22}(OH)_2$	Monoclinic $\angle\beta \simeq 105°$	80°–15° $r > v$ strong	−	{110} perfect parting {010}	$b = Y$ $c \wedge X = +30°$ to $+55°$ $a \wedge Z = +45°$ to $+70°$	X = yellow Y = yellow-brown Z = dark green	H = 5–6. Sp. Gr. = 3.3–3.0. Brown, green-brown
1.63–1.67	1.62–1.66	1.64–1.68	0.018–0.020	Tschermakite $Ca_2Mg_3Al_2[(Si_3Al)O_{11}]_2(OH)_2$	Monoclinic $\angle\beta \simeq 105°$	90°–73° $r \gtrless v$ moderate	− +	{110} perfect parting {100} or {001}	$b = Y$ $c \wedge Z \simeq -20°$ $a \wedge X \simeq -5°$	X = colorl Y = yellow green Z = green	H = 5–6. Sp. Gr. ≃ 3.1. Green, brown

1.63–1.68	1.62–1.67	1.65–1.69	0.022–0.017	Actinolite $Ca_2Fe_5^{2+}(Si_4O_{11})_2(OH)_2$	Monoclinic $\angle\beta \simeq 105°$	84°–75° $v > r$ weak	–	{110} perfect parting {100}	$b = Y$ $c \wedge Z = -17°$ to $-13°$ $a \wedge X = -2°$ to $+2°$	X = colorless Y = pale yellow-green Z = green	H = 5–6. Sp. Gr. = 3.2–3.4. Bright green, dark green
1.64–1.66	1.62–1.64	1.65–1.67	0.023–0.029	Holmquistite $Li(Mg,Fe^{2+})_3(Al,Fe^{3+})_2(Si_4O_{11})_2(OH)_2$	Orthorhombic	44°–52° $r > v$ weak	–	{210} perfect	$b = Y$ $c = Z$ $a = X$	X = pale yellow Y = violet Z = blue-violet	H = 5–6. Sp. Gr. = 3.1. Light blue, blue-violet, black
1.64–1.68	1.63–1.66	1.66–1.70	0.022–0.035	Cummingtonite $Mg_7(Si_4O_{11})_2(OH)_2$	Monoclinic $\angle\beta \simeq 102°$	60°–90° $v > r$ weak	+	{110} perfect	$b = Y$ $c \wedge Z \simeq -21°$ to $-15°$ $a \wedge X \simeq -9°$ to $-3°$	X = colorless Y = pale yellow Z = pale green, brown	H = 5–6. Sp. Gr. = 3.3. Dark green, brown
1.65–1.69	1.66–1.69	1.66–1.70	0.016–0.012	Crossite $Na_2Mg_3Fe_2^{2+}(Si_4O_{11})_2(OH)_2$	Monoclinic $\angle\beta \simeq 104°$	90°–0° $r > v$ extreme	– +	{110} perfect	$b = Z$ $c \wedge Y = -10°$ to $-8°$ $a \wedge X = +4°$ to $+6°$	X = pale yellow Y = deep blue Z = pale violet	H = 6. Sp. Gr. = 3.1–3.3. Dark blue, gray
1.66–1.69	1.64–1.68	1.66–1.69	0.009–0.021	Katophorite $NaCa(Mg,Fe^{2+})_4Fe^{3+}(Si_7Al)O_{22}(OH)_2$	Monoclinic $\angle\beta \simeq 105°$	0°–50° $r > v$ strong	–	{110} perfect parting {010}	$b = Z$ $c \wedge X = +36°$ to $+70°$ $a \wedge Y = +51°$ to $+85°$	X = pale yellow Y = green, brown Z = green, brown	H = 5. Sp. Gr. = 3.2. Dark green, black, brown
1.66–1.70	1.65–1.68	1.67–1.70	0.020–0.021	Hastingsite $NaCa(Mg,Fe^{2+})_4(Al,Fe)[(Si_3Al)O_{11}]_2(OH)_2$	Monoclinic $\angle\beta \simeq 105°$	90°–45° $v > r$ moderate	–	{110} perfect parting {100} or {001}	$b = Y$ $c \wedge Z \simeq -19°$ $a \wedge X \simeq -4°$	X = yellow Y = green Z = dark green	H = 5–6. Sp. Gr. ≃ 3.30. Dark green
1.66–1.71	1.65–1.70	1.67–1.73	0.02	Ferroedenite $NaCa_2Fe_5^{2+}(Si_7Al)O_{22}(OH)_2$	Monoclinic $\angle\beta \simeq 105°$	90°–20° $r \gtrless v$ weak	–	{110} perfect parting {100} or {001}	$b = Y$ $c \wedge Z \simeq -15°$ $a \wedge X \simeq 0°$	X = yellow Y = green Z = deep green	H = 5–6. Sp. Gr. ≃ 3.4. Dark green, black
1.67–1.77	1.65–1.70	1.68–1.80	0.018–0.083	Oxyhornblende $Ca_2Na(Mg,Fe^{3+},Al,Ti)_5[(Si_3Al)O_{11}]_2(OH)_2$	Monoclinic $\angle\beta \simeq 106°$	56°–88° $v > r$ weak $r > v$ strong	–	{110} perfect parting {100} or {001}	$b = Y$ $c \wedge Z = -19°$ to 0° $a \wedge X = -3°$ to $+16°$	X = pale yellow Y = brown Z = dark brown	H = 5–6. Sp. Gr. = 3.2–3.3. Brown, black
1.67–1.71	1.66–1.70	1.68–1.72	0.020	Ferrotschermakite $Ca_2Fe_3^{2+},Al_2[(Si_3Al)O_{11}]_2(OH)_2$	Monoclinic $\angle\beta \simeq 105°$	73°–50° $r \gtrless v$ moderate	–	{110} perfect parting {100} or {001}	$b = Y$ $c \wedge Z \simeq -18°$ $a \wedge X \simeq -3°$	X = yellow Y = yellow-green Z = blue-green	H = 5–6. Sp. Gr. ≃ 3.4. Dark green, black

(continued)

TABLE 5-3. Optical Properties of the Amphibole Group (*continued*)

n_β	n_α	n_γ	$n_\gamma - n_\alpha$	Mineral	System	$2V$	Optic Sign	Cleavage	Optical Orientation	Color in Sections	Physical Properties
1.68–1.71	1.66–1.69	1.70–1.73	0.035–0.045	Grunerite $Fe_7^{2+}(Si_4O_{11})_2(OH)_2$	Monoclinic $\angle\beta \simeq 102°$	90°–80° $r>v$ weak inclined	−	{110} perfect	$b=Y$ $c \wedge Z \simeq -15°$ to $-10°$ $a \wedge X \simeq -3°$ to $+2°$	X = pale yellow Y = pale yellow Z = green, brown	H = 5–6. Sp. Gr. = 3.3–3.6. Dark green, brown
1.68–1.71	1.67–1.70	1.68–1.72	0.015–0.012	Arfvedsonite $Na_2CaFe_4^{2+}Fe^{3+}(Si_7Al)O_{22}(OH)_2$	Monoclinic $\angle\beta \simeq 105°$	0°–70° $v>r$ strong	−	{110} perfect parting {010}	$b=Z$ $c \wedge X = +5°$ to $+30°$ $a \wedge Y = +20°$ to $+45°$	X = deep blue-green Y = pale blue-green Z = yellow-green	H = 5–6. Sp. Gr. = 3.5–3.2. Dark green, black
1.69–1.70	1.68–1.69	1.70–1.71	0.012–0.018	Barkevikite $NaCa_2(Fe^{2+},Mg,Mn)_5(Si_7Al)O_{22}(OH)_2$	Monoclinic $\angle\beta \simeq 105°$	40°–50° $r>v$ moderate	−	{110} perfect parting {100} or {001}	$b=Y$ $c \wedge Z \simeq -11°$ to $-18°$ $a \wedge X \simeq +4°$ to $-3°$	X = pale yellow-brown Y = red-brown Z = dark brown	H = 5–6. Sp. Gr. = 3.3–3.4. Dark brown, black
1.69–1.70	1.69–1.71	1.70–1.72	0.012–0.007	Riebeckite $Na_2Fe_3^{2+}Fe_2^{3+}(Si_4O_{11})_2(OH)_2$	Monoclinic $\angle\beta \simeq 103°$	50°–90° $r>v$ strong 90°–70° $v>r$ strong	− +	{110} perfect parting {010} or {001}	$b=Y$ $c \wedge X = -8°$ to $-7°$ $a \wedge Z = +6°$ to $+7°$	X = deep blue Y = yellow Z = yellow-green	H = 6. Sp. Gr. = 3.0–3.4. Dark blue-green, black
1.69–1.74	1.67–1.69	1.70–1.77	0.019–0.083	Kaersutite $NaCa_2(Mg,Fe^{2+})_4Ti[(Si_3Al)O_{11}]_2(OH)_2$	Monoclinic $\angle\beta \simeq 106°$	66°–82° $r>v$ strong	−	{110} perfect parting {100} or {001}	$b=Y$ $c \wedge Z = 0°$ to $-19°$ $a \wedge X = +16°$ to $-3°$	X = pale yellow Y = red-brown Z = dark red-brown	H = 5–6. Sp. Gr. = 3.2. Dark brown, black
1.70–1.73	1.69–1.70	1.71–1.73	0.022–0.024	Ferrohastingsite $NaCaFe_4^{2+}(Al,Fe^{3+})[(Si_3Al)O_{11}]_2(OH)_2$	Monoclinic $\angle\beta \simeq 105°$	45°–10° $v>r$ moderate inclined	−	{110} perfect parting {100} or {001}	$b=Y$ $c \wedge Z \simeq -12°$ $a \wedge X \simeq +3°$	X = yellow Y = dark green Z = very dark green	H = 5–6. Sp. Gr. ≃ 3.50. Dark green, black

Anthophyllite

$(Mg,Fe^{2+})_7(Si_4O_{11})_2(OH)_2$ ORTHORHOMBIC

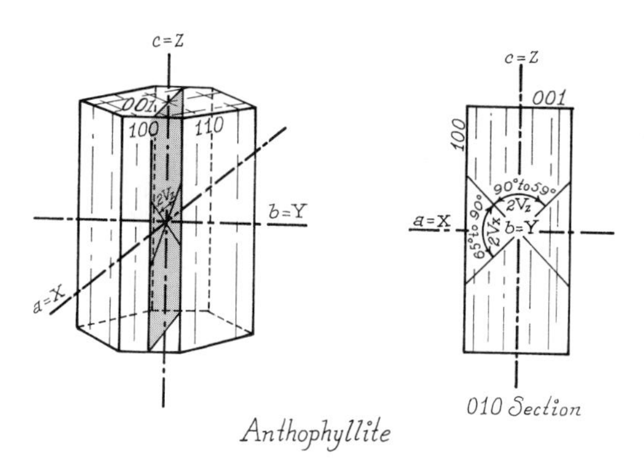

$n_\alpha = 1.588–1.694$
$n_\beta = 1.602–1.710$
$n_\gamma = 1.613–1.722$
$n_\gamma - n_\alpha = 0.025–0.013$
Biaxial negative, $2V_x = 75°–90°$ or
Biaxial positive, $2V_z = 90°–70°$
$a = X, b = Y, c = Z$
$r > v$ or $v > r$ weak to moderate
Colorless to pale brown or green, in section, with weak to moderate pleochroism
$Z > Y = X$ or $Z = Y > X$

COMPOSITION. In its broadest sense, anthothyllite encompasses a range of compositions where Mg and Fe replacement may be complete and where Al^{3+} may replace Si^{4+} up to Si_3AlO_{11}, with equal substitution of Al^{3+} for (Mg^{2+}-Fe^{2+}). The resulting complex may be described in terms of four theoretical end members:

Anthophyllite	$Mg_7(Si_4O_{11})_2(OH)_2$
Ferroanthophyllite	$Fe_7^{2+}(Si_4O_{11})_2(OH)_2$
Gedrite	$Mg_5Al_2(Si_3AlO_{11})_2(OH)_2$
Ferrogedrite	$Fe_5^{2+}Al_2(Si_3AlO_{11})_2 \cdot (OH)_2$

Replacement Al^{3+} for Si^{4+} normally produces compositions very near either Si_4O_{11} or Si_3AlO_{11} (intermediate Si:Al ratios are uncommon), giving rise to two dominant series:

Anthophyllite-ferroanthophyllite (anthophyllites)—see Fig. 5-29
Gedrite-ferrogedrite (gedrites)—see Fig. 5-30

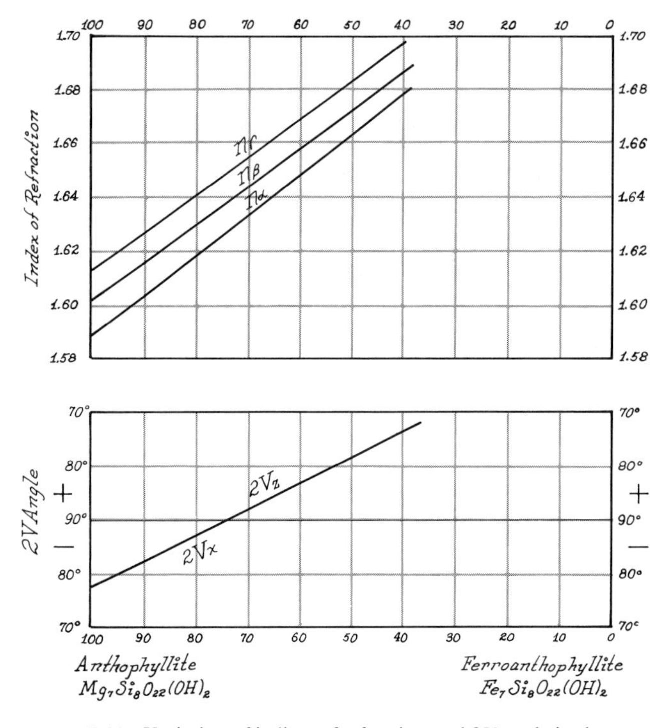

FIGURE 5-29. Variation of indices of refraction and $2V$ angle in the anthophyllite-ferroanthophyllite series.

* Indices increase with both Fe^{2+} and Al^{3+} to maximum values for ferrogedrite, $Fe_5^{2+}Al_2(Si_3AlO_{11})_2(OH)_2$, as shown in Fig. 5-29. Minimum refraction indices represent Mg-anthophyllite $Mg_7(Si_4O_{11})_2(OH)_2$. Birefringence is maximum for Mg-anthophyllite, decreasing with Fe^{2+}, and $2V$ ranges from about 75° (−) for Mg-anthophyllite through 90° to about 78° (+) at ferrogedrite.

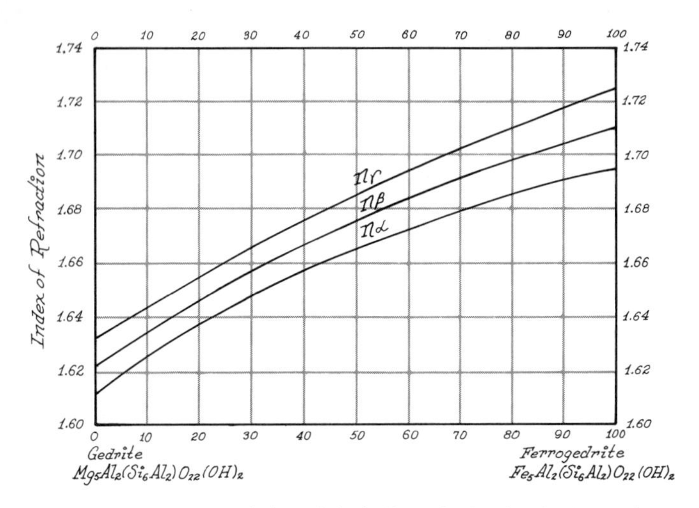

FIGURE 5-30. Variation of the indices of refraction in the gedrite-ferrogedrite series.

Ferroanthophyllite is theoretical, and the former series exists, in nature, only from $Mg_7(Si_4O_{11})_2(OH)_2$ to about 40 percent $Fe_7^{2+}(Si_4O_{11})_2(OH)_2$. Ideal gedrite $Mg_5Al_2(Si_3AlO_{11})_2 \cdot (OH)_2$ is also rare.

Minor F^- may replace $(OH)^-$, and Fe^{3+}, Mn^{2+}, Ti^{4+}, Ca^{2+}, and Na^+ are often present as minor impurities.

PHYSICAL PROPERTIES. H = $5\frac{1}{2}$–5. Sp. Gr. = 2.9–3.6. Color, in hand sample, is pale to dark brown, less commonly, green to gray. Fibrous, bladed, or columnar.

COLOR AND PLEOCHROISM. Anthophyllite is normally colorless in section; however, color increases with Fe^{2+}, and sections of iron-rich varieties may show weak color, moderately pleochroic in browns $Z > Y = X$ or $Z = Y > X$. Fragments commonly show color and weak to moderate pleochroism. X = pale brown, pale yellow, pale greenish yellow; Y = clove-brown, yellow-brown, yellow-green, blue-green; and Z = dark brown, gray-brown, gray-green, blue-green.

FORM. Crystal habits range from asbestiform, or fibrous, to bladed, columnar, and prismatic, and crystal aggregates are parallel to radiated. Prismatic sections are rectangular, and basal sections are diamond-shaped, with corners blunted by small {100} or {010} faces.

CLEAVAGE. Characteristic amphibole cleavages are perfect on {110}. In anthophyllite, these prismatic cleavages intersect at 54°30′ and 125°30′. Poor cleavages are also reported on {100} and {010}.

BIREFRINGENCE. Maximum birefringence, seen in {010} sections, is moderate to mild (0.025–0.013) decreasing as Fe^{2+} replaces Mg. Interference colors range to vivid colors of low second order, in standard section.

TWINNING. Twinning appears absent.

INTERFERENCE FIGURE. Anthophyllites and gedrites are commonly biaxial positive, with large $2V$. Pure anthophyllite $Mg_7(Si_4O_{11})_2(OH)_2$ is optically negative ($2V_x \simeq 65°$) but, with increasing Fe^{2+}, $2V$ opens to 90°, at about 10 percent $Fe_7(Si_4O_{11})_2(OH)_2$, and closes on Z, making most compositions optically positive. Ferrogedrite is optically negative, but other gedrite compositions are positive.

Sections parallel to {010} show flash figures and basal sections, showing both cleavages normally yield an acute bisectrix. Fragments on cleavages yield near-flash figures. Optic axis dispersion is weak to moderate $r > v$ or $v > r$, with possible inclined bisectrix dispersion.

OPTICAL ORIENTATION. Anthophyllites and gedrites are orthorhombic with $a = X$, $b = Y$, $c = Z$. Optic plane is {010}. Extinction angles are parallel to cleavages on cleavages and {$hk0$} sections and symmetrical on basal sections. Elongation is positive (length-slow), since Z and elongation parallel c.

DISTINGUISHING FEATURES. Anthophyllites and gedrites are the only common orthorhombic amphiboles and are distinguished by typical amphibole cleavages combined with strictly parallel extinction in sections containing c. Mg-rich anthophyllite and ferrogedrite are optically negative and are separated by refraction index. Common anthophyllites and gedrites are optically separated with great difficulty. Sillimanite and zoisite have smaller $2V$ and {010} cleavage.

ALTERATION. Anthophyllite alters to talc or serpentine at low temperatures.

OCCURRENCE. Anthophyllites and gedrites are metamorphic minerals formed by high-grade dynamothermal metamorphism of ultramafic or impure argillaceous rocks or by Mg and Fe metasomatism of argillaceous sediments. They appear in amphibolites, gneisses, and granulites of regional metamorphism and hornfels, very commonly associated with cordierite. Hornblende, plagioclase, sillimanite, talc, serpentine, and other Mg-minerals and Al-minerals are also common associates.

REFERENCES: ANTHOPHYLLITE

Cerny, P. 1968. Some properties of natural, synthetic, and extrapolated magnesium anthophyllite. *Neues Jahrb. Mineral. Monatsh., 1968*, 57–61.

Finger, L. W. 1970. Refinement of the crystal structure of anthophyllite. *Carnegie Inst. Wash. Year Book, 68*, 283–288.

Fyfe, W. S. 1962. Relative stability of talc, anthophyllite, and enstatite. *Amer. Jour. Sci., 260*, 460–466.

Jaffe, H. W., P. Robinson, and C. Klein. 1968. Exsolution lamellas and optic orientation of clinoamphiboles. *Science, 160*, 776-778.

Rabbitt, J. C. 1948. A new study of the anthophyllite series. *Amer. Mineral., 33*, 263–323.

Winchell, A. N. 1938. The anthophyllite and cummingtonite-grunerite series. *Amer. Mineral., 23*, 329–333.

The Cummingtonite-Grunerite Series

$(Mg,Fe^{2+})_7(Si_4O_{11})_2(OH)_2$

MONOCLINIC
$\angle\beta \simeq 102°$

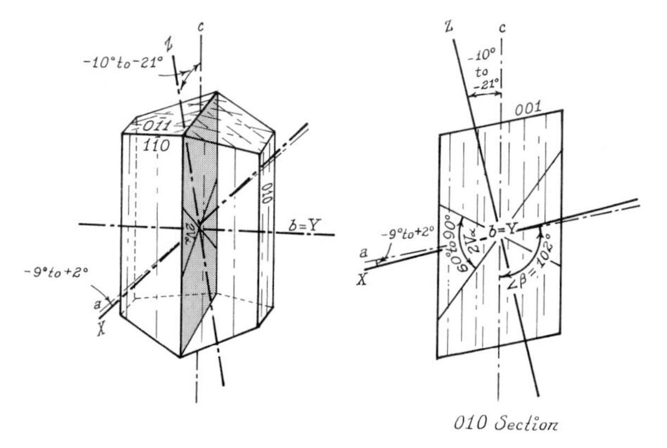

Cummingtonite-Grunerite

Cummingtonite	**Grunerite**
$n_\alpha = 1.632–1.663$*	$n_\alpha = 1.663–1.688$
$n_\beta = 1.638–1.677$	$n_\beta = 1.677–1.709$
$n_\gamma = 1.655–1.697$	$n_\gamma = 1.697–1.729$
$n_\gamma - n_\alpha = 0.022–0.035$	$n_\gamma - n_\alpha = 0.035–0.045$
Biaxial positive	Biaxial negative
$2V_z = 70°–90°$	$2V_x = 90°–80°$
$b = Y, c \wedge Z = -21°$ to $-16°$	$b = Y, c \wedge Z = -16°$ to $-12°$
$a \wedge X = -9°$ to $-3°$	$a \wedge X = -3°$ to $+2°$
$v > r$ weak	$r > v$, inclined Bxa weak

Colorless to pale brown or green, with weak pleochroism in pale yellow, browns and greens

$Z > Y \geqslant X$

* Refraction indices increase uniformly with Fe^{2+} and increase somewhat with Al and Mn. Birefringence and $2V_z$ also increase with Fe^{2+} and $c \wedge Z$ decreases (Fig. 5-31).

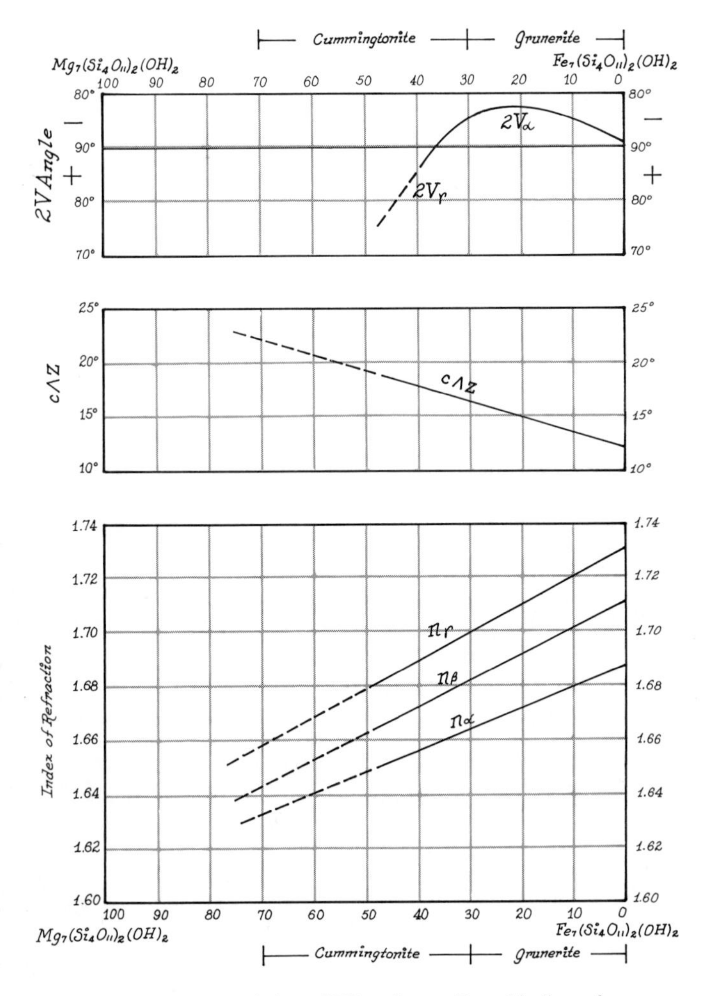

FIGURE 5-31. Variation of $2V$ angle, $c \wedge Z$, and indices of refraction in the cummingtonite-grunerite series. (Klein, Cummingtonite-grunerite series: A chemical, optical and x-ray study, *Amer. Mineral.*, p. 971, Fig. 3, 1964, © by the Mineralogical Society of America)

COMPOSITION. The cummingtonite-grunerite series is basically the monoclinic analog of anthophyllite-gedrite from $Mg_7(Si_4O_{11})_2(OH)_2$ to $Fe_7(Si_4O_{11})_2(OH)_2$. Neither series is continuous from Mg to Fe^{2+} end member, and the cummingtonite-grunerite series is represented, in nature, by compositions continuous from about 30 to 100 percent $Fe_7(Si_4O_{11})_2(OH)_2$. Cummingtonite is defined by the range 30 to 70 percent Fe^{2+} end member $Fe_7(Si_4O_{11})_2(OH)_2$ and grunerite by the range 70 to 100 percent. Compositions more Mg-rich than 30 percent $Fe_7(Si_4O_{11})_2(OH)_2$ are not reported but have been named *kupfferite*, despite nondiscovery. Correlation of optical properties with composition is shown in Fig. 5-32 and 5-33.

Replacement $Al \rightleftharpoons Si$ is usually minor but more important in Mg-rich varieties. Ca content is also small, and a continuous series with hornblende does not exist. Cummingtonite commonly occurs in direct association with either anthophyllite or hornblende as a definite, distinct phase. A Mn-rich variety is known (*dannemorite*), and minor amounts of Fe^{3+}, Ti^{4+}, Zn^{2+}, and Na^+ may appear in cummingtonite-grunerite compositions.

PHYSICAL PROPERTIES. H = 5–6. Sp. Gr. = 3.1–3.6. Color, in hand sample, is dark green to brown, and crystals are bladed, columnar, to fibrous.

COLOR AND PLEOCHROISM. Cummingtonite-grunerite is colorless to pale green or brown, in standard section, with weak pleochroism $Z > Y \geqslant X$. X = colorless, pale yellow; Y = pale yellow, pale yellow-brown; and Z = pale green, pale brown. Color and pleochroism tend to increase with iron.

FORM. Crystals are columnar, bladed, or acicular to fibrous, or asbestiform, in parallel or radiated aggregates. Asbestiform varieties, are called *amosite* or *montasite* (soft, silky fibers). Prismatic sections are rectangular, and cross sections are rhombic, showing prismatic cleavages.

CLEAVAGE. Cleavages are distinct, prismatic {110}, amphibole cleavages intersecting at 56° and 124°.

BIREFRINGENCE. Maximum birefringence, seen in {010} sections, is moderate, increasing with Fe^{2+} from 0.022 to 0.045, with the cummingtonite-grunerite junction at about 0.035. Cummingtonites, in standard section, show interference colors in first order, but grunerite may show colors to mid third order.

INTERFERENCE FIGURE. The arbitrary division between cummingtonite and grunerite is logically fixed by the change in

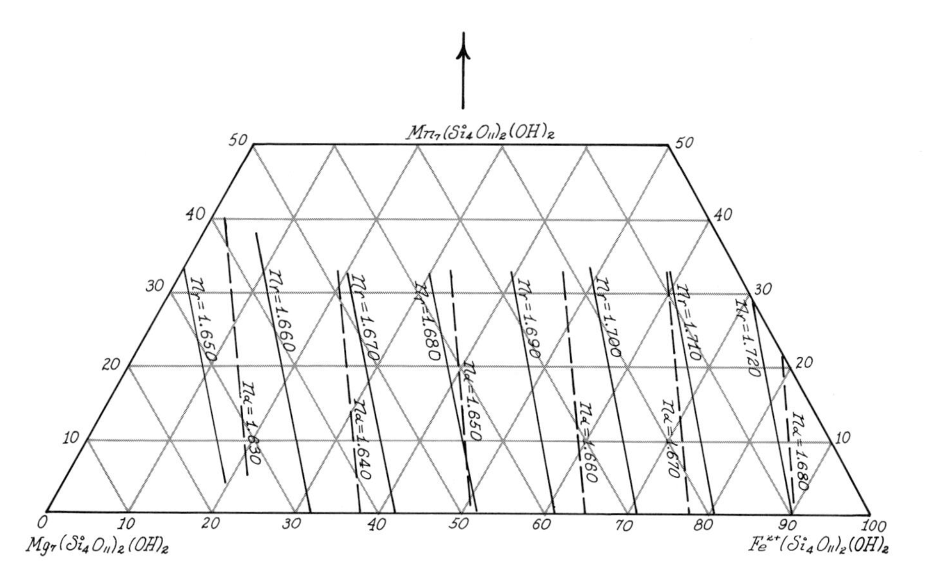

FIGURE 5-32. Relationship of indices of refraction (n_γ and n_α) with variation in Mg, Fe, and Mn in the cummingtonite-grunerite series. (Klein, Cummington-grunerite series: A chemical optical and x-ray study, *Amer. Mineral.*, p. 972, Fig. 4, 1964, © by the Mineralogical Society of America)

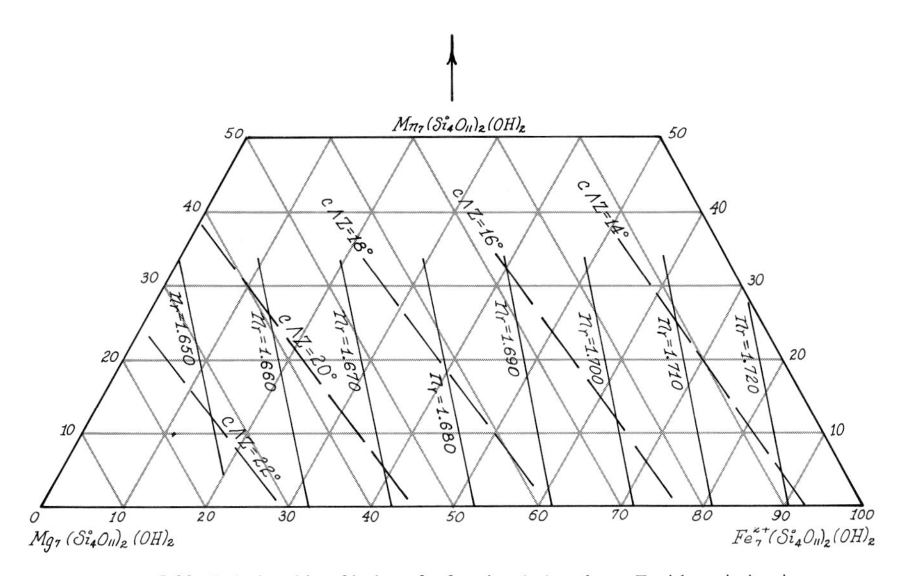

FIGURE 5-33. Relationship of index of refraction (n_γ) and $c \wedge Z$ with variation in Mg, Fe, and Mn in the cummingtonite-grunerite series. (Klein, Cummington-grunerite series: A chemical, optical and x-ray study, *Amer. Mineral.*, p. 972, Fig. 5, 1964, © by the Mineralogical Society of America)

optic sign. Cummingtonite is biaxial positive, with $2V_z$ increasing with iron from about 70° to 90°. At about 70 percent $Fe_7^{2+}(Si_4O_{11})_2(OH)_2$, $2V = 90°$ and with increasing iron grunerite becomes biaxial negative, with $2V_x$ decreasing from 90° to about 80° at the pure Fe^{2+} end member.

Sections parallel to {010} show flash figures, and cleavage fragments produce near-flash figures. Basal sections showing characteristic cleavages yield near-center Z bisectrix figures.

Dispersion is $v > r$ weak (cummingtonite) and $r > v$ weak, with weak inclined bisectrix dispersion (grunerite).

TWINNING. Simple contact or repeated twinning on {100} is very common, and twin lamellae are often very narrow.

ZONING. Cummingtonite or grunerite may be zoned by a reaction rim of hornblende or may form as reaction rims on pyroxenes. Intergrowths with anthophyllite or hornblende emphasize immiscibility zones between cummingtonite and related amphiboles.

OPTICAL ORIENTATION. The cummingtonite-grunerite series is monoclinic, with $b = Y$, $c \wedge Z = -21°$ to $-12°$ and $a \wedge X = -9°$ to $+2°$.

Maximum extinction angle, in {010} sections, decreases from 21° to 12° as Fe^{2+} increases, with the grunerite range beginning at about 15°. Extinction is parallel to cleavage, in {100} sections, and parallel, in basal sections.

Elongation is positive (length-slow), as Z is nearer elongation.

DISTINGUISHING FEATURES. Cummingtonite is biaxial positive and shows lower refraction index ($n_\beta < 1.677$), lower birefringence ($n_\gamma - n_\alpha < 0.035$), and higher maximum extinction ($c \wedge Z > 15°$) than grunerite. Anthophyllite-gedrite minerals show parallel extinction and no multiple twinning. Tremolite-actinolite minerals have low refractive indices, and tremolite is optically negative. Hornblende is optically negative and shows pronounced pleochroism.

ALTERATION. Cummingtonite or grunerite commonly inverts to hornblende by rim reaction and may be expected to alter to talc or serpentine with an iron oxide by-product.

OCCURRENCE. Cummingtonite is essentially confined to metamorphic rocks, where it appears in amphibolites, schists, gneisses, and granulites of moderately intense regional metamorphism. It is usually associated with hornblende or anthophyllite and appears commonly with cordierite, plagioclase, biotite, and garnet. It is sometimes present in hornfels or mafic xenoliths. Cummingtonite has been reported as a primary mineral in certain diorites, gabbros, and norites and occurs rarely in intermediate volcanics, often with brown hornblende.

Grunerite is most characteristic of dynamothermal metamorphism of iron-rich siliceous sediments, where it may appear in association with magnetite, hematite, quartz, fayalite, hedenbergite, or garnet.

REFERENCES: THE CUMMINGTONITE-GRUNERITE SERIES

Cooper, A. F., and J. F. Lovering. 1970. Greenschist amphiboles from Haast River, New Zealand. *Contrib. Mineral. Petrol.*, *27*, 11–24.

Finger, L. W. 1969. The crystal structure and cation distribution of a grunerite. *Mineral. Soc. Amer. Spec. Paper*, *2*, 95–100.

Gittos, M. F., G. W. Lorimer, and P. E. Champness. 1974. Electron microscopic study of precipitation (exsolution) in an amphibole (the hornblende-grunerite system). *Jour. Mater. Sci.*, *9*, 184–192.

Jaffe, H. W., P. Robinson, and C. Klein. 1968. Exsolution lamellas and optic orientation of clinoamphiboles. *Science*, *160*, 776–778.

Klein, C. 1964. Cummingtonite-grunerite series: A chemical, optical and x-ray study. *Amer. Mineral.*, *49*, 963–982.

Papike, J. J., M. Ross, and J. R. Clark. 1969. Crystal-chemical characterization of clinoamphiboles based on five new structure refinements. *Mineral. Soc. Amer. Spec. Paper*, *2*, 117–136.

Winchell, A. N. 1938. The anthophyllite and cummingtonite-grunerite series. *Amer. Mineral.*, *23*, 329–333.

The Tremolite-Actinolite Series $Ca_2(Mg,Fe^{2+})_5(Si_4O_{11})_2(OH)_2$

MONOCLINIC
$\angle\beta \approx 105°$

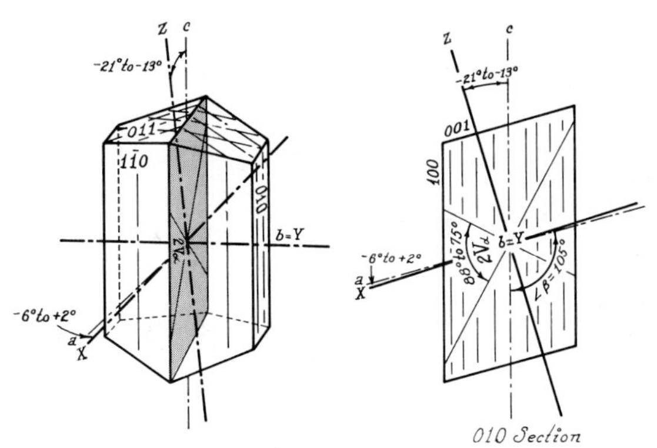

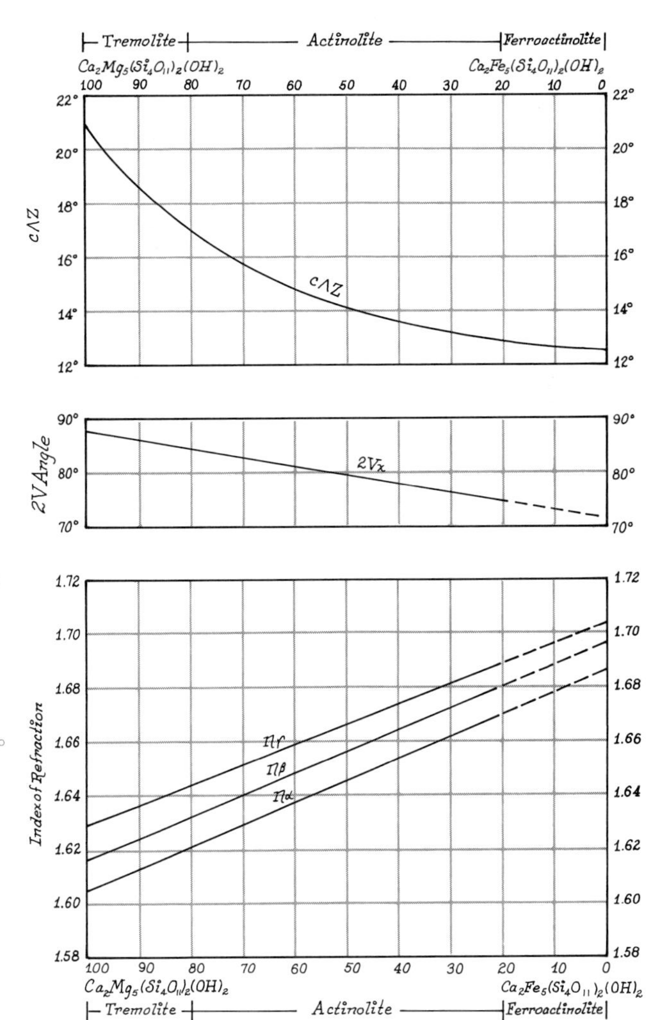

FIGURE 5-34. Variation of $c \wedge Z$, $2V$ angle, and indices of refraction in the tremolite-ferroactinolite series.

Tremolite	Actinolite
$n_\alpha = 1.600–1.620$*	$n_\alpha = 1.620–1.670$
$n_\beta = 1.612–1.630$	$n_\beta = 1.630–1.675$
$n_\gamma = 1.625–1.645$	$n_\gamma = 1.645–1.688$
$n_\gamma - n_\alpha = 0.027–0.022$	$n_\gamma - n_\alpha = 0.022–0.017$
Biaxial negative	Biaxial negative
$2V_x = 88° - 84°$	$2V_x = 84° - 75°$
Optic plane = {010}	Optic plane = {010}
$b = Y$	$b = Y$
$c \wedge Z = -21°$ to $-17°$	$c \wedge Z = -17°$ to $-13°$
$a \wedge X = -6°$ to $-2°$	$a \wedge X = -2°$ to $+2°$
$v > r$ weak	$v > r$ weak

Colorless to pale yellow green or blue green, in section, with weak pleochroism, increasing with Fe^{2+}

$Z > Y \geq X$

* Refraction index increases with increasing Fe^{2+} and is elevated slightly by minor Fe^{3+}, Al^{3+}, and Mn^{2+} and lowered by Na^+ and F^-. Birefringence is lowered by increasing Fe^{2+}, also by Na^+ and $Fe.^-$ $2V_x$ decreases with increasing Fe^{2+} (see Fig. 5-34).

COMPOSITION. The series tremolite, $Ca_2Mg_5(Si_4O_{11})_2(OH)_2$, to ferroactinolite, $Ca_2Fe_5^{2+}(Si_4O_{11})_2(OH)_2$, is not complete, and iron-rich compositions beyond the range 0–50 percent ferroactinolite are rare. Tremolite is defined by the range 0–20 percent; actinolite, 20–80 percent; and ferroactinolite, 80–100 percent Fe^{2+} end member, $Ca_2Fe_5^{2+}(Si_4O_{11})_2 \cdot (OH)_2$. Ferroactinolite is not reported in nature.

The substitution of Al^{3+} for Si^{4+} accompanied by Al^{3+} for (Mg^{2+},Fe^{2+}), forms a continuous series between tremolite-ferroactinolite, $Ca_2(Mg,Fe^{2+})_5(Si_4O_{11})_2(OH)_2$, and *tschermakite-ferrotschermakite*, $Ca_2[(Mg,Fe^{2+})_4Al](Si_7Al) \cdot O_{22}(OH)_2$, and it is necessary to arbitrarily define tremolite-ferroactinolite by compositions of low Al content $(<(Si_{7.5}Al_{0.5})O_{22})$.

The substitution of Na^+ for Ca^{2+}, accompanied by Al^{3+} for Mg^{2+} or Fe^{3+} for Fe^{2+}, causes tremolite-ferroactinolite to form a continuous series with glaucophane, $Na_2(Mg_3Al_2)(Si_4O_{11})_2(OH)_2$, or riebeckite, $Na_2(Fe_3^{+2}Fe_2^{+3})(Si_4O_{11})_2(OH)_2$, and again, tremolite-ferroactinolite must be defined by low Na, Al, and Fe^{3+}.

High Mn content is rare but Mn varieties of both tremolite (*hexagonite*—pale lilac in hand sample) and actinolite (*manganactinolite*) are known. Significant F^- may replace $(OH)^-$, and a Cr-tremolite has been reported.

PHYSICAL PROPERTIES. H = 5–6. Sp. Gr. = 3.0–3.4. Color, in hand sample, is white to gray (tremolite), ranging through bright green to dark green with increasing Fe^{2+} (actinolite).

COLOR AND PLEOCHROISM. In standard section, color and pleochroism increase with Fe^{2+} from colorless (tremolite) to pale green or green (actinolite), pleochroic in yellow and green $Z > Y \geqslant X$. X = colorless, pale yellow-green; Y = pale yellow-green, pale blue-green; and Z = pale green, green, blue-green.

FORM. Tremolite-actinolite ranges from columnar, bladed, to fibrous or asbestiform, normally appearing as parallel aggregates. Prismatic sections are rectangular, and basal sections are rhombic, showing typical amphibole cleavages {110} (Fig. 5-35).

The variety *nephrite* is a tough, compact variety of tremolite-actinolite, showing an interlaced aggregate of fibers. Precious jade may be either nephrite (amphibole) or jadeite (pyroxene).

CLEAVAGE. Typical amphibole cleavages, distinct prismatic {110}, intersecting at 56° and 124° are present. Parting may be present on {100} and, rarely, on {010}.

BIREFRINGENCE. Maximum birefringence, seen in {010} sections, is moderate (0.027) to mild (0.017), decreasing with increasing Fe^{+2}. Interference colors, in standard section, range to middle second order, for tremolite, but do not exceed low second order for actinolite.

TWINNING. Normal twinning on {100} is common and may be simple or repeated. Fine multiple twinning on {001} is rare.

ZONING. Pale green actinolite cores may grade to hornblende rims, and parallel growths of actinolite and hornblende fibers are reported.

INTERFERENCE FIGURE. The entire tremolite-ferroactinolite series is biaxial negative, with $2V_x$ decreasing from about 88° to about 75° with increasing iron. Fragments of deepest color yield best figures, and inherent color may obscure isochromes. {100} sections show off-center Bxa figures symmetrical on the optic plane, {010} sections show flash figures, and cleavage fragments show near-flash figures. Dispersion is weak $v > r$, with possible inclined bisectrix effects.

OPTICAL ORIENTATION. The series is monoclinic, with $b = Y$, $c \wedge Z = -21°$ to $-13°$, and $a \wedge X = -6°$ to $+2°$. Maximum extinction angles, on {010} sections, range from about 21° to 13°, decreasing with Fe^{2+} (tremolite, 21°–17°; actinolite, 17°–13°). Extinction angles of elongated fragments, on {110} cleavages, ranges from about 15° to 11°. Elongation is positive (length-slow), since Z lies nearer c.

DISTINGUISHING FEATURES. Tremolite shows lower refraction index, higher birefringence, larger $2V_x$, and larger extinction angle than does actinolite. Anthophyllite and cummingtonite are biaxial positive, and grunerite shows higher birefringence and refractive index. Hornblende is highly colored and pleochroic, with extinction angles normally above those for actinolite (>17°). Sillimanite and wollastonite resemble tremolite, but sillimanite is orthorhombic with parallel extinction, and wollastonite shows near-parallel extinction, moderate $2V$, and both positive and negative elongation. Nephrite is distinguished by felty, interwoven fibers and shows smaller extinction and lower refraction indices than does jadite.

ALTERATION. Tremolite-actinolite may alter to talc, chlorite, or calcite-dolomite. As *uralite*, actinolite is itself an alteration product, resulting from deuteric or hydrothermal alteration of many pyroxenes.

OCCURRENCE. Tremolite forms most commonly in carbonate contact metamorphic zones by the reaction of dolomite and quartz:

$$5(Ca,Mg)CO_3 + 8SiO_2 + H_2O \longrightarrow Ca_2Mg_5(Si_4O_{11})_2(OH)_2 + 3CaCO_3 + 2CO_2$$

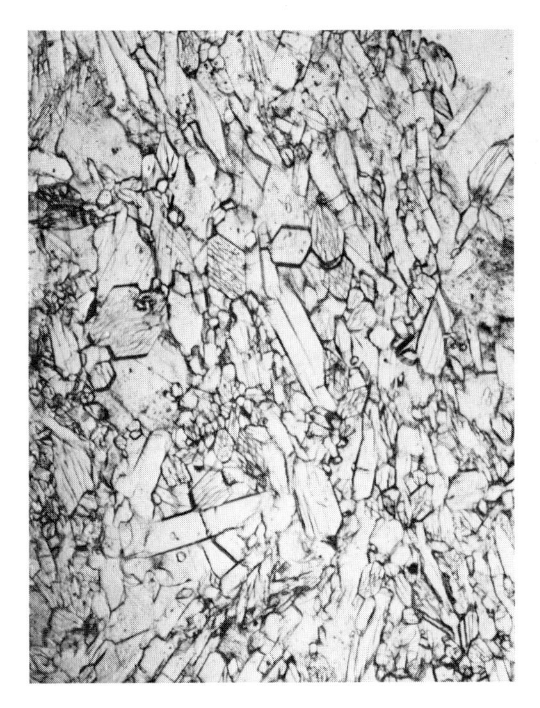

FIGURE 5-35. Tremolite crystals in a talc-tremolite schist show the characteristic amphibole sections and prismatic cleavages intersecting at ~56° and ~124°.

In contact zones, it is associated with calcite, Ca-garnet, wollastonite, phlogopite, and several Ca and Mg-silicates. With increasing metamorphic grade, tremolite reacts with calcite and remaining quartz to yield diopside or forsterite.

In rocks of relatively low-grade regional metamorphism (greenschist facies), tremolite and actinolite appear in schists derived from ultramafic or carbonate rocks. Talc is a very common associate, as are carbonates, chlorites, albite, and epidotes. As the grade of metamorphism increases to the amphibolite facies, actinolite accommodates more Al to become hornblende. At low grades, actinolite and hornblende may coexist, but at high temperatures the immiscibility zone disappears and only Al-poor hornblende exists. Actinolite may appear in high-pressure rocks with glaucophane and its associates.

Fine-grained fibrous actinolite (uralite) appears as the alteration product of pyroxenes, and a Cr-variety of tremolite is reported in pegmatites.

REFERENCES: THE TREMOLITE-ACTINOLITE SERIES

Grigor'ev, D. P. 1939. Experimental investigation on the effect of alumina on the optical properties of tremolite. *Compt. Rend. Acad. Sci. USSR*, *23*, 71–73.

Papike, J. J., M. Ross, and J. R. Clark. 1969. Crystal chemical characterization of clinoamphiboles based on five new structure refinements. *Mineral. Soc. Amer. Spec. Paper*, *2*, 117–136.

Posnjak, E., and N. L. Bowen. 1931. The role of water in tremolite. *Amer. Jour. Sci.*, *22*, 203–214.

Winchell, A. N. 1945. Variations in composition and properties of the calciferous amphiboles. *Amer. Mineral.*, *30*, 27–50.

Zussman, J. 1955. The crystal structure of anthophyllite. *Acta Crystallogr.*, *8*, 301–308.

The Hornblende Series ("Common" Hornblende) $Ca_2(Mg,Fe^{2+})_4(Al,Fe^{3+})(Si_7Al)O_{22}(OH)_2$

MONOCLINIC
$\angle\beta = 105°30'$

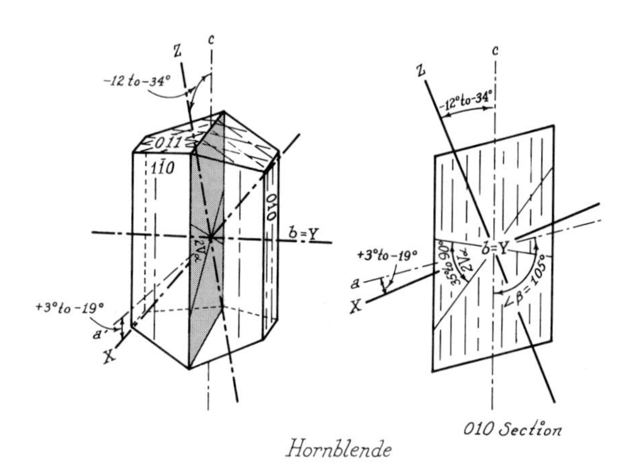

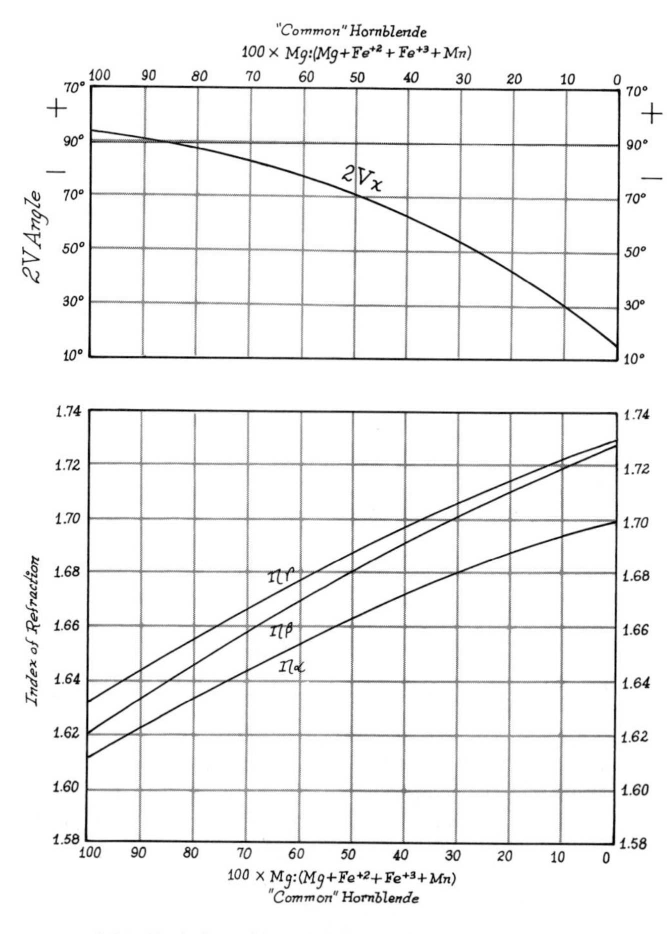

FIGURE 5-36. Variation of $2V$ angle and indices of refraction in the "common" hornblende series. (After Deer, Howie, and Zussman, 1962)

$n_\alpha = 1.61–1.69$*
$n_\beta = 1.62–1.70$
$n_\gamma = 1.63–1.73$
$n_\gamma - n_\alpha = 0.015–0.034$
Biaxial negative, $2V_x = 35°–90°$
$b = Y$, $c \wedge Z = -12°$ to $-34°$, $a \wedge X = +3°$ to $-19°$
$v > r$ or $r > v$ moderate
Numerous shades of green to brown with distinct pleochroism $Z \geq Y > X$

COMPOSITION. Hornblendes are Ca-rich amphiboles with appreciable $Si \rightleftharpoons Al$ substitution and enough iron to cause dark color. $Si \rightleftharpoons Al$ never exceeds Al_2Si_6 and is balanced by Na (or K) in the *W* site, by Al or Fe^{3+} in six-fold sites, or both, resulting in three hornblende series where complete interchange of $Mg^{2+} \rightleftharpoons Fe^{2+}$ provides series end members:

Edenite-ferroedenite	$NaCa_2(Mg,Fe^{2+})_5(Si_7Al)O_{22}(OH)_2$
Tschermakite-ferrotschermakite	$Ca_2(Mg,Fe^{2+})_4(Al,Fe^{3+})(Si_7Al)O_{22}(OH)_2$
Pargasite-ferrohastingsite	$NaCa_2(Mg,Fe^{2+})_4(Al,Fe^{3+})(Si_6Al_2)O_{22}(OH)_2$

* Refraction index and birefringence increase with iron, and $2V_x$ becomes less (Fig. 5-36). Substitution of F^- for $(OH)^-$ lowers indices, and Na for Ca has little effect.

K, Mn, and Ti are usually present as minor constituents, and a dozen or more elements may appear in traces. Optical properties of the tschermakite-ferrochermakite series are shown in Fig. 5-37.

PHYSICAL PROPERTIES. H = 5–6. Sp. Gr. = 3.02–3.45. Color in hand sample is dark green to black with vitreous to pearly luster.

COLOR AND PLEOCHROISM. Color and pleochroism are pronounced in both fragments and thin section and extremely variable in greens, blue-greens, and yellow-browns. X = yellow, pale green, or pale yellow-brown, Y = green, yellow-green, or brown; and Z = blue-green, green, or brown. Unfortunately, the darkest fragments are properly oriented for best interference figures.

FORM. Crystals are commonly prismatic, yielding typical amphibole cross sections (Fig. 5-26). Intermediate plutonic rocks often show hornblende as reaction rims around pyroxene crystals. As a secondary mineral (uralite), hornblende is fine-grained and fibrous, usually as reaction rims. Fragments show cleavage and tend to be splintery.

CLEAVAGE. Prismatic cleavages {110} are perfect, intersecting at 56° and 124°, and parting is possible on {100} or {001}.

BIREFRINGENCE. Moderate maximum birefringence, 0.015 (high Mg) to 0.034 (high Fe), is seen in {010} sections. Most hornblendes show maximum birefringence near 0.022. In section, interference colors are first order and low second order.

Fragments are often dark, with inherent mineral color, and first-order colors appear light to dark green. Dispersion may cause small fragments to appear bluish near extinction position.

TWINNING. Hornblende may twin on {100}, but this twin is more characteristic of the cummingtonite-grunerite series.

ZONING. Color zoning in greens or brown to green is likely with dark zones at either core or margin. Hornblende may grade to edenite, and inclusions may be confined to specific zones.

INTERFERENCE FIGURE. Intense mineral color often makes good figures dark and difficult to find. Broad isogyres appear on a greenish field without conspicuous isochromes. Most hornblende compositions are optically (−); Mg-rich varieties, however, may be (positive) (pargasite and tschermakite). Rather large $2V_x$ angles near 70° (negative) are most characteristic; however, increasing Fe^{2+} decreases $2V_x$ to as low as 10° (negative), and decreasing Fe^{2+} causes the $2V$ to converge on Z and become positive (Fig. 5-36).

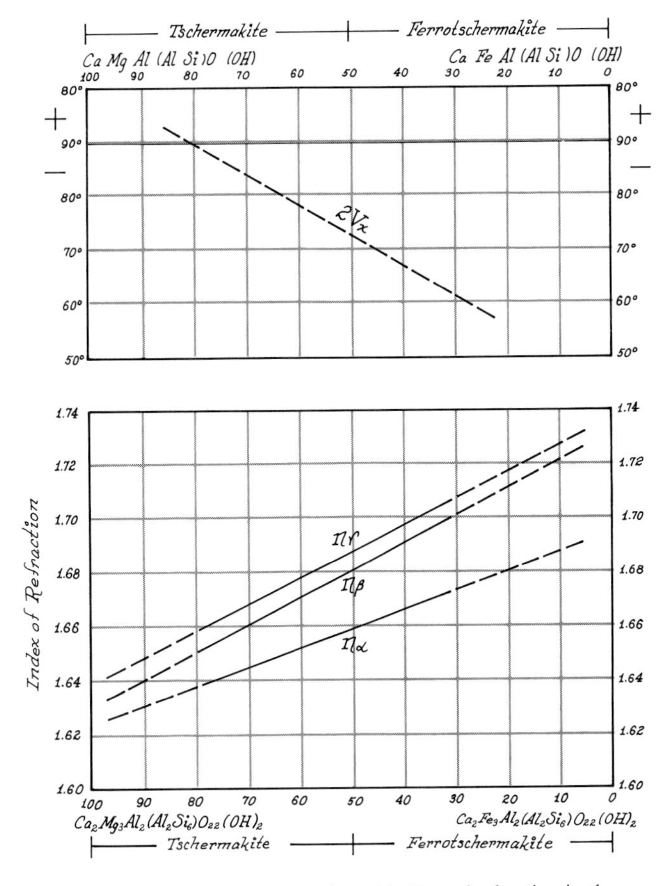

FIGURE 5-37. Variation of $2V$ angle and indices of refraction in the tschermakite-ferrotschermakite series. (After Deer, Howie, and Zussman, 1962)

Most compositions show moderate optic axis dispersion $v > r$ or, more rarely, $r > v$, and a few varieties (such as edenite and ferrohastingite) show pronounced inclined bisectrix dispersion.

OPTICAL ORIENTATION. The optic plane parallels {010} and $b = Y$, $c \wedge Z = -12°$ to $-34°$, and $a \wedge X = +3°$ to $-19°$. Angle $c \wedge Z$ is not very helpful in distinguishing hornblende varieties.

The extinction angle, in prismatic sections, ranges from a maximum of 12° to 34° in {010} sections to 0° in {100} sections. Basal sections show symmetrical extinction to intersecting {110} cleavages. The sign of elongation is positive (length-slow) in prismatic sections, as Z lies near c.

DISTINGUISHING FEATURES. Intense color and high refraction index distinguishes hornblende from iron-free amphiboles (such as tremolite and cummingtonite). Soda-amphiboles tend to be blue; basaltic hornblende, barkevekite, and kaersutite show intense absorption and pleochroism; and basaltic hornblende shows very high birefringence. Members of the cummingtonite-grunerite series are usually twinned on {100}.

ALTERATION. Hornblende normally alters to chlorite with associated calcite and epidote. Hydrothermal solutions may "bleach" hornblende on basal planes.

OCCURRENCE. Hornblende is one of the most widely distributed minerals in nature. It appears in igneous rocks from ultramafic to salic, and is the most characteristic ferromagnesian mineral of intermediate plutonics. Hornblende in ultramafic rocks tends to be Mg-rich and Al-poor (edenite or tschermakite), and in pegmatites, alkali granites, and feldspathoidal rocks, it tends to be Fe-rich and Al-rich (ferrohastingsite). In many igneous rocks, hornblende has been derived from primary pyroxene by hydration, yielding reaction rims of hornblende if the process is arrested short of completion. Basaltic hornblende is the characteristic amphibole of volcanics, but common green hornblende also appears.

A secondary, fibrous, light green amphibole of undetermined composition, formed at the expense of pyroxene by hydrothermal solutions, is called *uralite* and may be a variety of hornblende. It is often spatially related to fractures that allowed access to altering solutions.

Hornblende is one of the most common minerals of regional metamorphism, where it appears as a stress mineral characteristic of schists, gneisses, and amphibolites. It appears in rocks from the greenschist to granulite facies, forming the characteristic mineral of the amphibolite facies. It tends to become richer in Al and Na with increasing grade.

REFERENCES: THE HORNBLENDE SERIES ("COMMON" HORNBLENDE)

Deer, W. A., R. A. Howie, and A. Zussman. 1962. *Rock-Forming Minerals*. Vol. 2: *Chain Silicates*. New York: Wiley.

Gittos, M. F., G. W. Lorimer, and P. E. Champness. 1974. Electron microscopic study of precipitation (exsolution) in an amphibole (the hornblende-grunerite system). *Jour. Mater. Sci.*, *9*, 184–192.

Jaffe, H. W., P. Robinson, and C. Klein. 1968. Exsolution lamellas and optic orientation of clinoamphiboles. *Science*, *160*, 776–778.

Kalinin, E. P. 1967. The correlation between the principle ions of amphiboles and their influence on the optical properties of minerals. *Zap. Vses. Mineral. Obshchest.*, *96*, 170–182.

Papike, J. J., M. Ross, and J. R. Clark. 1969. Crystal-chemical characterization of clinoamphiboles based on five new structure refinements. *Mineral. Soc. Amer. Spec. Paper*, *3*, 117–136.

Profi, S., C. Sideris, and S. E. Filippakis. 1974. Relations of chemical composition to optical and structural properties in some igneous hornblendes. *Neues Jahrb. Mineral., Monatsh., 1974*, 68–82.

Oxyhornblende (Basaltic Hornblende)

$Ca_2Na(Mg,Fe^{2+},Fe^{3+},Al,Ti)_5[(Si_3Al)O_{11}]_2(O,OH)_2$

MONOCLINIC
$\angle\beta \approx 106°$

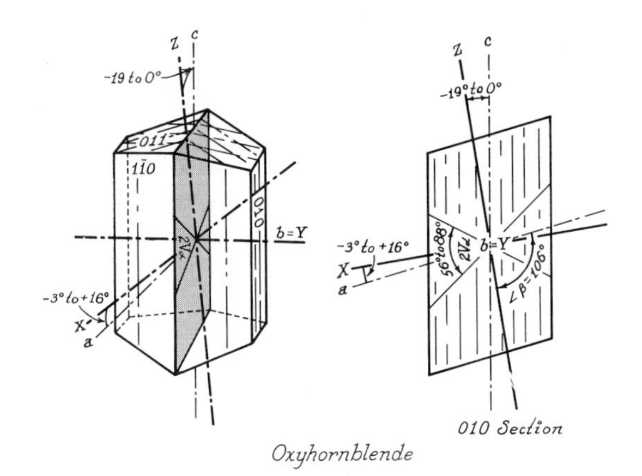

Oxyhornblende

$n_\alpha = 1.650\text{–}1.700$*
$n_\beta = 1.670\text{–}1.770$
$n_\gamma = 1.680\text{–}1.800$
$n_\gamma - n_\alpha = 0.018\text{–}0.083$
Biaxial negative, $2V_x = 56°\text{–}88°$
$b = Y, c \wedge Z = -19°$ to $0°$, $a \wedge X = -3°$ to $+16°$
$v > r$ weak to $r > v$ strong
Intense color and extreme pleochroism in deep shades of brown, red-brown, or green-brown $Z > Y > X$

COMPOSITION. Oxyhornblende seems no more than oxidized, common hornblende produced either naturally, in volcanic rocks, or by artificial heating. Hydrogen is driven from $(OH)^-$ radicals yielding extra O^{2-} with excess negative charge which, in turn, is balanced by the electron loss when Fe^{2+} becomes Fe^{3+}. Oxyhornblende may represent virtually any hornblende composition (for example, edenite, tschermakite, and hastingsite) where appreciable Fe^{2+} has been oxidized to Fe^{3+} with accompanying replacement of O^{2-} for $(OH)^-$. The degree of oxidation may range from none (hornblende) to complete (only Fe^{3+} or no $(OH)^-$). Most oxyhornblendes, however, show $Fe^{3+}:Fe^{2+}$ greater than 1:1, with Mg predominating over ($Fe^{3+} + Fe^{2+}$). Ferric end members have been proposed (Winchell, 1945) for most end-member hornblende compositions. Hydroxyl content is low, and Al ⇌ Si replacement is normally near maximum (Si_3AlO_{11}). *Kaersutite* is a Ti-rich oxyhornblende, usually iron-poor ($Fe^{2+} + Fe^{3+}$), with low $(OH)^-$ or high (Na,K) content.

* Composition and optical constants grade continuously to those of common hornblende. As Fe^{3+} or Ti^{4+} replaces Fe^{2+} or Mg^{2+}, we note increase in refractive indices, birefringence, color intensity, and pleochroism and decrease in $2V_x$ and extinction angle ($c \wedge Z$).

PHYSICAL PROPERTIES. H = 5–6. Sp. Gr. = 3.2–3.3. Color in hand sample is brown to black.

COLOR AND PLEOCHROISM. Oxyhornblendes show intense color and extreme absorption. $Z > Y > X$, increasing with Fe^{3+} and Ti^{4+}. In standard section, pleochroism is strong in yellow or reddish or greenish brown. X = pale yellow, yellow-brown; Y = brown, red-brown; and Z = dark red-brown, dark green-brown.

FORM. Oxyhornblende usually appears as euhedral phenocrysts in lavas. Prismatic sections are rectangular, and basal sections are rhombic (diamond-shaped) with small {010} faces. Because of easy alteration, rhombic sections alone may betray the former existence of oxyhornblende.

CLEAVAGE. Oxyhornblende shows normal amphibole cleavages (distinct prismatic cleavages {110} intersecting at 56° and 124°), and parting may appear on {100} or {001}.

BIREFRINGENCE. High birefringence is one of the distinctive features of oxyhornblendes. Maximum birefringence, seen in {010} sections, is moderate to strong (0.018–0.083), yielding colors to fifth order, in standard section. Birefringence increases rapidly with oxidation, and pale tints of third- or fourth-order colors are usually visible in section, when not obscured by intense mineral color. Both extinction ($c \wedge Z$) and $2V_x$ tend to be small, so {100} sections show low and {001} sections high birefringence.

TWINNING. As in hornblende, simple or multiple normal twinning may be present on {100}.

ZONING. Zoning is common, with outer zones showing higher oxidation (higher refractive indices, birefringence, and color, with lower $2V_x$ and extinction angles). Resorption is common on exposed surfaces, and alteration rims are very common.

INTERFERENCE FIGURE. Oxyhornblendes are biaxial negative, with a wide range of $2V_x$ values (56° to 88°), which become low with increased oxidation (Fe^{3+}). As X is near a, {100} sections show off-center Bxa figures symmetrical on the optic plane. Isogyres may remain in the visible field, with numerous isochromatic rings. Basal sections show near-Bxa figures, and {010} sections show flash figures.

Dispersion is normally weak to moderate $v > r$, with possible inclined bisectrix; however, kaersutite shows strong $r > v$.

OPTICAL ORIENTATION. Oxyhornblende is monoclinic, with $b = Y$, $c \wedge Z = -19°$ to $0°$ and $a \wedge X = -3°$ to $+16°$. Maximum extinction angle, seen in {010} sections, ranges from 19° to 0° and is near 0° for high oxidation states. Extinction about c decreases to 0° on {100} sections, and basal sections show symmetrical extinction to prismatic {110} cleavages. Elongation is positive (length-slow), since Z is near c.

DISTINGUISHING FEATURES. Oxyhornblende appears in volcanic rocks as phenocrysts showing strong birefringence, deep brown color, strong pleochroism, and, commonly, alteration rims. Common hornblende is green, with low birefringence and large extinction angle and $2V_x$. Barkevikite appears in high-alkali plutonic rocks and shows small $2V_x$ and high refraction index. Biotite shows single cleavage and very small $2V_x$.

ALTERATION. Oxyhornblende phenocrysts alter easily and are unstable in their lavas, showing resorption effects and alteration rims. Peripheral rims consist of fine granular aggregates of pyroxene (augite, pigeonite, titanaugite, and so on), plagioclase, biotite, and dusty magnetite or hematite (see Fig. 5-26). Kaersutite commonly alters to chlorite and granular sphene.

OCCURRENCE. Oxyhornblende is essentially confined to volcanic, or hypabyssal, rocks where it occurs as phenocrysts (rarely in groundmass). It is particularly characteristic of intermediate volcanics (andesites, latites, basalt, trachytes, and so on), and kaersutite is commonly found in alkaline volcanics (trachyandesites, camptonites, and so on). Common hornblende may appear in similar rocks but not in association with oxyhornblende. Kaersutite has been reported in syenites.

REFERENCES: OXYHORNBLENDE (BASALTIC HORNBLENDE)

Barnes, V. E. 1930. Changes in hornblende at about 800°. *Amer. Mineral.*, *15*, 393–417.

Hallimond, A. F. 1943. The graphical representation of the calciferous amphiboles. *Amer. Mineral.*, *28*, 65–89.

Kitamura, M., and M. Tokonami. 1971. Crystal structure of kaersutite. *Sci. Rep. Tohoku Univ.*, Series 3, *11*, 125–141.

Wilkinson, J. F. G. 1961. Some aspects of the calciferous amphiboles oxyhornblende, kaersutite, and barkevikite. *Amer. Mineral.*, *46*, 340–354.

Winchell, A. N. 1945. Variations in composition and properties of the calciferous amphiboles. *Amer. Mineral.*, *30*, 27–50.

The Glaucophane-Crossite-Riebeckite Series

$Na_2(Mg,Fe^{2+})_3(Al,Fe^{3+})_2(Si_4O_{11})_2(OH)_2$

MONOCLINIC
$\angle\beta \simeq 104^\circ$

Glaucophane *010 Section*

Glaucophane

	Biaxial negative
$n_\alpha = 1.594$–1.647*	$2V_x = 50°$–$0°$
$n_\beta = 1.612$–1.663	$b = Y$
$n_\gamma = 1.618$–1.663	$c \wedge Z = -6°$ to $-9°$
$n_\gamma - n_\alpha = 0.024$–$0.016$	$a \wedge X = +8°$ to $+5°$
	$v > r$ weak
	X = yellow, Y = violet, Z = blue

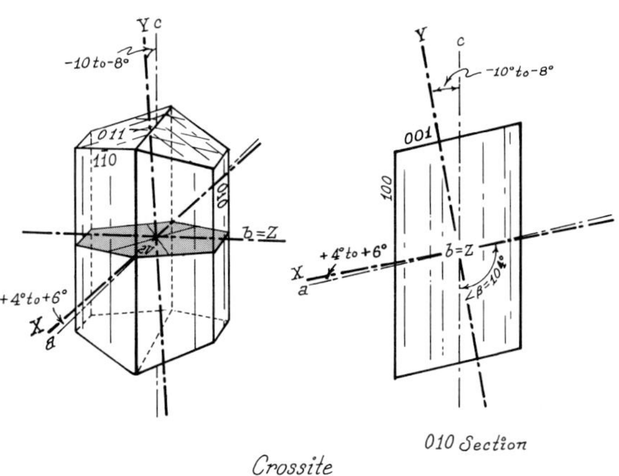

Crossite *010 Section*

Crossite

	Biaxial negative or positive
$n_\alpha = 1.647$–1.690	$2V = 0°$–$90°$
$n_\beta = 1.663$–1.690	$b = Z$
$n_\gamma = 1.663$–1.702	$c \wedge Y = -10°$ to $-8°$
$n_\gamma - n_\alpha = 0.016$–$0.012$	$a \wedge X = +4°$ to $+6°$
	$r > v$ extreme
	X = yellow, Y = blue, Z = violet

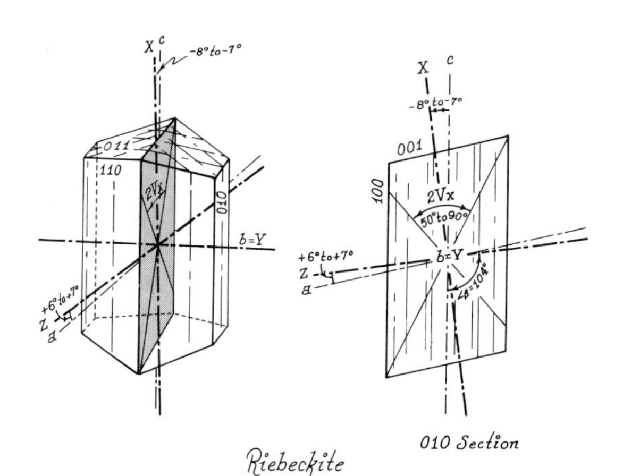

Riebeckite *010 Section*

Riebeckite

	Biaxial positive
	$2V_z = 0°$–$90°$
	Biaxial negative
$n_\alpha = 1.690$–1.702	$2V_x = 90°$–$50°$
$n_\beta = 1.690$–1.712	$b = Y$
$n_\gamma = 1.702$–1.719	$c \wedge X = -8°$ to $-7°$
$n_\gamma - n_\alpha = 0.012$–$0.017$	$a \wedge Z = +6°$ to $+7°$
	$v > r$ strong (opt. +)
	$r > v$ strong (opt. −)
	X = dark blue, Y = dark gray-blue, Z = yellow-brown

* Indices of refraction increase with iron substitution for magnesium or aluminum (Fig. 5-39).

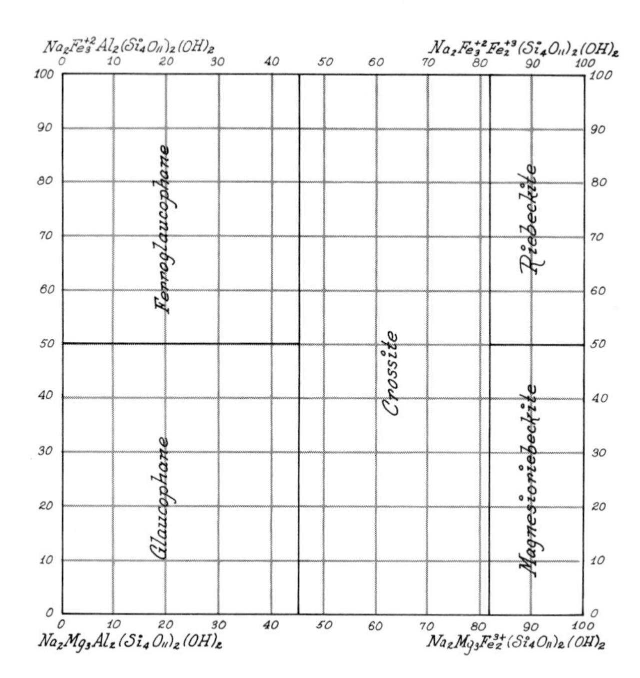

FIGURE 5-38. Composition fields of the blue amphiboles. (Modified from Miyashiro, 1957)

COMPOSITION. Ideal glaucophane, $Na_2Mg_3Al_2(Si_4O_{11})_2$-$(OH)_2$, forms at least major solid solution, with magnesioriebeckite, $Na_2Mg_3Fe_2{}^{3+}(Si_4O_{11})_2(OH)_2$, and riebeckite, $Na_2Fe_3{}^{2+}Fe_2{}^{3+}(Si_4O_{11})_2(OH)_2$, with crossite as an intermediate range (Fig. 5-38). Optically, it is advantageous to define glaucophane over the range 0–45 weight percent magnesioriebeckite and crossite 45–82 weight percent magnesioriebeckite, where $2V$ closes to 0° and optical orientation changes (Fig. 5-39). Glaucophane also forms a compositional series with ferroglaucophane, $Na_2Fe_3{}^{2+}Al_2(Si_4O_{11})_2(OH)_2$, and at least partial solution with tremolite, $Ca_2Mg_5(Si_4O_{11})_2$-$(OH)_2$, with soda tremolite as an intermediate member. Riebeckite $Na_2Fe_3{}^{2+}Fe_2{}^{3+}(Si_4O_{11})_2(OH)_2$ may well enter into solid solution with arfvedsonite, $Na_3Ca(Fe^{2+},Mg,Fe^{3+},Al)_{10}(Si_{15}AlO_{44})(OH)_4$, and possibly other rare soda amphiboles.

Minor Mn^{2+} and Ti^{4+} nearly always replace some Mg^{2+} or Al^{3+}, and some Ca^{2+} and K^+ usually replace Na^+.

Plots of optical properties against composition (Figs. 5-39 and 5-40) must be considered tentative, as deep color and dispersion make optical constants difficult to measure.

PHYSICAL PROPERTIES. H = 6. Sp. Gr. = 3.0 (Mg) to 3.4 (Fe). Color in hand sample is inky blue to gray (glaucophane); dark blue-green to black (riebeckite).

COLOR AND PLEOCHROISM. In thin section, both color and pleochroism are vivid in blue and violet. The pleochroic color scheme changes with optical orientation, as already noted. Glaucophane tends to show more pale lavender, and riebeckite shows more dark blue-green. In soda tremolite, pleochroism colors are pale and more greenish.

FORM. Parallel (rarely radiating) aggregates of acicular to columnar crystals, elongated on *c*, are most common. Typical, diamond-shaped, amphibole cross sections showing prismatic cleavages are common in thin section. Large riebeckite crystals are often poikilitic and a Mg-rich variety, *crocidolite*, is asbestiform.

CLEAVAGE. Typical prismatic {110} amphibole cleavages intersect at about 58° and 122° in glaucophane. Partings on {010} or {001} are possible.

BIREFRINGENCE. Maximum birefringence is moderate (0.024–0.012), with Mg-rich varieties showing greater birefringence (Fig. 5-39). Weaker birefringence is also commonly reported (~0.006–0.003). Interference colors are typically first order and low second order which may well be obscured by deep blue mineral color and strong dispersion.

TWINNING. Twins are not common but may appear as either simple or multiple forms on {100}.

ZONING. Composition zoning is very common and appears as concentric layers or irregular patches of different color or extinction position. Single, euhedral crystals commonly show zones of glaucophane and crossite optical orientation. Compositions may be Fe^{3+}-rich at either core or rim, and zoning may proceed either toward or from ideal glaucophane. Darker color, smaller $2V$, and smaller extinction angles usually denote Fe^{3+}-rich zones. Glaucophane may also form zoned crystals through tremolite enrichment. It may also appear as reaction rims on hornblende or jadeite.

INTERFERENCE FIGURE. Sections of low birefringence and strong absorption yield dark bisectrix figures with strong dispersion. Basal sections {001} show obtuse bisectrix figures for glaucophane, flash figures for crossite, and either acute or bisectrix figures for riebeckite. Broad, colored (by dispersion) isogyres may be difficult to see on a deep blue background. Glaucophane, crossite, and most riebeckite are biaxial negative; however, Mg-rich riebeckites may be positive.

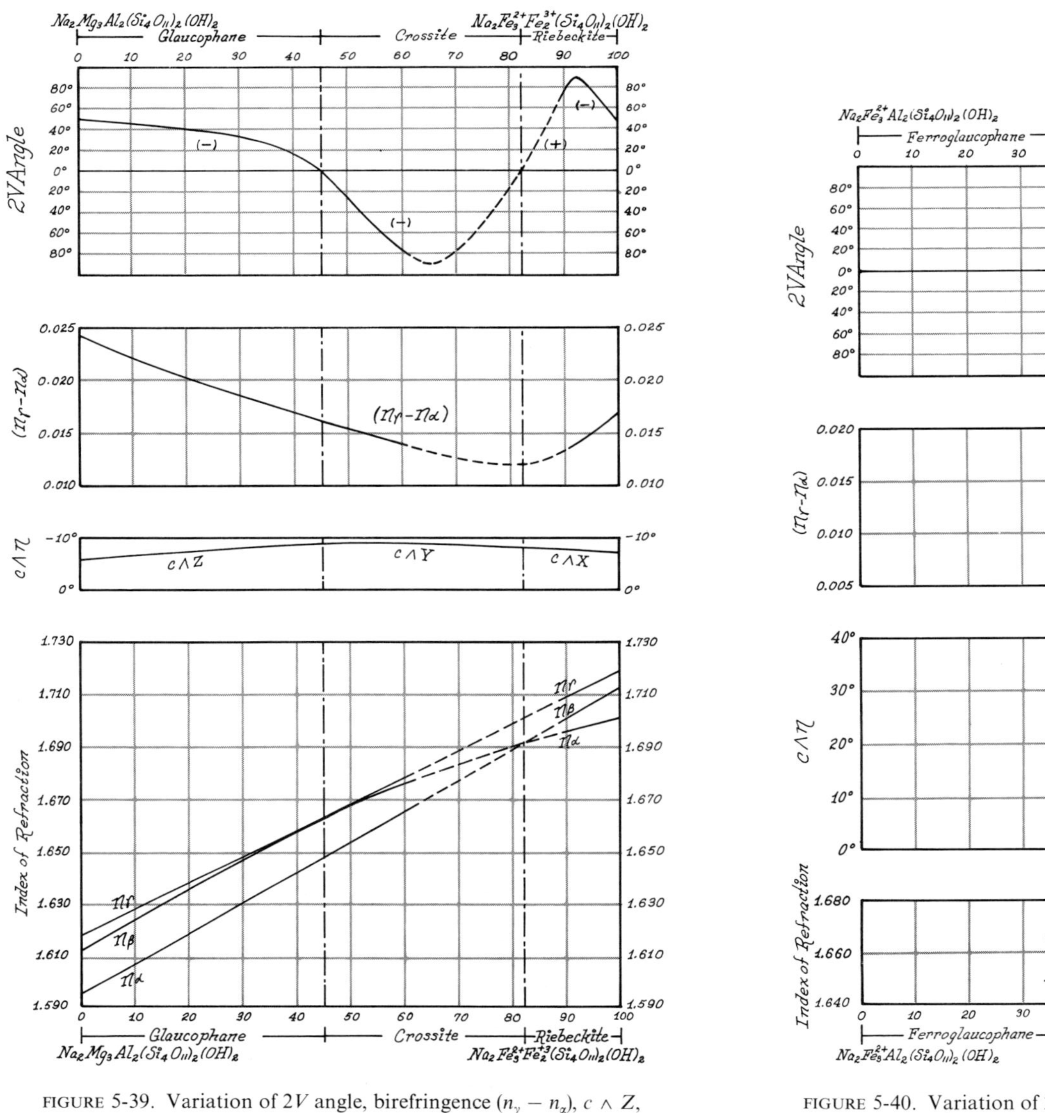

FIGURE 5-39. Variation of $2V$ angle, birefringence ($n_\gamma - n_\alpha$), $c \wedge Z$, or $c \wedge Y$ and indices of refraction in the glaucophane-crossite-riebeckite series. (After Borg, 1967)

FIGURE 5-40. Variation of $2V$ angle, birefringence ($n_\gamma - n_\alpha$), $c \wedge Y$, or $c \wedge X$ and indices of refraction in the ferroglaucophane-magnesioriebeckite series. (After Borg, 1967)

The $2V$ angle varies widely from 0° to 90° (Fig. 5-39). Glaucophane compositions show a relatively stable $2V_x$ at 50° to 30°; however, $2V_x$ closes quickly in the {010} plane to 0° at about 45 percent riebeckite, where the mineral is uniaxial negative. In the crossite range, the $2V_x$ opens again about X but now in the optic plane normal to {010}; $2V$ opens to 90° at 65 percent riebeckite and closes again to 0° on Z at about 82 percent riebeckite to yield a uniaxial positive mineral. In the riebeckite range, $2V_z$ appears to open on Z in {010}, yielding positive minerals to about 90 percent riebeckite, beyond which $2V$ continues to open, yielding biaxial negative minerals to 100 percent riebeckite, where $2V_x \sim 50°$ (see Fig. 5-39).

Dispersion is rather weak $v > r$ (inclined) for glaucophane, becoming very strong $r > v$ (horizontal) for the crossite range, $v > r$ (inclined) for positive riebeckite, and $r > v$ (inclined) for negative riebeckite.

OPTICAL ORIENTATION. From optical data, it seems most reasonable to define the breaks between glaucophane-crossite and crossite-riebeckite where $2V$ becomes zero and the optical orientation changes. Glaucophane, most riebeckite, and soda tremolite will then show the normal amphibole orientation, with $b = Y$, {010} optic plane, and small extinction angles.

For glaucophane, maximum extinction measured to cleavage traces in {010} sections is usually less than 10° ($c \wedge Z = -6°$ to $-9°$), although somewhat larger values have been reported. Cleavage traces are nearer the slow wave, and cleavage fragments are length-slow (positive elongation).

Most riebeckite shows a normal amphibole orientation $b = Y$, $c \wedge X = -8°$ to $-7°$, $a \wedge Z = +6°$ to $+7°$, which yields a maximum extinction of 7°–8° in {010} sections and negative elongation (length-fast). Some riebeckite is reported to have an odd orientation where $b = Z$, and the optic plane is normal to {010} and very near c. This orientation also shows length-fast fragments with small extinction angles.

Crossite has an orientation where $b = Z$, and the optic plane is normal to {010}, $c \wedge Y = -10°$ to $-8°$, $a \wedge X = +4°$ to $+6°$. The maximum extinction angle in {010} section is about 10° to cleavage traces. Fragments may show positive or negative elongation and small extinction angles.

DISTINGUISHING FEATURES. Glaucophane, crossite, and riebeckite all show typical amphibole cross sections, with prismatic cleavages at about 56° and 124° and pleochroism in blue characteristic of Na-amphiboles. Blue tourmaline is uniaxial, with parallel extinction and maximum absorption parallel to elongation. Chloritoid has larger indices of refraction and perfect basal cleavage, and dumortierite is orthorhombic, with parallel extinction. Arfvedsonite occurs in Na-rich plutonic igneous rocks, where it might easily be mistaken for riebeckite; however, arfvedsonite is always negative and has a crossite-like orientation ($b = Z$).

ALTERATION. Glaucophane or crossite may be replaced by green amphibole, presumably actinolite. Riebeckite commonly alters to ferric oxides or siderite with the release of free silica. The fibrous silica "tiger-eye" results from the alteration of fibrous riebeckite (*crocidolite*).

OCCURRENCE. Glaucophane and crossite are metamorphic minerals obviously characteristic of, but not strictly restricted to, the glaucophane schist facies (blueschist facies), which, in turn, implies low-temperature, high-pressure metamorphism. Glaucophane-crossite is essentially restricted to blueschists or greenschists of Cenozoic and Mesozoic plate subduction zones, where it may be associated with chlorite, albite, epidote, pumpellyite, lawsonite, aragonite, and actinolite. Glaucophane-crossite may also form as retrograde metamorphism of eclogite, where it normally appears with garnet, omphacite, and sphene or rutile.

Riebeckite is the most common Na-amphibole in igneous rocks. It appears in alkali granites, syenites, nepheline syenites, and similar Na-rich plutonic and hypabyssal igneous rocks. Its most common associate is quartz, but it also appears with other Na-pyriboles (aegirine, aegirine-augite, and arfvedsonite) and albite. Riebeckite is not common in volcanic rocks but is reported in rhyolite and trachyte. It is also a rare, late-crystallizing mineral in pegmatites or quartz veins.

Metamorphism produces riebeckite from iron-rich, siliceous sediments or igneous rocks. It is reported in granite gneiss, quartzite, and granulite, in association with quartz and other amphiboles and pyroxenes. Crocidolite (fibrous riebeckite) forms as veins in ironstones by sodium metasomatism.

Soda-tremolite is formed by sodium metasomatism of carbonate sediments.

REFERENCES: THE GLAUCOPHANE-CROSSITE-RIEBECKITE SERIES

Bocquet, J. 1974. Blue amphiboles of the Westerm Alps: Chemistry and physical characters. *Schweiz. Mineral. Petrol. Mitt.*, *54*, 425–448.

Borg, I. Y. 1967. Optical properties and cell parameters in the glaucophane-riebeckite series. *Contrib. Mineral. Petrol.*, *15*, 67–92.

Coleman, R. G., and J. J. Papike. 1968. Alkali amphiboles of the blueschists of Cazadero, California. *Jour. Petrol.*, *9*, 105–122.
Faye, G. H., and E. H. Nickel. 1970. Effect of charge-transfer processes on the color and pleochroism of amphiboles. *Can. Mineral.*, *10*, 616–635.
Makanjuola, A. A., and R. A. Howie. 1972. The mineralogy of the glaucophane schists and associated rocks from Île de Groix, Brittany, France. *Contrib. Mineral. Petrol.*, *35*, 83–118.
Miyashiro, A. 1957. The chemistry, optics and genesis of the alkali-amphiboles. *Jour. Fac. Sci. Univ. Tokyo*, Sect. 2, *11*, 57–83.
Sutherland, D. S. 1969. Sodic amphiboles and pyroxenes from fenites in East Africa. *Contrib. Mineral. Petrol.*, *24*, 114–135.

Katophorite

$Na(NaCa)(Mg,Fe^{2+})_4Fe^{3+}(Si_7Al)O_{22}(OH)_2$

MONOCLINIC
$\angle\beta \simeq 105°$

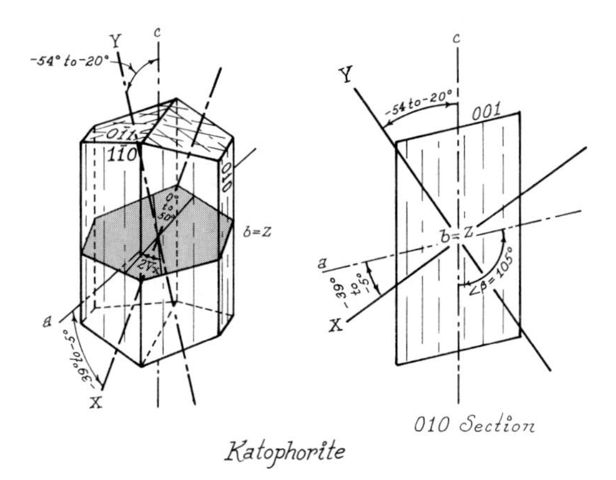

$n_\alpha = 1.681–1.639$*

$n_\beta = 1.688–1.658$

$n_\gamma = 1.690–1.660$

$n_\gamma - n_\alpha = 0.009–0.021$*

Biaxial negative, $2V_x \simeq 0°$ to $50°$

$b = Z$, $c \wedge Y \simeq -54°$ to $-20°$, $a \wedge X \simeq -39°$ to $-5°$†

$r > v$ strong†

Numerous shades of yellow, red-brown, or blue-green, in standard section, with strong pleochroism

$Z > Y > X$ or $Y > Z \simeq X$

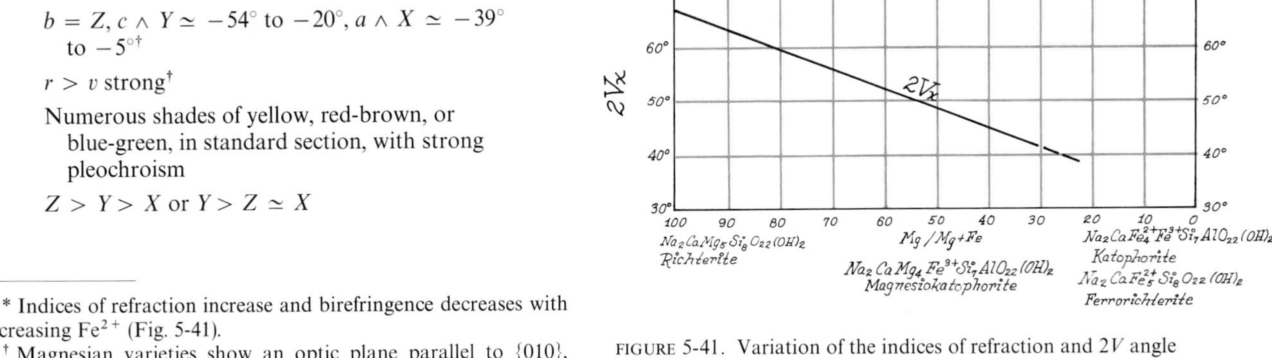

FIGURE 5-41. Variation of the indices of refraction and $2V$ angle in the richterite-katophorite-ferrorichterite series.

* Indices of refraction increase and birefringence decreases with increasing Fe^{2+} (Fig. 5-41).

† Magnesian varieties show an optic plane parallel to {010}, $b = Y$, and dispersion is $v > r$.

COMPOSITION. Most natural katophorite is Fe^{2+}-rich, and a complete solid solution probably exists between $Na(NaCa)Fe_4^{2+}Fe^{3+}(Si_7Al)O_{22}(OH)_2$ and the magnesian analog $Na(NaCa)Mg_4Fe^{3+}(Si_7Al)O_{22}(OH)_2$(magnesiokatophorite). The katophorite series contains more Al and Ca than the arfvedsonite, and there is probably some solid solution between the two series.

PHYSICAL PROPERTIES. H = 5. Sp. Gr. = 3.2–3.5. Color, in hand sample, is black or dark blue-green. Mg-varieties may be red-brown.

COLOR AND PLEOCHROISM. Fe-rich katophorite is deeply colored and highly pleochroic in yellow and green with absorption $Z > Y > X$. Mg-rich varieties tend to show pleochroism in yellow and red-brown and absorption $Y > Z \simeq X$. X = pale yellow, pale yellow-brown; Y = blue-green, deep brown; and Z = dark green, red-brown.

FORM. Euhedral crystals show characteristic diamond-shaped amphibole cross sections and rectangular prismatic sections. Katophorite may occur as interstitial grains or fibrous aggregates in the groundmass of alkali volcanics.

CLEAVAGE. Characteristic, perfect prismatic cleavages {110} intersect at 56° and 124°. Parting on {010} may also be apparent.

BIREFRINGENCE. Birefringence is rather low. Interference colors, in standard section, are first-order, usually gray and white. Strong dispersion may introduce shades of pale pink or blue, and grains may fail to extinguish completely with stage rotation.

TWINNING. Twinning on {100} is reported.

ZONING. Composition zoning is common, and katophorite commonly forms around cores of hornblende or aegirine and may be rimmed by arfvedsonite or aegirine.

INTERFERENCE FIGURE. The range of $2V_x$ is not well defined, and its relationship to composition is undetermined. Dispersion is strong $r > v$, with possible horizontal bisectrix dispersion for Fe-rich compositions, and strong $v > r$ and inclined bisectrix dispersion for magnesian varieties.

OPTICAL ORIENTATION. For most compositions, the optic plane is perpendicular to {010} and $b = Z$, $c \wedge Y \simeq -54°$ to $-20°$ and $a \wedge X \simeq -39°$ to $-5°$. For magnesian varieties, the optic plane becomes {010}, and $b = Y$. Maximum extinction angle, seen in {010} sections, ranges from about 20° to 45°, and either the fast or slow wave may lie nearer to cleavage traces.

DISTINGUISHING FEATURES. Katophorite, especially magnesiokatophorite, is typically pleochroic in red-brown and is most commonly confused with oxyhornblende. The latter mineral, however, shows greater birefringence, small extinction angles (<20°) and large $2V_x$ (56°–88°).

Large extinction angle and unusual orientation (optic plane $\perp$ {010}) are distinctive of katophorite and arfvedsonite. Arfvedsonite shows strong pleochroism in blue and green $X > Y > Z$ and dispersion $v > r$.

ALTERATION. Katophorite probably alters to fibrous amphibole (uralite), limonite, and siderite.

OCCURRENCE. Katophorite is a rare amphibole most characteristic of dark-colored, alkali intrusives where it may be associated with nepheline, arfvedsonite, and aegirine. It is also present in some mafic dike rocks and nepheline volcanics.

REFERENCE: KATOPHORITE

Kempe, D. R. C., and W. A. Deer 1970. Geological investigations in East Greenland. IX: The mineralogy of the Kangerdlugssuaq alkaline intrusion, East Greenland. *Medd. om Grønland, 190* (3), 1–95.

The Arfvedsonite-Eckermannite Series

$Na(Na_{1.5}Ca_{0.5})(Fe^{2+},Mg)_4Fe^{3+}(Si_{7.5}Al_{0.5})O_{22}(OH)_2$

MONOCLINIC
$\angle\beta \simeq 105°$

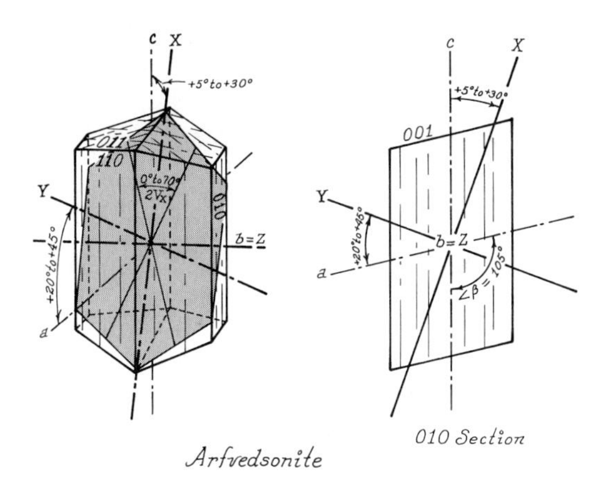

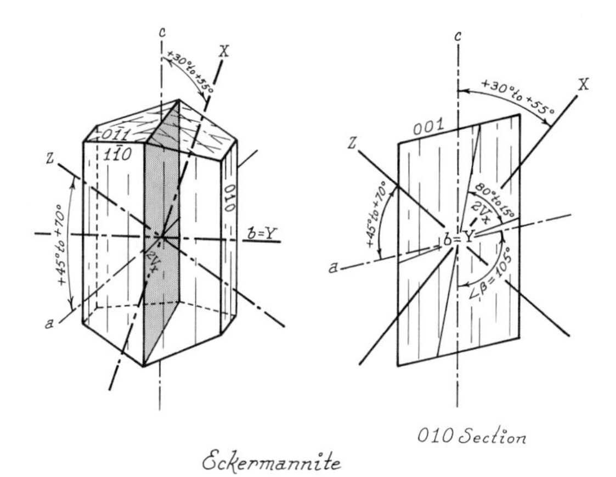

Arfvedsonite

$n_\alpha = 1.700–1.672$*
$n_\beta = 1.710–1.677$
$n_\gamma = 1.715–1.684$
$n_\gamma - n_\alpha = 0.015–0.012$*
Biaxial negative
$2V_x = 0°–70°$*
$b = Z$
$c \wedge X = +5°$ to $+30°$
$a \wedge Y = +20°$ to $+45°$
$r > v$ very strong
Strong pleochroism in blue-green
$X > Y > Z$

Eckermannite

$n_\alpha = 1.638 - 1.610$*
$n_\beta = 1.652 - 1.625$
$n_\gamma = 1.654 - 1.630$
$n_\gamma - n_\alpha = 0.016 - 0.020$*
Biaxial negative
$2V_x = 80°–15°$*
$b = Y$
$c \wedge X = +30°$ to $+55°$
$a \wedge Z = +45°$ to $+70°$
$r > v$ strong
Strong pleochroism in blue-green and yellow-green
Variable absorption

COMPOSITION. Complete solid solution may exist between Fe-rich arfvedsonite, $Na(Na_{1.5}Ca_{0.5})Fe_4^{2+}Fe^{3+}(Si_{7.5}Al_{0.5})$-$O_{22}(OH)_2$, and Mg-rich eckermannite, $Na(Na_{1.5}Ca_{0.5})$-$Mg_4Fe^{3+}(Si_{7.5}Al_{0.5})O_{22}(OH)_2$, and magnesioarfvedsonite is defined as an intermediate member (see Fig. 5-42). With increasing Fe^{2+}, eckermannite becomes magnesioarfvedsonite when the optic plane changes from {010} to perpendicular to {010}. Solid solution with the glaucophane-riebeckite series and katophorite series is probable.

Some K may replace Na, minor Al^{3+} or Ti^{4+} may replace Fe^{3+}, Mn^{2+} or Li^{+} may replace some (Mg,Fe^{2+}), and F^- may partially replace $(OH)^-$.

PHYSICAL PROPERTIES. H = 5–6. Sp. Gr. = 3.5–3.0 (high values represent Fe-rich compositions). Color in hand sample is black to dark green or blue-green. Mg-rich varieties may be brown or greenish brown.

* Optical constants of the arfvedsonite series are difficult to measure, because of strong absorption and dispersion and variation in optical orientation. Indices of refraction increase with Fe^{2+}, as expected; however, birefringence seems to be less in iron-rich varieties. Values of $2V_x$ are highly variable and very uncertain (see Fig. 5-42).

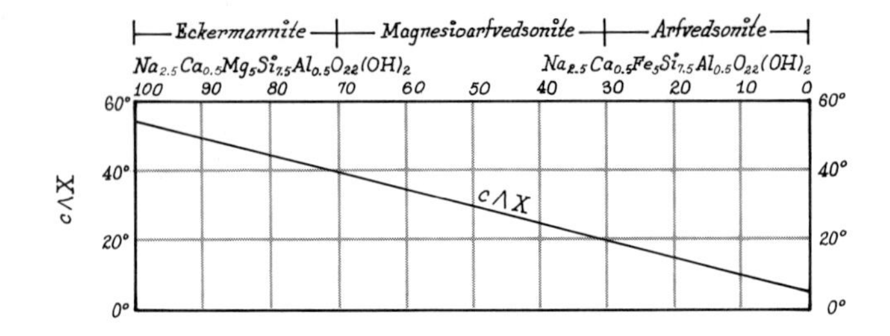

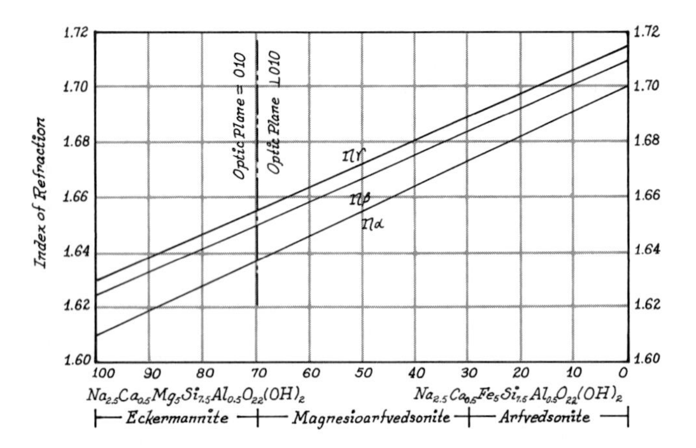

FIGURE 5-42. Variation of $c \wedge X$ and indices of refraction in the eckermannite-afrvedsonite series. (After Deer, Howie, and Zussman, 1962)

COLOR AND PLEOCHROISM. Arfvedsonite-eckermannite is highly colored in section or fragments and highly pleochroic, with variable absorption (usually $X > Y > Z$). X = deep blue-green, yellow, indigo; Y = pale blue-green, yellow-brown, gray-violet; and Z = pale yellow-green, deep green, pale brownish green.

FORM. Prismatic crystals are elongated on c and are euhedral to subhedral with typical amphibole cross sections. Fine, fibrous aggregates may form as interstitial masses in volcanic rocks.

CLEAVAGE. Typical perfect prismatic amphibole cleavages ({110} intersecting at 56° and 124°) may be accompanied by parting on {010}.

BIREFRINGENCE. Maximum birefringence is mild ($n_\gamma - n_\alpha \simeq$ 0.012–0.020), seen in {010} sections for eckermannite and in sections near {100} for arfvedsonite. Contrary to expectation, low birefringence values are indicative of Fe-rich compositions (arfvedsonite). Interference colors are first order in standard section but may be obscured by deep mineral color and dispersion.

TWINNING. Twinning on {100} may be simple or repeated.

ZONING. Composition zoning may grade from hornblende or barkevikite core to arfvedsonite rim.

INTERFERENCE FIGURE. The arfvedsonite-eckermannite series is biaxial negative, with $2V_x$ highly variable from essentially 0° to almost 90°. Interference figures may be dark and diffuse, due to strong absorption and mineral color. Fragments on cleavage planes yield highly eccentric figures.

Optic axis dispersion in arfvedsonite is very strong $v > r$ ($r > v$ is also reported), and bisectrix dispersion is horizontal. Dispersion in eckermannite is strong $r > v$ and inclined.

OPTICAL ORIENTATION. The optic plane is perpendicular to {010} for arfvedsonite and magnesioarfvedsonite compositions, and $b = Z$, $c \wedge X = +5°$ to $+30°$, and $a \wedge Y = +20°$ to $+45°$. Maximum extinction angle to cleavage traces, seen in {010} sections is 5° to 30°, and the faster wave lies nearer to elongation (negative elongation). With increasing magnesium, the optic plane becomes parallel to {010} to define eckermannite. For eckermannite compositions, $b = Y$, $c \wedge X = +30°$ to $+55°$, and $a \wedge Z = +45°$ to $+70°$. Maximum extinction angle to cleavage traces, seen in {010} sections, is 30° to 55°, and either the fast or slow wave may lie nearer elongation.

DISTINGUISHING FEATURES. Arfvedsonite-eckermannite is recognized as a Na-amphibole by characteristic amphibole sections and strong pleochroism in blue-green. Riebeckite, glaucophane, barkevikite, kaersutite, and most other amphiboles show the same optical orientation as eckermannite. Riebeckite may be optically positive; glaucophane shows pleochroism in blue and violet, and barkevikite and kaersutite in yellow, red, and brown; and all show maximum extinction angles less than 25°.

Crossite shares the unusual optical orientation of arfvedsonite, but shows pleochroism in blue and violet, dispersion $r > v$, and is length-slow in {010} sections.

ALTERATION. Arfvedsonite-eckermannite alters to fine, fibrous amphibole (uralite) with limonite or siderite by-product.

It may be replaced by aegirine and Fe-biotite in late stages of crystallization.

OCCURRENCE. Arfvedsonite-eckermannite is most common in Na-rich, plutonic igneous rocks, where it may appear with quartz (for example, alkali granite) or nepheline (for example, nepheline syenite). It is usually associated with albite and aegirine, aegirine-augite, or hastingsite hornblende. It is a rare mineral in alkali-rich volcanics and nepheline pegmatites.

REFERENCES: THE ARFVEDSONITE-ECKERMANITE SERIES

Deer, W. A., R. A. Howie, and J. Zussman. 1962. *Rock-Forming Minerals*. Vol. 2: *Chain Silicates*. New York: Wiley.

Hawthorne, F. C. 1976. The crystal chemistry of the amphiboles. V. The structure and chemistry of arfvedsonite. *Can. Mineral.*, *14*, 346–356.

Miyashiro, A. 1957. The chemistry, optics, and genesis of the alkali-amphiboles. *Jour. Fac. Sci. Univ. Tokyo*, Sect. 2, *11*, 57–83.

Sahama, T. G. 1956. Optical anomalies in arfvedsonite from Greenland. *Amer. Mineral.*, *41*, 509–512.

Barkevikite

$(Na,K)Ca_2(Fe^{2+},Mg,Fe^{3+},Mn)_5(Si_7Al)O_{22}(OH)_2$

MONOCLINIC
$\angle\beta \simeq 105^\circ$

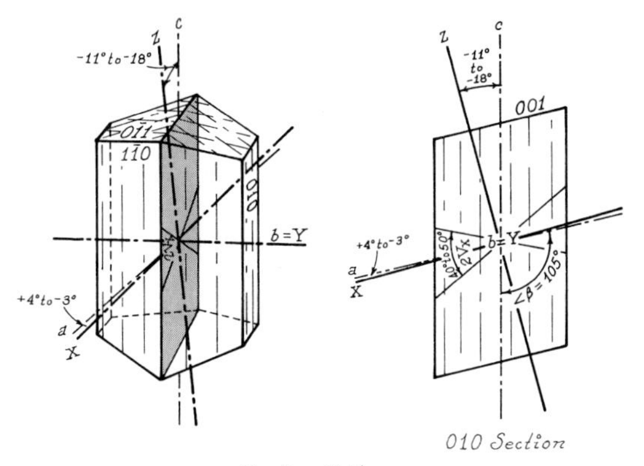

$n_\alpha = 1.691–1.685$
$n_\beta = 1.700–1.696$
$n_\gamma = 1.707–1.701$
$n_\gamma - n_\alpha \simeq 0.012–0.018$
Biaxial negative, $2V_x \simeq 40^\circ–50^\circ$
$b = Y, c \wedge Z \simeq -11^\circ$ to -18°, $a \wedge X \simeq +4^\circ$ to -3°
$r > v$ weak to strong
Deep color and strong pleochroism in yellow to deep red-brown $Z > Y > X$

COMPOSITION. Barkevikite is an iron-rich hornblende containing more ferrous iron (Fe^{2+}) and less ferric (Fe^{3+}) than normal oxyhornblende. Some varieties are Mn-rich.

PHYSICAL PROPERTIES. H = 5–6. Sp. Gr. = 3.35–3.44. Color in hand sample is black to deep brown.

COLOR AND PLEOCHROISM. In standard section, barkevikite is highly colored and strongly pleochroic in yellow and red-brown $Z > Y > X$. X = pale yellow-brown, Y = red-brown, and Z = dark brown.

FORM. Barkevikite normally appears as rather large, euhedral, prismatic crystals elongated on c; crystals may be poikilitic.

CLEAVAGE. Perfect prismatic cleavages {110} intersect at 56° and 124°, as with other amphiboles. Parting on {100} or {001} may accompany cleavages.

BIREFRINGENCE. Maximum birefringence, seen in {010} sections, is mild ($n_\gamma - n_\alpha = 0.012–0.018$). Interference colors, in standard section, are first-order, usually white or yellow and often tinted pale pink or blue by strong dispersion or obscured by deep mineral color.

TWINNING. Simple twinning on {100} is reported.

ZONING. Barkevikite has been reported as reaction rims about pyroxene.

INTERFERENCE FIGURE. Barkevikite is biaxial negative, with moderate $2V_x \simeq 40°$ to $50°$. Optic axis dispersion may be weak but is usually strong $r > v$, and bisectrix dispersion is inclined. Cleavage fragments yield highly eccentric figures, but fragments on {100} parting yield essentially centered acute bisectrix figures.

OPTICAL ORIENTATION. The optic plane is {010}, and $b = Y$, $c \wedge Z \simeq -11°$ to $-18°$ and $a \wedge X \simeq +4°$ to $-3°$. Maximum extinction angle, seen in {010} sections, is 11° to 18°, and the slow wave is nearest to elongation as defined by cleavage traces (negative elongation).

DISTINGUISHING FEATURES. Barkevikite is strongly pleochroic from yellow to deep red-brown and is most likely to be mistaken for oxyhornblende. The latter shows lower indices of refraction, greater birefringence, and distinctly different occurrence (volcanic rocks). Katophorite is also yellow to red-brown, but it shows a large maximum extinction angle ($>20°$), and its optic plane is perpendicular to {010}. Magnesiokatophorite has much lower indices of refraction than barkevikite and has reverse dispersion.

ALTERATION. Barkevikite alters to uralite or Na-pyroxene and iron oxides.

OCCURRENCE. The major occurrence of barkevikite is in alkaline plutonic igneous rocks (such as nepheline syenite and essexite). It has been reported from nepheline dike rocks and possibly alkaline volcanics.

REFERENCES: BARKEVIKITE

Faye, G. H., and E. H. Nickel. 1970. Effect of charge-transfer processes on the color and pleochroism of amphiboles. *Can. Mineral.*, *10*, 616–635.

Hallimond, A. F. 1943. The graphical representation of the calciferous amphiboles. *Amer. Mineral.*, *28*, 65–89.

CHAPTER 6

Phyllosilicates

Phyllosilicates derive their name from the Greek word for *leaf*, a term suggestive of the thin sheets stacked one atop another that several minerals of this group exhibit as their typical morphology. This feature arises from the arrangement of silicate tetrahedra into sheets, and phyllosilicates are thus often called *sheet silicates* or, less commonly, *layer silicates*. A portion of a "two-dimensionally infinite" tetrahedral sheet is shown in Fig. 6–1. There are two kinds of oxygen anions. Those of one kind, the *basal oxygens*, are all located in a layer at the "bases" of the tetrahedra and are shared by two silicon atoms (or substituent cations). Those of the other kind, called *apical oxygens*, are bonded to only one silicon atom and are thus left with an unsatisfied negative charge with which to form strong bonds to nontetrahedral cations. The common substitution of Al^{3+} for some of the Si^{4+} in the tetrahedral sheet allows relatively weak bonds to be formed between basal oxygens and nontetrahedral cations, but some kinds of sheet silicates exhibit only hydrogen bonds or Van der Waals bonds between layers. These weak interlayer bonds produce perfect one-directional cleavage and low hardness.

Sheet silicates contain not only tetrahedral layers but also layers of octahedrally coordinated cations to which the apical oxygens of the tetrahedral sheets are bonded. Octahedral cations may be sandwiched between apical oxygens of two tetrahedral sheets (as in micas), attached to the apical oxygens of a single tetrahedral sheet (as in kaolinite), or included in an independent layer between micalike sheets (as in chlorites).

The marked structural anisotropy of phyllosilicates makes it tempting to suggest at least qualitative correlations between crystal structure and optical properties. If the plane of a sheet be taken as "horizontal," then most of the cation-anion bonds within the sheet are oriented closer to horizontal than vertical. Thus the atoms experience more "horizontal"

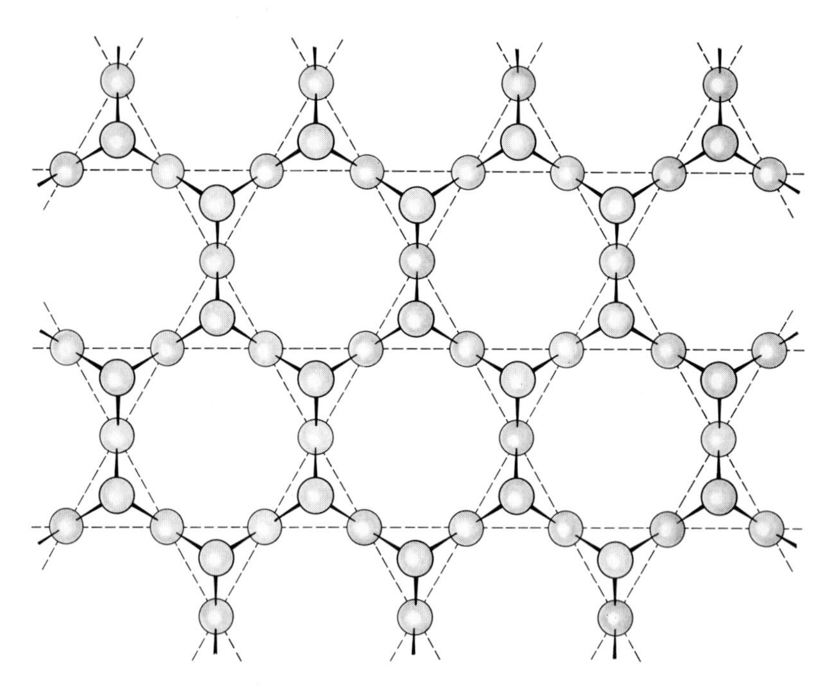

FIGURE 6-1. The idealized tetrahedral layer of the sheet silicates. Basal oxygen ions, shown at the corners of dashed triangles, are bonded to two silicon cations at the center of each tetrahedron. Apical oxygens, at the top of each tetrahedron, are bonded to only one silicon ion below it (not shown) and to metal cations in octahedral coordination above (not shown).

polarization than "vertical" polarization, and light vibrating in the horizontal plane (the plane of the sheet) may be expected to travel more slowly than light vibrating in other directions. In general, this conclusion is vindicated by observation. In the micas, the clay minerals, talc, pyrophyllite, the serpentines, and some chlorites, the *X* optical direction is within a few degrees of the normal to the sheets; moreover, they are all biaxial negative, so that n_β is numerically closer to n_γ than it is to n_α (the indicatrix is "flattened" nearly in the plane of the sheets).

There are two notable exceptions to this correlation; namely, apophyllite and some of the chlorites. Some apophyllite is uniaxial negative, but most is uniaxial positive with very low birefringence. The crystal structure of this mineral resembles that of feldspars in that it consists of four- and eight-membered rings of silicate tetrahedra, although in apophyllite these rings are arranged into sheets perpendicular to the unique crystallographic (and optic) axis. This sheet configuration possesses a lower bond density than does that of the other phyllosilicates, and thus the structure lacks strong net polarization in the plane of the sheets.

Of the iron chlorites, those rich in aluminum (poor in silicon and total iron) tend to be optically negative, with $c \wedge X$ small, which is consistent with the micas, clays, and so on; as aluminum is replaced by Fe^{2+} and Si^{4+}, the iron chlorites become positive, with $c \wedge Z$ small. Magnesium chlorites tend to exhibit the opposite behavior with respect to aluminum.

The Clays

The clay minerals are an ill-defined group of secondary minerals formed at or near the earth's surface by weathering or hydrothermal alteration of feldspars and other aluminous silicates. They are characteristically fine-grained, often below the resolution of a light microscope, making them difficult to identify by optical means. Most clay minerals are phyllosilicates with sheet structures based on combinations of brucite-type layers of octahedrally coordinated cations and Si_4O_{10} layers of tetrahedrally coordinated cations (Si^{4+} or Al^{3+}). Structurally and chemically, certain clay minerals are more closely related to micas and septechlorites than to other clays, and we can recognize several distinct clay mineral groups:

The Kaolin Group (Kandites)

Mineral	Formula
Kaolinite	$Al_2Si_2O_5(OH)_4$
Dickite	$Al_2Si_2O_5(OH)_4$
Nacrite	$Al_2Si_2O_5(OH)_4$
Halloysite (endellite)	$Al_2Si_2O_5(OH)_4 \cdot 2H_2O$
Meta-halloysite (halloysite)	$Al_2Si_2O_5(OH)_4$
Anauxite	$Al_{2-1.5}Si_{2-2.5}O_5(OH)_4$
Allophane	$Al_2Si_2O_5(OH)_4 \cdot nH_2O$

The kaolin structure is an asymmetrical, two-layer structure of one "gibbsite" layer of octahedral coordination and one Si_4O_{10} layer of tetrahedral coordination and is the dioctrahedral analog of the septechlorites. As electrostatic neutrality is achieved within the double layer, no interlayer cations are present, and the kaolin minerals have a soapy feel. Stacking of successive kaolin sheets with repetition on c at one, two, and six layers distinguishes the kaolinite, dickite, and nacrite polymorphs, respectively. Halloysite (endellite) contains a single layer of water molecules that separates kaolin layers, allowing them to curl to tubular forms. The axis of halloysite tubes is usually a or b, and the octahedral layer is on the inside of the tube, as it is too small to "fit" the tetrahedral layer. The curl direction is opposite to that of chrysotile serpentine where the octahedral layer is "large." Meta-halloysite is dehydrated halloysite (the water layer is lost), and united kaolin sheets tend to unroll, yielding split tubes and canoe-shaped sheets. Anauxite shows higher Si:Al ratios than kaolinite, probably due to soluble amorphous silica, and allophane is the amorphous form of what is essentially the kaolinite composition.

The Montmorillonite Group (Smectites)

Montmorillonite	$(\frac{1}{2}\text{Ca,Na})_{0.67}(Al_{3.33}Mg_{0.67})Si_8O_{20}(OH)_4 \cdot nH_2O$
Beidellite	$(\frac{1}{2}\text{Ca,Na})_{0.67}Al_4(Si_{7.33}Al_{0.67})O_{20}(OH)_4 \cdot nH_2O$
Nontronite	$(\frac{1}{2}\text{Ca,Na})_{0.67}Fe_4{}^{3+}(Si_{7.33}Al_{0.67})O_{20}(OH)_4 \cdot nH_2O$
Saponite	$(\frac{1}{2}\text{Ca,Na})_{0.67}Mg_6(Si_{7.33}Al_{0.67})O_{20}(OH)_4 \cdot nH_2O$
Hectorite	$(\frac{1}{2}\text{Ca,Na})_{0.67}(Mg_{5.33}Li_{0.67})Si_8O_{20}(OH)_4 \cdot nH_2O$

The montmorillonite structure is a three-layer arrangement of one "brucite" or "gibbsite" octahedral layer enclosed between two inward-pointing Si_4O_{10} layers of tetrahedral coordination. Montmorillonite clays are the "swelling" clays, and interlayer cations and water molecules are required to neutralize cation substitutions in both tetrahedral and octahedral sites.

Beidellite, nontronite, and saponite show cation substitution largely in tetrahedral sites where Al^{3+} replaces Si^{4+} to about $Si_{7.33}Al_{0.67}$; montmorillonite and hectorite are largely without tetrahedral substitution (Si_8). Beidellite, nontronite, and saponite show little cation substitution in octahedral sites and are distinguished by octahedral cations Al_4, $Fe_4{}^{3+}$, and Mg_6, respectively; montmorillonite and hectorite show cation substitution largely in octahedral sites, where a great variety of substitutions are possible. Montmorillonite usually shows significant $Mg^{2+} \rightarrow Al^{3+}$ replacement to about $Al_{3.33}Mg_{0.67}$, and hectorite is a rare Li-montmorillonite, with substitution $Li^+ \rightarrow Mg^{2+}$ to about $Mg_{5.33}Li_{0.67}$. Each cation substitution, tetrahedral or octahedral, causes positive charge deficiency to be balanced by interlayer cations of equal positive charge, usually Ca or Na, although many other ions are possible. Interlayer cations are enclosed in complete layers of water molecules—usually two water molecule layers for Ca-montmorillonites and one or more layers for Na-montmorillonites. Interlayer cations are readily exchangeable, and interlayer spaces swell and contract with the addition or removal of water molecules.

Because of interlayer separations, stacking of successive montmorillonite layers is largely random, and interlayering with chlorite, vermiculite, and micas is common.

The Illite Group (Hydromicas)

Illite	$K_{1-1.5}Al_4(Si_{7-6.5}Al_{1-1.5})O_{20}(OH)_4$
Brammallite	$Na_{1-1.5}Al_4(Si_{7-6.5}Al_{1-1.5})O_{20}(OH)_4$
Hydromuscovite	$(K,H_3O)_2Al_4(Si_6Al_2)O_{20}(OH)_4$
Glauconite	$(K,H_3O)_2(Fe^{3+},Al,Fe^{2+},Mg)_4(Si_{7-7.5}Al_{1-0.5})O_{20}(OH)_4$

The illites are secondary micas, usually very fine-grained, formed by processes of near-surface alteration. They display the typical three-layer structure of mica, an octahedral layer sandwiched between tetrahedral layers with intersheet cations required to balance the substitution of Al^{3+} for Si^{4+}. Illite and brammallite differ from muscovite and paragonite only by slightly less $Al^{3+} \rightarrow Si^{4+}$ substitution and correspondingly less interlayer K^+ or Na^+.

Hydromuscovite shows little silica excess, but K^+ is deficient, presumably partly replaced by $(H_3O)^+$. Hydrobiotite is usually interlayered biotite and vermiculite. Glauconite is also silica-rich and is essentially the ferric equivalent of illite bridging the composition gap between muscovite $KAl_2(Si_3Al)O_{10}(OH)_2$ and celadonite $K(Mg,Fe^{3+},Al)_2Si_4O_{10}(OH)_2$.

The Vermiculite Group

Vermiculite $(Mg,Ca)[(Mg,Fe^{2+})_5(Fe^{3+},Al)](Si_5Al_3)O_{20}(OH)_4 \cdot 8H_2O$

Vermiculite may represent a wide range of compositions and is closely related to both montmorillonites and chlorites. Its structure is again the three-layer mica structure with interlayer cations balancing charge deficiency largely in tetrahedral sites. Interlayer cations are largely Mg^{2+} sandwiched between two water-molecule layers to yield an interlayer configuration much like the "brucite" layer of chlorites. Rapid heating vaporizes the water molecules causing vermiculite layers to expand. Mixed-layer association with chlorite, montmorillonite, or micas is very common.

The Palygorskite Group

Palygorskite $(Mg,Al)_4Si_8O_{20}(OH)_2(OH_2)_4 \cdot 4H_2O$

Sepiolite $(Mg,Al,Fe^{3+})_8Si_{12}O_{30}(OH)_4(OH_2)_4 \cdot 8H_2O$

Palygorskite and sepiolite are fibrous clays. The structure of palygorskite contains amphibolelike chains (Si_4O_{11}) elongated on *c* and side-by-side along *b*, with tetrahedra of adjoining chains pointing alternately up and down. Adjoining chains are linked through oxygen ions so that each silicon ion shares three of four oxygen ions with other silicons (Si_4O_{10}), and continuous sheets of oxygen ions form parallel to $\{100\}$. These corrugated SiO_4 sheets are joined by cations, usually Mg^{2+} or Al^{3+}, in octahedral coordination to form narrow strips of "talc" alternating on opposite sides of the continuous oxygen layer. In the narrow, octahedral ribbons, $(OH)^-$ ions complete the octahedral coordination about Mg^{2+}, as in talc, and at the discontinuous edges $(OH)^-$ ions are neutralized by protons (H^+) to form bound water molecules, designated as (OH_2) to differentiate them from water molecules (H_2O), which occupy channels between chains, as in zeolites.

The structure of sepiolite differs only in the width of the "talc" ribbons (the amphibole-type chains), which are wider by one silicate tetrahedron.

Palygorskite and sepiolite represent middle segments of a solid solution series between an Mg end member and an Al end member that probably do not exist with palygorskite or sepiolite structures. It has been proposed that palygorskite and sepiolite are intermediate-composition series between talc and pyrophyllite, forming standard sheet structures near the series end members. A ferric variety of sepiolite (*xylotile*) and Ni-and Mn-rich compositions are known.

Attapulgite and *pilolite* are variety names for palygorskite.

Table 6-1 summarizes the optical and physical properties of the clay minerals.

TABLE 6-1. Optical Properties of the Common Clay Minerals

n_β	n_α	n_γ	$n_\gamma - n_\alpha$	Mineral	System	2V	Optic Sign	Cleavage	Optical Orientation	Color in Sections	Physical Properties
1.47–1.50	—	—	0.000	Allophane $Al_2Si_2O_5(OH)_4$	Amorphous	—		None	—	Colorless, bluish, brown	H = 2–3. Sp. Gr. = 1.8–1.9. White, iron-stained
1.50–1.53	1.49–1.52	1.50–1.53	0.009–0.020	Sepiolite $(Mg,Al,Fe^{+3})_8Si_{12}\cdot O_{30}(OH)_4(OH_2)_4\cdot 8H_2O$	Monoclinic $\angle\beta = 96°$	20°–70°	–	{110} distinct	$c \wedge Z$ = small length-slow	Colorless, pale yellow	H = 2–2½. Sp. Gr. = 1.0–2.6. White, gray, yellow, green
1.49–1.57	1.48–1.53	1.50–1.59	0.01–0.04	Saponite $(\frac{1}{2}Ca,Na)_{0.67}Mg_6(Si_{7.33}Al_{0.67})\cdot O_{20}(OH)_4\cdot nH_2O$	Monoclinic $\angle\beta \simeq 90°$	Moderate	–	{001} perfect	$b = Y$ $c \wedge X$ = small $a \wedge Z$ = small	Colorless ("muddy")	H = 1–2. Sp. Gr. = 2.0–2.7. White, yellow, red, brown, green, black
1.50–1.59	1.48–1.57	1.50–1.60	0.02–0.03	Montmorillonite $(\frac{1}{2}Ca,Na)_{0.67}(Al_{3.33}Mg_{0.67})\cdot Si_8O_{20}(OH)_4\cdot nH_2O$	Monoclinic $\angle\beta \simeq 90°$	0°–30°	–	{001} perfect	$b = Y$ $c \wedge X$ = small $a \wedge Z$ = small	Colorless ("muddy")	H = 1–2. Sp. Gr. = 2.0–2.7. White, yellow, red, brown, green, black
1.50–1.59	1.48–1.57	1.50–1.60	0.02–0.03	Beidellite $(\frac{1}{2}Ca,Na)_{0.67}Al_4(Si_{7.33}Al_{0.67})\cdot O_{20}(OH)_4\cdot nH_2O$	Monoclinic $\angle\beta \simeq 90°$	0°–30°	–	{001} perfect	$b = Y$ $c \wedge X$ = small $a \wedge Z$ = small	Colorless ("muddy")	H = 1–2. Sp. Gr. = 2.0–2.7. White, yellow, red, brown, green, black
1.52	1.49	1.52	0.03	Hectorite $(\frac{1}{2}Ca,Na)_{0.67}(Mg_{5.33}Li_{0.67})\cdot Si_8O_{20}(OH)_4\cdot nH_2O$	Monoclinic $\angle\beta \simeq 90°$	Small	–	{001} perfect	$b = Y$ $c \wedge X$ = small $a \wedge Z$ = small	Colorless ("muddy")	H = 1–2. Sp. Gr. = 2.0–2.7. White, yellow, red, brown, green, black
1.53–1.54	1.53–1.54	1.53–1.54	0.000–0.004	Halloysite (endellite) $Al_2Si_2O_5(OH)_4\cdot 2H_2O$	Monoclinic	Small	–	{001} perfect	$b = Z$	Colorless ("muddy")	H = 1–2. Sp. Gr. = 2.0–2.2. White, bluish, yellow, gray
1.53–1.56	1.50–1.52	1.54–1.56	0.020–0.035	Palygorskite $(Mg,Al)_4Si_8O_{20}(OH)_2(OH_2)_4\cdot 4H_2O$	Monoclinic $\angle\beta = 96°$	0°–60°	–	{110} distinct	$c \wedge Z$ = small length-slow	Colorless	H = 2–2½. Sp. Gr. = 1.0–2.6. White, gray, yellowish
1.54–1.58	1.52–1.56	1.54–1.58	0.02–0.03	Vermiculite $(Mg,Ca)[(Mg,Fe^{+2})_5(Fe^{+3},Al)]\cdot (Si_5Al_3)O_{20}(OH)_4\cdot 8H_2O$	Monoclinic $\angle\beta = 97°$	0°–8° $v > r$ weak	–	{001} perfect	$b = Y$ $c \wedge X = -3°$ to $-6°$ $a \wedge Z = +4°$ to $+1°$	Colorless, pale brown, pale green	H ≃ 1.5. Sp. Gr. ≃ 2.4. Pale brown (bronze), dark brown
1.55–1.56	1.55–1.56	1.55–1.56	0.002	Meta-halloysite $Al_2Si_2O_5(OH)_4$	Monoclinic			{001} perfect		Colorless, nearly opaque	H = 2–3. Sp. Gr. = 2.5–2.7. White, iron-stained

1.56–1.61	1.54–1.57	1.57–1.61	0.03	Hydromuscovite $(K,H_3O)Al_2(Si_3Al)O_{10}(OH)_2$	Monoclinic $\angle\beta \simeq 90°$	0°–5°	–	{001} perfect	$b=Z$ $c \wedge X=$ small $a \wedge Y=$ small	Colorless	H = 1–2. Sp. Gr. = 2.6–2.9. Colorless, iron-stained
1.56	1.56	1.56–1.57	0.006	Nacrite $Al_2Si_2O_5(OH)_4$	Monoclinic $\angle\beta \simeq 90°$	40°–90° $r \gtrless v$ weak	∓	{001} perfect	$b=Z$ $c \wedge X=+7°$ to $+10°$ $a \wedge Y=+7°$ to $+10°$	Colorless, pale brown, nearly opaque	H = 2–3. Sp. Gr. = 2.5–2.7. White, iron-stained
1.56–1.57	1.56	1.56–1.57	0.004–0.008	Dickite $Al_2Si_2O_5(OH)_4$	Monoclinic $\angle\beta \simeq 97°$	50°–80° $v > r$ weak	+	{001} perfect	$b=Z$ $c \wedge X=+7°$ to $+13°$ $a \wedge Y=+14°$ to $+20°$	Colorless, pale brown, nearly opaque	H = 2–3. Sp. Gr. = 2.5–2.7. White, iron-stained
1.56–1.63	1.53–1.61	1.56–1.64	0.03–0.04	Nontronite $(\frac{1}{2}Ca,Na)_{0.67}Fe_4^{+3}(Si_{7.33}Al_{0.67})\cdot O_{20}(OH)_4\cdot nH_2O$	Monoclinic $\angle\beta \simeq 90°$	25°–70°	–	{001} perfect	$b=Y$ $c \wedge X=$ small $a \wedge Z=$ small	$X=$ yellow-green $Y=Z=$ green	H = 1–2. Sp. Gr. = 2.0–2.7. Green, brown, black
1.56–1.57	1.55–1.57	1.56–1.57	0.008–0.005	Kaolinite $Al_2Si_2O_5(OH)_4$	Monoclinic $\angle\beta =$ 104°30′	23°–60° $r > v$ weak	–	{001} perfect	$b=Z$ $c \wedge X=-13°$ to $-10°$ $a \wedge Y=+1°$ to $+4°$	Colorless, pale brown, nearly opaque	H = 2–3. Sp. Gr. = 2.5–2.7. White, iron-stained
1.56–1.57	1.55–1.57	1.56–1.57	0.008–0.005	Anauxite $Al_{2-1.5}Si_{2-2.5}O_5(OH)_4$	Monoclinic $\angle\beta \simeq 104°$	23°–60° $r > v$ weak	–	{001} perfect	$b=Z$ $c \wedge X=-13°$ to $-10°$ $a \wedge Z=+1°$ to $+4°$	Colorless, pale brown, nearly opaque	H = 2–3. Sp. Gr. = 2.5–2.7. White, iron-stained
1.57–1.61	1.54–1.57	1.57–1.61	0.03	Illite $K_{1-1.5}Al_4(Si_{7-6.5}Al_{1-1.5})\cdot O_{20}(OH)_4$	Monoclinic $\angle\beta \simeq 90°$	0°–10°	–	{001} perfect	$b=Z$ $c \wedge X=$ small $a \wedge Y=$ small	Colorless	H = 1–2. Sp. Gr. = 2.6–2.9. Colorless, iron-stained
1.58–1.61	1.55–1.57	1.59–1.62	0.036–0.049	Sericite $KAl_2(Si_3Al)O_{10}(OH)_2$	Monoclinic $\angle\beta = 95°30'$	30°–47° $r > v$ distinct horizontal Bxa	–	{001} perfect	$b=Z$ $c \wedge X=0°$ to $-5°$ $a \wedge Y=+1°$ to $+3°$	Colorless	H = $2\frac{1}{2}$–4. Sp. Gr. = 2.8–2.9. Colorless, pale brown, red, green
1.58–1.61	1.55–1.58	1.59–1.61	0.03	Brammallite $Na_{1-1.5}Al_4(Si_{7-6.5}Al_{1-1.5})\cdot O_{20}(OH)_4$	Monoclinic $\angle\beta \simeq 90°$	0°–30°	–	{001} perfect	$b=Z$ $c \wedge X=$ small $a \wedge Y=$ small	Colorless	H = 1–2. Sp. Gr. = 2.6–2.9. Colorless, iron-stained
1.61–1.65	1.56–1.61	1.61–1.65	0.014–0.032	Glauconite (celadonite) $(K,H_3O)_2(Fe^{+3},Al,Fe^{+2},Mg)_4\cdot (Si_{7-7.5}Al_{1-0.5})O_{20}(OH)_4$	Monoclinic $\angle\beta \simeq 100°$	0°–20° $r > v$ weak inclined Bxa	–	{001} perfect	$b=Y$ $c \wedge X \simeq -10°$ $a \simeq Z$	$X=$ yellow-green $Y=Z=$ green	H = 2. Sp. Gr. = 2.4–3.0. Blue-green, olive-green, iron-stained

The Kaolin (Kandite) Group

$Al_2Si_2O_5(OH)_4$

MONOCLINIC
$\angle\beta = 104°–90°$

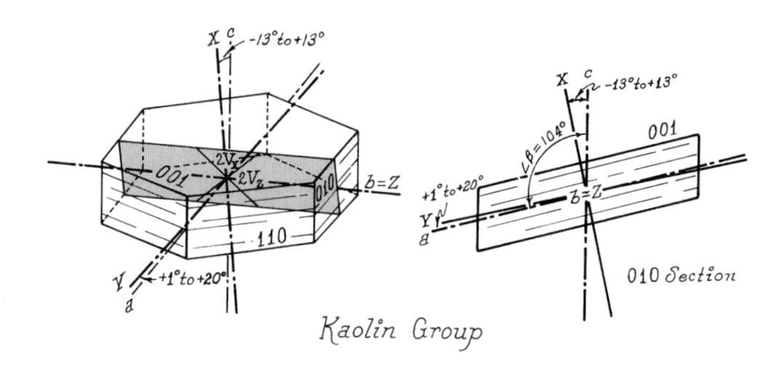

	n_α	n_β	n_γ	$n_\gamma - n_\alpha$	$2V$	$a \wedge Y$
Kaolinite / Anauxite	1.553–1.565	1.559–1.569	1.560–1.570	0.008–0.005	23°–60° (−)	1°–4°
Dickite	1.558–1.564	1.560–1.566	1.563–1.571	0.004–0.008	50°–80° (+)	14°–20°
Nacrite	1.557–1.560	1.562–1.563	1.563–1.566	0.006	40°–90° (−)	7°–10°
Halloysite	—	1.526–1.542	—	0.000–0.004	—	Small
Meta-halloysite	—	1.549–1.561	—	0.002	—	—

$b = Z$ (optic plane $\perp$ {010} near {100})

Kaolinite $c \wedge X = -13$ to $-10°$, $a \wedge Y = +1°$ to $+4°$

Dickite $c \wedge X = +7°$ to $+13°$, $a \wedge Y = +14°$ to $+20°$

Nacrite $c \wedge X = +7°$ to $+10°$, $a \wedge Y = +7°$ to $+10°$

$r > v$ weak on X ($v > r$ for positive varieties)

Colorless to very pale brown or yellow in section, with possible weak pleochroism $Z = Y > X$

COMPOSITION AND STRUCTURE. Kaolinite, dickite, and nacrite are identical in composition $Al_2Si_2O_5(OH)_4$ and are distinguished by stacking sequences of basic kaolin sheets (Fig. 6-2). Anauxite shows a slightly higher Si:Al ratio due to the presence of soluble amorphous silica between kaolinite flakes; thus anauxite is not a distinct mineral species. Halloysite (endellite) is a hydrated form with a single layer of water molecules between kaolin layers and meta-halloysite is dehydrated or partially dehydrated halloysite. Allophane is an amorphous mineral with essentially the kaolinite composition.

Kaolinite	$Al_2Si_2O_5(OH)_4$
Dickite	$Al_2Si_2O_5(OH)_4$
Nacrite	$Al_2Si_2O_5(OH)_4$
Halloysite (endellite)	$Al_2Si_2O_5(OH)_4 \cdot 2H_2O$
Meta-halloysite	$Al_2Si_2O_5(OH)_4$
Anauxite	$Al_{2-1.5}Si_{2-2.5}O_5(OH)_4$

Little chemical variation from the ideal composition is observed. Many cations, especially Na, K, and Ca, are ad-

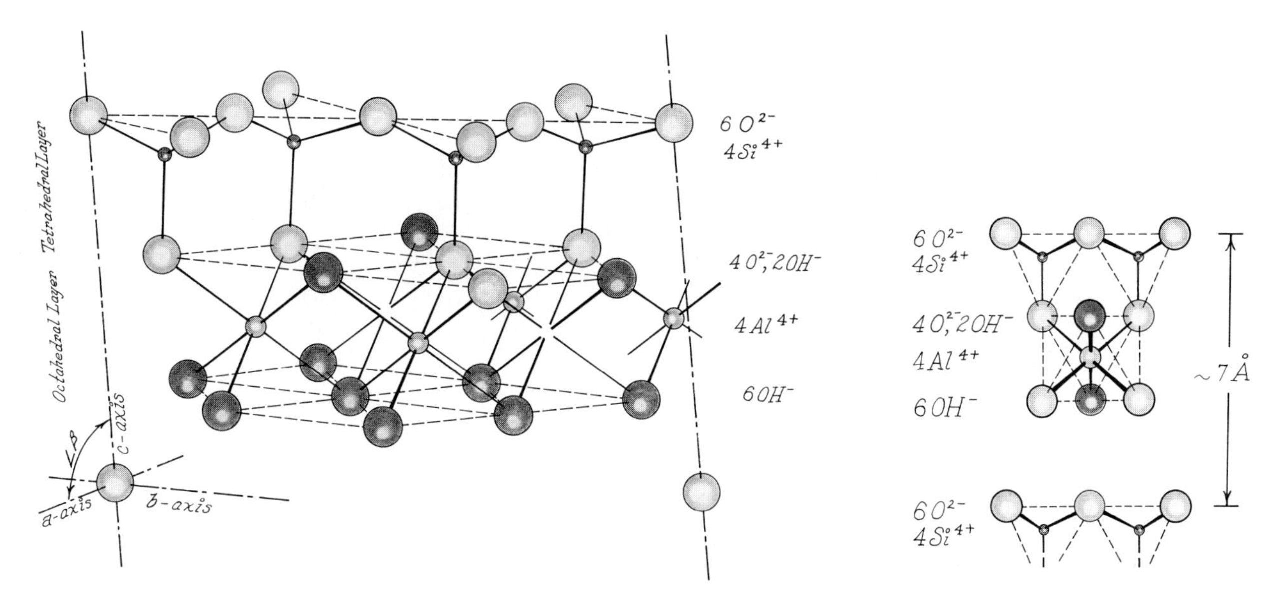

FIGURE 6-2. The idealized structure of kaolinite: Repeat unit on c is one tetrahedral and one octahedral (dioctahedral) layer (~7 Å).

sorbed by clay layers with broken bonds, and minor Mg, Fe, or Ti may replace Al in octahedral sites.

PHYSICAL PROPERTIES. H = 2–3. Sp. Gr. = 2.5–2.7. In hand sample, kaolin clays are commonly white to strongly iron-stained, with dull luster and greasy feel. Kaolins are non-expanding clays.

COLOR AND PLEOCHROISM. The kaolin clays are normally colorless to weakly colored as fragments or in section. Fine-grained aggregates are strongly light-absorbing and may appear dark as fragment clumps or dusty alteration in feldspars. Weak pleochroism is sometimes present as $Z = Y > X$. $Z = Y =$ pale yellow, $X =$ pale brown to buff.

FORM. Individual crystals of the kaolin clays are commonly below the resolution limit of light microscopes and appear as a cloudy aggregate mass of low birefringence. Coarser aggregates appear as shredded masses, veinlets, or scales. Large crystals may exceed 1 mm and appear as twisted, wormlike masses elongated on c with individual plates seeming to "float" apart like an accordion (Fig. 6-3). Halloysite is extremely fine grained, appearing as a uniform mass often cut by a random network of contraction cracks.

The true morphology of the kaolin clays is seldom recognized by optical means, but the electron microscope reveals the foliated structure of kaolinite, dickite, nacrite, and anauxite, commonly as tiny pseudohexagonal plates, and the tubular forms of halloysite. Meta-halloysite, as dehydrated halloysite, reveals partially unrolled tubes and scooplike forms, and allophane is truly amorphous, with colloform structures.

CLEAVAGE. Perfect basal cleavage {001} is not always visible but may control fragment orientation of larger flakes.

BIREFRINGENCE. Maximum birefringence, nil to weak (0.000–0.008), is seen in sections nearly normal to {001}, and basal sections show very weak birefringence. Interference colors are

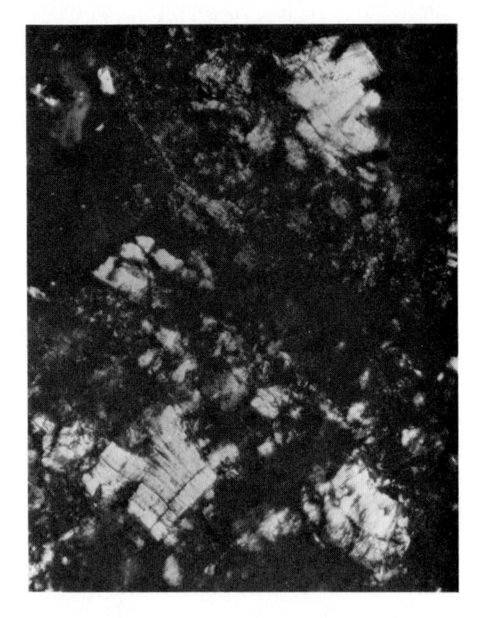

FIGURE 6-3. Kaolin clays occasionally are coarse enough to show vermicular forms and the perfect basal cleavage of the sheet silicates.

grays below first-order white and some varieties (for example, halloysite and allophane) are sensibly isotropic.

TWINNING. Twinning, probably on {001}, is reported but is rarely visible.

INTERFERENCE FIGURE. The task of finding a usable interference figure from kaolin clays is usually a hopeless one; however, unusually large plates or aggregated plates of similar orientation of kaolinite, dickite, or nacrite, may yield crude figures of diffuse isogyres on a gray field. X, essentially perpendicular to basal plates, is the acute bisectrix for kaolinite and nacrite (biaxial negative) and the obtuse bisectrix of dickite (biaxial positive). Dispersion is weak $r > v$ on X. $2V$ is small to moderate for kaolinite ($2V_x = 20°–60°$) and moderate to large for dickite ($2V_z = 50°–80°$) and nacrite ($2V_x = 40°–90°$).

OPTICAL ORIENTATION. For all crystallized forms, $b = Z$ and the optic plane is $\perp$ {010} a few degrees from {100}. For kaolinite, $c \wedge X = -13°$ to $-10°$ and $a \wedge Y = +1°$ to $-4°$, for dickite, $c \wedge X = +7°$ to $+13°$ and $a \wedge Y = +14°$ to $+20°$, for nacrite, $c \wedge X = +7°$ to $+10°$ and $a \wedge Y = +7°$ to $+10°$.

Maximum extinction to cleavage traces is $a \wedge Y$ (near parallel for kaolinite, 14° to 20° for dickite, and 7° to 10° for nacrite). Elongation of cleavage traces is positive (length-slow), as X lies almost normal to cleavage.

DISTINGUISHING FEATURES. Kaolin clays are the fine-grained alteration products of aluminous silicates, usually feldspars, showing weak to nil birefringence (white or gray interference colors) and positive (length-slow) elongation. Illites, montmorillonites, vermiculite, talc, and most other secondary minerals show higher birefringence. Chlorites and serpentine have weak birefringence but are commonly colored and pleochroic. Some chlorites are length-fast, and serpentine is derived from olivine and other ferromagnesian silicates.

Dickite and nacrite are distinguished from kaolinite by inclined extinction (7°–20°), halloysite is very fine-grained and essentially isotropic, dickite is optically positive, and allophane is truly isotropic.

ALTERATION. The kaolin clays are, themselves, stable alteration products of feldspars and other aluminous silicates. Excessive leaching in humid climates may form aluminous laterites (diaspore, gibbsite, and so on) at the expense of the kaolin clays.

OCCURRENCE. The kaolin clays form from the alteration of aluminous silicates, largely feldspars, at low temperatures and pressures. Kaolin clays are favored by acid environments and the illites and montmorillonites by alkaline environments. Kaolinite is perhaps the most common and important clay mineral. It forms by weathering of granites and similar Ca-poor igneous and metamorphic rocks and is a major constituent of shales and residual or transported soils in association with illites, quartz, chlorite, limonite, micas, and carbonates. Kaolinite is also a product of hydrothermal alteration and is common in wall rock alteration of sulfide ore veins between inner sericite zones and outer chlorite zones.

Dickite, nacrite, and halloysite are largely hydrothermal minerals produced by the action of acid solutions on aluminous silicates. Dickite and nacrite appear with quartz, sulfides, chalcedony, and limonite. Halloysite and meta-halloysite appear with kaolinite, alunite, and sulfide ores.

REFERENCES: THE KAOLIN (KANDITE) GROUP

Bailey, S. W., and R. B. Langston. 1969. Anauxite and kaolinite structures identical. *Clays and Clay Minerals.*, *17*, 241–243.

Bates, T. F. 1959. Morphology and crystal chemistry of 1:1 layer lattice silicates. *Amer. Mineral.*, *44*, 78–114.

Bates, T. F., F. A. Hildebrand, and A. Swineford. 1950. Morphology and structure of endellite and halloysite. *Amer. Mineral.*, *35*, 463–484.

Hendricks, S. B. 1939. Crystal structure of nacrite and the polymorphism of the kaolin minerals. *Z. Kristallogr.*, *100*, 509–518.

Langston, R. B., and J. A. Pask. 1969. The nature of anauxite. *Clays and Clay Minerals.*, *16*, 425–436.

Newnham, R. E., and G. W. Brindley. 1956. The crystal structure of dickite. *Acta Crystallogr.*, *9*, 759–764.

Ross, C. R., and P. F. Kerr. 1931. The kaolin minerals. *U.S. Geol. Surv., Prof. Paper*, *165-E*, 151–176.

Weaver, C. E., and L. D. Pollard. 1973. *The Chemistry of Clay Minerals*. Amsterdam: Elsevier.

The Montmorillonite (Smectite) Group

$(\frac{1}{2}Ca,Na)_{0.67}(Al,Mg,Fe)_{4-6}(Si,Al)_8O_{20}(OH_4)\cdot nH_2O$

MONOCLINIC
$\angle\beta \simeq 90^\circ$

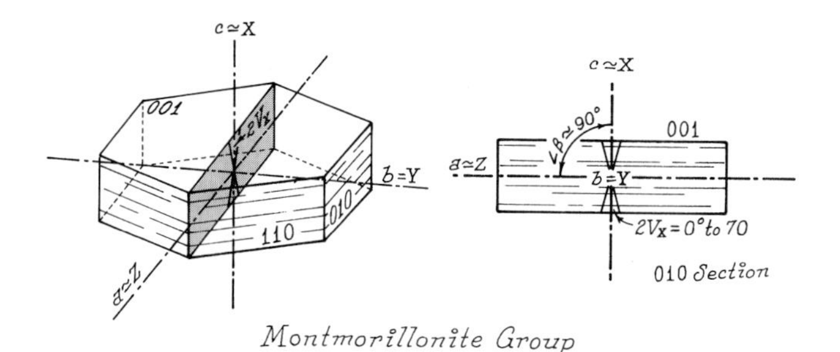

Montmorillonite Group

	n_α^*	n_β	n_γ	$n_\gamma - n_\alpha$	$2V_x$
Montmorillonite / Beidellite	1.48–1.57	1.50–1.59	1.50–1.60	0.02–0.03	0°–30° (−)
Nontronite	1.53–1.61	1.56–1.63	1.56–1.64	0.03–0.04	25°–70° (−)
Saponite	1.48–1.53	1.49–1.57	1.50–1.59	0.01–0.04	Moderate (−)
Hectorite	1.49	1.52	1.52	0.03	Small (−)

Biaxial negative, $2V_x = 0^\circ$ to moderately large

$b = Y$, $c \wedge X$ = small, $a \wedge Z$ = small

Colorless to pale yellow, brown, or pink in standard section

Nontronite is bright green and distinctly pleochroic from green to yellow $Z = Y > X$

* Refraction indices increase with increasing iron and with dehydration but are changed little by $Al \rightleftharpoons Si$ or $Al \rightleftharpoons Mg$.

COMPOSITION AND STRUCTURE. The montmorillonites (smectites) have a three-layer, talclike structure (see p. 302) and compositions related to that of pyrophyllite, $Al_4Si_8O_{20}(OH)_4$, and talc, $Mg_6Si_8O_{20}(OH)_4$. Beidellite, nontronite, and saponite are formed largely by substitution in tetrahedral sites ($Al \rightleftharpoons Si$) with largely a single element in octahedral sites to distinguish beidellite (Al), nontronite (Fe^{3+}), and saponite (Mg). Montmorillonite and hectorite show little substitution in tetrahedral sites and are distinguished by substitution in octahedral sites. Substitution in either tetrahedral (Al^{3+} for Si^{4+}) or octahedral (Mg^{2+} for Al^{3+} or Li^{+} for Mg^{2+}) sites requires interlayer cations for charge balance.

Beidellite	$(\frac{1}{2}Ca,Na)_{0.67}Al_4(Si_{7.33}Al_{0.67})O_{20}(OH)_4 \cdot nH_2O$
Nontronite	$(\frac{1}{2}Ca,Na)_{0.67}Fe_4{}^{3+}(Si_{7.33}Al_{0.67})O_{20}(OH)_4 \cdot nH_2O$
Saponite	$(\frac{1}{2}Ca,Na)_{0.67}Mg_6(Si_{7.33}Al_{0.67})O_{20}(OH)_4 \cdot nH_2O$
Montmorillonite	$(\frac{1}{2}Ca,Na)_{0.67}(Al_{3.33}Mg_{0.67})Si_8O_{20}(OH)_4 \cdot nH_2O$
Hectorite	$(\frac{1}{2}Ca,Na)_{0.67}(Mg_{5.33}Li_{0.67})Si_8O_{20}(OH)_4 \cdot nH_2O$

Interlayer cations, surrounded by one or more layers of water molecules, are usually Ca^{2+} or, less commonly, Na^{+}, but may be represented by almost any large cation.

Undoubtedly, much solid solution exists in the montmorillonite group. A montmorillonite-beidellite series appears largely continuous, and other series between ideal end members seem reasonable.

PHYSICAL PROPERTIES. H = 1–2. Sp. Gr. = 2.0–2.7, increasing with dehydration. In hand sample, the montmorillonites, except nontronite, are white, ideally, but as they are normally the alteration product of basic igneous rocks they are commonly impure and iron-stained to appear yellow, red, brown, or green from pale to almost black. The montmorillonites are all expanding clays that may enlarge in water and develop characteristic shrinkage cracks when dried.

COLOR AND PLEOCHROISM. In section or as fragments, montmorillonites are commonly colorless but tend to be opaque and "muddy". Colored varieties are pleochroic $Z = Y > X$, and iron-bearing compositions, notably nontronite, are highly pleochroic. X = yellow-green, and $Y = Z$ = bright green to brownish green. Basal sections are most deeply colored and without pleochroism.

FORM. Microscopic to submicroscopic aggregates of flakes or granules are most common. Aggregates commonly reflect the outline and structures of minerals they replace. Saponite, nontronite, and hectorite may be finely fibrous or bladed, with crystals elongated on *c*. An electron microscope reveals thin plates, irregular or pseudohexagonal in outline.

CLEAVAGE. Perfect basal {001} cleavage is seldom visible but will tend to control fragment orientation.

BIREFRINGENCE. Maximum birefringence, weak to moderate (0.01–0.04), is seen in sections normal to basal cleavage and may reveal bright colors of first and second order at standard thickness. Particles with structural continuity may be less than standard thickness, but bright colors are usually apparent. Basal sections commonly show low birefringence.

INTERFERENCE FIGURE. "Muddy," submicroscopic aggregates of most clays offer little hope of an interference figure. Unusually large plates or aggregates of oriented flakes may yield an acute bisectrix figure on basal sections. All montmorillonites are biaxial negative, and most show small to moderate $2V_x$. Nontronite may show rather large $2V_x$ to 70° and strong dispersion $r > v$.

OPTICAL ORIENTATION. The optic plane is {010} normal to basal cleavage, and X is essentially normal to {001}. Crystals or fibers elongated on *a* are length-fast.

DISTINGUISHING FEATURES. Montmorillonite clays show higher birefringence than kaolin clays and lower birefringence than illites and talc. Some kaolin clays (dickite and nacrite, for example) show inclined extinction, and vermiculite is commonly coarse grained. Nontronite is green, with higher refractive indices than other members of the group. Optical methods are inadequate to separate other members of the montmorillonite group and may, indeed, prove inadequate to distinguish major clay groups.

ALTERATION. The montmorillonite clays are themselves alteration products, largely of basic rocks where Mg and Ca are available. They are highly variable, however, and not the most stable alteration minerals. Water leaching may form kaolinite from montmorillonite, and chlorite or illite can be produced.

OCCURRENCE. Montmorillonite and beidellite form largely from the weathering of igneous rocks in semiarid climates where Mg and Ca are abundant and leaching does not remove them. Siliceous rocks tend to yield illite, mafic rocks yield montmorillonite, and sufficient leaching may produce kaolin from either. Montmorillonite and beidellite are the principal minerals of bentonite clays that result from the weathering of volcanic ash deposits. Quartz fragments, high-temperature silicas, zeolites, micas, and partially altered glass are common associates.

Hydrothermal alteration produces montmorillonite clays in outer wall rock zones adjoining ore veins and near thermal springs. Their formation is favored by alkaline solutions, although montmorillonites may appear with kaolin or chlorites. Saponite and nontronite are formed in hydrothermal veins and vesicles.

REFERENCES: THE MONTMORILLONITE (SMECTITE) GROUP

Greene-Kelly, R. 1955. Dehydration of the montmorillonite minerals. *Mineral. Mag.*, *30*, 604–615.

Grim, R. E., and G. Kublicki. 1961. Montmorillonite: High-temperature reactions and classification. *Amer. Mineral.*, *46*, 1329–1369.

Kerr, P. F. 1950. *Analytical Data on Reference Clay Minerals.* New York: Amer. Petroleum Inst., Columbia University.

Ross, C. S., and S. B. Hendricks. 1945. Minerals of the montmorillonite group. *U.S. Geol. Surv. Prof. Paper*, *205-B*.

Sawhrey, B. L., and M. L. Jackson. 1958. Soil montmorillonite formulas. *Proc. Soil Sci. Soc. Amer.*, *22*, 115–118.

Weaver, C. E., and L. D. Polland. 1973. *The Chemistry of Clay Minerals.* Amsterdam: Elsevier.

Weir, A. H., and R. Greene-Kelly. 1962. Beidellite. *Amer. Mineral.*, *47*, 137–146.

The Illite Group

$(K,Na,H_3O)_{1-2}Al_4(Si_{7-6}Al_{1-2})O_{20}(OH)_4$

MONOCLINIC
$\angle\beta \simeq 90°$

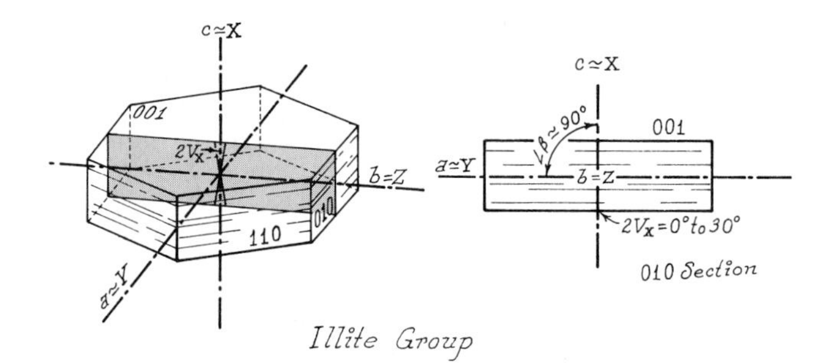

Illite Group

	n_α	n_β	n_γ	$n_\gamma - n_\alpha$	$2V_x$
Illite	1.54–1.57	1.57–1.61	1.57–1.61	≃0.03	<10° (−)
Hydromuscovite	1.53–1.57	1.56–1.60	1.56–1.61	≃0.03	< 5° (−)

Biaxial negative, $2V_x = 0° - 30°$ (usually <10°)
$b = Z$, $c \wedge X$ = small, $a \wedge Y$ = small
Colorless in standard section

COMPOSITION AND STRUCTURE. Illite clays have the three-layer structure of mica and essentially the composition of muscovite. Illite and brammallite differ from muscovite and paragonite by being slightly Si-rich, due to less $Al^{3+} \rightarrow Si^{4+}$ substitution and to having correspondingly less interlayer K^+ or Na^+. Hydromuscovite is K-deficient, due to the presence of $(H_3O)^+$.

Illite	$K_{1-1.5}Al_4(Si_{7-6.5}Al_{1-1.5})O_{20}(OH)_4$
Brammallite	$Na_{1-1.5}Al_4(Si_{7-6.5}Al_{1-1.5})O_{20}(OH)_4$
Hydromuscovite	$(K,H_3O)_2Al_4(Si_6Al_2)O_{20}(OH)_4$

Extensive substitution is possible for Al_4 in octahedral sites to form extensive solid solution with possible end-member compositions (for example, celadonite, $K_2Mg_2(Fe^{3+},Al)_2$-$Si_8O_{20}(OH)_4$), and trioctahedral illites are reported. Glau-

* Trioctahedral illites containing Mg and Fe^{2+} are reported to show higher indices of refraction (n_α = 1.58–1.63, n_γ = 1.61–1.67).

conite is an important dioctahedral illite considered separately in the following section.

PHYSICAL PROPERTIES. H = 1–2. Sp. Gr. = 2.6–2.9. In hand sample, the illites are a white or colorless aggregate of fine scales unless stained by iron oxides (yellow, red, or brown) or graphite (gray or black). Illites are not expanding clays.

COLOR AND PLEOCHROISM. Illites are mostly colorless in section or as fragments, rarely pale yellow, brown, pink, or green, with weak pleochroism.

FORM. Illites are characteristically fine-grained to submicroscopic aggregates of scales or flakes. Illite is often interlayered with chlorites, montmorillonite, and other clays.

CLEAVAGE. Perfect basal cleavage $\{001\}$ is characteristic of illites and other phyllosilicates. Cleavage may not be visible in section but may control fragment orientation.

BIREFRINGENCE. Maximum birefringence, seen in sections normal to basal sections (near $\{100\}$), is moderate (0.03), yielding bright interference colors to mid second order. Individual mineral grains may be less than standard thickness (0.03 mm), and maximum interference colors are often first-order yellow to red. Cleavage plates show very weak birefringence.

INTERFERENCE FIGURE. Usable interference figures are rare, owing to very fine grain size. Large cleavage plates or aggregates of oriented flakes show favorable orientation, however, as the acute bisectrix (X) is essentially normal to basal cleavage. The illites are biaxial negative, with $2V_x = 0°–30°$, usually about 5°, making the figure almost uniaxial.

OPTICAL ORIENTATION. The illites are monoclinic ($\angle\beta$ near 90°) with $b = Z$; $c \wedge X$ and $a \wedge Y$ are small. Extinction is near parallel, and elongation is positive (length-slow), as X is essentially normal to cleavage.

DISTINGUISHING FEATURES. Birefringence is lower and $2V_x$ smaller than that of muscovite. Birefringence of the illites is greater than that of kaolin clays but about the same as the montmorillonite clays. Members of the illite group are indistinguishable by optical means. X-ray diffraction, differential thermal analysis (dta) or even chemical analysis may prove necessary to separate illites from montmorillonite clays.

ALTERATION. The illite clays are themselves relatively stable alteration products, largely of acidic rock types. Intense leaching in humid climates, may yield kaolin clays, Al-hydroxides, or montmorillonite (by hydration).

OCCURENCE. Illite forms largely from the alteration of K-feldspars or muscovite and is a common product of the weathering of siliceous igneous or metamorphic rocks. Illites are the most abundant clays in most shales and argillaceous sediments, appearing in association with kaolin and montmorillonite clays, chlorites, quartz, calcite, organic carbon, and iron oxides. Illites are less abundant in soils. They are also formed by hydrothermal alteration and are common in the wall rock of ore veins, usually the inner zone, and near the channels of thermal springs. The formation of illites is favored by alkaline solutions and the abundance of K-silicates.

REFERENCES: THE ILLITE GROUP

Gaudette, H. E. 1965. Illite from Fond du Lac County, Wisconsin. *Amer. Mineral.*, *50*, 411–417.

Gaudette, H. E., J. L. Eades, and R. E. Grim. 1966. The nature of illite. In *Clays and Clay Minerals, Proceedings of the 13th National Conference*, W. F. Bradley and S. W. Bailey (eds.). London: Pergamon Press.

Grim, R. E., R. H. Bray, and W. F. Bradley. 1937. The mica of argillaceous sediments. *Amer. Mineral.*, *22*, 813–829.

Weaver, C. E. 1965. Potassium content of illite. *Science*. *147*, 603–605.

Weaver, C. E., and L. D. Pollard. 1973. *The Chemistry of Clay Minerals*. Amsterdam: Elsevier.

Glauconite

$(K,H_3O)_2(Fe^{3+},Al,Fe^{2+},Mg)_4(Si_{7-7.5}Al_{1-0.5})O_{20}(OH)_4$

MONOCLINIC
$\angle\beta \simeq 100°$

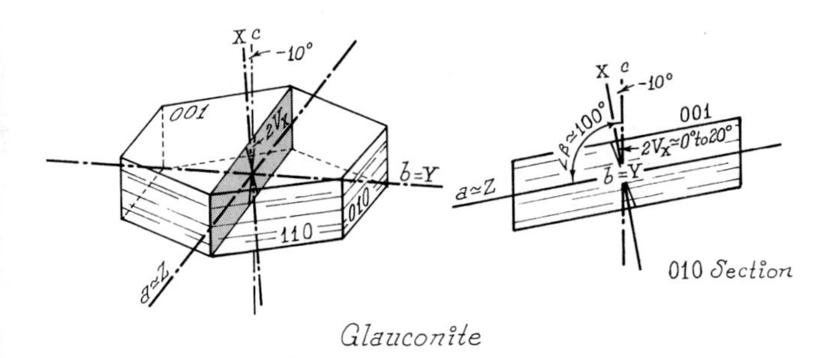

$n_\alpha = 1.56–1.61$*

$n_\gamma = n_\beta = 1.61–1.65$

$n_\gamma - n_\alpha = 0.014–0.032$

Biaxial negative, $2V_x = 0°–20°$

$b = Y, a \simeq Z, c \wedge X \simeq -10°$

$r > v$, inclined bisectrix dispersion

Yellow-green or olive-green in section

$Z = Y > X$; X = pale yellow, yellow-green, green; and $Y = Z$ = yellow-green, dark green, olive-green, blue-green.

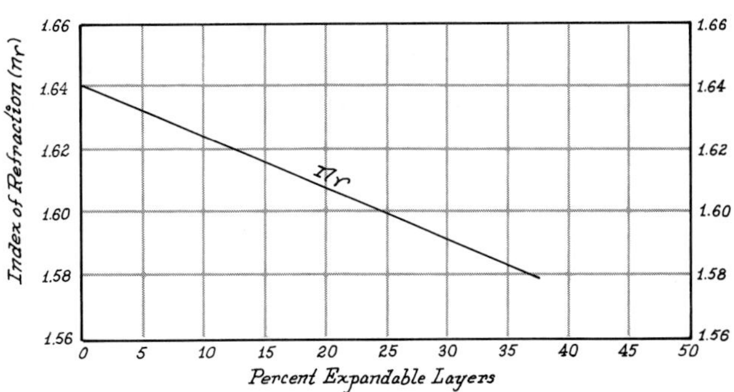

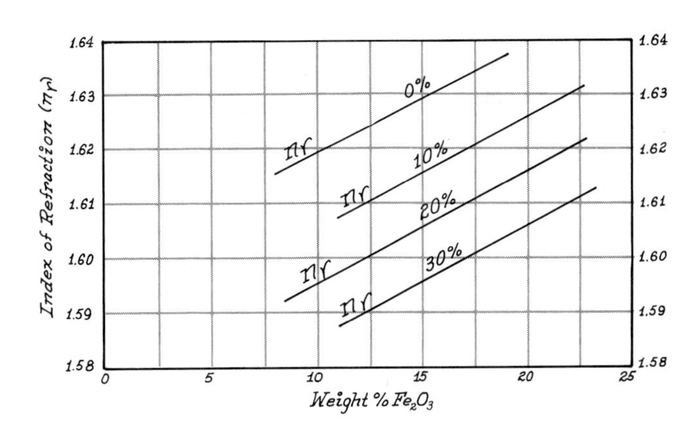

FIGURE 6-4. Variation of the index of refraction (n_γ) with the percent of expandable layers in glauconite; also, the influence of Fe_2O_3 and percentage expandable layers on the index of refraction of glauconite. (After Toler and Hower, 1959)

COMPOSITION AND STRUCTURE. Glauconite is a secondary mica and essentially the ferric equivalent of illite. Well-crystallized glauconite shows single-layer (1**M**) stacking repetition but disordered stacking (1**M**d) is common, and mixed-layer structures with montmorillonites are often encountered. Interlayer cations are largely K, with appreciable water molecules or $(H_3O)^+$ ions and lesser Na and Ca. Gauconite is essentially dioctahedral (only four out of six octahedral sites are filled) with ferric iron (Fe^{3+}) as the major cation in octahedral coordination, with lesser Al, Fe^{2+}, and Mg. Aluminous glauconites (Fe^{3+}:Al < 3) are called *skolite*. *Celadonite* is a variety enriched in (Mg + Fe^{2+}) and depleted in (Al + Fe^{3+}). Silicon ions dominate tetrahedral sites with only minor Al replacement, and the resulting excess plus charge is balanced by bivalent for trivalent ions in octahedral sites.

PHYSICAL PROPERTIES. H = 2. Sp. Gr. = 2.4–3.0. In hand sample, glauconite is blue-green to olive-green, often stained by limonitic alteration. It is usually mixed with clay minerals in tiny, rounded pellets or granules.

* Index of refraction and birefringence increase rapidly with ferric iron (Fe^{3+}) content and decrease with interlayer water. Titanium, manganese, and fluorine are seldom present in quantities necessary to influence optical constants. (Fig. 6-4.)

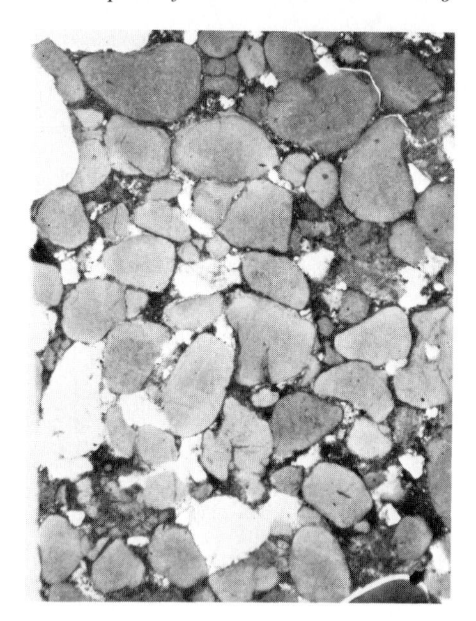

FIGURE 6-5. Glauconite occurs as rounded green granules in marine sediments. Crossed nicols reveal a fine aggregate structure in each pellet.

COLOR AND PLEOCHROISM. In section or fragments, glauconite is colored and pleochroic in yellows and greens. Color intensity increases with Fe^{3+}, and aluminous varieties may be nearly colorless. Absorption is $Z = Y > X$. X = pale yellow or yellow-green to green, $Y = Z$ = pale to dark green, olive-green or blue-green.

FORM. Glauconite is micaceous, but poorly formed. Small flakes or aggregates are most common, associated with clay minerals in small, rounded pellets, granules, or casts of micro fossils (Fig. 6-5). Disordered glauconites are without foliated form even on submicroscopic scale. Celadonite appears most commonly as radiating aggregates in vesicles.

CLEAVAGE. Cleavage is perfect on {001}, as with all micas.

BIREFRINGENCE. Birefringence is maximum in all sections normal to cleavage and ranges from mild to moderate (0.014–0.032). In standard section, interference colors are upper first- or lower second-order, often dulled by inherent mineral color. Basal sections are almost isotropic.

INTERFERENCE FIGURE. Fine grain size and poor crystallization suggest poor figures, but large cleavage flakes or basal sections should yield essentially centered, biaxial negative, acute bisectrix figures with few isochromes and small $2V$. $2V_x = 0°–20°$, usually $>10°$. Dispersion is perceptiable $r > v$, with possible inclined bisectrix dispersion.

OPTICAL ORIENTATION. The optic plane lies parallel to {010}, with $a \simeq Z$ and the acute bisectrix X essentially normal to basal cleavage. $b = Y$ and $c \wedge X \simeq 10°$. Extinction is essentially parallel to cleavage in all sections, and cleavage elongation is always positive (length-slow), since X is normal to cleavage. Basal sections are almost isotropic, without cleavage traces.

DISTINGUISHING FEATURES. Limited occurrence, color, and typical pellet form usually distinguish glauconite. Chlorites and septechlorites may appear as green, micaceous aggregates, and the chamosite variety commonly appears in oolitic form, but most chlorites show birefringence below 0.01.

ALTERATION. Limonite and goethite are the common weathering products of glauconite. It often appears as mixed layers with expandable montmorillonite, and becomes montmorillonite with loss of potassium.

OCCURRENCE. The characteristic occurrence of glauconite is as rounded, diagenetic pellets, or granules, in detrital sediments of marine origin ("greensands"). It may be intimately associated with clays and chlorites in pellets, which are, in turn, associated with detrital quartz, feldspar, mica, and so on. Glauconite may be associated with carbonates, clays, and collophane in glauconitic limestones and marls.

Glauconite suggests slow deposition and may be related to the action of sulfate-reducing bacteria and decaying organic matter on volcanic glass. Biotite and many other silicates may represent parent material for glauconite formation, and K may be derived from the sea. Older sediments seem to contain glauconite that is poor in ferric iron.

Celadonite occurs as radiating aggregates in vesicles of basaltic rocks.

REFERENCES: GLAUCONITE

Bailey, R. J., and M. P. Atherton. 1969. The petrology of a glauconitic sandy chalk. *Jour. Sed. Petrol.*, *39*, 1420–1431.

Burst, J. F. 1958. Mineral heterogeneity in "glauconite" pellets. *Amer. Mineral.*, *43*, 481–497.

Foster, M. D. 1969. Studies of celadonite and glauconite. *U.S. Geol. Surv., Prof. Paper, 614-F.*

Gruner, J. W. 1935. The structural relationship of glauconite and mica. *Amer. Mineral.*, *20*, 699–714.

Hendricks, S. B., and C. S. Ross. 1941. The chemical composition and genesis of glauconite and celadonite. *Amer. Mineral.*, *26*, 683–708.

Schneider, H. 1927. A study of glauconite. *Jour. Geol.*, *35*, 289.

Toler, L. G., and J. Hower. 1959. Determination of mixed layering in glauconites by index of refraction. *Amer. Mineral.*, *44*, 1314–1318.

Weaver, C. E., and L. D. Polland. 1973. *The Chemistry of Clay Minerals*. Amsterdam: Elsevier.

Wermund, E. G. 1961. Glauconite in Early Tertiary Sediments of Gulf Coast Province. *Amer. Assoc. Petrol. Geol. Bull.*, *45*, 1667–1697.

Vermiculite

$(Mg,Ca)[(Mg,Fe^{2+})_5(Fe^{3+},Al)](Si_5Al_3)O_{20}(OH)_4 \cdot 8H_2O$

MONOCLINIC
$\angle\beta \simeq 97°$

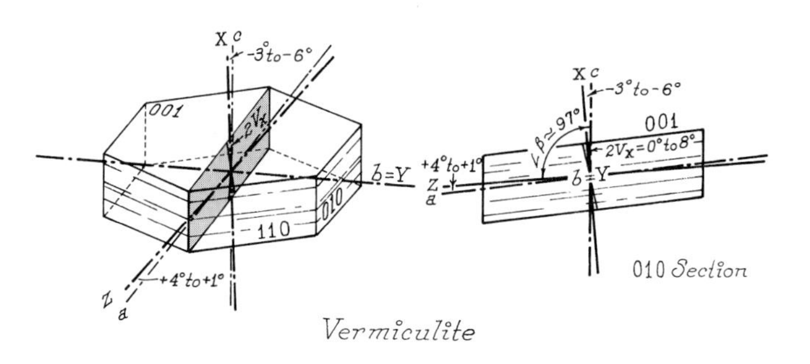

$n_\alpha = 1.525–1.564$*
$n_\beta = 1.545–1.583$
$n_\gamma = 1.545–1.583$
$n_\gamma - n_\alpha = 0.02–0.03$
Biaxial negative, $2V_x = 0°–8°$
$b = Y, c \wedge X = -3°$ to $-6°, a \wedge Z = +4°$ to $+1°$
$v > r$ weak
Colorless to pale yellow, brown, or green, with weak pleochroism $Z = Y > X$

* Iron-rich vermiculites show the higher refraction indices.

COMPOSITION AND STRUCTURE. Vermiculite has a trioctahedral, mica like structure with inter-layer cations enclosed between two sheets of water molecules (Fig. 6-6). Substitution in tetrahedral sites (Al^{3+} for Si^{4+}) is balanced by interlayer cations (usually Mg^{2+}) and trivalent ions in octahedral sites (Al^{3+} or Fe^{3+} for Mg^{2+}). Several solid solution series are suggested merely by diadochy of Mg^{2+}, Fe^{2+}, Al^{3+} and Fe^{3+} in octahedral coordination; Ni^{2+}, Cr^{3+}, Ti^{4+}, Li^{+}, and other ions may appear as major or minor constituents. In addition, Al $\rightleftharpoons$ Si substitution in tetrahedral sites is variable, and a wide variety of interlayer cations is possible. Polar water molecules surround interlayer cations in variable numbers, and hydration is highly variable.

PHYSICAL PROPERTIES. H $\simeq$ 1.5. Sp. Gr. $\simeq$ 2.4. In hand sample, vermiculite is light to dark brown, with bronze luster; rarely, green. Vermiculite is micaceous with flexible, inelastic sheets that separate or exfoliate, with great volume increase, when heated (expanding water layers).

COLOR AND PLEOCHROISM. Vermiculite is commonly colorless in section but may be weakly colored, with moderate pleochroism $Z = Y > X$: X = colorless, pale green; and $Y = Z$ = yellow-brown, green-brown.

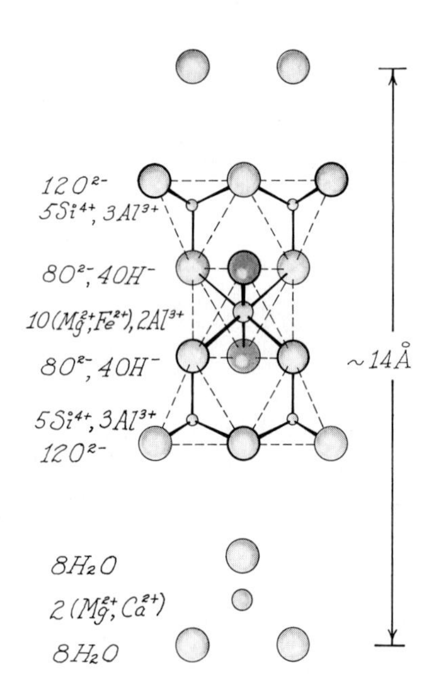

FIGURE 6-6. The idealized structure of vermiculite: Repeat unit on c is two tetrahedral layers and one octahedral layer, with interlayer cations and water molecules (~14 Å).

FORM. Vermiculite is a micaceous mineral ranging in form from tiny, submicroscopic flakes to large "books" strongly resembling biotite. It is very commonly interlayered with biotite (hydrobiotite) and chlorite. Fibrous vermiculite has been reported.

CLEAVAGE. Perfect basal {001} cleavage, typical of layered silicates, controls fragment orientation.

BIREFRINGENCE. Maximum birefringence, in sections normal to cleavage, is moderate (0.02–0.03), showing bright colors to middle second order in standard section. Birefringence of basal sections and cleavage plates is almost nil.

INTERFERENCE FIGURE. Vermiculite is biaxial negative, but very small $2V_x = 0°$ to $8°$ makes it sensibly uniaxial negative. The acute bisectrix (X) lies almost normal to basal cleavage, and thin cleavage plates yield good optic axis figures with color rings. Dispersion is weak $v > r$.

OPTICAL ORIENTATION. Vermiculite is monoclinic $b = Y$, $c \wedge X = -3°$ to $-6°$ and $a \wedge Z = +4°$ to $+1°$. Maximum extinction, in {010} sections, is near parallel (1° to 4°), and elongation is positive (length-slow), as X is essentially perpendicular to basal cleavage.

DISTINGUISHING FEATURES. Vermiculite shows little color and shows weaker pleochroism and lower birefringence than biotite. Birefringence of vermiculite is less than that of muscovite, phlogopite, or talc and greater than that of kaolin and most chlorites. Fine-grained vermiculite is not easily distinguished from montmorillonite or illite clays and may well require more than optical procedures for recognition.

ALTERATION. Vermiculite, itself, is an alteration product, usually of biotite or phlogopite. It shows an extensive range of hydration with corresponding range of physical properties and may alter to other clays with proper environment.

OCCURRENCE. Vermiculite forms from the weathering or hydrothermal alteration of biotite, phlogopite, and, rarely, other ferromagnesian minerals, and is commonly associated with serpentine, chlorite, and talc in altered peridotite, dunite, and other ultramafic intrusives. With this paragenesis, it commonly appears as large "books" pseudomorphic after mica. Hydrobiotite (interlayered vermiculite and biotite) represents the partial alteration of biotite, usually by hydrothermal or deuteric solutions.

Microscopic and submicroscopic flakes are common constituents of some soils and less common in marine sediments.

In metamorphic rocks, vermiculite appears in contact zones between acid and ultramafic intrusions and in carbonate rocks by the reaction of argillaceous impurities with Mg-carbonates.

REFERENCES: VERMICULITE

Barshad, I., and F. M. Kishk. 1969. Chemical composition of soil vermiculite clays as related to their genesis. *Contrib. Mineral. Petrol.*, *24*, 136–155.

Basset, W. A. 1959. Origin of the vermiculite deposit at Libby, Montana. *Amer. Mineral.*, *44*, 282–299.

Basset, W. A. 1963. The geology of vermiculite occurrences. *Clays and Clay Minerals, Proceedings of the 10th National Conference*, W. F. Bradley and S. W. Bailey (eds.). London: Pergamon Press.

De La Calle, C., J. Dubernat, H. Suguet, H. Pezerat, J. Gaultier, and J. Mamy. 1976. Crystal structure of two-layer Mg-vermiculites and Na, Ca-vermiculites. In *Proceedings of the International Clay Conference*, S. W. Bailey (ed.). Willamette, Ill.: Applied Publishing.

D'yakonov, Yu. S., and I. A. L'vova. 1967. Transformation of trioctahedral micas into vermiculite. *Dokl. Akad. Nauk SSSR*, *175*, 127–129.

Foster, M. D. 1963. Interpretation of the composition of vermiculites and hydrobiotites. In *Clays and Clay Minerals, Proceedings of the 10th National Conference*, W. F. Bradley and S. W. Bailey (eds.) London: Pergamon Press.

Weaver, C. E., and L. D. Pollard. 1973. *The Chemistry of Clay Minerals*. Amsterdam: Elsevier.

The Palygorskite-Sepiolite Series

Polygorskite	$(Mg,Al)_4Si_8O_{20}(OH)_2(OH_2)_4 \cdot 4H_2O$	MONOCLINIC $\angle\beta = 96^\circ$
Sepiolite	$(Mg,Al,Fe^{3+})_8Si_{12}O_{30}(OH)_4(OH_2)_4 \cdot 8H_2O$	ORTHORHOMBIC

	n_α*	n_γ	$n_\gamma - n_\alpha$	$2V_x$
Palygorskite	1.500–1.520	1.540–1.555	0.020–0.035	0°–60° (−)
Sepiolite	1.490–1.522	1.505–1.530	0.009–0.020	20°–70° (−)

Biaxial negative, $2V_x = 0^\circ$ to large (highly variable)

$c \wedge Z$ = small

Colorless to pale yellow or yellow-green in section, with weak pleochroism $Z = Y > X$

COMPOSITION AND STRUCTURE. Both palygorskite and sepiolite are fibrous clays with a structure of amphibolelike silica chains parallel to *c*, successively inverted and joined through oxygen ions along *b* and joined by cations, usually Mg^{2+} or Al^{3+}, in octahedral coordination on *a*. Water molecules (H_2O) occupy channels between chains, as in zeolites; some water molecules (OH_2) are bound to the free edge of octahedral sheets; and some $(OH)^-$ appears in the anion layers of octahedral sheets, as in most phyllosilicates. Palygorskite has two SiO_3 chains in the amphibolelike chain, and sepiolite has three.

Both minerals represent undefined segments of solid solution series between Mg-rich (trioctahedral) and Al-rich (dioctahedral) compositions. Significant Al^{3+}, or even Fe^{3+}, may replace Si^{4+} in tetrahedral coordination and the positive-charge deficiency is balanced by additional trivalent cations (Al^{3+} or Fe^{3+}) in octahedral sites or minor, exchangeable cations (for example, Ca^{2+} or Mg^{2+}) in open channels.

Octahedral cations are ideally Mg^{2+}, but much Al^{3+} is usually present and a Fe^{3+}-rich variety of sepiolite is recognized, as are Ni-rich and Mn-rich compositions.

* Refraction indices increase with increasing Mg and increasing dehydration.

PHYSICAL PROPERTIES. H = 2–2.5. Specific gravity ranges widely, from less than 1.0 to 2.6, varying with dehydration and porosity. In hand sample, both minerals are white to grayish, yellowish, or gray-green with earthy to waxy luster. Ni-sepiolites are apple-green.

COLOR AND PLEOCHROISM. Palygorskite and sepiolite are normally colorless as fragments or in section. Ferric varieties are colored and pleochroic, $Z = Y > X$; X = pale yellow; and $Y = Z$ = pale yellow-green.

FORM. Minerals of the palygorskite group are basically fibrous, from matted, tangled megascopic fibers to compact or aerated masses of felted submicroscopic fibers. Palygorskite has been described as cork, leather, cardboard, or skin; it is lightweight, supple, and tough. The electron microscope reveals laths or bundles of laths flattened on {100} and elongated on *c* and suggests a foliated, flaky β form.

CLEAVAGE. Good {110} cleavages augment a fibrous habit.

BIREFRINGENCE. Maximum birefringence ranges from mild (0.009–0.020) for sepiolite to moderate (0.020–0.035) for palygorskite and is seen in sections parallel to fiber length. Interference colors should range from mid-second-order to first-order gray; however, fiber thickness is commonly far below standard thin section thickness.

INTERFERENCE FIGURE. Palygorskite or sepiolite fibers are often much too small for interference figures. They are biaxial negative, with highly variable $2V_x = 0°–70°$. Some ferric sepiolites are biaxial positive. As Z lies near fiber length, fibers yield figures from acute bisectrix to flash figures, all symmetrical around the optic normal.

OPTICAL ORIENTATION. The Z optical direction lies on or near *c* yielding parallel extinction and positive (length-slow) elongation.

DISTINGUISHING FEATURES. Sepiolite shows lower birefringence than palygorskite and lower refractive index parallel to fiber length (n_γ). Palygorskite and sepiolite may show tangled masses of fibers, in contrast to foliated clays and other phyllosilicates. Most clays show positive relief, as does fibrous serpentine.

ALTERATION. Palygorskite and sepiolite are alteration products of Mg-carbonates and Mg-silicates (for example, serpentine, vermiculite, amphibole, and pyroxene). They may often be unstable with respect to the formation of montmorillonite; intense leaching may yield kaolin. With increased temperature, palygorskite and sepiolite become enstatite, cristobalite, and sillimanite.

OCCURRENCE. In soils and sediments, palygorskite and sepiolite appear with carbonates in lacustrine marls and dolomites and desert sediments. They seem to favor arid environments and are common in deposits of playa or saline lakes, where they appear with montmorillonites and illites but not kaolin or leached clays. They are very soluble in acid environments and rarely occur together.

Palygorskite and sepiolite also commonly appear in veins or cavities in carbonate rocks or mafic igneous rocks, due to hydrothermal alteration or weathering of serpentine, vermiculite, pyroxene, or amphibole. They may result from the reaction of Mg-solutions with colloidal silica liberated by surface weathering. Most common associates are calcite, dolomite, talc, chlorite, quartz, chalcedony, and opal.

REFERENCES: THE PALYGORSKITE-SEPIOLITE SERIES

Bradley, W. F. 1940. Structure of attapulgite. *Amer. Mineral.*, *25*, 405–410.

Brown, G. 1961. *The X-ray Identification and Crystal Structures of Clay Minerals*. London: Mineralogical Society.

Christ, G. L., J. C. Hathaway, P. B. Hostetler, and A. O. Shephard. 1969. Palygorskite: New X-ray data. *Amer. Mineral.*, *54*, 198–205.

Drits, V. A., and V. A. Aleksandrova. 1966. The crystallochemical nature of palygorskite. *Zap. Vses. Mineral. Obshchest.*, *95*, 551–560.

Grim, R. E. 1953. *Clay Mineralogy*. New York: McGraw-Hill.

Lowry, D. C. 1964. Palygorskite in a cave in New Zealand. *New Zealand Jour. Geol. Geophys.*, *7*, 917.

Nagy, B., and W. F. Bradley. 1955. The structure scheme of sepiolite. *Amer. Mineral.*, *40*, 885–892.

Serna, C., G. E. Van Scoyoc, and J. L. Ahlrichs. 1977. Hydroxyl groups and water in palygorskite. *Amer. Mineral.*, *62*, 784–792.

Stephen, I. 1954. An occurrence of palygorskite in the Shetland Isles, *Mineral. Mag.*, *30*, 471–480.

Weaver, C. E., and L. D. Polland. 1973. *The Chemistry of Clay Minerals*. Amsterdam: Elsevier.

The Mica Group

Micas are a diverse mineral group of wide geologic occurrence, but they are unified by certain common structural characteristics. They are sheet structures, in which each silicate tetrahedron shares the oxygen ions of one of its triangular bases with other tetrahedra, forming a layer as shown in Fig. 6-1. Octahedrally coordinated cations are "sandwiched" between the opposing apical oxygen ions of two such tetrahedral layers. The hydroxyl groups $[(OH)^-]$ that occupy anion sites at the centers of the rings of apical oxygens are also bonded to the cations of the octahedral layer, so that each octahedron consists of the central cation, four O^{2-} and two $(OH)^-$. These three-layer sheets are the "mica layers" and are bound to one another by alkali or alkaline earth ions known as *interlayer cations.* The presence of the interlayer cations distinguishes micas from some kinds of clays. Fig. 6-7 depicts the arrangement.

The general chemical formula for the mica group is $W_2Y_{4-6}Z_8O_{20}(OH)_4$, where W represents the interlayer cations, Y the octahedral cations, and Z the tetrahedral cations. Signifi-

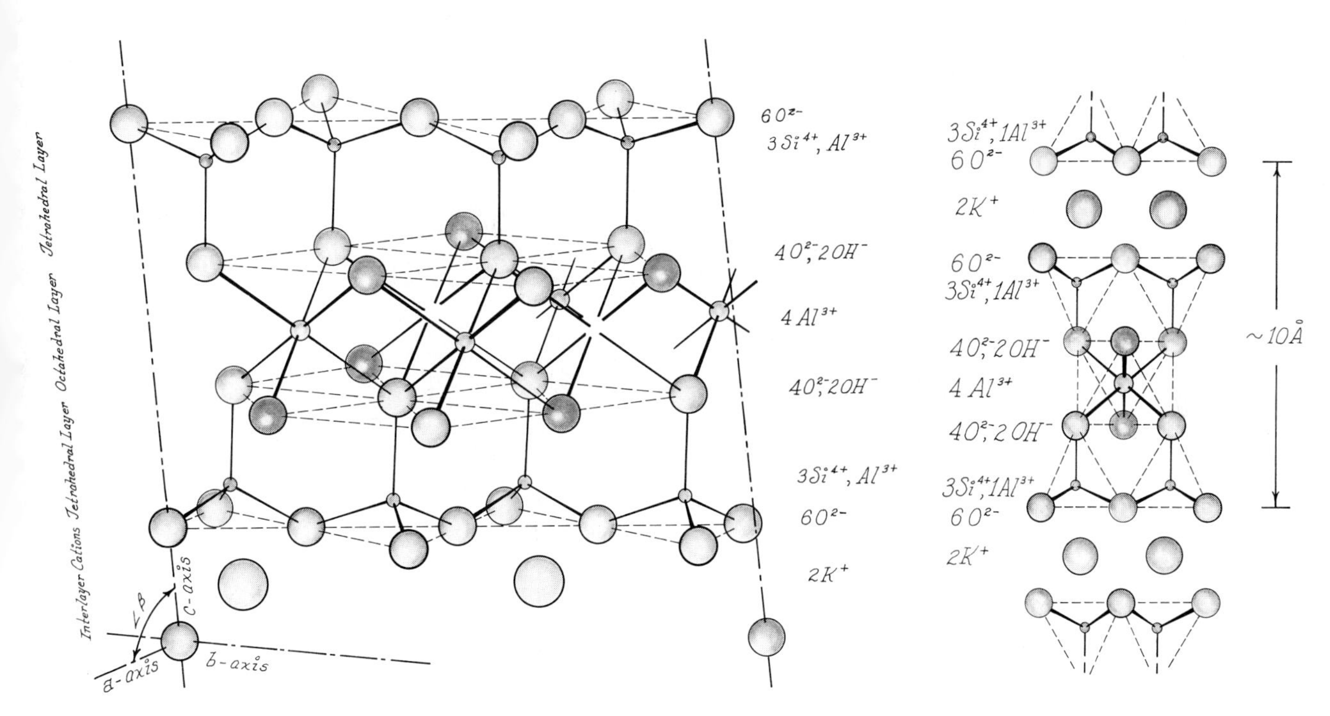

FIGURE 6-7. The idealized structure of muscovite: Repeat unit on c is two tetrahedral layers and one octahedral (dioctahedral) layer with interlayer cations (~ 10 Å).

cant fluorine (F^-) may replace hydroxyl ions in natural micas. As is typical of many silicate minerals, some of the "silicate" tetrahedra contain Al^{3+} instead of Si^{4+}. In most micas, the Al:Si ratio is 2:6, requiring a monovalent interlayer cation; if that cation is K^+, we have a "*normal mica*," and if Na^+, a "*paragonite mica*." Ratios of 4Al:4Si require a divalent interlayer cation; the "*brittle micas*" have Ca^{2+} between the sheets. Ratios between 4:4 and 2:6 are fairly common, with charge balance being effected by adjustments in the *W*- and/or *Y*-site chemistry. Micas more rich in silicon than 2Al:6Si are called *phengites*. Rarely, minor Fe^{3+} or Ti^{4+} may substitute for Si^{4+} in tetrahedral coordination.

The greatest chemical variation is found in the octahedral sites, where Li^+, Fe^{2+}, Mg^{2+}, Mn^{2+}, Fe^{3+}, Al^{3+}, Cr^{3+}, and V^{3+} may be present in major amounts. If all three of the available octahedral sites are filled, the mineral is called a *trioctahedral* mica; if only two of the three available sites are filled, the mineral is a *dioctahedral* mica. Although most micas fall into one or the other of these groups, those intermediate between them are not unknown.

The common interlayer cations are K^+, Na^+, and Ca^{2+}, but other alkalis and alkaline earths (Rb^+, Cs^+, Sr^{2+}, and Ba^{2+}) may be rare to minor substituents. The distinctive, perfect basal cleavage of all micas results from these weak bonds between mica layers and interlayer cations, and brittle micas are characterized by more difficult cleavage. Submicroscopic crinkling of the sheets causes the typical "bird's-eye maple" extinction seen in micas and other sheet silicates (Fig. 6-8).

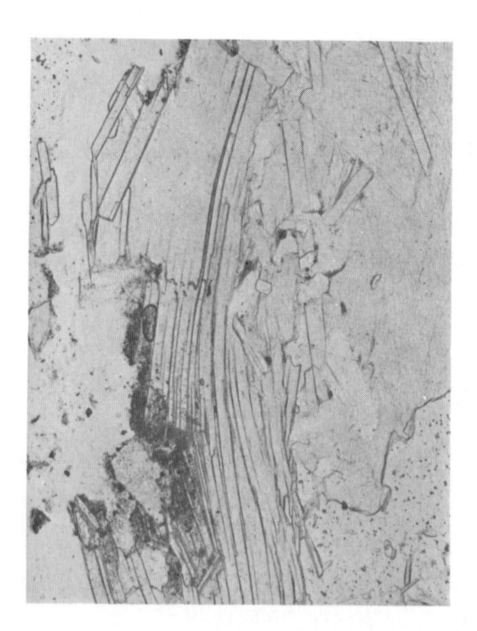

FIGURE 6-8. Muscovite is colorless and shows a single cleavage and "bird's-eye maple" extinction between crossed nicols.

The thickness of a single mica layer, as shown schematically in Fig. 6-7, is approximately 10 Å, but the structural repeat distance—the (001) *d*-spacing—may be 10 Å, 20 Å, 30 Å, and so on, depending on the stacking sequence of successive layers. The stacking sequence may be perfectly ordered (that is, a constant angular rotation of each layer relative to the one below it) to completely disordered, allowing numerous mica polytypes, which are optically indistinguishable.* Although most mica polytypes are monoclinic (the symmetry of a single mica layer), appropriate stacking can produce hexagonal, trigonal, orthorhombic, or even triclinic symmetry. Most biotites and lithia micas appear as simple 1**M** polytypes (*c* repeat distance is one layer, or 10 Å, resulting in monoclinic symmetry) with occasional 2**M** (two-layer monoclinic), 2**Or** (two-layer orthorhombic), and 3**T** (three-layer trigonal) polytypes. Muscovites and paragonites are usually 2**M** polytypes. The simple hexagonal (6**H**) polytype has not been reported.

The following list summarizes the common chemical constituents of micas. Table 6-2 shows the compositions of specific micas of the major groups, and Table 6-3 gives the optical properties of micas.

Interlayer Cations	
Normal micas	K_2
Paragonite micas	Na_2
Brittle micas	Ca_2
Cations in Tetrahedral Coordination	
Talcs	Si_8
Phengite micas	Si_7Al
Normal micas	Si_6Al_2
Intermediate micas	Si_5Al_3
Brittle micas	Si_4Al_4
Cations in Octahedral Coordination	
Dioctahedral micas	R_4^{3+}
Trioctahedral micas	R_6^{2+}

Phengites are micas richer in silicon than Si_6Al_2, and intermediate micas contain less silicon than the normal ratio (Si_6Al_2) and more than the brittle micas (Si_4Al_4).

* Polytypes are a special class of polymorphs in which the structural differences arise from the stacking of essentially identical layers in different ways.

TABLE 6-2. Composition of Specific Micas

	Interlayer Cations	Octahedral Cations	Tetrahedral Cations	Anions
Muscovite Micas (Dioctahedral)				
Muscovite	K_2	Al_4	Si_6Al_2	$O_{20}(OH,F)_4$
Roscoelite	K_2	$(Al,V)_4$	Si_6Al_2	$O_{20}(OH,F)_4$
Oellacherite	$(K,Ba)_2$	$(Al,Mg)_4$	Si_6Al_2	$O_{20}(OH,F)_4$
Fuchsite	K_2	$(Al,Cr)_4$	Si_6Al_2	$O_{20}(OH,F)_4$
Manganoan muscovite	K_2	$Al_2(Mn,Mg,Fe^{2+})_3$	Si_6Al_2	$O_{20}(OH,F)_4$
Phengite	K_2	$Al_3(Mg,Fe^{2+})$	Si_7Al	$O_{20}(OH,F)_4$
Alurgite	K_2	$Al_3(Mn,Mg,Fe^{2+})$	Si_7Al	$O_{20}(OH,F)_4$
Mariposite	K_2	$(Al,Cr)_3(Mg,Fe^{2+})$	Si_7Al	$O_{20}(OH,F)_4$
Biotite Micas (Trioctahedral)				
Phlogopite	K_2	Mg_6	Si_6Al_2	$O_{20}(OH,F)_4$
Annite	K_2	Fe_6^{2+}	Si_6Al_2	$O_{20}(OH,F)_4$
Eastonite	K_2	Mg_5Al	Si_5Al_3	$O_{20}(OH,F)_4$
Siderophyllite	K_2	$Fe_5^{2+}Al$	Si_5Al_3	$O_{20}(OH,F)_4$
Lepidomelane	K_2	$(Fe^{2+},Fe^{3+})_6$	$(Si,Al)_8$	$O_{20}(OH,F)_4$
Lithian Micas (Trioctahedral)				
Lepidolite	K_2	$(Li_{4-3}Al_{2-3})$	$Si_{8-6}Al_{0-2}$	$O_{20}(OH,F)_4$
Polylithionite	K_2	Li_4Al_2	Si_8	$O_{20}(OH,F)_4$
Paucilithionite	K_2	Li_3Al_3	Si_6Al_2	$O_{20}(OH,F)_4$
Lithian muscovite	K_2	Li_2Al_3	Si_7Al	$O_{20}(OH,F)_4$
Taeniolite	K_2	Li_2Mg_4	Si_8	$O_{20}(OH,F)_4$
Zinnwaldite	K_2	$Fe_{1-2}^{2+}Li_{2-3}Al_2$	$Si_{6-7}Al_{2-1}$	$O_{20}(OH,F)_4$
Brittle Micas				
Margarite (Dioctahedral)	Ca_2	Al_4	Si_4Al_4	$O_{20}(OH,F)_4$
Clintonite-xanthophyllite (Trioctahedral)	Ca_2	$Mg_{4.6}Al_{1.4}$	$Si_{2.5}Al_{5.5}$	$O_{20}(OH,F)_4$
Paragonite Micas (Dioctahedral)				
Paragonite	Na_2	Al_4	Si_6Al_2	$O_{20}(OH,F)_4$

TABLE 6-3. Optical Properties of the Micas

n_β	n_α	n_γ	$n_\gamma - n_\alpha$	Mineral	System	$2V$	Optic Sign	Cleavage	Optical Orientation	Color in Sections	Physical Properties
1.55–1.59	1.52–1.55	1.55–1.59	0.018–0.038	Lepidolite $K_2(Li_{4-2}Al_{2-3})(Si_{8-6}Al_{0-2})O_{20}(OH,F)_2$	Monoclinic $\angle\beta = 100°$	0°–58° $r > v$ weak	–	{001} very perfect	$b = Y$; $a \wedge Z = 0°$ to $+7°$; $c \wedge X = -10°$ to $-3°$	Colorless	H = $2\frac{1}{2}$–4. Sp. Gr. = 2.2–3.3. Colorless, gray, violet, rose-red, yellow-green
1.57–1.59	1.53–1.56	1.57–1.59	0.035	Zinnwaldite $K_2(Fe_{1-2}^{2+}Li_{2-3}Al_2)\cdot(Si_{6-7}Al_{2-1})\cdot O_{20}(F,OH)_4$	Monoclinic $\angle\beta = 100°$	0°–40° $r > v$ weak	–	{001} very perfect	$b = Y$; $a \wedge Z = 0°$ to $+2°$; $c \wedge X = -10°$ to $-8°$	X = colorless; Y = gray brown; Z = gray brown	H ≃ 3. Sp. Gr. ≃ 3.0. Yellow, dark brown, gray
1.56–1.64	1.53–1.59	1.56–1.64	0.028–0.049	Phlogopite $KMg_3(Si_3Al)\cdot O_{10}(OH)_2$	Monoclinic $\angle\beta = 100°$	0°–15° $v > r$	–	{001} very perfect	$b = Y$; $c \wedge X = -10°$ to $-5°$; $a \wedge Z = 0°$ to $+5°$	X = colorless; $Y = Z$ = yellow, pale green, red-brown	H = 2–$2\frac{1}{2}$. Sp. Gr. = 2.7–2.9. Golden brown, red-brown, green
1.58–1.61	1.55–1.57	1.59–1.62	0.036–0.049	Muscovite $KAl_2(Si_3Al)\cdot O_{10}(OH)_2$	Monoclinic $\angle\beta = 95°30'$	30°–47° $r > v$ moderate	–	{001} very perfect	$b = Z$; $c \wedge X = 0°$ to $-5°$; $a \wedge Y = +1°$ to $+3°$	Colorless	H = $2\frac{1}{2}$. Sp. Gr. = 2.77–2.88. Colorless to pale brown, green, or red
1.59–1.61	1.56–1.58	1.60–1.61	0.028–0.038	Paragonite $NaAl_2(Si_3Al)\cdot O_{10}(OH)_2$	Monoclinic $\angle\beta = 95°$	0°–40° $r > v$ moderate	–	{001} very perfect	$b = Z$; $a \simeq Y$; $c \wedge X \simeq -5°$	Colorless	H = $2\frac{1}{2}$. Sp. Gr. = 2.85. Colorless, pale yellow, and so on
1.61–1.65	1.56–1.61	1.61–1.65	0.014–0.032	Glauconite $(K,H_3O)_2\cdot(Fe^{+3},Al,Fe^{+2},Mg)_4\cdot(Si_{7-7.5}Al_{1-0.5})\cdot O_{20}OH)_4$	Monoclinic $\angle\beta \simeq 100°$	0°–20° $r > v$ weak	–	{001} very perfect	$b = Y$; $a \simeq Z$; $c \wedge X \simeq -10°$	X = yellow-green; $Y = Z$ = green	H = 2. Sp. Gr. = 2.4–3.0. Blue-green, olive-green
1.61–1.70	1.56–1.63	1.61–1.70	0.04–0.07	Biotite $K_2(Mg,Fe)_{6-5}Al_{0-1}\cdot(Si_{6-5}Al_{2-3})\cdot O_{20}(OH,F)_2$	Monoclinic $\angle\beta = 90°–100°$	0°–25° $v > r$ moderate	–	{001} very perfect	$b = Y$; $c \wedge X = -10°$ to $+9°$; $a \wedge Z = 0°$ to $+9°$	X = pale green, tan; Y = green, brown; Z = dark green, brown	H = $2\frac{1}{2}$–3. Sp. Gr. = 2.7–3.3. Black, dark brown, dark green
1.62–1.65	1.59–1.64	1.63–1.65	0.010–0.032	Margarite $CaAl_2(Si_2Al_2)\cdot O_{10}(OH)_2$	Monoclinic $\angle\beta = 95°$	26°–67° $v > r$ moderate	–	{001} perfect	$b = Z$; $a \wedge Y = -6°$ to $-8°$; $c \wedge X = -11°$ to $-13°$	Colorless	H = $3\frac{1}{2}$–$4\frac{1}{2}$. Sp. Gr. = 3.0–3.1. Flesh pink, pale yellow, pale green.
1.65–1.66	1.64–1.65	1.65–1.66	0.012	Clintonite $Ca_2(Mg_{4.6}Al_{1.4})\cdot(Si_{2.5}Al_{5.5})\cdot O_{20}(OH)_4$	Monoclinic $\angle\beta = 100°$	32° $v > r$ weak horizontal bxa	–	{001} perfect	$b = Z$; $a \simeq Y$; $c \wedge X \simeq -5°$	X = colorless; $Y = Z$ = pale green, brown	H = $3\frac{1}{2}$–6. Sp. Gr. = 3.0–3.1. Colorless, yellow, green, red-brown
1.65–1.66	1.64–1.65	1.65–1.66	0.012	Xanthophyllite $Ca_2(Mg_{4.6}Al_{1.4})\cdot(Si_{2.5}Al_{5.5})\cdot O_{20}(OH)_4$	Monoclinic $\angle\beta = 100°$	0°–23° $v > r$ weak inclined bxa	–	{001} perfect	$b = Y$; $a \simeq Z$; $c \wedge X \simeq -10°$	X = colorless; $Y = Z$ = pale green, brown	H = $3\frac{1}{2}$–6. Sp. Gr. = 3.0–3.1. Colorless, yellow, green, red-brown

Muscovite

$KAl_2(Si_3Al)O_{10}(OH)_2$

MONOCLINIC
$\angle\beta = 95°30'$

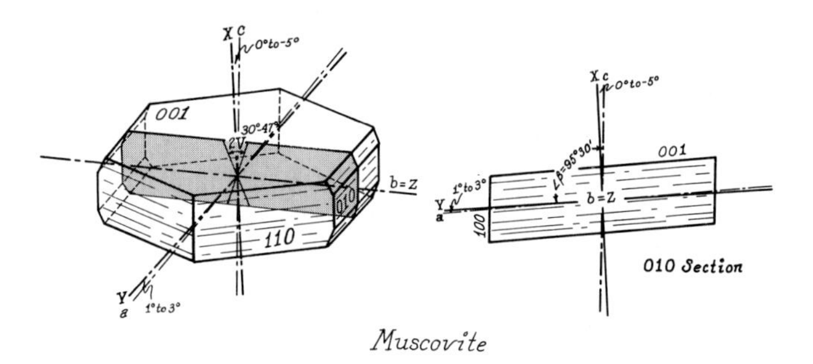

Muscovite

$n_\alpha = 1.552$–1.574*
$n_\beta = 1.582$–1.610
$n_\gamma = 1.587$–1.616
$n_\gamma - n_\alpha = 0.036$–$0.049$*
Biaxial negative, $2V_x = 30°$–$47°$
$b = Z, c \wedge X = 0°$ to $-5°$, $a \wedge Y = +1°$ to $+3°$
$r > v$ distinct, horizontal bisectrix dispersion
Colorless in section

COMPOSITION. Muscovite is the basic aluminum-potash mica $KAl_2(Si_3Al)O_{10}(OH)_2$, and many specimens closely approach ideal composition. Sodium replaces potassium in muscovite to form a distinct mineral *paragonite* with similar optical properties, but sodium is a minor impurity in muscovite. Large alkalis (Rb and Cs) and alkaline earths (Ca, Sr, and Ba) may appear as minor impurities replacing potassium, and a Ba-rich muscovite has been called *oellacherite.*

In the fine-grained, secondary muscovites of sedimentary and metamorphic rocks (hydromuscovite), $(H_3O)^+$ ions may replace K^+ as interlayer cations.

In octahedral coordination, vanadium (V^{3+}) may appear as a very major constituent (>15 percent V_2O_5 in complete analysis) to form *roscoelite,* which is probably a mineral species distinct from muscovite (no complete solid solution series). Lithium (Li^+) and chromium (Cr^{3+}) are often significant, forming *lithian muscovite* and *fuchsite,* respectively, and Fe^{2+}, Fe^{3+}, and Mn^{2+} (*manganoan muscovite*) may appear as minor constituents. *Mariposite* and *alurgite* appear to be phengite (see p. 268) equivalents of fuchsite and manganoan muscovite respectively.

Inclusions of accessory minerals (such as apatite, zircon, opaques, and garnet) are common in muscovite, and magnetite or hematite appears by exsolution at low temperatures. Fluorine, F^-, may replace some hydroxyl ions, $(OH)^-$.

PHYSICAL PROPERTIES. $H = 2\frac{1}{2}$–4, Sp. Gr. = 2.77–2.88. Hardness varies with composition and crystallographic direction (for example, hardness is greater normal to cleavage). Hand samples are generally colorless to pale brown, green, or red and is greatly influenced by minor impurities. Ferric iron is commonly responsible for brown colors, ferrous iron for greens, and manganese for pinks and reds. Fuchsite (Cr) is emerald-green, and manganoan muscovite is blue or violet. Color zoning is quite common as concentric hexagonal bands on cleavage surfaces.

COLOR AND PLEOCHROISM. In section or as fragments, muscovite is normally colorless. Fuchsite is pleochroic in greens $Z > Y > X$.

FORM. Muscovite is, of course, micaceous, showing one set of cleavage planes often bent by compression, in section, and highly oriented flakes as fragments. Secondary muscovite, or *sericite,* usually appears as fine scales or fibrous aggregates.

CLEAVAGE. Perfect basal {001} cleavage controls fragment orientation.

BIREFRINGENCE. Maximum birefringence, seen in {100} sections is moderate to strong (0.036–0.049), and all sections normal to cleavage show birefringence near maximum. Interference colors, in standard section are mostly brilliant second- and third-order colors. Basal sections and cleavage fragments show only weak birefringence.

---

* Ideal composition yields minimum refractive index and maximum $2V_x$. Refractive index increases and $2V_x$ decreases with Fe^{2+}, Fe^{3+}, Mn^{2+}, Cr^{3+}, V^{3+}, and Ti^{4+}. Phengites and hydromuscovites also tend to have small $2V$ and large indices of refraction.

Note that all micas in all sections show a characteristic granular or pebbled surface, between crossed nicols, which is especially apparent 1° or 2° from extinction.

TWINNING. The complex mica twin is formed about a twin axis lying in the {001} composition plane, and normal twinning on {110} is occasionally noted.

INTERFERENCE FIGURE. Basal sections or simple sheets or flakes of muscovite yield excellent, essentially centered, biaxial negative, acute bisectrix figures. Isogyres are distinct in a nearly white to colorful field of encircling isochromes, depending on thickness; melatopes are usually well separated yet well within the visible field, because of moderate to small $2V$. Most muscovite shows $2V_x$ between 30° and 47°, with values near 47 suggesting near theoretical composition. Phengite varieties favor small $2V_x$, and some structural forms are truly uniaxial ($2V_x = 0$). Sections showing distinct cleavage show moderate to high birefringence and poor interference figures. Optic axis dispersion is rather distinct $r > v$, with mostly theoretical, horizontal bisectrix dispersion.

OPTICAL ORIENTATION. Since $b = Z$ and $c \wedge X = 0°$ to $-5°$, the XZ (optic) plane contains b, lies no more than 5° from the cb {100} plane, and the acute bisectrix X is within 5° of being normal to cleavage surfaces {001}. Maximum extinction angle to cleavage traces is 0° to 3°, seen on {010} sections, and other sections show essentially parallel extinction. Bent or wrinkled sheets cause wavy extinction. Elongation is positive (length-slow) to cleavage traces, since X lies essentially perpendicular to cleavage.

DISTINGUISHING FEATURES. All micas show a characteristic pebble surface texture ("bird's-eye maple" structure), which is, unfortunately, shared by other sheet silicates (such as talc, pyrophyllite, vermiculite, and chlorite). Lack of color, single cleavage, and brilliant interference colors are characteristic, but not distinctive, of muscovite, and muscovite is easily confused with nearly any colorless phyllosilicate. Talc, vermiculite, colorless phlogopite, and clintonite show $2V_x <$ 30°, pyrophyllite $2V_x > 50°$, colorless chlorites show low birefringence, and lepidolite has refractive indices below those of muscovite. Paragonite is optically indistinguishable from muscovite and is more common than was once thought; X-ray studies may be required to distinguish sericite from fine aggregates of talc or pyrophyllite. Talc is associated with Mg-rich minerals, and pyrophyllite and paragonite are most commonly associated with metamorphic rocks, sediments, or alteration of K-poor minerals.

ALTERATION. Muscovite is one of the minerals most resistant to weathering and general alteration and, as sericite, is itself a stable alteration product of other minerals. Hydration or weathering may convert muscovite to illites and hydromuscovite, which are recognized by lower birefringence and which in turn may weather to montmorillonite or kaolin clays.

OCCURRENCE. Muscovite is more characteristic of metamorphic rocks than of igneous environments, where alumina content seldom exceeds that required by feldspar. It appears, however, in granites, usually associated with biotite and microcline, where it may crystallize as large crystals at high pressures directly from magma or may form in the solid state as small flakes in feldspars by late deuteric or metasomatic reactions. Greisen is a fine-grained, quartz-mica rock characteristic of fluorine metasomatism of granites.

Muscovite, with quartz and alkali feldspars, is a universal constituent of pegmatites. It appears both as large books near the pegmatite core and as fine, scaley aggregates in the outer zones, where it may have replaced aluminous minerals. Muscovite is the common mica of aplite dikes and appears in hydrothermal veins.

Primary muscovite is a "plutonic" mineral favored by high pressure, especially directed pressure, and is not found in volcanic rocks.

Metamorphic rocks of nearly all kinds and grades contain muscovite. In zones of regional metamorphism, it is one of the first minerals to appear, as visible scales in phyllites, through recrystallization of existing sericite. In chlorite, biotite, garnet, staurolite, and kyanite grade rocks (greenschist and amphibolite facies), muscovite is typical of schists and gneisses, but at higher grades muscovite tends to dissociate in favor of microcline and sillimanite. Metamorphism of siliceous sediments and quartz-feldspar rocks yield quartz-muscovite schists at low grades returning to quartz-feldspar rocks at higher grades. Contact metamorphism of argillaceous sediments yields muscovite, or sericite, at medium grades, which tend to dissociate to K-feldspar and andalusite with increasing metamorphism.

Sericite is characteristic of retrograde metamorphism and forms abundantly by hydrothermal alteration of feldspars in igneous rocks. Ca-feldspar is normally replaced before Na-feldspar, and K-feldspar is most resistant to sericitization. Sericitization of siliceous wall rock commonly accompanies introduction of hydrothermal veins in the mesothermal temperature range.

In argillaceous sedimentary rocks, muscovite occurs abundantly as sericite interlayered with illite, montmorillonite, and kaolinite clays.

REFERENCES: MUSCOVITE

Bairakov, V. V. 1967. Chromium-bearing muscovite from a biotite gneiss xenolith. *Dopov. Akad. Nauk Ukr. RSR.*, Series B29, 193–196.

Burnham, C. W., and E. W. Radoslovich. 1964. Refinement of the crystal structures of co-existing muscovite and paragonite. *Carnegie Inst. Wash. Year Book*, *63*, 232–236.

Emiliani, F. 1956. Crystal chemical studies of micas. Part I. Chemical and optical study of some muscovites of the pegmatitic orthogneiss of Val Venosta. *Acta Geol. Alpina*, *6*, 79–104.

Frey, M. 1969. Metamorphism of the Keuper from the Tafeljura to the Lukmanier region. Alteration of the clayey-marly rocks from the region of diagenesis to the staurolite zone. *Beitr. Geol. Karte. Schweiz.*, *137*, 1–160.

Guven, N. 1971. Structural factors controlling stacking sequences in dioctahedral micas. *Clays and Clay Minerals*, *19*, 159–165.

Guven, N., and C. W. Burnham. 1967. The crystal structure of 3T muscovite. *Carnegie Inst. Wash. Year Book*, *65*, 290–293.

Müller, G. 1966. The relationships among the chemical composition, refractive indices, and density of co-existing biotite, muscovite, and chlorite in granitic rocks. *Contrib. Mineral. Petrol.*, *12*, 173–191.

Radoslovich, E. W. 1960. The structure of muscovite, $KAl_2(Si_3Al)O_{10}(OH)_2$. *Acta Crystallogr.*, *13*, 919–932.

Snetsinger, K. G. 1966. Barium-vanadium muscovite and vanadium tourmaline from Mariposa County, California. *Amer. Mineral.*, *51*, 1623–1639.

Biotite

$K_2(Mg,Fe^{2+})_{6-5}Al_{0-1}(Si_{6-5}Al_{2-3})O_{20}(OH,F)_4$

MONOCLINIC
$\angle \beta = 90°–100°$

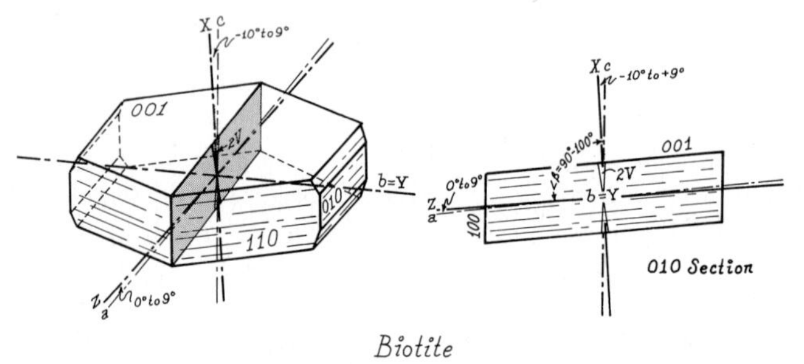

$n_\alpha = 1.565–1.625$*

$n_\beta = 1.605–1.696$

$n_\gamma = 1.605–1.696$

$n_\gamma - n_\alpha = 0.04–0.07$*

Biaxial negative, $2V_x = 0°–25°$*

$b = Y$, $c \wedge X = -10°$ to $-9°$, $a \wedge Z = 0°$ to $-9°$

$v > r$ weak (rarely $r > v$), inclined bisectrix dispersion

Strong color and absorption, $Z \simeq Y \gg X$

X = pale yellow, pale green, pale brown;
Y ≃ Z = dark brown, dark green, dark red-brown

* Index of refraction increases with Fe^{2+} and Mn, at a more accelerated rate with Ti and Fe^{3+} and is lowered by F. Complete replacement of Mg^{2+} by Fe^{2+} raises n_γ about 0.15 (see Fig. 6-9). Fe^{3+} and Ti^{4+} increase birefringence and $2V$.

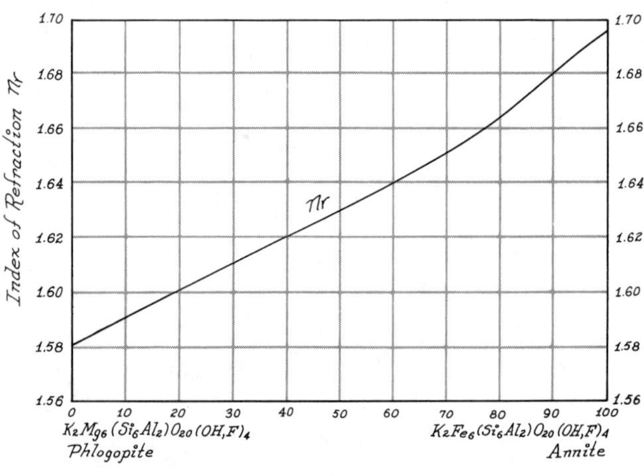

FIGURE 6-9. Variation of the index of refraction (n_γ) in the phlogopite-annite series (biotite).

COMPOSITION. Four ideal end members appear to join in complete mutual solid solution:

Phlogopite	$K_2Mg_6(Si_6Al_2)O_{20}(OH,F)_4$
Annite	$K_2Fe_6{}^{2+}(Si_6Al_2)O_{20}(OH,F)_4$
Eastonite	$K_2(Mg_5Al)(Si_5Al_3)O_{20}(OH,F)_4$
Siderophyllite	$K_2(Fe_5Al)(Si_5Al_3)O_{20}(OH,F)_4$

The four-end-member system is arbitrarily divided into magnesium-rich compositions (Mg:Fe > 2:1), collectively called *phlogopites* and iron-rich compositions (Mg:Fe < 2:1), collectively called *biotites*. Unusually iron-rich (ferrous plus ferric) biotites are called *lepidomelanes*.

Interlayer cations are largely K^+, with significant Na^+ and very minor Rb^+, Ca^{2+}, Ba^{2+}, and Cs^+.

Cations in tetrahedral coordination are usually silicon and aluminum near the ratio Si_6Al_2, but additional aluminum may replace silicon to Si_5Al_3.

In octahedral coordination, Mg^{2+} and Fe^{2+} are the common cations but Al^{3+}, Fe^{3+}, Ti^{4+}, or Mn^{2+} may be present in very significant amounts and minor Li^+ appears. Fluorine, or even chlorine, may replace much $(OH)^-$, especially in Ti-rich biotites, and biotites commonly contain less than the ideal $(OH,F)_4$ requiring additional oxygen cations and additional balancing cation charge.

Inclusions of accessory minerals (for example, apatite, sphene, zircon, and opaques) are to be expected in biotite grains. Tiny acicular crystals of rutile may produce six-ray asterism in phlogopite.

PHYSICAL PROPERTIES. H = $2\frac{1}{2}$–3, Sp. Gr. = 2.7–3.3. Maximum hardness is normal to cleavage and minimum hardness in {001} parallel to *a*. Color, in hand sample, is black to deep brown or green. Brown is favored by Fe^{3+} and Ti, and green results from Fe^{2+} content. Color zoning is uncommon in biotite.

COLOR AND PLEOCHROISM. Biotite is highly colored and strongly pleochroic in shades of brown or green. Absorption is strong (extreme with high iron) to light waves vibrating parallel to cleavage {001} and weak normal to cleavage ($Z \simeq Y \gg X$)—sections are dark parallel to vibration plane of lower nicol (see Fig. 6-10). Basal sheets show high absorption. In the absence of Fe^{3+}, Ti, and Mn, biotite is pleochroic in greens, darkening with increasing Fe^{2+}. Most commonly, sufficient Fe^{3+} is present to produce olive-green or brown shades, and titaniferous varieties are pleochroic in red-brown. X = pale green, pale tan, pale yellow; Y = blue-green, deep brown, red-brown; and Z = dark olive-green, dark brown, red-brown. Lepidomelane shows extreme absorption from nearly colorless to opaque.

FORM. Perfect basal cleavage gives biotite pronounced foliated form and dictates fragment orientation. Cleavage flakes and basal crystal sections commonly show pseudohexagonal outline. Sections may show wavy extinction of cleavages distorted by stress or ragged aggregates of fine-grained biotite.

CLEAVAGE. Perfect basal {001} cleavage controls fragment orientation.

BIREFRINGENCE. Maximum birefringence, seen in sections normal to cleavage, is strong (0.04–0.07) making bright, third-order colors most characteristic of sections showing sharp cleavage traces. Ti^{4+} and Fe^{3+} are responsible for upper limits of birefringence and fourth-order interference colors. Birefringence of cleavage plates and basal sections is almost nil. All micas show characteristic granular or pebbled ("bird's-eye maple") surface, between crossed nicols near the extinction position.

TWINNING. Twinning is rarely obvious in biotite micas, but normal twinning on {110} and complex twinning on {001} (mica law) are not uncommon.

ZONING. Color zoning may be seen in phlogopite as either alternating light and dark bands or darker central core.

INTERFERENCE FIGURE. Cleavage plates and basal sections of biotite and phlogopite micas yield centered, negative, acute bisectrix figures easily mistaken for uniaxial. Ti^{4+} and Fe^{3+} varieties may yield $2V_x$ near 20°, but common varieties show little isogyre separation. Inherent color absorption may darken the figure background and associated isochromatic rings.

Optic axis dispersion is generally weak, and usually $v > r$; however, some Mg-rich biotites show $r > v$. Weak bisectrix dispersion should be inclined. Phlogopite shows $v > r$, which may be very pronounced.

OPTICAL ORIENTATION. In the common monoclinic polytype (1**M**), $Y = b$ and $\angle\beta$ is near 100°. Z is very near a ($a \wedge Z$ = 0° to 9°) and X ranges near c ($c \wedge X$ = 0° to −10°). Most biotites are 1**M** polytypes with the optic plane parallel to {010}, but a few are 2**M** polytypes with the optic plane normal to {010}, and rare trigonal forms have no optic plane. Maximum extinction angle is 0°–9° from cleavage in {010} sections. Wavy extinction results from bent cleavages. Elongation is positive (length-slow), since the fast wave (X) vibrates essentially normal to cleavage.

DISTINGUISHING FEATURES. Biotite has greater absorption, darker color, and higher refractive index than other micas. Phlogopite varieties may show weak color and low index, in section, and are distinguished from muscovite by small $2V$ and from lepidolite by occurrence. Vermiculite shows lower refraction index and birefringence, and chlorites show much

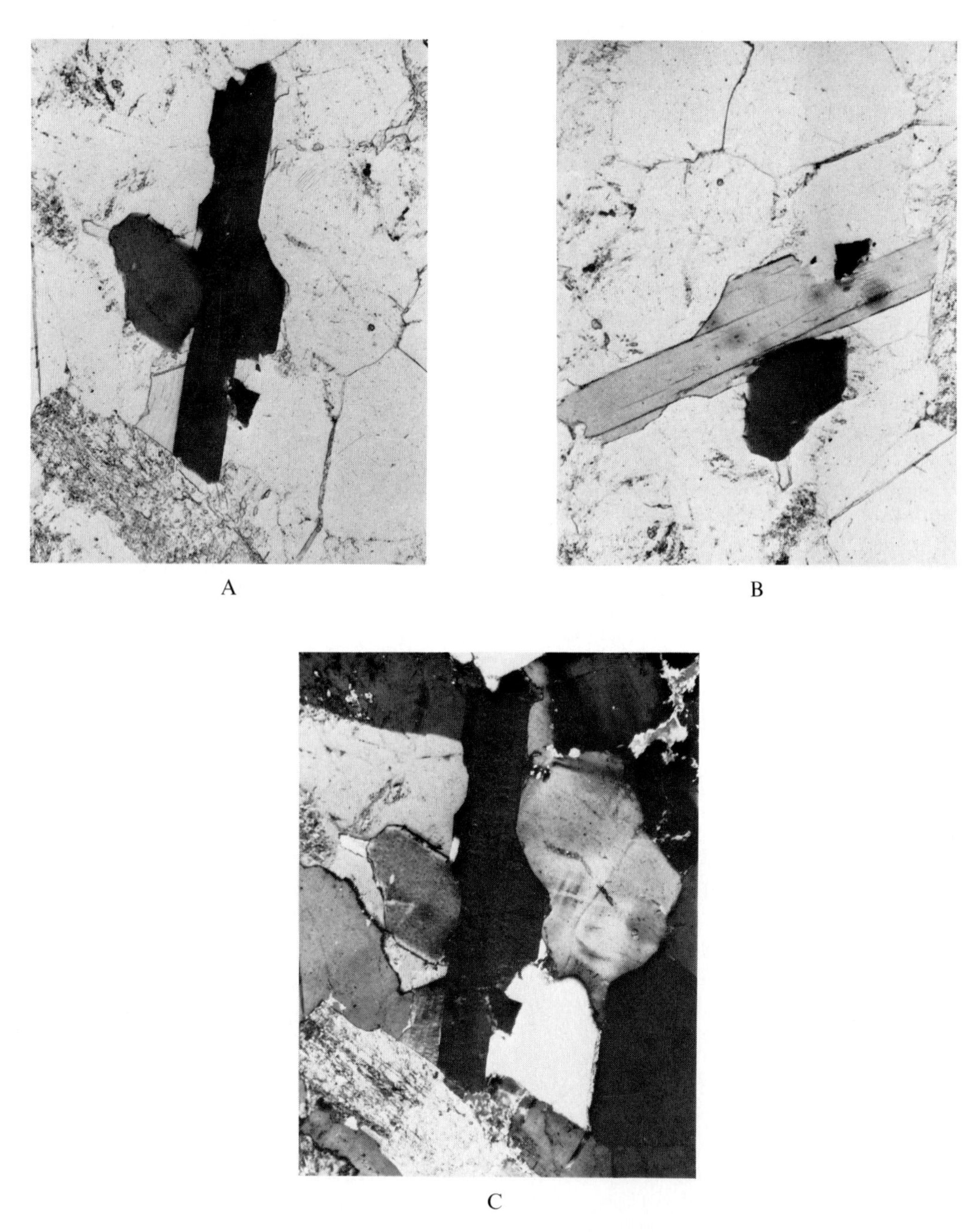

FIGURE 6-10. (A and B) Biotite is characterized by strong pleochroism. (C) All micas and many other sheet silicates show the "bird's-eye maple" extinction.

weaker birefringence than biotites. Stilpnomelane may be optically undistinguishable from biotite, but one may look for a second cleavage normal to the major basal cleavage and a much weaker bird's-eye maple appearance near extinction. Amphiboles commonly associated with biotite may show characteristic cross sections, two perfect cleavage directions, and large extinction angles. Tourmaline has no cleavage and shows maximum absorption when crystal length is elongated normal to the vibration direction of the lower nicol.

ALTERATION. Biotites are relatively resistant to weathering but less so than muscovite. Vermiculite is the common end product of biotite weathering with a "clay mica" (hydrobiotite), analogous to trioctahedral illite, as an intermediate step. Hydration replaces K^+ by $(H_3O)^+$ to form hydrobiotite. Oxidation may convert ferrous iron to ferric iron, giving a red-brown biotite, which, in turn, may be leached away and be replaced by magnesium to form bleached, pale brown vermiculite. Extreme weathering may proceed from vermiculite to kaolin or montmorillonite clays.

Hydrothermal alteration of biotite most commonly produces chlorites, with green biotite as an intermediate stage and magnetite and rutile as by-products. Biotite commonly recrystallizes before chlorite begins to form as mixed layers, and titanium and iron are excluded early as grains of rutile or leucoxene and magnetite. Calcite, epidote-zoisite, or sericite may be produced from biotite by hydrothermal solutions and H_2S may unite with ferrous iron to yield iron sulfides.

OCCURRENCE. Biotite is the most ubiquitous ferromagnesian mineral occurring in most igneous and metamorphic rocks and as partially weathered flakes in detrital sediments. Biotite is most characteristic of silicic and alkalic plutonic rocks and low- to moderate-grade metamorphic rocks. Phlogopite is largely confined to metamorphosed carbonate sediments, peridotite-type ultramafic rocks, lamprophyres, and leucite-bearing volcanic rocks.

Plutonic igneous rocks contain more biotite than volcanic ones, light-colored rocks are more likely to contain biotite than dark rock types; and few granites, granodiorites, and similar rock types are without it. The Fe:Mg ratio increases with late crystallization, and Fe^{3+}, Ti^{4+}, and Mn^{2+} are most abundant in biotite of silicic and alkalic rocks. Fe-rich biotite (dark colored and highly pleochroic) is characteristic of pegmatites and nepheline syenites and Mg-rich phlogopites (light color and weak absorption) are typical of lamprophyres, peridotites, and similar ultramafic rocks.

Volcanic conditions favor anhydrous minerals, and biotite in volcanic rocks is commonly highly altered or completely resorbed. Altered biotite phenocrysts are the most characteristic ferromagnesian minerals of rhyolites and quartz latites, and fresh phlogopite is characteristic of leucite-bearing volcanics.

Schists and gneisses of regional metamorphism commonly contain biotite formed from chlorite and muscovite. Biotite is characteristic of the greenschist and amphibolite facies associated with index minerals chlorite, almandine, and staurolite. At higher grades, biotite becomes unstable with respect to the formation of sillimanite or microcline and cordierite. Less Al replaces Si with increasing grade, and less Fe^{2+}, Fe^{3+}, and Mn appear in octahedral coordination. Ti, Mg, and Cr increase with grade, and color progresses from green to dark red-brown with grade. Impure dolomitic marble may yield phlogopite.

Contact metamorphism yields fine-grained biotite, from chlorite and sericite, in hornfels often associated with andalusite or cordierite. At higher grades, biotite yields to K-feldspar with pyroxene and sillimanite or cordierite, and retrograde processes may return chlorite. Contact Mg-marbles often contain phlogopite in association with calc-magnesium contact silicates.

REFERENCES: BIOTITE

Barshad, I. 1948. Vermiculite and its relation to biotite as revealed by base exchange reactions, x-ray analyses, differential thermal curves, and water content. *Amer. Mineral.*, *33*, 655–678.

Deer, W. A. 1937. The composition and paragenesis of biotites of the Carsphairn igneous complex. *Mineral. Mag.*, *24*, 495–502.

Eugster, H. P., and Wones, D. R. 1958. Phase relations of hydrous silicates with intermediate Mg:Fe ratios. *Carnegie Inst. Washington Year Book*, *57*, 193.

Grigoriev, D. P. 1934. The preparation of artifical magnesian mica. *Centr. Min. Geol.*, *1934-A*, 219–223.

Hall, A. J. 1941a. The relation between chemical composition and refractive index in the biotites. *Amer. Mineral.*, *26*, 34–41.

Hall, A. J. 1941b. The relation between color and chemical composition in the biotites. *Amer. Mineral.*, *26*, 29–33.

Heinrich, E. W. 1946. Studies in the mica group: The biotite-phlogopite series. *Amer. Jour. Sci.*, *244*, 836–848.

Hendricks, S. B., and M. E. Jefferson. 1939. Polymorphism of the micas. *Amer. Mineral.*, *24*, 729–753.

Kohn, J. A., and R. A. Hatch. 1955. Synthethic mica investigations. Part VI. X-ray and optical data on synthetic fluorphlogopite. *Amer. Mineral.*, *40*, 10–21.

Muller, G. 1966. The relationships among the chemical composition, refractive indices, and density of co-existing biotite, muscovite, and chlorite in granitic rocks. *Contrib. Mineral. Petrol.*, *12*, 173–191.

Nockolds, S. R. 1947. The relation between chemical composition and paragenesis in the biotite micas of igneous rocks. *Amer. Jour. Sci.*, *245*, 401–420.
Schwartz, G. M. 1958. Alteration of biotite under mesothermal conditions. *Econ. Geol.*, *53*, 164–177.
Walker, G. F. 1949. The decomposition of biotite in the soil. *Mineral. Mag.*, *28*, 693–703.
Wones, D. R. 1958a. Ferrous-ferric biotites. *Carnegie Inst. Wash. Year Book*, *57*, 195.
Wones, D. R. 1958b. The phlogopite-annite join. *Carnegie Inst. Wash. Year Book*, *57*, 194.

Lepidolite

$K_2(Li_{4-2}Al_{2-3})(Si_{8-6}Al_{0-2})O_{20}(OH,F)_4$

MONOCLINIC
$\angle\beta = 100°$

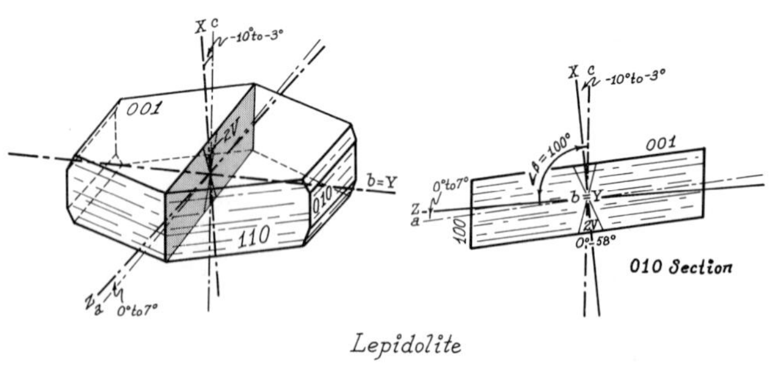

$n_\alpha = 1.525–1.548$*
$n_\beta = 1.551–1.585$
$n_\gamma = 1.554–1.587$
$n_\gamma - n_\alpha = 0.018–0.038$
Biaxial negative $2V_x = 0°–58°$ (usually 30°–50°)*
$b = Y, a \wedge Z = 0°$ to $+7°, c \wedge X = -10°$ to $-3°$
$r > v$ weak, inclined bisectrix dispersion
Colorless in standard section

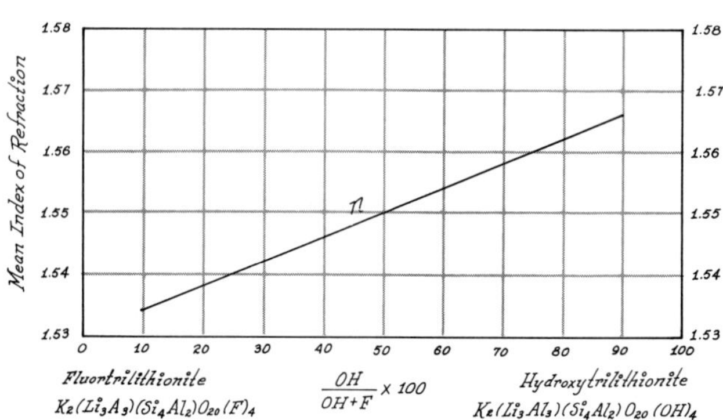

FIGURE 6-11. Variation of the mean index of refraction with the replacement of $(OH)^-$ by F^- in the fluortrilithionite-hydroxy-trilithionite series (lepidolite). (Data from Munoz, 1968)

COMPOSITION. Lepidolite is a composition range of lithian micas defined by the following end members:

Polylithionite	$K_2(Li_4Al_2)Si_8O_{20}(OH,F)_4$
Paucilithionite (trilithionite)	$K_2(Li_3Al_3)(Si_6Al_2)O_{20}(OH,F)_4$
Lithian muscovite	$K_2(Li_2Al_3)(Si_7Al)O_{20}(OH,F)_4$

Sites of octahedral coordination tend to be all filled in lepidolites, and only two-thirds filled in muscovite. Compositions requiring five of six octahedral sites filled are apparently rare, and the lepidolite-muscovite series may be discontinuous. Natural lepidolites may contain up to 98 percent polylithionite, 60 percent paucilithionite, and 35 percent muscovite. Optical constants of these natural lepidolites are shown in Fig. 6-12. Mn^{2+}, Fe^{2+}, Fe^{3+}, and Mg^{2+} may enter octahedral coordination with Li^+ and Al^{3+}, and manganese may be present in very significant amounts. Lepidolite often contains significant amounts of Rb^+, Cs^+, and Na^+ as substitutes for K^+, and traces of many other elements have been noted as octahedral or interlayer cations. Lepidolite commonly contains very appreciable fluorine in hydroxyl sites.

* Indices of refraction seem unrelated to lithium content but increase with Mn^{2+} and Fe^{3+}, and decrease with F^- (Fig. 6-11). $2V$ becomes smaller as manganese and iron increase.

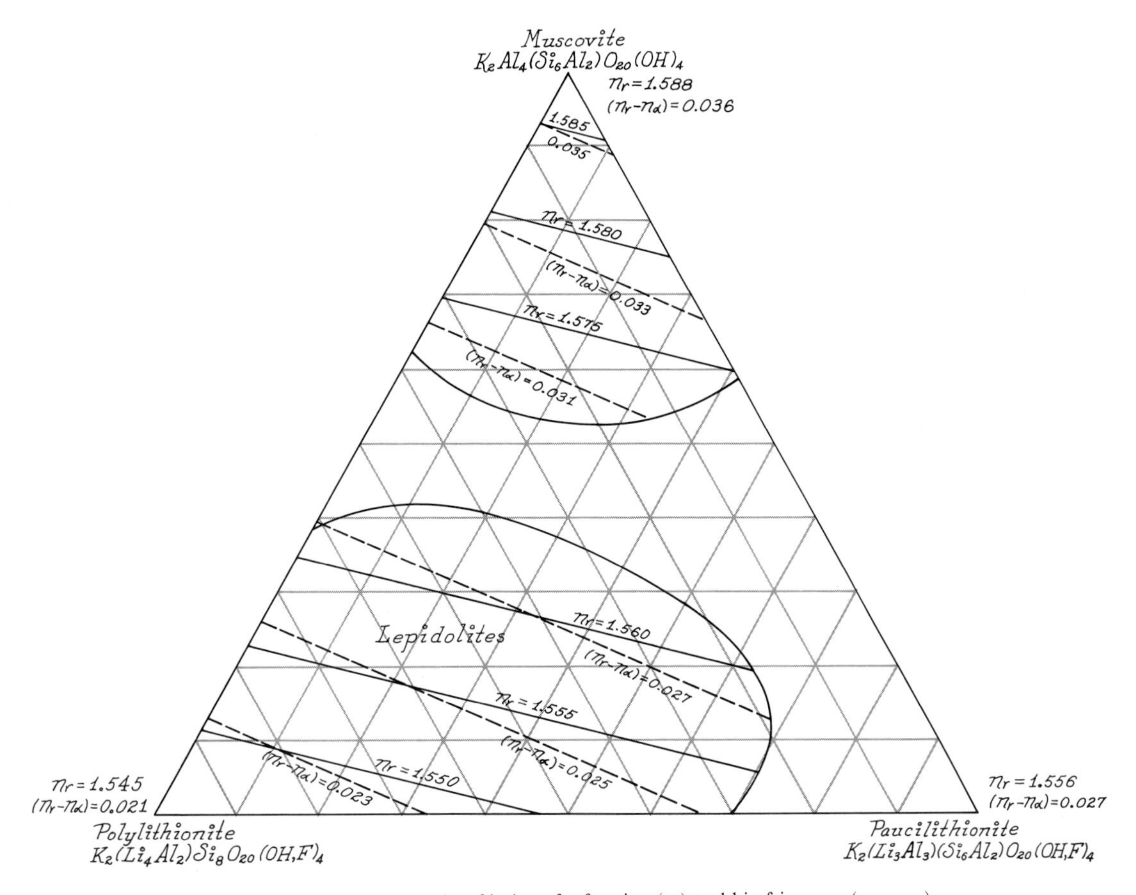

FIGURE 6-12. Relationship of index of refraction (n_γ) and birefringence ($n_\gamma - n_\alpha$) in the lithium micas (lepiodolites and lithian muscovites). (Tröger, 1962)

PHYSICAL PROPERTIES. H = $2\frac{1}{2}$–4, Sp. Gr. = 2.18–3.3. Hand sample is colorless or gray, to violet, rose-red, or yellow-green. Violet shades result from Mn^{2+} content and greens, presumably, from Fe^{2+}.

COLOR AND PLEOCHROISM. Lepidolite is commonly colorless, in section, but thick sections or fragments of highly colored forms may show pleochroism in pale violets or green with $Y = Z > X$.

FORM. Lepidolite appears most commonly as aggregates of small scales or flakes; however, large "books" are not uncommon. Fragments lie on basal cleavages. Structurally, four polytypes are known (1**M**, 2**M**, 2**Or**, and 3**T**), 2**M** being the commonest form.

CLEAVAGE. Perfect basal {001} cleavage controls fragment orientation.

BIREFRINGENCE. Maximum birefringence, seen in {010} sections, is moderate (0.018–0.038), yielding brilliant second-order interference colors in standard section. Basal sections and cleavage fragments show very weak birefringence and shades of first-order gray.

TWINNING. Lepidolite twins by the complex mica law with {001} composition plane.

INTERFERENCE FIGURE. Cleavage flakes and basal sections yield excellent acute bisectrix figures; biaxial negative. One or more orders of isochromes are expected at standard thickness, and melatopes are well separated within the visible field. $2V_x$ is usually between 30° and 50° but may range from 58° to 0° (rare trigonal forms). Optic axis dispersion is weak $r > v$, and possible bisectrix dispersion would be inclined, since $b = Y$.

OPTICAL ORIENTATION. Since $b = Y$, the optic plane parallels {010}, $a \wedge Z = 0°–7°$, and X must lie within 7° of the cleavage {001} normal. Maximum extinction angle, seen on {010} sections, should measure 0°–7° to cleavage traces. Elongation is positive (length-slow), since the fast wave always vibrates essentially normal to cleavage traces.

DISTINGUISHING FEATURES. In section, lepidolite is easily confused with other colorless micas. It usually has lower refractive indices than muscovite or paragonite, and its occurrence is somewhat distinctive.

ALTERATION. Lepidolite, like other micas, should be resistant to weathering but may be expected to hydrate to a hydromica.

OCCURRENCE. Lepidolite is essentially confined to granite pegmatites, where it is associated with quartz, microcline, albite, and the common pegmatite accessories tourmaline, topaz, beryl, and lithium minerals (such as spodumene and amblygonite). Lepidolite is occasionally reported in granites and hypothermal veins with cassiterite.

REFERENCES: LEPIDOLITE

Foster, M. D. 1960. Interpretation of the composition of lithium micas. *U.S. Geol. Surv., Prof. Paper, 354-E*, 115–147.

Franzini, M., and F. Sartori. 1969. Crystal data on 1M and $2M_2$ lepidolites. *Contrib. Mineral. Petrol, 23*, 257–270.

Hendricks, S. B., and M. E. Jefferson. 1939. Polymorphism of the micas. *Amer. Mineral., 24*, 729–753.

Hosking, K. F. G. 1957. Identification of lithium minerals. *Mining Mag., 96*, 271–276.

Levinson, A. A. 1953. Studies in the mica group: Relationship between polymorphism and composition in the muscovite-lepidolite series. *Amer. Mineral., 38*, 88–107.

Miser, H. D., and R. E. Stevens. 1938. Taeniolite from Magnet Cove, Arkansas. *Amer. Mineral., 23*, 104–110.

Munoz, J. L. 1968. Physical properties of synthetic lepidolites. *Amer. Mineral., 53*, 1490–1512.

Olav, H. J. C. 1961. The occurrence of a two-layer orthorhombic stacking polymorph of lepidolite. *Z. Kristallogr., 115*, 464–467.

Smith, J. V., and Yoder, H. S. 1956. Experimental and theoretical studies of the mica polymorphs. *Mineral. Mag., 31*, 209–225.

Tröger, W. E. 1962. Protolithionite and zinnwaldite, the chemistry and optics of lithium micas. *Beitr. Mineral. Petrogr., 8*, 418–432.

Winchell, A. N. 1942. Further studies of the lepidolite system. *Amer. Mineral., 27*, 114–130.

Zinnwaldite

$K_2(Fe_{1-2}^{2+}Li_{2-3}Al_2)(Si_{6-7}Al_{2-1})O_{20}(F,OH)_4$

MONOCLINIC
$\angle\beta = 100°$

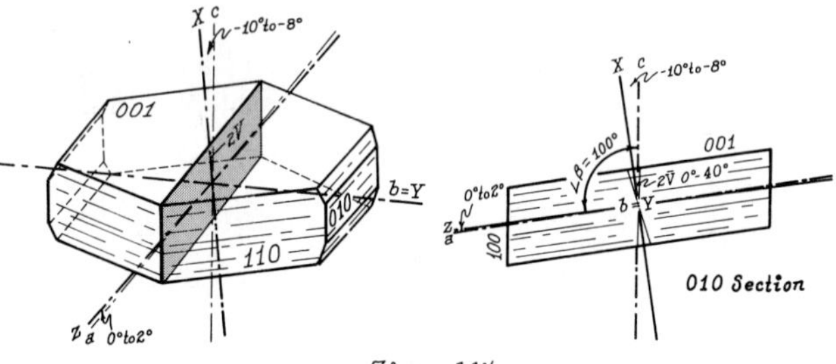

Zinnwaldite

$n_\alpha = 1.535–1.558$
$n_\beta = 1.570–1.589$
$n_\gamma = 1.572–1.590$*
$n_\gamma - n_\alpha \simeq 0.035$*

Biaxial negative, $2V_x = 0°–40°$
$b = Y, a \wedge Z = 0°$ to $-2°$, $c \wedge X = -10°$ to $-8°$
$r >$ weak, inclined bisectrix dispersion
Colorless to pale brown in section
X = colorless to pale yellow-brown,
Y = gray brown, Z = gray-brown

COMPOSITION. Zinnwaldite is a lithian-ferrous iron mica. Interlayer cations are largely K, with minor replacement by other large alkalis, and the cations of octahedral coordination are commonly Fe^{2+}, Li^{+}, and Al^{3+}, in essentially equal numbers, with minor Ti^{4+}, Fe^{3+}, Mg^{2+}, Mn^{2+}, and many other trace elements. F^- ions are commonly more abundant than $(OH)^-$ in hydroxyl sites.

* Optical constants of some L-Fe micas are shown in Fig. 6-13.

PHYSICAL PROPERTIES. H ≃ 3. Sp. Gr. ≃ 3.0. Color in hand sample ranges from yellow to dark brown or gray.

COLOR AND PLEOCHROISM. Zinnwaldite is usually weakly colored and pleochroic in pale yellow and brown, with greatest absorption when cleavage traces parallel the polarization direction of the lower nicol. $Z \geq Y > X$.

FORM. Zinnwaldite shows the characteristic foliated form of all micas. Most zinnwaldites are based on a single-layer, monoclinic (1**M**) structure, although 3**T** and 2**M** polytypes are reported.

CLEAVAGE. Perfect basal cleavage controls fragment orientation.

BIREFRINGENCE. Maximum birefringence, seen in {010} sections, is moderately strong. Bright, second-order colors should predominate in standard sections. Basal sections show very weak birefringence.

INTERFERENCE FIGURE. Cleavage fragments and basal sections yield excellent, centered, acute bisectrix figures with several isochromes and, usually, moderate $2V_x$.

OPTICAL ORIENTATION. $b = Y$, making {010} the optic plane, within which Z essentially parallels cleavage traces and X is within 10° of the cleavage normal. Extinction is essentially parallel to cleavage traces in all sections and elongation is positive (length-slow), since X is essentially normal to cleavage traces.

DISTINGUISHING FEATURES. Zinnwaldite occurs only in granite pegmatites and hydrothermal veins, where it is distinguished from biotite by the latter's high index and strong color.

OCCURRENCE. Zinnwaldite is essentially confined to granite pegmatites and high-temperature hydrothermal veins, where it is generally associated with cassiterite.

REFERENCES: ZINNWALDITE

Rieder, M. 1968. Zinnwaldite: Octahedral ordering in lithium-iron micas. *Science*, *160*, 1338–1340.

Tröger, W. E. 1962. Protolithionite and zinnwaldite, the chemistry and optics of lithium micas. *Beitr. Mineral. Petrogr.*, *8*, 418–431.

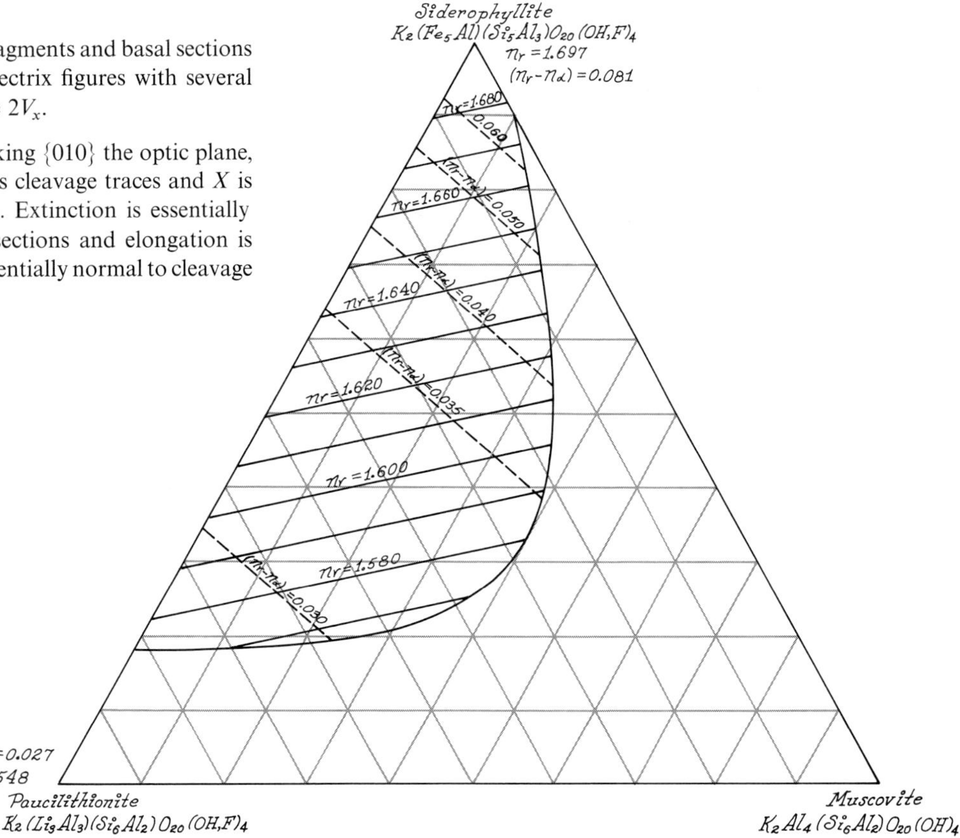

FIGURE 6-13. Relationship of index of refraction (n_γ) and birefringence ($n_\gamma - n_\alpha$) in the Li-Fe^{2+} micas. (Tröger, 1962)

BRITTLE MICAS

Margarite	$Ca_2Al_4(Si_4Al_4)O_{20}(OH)_4$	MONOCLINIC
Clintonite and Xanthophyllite	$Ca_2(Mg_{4.6}Al_{1.4})(Si_{2.5}Al_{5.5})O_{20}(OH)_4$	$\angle\beta = 95°–100°$

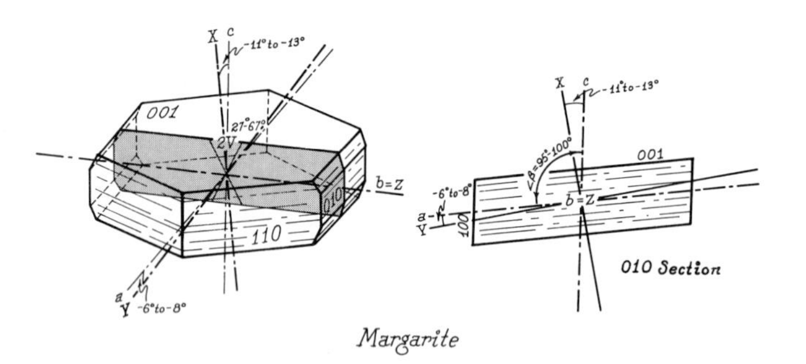

Margarite

Margarite

$n_\alpha = 1.595–1.638$*
$n_\beta = 1.625–1.648$
$n_\gamma = 1.627–1.650$
$n_\gamma - n_\alpha = 0.010–0.032$*
Biaxial negative
$2V_x = 26°–67°$

$b = Z, a \wedge Y = -6°$ to $-8°$
$c \wedge X = -11°$ to $-13°$
$v > r$ distinct
Horizontal Bxa dispersion
Colorless in section

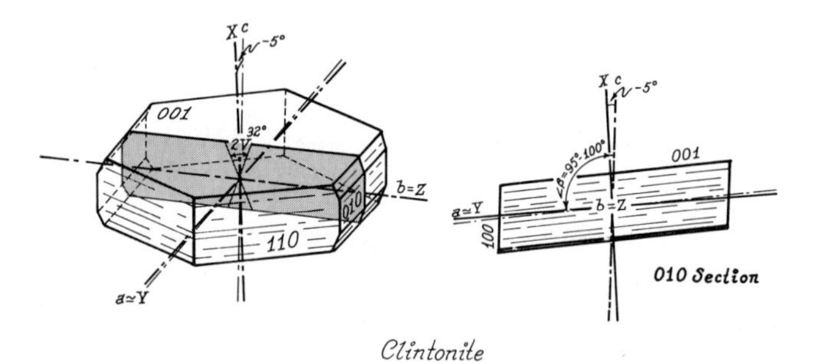

Clintonite

Clintonite

$n_\alpha = 1.643–1.648$
$n_\beta = 1.655–1.662$
$n_\gamma = 1.655–1.663$
$n_\gamma - n_\alpha \simeq 0.012$
Biaxial negative
$2V_x \simeq 32°$
$b = Z, a \simeq Y,$
$c \wedge X \simeq -5$

$v > r$ weak
Horizontal Bxa dispersion
Colorless or weakly colored in section
X = colorless to pale yellow, orange, or red-brown
$Y = Z$ = pale green or brown

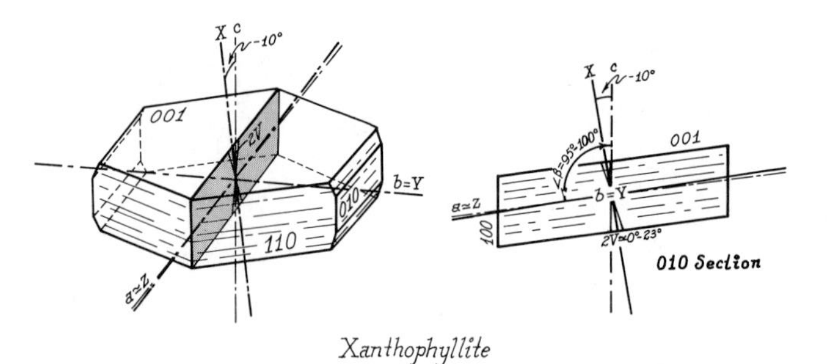

Xanthophyllite

Xanthophyllite

$n_\alpha = 1.643–1.648$
$n_\beta = 1.655–1.662$
$n_\gamma = 1.655–1.663$
$n_\gamma - n_\alpha \simeq 0.012$
Biaxial negative

$2V_x = 0°–23°$
$b = Y, a \simeq Z$
$c \wedge X \simeq -10°$
$v > r$ weak
Inclined Bxa dispersion

* Substitution of Na^+ for Ca^{2+} moves optical constants toward those of paragonite by decreasing refractive indices and increasing birefringence.

COMPOSITION. Margarite is the calcium analog of muscovite, or paragonite (dioctahedral), and clintonite-xanthophyllite is the calcium analog of the biotites (trioctahedral). Minor Sr^{2+} and Ba^{2+} may replace Ca^{2+} as interlayer cations, and appreciable Na^{+} moves optical constants toward those of paragonite, or muscovite. Ions in octahedral sites are reasonably uniform; two-thirds filled by Al^{3+} in margarite, and completely filled by Mg^{2+} and Al^{3+} in clintonite-xanthophyllite in a ratio near 4.5:1.5. Minor Fe^{2+}, Fe^{3+}, Mn^{2+}, and other ions appear, especially in clintonite-xanthophyllite.

Margarite frequently contains excess $(OH)^{-}$ for O^{2-}, balanced by Na^{+} for Ca^{2+}, and clintonite-xanthophyllite is one of the few silicates containing more Al^{3+} than Si^{4+} in tetrahedral coordination, balanced by Al^{3+} for Mg^{2+}, in octahedral sites.

PHYSICAL PROPERTIES. The Ca bond between layers gives brittle micas greater hardness and less elasticity than their K analog. H = $3\frac{1}{2}$–$4\frac{1}{2}$, on cleavage surfaces and normal to {001}. Sp. Gr. $\simeq$ 3.1. In hand sample, margarite is commonly gray, salmon-pink, yellow, or pale green, with distinctive pearly luster. Clintonite-xanthophyllite tends to be darker, in greens and rust brown.

COLOR AND PLEOCHROISM. Most brittle micas are colorless in section, but clintonite-xanthophyllite may show pronounced pleochroism in pale colors $Z = Y > X$. X = colorless, pale yellow, orange, or red-brown, $Y = Z$ = pale green or brown.

FORM. Brittle micas appear most frequently as coarse, platy aggregates. Margarite appears as a 1**M** polytype; clintonite appears generally as 3**T**; and xanthophyllite is normally 1**M**.

CLEAVAGE. Perfect basal cleavage {001}, more difficult than cleavage of normal micas, still controls fragment orientation.

BIREFRINGENCE. Birefringence is low compared to alkali micas and is maximum near {100} sections of margarite ($n_\gamma - n_\alpha$ = 0.010–0.032), and clintonite ($n_\gamma - n_\alpha \simeq 0.012$) and on {010} sections of xanthophyllite ($n_\gamma - n_\alpha \simeq 0.012$). Interference colors, in standard section, seldom exceed first order, and basal sections are nearly dark.

TWINNING. Complex, multiple twinning on {001} may be observed.

INTERFERENCE FIGURE. Cleavage plates and basal sections yield centered, negative, acute bisectrix figures with few isochromes and, usually, moderate $2V_x$. Margarite may show $2V_x$ ranging from about 26° to 67°, usually about 40°. Xanthophyllite commonly shows small $2V_x$ from 0° to about 23°, and clintonite $2V_x$ lies between those of margarite and xanthophyllite, about 32°. Dispersion is rather weak $v > r$, with possible horizontal bisectrix dispersion for margarite and clintonite and inclined for xanthophyllite.

OPTICAL ORIENTATION. Clintonite and xanthophyllite are defined by optical orientation. The optic plane of xanthophyllite lies parallel to {010} and the optic plane of clintonite and margarite is normal to {010}. For margarite and clintonite, $b = Z$, Y lies within a few degrees of a ($a \wedge Y = -6°$ to $-8°$ for margarite), and X is near the cleavage normal ($c \wedge X = -11°$ to $-13°$ for margarite and $c \wedge X \simeq -5°$ for clintonite). In xanthophyllite, $b = Y$, $a \simeq Z$, and X is normal to {001} cleavage. Extinction is essentially parallel to cleavage traces in all sections of clintonite and xanthophyllite and no more than 8° in sections of margarite. Elongation is positive (length-slow) in brittle micas, since X is essentially normal to cleavage.

DISTINGUISHING FEATURES. Xanthophyllite is distinguished from clintonite and margarite by optical orientation, small $2V$, and possible color and inclined dispersion. Clintonite has higher refractive index, generally smaller $2V$ and smaller extinction angle than margarite. Brittle micas, as a group, have higher refractive indices and lower birefringence than normal micas and lack the dark green color of most chlorites. Chloritoid is optically positive, and stilpnomelane is highly colored.

ALTERATION. Brittle micas alter to a yellow-brown, micaceous mineral, probably vermiculite.

OCCURRENCE. Brittle micas are formed in mica or chlorite schists and metasomatic zones in limestone. Margarite is associated with corundum or other aluminous minerals (such as diaspore, staurolite, and tourmaline). Clintonite-xanthophyllite is associated with talc, serpentine, and contact calc-silicates (such as grossular, vesuvianite, and phlogopite).

REFERENCES: BRITTLE MICAS

Foreman, S. A. 1951. Xanthophyllite. *Amer. Mineral.*, *36*, 450–457.

Foreman, S. A., H. Kodama, and S. Abbey. 1967. A reexamination of xanthophyllite (clintonite) from the type locality. *Can. Mineral.*, *9*, 25–30.

Knopf, A. 1953. Clintonite as contact-metasomatic product of the Boulder batholith, Montana. *Amer. Mineral.*, *38*, 1113–1117.

Manning, P. G. 1969. Origin of color and pleochroism of astrophyllite and brown clintonite. *Can. Mineral.*, *9*, 663–677.

Switzer, G. 1941. Hardness of micaceous minerals. *Amer. Jour. Sci. 239*, 316.

Takeuchi, Y. 1966. Structures of brittle micas. In *Clays and Clay Minerals*, *Proceedings of the 13th National Conference*, W. F. Bradley and S. W. Bailey (eds.). London: Pergamon Press.

Takeuchi, Y., and R. Sadanaga. 1967. Structural studies of brittle micas. I: The structure of xanthophyllite refined. *Mineral. J.*, *4*, 424–437.

The Chlorite Group

Chlorites are secondary layer silicates (phyllosilicates) based on a dual structural pattern. The structure of normal chlorites consists of one "talc" layer (an octahedral layer sandwiched between tetrahedral layers) and one "brucite" layer (a single octahedral layer) repeating on *c* and forming a single chlorite sheet about 14 Å thick (Fig. 6-14). Septechlorites possess an

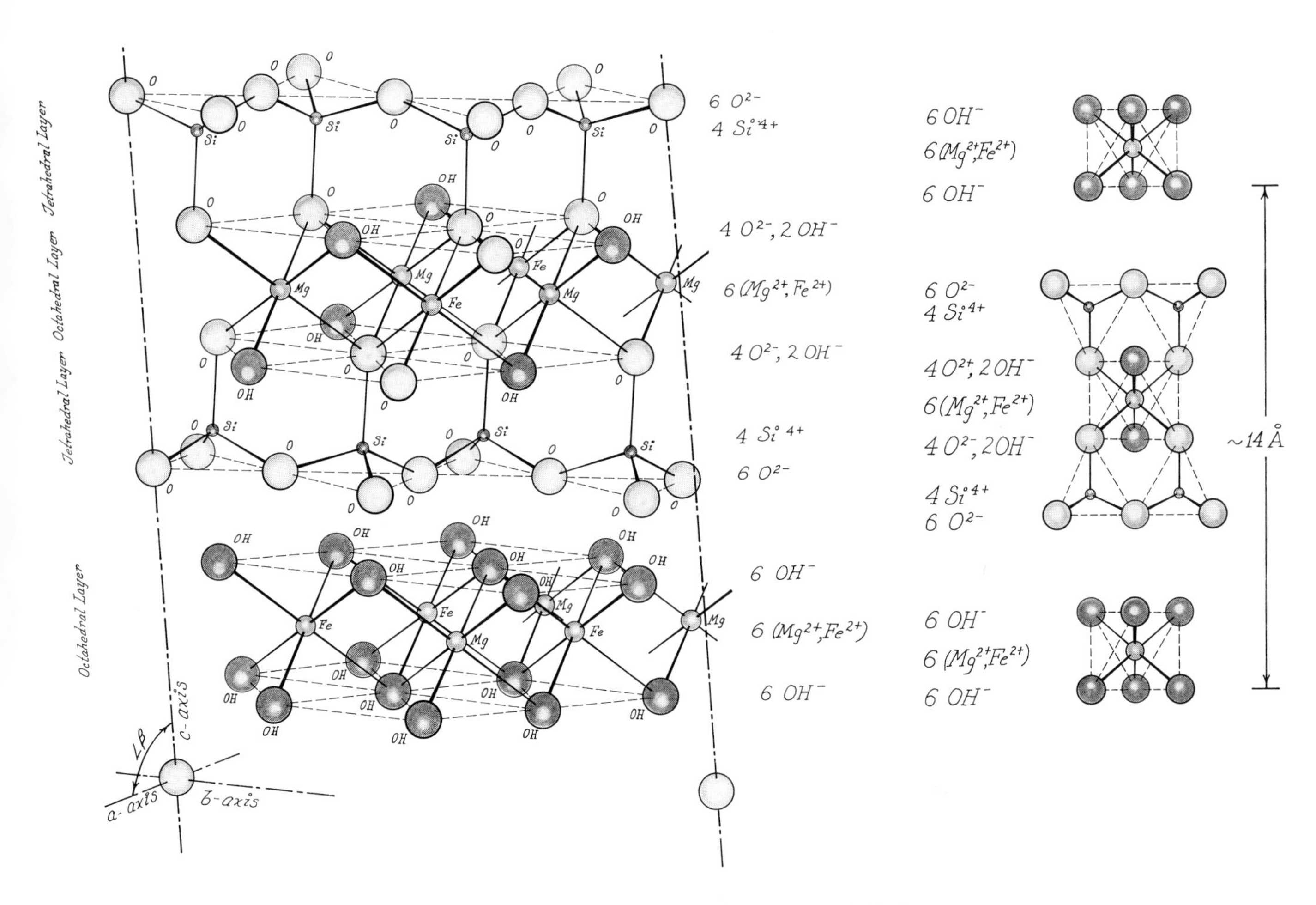

FIGURE 6-14. The idealized structure of the chlorites: The normal chlorite structure repeats on *c* at two tetrahedral and two octahedral layers (~14 Å).

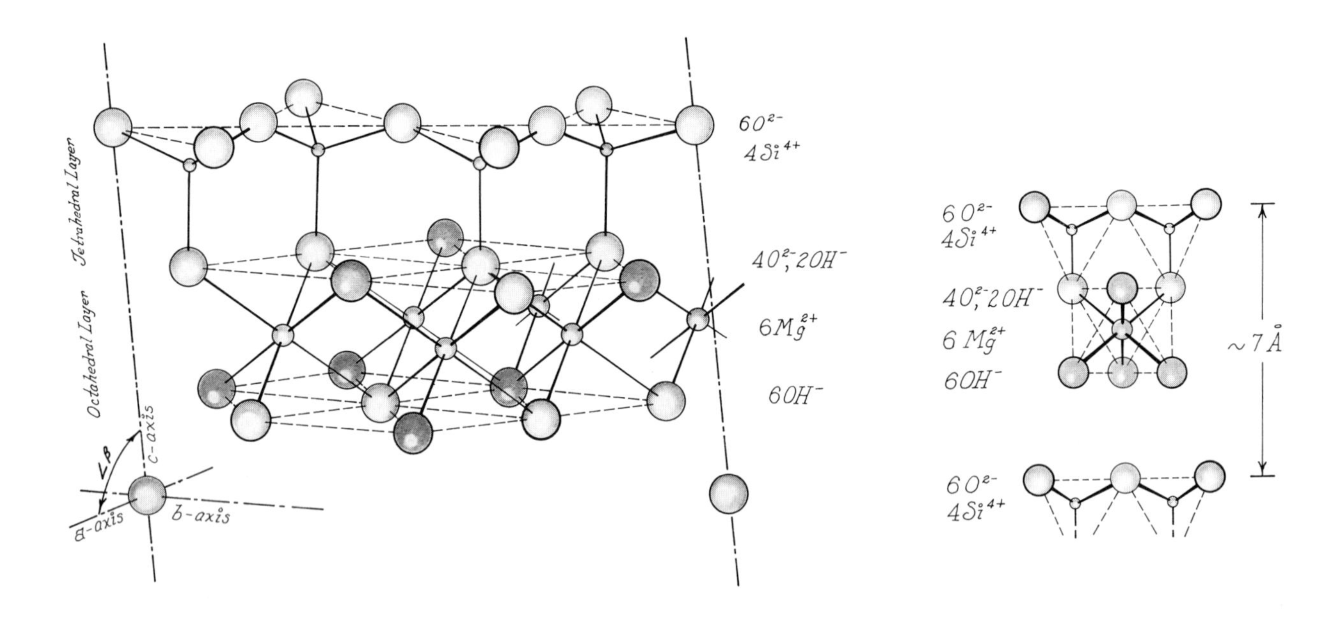

FIGURE 6-15. The idealized structure of antigorite: Repeat unit on *c* is one tetrahedral and one octahedral (trioctahedral) layer (~7 Å).

asymmetrical "kaolin" structure of one tetrahedral and one octahedral layer repeating on *c* and forming a single septechlorite layer about 7 Å thick (Fig. 6-15), hence "septe-". Many chlorite compositions, if not all, are polymorphic, appearing in both 14 Å and 7 Å sheets; high temperature favors 14 Å forms. Several stacking sequences are possible for both 14 Å and 7 Å layers, leading to several polytypes of each form, with different cell dimensions on *c*.

The most common chlorite compositions lie along the series between antigorite, $Mg_6Si_4O_{10}(OH)_8$, and amesite, $(Mg_4Al_2)(Si_2Al_2)O_{10}(OH)_8$, with more or less substitution of Fe^{2+} for Mg^{2+}. In tetrahedral layers, Al^{3+} may replace Si^{4+} to (Al_2Si_2) and, in rare cases, Fe^{3+} or Cr^{3+} are known to replace some Si^{4+} (for example, *cronstedtite*, $Fe_5^{2+}Fe^{3+}$-$(Si_3Fe^{3+})O_{10}(OH)_8$. In octahedral layers, one trivalent ion (Al^{3+},Fe^{3+}, or Cr^{3+}) must appear to balance each trivalent ion in tetrahedral coordination for Si^{4+}, and bivalent ions (Mg^{2+},Fe^{2+},Mn^{2+}, or Ni^{2+}) fill remaining octahedral sites. End-member compositions are rare but serve to limit chlorite series:

Antigorite	$Mg_6Si_4O_{10}(OH)_8$
Amesite	$(Mg_4Al_2)(Si_2Al_2)O_{10}(OH)_8$
Clinochlore	$(Mg_5Al)(Si_3Al)O_{10}(OH)_8$
Chamosite	$(Fe_5{}^{2+}Al)(Si_3Al)O_{10}(OH)_8$
Pennantite	$(Mn_5Al)(Si_3Al)O_{10}(OH)_8$
Nepouite	$(Ni_5Al)(Si_3Al)O_{10}(OH)_8$
Delessite	$(Mg_5Fe^{3+})(Si_3Al)O_{10}(OH)_8$
Kotschubeite	$(Mg_5Cr^{3+})(Si_3Al)O_{10}(OH)_8$
Strigovite	$(Fe_5{}^{2+}Fe^{3+})(Si_3Al)O_{10}(OH)_8$
Cronstedtite	$(Fe_5{}^{2+}Fe^{3+})(Si_3Fe^{3+})O_{10}(OH)_8$

The Al:Si ratio, in tetrahedral coordination, ranges considerably from compositions cited above but only Mg-chlorites range from Si_4O_{10} to $(Si_2Al_2)O_{10}$ to form the basic antigorite-amesite series.

Although most chlorite compositions are apparently possible as 14 Å or 7 Å structures, intermediate compositions are normally 14 Å chlorites, and compositional extremes tend to be 7 Å septechlorites. Amesite, chamosite, cronstedtite, and some Cr-chlorites (such as kämmererite) are commonly septechlorites, and antigorite appears only as the 7 Å polymorph. The prefix "septe-" has been proposed for all 7 Å chlorites (septeamesite, septecronstedtite, and so on); however, this structural distinction cannot be made by optical means.

Mg-chlorites are serpentines showing the 7 Å sheet structure, which tends to warp or curl, as the sheets are asymmetrical and the octahedral sheet is "too large" for the tetrahedral one (see p. 298). This structural strain is somewhat relieved by the substitution of Al^{3+} for Si^{4+} in tetrahedral layers and accompanying Al^{3+} for Mg^{2+} in octahedral layers, and amesite sheets are commonly large and flat. Chamosite, however, commonly shows warped sheets as small pellets, and kämmererite may be somewhat fibrous.

Chlorites are characteristically green, as implied by the name: Greek *chloros*, 'green'. Ferrous iron is usually present to supply the green color; however, Fe-poor, Mg-chlorites may be white, Cr-chlorites are usually lavender or lilac, Mn-chlorites are orange or brown-orange, and ferric iron tends to supply a brownish color or very dark green or brown in combination with ferrous iron.

Optical constants for the Mg-Fe^{2+} chlorites are summarized in Fig. 6-16.

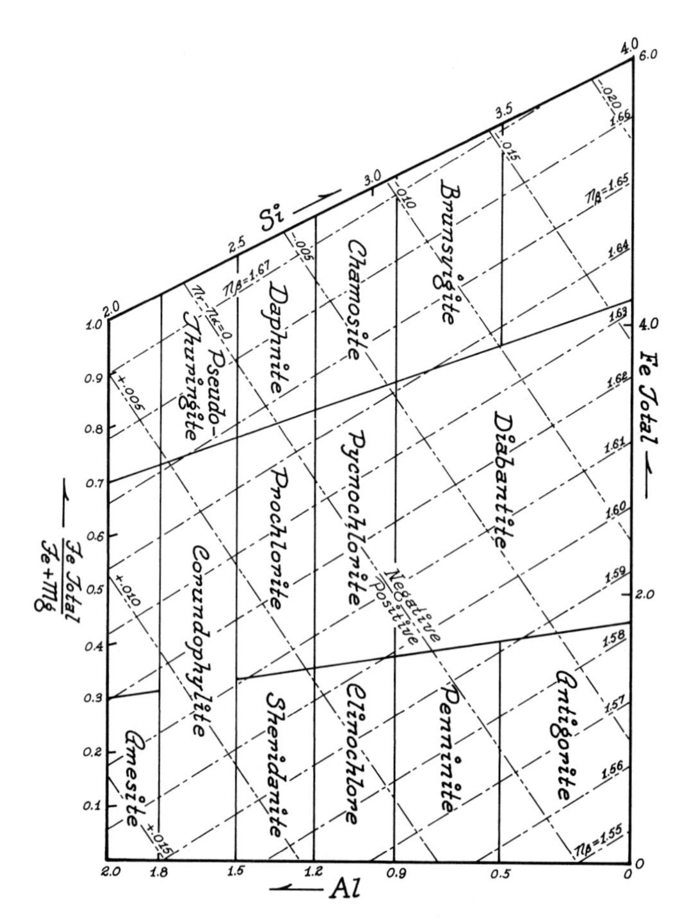

FIGURE 6-16. Optical characteristics of the Mg-Fe chlorites: This diagram is patterned after one presented by Hey (1954), and the lines representing index of refraction (n_{β}) and birefringence ($n_{\gamma} - n_{\alpha}$) are exactly as presented by him. Horizontally, Al^{3+} replaces Si^{4+} from $(Mg_4Al_2)(Si_2Al_2)O_{10}(OH)_8$ to $Mg_6Si_4O_{10}(OH)_8$, and, vertically, Fe^{2+} replaces Mg^{2+} from $Mg_6Si_4O_{10}(OH)_8$ to $Fe_6Si_4O_{10}(OH)_8$, right side, or from $(Mg_4Al_2)(Si_2Al_2)O_{10}(OH)_8$ to $(Fe_4Al_2)(Si_2Al_2)O_{10}(OH)_8$, left side. (Areas of chlorite species and nomenclature, from Phillips, 1964)

The Amesite-Antigorite Series

$(Mg,Al)_6(Si,Al)_4O_{10}(OH)_8$

MONOCLINIC
$\angle\beta = 97°$

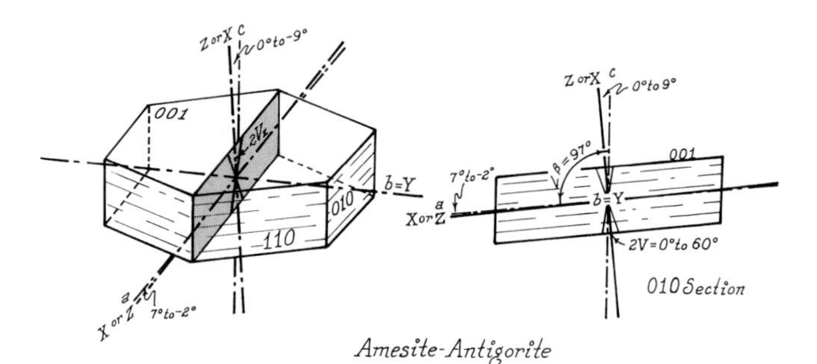

Amesite-Antigorite

	n_β*	$n_\gamma - n_\alpha$*	$2V$*	Dispersion
Amesite	1.59–1.62	0.010–0.017	10°–30° (+)	$v > r$
Corundophyllite	1.58–1.65	0.005–0.015	0°–40° (+)	$v > r$
Sheridanite	1.58–1.61	0.005–0.012	0°–50° (+)	$v > r$
Clinochlore	1.57–1.60	0.002–0.009	0°–40° (+)	$v > r$
Penninite	1.56–1.60	0.000–0.006	0°–20° (+)	$v > r$
Antigorite	1.55–1.59	0.000–0.008	20°–60° (−)	$r > v$

Biaxial positive or negative

$2V$ is small

$b = Y, c \wedge Z = 0°$ to $-9°$, $a \wedge X = 7°$ to $-2°$ (positive varieties)

$b = Y, c \wedge X = 0°$ to $-7°$, $a \simeq Z$ (negative varieties)

$v > r$ moderate on Z

Colorless to deeply colored, in standard section, with distinct pleochroism in green and yellow $X = Y > Z$

COMPOSITION. The Mg-chlorites form a solid solution series from antigorite (serpentine), $Mg_6Si_4O_{10}(OH)_8$, to amesite, $(Mg_4Al_2)(Si_2Al_2)O_{10}(OH)_8$(Am), by the replacement of Si^{4+} and Mg^{2+} by Al^{3+} and Al^{3+}.

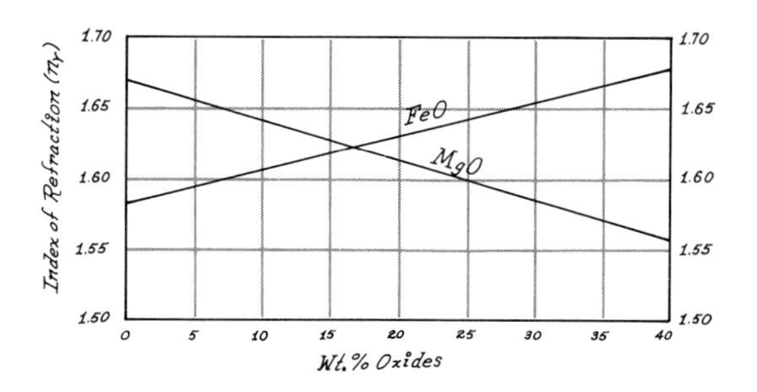

FIGURE 6-17. The effect of MgO and FeO on the index of refraction (n_γ) in the common chlorites. (Data from Il'vitskii and Tonatar-Barash, 1966)

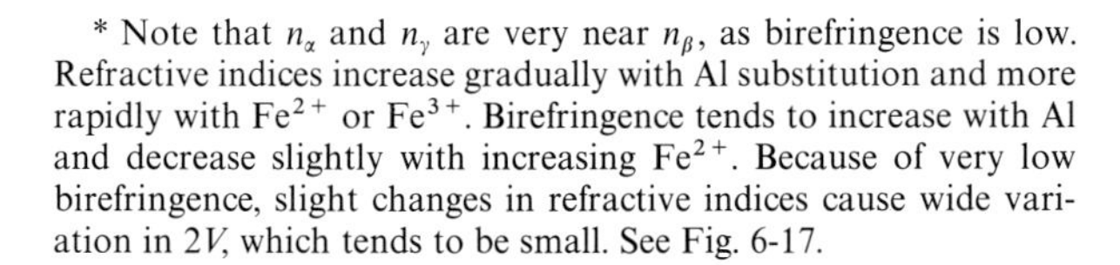
* Note that n_α and n_γ are very near n_β, as birefringence is low. Refractive indices increase gradually with Al substitution and more rapidly with Fe^{2+} or Fe^{3+}. Birefringence tends to increase with Al and decrease slightly with increasing Fe^{2+}. Because of very low birefringence, slight changes in refractive indices cause wide variation in $2V$, which tends to be small. See Fig. 6-17.

Amesite	Am_{100}–Am_{90}
Corundophyllite	Am_{90}–Am_{75}
Sheridanite	Am_{75}–Am_{60}
Clinochlore	Am_{60}–Am_{45}
Penninite	Am_{45}–Am_{25}
Antigorite	Am_{25}–Am_{0}

Antigorite (serpentine) seems to exist only as a septechlorite (7 Å), amesite is normally a 7 Å chlorite (see p. 286), and the series is known to be complete for the septechlorites. Natural chlorites most commonly show intermediate compositions and the normal chlorite (14 Å) structure. The series, apparently, is not complete in the 14 Å polymorph, and compositions between Am_0 (antigorite) and Am_{30} are very rare.

Iron is usually present as $Fe^{2+} \rightleftharpoons Mg^{2+}$, and ferrous analogs are known for intermediate members of the series. New mineral names are assigned, however, when Fe:(Mg + Fe) exceeds 0.3 (more Fe-rich than about Mg_4Fe_2).

In octahedral sites, minor Fe^{3+}, Cr^{3+}, Mn^{2+}, Ni^{2+}, and Ti^{4+} may replace Mg^{2+}; major replacement forms chlorite varieties beyond the amesite-antigorite series. (For example, *rhodophyllite-kotschubeite-kämmererite* are the chromium analogs of sheridanite-clinochlore-penninite, where Cr^{3+} replaces Al^{3+} in octahedral sites).

PHYSICAL PROPERTIES. H = 2–3 (hardness increases with Al for Si replacement). Sp. Gr. = 2.6–2.8. In hand sample, the Mg-chlorites are white to light green or bluish green and may be other colors with minor impurities. Antigorite tends to be soft with a soapy feel, and amesite is noticeably harder, with "brittle" sheets.

COLOR AND PLEOCHROISM. Mg-chlorites are usually colorless to pale green in standard section, with moderate pleochroism in greens and yellows $X = Y > Z$ (positive varieties). $X = Y$ = colorless, pale green to dark green or blue-green. Z = colorless, pale yellow-green, or yellow.

FORM. The amesite-antigorite minerals appear as large, blocky crystals in broad, flat sheets to scaley or shredded aggregates of tiny flakes (Fig. 6-18). Antigorite compositions commonly are fibrous or scaley as curled microsheets.

CLEAVAGE. Perfect basal cleavage {001} becomes slightly more difficult toward amesite.

BIREFRINGENCE. Maximum birefringence is seen in sections perpendicular to cleavage and ranges from zero to mild (0.000–0.017), showing interference colors of mostly gray and white ranging to first-order yellow and red. Mineral color may discolor interference white and grays, and anomalous interference colors such as pale olive or indigo are common for iron-bearing compositions near penninite. Basal sections and cleavage plates show near-zero birefringence.

TWINNING. Multiple twinning on {001} is common, and twins are reported with twin planes normal to {001}.

INTERFERENCE FIGURE. Basal sections and cleavage plates yield an essentially centered, acute bisectrix figure of black isogyres on a white field. All compositions, except near antigorite, are biaxial positive, with small $2V_z$ ranging from 0° (uniaxial) to 50° and moderate dispersion $v > r$.

OPTICAL ORIENTATION. The optic plane is {010}, as $b = Y$, and the acute bisectrix is essentially normal to basal cleavage, $c \wedge Z = 0°$ to $-9°$ (positive varieties) or $c \wedge X = 0°$ to $-7°$ (negative varieties), and the obtuse bisectrix X or Z essentially parallels a. Sections perpendicular to cleavage show near-parallel extinction (0° to about 7°) and negative elongation (length-fast), as Z lies essentially normal to cleavage. Negative varieties are length-slow.

DISTINGUISHING FEATURES. The chlorites are foliated minerals, usually as fine-grained aggregates, appearing in low-grade schists, argillaceous sediments, and altered igneous rocks. Chlorite is usually pale green and pleochroic, in section, with weak birefringence, often anomalous olive or indigo. Chlorite is distinguished from the micas, montmorillonites, and illites by low birefringence and small $2V$ and from kaolin clays by higher refraction index. Chlorites of the amesite-antigorite series are positive, except antigorite, which distinguishes the series from most other chlorites, micas, and clay minerals.

ALTERATION. Chlorite itself is a product of hydrothermal or metamorphic alteration of biotite, amphiboles, pyroxenes, olivine, garnet, staurolite, and perhaps other Al-bearing, ferromagnesian minerals. Members of the amesite-antigorite series are derived from Mg-rich, Fe-poor parent minerals. Chlorites are relatively stable at surface conditions and may even form in sediments at the expense of some clay minerals. Increasing metamorphism, however, returns chlorites to amphiboles and pyroxenes.

OCCURRENCE. Chlorite is widely distributed in sediments, low-grade metamorphic rocks, and hydrothermal environments.

In metamorphic rocks, chlorite is found in schists, phyllites, and amphibolites and is the characteristic mineral of the greenschist facies. Chlorite forms by the low-grade metamor-

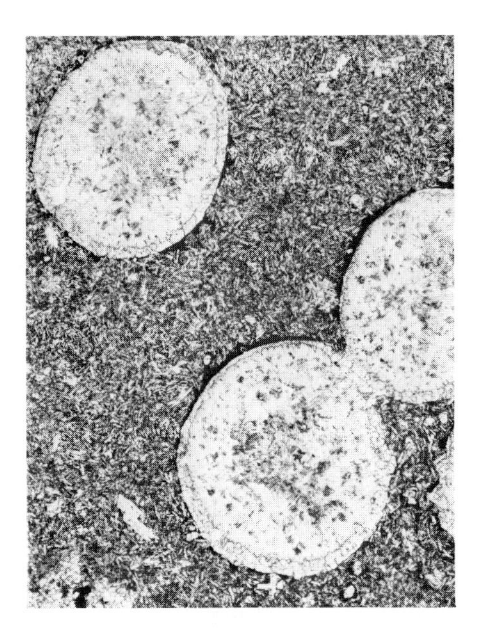
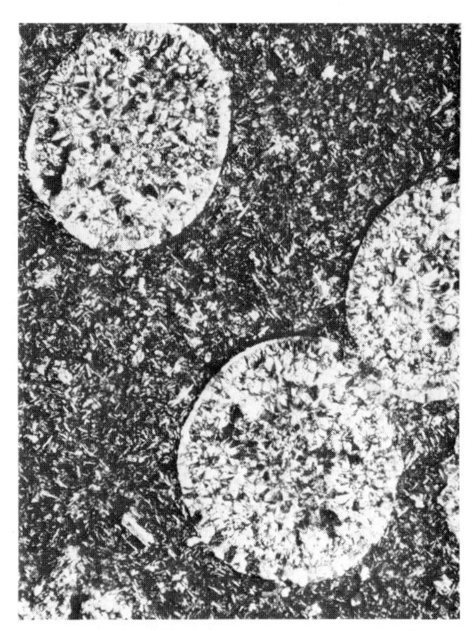

FIGURE 6-18. Chlorite aggregates fill amygdules in basalt. Interference colors are low first order and often anomalous, as indigo blue or drab olive.

phism of argillaceous sediments and other Al-rich rocks in association with albite, epidote, actinolite, hornblende, anthophyllite, biotite, chloritoid, garnet, sericite, and quartz. Chlorite becomes biotite, amphibole, or pyroxene with increasing metamorphism and may be restored by retrograde metamorphism.

In igneous rocks, chlorite is normally the product of hydrothermal or deuteric alteration of biotite, amphiboles, or pyroxenes. In granites and similar plutonic rocks, chlorite, near penninite, is commonly seen in partial replacement of biotite. Antigorite and penninite appear in hydrothermal veins in ultramafic rocks and serpentinites. In basalts, andesites, and similar volcanic rocks, chlorite appears in amygdules, cavities, and on fracture surfaces or may replace ferromagnesian phenocrysts or groundmass. Chlorite may also form in hydrothermal quartz veins.

In sedimentary rocks, chlorite appears with clays as detrital mineral flakes or as authigenic grains. It may form from unstable clay minerals (such as celadonite) or from the breakdown of detrital ferromagnesian grains and often appears in mixed-layer association with vermiculite or montmorillonite.

REFERENCES: THE AMESITE-ANTIGORITE SERIES

Alekandrova, V. A., V. A. Drits, and G. V. Sokolova. 1973. Crystal structure of dioctahedral chlorite. *Soviet Physics—Crystallogr.*, *18*, 50–53.

Balasubramaniam, K. S., and G. Viswanathan. 1971. Mineralogy of the greenschists near Jambughoda, Gujarat State. *Indian Mineral.*, *12*, 75–77.

Carroll, D. 1970. Clay minerals: A guide to their x-ray identification. Geol. Soc. Am., Spec. Paper, 126.

Gruner, J. W. 1944. Kaolinitic structure of amesite. *Amer. Mineral.*, *29*, 422–430.

Hey, M. H. 1954. A new review of the chlorites. *Mineral. Mag.*, *30*, 277–292.

Il'vitskii, M. M., and Z. I. Tanatar-Barash. 1966. Chemical constitution and correlation of optical properties with chemical composition of chlorites by statistical correlation analysis. *Zap. Vses. Mineral. Obshchest.*, *95*, 224–229.

Laajoki, K., and P. Ojanperä. 1973. Magnesioriebeckite and penninite from a shear zone in Puolanka, NE Finland. *Bull. Geol. Soc. Finland*, *45*, 143–153.

Marakis, G. 1972. Remarks on the mineral clinochlore from

rodingites of Kymi, Euboea Island, Greece. *Neues Jahrb. Min., Mh.*, 1972, 345–349.

Müller, G. 1966. The relationships among the chemical composition, refractive indices, and density of coexisting biotite, muscovite, and chlorite from granitic rocks. *Contrib. Mineral. Petrol.*, *12*, 173–191.

Nelson, B. W., and R. Roy. 1953. The serpentine-amesite join in the system $MgO\text{-}Al_2O_3\text{-}SiO_2\text{-}H_2O$ and classification of the chlorite minerals. *Ninth Quarterly Progress Report* (School of Mineral Industries, Pennsylvania State College).

Oinuma, K., S. Shimoda, and T. Sudo. 1973. Triangular diagrams in use of a survey of crystal chemistry of chlorites. In *Proceedings, International Clay Conference*, J. M. Serratosa (ed.). Madrid: Division de Ciencias C. S. I. C.

Orcel, J. 1927. Chemical classification of chlorites. *Bull. Soc. Franc. Min.*, *50*, 75–456.

Phillips, W. R. 1964. A numerical system of classification for chlorites and septechlorites. *Mineral. Mag.*, *33*, 1114–1124.

Tschermak, G. 1891. Chlorite group. *Sitz. Akad. Wiss. Wien, Math-nat. Kl.*, *100*, Abt. 1, 29.

The Pseudothuringite-Brunsvigite Series $(Fe,Al)_6(Si,Al)_4O_{10}(OH)_8$

MONOCLINIC
$\angle\beta$ 97°

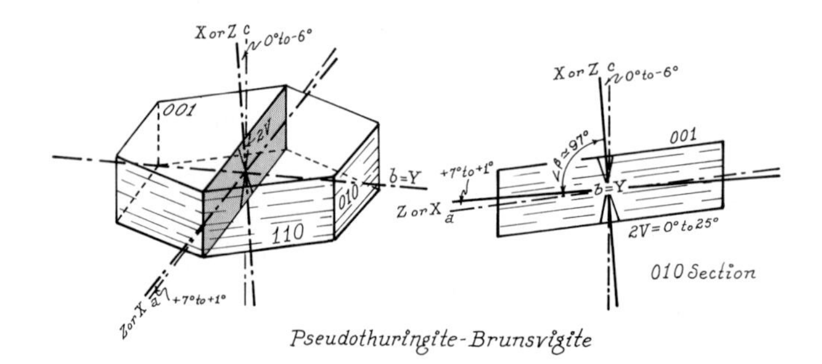

Pseudothuringite-Brunsvigite

	n_β*	$n_\gamma - n_\alpha$*	$2V$*	Dispersion
Pseudothuringite	1.65–1.68	0.005–0.000	Small + or −	$v > r$
Daphnite	1.64–1.68	0.000–0.006	Small − or +	$r > v$
Chamosite	1.64–1.67	0.002–0.010	Small −	$r > v$
Brunsvigite	1.64–1.67	0.005–0.015	Small −	$r > v$

Biaxial negative or positive; small $2V$ to 25°

$b = Y$, $c \wedge Z = 0°$ to $-6°$, $a \wedge X = +7°$ to $+1°$ (positive varieties)

$b = Y$, $c \wedge X = 0°$ to $-5°$, $a \wedge Z = +7°$ to $+2°$ (negative varieties)

$v > r$ strong (on Z)

Strong color and pleochroism in green and green-brown $Z = Y > X$

COMPOSITION. The Fe^{2+}-chlorites form a solid solution series analogous to that of the Mg-chlorites (amesite-antigorite), although the end-member compositions $Fe_6Si_4O_{10}(OH)_8$ and $(Fe_4Al_2)(Si_2Al_2)O_{10}(OH)_8$ either do not exist as natural minerals or are very rare. The ferrous analog of antigorite has been named *greenalite* (Gr), $Fe_6Si_4O_{10}(OH)_8$. It can be synthesized only as a septechlorite (7 Å) (see p. 286) and is reported in the iron deposits of the Mesabi Range. The Fe^{2+}-chlorite series is not well defined or extensively studied. The series seems incomplete, and compositions near the center

* Note that n_α and n_γ are near n_β, as birefringence is small. $2V$ varies widely with very small changes in refractive index. Refractive indices decrease with increasing Mg.

of the series are most common:

Greenalite	Gr_{100}–Gr_{90}
Pseudothuringite	Gr_{90}–Gr_{75}
Daphnite	Gr_{75}–Gr_{60}
Chamosite	Gr_{60}–Gr_{45}
Brunsvigite	Gr_{45}–Gr_{25}

Greenalite appears only as a septechlorite, and some other compositions may exist in the 7 Å polymorph (chamosite), but most natural chlorites are the normal 14 Å structure.

Besides Fe^{2+} and Al^{3+}, several ions appear in octahedral coordination (Mg^{2+}, Mn^{2+}, Ni^{2+}, Fe^{3+}, Cr^{3+}, and Ti^{4+}) and analogous Mn-chlorite and Ni-chlorite series exist through at least a limited range. Very significant Mg is usually present in Fe^{2+}-chlorites, and complete solid solution exists between the amesite-antigorite series and pseudothuringite-brunsvigite series, with intermediate members of intermediate optical properties, as shown in Fig. 6-16.

PHYSICAL PROPERTIES. H = 1.5–2.5. Sp. Gr. = 2.8–3.3. In hand sample, ferrous-chlorites are dark green.

COLOR AND PLEOCHROISM. Ferrous-chlorites are deeply colored and strongly pleochroic as fragments or in standard section. $Z = Y > X$ (negative varieties) or $X = Y > Z$ (positive varieties). X = pale green or green-brown; $Y = Z$ = dark green.

FORM. Minerals of the series usually appear as massive aggregates of small flakes or scales. Chamosite compositions commonly form as oolitic pellets, with concentric layers of discontinuous, fine-grained flakes formed on a core of coarse scales or some foreign mineral.

CLEAVAGE. Perfect basal cleavage {001}.

BIREFRINGENCE. Maximum birefringence is seen in sections perpendicular to cleavage and ranges from zero to mild (0.000–0.015), showing interference colors of mostly gray or white. Strong inherent mineral color is present, and anomalous interference colors are characteristic of Fe-chlorites. Basal sections and cleavage plates show near-zero birefringence.

TWINNING. Multiple twinning on {001} is common.

INTERFERENCE FIGURE. Basal sections and cleavage plates yield essentially centered, acute bisectrix figures of black isogyres on a white field. Pseudothuringite is usually biaxial positive, daphnite usually biaxial negative, and chamosite and brunsvigite are consistently negative. $2V$ is small, ranging from 0° to at least 25°, and dispersion is strong $r > v$ on X ($v > r$ for positive varieties) with possible inclined bisectrix dispersion.

OPTICAL ORIENTATION. The optic plane is {010}, as $b = Y$, and the acute bisectrix is essentially normal to the cleavage plane {001}, $c \wedge X = 0°$ to $-5°$ (negative varieties) or $c \wedge Z = 0°$ to 6° (positive varieties). The obtuse bisectrix Z or X lies near a. Sections normal to cleavage show near-parallel extinction (0° to about 7°) and positive elongation (length-slow for negative varieties) or negative elongation (length-fast for positive varieties).

DISTINGUISHING FEATURES. Fe-chlorites are deeply colored and strongly pleochroic with weak birefringence, commonly anomalous. Micas show strong birefringence; montmorillonite, illite, and kaolins are colorless; and Mg-chlorites are biaxial positive.

ALTERATION. Ferrous iron in chlorite structures may oxidize by weathering, and some ferric chlorites may be derived in this way. Chamosite is reported to weather to kaolin and goethite or limonite, and all Fe-chlorites commonly show limonite stains. Increasing metamorphism converts chlorite to biotite, stilpnomelane, amphiboles, or pyroxenes.

OCCURRENCE. Fe-chlorites should be most characteristic of late, hydrothermal quartz veins and cavity filling in mafic volcanics. Fe-chlorites form a major constituent of some iron-rich sediments, and chamozite oolites and flakes with siderite form some sedimentary iron ores.

REFERENCES: THE PSEUDOTHURINGITE-BRUNSVIGITE SERIES

Hallimond, A. F., F. A. Bannister, and C. O. Harvey. 1939. Chamosite and daphnite. *Mineral. Mag.*, *25*, 441.

Hey, M. H. 1954. A new review of the chlorites. *Mineral. Mag.*, *30*, 277–292.

Il'vitskii, M. M., and Z. I. Tanatar-Barash. 1966. Chemical constitution and correlation of optical properties with chemical composition of chlorites by statistical correlation analysis. *Zap. Vses. Mineral. Obshchest.*, *95*, 224–229.

Müller, G. 1966. The relationships among the chemical composition, refractive indices, and density of coexisting biotite, muscovite, and chlorite from granitic rocks. *Contrib. Mineral. Petrol.*, *12*, 173–191.

Orcel; J. 1927. Chemical classification of chlorites. *Bull. Soc. Franc. Min.*, *50*, 75–456.
Phillips, W. R. 1964. A numerical system of classification for chlorites and septechlorites. *Mineral. Mag.*, *33*, 1114–1124.
Sutherland, J. K. 1967. Chlorites of the Anaconda-Caribou deposit, New Brunswick. New Brunswick Res. and Productivity Council, Res. Note 8.
Tschermak, G. 1891. Chlorite group. *Sitz. Akad. Wiss. Wien, Math-nat. Kl.*, *100*, Abt. 1, 29.

The Klementite-Delessite Series

$(Mg,Fe^{3+})_6(Si,Al)_4O_{10}(OH)_8$

MONOCLINIC
$\angle\beta \simeq 97°$

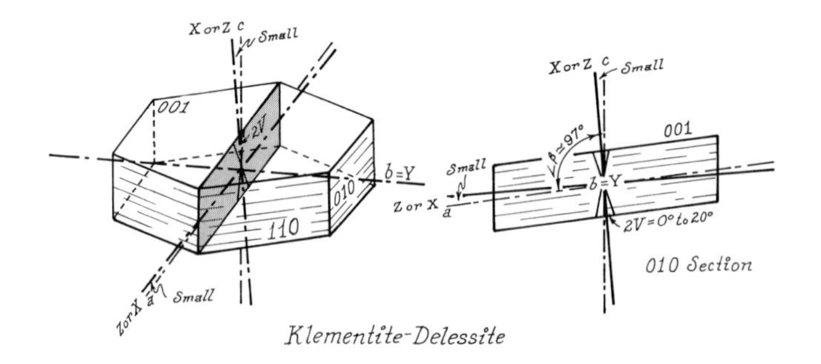

Klementite-Delessite

	n_β*	$n_\gamma - n_\alpha$*	2V	Dispersion
Klementite	1.63–1.60	0.000–0.001	Small + or −	$r \lesseqgtr v$
Delessite	1.60–1.57	0.001–0.002	Small −	$r > v$

Biaxial negative or positive; $2V$ = small
$b = Y$, small $c \wedge X$ or Z, small $a \wedge Z$ or X
$r > v$ (on X)
Colorless to weakly colored and pleochroic in pale yellow and green $Z = Y > X$

COMPOSITION. The klementite-delessite chlorite series lies between ideal end members $(Mg_4Fe_2{}^{3+})(Si_2Al_2)O_{10}(OH)_8$ and $Mg_6Si_4O_{10}(OH)_8$, antigorite (At) making the series analogous to the amesite-antigorite series, with Fe^{3+} replacing Al in octahedral sites.

Klementite	At_{10}–At_{40}
Delessite	At_{40}–At_{75}
Antigorite	At_{75}–At_{100}

The klementite-delessite series forms continuous solid solution with its ferrous analog series thuringite-strigovite and probably with other chlorite series. In octahedral sites, therefore, major amounts of Fe^{2+} may be expected, with smaller impurities of Cr^{3+}, Mn^{2+}, and Ti^{4+}.

Normal chlorite structure is most common (14 Å), although septe-analogs probably exist.

* Note that n_α and n_γ are very near n_β, as birefringence is very small. $2V$ is commonly small but may vary widely with very small changes in refraction index. Refractive indices increase and birefringence decreases as Al^{3+} and Fe^{3+} replace Si^{4+} and Mg^{2+}. Optical constants are not greatly affected by the subsitution $Al \rightleftharpoons Fe^{3+}$.

PHYSICAL PROPERTIES. H = 2–3. Sp. Gr. = 2.6–3.0. In hand sample, ferric-rich chlorites tend to be brownish and may range from nearly white to green-brown.

COLOR AND PLEOCHROISM. Sections are normally colorless to pale green or brownish with moderate pleochroism $Z = Y > X$ (negative varieties).

FORM. Foliated as sheets or aggregates of fine-grained scales.

CLEAVAGE. Perfect basal cleavage {001}.

BIREFRINGENCE. Maximum birefringence is seen in sections perpendicular to cleavage and is very weak (0.000–0.002), showing no more than first-order gray in standard section.

TWINNING. Multiple twinning on {001} is common.

INTERFERENCE FIGURE. Basal sections and cleavage plates yield essentially centered, acute bisectrix figures of broad, dark isogyres on a white or gray field. Most of the series is biaxial negative; however, high Al^{3+} and Fe^{3+} replacement of Si^{4+} and Mg^{2+} produces positive varieties.

$2V$ is small, ranging to about 20°, and dispersion is moderate $r > v$ on X.

OPTICAL ORIENTATION. The optic plane is {010}, as $b = Y$. The acute bisectrix is essentially normal to basal cleavage, $c \wedge X$ (negative varieties) or $c \wedge Z$ (positive varieties) is small, and the obtuse bisectrix Z or X lies near a. Sections normal to cleavage show near-parallel extinction and positive elongation (length-slow for negative varieties) or negative elongation (length-fast for positive varieties).

DISTINGUISHING FEATURES. Micas, illites, and montmorillonites show stronger birefringence, and kaolins are colorless, with lower refraction index. Mg-chlorites are usually biaxial positive, and Fe-chlorites are deeply colored.

ALTERATION. Chlorites are themselves secondary minerals derived from Al-bearing ferromagnesian minerals by the processes of low-grade metamorphism or hydrothermal alteration. Fe-chlorites tend to oxidize easily, and some ferric chlorites may be derived from their ferrous analogs. Increasing metamorphism returns chlorite to biotite, amphiboles, and pyroxenes.

OCCURRENCE. The klementite-delessite series occurs with other chlorites in low-grade metamorphic rocks. Some ferric chlorites appear to result from the oxidation of ferrous varieties, and we may expect klementite and delessite in sediments and other weathered environments.

REFERENCES: THE KLEMENTITE-DELESITE SERIES

Hey, M. H. 1954. A new review of the chlorites. *Mineral. Mag.*, *30*, 277–292.

Il'vitskii, M. M., and Z. I. Tanatar-Barash. 1966. Chemical constitution and correlation of optical properties with chemical composition of chlorites by statistical correlation analysis. *Zap. Vses. Mineral. Obschest.*, *95*, 224–229.

Müller, G. 1966. The relationships among the chemical composition, refractive indices, and density of coexisting biotite, muscovite, and chlorite from granitic rocks. *Contrib. Mineral. Petrol.*, *12*, 173–191.

Oinuma, K., S. Shimoda, and T. Sudo. 1973. Triangular diagrams in use of a survey of crystal chemistry of chlorites. In *Proceedings, International Clay Conference*, J. M. Serratosa (ed.). Madrid: Division de Ciencias, C.S.I.C.

Orcel, J. 1927. Chemical classification of chlorites. *Bull. Soc. Franc. Min.*, *50*, 75–456.

Pande, I. C., and P. K. Verma. 1968. A klementite from chlorite-sericite schist, Wajula, Distr. Almora, U.P., India. *Mineral. Mag.*, *36*, 752–753.

Phillips, W. R. 1964. A numerical system of classification for chlorites and septechlorites. *Mineral. Mag.*, *33*, 1114–1124.

Tschermak, G. 1891. Chlorite group. *Sitz. Akad. Wiss. Wien, Math-nat. Kl.*, *100*, Abt. 1, 29.

The Thuringite-Strigovite Series

$(Fe^{2+},Fe^{3+})_6(Si,Al)_4O_{10}(OH)_8$

MONOCLINIC
$\angle\beta \simeq 97°$

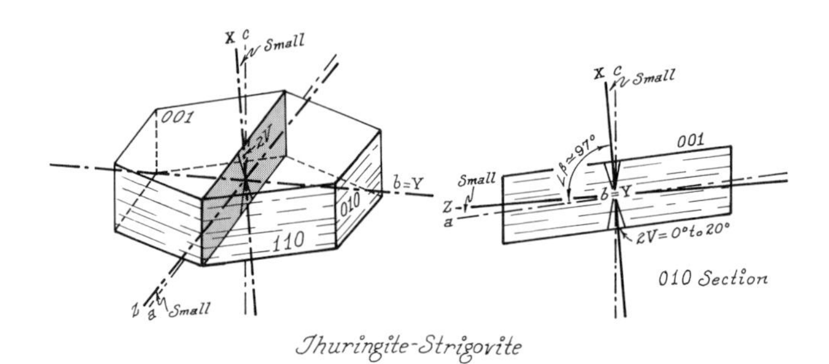

Thuringite-Strigovite

	n_β*	$n_\gamma - n_\alpha$*	$2V_x$*	Dispersion
Thuringite	1.73–1.70	0.014–0.017	Small –	$r > v$
Strigovite	1.70–1.68	0.017–0.020	Small –	$r > v$

Biaxial negative; $2V_x = 0°–20°$
$b = Y$, small $c \wedge X$, small $a \wedge Z$
$r > v$ strong
Deeply colored and strongly pleochroic in yellow-brown and deep green $Z = Y > X$

COMPOSITION. The thuringite-strigovite chlorite series lies between ideal end members $(Fe_4^{2+}Fe_2^{3+})(Si_2Al_2)O_{10}(OH)_8$ and $Fe_6Si_4O_{10}(OH)_8$, "greenalite" (Gr), making the series the analog of pseudothuringite-brunsvigite, where Fe^{3+} replaces Al in octahedral sites.

Thuringite	Gr_{10}–Gr_{40}
Strigovite	Gr_{40}–Gr_{75}
"Greenalite"	Gr_{75}–Gr_{100}

The thuringite-strigovite series forms continuous solid solution with its magnesian analog series klementite-delessite and with other chlorite series. In octahedral sites, therefore, major amounts of Mg^{2+} may be present with lesser Al^{3+}, Cr^{3+}, Mn^{2+}, and Ti^{4+}. In the rare mineral *cronstedtite* $(Fe_5^{2+}Fe^{3+})(Si_3Fe^{3+})O_{10}(OH)_8$, ferric iron replaces silicon in tetrahedral sites.

The normal chlorite structure (14 Å) is common, and septechlorites may exist in this composition range.

PHYSICAL PROPERTIES. H = 1.5–2.5. Sp. Gr. = 2.9–3.3. In hand sample, iron-rich chlorites tend to be dark green to dark brown or nearly black.

COLOR AND PLEOCHROISM. Chlorites containing both ferrous and ferric iron are unusually dark and strongly pleochroic in greens or browns $Z = Y > X$. X = yellow-brown, $Y = Z$ = dark green or dark greenish brown.

FORM. Commonly as fine-grained aggregates of scales or flakes.

CLEAVAGE. Perfect basal cleavage {001}.

* $2V_x$ is commonly small but may range to 20°. Refractive indices increase and birefringence decreases as Al^{3+} and Fe^{3+} replace Si^{4+} and Fe^{2+}. Refractive indices and birefringence decrease with Mg for Fe^{2+} or Al for Fe^{3+}.

BIREFRINGENCE. Maximum birefringence is seen in sections perpendicular to cleavage and is mild (0.014–0.020) showing first-order colors in standard section. Interference colors are commonly dulled and masked by deep, inherent mineral color, and anomalous colors may appear. Basal sections and cleavage plates show near-zero birefringence.

TWINNING. Multiple twinning on {001} is common.

INTERFERENCE FIGURE. Basal sections and cleavage plates yield essentially centered, acute bisectrix figures with few isochromes. The series is biaxial negative, $2V_x$ is small, and dispersion strong $r > v$, with possible inclined bisectrix dispersion.

OPTICAL ORIENTATION. Sections normal to cleavage show near-parallel extinction and positive elongation (length-slow).

DISTINGUISHING FEATURES. Fe-chlorites are strongly colored and pleochroic, biaxial negative, with small $2V_x$, and moderately birefringent, often showing anomalous interference colors. Colored micas show stronger birefringence, and most clay minerals are colorless, with much lower relief. Mg-chlorites are largely positive, and ferrous chlorites tend to show weaker color and birefringence and lower refraction index.

ALTERATION. Ferrous iron may be partially oxidized by weathering to yield some ferric chlorites. Increasing metamorphism returns chlorites to biotite, amphiboles, or pyroxenes.

OCCURRENCE. Thuringite and strigovite are Fe-rich chlorites suggestive of Fe-rich sediments or late hydrothermal deposition, and since at least some ferric chlorites are oxidation products of ferrous chlorites we may expect them in sedimentary iron formations and other environments subject to weathering.

REFERENCES: THE THURINGITE-STRIGOVITE SERIES

Abbona, F., and R. Compagnoni. 1969. Sulle cloriti ferrifere di Morgex (Valle d'Aosta). *Atti Accad. Sci. Torino, 103*, 533–549.

Hey, M. H. 1954. A new review of the chlorites. *Mineral. Mag., 30*, 277–292.

Il'vitskii, M. M., and Z. I. Tanatar-Barash. 1966. Chemical constitution and correlation of optical properties with chemical composition of chlorites by statistical correlation analysis. *Zap. Vses. Mineral. Obshchest., 95*, 224–229.

Kepezhinskas, K. B., and V. S. Sobolev. 1965. Paragenetic types of chlorite. *Dokl. Acad. Sci. USSR, Earth Sci. Sect., 161*, 436–439.

Müller, G. 1966. The relationships among the chemical composition, refractive indices, and density of coexisting biotite, muscovite, and chlorite from granitic rocks. *Contrib. Mineral. Petrol., 12*, 173–191.

Orcel, J. 1927. Chemical classification of chlorites. *Bull. Soc. Franc. Min., 50*, 75–456.

Phillips, W. R. 1964. A numerical system of classification for chlorites and septechlorites. *Mineral. Mag., 33*, 1114–1124.

Tschermak, G. 1891. Chlorite group. *Sitz. Akad. Wiss. Wien, Math-nat. Kl., 100*, Abt. 1, 29.

OTHER PHYLLOSILICATES

Serpentine

$Mg_3Si_2O_5(OH)_4$

MONOCLINIC
$\angle\beta = 93^\circ$ or 90°

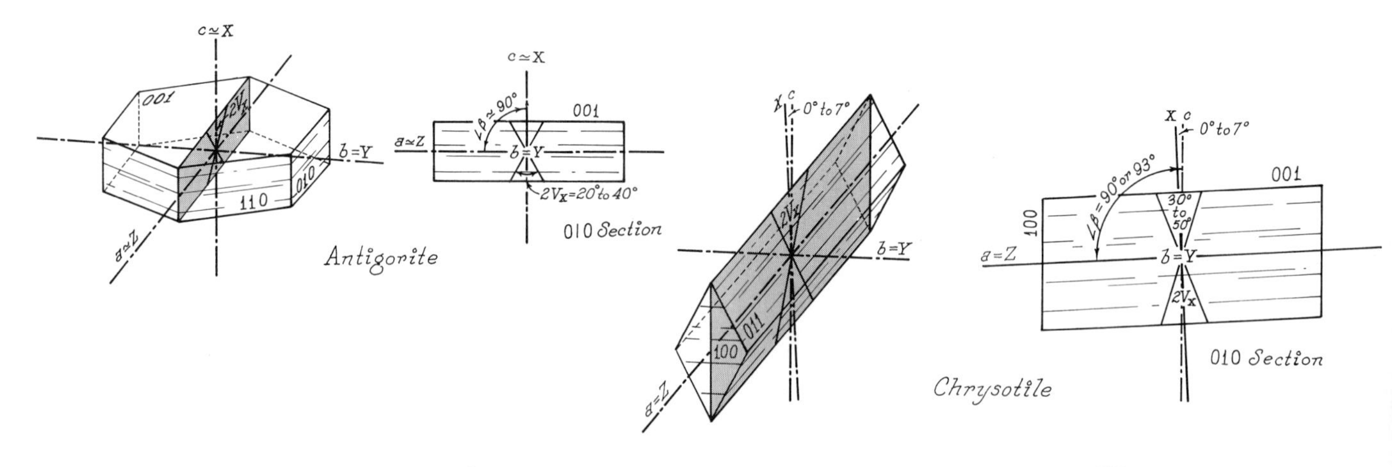

	n_α*	n_β	n_γ	$n_\gamma - n_\alpha$	$2V_x$
Antigorite	1.555–1.567	1.560–1.573	1.560–1.574	0.004–0.009	20°–60° (−)
Lizardite	1.538–1.561	—	1.546–1.567	0.007	—
Chrysotile	1.529–1.559	1.530–1.564	1.537–1.567	0.008	30°–50° (−)

Biaxial negative; $2V_x$ varies widely (20°–80°); may appear uniaxial

$b = Y, a \simeq Z, c \wedge X = 0^\circ$ to -7°

$r > v$ weak (antigorite)

Colorless to pale green or yellow in section, with weak pleochroism $Z > Y = X$

COMPOSITION AND STRUCTURE. All serpentine varieties have basically the same structure, a repeated two-layer arrangement of one tetrahedral (SiO_4) and one octahedral ($Mg(OH)_2$) layer (Fig. 6-15). Serpentine, $Mg_3Si_2O_5(OH)_4$, is the trioctahedral equivalent of kaolinite, $Al_2Si_2O_5(OH)_4$, and, by definition, the Mg-septechlorite. The basic two-layer unit is monoclinic, but polytypes may show different symmetry.

Three varieties of serpentine (antigorite, lizardite, and chrysotile) are distinguished by physical deformation, disposition of the basic two-layer, 7 Å sheet structure and concentration ranges in certain chemical components. One oxygen ion of each SiO_4 tetrahedron in the tetrahedral layer is bonded to

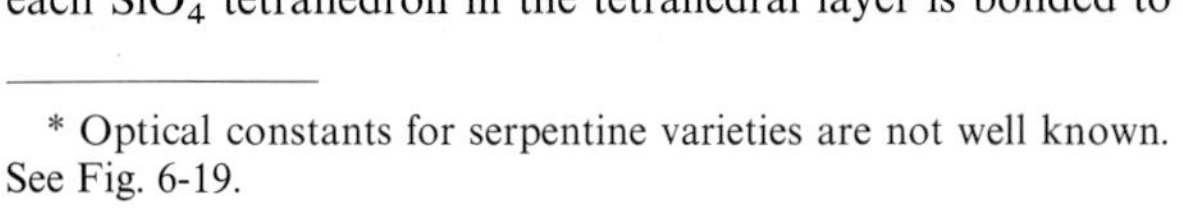

* Optical constants for serpentine varieties are not well known. See Fig. 6-19.

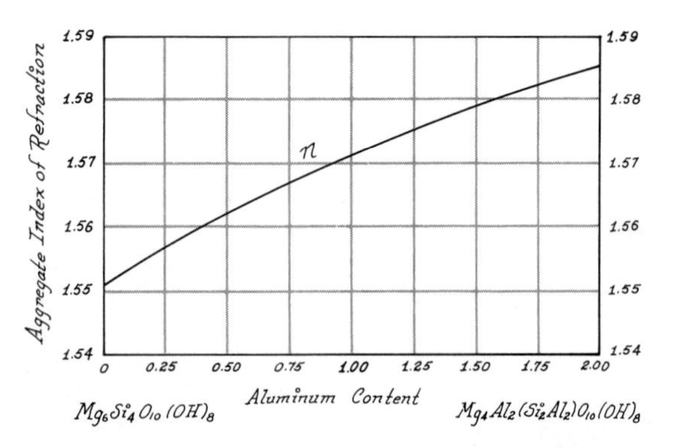

FIGURE 6-19. Variation of the aggregate refractive index with aluminum in tetrahedral coordination in synthetic Mg-serpentines. (After Chernosky, 1975)

Mg^{2+} ions and replaces an $(OH)^-$ at one corner of each anion octahedron in the octahedral layer to unite the two basic layers. The serpentine structure is strained, because tetrahedral and octahedral layers do not "fit." The tetrahedral layer is contracted about small Si ions, and the octahedral layer expanded about larger Mg ions. In three-layer, symmetrical structures (for example, talc, mica, and chlorite), where two tetrahedral layers sandwich the octahedral layer, this minor strain is balanced, and the mineral habit is foliated as large sheets. In two-layer structures (such as serpentine, kaolinite, and septechlorite), the strain is unbalanced, and the sheets tend to curl or form as very small flakes.

Chrysotile is normally fibrous (asbestos) and apparently consists of rolls or tubes of the basic, two-layer serpentine sheet, usually rolled, or elongated, on the *a* axis.

Lizardite is normally very fine grained and massive, resembling soapstone. It apparently consists of tiny flakes only a few tens of ångstroms in extent, over which distance, the layer strain may be small. Substitution of $2Al^{3+}$ for Si^{4+} and Mg^{2+} causes the tetrahedral layer to expand and the octahedral layer to contract, thereby reducing strain and allowing layers of Al-rich lizardite to grow much larger.

Antigorite is normally a foliated mineral, often occurring as large sheets. The layer structure of antigorite is apparently inverted over few tens of angstroms as structural "ribbons" parallel to *b*. The individual strips tend to curl, and antigorite sheets are submicroscopically corrugated, like sheets of corrugated iron.

The composition of natural serpentines is very near the ideal, but varies within limits characteristic of the particular variety (chrysotile, antigorite, or lizardite). Much Al^{3+} may replace Si^{4+} and Mg^{2+}, and Fe^{2+} or Fe^{3+} may replace Mg^{2+}, although significant substitutions form septechlorites. Pure Ni-serpentines can be synthesized, and *garnierite* $(Ni,Mg)_3Si_2O_5(OH)_4$ is a natural Ni-rich serpentine.

PHYSICAL PROPERTIES. $H = 2\frac{1}{2}–3\frac{1}{2}$. Sp. Gr. = 2.5–2.6. Serpentine, in hand sample, commonly displays a smooth or soapy feel and a silky, pearly, or waxy luster, depending on physical state. It is most commonly yellow-green to very dark green but may be white, gray, or brown.

COLOR AND PLEOCHROISM. Serpentine is colorless to pale green, in section or as fragments, with weak pleochroism $Z > Y = X$. X = colorless to pale yellow-green, Y = colorless to pale yellow-green or pale green, and Z = yellow to light green.

FORM. Chrysotile is commonly asbestiform, as parallel fibers or as aggregates with cross fibers, commonly as distinct veins. Antigorite is foliated as large sheets or as scaley or shredded aggregates. Lizardite is largely massive with netlike mesh structure and undulatory extinction (Fig. 6-20).

CLEAVAGE. Perfect basal cleavage {001} is characteristic of all sheet silicates. Antigorite, commonly shows characteristic foliation.

BIREFRINGENCE. Maximum birefringence of serpentine is weak to very weak (<0.01) and may be near zero on cleavage flakes. Interference colors, in standard section or as fragments, are grays to white of first order and may be anomalous, as pale yellow or pale olive. Extinction is often uneven, as netlike or mesh patterns.

TWINNING. Microscopic or megascopic twinning is rare, although submicroscopic, repeated twinning is probably very common and may be responsible for antigorite sheets.

INTERFERENCE FIGURE. Fine grain size, curled sheets, generally disrupted crystal structure, and very low birefringence favor poor interference figures. Antigorite sheets, like cryptoperthites, form a composite acute bisectrix figure of very broad and diffuse isogyres on a gray field. $2V_x$ may appear near zero, due to disordered sheet stacking. Dispersion is weak $r > v$.

OPTICAL ORIENTATION. The relation of crystal and optical directions is subject to dispute, although X is essentially normal to basal cleavage ($c \wedge X = 0°–7°$), and Y and Z are nearly indistinguishable and essentially parallel to crystallographic axes. Chrysotile fibers are elongated on $a = Z$, rarely on $b = Y$, and fibers are normally length-slow, with parallel extinction. Sections of antigorite sheets also show essentially parallel extinction and positive elongation (length-slow).

DISTINGUISHING FEATURES. Chrysotile fibers show parallel extinction, lower refraction index, and lower birefringence than asbestiform amphiboles. Antigorite shows much lower birefringence than micas and has the lowest refraction index among the chlorites.

Fe-bearing chlorites show stronger color and pleochroism, and most chlorites show stronger birefringence or anomalous interference colors. Brucite also commonly shows anomalous indigo colors and is uniaxial.

ALTERATION. Serpentine itself is the hydrothermal alteration product of olivine and Mg-pyroxenes, commonly with mag-

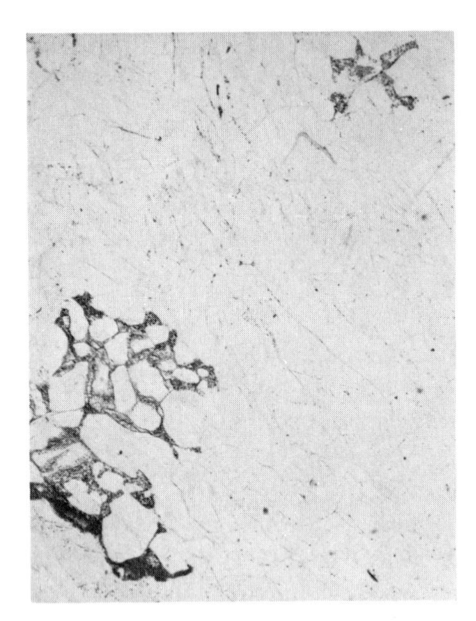

FIGURE 6-20. Antigorite commonly results form the low-temperature autometamorphism of olivine in peridotite-type rocks. The irregular fracture pattern of olivine is commonly preserved in the serpentinized rock.

netite as a by-product. It rarely alters to clinochlore and may be replaced by silica.

OCCURRENCE. The serpentine minerals are always secondary. They often occur together as the alteration products of Al-poor, magnesian minerals, especially olivine, Mg-pyroxenes, and Mg-amphiboles; less commonly, garnet, chondrodite, idocrase, and so on. The most characteristic occurrence of serpentine is that derived from dunites, peridotites, and pyroxenites, where it appears to form in the final cooling stages by reaction with residual water. Chrysotile is derived largely from olivine, by this low-grade hydrothermal or metamorphic process, and antigorite may form from chrysotile or enstatite and is favored by shearing stress. Serpentine formed in this way is commonly associated with chromite, magnetite, garnet, pyroxenes, amphiboles, chlorites, and talc.

A second occurrence of serpentine is in metamorphic contact zones in carbonate rocks, where it results from the alteration of forsterite marble as ophicalcite. Calcite, dolomite, magnesite, and the calc-silicates are common associates.

REFERENCES: SERPENTINE

Chernosky, J. V., Jr. 1975. Aggregate refractive indices and unit cell parameters of synthetic serpentine in the system MgO-Al_2O_3-SiO_2-H_2O. *Amer. Mineral.*, *60*, 200–208.

Jahanbagloo, I. C., and T. Zoltai. 1968. The crystal structure of a hexagonal aluminum serpentine. *Amer. Mineral.*, *53*, 14–24.

Krstanovic, I., and S. Pavlovic. 1967. X-ray study of six-layer orthoserpentine. *Amer. Mineral.*, *52*, 871–876.

Nagy, B., and G. T. Faust. 1956. Serpentines: Natural mixtures of chrysolite antigorite. *Amer. Mineral.*, *41*, 817–838.

Page, N. J. 1966. Mineralogy and chemistry of the serpentine group minerals and the serpentinization process. Unpublished doctoral dissertation, University of California at Berkeley.

Page, N. J., and R. G. Coleman. 1967. Serpentine-mineral analyses and physical properties. *U.S. Geol. Surv.*, *Prof. Paper*, *575-B*, 103–107.

Wicks, F. J., and E. J. W. Whittaker. 1975. A reappraisal of the structures of the serpentine minerals. *Can. Mineral.*, *13*, 227–243.

Talc $Mg_3Si_4O_{10}(OH)_2$

TRICLINIC
$\angle\alpha \simeq 90°30'$
$\angle\beta = 98°55'$
$\angle\gamma \simeq 90°$

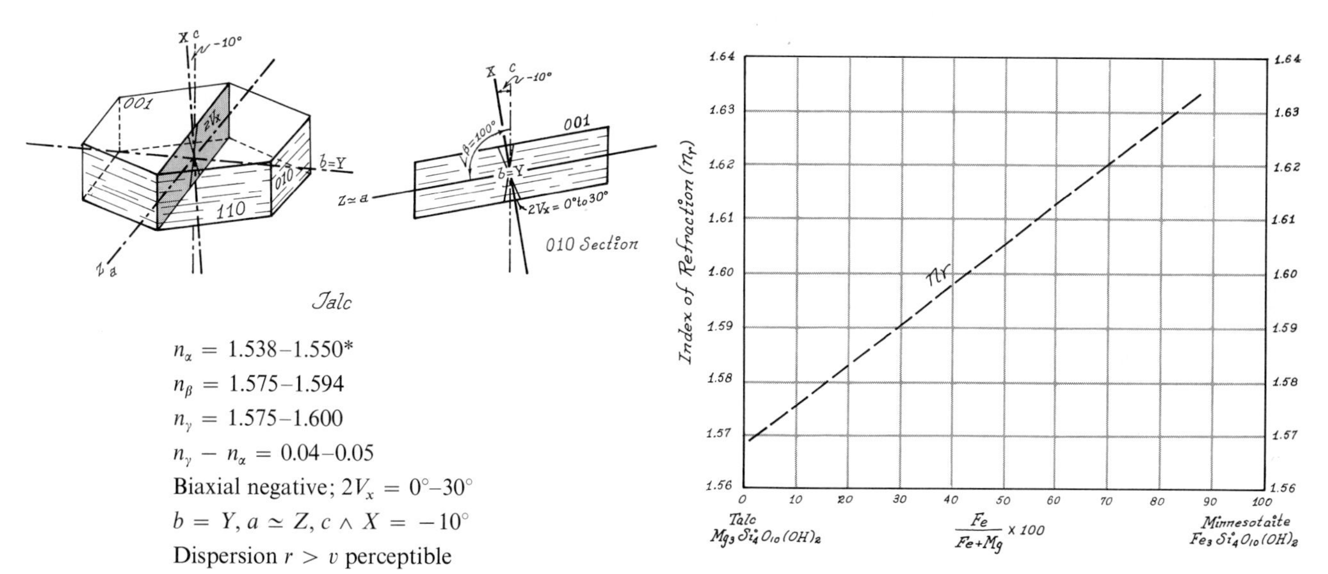

$n_\alpha = 1.538\text{–}1.550$*
$n_\beta = 1.575\text{–}1.594$
$n_\gamma = 1.575\text{–}1.600$
$n_\gamma - n_\alpha = 0.04\text{–}0.05$
Biaxial negative; $2V_x = 0°\text{–}30°$
$b = Y, a \simeq Z, c \wedge X = -10°$
Dispersion $r > v$ perceptible
Colorless in section

FIGURE 6-21. Variation of index of refraction (n_γ) in the talc-minnesotaite series.

COMPOSITION AND STRUCTURE. Talc is a three-layer, pseudo-monoclinic sheet-silicate (phyllosilicate); an octahedral layer is sandwiched between tetrahedral silicate layers (Fig. 6-22). Talc differs from the micas in that the three-layer sheet is electrically balanced and no interlayer cations are present to tie adjoining sheets; thus, talc has a soapy feel.

Very minor Al^{3+} may replace Si^{4+} in tetrahedral sites, and a few interlayer alkalis may balance the charge. All octahedral sites are filled by Mg^{2+}, with minor replacement by Fe^{2+}, Mn^{2+}, or Ni^{2+}. Nickel talc, $Ni_3Si_4O_{10}(OH)_2$, and *iron talc*, $Fe_3Si_4O_{10}(OH)_2$, are produced synthetically, and compositions near the iron analog appear in nature as *minnesotaite*. Dioctahedral talc, where $2Al^{3+}$ replaces $3Mg^{2+}$, is pyrophyllite, $Al_2Si_4O_{10}(OH)_2$.

PHYSICAL PROPERTIES. H = 1. Sp. Gr. = 2.6–2.8. Talc, in hand sample, displays a soapy feel and pearly luster and is white, colorless, or pale to dark green or gray and yellowish to red or brown by alteration.

COLOR AND PLEOCHROISM. Talc is colorless as fragments or in section. Minnesotaite is slightly colored and weakly pleochroic in pale green and yellow.

FORM. Talc is a foliated mineral appearing as an aggregate of flexible, inelastic sheets. It is commonly a fine-grained aggregate of fibers, shreds, or flakes strongly resembling colorless mica.

---

* Refraction indices increase with increasing Fe^{2+} or Ni^{2+}. Indices of pure synthetic talc are lower than those shown here, as shown in Fig. 6-21.

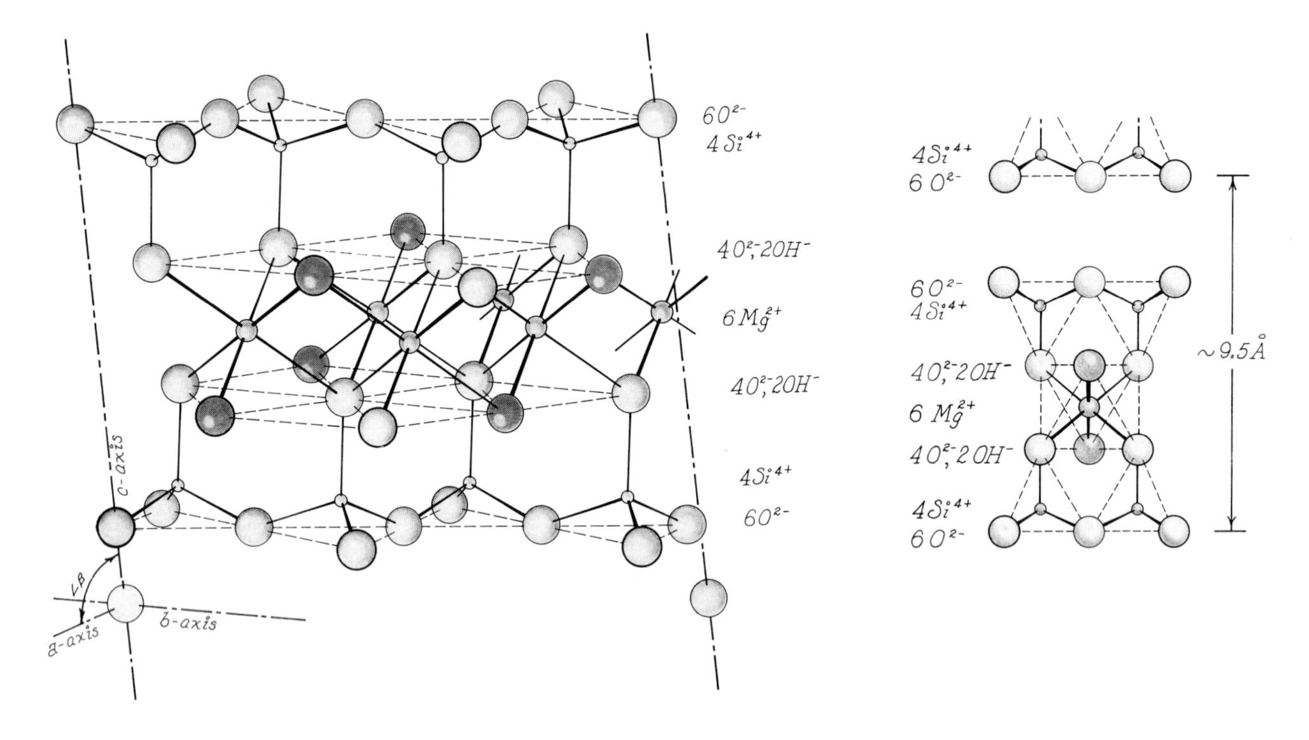

FIGURE 6-22. The idealized structure of talc: Repeat unit on c is two tetrahedral layers and one octahedral (trioctahedral) layer (~9.5 Å).

CLEAVAGE. Perfect basal cleavage {001} fails to produce extensive flat surfaces, due to bent and contorted sheets.

BIREFRINGENCE. Maximum birefringence, seen in {010} sections or essentially any section normal to cleavage, is strong, and low third-order colors are visible in standard section. Basal section and cleavage flakes show very low birefringence ($n_\gamma - n_\beta = 0.006$ or less).

TWINNING. Twinning is rare in talc but may produce pseudocubic crystals.

INTERFERENCE FIGURE. Offset stacking of adjacent sheets or twinning causes $2V_x$ to appear small, and most figures are sensibly uniaxial. Basal sections yield acute bisectrix (optic axis) figures with very minor isogyre separation, few color rings, and perceptible dispersion $r > v$.

OPTICAL ORIENTATION. Extinction is essentially parallel to cleavages, and elongation is positive (length-slow), as X is normal to cleavages.

DISTINGUISHING FEATURES. Talc closely resembles colorless mica and pyrophyllite. Muscovite and pyrophyllite have larger $2V_x$ and phlogopite and phengites show higher refractive indices. Brucite often shows anomalous blue interference colors and is uniaxial positive. Fine-grained talc and sericite may be impossible to distinguish optically.

ALTERATION. Talc, itself, is an uncommon alteration product of magnesium silicates (olivine, enstatite, and so on), and hydrothermal alteration of ultramafic rocks produces talc. CO_2 released by carbonate metamorphism combines with serpentine to yield talc and magnesite, and additional metamorphism and CO_2 may cause talc to decompose in favor of magnesite and quartz.

OCCURRENCE. The most common appearance of talc is in low-grade talc and talc-tremolite schists. It forms as the first new silicate mineral in the contact zones of impure dolomites. Talc is stable at slightly higher temperatures than chlorite or tremolite and may form at their expense by low-grade greenschist metamorphism. With progressive decarbonation talc disappears in favor of the higher-temperature calc-magnesium silicates. In this paragenesis, talc is commonly associated with calcite, dolomite, tremolite, quartz, serpentine, chlorite, and anthophyllite.

Hydrothermal alteration of peridotites and similar ultramafic rocks may yield massive impure talc rock (soapstone or steatite) by "steatitization," which is closely related to serpentinization. CO_2 metasomatism of serpentinite yields talc and magnesite on joints and shear planes.

REFERENCES: TALC

Blake, R. L. 1965. Iron phyllosilicates of the Cayuna district in Minnesota. *Amer. Mineral., 50*, 148–169.

Forbes, W. C. 1969. Unit-cell parameters and optical properties of talc on the join $Mg_3Si_4O_{10}(OH)_2$–$Fe_3Si_4O_{10}(OH)_2$. *Amer. Mineral., 54*, 1399–1408.

Gruner, J. W. 1934. The crystal structure of talc and pyrophyllite. *Z. Kristallogr., 88*, 412.

Gruner, J. W. 1944. The composition and structure of minnesotaite, a common iron silicate in iron formations. *Amer. Mineral., 29*, 363–372.

Hendricks, S. B. 1938. On the crystal structure of talc and pyrophyllite. *Z. Kristallogr., 99*, 264–274.

Rayner, J. H., and G. Brown. 1973. Crystal structure of talc. *Clays and Clay Minerals., 21*, 103–114.

Ross, M., W. L. Smith, and W. H. Ashton. 1968. Triclinic talc and associated amphiboles from Gouverneur mining district, New York. *Amer. Mineral., 53*, 751–769.

Smolin, P. P. 1967. Limits and nature of variations in the composition of talc minerals. *Dokl. Akad. Nauk SSSR, 172*, 187–190.

Pyrophyllite

$Al_2Si_4O_{10}(OH)_2$

MONOCLINIC
$\angle\beta = 100°$

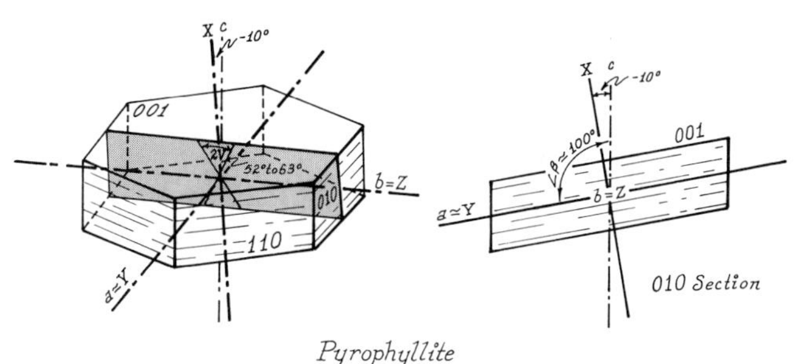

Pyrophyllite

$n_\alpha = 1.534–1.556$
$n_\beta = 1.586–1.589$
$n_\gamma = 1.596–1.601$
$n_\gamma - n_\alpha \simeq 0.050$
Biaxial negative; $2V_x = 53°–62°$
$b = Z, a \simeq Y, c \wedge X = -10°$
$r > v$ weak
Colorless in standard section

COMPOSITION. Pyrophyllite, $Al_2Si_4O_{10}(OH)_2$, is the dioctahedral analog of talc, $Mg_3Si_4O_{10}(OH)_2$, and has physical and optical properties similar to talc.

FORM. Pyrophyllite may appear as aggregates of sheets or flakes, rosettes of needles or fibers, or in a fine-grained, massive form resembling soapstone (Fig. 6-23).

OPTICAL ORIENTATION. The optic plane of pyrophyllite lies normal to {010} and nearly parallel to {100}. Extinction is

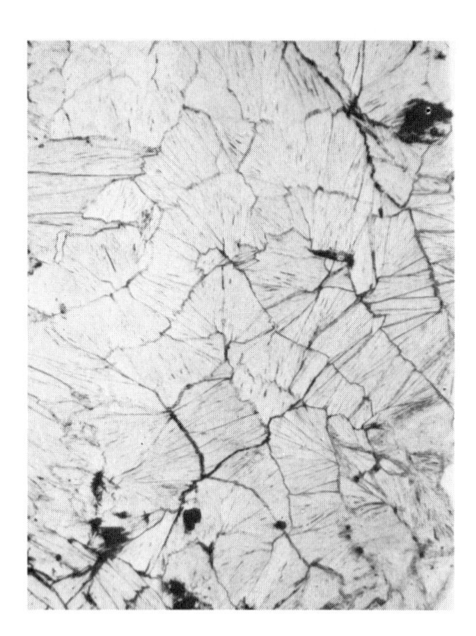

FIGURE 6-23. Pyrophyllite commonly appears as rosettes of flakes or sheets.

essentially parallel to cleavage traces, and elongation is always positive (length-slow), as X is normal to cleavages.

DISTINGUISHING FEATURES. Pyrophyllite shows larger $2V_x$ than talc or muscovite.

OCCURRENCE. Pyrophyllite is a relatively uncommon mineral that forms largely from hydrothermal alteration of feldspars and other aluminous silicates. It appears in schistose rocks and hydrothermal veins, in association with quartz, muscovite, and kyanite.

REFERENCES: PYROPHYLLITE

Haas, H., and M. J. Holdaway. 1973. Equilibria in the system Al_2O_3-SiO_2-H_2O involving the stability limits of pyrophyllite and thermodynamic data of pyrophyllite. *Amer. Jour. Sci.*, *273*, 449–464.

Rayner, J. H., and G. Brown. 1966. Structure of pyrophyllite. In *Clays and Clay Minerals, Proceedings of the 13th National Conference*, W. F. Bradley and S. W. Bailey (eds.). London: Pergamon Press.

Wardle, R., and G. W. Brindley, 1972. The crystal structures of pyrophyllite, 1Tc, and of its dehydroxylate. *Amer. Mineral.*, *57*, 732–750.

TRICLINIC

Stilpnomelane $(K,Na,Ca)_{0-1}(Fe^{3+},Fe^{2+},Mg,Mn,Al)_2(O,OH,H_2O)_{6-7}(Fe^{3+},Fe^{2+},Mg,Mn,Al)_{5-6}(Si_8O_{20})(OH)_4$

$\angle\alpha = 124°$
$\angle\beta = 96°$
$\angle\gamma = 120°$

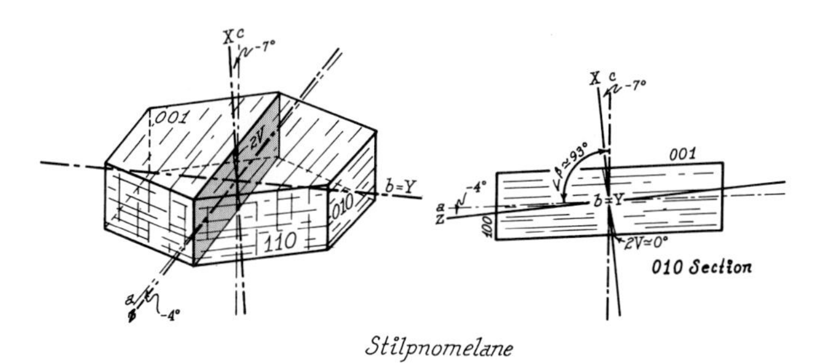

Stilpnomelane

$n_\alpha = 1.543–1.634$*
$n_\gamma = n_\beta = 1.576–1.745$*
$n_\gamma - n_\alpha = 0.030–0.110$*
Biaxial (uniaxial) negative, $2V_x \simeq 0°$
$b = Y, a \simeq Z$, and $c \wedge X \simeq -7°$
Highly colored in section and intensely pleochroic from yellow to dark brown or green
$Z = Y \gg X$

COMPOSITION. Stilpnomelane is a trioctahedral sheet silicate with an unusual structure. It contains five planes of cations parallel to (001). The central plane of each layer consists of octahedrally coordinated cations, flanked on each side by a tetrahedral sheet. These tetrahedral sheets consist of "islands" of 24 tetrahedra whose apices coordinate the octahedral cations, linked by trigonal rings of 6 tetrahedra whose apices point away from the octahedral sheet (see Fig. 6-25). Tetrahedra of the trigonal rings share apices with similar rings in the next layer. Thus there are no essential interlayer cations, although occasional K^+, Na^+, or Ca^{2+} appear as nonessential constituents.

Cations in octahedral coordination are dominantly Fe^{3+} (stilpnomelane or *ferristilpnomelane*), Fe^{2+} (*ferrostilpnome-*

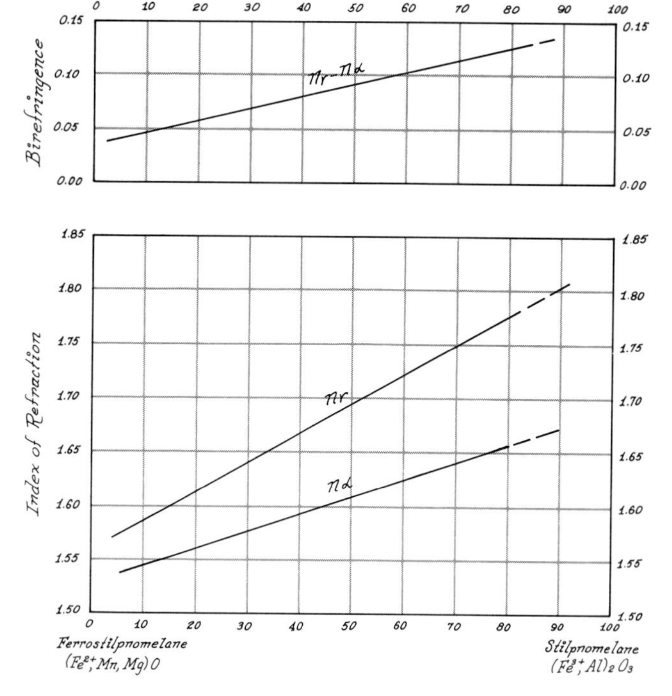

FIGURE 6-24. Variation of birefringence ($n_\gamma - n_\alpha$) and indices of refraction in the stilpnomelane-ferrostilpnomelane series. (After Blake, 1964)

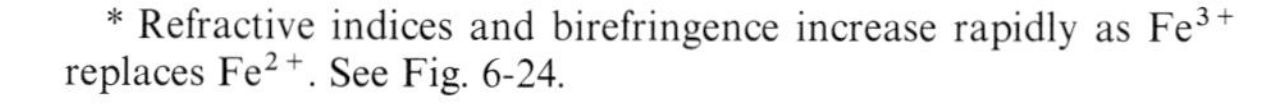

* Refractive indices and birefringence increase rapidly as Fe^{3+} replaces Fe^{2+}. See Fig. 6-24.

FIGURE 6-25. The tetrahedral sheet in stilpnomelane. (After Eggleton, 1970)

lane), or Mn^{2+} (*parsettensite*), with minor Mg^{2+}, Ti^{4+}, and Al^{3+}. Tetrahedral cations are largely Si^{4+}, with only minor Al^{3+} replacement.

PHYSICAL PROPERTIES. H = 3–4. Sp. Gr. = 2.59–2.96. In hand sample, ferric-rich varieties (stilpnomelane) are dark, reddish brown to black, ferrous-rich varieties (ferrostilphomelane) are deep green, and the manganostilpnomelane (parsettensite) is coppery red. Cleavage plates of the stilpnomelane minerals tend to be brittle and neither elastic nor flexible, compared to biotite and chlorite, respectively.

COLOR AND PLEOCHROISM. Stilpnomelane minerals (see Table 6-4) are intensely colored and highly pleochroic in browns and greens. $Z = Y > X$. Like biotite, stilpnomelane sections are dark when cleavage traces are parallel to polarization of the lower nicol.

TABLE 6-4. Stilpnomelane Color and Pleochroism

	X	$Y = Z$
Stilpnomelane	Golden yellow	Dark brown to black
Ferrostilpnomelane	Pale yellow	Dark green
Parsettensite	Colorless to pale yellow	Pale yellow-green

FORM. Micaceous plates of stilpnomelane are often arranged as radiating or sheaflike aggregates.

CLEAVAGE. Perfect basal {001} cleavage is less perfect than that of mica or chlorite, and an imperfect cleavage lies parallel to {010}.

BIREFRINGENCE. Maximum birefringence is seen on {010} sections and is highly variable, ranging from 0.03, for ferrous varieties (ferrostilpnomelane), to 0.110, for ferric composition (stilpnomelane). Interference colors in standard section range from brilliant second-order colors to fifth-order pastels. Basal sections are sensibly isotropic. Stilpnomelane does not show the pebbled surface typical of micas under crossed nicols.

ZONING. Color zoning often shows oxidation of outer zones (lighter Fe^{2+} core).

INTERFERENCE FIGURE. Basal sections or cleavage plates are highly absorbing and show essentially optic axis-centered, uniaxial, negative figures with numerous darkened isochromes. $2V_x$ is essentially 0°; however, minor isogyre separation may be observed, and angles as large as 40° are reported.

OPTICAL ORIENTATION. $b = Y$, $a \simeq Z$, and X lies normal to cleavage plates within 7° of c. Extinction is parallel to cleavage in all sections, and elongation is positive (length-slow), since the fast wave vibrates normal to basal cleavage.

DISTINGUISHING FEATURES. Stilpnomelane is commonly mistaken for biotite or chlorite. It may show a second minor cleavage, however, and does not show the "bird's-eye maple" surface of micas, under crossed nicols. Birefringence is much greater than that of chlorite, and ferric varieties show birefringence too great for biotite.

ALTERATION. Ferrostilpnomelane apparently oxidizes to stilpnomelane, and the latter should alter to iron oxides and clays.

OCCURRENCE. Stilpnomelane is most characteristic of low-grade, regional metamorphic rocks, where it appears with chlorite, sericite, epidote, albite, almandine, and actinolite in slates and schists of the greenschist facies, derived from graywackes and other iron-rich sediments. It also forms under high pressure and low temperature with glaucophane, lawsonite, and other minerals of the glaucophane schist facies. Stilpnomelane is a major constituent of the mildly metamorphosed iron-silicate formations of the Lake Superior region, in association with Fe-talc, Fe-chlorite, and hematite, and it appears in veins that cut these formations.

REFERENCES: STILPNOMELANE

Blake, R. L. 1964. Some iron phyllosilicates of the Cayuna and Mesabi districts of Minnesota. U.S. Bureau of Mines Report, Investigation 6394.

Eggleton, R. A. 1970. Silicon tetrahedral sheet in stilpnomelane. *Nature*, *225*, 625–626.

Grout, E. F., and Thiel, G. A. 1924. Notes on stilpnomelane. *Amer. Mineral.*, *9*, 228–231.

Gruner, J. W. 1944. The structure of stilpnomelane re-examined. *Amer. Mineral.*, *29*, 291–298.

Hashimoto, M. 1969. A note on stilpnomelane mineralogy. *Contrib. Mineral. Petrol*, *23*, 86–88.

Hutton, C. O. 1938. The stilpnomelane group of minerals. *Mineral. Mag.*, *25*, 172–206.

Hutton, C. O. 1945. Additional optical and chemical data on the stilpnomelane group of minerals. *Amer. Mineral.*, *30*, 714–718.

Hutton, C. O. 1956. Further data on the stilpnomelane mineral group. *Amer. Mineral.*, *41*, 608–615.

Smith, M. L., and C. Frondel. 1968. Related layered minerals granophyllite, bannisterite, and stilpnomelane. *Mineral. Mag.*, *36*, 893–913.

Turner, F. J., and Hutton, C. O. 1935. Stilpnomelane and related minerals as constituents of schists of western Otago, New Zealand. *Geol. Mag.*, *72*, 1–8.

Prehnite

$Ca_2Al(AlSi_3)O_{10}(OH)_2$ ORTHORHOMBIC

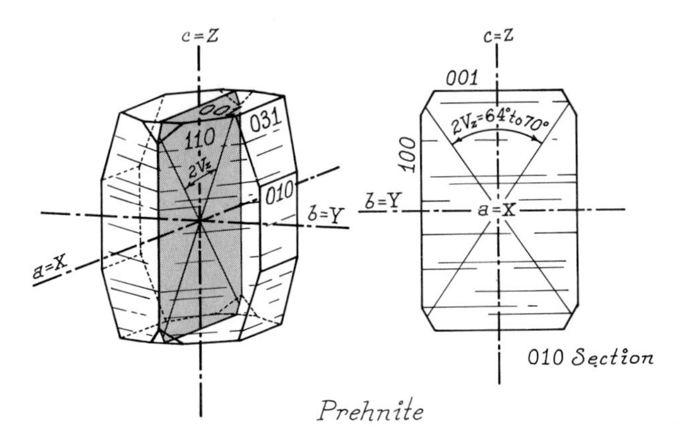

n_α = 1.610–1.637*
n_β = 1.615–1.647
n_γ = 1.632–1.673
$n_\gamma - n_\alpha$ = 0.020–0.035
Biaxial positive, $2V_z$ = 60°–70°*
$a = X, b = Y, c = Z$
Usually $r > v$ weak, sometimes $v > r$ weak to strong
Colorless in standard thin section

* Indices of refraction increase with substitution of Fe^{3+} for Al^{3+}. See Fig. 6-26. $2V_z$ often appears less than indicated values. Anomalous crystal sectors are rather common and may show abnormal interference colors, nonextinction, abnormally small $2V$ with crossed dispersion, and twinning.

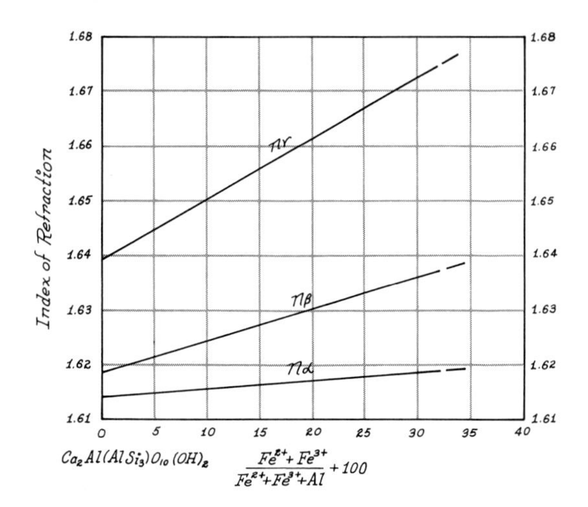

FIGURE 6-26. Variation of indices of refraction with iron (Fe^{2+} + Fe^{3+}) in prehnite. (After Zolotukhin, Vasilév, and Zyuzin, 1965)

COMPOSITION AND STRUCTURE. Prehnite is probably a sheet-type silicate and may be related to the brittle micas. Its composition shows surprisingly little variation, although significant (~30 weight percent) Fe^{3+} may replace Al^{3+}, and the Si:Al ratio may vary slightly.

PHYSICAL PROPERTIES. H = 6–6½. Sp. Gr. = 2.80–2.95. In hand sample, prehnite is pale green to yellow, sometimes white to gray.

COLOR AND PLEOCHROISM. Prehnite is colorless in section or as fragments.

FORM. Prehnite usually occurs as globular or botryoidal masses of radiating crystals. Columnar or platy aggregates are often sheaflike or fan-shaped ("bow-tie" structure). Single crystals are rare as tabular or prismatic forms. Individual crystals often show radiating sectors or "hourglass" segments of differing optical orientation.

CLEAVAGE. Distinct basal cleavage {001} dominates fragment orientation. Prismatic cleavages {110} are poor.

BIREFRINGENCE. Maximum birefringence, seen in {010} sections, is moderate ($n_\gamma - n_\alpha$ = 0.020–0.035). Interference colors are brilliant, ranging to middle second-order, and may be anomalous.

TWINNING. Two sets of fine lamellar twinning, intersecting at high angles and resembling microline twinning, may be present.

ZONING. Color zoning is rather common and usually grades from a darker core.

INTERFERENCE FIGURE. Prehnite is biaxial positive, with rather large $2V_z = 60°–70°$. Dispersion is usually weak $r > v$; however, abnormal crystal sectors may show smaller $2V$, possibly reversed dispersion ($v > r$ weak to strong) and even crossed bisectrix dispersion.

Basal sections and cleavage plates show centered acute bisectrix figures.

OPTICAL ORIENTATION. Extinction is parallel to cleavage traces or crystal elongation. Prismatic crystals, elongated on $c = Z$, are length-slow (positive elongation); however, the faster wave vibrates parallel to basal cleavage and tabular crystals (negative elongation). Some crystals may be composed of submicroscopic layers normal to {001} and having opposite crystallographic orientations.

DISTINGUISHING FEATURES. Prehnite may be distinguished from minerals of similar occurrence (secondary cavity filling) by "hourglass" or "bow-tie" structure and anomalous optics (incomplete extinction, abnormal interference colors, and so on). Prehnite has greater birefringence than zeolites, topaz, danburite, lawsonite, wollastonite, or andalusite. Datolite has higher birefringence and is biaxial negative.

ALTERATION. Prehnite may alter to zeolites (scolecite) or chlorite and may be transformed to epidote and grossular by mild metamorphism. Prehnite may be a constituent of sausserite alteration derived from plagioclase or pyroxene.

OCCURRENCE. Prehnite occurs most commonly in amygdules, cavities, or veins in basalt, andesite, diabase, and similar rocks with zeolites, epidote, chlorite, datolite, calcite, chalcedony, and pectolite. It is less common in veins and cavities in granite, monzonite, or pegmatite. Prehnite may occur in contact metamorphic marbles with calcite, wollastonite, grossular, diopside, and other calc-silicate minerals. It may be a constituent of saussuritic alteration of plagioclase feldspar, and zoisite-prehnite rocks may result from alteration of mafic igneous rocks. It has been observed in sulfide-bearing trap intrusions in a relatively high-temperature paragenesis.

REFERENCES: PREHNITE

Aumento, F. 1968. Space group of prehnite. *Can. Mineral., 9*, 485–492.

Liou, J. G. 1971. Synthesis and stability relations of prehnite, $Ca_2Al_2Si_3O_{10}(OH)_2$. *Amer. Mineral., 56*, 507–531.

Matsuedo, H. 1975. Iron-rich prehnite from the skarn of Sampo mine, Okayama Prefecture, Japan. *Kyu. Daig. Rigak. Ken. Hokoku, Chish., 12*, 91–100.

Papike, J. J., and T. Zoltai. 1967. Ordering of tetrahedral aluminum in prehnite, $Ca_2(Al,Fe^{3+})[Si_3AlO_{10}](OH)_2$. *Amer. Mineral., 54*, 974–984.

Zolotukhin, V. V., Yu. R. Vasil'er, and N. I. Zyuzin. 1965. High-iron variety of prehnite and a new diagram for prehnites. *Dokl. Akad. Nauk SSSR, 164*, 138–141.

Apophyllite

$KCa_4(Si_4O_{10})_2F \cdot 8H_2O$

TETRAGONAL

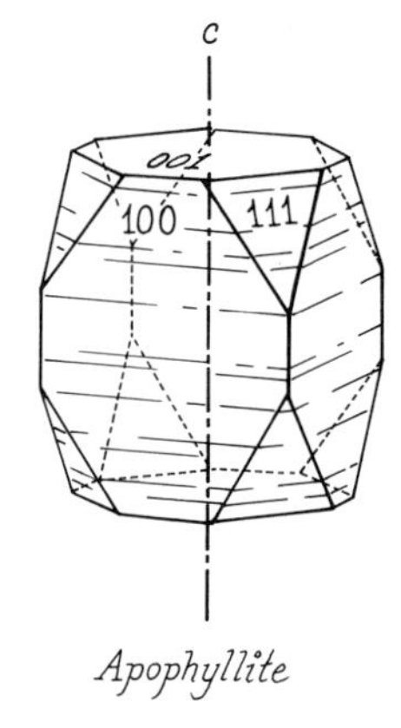

Apophyllite

$n_\omega = 1.531–1.536$*

$n_\varepsilon = 1.533–1.538$

$n_\varepsilon - n_\omega = 0.000–0.003$*

Uniaxial positive, rarely negative*

Colorless in standard section

* Most apophyllite is uniaxial positive; however, some varieties are essentially isotropic ($n \simeq 1.542$), and others are uniaxial negative, with greater indices of refraction ($n_\omega = 1.537–1.545$ and $n_\varepsilon = 1.537–1.544$).

COMPOSITION AND STRUCTURE. Apophyllite is a sheet silicate, each silicate tetrahedron sharing three of four oxygen ions to form Si_4O_{10} sheets parallel to {001}. The hexagonal symmetry of most silica sheets is replaced by unusual fourfold and eightfold rings yielding tetragonal symmetry. Ions of K^+, Ca^{2+}, and F^- link successive sheets, and water molecules lie between sheets coordinated with the cations.

Apophyllite composition is usually near the ideal, but some Na may replace K, and the Si:Al ratio may vary somewhat. Some $(OH)^-$ apparently replaces F^- as F-poor apophyllite is water-rich.

PHYSICAL PROPERTIES. H = $4\frac{1}{2}$–5. Sp. Gr. = 2.33–2.37. Apophyllite is colorless or white in hand sample; rarely, pink, yellow, or green. Luster is vitreous to distinctly pearly on {001}.

COLOR AND PLEOCHROISM. Apophyllite is colorless in section or as fragments.

FORM. Apophyllite is platy; basal {001} plates may build to prismatic crystals elongated on *c*.

CLEAVAGE. Very perfect basal cleavage {001} controls fragment orientation. Poor prismatic cleavages {110} are seldom apparent.

BIREFRINGENCE. Prismatic sections show maximum birefringence, which is nil to very weak ($n_\varepsilon - n_\omega = 0.000–0.003$) and commonly anomalous. Apophyllite may be isotropic for any wavelength, and interference colors are first-order gray with anomalous tints.

TWINNING. Rare twinning may be observed on {111}; penetration twinning also exists.

INTERFERENCE FIGURE. Apophyllite is normally uniaxial positive; however, some varieties are isotropic, others uniaxial negative, and some are even biaxial, with $2V$ up to 60° and crossed dispersion of optic axis planes (see brookite, p. 22). Basal sections and cleavage plates yield centered optic axis figures showing a broad isogyre cross and no isochromes.

OPTICAL ORIENTATION. Extinction is parallel to cleavage or crystal elongation, and the fast wave normally vibrates parallel to cleavage traces (negative elongation to basal cleavage).

DISTINGUISHING FEATURES. Apophyllite is tetragonal, with perfect basal cleavage, negative relief, and very low anomalous birefringence. Zeolites normally show lower indices of refraction and different habit or cleavage.

ALTERATION. Apophyllite usually alters to opal, chert, kaolin clays, and calcite.

OCCURRENCE. The major occurrence of apophyllite is in amygdules, vugs, and fissures in basalt, diabase, and similar volcanics, where it is deposited in association with zeolites, prehnite, pectolite, calcite, and datolite. It is much less common in cavities and veins in granite, aplite, or granite gneiss and appears rarely in contact-metamorphic marbles and nepheline syenite veins with natrolite, arfvedsonite, aegirine, and apatite. Apophyllite is a minor alteration product of wollastonite.

REFERENCES: APOPHYLLITE

Coleville, A. A., C. P. Anderson, and P. M. Black. 1971. Refinement of the crystal structure of apophyllite. *Amer. Mineral.*, *56*, 1220–1231.

Sowani, P. V., and A. V. Pahdke. 1964. A note on apophyllite occurring around Poona. *J. Univ. Poona*, *28*, 81–83.

Taylor, W. H., and St. Naray-Szabo. 1931. The structure of apophyllite. *Z. Kristallogr.*, *77*, 146–158.

CHAPTER 7

Tektosilicates

Tektosilicates, or framework silicates, possess crystal structures in which each oxygen ion is shared by two tetrahedra; thus each anion is a bridging oxygen. Since the electrostatic bond strength of the silicon-oxygen bond

$$\frac{\text{charge on the Si ion}}{\text{coordination number of the Si ion}} = \frac{4}{4} = 1$$

is exactly equal to one-half the anion charge, each divalent oxygen ion is electrically neutralized by bonds to two cations, if there are no substituents replacing some of the silicon. Likewise, bonds from four anions satisfy the charge on the silicon ions. Each cation effectively "owns" one-half of each anion of its tetrahedron, and the silicon-oxygen ratio over the infinite structure is, therefore, Si:O = 1:2. Consequently, the only possible electrostatically neutral framework silicates would be the silica poymorphs, with composition SiO_2, were it not for the common substitution of other cations (principally Al^{3+}) for silicon in some of the tetrahedra. In the tektosilicates, the tetrahedral Si:Al ratio is usually a simple one, commonly ranging between Si:Al = 3:1 and Si:Al = 1:1. Such substitutions in the tetrahedral framework result in a deficiency of positive charge that is balanced by the introduction of large monovalent or divalent cations (such as Na^+, K^+, and Ca^{2+}) into voids in the structure. Thus we have minerals such as nepheline, $Na(SiAl)O_4$; orthoclase, $K(Si_3Al)O_8$; and anorthite, $Ca(Si_2Al_2)O_8$.

Because the tektosilicates, with four bridging oxygens per tetrahedron, constitute three-dimensional tetrahedral networks, there are no strongly anisotropic structural units such as chains or sheets. Consequently, there is little consistent planar or linear polarization of the electron distribution, and a similar ion polarization is presented to light waves irrespective of propagation or vibration direction. Thus all light rays travel with similar velocities, and the maximum birefringence of tektosilicates is small.

Silica

Although quartz (α-quartz or low quartz) is the only form of primary free silica to occur in plutonic rocks, uncombined SiO_2 is known to exist in a dozen or more structural modifications, each requiring some combination of physical conditions (Table 7-1). About half the known forms are synthetic, without known representation in nature. All silica structures consist of silicate tetrahedra (SiO_4^{4-}) linked together on all four corners to form a framework structure, wherein every silicon ion occurs in four fold (tetrahedral) oxygen coordination and every oxygen is coordinated between two silicons. The correlation between density and refractive index is well illustrated by the silica polymorphs, as shown in Fig. 7-1.

The three major polymorphs of natural SiO_2 are quartz, tridymite, and cristobalite, each having a distinctive crystal structure stable over a specific temperature-pressure range, and each showing a high-temperature (β) and low-temperature (α) form of closely related structure.* High-temperature modifications tend to be open structures of higher symmetry (Fig. 7-2). Both cristobalite and tridymite may persist, and may even crystallize directly, below their thermodynamic stability ranges. Their low-temperature forms are found in volcanic rocks of relatively recent geologic origin, in which they may represent a major percentage of a given flow. They exist metastably in nature, because of sluggish (reconstructive) transitions to other polymorphs, which require breaking and reforming of bonds. The high- and low-temperature (β and α) modifications of each major form represent only slightly structural adjustments (displacive), and inversion to α forms is therefore rapid.

Low quartz is the most common form of uncombined silica in volcanic rocks. β-quartz apparently crystallizes in some flows but inverts to α-quartz on cooling below 573°C, and only the α form exists at surface conditions. Opal and the cryptocrystalline forms of quartz introduced as colloids by meteoric waters are very common in veins, vesicles, vugs, and other openings in tuffs and volcanic rocks.

Coesite and stishovite are high-pressure polymorphs of SiO_2 found at meteor impact sites. Unlike the other silica polymorphs, stishovite is not a tetrahedral framework structure, but has the rutile (TiO_2) structure (Fig. 1-8). This is the only known mineral in which Si^{4+} is in other than tetrahedral coordination.

---

* Tridymite has high-, middle-, and low-temperature forms. Middle tridymite is never stable but may exist metastably between 117° and 163°C. There appear to be significant differences in symmetry between terrestrial and meteoritic low tridymites.

TABLE 7-1. Structural Forms of Silica

α-Quartz	Stable below 573°C	Trigonal
β-Quartz	Stable from 573° to 870°C	Hexagonal
α-Tridymite	Never stable; exists metastably below 117°C	Orthorhombic (terrestrial) Monoclinic (meteoritic)
β_1-Tridymite	Never stable; exists metastably from 117° to 163°C	Orthorhombic
β_2-Tridymite	Stable from 867° to 1470°C; exists metastably from 163° to 867°C	Hexagonal
α-Cristobalite	Never stable; exists metastably below 267°C	Tetragonal (pseudoisometric)
β-Cristobalite	Stable from 1470° to 1723°C (MP); exists metastably from 267° to 1470°C[a]	Isometric
Coesite	High-pressure phase found in meteor impact craters; stable at pressures above $P = 19.5 + 0.0112\ T$, from 700–1700[a]	Monoclinic
Stishovite	High-pressure phase found in meteor impact craters; stability field unknown	Tetragonal
Melanophlogite	Crystallizes at low T and P, stability field unknown; inverts to cristobalite at high T (~1000°C) and to quartz under mechanical stress[a]	Isometric
Bobkovite	Stability field unknown; inverts to cristobalite on heating	Isometric
Keatite	Synthetic phase produced at 380–585°C, 330–1200 atm: stability range unknown	Tetragonal

[a] MP = melting point; T = temperature (°C); P = pressure (kb, kilobars)

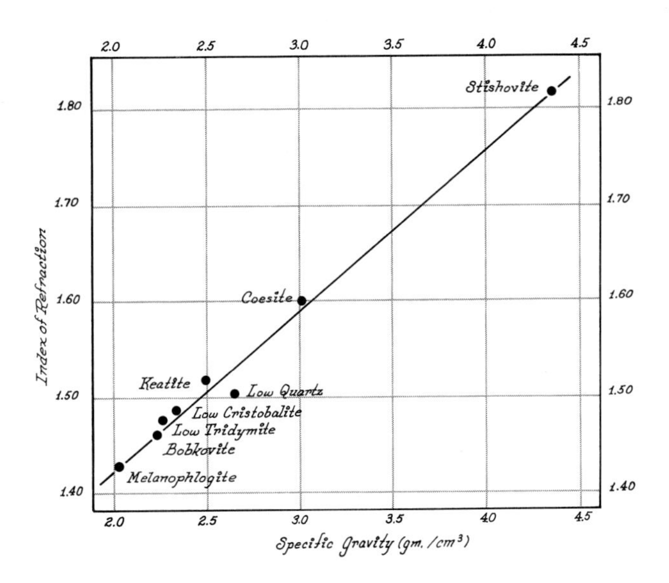

FIGURE 7-1. Index of refraction and specific gravity of the silica polymorphs.

α-Quartz SiO_2 TRIGONAL

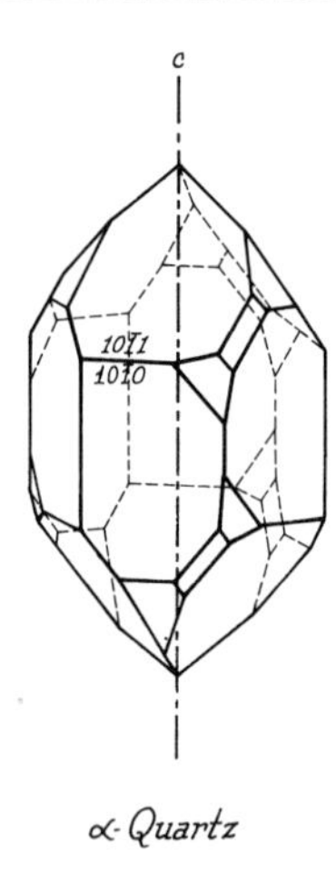

α-Quartz

$n_\omega = 1.544$
$n_\varepsilon = 1.553$
$n_\varepsilon - n_\omega = 0.009$
Uniaxial positive
Colorless in standard section

COMPOSITION AND STRUCTURE. Quartz is essentially pure SiO_2 without major or consistent impurities, and its optical properties are equally consistent. The quartz crystal structure is shown in Fig. 7-2. Of the common natural substances, only diamond, graphite, and ice are less variable in composition. A few Al^{3+} ions may replace Si^{4+} ions, usually coupled with the introduction of small alkali ions (Li^+ or Na^+). This substitution is much more important in tridymite and cristobalite, where the structure is sufficiently open to allow significant introduction of interstitial ions. Amethyst and citrine are apparently colored by minor ferric iron. The colors of rose quartz and blue quartz are attributed to minor titanium as a silicon substituent and usually contain very fine needles of rutile as an exsolution phase. Germanium is a possible substitute for some silicon but is very minor in quartz analyses and is much more abundant in some silicates.

PHYSICAL PROPERTIES. H = 7. Sp. Gr. = 2.65. In hand sample, quartz is most commonly colorless and transparent, with vitreous luster and classical conchoidal fracture. Color, however, is highly variable, and many varieties of coarse crystalline quartz are defined by color—*amethyst* (violet), *citrine* (yellow-brown), *smoky quartz* (black-brown), and *rose quartz* (pink).

COLOR AND PLEOCHROISM. Quartz is normally colorless and free of alteration as fragments or in thin section. It may, however, show numerous tiny, liquid inclusions or inclusions of other minerals (such as rutile, chlorite, hematite, tourmaline, sillimanite, and apatite). Colored quartz varieties show very uneven color distribution, which is seldom intense enough to show in fragments or section. Color, if visible, is pleochroic $E > O$.

FORM. In plutonic rocks, quartz is typically anhedral and interstitial as a late-forming mineral. Vermicular blebs of quartz as graphic intergrowths with K-feldspar (micropegmatite) or plagioclase (myrmekite) are also common in siliceous, plutonic igneous rocks (such as pegmatite and granite). Vein quartz may show crude lamellar structure. Euhedral crystals as wall lining in veins may show systematic malformation due to consistent direction of solution flow; a lesser thickness being deposited on the leeward side as revealed by concentric growth zones of equivalent faces (phantoms).

In rhyolite and related volcanic rocks, quartz phenocrysts are often euhedral as stubby, doubly terminated, trigonal crystals, often showing rounding or embayments of resorption. Although natural quartz is always the low-temperature form, hexagonal bipyramids with negligible prism faces suggest original crystallization above 573°C as β-quartz, especially when crystals are highly fractured. Quartz may also be anticipated as part of a fine, granular groundmass or in fibrous form as spherulites.

Detrital grains are often rounded or finely pitted by transportation and may show optically continuous, authigenic overgrowths. Euhedral, doubly terminated, authigenic crystals occur in some sediments.

Metamorphic quartz commonly appears as elongated, interlocking grains and may show lamellar structure (Boehm lamellae) due to optical disorientation on closely spaced, parallel planes.

Undulatory, or patchy, extinction is one of the most characteristic features of plutonic quartz, especially in metamorphic rocks, and in veins and pegmatites (Fig. 7-3). It results

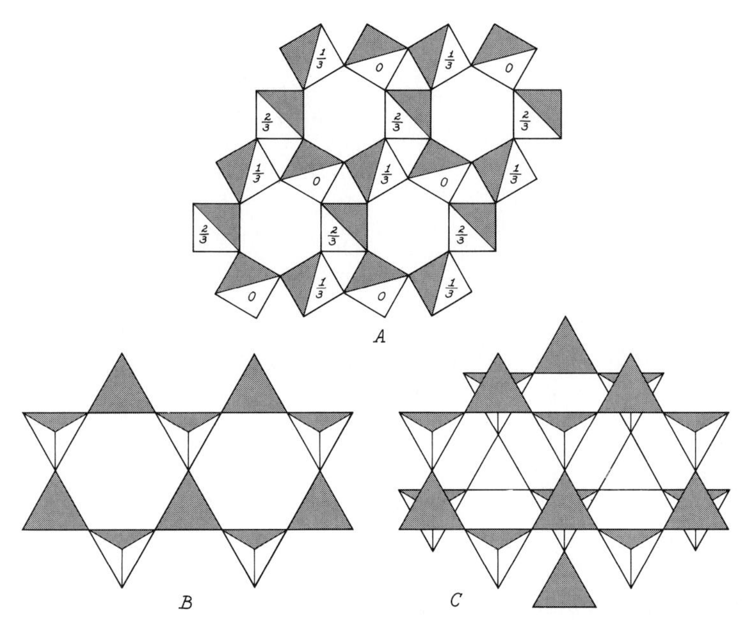

FIGURE 7-2. Idealized structures of major SiO_2 polymorphs to the same scale. Each tetrahedron represents an SiO^{4-} group. Minerals represented are (A) quartz (fractions indicate height on the *c*-axis); (B) tridymite (the sheet extends indefinitely in two dimensions and is attached to identical sheets above and below; (C) cristobalite (sheets are identical but "point" in opposite directions).

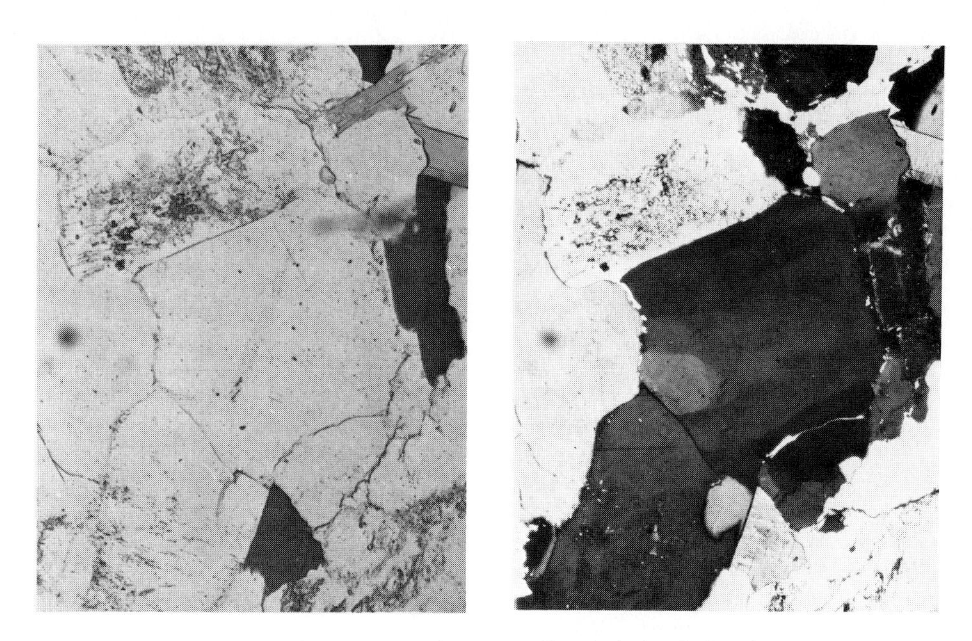

FIGURE 7-3. α-Quartz grains are usually anhedral and often show nonuniform (undulatory) extinction between crossed nicols.

from mechanical strain caused by external stresses of metamorphism or of confined growth and produces disorientation of irregular, adjacent, microscopic patches that may dislocate the optic axis by several degrees.

CLEAVAGE. Quartz seldom shows cleavage in section, and fragments show essentially random orientation. Several difficult cleavages are known to exist in quartz, and massive pegmatite or vein quartz may show rather distinct rhombohedral $\{10\bar{1}1\}$ cleavage, giving rise to crude lamellar form.

BIREFRINGENCE. Weak maximum birefringence (0.009) is seen on sections parallel to *c*. Interference colors, in thin section, are first-order grays and white to pale "straw"-yellow, and fragments usually show first-order and low second-order colors. Quartz is commonly used as the standard for determining section thickness, as its maximum birefringence is invariant.

TWINNING. Although twinning is rarely observed in sections or fragments of quartz, at least one twin law is almost universally present. The Dauphine law is by far the most abundant type, but goes unnoticed as a parallel twin with *c* as the twin axis and an irregular, vertical composition plane. Twin parts are congruent (both either right- or left-hand), and indicatrix orientation is identical in both parts. The Brazil law is the less common parallel twin, and twin parts are related by reflection across the vertical prism $\{11\bar{2}0\}$. Again, indicatrix orientation is unchanged, and the twin goes unnoticed in normal thin sections. Twin parts, however, are enantiomorphous (one right- and one left-hand), and optical rotation is reversed on adjoining parts and may be detected in very thick sections.

Several rare, simple, normal twin laws, of which the Japan laws are best documented, are known in quartz and may be observed in fragments and section. They are simple, knee-type twins with pyramidal twin planes, and indicatrix orientation differs appreciably on adjoining twin parts.

INTERFERENCE FIGURE. Random orientation of fragments and sections favors no figure type. Sections showing negligible birefringence yield a figure of broad black isogyres, usually somewhat off-center, on a white field. Biaxial figures with $2V$ as large as 10–20° may accompany distinct undulatory extinction as the result of mechanical strain, and amethyst and smoky varieties may display similar or even greater biaxial character.

Quartz is said to be *optically active*, meaning that the plane of polarized waves moving parallel to the optic axis is rotated to the right or left by the spiral crystal structure of right-or left-hand quartz. The angle of rotation of the vibration plane per mm of thickness is called the *rotatory power* and is negligible in sections of normal thickness. A very thick (1 mm or more) basal section of quartz, however, yields a field of many isochromes with distinct black isogyre cross except near the intersection, where it is obscured by a circle of yellow light as the result of this rotatory effect. Right-hand quartz rotates the vibration plane to the right (clockwise), and clockwise rotation of the *E-W* analyzer nicol causes the central color to ascend the color scale to colors of higher order and causes isochromes to move toward the center.

OPTICAL ORIENTATION. Crystals elongated on *c* show parallel extinction to prism faces and positive elongation (length-slow), since the extraordinary ray is the slower ray. It is often apparent that optical, and hence crystallographic, orientation is not exactly uniform throughout a given crystal, as evidenced by a nonuniform, or undulatory, extinction representing an optic axis displacement of several degress in adjoining extinction areas.

DISTINGUISHING FEATURES. Quartz is distinguished from alkali feldspar by its positive relief in balsam, lack of alteration and cleavage, and uniaxial figure. Quartz lacks the multiple twinning of most plagioclase and differs from untwinned oligoclase by uniaxial figure and lack of cleavage. Cordierite, beryl, and nepheline resemble quartz in hand sample, as fragments or in thin section where cordierite and beryl consistently associate with quartz. Cordierite is distinctly biaxial, is frequently twinned (polysynthetic), and alters readily; beryl and nepheline are optically negative, with somewhat higher and lower refractive indices respectively.

ALTERATION. α-Quartz is one of the most chemically stable minerals in nature at surface conditions and is characteristically free from alteration. Under exceptional conditions, it may be partially replaced by sericite, talc, chlorite, pyrite, magnetite, and certain clays.

OCCURRENCE. α-Quartz is probably the most abundant single mineral variety making up about 12 percent of the earth's crust, mainly as an essential constituent in acidic igneous rocks. The student may expect to observe quartz in all rocks except basic and undersaturated igneous varieties.

Quartz is essential in alaskites, granites, adamellites, granodiorites, quartz diorites, and similar plutonic igneous rocks. Quartz represents free, uncombined silica and is therefore incompatible with nepheline, leucite, and other feldspathoids. It is, similarly, not associated with forsteritic olivine, corundum, perovskite, and other silica-deficient minerals. Quartz is stable with high-iron olivines (fayalite) in vugs and cavities of hydrothermal or pegmatitic origin. Iron-rich nonsilicates (such as magnetite and hematite) commonly occur in association with quartz. In pegmatites, quartz is ubiquitous as huge masses in the central core or intergrown with alkali feldspar as graphic granite in outer zones. Massive veins of quartz are deposited by ascending hydrothermal solutions, making quartz a common gangue mineral of many metalliferous deposits.

In volcanic rocks, quartz is common in rhyolites, dellenites, and dacites, both in granular form in the aphanitic groundmass and as euhedral phenocrysts, often highly resorbed. Although quartz is not common in mafic flows, xenocrysts of quartz may occur in basalts and similar rocks, often showing reaction rims of pyroxene. Quartz pseudomorphs after tridymite and cristobalite are common in siliceous volcanics.

Some sedimentary rocks and unconsolidated sediments are almost wholly quartz sand and quartz pebbles, and detrital quartz grains are common in nearly all sediments. Crystallization of authigenic quartz in crystallographic continuity with detrital grains is an important cementation process and euhedral, authigenic quartz crystals may appear in most detrital and chemical sediments. Colloidal silica from silicate weathering is deposited in all rock types as veins or cavities of cryptocrystalline quartz by descending meteoric waters. Cavities, vugs, and similar openings in veins, sediments, and pyroclastics are often lined with secondary quartz crystals, forming geodes and similar deposits of great beauty and interest.

Most metamorphic rocks contain quartz either unchanged from the original rock or coarsely recrystallized by high metamorphic intensity. Certain metamorphic rocks, derived from sedimentary rocks are largely quartz. Quartz is also formed by metamorphic processes such as the conversion of serpentine to magnesite with release of silica. Quartz veins and dikes in many kinds of metamorphic rocks may represent similar metamorphic differentiation.

REFERENCES: QUARTZ

Bloss, F. D., and G. V. Gibbs. 1963. Cleavage in quartz. *Amer. Mineral.*, *48*, 821–838.

Konno, H. 1967. Relations between colors and trace elements of smoky quartz and amethyst. II. *Sci. Rep. Tohuku Univ.*, Series 3, *10*, 21–39.

Lehmann, G. 1969. Color of rose quartz. *Neues Jahrb. Mineral., Monatsh.*, *1969*, 222–225.

Nakashiro, F. M. 1966. Undulatory range and crystal size of quartz. *Can. Mineral.*, *8*, 640–643.

Nitayananda, R., and S. Ramaseshon. 1970. Dispersion of the birefringence of quartz. *Jour. Opt. Soc. Amer.*, *60*, 1531–1532.

Tatarskii, V. B., and V. F. Chernyshova. 1968. An investigation of natural and synthetic quartz by the Hilger-Chance refractometer. *Pap. Proc. Gen. Meet., Int. Mineral. Assoc., Fifth*, 123–130.

Tullis, J. 1970. Quartz: preferred orientation in rocks produced by Dauphine twinning. *Science*, *168*, 1342–1344.

Young, R. A., and B. Post. 1962. Electron density and thermal effects in α-quartz. *Acta Crystallogr.*, *15*, 337–346.

β-Quartz SiO_2 HEXAGONAL

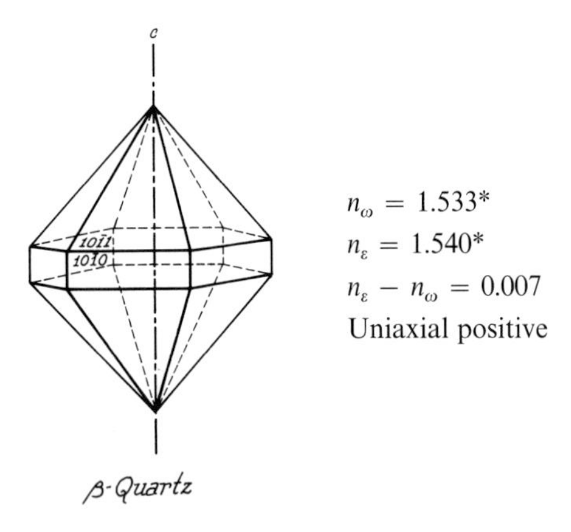

$n_\omega = 1.533$*
$n_\varepsilon = 1.540$*
$n_\varepsilon - n_\omega = 0.007$
Uniaxial positive

COMPOSITION. β-Quartz is nearly pure SiO_2 but should allow slightly more substitution of Al^{3+} and accompanying alkalis for Si^{4+} than α-quartz because of its more open structure.

PHYSICAL PROPERTIES. H ≈ 7. Sp. Gr. = 2.53. Color of α-quartz paramorphs after β-quartz is usually opaque milky white to gray, due to extensive, minute fractures.

* Refractive indices measured at 580°C. Index decreases slightly with increasing temperature.

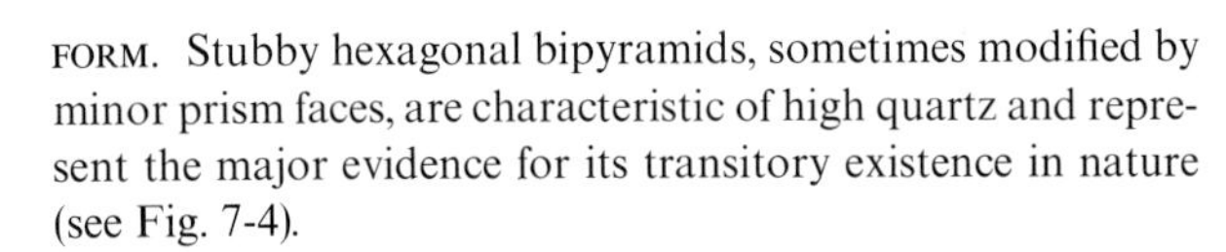

FORM. Stubby hexagonal bipyramids, sometimes modified by minor prism faces, are characteristic of high quartz and represent the major evidence for its transitory existence in nature (see Fig. 7-4).

CLEAVAGE. Heating studies suggest good cleavage on $\{10\bar{1}1\}$ and lesser cleavage on $\{10\bar{1}0\}$.

TWINNING. β-Quartz forms several kinds of normal twins with pyramidal twin planes.

INTERFERENCE FIGURE. Both forms of quartz are enantiomorphous, and β-quartz should show "optical activity" similar to its α form (see p. 316).

OCCURRENCE. β-Quartz exists only briefly after its crystallization as phenocrysts in siliceous volcanic lava and yields its identity to the stable α form before the lava cools. Only the stubby bipyramidal habit and intense fracturing remain to tell of the existence of β-quartz. The student may expect not to encounter β-quartz either in section or as fragments but may see α-paramorphs in rhyolites, dellenites, dacites, and similar volcanic rocks. Bipyramidal crystals of α-quartz have been reported in granites and even pegmatites.

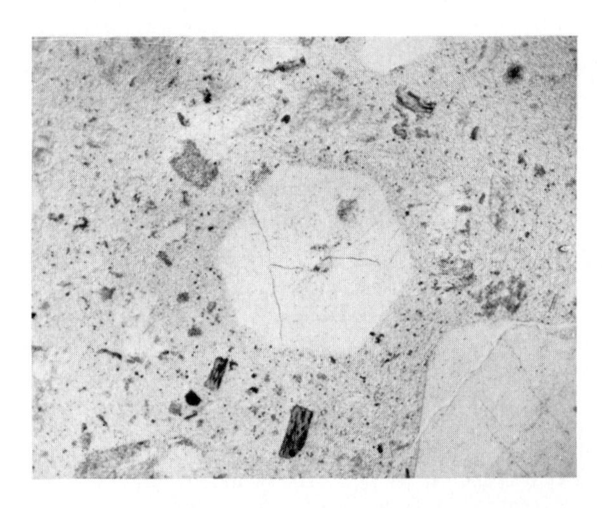

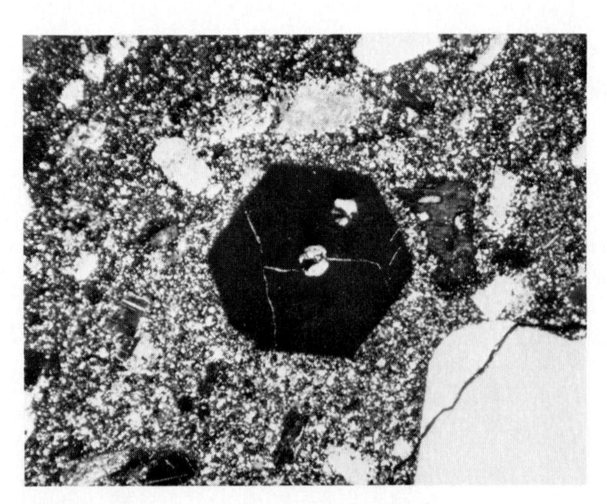

FIGURE 7-4. β-Quartz is often implied by stubby, hexagonal, bipyramidal habit and fractures, indicating volume loss. Although all quartz found in nature is α-quartz, or low-quartz, crystals of high-temperature β-quartz form in some lavas and tuffs and invert to the α-structure as the lava cools below 573°C.

α-Tridymite

SiO_2

ORTHORHOMBIC*
(PSEUDOHEXAGONAL)

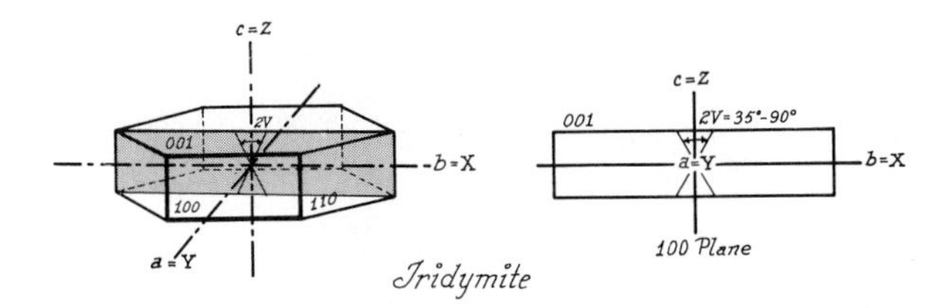

n_α = 1.469–1.479*
n_β = 1.469–1.480
n_γ = 1.473–1.483
$n_\gamma - n_\alpha$ = 0.002–0.004†
Biaxial positive, $2V_z$ = 35°–90°
$a = Y, b = X, c = Z$
Colorless in standard section

COMPOSITION AND STRUCTURE. In contrast to the quartz structure, the very open structure of tridymite (see Fig. 7-2) allows introduction of large ions, and appreciable Si^{4+} may be replaced by Al^{3+}, with the introduction of alkalis (Na^+ and K^+) or calcium into interstitial positions maintaining electrostatic neutrality. Inversion temperatures of high-low forms are significantly affected by ionic substitution, and interstitial ions may well tend to stabilize open structures.

Nepheline ($NaAlSiO_4$, hexagonal) has a structure much like that of tridymite (SiO_2, hexagonal above 117°C), and substitution of Al^{3+} for half the Si^{4+} in tridymite with Na^+ introduction into half the structural voids would essentially yield nepheline. A nepheline-tridymite series has been proposed but is probably far from a continuous series. Nepheline has low-, middle- and high-temperature forms.

PHYSICAL PROPERTIES. H = 7, Sp. Gr. = 2.27. In hand sample, tridymite is transparent, colorless to white, very brittle, and has vitreous luster.

COLOR AND PLEOCHROISM. Colorless as fragments or in section.

FORM. High β_2 tridymite is the stable form of silica between 870° and 1470°C. Inversion to stable quartz below 870°C or cristobalite above 1470°C requires structural adjustments of the reconstructive type (bonds are broken and reformed), which are very sluggish, and tridymite may exist metastably at normal temperatures or be heated to its melting point (1670°C) without inversion. On cooling, two simple displacive inversions (only slight structural adjustments) occur rapidly in the lower metastable range. High-tridymite (β_2) is the existing form of tridymite at temperatures above 163°C. Middle-tridymite (β_1) exists metastably from 163° to 117°C. Low-tridymite (α), like the β_1 form is never stable but exists below 117°C and, like α-quartz, α-tridymite is the only form encountered at normal temperatures. Cation substitution tends to alter ideal inversion temperatures, and the β_2 form may invert directly to α-tridymite.

Crystals are seldom larger than 1–2 mm and are most commonly delicate basal plates bounded by six prism faces {10$\bar{1}$0}.

CLEAVAGE. Cleavage is lacking or poor in α-tridymite. Occasional basal or prismatic parting may suggest cleavages in the high-temperature form.

BIREFRINGENCE. Very weak maximum birefringence, 0.002–0.004, is seen in {100} sections. Interference colors in thin section are no more than dark, first-order gray, and use of the $\frac{1}{4}\lambda$ (sensitive) plate may be necessary to detect birefringence.

TWINNING. Paramorph crystals are commonly twinned on a pyramid face, usually {10$\bar{1}$6}, producing wedges or pie-shaped individuals as contact twins characteristic of tridymite (Fig. 7-5). Twinning is commonly repeated, producing three wedges in contact (trilling) from which tridymite derives its name (Greek *tridumos*, "threefold"). More complex groups are possible, and other twin laws are occasionally represented as simple or multiple, contact or penetration twins.

INTERFERENCE FIGURE. Small crystal size and very weak birefringence combine to make the task of observing a useful interference figure on fragments or sections of natural tridymite an almost hopeless one. Thick basal sections {001} of large crystals should show a positive acute bisectrix figure of broad, indistinct isogyres on a gray field.

* All natural tridymite crystals are pseudomorphs after the high-temperature form, which is hexagonal.
† Lower refractive index and $2V$ values are recorded for pure synthetic tridymite. Larger values are more characteristic of the natural mineral and increase with Al-Na substitution for silicon.

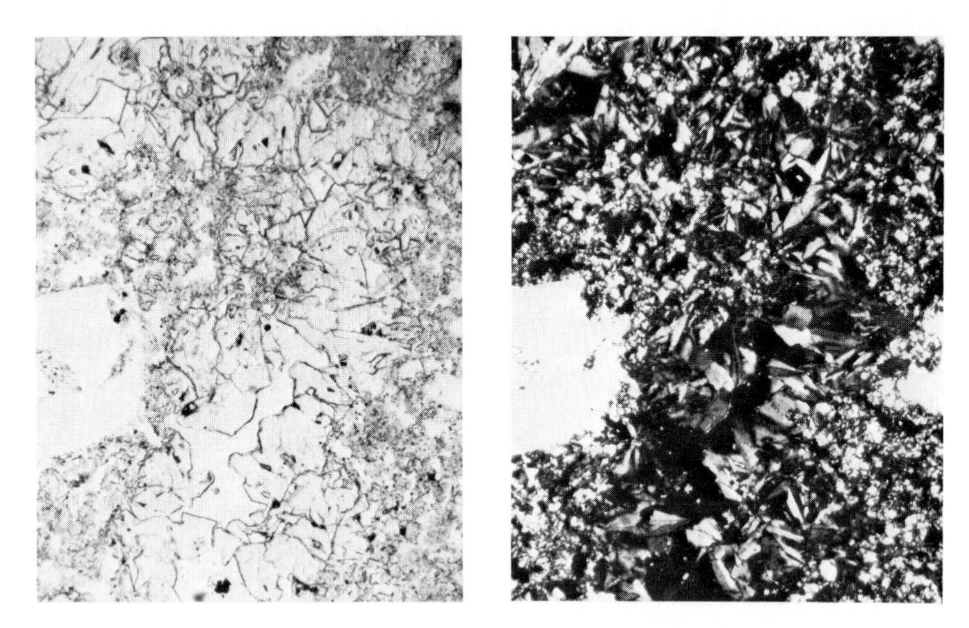

FIGURE 7-5. Tridymite derives its name from triangular (pie-shaped) crystals, commonly twinned, which appear as aggregates in the voids of rhyolitic flows and tuffs. Strong negative relief and gray interference colors are distinctive.

OPTICAL ORIENTATION. Extinction is parallel to basal or prismatic parting and symmetrical to pyramidal twin planes. Since Z lies normal to $\{001\}$, basal parting produces negative elongation (length-fast).

DISTINGUISHING FEATURES. The most distinctive characteristics of tridymite are its low refractive index (moderately high negative relief in balsam), very weak birefringence, wedge-shaped twin individuals, and tendency to occur in open cavities in acidic volcanic rocks. Tridymite and cristobalite are most similar in appearance and occurrence, and are distinguished with great difficulty. Cristobalite shows slightly higher refractive index (>1.48) and is uniaxial negative. Tridymite resembles certain zeolites in cavities of volcanic rocks but zeolites are most common in mafic flows, especially basalt, where tridymite and cristobalite are rare.

ALTERATION. Being an unstable form of silica, tridymite tends to invert to fine-grained α-quartz, particularly in older volcanics.

OCCURRENCE. The most typical occurrence of tridymite is as relatively coarse crystals in vugs, vesicles, lithophysae, and other cavities in fine-grained silicic and intermediate volcanic rocks such as obsidian, acid tuff, rhyolite, trachyte, dellenite, dacite, and andesite (less commonly in basalts) where it has been deposited by late, hot gases. Tridymite may also occur as tiny, disseminated grains in the groundmass of these rocks and, occasionally, as distinct phenocrysts. It represents the major silica mineral in some rocks and rarely constitutes as much as 20–30 percent of a flow. Tridymite may occupy a cavity alone or in association with cristobalite, sanidine, quartz, chalcedony, specular hematite, fayalite, augite, hornblende, topaz, or cassiterite. Reaction rims on pyroxene crystals in intermediate volcanics may contain fine-grained tridymite. Tridymite and cristobalite are rare in flows older than Tertiary, because of inversion to stable quartz, and tridymite is usually much more abundant than cristobalite. Tridymite has been reported in siliceous xenoliths in lavas and in contact metamorphic zones adjoining siliceous sediments. Tridymite is rather common in stony meteorites. Industrial silica bricks contain synthetic tridymite as their major constituent, with cristobalite, quartz, and pseudowollastonite.

REFERENCES: TRIDYMITE

Dollase, W. A. 1967. Crystal structure at 220° of orthorhombic high tridymite from the Steinbach meteorite. *Acta Crystallogr.*, *23*, 617–623.

Dollase, W. A., and W. H. Baur. 1976. The superstructure of meteoritic low tridymite solved by computer simulation. *Amer. Mineral.*, *61*, 971–978.

Grant, R. W. 1967. New data on tridymite. *Amer. Mineral.*, *52*, 536–540.

Hirota, K., and A. Ono. 1977. On the stability of tridymite. *Naturwissenschaften*, *64*, 39–40.

Hoffmann, W. 1967. Lattice constants and space groups of tridymite at 20°. *Naturwissenschaften*, *54*, 114.

α-Cristobalite

SiO_2

TETRAGONAL (PSEUDOISOMETRIC)

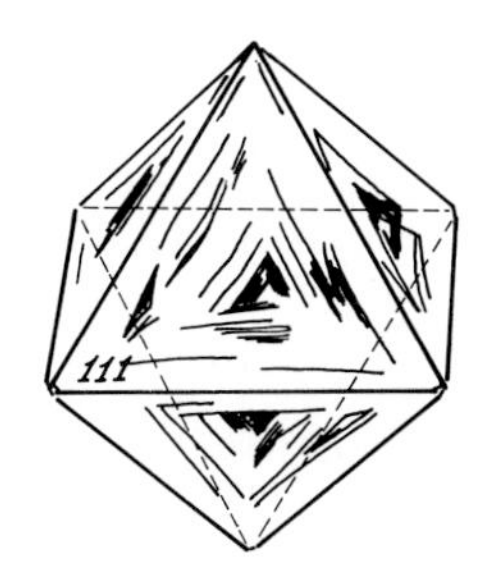

Cristobalite

$n_\omega = 1.487$
$n_\varepsilon = 1.484$
$n_\omega - n_\varepsilon = 0.003$
Uniaxial negative
Colorless in standard section

COMPOSITION AND STRUCTURE. Cristobalite (see Fig. 7-2) has an even more open structure than tridymite and should, therefore, allow extensive introduction of interstitial ions. Al^{3+}, Ca^{2+}, Fe^{2+}, and Na^+ are commonly present. Although supporting analyses are not numerous, we may expect appreciable substitution of Al^{3+} for Si^{4+} in tetrahedral coordination with accompanying introduction of interstitial ions (usually Na^+) for electrostatic neutrality. Indeed, a SiO_2-$NaAlSiO_4$ series has been proposed. The cristobalite structure is analogous to carnegieite, the high-temperature polymorph of nepheline, and at least a partial cristobalite (SiO_2)-carnegieite ($NaAlSiO_4$) series may exist. Both cristobalite and carnegieite have high-temperature forms, which are isometric. Ionic substitution, if significant, does not appear to influence optical constants, as is the case with tridymite.

PHYSICAL PROPERTIES. H = $6\frac{1}{2}$. Sp. Gr. = 2.33. In hand sample, cristobalite is colorless to milky white or yellowish and vitreous.

COLOR AND PLEOCHROISM. Cristobalite is colorless as fragments or in section.

FORM. Cristobalite is the stable form of silica between 1470° and 1728°C (melting point) but may exist metastably and even crystallize at surface temperatures. High-temperature (β) cristobalite is truly isometric and inverts to its low-temperature (α) form at some temperature below 268°C, depending on structural order and ionic substitution. β-cristobalite may persist near normal temperatures. α-cristobalite is never stable but is the polymorph normally encountered at surface conditions. Paragenesis and morphological habits have led to the definition of cristobalite varieties. Crystals of cristobalite in vugs and other cavities in lavas and tuffs are small (usually <1 mm) pseudomorphs of α-cristobalite after its β form. Crystals are usually octahedrons, rarely cubes, tending toward skeletal (indented faces) or dendritic forms and frequently showing twinning. Spherulites in volcanic glass contain a distinctly fibrous cristobalite, often intergrown with sanidine or tridymite as radial needles. The variety name *lussatite* has been applied to microscopically fibrous cristobalite, which in intimate association with cryptocrystalline fibrous quartz forms the well-known but poorly understood chalcedony. Lussatite may contain both β- and α-cristobalite, although it apparently forms from the crystallization of silica gel deposited by near-surface meteoric or hydrothermal waters. A vitreous, massive, submicrocrystalline form of cristobalite, forming from gel at surface temperatures, may constitute much of what is common opal.

CLEAVAGE. No cleavage is observed in cristobalite, but a curved fracture pattern seems characteristic.

BIREFRINGENCE. Birefringence is extremely weak at 0.002 or slightly higher and may appear essentially zero, even with the aid of the $\frac{1}{4}\lambda$ (sensitive) plate. It is usually sufficient, however, to determine elongation of fibers.

TWINNING. Normal, spinel twinning on $\{111\}$ is quite common, either as simple or repeated groups. One set of polysynthetic twins may be apparent, or two intersecting sets may be observed.

INTERFERENCE FIGURE. Small crystals and negligible birefringence virtually preclude usable interference figures. Thick basal sections may show a broad black cross on a gray or white field.

OPTICAL ORIENTATION. Inversion from β to α forms causes one of the octahedron axes (a) to become the optic axis (c). Octahedron faces should show symmetrical extinction. Cristobalite fibers are usually length-slow (positive elongation), with parallel extinction, but are occasionally length-fast. Lussatite has been defined as length-slow, and the name *lussatine* has been applied to length-fast fibers.

DISTINGUISHING FEATURES. Tiny, equant crystals showing very weak birefringence, moderately high negative relief in openings of silicic volcanics, are characteristic of cristobalite. Tridymite in the same environment may be indistinguishable without special effort. Tridymite usually has lower refractive index (< 1.48) and is biaxial positive and much more common. Fibrous cristobalite has lower refractive index than either quartz (chalcedony) or sanidine. Zeolites resemble cristobalite and tridymite in appearance and occurrence but are most characteristic of basalts.

ALTERATION. Cristobalite, like tridymite, may invert to fine-grained quartz in older volcanics.

OCCURRENCE. Cristobalite appears to form in nature usually, if not always, below its true stability range. It has a distinct high-temperature origin in lavas, metamorphic contact zones and meteorites and a low-temperature crystallization in chalcedony, opaline masses, biochemical deposits, and siliceous sinter. Many occurrences have been recorded in recent years through the evidence of x-ray diffraction. The student of optical petrography may expect to observe α-cristobalite most often as tiny crystals lining gas cavities and fractures in acidic volcanic rocks, where it was deposited as original β-cristobalite by late volatile activity. Here it is most commonly associated with tridymite (more common than cristobalite), quartz, sanidine, anorthoclase, magnetite, or fayalite. Cristobalite is most typical of lithophysae or vugs in obsidian or rhyolites, but may appear in most any common, saturated volcanic rock of Quaternary or Tertiary age, including mafic varieties, and is probably the most common form of silica in gas cavities of basalts. X-ray diffraction may detect cristobalite in the submicroscopic groundmass of siliceous flows, and it may be found to constitute 10 percent or more of the rock. Spherulites in obsidian are radiating fibers of cristobalite and sanidine with possible tridymite.

Cristobalite appears in xenoliths and metamorphic contact zones (buchites) where high-temperature lavas contact siliceous sediments. Lussatite appears to form at near-surface conditions in many chalcedony occurrences as crusts and colloform masses, especially in association with serpentines. Much opal appears to be, or to contain, submicrocrystalline cristobalite, which, strangely, may be largely β-cristobalite. Synthetic silica bricks contain cristobalite in association with tridymite.

REFERENCES: CRISTOBALITE

Floerke, O. W. 1962. Amorphous and microcrystalline SiO_2. *Chem. Erde*, *22*, 91–110.

Leadbetter, A. J., T. W. Smith, and A. F. Wright. 1973. Structures of high cristobalite. *Nature*, *Phys. Sci*, *244*, 125–126.

Wilson, M. J., J. D. Russell, and J. M. Tait. 1974. New interpretation of the structure of disordered α-cristobalite. *Contrib. Mineral. Petrol.*, *47*, 1–6.

Chalcedony

SiO_2

TRIGONAL (MICROCRYSTALLINE)

$n_\omega = 1.526–1.544$*
$n_\varepsilon = 1.531–1.553$
$n_\varepsilon - n_\omega = 0.005–0.009$*
Uniaxial positive*
Colorless or pale yellow, brown, without pleochroism

COMPOSITION AND STRUCTURE. Chalcedony and its many subvarieties are cryptocrystalline quartz, either as fine-granular masses or as fibrous aggregates; it may also contain lesser amounts of microfibrous cristobalite.

Chalcedony is largely SiO_2 with variable amounts of H_2O (up to 10 percent) trapped in pores, between fibers, and in closed inclusions. Very minor Fe^{3+}, Al^{3+}, and alkalis may be present as impurities, and some highly colored varieties contain appreciable amounts of other minerals (such as hematite, goethite, and graphite) as finely divided coloring matter. Tiny calcite inclusions in chert or flint may be remnants of incomplete fossil replacement.

Numerous variety names are somewhat poorly defined: *Chalcedony* commonly implies microfibrous structure, normal to fine layering formed as rounded, botryoidal masses in open seams or cavities. Deposition as colloidal gel is most likely. The rock name *chert* seems to imply microgranular structure and massive, structureless beds, seams, or nodules often resulting from replacement of carbonate rock, although some extensive deposits are direct precipitates.

PHYSICAL PROPERTIES. H = $6\frac{1}{2}$–7. Sp. Gr. = 2.57–2.64. Color, in hand sample, is highly variable due to finely disseminated mineral impurities. Many varieties of chalcedony and chert are defined by color or color patterns.

Sard: brown, translucent chalcedony colored by colloidal iron oxides.

Carnelian: red, translucent chalcedony colored by colloidal hematite.

Prase and Plasma: dark green, translucent chalcedony colored by disseminated ferrous silicates (such as chlorite and hornblende).

Chrysoprase: apple-green, translucent chalcedony colored by hydrated nickel silicate.

Agate: banded chalcedony, usually as cavity filling (geodes).

Moss Agate: dendritic patterns in translucent chalcedony. Dendrites are oxides of Mn, Fe, and so on, representing diffusion into gelatinous silica.

Flint: gray to black nodules of chert from chalk formations

Novaculite: massive, white, coarse chert probably resulting from the mild metamorphism of chert beds.

Jasper: red, brown, or yellow chert colored by finely disseminated hematite or geothite.

Luster is dull vitreous to greasy, because of the tiny pores associated with microfibrous or microgranular structure. Fracture is essentially conchoidal, in microgranular types, to almost splintery, in microfibrous varieties.

COLOR AND PLEOCHROISM. Chalcedony and chert are usually colorless in standard section or as small fragments; however, highly colored forms may show weak color, without pleochroism, and some chert may be almost opaque due to finely disseminated oxides.

FORM. Under the microscope, chalcedony commonly shows distinct coarse banding, as part of megascopic botryoidal forms, with microscopic fibers perpendicular to planes separating successive bands (Fig. 7-6). Interlayered granular silica is common.

Chert and its varieties show extremely fine to moderately coarse granular quartz. Flint commonly shows irregular patches of microfibrous silica in random parallel or radial array, probably due to fossil replacement, and organic forms may be evident as sponge spicules, radiolaria, foraminifera, and so on.

* Refractive indices and birefringence decrease with increasing water content. With no water, chalcedony has the optical constants of α-quartz. Chalcedony may appear biaxial, usually positive.

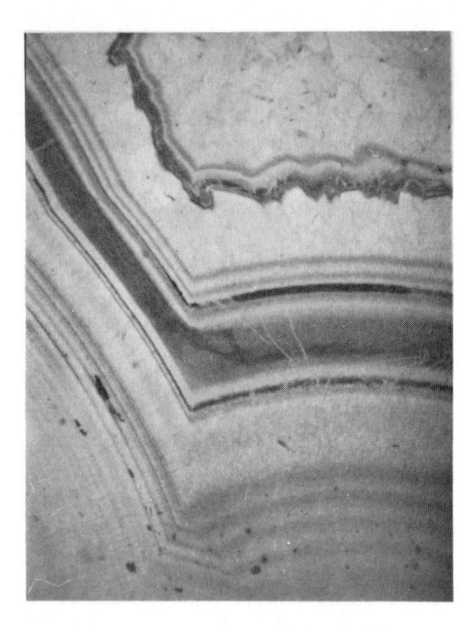

FIGURE 7-6. Chalcedony is often deposited in layered botryoidal forms that appear in thin section with microfibrous to microgranular structure.

CLEAVAGE. Microcrystalline silicas show no cleavage; however, parting may be prominent on layer boundaries, and splintery fracture may develop parallel to fiber orientation.

BIREFRINGENCE. Birefringence decreases from that of quartz (0.009) as water content increases. The low birefringence of chalcedony (0.005–0.009) yields first-order gray and white, and some chert may be sensibly isotropic, due to extremely fine grain size.

ZONING. Color banding and layers of dense inclusions are common.

INTERFERENCE FIGURE. Individual grains and fibers are too small to show separate interference figures, but their combined effect may show a biaxial type figure of highly variable $2V$, usually positive. Radiating fibers show a pseudouniaxial cross in crossed, polarized, orthoscopic illumination.

OPTICAL ORIENTATION. Fibers are usually not elongated on the *c* quartz axis but, strangely, are elongated perpendicular to *c* to show negative elongation (length-fast). Fibers of adjoining bands may show opposite elongation. Pseudomorphs after sulfate evaporite minerals are generally length-slow. Extinction is parallel, but fibers are sometimes twisted causing only sections of the fibers to be extinct at any one time.

DISTINGUISHING FEATURES. Microfibrous or microgranular structure, low relief, low birefringence, and parallel extinction are usually distinctive. Chalcedony is most likely mistaken for fibrous serpentine, which is biaxial negative and often pale green.

ALTERATION. Chalcedony and chert are relatively stable under surface conditions and are very common as pebbles and sand grains in detrital sediments. Although chert is well known in sediments as old as late Precambrian, chalcedony and chert tend to crystallize to coarse quartz with time and especially with the elevated temperatures of mild metamorphism. Cryptocrystalline silica may be partially dissolved by alkaline solutions, which are highly reactive, because of the large surface area afforded by microporosity.

OCCURRENCE. Chalcedony is deposited in open cavities and fissures at or near the earth's surface (low temperature and low pressure) by meteoric groundwater or late hydrothermal solutions (Fig. 7-7). It often results from the hardening and devitrification of gelatinous silica, which is a by-product of

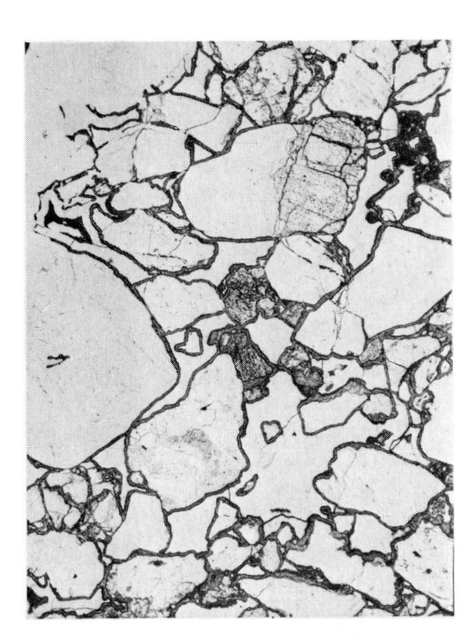

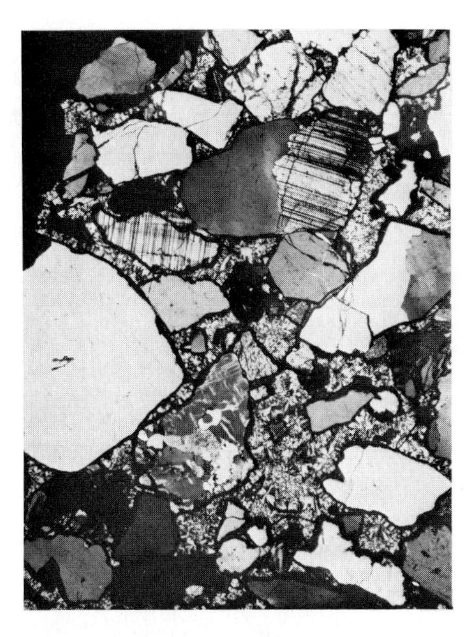

FIGURE 7-7. Chalcedony is the cementing agent in this arkosic breccia. A thin film of iron oxide surrounds each mineral grain.

the weathering of silicate minerals. Chalcedony is usually associated with siliceous volcanics, especially tuffs and breccias, but it may appear as veins in almost any igneous rock, or even sedimentary rocks, and as vesicle filling in basalt in association with carbonates, zeolites, and chlorite. Chalcedony may replace fossils (for example, corals and sponges) and is the common petrifying agent in fossil wood and bone, especially in volcanic ash deposits. It forms geodes as cavity fill and is a common cementing agent in detrital sediments.

Chert normally appears as microgranular silica owing to direct marine precipitation of silica gel or organic silica or to the replacement of carbonate sediments. In form, it ranges from irregular nodules in chalk beds (flint) to massive beds, commonly the result of submarine volcanics (pillow lavas), or recrystallization of organic opaline deposits (radiolaria, diatoms, and so on). Red chert, or jasper, is commonly associated with Precambrian hematite ores in large, interbedded deposits.

Grains and pebbles of microcrystalline silica are very common in coarse detrital sediments.

REFERENCES: CHALCEDONY

Jocelyn, J. 1971. Chalcedony and sulfate evaporite deposits. *Nature Phys. Sci.*, *232*, 23–24.

Morikawa, H., A. Nukui, and S. Iwai. 1973. Texture of chalcedony. *Bull. Tokyo Inst. Technol.*, *117*, 49–54.

Pittman, J. S., and R. L. Folk. 1971. Length-slow chalcedony after sulfate evaporite minerals in sedimentary rocks. *Nature Phys. Sci.*, *230*, 64–65.

Opal

$SiO \cdot nH_2O$

"AMORPHOUS"

$n = 1.435–1.460$*

Isotropic

Colorless, rarely pale yellow, brown, and so on, without pleochroism

COMPOSITION AND STRUCTURE. Opal has long been described as an amorphous gel. It is now known, however, to be a more-or-less close-packed aggregate of submicroscopic spheroids of tridymite and/or cristobalite. It is essentially silica with highly variable amounts of inessential water, usually 5 to 10 percent. Water is lost by continuous dehydration, often accompanied by shrinking and cracking. Opal may contain a few percent of Al, Fe^{3+}, Ca, Mg, Ti, Na, and K often in submicroscopic admixed mineral grains.

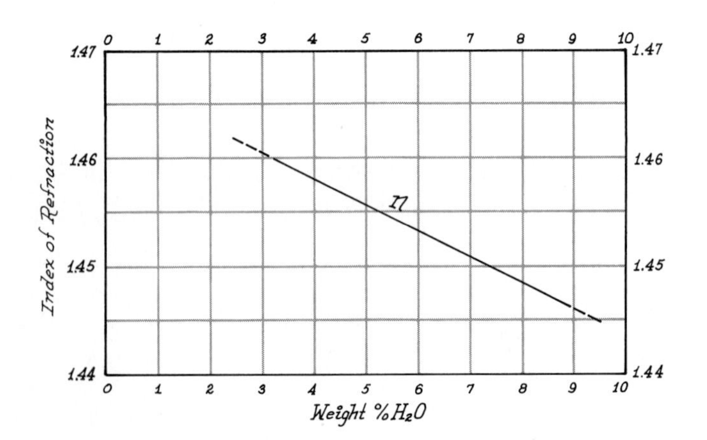

FIGURE 7-8. Variation of index of refraction with water content in opal.

* Index of refraction decreases as water content increases and may fall to 1.40 or less. Opal fragments may absorb index liquids. See Fig. 7-8.

Several opal varieties are easily distinguished:

Precious opal: shows brilliant display of spectral colors
- *Black opal:* largely short-wavelength spectral colors (violet, blue, green) and dark, transparent body color
- *Fire opal:* largely long-wavelength spectral colors (red, orange, yellow) and pale, transparent body color

Common opal: opaque to transparent without play of color; white or pale dull colors

Hyalite: colorless, transparent

Wood opal: petrifying material of wood, usually yellow to brown, showing woody structure

PHYSICAL PROPERTIES. $H = 5\frac{1}{2}–6\frac{1}{2}$. Sp. Gr. = 1.99–2.25. Color, in hand sample, is highly variable, usually white, colorless, or pale shades of yellow, red, or brown. Internal spectral colors define precious opal and appear to be a diffraction phenomenon caused by minute internal surfaces of debatable kind. Luster is highly vitreous to greasy, and fracture is distinctly conchoidal.

COLOR AND PLEOCHROISM. Sections or fragments of opal are normally colorless; however, nearly any pale color is possible, without pleochroism.

FORM. Opal is optically amorphous; colloform structures are characteristic (botryoidal, reniform, and so on). Patches of tiny chalcedony fibers are common in opal. Microscopic tests of radiolaria, diatomite, and sponge spicules are common in many opaline rocks.

CLEAVAGE. None.

BIREFRINGENCE. Opal is sensibly isotropic, but very weak strain birefringence is sometimes observed.

ZONING. Color distribution is commonly highly irregular.

DISTINGUISHING FEATURES. Moderately high negative relief, structureless form and unique occurrence are usually distinctive. Volcanic glasses ($n > 1.48$) and silica glass fused by lightning (*fulgurites*, $n \simeq 1.46$) show higher refractive index and distinctly different occurrence. Analcime ($n = 1.49$) and fluorite ($n = 1.43$) commonly occur in veins, but analcime may show weak birefringence, complex lamellar twinning and cubic cleavage. Fluorite has perfect octahedral cleavage.

ALTERATION. Opal tends to crystallize to chalcedony, and pre-Tertiary opal is rare. It is highly reactive to alkaline solutions and is completely soluble in hot, strong solutions.

OCCURRENCE. Most opaline silica in nature is the direct or indirect result of silica-secreting organisms in the sea (such as radiolaria and diatoms), lakes (diatoms), or thermal springs (such as algae), and even in certain large plants (such as bamboo). Extensive sedimentary deposits may be largely opaline as the accumulation of countless microskeletons of diatoms, radiolaria, sponge spicules, and so on, and mild metamorphism of these sediments produces massive beds of microcrystalline quartz.

Colloidal silica dissolved from subterranean channels is deposited at the surface by hydrothermal waters as siliceous sinter, or geyserite, with or without the aid of microorganisms, in hot springs and geysers.

Precious opal and many other forms are apparently deposited in veins and cavities as gelatinous silica by circulating meteoric groundwaters, especially in volcanic tuffs, breccias, and rhyolitic flow rocks in arid climates. Opal is less common in veins or cavities in basalt or serpentinites deposited by hydrothermal solutions in association with zeolites, calcite, or magnesite. Colloidal silica is largely the by-product of the alteration of feldspars and other silicates to clay minerals.

Wood, shells, and other organic remains may be replaced or petrified by introduced opal in volcanic breccia, sandstone, or cherty limestones. Opal is a rather rare cementing agent in detrital sediments and is not uncommon as detrital fragments.

REFERENCES: OPAL

Cole, S. H., and E. A. Monroe. 1967. Electron microscope studies of the structure of opal. *Jour. Appl. Phys.*, *38*, 1872–1873.

Floerke, O. W., J. B. Jones, and E. R. Segnit. 1975. Opal-CT crystals. *Neues Jahrb. Mineral. Monatsh.*, *1975*, 369–377.

Jones, J. B. 1971. Nature of opal. I. Nomenclature and constituent phases. *Jour. Geol. Soc. Aust.*, *18*, 57–68.

Sanders, J. V. 1971. Microstructure of precious opal. *Mineral. Rec.*, *2*, 261–268.

Natural Glasses

AMORPHOUS

n = 1.48–1.65*
Rhyolitic glass ~1.49
Trachytic glass ~1.51
Andesitic glass ~1.52
Leucite tephrite glass ~1.55
Basaltic glass ~1.60

Color in section ranges from colorless (acidic glasses) through shades of gray, pale brown, or green to dark brown or completely opaque (basic glasses)

COMPOSITION AND STRUCTURE. All volcanic glasses derive from silicate melts and may vary in composition over the range of the common igneous rocks (silica ranges from about 40 to over 77 percent), excluding the ultramafic types. Only from chemical analysis or refractive index can we equate a glass to its crystalline analog and designate it as a rhyolitic glass, andesitic glass, basaltic glass, and so on. A given volcanic rock may be entirely glass, glass with crystallites, glass with phenocrysts, or largely crystalline, with only minor glass filling interstices. Most rocks that are largely glass are rhyolitic; however, interstitial basaltic glass is common in basalt flows. The field classification of glasses is based largely on structure, because chemical composition is only inferred, as follows.

Obsidian is usually a black, massive glass with a shiny, vitreous luster and conchoidal fracture. It is usually rhyolitic, with less than 0.50 percent water.

Perlite is a gray or pale brown glass, very brittle and highly fractured, with spherical cracks. Perlite is usually rhyolitic and probably results from the submagmatic hydration of obsidian; it normally contains 2 to 5 weight percent water.

Pumice is a highly vesiculated, cellular glass froth. It is usually rhyolitic, with 2 to 4 weight percent water.

Pitchstone is a massive dark glass that displays a resinous or pitchy luster and conchoidal fracture. It may be rhyolitic or slightly more basic (trachytic) and owes its pitchy luster to

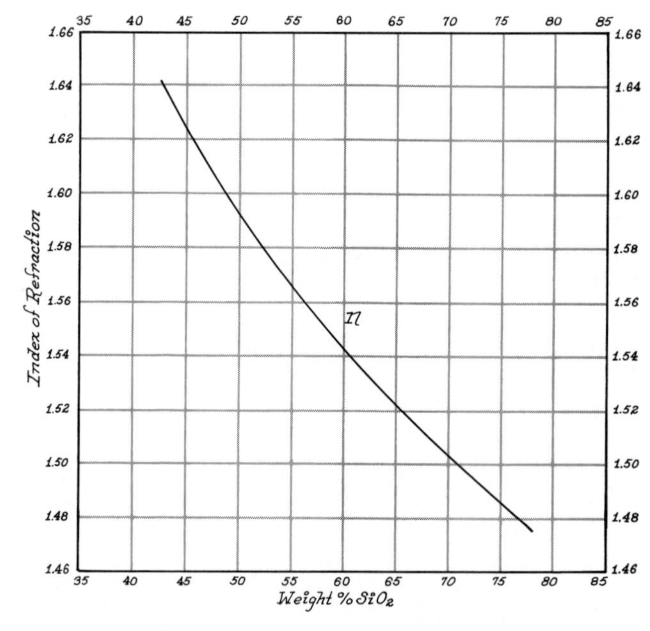

FIGURE 7-9. Variation of index of refraction with silica content in volcanic glasses. (Modified from Huber and Rinehart, 1966)

* The refractive index of a natural glass largely depends on the silica content of the glass, and Fig. 7-9 is a recent plot of the "average" silica content against the refractive index curve for volcanic rocks (Huber and Rinehart, 1966).

Other compositional variables influence the refractive index as follows:

1. Water of hydration significantly increases the refraction index: obsidian, n = 1.482 (0 weight percent H_2O), n = 1.486 (0.35 weight percent H_2O); perlite, n = 1.482 (0 weight percent H_2O), n = 1.496 (3.35 weight percent H_2O); and tachylite, $n \sim 1.60$. The refractive index of basaltic glasses appears to be largely independent of water content; however, increasing hydration is probably more or less balanced by oxidation of ferrous iron.

2. Different volcanic suites yield somewhat different IR (index of refraction) versus SiO_2 weight percent curves. Glasses from alkalic rocks yield low refractive indices (by as much as 0.02) for a given SiO_2 weight percent compared to the more normal calc-alkaline glasses. Alkalic glasses yield higher refractive indices than we would predict from their specific gravities.

either abundant crystallites or high water content (5 to 11 weight percent).

Vitrophyre is a porphyritic glass, usually obsidian or pitchstone, containing phenocrysts of sanidine, plagioclase, quartz, biotite, or, rarely, amphibole or pyroxene, which are often broken or corroded.

Tachylite or *sideromelane* is a dark basaltic glass common in the groundmass of basalts or as thin shells on pillow lavas or basaltic pyroclastics. It ranges in silica content between about 40 to 55 percent and commonly contains 0 to 5 weight percent water. Hyalomelane is a basaltic vitrophyre.

Although a glass is, by definition, amorphous, considerable short-range order undoubtedly exists, and complex silicate chains are responsible for the high viscosity of acidic lavas, which impedes crystallization.

PHYSICAL PROPERTIES. H $= 5\frac{1}{2}$ (basic glasses) to 7 (acidic glass). Sp. Gr. $\simeq$ 2.25–3.17 (obsidian ~2.32, perlite ~2.35, pumice floats on water, pitchstone ~2.34, alkalic glass ~2.56, tachylite ~2.76). Color in hand sample may be white to black: Obsidian is black, smoky transparent, brown, mahogany-red, or greenish. Obsidian is basically a transparent glass, colored dark by tiny microlites, or crystallites, of magnetite and pyroxene. Brown and red colors result from the oxidation of magnetite to hematite and often appear as uneven streaks or mottled areas in the black glass. Pyroxene crystallites aligned by flow may polarize light, adding to the dark color or causing a silky surface sheen, or play of colors, which may appear dominantly golden, silvery, greenish, or violet. Perlite is pearl gray, bluish gray, or pale brown; pumice is white, light gray, or brownish yellow; pitchstone is dark brown or greenish brown to pitch black; tachylite is dark brown to black; and vitric tuff is white to black.

COLOR AND PLEOCHROISM. Acidic glasses are usually colorless to pale gray or pale yellow-brown in standard thin section. They may be clouded by swarms of microlites in flow swirls. Basaltic glass is usually yellow-brown to dark brown and almost opaque. Isotropic substances, of course, show no pleochroism.

FORM. Obsidian and pitchstone are massive glasses containing swarms of crystallites or microlites aligned by flow banding or swirling flow structures. Spherulites ranging from a few millimeters to 1 or 2 centimeters in diameter are common in obsidian and are composed of radiating, acicular, or tabular crystals of feldspar and tridymite or cristobalite. Spherulites are usually pale gray in hand sample but yellowish in thin section, and flow banding can usually be traced without interruption through the spherulites, indicating an origin by early-stage devitrification after the glass solidified. Crystal-lined cavities (lithophysae) are also common.

Perlite is characterized by intense fracturing and spherical hydration cracks. Pearlike spheres (usually a few millimeters in diameter) may break from a perlite mass, and similar, but larger, nodules of normal black obsidian (*marekanite*) may remain unhydrated in some perlites.

Pumice is a glass froth showing tiny spherical to tubelike vesicles, stretched out by flow, and fragile chamber walls.

Vitric tuffs are composed largely of glass fragments and the thin, curved plates and Y-shaped shards of cellular walls from pumice vesicles (Fig. 7-10). Even welded tuffs retain pyroclastic structures in the massive glass formed by compression and heat welding at the base of vitric ash flows.

Tachylite is commonly clouded by numerous microlites and may form as hairlike (*Pele's hair*) masses with tiny glass spheres in basaltic tuffs. Tachylite is often highly scoriacious.

Devitrified glass yields a uniform, aphanitic aggregate of anhedral quartz and feldspar grains. Relict perlitic cracks and flow banding may persist in the cryptocrystalline mass.

BIREFRINGENCE. Glasses are, of course, isotropic; however, spherical obsidian masses in perlite and other natural glasses strained by flow or deformation may show weak strain birefringence.

DISTINGUISHING FEATURES. Analcite shows the moderate negative relief and weak birefringence of some glasses; however, it occurs only in the basic volcanics where glasses are dark with positive relief. Analcite may also show cleavage traces. Other zeolites also show negative relief and distinctive structures, often fibrous.

Opal is common in tuffs, where it fills cavities and fractures in contrast to glassy pyroclastic fragments. The refractive index of opal is always lower than that of the lowest rhyolitic glass.

ALTERATION. The most common form of alteration (hydrothermal or weathering) in any natural glass is late-stage devitrification, whereby the unstable amorphous state of an acidic glass crystallizes to a microcrystalline uniform aggregate of quartz and feldspar. Essentially all glass older than Mesozoic has been devitrified. An early devitrification state may produce spherulites in the supercooled glass. Basaltic glass devitrifies to a sandy-looking yellow or orange isotropic mineraloid having an index of refraction less than balsam and

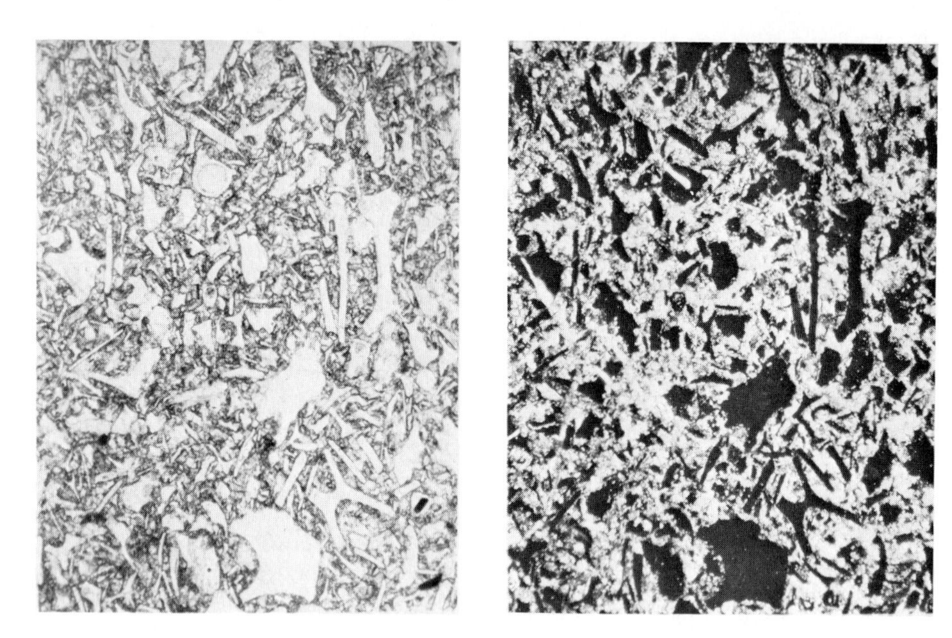

FIGURE 7-10. Glass shards cemented by calcite form a soft, tuffaceous rock.

called *palagonite*. Palagonite may form in amygdales in basalt.

Air-fall vitric tuffs lose alkalis and hydrate to clays of the montmorillonite type, which form widespread bentonite beds. Tuffs may also be silicified to a porcelainlike rock.

OCCURRENCE. Volcanic glass is common in acidic volcanic flows, where it may be the major, or the only, constituent. Obsidian or pitchstone results where the rapid cooling of an extremely viscous rhyolitic lava slows crystallization to a halt. Perlite is characteristic of rhyolite domes or the contacts of acidic sills or dikes. Pumice appears on the surface of rhyolite flows and in air-fall tuffs and vitric ash flows. Ignimbrites (welded tuffs) and ash-flow tuffs may contain numerous obsidian fragments, glassy shards, and collapsed pumice fragments, often compressed to a solid glassy mass.

Basaltic glass (tachylite or sideromelane) is much less common than the acidic glasses. It forms on the quenched edges of pillow lavas and the chilled margins of dikes and flows. It is often a major interstitial constituent of hyalophitic or intersertal basalts. Basaltic pyroclastics tend to be glassy and highly scoriaceous and may be stretched and twisted to hairlike forms and tiny glass spheres called *Pele's hair* and *Pele's tears*.

All volcanic glasses are geologically young; the oldest welded tuffs appear to be late Cretaceous ($\sim$70 million years) and the oldest obsidian, Miocene ($\sim$25 million years). Unusual glasses occur in meteorites and tektites and are very abundant in lunar soils.

Metamorphic glass (*buchite*) of the sanidinite facies is rare, but is formed by the partial fusion of country rock by high-temperature intrusions, usually diabasic. Fusion normally occurs in argillaceous or quartz-feldspar rocks (such as arkose and granite), where it begins on the quartz-feldspar contacts and may proceed for a few inches to a hundred or more feet from the igneous contact. The glass may be colorless, yellow, green, or brown to almost black and commonly encloses crystals of cordierite, mullite, sillimanite, corundum, or tridymite.

REFERENCES: NATURAL GLASSES

Duffield, W. A., G. H. Everett, and E. K. Gibson. 1977. Some characteristics of Pele's hair. *Jour. Res. U.S. Geol. Surv.*, *5*, 93–101.

George, W. O. 1924. The relation of physical properties of natural glasses to their chemical composition. *Jour. Geol.*, *32*, 353–372.
Huber, N. K., and C. D. Rinehart. 1966. Some relationships between the refractive index of fused glass beads and the petrologic affinity of volcanic rock suits. *Bull. Geol. Soc. Amer.*, *77*, 101–110.
Mathews, W. H. 1951. A useful method for determining approximate composition of fine-grained igneous rocks. *Amer. Mineral.*, *36*, 92–101.
Pollack, J. B., O. B. Toon and B. N. Khare. 1973. Optical properties of some terrestial rocks and glasses. *Icarus*, *19*, 372–389.
Ross, C. S. 1964. Volatiles in volcanic glasses and their stability relations. *Amer. Mineral.*, *49*, 258–271.
Ross, C. S., and R. L. Smith. 1955. Water and other volatiles in volcanic glasses. *Amer. Mineral.*, *40*, 1071–1089.

Feldspars

In the earth's crust, feldspar is more abundant than all other minerals combined. It is a major essential constituent of all but the most mafic or most alkaline igneous rocks: It is present in most metamorphic rock types; it appears as authigenic growths and detrital fragments in sedimentary rocks; and its alteration products make up the clays of soils and sediments. As the major mineral group of igneous rocks, the feldspars form a basis for igneous rock classification, and the student of igneous petrography must gain deeper intimacy with feldspar varieties than with any other mineral group. Unfortunately, the feldspars constitute one of the more complex mineral groups, and, in spite of detailed studies by many capable investigators, some important questions are still unanswered.

Composition and Structure

Feldspar is a tektosilicate, or framework structure of silicate tetrahedra. In tektosilicates, each tetravalent silicon ion (Si^{4+})* is surrounded by four divalent oxygen ions (O^{2-}) at the corners of a tetrahedron, and each oxygen ion is shared between two silicon ions (or their substituents) to provide a three-dimensional framework of corner-sharing tetrahedra. If silicon were the only cation present, the chemical formula would be SiO_2 (quartz), and the structure would be electrostatically neutral. If one of every four silicon ions is replaced by Al^{3+}, however, a deficiency of positive charge results, which is balanced by the introduction into the structure of an alkali ion (K^+ or Na^+) to form the *alkali feldspars*, ranging in composition from $K(AlSi_3)O_8$ (called *orthoclase*)† to $Na(AlSi_3)O_8$ (called *albite*). Two Al^{3+}

* The Si-O bond is about 50 percent covalent, but the ionic character is emphasized here in order to show electrostatic charge balance.

† Various structural polymorphs of $K(AlSi_3)O_8$ are called *sanidine*, *orthoclase*, and *microcline*, but the potassium feldspar chemical composition is called *orthoclase* regardless of the polymorph under consideration.

substitutions are balanced by an alkaline earth (Ca^+ or Ba^+) to yield the Ca-feldspar $Ca(Al_2Si_2)O_8$ (anorthite) and Ba-feldspar $Ba(Al_2Si_2)O_8$ (celsian).* A portion of a typical feldspar crystal structure is shown in Figs. 7-11 and 7-12.

Feldspar structures are based on monoclinic and triclinic lattices, as determined by the effective size of the large cations (called *M*-cations) and the Al-Si distribution. The larger *M*-cations (K^+ and Ba^{2+}) tend to support the more open monoclinic structures, while smaller cations (Na^+ and Ca^{2+}) allow the structure to "collapse" to triclinic symmetry. Preferential ordering of Al^{3+} into one of the tetrahedral sites eliminates symmetry elements present in a monoclinic feldspar, thereby altering the symmetry to triclinic. High temperature increases effective ion size and Al-Si disorder, so that high-temperature forms tend to be monoclinic even if the *M*-cation is small at room temperature (for example, $Na(AlSi_3)O_8$ above ~1000°C) and low-temperature forms are generally triclinic even if the *M*-cation is large (for example, triclinic $K(AlSi_3)O_8$). High-temperature and low-temperature polymorphs exist for each of several isomorphous series described hereafter.

The Alkali Feldspar Series

At high temperatures, the alkali feldspar series is a continuous solid solution series from $Na(AlSi_3)O_8$ to $K(AlSi_3)O_8$, but at low temperatures the solid solution range is limited, and, except for some K-rich sanidine and a small range near albite, the alkali feldspars are composed of a sodium-rich phase and a potassium-rich phase that have "unmixed," or exsolved, in the solid state. Nearly all alkali feldspars are, then, exsolution intergrowths of two phases on a megascopic (*perthite*), microscopic (*microperthite*), or submicroscopic (*cryptoperthite*) scale (Fig. 7-13). Perthite and its small-scale equivalents are lattice-type intergrowths of a K-rich alkali feldspar and a Na-rich variety, where exsolution layers and stringers show preference for certain crystallographic planes.

The alkali feldspars, defined by composition, have been further divided into four structural series, originally on the basis of optical properties and occurrence. A wealth of more recent work has identified the distribution of aluminum and silicon atoms among the tetrahedral sites as the distinguishing feature among these series, with variations in optical and physical properties merely reflections of variations in composition and ordering (see Fig. 7-14). These series are described as follows, along with a fifth series (adularia to pericline) which is based on crystal morphology and paragenesis.

---

* Calcium is the only alkaline earth commonly abundant in feldspars, but nearly pure barium varieties do occur in nature and strontium feldspars have been made synthetically. Magnesium feldspars do not exist, because the available structural opening is too large for the Mg^{2+} ion; similar reasoning eliminates an alkali lithium variety. Rubidium (Rb^+) and cesium (Cs^+) are found as significant impurities in alkali feldspars, some of which constitute a major natural source of rubidium; rubidium feldspars have been synthesized.

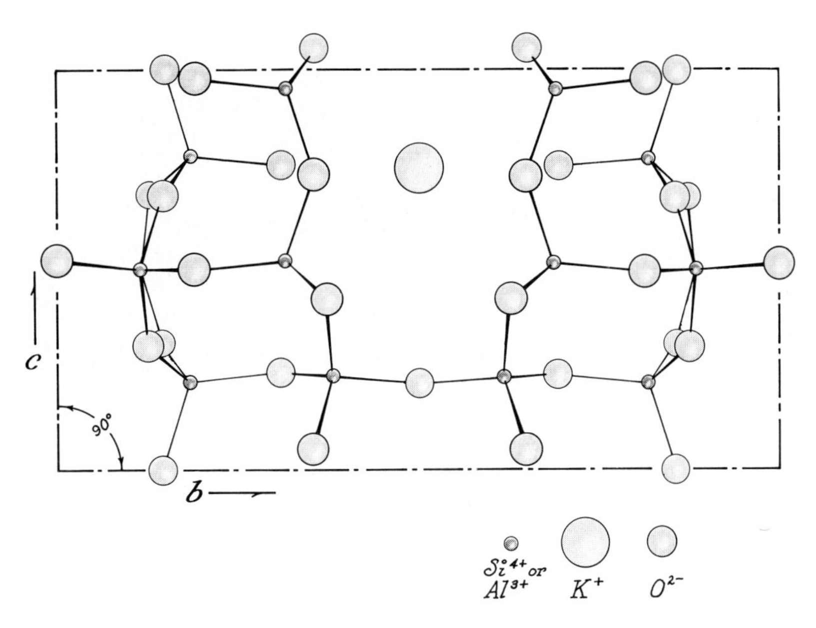

FIGURE 7-11. The structure of feldspar projected onto the *bc* (100) plane. This monoclinic feldspar shows a vertical symmetry plane and $\angle\alpha = 90°$. (After Stewart and Ribbe, 1969)

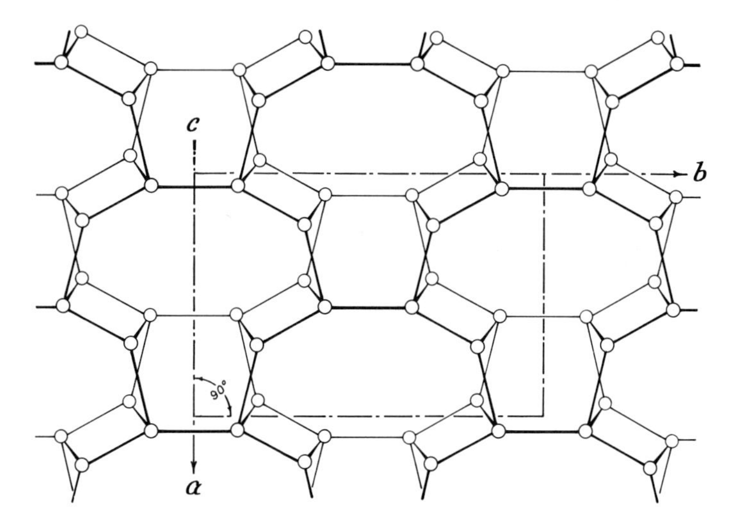

FIGURE 7-12. The "dog face" projection of the feldspar structure on the (001) plane. All small circles represent tetrahedral sites, which are occupied by silicon or aluminum ions. The structure shown represents a monoclinic feldspar ($\angle\gamma = 90°$) and shows a vertical symmetry plane. (After Laves, 1960)

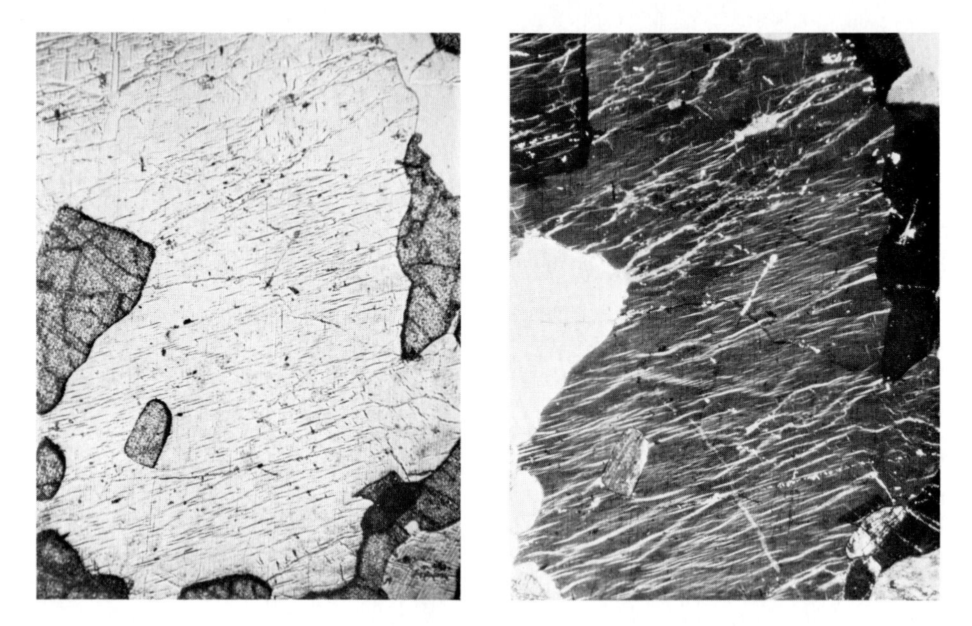

FIGURE 7-13. Microperthite displays irregular stringers of sodium-rich feldspar (light) in potassium-rich feldspar (dark) resulting from solid state exsolution of a single-phase alkali feldspar. Perthite and microperthite are favored by slow cooling and are characteristic of plutonic rocks, either igneous or metamorphic.

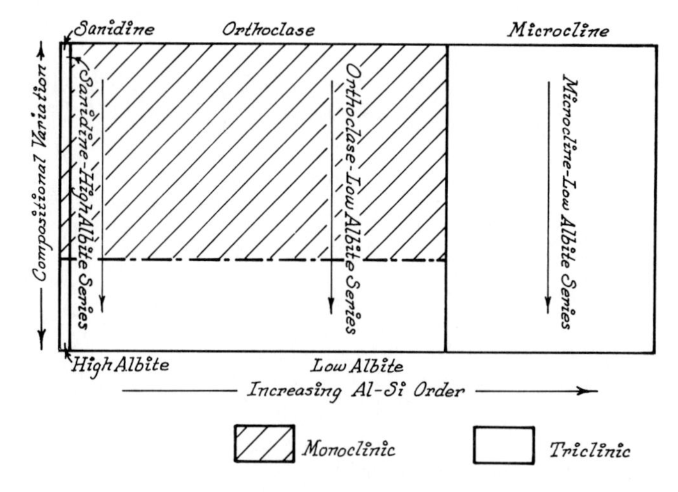

FIGURE 7-14. Alkali feldspar varieties, as determined by compositional variation and Al-Si ordering.

1. THE HIGH SANIDINE-HIGH ALBITE SERIES. This very high-temperature series appears to form complete solid solution from monoclinic high sanidine to triclinic high albite, with the symmetry change at Ab_{63}.* Only K-rich ($<Ab_{33}$) members of the series appear in nature. The only known distinction between high sanidine and sanidine (next series) is that the optic axial plane in high sanidine is parallel to (010), whereas it is normal to (010) in sanidine and the other K-feldspars.

2. THE SANIDINE-HIGH ALBITE SERIES (see Fig. 7-15). This series is characteristic of alkali feldspars of volcanic rocks formed above about 900°C for sanidine and about 1350°C for albite. Monoclinic sanidine and triclinic high albite form limited solid solution with unmixing to submicroscopic cryptoperthite in the range Ab_{40}–Ab_{75}, and monoclinic-triclinic symmetry change at Ab_{63}. All monoclinic members of the series are commonly designated *sanidine* (Ab_0–Ab_{63}) and triclinic varieties are called *anorthoclase* (Ab_{63}–Ab_{90}) and *albite* (Ab_{90}–Ab_{100}). The Al-Si distribution is completely random over the tetrahedral sites—that is, it is completely disordered.

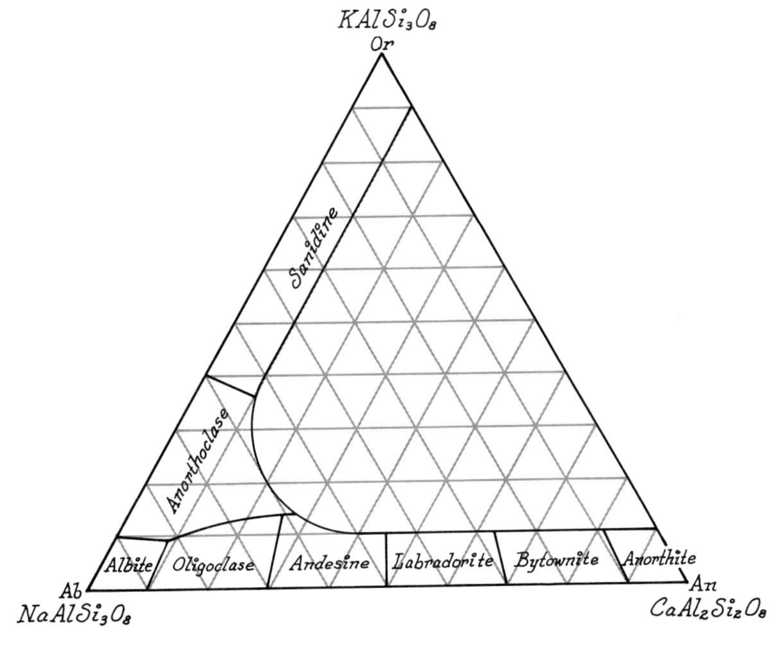

FIGURE 7-15. Solid solution in the high-temperature feldspars, showing composition ranges of the sanidine-high albite series and the high albite-anorthite plagioclase series. (After Deer, Howie and Zussman, 1963)

* Standard abbreviations for feldspar compositions are Ab (albite), Or (orthoclase), An (anorthite), and Cn (celsian).

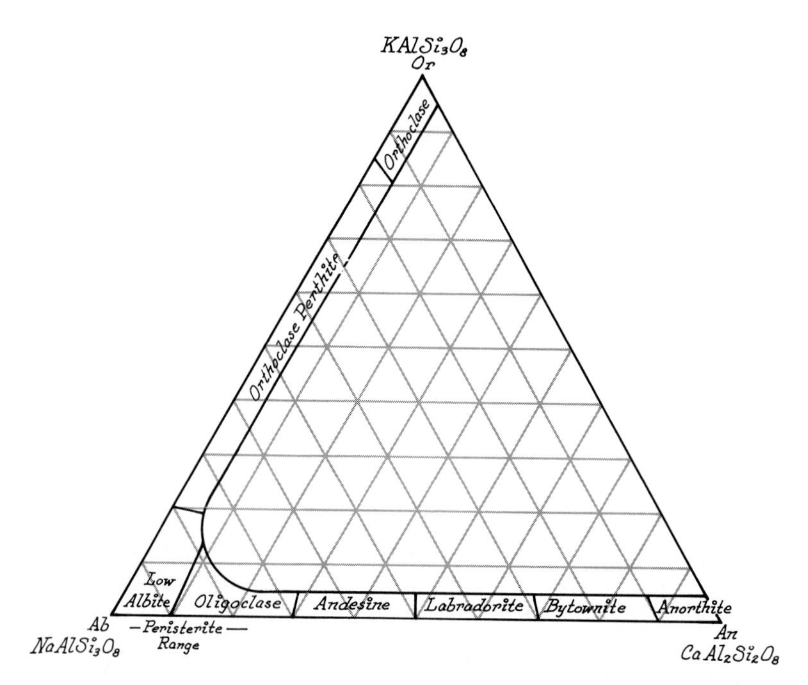

FIGURE 7-16. Solid solution in medium-temperature feldspars, showing composition ranges of the orthoclase-low albite series and the low albite-anorthite plagioclase series.

3. THE ORTHOCLASE-LOW ALBITE SERIES (see Fig. 7-16). Partial Al-Si ordering produces the common low-temperature series found in plutonic rocks. Unmixing appears over a wide range from Ab_{15} to Ab_{80} as microscopically visible microperthites, where the monoclinic phase is called *orthoclase* (Ab_0–Ab_{15}) and the triclinic phase *albite* (Ab_{80}–Ab_{100}). Albite may display multiple twinning.

4. THE MICROCLINE-LOW ALBITE SERIES (see Fig. 7-17). Although microcline often occurs in the same plutonic rocks with orthoclase, it appears to be the lower-temperature form, with a degree of Al-Si order sufficient to reduce the symmetry to triclinic. Its obliquity ("triclinicity") ranges from nearly monoclinic to a maximum obliquity for "maximum microcline." Solid solution is very limited, and members of the series more sodic than Ab_8 appear as microcline microperthites or even megascopic perthites. Microcline is very seldom untwinned by one or more multiple twin laws, and some investigators have proposed that orthoclase is no more than microcline, with submicroscopic twinning on {010}, producing monoclinic symmetry.

5. THE ADULARIA-PERICLINE SERIES. Adularia is a K-feldspar and pericline a Na-feldspar forming below the normal feldspar stability range in hydrothermal veins. Both minerals are unusually pure, with very minor K-Na substitution, and both are discernible only by char-

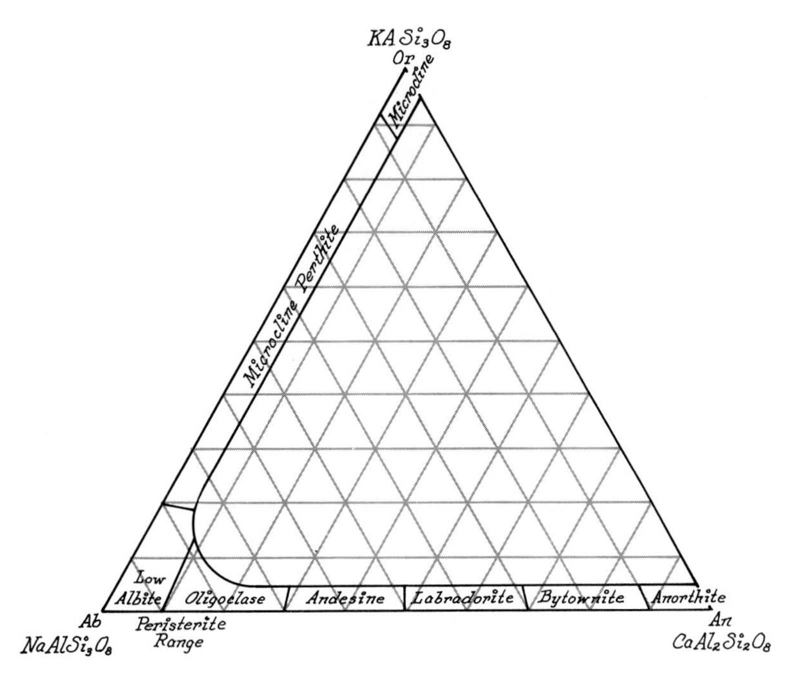

FIGURE 7-17. Solid solution in low-temperature feldspars, showing composition ranges of the microcline-low albite series and the low albite-anorthite plagioclase series.

acteristic crystal habit and occurrence. Although adularia forms at a temperature below that of microcline, it crystallizes rapidly, preventing extensive Al-Si ordering; thus adularia is essentially orthoclase of a special crystal habit and paragenesis.

For alkali feldspars, $2V$ is shown in Fig. 7-18 as a function of composition and Al-Si order.

The Plagioclase Feldspar Series

The albite-anorthite, or plagioclase, series is most commonly subdivided as

Albite	An_0–An_{10}
Oligoclase	An_{10}–An_{30}
Andesine	An_{30}–An_{50}
Labradorite	An_{50}–An_{70}
Bytownite	An_{70}–An_{90}
Anorthite	An_{90}–An_{100}

The compositional relationships among these and the alkali feldspars are shown in Figs. 7-15, 7-16, and 7-17. The plagioclases form high-temperature and low-temperature structural series from high albite and low albite, respectively, to anorthite, which is completely ordered,

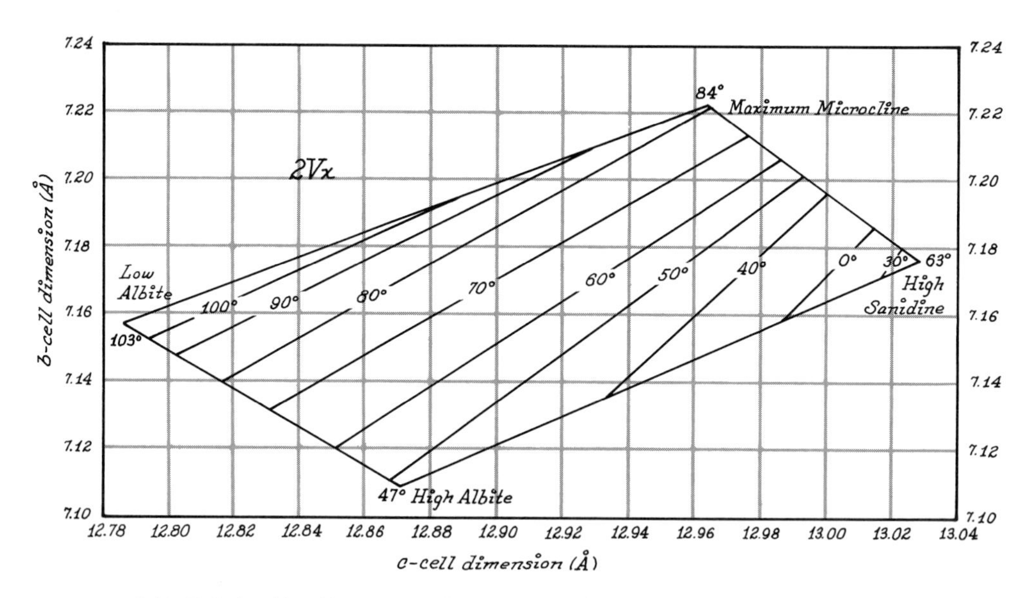

FIGURE 7-18. Relationship of $2V_x$ angle and the *b* and *c* lattice parameters in the alkali feldspars. The diagram is patterned after the *bc*-plot of Stewart and Wright (1974), which relates cell parameters and Si-Al ordering. (After Stewart, 1974)

but all forms are triclinic.* The distinction is again based on Al-Si ordering, with disorder characterizing high-temperature forms.

1. HIGH-TEMPERATURE PLAGIOCLASE SERIES. High albite appears to form a complete solid solution with anorthite. High plagioclases are found only in rapidly cooled volcanic rocks.

2. LOW-TEMPERATURE PLAGIOCLASE SERIES. The low albite-anorthite series represents plagioclase of most common igneous rocks. High albite appears to invert to the low form between 500°C and 700°C, and intermediate albites are not found in nature. Because of the variable Al:Si ratio of the plagioclases, those of intermediate compositions are necessarily partially disordered (unlike the end members, low albite and anorthite, which are fully ordered). This gives rise to unmixing, or exsolution, in compositional ranges where the exsolved phases are presumably more stable than the more disordered single phase. The three such ranges of submicroscopic intergrowth are An_{2-16}, which exsolves to $An_{\sim 0}$ and $An_{25 \pm 5}$ and is called *peristerite*; An_{48-58}, which exsolves to $An_{\sim 40}$ and $An_{\sim 60}$ and is called *Bøggild intergrowth*; and An_{70-90}, which exsolves to $An_{\sim 65}$ and $An_{\sim 95}$ and is called *Huttenlocher intergrowth*. The peristerites and Bøggild intergrowths are responsible for the irridescent play of colors seen in some plagioclase.

* Above ~1000°C, high albite inverts to "monalbite," which is monoclinic, because of the high thermal motion of the sodium atoms (see p. 332).

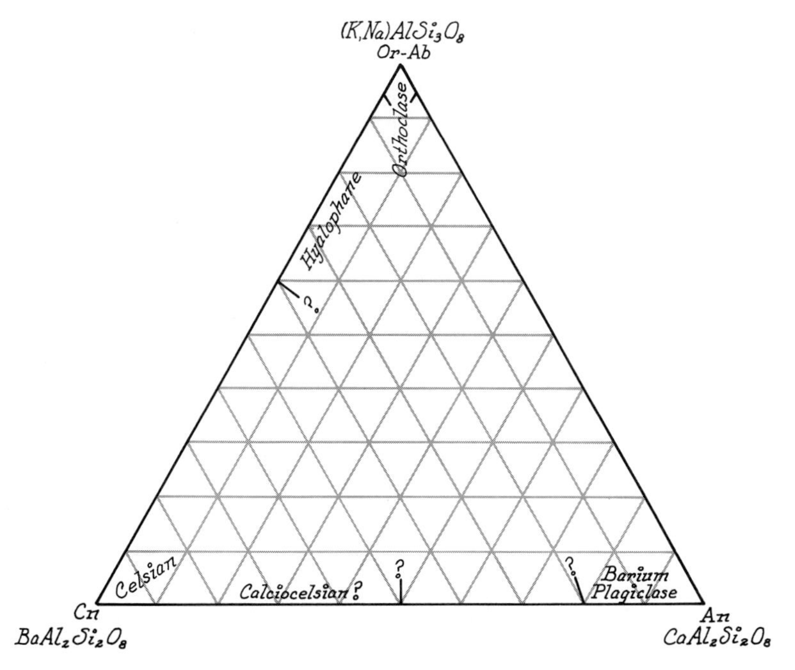

FIGURE 7-19. Solid solution in the barium feldspars shows indefinite composition ranges for the celsian-orthoclase series and a supposed barium-rich plagioclase series.

Megascopic unmixing of K-feldspar in plagioclase forms *antiperthite*, usually in the oligoclase-andesine range; less commonly, with albite. Replacement antiperthites are recognized by irregular distribution of large, odd-shaped blebs of K-feldspar.

The Barium Feldspar Series

The feldspars may well be a four-component system with a pure barium analog of anorthite called *celsian* as the fourth end member.

Celsian, $Ba(Al_2Si_2)O_8$, may form a complete, monoclinic series with the potassium feldspars, but natural feldspars in the range Cn_{40}–Cn_{80} are very uncommon. The name *celsian* applies to the nearly pure barium end member, Cn_{85-100}. *Hyalophane* is the only name given to intermediate members of the Ba-K feldspar series, and these are potassium-rich ($<Cn_{40}$). A possible, but as yet unknown, celsian-anorthite series has been proposed with a hypothetical intermediate member called *calciocelsian* (for Cn > An). Fig. 7-19 depicts the compositional relationships for barium-bearing feldspars. Table 7-2 summarizes the optical and physical properties of the feldspar minerals.

TABLE 7-2. Optical Properties of the Feldspars

n_β	n_α	n_γ	$n_\gamma - n_\alpha$	Mineral	System	$2V$	Optic Sign	Cleavage	Optical Orientation	Color in Sections	Physical Properties
1.518–1.519	1.514–1.516	1.521–1.522	0.007–0.008	Microcline $KAlSi_3O_8$ (Or_{100-92})[a]	Triclinic	66°–68° $r > v$ weak	−	{001} perfect {010} distinct {110} poor	Optic plane is near {001} $X \wedge \{001\} = 15°$	Colorless	H = 6–6½. Sp. Gr. ≃ 2.57. White, salmon-red, apple-green
1.520–1.531	1.515–1.527	1.523–1.536	0.008–0.009	Microcline cryptoperthite (Or_{92-20})[a]	Triclinic	66°–90° 90°—87° $r \gtrless v$ weak	− +	{001} perfect {010} distinct {110} poor	Optic plane is near {001} $X \wedge \{001\} \simeq 17°$	Colorless	H = 6–6½. Sp. Gr. ≃ 2.60. White, salmon-red, apple-green
1.522–1.524	1.518–1.520	1.522–1.525	0.005	Orthoclase $KAlSi_3O_8$ (Or_{100-85})[b]	Monoclinic ∠β = 116°	35°–50° $r > v$ moderate	−	{001} perfect {010} distinct {110} poor	$b = Z$ $c \wedge Y = -13°$ to −21° $a \wedge X = +14°$ to +6°	Colorless	H = 6–6½. Sp. Gr. ≃ 2.57. White, salmon-red, yellow
1.522–1.529	1.518–1.524	1.522–1.530	0.005–0.006	Sanidine (K,Na)$AlSi_3O_8$(Ab_{0-63})[b]	Monoclinic ∠β = 116°	18°–42° $r > v$ weak	−	{001} perfect {010} distinct possible parting {100}	$b = Z$ $c \wedge Y \simeq -21°$ $a \wedge X \simeq +5°$	Colorless	H = 6. Sp. Gr. ≃ 2.6. White (glassy)
1.523–1.525	1.518–1.521	1.524–1.526	0.006	High sanidine (K,Na)$AlSi_3O_8$(Ab_{0-33})[b]	Monoclinic ∠β = 116°	15°–63° $r > v$ weak	−	{001} perfect {010} distinct	$b = Y$ $c \wedge Z \simeq -20°$ $a \wedge X \simeq +5°$	Colorless	H = 6. Sp. Gr. ≃ 2.6 White (glassy)
1.524–1.531	1.520–1.527	1.525–1.536	0.006–0.008	Orthoclase cryptoperthite (Or_{85-20})[b]	Monoclinic ∠β ≃ 116°	50°–90° $r > v$ weak	−	{001} perfect {010} distinct {110} poor	$b = Z$ $c \wedge Y = -13°$ to −21° $a \wedge X = +14°$ to +6°	Colorless	H = 6–6½. Sp. Gr. ≃ 2.60. White, salmon-red, yellow
1.524–1.545	1.520–1.542	1.526–1.547	0.005–0.010	Hyalophane (Cn_{30-5})[b]	Monoclinic ∠β = 116°	40°–79° $r > v$ weak	−	{001} perfect {010} distinct {110} poor	$b = Z$ $c \wedge Y = -25°$ to −45° $a \wedge X = +1°$ to −19°	Colorless	H = 6–6½. Sp. Gr. ≃ 2.7 White, yellow
1.529–1.532	1.524–1.526	1.530–1.534	0.006–0.007	Anorthoclase (Ab_{63-90})[a]	Triclinic	42°–52° $r > v$ weak	−	{001} perfect {010} distinct possible parting {100}	$b \wedge Z \simeq 5°$ $a \wedge X \simeq 10°$ $c \wedge Y \simeq 20°$	Colorless	H = 6. Sp. Gr. ≃ 2.6. White, gray

1.532–1.533	1.526–1.527	1.534–1.541	0.007–0.008	High albite $NaAlSi_3O_8(Ab_{100-90})$[c]	Triclinic	52°–54° $r>v$ weak	−	{001} perfect {010} distinct	Optic plane almost ⊥ {010}	Colorless	H = 6. Sp. Gr. ≃ 2.6. White
1.532–1.538	1.528–1.534	1.538–1.542	0.010	Albite $NaAlSi_3O_8(An_{0-10})$[c]	Triclinic	78°–84° $v>r$ weak	+	{001} perfect {010} distinct {110} poor	See Fig. 7-28	Colorless	H = 6–6½. Sp. Gr. ≃ 2.63. White
1.538–1.548	1.534–1.544	1.542–1.552	0.008	Oligoclase (An_{10-30})[c]	Triclinic	84°–90° 90°–86° $v \gtrless r$ weak	+ −	{001} perfect {010} distinct {110} poor	See Fig. 7-28	Colorless	H = 6–6½. Sp. Gr. ≃ 2.65. White, salmon pink
1.548–1.558	1.544–1.555	1.552–1.562	0.008	Andesine (An_{30-50})[c]	Triclinic	90°–78° $r>v$ weak	+	{001} perfect {010} distinct {110} poor	See Fig. 7-28	Colorless	H = 6–6½. Sp. Gr. ≃ 2.68. Pale gray
1.558–1.569	1.555–1.564	1.562–1.573	0.008	Labradorite (An_{50-70})[c]	Triclinic	77°–86° $r>v$ weak	+	{001} perfect {010} distinct {110} poor	See Fig. 7-28	Colorless	H = 6–6½. Sp. Gr. ≃ 2.70. Gray, black
1.569–1.580	1.564–1.573	1.573–1.584	0.010	Bytownite (An_{70-90})[c]	Triclinic	87°–90° 90°–77° $v \gtrless r$ weak	+ −	{001} perfect {010} distinct {110} poor	See Fig. 7-28	Colorless	H = 6–6½. Sp. Gr. ≃ 2.73. Gray, yellow
1.580–1.585	1.573–1.577	1.584–1.590	0.013	Anorthite $CaAl_2Si_2O_8(An_{90-100})$[c]	Triclinic	77°–79° $v>r$ weak	−	{001} perfect {010} distinct {110} poor	See Fig. 7-28	Colorless	H = 6–6½. Sp. Gr. ≃ 2.75. White, gray
1.583–1.593	1.579–1.587	1.588–1.600	0.009–0.013	Celsian $BaAl_2Si_2O_8(Cn_{100})$[b]	Monoclinic $\angle\beta \simeq 115°$	83°–90°	+	{001} perfect {010} distinct {110} poor	$b=Y$ $c \wedge X = +3°$ to $+5°$ $a \wedge Z = +28°$ to $+30°$	Colorless	H = 6–6½. Sp. Gr. ≃ 3.3. White, yellow

[a] Combined albite and pericline yields a distinctive quadrille pattern
[b] No repeated twinning
[c] Distinct lamellar (albite) twinning is characteristic

ALKALI FELDSPARS

The Sanidine-High Albite Series

$(K,Na)AlSi_3O_8$

MONOCLINIC (Sanidine)
$\angle \ = 116°$
TRICLINIC (High Albite)
$\angle\alpha = 93°22'$
$\angle\beta = 116°18'$
$\angle\gamma = 90°17'$

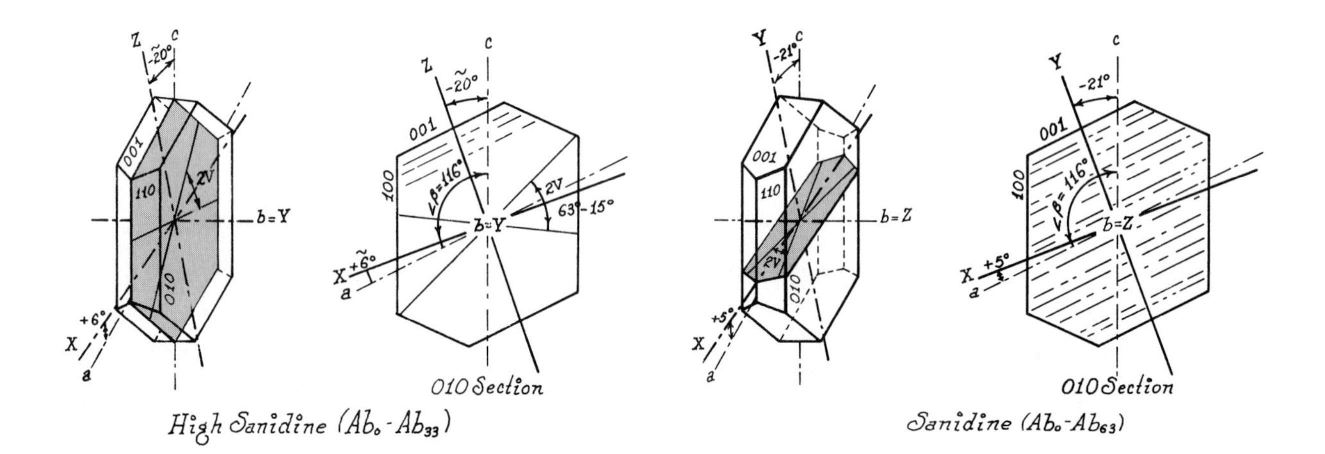

High Sanidine $(Ab_0\text{-}Ab_{33})$

Sanidine $(Ab_0\text{-}Ab_{63})$

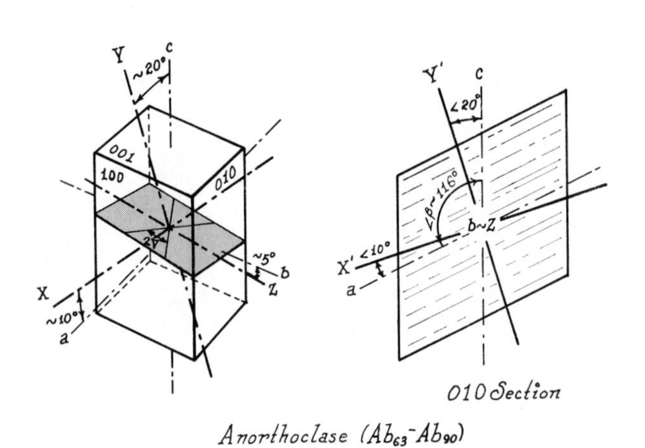

Anorthoclase $(Ab_{63}\text{-}Ab_{90})$

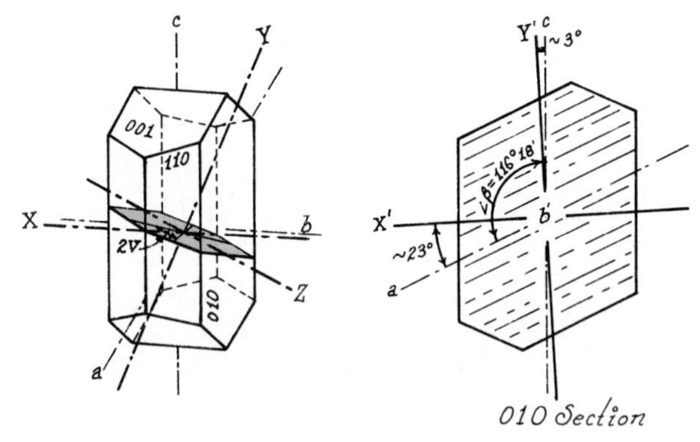

High Albite $(Ab_{90}\text{-}Ab_{100})$

High Sanidine (Ab_0–Ab_{33})
$n_\alpha = 1.518$–1.521
$n_\beta = 1.523$–1.525
$n_\gamma = 1.524$–1.526
$n_\gamma - n_\alpha = 0.006$
Biaxial negative
$2V_x = 63°$–$15°$*
$b = Y, c \wedge Z \simeq -20°$, optic plane ∥ 010
$v > r$ weak

Sanidine (Ab_0–Ab_{63})
$n_\alpha = 1.518$–1.524*
$n_\beta = 1.522$–1.529
$n_\gamma = 1.522$–1.530
$n_\gamma - n_\alpha = 0.005$–$0.006$
Biaxial negative
$2V_x = 18°$–$42°$*
$b = Z, c \wedge Y \simeq -21°$, optic plane ⊥010
$r > v$ weak

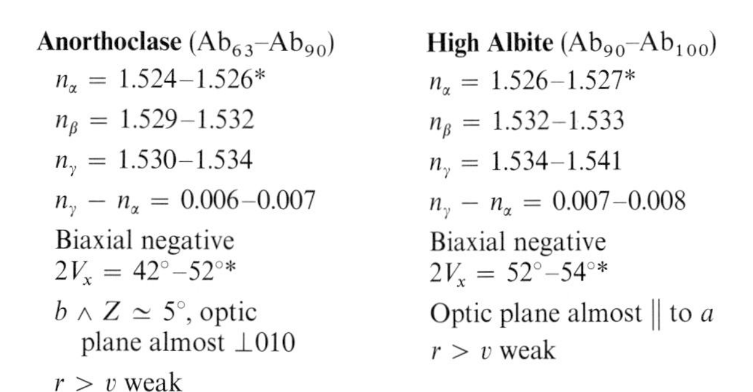

Anorthoclase (Ab_{63}–Ab_{90})
$n_\alpha = 1.524$–1.526*
$n_\beta = 1.529$–1.532
$n_\gamma = 1.530$–1.534
$n_\gamma - n_\alpha = 0.006$–$0.007$
Biaxial negative
$2V_x = 42°$–$52°$*
$b \wedge Z \simeq 5°$, optic plane almost ⊥010
$r > v$ weak

High Albite (Ab_{90}–Ab_{100})
$n_\alpha = 1.526$–1.527*
$n_\beta = 1.532$–1.533
$n_\gamma = 1.534$–1.541
$n_\gamma - n_\alpha = 0.007$–$0.008$
Biaxial negative
$2V_x = 52°$–$54°$*
Optic plane almost ∥ to a
$r > v$ weak

COMPOSITION. Sanidine is the high-temperature form of orthoclase, and sanidine itself has a higher-temperature form of different optical orientation (high sanidine). High sanidine appears to form a complete solid solution with high albite but only K-rich members of the series ($<Ab_{33}$) occur in natural rocks.

Sanidine and high albite form an optically continuous series, but submicroscopic unmixing is known to occur in the range Ab_{75}–Ab_{40}. Monoclinic symmetry becomes triclinic at Ab_{63} due to structural collapse about the small sodium ion, forming a natural division for monoclinic sanidine (Ab_0–Ab_{63}) and triclinic anorthoclase (Ab_{63}–Ab_{90}).

Up to 30 percent $BaSi_2Al_2O_8$ has been reported in sanidine, suggesting isomorphism with celsian.

PHYSICAL PROPERTIES. H = 6. Sp. Gr. = 2.56–2.62. The series is white, yellow, and so on, in hand sample and usually transparent and glassy.

COLOR. Minerals of the sanidine-albite series are clear and colorless as fragments or in thin section.

FORM. Sanidine crystals are commonly euhedral as phenocrysts, either essentially square in section or elongated parallel to a. Sanidine is occasionally acicular as spherulites in volcanic glass.

CLEAVAGE. Cleavages are {001} perfect and {010} distinct, intersecting at 90° in sanidine and 93° in high albite. Parting is sometimes present parallel to {100}.

BIREFRINGENCE. Low maximum birefringence (0.005–0.008) is seen in essentially {001} sections in sanidine and anorthoclase

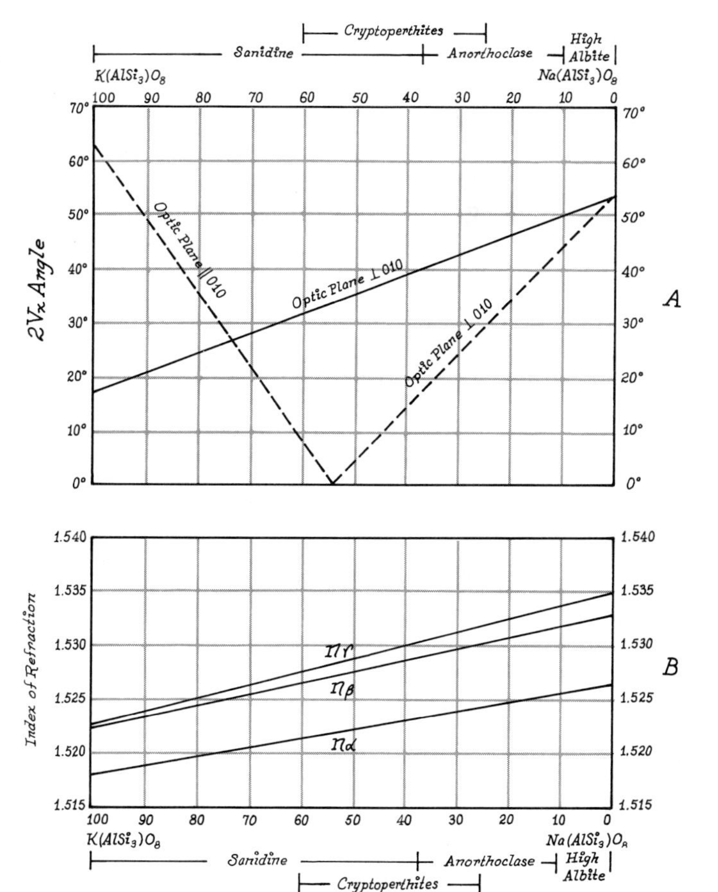

FIGURE 7-20. Variation of $2V$ angle and indices of refraction in the sanidine-high albite, alkali feldspar series. The dashed line in (A) shows variation in $2V$ angle for high sanidine.

* Indices of refraction and $2V_x$ are related to composition as shown in Fig. 7-20.

and {010} sections for high sanidine. Interference colors in standard section are first-order grays to white.

TWINNING. Only Carlsbad twinning is common in sanidine, bisecting phenocrysts or microlites parallel to their length. Baveno and Manebach twins are occasionally seen.

Triclinic anorthoclase is distinguished by combined albite and pericline twinning, which form the diffuse grid, or quadrille, pattern described for microcline but with different crystallographic orientation. The {010} sections show only pericline twinning, with the composition plane only a few degrees ($-2°$ to $-5°$) from {001}, {100} sections show both albite and pericline twinning (grid pattern), and {001} sections show only albite twinning (compare with microcline twinning, p. 350). Twin laminae are very narrow compared to those of microcline and become imperceptibly fine with increasing potash content. Baveno, Carlsbad, and Manebach may occur as simple twins.

INTERFERENCE FIGURE. Sections or fragments of sanidine, anorthoclase, or high albite on {001} show essentially flash figures, and those on {010} show a broad, diffuse obtuse bisectrix. Acute bisectrix figures require unlikely fragment orientation (near 100) and are obviously biaxial negative, with $2V_x$ ranging from 18° (sanidine Ab_0) to 54° (high albite Ab_{100}) (Fig. 7-20). Dispersion throughout the series is weak $r > v$, with possible horizontal bisectrix dispersion.

High sanidine shows obtuse bisectrix figures on {001}, and flash figures on {010}. {100} sections yield an optic axis figure with $2V_x$ of about 63° (Ab_0) to 15° (Ab_{33}). Dispersion $v > r$, with possible inclined bisectrix dispersion.

OPTICAL ORIENTATION. For sanidine, $b = Z$, $a \wedge X = +5°$, $c \wedge Y = -21°$. Anorthoclase is triclinic, with $b \wedge Z$ about 5°, $a \wedge X$ about 10°, and $c \wedge Y$ about 20°, these angles becoming somewhat larger at high albite. The optic plane for the sanidine-high albite series is perpendicular to {010} for sanidine, becoming only slightly inclined to {010} beyond Ab_{63}.

High sanidine shows an entirely different optical orientation with $b = Y$, $c \wedge Z$ about $-20°$ and $a \wedge X$ about $+5°$. The optic plane is, then, parallel to {010}. By heating low-temperature K-feldspar (optic plane normal to 010), we can cause $2V_x$ to close to 0° on X and, with prolonged heating at 1075°C, to open again in the {010} plane (Fig. 7-20), suggesting high-temperature formation and rapid chilling for high sanidine occurrences.

Maximum extinction angle, for both sanidine and high sanidine series, is about 5° to {001} cleavage, as seen in {010} sections, increasing slightly at high albite. Basal {001} sections or fragments on {001} show essentially parallel extinction to {010} cleavages.

Sign of elongation is negative (length-fast) for both series, since X lies only a few degrees from a and cleavages ({010} and {001}) parallel a.

DISTINGUISHING FEATURES. The sanidine-high albite series is most conveniently distinguished from low-temperature feldspars by its small $2V_x$ and extinction angles and by its characteristic occurrences. Anorthoclase shows characteristic double polysynthetic twinning yielding the grid, or quadrille, pattern of microcline. Anorthoclase, however, is confined to volcanic rocks and microcline to plutonic ones, and the twin laminae, in anorthoclase, are very narrow and have different crystallographic orientation (see Section on twinning).

High albite has a much smaller $2V$ than low albite, with reverse optic sign and dispersion. High sanidine is distinguished from sanidine by different optical orientation ($b = Y$) and reverse dispersion.

ALTERATION. High-temperature feldspars are less stable, at surface conditions, than low-temperature varieties and alter to essentially the same products as other alkali feldspars (see p. 347). Kaolin-group clays or sericite with cryptocrystalline silica are the most common alteration products.

OCCURRENCE. Anorthoclase and high albite are rare minerals occurring only in high-temperature, soda-rich lavas. Sanidine is the common alkali feldspar in siliceous volcanic rocks as phenocrysts and microlites or as needles in spherulites of glassy rocks. It often forms large, glassy, euhedral crystals in dike rocks and is the characteristic mineral of the sanidinite facies of contact metamorphism. In volcanic rocks, sanidine commonly ranges between Ab_{20} and Ab_{62} and is most soda-rich in the absence of associated plagioclase.

High sanidine, of course, suggests unusually high temperature and rapid chilling, and $2V$ angle and optic plane orientation may serve to imply temperature of formation. In metamorphic rocks, sanidine appears in xenoliths in lavas or dike rocks and in contact zones with near-surface intrusions where temperature is maximum and pressure minimum.

MONOCLINIC (ORTHOCLASE)
$\angle\beta = 116°$
TRICLINIC (LOW ALBITE)
$\angle\alpha = 94°20'$
$\angle\beta = 116°34'$
$\angle\gamma = 87°39'$

The Orthoclase-Low Albite Series

$(K,Na)AlSi_3O_8$

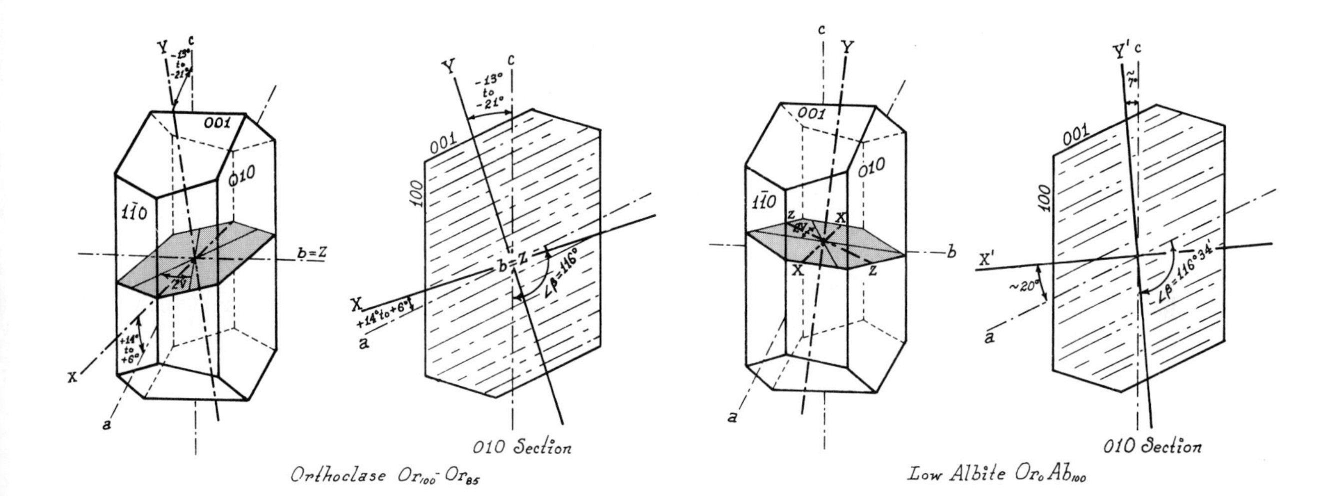

Orthoclase (Or_{100}–Or_{85})

$n_\alpha = 1.518–1.520$*

$n_\beta = 1.522–1.524$

$n_\gamma = 1.522–1.525$

$n_\gamma - n_\alpha = 0.005$

$2V_x = 35°–50°\ (-)$

$b = Z, c \wedge Y = -13°$ to $-21°$, $a \wedge X = +14°$ to $+6°$

$r > v$ distinct, horizontal bisectrix dispersion

Colorless in standard section

Orthoclase Cryptoperthites (Or_{85}–Or_{20})

Intermediate Values for Cryptoperthites

Low Albite (Or_{20}–Or_0)

$n_\alpha = 1.527–1.530$

$n_\beta = 1.531–1.533$

$n_\gamma = 1.536–1.540$

$n_\gamma - n_\alpha = 0.009–0.010$

$2V_z = 90°–85°\ (+)$

Optic plane approx. normal to $\{010\}$
$X \wedge \{001\} = 19°$

$v > r$ weak

Colorless in standard section

* Indices of refraction increase progressively with Na content and are strongly influenced by foreign elements (Fig. 7-21). Most natural "orthoclase" is rich in sodium ($<Or_{85}$) appearing as microperthite or cryptoperthite, with optical constants between those shown for orthoclase and low albite.

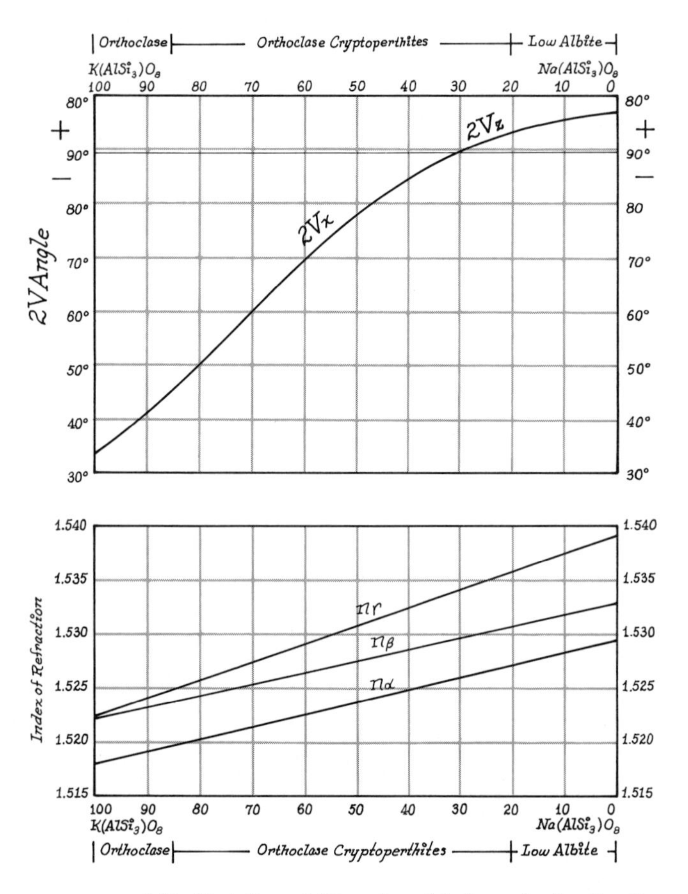

FIGURE 7-21. Variation of $2V$ angle and indices of refraction for the orthoclase-low albite, alkali feldspar series.

COMPOSITION. In the ranges Or_{100}–Or_{85} (*orthoclase*) and Or_{20}–Or_0 (*albite*), a single phase exists with optical properties as indicated above. In the intermediate range Or_{85}–Or_{20}, however, unmixing occurs, producing K-rich and Na-rich phases interlayered as visible microperthites (Fig. 7-13) or submicroscopic cryptoperthites, with optical constants intermediate between orthoclase and low albite, as shown in Fig. 7-21. Adularia and pericline are unusually pure K-feldspar and Na-feldspar respectively. The anorthite "molecule" (An) is usually $<An_5$, except near albite. Significant barium may be present, with minor Sr^{2+}, Mg^{2+}, Fe^{2+}, and Mn^{2+} as replacement for K^+ and Na^+. Rarely, large amounts of Fe^{3+} replace Al^{3+} in tetrahedral coordination.

PHYSICAL PROPERTIES. H = 6–6½. Sp. Gr. = 2.55–2.63. Color, in hand sample, is normally colorless to white or salmon, more rarely yellow, gray, or green. Microperthite and cryptoperthite interlayering often produces an iridescent luster (moonstone). Luster is vitreous to pearly. "Common orthoclase" is not common as hand specimens.

COLOR AND PLEOCHROISM. Colorless as fragments or in section. Often clouded by argillaceous alteration.

FORM. Crystals are usually anhedral. Exsolution lattice intergrowths of albite are almost universal, as visible perthite and microperthite or submicroscopic cryptoperthite. Adularia frequently forms euhedral crystals.

CLEAVAGE. Characteristic cleavages are {001} perfect and {010} distinct intersecting at 90°; {110} poor. Parting is possible parallel to {100}.

BIREFRINGENCE. Low maximum birefringence (0.005–0.010), seen on {001} sections, increases with Na content. In section, interference colors are first-order grays and white, usually less than quartz.

TWINNING. No repeated twinning. Carlsbad twins are most common and are best seen in sections containing *b*. Both individuals show extinction when the composition plane is north-south or east-west but different birefringence in other positions, unless the section is parallel or normal to *c*. Baveno twins produce nearly square prisms, with the composition plane at 45°. Both individuals extinguish simultaneously, although fast and slow rays are reversed. Manebach twinning produces individuals of different birefringence except in sections parallel or normal to {001}. Other twin laws are occasionally represented.

ZONING. Zoning is not common in alkali feldspars of the low-temperature series and shows very limited composition range.

INTERFERENCE FIGURE. Diffuse isogyres on a white field usually suggest a very large $2V$ angle, which increases with Na-content from about 35° at Or_{100} to 90° at about Or_{30} (Fig. 7-21). Orthoclase is optically negative, becoming positive for albite beyond Or_{30}. Y is nearly normal to {001}, and Z is normal to {010}, so cleavage fragments usually show flash and obtuse bisectrix figures. Dispersion is quite distinct, with $r > v$ for orthoclase compositions, reversing to $v > r$ near albite.

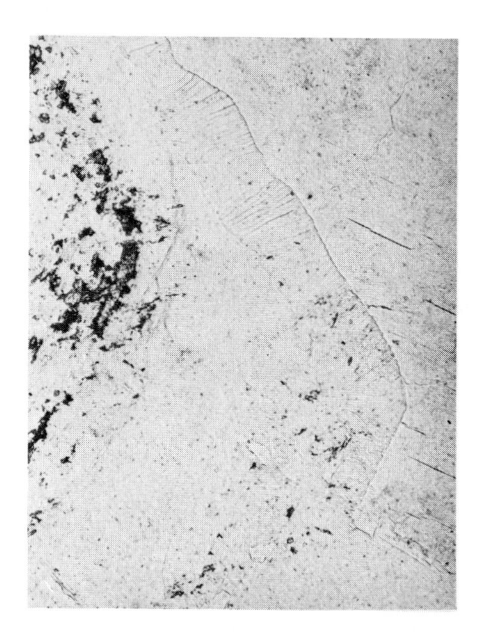

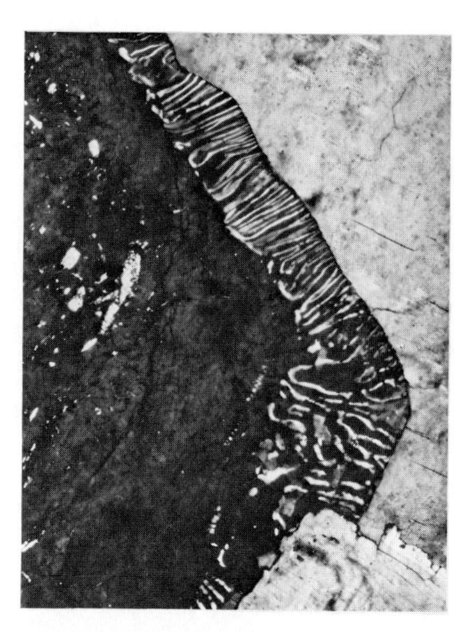

FIGURE 7-22. Myrmekite intergrowths of quartz and plagioclase rim a large plagioclase crystal in an intermediate composition, plutonic igneous rock.

OPTICAL ORIENTATION. The optic plane in orthoclase lies normal to {010} and $b = Z$, $c \wedge Y = -13°$ to $-21°$, $a \wedge X = +14°$ to $+6°$. In low albite, the optic plane becomes almost parallel to *b*, making an angle of about +20° to {001}.

Maximum extinction angle for orthoclase increases with Na content from 6° to 14° measured to {001} cleavage in {010} sections or fragments lying on {010}, and extinction is parallel to {010} cleavage in sections parallel to *b*. Albite extinction shows maximum value of about 19° to {010} in sections normal to {010} as described for plagioclase orientation. Sign of elongation is negative (length-fast) for orthoclase, since X lies near *a*, and {001} and {010} cleavages produce elongation on *a*.

DISTINGUISHING FEATURES. Because it is colorless and anhedral with low relief and low birefringence and usually untwinned, orthoclase is not conspicuous and may be overlooked or mistaken for quartz, so one must make a conscious effort to find and identify it. Orthoclase is normally distinguished from microcline or plagioclase by its lack of multiple twinning and from quartz by its negative relief in balsam, biaxial figure, and its tendency to be turbid with alteration. It has a greater negative relief than either nepheline or untwinned albite and is biaxial negative. Large $2V$ angle distinguishes it from its high-temperature form, sanidine.

ALTERATION. Orthoclase alters most commonly to sericite, illite, or kaolin by either hydrothermal alteration or weathering. Kaolin minerals (such as kaolinite and halloysite) imply acid solutions, possibly near-surface weathering, and sericite suggests neutral or alkaline solutions, often of hydrothermal origin. The brownish, turbid appearance of most orthoclase may be due to dispersed kaolin or tiny liquid inclusions of deuteric origin. Quartz, gibbsite, and pyrophyllite are common alteration products of orthoclase, and alunite may form in sulfate solutions. Orthoclase may be a source of glauconite in clastic sediments. *Myrmekite* (quartz) intergrowth patches penetrating crystals of alkali feldspar that abut plagioclase appear to represent replacement of K-feldspar by plagioclase with release of silica (Fig. 7-22).

OCCURRENCE. Orthoclase is common in felsic plutonic rocks, usually as an associate of either quartz or nepheline. Orthoclase commonly occurs with microcline, although microcline, apparently the more stable low-temperature phase, is favored in older intrusives (Precambrian), and orthoclase is most common in younger plutonics. In silicic rock types, orthoclase averages about Or_{70} in the orthoclase microperthite range; hence orthoclase is normally perthitic. Coarse perthites are characteristic of large batholiths, microperthites of small intrusions, and cryptoperthites of hypabyssal occurrences and flows. Orthoclase is the distinctive mineral of syenites, where it occurs as Na-rich (Or_{50}–Or_{20}) microperthite.

In volcanic rocks, sanidine is the common alkali feldspar but orthoclase may appear in some low-temperature silicic rocks.

In metamorphic rocks, microcline is more common than orthoclase, particularly in lower-grade rocks. Orthoclase microperthite is most characteristic of the higher-grade gneisses and granulites of the katazone, where it may show a dark color. Progressive metamorphism beyond the kyanite zone exceeds the stability range of muscovite, and orthoclase plus sillimanite appears in its stead.

In sedimentary rocks, orthoclase is relatively stable as detrital grains, and authigenic orthoclase (probably adularia) may form as clear growths on orthoclase or microcline nuclei during sedimentation.

The Microcline-Low Albite Series

$(K,Na)AlSi_3O_8$

TRICLINIC (MAXIMUM MICROCLINE)
$\angle\alpha = 90°41'$
$\angle\beta = 115°59'$
$\angle\gamma = 87°30'$

TRICLINIC (LOW ALBITE)
$\angle\alpha = 94°20'$
$\angle\beta = 116°34'$
$\angle\gamma = 87°39'$

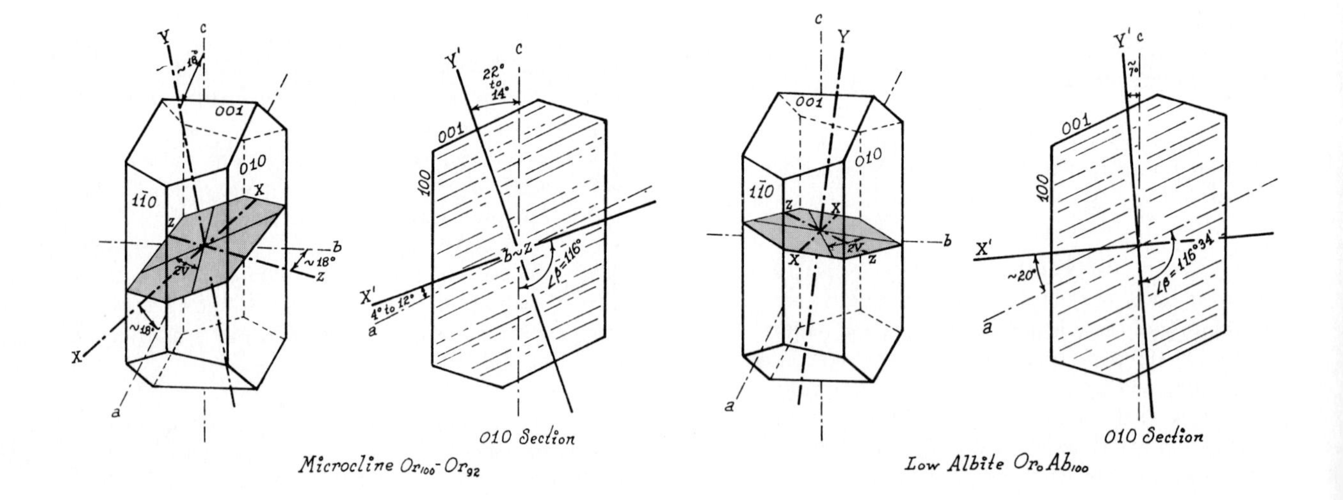

Microcline (Or_{100}–Or_{92})

$n_\alpha = 1.514–1.516$*
$n_\beta = 1.518–1.519$
$n_\gamma = 1.521–1.522$
$n_\gamma - n_\alpha = 0.007$
$2V_x = 66°–68° (-)$*
Optic plane near {001}
$X \wedge \{001\} = 15°$
$r > v$ perceptible, horizontal bisectrix dispersion
Colorless in standard section

Microline

Cryptoperthites (Or_{92}–Or_{20})

Intermediate Values for Microperthites

Low Albite (Or_{20}–Or_0)

$n_\alpha = 1.527–1.530$
$n_\beta = 1.531–1.533$
$n_\gamma = 1.536–1.540$
$n_\gamma - n_\alpha = 0.009–0.010$
$2V_z = 90°–85° (+)$
Optic plane approx. normal to {010}
$X \wedge \{001\} = 19°$
$v > r$ weak

Colorless in standard section

COMPOSITION. Very little solid solution exists in the microcline-low albite series, and differences between it and the orthoclase-low albite series are significant only at the K end of the series, where microcline is contrasted with orthoclase. Microcline perthites, microperthites, and cryptoperthites, showing intermediate optical properties, represent almost the entire series (see Fig. 7-23). Microcline contains much the same impurities as orthoclase and in about the same amounts ($<An_5$).

PHYSICAL PROPERTIES. H = 6–6½. Sp. Gr. = 2.56–2.63. Color, in hand sample, is most commonly white or salmon, more rarely apple-green (amazonstone) or yellow. Luster is vitreous to pearly.

COLOR AND PLEOCHROISM. Colorless as fragments or in section. May be clouded by alteration.

FORM. Crystals are normally anhedral in section. Lattice intergrowths are almost universal, as coarse perthites, microperthites, or cryptoperthites.

CLEAVAGE. Characteristic cleavages are {001} perfect and {010} distinct, intersecting at 90°41′. {110} cleavage is poor.

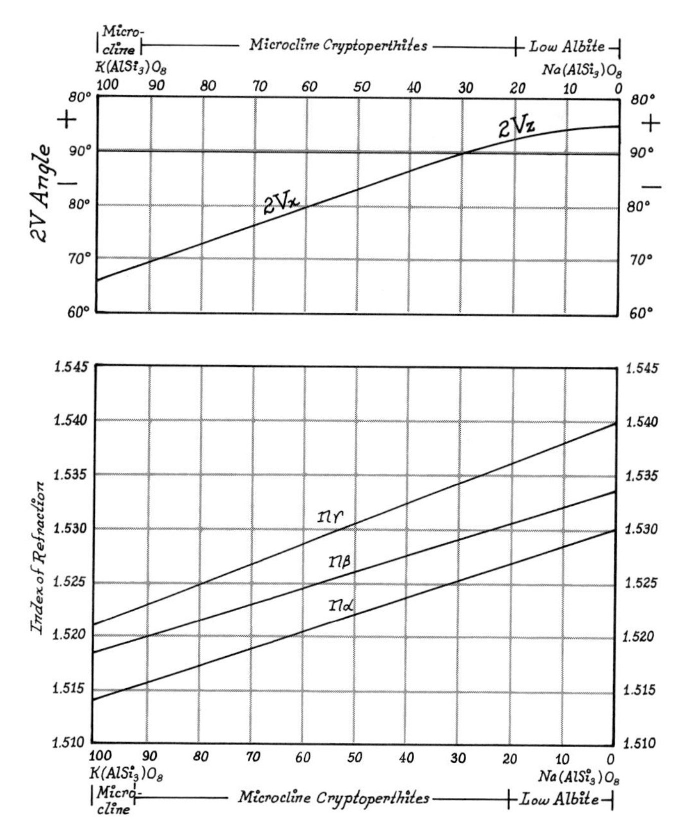

FIGURE 7-23. Variation of $2V$ angle and indices of refraction for the microcline-low albite, alkali feldspar series.

* Refractive index and $2V_x$ of the cryptoperthites increase steadily with Na content (Fig. 7-23). One cannot, however, reliably determine Na-K composition of cryptoperthites by refractive index measurements, because impurities greatly affect optical constants, and several structural states are possible.

Single-phase microcline seldom contains more than 5 percent Na-feldspar, and most natural microcline is perthitic, microperthitic, or cryptoperthitic with optical constants between those shown above for the pure microcline and albite end members.

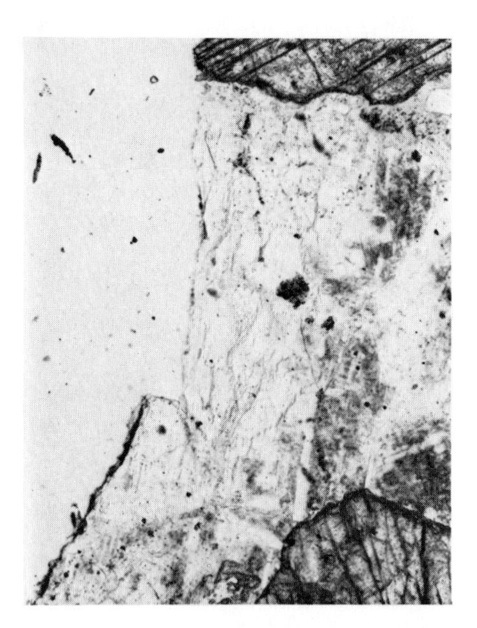

FIGURE 7-24. Microcline twinning appears as a double set of polysynthetic twin laminae, at almost 90°, which pinch out and swell in lenticular forms. Microcline is characteristic of plutonic rocks, both igneous and metamorphic, and is often perthitic.

BIREFRINGENCE. Weak maximum birefringence (0.007–0.008) seen on {001} sections. Interference colors are first-order gray to white.

TWINNING. The most characteristic feature of microcline is its double set of polysynthetic twins according to albite and pericline laws (Fig. 7-24). Diffuse twin laminae, intersecting at essentially 90°, pinch and swell, forming lenticular or spindle-shaped bands of this characteristic grid or quadrille pattern. Composition plane of the pericline twins contains b and lies about $-35°$ from {001}. Only albite twinning is visible on {100} sections, and only pericline twinning on {010} at 83° to the basal cleavage. Grid twinning indicates that the mineral originally crystallized at high temperature, with monoclinic symmetry, and subsequently inverted to the triclinic polymorph.

Simple Baveno, Carslbad, or Manebach twinning are also reported.

INTERFERENCE FIGURE. Both fragments and sections of microcline yield poor interference figures, because of compound twinning. Figures are diffuse and show large $2V$ ranging from 66° (−) through 90° to about 70° (+). Common microcline microperthites usually show a $2V_x$ about 80°–85° (−). $2V_x$ increases with Na and Ca content (Fig. 7-23). Fragments lying on {001} and {001} sections show flash figures and sections, and fragments on {010} give an off-center obtuse bisectrix.

Dispersion is usually quite distinct $r > v$, with horizontal bisectrix dispersion.

OPTICAL ORIENTATION. Z lies near b, $c \wedge Y$ is about $-18°$ and $a \wedge X$ is about $+18°$. Maximum extinction angle is seen on {001} sections and measures about 15° to the {010} cleavage. Sections parallel to {010} give about 5° extinction to {001} cleavage traces. Sign of elongation is negative (length-fast), since X is near a and fragments tend to be elongated on a.

DISTINGUISHING FEATURES. A gridlike, or quadrille, twinning pattern is, undoubtedly, the most characteristic feature of microcline (Fig. 7-24). Individual crystals of plagioclase showing combined albite and pericline twinning may be confused with microcline. Twin laminae, however, are lenticular and diffuse in microcline, and the pericline composition plane lies nearly normal to {001} cleavage.

ALTERATION. Kaolin clays, illite, and sericite are the most common alteration products of microcline. Weathering commonly yields kaolin, and either kaolin or sericite most often result from hydrothermal alteration. Quartz, gibbsite, pyrophyllite, and alunite are additional alteration possibilities

OCCURRENCE. Microcline is a low-temperature form of K-feldspar characteristic of moderately deep-seated rocks, either igneous or metamorphic. It is very common in all silicic, plutonic, igneous rocks such as granite and granodiorite. Microcline seems more stable than orthoclase, and its formation is favored by slight shearing pressure. Precambrian rocks usually contain microcline in preference to orthoclase. Microcline is especially characteristic of pegmatites, where it often appears as the feldspar phase in graphic granites.

In metamorphic rocks, microcline appears in low-grade schists of the amphibolite and greenschist facies, yielding to orthoclase in the deeper zones. The "eyes" in augen schists and gneisses are commonly microcline. Porphyroblasts, or phenocrysts, of microcline perthite or microcline mantled by sodic plagioclase are said to show *rapakivi structure.*

In sedimentary rocks, microcline is a rather stable detrital mineral, and new authigenic overgrowths of microcline may form on microcline nuclei during sedimentation. Authigenic K-feldspar is unusually pure $K(AlSi_3)O_8(Or_{99+})$.

REFERENCES: ALKALI FELDSPARS

Bailey, E. M., and E. S. Rollin. 1960. Selective staining of plagioclase and K-feldspar on rock slabs and thin sections (abs.). *Bull. Geol. Soc. Amer., 71*, 2047.

Baskin, Y. 1956. A study of authigenic feldspars, *Jour. Geol., 64*, 132–155.

Chaisson, V. 1950. The optics of triclinic adularia, *Jour. Geol., 58*, 537–547.

Deer, W. A., R. A. Howie, and J. Zussman. 1962. *Rock-Forming Minerals.* Vol. 4: *Framework Silicates.* New York: Wiley.

Donnay, G., and J. D. H. Donnay. 1952. The symmetry change in the high-temperature alkali-feldspar series. *Amer. Jour. Sci., Bowen Vol.*, 115–132.

Goldsmith, J. R., and F. Laves. 1954. Potassium feldspars structurally intermediate between microcline and sanidine. *Geochim. Cosmochim. Acta, 7*, 212–230.

Hsu, K. J. 1954. A study of the optic properties and petrologic significance of zoned sanidines, *Amer. Jour. Sci., 252*, 441–443.

Kennedy, G. C. 1947. Charts for correlation of optical properties with chemical composition of some common rock-forming minerals. *Amer. Mineral., 32*, 561–573.

Köhler, A. 1948. Zur Optik des Adulars. *Neues Jahrb. Min., Abh.*, 1948, 49–55.

Köhler, A. 1949. Recent results of investigations of the feldspars. *Jour. Geol., 57*, 592–599.

Laves, F. 1960. Al/Si-Verteilungen, Phasen-Transformationen und Namen der Alkalifeldspäte. *Z. Kristallogr., 113*, 265–296.

MacKenzie, W. S. 1952. The effect of temperature on the symmetry of high-temperature and soda-rich feldspars, *Amer. Jour. Sci., Bowen Vol.*, 319–342.

MacKenzie, W. S., and J. V. Smith. 1956. The alkali feldspars. III. An optical and x-ray study of high-temperature feldspars. *Amer. Mineral., 41*, 405–427.

Ribbe, P. H., ed. 1975. *Feldspar Mineralogy: Mineral. Soc. Amer. Short Course Notes*, Vol. 2. Blacksburg, Va.: Southern Printing Co.

Smith, J. V. 1974. *Feldspar Minerals.* Vol. 1: *Crystal Structure and Physical Properties.* Heidelberg: Springer-Verlag.

Smith, J. V. 1974. *Feldspar Minerals.* Vol. 2: *Chemical and Textural Properties.* Heidelberg: Springer-Verlag.

Stewart, D. B. 1974. Optic axial angle and extinction angles of alkali feldspars related by cell parameters to Al/Si order and composition. In W. S. MacKenzie and J. Zussman, eds., *The Feldspars.* Manchester, England: Manchester University Press. pp. 145–161.

Stewart, D. B., and P. H. Ribbe. 1967. Structural explanation for variations in cell parameters of alkali feldspar with Al/Si ordering. *Amer. Jour. Sci., 276A*, 444–462.

Tuttle, O. F. 1952. Optical studies on alkali feldspars. *Amer. Jour. Sci., Bowen Vol.*, 553–567.

PLAGIOCLASE FELDSPARS

TRICLINIC

Plagioclase	$(NaSi,CaAl)AlSi_2O_8$	Ab An
		$\angle\alpha = 94°20'–93°10'$
		$\angle\beta = 116°34'–115°51'$
		$\angle\gamma = 87°39'–91°13'$

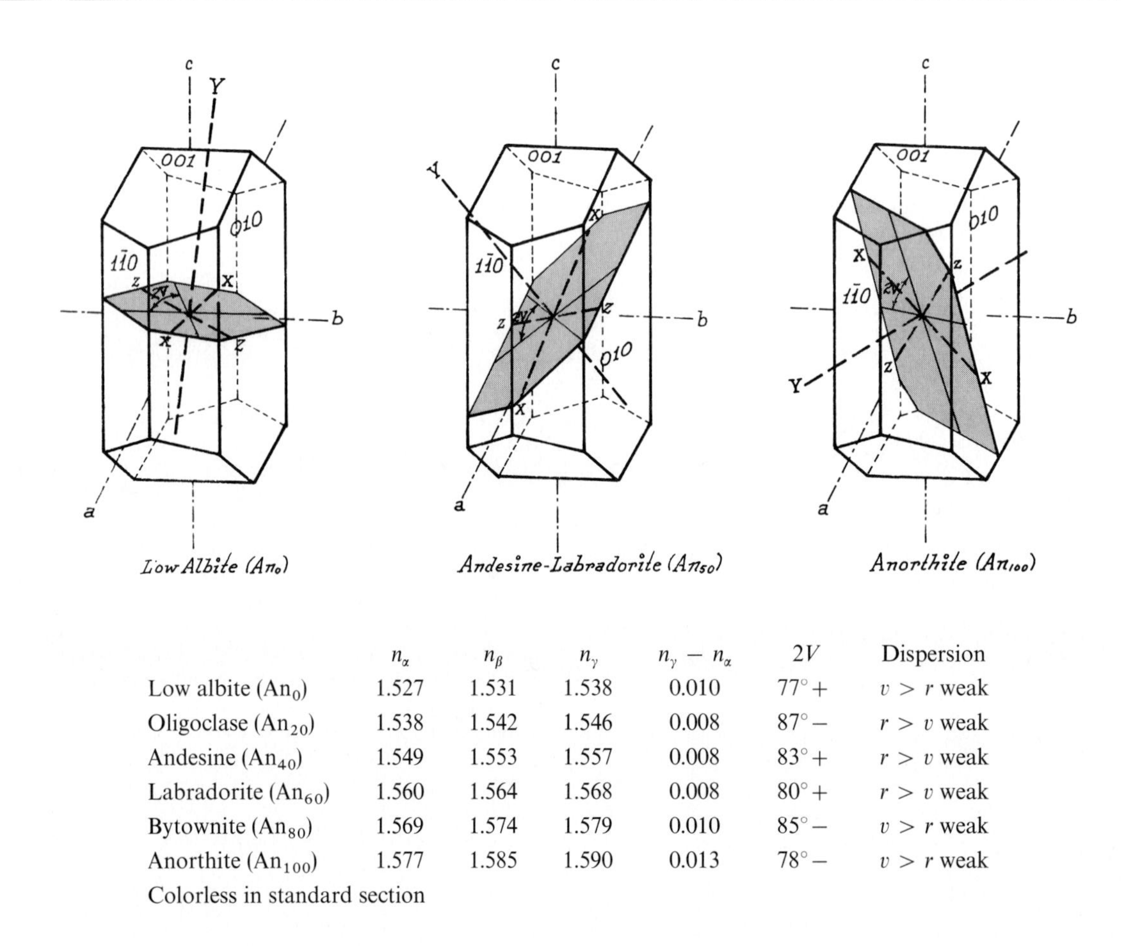

	n_α	n_β	n_γ	$n_\gamma - n_\alpha$	$2V$	Dispersion
Low albite (An_0)	1.527	1.531	1.538	0.010	77°+	$v > r$ weak
Oligoclase (An_{20})	1.538	1.542	1.546	0.008	87°−	$r > v$ weak
Andesine (An_{40})	1.549	1.553	1.557	0.008	83°+	$r > v$ weak
Labradorite (An_{60})	1.560	1.564	1.568	0.008	80°+	$r > v$ weak
Bytownite (An_{80})	1.569	1.574	1.579	0.010	85°−	$v > r$ weak
Anorthite (An_{100})	1.577	1.585	1.590	0.013	78°−	$v > r$ weak

Colorless in standard section

* Albite twinning is recognized as repeated twinning, usually running parallel to crystal length. It is the only repeated, normal plagioclase twinning, which requires that, when the composition planes are vertical and north-south, both sets of twin laminae show equal interference colors.

COMPOSITION. The low-temperature plagioclase series is chemically continuous from pure albite, $Na(AlSi_3)O_8$, to pure anorthite, $Ca(Al_2Si_2)O_8$, with submicroscopic unmixing in the range An_2–An_{16} (peristerites). An_{48}–An_{58} (Bøggild intergrowths), and An_{70}–An_{90} (Huttenlocher intergrowths). Small amounts of K-feldspar may enter into the plagioclase composition, reaching a maximum of about 8 percent $K(AlSi_3)O_8$ (Or_8) near the albite end of the series. Potassium does not appear to have significant influence on optical properties of the plagioclase series.

Composition is best determined by refractive index measurements on fragments (Fig. 7-25), where an accuracy of ± 0.001 yields composition within about 2 percent An. An equally accurate but simpler method utilizes fragments lying on the $\{001\}$ cleavage, which yield a low refractive index very near n_α for both high- and low-temperature varieties. This value may be substituted into the following equations to obtain composition to within ± 1–2 percent An if Or content is low:

$$An_{0-24} \quad An = 1936(n - 1.5287)$$
$$An_{24-31} \quad An = 1790(n - 1.5277)$$
$$An_{31-84} \quad An = 1944(n - 1.5290)$$
$$An_{84-100} \quad An = 2133(n - 1.5328)$$

The appropriate An range (and hence the appropriate equation to use) can be found by first using the equation for the range An_{31-84} to obtain an approximate composition. In thin section, composition is determined by optical orientation as described later.

PHYSICAL PROPERTIES. H = 6–6½. Sp. Gr. = 2.63 (albite) to 2.76 (anorthite). Color, in hand sample, ranges from white to colorless for albite, to dark gray for labradorite, and lighter again for anorthite. Luster is vitreous to pearly.

COLOR AND PLEOCHROISM. Colorless as fragments or in section, often clouded by alteration.

FORM. Crystals may be euhedral to anhedral. Sodic varieties may appear as plates parallel to $\{010\}$ (*cleavelandite*) or as granular aggregates ("sugary"). Calcic varieties tend to be euhedral in plutonic rocks, and all varieties may be euhedral as phenocrysts or microlites, usually elongated parallel to a, in volcanic rocks.

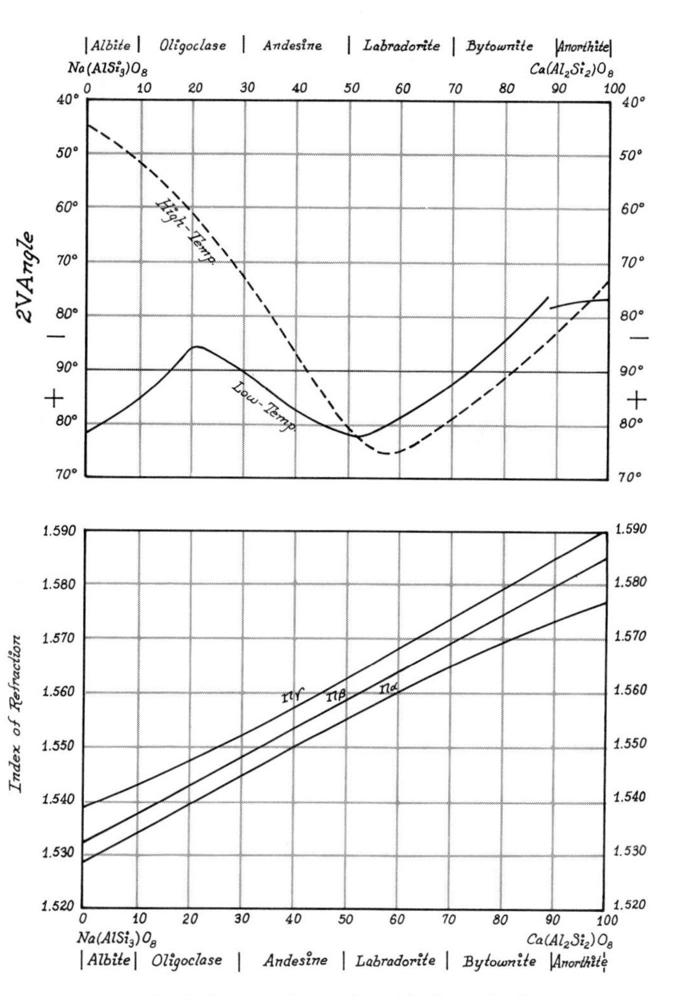

FIGURE 7-25. Variation of $2V$ angle and indices of refraction in the plagioclase series (Smith, The optical properties of heated plagioclases, *Amer. Mineral.*, pp. 1188, 1189, Figs. 2 and 3, 1958, © by the American Mineralogical Society of America). In the $2V$ angle diagram, the solid curve shows $2V$ variation in the natural, low-temperature plagioclase series. The dashed line shows $2V$ angles of high-temperature modifications.

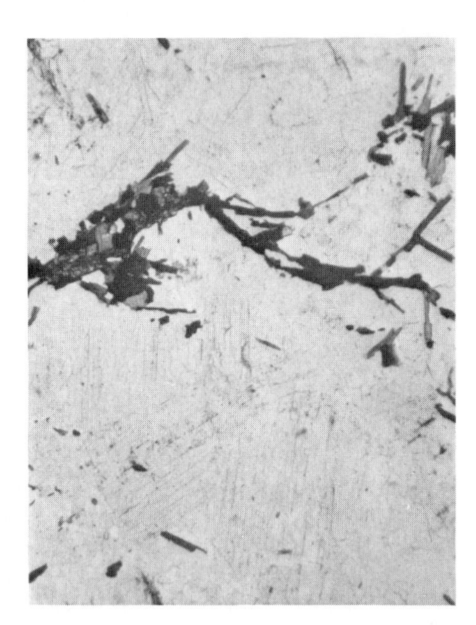

FIGURE 7-26. Plagioclase that is highly zoned suggests considerable disequilibrium during crystallization.

CLEAVAGE. Characteristic cleavages are {001} perfect and {010} distinct, intersecting essentially at right angles; {110} poor.

BIREFRINGENCE. Weak maximum birefringence ranges from 0.008 (andesine) to 0.010 (albite) and 0.013 (anorthite). In section, interference colors are mostly first-order grays and white to a maximum of deep yellow (anorthite).

TWINNING. Polysynthetic, normal twinning by the albite law is almost universal in all plagioclase varieties and is its most characteristic feature. Albite twin laminae, usually parallel to the long crystal dimension, tend to be very narrow in the oligoclase range, becoming broad for all calcic varieties and albite. Carlsbad (simple, parallel) and pericline (repeated, parallel) twinning are common in combination with albite twins, and other laws are not rare.

Authigenic albite in sedimentary rocks and sodic plagioclase in low-grade metamorphic rocks show only simple twinning, mostly albite twinning, or no twinning at all.

ZONING. Composition zoning is an indication of disequilibrium conditions during the cooling history and is especially characteristic of plagioclase in intermediate igneous rock types (Fig. 7-26). Normal zoning produces a calcic core with sodic exterior so that extinction angles are usually greater for central zones. Zoning may be essentially continuous, suggesting uniform cooling history or may show sharp discontinuities. Reverse zoning implies a sodic core becoming calcic with progressive cooling, and may result from a decrease of dissolved H_2O in the magma or magmatic assimilation of carbonate or other high-calcium rocks. Oscillatory zoning showing separated zones of equal extinction, may suggest a turbulent history of rising and falling crystals, complex assimilation, or variable water pressure.

INTERFERENCE FIGURE. Useful interference figures are often difficult to obtain, especially from fragments, owing to ubiquitous twinning. In sections, untwinned areas may show biaxial figures of diffuse, illusive isogyres on a white field suggesting a large $2V$ angle. Difficulty is also encountered in sign determination, as the $2V$ is always large and varies on either side of 90° almost sinusoidally with changing composition (see Fig. 7-25). Dispersion is always rather weak, especially near albite, with $v > r$ or $r > v$, as shown earlier. Bisectrix dispersion in theory shows triclinic symmetry as combined inclined and either horizontal or crossed dispersion.

OPTICAL ORIENTATION. Systematic changes in optical-crystallographic orientation reflect changes in composition, and herein lies the key to plagioclase composition in thin section. The optic plane lies almost perpendicular to *c* for low albite, becoming almost parallel to *c* in anorthite.

The maximum extinction angle, measured from fast ray to {010}, varies continuously with composition and is suitable for composition determination for the student who may yet be unfamiliar with universal stage procedures. The basic principle of the *Michel-Lévy method* is to measure the maximum extinction angle from the {010} crystallographic plane, as follows (see Fig. 7-27).

1. Select a plagioclase crystal showing distinct albite twinning* where {010} is normal to the plane of the section as indicated by the following criteria:

   a. Composition planes between adjacent twin laminae appear as distinct, narrow lines.
   b. When composition planes are north-south (N-S), both sets of twin laminae show uniform birefringence (equal interference colors).
   c. Extinction angles to be measured will be essentially equal (within four or five degrees) on either side of the N-S position.

2. From the N-S position (twin laminae are N-S), the stage is rotated to the right, and the angle is measured to where the first set of twin laminae becomes extinct. Again from the N-S position, the stage is rotated left and the angle recorded to where the alternate twin set becomes extinct. The two extinction angles should be equal and must measure so within four or five degrees, and the average of the two measurements is taken as the correct value.

3. The measured angle must be the angle between {010} (albite composition plane) and the vibration direction of the faster wave. For compositions up to 80 percent anorthite, the first set of twin laminae to become extinct on either side of the N-S position will do so when the vibration direction

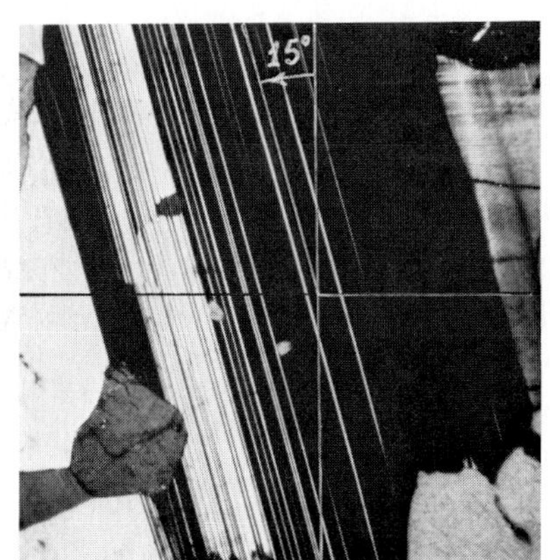

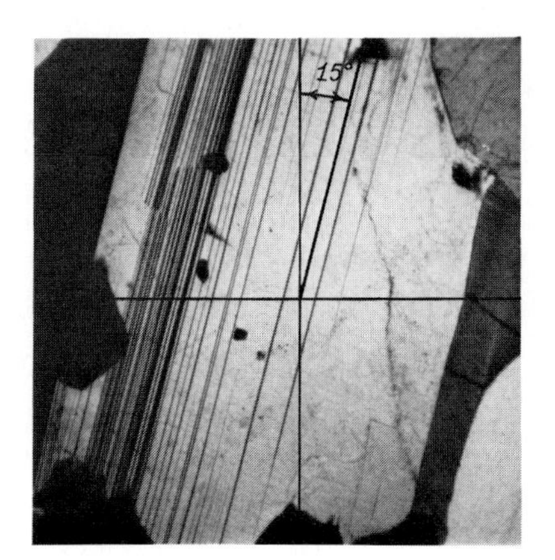

FIGURE 7-27. The Michel-Lévy method of determining plagioclase composition. Extinction angles are measured to albite twins (the {010} plane) for several plagioclase crystals until a maximum extinction angle is determined, as described in the text. This maximum extinction angle plotted on Fig. 7-28(A) reveals the plagioclase composition. In the example shown, an extinction angle of 15° is measured to the albite twin lamellae ({010} trace). If this proves to be the maximum angle after measuring similar extinction angles on a representative number of twinned plagioclase crystals in the same thin section, reference to Fig. 7-28 shows the plagioclase to have a composition near An_{33}, if its index of refraction is >1.54, or An_7, if its index of refraction is <1.54.

of its faster wave is N-S; for a composition of 80 percent anorthite, both sets of twin laminae become extinct simultaneously at 45°; and for compositions more calcic than 80 percent anorthite the first set of twin laminae becomes extinct when the vibration direction of its slower wave is N-S. In the latter case, one must record the complement of the angle measured to the first extinction or turn past the first extinction to measure the angle where the second set of twin laminae becomes extinct. Although one seldom encounters plagioclase more calcic than 80 percent anorthite, one should always check to make sure measurements are being made to the faster wave. At extinction, one vibration direction lies N-S, and from this position a stage rotation of 45° clockwise and use of the gypsum plate will confirm this vibration direction as either fast or slow.

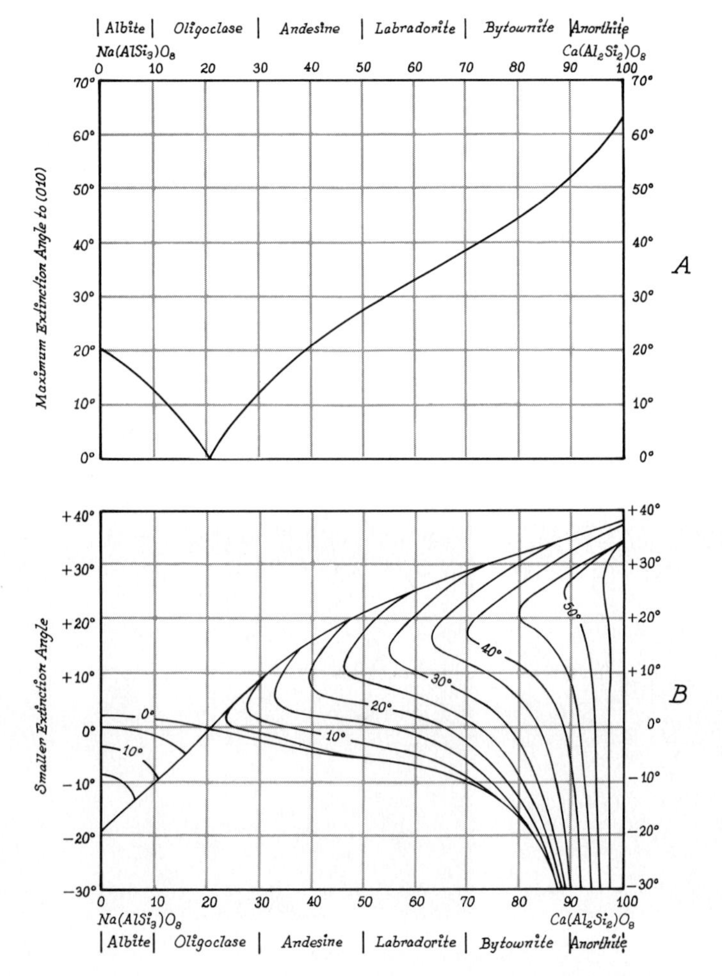

FIGURE 7-28. Variation of extinction angles measured to (010) traces in sections normal to (010) with variation in plagioclase composition (see text): (A) Variation in the maximum extinction angle of albite twins; (B) extinction angles of albite twins on the two sides of a Carlsbad twin. (From *Microscopic Identification of Minerals* by Heinrich. Copyright © 1965. Used with permission of McGraw-Hill Book Company.)

4. The only useful angle is the maximum angle between {010} and the vibration direction of the faster wave. Note that this extinction angle ranges from minimum to maximum, with orientation of the section about the twin axis normal to {010}. This requirement necessitates measurement of a sufficient number of extinction angles, as outlined, to be assured of a maximum value.* The student will learn to quickly discard, with very cursory measurement, sections oriented to give small extinction angles, and only a few formal measurements are usually necessary to arrive at a maximum value.

5. By referring the maximum extinction angle to Fig. 7-28(A), one quickly arrives at an approximate plagioclase composition. For extinction angles of less than 20°, the curve indicates two possible compositions, and the refractive index of the mineral is compared to that of balsam to determine which possibility is the correct one. Negative relief indicates a composition more sodic than 20 percent anorthite and positive relief a composition more calcic.

The albite-Carlsbad method assumes a section normal to {010} divided by a Carlsbad twin with multiple albite twins on both sides of the Carlsbad composition plane. This fortunate occurrence may allow the approximate plagioclase composition to be determined with this single crystal section as follows (see Fig. 7-29):

1. Locate a crystal section normal to {010}, as described in the previous section, showing a Carlsbad twin with albite twinning on both sides. Note that when twin traces are N-S crystal halves on either side of the Carlsbad trace may show different interference colors, but albite twins show similar colors on a given half.

2. On each side of the Carlsbad trace, measure extinction angles from the fast wave to {010} albite traces, as in the

* The method is, of course, based on the assumption that all plagioclase crystals in a given section are of like composition.

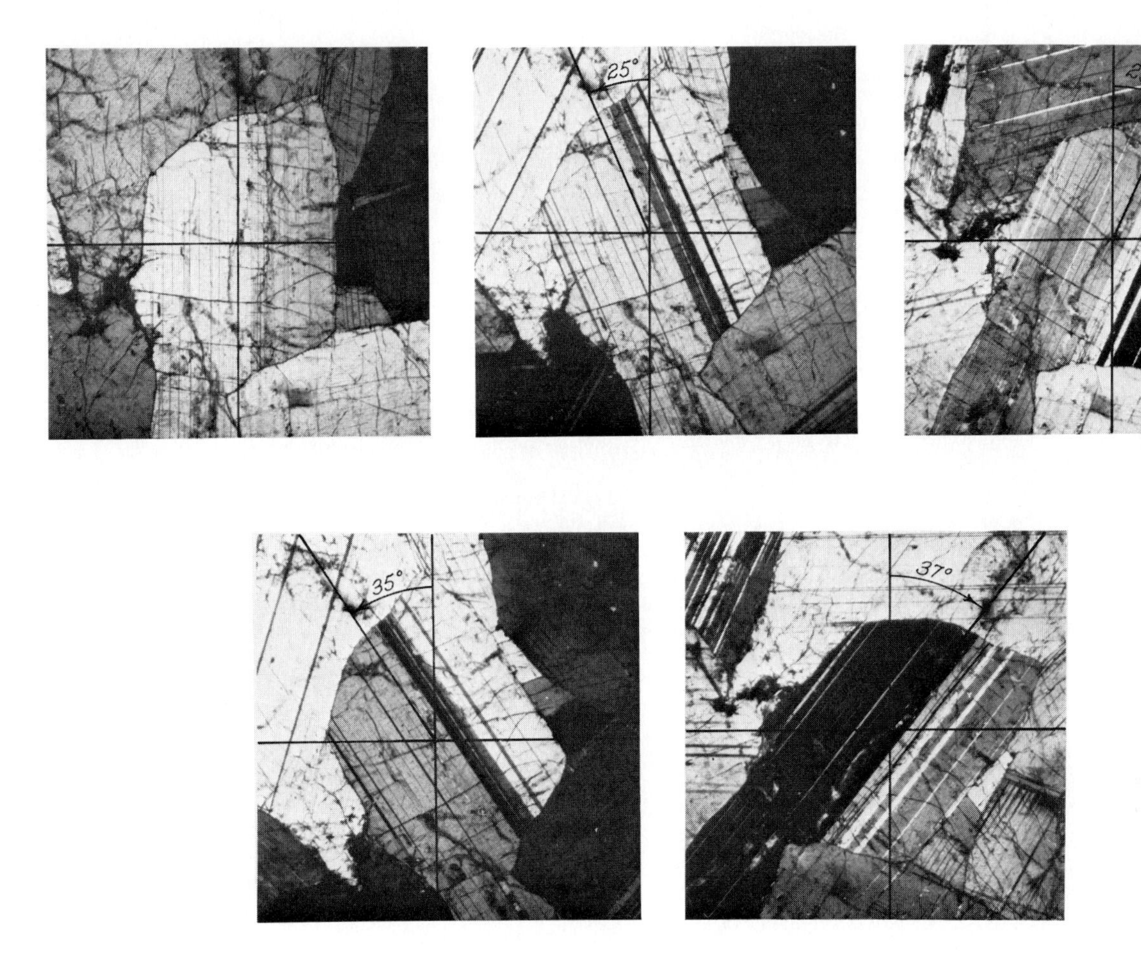

FIGURE 7-29. The albite-Carlsbad method of determining plagioclase composition. Extinction angles are measured for albite twin sets on both sides of a Carlsbad twin, as described in the text. These angles plotted on Fig. 7-28(B) reveal the plagioclase composition. In the example shown, the smaller extinction angle, 25°, is seen on the albite twin lamellae on the right side of the Carlsbad composition plane. The larger extinction angle, 36° (average), is seen on albite lamellae to the left of the Carlsbad twin. Reference to Fig. 7-28(B) shows this plagioclase to have a composition about An_{75} (bytownite). Note that An_{91} is also possible from Fig. 7-28, and measurement of another optical constant (for example, index of refraction or $2V$) may be necessary to eliminate one possibility.

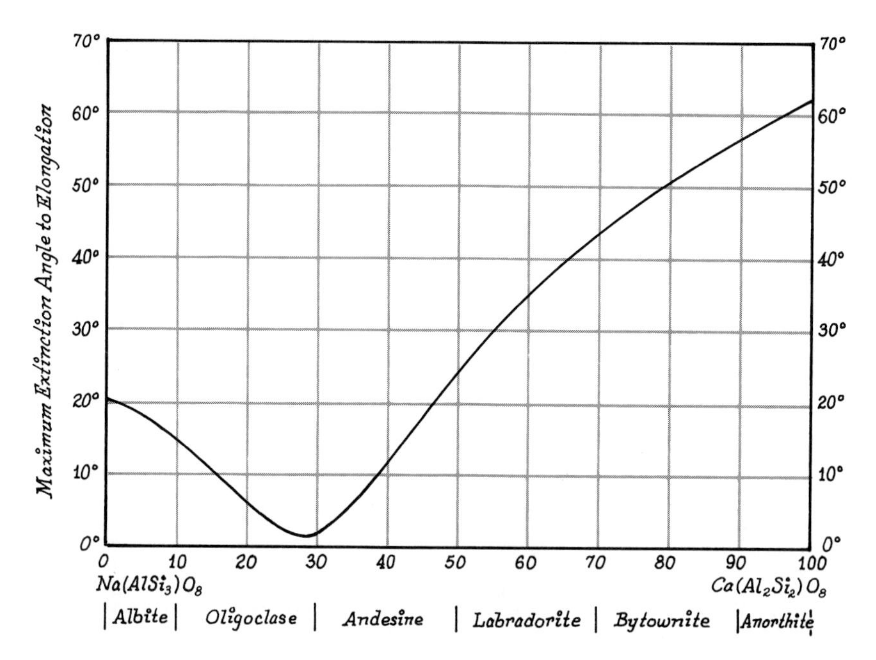

FIGURE 7-30. The plagioclase microlite method of determining approximate composition of very small plagioclase crystals in volcanic rocks. Extinction angles of many microlites are measured, and the maximum angle found is referred to this curve. Note that two compositions are possible at extinction angles below 20°. (From *Microscopic Identification of Minerals* by Heinrich. Copyright © 1965. Used with permission of McGraw-Hill Book Company.)

previous section. On a given Carlsbad half, the right and left extinction angles to albite traces must agree within four or five degrees, as previously required; however, the extinction angle for one Carlsbad half may differ greatly from the extinction angle for the other Carlsbad half.

3. Locate the smaller of these two extinction angles on the ordinate of Fig. 7-28(B) and the larger angle on curved lines in the body of the diagram. Their intersection, projected to the abscissa, indicates the albite-anorthite composition. Since one may not know the sign of these extinction angles, one must assume either sign, and as many as four intersections may result, perhaps making other optical data (such as refractive index) necessary to limit the composition.

Plagioclase microlites in volcanic rocks often present a special problem, as they may be too small for the preceding methods to be practical. They are usually elongated on a; if so, a good approximation of composition can be obtained by measuring extinction angles of many complete microlites and referring the maximum measured angle to Fig. 7-30.

DISTINGUISHING FEATURES. Undoubtedly, the most distinctive feature of plagioclase, in section or fragments, is polysynthetic twinning combined with low birefringence and low relief. Note that albite twinning is not seen on {010} sections, and rarely, plagioclase is untwinned. In metamorphic gneiss and hornfels, rarely in granite and pegmatite, cordierite may be mistaken for plagioclase, when the former shows polysynthetic twinning. Cordierite is usually highly contaminated with mineral inclusions or yellowish alteration and may show cyclic, pseudohexagonal sector twinning.

ALTERATION. Plagioclase is normally altered before alkali feldspar, and in zoned crystals higher calcic regions are altered first. Alteration products of plagioclase are most commonly calcite, sericite (muscovite or paragonite), or some clay of the kaolinite group, as determined by conditions of alteration. Surface or near-surface weathering commonly alters plagioclase to calcite and kaolin clays or to a montmorillonite clay in the presence of Mg-bearing waters. Zeolites (for example, analcime, natrolite, heulandite, scolecite, and thomsonite) may form from microlites of basic volcanics, often suggesting partial solution and redeposition in amygdales. Hydrothermal alteration leads to the formation of sericite, boehmite, kaolin clays, and secondary albite or orthoclase (along cracks), with accompanying calcite. A fine-grained aggregate of epidote minerals, notably zoisite, with secondary albite, sericite, and possibly other common alteration products is called *saussurite* and may be produced from calcic plagioclase by deuteric or hydrothermal alteration. Scapolite, prehnite, and pyrophyllite (from sericite) are additional alteration products of plagioclase.

OCCURRENCE. The plagioclase of plutonic rocks may have any composition in the plagioclase range. Oligoclase and andesine are characteristic of the common felsic and intermediate igneous rocks (for example, granite and granodiorite). Albite, or possibly oligoclase, may be anticipated in pegmatites or light-colored nepheline-bearing rocks, and albite, as pericline, may occur with adularia in hydrothermal veins. Labradorite and, less commonly, bytownite are the varieties characteristic of gabbro, anorthosite, and other mafic igneous rocks. Anorthite is not common in plutonic rocks; however, it is reported in olivine-bearing norites.

Stony meteorites often contain a Ca-rich plagioclase, usually in the bytownite range.

Volcanic rocks may contain representatives of the high-temperature plagioclase series. Rhyolites usually contain oligoclase or Na-rich andesine, and Ca-rich labradorite is most typical of basalts. Intermediate volcanics commonly show complexly zoned plagioclase in the andesine range. Neither albite nor anorthite are common in volcanic rocks, but anorthite occurs rarely in basalts, and albite is the distinctive mineral of spilites (that is, basic volcanics altered by Na-metasomatism) and keratophyres.

Metamorphic rocks seldom yield plagioclase in the labradorite-bytownite range or sodic-oligoclase (An_{10}–An_{20}). Albite (commonly untwinned), in combination with epidote, is characteristic of low-grade regional or contact metamorphism (greenschist, zeolite, and albite-epidote hornfels facies). Increasing metamorphism moves calcium from epidote to plagioclase, forming hornblende-oligoclase or hornblende-andesine associations of the amphibolite facies, and when hornblende becomes pyroxene the by-product plagioclase appears. Andesine, frequently as antiperthite, is characteristic of high-grade metamorphism in the granulite facies. Anorthite appears commonly with calc-silicates and carbonates in metamorphosed limestones and dolomites. Metasomatic albite showing "chessboard" twinning, forms in contact zones in argillaceous sediments through the action of Na-rich fluids.

Sedimentary rocks often contain detrital fragments of more-or-less altered plagioclase, sodic varieties being most stable. Authigenic albite commonly forms during diagenesis or by replacement of detrital grains. Authigenic albite contains very minor potash and lime and is commonly untwinned.

REFERENCES: PLAGIOCLASE

Bailey, E. M., and E. S. Rollin. 1960. Selective staining of plagioclase and K-feldspar on rock slabs and thin sections (abs.). *Bull. Geol. Soc. Amer., 71*, 2047.

Barber, C. T. 1936. The effect of heat on the optical orientation of plagioclase feldspar, *Mineral. Mag., 24*, 343–352.

Baskin, Y. 1956. A study of authigenic feldspars, *Jour. Geol., 64*, 132–155.

Brown, W. L. 1960. The crystallographic and petrologic significance of peristerite unmixing in the acid plagioclases. *Z. Kristallogr., 113*, 330–344.

Chayes, F. 1952. Relations between composition and indices of refraction in natural plagioclases. *Am. Jour. Sci., Bowen Vol.*, 85–105.

Chao, S. H., and W. H. Taylor, 1940. Isomorphous replacement and superlattice structures in the plagioclase feldspars. *Proc. Roy. Soc., 176A*, 76–87.

Dodge, T. A. 1936. A rapid microscopic method for distinguishing quartz from untwinned oligoclase-andesine. *Amer. Mineral., 21*, 531–532.

Doeglas, D. J. 1940. Reliable and rapid method for distinguishing quartz and untwinned feldspar with the universal stage. *Amer. Mineral., 25*, 286.

Donnay, J. D. H. 1943. Plagioclase twinning. *Bull. Geol. Soc. Amer., 54*, 1645–1652.

Emmons, R. C., and others, eds. 1953. Selected petrogenetic relationships of plagioclase. *Geol. Soc. Am. Mem., 52*.

Foster, W. R. 1955. Simple method for the determination of the plagioclase feldspars. *Amer. Mineral., 40*, 179–185.

Gorai, M. 1951. Petrological studies on plagioclase twins. *Amer. Mineral, 36*, 884–901.

Heinrich, E. W. 1965. *Microscopic Identification of Minerals.* New York: McGraw-Hill.

Kennedy, G. C. 1947. Charts for correlation of optical properties with chemical composition of some common rock-forming minerals. *Amer. Mineral., 32*, 561–573.

Köhler, A. 1949. Recent results of investigations of the feldspars. *Jour. Geol.*, *57*, 592–599.

Michel-Lévy, A. 1904. Étude sur la dètermination des feldspathes. *Trois. Fasc.*, Paris.

Morse, S. A. 1968. Revised dispersion method for low plagioclase. *Amer. Mineral.*, *53*, 105–115.

Morse, S. A. 1978. Test of plagioclase dispersion method and rapid probe analysis. *Amer. Mineral.*, *63*, 768–770.

Muir, I. D. 1955. Transitional optics of some andesines and labradorites. *Mineral. Mag.*, *30*, 545–568.

Poldervaart, A. 1950. Correlation of physical properties and chemical composition in the plagioclase, olivine and orthopyroxene series. *Amer. Mineral.*, *35*, 1067–1079.

Reynolds, D. L. 1952. The difference in optics between volcanic and plutonic plagioclases, and its bearing on the granite problem. *Geol. Mag.*, *89*, 233–250.

Ribbe, P. H., ed. 1975. *Feldspar Mineralogy: Mineral. Soc. Amer. Short Course Notes*, Vol. 2. Blacksburg, Va.: Southern Printing Co.

Slemmons, D. B. 1962. Determination of volcanic and plutonic plagioclases using a three- or four-axis universal stage. *Geol. Soc. Am., Spec. Paper 69*.

Smith, J. R. 1958. The optical properties of heated plagioclases. *Amer. Mineral.*, *43*, 1179–1194.

Smith, J. V. 1958. Effect of temperature, structural state and composition of the albite, pericline and acline-A twins of plagioclase feldspars. *Amer. Minerals.*, *43*, 546–551.

Smith, J. V. 1974a. *Feldspar Minerals.* Vol. 1: *Crystal Structure and Physical Properties.* Heidelberg: Springer-Verlag.

Smith, J. V. 1974b. *Feldspar Minerals.* Vol. 2: *Chemical and Textural Properties.* Heidelberg: Springer-Verlag.

Tunell, G. 1952. The angle between the *a*-axis and the trace of the rhombic section on the (010) pinacoid in the plagioclases, *Amer. Jour. Sci., Bowen Vol.*, 547–551.

Tunell, G. 1953. Two definitions of positive and negative extinction angles in the plagioclase feldspars: one leading to consistency and clarity, the other to inconsistency and confusion. *Amer. Mineral.*, *38*, 404–411.

Turner, F. J. 1947. Determination of plagioclase with the four-axis universal stage, *Amer. Mineral.*, *32*, 389–410.

Van der Kaaden, G. 1951. Optical studies on natural plagioclase feldspars with high- and low-temperature optics. Unpublished doctoral dissertation Utrecht Ryksuniversitat.

BARIUM FELDSPARS

MONOCLINIC

The Celsian-Orthoclase Series

$(BaAl,KSi)AlSi_2O_8$

	Cn	Hy	Or
$\angle\beta$ =	115°12′	116°	116°

Celsian (Cn_{100})	**Hyalophane** (Cn_{30}–Cn_5)	**Orthoclase** (Cn_5–Cn_0)
$n_\alpha = 1.587–1.579$	$n_\alpha = 1.542–1.520$	$n_\alpha = 1.520–1.518$
$n_\beta = 1.593–1.583$	$n_\beta = 1.545–1.524$	$n_\beta = 1.524–1.522$
$n_\gamma = 1.600–1.588$	$n_\gamma = 1.547–1.526$	$n_\gamma = 1.525–1.522$
$n_\gamma - n_\alpha = 0.013–0.009$	$n_\gamma - n_\alpha = 0.010–0.005$	$n_\gamma - n_\alpha = 0.005$
$2V_z = 83°–90°(+)$	$2V_x = 48°–79°(-)$	$2V_x = 50°–35°(-)$
$b = Y$	$b = Z$	$b = Z$
$c \wedge X = +3°$ to $+5°$	$c \wedge Y = -25°$ to $-45°$	$c \wedge Y = -13°$ to $-21°$
$a \wedge Z = +28°$ to $+30°$	$a \wedge X = +1°$ to $-19°$	$a \wedge X = +14°$ to $+6°$
Dispersion imperceptible	$r > v$ perceptible	$r > v$ distinct
	Horizontal bisectrix	Horizontal bisectrix
Colorless	Colorless	Colorless

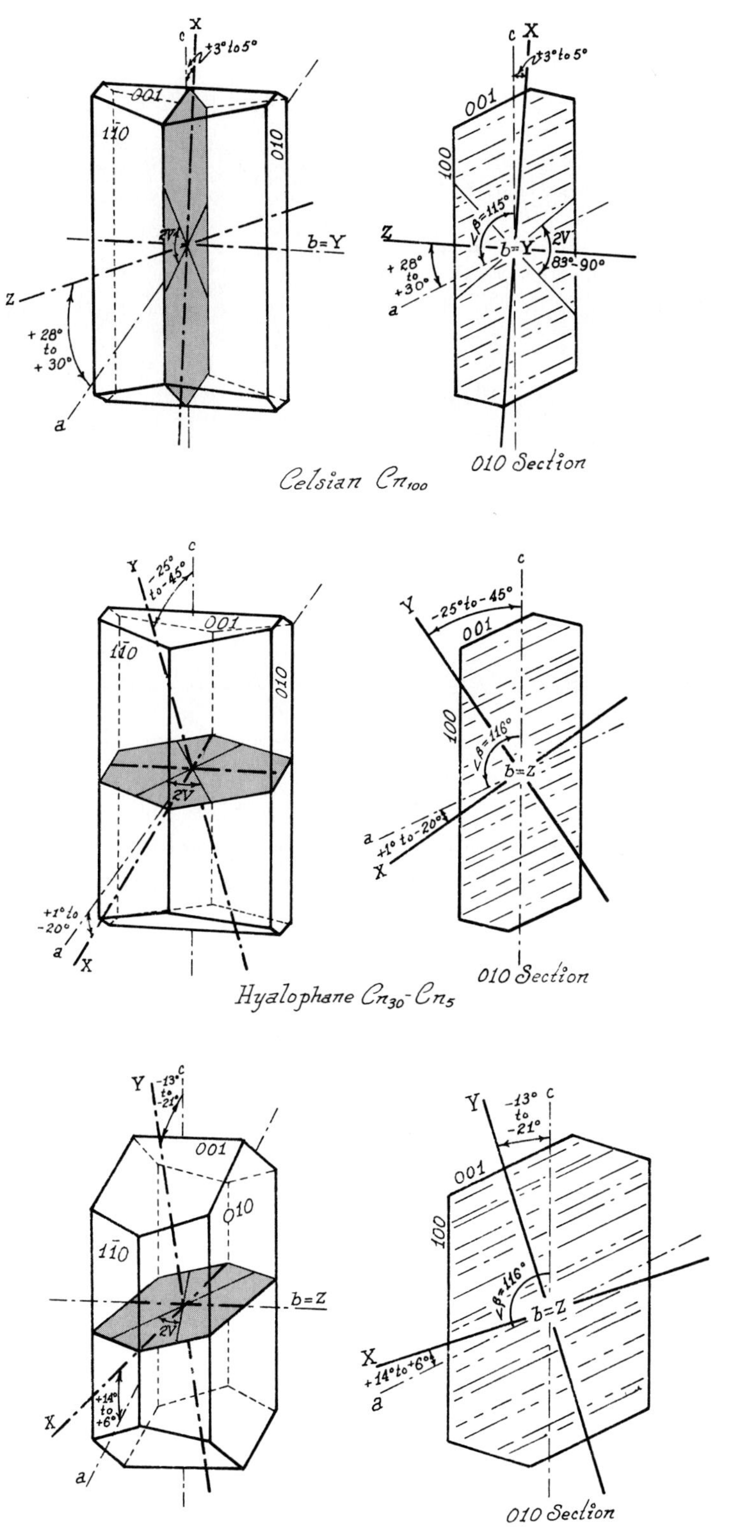

c X
+3° to 5°
001
1$\bar{1}$0
010
2V
b=Y
Z
+ 28° to + 30°
a
100
β=115°
b=Y
2V
83°-90°
010 Section
Celsian Cn_{100}
Y
-25° to -45°
c
001
1$\bar{1}$0
010
2V
+1° to -20°
a
X
100
β=116°
b=Z
010 Section
Hyalophane Cn_{30}-Cn_5
-13° to -21°
001
010
1$\bar{1}$0
b=Z
2V
+14° to +6°
β=116°
010 Section
Orthoclase Cn_5-Cn_0

COMPOSITION. Although a continuous series may exist between celsian ($BaAl_2Si_2O_8$) and orthoclase ($KAlSi_3O_8$), compositions in the range Cn_{100}–Cn_{40} are exceedingly rare, except for nearly pure celsian (Cn_{100}), and a break may well exist in the series at about Cn_{40}. As seen in Fig. 7-30, current data suggest a discontinuity in optical properties. Hyalophane is the only name applied to intermediate members of the series and is usually restricted to the range Cn_{30}–Cn_5. A celsian-anorthite series may exist, but optical data are too scarce to warrant consideration here.

PHYSICAL PROPERTIES. H = 6–6½. Sp. Gr. = 3.39–3.10 (celsian), 2.82–2.55 (hyalophane-orthoclase). Color, in hand sample, is white to yellow. Luster is vitreous.

COLOR AND PLEOCHROISM. Colorless as fragments or in section.

CLEAVAGE. Characteristic feldspar cleavages are {001} perfect and {010} distinct, intersecting at 90°; {110} cleavage is poor.

BIREFRINGENCE. Maximum birefringence for celsian is seen on {010} sections and is mild (0.009–0.013), yielding first-order gray, white, or yellow in thin section. Maximum birefringence for hyalophane is somewhat weaker (0.005–0.010), showing gray, white, or possibly pale yellow in sections near {001}.

TWINNING. Simple twinning by Carlsbad, Manebach, and Baveno laws. No repeated twinning in the barium feldspars.

INTERFERENCE FIGURE. Celsian yields a figure of broad, diffuse isogyres on a white field. $2V_z$ is large (83°–90°) and positive (Fig. 7-31), and dispersion is imperceptible. Cleavage fragments of celsian show flash figures on {010} and off-center optic axis figures on {001}.

Hyalophane is optically negative with a smaller $2V_x$(48°–79°) and its dispersion is perceptible $r > v$, with horizontal bisectrix dispersion. Cleavage fragments lying on {010} yield obtuse bisectrix figures, and those on {001} show essentially flash figures.

OPTICAL ORIENTATION. The optic plane of celsian is parallel to {010}, and $b = Y$. X is very near c (+3 to +5°), and $a \wedge Z = +28°$ to $+30°$ (Fig. 7-31). The acute bisectrix Z is almost normal to {100}. The maximum extinction angle for celsian seen on {010} sections, is about 29° to {001} cleavages, and, since the slow wave vibration (Z) lies nearest both cleavages, elongation is positive (length-slow). The optic plane of hyalophane is near {001} with the obtuse bisectrix $b = Z$, $a \wedge X = +1°$ to $-19°$, and $c \wedge Y = -25°$ to $-45°$.

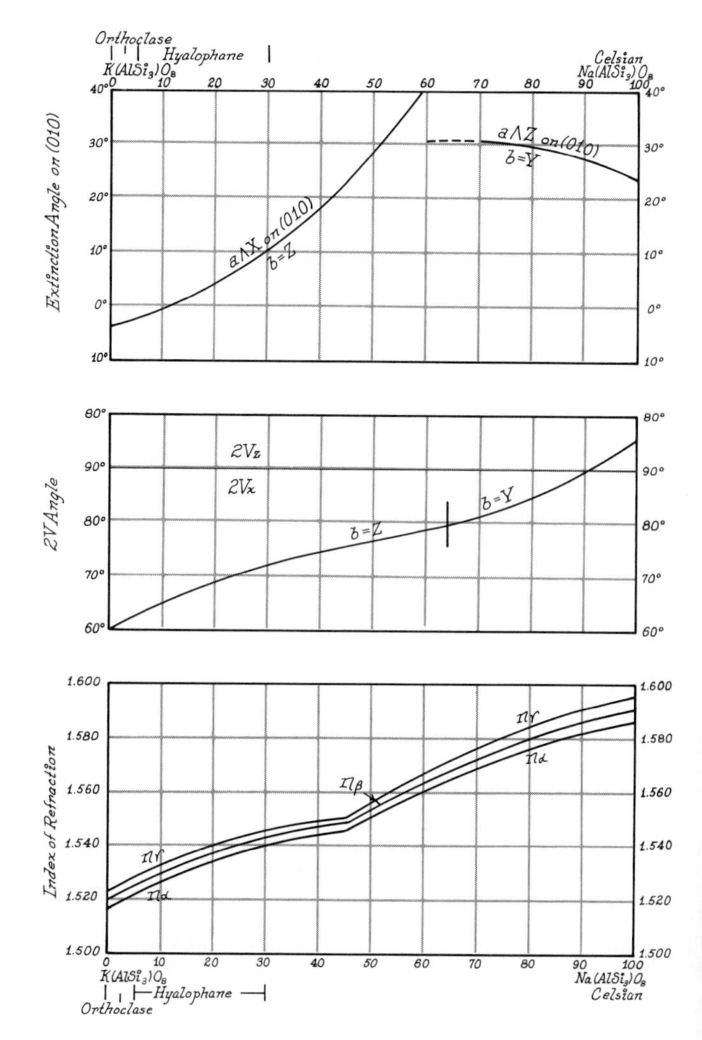

FIGURE 7-31. Variation of $a \wedge X$ or $a \wedge Z$, $2V$ angle, and indices of refraction in the orthoclase-celsian series. (After Smith, 1974)

The maximum extinction angle for hyalophane, seen on {010} sections, is 0° to 19° to the {001} cleavage. Since X is near a (X essentially lies in both cleavages), cleavage elongation is always negative (length-fast).

DISTINGUISHING FEATURES. Celsian is distinguished from alkali feldspars by its much higher refraction index and from plagioclase by the absence of multiple twinning and higher index of refraction. High barium content gives celsian maximum specific gravity for a feldspar.

Hyalophane may be distinguished from triclinic feldspars by its lack of multiple twinning, but it may prove optically indistinguishable from orthoclase or untwinned plagioclase. Ba-rich hyalophane shows a higher refractive index than orthoclase but grades chemically and optically into K-feldspar with decreasing Ba content.

OCCURRENCE. Barium-rich feldspars are very rare and seem restricted to rocks rich in manganese. Celsian and hyalophane occur most commonly in hydrothermal veins in association with manganese minerals (for example, rhodonite, rhodochrosite, spessartine and Mn-rich ferromagnesian minerals) and may occur with normal plagioclase. Celsian, hyalophane, and barium-rich plagioclase have been reported in granite gneisses and barium-rich sanidine in undersaturated volcanics.

REFERENCES: BARIUM FELDSPARS

Köhler, A. 1949. Recent results of investigations of the feldspars. *Jour. Geol.*, *57*, 592–599.

Kroll, H., and M. W. Phillips. 1976. Comparison of feldspar and paracelsian-type structures. *Z. Kristallogr.*, *143* (Fritz Laves Festband), 285–299.

Lin, H. C., and W. R. Foster. 1968. Studies in the system BaO-Al_2O_3-SiO_2. Part I. The polymorphism of celsian. *Amer. Mineral.*, *53*, 134–144.

Ribbe, P. H., ed. 1975. Feldspar Mineralogy, *Mineral. Soc. Amer. Short Course Notes*, Vol. 2. Blacksburg, Va.: Southern Printing Co.

Smith, J. V. 1974. *Feldspar Minerals*. Vol. 1: *Crystal Structure and Physical Properties* Heidelberg: Springer-Verlag.

Smith, J. V. 1974. *Feldspar Minerals*. Vol. 2: *Chemical and Textural Properties*. Heidelberg: Springer-Verlag.

Souček, J., and E. Jelínek. 1973. Celsian in the quartzite from Zlaté Hory (Selesia, Czechoslovokia). *Acta Univ. Carolinae, Geol., Rost Vol.*, 97–109.

The Feldspathoid Group

The feldspathoid group consists of tektosilicates that have feldsparlike chemical compositions but that are relatively poorer in silica than the feldspars. All minerals of this group occur as primary phases in alkaline igneous rocks (that is, igneous rocks with high alkali and low silica contents). Although volumetrically small, this group of rocks is petrologically important, and the feldspathoids are major and diagnostic constituents of them. Table 7-3 contains a summary of the optical and physical properties of the feldspathoids.

TABLE 7-3. Optical Properties of the Feldspathoid Minerals

n, n_ω, n_β	n_ε, n_α	n_γ	$n_\gamma - n_\alpha$	Mineral	System	$2V$	Optic Sign	Cleavage	Optical Orientation	Color in Sections	Physical Properties
1.479–1.493			0.000–0.001	Analcime $NaAlSi_2O_6 \cdot H_2O$	Isometric			{100} poor		Colorless	H = 5–5½. Sp. Gr. = 2.2–2.3. White, rarely pink, green
1.483–1.487			0.000–0.001	Sodalite $Na_8Al_6Si_6O_{24}Cl_2$	Isometric			{110} imperfect		Colorless, pale blue, pale pink	H = 5½–6. Sp. Gr. = 2.3. Blue, gray, salmon, green, yellow.
1.487			0.000	Hackmanite $Na_8Al_6Si_6O_{24}(Cl,S)_2$	Isometric			{110} poor		Colorless	H = 5. Sp. Gr. = 2.3. Red-violet, pink
1.485–1.495			0.000–0.001	Noselite $Na_8Al_6Si_6O_{24}SO_4$	Isometric			{110} poor		Colorless, pale gray, blue, brownish	H = 5½–6. Sp. Gr. = 2.3–2.4. Blue, gray, brown
1.496–1.505			0.000–0.001	Haüynite $(Na,Ca)_{8-4}Al_6Si_6O_{24}(SO_4)_{1-2}$	Isometric			{110} poor		Colorless, blue	H = 5½–6. Sp. Gr. = 2.4–2.5. Blue, gray, white, green
1.500–1.514			0.000–0.010	Lazurite $(Na,Ca)_8Al_6Si_6O_{24}(SO_4,S,Cl)_2$	Isometric			{110} imperfect		Blue	H = 5–5½. Sp. Gr. = 2.4. Blue
1.490–1.507	1.488–1.495		0.012–0.002	Vishnevite $(Na,Ca,K)_{6-7}(AlSiO_4)_6(SO_4,CO_3,Cl) \cdot 1\text{–}5\ H_2O$	Hexagonal		–	{10$\bar{1}$0} perfect {0001} poor	length-fast	Colorless	H = 5–6. Sp. Gr. = 2.4. White, pale blue, yellow
1.507–1.530	1.495–1.503		0.028–0.012	Cancrinite $(Ca,Na)_{7-8}(AlSiO_4)_6(CO_3,SO_4,Cl)_{1-2}$ $1\text{–}5H_2O$	Hexagonal		–	{10$\bar{1}$0} perfect {0001} poor	length-fast	Colorless	H = 5–6. Sp. Gr. = 2.5. Yellow, reddish, white, blue
1.508–1.511	1.509–1.511		0.001	Leucite $KAlSi_2O_6$	Pseudoisometric (Tetragonal)		–	{110} very poor		Colorless (cloudy)	H = 5½–6. Sp. Gr. = 2.5. Colorless, light gray, white
1.514	1.509	1.514	0.005	Carnegieite $Na_3KAl_4Si_4O_{16}$	Triclinic (?)	12°–15°	–	poor		Colorless	H = ? Sp. Gr. = 2.5. White
1.532–1.537	1.527–1.533		0.004–0.005	Kaliophilite $KAlSiO_4$	Hexagonal		–	{0001} poor, {10$\bar{1}$1} poor	length-fast	Colorless	H = 6. Sp. Gr. = 2.6. White
1.529–1.549	1.526–1.544		0.003–0.005	Nepheline $Na_3KAl_4Si_4O_{16}$	Hexagonal		–	{0001} poor {10$\bar{1}$0} poor	length-fast	Colorless	H = 5½–6. Sp. Gr. = 2.5–2.7. White, gray
1.538–1.543	1.532–1.537		0.005–0.007	Kalsilite $KAlSiO_4$	Hexagonal		–	{10$\bar{1}$0} poor, {0001} poor	length-fast	Colorless	H = 6. Sp. Gr. = 2.6. Colorless, white, gray
1.540–1.600	1.535–1.565		0.004–0.037	Scapolite $(Ca,Na)_4[(Al,Si)_3Al_3Si_6O_{24}](Cl,CO_3)$	Tetragonal		–	{100} imperfect, {110} imperfect	length-fast	Colorless	H = 5–6. Sp. Gr. = 2.5–2.8. Colorless, gray, yellow, green, violet

Nepheline $Na_3KAl_4Si_4O_{16}$ HEXAGONAL

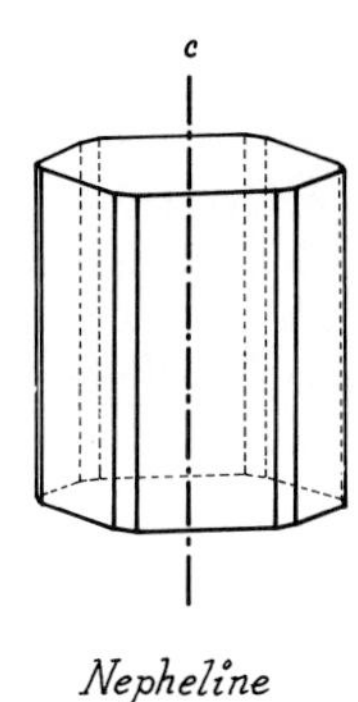

Nepheline

$n_\omega = 1.529–1.549$*
$n_\varepsilon = 1.526–1.544$
$n_\omega - n_\varepsilon = 0.003–0.005$*
Uniaxial negative
Colorless in standard section

COMPOSITION AND STRUCTURE. Nepheline is the most common mineral variety in a very complex group of polymorphs ranging from $NaAlSiO_4$ (Ne) to $KAlSiO_4$ (Ks). At high temperatures (>1150°C) solid solution is complete, but at low temperatures Na-rich phases (nepheline, Ne) and K-rich phases (*kalsilite*, Ks) are distinct. Kalsilite normally contains little Ne, but nepheline commonly ranges to 27 percent Ks, and it is quite possible that $Na_3KAl_4Si_4O_{16}$ is a unique intermediate compound, with immiscibility gaps between it and both Ne and Ks. The nepheline structure (hexagonal) is based on a tridymite-type framework in which one-half the Si^{4+} are replaced by Al^{3+} and an equal number of alkali ions appear in voids in the open structure to restore charge balance. One-fourth of the alkali sites are somewhat larger, favoring K^+ over Na^+ and leading to the ideal composition $Na_3KAl_4Si_4O_{16}$. Either site, however, may accommodate either alkali species.

Partial solid solution with both albite $NaAlSi_3O_8$(Ab) and anorthite $CaAl_2Si_2O_8$(An) is possible, to about Ab_{33} or An_{35}, and, although most natural nepheline is Ca-poor, solid solution with albite is usually evidenced by excess silicon.

Several polymorphs exist for both Na-rich and K-rich phases. Pure $NaAlSiO_4$(Ne) inverts to a high-temperature Ne at 900°C and to *carnegietie* (isometric) at 1254°C. Ideal nepheline ($Na_3KAl_4Si_4O_{16}$) changes directly to carnegieite, with increasing temperature, and carnegieite also exhibits two structural forms. Kalsilite (hexagonal), not really isostructural with nepheline, shows two high-temperature, orthorhombic polymorphs. *Trikalsilite* and *tetrakalsilite* (hexagonal) are additional K-rich polymorphs, naturally occurring, and roughly in the composition range Ks_{95}–Ks_{70}, and *kaliophilite* represents yet another rare, K-rich polymorph of questionable relationship.

PHYSICAL PROPERTIES. H = $5\frac{1}{2}$–6. Sp. Gr. = 2.56–2.67. Nepheline is commonly white to gray, in hand sample but impurities or alteration may yield darker colors. Nepheline has been called *eleolite* (Greek *elaion*, "oil") and commonly shows an oily or greasy luster.

COLOR. Nepheline is colorless to cloudy in section or as fragments. In volcanic rocks, nepheline crystals are commonly clear and glassy, while plutonic rocks normally yield nepheline clouded by inclusions or alteration.

FORM. Nepheline tends to form early in the crystallization and commonly appears as euhedral crystals showing hexagonal and stubby rectangular sections. In plutonic rocks, nepheline may be subhedral to anhedral, and in volcanic rocks it may appear as euhedral phenocrysts and as part of the fine granular groundmass.

CLEAVAGE. Poor basal {0001} and prismatic $\{10\bar{1}0\}$ cleavages are seldom evident in section, and irregular fractures are more characteristic.

BIREFRINGENCE. Maximum birefringence is very weak (0.003–0.005), increasing slightly with K content. Basal sections are dark, between crossed nicols, and prismatic sections show first-order gray to white.

* Refractive indices and birefringence increase slightly with K-content. Indices increase with anorthite (An) and decrease with increasing albite (Ab) solid solution.

TWINNING. Twinning is not common in nepheline.

ZONING. Composition zoning and the zonal arrangement of inclusions are commonly observed in crystals of nepheline. Normal zoning favors albite-rich (silica-rich) cores. Oscillatory zoning testifies to varying magma composition.

INTERGROWTHS. Inasmuch as the tolerance of nepheline for both kalsilite and albite decreases with lowering temperature, one may well expect to observe "microperthite" type lattice intergrowths involving nepheline and K-rich or Ab-rich phases. Graphic intergrowths with alkali feldspars are also known.

INTERFERENCE FIGURE. Basal sections and fragments on {0001} yield a very broad, diffuse cross on a gray to white field. The figure proves negative by the sensitive colors of the first-order plate and may show slight biaxial separation on stage rotation.

OPTICAL ORIENTATION. Longitudinal crystal sections are elongated on c and show parallel extinction and negative elongation (length-fast).

DISTINGUISHING FEATURES. Nepheline resembles alkali feldspars but is uniaxial negative, without twinning. Sodalite, haüynite, noselite, and analcime are isotropic, and leucite is nearly so, with multiple twinning. Melilite crystals are length-slow, with high refractive index.

ALTERATION. Nepheline may form in the early stages of crystallization and may alter in the early stages of alteration. It may be mildly turbid to completely replaced by a fine-grained, fibrous, or micaceous aggregate called *hydronepheline* or *gieseckite*, consisting largely of secondary micas, probably paragonite, and fibrous zeolites (probably natrolite). Nepheline also alters to other feldspathoids (such as cancrinite, sodalite, noselite, and haüynite) and Na-rich zeolites (such as analcime and natrolite).

OCCURRENCE. Nepheline appears only in quartz-free alkaline rocks in association with other alkali-rich or silica-poor minerals (such as alkali feldspars, other feldspathoids, Na-pyroxenes, and Na-amphiboles). It forms as a primary, magmatic mineral through the crystallization of Na-rich, Si-poor magmas and lavas; it may result from the assimilation of limestone by basic or intermediate magmas; or it may appear in rocks of Na-metasomatism (nephelinization). Nepheline may appear in a great number of odd rock types of widely varying composition, with a very wide range of unusual mineral associates.

In plutonic igneous rocks, nepheline occurs with albite and microcline in nepheline syenites and similar light-colored rocks and with labradorite and olivine in a few rare, dark-colored rocks. Na-rich, Fe-rich, or Si-poor minerals are common accessories (for example, Fe-rich biotite, Na-pyroxenes, Na-amphiboles, titanaugite, perovskite, melanite, corundum, and so on). Cancrinite, sodalite, and melilite are related feldspathoids, and exsolution intergrowths of nepheline with kalsilite or albite are common.

In volcanic rocks, nepheline ranges widely in composition. It appears as phenocrysts or groundmass granules with anorthoclase, sanidine, and high albite, especially in phonolites and related rocks. Leucite, sodalite, haüynite or noselite are commonly also present.

In hybrid rocks, nepheline is formed by the assimilation of carbonate sediments by basic magmas and is characteristically associated with calcite, Ca-rich pyroxenes, and amphiboles and Ca-rich accessories (such as perovskite, pyrochlore, and niocalite).

In metamorphic rocks, nepheline results from alkali metasomatism (nephelinization). It appears in gneisses and migmatites with albite, microcline, biotite, and other plutonic associates. Nepheline borders on embayed feldspar crystals, and graphic intergrowths suggest metasomatic development.

Kalsilite. Kalsilite is a rare mineral occurring in volcanic rocks very poor in Si and Na. It appears as groundmass grains or phenocrysts complexly zoned or intergrown with nepheline. Leucite, nepheline, olivine, and pyroxenes are common associates. Kalsilite may contain minor Na and may be unusually rich in ferric iron, which replaces Al. Several high-temperature polymorphs are known from rare lavas: $n_\omega = 1.538–1.543$, $n_\varepsilon = 1.532–1.537$, and $n_\omega - n_\varepsilon = 0.005$.

Kaliophilite. Kaliophilite is a very rare mineral similar in composition to kalsilite. It appears in ejecta from Mt. Vesuvius in association with pyroxene, melilite, calcite, leucite, haüynite and garnet. $n_\omega = 1.532$ and $n_\varepsilon = 1.527$.

Carnegieite. Carnegieite is not known in nature but is a synthetic polymorph of nepheline. It is structurally related to nepheline as cristobalite is related to tridymite, and, like cristobalite, it shows high-temperature (isometric) and low-temperature (pseudoisometric) forms. Like leucite, it is sensibly isotropic with multiple twinning. $n_\alpha = 1.509$, $n_\beta = 1.514$, $n_\gamma = 1.514$, and $2V_x = 12°–15°$.

REFERENCES: NEPHELINE

Bannister, F. A. 1931. A chemical, optical and X-ray study of nepheline and kaliophilite. *Mineral. Mag.*, *22*, 569–608.

Barth. T. F. W. 1963. The composition of nepheline. *Schweiz. Mineral. Petrog. Mitt.*, *43*, 153–164.

Barth, T. F. W., and E. Posnjak. 1932. Silicate structures of the cristobalite type. I. The crystal structure of carnegieite ($NaAlSiO_4$). *Z. Kristallogr. 81*, 135–141.

Buerger, M. J., G. E. Kline, and G. Donnay. 1954. Determination of the crystal structure of nepheline. *Amer. Mineral.*, *39*, 805–818.

Bowen, N. L. 1917. The sodium-potassium nephelites. *Amer. Jour. Sci.*, *43*, 115–132.

Claringbull, G. F., and F. A. Bannister. 1948. The crystal structure of kalsilite. *Acta Crystallogr.*, *1*, 42–43.

Dollase, W. A., and D. R. Peacor. 1971. Silicon-aluminum ordering in nephelines. *Contrib. Mineral. Petrol.*, *30*, 129–134.

Donnay, G., J. F. Schairer, and J. D. H. Donnay. 1959. Nepheline solid solution. *Mineral. Mag.*, *32*, 93–109.

Hamilton, D. L., and W. S. MacKenzie. 1960. Nepheline solid solution in the system $NaAlSiO_4$-$KAlSiO_4$-SiO_2. *Jour. Petrol.*, *1*, 56–72.

Sahama, Th. G. 1962. Order-disorder in natural nepheline solid solutions. *Jour. Petrol.*, *3*, 65–81.

Smith, J. V., and O. F. Tuttle. 1957. The nepheline-kalsilite system. I. X-ray data for the crystalline phases. *Amer. Jour. Sci.*, *225*, 282–305.

Tuttle, O. F., and J. V. Smith. 1958. The nepheline-kalsilite system. II. Phase relations. *Amer. Jour. Sci.*, *256*, 571–589.

Winchell, A. N. 1941. Nepheline. *Amer. Mineral.*, *26*, 536–540.

Leucite

$KAlSi_2O_6$

TETRAGONAL (PSEUDOISOMETRIC)

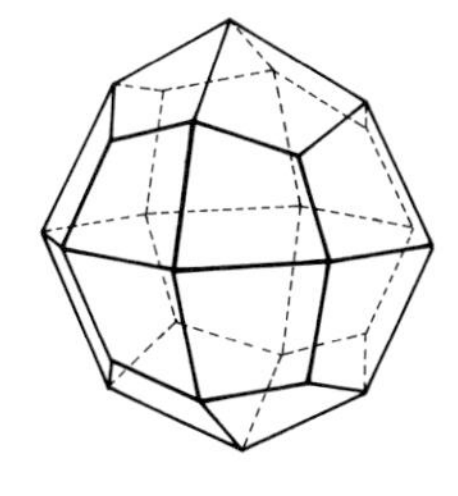

Leucite

$n_\omega = 1.508–1.511$
$n_\varepsilon = 1.509–1.511$
$n_\varepsilon - n_\omega = 0.001$
Colorless in standard section

COMPOSITION AND STRUCTURE. The composition of leucite approaches very nearly the ideal formula $KAlSi_2O_6$. Minor Na^+ may replace K^+, and Fe^{3+} may replace Al^{3+}. The structure of leucite is analogous to that of analcime and pollucite and is based on a cubic lattice containing rings of four and six tetrahedra oriented normal to the fourfold and threefold symmetry axis. Two distinctly different sites are available for large cations, and in leucite the larger sites contain K^+ ions and the smaller sites are vacant, yielding a very open, low-density structure. Above 605 ± 5°C leucite is isometric, but below that temperature the open structure collapses somewhat to tetragonal symmetry.

PHYSICAL PROPERTIES. H = $5\frac{1}{2}$–6. Sp. Gr. = 2.47–2.50. Color, in hand sample, is colorless to light gray, with vitreous luster.

COLOR AND PLEOCHROISM. Leucite is colorless as fragments or in section, often clouded by alteration or inclusions.

FORM. Large, euhedral, trapezohedron {112} phenocrysts in dark flow rocks represent the common habit of leucite. In section, trapezohedron crystals yield octahedral to roughly circular outline. Rarely, dodecahedron {110} or even cube {100} forms may appear on leucite crystals.

CLEAVAGE. Very poor dodecahedral {110} cleavage is seldom evident in section or on fragment surfaces.

BIREFRINGENCE. Maximum birefringence is very low (0.001), yielding no more than dark first-order gray, which may well go unnoticed, except for subtle polysynthetic twinning (Fig. 7-32).

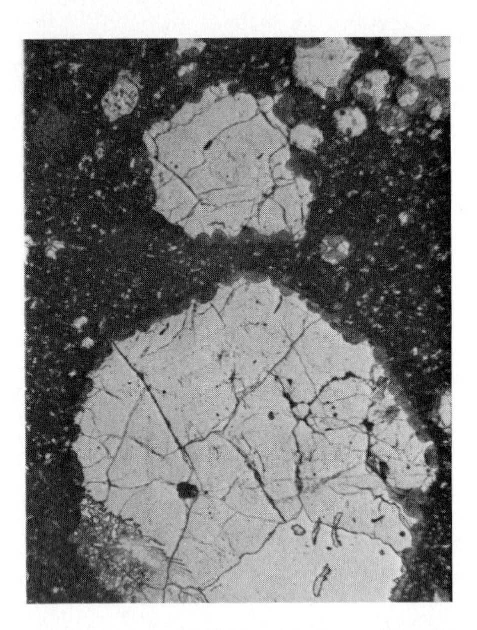

FIGURE 7-32. Leucite crystals appear only in K-rich lavas and tuffs. They are roughly euhedral (trapezohedral), with crude circular sections. Between crossed nicols, crystals show very weak birefringence (dark gray interference colors) and several sets of fine polysynthetic twins. Note also alteration advancing inward from crystal boundaries.

TWINNING. Several sets of diffuse, parallel twin lamellae are usually evident through subtle interference color differences in blacks and grays. Twin lamellae resemble those of microcline but tend to appear in three sets, intersecting at about 60° (Fig. 7-32).

ZONING. Composition zoning is not evident in leucite but inclusions of gas, liquid, glass, or numerous minerals (such as magnetite, pyroxenes, spinel, or feldspathoids) are commonly present in radial or concentric patterns. Leucite crystals may also be rimmed by fine alteration aggregates (pseudoleucite).

INTERFERENCE FIGURE. Leucite is sensibly isotropic, and dark, diffuse figures are further confused by twinning. Favorable sections without twinning may yield poor figures that appear optically negative and may show mild isogyre separation.

OPTICAL ORIENTATION. Leucite is sensibly isotropic without crystal elongation or cleavage. Extinction is often wavy.

DISTINGUISHING FEATURES. Very restricted occurrence, roughly circular phenocryst outline, very low birefringence, and polysynthetic twinning are usually sufficient to distinguish leucite. Microcline shows similar twinning but has higher birefringence and refractive index and is a mineral of plutonic enironments. Analcime may show similar outline and birefringence, but lacks twinning, and sodalite and haüynite are truly isotropic.

ALTERATION. The common alteration product of leucite is an aggregate or intergrowth of nepheline and K-feldspar called *pseudoleucite*. It may form through exsolution, Na-metasomatism, or reaction of leucite with Na-rich magma. Leucite may also alter to analcime or clays (sericite and kaolinite).

OCCURRENCE. Leucite is confined to volcanic or shallow dike rocks that are Si-poor, Na-poor, and K-rich. It is unknown in plutonic rocks and alters too quickly to appear in sediments or ancient flow rocks. Leucite occurrences are sufficiently rare to make practical the actual listing of specific localities. The Italian Peninsula (Mt. Vesuvius and so on) contains most leucite localities but leucite also appears in the Leucite Hills of Wyoming, in central Montana, and in several areas of Africa.

Leucite volcanics or tuffs may be almost pure leucite

(italite) or may contain leucite in association with a wide range of minerals (such as sanidine, olivine, phlogopite, diopside, nepheline, melilite, and other feldspathoids). Leucite appears in leucite basalts, leucite tephrites, leucite basanites, leucitites, and so on.

Euhedral aggregates of pseudoleucite show much wider occurrence and are even reported in plutonic rocks.

REFERENCES: LEUCITE

Barrer, R. M., and L. Hinds. 1953. Ion-exchange in crystals of analcite and leucite. *Jour. Chem. Soc.*, *1953*, 1879–1888.

Barrer, R. M., L. Hinds, and E. A. White 1953. The hydrothermal chemistry of silicates. III. Reactions of analcite and leucite. *Jour. Chem. Soc.*, *1953*, 1466.

Bowen, N. L., and R. B. Essestad. 1937. Leucite and pseudoleucite. *Amer. Mineral.*, *22*, 409–415.

Korekawa, M. 1969. Twinning of leucite. *Z. Kristallogr.*, *129*, 343–350.

Mazzi, F., E. Galli, and G. Gottardi. 1976. The crystal structure of tetragonal leucite. *Amer. Mineral.*, *61*, 108–115.

Peacor, D. R. 1968. High-temperature single-crystal diffractometer study of leucite, $(K,Na)AlSi_2O_6$. *Z. Kristallogr.*, *127*, 213–224.

Taylor, W. H. 1930. The structure of analcite ($NaAlSi_2O_6 \cdot H_2O$). *Z. Kristallogr.*, *74*, 1–19.

THE SODALITE GROUP

Sodalite	$Na_8Al_6Si_6O_{24}Cl_2$	ISOMETRIC
Noselite (Nosean)	$Na_8Al_6Si_6O_{24}SO_4$	ISOMETRIC
Haüynite (Haüyne)	$(Na,Ca)_{8-4}Al_6Si_6O_{24}(SO_4)_{1-2}$	ISOMETRIC

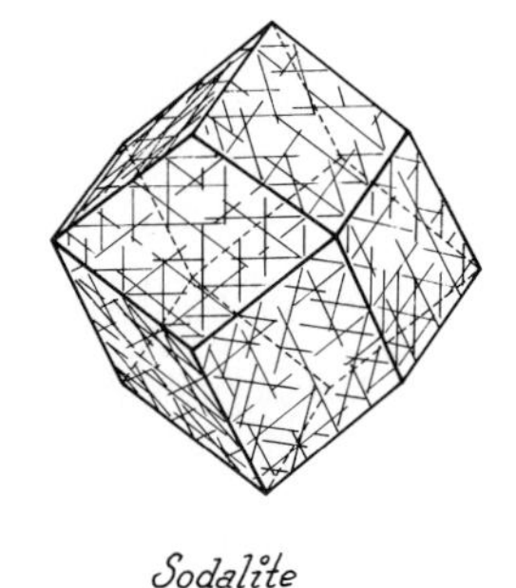

Sodalite

$n = 1.483–1.487$ (sodalite)
$n = 1.485–1.495$ (noselite)
$n = 1.496–1.505$ (haüynite)*
Colorless to blue in standard section

COMPOSITION AND STRUCTURE. The sodalite group feldspathoids are based on the cubic ultramarine structure, which is a tektosilicate framework. One-half the Si^{4+} are replaced by Al^{3+} requiring additional cations for charge balance. The open structure affords an equivalent of two very large cavities per unit cell, which accommodate large anions (Cl^-, S^{2-}, or $SO_4{}^{2-}$) surrounded by Na^+ or Ca^{2+} in tetrahedral coordination. The large anions and cations distinguish minerals of the sodalite group:

Sodalite	$Na_8Al_6Si_6O_{24}Cl_2$
Hackmanite	$Na_8Al_6Si_6O_{24}(Cl,S)_2$
Noselite	$Na_8Al_6Si_6O_{24}(SO_4)$
Haüynite	$(Na,Ca)_{8-4}Al_6Si_6O_{24}(SO_4)_{1-2}$
Lazurite	$(Na,Ca)_8Al_6Si_6O_{24}(SO_4,S,Cl)_2$

The composition of sodalite closely approaches the ideal value, with minor substitution of Ca^{2+} or K^+ for Na^+, Fe^{3+} for Al^{3+} and S^{2-} (hackmanite) or I^- for Cl^-. Replacement of Na^+ by Ca^{2+} allows bivalent anions ($SO_4{}^{2-}$ for Cl^-), and other members of the group result.

* Refractive index increases with Ca and SO_4.

An entire family of synthetic ultramarines are produced, with a wide variety of large anions (for example, Br^-, I^-, CO_3^{2-}, and WO_4^{2-}).

PHYSICAL PROPERTIES. H = $5\frac{1}{2}$–6. Sp. Gr. = 2.27–2.40 (increasing with Ca and SO_4). The characteristic color of the ultramarines is ultramarine blue; however, natural sodalite, noselite, and haüynite are often gray, green, pink, or yellow in hand sample.

COLOR AND PLEOCHROISM. Sodalite and noselite are colorless to pale gray-blue in section, while haüynite may be distinctly blue, without pleochroism. Color intensity appears related to sulfur content and may show irregular distribution.

FORM. Minerals of the sodalite group commonly form as euhedral phenocrysts of dodecahedron habit yielding hexagonal sections, often showing resorption. Sodalite often appears as anhedral masses or granular aggregates.

CLEAVAGE. Poor {110} cleavages are seldom obvious in thin section.

BIREFRINGENCE. The sodalite minerals are isotropic, but a minor structure collapse about anion groups, too small for the available cavity, may yield slight birefringence and dark gray interference colors, particularly near inclusions.

TWINNING. Twinning on {111} is possible but seldom evident in fragments or section.

ZONING. Tiny inclusions of gas or iron oxides are often distributed in concentric zones, and color intensity, related to inclusions, may be likewise zoned. Sharp compositional zoning is reported noselite and haüynite.

DISTINGUISHING CHARACTERISTICS. Sodalite minerals are isotropic, with moderate negative relief and possible blue color. They do not show the polysynthetic twinning of leucite or the good cleavage of fluorite. Analcime is difficult to distinguish from the sodalite group, and members of the group are distinguished with difficulty. Sodalite is commonly colorless without inclusions, noselite commonly shows zoned inclusions, and haüynite. is commonly blue. Analcime forms in many rock types, but in alkaline volcanics it is commonly confined to the groundmass. Only sodalite appears in limestone contact zones, and sodalite commonly shows red-orange fluorescence in ultraviolet light.

ALTERATION. Minerals of the sodalite group normally alter to fibrous zeolites (natrolite or thomsonite); to clays (kaolinite, diaspore, sericite, gibbsite); or to cancrinite. Calcite and limonite may accompany the alteration. Sodalite itself may be an alteration product of nepheline.

OCCURRENCE. The sodalite minerals are characteristic of alkaline, undersaturated igneous rocks. Noselite and haüynite are essentially confined to volcanic varieties, and sodalite appears in both intrusive and extrusive varieties.

Sodalite is most common in nepheline syenites and related rock types in association with nepheline, cancrinite, and alkali feldspars (albite, orthoclase). It appears in nepheline pegmatites, nepheline dike rocks and is common in alkaline volcanics (for example, phonolites and trachytes) with nepheline and sanidine. A unique occurrence of sodalite is in metasomatic zones in carbonate sediments with calcite and albite.

Noselite appears only in phonolites and similar flow rocks and in volcanic ejecta of similar composition in association with nepheline, leucite, and sanidine.

Haüynite is most characteristic of phonolites, haüynite, basalts, and similar volcanics and pyroclastics. It appears with nepheline, leucite, melilite, melanite, pyroxene, and sanidine. Haüynite may be found in a few rare dike rocks (such as okaite).

Lazurite is formed in metamorphosed limestone with forsterite and phlogopite. It commonly contains megascopic inclusions of calcite, pyrite, and so on, forming a brilliant blue ornamental stone, lapis lazuli.

REFERENCES: THE SODALITE GROUP

Barrer, R. M., and J. D. Falconer. 1956. Ion exchange in feldspathoids as a solid state reaction. *Proc. Roy. Soc., A, 236*, 227–249.

Barth, T. F. W. 1932a. The chemical composition of noselite and haüynite. *Amer. Mineral., 17*, 466–471.

Barth. T. F. W. 1932b. The structures of the minerals of the sodalite families. *Z. Kristallogr., 83*, 405–414.

Loehn, J., and H. Schulz. 1967. Structure refinement of sodalite, $Na_8Si_6Al_6O_{24}Cl_2$. *Acta Crystallogr., 23*, 434–436.

Loehn, J., and H. Schulz. 1968. Structural refinement of disordered haüyne, $(Na_5KCa_2)Al_6Si_6O_{24}(SO_4)_{1.5}$. *Neues Jahrb. Mineral., Abh., 109*, 201–210.

Prener, J. S., and R. Ward. 1950. The preparation of ultramarines. *Jour. Amer. Chem. Soc., 72*, 2780–2781.

Schulz, H., and H. Saalfeld. 1965. The crystal structure of nosean $Na_8(SO_4)(Si_6Al_6O_{24})$. *Mineral. Petrogr. Mitt., 10*, 225–232.

Taylor, D. 1967. The sodalite group of minerals. *Contrib. Mineral. Petrol., 16*, 172–188.

Cancrinite $(Ca,Na)_{7-8}(AlSiO_4)_6(CO_3,SO_4,Cl)_{1-2}\cdot 1\text{–}5H_2O$ HEXAGONAL

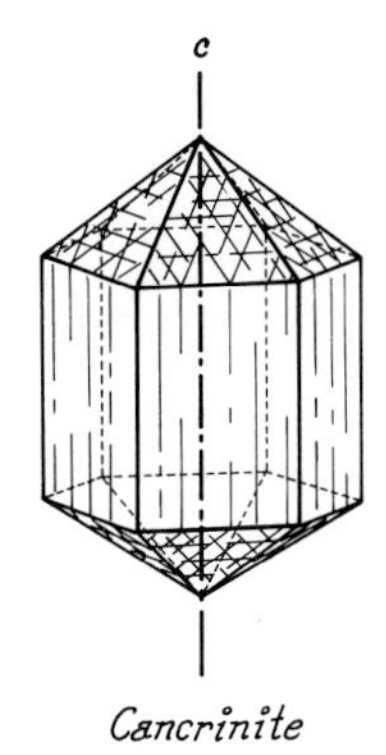

$n_\omega = 1.530\text{–}1.490$*
$n_\varepsilon = 1.503\text{–}1.488$
$n_\omega - n_\varepsilon = 0.028\text{–}0.002$*
Uniaxial negative
Colorless in standard section

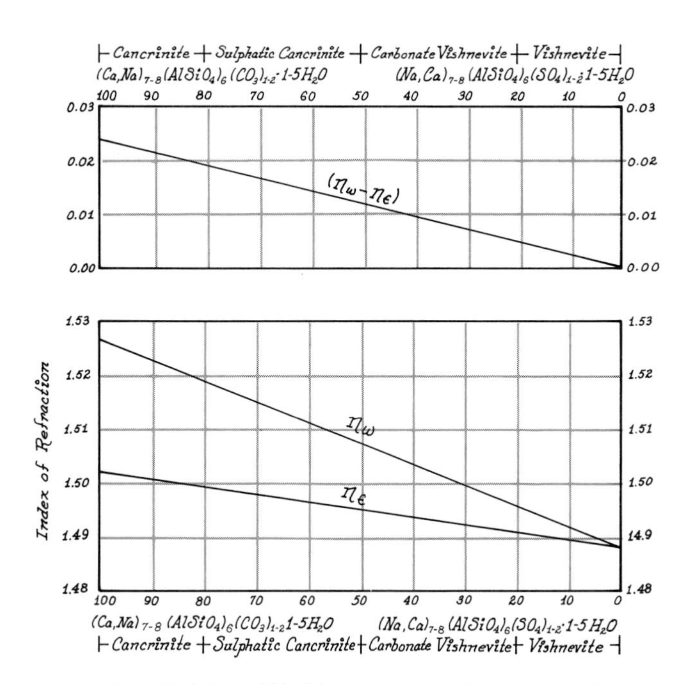

FIGURE 7-33. Variation of birefringence ($n_\omega - n_\varepsilon$) and indices of refraction in the cancrinite-vishnevite series. (After Deer, Howie, and Zussman, 1962)

COMPOSITION AND STRUCTURE. Cancrinite is an open framework (tektosilicate) structure with tetrahedra linked in complex rings. Large cations (Na^+,Ca^{2+},K^+) and complex anions (CO_3^{2-},SO_4^{2-},Cl^-) occupy large structural openings in channels parallel to *c*, and water molecules may fill structural tunnels, as in the zeolites. Cancrinite is commonly rich in Ca and CO_3; varieties rich in Na and SO_4 are called *vishnevite*; and a K-Cl variety is called *microsommite*. Water content may vary widely without affecting optical properties.

PHYSICAL PROPERTIES. H = 5–6. Sp. Gr. = 2.5–2.3 (higher values for Ca-CO_3 varieties). In hand sample, cancrinite is commonly yellow but may be white, gray, pale blue, green, or reddish.

COLOR. Cancrinite is colorless in thin section or as fragments.

FORM. Cancrinite tends to crystallize late in the crystallization sequence or to form by alteration of nepheline and is consequently subhedral to anhedral.

CLEAVAGE. Perfect prismatic cleavages $\{10\bar{1}0\}$ are evident in section and control fragment orientation.

BIREFRINGENCE. Maximum birefringence is seen in prismatic sections and fragments lying on prismatic cleavages. It ranges from moderate (0.028) to very weak (0.002). Vishnevite (Na-SO_4-rich) varieties show weak birefringence, which increases with Ca-CO_3 substitution (Fig. 7-33). Upper first-order colors are common in standard section.

TWINNING AND ZONING. Neither twinning nor zoning are characteristic of cancrinite, although repeated twinning is reported, and composition zoning seems probable.

---

* Refractive indices and birefringence decrease as Na-SO_4 replaces Ca-CO_3 (see Fig. 7-33).

INTERFERENCE FIGURE. Basal sections yield a uniaxial negative figure with few isochromatic rings. Isogyres sometimes separate with stage rotation, and microsommite compositions are uniaxial positive. As Na-SO_4 replaces Ca-CO_3, n_ω approaches n_ε. K-Cl substitution brings birefringence ($n_\omega - n_\varepsilon$) to zero, and microsommite is optically positive. Cleavage fragments on $\{10\bar{1}0\}$ yield only flash figures.

OPTICAL ORIENTATION. Cleavage produces elongation on *c*, with parallel extinction and negative elongation (length-fast). Microsommite is length-slow.

DISTINGUISHING FEATURES. Most cancrinite shows higher birefringence than nepheline and orthoclase; negative relief distinguishes it from basal mica sections. Scapolite shows higher refractive indices and imperfect cleavages. Cancrinite effervesces with hydrochloric acid.

ALTERATION. Cancrinite is itself a possible alteration product of nepheline. Natrolite is reported to be a rare alteration product of cancrinite.

OCCURRENCE. Cancrinite is characteristic of nepheline syenites, ijolites, and similar alkaline plutonic rocks. It may form as a primary igneous mineral near the end of the crystallization sequence or from the interaction of nepheline and calcite. Cancrinite is most commonly associated with nepheline and calcite and may form reaction zones between the two. Other common associates are sodalite, alkali feldspars, melanite, wollastonite, corundum, alkali pyroxenes, and amphiboles.

REFERENCES: CANCRINITE

Brown, W. L., and F. Cesbron. 1973. Cancrinite superstructure. *C. R. Acad. Sci. Paris*, Series D, *276*, 1–4.

Deer, W. A., R. A. Howie, and J. Zussman. 1962. *Rock-Forming Minerals. Vol. 4: Frame Work Silicates*. New York: Wiley.

Jarchow, O. 1966. Atom arrangement and structure refinement of cancrinite. *Z. Kristallogr.*, *122*, 407–422.

Novozhilov, A. I., M. I. Samoilovich, E. K. Mikul'skaya, and L. I. Parushnikova. 1966. Nature of the blue coloration of cancrinite crystals. *Zap. Vses. Mineral. Obshchest.*, *95*, 736–738.

Phoenix, R., and E. W. Nuffield. 1949. Cancrinite from Blue Mountain, Ontario. *Amer. Mineral.*, *34*, 452–455.

Rauff, H. 1878. Ueber die chemische Zusammensetzung des Nephelius, Cancrinits und Microsommits. *Z. Kristallogr.*, *2*, 445.

The Zeolite Group

The zeolites are a complex tektosilicate mineral group of 30 to 40 mineral varieties; that is, they possess a framework silicate structure in which each oxygen ion is shared by two Si^{4+}. Some number of Si^{4+}, depending on the zeolite variety, is replaced by Al^{3+}, and the charge deficiency is balanced by the introduction of Na^+, K^+, or Ca^{2+}. The structures are very open, as indicated by low densities, yielding large cavities that accommodate large cations and open channels or tunnels that commonly contain neutral water molecules. The general formula is

$$(Na^+,K^+,\tfrac{1}{2}Ca^{2+})_{XZ}[(Si_YAl_X)O_{2(X+Y)}]_Z \cdot nH_2O$$

Gentle heating drives out the water molecules, leaving open structural tunnels that will absorb other small molecules (such as NH_4, I_2, NO_2, and H_2S) or act as molecular sieves. The hydration of a natural zeolite is, therefore, quite variable, ranging from a maximum, limited by tunnel volume, to nothing, represented by dehydrated forms.

The large cations Na^+, K^+, and Ca^{2+} occupy large cavities and are easily replaced by one another or by other cations of comparable size and charge (such as Sr^{2+}, Ba^{2+}, Li^+, Rb^+, Cs^+, Ag^+, Cu^+, and Pb^{2+}). Because all available cavities are not filled, two Na^+ or K^+ may replace one Ca^{2+} ($2Na^+ \rightleftharpoons Ca^{2+}$) without altering the Al:Si ratio.

Two common, independent replacements form varieties of the zeolite group: $Ca^{2+} \rightleftharpoons 2Na^+$ or $2K^+$, as described earlier, and the coupled replacement $Na^+Si^{4+} \rightleftharpoons Ca^{2+}Al^{3+}$. Some of the more common varieties are

Analcime	$Na_2(Al_2Si_4)O_{12} \cdot 2H_2O$*
Natrolite	$Na_2(Al_2Si_3)O_{10} \cdot 2H_2O$
Laumontite	$Ca(Al_2Si_4)O_{12} \cdot 4H_2O$
Chabazite	$Ca(Al_2Si_4)O_{12} \cdot 6H_2O$
Scolecite	$Ca(Al_2Si_3)O_{10} \cdot 3H_2O$
Gismondite	$Ca(Al_2Si_2)O_8 \cdot 4H_2O$
Heulandite	$(Ca,Na_2)(Al_2Si_7)O_{18} \cdot 6H_2O$
Gmelinite	$(Na_2,Ca)(Al_2Si_4)O_{12} \cdot 6H_2O$
Mesolite	$Na_2Ca_2[(Al_2Si_3)O_{10}]_3 \cdot 8H_2O$
Thomsonite	$NaCa_2[(Al,Si)_5O_{10}]_2 \cdot 6H_2O$
Gonnardite	$Na_2Ca[(Al,Si)_5O_{10}]_2 \cdot 6H_2O$
Mordenite	$(Na_2,K_2,Ca)(Al_2Si_{10})O_{24} \cdot 7H_2O$
Stilbite	$(Ca,Na_2,K_2)(Al_2Si_7)O_{18} \cdot 7H_2O$
Phillipsite	$(Ca,K_2,Na_2)_6[(Al_3Si_5)O_{16}]_2 \cdot 12H_2O$
Ashcraftine	$KNaCa(Al_4Si_5)O_{18} \cdot 8H_2O$
Harmotome	$Ba(Al_2Si_6)O_{16} \cdot 6H_2O$
Edingtonite	$Ba(Al_2Si_3)O_{10} \cdot 4H_2O$

---

* Analcime is chemically a zeolite, but it is the only zeolite to crystallize as a primary mineral in igneous rocks, and its paragenesis approximates that of the feldspathoids (see p. 363).

Zeolites exhibit low refractive indices (lower than balsam), low birefringence, and low density.

They are characteristically found in vesicles, amygdales, and fissures in mafic volcanic rocks where formed by late hydrothermal processes. Zeolites may appear in many hydrothermal deposits; as authigenic, interstitial materials in detrital sediments; or in very low-grade metamorphic environments (zeolite facies). Less hydrated varieties (for example, analcime, natrolite, laumontite, and scolecite) suggest higher temperatures of formation. Silica-rich environments yield silica-rich zeolite varieties, and mordenite, heulandite, stilbite, and so on may appear with quartz or chalcedony. Silica-poor rocks favor the association of silica-poor zeolites (such as gismondite, scolecite, natrolite, phillipsite, and ashcroftine) with nepheline and other feldspathoids. Zeolites are relatively stable under surface conditions, although some varieties alter to beidellite or other clay minerals.

Three structural patterns, related to the linking of silicate tetrahedra, logically divide the zeolites as fibrous zeolites (for example, natrolite, scolecite, thomsonite, and gonnardite); layered, or foliated, zeolites (such as heulandite and stilbite); and equidimensional, or blocky, zeolites (such as analcime, chabazite, gmelinite, phillipsite, and harmotome).

Tetrahedra of the fibrous group are arranged in chains parallel to the *c* crystallographic direction. Few tetrahedral corners link adjacent chains causing cleavages parallel to *c* ({100}, {010}, {110}, and so on) and fibrous habit. Large cations and water molecules lie in cavities and channels between chains. Chains have fourfold symmetry on *c* to yield tetragonal or pseudotetragonal symmetry.

Tetrahedra of the layered zeolites share most oxygens within sheets parallel to {010}. Sheets are held together by relatively few shared oxygens, and prominent {010} cleavage results. Large cations and water molecules are located in cavities and channels between sheets and within the sheets in tetrahedron rings.

The blocky zeolites represent a framework of silicate tetrahedra with similar bond densities in all directions, yielding isometric, or at least blocky, habit. The tetrahedra are linked as rings and are arranged so that they surround open tunnels and cavities for large cations and water molecules.

Table 7-4 summarizes the optical and physical properties of the zeolites.

TABLE 7-4. Optical Properties of the Zeolites

n	Mineral	System	Cleavage	Color in Sections	Physical Properties
1.48	Faujasite $(Na_2,Ca)_{1.75}(Al_{3.5}Si_{8.5})\cdot O_{24}\cdot 16H_2O$	Isometric Octahedral	{111} distinct	Colorless	H = 5. Sp. Gr. ≃ 1.9. White
1.479–1.493	Analcme $Na_2(Al_2Si_4)O_{12}\cdot 2H_2O$	Isometric trapezohedral	{100} poor (cubic)	Colorless	H = 5–5½. Sp. Gr. = 2.2–2.3. White, rarely pink, green

n_ω	n_ε	$(n_\omega - n_\varepsilon)$ or $(n_\varepsilon - n_\omega)$	Mineral	System	Optic Sign	Cleavage	Color in Sections	Physical Properties
1.468–1.472	1.473–1.476	0.003	Erionite $(Na_2,K_2,Ca,Mg)_{4.5}(Al_9Si_{27})O_{72}\cdot 27H_2O$	Hexagonal fibrous (wooly)	+	—	Colorless	H = ? Sp. Gr. = 2.0 White
1.472–1.494	1.470–1.485	0.002–0.010	Chabazite $Ca(Al_2Si_4)O_{12}\cdot 6H_2O$	Trigonal rhombohedral	−	{10$\bar{1}$1} distinct	Colorless	H = 4½. Sp. Gr. = 2.1. White, salmon pink
1.476–1.494	1.474–1.480	0.002–0.015	Gmelinite $(Na_2,Ca)(Al_2Si_4)O_{12}\cdot 6H_2O$	Trigonal pyramidal	−	{10$\bar{1}$0} distinct {0001} parting	Colorless	H = 4½. Sp. Gr. = 2.0–2.2. Colorless, white, yellowish
1.496–1.505	1.491–1.500	0.005	Levynite $Ca(Al_2Si_4)O_{12}\cdot 6H_2O$	Trigonal tabular {0001}	−	—	Colorless	H = 4½. Sp. Gr. ≃ 2.1. White, yellowish, pink
1.536	1.545	0.009	Ashcroftine $KNaCa(Al_4Si_5)O_{18}\cdot 8H_2O$	Tetragonal prismatic	+	{100} perfect {001} distinct	Colorless	H = ? Sp. Gr. ≃ 2.6. White, pink

(continued)

TABLE 7-4. Optical Properties of the Zeolites *(continued)*

n_β	n_α	n_γ	$n_\gamma - n_\alpha$	Mineral	System	$2V$	Optic Sign	Cleavage	Optical Orientation	Color in Sections	Physical Properties
1.475–1.485	1.472–1.483	1.477–1.487	0.002–0.005	Mordenite $(Na_2,K_2,Ca)(Al_2Si_{10})O_{24} \cdot 7H_2O$	Orthorhombic acicular on c	76°–90° 90°–76°	− +	{010} perfect {100} perfect	$a = Y$ $b = Z$ $c = X$	Colorless	H = 3–4. Sp. Gr. ≃ 2.2. White, yellowish, pinkish
1.476–1.486	1.473–1.483	1.485–1.496	0.012–0.013	Natrolite $Na_2(Al_2Si_3)O_{10} \cdot 2H_2O$	Orthorhombic acicular on c	58°–64° $v > r$ weak	+	{110} perfect {010} parting	$a = X$ $b = Y$ $c = Z$	Colorless	H = 5. Sp. Gr. = 2.2–2.6. White, gray, yellow, pink
1.478	1.479	1.482	0.004	Ferrierite $(Na,K)_4Mg_2(Al_6Si_{30})O_{72}(OH)_2 \cdot 18H_2O$	Orthorhombic bladed on {100}	50°	+	{100} perfect	$a = X$ $b = Y$ $c = Z$	Colorless	H = 3–3½. Sp. Gr. ≃ 2.1. White
1.484–1.509	1.483–1.504	1.486–1.514	0.003–0.010	Phillipsite $(Ca,K_2,Na_2)_6[(Al_3Si_5)O_{16}]_2 12H_2O$	Monoclinic $\angle\beta = 125°40'$ blocky	60°–80° $v > r$ weak	+	{010} distinct {100} distinct	$b = X$ $c \wedge Z = +10°$ to $+30°$ $a \wedge Y = +46°$ to $+65°$	Colorless	H = 4–4½. Sp. Gr. = 2.2. White
1.487–1.505	1.487–1.505	1.488–1.512	0.001–0.007	Heulandite $(Ca,Na_2)(Al_2Si_7)O_{18}$ $6H_2O$	Monoclinic $\angle\beta = 91°26'$ tabular on {010}	30° $r > v$ moderate crossed bxa	+	{010} perfect	$b = Z$ $c \wedge Y = +8°$ to $+32°$ $a \wedge X = +9°$ to $+33°$	Colorless	H = 3½–4. Sp. Gr. = 2.2. White, yellow, brown, pink
1.491–1.507	1.482–1.500	1.493–1.513	0.006–0.013	Stilbite $(Ca,Na_2,K_2)(Al_2Si_7)O_{18} \cdot 7H_2O$	Monoclinic $\angle\beta = 129°10'$ platy on {010}	30° to 49° $v > r$ weak	−	{010} perfect {100} poor	$b = Y$ $c \wedge X = 0°$ to $-7°$ $a \wedge Z = +39°$ to $+32°$	Colorless	H = 3½–4. Sp. Gr. = 2.1–2.2. White, yellowish, brownish
1.496	1.491	1.499	0.008	Dachiardite $(Ca,K_2,Na_2)_3(Al_4Si_{18})O_{45}$ $14H_2O$	Monoclinic $\angle\beta = 108°$ platy	65°–73°	+	{100} perfect {001} perfect	$b = X$ $c \wedge Z \simeq -38°$ $a \wedge Y \simeq -20°$	Colorless	H = 4–4½. Sp. Gr. ≃ 2.2. White
1.498–1.507	1.497–1.506	1.499–1.508	0.002	Gonnardite $Na_2Ca[(Al,Si)_5O_{10}]_2 \cdot 6H_2O$	Orthorhombic acicular on c	50°	−	Good parallel to c	$c = Z$	Colorless	H = 4½–5. Sp. Gr. ≃ 2.3. White
1.497–1.515	1.485–1.505	1.497–1.519	0.010–0.014	Epistilbite $Ca(Al_2Si_6)O_{16} \cdot 5H_2O$	Monoclinic $\angle\beta = 124°20'$ platy on {010}	44° $v > r$	−	{010} distinct	$b = Y$ $c \wedge Z \simeq +10°$ $a \wedge X \simeq -24°$		$H = 4$ Sp. Gr. ≃ 2.2. White, yellowish, pink

1.504–1.508	1.504–1.508	1.505–1.509	0.001	Mesolite $Na_2Ca_2[(Al_2Si_3)O_{10}]_3 \cdot 8H_2O$	Monoclinic $\angle\beta \simeq 90°$ acicular on *b*	80° $v>r$ strong inclined bxa	+	{101} perfect, $\{10\bar{1}\}$ perfect	$b=Y$ $c \wedge X=-8°$ $a \wedge Z \simeq -8°$	Colorless	H = 5. Sp. Gr. = 2.3. White, yellowish
1.505–1.509	1.503–1.508	1.508–1.514	0.005–0.008	Harmotome $Ba(Al_2Si_6)O_{16} \cdot 6H_2O$	Monoclinic $\angle\beta=125°$ twin crystals	80° crossed bxa	+	{010} distinct {001} poor	$b=Z$ $c \wedge Y=+28°$ to $+32°$ $a \wedge X=+63°$ to $+67°$	Colorless	H = $4\frac{1}{2}$. Sp. Gr. = 2.4–2.5. White, yellowish, reddish
1.512	1.510	1.523	0.011	Brewsterite $Na(Sr,Ba,Ca)_5 \cdot (Al_{11}Si_{29})O_{80} \cdot 25H_2O$	Monoclinic $\angle\beta=93°$ platy on {010}	47°–65° $r>v$ weak crossed bxa	+	{010} perfect	$b=Z$ $c \wedge X \simeq -22°$ $a \wedge Y \simeq -19°$	Colorless	H = 5. Sp. Gr. ≃ 2.4. White, yellowish, greenish
1.512–1.522	1.502–1.514	1.514–1.525	0.008–0.016	Laumontite $Ca(Al_2Si_4)O_{12} \cdot 4H_2O$	Monoclinic $\angle\beta=111°30'$ prismatic on *c*	25°–47° $v>r$ strong	−	{010} perfect {110} perfect	$b=Y$ $c \wedge Z=$ $+8°$ to $+40°$ $a \wedge X=$ $+30°$ to $+62°$	Colorless	H = 3–4. Sp. Gr. = 2.2–2.4. White, yellow, gray, reddish
1.513–1.533	1.497–1.530	1.518–1.544	0.006–0.021	Thomsonite $NaCa_2[(Al,Si)_5O_{10}]_2 \cdot 6H_2O$	Orthorhombic acicular on *c*	42°–75° $r>v$ strong	+	{010} perfect, {100} distinct	$a=X$ $b=Z$ $c=Y$	Colorless	H = 5–$5\frac{1}{2}$. Sp. Gr. = 2.1–2.4. White, brown, red, green
1.516–1.520	1.507–1.513	1.517–1.521	0.010–0.007	Scolecite $Ca(Al_2Si_3)O_{10} \cdot 3H_2O$	Monoclinic $\angle\beta=90°39'$ fibrous on *c*	36°–56° $v>r$ strong	−	{110} perfect	$b=Z$ $c \wedge X \simeq -18°$ $a \wedge Y \simeq -17°$	Colorless	H = 5. Sp. Gr. = 2.2–2.3. White, yellowish
1.540–1.543	1.531–1.538	1.548	0.008–0.018	Gismondine $Ca(Al_2Si_2)O_8 \cdot 4H_2O$	Monoclinic $\angle\beta=92°30'$ bipyramidal	82°–86°	−	{101} distinct	$a \sim Z$ $b=X$ $c \sim Y$	Colorless	H = $4\frac{1}{2}$. Sp. Gr. ≃ 2.3. White, gray, reddish
1.553	1.541	1.557	0.016	Edingtonite $Ba(Al_2Si_3)O_{10} \cdot 4H_2O$	Orthorhombic fibrous on *c*	54° $v>r$ strong	−	{110} perfect	$a=Z$ $b=Y$ $c=X$	Colorless	H = 4–$4\frac{1}{2}$. Sp. Gr. = 2.7–2.8. White, gray, pink

Analcime (Analcite) $Na(AlSi_2)O_6 \cdot H_2O$ ISOMETRIC

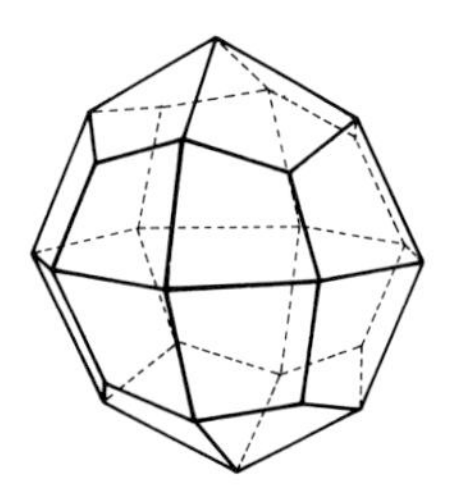

Analcime

$n = 1.479–1.493$*
Colorless in standard section

COMPOSITION AND STRUCTURE. The cubic framework structure of analcime comprises rings of four and six silicate tetrahedra normal to fourfold and threefold symmetry axes. Open tunnels parallel to threefold symmetry axes accommodate water molecules, and Na^+ ions occupy adjoining cavities to make up charge deficiencies when Al^{3+} replaces Si^{4+}.

Ideally, one of three Si^{4+} in tetrahedral coordination is replaced by Al^{3+} with the addition of one Na^+, and natural compositions are remarkably ideal, considering the fact that additional Na^+ sites are available. The Al:Si ratio may vary widely from less than 1:3 to greater than 1:2, and the water content appears to increase with silicon. Numerous, moderate-sized cations may replace some Na^+, and a pure Ca-analcime (*wairakite*) and K-analcime are known. Cesium analcime is known as *pollucite* (see Fig. 7-34).

PHYSICAL PROPERTIES. H = $5–5\frac{1}{2}$. Sp. Gr. = 2.22–2.29. In hand sample, analcime is normally colorless to white, with vitreous luster, rarely pink or greenish.

COLOR AND PLEOCHROISM. Analcime is colorless in thin section or as fragments.

FORM. The characteristic habit of analcime is well-formed isometric, usually trapezohedral, crystals that yield nearly round sections. Analcime is most common as vein or cavity filling but may appear as interstitial granules.

CLEAVAGE. Poor cubic {100} cleavages may show in section.

BIREFRINGENCE. Analcime is normally isotropic, but crystallization at low temperatures or high pressures favors anisotropism, shown by very weak birefringence (<0.001). Loss of water or Al-Si ordering may cause structural strain.

TWINNING. Analcime crystals are commonly penetration twins, and birefringent sections commonly show fine polysynthetic twinning on {100} or {110}.

INTERFERENCE FIGURE. Analcime is essentially isotropic; however, birefringent varieties may yield broad, diffuse isogyres on a gray field. The vague figure is usually biaxial negative, with $2V_x$ ranging from 0° to 90°.

DISTINGUISHING FEATURES. Analcime is an isotropic mineral with very low refractive index and is characteristic as vein and cavity filling. It is very difficult to distinguish from leucite, sodalite, and some other zeolites. Leucite has a somewhat higher refractive index and is essentially confined to volcanic rocks. Sodalite may show weak color, and most zeolites are distinctly anisotropic. Opal is amorphous, with very low refractive index.

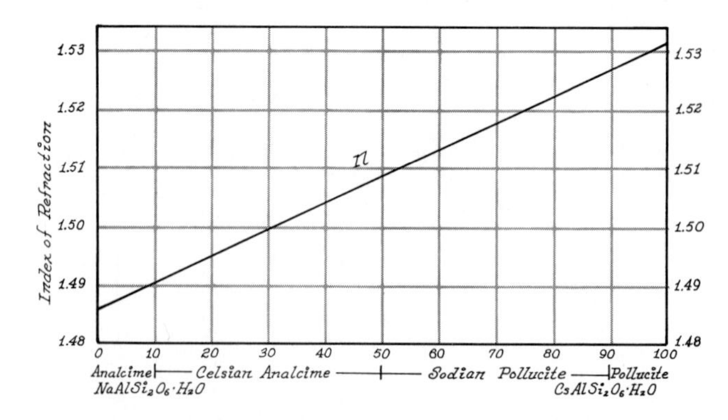

FIGURE 7-34. Variation of the single index of refraction in the analcime-pollucite series. (Data from Černy, 1974)

* For the ideal composition, $n = 1.488$. Refractive index increases with potash content and decreases with an increase in silica.

Chemical tests or staining techniques may be necessary to confirm analcime. Analcime gelatinizes with HCl to absorb malachite dye.

ALTERATION. Analcime is a stable mineral at low temperatures and pressures and may itself be an alteration product of alkali feldspars.

OCCURRENCE. Analcime may crystallize as a primary mineral in silica-deficient igneous rocks, either plutonic or volcanic; it is formed by hydrothermal or pneumatolitic fluids; it may appear as an authigenic mineral in sediments and may result from the alteration of alkali feldspars by Na-rich solutions.

In plutonic igneous rocks, analcime may constitute a large percentage of analcime syenite, teschenite, or alkali diabase with alkali feldspars, olivine, sphene, or melanite garnet. In silica-poor volcanic rocks, analcime appears as a primary constituent of the granular groundmass of some basalts or trachybasalts.

Hydrothermal crystallization of analcime occurs with calcite and zeolites in vesicles and fractures, usually in basalts and related flow rocks. Sedimentary rocks may show analcime as a major authigenic constituent of argillaceous or feldspathic sediments or pyroclastics related to alkaline lake deposits. Alteration of alkali feldspars by Na-rich solutions in near-surface environments may form analcime as a constituent of the "zeolite facies."

REFERENCES: ANALCIME

Beattie, I. R. 1954. The structure of analcite and ion-exchange forms of analcite. *Acta Crystallogr.*, *7*, 357–359.

Cerný, P. 1974. The present status of the analcime-pollucite series. *Can. Mineral.*, *12*, 334–341.

Pertsowsky, R. 1976. Optical and crystallographic properties of synthetic analcite: Effect of temperature and composition of the medium on formation. *Bull. Soc. Fr. Mineral. Cristallogr.*, *99*, 407–408.

Ross, C. S. 1941. Sedimentary analcite. *Amer. Mineral.*, *26*, 627–629.

Saha, P. 1959. Geochemical and X-ray investigation of natural and synthetic analcites. *Amer. Mineral.*, *44*, 300–313.

Steiner, A. 1955. Wairakite, the calcium analogue of analcime, a new zeolite mineral. *Mineral. Mag.*, *30*, 691–698.

Taylor, W. H. 1930. The structure of analcite ($NaAlSi_2O_6 \cdot H_2O$). *Z. Kristallogr.*, *74*, 1–19.

Wilkinson, J. F. G. 1963. Some natural analcime solid solutions. *Mineral. Mag.*, *33*, 498–505.

Yoder, H. S., and C. E. Weir. 1960. High-pressure form of analcite and free-energy change with pressure of analcite reactions. *Amer. Jour. Sci.*, *258-A*, 420–433.

Natrolite

$Na_2(Al_2Si_3)O_{10} \cdot 2H_2O$

ORTHORHOMBIC (PSEUDOTETRAGONAL)

$n_\alpha = 1.473–1.483$*
$n_\beta = 1.476–1.486$
$n_\gamma = 1.485–1.496$
$n_\gamma - n_\alpha = 0.012–0.013$
Biaxial positive, $2V_z = 58°–64°$
$a = X, b = Y, c = Z$
$v > r$ weak
Colorless in standard section

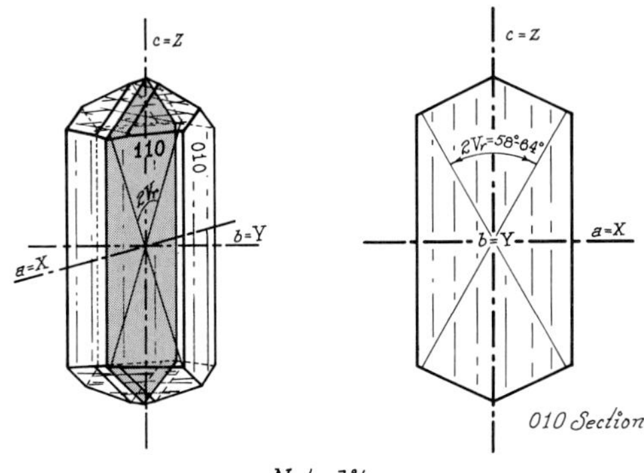

Natrolite

* Refractive indices are lowered by Ca substitution. Dehydrated zeolites show higher indices of refraction, but partial dehydration appears to lower indices.

COMPOSITION. Some K^+ and Ca^{2+} are always present, and the ideal Al:Si ratio varies somewhat. A tetragonal variety having an Si:Al ratio of 1.70 has been reported, with n_ω = 1.480 and n_ε = 1.494.

PHYSICAL PROPERTIES. H = 5. Sp. Gr. = 2.20–2.26. In hand sample, color is white, gray, yellow, or reddish, and luster may be silky.

COLOR AND PLEOCHROISM. Colorless in section or as fragments.

FORM. Natrolite is a fibrous zeolite appearing as acicular needles or blades elongated on *c*, often as radial aggregates.

CLEAVAGE. Perfect prismatic {110} cleavages and {010} parting favor the fibrous habit.

BIREFRINGENCE. Maximum birefringence, seen in {010} sections, is mild (0.012–0.013) yielding white and gray to first-order yellow in standard section. Cleavage fragments may show near-maximum birefringence.

TWINNING. Twinning is reported on {110}, {011}, and {031}, but is considered rare.

INTERFERENCE FIGURE. Figures show few isochromatic color rings and may well be difficult to obtain in usable form, as the acute bisectrix parallels fibrous elongation. Dispersion is $v > r$ weak, without bisectrix dispersion.

OPTICAL ORIENTATION. $a = X$, $b = Y$, $c = Z$. Extinction is parallel to elongation, and elongation is positive (length-slow) as $c = Z$.

DISTINGUISHING FEATURES. Natrolite is a fibrous zeolite with parallel extinction and positive elongation (length-slow). Scolecite shows inclined extinction, edingtonite is length-fast, and mesolite and thompsonite show both length-slow and length-fast elongation. Gonnardite is biaxial negative and shows very weak birefringence (0.002).

ALTERATION. Zeolites are relatively stable under surface conditions, but some are known to alter to montmorillonite-type clays (such as beidellite).

OCCURRENCE. Natrolite is a common zeolite in vesicles, amygdales, fractures, and other cavities in mafic volcanic rocks. It is deposited later than most other zeolites by hydrothermal solutions or even surface water. Natrolite commonly appears in association with other zeolites (such as analcime, stilbite, chabazite, heulandite, and laumontite). Natrolite may result from the alteration of nepheline or other feldspathoids, in undersaturated rock types, or from the alteration of alkali feldspars in mafic to quartz-bearing rock types.

REFERENCES: NATROLITE

Foster, M. D. 1965a. Zeolites: Composition relations among thomsonites, gonnardites, and natrolites. *U.S. Geol. Surv. Prof. Paper*, *504-E*.

Foster, M. D. 1965b. Zeolites: Composition of zeolites of the natrolite group. *U.S. Geol. Surv. Prof. Paper*, *504-D*.

Guseva, L. D., Yu. P. Men'shikov, T. S. Romanova, and I. V. Bussen. 1975. Tetragonal natrolite from the Lovozero alkaline massif. *Zap. Vses. Mineral. Obshchest.*, *104*, 66–99.

Torie, B. H., I. D. Brown, and H. E. Petch. 1964. Neutron diffraction determination of the hydrogen positions in natrolite. *Can. Jóur. Phys.*, *42*, 229–240.

Mesolite

$Na_2Ca_2[(Al_2Si_3)O_{10}]_3 \cdot 8H_2O$

MONOCLINIC (PSEUDOORTHORHOMBIC)
$\angle\beta \simeq 90°$

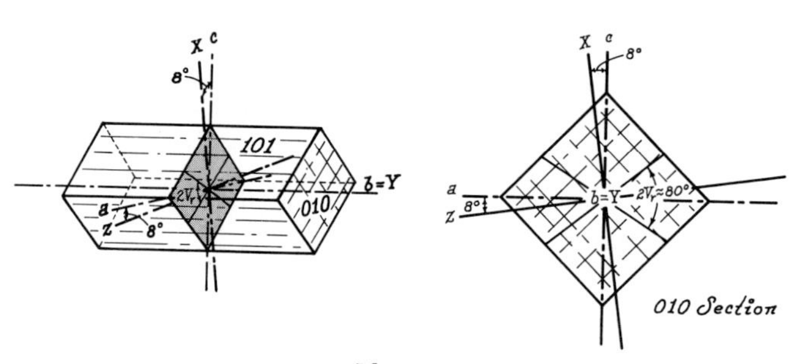

Mesolite

$n_\alpha = 1.504\text{–}1.508$
$n_\beta = 1.504\text{–}1.508$
$n_\gamma = 1.505\text{–}1.509$
$n_\gamma - n_\alpha \leq 0.001$
Biaxial positive, $2V_z \simeq 80°$*
$b = Y, c \wedge X = -8°, a \wedge Z \simeq -8$
$v > r$ strong, inclined bisectrix dispersion
Colorless in standard section

COMPOSITION. Mesolite represents intermediate compositions between natrolite, $Na_2(Al_2Si_3)O_{10} \cdot 2H_2O$, and scolecite, $Ca(Al_2Si_3)O_{10} \cdot 3H_2O$. Both Ca:Na and Al:Si ratios may vary from the ideal.

PHYSICAL PROPERTIES. H = 5. Sp. Gr. = 2.26. Hand sample color is usually white or yellowish with silky luster.

COLOR AND PLEOCHROISM. Colorless in section or as fragments.

FORM. Mesolite is a fibrous-type zeolite appearing as acicular needles elongated on *b* or as hairlike aggregates.

CLEAVAGE. Perfect cleavages on {101} and $\{10\bar{1}\}$ produce elongation on *b*.

BIREFRINGENCE. Maximum birefringence, seen in {010} sections, is very weak (0.001), yielding only dark gray interference colors. The first-order red accessory may be necessary to detect birefringence on the nearly isotropic fibers, which do not show maximum birefringence.

TWINNING. Twinning on {100} is always present.

INTERFERENCE FIGURE. Mesolite is essentially isotropic but is considered biaxial positive with large $2V_z$ which appears very sensitive to temperature change, $2V_z$ increases from about 70°, at 0°C, to exceed 90°, above 20°C (mesolite becomes biaxial negative), and $2V_x$ closes to 0°, at 59°C (uniaxial negative), and opens again in the {100} plane with increasing temperature. With very low birefringence, very slight changes in refractive indices greatly alter $2V$. Usable figures are almost impossible to obtain, because of fibrous habit and exceedingly low birefringence, and cleavage favors off-center figures. Dispersion is $v > r$ strong, with possible inclined bisectrix dispersion.

OPTICAL ORIENTATION. $b = Y, c \wedge X \simeq -8°, a \wedge Z \simeq -8°$. Extinction is parallel to cleavages and crystal elongation, as elongation is on *b*. Elongation may be either positive or negative, as $b = Y$.

DISTINGUISHING FEATURES. Mesolite is a fibrous zeolite showing twinning, very weak birefringence, and unique elongation on *b*, which is both length-fast and length-slow, as $b = Y$. Scolecite shows inclined extinction, natrolite and gonnardite are strictly length-slow, edingtonite is strictly length-fast, and thompsonite shows stronger birefringence.

ALTERATION. Zeolites are relatively stable under surface conditions, but some are known to alter to montmorillonite clays (such as beidellite).

OCCURRENCE. Mesolite is commonly formed in amygdales, vesicles, fissures, and other cavities in subsiliceous volcanic rocks by hydrothermal solutions. It is commonly associated with calcite and other zeolites (such as analcime, stilbite, and chabazite).

REFERENCE: MESOLITE

Harada, K., K. Nakao, and K. Nagashima. 1969. Ionic substitution in natural mesolites. *Gans. Kob. Kosho Gakk.*, *61*, 112–115.

* Optical constants change markedly with slight temperature change. $2V_z$ increases with temperature, and the mineral becomes biaxial negative at 20°C and uniaxial negative at 59°C.

Scolecite

$Ca(Al_2Si_3)O_{10} \cdot 3H_2O$

MONOCLINIC (PSEUDOTETRAGONAL)
$\angle = 90°39'$

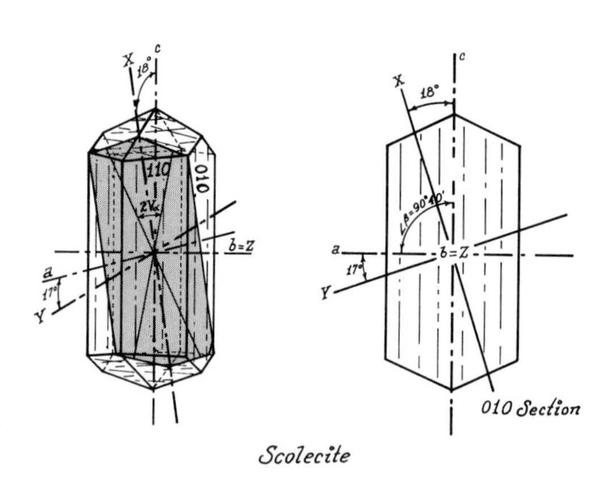

Scolecite

$n_\alpha = 1.507–1.513$
$n_\beta = 1.516–1.520$
$n_\gamma = 1.517–1.521$
$n_\gamma - n_\alpha = 0.010–0.007$
Biaxial negative, $2V_x = 36° - 56°$
$b = Z, c \wedge X \simeq -18°, a \wedge Y \simeq -17°$
$v > r$ strong, horizontal bisectrix dispersion
Colorless in standard section

COMPOSITION. Minor Na^+ or K^+ may replace Ca^{2+}.

PHYSICAL PROPERTIES. H = 5. Sp. Gr. = 2.25–2.29. Hand sample color is white to yellowish with vitreous to silky luster.

COLOR AND PLEOCHROISM. Colorless in section or as fragments.

FORM. Scolecite is a fibrous-type zeolite appearing as needles or hairlike fibers, often radiating and elongated on *c*.

CLEAVAGE. Nearly perfect, prismatic cleavages {110} produce fragment elongation on *c*.

BIREFRINGENCE. Maximum birefringence, seen in sections near {100}, is weak (0.007–0.010) yielding only gray to white interference colors in standard section. Fragments show less than maximum birefringence.

TWINNING. Parallel twins are common on {100} (twin axis parallel to *c*), producing multiple visible twins on {010}. Twinning is rare on {110} or {001}, and penetration twins are known.

INTERFERENCE FIGURE. Scolecite is biaxial negative, with $2V_x = 36°–56°$. Usable figures are rare, as the acute bisectrix is near fiber elongation, and birefringence is weak. Figures should show broad isogyres on a white field, with strong dispersion $v > r$ and possible horizontal bisectrix dispersion.

OPTICAL ORIENTATION. $b = Z, c \wedge X \simeq -18°, a \wedge Y \simeq -17°$. Scolecite is unique among the fibrous zeolites by showing inclined extinction to cleavages and crystal elongation. Maximum extinction angle is 18°, seen on {010} sections ($c \wedge X \simeq -18°$). Extinction is parallel on {100} and symmetrical on {001}. Elongation is negative (length-fast), as *X* lies nearest the elongation.

DISTINGUISHING FEATURES. Scolecite is the only prominent fibrous zeolite with inclined extinction and negative elongation.

ALTERATION. Zeolites are relatively stable under surface conditions but some are known to alter to montmorillonite clays (such as beidellite).

OCCURRENCE. Scolecite, like other zeolites, is characteristic of cavities in mafic volcanic rocks in association with other zeolites (such as stilbite) and calcite. It appears in the zeolite facies of some metamorphosed carbonate rocks with albite, epidote, chlorite, and so on.

REFERENCES: SCOLECITE

Ivleva, L. V., and S. P. Gabuda. 1971. Position of water molecules in scolecite. *Kristallografiya*, *16*, 825–826.
Smith, G. W., and R. Walls. 1971. Redetermination of the unit-cell geometry of scolecite. *Mineral. Mag.*, *38*, 72–75.

Thomsonite $NaCa_2[(Al,Si)_5O_{10}]_2 \cdot 6H_2O$ ORTHORHOMBIC (PSEUDOTETRAGONAL)

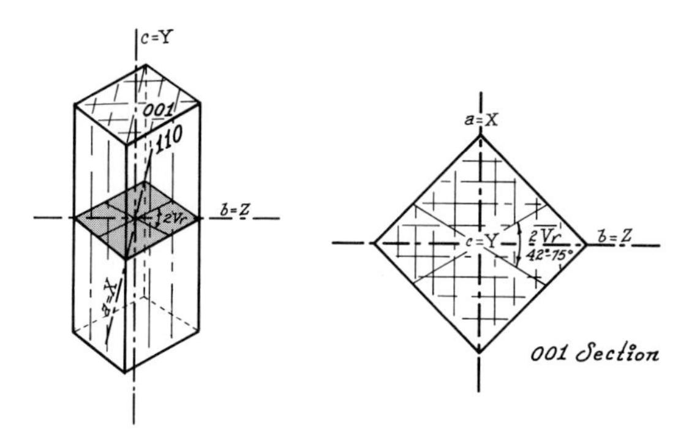

$n_\alpha = 1.497$–1.530*
$n_\beta = 1.513$–1.533
$n_\gamma = 1.518$–1.544
$n_\gamma - n_\alpha = 0.006$–$0.021$
Biaxial positive, $2V_z = 42°$–$75°$*
$a = X, b = Z, c = Y$
$r > v$ moderate to strong
Colorless in standard section

COMPOSITION. The Al:Si ratio tends to be about 2:3 but is somewhat variable, as is Ca:(Na,K).

PHYSICAL PROPERTIES. H = 5–5½. Sp. Gr. = 2.10–2.39. Hand sample color is white to brown, reddish, or greenish, with vitreous luster.

COLOR AND PLEOCHROISM. Colorless in section or as fragments.

FORM. Thomsonite is a fibrous-type zeolite appearing as columnar, acicular, or bladed crystals elongated on *c*.

CLEAVAGE. Cleavage is perfect on {010} and distinct on {100} to yield fragments elongated on *c*.

BIREFRINGENCE. Maximum birefringence, seen in basal {001} sections is weak to mild (0.006–0.021). Interference colors are normally white and gray but may range into first-order yellow or red. Fragments show less than maximum birefringence.

TWINNING. Twinning, on {110}, is apparently rare.

INTERFERENCE FIGURE. Fragments on {010} yield acute bisectrix figures, and those on {100} yield obtuse bisectrix figures. Figures show broad isogyres, few isochromes, and moderate to strong dispersion $r > v$ without bisectrix dispersion.

OPTICAL ORIENTATION. $a = X$, $b = Z$, $c = Y$. Extinction is strictly parallel and elongation is either positive or negative, as $c = Y$.

DISTINGUISHING FEATURES. Thomsonite is a fibrous zeolite showing parallel extinction and both length-fast and length-slow elongation. Mesolite also shows both elongations, but its birefringence is very weak (0.001).

ALTERATION. Zeolites tend to be stable at surface conditions but some are known to alter to montmorillonite clays (such as beidellite).

OCCURRENCE. Thomsonite appears in cavities and fractures in mafic volcanic and hypabyssal rocks in association with calcic plagioclase and other zeolites (such as chabazite, analcime, and natrolite). It is a possible alteration product of both nepheline and calcic-plagioclase and may form in metamorphic contact zones in limestone.

* Refractive indices and $2V_z$ increase with substitution of CaAl for NaSi.

REFERENCE: THOMSONITE

Foster, M. D. 1965. Zeolites: Composition relations among thomsonites, gonnardites, and natrolites. *U.S. Geol. Surv. Prof. Paper, 504-E.*

Gonnardite

$Na_2Ca[(Al,Si)_5O_{10}]_2 \cdot 6H_2O$

ORTHORHOMBIC (PSEUDOTETRAGONAL)

$n_\alpha = 1.497\text{–}1.506$
$n_\beta = 1.498\text{–}1.507$
$n_\gamma = 1.499\text{–}1.508$
$n_\gamma - n_\alpha \simeq 0.002$
Biaxial negative, $2V_x \simeq 50°$
$c = Z$
Colorless in standard section

COMPOSITION. The Al:Si ratio tends to be 1:1 but varies by the replacement CaAl $\rightleftharpoons$ NaSi.

PHYSICAL PROPERTIES. H = $4\frac{1}{2}$–5. Sp. Gr. $\simeq$ 2.3. Hand sample color is usually white, with vitreous to silky luster.

COLOR AND PLEOCHROISM. Colorless in section or as fragments.

FORM. Gonnardite is a fibrous-type zeolite, commonly appearing as radiating acicular crystals in spherulites.

CLEAVAGE. Cleavages are probably good parallel to *c*.

BIREFRINGENCE. Maximum birefringence is very weak (0.002), yielding only dark gray interference colors in standard section. Fragments show maximum birefringence.

TWINNING. Twinning is not characteristic of gonnardite.

INTERFERENCE FIGURE. As the birefringence is very weak, slight changes in refractive indices may cause a broad range for $2V_x$. Fragments should yield acute bisectrix to flash figures, as the obtuse bisectrix parallels elongation, but figures are poor, owing to fibrous habit and weak birefringence.

OPTICAL ORIENTATION. $c = Z$. Extinction should be parallel, and elongation is positive (length-slow).

DISTINGUISHING FEATURES. Gonnardite is a rare fibrous zeolite showing parallel extinction, positive (length-slow) elongation, and very weak birefringence. Mesolite is both length-fast and length-slow.

ALTERATION. Zeolites are relatively stable at surface conditions, but some are known to alter to montmorillonite clays (such as beidellite). Gonnardite may alter to thomsonite.

OCCURRENCE. Gonnardite appears in cavities in subsiliceous volcanics with other zeolites (such as thomsonite and phillipsite) and may represent an alteration product of nepheline.

REFERENCES: GONNARDITE

Foster, M. D. 1965. Zeolites: Composition relations among thomsonites, gonnardites and natrolites. *U.S. Geol. Surv. Prof. Paper*, *504-E*.

Harada, K., S. Iwamoto, and K. Kihara. 1967. Erionite, phillipsite, and gonnardite in the amygdales of altered basalt from Mazé, Niigata Prefecture, Japan. *Amer. Mineral.*, *52*, 1785–1794.

Mordenite (Ptilolite)

$(Na_2,K_2,Ca)(Al_2Si_{10})O_{24} \cdot 7H_2O$

ORTHORHOMBIC

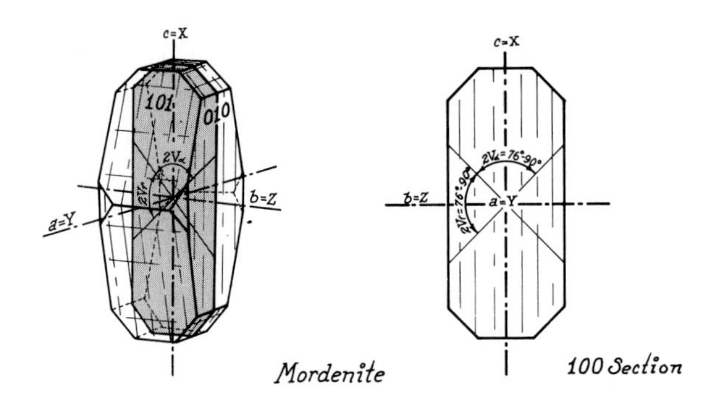

$n_\alpha = 1.472–1.483$
$n_\beta = 1.475–1.485$
$n_\gamma = 1.477–1.487$
$n_\gamma - n_\alpha = 0.002–0.005$
Biaxial negative to positive, $2V_x = 76°–90°$, $2V_z = 90°–76°$
$a = Y, b = Z, c = X$
Colorless in standard section

COMPOSITION. Alkalis are normally dominant over calcium, and Na^+ usually exceeds K^+.

PHYSICAL PROPERTIES. H = 3–4. Sp. Gr. = 2.12–2.22. Hand sample color is white to yellowish or pinkish.

COLOR AND PLEOCHROISM. Colorless in section or as fragments.

FORM. Tiny crystals, resembling heulandite, are commonly acicular or fibrous, often radiating as rosettes or as cottony masses.

CLEAVAGE. Perfect cleavages on {010} and {100} produce fragments elongated on *c*.

BIREFRINGENCE. Maximum birefringence, seen in {100} sections, is very weak (0.002–0.005), and interference colors are grays in standard section.

TWINNING. Twinning is not reported in mordenite.

INTERFERENCE FIGURE. Mordenite is either biaxial negative or positive, as $2V$ ranges about 15° on either side of 90°. Note that, with weak birefringence, very slight changes in refractive indices greatly change $2V$. Figures are usually poor, owing to weak birefringence and small crystals. Cleavage fragments should yield flash figures and bisectrix figures on Z.

OPTICAL ORIENTATION. $a = Y, b = Z, c = X$. Fragments and crystals are elongated on c to show parallel extinction and negative elongation (length-fast).

DISTINGUISHING FEATURES. Mordenite is a fibrous zeolite with parallel extinction, negative (length-fast) elongation, and very weak birefringence. Natrolite is length-slow, with higher birefringence; mesolite and thomsonite are length-fast and length-slow, with higher refractive index; scolecite shows inclined extinction; and gonnardite is length-slow.

ALTERATION. Zeolites tend to be stable at surface conditions.

OCCURRENCE. Mordenite occurs as fibrous or cottony aggregates in amygdales or fractures in mafic volcanics, where it may be associated with quartz. It results from the alteration of volcanic glass and may appear in altered tuffs or bentonite clays. Authigenic mordenite appears to form in detrital marine sediments and may be associated with illites and glauconite.

REFERENCES: MORDENITE

Gvakhariya, G. V., and T. V. Batiashvili. 1970. Properties of natural and synthetic mordenites. *Termoanal. Issled. Sovrem. Mineral., 1970*, 81–90.

Hay, R. L. 1966. Zeolites and zeolitic reactions in sedimentary rocks. *Geol. Soc. Amer., Spec. Paper, 85.*

Passaglia, E. 1975. Crystal chemistry of mordenites. *Contrib. Mineral. Petrol., 50*, 65–77.

Pobedimskaya, E. A., and N. V. Belov. 1963. The crystal structure of mordenite (ptilolite) $Na_8Al_8Si_{40}O_{96} \cdot 24H_2O = 8NaAlSi_5O_{12} \cdot 3H_2O$. *Kristallografiya, 8*, 919–921.

Laumontite

$Ca(Al_2Si_4)O_{12} \cdot 4H_2O$

MONOCLINIC
$\angle \beta = 111°30'$

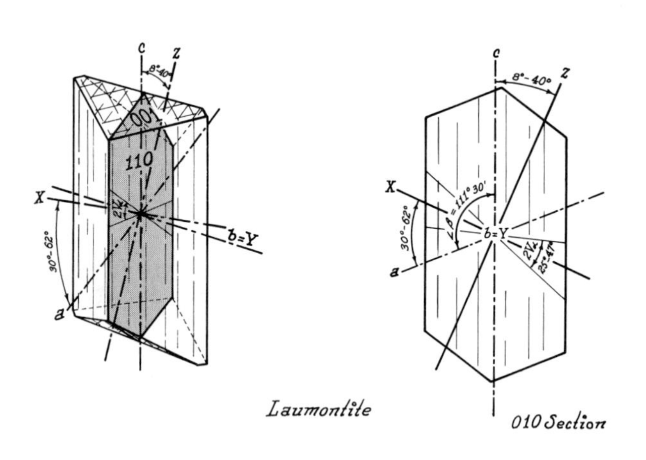

$n_\alpha = 1.502–1.514$*
$n_\beta = 1.512–1.522$
$n_\gamma = 1.514–1.525$
$n_\gamma - n_\alpha = 0.008–0.016$
Biaxial negative, $2V_x = 25°–47°$
$b = Y, c \wedge Z = +8°$ to $+40°, a \wedge X = +30°$ to $+62°$
$v > r$ strong, inclined bisectrix
Colorless in standard section

COMPOSITION. On exposure to air, laumonite loses one-eighth of its water to become *leonhardite*, $Ca(Al_2Si_4)O_{12} \cdot 3\frac{1}{2}H_2O$, with low refractive index, small $2V_x$, and large extinction angle ($c \wedge Z$). Some alkali may replace calcium with (for example, NaSi ⇌ CaAl) or without (for example, 2Na ⇌ Ca) changing the Si:Al ratio. Some Fe^{3+} may proxy for Al^{3+}.

PHYSICAL PROPERTIES. H = 3–4. Sp. Gr. = 2.23–2.41. Hand sample color is colorless or white to yellow, gray, or reddish. Luster is vitreous to pearly.

COLOR AND PLEOCHROISM. Colorless in section or as fragments.

FORM. Laumontite appears as columnar, prismatic crystals elongated on c or as radiating fibrous aggregates.

CLEAVAGE. Perfect cleavages parallel to {010} and {110} produce fragments elongated on c.

BIREFRINGENCE. Maximum birefringence, seen in {010} sections is weak to mild (0.008–0.016), and interference colors are gray and white, with possible first-order yellow or red, in standard section. Cleavage fragments showing maximum extinction angle also show maximum birefringence.

TWINNING. Simple twinning on {100} is common.

INTERFERENCE FIGURE. Laumontite is biaxial negative, with moderate $2V_x = 25°$ to $47°$. Acute bisectrix figures appear in sections near {100}, and fragments lie so as to yield flash figures and off-centered figures. Broad isogyres appear with very few isochromes, and dispersion is $v > r$ strong, with possible inclined bisectrix.

OPTICAL ORIENTATION. $b = Y$, $c \wedge Z = +8°$ to $+40°$, $a \wedge X = +30°$ to $+62°$. Maximum extinction angle, seen in {010} sections, is 8° to 40° measured to cleavage traces. All fragments on cleavages show inclined extinction, those on {010} showing maximum extinction angle. Crystals and cleavage fragments are elongated on c to show positive elongation (length-slow), as Z lies near c.

DISTINGUISHING FEATURES. Laumontite is a fibrous zeolite, biaxial negative with inclined extinction. Scolecite, heulandite, stilbite, and phillipsite show inclined extinction but scolecite is length-fast, heulandite is a layered zeolite (only one cleavage) and biaxial positive, stilbite also has one cleavage and is either length-fast or length-slow, and phillipsite is biaxial positive.

ALTERATION. Zeolites tend to be stable at surface conditions, but laumontite easily dehydrates to become leonhardite.

OCCURRENCE. Laumontite appears in cavities and veins in many types of igneous rocks from basalt to granite and is deposited in metalliferous hydrothermal veins. It forms in preference to other zeolites with increasing depth and results from mild metamorphic alteration of volcanic glass and feldspars to appear in tuffs and feldspathic detrital sediments.

* Refraction indices decrease as NaSi replaces CaAl and decrease with loss of water.

REFERENCES: LAUMONTITE

Bartl, H. 1970. Structure refinement of leonhardite by neutron diffraction. *Neues Jahrb. Mineral. Monatsh., 1970*, 298–310.
Bartl, H., and K. F. Fischer. 1967. Crystal structure of the zeolite laumontite. *Neues Jahrb. Mineral. Monatsh., 1967*, 33–42.
Pipping, F. 1964. The dehydration and chemical composition of laumontite. *Int. Mineral. Ass., Pap. Proc. Gen. Meet., 4th, New Delhi, India, 1964*, 159–166.

Edingtonite

$Ba(Al_2Si_3)O_{10} \cdot 4H_2O$

ORTHORHOMBIC (PSEUDOTETRAGONAL)

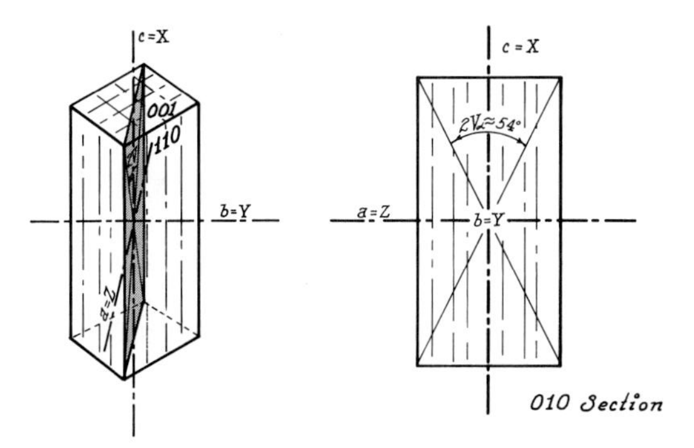

$n_\alpha = 1.541$
$n_\beta = 1.553$
$n_\gamma = 1.557$
$n_\gamma - n_\alpha = 0.016$
Biaxial negative, $2V_x \simeq 54°$
$a = Z, b = Y, c = X$
$v > r$ strong
Colorless in standard section

COMPOSITION. Edingtonite is one of the rare barium zeolites. Appreciable Ca^{2+} may replace Ba^{2+}.

PHYSICAL PROPERTIES. H = 4–4½. Sp. Gr. = 2.69–2.8. Hand sample color is white to grayish or pink.

COLOR AND PLEOCHROISM. Colorless in section or as fragments.

FORM. Edingtonite is a fibrous-type zeolite, appearing as tiny pyramidal crystals.

CLEAVAGE. Perfect prismatic {110} cleavages yield fragments elongated on *c*.

BIREFRINGENCE. Maximum birefringence, seen in {010} sections, is mild (0.016), yielding first-order colors in standard section. Cleavage fragments show near-maximum birefringence.

TWINNING. Twinning is not reported in edingtonite.

INTERFERENCE FIGURE. Edingtonite is biaxial negative, with moderate $2V_x$ near 54°. The acute bisectrix lies along fragment elongation, making figures difficult. Figures should show broad isogyres, few isochromes, and strong dispersion $v > r$ without bisectrix dispersion.

OPTICAL ORIENTATION. $a = Z, b = Y, c = X$. Extinction is parallel to elongation, and elongation is negative (length-fast), as $c = X$.

DISTINGUISHING FEATURES. Edingtonite is a rare, barium, fibrous-type zeolite showing parallel extinction and negative elongation. Thomsonite and mesolite show both positive and negative elongation, and scolecite shows inclined extinction.

ALTERATION. Zeolites tend to be stable at surface conditions, but some are known to alter to montmorillonite clays (such as beidellite).

OCCURRENCE. Edingtonite is a rare barium zeolite that tends to be associated with manganese mineralization. It appears in mafic volcanics with harmotome and other zeolites, calcite, and prehnite.

REFERENCE: EDINGTONITE

Galli, E. 1976. Crystal structure refinement of edingtonite. *Acta Crystallogr., B32*, 1623–1627.

Heulandite

$(Ca,Na_2)(Al_2Si_7)O_{18} \cdot 6H_2O$

MONOCLINIC
$\angle\beta = 116°25'$

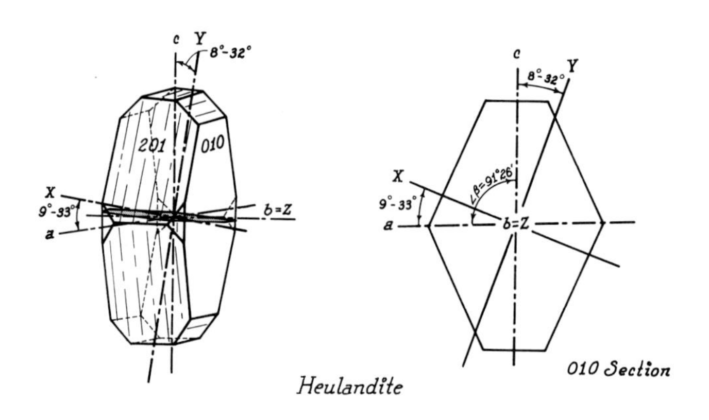

Heulandite

$n_\alpha = 1.487–1.505$*
$n_\beta = 1.487–1.505$
$n_\gamma = 1.488–1.512$
$n_\gamma – n_\alpha = 0.001–0.007$
Biaxial positive, $2V_z \simeq 30°$
$b = Z$, $c \wedge Y = +8°$ to $+32°$, $a \wedge X = +9°$ to $+33°$
$r > v$ and crossed bisectrix dispersion distinct
Colorless in standard section

COMPOSITION. Heulandite may contain significant K^+, and potash-rich varieties are known. A silica-rich variety (*clinoptilolite*) tends to be rich in alkalis via the substitution $Ca^{2+}Al^{3+} \rightleftharpoons Na^+Si^{4+}$.

PHYSICAL PROPERTIES. H = $3\frac{1}{2}$–4. Sp. Gr. = 2.2. Hand sample color is commonly white; rarely yellow, brown, or reddish. Luster is pearly on {010}.

COLOR AND PLEOCHROISM. Colorless in section or as fragments.

FORM. Heulandite is a layered zeolite commonly appearing as plates or tabular crystals flattened on {010}. Granular and crystal aggregates are less common.

CLEAVAGE. Perfect cleavage on {010} fosters a foliated habit.

* Index of refraction is less in silica-rich varieties.

BIREFRINGENCE. Maximum birefringence, seen in near basal sections, is very weak (0.001–0.007), and interference colors are no more than gray in standard section.

TWINNING. Twinning is not characteristic of heulandite.

INTERFERENCE FIGURE. Cleavage plates and {010} sections show acute bisectrix figures of broad isogyres and no isochromes. Dispersion is distinct $r > v$ with discernible crossed bisectrix dispersion.

OPTICAL ORIENTATION. $b = Z$, $c \wedge Y = +8°$ to $+32°$, and $a \wedge X = +9°$ to $+33°$. The optic plane is perpendicular to {010}. Extinction is parallel to {010} cleavages in sections normal to {010}, and maximum extinction angles to {100} or {001} planes range from 8° to 33°.

Elongation in sections perpendicular to {010} is always negative (length-fast), as Z is normal to cleavage. Silica-rich specimens (Si:Al > 3.57) have been observed to change to length-slow when heated during mounting of thin sections.

DISTINGUISHING FEATURES. Heulandite is a layered zeolite showing acute bisectrix figures on cleavage plates. Figures are biaxial positive, with small $2V$ and distinct crossed dispersion $r > v$. Stilbite is optically negative, shows higher birefringence, and is often twinned.

ALTERATION. Zeolites tend to be stable under surface conditions, but some are known to alter to montmorillonite clays (such as beidellite).

OCCURRENCE. Heulandite is most characteristic of cavities and fractures in basalts and other mafic rocks in association with stilbite and chabazite. It forms by devitrification of volcanic glass and the alteration of tuff or ash deposits and appears, consequently, in bentonite clays. Heulandite is reported in cavities in metamorphic schists, gneisses, and contact rocks and as an authigenic mineral in detrital sediments.

REFERENCES: HEULANDITE

Alberti, A. 1972. Crystal structure of the zeolite heulandite. *Tschermaks Mineral. Petrogr. Mitt.*, *18*, 129–146.

Alietti, A. 1972. Polymorphism and crystal chemistry of heulandites and clinoptilolites. *Amer. Mineral.*, *57*, 1448–1462.

Boles, J. R. 1972. Composition, optical properties, cell dimensions, and thermal stability of some heulandite group zeolites. *Amer. Mineral.*, *57*, 1463–1493.

Hawkins, D. B. 1974. Statistical analyses of the zeolites clinoptilolite and heulandite. *Contrib. Mineral. Petrol.*, *45*, 27–36.

Hay, R. L. 1966. Zeolites and zeolitic reactions in sedimentary rocks. *Geol. Soc. Amer. Spec. Paper*, *85*.

Stilbite (Desmine)

$(Ca,Na_2,K_2)(Al_2Si_7)O_{18} \cdot 7H_2O$

MONOCLINIC
$\angle\beta = 129°10'$

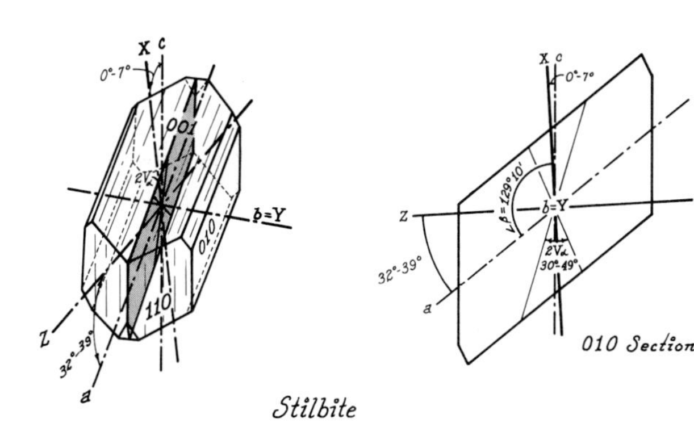

$n_\alpha = 1.482$–1.500*
$n_\beta = 1.491$–1.507
$n_\gamma = 1.493$–1.513
$n_\gamma - n_\alpha = 0.006$–$0.013$
Biaxial negative; $2V_x = 30°$–$49°$
$b = Y, c \wedge X = 0°$ to $-7°$, $a \wedge Z = +39°$ to $+32°$
$v > r$, inclined bisectrix dispersion
Colorless in standard section

COMPOSITION. Considerable variation is possible in the relative amounts of Ca, Na, and K. Sodium tends to be the dominant cation, but K-rich varieties are known and a Ca variety is called *epistilbite*, $Ca(Al_2Si_6)O_{16} \cdot 5H_2O$. Variation in the ratio Al:Si produces silica-poor (*hypostilbite*) and silica-rich (*stellerite*) varieties.

PHYSICAL PROPERTIES. H = $3\frac{1}{2}$–4. Sp. Gr. = 2.1–2.2 Hand sample color is white to yellowish or brownish. Luster is distinctly pearly on {010}.

* Refractive index increases somewhat with calcium content.

COLOR AND PLEOCHROISM. Colorless in section or as fragments.

FORM. Crystals are platy on {010} and elongated on *c*, often arranged as radiated, sheaflike aggregates (Fig. 7-35).

CLEAVAGE. Nearly perfect {010} cleavage ({100} poor) is responsible for layered habit and pearly luster.

BIREFRINGENCE. Maximum birefringence, seen on {010} sections and cleavage plates, is weak (0.006–0.013) and interference colors are gray, white, and possibly first-order yellow.

TWINNING. Stilbite crystals are consistently twinned on {001} as cruciform penetration twins that display pseudotrigonal symmetry.

INTERFERENCE FIGURE. Stilbite is biaxial negative, with small to moderate $2V_x = 30°$ to $49°$. Cleavage plates show flash figures, and near-basal sections show acute bisectrix figures of broad, diffuse isogyres and no isochromes. Dispersion is perceptible $v > r$, with possible inclined bisectrix dispersion.

OPTICAL ORIENTATION. $b = Y$, $c \wedge X = 0°$ to $-7°$, $a \wedge Z = +39°$ to $+32°$. The optic plane is {010}. Extinction is parallel to {010} cleavage in sections perpendicular to {010}, and maximum extinction, seen in {010} sections, is 32° to 39° measured to poor {001} cleavage. Elongation in sections perpendicular to {010} is either positive or negative, as *Y* is normal to cleavage.

DISTINGUISHING FEATURES. Stilbite is a layered zeolite showing twinning, higher birefringence, and opposite sign to heulandite.

ALTERATION. Zeolites tend to be stable at surface conditions but some are known to alter to montmorillonite clays.

OCCURRENCE. Stilbite is deposited by hydrothermal agents in amygdales and fractures in subsiliceous volcanics with heu-

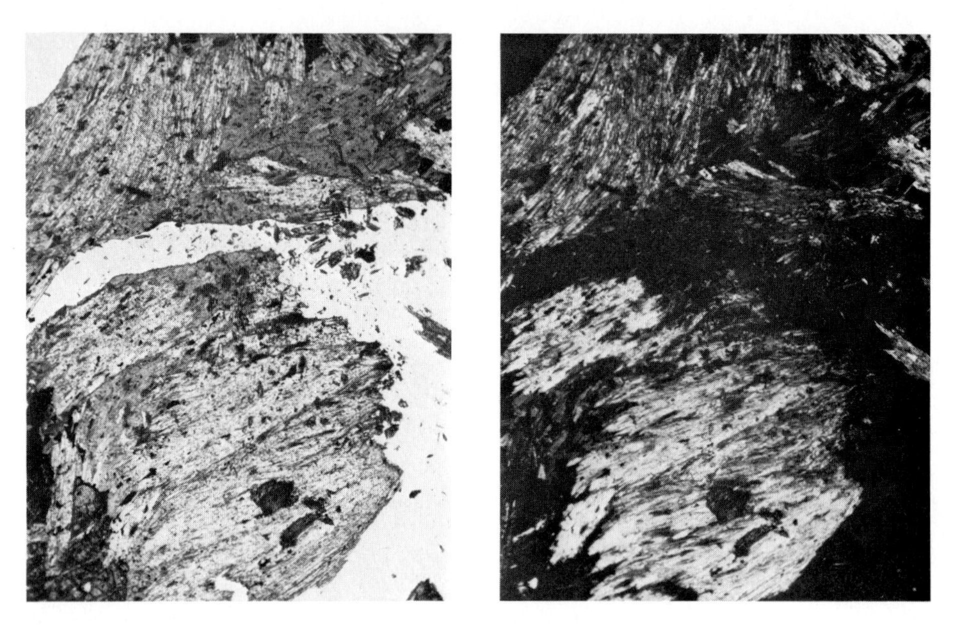

FIGURE 7-35. Stilbite appears as a coarse, fibrous aggregate in thin section.

landite, chabazite, natrolite, and other hydrothermal minerals. It may appear in cavities in almost any igneous rock, hydrothermal vein, and thermal spring deposit. Stilbite appears with epidote, quartz, adularia, and other hydrothermal minerals in cavities and fractures in some schists and detrital sediments.

REFERENCES: STILBITE

Galli, E. 1971. Refinement of the crystal structure of stilbite. *Acta Crystallogr.*, *B27*, 833–841.

Simonot-Grange, M. H., A. Cointot, and M. Lallemant. 1969. Stilbite-metastilbite crystalline transformation. *C. R. Acad. Sci. Paris*, Series C, *269*, 1098–1100.

Slaughter, M. 1970. Crystal structure of stilbite. *Amer. Mineral.*, *55*, 387–397.

Phillipsite

$(Ca,K_2,Na_2)_6[(Al_3Si_5)O_{16}]_2 \cdot 12H_2O$

MONOCLINIC
$\angle \beta = 124°12'$

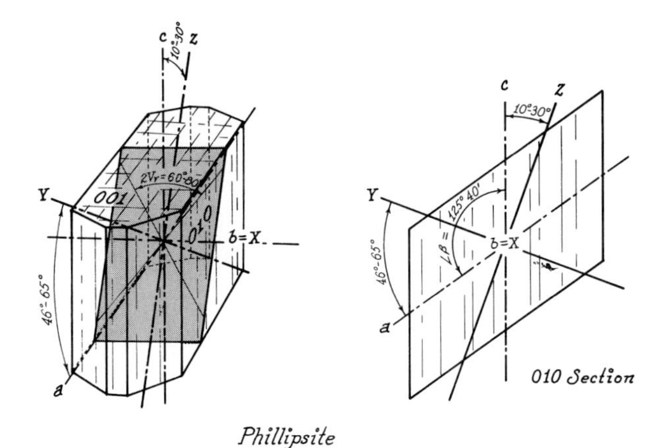

Phillipsite

$n_\alpha = 1.483–1.504$
$n_\beta = 1.484–1.509$
$n_\gamma = 1.486–1.514$
$n_\gamma - n_\alpha = 0.003–0.010$
Biaxial positive, $2V_z = 60°–80°$
$b = X, c \wedge Z = +10°$ to $+30°, a \wedge Y = +46°$ to $+65°$
$v > r$ weak, horizontal bisectrix
Colorless in standard section

COMPOSITION. Phillipsite tends to be a K-rich zeolite, but Ca may be dominant by the substitution $CaAl \rightleftharpoons KSi$, and Na-rich compositions are reported. *Wellsite* is a Ba-rich phillipsite.

PHYSICAL PROPERTIES. H = 4–4½. Sp. Gr. = 2.2. Hand sample color is white, rarely stained yellow or reddish. Luster is vitreous.

COLOR AND PLEOCHROISM. Colorless in section or as fragments.

FORM. Crystals are commonly euhedral as interpenetration twins simulating orthorhombic or tetragonal forms.

CLEAVAGE. Distinct cleavages parallel to {010} and {100} favor fragment elongation on *c*.

BIREFRINGENCE. Maximum birefringence, seen in sections near {100}, is weak (0.003–0.010), and interference colors are gray to white in standard section.

TWINNING. Phillipsite shows complex, multiple penetration twinning on {001}, {110}, and {021}, producing striated faces and crystal forms of pseudoorthorhombic, pseudotetragonal, and even pseudoisometric symmetry.

INTERFERENCE FIGURE. Phillipsite is biaxial positive, with moderate to large $2V_z = 60°–80°$. Acute bisectrix figures, seen in chance sections parallel to *b*, show broad isogyres without isochromes. Cleavage fragments tend to show the obtuse bisectrix. Dispersion is $v > r$, weak, with possible horizontal bisectrix dispersion.

OPTICAL ORIENTATION. $b = X$, $c \wedge Z = +10°$ to $+30°$, $a \wedge Y = +46°$ to $+65°$. The optic plane is perpendicular to {010}, 10° to 30° from {100}. Extinction is parallel to cleavages in sections perpendicular to {010}, and maximum extinction, seen in {010} sections, is 10° to 30° to {100} cleavage. Fragments tend to be elongated on *c* showing parallel or maximum extinction (10°–30°) and positive elongation.

DISTINGUISHING FEATURES. Phillipsite usually shows lower refractive index than harmotome and yields obtuse bisectrix figures on {010} cleavage plates. Harmotome and stilbite show penetration sector twinning, as does phillipsite, but show acute bisectrix and flash figure, respectively, on {010} cleavage plates. Tridymite and cristobalite are rare in basic volcanics.

ALTERATION. Zeolites tend to be stable at surface conditions, but some are known to alter to montmorillonite clays.

OCCURRENCE. Phillipsite is common in amygdales or fractures in mafic volcanics associated with chabazite. It is a silica-poor zeolite and consequently is characteristic of leucite rocks, phonolites, and other silica-deficient igneous rocks. Authigenic phillipsite is reported in thermal spring deposits and abounds in the red clays of deep-sea deposits.

REFERENCES: PHILLIPSITE

Galli, E., and A. G. L. Ghittoni. 1972. The crystal chemistry of phillipsites. *Amer. Mineral.*, *57*, 1125–1145.

Harada, K., S. Iwamoto, and K. Kihara. 1967. Erionite, phillipsite, and gonnardite in the amygdales of altered basalt from Maze, Niigata Prefecture, Japan. *Amer. Mineral.*, *52*, 1785–1794.

Hay, R. L. 1966. Zeolites and zeolitic reactions in sedimentary rocks. *Geol. Soc. Amer., Spec. Paper*, *85*.

Rinaldi, R., J. J. Pluth, and J. V. Smith. 1974. Zeolites of the phillipsite family: Refinement of the crystal structures of phillipsite and harmotome. *Acta Crystallogr.*, *B30*, 2426–2433.

Stonecipher, S. A. 1976. Origin, distribution and diagenesis of phillipsite and clinoptilolite in deep-sea sediments. *Chem. Geol.*, *17*, 307–318.

Harmotome

$Ba(Al_2Si_6)O_{16} \cdot 6H_2O$

MONOCLINIC

$\angle\beta = 125°$

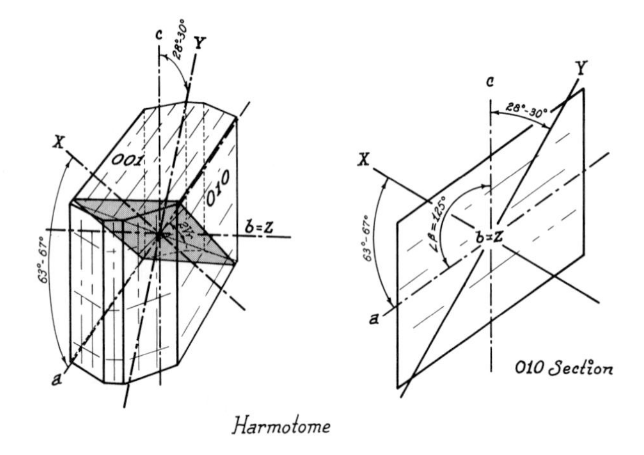

$n_\alpha = 1.503–1.508$

$n_\beta = 1.505–1.509$

$n_\gamma = 1.508–1.514$

$n_\gamma - n_\alpha = 0.005–0.008$

Biaxial positive, $2V_z \simeq 80°$

$b = Z$, $c \wedge Y = +28°$ to $+32°$, $a \wedge X = +63°$ to $+67°$

Crossed bisectrix dispersion weak

Colorless in standard section

COMPOSITION. Harmotome is a rare Ba-zeolite that may contain appreciable alkalis by the substitution $BaAl \rightleftharpoons (K,Na)Si$. Calcium is minor.

PHYSICAL PROPERTIES. H = $4\frac{1}{2}$. Sp. Gr. = 2.4–2.5. Hand sample color is white to yellow, brownish, or reddish. Luster is vitreous.

COLOR AND PLEOCHROISM. Section and fragments are colorless.

FORM. Harmotome appears most commonly as euhedral interpenetration (cruciform) twins, often repeated as fourlings.

CLEAVAGE. Cleavage is distinct on {010} and poor on {001}.

BIREFRINGENCE. Maximum birefringence, seen in unlikely sections between {100} and {001}, is weak, and interference colors are gray to white.

TWINNING. Harmotome, like phillipsite, shows complex, multiple penetration twinning on {001}, {110}, and {021}.

INTERFERENCE FIGURE. Cleavage fragments and {010} sections show acute bisectrix figures of diffuse isogyres on a white field. Dispersion is weak crossed bisectrix, without perceptible optic axis dispersion.

OPTICAL ORIENTATION. $b = Z$, $c \wedge Y = +28°$ to $+32°$, $a \wedge X = +63°$ to $+67°$. The optic plane is perpendicular to {010}. Maximum extinction, seen in {010} sections, is 27° to 32° to poor {001} cleavage. Elongation, seen in sections perpendicular to {010}, is negative (length-fast), as Z is normal to cleavage.

DISTINGUISHING FEATURES. Harmotome tends to be associated with manganese minerals. It usually shows higher refractive index than phillipsite. Penetration twinning is characteristic of harmotome, phillipsite, and stilbite; however, on {010} cleavage plates, harmotome shows an acute bisectrix, phillipsite an obtuse bisectrix, and stilbite a flash figure.

ALTERATION. Zeolites tend to be stable at surface conditions, but some are known to alter to montmorillonite clays.

OCCURRENCE. Harmotome is most commonly deposited by hydrothermal solutions in vesicles and fractures in basalt and silica-poor volcanics. It tends to be associated with manganese mineralization and appears in hydrothermal veins with strontianite, barite, calcite, pyrite, and galena. Harmotome may form in laterites with gibbsite or kaolin clays.

REFERENCES: HARMOTOME

Barrer, R. M., F. W. Bultitude, and I. S. Kerr. 1959. Some properties, and a structural scheme for, the harmotome zeolites. *Jour. Chem. Soc., 1959*, 1521–1528.

Rinaldi, R., J. J. Pluth, and J. V. Smith. 1974. Zeolites of the phillipsite family: Refinement of the crystal structures of phillipsite and harmotome. *Acta Crystallogr., B30*, 2426–2433.

Chabazite $Ca(Al_2Si_4)O_{12} \cdot 6H_2O$ TRIGONAL

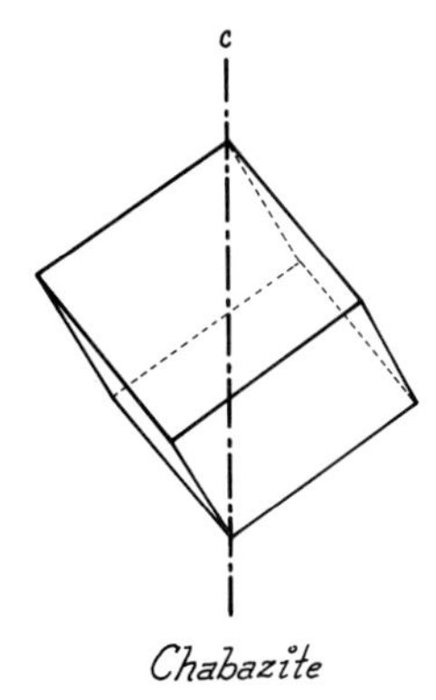

$n_\omega = 1.472–1.494$*
$n_\varepsilon = 1.470–1.485$
$n_\omega - n_\varepsilon = 0.002–0.010$
Uniaxial negative (may show small $2V_x$)
Colorless in standard section

COMPOSITION. Great variation is possible in the relative amounts of Na^+, K^+, and Ca^{2+}, the Al:Si ratio varies with $CaAl \rightleftharpoons (Na,K)Si$.

PHYSICAL PROPERTIES. H = $4\frac{1}{2}$. Sp. Gr. = 2.1. Hand sample color is white or salmon-pink. Vitreous luster.

COLOR AND PLEOCHROISM. Colorless in section or as fragments.

FORM. Chabazite crystals are rhombohedrons closely resembling cubes. Simple penetration and complex twins are common.

CLEAVAGE. Rather good rhombohedral $\{10\bar{1}1\}$ cleavages favor equant fragments.

BIREFRINGENCE. Maximum birefringence, seen in sections parallel to c is weak (0.002–0.010), and interference colors are gray to white.

TWINNING. Simple penetration twins on {0001} are common, and contact twins on $\{10\bar{1}1\}$ are less common.

ZONING. Chabazite is commonly nonuniform in composition and may show zones of different refractive index and even different optic sign.

INTERFERENCE FIGURE. Chabazite is commonly uniaxial negative, but may be positive or even biaxial with small to moderate $2V$. Basal sections may show areas of different orientation, and figures are indistinct, owing to twinning, zoning, and weak birefringence. Cleavage fragments yield off-center figures.

OPTICAL ORIENTATION. Crystals and cleavage fragments tend to be nearly "cubic" rhombohedrons showing symmetrical extinction.

DISTINGUISHING FEATURES. Chabazite is an equigranular (pseudocubic), uniaxial negative zeolite that cannot be distinguished from gmelinite or levynite by optical means unless gmelinite shows basal parting.

ALTERATION. Zeolites tend to be stable at surface conditions, but some are known to alter to montmorillonite clays (for example, beidellite).

* Refractive indices increase with Ca.

OCCURRENCE. Chabazite is a common zeolite, most common in cavities and fractures in basalts and other mafic volcanics in association with calcite and a wide variety of zeolites. It is deposited by hydrothermal agents in thermal springs and hydrothermal veins and is reported in a variety of metamorphic rocks with other zeolites.

REFERENCES: CHABAZITE

Passaglia, E. 1970. Crystal chemistry of chabazites. *Amer. Mineral.*, *55*, 1278–1301.

Smith, J. V., F. Rinaldi, and L. S. D. Glasser. 1963. Crystal structures with a chabazite framework. II. Hydrated Ca chabazite at room temperature. *Acta Crystallogr.*, *16*, 45–53.

Gmelinite

$(Na_2,Ca)(Al_2Si_4)O_{12} \cdot 6H_2O$ TRIGONAL

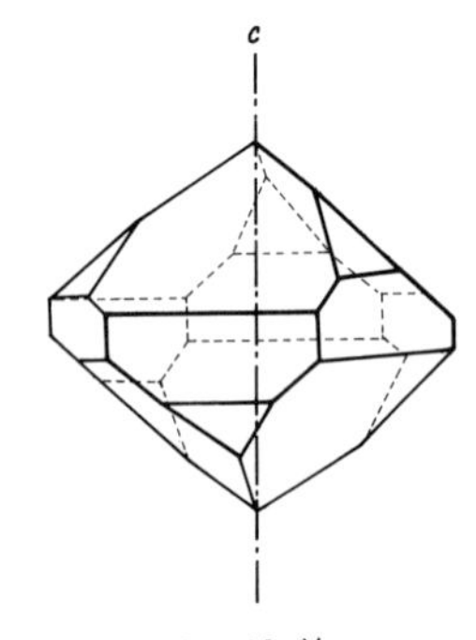

Gmelinite

$n_\omega = 1.476–1.494$
$n_\varepsilon = 1.474–1.480$
$n_\omega - n_\varepsilon = 0.002–0.015$
Uniaxial negative
Colorless in standard section

COMPOSITION. Considerable K^+ may replace Na^+, and some Sr^{2+} may replace Ca^{2+}.

PHYSICAL PROPERTIES. H = $4\frac{1}{2}$. Sp. Gr. = 2.0–2.2. Hand samples, are colorless or white, often stained yellow, brown, reddish, or greenish. Vitreous luster.

COLOR AND PLEOCHROISM. Colorless in section or as fragments.

FORM. Gmelinite crystals are commonly euhedral as rhombohedrons or pyramidal forms resembling quartz crystals; rarely as radiating aggregates.

CLEAVAGE. Distinct rhombohedral $\{10\bar{1}1\}$ cleavages may be accompanied by basal parting $\{0001\}$.

BIREFRINGENCE. Maximum birefringence, seen in sections parallel to *c*, is weak (0.002–0.015) and interference colors, usually gray to white, may range to first-order yellow.

TWINNING. Penetration twins on $\{0001\}$ are reported, and twinning on $\{10\bar{1}1\}$ may also exist.

INTERFERENCE FIGURE. Gmelinite is normally uniaxial negative but may be positive or even biaxial, with small to moderate $2V$. Rhombohedral cleavage, low birefringence, and twinning allow few usable figures.

OPTICAL ORIENTATION. Crystals and cleavage fragments tend to be rhombohedrons showing inclined or symmetrical extinction.

DISTINGUISHING FEATURES. Gmelinite is an equigranular (pseudocubic), uniaxial negative zeolite. It is distinguished from chabazite only by the basal parting of gmelinite.

ALTERATION. Zeolites tend to be stable at surface conditions, but some are known to alter to montmorillonite clays (such as beidellite).

OCCURRENCE. Gmelinite is associated with a wide variety of other zeolites in amygdales and fissures in basalts and similar volcanics.

REFERENCES: GMELINITE

Fedorova, G. M., and T. N. Shishakova. 1967. Structure of gmelinite. *Izv. Akad. Nauk. SSSR, Ser. Khim*, *1967*, 1856–1857.

Fischer, K. 1966. Crystal structure of gmelinite. *Neues Jahrb. Mineral., Monatsh.*, *1966*, 1–13.

OTHER TEKTOSILICATES

Scapolite

$(Ca,Na)_4[(Al,Si)_3Al_3Si_6O_{24}](Cl,CO_3)$ TETRAGONAL

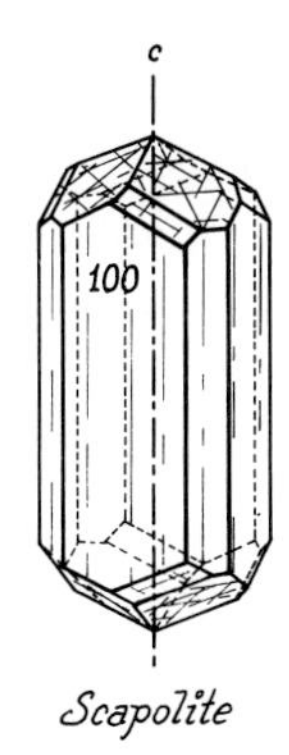

$n_\omega = 1.540–1.600$*
$n_\varepsilon = 1.535–1.565$
$n_\omega - n_\varepsilon = 0.004–0.037$*
Uniaxial negative
Colorless in standard section

COMPOSITION AND STRUCTURE. Scapolite is basically a solid solution series between $Na_4(Al_3Si_9O_{24})Cl$(*marialite*, Ma) and $Ca_4(Al_6Si_6O_{24})CO_3$(*meionite*, Me), divided as follows:

Marialite	$Me_0–Me_{20}$
Dipyre	$Me_{20}–Me_{50}$
Mizzonite	$Me_{50}–Me_{80}$
Meionite	$Me_{80}–Me_{100}$

Pure end-member compositions are unknown in nature, and compositions in the marialite or meionite ranges are rare.

* Indices of refraction and birefringence increase rapidly from marialite toward meionite. K^+, SO_4^{2-}, and $(OH)^-$ lower indices slightly. See Fig. 7-36.

Scapolite contains fourfold rings of silicate tetrahedra, in the basal plane, linked into columns along *c*, with columns joined together by other tetrahedra. Smaller structural cavities are filled by Na^+ or Ca^{2+}, and larger openings by Cl^- or CO_3^{2-} anions. In the range Ca:(Ca + Na) = 0–0.75, composition varies from pure marialite to $NaCa_3Al_5Si_7O_{24}CO_3$ by substitution of $Ca_3Al_2CO_3$ for Na_3Si_2Cl. At Ca:(Ca + Na) = 0.75 anion substitution is complete, and, in the range

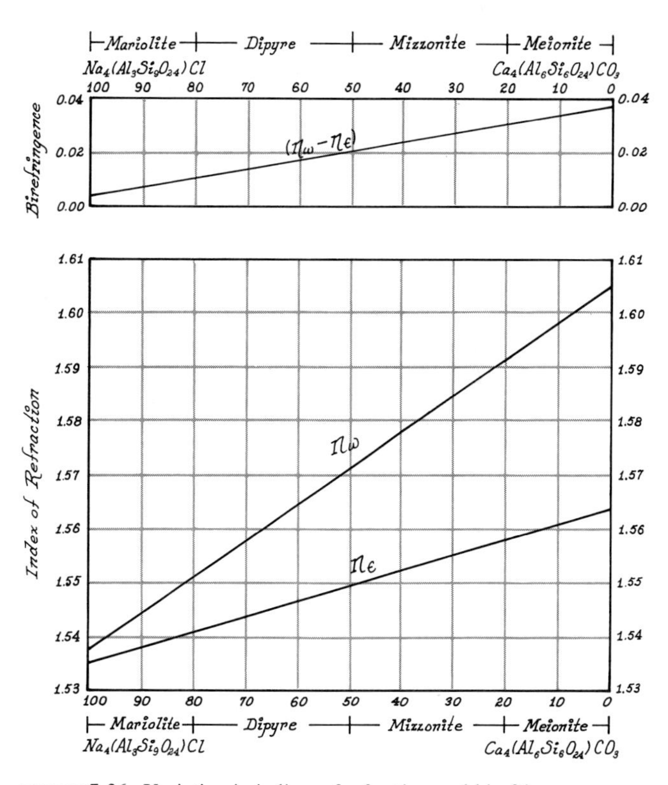

FIGURE 7-36. Variation in indices of refraction and birefringence in the scapolite minerals. (Data from Ulbrich, 1973)

Ca:(Ca + Na) = 0.75–1.00, composition changes from $NaCa_3Al_5Si_7O_{24}CO_3$ to $Ca_4Al_6Si_6O_{24}CO_3$ by replacement of NaSi by CaAl.

PHYSICAL PROPERTIES. H = 5–6. Sp. Gr. = 2.5–2.8 (increasing toward meionite). In hand sample, scapolite is colorless or white, ranging to pale shades of gray, yellow, green, violet, and so on.

COLOR AND PLEOCHROISM. Scapolite is colorless in section or as fragments.

FORM. Prismatic crystals, elongated on *c*, range from acicular aggregates to large euhedral crystals. Large crystals commonly show spongy, sieve structure. Granular masses also occur.

CLEAVAGE. Four directions of prismatic cleavage, {100} and {110}, are imperfect and intersect at 90° and 45°. Cleavage fragments are elongated on *c*.

BIREFRINGENCE. Maximum birefringence is very weak (0.004) for marialite ranging to moderately strong (0.037) at the ideal meionite composition. Most scapolite compositions show birefringence between 0.01 and 0.03 and interference colors from low first order to middle second order. Maximum birefringence is seen in prismatic sections and fragments on cleavages.

TWINNING AND ZONING. Twinning is not reported, but composition zoning is to be expected. Me-rich outer zones show higher birefringence and reflect $Ca\text{-}CO_3$ enrichment with crystal growth. Scapolite sometimes forms graphic intergrowths with quartz, microcline, or diopside.

INTERFERENCE FIGURE. Scapolite is uniaxial negative, and basal sections yield an optic axis figure, which may show slight, anomalous isogyre separation ($2V_x < 10°$). Isochromatic rings may be absent to moderately abundant, depending on composition and thickness. Fragments on prismatic cleavages yield only flash figures.

OPTICAL ORIENTATION. Prismatic crystals and cleavage fragments are elongated on *c* and show parallel extinction and negative elongation (length-fast).

DISTINGUISHING FEATURES. Scapolite near marialite shows low relief and low birefringence and may be mistaken for quartz, feldspar, nepheline, cordierite, or andalusite. Quartz is uniaxial positive without cleavage, plagioclase and cordierite are biaxial and commonly show polysynthetic twinning, K-feldspar is biaxial with negative relief, andalusite is biaxial with high relief, and nepheline usually shows slightly lower refraction indices and restricted occurrence. Cancrinite shows lower indices of refraction and perfect cleavage.

ALTERATION. Scapolite commonly replaces plagioclase and may also be replaced by it. Scapolite may alter to fine aggregates containing some combination of the minerals calcite, chlorite, sericite, epidote, plagioclase, idocrase, zeolites, or clays (kaolinite or montmorillonite).

OCCURRENCE. Scapolite shows a remarkably wide range of occurrences but is most characteristic of metamorphic and metasomatic environments and is most often related to carbonate sediments.

In calcareous rocks of regional metamorphism, scapolite may form from the greenschist to granulite or sanidinite facies. It is especially characteristic of marbles and the amphibolite facies. Scapolite may be a primary mineral of regional metamorphism or may represent regional metasomatism. Its common associates are calcite, sphene, hornblende, diopside, plagioclase, garnet, epidote, and apatite.

By contact metasomatism, scapolite is formed in carbonate sediments through the introduction of Na and Cl from hydrothermal sources. It may also form in contact zones with amphibolites, metagabbros, and other mafic rocks. Grossular, wollastonite, diopside, fluorite, and albite are common associates.

In pegmatites, scapolite may replace plagioclase or quartz to appear in association with apatite, fluorite, calcite, allanite, albite, microcline, and quartz. Scapolite aplites represent limestone assimilation.

With nepheline, scapolite appears in nepheline syenites and corundum nepheline syenites.

REFERENCES: SCAPOLITE

Burley, B. J., E. B. Freeman, and D. M. Shaw. 1961. Studies on scapolite, *Can. Mineral.*, *6*, 670.

Eugster, H. P., and H. J. Prostka. 1960. Synthetic scapolites (abstr.). *Bull. Geol. Soc. Amer.*, *71*, 1859.

Evans, B. W., D. M. Shaw, and D. R. Haughton. 1969. Scapolite stoichiometry. *Contrib. Mineral. Petrol.*, *24*, 293–305.

Ingamells, C. O., and J. Gittins. 1967. The stoichiometry of scapolite. *Can. Mineral.*, *9*, 214–236.

Phakey, P. P., and S. Ghose. 1972. Scapolite: Antiphase domain structure. *Nature. Phys. Sci.*, *238*, 78–80.

Shaw, D. M. 1960. The geochemistry of scapolite. Part I. Previous work and general mineralogy. Part II. Trace elements, petrology and general geochemistry. *Jour. Petrol.*, *1*, 218–261.

Ulbrich, H. H. 1973. Crystallographic data and refractive indices of scapolites. *Amer. Mineral.*, *58*, 81–92.

White, A. J. B. 1959. Scapolite-bearing marbles and calc-silicate rocks from Tungkillo and Milendella, South Australia. *Geol. Mag.*, *96*, 285.

Petalite

$Li(AlSi_4)O_{10}$

MONOCLINIC
$\angle\beta = 112°44'$

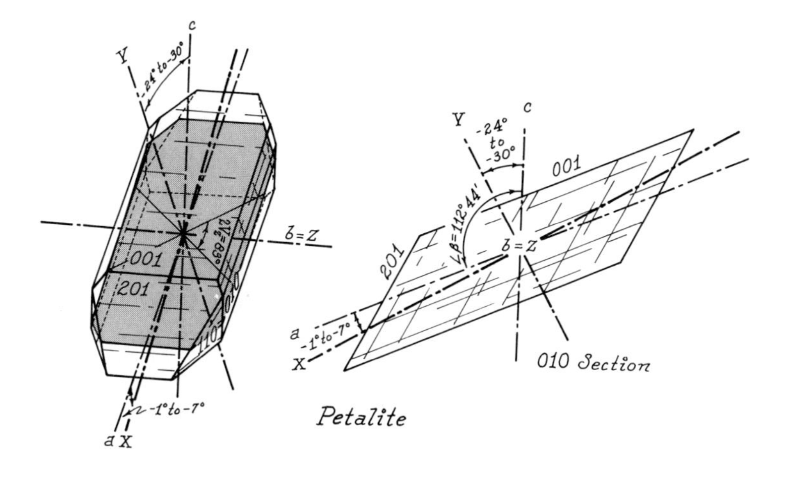

$n_\alpha = 1.504–1.507$
$n_\beta = 1.510–1.513$
$n_\gamma = 1.516–1.523$
$n_\gamma - n_\alpha = 0.011–0.017$
Biaxial positive, $2V_z = 83°$
$b = Z, c \wedge Y = -24°$ to $-30°, a \wedge X = -1°$ to $-7°$
$r > v$ weak, crossed bisectrix dispersion
Colorless in standard section

COMPOSITION AND STRUCTURE. Petalite is a framework silicate in which one of five Si^{4+} is replaced by Al^{3+}. Sheets of SiO_4 tetrahedra parallel to {001} are linked by sheets of AlO_4 tetrahedra to account for perfect basal cleavage. Lithium ions lie in a distorted tetrahedral coordination. Minor Na^+ may replace Li^+, and small amounts of Fe^{3+} are reported in some analyses.

PHYSICAL PROPERTIES. H = $6\frac{1}{2}$. Sp. Gr. = 2.41–2.42. Color in hand sample is white or pale gray; rarely pink or green.

COLOR AND PLEOCHROISM. Petalite is colorless in section or as fragments.

FORM. Crystals are often large and blocky, elongated somewhat on a; sometimes fibrous.

CLEAVAGE. Perfect basal cleavage {001} and distinct cleavage on {201} yield fragments elongated on b. (001) $\wedge$ (201) = $38\frac{1}{2}°$.

BIREFRINGENCE. Maximum birefringence, seen essentially on basal sections and cleavage plates, is mild ($n_\gamma - n_\alpha = 0.011$–0.017). Interference colors, in standard section, are gray and white, ranging to first-order yellow for basal sections.

TWINNING. Lamellar twinning on {001} is common.

INTERFERENCE FIGURE. Sections parallel to {010} yield acute bisectrix figures showing a white field and broad isogyres, which leave the visible field with stage rotation. Dispersion is weak $r > v$, with weak crossed bisectrix dispersion.

OPTICAL ORIENTATION. Petalite is monoclinic, with $b = Z$, $c \wedge Y = -24°$ to $-30°$, $a \wedge X = -1°$ to $-7°$. Maximum extinction, seen in {010} sections, is almost parallel (1° to 7°) to basal cleavage and about 35° to {201}. The fast ray is nearly parallel to crystal elongation a and to basal cleavage (negative elongation). Cleavage fragments elongated on b show parallel extinction and positive elongation (length-slow).

DISTINGUISHING FEATURES. Petalite resembles feldspars or quartz in thin section, and these minerals are likely associates of petalite in pegmatites. Quartz is uniaxial positive, with positive relief and neither cleavage nor polysynthetic twinning. All alkali feldspars have higher indices of refraction and slightly less birefringence than petalite, and their cleavages intersect at near 90°.

ALTERATION. Petalite normally alters to montmorillonite clays (hectorite); it may, however, also yield zeolites (heulandite or stilbite).

OCCURRENCE. Petalite is found in granite pegmatites associated with quartz, microcline, lepidolite, amblygonite, topaz, apatite, tourmaline, pollucite, and especially with spodumene.

REFERENCE: PETALITE

Zemann-Hedlik, A., and J. Zemann. 1955. The structure of petalite, $LiAlSi_4O_{10}$. *Acta Crystallogr.*, *8*, 781–787.

PART II

Optical and Physical Constants of the Nonopaque Minerals

Tables of Optical and Physical Constants of the Nonopaque Minerals

The nonopaque minerals are separated in the following tables as:

Isotropic
Uniaxial positive
Uniaxial negative
Biaxial positive
Biaxial negative

Minerals that show very small birefringence are listed both in the appropriate anisotropic group and as isotropic minerals; those with very small $2V$ are listed in both uniaxial and biaxial groups, and those with very large $2V$ are listed in both positive and negative biaxial groups.

Within each of these five groups, minerals are arranged in order of ascending index of refraction (n_ω for uniaxial minerals and n_β for biaxial minerals). Many mineral varieties show wide variation in their indices of refraction. Such a mineral is placed in the list at the mean value of its principal index, and the variation is shown in the left column. If the known variation exceeds 0.01, the mineral is designated by an asterisk (*) and, if the variation exceeds 0.10, by a double asterisk (**). When trying to assign a measured refraction index to a specific mineral in the tables, the user should be aware that his measured index may match that of a double-asterisk mineral several pages away from other minerals having that index of refraction.

Asterisks also tend to attract attention to the more common minerals. Refraction indices of many less well-known minerals undoubtedly have similar ranges; however, the extent of this variation has not been determined in most cases.

A few simple abbreviations are used in the tables, as follows:

acic—acicular
agg—aggregate
alter—alteration
anom—anomalous
biref—birefringence
bl—blue
blk—black
brn—brown
column—columnar
decomp—decomposed
diag—diagonal
dimorph—dimorphous
disp—dispersion
dist—distinct
dk—dark
effer—effervescent
elong—elongation
elong (*c*)—elongation on *c*-axis
+ elong—positive elongation
evap—evaporite
||ext—parallel extinction
fib—fibrous
fluor. U.V.—fluorescence in ultraviolet.
fluor. short λ-U.V.—fluorescence in short-wave ultraviolet
gelat—gelatinous
gran—granular
grn—green
hex—hexagonal
high T—high temperature
horiz—horizontal
imperf—imperfect
interfer—interference
low P—low pressure
lt—light
mass—massive
micro—microscopic
mono—monoclinic
ortho—orthorhombic
perc—perceptible
perf—perfect
prism—prismatic
pyram—pyramidal
rad—radiated
rhomb—rhombohedral
sed—sedimentary
sol—soluble
tab—tabular
U-deposits—uranium deposits
var—variation
xls—crystals
xline—crystalline
yel—yellow

One literature reference, in English, is given for each mineral, using the following abbreviations:

Dana I (II or III)—C. Palache, H. Berman, and C. Frondel, 1962, *Dana's System of Mineralogy*, 7th ed., Vol. I (II, III) (New York: Wiley).

D.H.Z *1* (*2*, *3*, *4*, *5*)—W. A. Deer, R. A. Howie, and J. Zussman, 1963, *Rock-Forming Minerals*, Vol. 1 (2, 3, 4, 5) (New York: Wiley).

Vlasov II—K. A. Vlasov, *Mineralogy of Rare Elements*, Vol. 2. Israel Program for Scientific Translations, 1966. (New York: Davey).

Win. and Win. II—A. N. Winchell and H. Winchell, 1951, *Elements of Optical Mineralogy*, Part II: *Descriptions of Minerals*, 4th ed. (New York: Wiley).

Those mineral varieties described in detail in this volume are referenced only by page number.

General References

Deer, W. A., R. A. Howie, and J. Zussman. 1963. *Rock-Forming Minerals*. New York: Wiley.
- Vol. 1: *Ortho- and Ring Silicates*
- Vol. 2: *Chain Silicates*
- Vol. 3: *Sheet Silicates*
- Vol. 4: *Framework Silicates*
- Vol. 5: *Nonsilicates*

Heinrich, W. E. 1965. *Microscopic Identification of Minerals*. New York: McGraw-Hill.

Larsen, E. S., and Berman, H. 1934. *The Microscopic Determination of the Nonopaque Minerals*. U.S. Geol. Survey Bulletin 848. Washington, D.C.: U.S. Gov. Printing Office.

Palache, C., H. Berman, and C. Frondel. *Dana's System of Mineralogy*. 7th ed. New York: Wiley.
- Vol. I (1944): *Elements, Sulfides, Sulfosalts, Oxides*
- Vol. II (1951): *Halides, Nitrates, Borates, Carbonates, Sulfates, Phosphates, Arsenates, Tungstates, Molybdates, etc.*
- Vol. III (1962): *Silica Minerals*

Roberts, W. L., G. R. Rapp, Jr., and J. Weber. 1974. *Encyclopedia of Minerals*. New York: Van Nostrand Reinhold.

Tröger, W. E. 1959. *Optische Bestimmung der Gesteinsbildenden Minerale. Teil 1: Bestimmungstabellen*. Stuttgart: E. Schweizerbart'sche Verlagsbuchhandlung.

Vlasov, K. A., ed. 1966. *Geochemistry and Mineralogy of Rare Elements and Genetic Types of Their Deposits*. Vol. 2: *Mineralogy of Rare Elements*. Israel Program for Scientific Translations. New York: Davey.

Winchell, A. N., and H. Winchell. 1951. *Elements of Optical Mineralogy*. Part II: *Descriptions of Minerals*. New York: Wiley.

Winchell, A. N., and H. Winchell. 1964. *The Microscopical Characteristics of Artificial Inorganic Solid Substances: Optical Properties of Artificial Minerals*. New York: Academic Press.

Isotropic Minerals

n Var.	Refractive Index	Mineral Name and Composition	Crystal System	Habit	Cleavage	Twinning
1.3270	n 1.3270	Villiaumite NaF	Isometric	Tiny xls, massive, granular	{100} perf	
1.340	n 1.340	Cryolithionite $Li_3Na_3Al_2F_{12}$	Isometric	Dodecahedral	{011} dist	
1.340	n 1.340	Hieratite K_2SiF_6	Isometric	Stalactitic concretions. Octahedral	{111} perf	
1.352 1.362	n 1.352-1.362	Carobbiite KF	Isometric	Cubic	{100} perf (cubic)	
1.369	n 1.369	Cryptohalite $(NH_4)_2SiF_6$	Isometric	Oct, massive crusts	{111} perf	
1.376	n 1.376	Elpasolite K_2NaAlF_6	Isometric	Massive	None	
1.339	n 1.399	Ralstonite (Hagemannite) $NaMgAl(F,OH)_6 \cdot H_2O$	Isometric	Octahedral, cubic	{111} imperf	Sector (birefringent)
1.425	n 1.425	Melanophlogite SiO_2	Tetragonal-Isometric	Cubic xls, thin crusts	None	
1.433 1.435	n 1.433-1.435	Fluorite (Fluorspar) CaF_2	Isometric	Cubic xls, granular, earthy	{111} perf	{111} penetration, contact
1.438	n 1.4388	Soda Alum $NaAl(SO_4)_2 \cdot 12H_2O$	Isometric	Octahedral	None	
1.440	n 1.440	Chukhrovite $Ca_3(Y,Ce)Al_2(SO_4)F_{13} \cdot 10H_2O$	Isometric	Cubic, octahedral	{111} dist	
*1.435 1.460	n 1.435-1.460	Opal $SiO_2 \cdot nH_2O$	Amorphous	Colliform structures	None	None
1.45 1.46	n 1.45-1.46	Bobkovite $(K,Ca,Mg,Fe)_{0.5}(Si_{29}Al)O_{60}$	Isometric	Massive		
1.455	n 1.455	Sulphohalite (Sulfohalite) $Na_6(SO_4)_2FCl$	Isometric	Dodecahedral, cubo-octahedral	None	
1.456	n 1.456	Alum (Potash) $KAl(SO_4)_2 \cdot 12H_2O$	Isometric	Massive, stalactitic, powdery	{111} poor	{111} very rare
1.457 1.459	n 1.457-1.459	Alum (Ammonium)(Tschermigite) $NH_4Al(SO_4)_2 \cdot 12H_2O$	Isometric	Octahedral crystals, fibrous	None	
1.460	n 1.460	Natrophosphate $Na_6H(PO_4)_2F \cdot 17H_2O$	Isometric	Massive	{111} imperf	
1.473	n 1.473	Paulingite $\sim K_2(Ca,Ba)_{1.3}(Si,Al)_{12}O_{24} \cdot 14H_2O$	Isometric	Dodecahedral	Poor	
1.475	n 1.475	Frankdicksonite BaF_2	Isometric	Cubic	{111} perf	
1.48	n ~1.48	Faujasite $(Na_2,Ca)Al_2Si_4O_{12} \cdot 6H_2O$	Isometric	Octahedral	{111} dist	{111} common
*1.47 1.49	n 1.47-1.49	Allophane $Al_2SiO_5 \cdot nH_2O$	Probably amorphous	Massive glassy crusts	None	None
1.483 1.487	n 1.483-1.487	Sodalite $Na_8AL_6Si_6O_{24}Cl_2$	Isometric	Dodecahedral	{110} poor	{111} contact, penetration, repeated
1.487	n 1.487	Hackmanite $Na_8Al_6Si_6O_{24}(Cl,S)_2$	Isometric	Dodecahedral	{110} poor	

* n variation exceeds 0.01

Color In Thin Section	Hardness Sp Gravity	Color Hand Sample	Alteration	Occurrence	Remarks	Reference
O=pink, E=yellow. Strong anomalous pleochroism	2-2½ 2.79-2.81	Carmine red, pink to lt orange		Alkalic intrusions in cavities. Saline lake deposits	Sol. H_2O. Weak anomalous birefringence	Am Min 55, 126-134
Colorless	2½-3 2.77	Colorless, white		Pegmatites with cryolite		Dana II, 99-100
Colorless	2½ 2.67	Colorless, white		Volcanic sublimate with sulfur, alum, realgar	Sol H_2O	Dana II, 103-104
Colorless	~2 2.52	Colorless		Cavities in lavas	Sol H_2O	Am Min 42, 117
Colorless	2½ 2.01-2.03	Colorless, white, gray		Sublimate from fumeroles, burning coal	Dimorph bararite. Sol H_2O, saline	Dana II, 104-105
Colorless	2½ 2.99-3.02	Colorless		Cavities in cryolite pegmatites, with pachnolite		Dana II, 114
Colorless	4½ 2.56-2.62	Colorless, white		Cavities in cryolite pegmatites with pachnolite, thomsenolite		Am Min 50, 1851-1864
Colorless	6½-7 2.005-2.052	Colorless, yellow, red-brown		With lussatite, sulfur in veins. Overgrowth on quartz		Am Min 57, 779-796
Colorless, violet, pale tints (often zoned)	4 3.180	Colorless, violet, green, yel etc.	Many minerals replace fluorite (e.g., carbonates)	Hydrothermal veins with ore minerals. Contact zones, pegmatites		p 13
Colorless	~3 1.68	Colorless		May occur in tunnels or springs. May not occur in nature	Sol H_2O	Dana II, 474
Colorless	~3 2.27-2.40	Colorless, white		Oxide zones Mo-W-deposits		Am Min 45, 1132-1133
Colorless	5½-6½ 1.99-2.25	Colorless, white, all tints	Devitrifies to chalcedony	Secreted by organisms. Hydrothermal & groundwater deposition	Seldom older than Tertiary	p 326
Colorless	2.238	Grayish white		Hydrothermal veins and metasomatic zones in leached carbonates		Am Min 42, 440
Colorless	3½ 2.505	Colorless, gray, lt. green-yellow		Lake salts with hanksite	Sol H_2O. Taste slightly saline	Dana II, 548-549
Colorless	2-2½ 1.75-1.76	Colorless, white		Efflorescence in clays, coal beds, fumeroles	Sol H_2O; astringent taste	Dana II, 472-474
Colorless	1½ 1.642-1.645	Colorless, white		In lignite coals, bituminous shales. Fumeroles	Sol H_2O; astringent taste	Dana II, 475-476
Colorless	2½ 1.71-1.72	Colorless	Powdery coating	Cavities in central pegmatite zone with aegirine, albite, natrolite	Fluor. U.V.-weak orange	Int Geol Rev 14, 984-989
Colorless	~5 ?	Colorless		Vesicles in basalt	Zeolite group	Am Min 45, 79-91
Colorless	2½ 4.89	Colorless		Hydrothermal mineral in quartz veinlets	Isomorph fluorite	Am Min 59, 885-888
Colorless	5 1.92	Colorless, white		Mafic volcanics with other zeolites	Zeolite group	DHZ 4, 392, 397
Colorless	2-3 1.85-1.89	Colorless, white, bluish, grn, etc.	Alumohydrocalcite	Veins and cavities in ore veins and coal beds	Kaolinite group	Win & Win II, 531
Colorless to pale blue	5½-6 2.14-2.4	Blue, white, yel, green, reddish	Zeolites, kaolin clays, diaspore, gibbsite, sericite, cancrinite	Nepheline-syenites with cancrinite, alkali feldspars. Phonolites	Feldspathoid (Sodalite-group)	p. 369
Colorless	5 2.3	White, pink, red-violet	Zeolite, sericite	Alkaline igneous rocks	Feldspathoid. Sodalite group	p. 369

Isotropic

n Var.	Refractive Index	Mineral Name and Composition	Crystal System	Habit	Cleavage	Twinning
* 1.479 1.493	n 1.479-1.493	Analcime (Analcite) $Na(AlSi_2)O_6 \cdot H_2O$	Isometric	Trapezohedral xls as crusts and veins	{100} poor	{100} or {110} polysynthetic
1.49	n~1.49	Rhyolitic glass	Amorphous	Massive, perlitic, pumaceous	None	None
1.485 1.495	n 1.485-1.495	Noselite (Nosean) $Na_8Al_6Si_6O_{24}SO_4$	Isometric	Dodecahedral xls, massive	{110} indist	{111} contact, multiple
1.4903	n 1.4903	Sylvite KCl	Isometric	Cubic or octahedral xls, granular	{100} perf	{111} possible (?)
1.495	n 1.495	Beryllosodalite (Tugtupite) $Na_4BeAlSi_4O_{12}Cl$	Isometric	Cryptocrystalline	{111} dist	
1.498	n 1.498	Zaherite $Al_{12}(SO_4)_5(OH)_{26} \cdot 20H_2O$	Poorly crystalline	Very fine-grained massive	Basal? good	
1.50	n~1.50	Lazurite (Lapis Lazuli) $(Na,Ca)_8(Al,Si)_{12}O_{24}(S,SO_4)$	Isometric	Dodecahedral, massive	{110} imperf	
1.496 1.505	n 1.496-1.505	Haüynite (Haüyne) $(Na,Ca)_{8-4}Al_6Si_6O_{24}(SO_4)_{1-2}$	Isometric	Dodecahedral, octahedral	{110} poor	{111} contact, penetration, repeated
1.50 1.51	n 1.50-1.51	Bolivarite $Al_2(PO_4)(OH)_3 \cdot 4\text{-}5H_2O$	Amorphous	Crusts, botryoidal	None	
*1.508 1.511	n 1.508-1.511	Leucite $KAlSi_2O_6$	Tetragonal (Pseudoisometric)	Trapezohedral xls	{110} very poor	Polysynthetic-several sets
1.51	n~1.51	Trachytic glass	Amorphous	Massive	None	None
1.51	n 1.51	Motukoreaite $NaMg_{19}Al_{12}(CO_3)_{6.5}(SO_4)_4(OH)_{54} \cdot 28H_2O$	Hexagonal	Clay-like		
1.510	n 1.510	Tychite $Na_6Mg_2SO_4(CO_3)_4$	Isometric	Minute octahedrons	None	
1.513	n 1.513	Cadwaladerite $Al(OH)_2Cl \cdot 4H_2O$	Isometric	Small masses, grains	None	
1.520	n 1.520	Pollucite $Cs_{1-x}Na_xAlSi_2O_6 \cdot xH_2O$ x~0.3	Isometric	Cubic or dodec xls, massive	None	
1.52	n~1.52	Andesitic glass	Amorphous	Massive	None	None
1.53	n 1.53	Viséite $NaCa_5Al_{10}Si_3P_5O_{30}(OH,F)_{18} \cdot 16H_2O$	Isometric or Pseudoisometric	Massive, minute nodes	None	
*1.526 1.542	n 1.526-1.542	Halloysite $Al_2Si_2O_5(OH)_4 \cdot 2H_2O$	Monoclinic	Uniform massive, shrinkage cracks	{001} perf	None visible
1.533 1.536	n 1.533-1.536	Langbeinite $K_2Mg_2(SO_4)_3$	Isometric	Massive, bedded, nodular	None	

* n variation exceeds 0.01

Color In Thin Section	Hardness Sp Gravity	Color Hand Sample	Alteration	Occurrence	Remarks	Reference
Colorless	5-5½ 2.22-2.29	Colorless, white		Si-poor igneous rocks. Low-T hydrothermal veins & vugs. Alter K-feldspar	Feldspathoid & Zeolite group	p. 378
Colorless, gray, lt yellow or brown	~7 2.32-2.35	Gray, brown, black, dk green	Devitrifies to fine aggregate of quartz and feldspar	Acid volcanic flows and tuffs		p. 328
Colorless to pale blue	5½ 2.20-2.40	Colorless, white, blue, gray, reddish	Zeolites, kaolin clays, sericite, gibbsite, diaspore, cancrinite	Phonolites & similar volcanics with nepheline, leucite and sanidine	Feldspathoid (Sodalite-group)	p. 369
Colorless	2 1.993	Colorless, reddish, bluish		Bedded evaporite deposits with halite & other evaporite salts	Highly sol H_2O. Saline-bitter taste	p. 12
Colorless	~4 2.28	White, pink, blue, green		Alkaline pegmatite alter	Strong red fluor U.V.	Am Min 46, 241
Colorless	3½ 2.01	White	Dehydrates quickly	Veinlets in kaolinite-boehmite rock	Low birefringence (0.001)	Am Min 62, 1125-1128
Blue	5-5½ 2.38-2.45	Dk blue, violet-blue, grn-blue		Contact limestone deposits with calcite, pyrite	Sodalite group	Win & Win II, 350-351
Colorless to pale blue	5½-6 2.44-2.50	Ultramarine blue, gray, green	Zeolites, kaolin clays, sericite, diaspore, gibbsite, cancrinite	Phonolites or haüynite basalts with nepheline, melilite, melanite	Feldspathoid (Sodalite-group)	p. 369
Colorless	2½-3½ 1.97-2.05	Yellowish-green		Weather zone pegmatite, granite	Opt (-) above 1050°C. Strong fluor U.V. green	Min Mag 38, 418-423
Colorless	5½-6 2.47-2.50	Colorless, pale gray	Pseudoleucite (fine aggregate of nepheline & K-feldspar), analcite	K-rich, Si- and Na-poor volcanic flows and tuffs	Feldspathoid	p. 367
Colorless	~6½ ~2.56	Gray, brown, black, etc.	Devitrification to fine-grained feldspar	Volcanic flows and tuffs		p. 328
Colorless	1-1½ 1.48-1.53	White		Fractures in basaltic tuff with quartz, calcite & goethite		Min Mag 41, 389-390
Colorless	3½-4 2.55-1.59	Colorless		Muds in saline lakes with northupite, gaylussite, thenardite	Sol dilute acids	Am Min 54, 302-305
Colorless	? 1.66	Lemon yellow		Evaporites		Dana II, 77
Colorless	6½-7 2.94	Colorless, white, pink, bluish		Granite pegmatites with lepidolite, spodumene, amblygonite	Zeolite group	Am Min 52, 1515-1518
Colorless, pale yellow or brown	~6 ~2.6	Gray, brown, black	Devitrification	Volcanic flows		p. 328
Colorless	3-4 2.17-2.2	White, yellowish, bluish		With delvauxite	Zeolite group	DHZ 4, 342, 345, 408
Colorless-opaque	2-3 2.5-2.7	White	Diaspore, gibbsite	Hydrothermal alter Al-silicates, with kaolinite, alunite, sulfides	Tubular forms (electron microscope)	p. 254
Colorless	3½-4 2.83-2.85	Colorless, pale colors	Picromerite, epsomite	Bedded salt deposits of marine origin	Slowly sol H_2O	Dana II, 434-435

Isotropic

n Var.	Refractive Index	Mineral Name and Composition	Crystal System	Habit	Cleavage	Twinning
1.542	n 1.542	Apophyllite $KCa_4(Si_4O_{10})F \cdot 8H_2O$	Tetragonal	Square prism xls, platy	{001} perf	{111} rare, penetration
1.544	n 1.544	Halite NaCl	Isometric	Cubic or octahedral xls, granular	{100} perf	{111} rare
1.55	n~1.55	Leucite Tephrite glass	Amorphous	Massive	None	None
*1.549 1.561	n 1.549-1.561	Meta-halloysite $Al_2Si_2O_5(OH)_4$	Monoclinic	Uniform massive, shrink-age cracks	{001} perf	Not visible
*1.55 1.59	n 1.55-1.59	Antigorite $Mg_6Si_4O_{10}(OH)_8$	Monoclinic	Massive, fine scaly, micaceous	{001} perf {010} & {100} dist	Rare
1.5714	n 1.5714	Nitrobarite $Ba(NO_3)_2$	Isometric	Octahedral	None	{111}
1.572	n 1.572	Manganolangbeinite $K_2Mn_2(SO_4)_3$	Isometric	Small tetrahedra		
*1.56 1.60	n 1.56-1.60	Penninite (Pennine) $(Mg,Al)_6(Si,Al)_4O_{10}(OH)_8$	Monoclinic	Micaceous, hex-tab xls, scaly	{001} perf	{001}
*1.56 1.61	n 1.56-1.61	Zaratite $Ni_3CO_3(OH)_4 \cdot 4H_2O$	Isometric	Massive, mammillary, fib (?)	None	
1.590	n 1.590	Schoenfliesite $MgSn(OH)_6$	Isometric	Fine-grained massive		
1.593	n 1.593	Georgeite $Cu_5(CO_3)_3(OH)_4 \cdot 6H_2O$	Amorphous			
1.600	n 1.600	Zunyite $Al_{13}Si_5O_{20}(OH,F)_{18}Cl$	Isometric	Tetrahedral, octahedral	{111} good	Common
1.60	n~1.60	Basaltic glass (Tachylite) (Sideromelane)	Amorphous	Massive, pumaceous, hair-like	None	None
1.593 1.608	n 1.593-1.608	Voltaite $K_2Fe_5^{2+}Fe_4^{3+}(SO_4)_{12} \cdot 18H_2O(?)$	Isometric	Octahedral-dodec, granular	None	
1.608	n 1.608	Monsmedite $K_2OTlO_3(SO_3)_8 \cdot 15H_2O$	Isometric	Cubic	{100}	
1.613	n 1.613	Hsianghualite $Ca_3Li_2Be_3(SiO_4)_3F_2$	Isometric	Trisoct or dodecahedral, granular	None	
*1.58 1.66	n 1.58-1.66	Diabantite $(Mg,Fe^{2+}Al)_6(Si,Al)_4O_{10}(OH)_8$	Monoclinic	Massive, foliated, scaly, rad	{001} perf	{001}
1.618	n 1.618	Metaborite HBO_2	Isometric	Tetrahedral	None	
*1.57 1.67	n 1.57-1.67	Bořickite $\sim CaFe_5(PO_4)_2(OH)_{11} \cdot 3H_2O$	Amorphous?	Reniform, opaline	None	
*1.60 1.63	n 1.60-1.63	Klementite $(Mg,Fe^{3+})_6(Si,Al)_4O_{10}(OH)_8$	Monoclinic β ~ 97°	Micaceous, scaly	{001} perfect	{001} multiple
*1.60 1.64	n 1.60-1.64	Pycnochlorite $(Mg,Fe^{2+},Al)_6(Si,Al)_4O_{10}(OH)_8$	Monoclinic	Massive, scaly, oolitic	{001} perf	{001}
*1.60 1.65	n 1.60-1.65	Zirfesite $\sim ZrO_2Fe_2O_3SiO_2 \cdot nH_2O$	Unknown (amorphous)	Massive, fine flakes	None	
*1.60 1.65	n 1.60-1.65	Prochlorite (Ripidolite) $(Mg,Fe^{2+},Al)_6(Si,Al)_4O_{10}(OH)_8$	Monoclinic	Micaceous-scaly, hex-tab xls	{001} perf	{001}

* n variation exceeds 0.01

Color In Thin Section	Hardness Sp Gravity	Color Hand Sample	Alteration	Occurrence	Remarks	Reference
Colorless	4½ 2.33-2.37	White, pink, yellow, green	Opal, chert, kaolin, calcite	Amygdules or veins in basalt with zeolites. Cavities in granite, gneiss		p. 309
Colorless	2-2½ 2.155	Colorless		Lake or marine evaporites with other evaporite salts	Sol H_2O. Saline taste	p. 11
Colorless, pale yellow or brown	~6 ~2.6	Gray, brown, black	Devitrification	Volcanic flows and tuffs		p. 328
Colorless, ~opaque	2-3 2.5-2.7	White	Diaspore, gibbsite	Hydrothermal alter. Al-silicates, with kaolinite, alunite, sulfides	Dehydrated halloysite	p. 254
Colorless, very pale green	2-3 2.58-2.61	White, yellowish, shades of green		Alter ultramatic rocks as serpentenite often with chrysotile	Chlorite & serpentine groups. Smooth-soapy feel	p. 298
Colorless	? 3.24-3.25	Colorless		Evaporite nitrate deposits	Possible birefringence in sectors. Easily sol H_2O	Dana II, 305-306
Colorless pink	? 3.02-3.06	Rose red		Cavities in alkali lavas with halite, sylvite, thenardite	Langbeinite-manganolangbeinite series	Dana II, 435
Colorless, pale yel, pale green	2-2½ 2.6-2.85	Emerald green, white, olive		Mg-rich schists, serpentenite	Chlorite group. Anomalous interference colors	p. 289
Green	3½ 2.57-2.69	Emerald green		Secondary mineral in ultramafic rocks, serpentenite, chromite	Easily sol warm HCL with effervescence	Dana II, 245-246
	? 3.483			Alter of hulsite in met ls with maghemite, goethite, fluorite		Am Min 57, 1557-1558
	Soft 2.55	Pale blue		Secondary Cu-ores with malochite, chalconatrolite		Am Min 64, 1330
Colorless	7 2.85-2.87	Colorless, gray, flesh color		Hydrothermal alteration of feldspar in veins, shales		Am Min 37, 960-965
Yellow-brown, dk brn to black without pleochroism	~5½ ~2.76	Dk brown to black	Yellow to orange mineraloid (palagonite)	Quenched or chilled basaltic flows or dikes		p. 328
Pale green to oil grn	3 2.7	Dk olive green to black	Yellow powder	Fumeroles & springs. Oxide zones ore-deposits with other sulfates	Sol acids	Dana II, 464-465
~opaque, dark green	2 3.00	Black		Cavities in oxide zones ore deposits with marcasite, barite		Am Min 54, 1496
Colorless	6½ 2.97-3.00	White		Metamorphosed limestone with phlogopite		Am Min 46, 244
Green-brown, dk green	2-2½ 2.77-3.03	Dk green		Altered diabase, mafic rocks	Chlorite group	p. 288
Colorless	5 2.47-2.49	Colorless, brownish		Rock salt deposits with anhydrite, boracite, aksaite, etc	Slowly sol H_2O	Am Min 50, 261-262
Red-brown, ~opaque	3½ ~2.70	Reddish-brown		Oxide zones(gossan) Fe-deposits		Dana II, 915-916
Colorless, pale green, pale brown. O > E	2-3 2.6-3.0	Green-brown		Low grade schists. Oxidation of ferrous chlorites	Chlorite group	p. 294
Lt green-brown to dk green	2-3 2.8-3.1	Dk green, dk olive-green		Schists, altered mafic rocks	Chlorite group	p. 288
Colorless, pale yel	? 2.70	Light yellow		Pseudomorphs after eudialyte in nepheline-syenite	Decomp acids	Vlasov II, 392-393
Pale yellow-green, green	2-3 2.88-3.08	Green to dk green, olive		Schists & similar metamorphic rocks. Ore veins	Chlorite group. Anom interference colors	p. 288

Isotropic

n Var.	Refractive Index	Mineral Name and Composition	Crystal System	Habit	Cleavage	Twinning
1.625 1.628	n 1.625-1.628	Bicchulite $Ca_2Al(SiAl)O_7 \cdot H_2O$	Isometric			
1.635	n 1.635	Brindleyite (Nimesite) $(Ni_{1.75}Al_{1.0})(Si_{1.5}Al_{0.5})O_5(OH)_4$	Monoclinic ?	Clay-like. Micro platy	{001}?	
1.639	n 1.639	Sal-ammoniac NH_4Cl	Isometric	Iso xls, fib masses, stalac, earthy	{111} imperf	{111} common
1.638 1.641	n 1.638-1.641	Sakhaite $Ca_{12}Mg_4(CO_3)_4(BO_3)_7Cl(OH)_2 \cdot H_2O$	Isometric	Massive in lense-like bodies	None	
1.640	n 1.640	Umbozerite $Na_3Sr_4ThSi_8(O,OH)_{24}$	Amorphous (Metamict)	Massive, poor tetragonal xls	None	
1.643	n 1.643	Mayenite $Ca_{12}Al_{14}O_{33}$	Isometric	Rounded grains		
*1.63 1.66	n 1.63-1.66	Griphite $(Na,Al,Ca,Fe)_3Mn_2(PO_4)_{2.5}(OH)_2$	Isometric	Massive, reniform	None	
1.653	n 1.653	Harkerite $Ca_{48}Mg_{16}Al_3(BO_3)_{15}(CO_3)_{18}(SiO_4)_{12}$ $Cl_2(OH)_6 \cdot 3H_2O$	Isometric	Octahedral	None	
*1.64 1.68	n 1.64-1.68	Daphnite $(Fe^{2+},Mg,Al)_6(Si,Al)_4O_{10}(OH)_8$	Monoclinic	Massive, foliated, botryoidal	{001} perf	{001}
*1.65 1.68	n 1.65-1.68	Pseudothuringite $(Fe^{2+},Mg,Al)_6(Si,Al)_4O_{10}(OH)_8$	Monoclinic	Massive, foliated	{001} perf	
1.674	n~1.674	Greenalite $(Fe^{2+},Fe^{3+})_{5-6}Si_4O_{10}(OH)_8$	Orthorhombic	Micro round granules	{001} perf (not observed)	
*1.676 1.704	n 1.676-1.704	Pharmacosiderite $Fe_3(AsO_4)_2(OH)_3 \cdot 5H_2O$	Isometric	Cubic, tetrahedral xls, earthy	{100} fair	Sector
1.6935	n 1.6935	Rhodizite $C_5B_{12}Be_4Al_4O_{28}$	Isometric	Tetrahedral or dodecahedral	{111} and {$\bar{1}$11} difficult	
**1.630 1.770	n 1.630-1.770	Tritomite-(Y) $(Y,Ca,La,Fe)_5(Si,B,Al)_3(O,OH,F)_{13}$	Hexagonal (metamict)	Massive compact	None	
**1.650 1.757	n 1.650-1.757	Tritomite $(Ce,La,Y,Th)_5(Si,B)_3(O,OH,F)_{13}$	Hexagonal (metamict)	Triangular pyram xls	None	
*1.675 1.735	n 1.675-1.735	Hydrogrossular (Hibschite) (Plazolite) $Ca_3Al_2(SiO_4)_2(OH)_4$	Isometric	Octahedral xls, massive	None	
1.705	n 1.705	Wickmanite $(Mn,Ca)Sn(OH)_6$	Isometric	Octahedral		
**1.61 1.82	n 1.61-1.82	Partzite $Cu_2Sb_2(O,OH)_7$(?)	Isometric	Massive	None	
1.713	n 1.713	Pyrope $Mg_3Al_3(SiO_4)_3$	Isometric	Dodecahedral, trapezohedral	None, {100} parting	
1.719	n 1.719	Spinel $MgAl_2O_4$	Isometric	Octahedral xls, grains	{111} parting	{111} usually simple contact
1.716 1.725	n 1.716-1.725	Dzhalindite $In(OH)_3$	Isometric	Small grains		
*1.728 1.747	n 1.728-1.747	Helvite (Helvine) $Mn_4(BeSiO_4)_3S$	Isometric	Tetrahedral, rounded agg	{111} dist	

* n variation exceeds 0.01

** n variation exceeds 0.10

Color In Thin Section	Hardness Sp Gravity	Color Hand Sample	Alteration	Occurrence	Remarks	Reference
Colorless	2.75-2.81	Colorless		Skarns with vesuvianite, hydrogrossular, gehlenite	"Gehlenite hydrate"	Am Min 59, 1330
Colorless, pale green	2½-3 3.16	Dk yellowish-grn		Bauxite	Ni-serpentine	Am Min 63, 484-489
Colorless	1-2 1.519	Colorless, white, yellowish		Sublimate in coal seams, oil-shales, fumeroles	Sol H_2O. Taste saline-stinging. Sectile	Dana II, 15-18
Colorless	5 2.78-2.83	Colorless, white, grayish		Mg-skarn with kotoite, clino-humite, forsterite, saunite, spinel		Am Min 51, 1817
Pale greenish or brownish	5 3.60	Bottle-green to green-brown		Veinlets in alkali rocks with ussingite, sphalerite, belovite, schizolite		Am Min 60, 341
Colorless	? 2.85	Colorless		Contact ls deposits. Ls. inclusions in volcanics		Am Min 50, 2106-2107
Yellowish-brown, brown	5½ 3.40	Dk brown, brownish-black		Granite pegmatites	Easily sol acids	Am Min 57, 269-272
Colorless	? 2.96	Colorless		Skarns-dolomitic contact zones		Min Mag 29, 621-666
Lt green-brown, dk green	2-3 3.20	Dk green		Ore veins with quartz, arsenopyrite	Chlorite group	p. 292
Brownish-green	2-3 ~3.2	Dk green		Hydrothermal veins	Chlorite group	p. 292
Lt yellowish-green to blue-green	2.85-3.61	Dk green, black		Primary mineral in iron formations	Septechlorite group. ~isotropic	Am Min 21, 449-455
Green, yellow-brown	2½ 2.80-2.87	Green, amber to dk brown		Oxide zones As-deposits, pegmatites, hydrothermal ores	Sectile. Sol HCl	Dana II, 995-997
Colorless	8½ 3.44-3.48	Colorless, white, yellowish		Pegmatites with rubellite tourmaline	Anomalous double refraction. Insol acids	Science 152, 500-502
Dk brown	3½, 6½ 3.05, 3.40	Reddish-brown to green-black		Granite pegmatites with apatite, fergusonite, zircon, fluorite		Can Min 6, 576-581
Brown	5½ 4.2	Dk brown		Nepheline syenite pegmatites		Am Min 47, 9-25
Colorless	6-6½ 3.0-3.6	White, lt brown, lt green, pink		Calc-silicate skarns. With prehnite, jadeite	Garnet group	p. 117
Pale yellow-brown	3.82-3.89	Brownish yellow		Low-T, late stage hydrothermal vein mineral		Am Min 53, 1063
Yellowish-green	3-4 2.96-3.96	Olive-green, blackish-green		Oxide zones Sb-sulfide deposits	Cu-bindheimite	Min Mag 30, 100-112
Colorless, pale red	7½ 3.535	Pink, deep red, purplish	Chlorite, epidote, hornblende	High pressure rocks (e.g., peridotites, eclogites, anorthosites)	Garnet group	p. 115
Colorless, pale colors	7½-8 3.55-4.04	Colorless, red, blue, green, etc.	Talc, mica, chlorite, corundum, diaspore	High-T marbles with chondrodite, forsterite. Al-xenoliths in mafic plutons	Spinel series	p. 34
Pale yellow	4.34	Yellow-brown		Alteration of indrite, in cassiterite		Am Min 49, 439-440
Pale yellow, brown	6 3.17-3.37	Brown, reddish, yel, yel-green		Granite pegmatites, alkali-pegmatites, skarns, hydro veins	Helvite-Danalite-Genthelvite Group	Vlasov II, 119-127

Isotropic

n Var.	Refractive Index	Mineral Name and Composition	Crystal System	Habit	Cleavage	Twinning
1.734	n 1.734	Grossular (Grossularite) $Ca_3Al_2(SiO_4)_3$	Isometric	Dodecahedral, massive	None	Rad wedges
1.736	n 1.736	Söhngeite $Ga(OH)_3$	Isometric	Xline agg		
1.735 1.745	n 1.735-1.745	Periclase MgO	Isometric	Cubic xls or fragments	{100} perf	{111}
	n 1.74	Tsavolite $(Ca,Mn)_3(Al,Cr,V)_2(SiO_4)_3$	Isometric		None	
1.742 1.745	n 1.742-1.745	Genthelvite $(Zn,Fe,Mn)_4Be_3Si_3O_{12}S$	Isometric	Tetrahedral	None	
*1.71 1.79	n 1.71(Mg)- 1.79(Mn)	Berzeliite $(Ca,Na)_3(Mg,Mn)_2(AsO_4)_3$	Isometric	Massive, rounded grains	None	
1.755	n 1.755	Arsenolite As_2O_3	Isometric	Octahedral, as crusts, earthy	{111} distinct	
1.760	n 1.760	Rhodolite $(Mg,Fe)_3Al_2(SiO_4)_3$	Isometric	Dodecahedral-trapezohedral	None	
*1.753 1.771	n 1.753-1.771	Danalite $(Fe,Mn,Zn)_4Be_3Si_3O_{12}S$	Isometric	Octahedral or dodecahedral	{111} in traces	
**1.664 1.87	n 1.664-1.87	Thorite $ThSiO_4$	Tetragonal (Metamict)	Short prism, pyram xls, massive	None if metamict	
1.768	n 1.768	Ringwoodite $(Mg,Fe)_2SiO_4$	Isometric	Rounded grains		
1.77	n 1.77	Pleonaste $(Mg,Fe)Al_2O_4$	Isometric	Octahedral	None	{111} repeated
**1.605 2.05	n 1.605-2.05	Stibiconite (Hydroromeite, Stibianite, Volgerite) $Sb_3O_6(OH)$	Isometric	Massive, crusts, botryoidal		
1.790	n 1.790	Manganberzeliite $(Ca,Na)_3(Mn,Mg)_2(AsO_4)_3$	Isometric	Massive, minute trapezohedrons	None	
*1.790 1.805	n 1.790-1.805	Gahnite $ZnAl_2O_4$	Isometric	Octahedral, massive	{111} parting indist	{111} simple, sixlings
1.799	n 1.799	Spessartine (Spessartite) $Mn_3Al_2(SiO_4)_3$	Isometric	Trapezohedral, dodecahedral	None	
1.803	n 1.803	Knorringite $Mg_3Cr_2(SiO_4)_3$	Isometric	Massive, minute grains	None	
*1.80 1.83	n 1.80-1.83	Hercynite $FeAl_2O_4$	Isometric	Massive, fine-granular	None	
1.81 1.82	n 1.81-1.82	Anthoinite $Al_2W_2O_9 \cdot 3H_2O$	Mono or Triclinic ($<\beta=96°$)	Chalky masses		
1.825 1.830	n 1.825-1.830	Hydrougrandite $(Ca,Mg,Fe^{2+})_3(Fe^{3+},Al)_2(SiO_4)_{3-x}(OH)_{4x}$	Isometric	Small grains	None?	
1.83	n 1.83	Liandratite $U(Nb,Ta)_2O_8$	Trigonal (metamict)	Amorphous coatings	None	

* n variation exceeds 0.01

** n variation exceeds 0.10

Color In Thin Section	Hardness Sp Gravity	Color Hand Sample	Alteration	Occurrence	Remarks	Reference
Colorless	6½-7 3.560	Colorless, yel, gray green, brn	Chlorite, calcite, epidote, feldspar	Skarns & marble with Ca-Mg-silicates. Nepheline syenites. Pegmatites	Garnet group. May be anisotropic. Color zoning	p. 117
Colorless, pale brn	4-4½ 3.84-3.85	Light brown		Altered germanite		Am Min 51, 1815
Colorless	5½-6 3.58-3.68	White, gray, yellow, brown	Brucite	High-T contact deposits in dolomite or magnesite with brucite	Easily sol dilute HCl or HNO_3	p. 18
	3.68	Bright green-dk green			Grossularite var	Am Min 61, 178
Colorless, pinkish, yellowish-green	6-6½ 3.55-3.70	Colorless, white, yellow, green, pink		Carbonatite, alkaline pegmatites, alkali granite	Helvite group. Sol HCl	Vlasov II, 119-127
Colorless to orange	4½-5 4.08(Mg)-4.46(Mn)	Yellow, orange-yellow		Skarn in limestone	Berzeliite-Manganberzeliite series	Dana II, 681-683
Colorless	1½ 3.87	White, yellowish, bluish, etc.		Oxidation of As minerals mine sublimation	Soluble in H_2O. Taste biting, sweet. Anomalous birefringence	Dana I, 543-544
None	7-7½ 3.84	Rose pink, purple	Chlorite	Schists and gneisses	Garnet group	p. 112
Colorless, pale yellow, pink	5½-6 3.31-3.46	Gray, yellow, pink, red, brown		Granite pegmatites, gneiss, contact met, hydro veins	Helvite group	Vlasov II, 119-127
Brown, yellow-green	~4½ 4.1-6.7	Dk brown, yel, orange, grn, blk	Thorogummite	Pegmatites, hydrothermal veins, metasomatic deposits	Not fluor. U.V.	USGS Bull 1064, 265-276
Colorless, purplish, gray-blue	3.90	Purple, bluish-gray		Pseudomorphs after olivine in meteorites	Dimorph olivine	Nature 221, 943-944
Green	7½-8 3.8	Dk. green, black		Peridotites with magnetite. Contact meta. with corundum	Spinel group	p. 34
Colorless, cloudy	3-7 3.3-5.5	White, gray, yel, brown, black		Alteration of stibnite with valentinite		Dana I, 597-598
Colorless to orange	4½-5 4.21-4.46	Yellow, orange-yellow		Veins in Mn-O ores, limestone skarns with Mn-minerals	Berzeliite-Manganberzeliite series. Easily sol acids	Dana II, 681-683
Green, blue	7½-8 4.607	Dk green, green-blue, brown		Contact limestone zones, schists, granite pegmatites	Spinel group. Possible pleochroism	Dana I, 689-697
Pink, pale brown	7-7½ 4.190	Dk red, violet, black, brown-red	Chlorite, epidote, hornblende, Mn-oxides	Skarns with rhodonite, tephroite. Granite pegmatites	Garnet group. May be anisotropic	p. 115
Pale green	~7 3.75-3.85	Green		Kimberlite pipes	Garnet group	Am Min 53, 1833-1840
~opaque green	7½-8 4.32	Black		Emery deposits	Spinel group	p. 34
Colorless	1 4.6-5.0	White		High-T, ore veins	Essentially isotropic	Am Min 43, 384
Colorless, pale green	>5 3.45	Green		Peridotites with olivine, augite	Garnet group	Am Min 50, 2100
Yellow	3½ ~6.8	Yellow to yellow-brown		Alteration crusts on petscheckite in complex pegmatites		Am Min 63, 941-946

n Var.	Refractive Index	Mineral Name and Composition	Crystal System	Habit	Cleavage	Twinning
1.830	n 1.830	Almandine (Almandite) $Fe_3Al_2(SiO_4)_3$	Isometric	Dodecahedral, trapezohedral	None	
1.830 1.835	n 1.830-1.835	Obruchevite $(Y,Na,Ca,U)(Nb,Ta,Ti,Fe)_2(O,OH)_7$	Isometric	Dense masses	None	
*1.821 1.855	n 1.821-1.855	Goldmanite $Ca_3V_2(SiO_4)_3$	Isometric	Dodecahedral	None	
1.838	n 1.838	Lime CaO	Isometric	Massive	{100} perf {011} parting	
*1.817 1.870	n 1.817-1.870	Roméite $(Ca,Fe,Na)_2(Sb,Ti)_2(O,OH)_7$	Isometric	Small octahedral xls, massive	{111} imperf	
1.863	n 1.863	Uvarovite $Ca_3Cr_2(SiO_4)_3$	Isometric	Dodecahedral, massive	None	
>1.86	n>1.86	Arsenobismite $Bi_2AsO_4(OH)_3$	Isometric	Friable microcrystalline masses		
1.887	n 1.887	Andradite $Ca_3Fe_2(SiO_4)_3$	Isometric	Trapezohedral, dodecahedral, massive	None. Possible parting {110}	Rad wedges
*1.84 1.93	n 1.84-1.93	Bindheimite $Pb_2Sb_2O_6(O,OH)$	Isometric	Cryptocrystalline, earthy	None	
1.923	n 1.923	Galaxite $MnAl_2O_4$	Isometric	Massive, fine granular	None	{111} possible
1.930	n 1.930	Nantokite CuCl	Isometric	Massive, granular	{110}	
*1.915 1.96	n 1.915-1.96	Betafite $(Ca,Fe,U)_{2-x}(Nb,Ti,Ta)_2O_6(OH,F)_{1-z}$	Isometric	Octahedral	None	
1.94	n 1.94	Kimzeyite $Ca_3(Zr,Ti)_2(Al,Si)_3O_{12}$	Isometric	Dodecahedral	None	
1.94	n 1.94	Melanite $Ca_3(Fe^{3+},Ti)_2[(Si,Ti)O_4]_3$	Isometric	Trapezohedral	None	
1.95	n 1.95	Stetefeldtite (Stetefeldite) $Ag_{2-y}Sb_{2-x}(O,OH,H_2O)_{6-7}$	Isometric	Massive		
1.9 2.0	n 1.9-2.0	Schorlomite $Ca_3(Fe^{3+},Ti)_2[(Si,Ti)O_4]_3$	Isometric	Massive	None	
>1.95	n>1.95	Mosesite $Hg_2N(Cl,SO_4,MoO_4,CO_3)\cdot H_2O$	Isometric	Octahedral or cubic	{111} imperf	{111} contact (Spinel-twins)
*1.93 2.02	n 1.93-2.02	Microlite $(Ca,Na,Fe)_2(Ta,Nb)_2(O,OH,F)_7$	Isometric	Octahedral xls, irregular mass	{111} dist, may be parting	{111} uncommon
*1.96 2.01	n 1.96-2.01	Pyrochlore (Ellsworthite) (Hatchettolite) $(Na,Ca,U)_2(Nb,Ta,Ti)_2O_6(OH,F)$	Isometric	Octahedral	{111} sometimes dist	{111} rare
>2.0	n>2.0	Cerianite CeO_2	Isometric	Octahedral		
>2.00	n>2.00	Westgrenite $(Bi,Ca)(Ta,Nb)_2O_6(OH)$	Isometric	Massive, veinlets, crusts	None	
*2.0 2.1	n 2.0-2.1	Limonite $FeO(OH)\cdot nH_2O$	Amorphous	Colloform structures	None	None
2.05	n 2.05	Percylite $PbCuCl_2(OH)_2$	Isometric?	Cubic or dodec xls, massive		
2.05	n 2.05	Eulytite (Eulytine)(Agricolite) $Be_4Si_3O_{12}$	Isometric	Tetrahedral	{110} indist	

* n variation exceeds 0.01

Color In Thin Section	Hardness Sp Gravity	Color Hand Sample	Alteration	Occurrence	Remarks	Reference
Pale pink, lt brown	7-7½ 4.325	Dk red, brown-red, black	Chlorite, epidote, hornblende	Regional schists & gneisses with mica, staurolite. Granites, andesites	Garnet group	p. 115
Brown	4½-5 3.60-3.80	Shades of brown		Pegmatites with garnet, fergusonite, columbite	Pyrochlore group	Am Min 43, 797
Green, brown	7 3.74	Dk green, brownish-green		Metamorphosed U-V-deposits	Garnet group. Weakly anisotropic	Am Min 49, 644-655
Colorless	3½ 3.35	White		Carbonate rock inclusions in lavas		Dana I, 503
Pale yellow, colorless	5½-6½ 4.7-5.4	Lt yellow to dk brown		Mn-ores		Dana II, 1020-1022
Pale emerald green	7½ 3.895	Emerald, dk green		Seams in chromite ores, serpentines, peridotites	Garnet group. Maybe anisotropic	p. 117
	~5.7	Yellow-brown, yellow-green		Oxide zones of ore deposits		Am Min 28, 536
Pale yellow, lt to dk brown	6½-7 3.835	Brown, grn, yel, black	Chlorite, calcite, epidote, limonite, nontronite	Skarns & marbles with Ca-Mg-silicates. Nepheline syenites, etc.	Garnet group. Maybe anisotropic. Color zoning	p. 117
Colorless, yellow, brownish	4-4½ 4.6-7.3	Yellow, brown, gray, white		Oxide zones Pb-Sb deposits		Dana II, 1018-1020
Dk brown	7½-8 4.077	Black, brown		Hydrothermal veins with Mn-silicates	Spinel group	Dana I, 689-697
Colorless	2½ 4.13-4.22	Colorless, white, greenish	Paratacamite	Oxidezones Cu-deposits with native Cu, cuprite, atacamite, etc.	Easily sol HCl, HNO_3, NH_4OH	Dana II, 18-19
Colorless, brownish	3-5½ 4.15-5.2	Black, brown, grn-brn, yel		Granite pegmatites	Pyrochlore group	USGS Bull 1064, 320-325
Brown	~7 4.03	Dark brown		Carbonatite with apatite, monticellite, perovskite, magnetite	Garnet group	Am Min 46, 533-548
Brown	7-7½ 3.7	Black		Volcanics	Garnet group. Ti-andradite	pp. 112, 117
Dk brown	3½-4½ 4.6-5.38	Black to brown		Cu-Ag-deposits with chalcocite, pyrite	Ag-bindheimite	Min Mag 30, 100-112
Dk. reddish brown	7-7½ 3.81-3.88	Black		Nepheline syenite	Garnet group. Ti-andradite	pp. 112, 117
Colorless, pale yel	3½ 7.53-7.72	Yellow, lt olive-green		Hg-deposits with montroydite, native Hg, cinnabar		Am Min 38, 1225-1234
Yellow-brown, reddish, green	5-5½ 4.3-6.31	Yellow to brown, reddish, green		Pegmatites with lepidolite, spodumene	Pyrochlore group. Not fluor. U.V.	USGS Bull 1060, 326-333
Yellow-brown, red-brown	5-5½ 4.31-4.48	Yellow-brown, red-brown, black		Carbonatites, pegmatites, alkali intrusives	Pyrochlore-Microlite series. Not fluor. U.V.	USGS Bull 1064, 326-333
Yellowish	? 7.19-7.22	Dk green, amber		Carbonatites, alteration of bastnäsite		USGS Bull 1064, 53-55
~Colorless	5 6.5-6.83	Yellow, pink, brown		Replacement of bismutotantalite in Li-pegmatites		GSA Special Paper 73, 256
Yellow to yellow-brown to red-brown	4-5½ 2.7-4.3	Yellow-brown to black-brown		Weathering product of Fe-minerals with goethite, hematite		p. 48
Pale blue	2½	Sky blue		Oxide zones Cu-Pb-deposits		Dana II, 81-82
Colorless, pale brn	4½ 6.6-6.76	Colorless, yel, dk brown		With native Bi	Gelatinizes HCl. Anomalous biref	Win & Win II, 494

n Var.	Refractive Index	Mineral Name and Composition	Crystal System	Habit	Cleavage	Twinning
2.05	n 2.05	Picotite $(Mg,Fe)(Al,Cr)_2O_4$	Isometric	Octahedral	None	{111} repeated
2.071	n 2.071	Cerargyrite (Chlorargyrite) AgCl	Isometric	Cubic, waxy coatings	None (sectile)	
2.08	n 2.08	Plumbopyrochlore $(Pb,Y,U,Ca)_{2-x}Nb_2O_6(OH)$	Isometric	Octahedral, grains	None	
2.085	n 2.085	Gianellaite $(NHg_2)_2(SO_4)$	Isometric	Octahedral, rosettes		
2.087	n 2.087	Senarmontite Sb_2O_3	Isometric	Octahedral xls, crusts	{111} traces	
>2.11	n>2.11	Cliffordite UTe_3O_8	Isometric	Octahedral		
**2.05 2.19	n 2.05-2.19	Fergusonite $(Y,Er,Ce,Fe)(Nb,Ta,Ti)O_4$	Tetragonal (metamict)	Prism, pyram, massive	{111} traces	
**2.05 2.19	n 2.05-2.19	Formanite $Y(Ta,Nb)O_4$	Tetragonal (metamict)	Granular, massive	{111} traces	
*2.12 2.15	n 2.12-2.15	Yttrocrasite $(Y,Th)Ti_2(O,OH)_6$(?)	Orthorhombic (metamict)	Prism tab xls		
2.137	n 2.137	Oldhamite (Ca,Mn)S	Isometric	Small spherules	{001}	
2.15	n 2.15	Embolite Ag(Cl,Br)	Isometric	Cubic, waxy coatings	None	{111} common
2.15	n 2.15	Yttrotantalite $(Y,U,Fe)(Ta,Nb)O_4$	Monoclinic (metamict)	Prism (c), tab {010} xls, massive	{010} indist	
**2.06 2.24	n 2.06-2.24	Euxenite $(Y,Er,Ce,La,U)(Nb,Ti,Ta)_2(O,OH)_6$	Orthorhombic (metamict)	Prism, diverg agg, massive	None	{201} common
2.18	n 2.18	Manganosite MnO	Isometric	Octahedral xls, massive	{001} fair, {111} parting	
2.19	n 2.19	Zirkelite (Zirconolite) $(Ca,Th,La)Zr(Ti,Nb,Fe)_2O_7$	Isometric	Flattened octahedrons	None	{111} polysynthetic, fourlings
2.192	n 2.192	Bideauxite $Pb_2AgCl_3(F,OH)_2$	Isometric	Small xls	None	
2.20	n 2.20	Samarskite (Ampangabeite)(Neuvite) $(Fe,Y,U)_2(Nb,Ti,Ta)_2O_7$	Orthorhombic (metamict)	Rough prism or tab xls, massive	{010} indist	
2.215	n 2.215	Polymignite $(Ca,Fe,Y,Th)(Nb,Ti,Ta)O_4$	Orthorhombic (metamict)	Prism xls	{100} & {010} traces	
2.23	n 2.23	Bunsenite NiO	Isometric	Octahedral		
2.248	n 2.248	Polycrase $(Y,Er,Ce,La,U)(Ti,Nb,Ta)_2(O,OH)_6$	Orthorhombic metamict	Short prism xls, divergent, massive	None	{201} common
2.253	n 2.253	Bromargyrite (Bromyrite) AgBr	Isometric	Cubic, waxy coatings	None	{111} rare
*2.262 2.315	n 2.262-2.315	Jixianite $(Pb(W,Fe^{3+})_2(O,OH)_7$	Isometric	Crust agg., octahedral	None	
~2.30	n~2.30	Lueshite (Igdloite) $NaNbO_3$	Ortho (Pseudo-isometric)	Cubic	{001} imperf	
2.26 2.34	n 2.26-2.34	Yttromicrolite (Hjelmite) $(Y,RE,Ca,Mn)(Ta,Nb,Fe,Sn)_2O_6(OH)$	Isometric (metamict)	Octahedral, massive	None	
2.3	n 2.3	Jacobsite $(Mn,Fe,Mg)(Fe,Mn)_2O_4$	Isometric	Octahedral, massive, gran	None	

* n variation exceeds 0.01 ** n variation exceeds 0.10

Color In Thin Section	Hardness Sp Gravity	Color Hand Sample	Alteration	Occurrence	Remarks	Reference
Dk. brown	7½-8 4.08	Yellow-brown, green-brown		Peridotites and dunites. Limestone contact deposits	Spinel group	p. 34
Colorless	2½ 5.55	Gray, greenish, brownish		Oxide zones Ag-deposits		Dana II, 11-14
Dk brn at center of xls, greenish-yellow to red on periphery	~5.0	Dark brown		Metasomatically altered granitic rocks with aegirine, albite	Pyrochlore group	Am Min 55, 1068
Pale yellow, colorless	3 7.13-7.19	Straw yellow		Veinlets with terlinguaite, calomel, montroydite, cinnabar		Am Min 62, 1057
Colorless	2-2½ 5.50-5.58	Colorless, white, grayish		Alter Sb-minerals with cervantite, stibnite, boulangerite	Dimorph valentinite. Easily sol HCl	Dana I, 544-545
Pale yellow	4 6.57-6.77	Sulfur yellow		Oxide zones Te-deposits		Am Min 54, 697-701
~Opaque, dk brown in splinters	5½-6½ ~5.38	Brownish-black, black		Rare-earth pegmatites	Fergusonite series. β-Fergusonite (Mono., Lt yellow)	Dana I, 757-763
~Opaque, dk brown in thin splinters	5½-6½ 7.03	Brownish-black, black		High-T veins with cassiterite, monazite, etc.		Dana I, 757-763
Amber to pale yellow Weak pleochroism	5½-6 4.80	Black		Pegmatites	Radioactive	Dana I, 793
Pale brown	4 2.58-2.60	Light brown		Meteorites	Readily sol HCl. Decomp boiling H_2O	Dana I, 208-209
Colorless, yellow, pale green	2½ 5.7-5.8	Gray, yellowish, greenish		Oxide zones Ag-deposits	Ductile, very plastic, anomalous biref	Dana II, 11-15
Red-brown	5-5½ 5.7	Dk brown to black		Pegmatites with monazite, samarskite, columbite, fergusonite		Dana I, 763-764
Opaque, dk brown in thin splinters	5½-6½ 4.30-5.87	Black, brownish		Granite pegmatites	Euxenite-Polycrase series	Dana I, 787-792
Emerald green	5½ 5.36	Emerald green-blk		With franklinite, zincite, willemite in Zn-O ores		Dana I, 501-502
Dk brown, reddish brown	5½ 4.70-4.74	Black		Magnetite pyroxenites with perovskite & baddeleyite		Dana I, 740
Colorless	3 6.26-6.27	Colorless, pale lilac		Oxide zones ore deposits		Min Mag 37, 637-640
Lt brown to dk brown	5-6 5.25-5.69	Brownish-black, black		Granite pegmatites with columbite, monazite, zircon		Dana I, 797-800
Lt brown, red-brown	6½ 4.77-4.85	Black		Pegmatites		Vlasov II, 496-499
Pale yellow-green	3½-5½ 6.7-6.9	Dk yellow-green		Oxide zones Ni-deposits. Slags		Dana I, 500-501
Yellow-brown to red-brown	5½-6½ 4.30-5.87	Black		Granite pegmatites with monazite, rare-earth minerals	Euxenite-Polycrase series	Dana I, 787-792
Colorless	2½ 6.47-6.50	Gray, greenish, brown, yellow		Oxide zones Ag-deposits		Dana II, 11-14
Yellow-brown, greenish yellow	6.04-7.22	Red, red-brown		Oxide zones W quartz veins with bismuthite, stolzite	Weak birefringence	Am Min 64, 1330
Reddish-brown in thin splinters	5½ 4.44-4.55	Black		Contact zone between cancrinite syenite and carbonatite	Perovskite group. 2V ~ 45°	Am Min 46, 1004
Very dk green, amber, brown	5-5½ 5.57-6.53	Black		Granite pegmatites with allanite, gadolinite,	Weak anomalous biref. Pyrochlore group	Am Min 64, 890-892
Brown in very thin splinters	5½-6½ 4.76-5.03	Black		Crystalline limestone, gneiss, contact Mn-deposits	Spinel group, magnetite series	Dana I, 698-707

Isotropic

n Var.	Refractive Index	Mineral Name and Composition	Crystal System	Habit	Cleavage	Twinning
~2.3	n~2.3	Trevorite $NiFe_2O_4$	Isometric	Octahedral xls, massive	None	
*2.30 2.38	n 2.30-2.38	Perovskite $CaTiO_3$	Pseudoisometric (Mono or Ortho)	Tiny cubic or octahedral xls	{100} poor	{111} complex lamellar
2.346 (Na)	n 2.346 (Na)	Marshite CuI	Isometric	Tetrahedral, cubo-octahedral	{110} perf	{111} sometimes repeated
2.35	n 2.35	Tazheranite $(Zr,Ca,Ti)O_2$	Isometric	Irregular grains, thick tab xls	None	
2.38	n 2.38	Magnesioferrite $(Mg,Fe^{2+})Fe^{3+}_2O_4$	Isometric	Oct, minute xls, massive	None	{111} common
	n>2.4(Na)	Maghemite $\gamma\text{-}Fe_2O_3$	Isometric	Massive		
2.4195	n 2.4195	Diamond C	Isometric	Octahedral-tetrahedral xls, gran	{111} perf	{111} simple or lamellar, {001}
>2.42	n>2.42	Sillénite $\gamma\text{-}Bi_2O_3$	Isometric	Fine-grained massive, earthy		
**2.37 2.50	n 2.37-2.50	Sphalerite ZnS	Isometric	Tetrahedral xls, granular	{110} perf	{111} simple or multiple
2.49 (Li)	n 2.49 (Li)	Eglestonite $Hg_6Cl_3O_2H$	Isometric	Dodecahedral, cubic, oct, massive	None ?	
2.49 (Li)	n 2.49 (Li)	Monteponite CdO	Isometric	Octahedral, powdery	{111} dist	Penetration
2.61 (Li)	n 2.61 (Li)	Blakeite $Fe_2(TeO_3)_3$		Crusts, microcrystalline	Friable	
2.70	n 2.70	Alabandite MnS	Isometric	Cubic or octahedral crystals	{100} perfect	{111}
>2.72 (Li)	n>2.72 (Li)	Tennantite $(Cu,Fe)_{12}As_4S_{13}$	Isometric	Tetrahedral, massive, granular	None	Twin axis [111], contact or penetration
>2.72 (Li)	n>2.72 (Li)	Tetrahedrite (Falkenhaynite) (Stylotypite) $Cu_{12}Sb_4S_{13}$	Isometric	Tetrahedral, massive, granular	None	Twin axis [111], contact or penetration
2.849	n 2.849	Cuprite Cu_2O	Isometric	Cubic, octahedral, massive	{111} interrupted {001} rare	
2.85	n 2.85	Pandaite $(Ba,Sr)_2(Nb,Ti)_2(O,OH)_7$	Isometric	Small octahedral xls	{111} very poor	

** n variation exceeds 0.10

Color In Thin Section	Hardness Sp Gravity	Color Hand Sample	Alteration	Occurrence	Remarks	Reference
Dk brown in thin splinters	5+ 5.16-5.24	Black		Talc rocks	Spinel group, magnetite series. Strongly magnetic	Dana I, 698-707
Colorless, pale yel to deep brown (often zoned)	5½ 3.98-4.26	Yellow to brown or black	Lucoxene	Alkali or ultra mafic igneous rocks with nepheline, melilite, olivine	Very weak birefringence	p. 30
Colorless	2½ 5.60-5.68	Colorless, lt yel, pink, red-brown		Oxide zones Cu-deposits in arid regions	Fluor U.V. dark red. Anomalous biref	Dana II, 20-22
Pale yellow, sometimes reddish tint	7.5 5.01	Yellow-orange to orange-red		Alkali syenites, periclase-brucite marble		Am Min 55, 318
Dk brown ~opaque	5½-6½ 4.43-4.55	Black, brownish-black		Xline limestones, chlorite schist, fumeroles	Magnetite series of spinel group. Strongly magnetic	Dana I, 698-707
Brown ~ opaque	5 5.0-5.2	Brown, black		Formed in gossans	Strongly magnetic	Dana I, 708-709
Colorless	10 3.511	Colorless, yellowish, black, etc.	None	Olivine peridotites with pyroxene, garnet, magnetite, phlogopite		p. 4
Yellow to golden-yel	Soft 8.80	Green, yel-green, brn-green		Secondary mineral in Bi-deposits with bismutite, pucherite	Dimorph bismite	Dana I, 601
Colorless, pale yel, pale brown	3½-4 3.9-4.1	Honey yellow, brown, black	Smithsonite, hemimorphite, goslarite, limonite	Hydrothermal sulfide veins with galena, chalcopyrite, pyrite		p. 6
Pale yellow, yellow-brown	2½ 8.33-8.56	Yellow, orange-yel, brown-yel		Oxide zones Hg-deposits	Blackened by NH_4OH, decomposed by acids	Dana II, 51-52
Red to orange-brown	3 8.1-8.2	Black		Oxide zones Zn-deposits		Dana I, 502-503
Yellow, yellow-brown	2+ 3.1+	Dk brown, red-brown		Oxide zones Te-deposits	Essentially isotropic	Dana II, 643
Almost opaque	3½-4 3.95-4.05	Iron black		Sulfide vein deposits with pyrite, sphalerite, galena, etc.		Dana I, 207-208
Deep red in very thin splinters	3-4½ 4.59-4.75	Steel gray to iron black		Hydrothermal ore veins, contact metasomatic deposits, pegmatites	Tetrahedrite-Tennantite series. Decomp. HNO_3	Dana I, 374-384
Cherry-red in very thin splinters	3-4½ 4.6-5.1	Steel gray to iron black		Hydrothermal ore veins, contact metasomatic ores, pegmatites	Tetrahedrite-Tennantite series. Decomp. HNO_3	Dana I, 374-384
Red, orange-yellow	3½-4 6.14-6.15	Red, red-brown, ~black		Oxide zones Cu-deposits	Anomalous anisotropism & pleochroism. Sol HCl, NH_4OH	Dana I, 491-494
Colorless, yellowish	4½-5 4.00-4.01	Yellowish-gray to olive-gray		Biotite-rich contact zone near carbonatite	Pyrochlore group	Min Mag 32, 10-25

n_ω Var.	Refractive Index	Mineral Name and Composition	$(n_\varepsilon-n_\omega)$	Crystal System	Habit	Cleavage	Twinning
1.309	n_ω 1.30907 n_ε 1.31052	Ice H_2O	0.0145	Hexagonal	Snow xls, lath xls	None	{0001} common
1.378	n_ω 1.378 n_ε 1.390	Sellaite MgF_2	0.012	Tetragonal	Prism-acic xls (c), fib. agg	{010} & {110} perf	{110}
1.440	n_ω 1.440 n_ε 1.445	Schairerite $Na_3SO_4(F,Cl)$	0.005	Trigonal	Minute trigonal xls, rhombs	None	{0001} common
1.447	n_ω 1.447 n_ε 1.449	Galeite $Na_{15}(SO_4)_5F_4Cl$	0.002	Trigonal	Rhomb, hex prisms		
1.461	n_ω 1.461 n_ε 1.477	Tincalconite $Na_2B_4O_5(OH)_4 \cdot 3H_2O$	0.016	Trigonal	Powdery, pseudo-cubic xls		
1.471	n_ω 1.471 n_ε 1.474	Erionite (Offretite) $(Ca,Na_2,K_2)_{1.5}Al_9Si_{27}O_{72} \cdot 27H_2O$	0.003	Hexagonal	Rad fib, wool-like		
1.472	n_ω 1.472 n_ε 1.492	Gagarinite $NaCaY(F,Cl)_6$	0.020	Hexagonal	Massive, crypto-xline	{10$\bar{1}$0} fair	
1.479	n_ω 1.479 n_ε 1.481	Herschelite $(Na,Ca,K)AlSi_2O_6 \cdot 3H_2O$	0.002	Trigonal	Agg. hex. plates	{10$\bar{1}$1}	
1.484	n_ω 1.484 n_ε 1.603	Urea $CO(NH_2)_2$	0.119	Tetragonal	Elong. pyram. xls.		
1.488	n_ω 1.488 n_ε 1.500	Douglasite $K_2FeCl_4 \cdot 2H_2O$	0.012	Mono β=105°	Massive, coarse granular	{$\bar{2}$01}	
1.487 1.491	n_ω 1.487-1.491 n_ε 1.492-1.499	Aphthitalite (Glaserite) $(K,Na)_3Na(SO_4)_2$	0.005 to 0.008	Trigonal	Tabular, pseudo-hexagonal twins	{10$\bar{1}$0} fair {0001} poor	{11$\bar{2}$0} cyclic, {0001}
1.496	n_ω 1.496 n_ε 1.502	Tugtupite (Beryllosodalite) $Na_4BeAl(SiO_3)_4Cl$	0.006	Tetragonal	Massive, compact, cryptocrystalline	{hkℓ} dist	{101}
1.508 1.511	n_ω 1.508-1.511 n_ε 1.509-1.511	Leucite $KAlSi_2O_6$	~0.001	Tetragonal (Pseudoiso-metric)	Trapezohedral xls.	{110} very poor	Several sets polysynthetic
1.510	n_ω 1.510 n_ε 1.512	Franzinite $\sim(Na,Ca)_7Al_6Si_6O_{24}[(SO_4),(OH),(CO_3),Cl]_3 \cdot H_2O$	0.002	Trigonal	Stubby prism xls		
1.510	n_ω 1.510 n_ε 1.515	Francoanellite $H_6K_3Al_5(PO_4)_8 \cdot 13H_2O$	0.005	Trigonal	Nodular agg, earthy		
1.510	n_ω 1.510 n_ε 1.545	Solongoite $Ca_4B_6O_8Cl(OH)_9$	0.035	Mono	Prism xls		
*1.500 1.515	n_ω 1.500-1.515 n_ε 1.502-1.512	Garronite $Na_2Ca_5Al_{12}Si_{20}O_{64} \cdot 27H_2O$	0.002	Tetragonal	Rad-acic in amygdules	{110} dist	
1.516	n_ω 1.516 n_ε 1.520	Leifite $(Na,H_3O)_2(Si,Al,Be,B)_7(O,OH,F)_{14}$	0.004	Trigonal	Acic. xls	{10$\bar{1}$0} dist	

* n_ω variation exceeds 0.01

Color In Thin Section	Hardness Sp Gravity	Color Hand Sample	Alteration	Occurrence	Remarks	Reference
Colorless	1½ 0.9167	Colorless, white		Snowfields, glaciers, ice caves	Liquid above 0°C	Dana I, 494-498
Colorless	5 3.14-3.15	Colorless, white		Fumeroles or evaporites with sulfur, gypsum, anhydrite, fluorite	Slightly sol H_2O. Decomp H_2SO_4	Dana II, 37-39
Colorless	3½ 2.60-2.63	Colorless		Evaporite lake deposits with tychite, pirssonite, gaylussite, etc.	Slowly sol H_2O	Dana II, 547-548
Colorless	2.605-2.61	White		Xls in saline clays with gaylussite and northupite		Am Min 48, 485-510
Colorless	1.88-1.94	White		Dehydration of borate minerals in playa-lake evaporites	Sol H_2O. Alkaline taste	Am Min 58, 523-530
Colorless	~2.02	White		Fractures in acid tuffs, basalts		Am Min 54, 875-886
Colorless	4½ 4.21	White, yellowish, pink	Tengerite, synchisite, yttrofluorite	Alkali granites	Decomposed by acids. Small 2V.	Am Min 47, 805
Colorless	4-5 2.05-2.16	White		Cavities and veins in lavas	Zeolite group	Am Min 47, 985-987
Colorless	1.33	White, yellowish, brownish		Cave guano with weddellite, phosphammite, NH_4-aphthitalite	Sol H_2O	Am Min 59, 874
Colorless, pale grn	2.16	Light green		Evaporite salts	Sol H_2O	Dana II, 100
Colorless	3 2.66-2.72	Colorless, white, bluish, greenish		Evaporite sediments, efflorescences, fumerole deposits	Soluble in H_2O and acids	Dana II, 400-403
Colorless	~4 2.28-2.36	White, rose, bluish, greenish		Alkali veins and pegmatites with analcime, chkalovite	Possible small 2V=0°-10°. Fluor. U.V-bright rose	Am Min 48, 1178
Colorless	5½-6 2.47-2.50	Colorless, pale gray	Pseudoleucite (fine aggregate of nepheline & K-feldspar), analcite	K-rich, Si- and Na-poor volcanic flows and tuffs	Feldspathoid	p. 367
Colorless	5 2.49-2.52	White		Ejecta in pumice deposits with afghanite and liottite	Cancrinite group	Am Min 62, 1259
Colorless	~1 2.26	Yellowish-white		With bat guano in caves	Sol dilute HCl or HNO_3	Am Min 61, 1054
Colorless	3.5 2.51-2.53	Colorless		Contact limestone with szaibelyite, grossular, kurchatovite	Readily sol cold HCl or H_2SO_4	Am Min 60, 162-163
Colorless	~4 2.13-2.20	White		Amygdules in basalts with other zeolites	Zeolite group	Min Mag 33, 173-186
Colorless	6 2.49-2.57	Colorless		Cavities in alkali-pegmatites with acmite, calcite		Win & Win II, 355

Uniaxial +

n_ω Var.	Refractive Index	Mineral Name and Composition	$(n_\varepsilon - n_\omega)$	Crystal System	Habit	Cleavage	Twinning
1.519	n_ω ~1.519 n_ε ~1.520	Davyne (Microsommite) $(Na,Ca,K)_8Al_6Si_6O_{24}(Cl,SO_4,CO_3)_{2-3}$	0.001	Hexagonal	Prismatic, groups	{10$\bar{1}$0} perf {0001} poor	
1.523	n_ω 1.523 n_ε 1.529	Afghanite $(Na,Ca,K)_{12}(Si,Al)_{16}O_{34}(Cl,SO_4,CO_3)_4 \cdot O \cdot 6H_2O$	0.006	Hexagonal	Massive	{10$\bar{1}$0}	
1.523	n_ω 1.523 n_ε 1.544	Weddellite $CaC_2O_4 \cdot 2\text{-}3H_2O$	0.021	Tetragonal	Minute pyramidal xls	None	
1.524	n_ω 1.524 n_ε 1.530	Berlinite $AlPO_4$	0.006	Trigonal	Mass, granular	None	
1.527	n_ω 1.527 n_ε 1.535	Roggianite $NaCa_6Al_9Si_{13}O_{46} \cdot 20H_2O$	0.008	Tetragonal	Fib. agg (c)	{110} perf	
1.52	n_ω 1.52 n_ε 1.55	Koenenite $Na_4Mg_9Al_4Cl_{12}(OH)_{22}$	0.03	Trigonal	Rosettes, intergrown tab	{0001} perf	
1.532	n_ω 1.532 n_ε 1.570	Fibroferrite $Fe^{3+}SO_4OH \cdot 5H_2O$	0.038	Trigonal	Rad fib, botryoidal, crusts	{0001} perf	Trilling
1.533	n_ω 1.533 n_ε 1.540	β-Quartz SiO_2	0.007	Hexagonal	Stubby bipyramids	{10$\bar{1}$1} good, {10$\bar{1}$0} fair	
1.531 1.536	n_ω 1.531-1.536 n_ε 1.533-1.538	Apophyllite $KCa_4(Si_4O_{10})_2F \cdot 8H_2O$	0.000 to 0.003	Tetragonal	Square prism xls, platy	{001} perf, {110} imperf	{111} rare, penetration
*1.526 1.544	n_ω 1.526-1.544 n_ε 1.531-1.553	Chalcedony SiO_2	0.005 to 0.009	Trigonal (microcrystal-line)	Colloform banding, fib, acic, granular	None	None
1.536	n_ω 1.536 n_ε 1.544	Yagiite $(Na,K)_3Mg_4Al_6(Si,Al)_{12}O_{60}$	0.008	Hexagonal	Massive		
1.536	n_ω 1.536 n_ε 1.545	Ashcroftine (Kalithomsonite) $KNaCaY_2Si_6O_{12}(OH)_{10} \cdot 4H_2O$	0.009	Tetragonal	Fibrous on c	{100} perf {001} dist	
1.537	n_ω 1.537 n_ε 1.542	Roedderite $(Na,K)_2Mg_5Si_{12}O_{30}$	0.005	Hexagonal	Platy xls	None	
1.540	n_ω 1.540 n_ε 1.560	Chlormanasseite $Mg_5Al_{2.7}(OH)_{16}Cl_2 \cdot 3H_2O$	0.020	Hexagonal	Platy, dipyramidal xls	{001} perfect	
1.544	n_ω 1.544 n_ε 1.553	α-Quartz SiO_2	0.009	Trigonal	Anhedral to euhedral	{10$\bar{1}$1} poor	Rarely observed
*1.536 1.552	n_ω 1.536-1.552 n_ε 1.548-1.558	Coquimbite $Fe_{2-x}Al_x(SO_4)_3 \cdot 9H_2O$	0.009 to 0.006	Trigonal	Short prism, pyra-midal, granular	{10$\bar{1}$1} imperf {10$\bar{1}$0} difficult	{0001}
1.545	n_ω 1.545 n_ε 1.590	Monohydrocalcite $CaCO_3 \cdot H_2O$	0.045	Trigonal	Massive, fine-grained		
1.545 1.547	n_ω 1.545-1.547 n_ε 1.549-1.551	Osumilite $(K,Na)(Mg,Fe^{2+})_2(Al,Fe^{3+})_3(Si,Al)_{12}O_{30} \cdot H_2O$	0.004	Hexagonal	Prism xls (c), tab xls {0001}	None	
1.550	n_ω 1.550 n_ε 1.555	Paracoquimbite $Fe_2(SO_4)_3 \cdot 9H_2O$	0.005	Trigonal	Rhomb, pseudo-cubic xls, granular	{01$\bar{1}$2} & {10$\bar{1}$4} imperf	{0001} common

* n_ω variation exceeds 0.01

Color In Thin Section	Hardness Sp Gravity	Color Hand Sample	Alteration	Occurrence	Remarks	Reference
Colorless	6 2.42-2.52	Colorless, white		Leucite-rich lavas	Cancrinite group	DHZ 4, 310-320
Colorless	5½-6 2.55-2.65	Colorless, bluish		Associated with lazurite, sodalite, nepheline	Cancrinite group	Am Min 53, 2105
Colorless	~4 1.94	Colorless, white, yellowish		Antarctic ocean muds	Insol H_2O	Dana II, 1101-1102
Colorless	6½ 2.62-2.64	Colorless, gray, pink		Contact met., pegmatites	Isotypic with quartz	Am Min 26, 631
Colorless	 2.02	White, yellowish		Fractures in albite dike		Am Min 55, 322-323
O = red-brown, E = colorless	~1½ 1.98	Colorless, yellow, red		Potash evaporite salts with halite, sylvite, carnallite	Decomp. hot HCl	Dana II, 86-87
O = colorless, E = pale amber	2½ 1.92-2.01	White, pale yellow, greenish		Oxidation of pyrite in ore-deposits	Decomposed H_2O	Dana II, 614-616
Colorless to opaque	 2.53	White, gray	β-quartz inverts to α-quartz below 573°C	Exists only briefly in siliceous volcanic lavas	Not found in nature below 573°C	p. 318
Colorless	4½-5 2.33-2.37	White, pink, yellow, green	Opal, chert, kaolin, calcite	Amygdules or veins in basalt with zeolites. Cavities in granite, gneiss	Decomp. HCl. Optically anomalous	p. 309
Colorless, pale yel, brown etc. without pleochroism	6½-7 2.57-2.64	Colorless, many colors & patterns	Rather stable	Devitrification of gelatinous silica in near-surface veins and voids		p. 323
Colorless	 2.70	Colorless		Fe-Ni meteorites		Am Min 54, 14-18
Colorless	 2.51-2.61	White, pink		Cavities in pegmatites or syenites	No longer considered a zeolite	Min Mag 37, 515-517
Colorless	 2.63	Colorless		Meteorites		Am Min 51, 949-955
Colorless	Soft 1.98-2.09	Colorless, greenish-brown		Cavities in magnetite with chlorite		Am Min 64, 1329
Colorless	7 2.65	White, black, violet, yellow, etc.	Usually unaltered	Siliceous igneous rocks (e.g., granite, rhyolite, etc.), pegmatites, detrital sediments, veins and cavities, schists, gneisses, etc.		p. 314
Colorless	2½ 2.11	Colorless, lavender, yel, green		Oxide zones in ore deposits	Dimorph. paracoquimbite, Sol H_2O, astringent. Abnormal interference colors.	Am Min 55, 1534-1540
Colorless	 2.38	White, gray		Lake bottom sediments with calcite	Synthetic compound is optically (-)	Am Min 49, 1151
O = pale blue, E = colorless	 2.64	Dk blue to black		Cavities in andesite with cristobalite, tridymite, hortonolite		Am Min 54, 101-116
Colorless	2½ 2.11	Pale violet		Secondary sulfate with coquimbite	Dimorph coquimbite. Sol H_2O. Astringent taste	Am Min 56, 1567-1572

Uniaxial +

n_ω Var.	Refractive Index	Mineral Name and Composition	$(n_\varepsilon - n_\omega)$	Crystal System	Habit	Cleavage	Twinning
1.550	n_ω 1.550 n_ε 1.650	Vaterite μ-$CaCO_3$	0.100	Hexagonal	Fine Fib, platy		
1.552	n_ω 1.552 n_ε~1.56	Zirklerite $(Fe,Mg,Ca)_9Al_4Cl_{18}(OH)_{12}\cdot14H_2O$ (?)	~0.008	Hexagonal (?)	Massive, granular to fibrous	$\{10\bar{1}1\}$	
1.556	n_ω 1.556 n_ε 1.645	Juliënite $Na_2Co(SCN)_4\cdot8H_2O$	0.089	Tetragonal	Crusts acic xls (c)	{001}	
1.556 1.558	n_ω 1.556-1.558 n_ε 1.610-1.614	Ferrinatrite $Na_3Fe^{3+}(SO_4)_3\cdot3H_2O$	0.055	Trigonal	Prism, rad fib agg	$\{10\bar{1}0\}$ perf $\{11\bar{2}0\}$ less perf	
1.565	n_ω 1.565 n_ε 1.575	Pinnoite $MgB_2O_4\cdot3H_2O$	0.010	Tetragonal	Rad fib, prism (c) nodular		
1.568	n_ω 1.568 n_ε 1.590	Natroalunite $NaAl_3(SO_4)_2(OH)_6$	0.022	Trigonal	Rhomb, tab xls, fib, massive	{0001} dist $\{01\bar{1}2\}$ traces	
1.569	n_ω 1.569 n_ε 1.581	Schaurteite $Ca_3Ge^{4+}[(OH)_6(SO_4)_2]\cdot3H_2O$	0.012	Hexagonal	Fine fib - acic		
1.571	n_ω 1.571 n_ε 1.600	Slavyanskite $CaAl_2O_4\cdot nH_2O$		Tetragonal	Bipyramidal xls	{110} & {100} dist	
1.572	n_ω 1.572 n_ε 1.586	Eucryptite $LiAlSiO_4$	0.014	Trigonal	Prism, massive-granular	$\{10\bar{1}1\}$ indist	
1.573	n_ω 1.573 n_ε 1.599	Tunisite $NaHCa_2Al_4(CO_3)_4(OH)_{10}$	0.026	Tetragonal	Fine agg small tab xls	{001} & {110} dist	
*1.559 1.590	n_ω 1.559-1.590 n_ε 1.580-1.600	Brucite $Mg(OH)_2$	0.02 to 0.01	Trigonal	Fine scales, tab xls, fib (c)	{0001} perf	None
1.573 1.576	n_ω 1.573-1.576 n_ε 1.594-1.596	Preobrazhenskite $Mg_3B_{11}O_{15}(OH)_9$	0.020	Ortho (Pseudohex)	Massive, fine-grained nodular		
*1.55 1.583	n_ω 1.55 -1.583 n_ε	Bayerite $Al(OH)_3$	Small	Hexagonal	Micro fib		
*1.568 1.585	n_ω 1.568-1.585 n_ε 1.590-1.601	Alunite $KAl_3(SO_4)_2(OH)_6$	0.010 to 0.023	Trigonal	Rhomb-tab xls, fib-column, gran	{0001}	None
1.580	n_ω 1.580 n_ε 1.588	Coeruleolactite $(Ca,Cu)Al_6(PO_4)_4(OH)_8\cdot4\text{-}5H_2O$	0.008	Triclinic	Micro-xline, fib, botryoidal	None	
1.580 1.581	n_ω 1.580-1.581 n_ε 1.596-1.631	Nordstrandite $Al(OH)_3$	0.016 to 0.050	Triclinic	Tab {001} xls, rhomb outline	{001} perf	
1.583	n_ω 1.583 n_ε 1.593	Xonotlite (Jurupaite) $Ca_3Si_3O_8(OH)_2$	0.010	Mono $\beta \sim 90°$	Acic-fib xls, matted fib, massive	One good	
1.5886	n_ω 1.5886 n_ε 1.5894	Rinneite $K_3NaFeCl_6$	~0.001	Trigonal	Massive granular, prism, tab xls	$\{11\bar{2}0\}$ dist	{0001}
1.59	n_ω 1.59 n_ε	Chlormanganokalite K_4MnCl_6	Small	Trigonal	Rhombohedral	None	

* n_ω variation exceeds 0.01

Color In Thin Section	Hardness Sp Gravity	Color Hand Sample	Alteration	Occurrence	Remarks	Reference
Colorless	~3 2.54-2.65	Colorless		Hydrogel pseudomorphs after larnite	Trimorph calcite, aragonite. Sol HCl with effervescence	Min Mag 32, 535-544
Colorless	~3½ ~2.6	Lt gray		Marine evaporite salts with halite, K-salts, anhydrite		Dana II, 87
Blue	 1.64-1.65	Blue		Co-wad	Sol H_2O → red solution	Dana II, 1106-1107
Colorless	2½ 2.55-2.61	White, lt green, blue, purplish		Arid sediments with other sulfates. Fumeroles		Dana II, 456-458
Pale yellow	3½ 2.29	Yellow, yellow-green		Evaporite salt deposits with kainite, boracite	Solution dilute acids, boiling H_2O	Dana II, 334-335
Colorless	3½-4 2.6-2.9	White, yellowish, brownish		Fumeroles, metamorphic rocks, veins in ore deposits	Alunite group	Dana II, 556-560
Colorless	 2.64	White		Alteration Ge-bearing ores		Am Min 53, 507
Colorless	4-5 2.49-2.52	Colorless		Salt deposits		Am Min 63, 599
Colorless	6½ 2.66	Colorless, white		Li-pegmatites	Fluor U.V.-pink	Am Min 47, 557-561
Colorless	3½-4½ 2.48-2.51	Colorless, white		Pb-Zn hydrothermal veins with marcasite, pyrite, sphalerite		Am Min 54, 1-13
Colorless	2½ 2.40	White, pale grn, blue, brown	Hydromagnesite, serpentine, periclase	Low-T hydrothermal veins with periclase, magnesite, chlorite	Easily sol acids	p. 41
Colorless	4½-5	Colorless, yellow dk gray	Inyoite	Evaporite beds with halite, polyhalite, kaliborite, boracite, inyoite		Am Min 55, 1071
Colorless	? 2.53-2.54	White		Argillaceous sediments	Dimorph. of Gibbsite	Am Min 49, 819
Colorless	3½-4 2.6-2.9	White, gray, reddish, yellow		Felsic volcanics altered by sulfur gases and solutions, with clays	Slowly sol dilute H_2SO_4	p. 94
Colorless	5 2.57-2.70	White, pale blue		Oxide zones Cu-deposits	Ca-turquoise, ~uniaxial	Dana II, 961
Colorless	3 2.44-2.51	Colorless, pink		Solution cavities in limestone	Trimorph gibbsite and bayerite	Nature, 196, 264-266
Colorless	6-6½ 2.71	White, colorless, gray, pink		Veinlets in serpentine, limestone contact zones	Small 2V, Sol. HCl	Win & Win II, 455
Colorless, yellowish	3 2.35-2.41	Colorless, yellow, rose, violet		Marine salt beds with halite, anhydrite, sylvite, kieserite	Sol H_2O. Taste astringent	Dana II, 107-108
Colorless	2½ 2.31	Pale yellow		Amygdules in alkaline lavas	Sol H_2O	Dana II, 109

Uniaxial +

n_ω Var.	Refractive Index	Mineral Name and Composition	$(n_\varepsilon - n_\omega)$	Crystal System	Habit	Cleavage	Twinning
1.586 1.594	n_ω 1.586-1.594 n_ε 1.595-1.604	Wardite (Soumansite) $NaAl_3(PO_4)_2(OH)_4 \cdot 2H_2O$	0.009 to 0.010	Tetragonal	Pyram xls, rad fib, gran crusts	{001} perf	
*1.580 1.600	n_ω 1.580-1.600 n_ε 1.640-1.680	Cacoxenite $Fe_4^{3+}(PO_4)_3(OH)_3 \cdot 12H_2O$	0.060 to 0.080	Hexagonal	Rad tufted, fib (c)	None?	
*1.567 1.623	n_ω 1.567-1.623 n_ε 1.572-1.633	Eudialyte $Na_4(Ca,Fe,Ce,Mn)_2ZrSi_6O_{17}(OH,Cl)_2$	0.002 to 0.008	Trigonal	Hex tab, prism, rhomb	{0001} dist, {10$\bar{1}$0} & {11$\bar{2}$0} imperf	
1.596	n_ω 1.596 n_ε 1.624	Catapleiite $(Na_2,Ca)ZrSi_3O_9 \cdot 2H_2O$	0.028	Hexagonal	Thin hex plates	{10$\bar{1}$0} perf, {10$\bar{1}$1} & {10$\bar{1}$2} imperf {000$\bar{1}$} parting	{10$\bar{1}$0} or {000$\bar{1}$} polysynthetic
1.598	n_ω 1.598 n_ε 1.598	Combeite $Na_4Ca_3Si_6O_{16}(OH,F)_2$	small	Trigonal	Hex prisms	None	
1.603	n_ω 1.603 n_ε 1.639	Calcium Catapleiite $CaZrSi_3O_9 \cdot 2H_2O$	0.036	Hexagonal	Massive		
1.606	n_ω 1.606 n_ε 1.620	Saryarkite $(Ca,Y,Th)_2Al_4(SiO_4,PO_4)_4(OH) \cdot 9H_2O$	0.014	Tetragonal	Massive	None	
1.604 1.609	n_ω 1.604-1.609 n_ε 1.625-1.653	Narsarsukite (Goureite) $Na_2(Ti,Fe)Si_4(O,F)_{11}$	0.021 to 0.044	Tetragonal	Short prism or tab {001} xls	{100} & {110} perf	
*1.588 1.635	n_ω 1.588-1.635 n_ε 1.560-1.590	Coalingite $Mg_{10}Fe_2(CO_3)(OH)_{24} \cdot 2H_2O$	0.022 to 0.045	Hexagonal	Platelets, fibrous	Primatic and 45° to elong	
*1.60 1.622	n_ω 1.60-1.622 n_ε 1.61-1.631	Crandallite (Pseudowavellite) $CaAl_3(PO_4)_2(OH)_5 \cdot H_2O$	0.009	Hexagonal	Trig prisms, rad fib massive	{0001} perf	
*1.625	n_ω 1.625 n_ε 1.655	Wadeite $K_2CaZr(SiO_3)_4$	0.030	Hexagonal	Hex prisms	None	
*1.604 1.640	n_ω 1.604-1.640 n_ε 1.615-1.657	Sarcolite $NaCa_4Al_3Si_5O_{19}$	0.011 to 0.017	Tetragonal	Equant xls		
1.625	n_ω 1.625 n_ε 1.63	Gorceixite $BaAl_3(PO_4)_2(OH)_5 \cdot H_2O$	small	Trigonal	Rad fib agg, botryoidal	None	
*1.618 1.649	n_ω 1.618-1.649 n_ε 1.622-1.646	Metatorbernite $Cu(UO_2)_2(PO_4)_2 \cdot 8H_2O$	0.002	Tetragonal	Platy {001}, rosettes, sheaflike	{001} perf	
1.629	n_ω 1.629 n_ε 1.632	Babefphite $BaBe(PO_4)(O,F)$	0.003	Tetragonal	Grains, tab	None	
1.631	n_ω 1.631 n_ε 1.652	Faheyite $(Mn,Mg,Na)Be_2Fe_2^{3+}(PO_4)_4 \cdot 6H_2O$	0.021	Hexagonal	Fib (c), botryoidal, rosettes	Perf prismatic	
1.631 1.635	n_ω 1.631-1.635 n_ε 1.646-1.649	Svanbergite $SrAl_3PO_4SO_4(OH)_6$	0.015	Trigonal	Rhomb-pseudocubic xls granular	{0001} dist	
*1.620 1.653	n_ω 1.620-1.653 n_ε 1.630-1.661	Goyazite (Strontiohitchcockite) (Hamlinite) $SrAl_3(PO_4)_2(OH)_5 \cdot H_2O$	0.010 to 0.008	Trigonal	Rhomb, pseudocubic	{0001} perf	
*1.651 1.632	n_ω 1.651-1.632 n_ε 1.651-1.640	Åkermanite $Ca_2MgSi_2O_7$	0.000 to 0.008	Tetragonal	Prism. xls	{001} dist., {110} poor	None

* n_ω variation exceeds 0.01

Color In Thin Section	Hardness Sp Gravity	Color Hand Sample	Alteration	Occurrence	Remarks	Reference
Colorless	5 2.81	Colorless, white, lt green, bluish		Variscite nodules. Pegmatites with amblygonite, apatite, morinite	Sol in acids with difficulty	Dana II 940-941
O = Pale yellow, E = orange-yellow	3-4 2.25-2.4	Golden yellow, brown-yellow		Oxide zones Fe-deposits	Sol acids	Am Min 51, 1811-1814
Pink or yellow E > O	5-5½ 2.74-2.98	Yellow-brown, brown-red, pink		Nepheline syenite, nepheline pegmatites	Var. Eucolite, Sol. HCl, usually zoned	Vlasov II, 355-364
Colorless	5-6 2.65-2.79	Yellow, brown, orange, skyblue		Alkali igneous rocks and pegmatites	May be biaxial	Vlasov II, 367-370
Colorless	 2.84	Colorless		Melilite-nephelinite volcanics		Min Mag 31, 503-510
~Opaque	4½-5 2.77-2.78	Pale yellow		Cavities in syenite pegmatite		Am Min 49, 1153
Colorless	3½-4 3.07-3.35	White		Altered felsic rocks with barite, molybdenite, pyrite, Fe-oxides	Partly sol HCl, HNO_3, H_2SO_4(1:1)	Am Min 49, 1775-1776
O = colorless E = pale yellow	6-7 2.78	Colorless, yellow, green, brown		Alkali rocks, carbonatites with aegirine		Am Min 47, 539-556
O = golden brown, E = colorless	1-2 2.33-2.42	Red-brown, lt. brown, yellow		Veins in serpentenite	Small 2V	Am Min 54, 437-447
Colorless	5 2.78-2.92	Yellow, white, gray		Nodules in sediments, with variscite	Triclinic dimorph	Am Min 48, 1144-1153
Colorless	5½-6 3.10-3.16	Colorless, pink, lilac		K-rich igneous rocks with zeolites, leucite, perovskite, barite		Vlasov II, 371-372
Colorless	6 2.92-2.96	Flesh red		Potassic volcanics		Win & Win II, 496
Colorless	6 3.30-3.32	White, brown		Spherical masses in detrital sediments or novaculite		Dana II, 833
Pale green, O > E	2½ 3.4-3.8	Pale green to dk green		Oxide zones Cu-U-veins containing uraninite	Not fluor U.V., Anomalous interference colors	USGS Bull 1064, 208-211
Colorless	 4.31-4.44	White		Fluorite deposit in subalkalic syenite with zircon, phenakite, scheelite		Am Min 51, 1547
Colorless	 2.66-2.67	White, pale blue, brownish		Coatings on pegmatite minerals		Am Min 38, 263-270
Colorless	5 3.22-3.24	Colorless, yellow, rose, red-brown		Al-rich metamorphic rocks with pyrophyllite, kyanite, diaspore	Insol acids	Dana II, 1005-1006
Colorless	4½-5 3.26-3.29	Colorless, yellowish, pink		Hydrothermal veins, pegmatites	Possible small 2V	Dana II, 834-835
Colorless	5-6 2.99-2.94	Colorless, yellow, gray-green	Cebollite, calcite, zeolite, garnet, diopside	High grade contact metamorphism of carbonate rocks	Melilite group	p. 159

n_ω	Refractive Index	Mineral Name and Composition	$(n_\varepsilon - n_\omega)$	Crystal System	Habit	Cleavage	Twinning
1.636	n_ω 1.636 n_ε 1.647	Woodhouseite $CaAl_3PO_4SO_4(OH)_6$	0.011	Trigonal	Pseudocubic to tab xls	{0001} perf	
1.643	n_ω 1.643 n_ε 1.73	Synchysite-(Y), (Doverite) $(Y,Ce)Ca(CO_3)_2F$	0.087	Ortho	Massive, fine-grained masses		
1.646	n_ω 1.646 n_ε 1.647	Braitschite $(Ca,Na_2)_7(Ce,La)_2B_{22}O_{43}\cdot 7H_2O$	0.001	Hexagonal	Hex plates		
1.641 1.650	n_ω 1.641-1.650 n_ε 1.740-1.745	Synchysite (Synchisite) $(Ce,La)Ca(CO_3)_2F$	0.099 to 0.091	Ortho	Minute tab xls, pseudohex	{0001} dist (parting?)	Very common
*1.644 1.658	n_ω 1.644-1.658 n_ε 1.697-1.709	Dioptase $CuSiO_2(OH)_2$	0.053 to 0.051	Trigonal	Prism rhomb, massive	{10$\bar{1}$1} perf	
1.653	n_ω 1.653 n_ε 1.675	Plumbogummite $PbAl_3(PO_4)_2(OH)_5\cdot H_2O$	0.022	Trigonal	Minute hex xls, botryoidal, crusts		
1.654	n_ω 1.654 n_ε 1.670	Phenacite (Phenakite) Be_2SiO_4	0.016	Trigonal	Rhomb, acic, prism xls, rad fib	{11$\bar{2}$0} dist, {10$\bar{1}$1} imperf	{10$\bar{1}$0} common
1.654	n_ω 1.654 n_ε 1.703	Rhabdophane $(Ce,Y,La,Di)(PO_4)\cdot H_2O$	0.049	Hexagonal	Rad fib, botryoidal-stalactitic	None	
1.656	n_ω 1.656 n_ε 1.682	Despujolsite $Ca_3Mn^{4+}(SO_4)_2(OH)_6\cdot 3H_2O$	0.026	Hexagonal	Hex prisms	None	
1.662	n_ω 1.662 n_ε 1.663	Cahnite $Ca_2BAsO_4(OH)_4$	0.001	Tetragonal	Pseudotetrahedral	{110} perf	{110} common
1.662	n_ω 1.662 n_ε 1.756	Roentgenite (Röntgenite) $Ca_2(Ce,La)_3(CO_3)_5F_3$	0.094	Trigonal	Rhomb or bipyram xls	None	
1.671	n_ω 1.671 n_ε 1.676	Fillowite $H_2Na_6(Mn,Fe,Ca)_{14}(PO_4)_{12}\cdot H_2O$ (?)	0.005	Trigonal	Rhomb, granular	{0001} perf	
1.671	n_ω 1.671 n_ε 1.689	Hinsdalite $(Pb,Sr)Al_3PO_4SO_4(OH)_6$	0.018	Trigonal	Rhomb (pseudocubic), tab, gran	{0001} perf	
1.6751	n_ω 1.6751 n_ε 1.6850	Bazirite $BaZrSi_3O_9$	0.01	Hexagonal	Anhedral grains		
1.676	n_ω 1.676 n_ε 1.757	Parisite $Ca(Ce,La)_2(CO_3)_3F_2$	0.081	Trigonal	Rhomb or hex dipyramid xls	{0001} dist (parting?)	
1.680	n_ω 1.680 n_ε 1.695	Brockite $(Ca,Th,Ce)(PO_4,CO_3)\cdot H_2O$	0.015	Hexagonal	Rad fib agg	None	
1.6813	n_ω 1.6813 n_ε 1.691	Bazirite $BaZrSi_3O_9$	0.01	Hexagonal	Anhedral grains		
*1.695 1.680	n_ω 1.695-1.680 n_ε 1.705	Florencite $CeAl_3(PO_4)_2(OH)_6$	0.010	Trigonal	Rhomb, pseudocubic	{0001} dist, {11$\bar{2}$0} traces	
1.689	n_ω 1.689 n_ε 1.695	Ilimaussite $Ba_2Na_4CeFeNb_2Si_8O_{28}\cdot 5H_2O$	0.006	Hexagonal	Lamellar agg		Polysynthetic (rare)
1.689	n_ω 1.689 n_ε 1.696	Lithiophilite-Triphylite $Li(Mn,Fe)PO_4$	0.007	Ortho	Cleavage masses	{001} perf, {010} dist, {110} imperf	None

* n_ω variation exceeds 0.01

Color In Thin Section	Hardness Sp Gravity	Color Hand Sample	Alteration	Occurrence	Remarks	Reference
Colorless	4½ 3.00-3.01	White to flesh pink		Cavities in quartz veins with topaz, augelite, lazulite, pyrophyllite		Dana II, 1006-1007
Brownish-red	~6½ 3.61-3.72	Reddish brown		Pegmatites with xenotime, cenosite, hematite		Am Min 47, 337-343
Colorless	 2.84-2.90	Colorless, white		Evaporite salt deposits		Am Min 53, 1081-1095
Colorless, pale yellow-brown	4½ 3.90-4.15	Gray, yellowish to brown		Alkali pegmatites, carbonatites, alpine veins-with thorite, parisite, etc.		Am Min 38, 932-963
Pale green O > E	5 2.28-3.35	Green, dk blue-green		Oxide zones Cu-deposits	Gelat HCl	Win & Win II, 453
Colorless	4½-5 4.01-4.08	White, gray, yellow, brown		Oxide zone Pb-deposits	Sol hot acids	Dana II, 831-832
Colorless	7½-8 2.93-3.00	Colorless, yellow, pink, brown		Granite pegmatites, greisens, hydrothermal and Alpine veins		Vlasov II, 81-85
Colorless	3½ 3.94-4.01	Brown, pink, white		Thin crusts on limonite	Easily sol HCl	Dana II, 774
O ~ colorless E - pale yellow	~2½ 2.46-2.54	Lemon yellow		Cavities with gaudefroyite	Abnormal interference colors	Am Min 54, 326
Colorless	3 3.16-3.18	Colorless, white		Contact meta skarns, pegmatites	Abnormal interference colors	Am Min 46, 1077-1085
Pale yellow	~4½ 4.19	Yellow to brown		Alkali pegmatites with synchysite, parisite, bastnaesite		Am Min 38, 932-962
Colorless, yellow	4½ 3.42-3.43	Yellow-brown, red-brown		Granite pegmatites	Small 2V. Sol acids	Am Min 50, 1647-1669
Colorless	4½ 3.65-4.07	Colorless, yellowish, greenish		Late mineral in sulfide ore veins	Insol acids. Possible biaxial sectors	Dana II, 1004
Colorless	 3.82	Colorless		Alkali granite with elpidite	Fluor. short λ U.V.-bluish-white	Min Mag 42, 35
Colorless to yellow E > O	4½ 4.36-4.38	Yellow to brown		Alkali pegmatites and plutons with aegirine, riebeckite	Sol hot acids	Dana II, 282-285
Pale yellow on thin edges	? 3.9-4.0	Red-brown to pale yellow		Hydrothermal veins in granitic rocks	Hematite inclusions	Am Min 47, 1346-1355
Colorless	 3.82	Colorless		Alkali granite with elpidite	Fluor. short λ U.V.-bluish-white	Min Mag 42, 35
Colorless	5-6 3.46-3.71	Pink, pale yellow		Mica schists, pegmatites, carbonatites		Dana II 838-839
Pale yellow	~4 3.6	Brownish yellow		Hydrothermal veins in alkali plutons		Am Min 54, 992-993
Colorless, pale pink	4½-5 3.34-3.48	Salmon-pink, blue-gray	Rare phosphates	Granite pegmatites with Li-minerals and other phosphates	Lithiophilite-Triphylite series	p. 79

Uniaxial +

n_ω Var.	Refractive Index	Mineral Name and Composition	$(n_\varepsilon - n_\omega)$	Crystal System	Habit	Cleavage	Twinning
1.690	n_ω 1.690 n_ε 1.673	Iron Åkermanite $Ca_2Fe^{2+}Si_2O_7$	0.017	Tetragonal	Prism. xls	{001} dist., {110} poor	None
1.691	n_ω 1.691 n_ε 1.719	Willemite Zn_2SiO_4	0.028	Trigonal	Prism hex xls, massive, fib	{0001} & {11$\bar{2}$0} poor	
1.697	n_ω 1.697 n_ε 1.704	Muirite $Ba_{10}Ca_2MnTiSi_{10}O_{30}(OH,Cl,F)_{10}$	0.007	Tetragonal	Tet prisms	{001} & {100} indist	
1.701	n_ω 1.701 n_ε 1.707	Kemmlitzite $(Sr,Ce)Al_3(AsO_4)(P,S)O_4(OH)_6$	0.006	Trigonal	Pseudocubic rhomb xls	{0001} indist	
1.704	n_ω 1.704 n_ε 1.765	Goudeyite $Cu_6Al(AsO_4)_3(OH)_6\cdot3H_2O$	0.061	Hexagonal	Hair-like tufts. Hex prisms		
*1.707 1.723	n_ω 1.707 n_ε 1.722	Amakinite $(Fe,Mg)(OH)_2$	0.015	Trigonal	Irregular grains	Poor	
*1.701 1.723	n_ω 1.701-1.723 n_ε 1.777-1.815	Agardite $(Y,Ca)_2Cu_{12}(AsO_4)_6(OH)_{12}\cdot6H_2O$	0.072 to 0.082	Hexagonal	Acicular		
1.714	n_ω 1.714 n_ε 1.731	Osarizawaite $PbCuAl_2(SO_4)_2(OH)_6$	0.017	Trigonal	Tiny Hex or rhomb xls		
1.719	n_ω 1.719 n_ε 1.733	Bromellite BeO	0.014	Hexagonal	Hex prisms	{10$\bar{1}$0} dist	
1.72	n_ω 1.72 n_ε 1.82	Bastnaesite (Bastnasite) $(Ce,La)CO_3F$	0.10	Hexagonal	Platy {0001}, granular	{10$\bar{1}$0} poor, {0001} parting	
1.720 1.724	n_ω 1.720-1.724 n_ε 1.816-1.827	Xenotime YPO_4	0.095 to 0.170	Tetragonal	Bipyram-prism xls, rad agg	{110} dist	{101} simple, rare
*1.724 1.738	n_ω 1.724-1.738 n_ε 1.746-1.758	Connellite $Cu_{19}Cl_4SO_4(OH)_{32}\cdot3H_2O$	~0.020	Hexagonal	Rad acic		
*1.724 1.738 v	n_ω 1.738-1.724 n_ε 1.752-1.746	Buttgenbachite $Cu_{19}Cl_4(NO_3)_2(OH)_{32}\cdot2\text{-}3H_2O$	0.014 to 0.022	Hexagonal	Rad. acic		
>1.74	n_ω >1.74 n_ε ?	Schafarzikite $FeSb_2O_4$	Weak	Tetragonal	Prism xls (c)	{110} perf, {100} ~ perf, {001} trace	
*1.73 1.75	n_ω 1.73-1.75 n_ε ?	Coffinite $U(OH)_{4x}(SiO_4)1\text{-}x$	Small	Tetragonal	Earthy agg		
1.747	n_ω 1.747 n_ε 1.776	Fleischerite $Pb_3Ge(SO_4)_2(OH)_6\cdot3H_2O$	0.029	Hexagonal	Fibrous agg	None	
1.730 1.750	n_ω 1.730-1.750 n_ε 1.810-1.830	Mixite $Cu_{12}Bi_2(AsO_4)_6(OH)_{12}\cdot6H_2O$	~0.080	Hexagonal	Fib acic xls (c), rad, concentric		
1.757	n_ω 1.757 n_ε 1.804	Benitoite $BaTiSi_3O_9$	0.047	Hexagonal	Tabular, triangular crystals	{10$\bar{1}$1} poor	
1.760	n_ω 1.760 n_ε 1.870	Hydroxyl-bastnaesite $(Ce,La)(CO_3)(OH,F)$	0.110	Hexagonal	Reniform aggregates	{11$\bar{2}$0} imperf.	
1.761	n_ω 1.761 n_ε ~1.762	Macgovernite (McGovernite) $Mn_9Mg_4Zn_2As_2Si_2O_{17}(OH)_{14}$	Very small	Trigonal	Micaceous "books," massive	{0001} perf (micaceous)	

* n_ω variation exceeds 0.01

Color In Thin Section	Hardness Sp Gravity	Color Hand Sample	Alteration	Occurrence	Remarks	Reference
Yellow-brown O > E	~5			High grade contact metamorphism	Melilite group	p. 159
Colorless	5½ 3.89-4.19	Colorless, white, grn, yel, red, brn		Zn-oxide ores with zincite, franklinite	Fluor U.V. intense yellow-green. Often phosphorescent	Win & Win II, 497
O = orange, E = colorless	~2½ 3.86-3.88	Orange		Quartz-sanbornite contact rocks	Decomp weak acids	Am Min 50, 1500-1503
Colorless, brownish zones	5.5 3.60-3.63	Colorless-brownish (zonal)		Acidic plutons with apatite, anatase, zircon		Am Min 55, 320-321
O = pale yellow-green E = green	3-4 3.50	Pale yellow-green		Secondary coatings on altered Cu-ores	Slowly sol dilute HCl	Am Min 63, 704-708
Colorless	3½-4 2.98	Pale yellow-green	Limonite goethite	Veins in kimberlite	Weakly magnetic	Am Min 47, 1218
Colorless, pale blue-green	3-4 3.66-3.72	Blue-green		Oxide zones of ore deposits	Mixite group	Am Min 55, 1447-1448
Pale green	4.01-4.11	Green, yellow-green		Oxide zone Pb-Zn-Cu deposits as earthy friable incrustations	Al-beaverite	Am Min 47, 1079-1093
Colorless	~9 3.01-3.02	White		Calcite veins in skarn rocks		Dana I, 506-507
Colorless, pale yellow, E > O	4-4½ 4.78-5.2	Yellow to reddish-brown		Contact metamorphic, pegmatite		Am Min 38, 932-963
O = colorless, pale pink, pale yellow. E = yellow, gray-brn, yellow-green	4-5 4.3-5.1	Yellow-brown to red-brown		Granites, syenites, granite pegmatites with zircon, monazite	Not sol acids	p. 74
Pale blue, not pleochroic	3 3.39-3.41	Azure blue, blue-green		Oxide zones Cu-deposits	Isomorph buttgenbachite, sol acids, NH_4OH	Dana II, 572-573
Pale blue, not pleochroic	3 3.42-3.44	Azure blue		Oxide zones Cu-deposits	Isomorph connellite, sol acids or NH_4OH	Dana II, 572-573
O = straw-yellow, E = brownish-yellow	3½ 4.3	Red, reddish-brown		Stibnite ores with senarmontite, valentinite and kermesite		Dana II, 1035-1036
Pale to dark brown	5-6 5.1	Black		Oxide zones U-deposits	Not fluor U.V.	Am Min 47, 26-33
Colorless	Soft 4.2-4.6	White, pink		Oxide zones Pb-deposits		Am Min 45, 1313
O ~ colorless, E = green	3-4 3.77-3.79	Green, blue-green whitish		Oxide zones Cu-deposits		Dana II, 943-944
Colorless, pale blue E > O	6-6½ 3.64-3.69	Blue, colorless, pink		Veinlets with neptunite, natrolite in serpentine		Win & Win II, 453
Colorless, pale brown	~4 4.75	Yellow, dk grown, colorless		Carbonatite veins with barite, monazite, sulfides	Sol HCl or H_2SO_4	Am Min 50, 805
Red-brown	3.72	Lt brown to dk red-brown		Hydrothermal Zn-O deposits with franklinite, zincite, willemite		Am Min 45, 937-945

Uniaxial +

n_ω Var.	Refractive Index	Mineral Name and Composition	$(n_\varepsilon - n_\omega)$	Crystal System	Habit	Cleavage	Twinning
1.765	n_ω 1.765 n_ε 1.780	Stillwellite $(Ce,La,Ca)BSiO_5$	0.015	Trigonal	Rhomb xls, massive, compact		
1.765	n_ω 1.765 n_ε 1.800	Henritermierite $Ca_3(Mn^{3+}_{1.5}Al_{0.5})(SiO_4)_2(OH)_4$	0.035	Tetragonal	Fine granular	None	{101} 4 sectors
1.78	n_ω 1.78 n_ε 1.82	Britholite (Beckelite, Lessingite) $(Ca,Ce)_5(SiO_4,PO_4)_3(OH,F)$	0.04	Hexagonal?	Prism xls, massive	None	
1.783	n_ω 1.783 n_ε 1.879	Chernovite $YAsO_4$	0.096	Tetragonal	Prism	{010} perf	
1.799	n_ω 1.799 n_ε 1.826	Stishovite SiO_2	0.027	Tetragonal	Massive, micro-granular		
1.80	n_ω 1.80 n_ε 1.81	Nigerite $(Zn,Fe,Mg)(Sn,Zn)_2(Al,Fe)_{12}O_{22}(OH)_2$	0.01	Trigonal	Hex plates	None	
*1.78 1.825	n_ω 1.78-1.825 n_ε 1.79-1.840	Thorite $ThSiO_4$	0.01 to 0.015	Tetragonal	Short prism, pyram xls, massive	{100} dist	
>1.80	n_ω >1.80 n_ε	Downeyite (Selenolite) SeO_2	Mod	Tetragonal	Acicular xls (c)		Contact
1.805	n_ω 1.805 n_ε	Kraisslite $Mn\ Zn(AsO_4)(SiO_2)_2(OH)_3$	Weak	Hexagonal	Scaly micaceous	{0001} perfect	
1.806 1.815	n_ω 1.806-1.815 n_ε 1.808-1.820	Cerite $(Ca,Mg)_2(RE)_8(SiO_4,FCO_3)_7(OH,H_2O)_3$	0.002 to 0.005	Trigonal	Pseudo-octahedral, massive	None	
*1.77 1.855	n_ω 1.77-1.855 n_ε	Lusungite $(Sr,Pb)Fe_3(PO_4)_3(OH)_5 \cdot H_2O$	0.03 to 0.04	Trigonal	Massive		
1.81	n_ω 1.81 n_ε 2.02	Gaudefroyite $Ca_4Mn_{3-x}[(BO_3)_3/(CO_3)/(O,OH)_3]$ $x \sim .17$	0.21	Hexagonal	Acic hex prisms	None ?	
*1.81 1.85	n_ω 1.81-1.85 n_ε 1.84-1.88	Chromatite $CaCrO_4$	0.03	Tetragonal	Fine xline	None	
1.842	n_ω 1.842 n_ε 1.848	Paratacamite $Cu_2(OH)_3Cl$	0.006	Trigonal	Rhomb xls, massive granular	{10$\bar{1}$1} dist	{10$\bar{1}$1} common
1.85	n_ω 1.85 n_ε 1.93	Zemannite $(Na,Fe)_2(TeO_3)_3Na_xH_{2-x} \cdot nH_2O$	0.08	Hexagonal	Hex prism xls		
1.864	n_ω 1.864 n_ε 1.88	Welinite $Mn^{4+}Mn^{2+}_3O_3(SiO_4)$	0.016	Hexagonal	Crude xls	{001} poor to dist	
1.880	n_ω 1.884 n_ε 1.880	Morelandite $Ba_5(AsO_4)_3Cl$	0.004	Hexagonal	Massive	{0001} v. poor	
1.898	n_ω 1.898 n_ε 1.915	Hugelite $Pb_2(UO_2)_3(AsO_4)_2(OH)_4 \cdot 3H_2O$	0.017	Monoclinic	Tab {100}, elong (c)	{100} good	
1.89	n_ω 1.89 n_ε 2.00	Denningite $(Mn,Ca,Zn)Te_2O_5$	0.011	Tetragonal	Thin octagonal plates	{001} perf	

* n_ω variation exceeds 0.01

Color In Thin Section	Hardness Sp Gravity	Color Hand Sample	Alteration	Occurrence	Remarks	Reference
Colorless	4.612	Lavender-gray, yellow-brown		Metasomatic limestone deposits with allanite, garnet, uraninite		Nature 176, 509-510
O = very pale yellow, E = lemon-yellow	~7 3.34-3.40	Red-brown		Fractures in Mn-ores	Garnet group. Sol warm HCl	Am Min 54, 1739
O = colorless, pale brown, E = colorless	5 3.86-4.69	Yellow, green, brown, black		Alkali igneous rocks, nepheline pegmatites, contact met	Apatite group. Usually (-)	Vlasov II, 297-300
Colorless	4½ 4.87	Colorless, pale yellow		Hydrothermal veins	Xenotime group	Am Min 53, 1777
Colorless	4.28-4.35	Colorless		High pressure silica in meteorite craters with coesite, silica glass		Dana III, 317-318
Dk brown. ~opaque	8½ 4.51	Dk brown, black		Sn-bearing pegmatites with cassiterite, columbite, sillimanite	Weakly magnetic	Am Min 52, 864-866
Shades of green, brn E > O	~4½ 4.1-6.7	Dk brown, yel, orange, grn, blk	Thorogummite	Pegmatites, hydrothermal veins, metasomatic deposits	Usually metamict. Not fluor U.V.	USGS Bull, 1064, 265-276
Colorless	4.15	Colorless, red		Sublimate of gases from burning coal	Readily sol H_2O, acetone. Extremely hygroscopic	Am Min, 62, 316-320
Brown	3-4 3.88-3.90	Coppery-brown		Coatings and fissures in Zn-oxide ores		Am Min, 63, 938-940
Pale brown	5½ 4.78-4.86	Brown, red		Rare earth veins	~isotropic, small 2V	Am Min 43, 460-475
Yellow-brown. Slightly pleochroic		Dark-brown		Pegmatites with goethite, quartz, phosphates		Am Min 44, 906-907
O = Lt russet, E = orange-red	6 3.35-3.50	Black	Polianite	Calcite veins with Mn-minerals		Am Min 50, 806-807
Pale yellow	3.14	Yellow		Limestone with gypsum		Am Min 49, 439
Deep green. Not pleochroic	3 3.74-3.75	Green, greenish-black		Oxide zones Cu-deposits as xl coatings	Dimorph atacamite. Patchy ext, small 2V	Dana II, 74-76
Colorless to brown	>4.05	Dk to lt brown		Minute xls in telluride ores		Can Min, 10, 139-140
Red-brown	4 4.41-4.47	Dk red-brown to reddish black		Mn-ores with hausmannite, calcite, barite, sarkinite,	Possible small 2V	Am Min, 53, 1064
Colorless	4½ 5.33	Lt. yellow to gray		With calcite and hausmannite from Jakobsberg, Sweden	Apatite group	Am Min 65, 207
O = yellow, E = colorless	~5.1	Brown, orange-yellow		Secondary U-mineral	Anomalous interference colors	Am Min, 47, 418-419
Colorless	4 5.05-5.07	Colorless, pale green		With other tellurites	Small 2V < 15° possible	Can Min, 7, 443-452

Uniaxial +

n_ω Var.	Refractive Index	Mineral Name and Composition	$(n_\varepsilon - n_\omega)$	Crystal System	Habit	Cleavage	Twinning
1.90	n_ω 1.90 n_ε 2.12	Trippkeite $CuAs_2O_4$	0.22	Tetragonal	Short prism xls (c), fib	{100} perf {110} good	
1.910	n_ω 1.910 n_ε 1.945	Ganomalite $Pb_6Ca_4(OH)_2(Si_2O_7)_3$	0.035	Hexagonal	Prism, massive, granular	Prismatic	
1.921	n_ω 1.9208 n_ε 1.9375	Scheelite $CaWO_4$	0.017	Tetragonal	Dipyram or tab xls, massive, granular	{101} dist,{112} & {001} indist	{110} contact or penetration
*1.920 1.960	n_ω 1.920-1.960 n_ε 1.967-2.015	Zircon $ZrSiO_4$	0.042 to 0.065	Tetragonal	Square prism-pyram xls	{110} poor	{111} "knee" twins
1.944	n_ω 1.944 n_ε 2.16	Baotite $Ba_4(Ti,Nb)_8Si_4O_{28}Cl$	0.216	Tetragonal	Granular	{110} at 90°	
1.945	n_ω 1.945 n_ε 1.971	Nasonite $Ca_4Pb_6Si_6O_{21}Cl_2$	0.026	Hexagonal	Prism xls, massive granular	{0001} dist, prismatic indist	
>1.95	n_ω>>1.95 n_ε>>1.95	Ordoñezite $ZnSb_2O_6$		Tetragonal	Tetragonal xls, stalactitic	None	{013} common
1.96	n_ω 1.96 n_ε	Dixenite $Mn_6(OH)_2Si_2O_4(AsO_3)_2$	Small	Trigonal	Massive, micaceous agg	{0001} perf	
1.967	n_ω 1.967 n_ε 1.978	Powellite $Ca(Mo,W)O_4$	0.011	Tetragonal	Pyram or tab xls, massive	{112}, {011} & {001} indist	
1.973	n_ω 1.973 n_ε 2.656	Calomel $HgCl$	0.683	Tetragonal	Tab {001}, prism (c), earthy	{110} dist, {011} imperf	{110} cont, penetr, repeated
~2.00	n_ω~2.00 n_ε 2.01	Simpsonite (Calogerasite) $Al_4Ta_3O_{13}OH$	0.01	Hexagonal	Tab or short prism (c) xls	None	
*1.990 2.010	n_ω 1.990-2.010 n_ε 2.091-2.100	Cassiterite SnO_2	0.10 to 0.09	Tetragonal	Bipyram, acic xls, rad, colloform	{100}, {110} & {111} poor	{101} simple, polysynthetic, cyclic
2.00	n_ω 2.00 n_ε 2.14	Wakefieldite YVO_4	0.14	Tetragonal	Massive, pulverulent		
2.013	n_ω 2.013 n_ε 2.029	Zincite $(Zn,Mn)O$	0.016	Hexagonal	Massive, platy, gran. Hemimorphic	{10$\bar{1}$0} perf, {0001} parting	{0001}
>2.10	n_ω>2.10 n_ε	Priderite $(K,Ba)(Ti,Fe^{3+})_8O_{16}$		Tetragonal	Stout prism xls, columnar	{001} perf, prism fair	
2.118	n_ω 2.1181 n_ε 2.1446	Phosgenite $Pb_2CO_3Cl_2$	0.026	Tetragonal	Prism, thick tab, massive, granular	{001} & {110} dist,{100} indist	
2.12	n_ω~2.12 n_ε>2.12	Behierite $(Ta,Nb)BO_4$	Large	Tetragonal	Octahedral	{110} & {010} dist	
2.13	n_ω 2.13 n_ε 2.21	Penfieldite Pb_2Cl_3OH	0.08	Hexagonal	Prism (c), tab {0001} xls	{0001} dist	{0001} common
2.19 (Li)	n_ω 2.19 (Li) n_ε 2.21	Mackayite $FeTe_2O_5(OH)$?	0.02	Tetragonal	Prism, pyramidal xls	None	
2.2	n_ω 2.2 n_ε	Russellite Bi_2WO_6		Tetragonal	Massive, compact, fine-grained		

* n_ω variation exceeds 0.01

Color In Thin Section	Hardness Sp Gravity	Color Hand Sample	Alteration	Occurrence	Remarks	Reference
Pale blue-green, Not pleochroic	Soft 4.8	Greenish-blue		Oxide zones Cu-deposits with olivenite	Easily sol HCl or HNO_3	Dana II, 1034
Colorless	3 5.74	Colorless, white, gray		Skarns	Nasonite Cl-analogue	Win & Win II, 478-479
Colorless	4½-5 6.10-6.12	White, yellow, brownish, gray		Contact met deposits, hydrothermal veins, pegmatites	Decomp HCl or HNO_3. Fluor short λ U.V.white or yel.	Dana II, 1074-1079
Colorless. Patchy brown	7½ 4.67	Colorless, red-brown, yel-brown	Resistant	Accessory in siliceous and alkaline igneous rocks. Pegmatites. Heavy sands	Often metamict	p. 131
O = colorless, brown, greenish-yellow E = Dk brown, greenish-yellow	6 4.42-4.71	Lt brown-black		Carbonate veins in alkalic rocks		Am Min, 47, 987-993
Colorless	4 5.55-5.63	White		Hydrothermal ore veins with barysilite, willemite, datolite, prehnite		USGS,PP 180, 92-93
Colorless, pale brown	6½ 6.63-6.66	Lt to dk brown, olive, colorless		Veinlets in rhyolite with cassiterite, tridymite, sanidine, topaz		Am Min, 40, 64-69
Deep red	3-4 4.2	Bronze, ~black		With adelite	Decpd HCl	Geol. Fören. Stockh. Förhandl., 72, 64
Yellow, brown, green, blue, colorless	3½-4 4.23-4.26	Yellow, brown, green, blue, gray	Ferrimolybdite	Oxide zones ore deposits	Fluor U.V. yellow. Decomp HCl or HNO_3	Dana II, 1079-1081
Colorless	1½ 7.15-7.23	Colorless, white, yellow, brown		Oxide zones Hg-deposits	Fluor U.V. dark red	Dana II, 25-28
Colorless	7-7½ 5.92-6.84	Colorless, lt yellow, lt brown		Granite pegmatites with tantalite, microlite, tapiolite, beryl	Fluor U.V. pale yellow, bluish-white, yellowish-blue	Vlasov II, 446-449
Shades of yel, red, brown or greenish. E > O (often zoned)	6-7 6.9-7.1	Dk brown to black		Granites, hydrothermal veins, pegmatites with topaz, W-Mo minerals		p. 23
Colorless, pale yel	~5 4.21-4.25	Pale tan to yel		Pegmatite with allanite, biotite, chlorite, bismutite, quartz	Xenotime group	Am Min, 56, 395-410
Deep red, yellow. Not pleochroic	4 5.68	Deep yellow, orange, dk red		Zn-deposits with franklinite, willemite, calcite	Sol acids	Dana I, 504-506
O = reddish-brown, E = deep reddish-brn to black	3.68	Black, reddish		Accessory in leucite-bearing rocks		Min Mag, 29, 496-501
Colorless	2-3 6.13	Colorless, white, brown		Oxide zones Pb-deposits	Fluor U.V.shades of yel Sol dilute HNO_3	Dana II, 256-259
Colorless	7-7.5 7.86-7.91	Pinkish-gray		Pegmatites		Am Min, 47, 414
Colorless	5.82-6.00	Colorless, white, yellowish		Secondary Pb-mineral with percylite, laurionite, fiedlerite	Sol dilute HNO_3	Dana II, 66-67
O = green, E = yellow-green	4½ 4.79-4.86	Yellow-green, brown-green		Oxide zones ore deposits with native Te, emmonsite		Am Min, 55, 1072
Pale yellow	3½ 7.35-7.40	Pale yellow to greenish		High-T hydro veins with Bi, bismuthite, wolframite, cassiterite		Dana I, 604-605

Uniaxial +

n_ω Var.	Refractive Index	Mineral Name and Composition	$(n_\omega - n_\varepsilon)$	Crystal System	Habit	Cleavage	Twinning
2.21	n_ω 2.21 n_ε 2.22	Iodargyrite (Iodyrite) AgI	0.01	Hexagonal	Prism (c), tab {0001} scaly	{0001} perf	{30$\bar{3}$4} cyclic (fourlings)
*2.12 2.30	n_ω 2.12-2.30 n_ε	Bismutite (Basobisutite) (Bismutosphaerite) $(BiO)_2CO_3$	Mod	Tetragonal	Massive, fib, scaly agg	{001} dist	
2.19 2.27	n_ω 2.19-2.27 n_ε 2.30-2.36	Calzirtite $CaZr_3TiO_9$	0.11 to 0.07	Tetragonal	Prism	Non-?	
2.27	n_ω 2.27 n_ε 2.30	Raspite $PbWO_4$	0.03	Monoclinic	Tab elong (b)	{100} perf	{100} common, {$\bar{1}$02} rare
2.27 (Li)	n_ω 2.27 (Li) n_ε 2.42	Tapiolite $(Fe,Mn)(Ta,Nb)_2O_6$		Tetragonal	Short prism or equant xls	None	{013} simple or polysynthetic
2.27 (Li)	n_ω 2.27 (Li) n_ε 2.42	Mossite $Fe(Nb,Ta)_2O_6$		Tetragonal	Short prism or equant xls	None	{013} simple or polysynthetic
2.356 (Na)	n_ω 2.356 (Na) n_ε 2.378	Wurtzite ZnS	0.022	Hexagonal	Short prism to tab xls, fib, crusts	{11$\bar{2}$0} dist, {0001} imperf	
2.37 (Li)	n_ω 2.37 (Li) n_ε 2.42	Freundenbergite $Na_2(Ti,Nb)_6Fe^{3+}_2(O,OH)_{18}$	0.05	Hexagonal or Monoclinic	Granular	{0001} and {10$\bar{1}$0} dist	
2.476	n_ω 2.476 n_ε 2.485	Changbaiite $PbNb_2O_6$	0.009	Trigonal	Tabular crystals, spherules	{0001} perfect, {10$\bar{1}$1} distinct	
2.5 (Li)	n_ω ~2.5(Li) n_ε ?	Eskolaite Cr_2O_3		Trigonal	Prism, tab xls	{10$\bar{1}$1} dist	{0001} common
2.506	n_ω 2.506 n_ε 2.529	Greenockite (Xanthochroite) CdS	0.023	Hexagonal	Earthy. Hemimorphic pyram xls	{11$\bar{2}$2} dist, {0001} imperf	{11$\bar{2}$2} rare
2.583	n_ω 2.583 n_ε 2.700	Brookite (Pyromelane) TiO_2	0.117	Orthorhombic	Tab {010}, prism (c) xls	{120} indist, {001} poor	{120} rare
*2.605 2.616	n_ω 2.605-2.616 n_ε 2.890-2.903	Rutile TiO_2	0.285 to 0.296	Tetragonal	Short prism to acic xls	{110} dist, {100} fair	{011} knee-twins, cyclic
2.654	n_ω 2.654 n_ε 2.697	Carborundum (Moissanite) SiC	0.043	Hexagonal	Hex plates		
2.905	n_ω 2.905 n_ε 3.256	Cinnabar HgS	0.351	Trigonal	Tab, prism, massive, earthy	{10$\bar{1}$0} perf	{0001} penetration
3.0 (Na)	n_ω 3.0 (Na) n_ε 4.04	Selenium Se	~1.0	Trigonal	Hollow acic xls (c), felty, globules	{01$\bar{1}$2} dist	

* n_ω variation exceeds 0.01

Color In Thin Section	Hardness Sp Gravity	Color Hand Sample	Alteration	Occurrence	Remarks	Reference
Colorless	1½ 5.69-5.70	Colorless, yel, green, brown	Miersite, native Ag	Oxide zones Ag-deposits	Sectile. Possible biref. Anom green interfer colors	Dana II, 22-25
Colorless (~opaque)	2½-3½ 6.1-8.3	White, grayish, yellowish		Oxide zones Bi-deposits	Sol with efforvescence in acids	Dana II, 259-262
Brown, not pleochroic	6-7 5.01	Lt-dk brown, greenish		Carbonatites, metasomatic rocks		Am Min, 52, 1880-1885
Pale yellow	2½-3 8.46-8.52	Yellow, yellow-brown, gray		Oxide zones Pb-deposits with stolzite	Dimorph. stolzite	Dana II, 1089-1090
O = pale yellow to reddish-brown, E ~ opaque	6-6½ 7.82-8.13	Black		Albitized granite pegmatites	Tapiolite series. Dimorph tantalite	Vlasov II, 449-452
O = yellowish or reddish-brown, E ~ opaque	6-6½ 6.93	Black		Granite pegmatites	Tapiolite series	Dana I, 775-778
	3½-4 3.98-4.09	Orange-brown to brown-black	Sphalerite	Hydrothermal veins with sphalerite. Clay-ironstone concretions	Trimorph sphalerite and matraite. Sol HCl	Dana I, 227-228
O = dk brown, E = yellow-brown	~5 4.38	Black		Alkali syenite with apatite. Late magmatic.		Am Min, 46, 765-766
	~5 6.48-6.51	Colorless, cream, lt brown		Kaolinite veins in K-rich granite	May show small 2V (r > v)	Am Min 64, 242
Green	5.2	Green to black		Cr-bearing skarn with tremolite. Refractories		Am Min 46, 998-999
Pale yellow. Weak pleochroism	3-3½ 4.77-4.9	Yellow, orange		Earthy coatings on Zn-minerals, with zeolites in cavities in basic rocks	Dimorph hawleyite. Sol conc HCl	Dana I, 228-230
Yellow-brown or red-brown to deep brown. E > O	5½-6 4.12	Yellow-brown to red-brown, black	Rutile	Gneisses, hydrothermal veins, pegmatites with rutile and anatase	Trimorph rutile, anatase. Properties vary greatly with λ.	p. 22
Red-brown to yellow-brown, ~opaque E > O	6-6½ 4.2-5.6	Brick red to black	Sphene, leucoxene, anatase	High-T & P plutonic rocks, igneous and met. Pegmatites, quartz veins	Trimorph anatase, brookite	p. 19
Bluish	9½ 3.22	Green, bluish, black		Meteorites	Grinding compound for thin sections	Dana I, 123-124
Deep red	2-2½ 8.09-8.19	Bright red		Low-T hydrothermal veins, hot springs, volcanic sublimate	Optically active	Dana I, 251-255
Red in thin splinter	2 4.80	Gray, violet-gray, reddish		Veins and coatings in sandstone with secondary U-V-ores	Metallic	Dana I, 136-137

Uniaxial Negative Minerals

n_ω Var.	Refractive Index	Mineral Name and Composition	$(n_\omega - n_\varepsilon)$	Crystal System	Habit	Cleavage	Twinning
1.313	n_ω 1.313 n_ε 1.309	Malladrite Na_2SiF_6	0.004	Trigonal	Crusts, prism xls(c)		
1.349	n_ω 1.349 n_ε 1.342	Chiolite $Na_5Al_3F_{14}$	0.007	Tetragonal	Dipyram	{001} perfect {011} distinct	{011}
1.406	n_ω 1.406 n_ε 1.391	Bararite $(NH_4)_2SiF_6$	0.015	Trigonal	Crusts, tab	{0001} perfect	"paddlewheel"
1.474	n_ω 1.474 n_ε 1.436	Humberstonite (Chili-Loweite) $Na_7K_3Mg_2(SO_4)_6(NO_3)_2 \cdot 6H_2O$	0.038	Trigonal	Thin tab {0001}	{0001} perfect	None ?
1.479	n_ω 1.479 n_ε 1.408	Zhemchuzhnikovite $NaMg(Al,Fe^{3+})(C_2O_4)_3 \cdot 8H_2O$	0.071	Hexagonal	Acic-fib xls, hex prism xls	{0001} fair	
1.481	n_ω 1.481 n_ε 1.461	Hanksite $Na_{22}K(SO_4)_9(CO_3)_2Cl$	0.020	Hexagonal	Hex prism, tab	{0001} distinct	Interpenetrant
*1.472 1.494	n_ω 1.472-1.494 n_ε 1.470-1.485	Chabazite $Ca(Al_2Si_4)O_{12} \cdot 6H_2O$	0.002 to 0.010	Trigonal	Rhombohedral xls	{10$\bar{1}$1} good	{0001} penetration, {10$\bar{1}$1} contact
*1.476 1.494	n_ω 1.476-1.494 n_ε 1.474-1.480	Gmelinite $(Na_2,Ca)(Al_2Si_4)O_{12} \cdot 6H_2O$	0.002 to 0.015	Trigonal	Rhombohedral, prism, Rad agg	{10$\bar{1}$1} distinct {0001} parting	{0001} penetration, {10$\bar{1}$1} rare
1.487	n_ω 1.487 n_ε 1.484	α-Cristobalite SiO_2	0.003	Tetragonal (pseudoisometric)	Octahedral or cubic xls, Rad acic xls.	None	{111} simple or polysynthetic
1.489	n_ω 1.489 n_ε 1.486	Offretite $(K,Ca)_3(Al_5Si_{13})_{36} \cdot 14H_2O$	0.003	Hexagonal	Tiny prism xls, radiated	{0001} distinct	
1.490	n_ω 1.490 n_ε 1.471	Loeweite (Löweite) $Na_{12}Mg_7(SO_4)_{13} \cdot 15H_2O$	0.019	Trigonal	Massive, grains		
1.490	n_ω 1.490 n_ε 1.476	Wardsmithite $Ca_5MgB_{24}O_{42} \cdot 30H_2O$	0.014	Hexagonal or pseudohex	Platy aggregates	{0001} ?	
1.491	n_ω 1.491 n_ε 1.470	Ettringite $Ca_6Al_2(SO_4)_3(OH)_{12} \cdot 24H_2O$	0.021	Hexagonal	Flat hex dipyram, fib	{10$\bar{1}$0} perfect	
1.493	n_ω 1.493 n_ε 1.482	Wermlandite $Ca_2Mg_{14}Al_4(CO_3)(OH)_{42} \cdot 30H_2O$	0.011	Trigonal	Thin hex plates	{0001} perfect	
1.495	n_ω 1.495 n_ε 1.460	Acetamide CH_3CONH_2	0.035	Hexagonal	Gran agg hex xls	None ?	
*1.490 1.507	n_ω 1.490-1.507 n_ε 1.488-1.495	Vishnevite $(Na,Ca,K)_{6-7}Al_6Si_6O_{24}(SO_4,CO_3,Cl)_{1-1.5} \cdot 1\text{-}5H_2O$	0.002 to 0.012	Hexagonal	Massive, prism xls rare	{10$\bar{1}$0} perfect {0001} poor	Lamellar rare
1.496 1.505	n_ω 1.496-1.505 n_ε 1.491-1.500	Levyne (Levynite) $(Na,Ca)_2(Al,Si)_9O_{18} \cdot 8H_2O$	0.005	Trigonal	Thin tab {0001}, sheaf-like agg	None	
1.502	n_ω 1.502 n_ε 1.449	Ungemachite $Na_8K_3Fe^{3+}(SO_4)_6(OH)_2 \cdot 10H_2O$		Trigonal	Thick tab xls {0001}, pseudorhomb	{0001} perfect	
1.500 1.507	n_ω 1.500-1.507 n_ε 1.464-1.468	Thaumasite $Ca_3Si(OH)_6(CO_3)(SO_4) \cdot 12H_2O$	0.036 to 0.039	Hexagonal	Acic-filiform, massive	{10$\bar{1}$1} traces	

* n_ω variation exceeds 0.01

Color In Thin Section	Hardness Sp Gravity	Color Hand Sample	Alteration	Occurrence	Remarks	Reference
Colorless	? 2.714	Pale rose, white		Crusts on lavas with salammoniac, feruccite, avogadrite	Slightly sol H_2O	Dana II, 105-106
Colorless	3½-4 2.99-3.00	Colorless, white		Pegmatites with cryolite		Dana II, 123-124
Colorless	2½ 2.14-2.15	White		Sublimate in coal beds, fumeroles	Dimorph Cryplohalite, sol H_2O, saline	Dana II, 106-107
Colorless	~2½ 2.25	Colorless		Evaporite nitrate deposits	Sol H_2O	Am Min 55, 1518-1533
O = greenish-yellow, E = reddish-violet	2 1.66-1.69	Smoky green, violet		Veinlets in coal.	Easily sol H_2O	Am Min, 49, 442-443
Colorless	3-3½ 2.56-2.57	Colorless, yellowish		Saline lake deposits with borates	Sol H_2O. Saline taste. U.V. (ℓλ) fluor yellow	Dana II, 628-629
Colorless	4½ 2.1	White, salmon pink	Montmorillonite clays	Cavities and veins in basalts. Hydrothermal veins and springs	Zeolite group	p. 393
Colorless	4½ 2.0-2.2	Colorless, white, yellow, brown	Montmorillonite clays	Amygdules in basalt with other zeolites.	Zeolite group	p. 394
Colorless	6½ 2.33	Colorless, white, yellowish	Fine-grained α-quartz	Cavities and spherulites in glasses and siliceous volcanics. Opalline masses	Metastable below 1470°C. Seldom older than Tertiary	p. 321
White	? 2.13	Colorless, white		Amygdules in basalt with levynite	Zeolite group	Am Min, 54, 875-886
Colorless	2½-3 2.36-2.38	Colorless, gray, lt yellow	White crust	Bedded marine salt deposits	Sol H_2O. Bitter taste	Am Min, 55, 378-386
Colorless	2½ 1.88	Colorless		Borate lake deposits with colemanite, priceite, gowerite, ulexite		Am Min, 55, 349-357
Colorless	2-2½ 1.77-1.79	Colorless, white		Contact limestone metamorphism		Am Min, 45, 1137-1143
Colorless	1½ 1.932	Greenish gray		Fe-ores with magnetite, calcite	Possible small 2V	Lithos, 4, 213-217
Colorless	1-1½ 1.17	Colorless, gray		Coal waste piles, only in dry weather	Easily sol H_2O. Very bitter taste. Volatilizes in sun	Am Min, 61, 338
Colorless	5-6 2.32-2.42	Colorless, white, pale tints		Primary in alkali intrusives. Alter nepheline	Sulfate Cancrinite	DHZ, 4, 310
Colorless	4-4½ 2.09-2.16	Colorless, white, yellowish		Cavities in basalt	Zeolite group	DHZ, 4, 387-400
Colorless	2½ 2.29	Colorless, pale yellow		Oxide zones ore deposits with clino-ungemachite, sideronatrite	Easily sol dilute HCl	Dana II, 596-597
Colorless	3½ 1.91	Colorless, white		Metasomatic deposits with spurrite, ettringite, prehnite		Win & Win II, 179

n_ω Var.	Refractive Index	Mineral Name and Composition	$(n_\omega - n_\varepsilon)$	Crystal System	Habit	Cleavage	Twinning
1.504	n_ω 1.504 n_ε 1.459	Macallisterite (Trigonomagneborite) $Mg_2B_{12}O_{20}\cdot 15H_2O$	0.045	Trigonal	Rhomb xls, nodular	{0001} & {01$\bar{1}$2} dist	
1.506	n_ω 1.506 n_ε 1.499	Mazzite $K_2CaMg_2(AlSi_2)_{36}O_{72}\cdot 28H_2O$	0.007	Hexagonal	Acic bundles		
*1.500 1.515	n_ω 1.500-1.515 n_ε 1.502-1.512	Garronite $Na_2Ca_5Al_{12}Si_{20}O_{64}\cdot 27H_2O$	0.003	Tetragonal	Rad acic in amyg-dules	{110} distinct	
*1.490 1.530	n_ω 1.490-1.530 n_ε 1.488-1.503	Cancrinite $(Ca,Na)_{7-8}(AlSiO_4)_6(CO_3,SO_4,Cl)_{1-2}\cdot 1\text{-}5H_2O$	0.002 to 0.028	Hexagonal	Massive	{10$\bar{1}$0} perfect	
1.511	n_ω 1.5109 n_ε 1.4873	Retgersite $NiSO_4\cdot 6H_2O$	0.024	Tetragonal	Tab, prism (c), fib veinlets	{001} perfect {110} traces	
1.513	n_ω 1.513 n_ε 1.470	Archerite $(K,NH_4)H_2PO_4$	0.043	Tetragonal	Stalactitic	None	
1.515	n_ω 1.515 n_ε 1.417	Stepanovite $NaMg[Fe(C_2O_4)_3]\cdot 8\text{-}9H_2O$	0.098	Trigonal	Rhomb or hex prism xls	None	{0001}
1.515 1.519	n_ω 1.515-1.519 n_ε	Basaluminite $Al_4SO_4(OH)_{10}\cdot 5H_2O$		Hexagonal (?)	Microcrystalline, elongated		
1.519	n_ω 1.519 n_ε 1.503	Teepleite $Na_2B(OH)_4Cl$	0.016	Tetragonal	Tab xls {001}, xl groups	None	
*1.511 1.531	n_ω 1.511-1.531 n_ε 1.495-1.529	Hydrotalcite $Mg_6Al_2CO_3(OH)_{16}\cdot 4H_2O$	0.016 to 0.002	Trigonal	Flexible laminae, fib, massive	{0001} perfect	
1.520	n_ω 1.520 n_ε 1.512	Tachhydrite (Tachyhydrite) $CaMg_2Cl_6\cdot 12H_2O$	0.008	Hexagonal	Massive, rounded masses	{10$\bar{1}$1} perfect	
1.520	n_ω 1.520 n_ε~1.515	Virgilite $Li_{0.61}(Si_{2.37}Al_{0.60}Fe_{0.02}P_{0.01})O_6$	0.005 to 0.006	Hexagonal	Tiny hex bipyramids, Fib rosettes		
1.521	n_ω 1.521 n_ε 1.517	Carletonite $KNa_4Ca_4Si_8O_{18}(CO_3)_4(F,OH)\cdot H_2O$	0.004	Tetragonal	Massive	{001} perfect {110} distinct	
1.522 1.524	n_ω 1.522-1.524 n_ε 1.553-1.554	Taeniolite (Tainiolite) $KLiMg_2Si_4O_{10}F_2$	~0.030	Monoclinic β~100°	Micaceous, pseudohex xls	{001} perfect	Cyclic trillings rare
1.524	n_ω 1.524 n_ε 1.510	Manasseite $Mg_6Al_2CO_3(OH)_{16}\cdot 4H_2O$	0.014	Hexagonal	Massive foliated {0001}	{0001} perfect	
1.525	n_ω 1.525 n_ε 1.480	Biphosphammite $NH_4H_2PO_4$	0.045	Tetragonal	Radiated prism	None	
1.526	n_ω 1.526 n_ε 1.602	Anthonyite $Cu(OH,Cl)_2\cdot 3H_2O$	0.076	Monoclinic β = 112.5°	Tiny bent prisms on c	{100} distinct	
*1.519 1.534	n_ω 1.519-1.534 n_ε	Strätlingite $2CaO\cdot Al_2O_3\cdot SiO_2\cdot 8H_2O$		Trigonal	Platy	{0001} perfect	
1.530	n_ω 1.530 n_ε 1.48	Fairchildite $K_2Ca(CO_3)_2$	0.050	Hexagonal	Micro hex plates {0001}	{0001} distinct	

* n_ω variation exceeds 0.01

Color In Thin Section	Hardness Sp Gravity	Color Hand Sample	Alteration	Occurrence	Remarks	Reference
Colorless	2½ 1.867	Colorless, lt amber		Borax lakes and lake deposits with ginorite, sassolite, rivadite	Sol H_2O	Am Min, 52, 1776-1784
Colorless	 2.108	White		Cavities in olivine basalt with phillipsite, offertite, chabazite		Am Min, 60, 340
Colorless	~4 2.13-2.20	White		Amygdules in basalts with other zeolites	Zeolite group	Min Mag 33, 173-186
Colorless	5-6 2.3-2.5	Yellow, white, gray, reddish	Natrolite	Alkaline plutonic rocks with nepheline, calcite. Alteration of nepheline	Feldspathoid (cancrinite group)	p. 371
Pale green	2½ 2.07	Dk emerald green		Oxide zone Ni-deposits with patronite	Sol H_2O. Taste bitter-metallic	Dana II, 497-498
Colorless	soft 2.23	White		Cave stalactites with saline minerals and other phosphates	Sol H_2O	Min Mag, 41, 33-35
O = yellow-green, E = colorless	2 1.69	Greenish		Veinlets in coal	Easily sol H_2O	Am Min, 49, 442-443
Colorless	 2.12-2.24	White		Veins, coatings, concretions		Am Min, 33, 787
Colorless	3-3½ 2.07	White		Lake evaporites with trona, halite	Easily sol H_2O	Am Min, 44, 875
Colorless	2 2.06-2.16	White, brownish		Serpentenite, talc schist, alter spinel	Dimorph manasseite. Greasy feel. Sol HCl-effer	Dana I, 653-655
Colorless to pale yel Not pleochroic	2 1.667	Colorless, yellow		K-rich, marine salt beds with kainite, carnallite, sylvite	Sol H_2O. Bitter taste, Deliquescent	Dana II, 95-96
Colorless	 2.46	Colorless		High-Al volcanic glass. Overgrowths on quartz, K-feldspar, biotite, spinel	β-quartz-β-eucryptite series	Am Min, 63, 461-465
Colorless	4-4½ 2.43-2.45	Pink, lt blue, colorless		Contact metamorphism		Am Min, 56, 1855-1866
Colorless	2½-3 2.82-2.90	Colorless, brownish		Nepheline-syenite pegmatites with natrolite, neptunite, apatite	Mica group	Vlasov II, 29-31
Colorless	2 2.00-2.05	White, brownish, bluish		With hydrotalcite in serpentenite	Dimorph hydrotalcite. Greasy feel. Sol HCl-effer	Dana I, 658-659
Colorless	1 2.04	Colorless, lt buff		Guano	Sol. H_2O	Min Mag, 38, 965-967
X = lavender, Y = Z = deep smoky blue	2	Lavender		In cavities and incrustations in basalt	Easily sol in cold dilute acids. Small 2V~3°	Am Min, 48, 614-619
Colorless	 1.95	Colorless, lt green		Linestone inclusionsin basalt with nepheline, zeolites	Decomp dilute HCl	Am Min, 62, 395
Colorless	 2.45	Colorless		Wood ash		Am Min, 32, 607

n_ω Var.	Refractive Index	Mineral Name and Composition	$(n_\omega - n_\varepsilon)$	Crystal System	Habit	Cleavage	Twinning
1.530	n_ω 1.530 n_ε 1.528	Liottite $(Na,Ca,K)_{6-8}Al_6Si_6O_{24}(CO_3,SO_4,Cl,OH)_{1-2} \cdot 1\text{-}5H_2O$	0.002	Hexagonal	Hexagonal prisms	None ?	
1.530 1.537	n_ω 1.530-1.537 n_ε 1.495-1.506	Slavikite $MgFe_3^{3+}(SO_4)_4(OH)_3 \cdot 18H_2O$	0.035 to 0.031	Trigonal ?	Minute tab-scales {0001}, crusts		
1.534	n_ω 1.534 n_ε 1.514	Zincaluminite $Zn_6Al_6(SO_4)_2(OH)_{26} \cdot 5H_2O$	0.020	Orthorhombic ?	Tiny hex plates, Crusts and tufts		
1.537	n_ω 1.537 n_ε 1.533	Kaliophilite $KAlSiO_4$	0.004	Hexagonal	Prism, acicular	{0001} & {10$\bar{1}$0} indistinct	
1.539	n_ω 1.539 n_ε 1.511	Mellite $C_{12}Al_2O_{12} \cdot 18H_2O$	0.028	Tetragonal	Prism or pyramidal crystals, massive	{011} indistinct	
*1.529 1.549	n_ω 1.529-1.549 n_ε 1.526-1.544	Nepheline (Nephelite)(Eleolite) $Na_3KAl_4Si_4O_{16}$	0.003 to 0.005	Hexagonal	Stubby hexagonal prisms. Massive	{0001} & {10$\bar{1}$0} poor	Rare
1.540	n_ω 1.540 n_ε 1.510	Brugnatellite $Mg_6FeCO_3(OH)_{13} \cdot 4H_2O$	0.030	Hexagonal	Foliated {0001}	{0001} perfect	
1.540	n_ω 1.540 n_ε 1.535	Tetrakalsilite $K_3NaAl_3Si_3O_{12}$	0.005	Hexagonal			
1.538 1.543	n_ω 1.538-1.543 n_ε 1.532-1.537	Kalsilite $KAlSiO_4$	0.006	Hexagonal	Massive, embedded grains	{10$\bar{1}$0} & {0001} poor	
*1.532 1.551	n_ω 1.532-1.551 n_ε 1.529-1.548	Milarite $KCa_2AlBe_2(Si_{12}O_{30}) \cdot H_2O$	0.003	Hexagonal	Prismatic crystals	None	Crosstwinning
1.537 1.545	n_ω 1.537-1.545 n_ε 1.537-1.544	Apophyllite $KCa_4(Si_4O_{10})_2F \cdot 8H_2O$	0.000 to 0.001	Tetragonal	Square prism xls. platy	{001} perfect {110} imperfect	{111} rare, penetration
1.543	n_ω 1.543 n_ε 1.533	Iowaite $Mg_4Fe^{3+}(OH)_8OCl \cdot 2\text{-}4H_2O$	0.010	Hexagonal	Platy crystals	{0001} perfect	
*1.538 1.550	n_ω 1.538-1.550 n_ε 1.536-1.541	Marialite $Na_4(Al_3Si_9)O_{24}Cl$	~0.002 to 0.009	Tetragonal	Prism. xls., granular	{100} & {110} imperfect	None
1.545	n_ω 1.545 n_ε 1.518	Stichtite $Mg_6Cr_2CO_3(OH)_{16} \cdot 4H_2O$	0.027	Trigonal	Massive, foliated-scaly, fibrous	{0001} perfect	
*1.540 1.555	n_ω 1.540-1.555 n_ε 1.500-1.520	Palygorskite (Attapulgite) (Pilolite) $(Mg,Al)_4Si_8O_{20}(OH)_2(OH_2)_4 \cdot 4H_2O$	0.020 to 0.035	Monoclinic	Matted fibers (c), felted	{110} distinct	None
1.549	n_ω 1.549 n_ε 1.536	Gyrolite $Ca_2Si_3O_7(OH)_2 \cdot H_2O$	0.013	Hexagonal	Massive, lamellar-rad	{0001} perfect	
1.550	n_ω 1.550 n_ε 1.495	Antarcticite $CaCl_2 \cdot 6H_2O$	0.055	Trigonal	Acicular groups	{0001} perfect, {10$\bar{1}$0} distinct	
1.550	n_ω 1.550 n_ε 1.546	Osumilite $(K,Na)(Mg,Fe^{2+})_2(Al,Fe^{3+})_3(Si,Al)_{12}O_{30} \cdot H_2O$	0.004	Hexagonal	Prism crystals (c), tabular crystals {0001}	None	

* n_ω variation exceeds 0.01

Color In Thin Section	Hardness Sp Gravity	Color Hand Sample	Alteration	Occurrence	Remarks	Reference
Colorless	5 2.56	Colorless		Cavities in volcanic ejecta	Cancrinite-davyne group	Am Min, 62, 321-326
O = Lt yellow, E ~ colorless	1.89-1.99	Greenish-yellow		Alteration of pyrite in ore deposits and shales		Dana II, 621-622
Colorless	2½-3 2.26	White, pale blue		Oxide zones Zn-deposits with smithsonite, serpierite	Possible small 2V	Dana II, 579-580
Colorless	6 2.49-2.67	Colorless		Feldspathoid volcanic ejecta with leucite, haugite, melilite	Dimorph kalsilite	DHZ, 4, 231-270
O = yellow-brown, E = pale yellow, colorless	2-2½ 1.64-1.70	Golden-brown, reddish, white		Fractures and cavities in lignite coal	Fluor short λ U.V.-blue. Sol HNO_3	Dana II, 1104-1105
Colorless	5½-6 2.56-2.67	White, gray	Alters readily to fibrous micas, cancrinite minerals, Na-zeolites	Quartz-free, Na-rich plutons and volcanics with feldspar, calcite	Feldspathoid	p. 365
O = pale orange, E = colorless	~2 2.14-2.21	Pink, yellow, lt brown		Coatings on altered serpentine	Sol and efforvescence in dilute HCl	Dana I, 660-661
Colorless	2.59-2.62	White		Volcanic ejecta blocks (Mt. Vesuvius) with kalsilite, melilite		Am Min 64, 658
Colorless	6 2.59-2.63	Colorless, white, gray		Phenocrysts or groundmass in K-rich lavas	Dimorph kaliophilite	DHZ, 4, 231-270
Colorless, pale yel	5½-6 2.46-2.61	Colorless, pale yellow-green		Pegmatites, aplite and low-T hydrothermal veins	Possible 2V	Vlasov II, 108-111
Colorless	4½-5 2.33-2.37	White, pink, yellow, green	Opal, chert, kaolin, calcite	Amygdules or veins in basalt with zeolites. Cavities in granite-gneiss	Usually uniaxial +. Decomp HCl	p. 309
Colorless	1.5 2.11	Bluish-green, pale green	Pyroaurite	Serpentinite with chrysotile, dolomite, brucite, magnesite	Soapy feel. Possible small 2V	Am Min, 54, 296-299
Colorless	5-6 2.5-2.8	White, yellow, lt. green, violet	Plagioclase, calcite, chlorite, sericite, epidote, zeolite	Contact or regional met. of carbonate rocks	Scapolite group	p. 395
Rose-pink to lilac. O > E	1½-2 2.11-2.16	Lilac to rose-pink		Serpentinite with chromite	Dimorph barbertonite. Soapy feel. Sol HCl with efforv	Dana I, 655-656
Colorless	2-2½ 1.0-2.6	White, gray, yellow, gray-green	Montmorillonite, kaolin	Desert (playa) soils, marls. Veins and cavities in altered mafic rocks	Often called mineral cork, leather, cardboard or skin	p. 265
Colorless	3-4 2.34-2.45	Colorless, white		Alteration of lime-silicate minerals		Am Min, 46, 913-933
Colorless	2-3 1.68-1.72	Colorless		Evaporite	Highly soluble. Absorbs H_2O from air, melts	Am Min, 54, 1018-1025
O = pale blue, E = colorless	2.64	Dk blue to black		Cavities in andesite with cristobalite, tridymite, hortonolite	Cordierite group	Am Min, 54, 101-116

n_ω Var.	Refractive Index	Mineral Name and Composition	$(n_\omega - n_\varepsilon)$	Crystal System	Habit	Cleavage	Twinning
1.552	n_ω 1.552 n_ε 1.530	Truscottite $(Ca,Mn)_2Si_4O_9(OH)_2$	0.022	Trigonal	Massive, spherical aggregate	{0001} perfect	
1.552	n_ω 1.552 n_ε 1.535	Satimolite $KNa_2Al_4B_6O_{15}Cl_3 \cdot 13H_2O$	0.017	Orthorombic	Earthy aggregates, tab-rhomb crystals	None ?	
*1.54 1.56	n_ω 1.54-1.56 n_ε	Carrboydite $Ni_7Al_{4.5}(SO_4,CO_3)_{2.8}(OH)_{22} \cdot 3.7H_2O$	Low to mod	Hexagonal	Microscopic nodular, platy	{0001} ?	
1.556	n_ω 1.556 n_ε 1.546	Chrompyroaurite $Mg_6(Fe^{3+},Cr)_2(CO_3)(OH)_{16} \cdot 4H_2O$	0.010	Trigonal			
1.556	n_ω 1.556 n_ε 1.540	Jouravskite $Ca_6Mn_2(SO_4)_2(CO_3)_2(OH)_{12} \cdot 24H_2O$	0.016	Hexagonal	Tiny embedded grains	{100} distinct	
1.557	n_ω 1.557 n_ε 1.529	Barbertonite $Mg_6Cr_2CO_3(OH)_{16} \cdot 4H_2O$	0.028	Hexagonal	Fibrous, plates	{0001} perfect	
1.560	n_ω 1.560 n_ε 1.34	Nitromagnesite $Mg(NO_3)_2 \cdot 6H_2O$	0.22	Monoclinic	Efflorescence, long prism crystals (c)	{110} perfect	
1.560	n_ω 1.560 n_ε 1.507	Chloraluminite $AlCl_3 \cdot 6H_2O$	0.053	Trigonal	Rhombohedral		
1.559 1.562	n_ω 1.559-1.562 n_ε 1.514	Zincsilite $\sim Zn_3Si_4O_{10}(OH)_2 \cdot nH_2O$	~0.046	Monoclinic ?	Fine foliae or lamellae	{001} perfect	
*1.550 1.572	n_ω 1.550-1.572 n_ε 1.541-1.550	Dipyre $(Na,Ca)_4(Al,Si)_{12}O_{24}(Cl,CO_3)$	0.010 to 0.022	Tetragonal	Prism. xls., granular	{100} & {110} imperfect	None
1.561	n_ω 1.561 n_ε 1.549	Lovozerite $(Na,Ca)_2(Zr,Ti)Si_6O_{13}(OH)_6 \cdot 3H_2O$	0.012	Monoclinic β=92.5°			Common, polysynthetic
1.563	n_ω 1.563 n_ε 1.558	Reyerite $(Na,K)_4Ca_{14}Si_{24}O_{60}(OH)_5 \cdot 5H_2O$	0.005	Trigonal	Massive, micaceous spherulites	{0001} perfect	
1.564	n_ω 1.564 n_ε 1.543	Pyroaurite $Mg_6Fe^{3+}CO_3(OH)_{16} \cdot 4H_2O$	0.021	Trigonal	Thick or thin tab {0001}, fibrous	{0001} perfect	
*1.545 1.583	n_ω 1.545-1.583 n_ε 1.525-1.564	Vermiculite $(Mg,Ca)[(Mg,Fe^{2+})_5(Fe^{3+},Al)](Si_5Al_3)O_{20}(OH)_4 \cdot 8H_2O$	0.02 to 0.03	Monoclinic	Micaceous, scaly to large books	{001} perfect	Mica laws?
1.567	n_ω 1.567 n_ε	Lawrencite $FeCl_2$	Weak	Trigonal	Massive	{0001} perfect	
1.565 1.570	n_ω ~1.565-1.570 n_ε ~1.560	Zeophyllite (Radiophyllite) $Ca_4(Si_3O_7)(OH)_4F_2$	~0.005	Triclinic	Platy crystals, rad foliated, spherical	{001} perfect	
1.567	n_ω 1.567 n_ε 1.566	Brannockite $KLi_3Sn_2Si_{12}O_{30}$	0.001	Hexagonal	Thin plates {001}	None	
1.569	n_ω 1.569 n_ε 1.547	Desautelsite $Mg_6Mn_2^{3+}(CO_3)(OH)_{16} \cdot 4H_2O$	0.022	Hexagonal	Hexagonal plates	{0001} perfect	
1.569	n_ω 1.569 n_ε 1.549	Kalistrontite $K_2Sr(SO_4)_2$	0.020	Trigonal	Long prism crystals, Hexagonal tabular	{0001} perfect	

* n_ω variation exceeds 0.01

Color In Thin Section	Hardness Sp Gravity	Color Hand Sample	Alteration	Occurrence	Remarks	Reference
Colorless	2.35	White		Ore deposits (Gold mine)	Possible small 2V	Min Mag, 30, 450-457
Colorless		White		Clays and borate evaporites with polyhalite, boracite, kieserite		Am Min, 55, 1069
Pale yellow-green to lt blue-green	2.50	Yellow-green to blue-green		Weathered zones of Ni-Cu deposits with malachite, gypsum		Am Min, 61, 366-372
O = colorless, E = pale pink	1-1½ 2.12	Violet		In serpentine	Pyroaurite-stichtite series	Am Min 64, 1329
Bright yellow, weak pleochroism	2½ 1.93-1.95	Yellow-green, orange-green		Oxide zones Mn-deposits with calcite	Abnormal interference colors	Am Min, 50, 2102
O = dk rose, E = lt pink	1½-2 2.10-2.11	Dk lilac, pink		Veins in serpentenite	Greasy feel	Dana I, 659
Colorless	1.58-1.64	Colorless		Efflorescence in limestone caves	Easily sol H_2O. Taste bitter	Dana II, 307
Colorless		Colorless, white, yellow		Fumeroles, volcanics	Deliquescent. Sol H_2O	Dana II, 50
Colorless	1½-2 2.61-2.71	White to bluish		Oxide zones galena-sphalerite-chalcopyrite skarn deposits	Montmorillonite group	Am Min, 46, 241
Colorless	5-6 2.5-2.8	White, yellow, lt. green, violet	Plagioclase, calcite, chlorite, sericite, epidote, zeolite	Contact or regional met. of carbonate rocks	Scapolite group	p. 395
Yellow-pink, E > O	5 2.3-2.7	Red-brown, dk brown, black		Alkali plutons and pegmatites with nepheline, aegirine, eudialyte		Vlasov II, 364-365
Colorless	3-4 2.54-2.59	Colorless, white		Amygdules in diabase, basalt with chlorite		Am Min, 58, 517-522
O = pale yellow, reddish, brownish, E = colorless	2½ 2.12	White, yellowish, brownish		Low-T hydrothermal veins in serpentine or dolomite	Dimorph sjogrenite. Easily sol HCl with efferv	Am Min, 54, 296-299
O = yellow brown, green brown, E = colorless, pale green	1.5 ~2.4	Dk brown, green		Alter biotite in ultramafic rocks with serpentine, chlorite, talc.	Volume increase when heated. Interlayered with biotite	p. 263
Green to brown	Soft 3.16-3.22	Green to brown	Ferric-chloride	Inclusions in native iron. Fissures in iron meteorites	Readily sol H_2O. Deliquescent	Dana II, 40
Colorless	3 2.61-2.76	White		Cavities in basalt with apophyllite and zeolites	Zeolite group	Min Mag, 31, 726-735
Colorless	2.98-3.08	Colorless		Fractures in pegmatites	Li-Sn Osumilite, Fluor U.V. blue-white	Min Rec, 4, 73-76
O = deep orange, E = lt orange	~2 2.13	Bright orange		Altered serpentine		Am Min 64, 127
Colorless	2 3.20-3.32	Colorless		Evaporite salts with anhydrite, sylvite, dolomite	Sol hot HCl	Am Min, 48, 708-709

n_ω Var.	Refractive Index	Mineral and Name Composition	$(n_\omega - n_\varepsilon)$	Crystal System	Habit	Cleavage	Twinning
1.568	n_ω 1.568 n_ε 1.548-1.553	Tungusite $Ca_4Fe_2Si_6O_{15}(OH)_6$	0.015 to 0.020	Unknown	Fine platelets, rad fibrous, crusts	One perfect	
*1.55 1.583	n_ω 1.55-1.583 n_ε	Bayerite $Al(OH)_3$	Small	Hexagonal	Microscopic fibrous		
1.570	n_ω 1.570 n_ε 1.534	Fluoborite (Nocerite) $Mg_3(BO_3)(F,OH)_3$	0.036	Hexagonal	Radiated fibrous aggregate, prism, felted	{0001} indist	
*1.540 1.600	n_ω 1.540-1.600 n_ε 1.535-1.565	Scapolite (Wernerite) $(Ca,Na)_4[(Al,Si)_3Al_3Si_6O_{24}](Cl,CO_3)$	0.004 to 0.037	Tetragonal	Prism crystals (c), acicular aggregate, granular	{100} & {110} imperf	None
1.572	n_ω 1.572 n_ε 1.570	Englishite $K_2Ca_4Al_8(PO_4)_8(OH)_{10}\cdot 9H_2O$	0.002	Monoclinic ?	Micaceous, scaly agg	{001} perfect	
1.573	n_ω 1.573 n_ε 1.550	Sjögrenite $Mg_6Fe_2^{3+}CO_3(OH)_{16}\cdot 4H_2O$	0.023	Hexagonal	Thin hex plates {0001}	{0001} perfect	
1.573	n_ω 1.573 n_ε 1.572	Ekanite $(Ca,Na,K,Th)_2Si_4O_{10}$	0.001	Tetragonal	Prism (c), massive	None	Common
1.574	n_ω 1.574 n_ε 1.547	Portlandite $Ca(OH)_2$	0.027	Trigonal	Minute hex plates	{0001} perfect	
1.57 1.58	n_ω 1.57-1.58 $n_\varepsilon \sim n_\omega$	Calcioferrite $Ca_2Fe_2(PO_4)_3(OH)\cdot 7H_2O$	Small	Hexagonal	Foliated nodular masses	One perfect (foliation)	
1.576	n_ω 1.576 n_ε 1.546	Parsettensite $KMn_{10}Si_{12}O_{30}(OH)_{12}$	0.030	Hexagonal	Massive to micaceous	{0001} perfect	
*1.571 1.585	n_ω 1.571-1.585 n_ε 1.544-1.565	Saléeite $Mg(UO_2)_2(PO_4)_2\cdot 8\text{-}10H_2O$	0.027 to 0.020	Tetragonal	Platy rect crystals {001}, scaly, rad	{001} perfect {010} & {110} indist	
1.577 1.578	n_ω 1.577-1.578 n_ε 1.553-1.555	Autunite $Ca(UO_2)_2(PO_4)_2\cdot 10\text{-}12H_2O$	0.023 to 0.025	Tetragonal	Scaly aggregates, earthy	{001} perfect, {100} poor	
1.578	n_ω 1.578 n_ε 1.542	Eastonite $K_2(Mg_5Al)(Si_5Al_3)O_{20}(OH,F)_4$	0.036	Monoclinic β ~ 100°	Micaceous	{001} perfect	{110} normal, {001} complex
1.578	n_ω 1.578 n_ε 1.560	Indialite (α-Cordierite) $Mg_2Al_3(Si_5Al)O_{18}$	0.018	Hexagonal	Hex. prism., granular		None
*1.570 1.589	n_ω 1.570-1.589 n_ε 1.535-1.558	Zinnwaldite $K_2(Fe_{1-2}^{2+}Li_{2-3}Al_2)(Si_{6-7}Al_{2-1})O_{20}(OH,F)_4$	~0.035	Monoclinic	Micaceous	{001} perfect	
1.580	n_ω 1.580 n_ε 1.485	Berborite $Be_2(BO_3)(OH,F)\cdot H_2O$	0.095	Trigonal	Rhombohedral	{0001} perfect	Obs
1.580	n_ω ~1.580 n_ε	Meta-ankoleite $K_2(UO_2)_2(PO_4)_2\cdot 6H_2O$		Tetragonal	Platy, micaceous	{001} perfect {100} distinct	
1.580	n_ω 1.580 n_ε 1.575	Sodium Melilite $NaCaAlSi_2O_7$	0.005	Tetragonal	Prism. xls	{001} distinct {110} poor	None
*1.56 1.60	n_ω 1.56-1.60 n_ε 1.53-1.57	Hydromuscovite $(K,H_3O)_2Al_4(Si_6Al_2)O_{20}(OH)_4$	~0.03	Monoclinic	Fine scales	{001} perfect	Not visible
1.580	n_ω 1.580 n_ε 1.575	Darapiosite $(K,Na)_3Li(Mn,Zn)_2ZrSi_{12}O_{30}$	0.005	Hexagonal			

* n_ω variation exceeds 0.01

Color In Thin Section	Hardness Sp Gravity	Color Hand Sample	Alteration	Occurrence	Remarks	Reference
Pale yel, pale grn. Slight pleochroism	~2 2.59	Grass green, yellow-green		Amygdules in lavas with zeolites, gyrolite, calcite, apophyllite		Am Min, 52, 927-928
Colorless	 2.53-2.54	White		Argillaceous sediments	Dimorph of Gibbsite	Am Min, 49, 819
Colorless	3½ 2.98-3.01	Colorless, white		Contact carbonate zones, Hydrothermal veins		Dana II, 369-370
Colorless	5-6 2.5-2.8	Colorless, white, yel, grn, violet	Plagioclase, fine calcite, sericite, epidote zeolites, clays	Metasomatic rocks with calcite, sphene, garnet, epidote, apatite. Pegmatites		p. 395
Colorless	~3 ~2.65	Colorless		Phosphate nodules with variscite		Dana II, 957-958
O = pale yellow, brownish, E = colorless	2½ 2.11	White, yellowish, brownish		Low-T hydrothermal mineral with calcite and pyroaurite	Dimorph pyroaurite. Easily sol dilute acids with efforvescence	Dana I, 659-660
	5 3.32	Dk brown, green		Gem gravels		Can Min, 11, 913-927
Colorless	2 2.23-2.24	Colorless		Contact carbonate rocks with larnite, spurrite, afwillite	Sectile. Sol H_2O	Dana I, 641-642
Yellow, ~opaque	2½ 2.53	White, yellow, yellow-green		Nodules in clay		Dana II, 976-977
O = colorless to pale yellow, E = greenish-yellow	~1.5 2.59	Copper red		Mn-deposits	Sol HCl	Am Min, 10, 107
Pale yellow	2½ 3.27	Yellow, yellow-green		Secondary U-mineral with carnotite, autunite, etc.	Alunite group. Fluor. U.V.-bright yellow-green	USGS Bull, 1064, 177-183
Pale yellow, greenish	2-2½ 3.05-3.20	Yellow, green		Secondary U-mineral in pegmatites, granite, sediments, etc	Strong green fluor U.V. Meta-alunite may be biaxial	Am Min, 46, 812-822
Colorless, yellow brown. O > E	2½-3 2.86	Colorless, brown	Chlorite	Ultramafic igneous rocks	Mica group. Biotite series	pp. 270-274
Colorless, pale purple. Pleochroic	7-7½	Colorless	Pinite (chlorite, muscovite, biotite)	Buchites with sillimanite, spinel, anorthite	Cordierite polymorph	p. 167
O = gray-brown, E = colorless, pale yellow	~3 ~3.0	Yellow to dk brown, gray		Granite pegmatites, high-T hydrothermal veins with cassiterite	Mica group	p. 281
Colorless	3 2.20	Colorless		Skarns		Am Min, 53, 348-349
Colorless, yellow	? 3.54	Yellow		Granite pegmatites	Fluor. U.V.-yellow-green	Bull Geol Surv Great Brit, 25, 49
Colorless	~5 ~3.0	Colorless	Cebollite	High grade contact metamorphism of carbonate rocks	Melilite group	p. 159
Colorless	1-2 2.6-2.9	White, Fe-stained		Alter K-feldspar or muscovite. Shale, soils		p. 259
O = colorless, violet E = colorless, blue	2.92	Colorless, white, brownish, blue		Alkalic plutons with aegirine, quartz, eudialyte, sogdianite		Am Min, 61, 1053-1054

n_ω Var.	Refractive Index	Mineral Name and Composition	$(n_\omega - n_\varepsilon)$	Crystal System	Habit	Cleavage	Twinning
1.578 1.585	n_ω 1.578-1.585 n_ε 1.559-1.564	Sodium Autunite $Na(UO_2)(PO_4)\cdot 4H_2O$	0.019 to 0.021	Tetragonal	Foliated-platy, rad. masses	{001} perfect {100} less perf	
*1.572 1.592	n_ω 1.572-1.592 n_ε 1.550-1.558	Mizzonite $(Ca,Na)_4(Si,Al)_{12}O_{24}(CO_3,Cl)$	0.022 to 0.034	Tetragonal	Prism. xls., granular	{100} & {110} imperfect	None
1.582 1.584	n_ω 1.582-1.584 n_ε 1.564-1.565	Sabugalite $HAl(UO_2)_4(PO_4)_4\cdot 16H_2O$	0.018 to 0.019	Tetragonal	Thin plates or laths (Square-rect.)	{001} perfect	
*1.568 1.602	n_ω 1.568-1.602 n_ε 1.563-1.594	Beryl $Be_3Al_2(SiO_3)_6$	0.004 to 0.008	Hexagonal	Large hex. prism. xls	{0001} poor	Rare
1.585	n_ω 1.585 n_ε 1.564	Uramphite $NH_4UO_2PO_4\cdot 3H_2O$	0.021	Unknown	Square tab, rosettes	Two distinct	
1.583 1.588	n_ω 1.583-1.588 n_ε 1.574	Uranocircite II $Ba(UO_2)_2(PO_4)_2\cdot 10H_2O$	0.014	Tetragonal ?	Thin rectangular plates, fan-like	{001} perfect {100} distinct	Polysynthetic (~microcline)
1.585	n_ω 1.585 n_ε 1.612	Sodium Uranospinite $(Na_2,Ca)(UO_2)_2(AsO_4)_2\cdot 5H_2O$	0.027	Tetragonal	Tab to prism xls rad-fibrous	{001} perfect {010} & {100} distinct	
1.587	n_ω 1.5874 n_ε 1.3361	Soda-niter (Soda-nitre) $NaNO_3$	0.251	Trigonal	Rhomb crystals, massive granular	{10$\bar{1}$1} perfect {01$\bar{1}$2} & {0001} imperfect	{01$\bar{1}$2}, {02$\bar{2}$1}, {0001}
1.590	n_ω 1.590 n_ε 1.585	Iraqite $(Ca_3K)(La,Ce,Y,Th)_3Si_{16}O_{40}$	0.005	Tetragonal			
1.588 1.595	n_ω 1.588-1.595 n_ε 1.573-1.581	Metavoltine $(K,Na)_5Fe_3(SO_4)_6(OH)_2\cdot 8H_2O$	0.015 to 0.014	Hexagonal	Minute hex plates, scaly granular	{0001} perfect	
1.590 1.592	n_ω 1.590-1.592 n_ε 1.581-1.582	Torbernite $Cu(UO_2)_2(PO_4)_2\cdot 8\text{-}12H_2O$	0.009 to 0.010	Tetragonal	Tabular crystals {001}, scaly, gran-earthy	{001} perfect {100} indistinct	{110} rare
1.593	n_ω 1.593 n_ε 1.585	Machatschkiite $Ca_3(AsO_4)_2\cdot 9H_2O$	0.008	Trigonal	Crusts	None	
1.594	n_ω 1.594 n_ε 1.55	Alurgite $K_2(Mg,Mn)Al_3(AlSi_7)O_{22}(OH)_2$	0.04	Monoclinic $\beta \sim 100°$	Micaceous	{001} perfect	
*1.572 1.621	n_ω 1.572-1.621 n_ε 1.55 -1.596	Uranospinite $Ca(UO_2)_2(AsO_4)_4\cdot 10H_2O$	0.022 to 0.025	Tetragonal	Thin rectangular plates {001}	{001} perfect {100} distinct	
1.596	n_ω 1.596 n_ε 1.589	Wenkite $(Ba,Ca)_9Al_9Si_{12}O_{42}(SO_4)_2(OH)_5$	0.007	Hexagonal	Prismatic crystals	Prismatic poor	
*1.557 1.637	n_ω 1.557-1.637 n_ε 1.530-1.590	Phlogopite $KMg_3(Si_3Al)O_{10}(OH)_2$	0.028 to 0.047	Monoclinic	Micaceous, pseudohex crystals	{001} perfect	{110} normal, {001} complex
1.597	n_ω 1.597 n_ε 1.570	Abernathyite $KUO_2AsO_4\cdot 3H_2O$	0.027	Tetragonal	Thin to thick tabular crystals	{001} perfect	
*1.592 1.605	n_ω 1.592-1.605 n_ε 1.558-1.564	Meionite $Ca_4(Al_6Si_6)O_{24}CO_3$	0.034 to 0.041	Tetragonal	Prism. xls., granular	{100} & {110} imperfect	None
1.6	n_ω 1.6 n_ε	Molysite $FeCl_3$	Very large	Trigonal	Massive coatings	{0001} perfect	
1.601	n_ω 1.601 n_ε 1.480	Grimselite $K_3Na(UO_2)(CO_3)_3\cdot H_2O$	0.121	Hexagonal	Prism crystals, crusts	None	

* n_ω variation exceeds 0.01

Color In Thin Section	Hardness Sp Gravity	Color Hand Sample	Alteration	Occurrence	Remarks	Reference
Pale yellow, weak pleochroism	2–2½ 3.58-3.89	Lemon yellow, greenish-yellow		Lignite deposits	Autunite group. Fluor U.V.-strong yellow-green	Am Min, 43, 383
Colorless	5-6 2.5-2.8	White, yellow, lt. green, violet	Plagioclase, calcite, chlorite, sericite, epidote, zeolite	Contact or regional met. of carbonate rocks with diopside, calcite	Scapolite group	p. 359
O = pale yellow, E = colorless	2½ 3.15-3.20	Bright yellow		Secondary U-mineral in sand-stone type deposits	Fluor U.V.-lemon yellow	USGS Bull, 1064, 196-200
Colorless	7½-8 2.68-2.80	White, pale blue, green, yellow	Kaolin clays, sericite, illite, bayenite, bertrandite	Granite pegmatites		p. 164
X-colorless, Y = Z = pale green	? 3.7	Pale green to bottle green		Oxide zone U-coal deposit.	Fluor U.V.-yellow green	Am Min, 44, 464
O = pale yellow, E = colorless	2-2½ 3.46-4.10	Yellow		Secondary U-mineral	Aulunite group. Fluor. U.V.-green	USGS Bull, 1064, 211-215 & 177
O = pale yellow, E = colorless	2½ 3.846	Yellow-green to yellow		Oxide zone hydrothermal veins, felsic volcanics	Small 2V. Fluor U.V.-strong yellow-green	Am Min, 43, 383-384
Colorless	1½-2 2.25-2.27	Colorless, white, yellowish		Soil efflorescence in arid regions	Easily sol H_2O	Dana II, 300-302
Colorless, pale yel	4½ 3.27-3.28	Greenish yellow		Dolomite contact zone		Am Min, 61, 1054
O = Dk yellow, brown E = Lt yellow	2½ 2.40-2.51	Yel-brn, grn-brn, orange-brown		Secondary sulfate with krausite, coquimbite, alunite	Partially sol H_2O. Decomposed acids	Dana II, 619-621
O = Pale-dk green, E = colorless, pale greenish blue	2-2½ 3.22-3.28	Emerald to grass green		Pegmatites, hydrothermal veins, sed U-deposits	Autunite group	USGS Bull, 1064, 170-177
Colorless	2-3 2.50	Colorless		Secondary crusts with gypsum, pharmacolite, sainfeldite	Readily sol dilute HCl or HNO_3. Small 2V	Am Min, 62, 1260
Pink. Faint pleochroism	2½-3 2.84	Copper red		Mn ore deposits	Mica group	pp. 270, 272
O = pale yellow, E ~ colorless	2-3 3.45-3.65	Lemon yellow, siskin green		Secondary U-mineral, alter uraninite and arsenides	Autunite group. Fluor. U.V.-bright yellow	USGS Bull, 1064, 183-187
Colorless	6 3.13-3.28	Lt gray		Calc-silicate marble with barite	Possible small 2V < 10°	Am Min, 48, 213
O = pale yellow, pale pink-brown, pale grn, E = colorless	2-2½ 2.76-2.90	Bronze brown, colorless, greenish		Metamorphosed dolomite, peri-dotites, lamprophyres, leucite volcanics	Mica (biotite) group	p. 274
Colorless	~2½ 3.32-3.57	Yellow		Secondary mineral as coatings of fracture filling		Am Min, 49, 1578-1602
Colorless	5-6 2.5-2.8	White, yellow, lt. green, violet	Plagioclase, calcite, chlorite, sericite, epidote, zeolite	Contact or regional met. of carbonate rocks with diopside, calcite	Scapolite group	p. 395
Yellow	Soft 2.90-3.04	Yellow to brownish-red	Readily hydro-lized by H_2O to hydrous ferric oxide	Volcanic sublimate	Deliquescent	Dana II, 47-48
O = pale yellow, E = colorless	2-2½ 3.27-3.30	Yellow		Secondary U-mineral in ore veins	Nonfluorescent U.V. Possible small 2V	Am Min, 58, 139

Uniaxial −

n_ω Var.	Refractive Index	Mineral Name and Composition	$(n_\omega - n_\varepsilon)$	Crystal System	Habit	Cleavage	Twinning
*1.594 1.609	n_ω 1.594-1.609 n_ε 1.564-1.580	Paragonite $NaAl_2(Si_3Al)O_{10}(OH)_2$	0.036 to 0.029	Monoclinic β=95°	Micaccous, fine scaly, massive	{001} perfect	{001} complex
*1.595 1.613	n_ω 1.595-1.613 n_ε 1.585-1.600	Meta-autunite $Ca(UO_2)_2(PO_4)_2 \cdot 2\text{-}6H_2O$	0.010 to 0.013	Tetragonal	Pseudomorphic after autunite	{001} perfect {100} indistinct	Sector
1.605	n_ω 1.605 n_ε 1.580	Montdorite $K(Fe,Mn,Mg,Ti)_5(Si,Al)_8O_{20}(F,OH)_4$	0.025	Monoclinic β = 99.9°			
1.605	n_ω 1.605 n_ε 1.450	Eitelite $Na_2Mg(CO_3)_2$	0.145	Trigonal	Rhombohedral crystals	{0001} distinct	
1.605	n_ω 1.605 n_ε 1.573	Heinrichite $Ba(UO_2)_2(AsO_4)_2 \cdot 10\text{-}12H_2O$	0.032	Tetragonal	Tabular {001}	{001} perfect {100} distinct	
1.608	n_ω 1.608 n_ε 1.606	Sogdianovite $(K,Na)_2Li_2(Li,Fe,Al,Ti)_2Zr_2(Si_2O_5)_6$	0.002	Hexagonal	Massive, platy	{0001} perfect	
1.610	n_ω 1.610 n_ε 1.582	Zeunerite $Cu(UO_2)_2(AsO_4)_2 \cdot 12H_2O$	0.028	Tetragonal	Tabular crystals {001}	{001} perfect {100} distinct	
1.610	n_ω 1.610 n_ε 1.605	Zirsinalite $Na_6(Ca,Mn,Fe)Zr(SiO_3)_6$	0.005	Trigonal	Granular	None	
1.610	n_ω 1.610 n_ε 1.607	Sugilite $(K,Na)(Fe^{3+},Al)_2(Li,Al,Fe^{3+})_3Si_{12}O_{30} \cdot H_2O$	0.003	Hexagonal	Granular aggregate	{0001} poor	
*1.58 1.65	n_ω 1.58-1.65 n_ε 1.57-1.62	Delessite (Melanolite) $(Mg,Al,Fe^{3+})_6(Si,Al)_4O_{10}(OH)_8$	0.001 to 0.020	Monoclinic	Massive, granular, foliated	{001} perfect	{001}
1.612	n_ω 1.612 n_ε 1.593	Meliphanite (Melinophane) $(Ca,Na)_2Be(Si,Al)_2(O,F)_7$	0.019	Tetragonal	Tabular, lamellar aggregate	{010} perfect {001} distinct {110} indistinct	
1.612	n_ω 1.612 n_ε 1.584	Troegerite $H_2(UO_2)_2(AsO_4)_2 \cdot 8H_2O$	0.028	Tetragonal	Micaceous {001}, thin tabular crystals	{001} perfect {100} good	
1.611 1.617	n_ω 1.611-1.617 n_ε 1.603-1.606	Cymrite $BaAl_2Si_2O_8 \cdot H_2O$	0.008 to 0.011	Hexagonal	Hexagonal prism or plates, massive	{0001} perfect, {10$\bar{1}$0} imperfect	
*1.603 1.628	n_ω 1.603-1.628 n_ε 1.598-1.619	Dahllite (Carbonate-apatite) (Francolite) $Ca_5(PO_4,CO_3)_3(OH,F)$	0.005 to 0.009	Hexagonal	Hexagonal prism	{0001} poor, {10$\bar{1}$0} trace	
1.612 1.618	n_ω 1.612-1.618 n_ε 1.607-1.611	Fluocerite (Tysonite) $(Ce,La)F_3$	0.005 to 0.007	Trigonal	Prism (c), tab {0001}, massive	{0001} distinct {11$\bar{2}$0} indist	
1.617	n_ω 1.617 n_ε 1.595	Karnasurtite (Kozhanovite) (Carnasurtite) $(La,Ce,Th)(Ti,Nb)(Al,Fe)(Si,P)_2O_7(OH)_4 \cdot 3H_2O$	0.022	Hexagonal ?	Granular, hexagonal tabular crystals	1 perfect, 1 imperfect	
1.618	n_ω 1.618 n_ε 1.552	Chalcophyllite $Cu_{18}Al_2(AsO_4)_3(SO_4)_3(OH)_{27} \cdot 36H_2O$	0.066	Trigonal	Hexagonal tabular, scaly	{0001} perfect {10$\bar{1}$1} traces	
1.621	n_ω 1.621 n_ε 1.619	Gillespite $BaFeSi_4O_{10}$	0.002	Tetragonal	Massive compact	{001} perfect {100} indistinct	
1.623	n_ω 1.623 n_ε 1.610	Meta-uranocircite $Ba(UO_2)_2(PO_4)_2 \cdot 6\text{-}8H_2O$	0.013	Tetragonal	Thin flexible plates {0001}, radiated	{001} perfect {100} distinct	{100} & {010} polysynthetic

* n_ω variation exceeds 0.01

Color In Thin Section	Hardness Sp Gravity	Color Hand Sample	Alteration	Occurrence	Remarks	Reference
Colorless	2½ 2.78-2.90	Colorless, pale yellow		Phyllite, schist, gneiss, quartz veins. Fine sediments	Mica group	p. 272
X = colorless to pale yellow, Y = Z = dk yellow	2-2½ 3.45-3.55	Yellow, greenish-yellow		Dehydration of autunite at weathered exposures	Fluor U.V.-pale yellow-green	Am Min, 48, 1389-1393
	3.15-3.16	Green, brownish green		Peralkaline rhyolite		Am Min 64, 1331
Colorless	3½ 2.74	Colorless, white		Evaporite salts	Small 2V	Am Min, 58, 211-217
O = pale yellow, E = colorless	2.5 3.61	Yellow, green		Secondary U-mineral, alter pitchblende	Fluor U.V.-bright green, yel-green. 2V = 0° to 20°	Am Min, 43, 1134-1143
Colorless	7 2.90	Pale red-violet		Alkali pegmatites with thorite stillwellite, aegirine		Am Min, 54, 1221-1222
O = blue-green, E = pale blue-green	2½ 3.39	Yellow-green to emerald green		Oxide zones U-deposits containing uraninite and As-minerals	Autunite group. Not fluor U.V.	USGS Bull, 1064, 191-194
Colorless	5½ 2.90-3.08	Colorless, yellowish-gray	Rapidly decomp in air to powdery Na_2CO_3. Eudialyte, lovozerite	Pegmatitic veinlets in alkalic rocks with aegirine, lomonosovite	Decomp. cold acids	Am Min, 60, 489
Colorless	6-6½ 2.74-2.80	Brownish-yellow		Aegirine syenite with albite, pectolite, sphene, allanite, andradite		Am Min, 62, 596
O = dk green, E = lt yellow-green	2-3 2.73	Lt yellow, dk green, black		Altered mafic rocks	Chlorite group	p. 294
Colorless	5-5½ 3.03	Yellow, colorless		Alkali pegmatites with nepheline. Skarns with melanite		Vlasov II, 129-134
Colorless, pale yel	2-3 3.55	Yellow		Secondary U-mineral with zeunerite, metazeunerite, walpurgite	Fluor U.V-lemon yellow	USGS Bull, 1064, 187-191
Colorless	2-3 3.41-3.44	Dk green, brown		High P-low T metamorphic rocks, ore veins	Small 2V	Am Min, 49, 158-165
Colorless	5 2.9-3.1	Yellow, green, brown, etc.		Granite pegmatites	Apatite group	Dana II, 879-889
Colorless, pale pink	4-5 5.93-6.14	Yellow-brown, red-brown		Rare earth pegmatites	Sol H_2SO_4	Dana II, 48-50
Pale yellow non-pleochroic	2 2.89-2.95	Honey yellow	Polylithionite	Alkali pegmatites with schizolite, natrolite, epididymite		Am Min, 45, 1133-1134
O = pale bluish-green E = colorless	2 2.64-2.67	Green, bluish-green		Oxide zone Cu-deposits	I.R. decreases with H_2O content. Sol acids, NH_4OH	Dana II, 1008, 1010
O = pale pink, E = deep rose	3 3.402	Red		Hydrothermal veins with sanbornite, celsian, witherite		Am Min, 14, 319-322
O = pale yellow, E ~ colorless	2-2½ 3.95-4.00	Yellow-green		Secondary U-mineral, also primary in low-T veins	Fluor U.V.-green. Small 2V common	USGS Bull, 1064, 211-215

n_ω Var.	Refractive Index	Mineral Name and Composition	$(n_\omega - n_\epsilon)$	Crystal System	Habit	Cleavage	Twinning
1.623 1.627	n_ω 1.623-1.627 n_ϵ 1.602-1.607	Bazzite $Be_3(Sc,Al)_2Si_6O_{18}$	0.020	Hexagonal	Acicular (c)		
1.627	n_ω 1.627 n_ϵ 1.615	Burbankite $(Na,Ca,Sr,Ba,Ce)_6(CO_3)_5$	0.012	Hexagonal	Anhedral crystals	{10$\bar{1}$0} distinct	
1.629	n_ω 1.629 n_ϵ 1.626	Whitlockite $(Ca,Mg)_3(PO_4)_2$	0.03	Trigonal	Rhomb, tabular crystals, granular earthy	None	
1.630	n_ω 1.630 n_ϵ 1.60	Mariposite $K_2(Al,Cr)_3(Mg,Fe^{2+})Si_7AlO_{20}(OH,F)_4$	0.03	Monoclinic β ~ 100°	Micaceous	{001} perfect	
*1.61 1.65	n_ω 1.61-1.65 n_ϵ 1.56-1.61	Glauconite $(K,H_3O)_2(Fe^{3+},Al,Fe^{2+},Mg)_4$ $(Si_{7-7.5}Al_{1-0.5})O_{20}(OH)_4$	0.014 to 0.032	Monoclinic β ~ 100°	Fine flakes in pellets or granules	{001} perfect	Not visible
1.632	n_ω 1.632 n_ϵ 1.595	Metanovacekite $Mg(UO_2)_2(AsO_4)_2 \cdot 4\text{-}8H_2O$	0.037	Tetragonal	Platy {001}	{001} perfect {100} & {110} indistinct	
*1.620 1.641	n_ω 1.620-1.641 n_ϵ 1.620-1.625	Novacekite $Mg(UO_2)_2(AsO_4)_2 \cdot 12H_2O$	Very small to 0.016	Tetragonal	Rectangular plates {001}, crusts	{001} perfect {010} & {110} indistinct	
1.634	n_ω 1.634 n_ϵ 1.632	Kahlerite $Fe^{2+}(UO_2)_2(AsO_4)_2 \cdot 12H_2O$	0.002	Tetragonal	Tabular crystals {001}	{001} perfect	
1.633 1.636	n_ω 1.633-1.636 n_ϵ 1.590-1.591	Cuprorivaite $CaCuSi_4O_{10}$	0.043	Tetragonal	Tabular crystals {001}	{001} perfect	
1.637	n_ω 1.637 n_ϵ 1.621	Liddicoatite $Ca(Li,Al)_3Al_6B_3Si_6O_{27}(O,OH)_3$ (OH,F)	0.016	Trigonal	Large prism crystals	{0001} poor	
1.637 1.641	n_ω 1.637-1.641 n_ϵ 1.609	Metaheinrichite $Ba(UO_2)_2(AsO_4)_2 \cdot 8H_2O$	0.028 to 0.032	Tetragonal	Tabular {001}	{001} perfect {100} distinct	
1.6365 1.6438	n_ω 1.6365-1.6438 n_ϵ 1.6148-1.6097	Mitscherlichite $K_2CuCl_4 \cdot 2H_2O$	0.0217 to 0.0341	Tetragonal	Short prism or pyram crystals	None ?	{101}
*1.633 1.650	n_ω 1.633-1.650 n_ϵ 1.629-1.646	Fluorapatite $Ca_5(PO_4)_3F$	0.003 to 0.005	Hexagonal	Small hexagonal prism crystals	{0001} poor, {10$\bar{1}$0} very poor	{11$\bar{2}$1} or {10$\bar{1}$3} simple, rare
1.642	n_ω 1.642 n_ϵ 1.608	Metakahlerite $Fe(UO_2)_2(AsO_4)_2 \cdot 8H_2O$	0.034	Tetragonal	Scaly aggregates	{001} perfect {100} distinct	
1.643	n_ω 1.643 n_ϵ 1.623	Zussmanite $K(Fe^{+2},Mg,Mn)_{13}(Si_{17}Al)O_{42}(OH)_{14}$	0.020	Trigonal	Tabular	{0001} perfect	
1.644	n_ω 1.644 n_ϵ 1.617	Metakirchheimerite $Co(UO_2)_2(AsO_4)_2 \cdot 8H_2O$	0.027	Tetragonal	Tabular crystals, crusts	{001} perfect	
1.650 1.640	n_ω 1.650-1.640 n_ϵ 1.646-1.636	Wilkeite $Ca_5(SiO_4,PO_4,SO_4)_3(O,OH,F)$	0.004	Hexagonal	Rounded crystals, massive granular	{0001} imperfect	
1.646	n_ω 1.646 n_ϵ 1.635	Zapatalite $Cu_3Al_4(PO_4)_3(OH)_9 \cdot 4H_2O$	0.011	Tetragonal	Massive	{001} good	
*1.635 1.658	n_ω 1.635-1.658 n_ϵ 1.615-1.633	Elbaite $Na(Li,Al)_3Al_6(SiO_3)_6(BO_3)_3(OH)_4$	0.020 to 0.025	Trigonal	Prism. xls., acic.-rad	{11$\bar{2}$0} & {10$\bar{1}$1} very poor	{10$\bar{1}$1} rare
1.647	n_ω 1.647 n_ϵ 1.637	Aminoffite $Ca_2(Be,Al)Si_2O_7(OH) \cdot H_2O$	0.010	Tetragonal	Pyramidal crystals	{001} imperfect	

170

* n_ω variation exceeds 0.01

Color In Thin Section	Hardness Sp Gravity	Color Hand Sample	Alteration	Occurrence	Remarks	Reference
O = lt greenish yel, E = dk blue	6½ 2.77	Blue		Granite pegmatites	Sc-Beryl	Min Abs, 18, 115 (1967)
Colorless	3½ 3.50	Grayish yellow		Hydrothermal veins		Am Min, 38, 1169-1183
Colorless	5 3.12	Colorless, white, yellow, pink		Secondary mineral in granite pegmatites. Guano phosphates	Easily sol dilute acids	Dana II, 684-686
Colorless, pale green E > O	2½-3 2.79	Apple green			Mica group. Cr-phengite	pp. 270-272
O = yellow-green, olive-green, blue-green, E = pale yellow, grn	2 2.4-3.0	Olive-green, blue-green	Limonite, goethite	With clays and chlorite in diagenetic pellets in marine sediments	Mica group	p. 261
Colorless, lt yellow	2½ 3.51-3.72	Pale yellow, yel		Secondary U-mineral	Autunite group. Fluor. U.V.-dull green	Min Mag, 31, 966
Pale yellow	2½ ~3.7	Yellow		Secondary U-mineral	Novacekite-Saléeite series Autunite group Fluor U.V. dull green. Small 2V	USGS Bull, 1064, 177-183
Pale yellow		Yellow, yellow-green		Secondary U-mineral, oxide zone Fe-deposits	Not fluor U.V. Possible small 2V	USGS Bull, 1064, 204-205
O = blue, E = pale rose	~5 3.08-3.09	Blue		Volcanic rocks (Mt Vesuvius)		Am Min, 47, 409-411
Strong pleochroism O > E	~7½ 3.02-3.05	Zoned, red, green, brown, blue		Pegmatites with elbaite	Tourmaline group. Caelbaite	Am Min, 62, 1121-1124
	2.5 4.04-4.09	Yellow, green		Secondary U-mineral, alter. pitchblende	Fluor U.V.-bright green, yellow-green. 2V = 0° to 18°	Am Min, 43, 1134-1143
Colorless to pale blue or green, O > E	2½ 2.41-2.42	Green, greenish-blue		Volcanic emanations associated with metavoltine, sylvite, gypsum	Deliquescent	Dana II, 100-101
Colorless, pale colors, E > O	5 2.9-3.5	Yellow, green, any color	Serpentine, kaolin clays, wavellite	Most igneous rocks. Pegmatites and hydrothermal veins. Many metamorphic rocks	Apatite group	p. 81
O = pale yellow, E ~ colorless	? 3.84	Sulfur yellow		Secondary U-mineral	2V = 0°-22°. Not fluor U.V.	Am Min, 45, 254
O = pale green, E = colorless	3.146	Pale green		Metamorphosed siliceous iron-stones, shale or limestone		Am Min, 50, 278
Colorless, greenish-yellow, not pleochroic	2-2½ <3.33	Pale rose		Secondary U-mineral with pitch blende, erythrite	Possible small 2V. Metatorbenite group	Am Min, 44, 466
Colorless	~5 3.12-3.23	Pale pink, yellowish	Crestmoreite	Contact marble with idocrase, diopside, garnet, tobermorite	Apatite group. Easily sol HCl or HNO_3	Dana II, 905
Pale green. E > O	1½ 3.016	Pale blue-green		Silicified limestone with libethenite, chenevixite, beaverite, alunite	May be biaxial	Min Mag 38, 541-544
Colorless, pale blue, green, rose. O > E	7-7½ 3.0-3.25	Colorless, rose, green, blue	Illite, sericite, biotite, chlorite	Granite pegmatites	Tourmaline group	p. 170
Colorless	5½ 2.94-3.12	Colorless		Veins and cavities in iron ores		Vlasov II, 129-134

n_ω Var.	Refractive Index	Mineral Name and Composition	$(n_\omega - n_\varepsilon)$	Crystal System	Habit	Cleavage	Twinning
1.648	n_ω 1.648 n_ε 1.625	Kazakovite $Na_6H_2Ti(SiO_3)_6$	0.023	Trigonal	Rhomb crystals	None	{11$\bar{2}$4} poly-synthetic
*1.605 1.696	n_ω 1.605–1.696 n_ε 1.565–1.625	Biotite $K(Mg,Fe^{2+})_{6-5}Al_{0-1}(Si_{6-5}Al_{2-3})O_{20}(OH,F)_4$	0.04 to 0.07	Monoclinic	Micaceous, pseudohex. crystals	{001} perfect	{110} normal, {001} complex
1.650	n_ω 1.650 n_ε 1.626	Metazeunerite $Cu(UO_2)_2(AsO_4)_2 \cdot 8H_2O$	0.024	Tetragonal	Tabular {001}, micaceous, platy	{001} perfect {100} distinct	
*1.643 1.658	n_ω 1.643–1.658 n_ε 1.637–1.654	Hydroxylapatite $Ca_5(PO_4)_3OH$	0.007 to 0.004	Hexagonal	Small hexagonal prism crystals	{0001} poor	{11$\bar{2}$1} or {10$\bar{1}$3} simple, rare
1.651	n_ω 1.651 n_ε 1.637	Strontium-apatite $(Sr,Ca)_5(PO_4)_3(OH,F)$	0.014	Hexagonal	Corroded prism crystals	Poor	
1.653	n_ω 1.653 n_ε 1.640	Jeremejevite (Eremeyevite) $Al_6B_5O_{15}(OH)_3$	0.013	Hexagonal	Elong hexagonal prisms	None	Sector
1.653	n_ω 1.653 n_ε 1.642	Chantalite $CaAl_2SiO_4(OH)_4$	0.011	Tetragonal	Anhedral grains		
1.654	n_ω 1.654 n_ε 1.650	Hydroxylellestadite $Ca_{10}(SiO_4)_3(SO_4)_3(OH,Cl,F)$ OH>Cl,F	0.004	Hexagonal	Massive	{10$\bar{1}$0} distinct	
*1.64 1.67	n_ω 1.64–1.67 n_ε 1.62–1.65	Brunsvigite $(Fe^{2+},Mg,Al)_6(Si,Al)_4O_{10}(OH)_8$	0.006 to 0.015	Monoclinic	Massive, scaly, foliated, hex-tab xls	{001} perfect	{001}
1.655	n_ω 1.655 n_ε 1.650	Ellestadite $Ca_5[(Si,S,P,C)O_4]_3(Cl,F,OH)$	0.005	Hexagonal	Massive, granular	{0001} & {10$\bar{1}$0} indistinct	
*1.631 1.658	n_ω 1.631–1.658 n_ε 1.610–1.633	Dravite $NaMg_3Al_6(SiO_3)_6(BO_3)_3(OH)_4$	0.021 to 0.025	Trigonal	Prism. xls., acic.-rad	{11$\bar{2}$0} & {10$\bar{1}$1} very poor	{10$\bar{1}$1} rare
1.655 1.658	n_ω 1.655–1.658 n_ε 1.633–1.636	Eudialyte $Na_4(Ca,Fe,Ce,Ma)_2ZrSi_6O_{17}(OH,Cl)_2$	0.004 to 0.022	Trigonal	Hexagonal tabular, prism, rhombohedral	{0001} distinct {10$\bar{1}$0} & {11$\bar{2}$0} imperfect	
1.658	n_ω 1.658 n_ε 1.486	Calcite $CaCO_3$	0.172	Trigonal	Rhomb-scalenohedral crystals	{10$\bar{1}$1} perfect {01$\bar{1}$2} parting	{01$\bar{1}$2} lammelar {0001} contact
*1.650 1.667	n_ω 1.650–1.667 n_ε 1.647–1.665	Chlorapatite $Ca_5(PO_4)_3Cl$	0.004 to 0.001	Hexagonal	Small hexagonal prism crystals	{0001} poor	Rare {11$\bar{2}$1} or {10$\bar{1}$3}
1.654 1.664	n_ω 1.654–1.664 n_ε 1.620–1.629	Friedelite (Ferroschallerite) $(Mn,Fe)_8Si_6O_{18}(OH,Cl)_4 \cdot 3H_2O$	0.029 to 0.035	Trigonal	Tabular, rarely acicular, fibrous, massive	{0001} perfect	
1.660	n_ω 1.660 n_ε	Fermorite $(Ca,Sr)_5[(As,P)O_4]_3(F,OH)$	Small	Hexagonal	Massive, granular	None	
**1.576 1.745	n_ω 1.576–1.745 n_ε 1.543–1.634	Stilpnomelane $(Fe^{3+},Fe^{2+},Mg,Mn,Al)_{5-6}(Si_4O_{10})_2(OH)_4$	0.030 to 0.110	Triclinic	Micaceous, radiated	{001} perfect	
*1.670 1.651	n_ω 1.670–1.651 n_ε 1.658–1.651	Gehlenite $Ca_2Al(SiAl)O_7$	0.012 to 0.000	Tetragonal	Prism. xls.	{001} distinct {110}	None
1.660	n_ω 1.660 n_ε 1.640	Belovite $(Sr,Ce,Na,Ca)(PO_4)_3(O,OH)$	0.020	Hexagonal	Prism	{10$\bar{1}$0} & {0001} imperfect	

* n_ω variation exceeds 0.01

** n_ω variation exceeds 0.10

Color In Thin Section	Hardness Sp Gravity	Color Hand Sample	Alteration	Occurrence	Remarks	Reference
Colorless, pale yellow	4 2.84	Pale yellow		Alkali syenites with sodalite, nordite, belovite, vuonnemite		Am Min 60, 161-162
O = dk brown, dk grn, dk red-brown, E = pale yellow, pale green, pale brown	2½-3 2.7-3.3	Black to dk brown or green	Chlorite, vermiculite, hydrobiotite	Granitic plutons, silicic volcanics, schists and gneisses	Mica group	p. 274
	2-2½ 3.64-3.79	Pale green to emerald green		Secondary U-mineral in oxide zones of As-bearing deposits	Fluor U.V.-dull yellow-green	USGS Bull, 1064, 215-220
Colorless, pale colors, E > O	5 2.9-3.5	Yellow, green, any color		Pegmatites and hydrothermal veins. Talc schists	Apatite group	p. 81
Colorless	5 3.84	Pale green, yellowish green		Alkalic pegmatites with batisite, innelite, ramsayite, eudialyte	Apatite group	Am Min, 47, 808
Colorless	6½ 3.27-3.28	Colorless, pale yellow-brown		Granitic plutons with quartz and orthoclase	Hexagonal crystals, biaxial cores	Dana II, 330-332
Colorless	2.8-2.97	Colorless, white		Rodingite dikes in ophiolites with prehnite, vuagnatite, chlorite, calcite		Am Min 63, 1282
Colorless	4½ 3.02	Pale purplish		Contact carbonate zones with diopside, wollastonite	Apatite group	Am Min, 56, 1507-1518
O = very dk green, E = pale green-brown	~2 2.99-3.08	Olive-green, dk green		Hydrothermal veins. Veins & voids in granite, gabbro, spilite	Chlorite group	p. 292
Colorless	~5 3.068	Pale rose		Veins in contact metamorphic marble	Apatite group	Dana II, 906
Colorless, yellow, brown O >> E	7-7½ 3.0-3.25	Dk. brown, yellow	Illite, sericite, biotite, chlorite	Mica schists. Pelitic contact rocks	Tourmaline group	p. 170
Pink or yellow, E > O	5-5½ 2.74-2.98	Yel-brown, brn-red, pink		Nepheline syenite, nepheline pegmatites	Var. Eucolite. Sol HCl. usually zoned	Vlasov II, 355-364
Colorless	3 2.710	Colorless, white, any color		Very widespread mineral in carbonate sediments, veins, etc.	Easily sol. cold dilute acids with effervescence	p. 53
Colorless, pale colors, E > O	5 2.9-3.5	Yellow, green, any color		Accessory in mafic igneous rocks. Skarns with scapolite	Apatite group	p. 81
Pale pink, brownish-red	4-5 3.04-3.06	Pink to dk red, brown, yellow		Hydrothermal Mn-ores	2V small	USGS pp 180, 88-90
Colorless	5 3.52	White, pinkish		Veinlets in Mn-ores	Apatite group. Sol acids	Dana II, 904
O = black, dk brown, dk green, E = colorless, yel	3-4 2.59-2.96	Black, red-brown, deep green	Fe-oxides and clays	Low grade or low T-high P metamorphic rocks		p. 305
Colorless	5-6 3.04-2.99	Colorless, yellow, gray green	Cebollite, calcite, zeolite, garnet, diopside	High grade contact metamorphism of carbonate rocks	Melilite group	p. 159
Colorless, yellow	5 4.19	Yellow		Alkaline Pegmatites	Sr-hydroxyapatite. Sol dilute HCl or HNO_3	Am Min, 40, 367-368

Uniaxial −

n_ω Var.	Refractive Index	Mineral Name and Composition	$(n_\omega - n_\varepsilon)$	Crystal System	Habit	Cleavage	Twinning
*1.643 1.679	n_ω 1.643-1.679 n_ε 1.681-1.704	Schallerite $(Mn,Fe)_8Si_6As(O,OH,Cl)_{26}$	0.038 to 0.025	Hexagonal	Massive, granular	{0001} distinct	
1.665	n_ω 1.665 n_ε 1.663	Steenstrupine $CeNaMn(SiO_3)_3$	0.002	Hexagonal	Rhombohedral crystals	None	
1.66	n_ω 1.66 n_ε 1.57	Mackelveyite (Mckelveyite) $(Na,Ca)(Ba,Y,U)_2(CO_3)_3 \cdot 1\text{-}2H_2O$	0.09	Trigonal	Hemimorphic xls, platy {0001}	None or poor {0001}	
1.666	n_ω 1.666 n_ε 1.653	Tienshanite $BaNa_2MnTiB_2Si_6O_{20}$	0.013	Hexagonal	Massive, fine crystalline	{0001} distinct	
1.669	n_ω 1.669 n_ε 1.631	Manganpyrosmalite $(Mn,Fe)_8(Si_6O_{15})(OH,Cl)_{10}$	0.038	Trigonal	Massive granular	{0001} perfect	
1.669	n_ω 1.669 n_ε 1.657	Hardystonite $Ca_2ZnSi_2O_7$	0.012	Tetragonal	Massive, granular	{001} distinct {100} & {110} indistinct	
1.670	n_ω 1.670 n_ε 1.616	Siderophyllite $K_2(Fe^{2+}Al)_2(Si_5Al_3)O_{20}(OH,F)_4$	0.064	Monoclinic β ~ 100°	Micaceous	{001} perfect	{110} normal, {001} complex
1.670	n_ω 1.670 n_ε 1.582	Ekmanite $(Fe,Mn,Mg)_6(Si,Al)_8O_{20}(OH)_8 \cdot 2H_2O$	0.088	Orthorhombic	Foliated, columnar, scaly, massive	{001} perfect	
1.675	n_ω 1.675 n_ε 1.59	Chloromagnesite (Chlormagnesite) $MgCl_2$	0.085	Trigonal	Micro hexagonal plates		
*1.658 1.698	n_ω 1.658-1.698 n_ε 1.633-1.675	Schorl $NaFe_3^{2+}Al_6(SiO_3)_6(BO_3)_3(OH)_4$	0.025 to 0.023	Trigonal	Prism. xls., acic.-rad	{11$\bar{2}$0} & {10$\bar{1}$1} very poor	{10$\bar{1}$1} rare
1.675 1.682	n_ω 1.675-1.682 n_ε 1.636-1.647	Pyrosmalite $(Mn,Fe)_{14}Si_3O_7(OH,Cl)_6$	0.039 to 0.035	Trigonal	Hexgonal prism or tabular, massive	{0001} perfect	
1.679	n_ω 1.679 n_ε 1.500	Dolomite $CaMg(CO_3)_2$	0.179	Trigonal	Rhomb crystals, fine-coarse cleavage agg	{10$\bar{1}$1} perfect	{02$\bar{2}$1} lamellar, {0001} simple
1.680	n_ω~1.680 n_ε~1.655	Sincosite $Ca(VO)_2(PO_4)_2 \cdot 5H_2O$	0.025	Tetragonal	Tab (square) crystals, scaly, rosettes	{001} perfect {100} & {110} distinct	{110} rare
1.681 1.686	n_ω 1.681-1.686 n_ε 1.627-1.638	Spangolite $Cu_6AlSO_4(OH)_{12}Cl \cdot 3H_2O$	0.054 to 0.048	Trigonal	Hexagonal prism or tabular crystals, holohedral	{0001} perfect {10$\bar{1}$1} distinct	{0001} rare
1.683	n_ω 1.683 n_ε 1.672	Verplanckite $Ba_2(Mn,Fe,Ti)Si_2O_6(O,OH,Cl,F)_2 \cdot 3H_2O$	0.011	Hexagonal	Hex prisms (c), radiated prism	{11$\bar{2}$0} distinct {0001} poor	
1.685	n_ω 1.685 n_ε 1.674	Pabstite $BaSnSi_3O_9$	0.011	Hexagonal	Anhedral grains		
1.690	n_ω 1.690 n_ε 1.527	Benstonite $MgCa_6(Ba,Sr)_6(CO_3)_{13}$	0.163	Trigonal	Flat rhomb xls, crusts	{10$\bar{1}$1} distinct	
1.690	n_ω 1.690 n_ε 1.666	Calumetite $Cu(OH,Cl)_2 \cdot 2H_2O$	0.024	Orthorhombic ?	Scaly, spherules	{001} distinct	
1.691	n_ω 1.691 n_ε 1.641	Bandylite $CuB(OH)_4Cl$	0.050	Tetragonal	Tabular crystals	{001} perfect	
1.692	n_ω 1.6918 n_ε 1.6480	Wickenburgite $CaPb_3Al_2Si_{10}O_{24}(OH)_6$	0.044	Hexagonal	Tabular crystals	{0001} indistinct	

* n_ω variation exceeds 0.01

Color In Thin Section	Hardness Sp Gravity	Color Hand Sample	Alteration	Occurrence	Remarks	Reference
Red-brown	~5 3.368	Reddish-brown		Mn-Zn-ores with rhodonite and calcite		USGS PP 180, 90
Brown tints, weak pleochroism	5 3.1-3.6	Brownish-red to black		Alkali pegmatites or plutons with nepheline, sodalite, natrolite		Vlasov II, 321-324
O = pale green E ~ colorless	? 3.47-3.62	Lime yellow to black		Lacustrine shales with labuntsovite, ewaldite, searlesite, etc.	Dimorph ewaldite	Am Min, 52, 860-864
Pale yellow-green	6-6½ 3.29	Pistachio green		Alkali pegmatites with aegirine, quartz, pyrochlore, datolite, etc.		Am Min, 53, 1426
Brown	4½ 3.13-3.14	Brown		Zn-O ores with willemite, friedelite, bementite, franklinite		Am Min, 38, 755-760
Colorless	3-4 3.39-3.44	White, pink, lt brown		Hydrothermal Zn-deposits with willemite, franklinite	Melilite group	USGS PP 180, 93-94
Colorless, yellow, brown O > E	2½-3 3.19	Brown, black	Chlorite	Pegmatites, Alkali igneous rocks	Mica group. Biotite series	pp. 270-274
O = pale green, dk greenish brown E = colorless	2-2½ 2.79	Gray to black, greenish		Magnetite ores, with pyrochroite		Am Min, 39, 946-956
Colorless	? 2.33-2.44	Colorless, white		Volcanic sublimate	Sol H_2O	Dana II, 41
O = black, blue-blk, dk. green, dk. brn, E = gray, lt. blue, yellow	7-7½ 3.0-3.25	Black, brown	Illite, sericite, biotite, chlorite	Granite pegmatites, quartz veins	Tourmaline group	p. 170
~Colorless, pale tints O > E	4-4½ 3.06-3.19	Dk green, lt brown, lt green		Fe-Mn-ores with magnetite, pyroxene, apophyllite	Gelat HCl	Win & Win II, 359
Colorless	3½-4 2.86	White, yellowish, brownish		Bedded carbonate sediments with calcite, evaporite salts. Hydro veins	Sol warm acids with efforvescence	p. 63
E ~ colorless, pale yellow O = gray-green	Soft ~2.84	Yellow-green to brown-green		Vugs, cavities, veins	Sol dilute acids	Dana II, 1057-1058
O = pale green E = pale blue-green	~3 3.14	Dk green, green, blue-green		Oxide zones Cu-deposits with malachite, cuprite, linarite, etc.	Easily sol acids	Dana II, 576-578
O = orange-yellow E = colorless	2½-3 3.46-3.52	Brownish-orange, brown-yellow		Quartz-sanbornite contact rocks with celsian, fresnoite	Slowly sol HCl	Am Min, 50, 1500-1503
Colorless	~6 4.03-4.07	Colorless, white, pinkish		Assoc with tremolite, witherite, phlogopite, diopside, forsterite	Fluor short λ U.V.-bluish-white	Am Min, 50, 1164-1169
Colorless	3-4 3.60-3.66	White, yellow, yellow-brown		Hydrothermal veins	Fluor U.V.	Am Min, 47, 585-598
Blue O > E	2	Azure blue		Incrustations oxide zones Cu-deposits	Small 2V, Sol dilute acids	Am Min, 48, 614-619
O = blue E = pale greenish-yel	2½ 2.81	Dk blue		Oxide zone ore deposits	Sol NH_4OH	Am Min, 23, 85
Colorless	5 3.85-3.88	Colorless, white, pink		Oxide zones Pb-deposits with phoenicochroite, mimetite, cerussite	Fluor short λ U.V.-dull orange	Am Min, 53, 1433-1438

n_ω Var.	Refractive Index	Mineral Name and Composition	$(n_\omega - n_\varepsilon)$	Crystal System	Habit	Cleavage	Twinning
*1.682 1.704	n_ω 1.682–1.704 n_ε 1.670–1.691	Orpheite $H_6Pb_{10}Al_{20}(PO_4)_{12}(SO_4)_5(OH)_{40}\cdot 11H_2O$	0.012 to 0.013	Hexagonal ?		{0001} poor (?)	
1.694	n_ω 1.694 n_ε 1.519	Narsethite $BaMg(CO_3)_2$	0.174	Trigonal	Tabular rhomb crystals, circular plates	{10$\bar{1}$0} distinct	
1.691 1.697	n_ω 1.691–1.697 n_ε 1.625–1.631	Annite $K_2Fe_6^{2+}(Si_6Al_2)O_{20}(OH,F)_4$	0.066	Monoclinic $\beta \sim 100°$	Micaceous	{001} perfect	{110} normal, {001} complex
1.700	n_ω 1.700 n_ε 1.509	Magnesite $MgCO_3$	0.191	Trigonal	Cleavage aggregates, earthy	{10$\bar{1}$1} perfect	Rare
1.706	n_ω 1.706 n_ε 1.698	Svabite $Ca_5(AsO_4)_3(F,Cl,OH)$	0.008	Hexagonal	Short prism xls, massive	{10$\bar{1}$0} indistinct	
*1.699 1.724	n_ω 1.699–1.724 n_ε 1.658–1.690	Phosphuranylite $Ca(UO_2)_4(PO_4)_2(OH)_4\cdot 7H_2O$	0.041 to 0.034	Orthorhombic	Tiny scales, plates, laths. Earthy	{100} perfect {010} indistinct	
1.712	n_ω 1.712 n_ε	Palmierite $(K,Na)_2Pb(SO_4)_2$	Large	Trigonal	Micro hexagonal plates {0001}		
1.714	n_ω 1.714 n_ε 1.702	Traskite $Ba_9Fe_2Ti_2Si_{12}O_{36}(OH,Cl,F)_6\cdot 6H_2O$	0.012	Hexagonal	Hexagonal prisms	None	
*1.690 1.750	n_ω 1.690–1.750 n_ε 1.510–1.548	Ankerite $CaFe(CO_3)_2$	0.180 to 0.202	Trigonal	Rhomb crystals, fine-coarse cleavage agg	{10$\bar{1}$1} perfect	{02$\bar{2}$1} lamellar, {0001} simple
1.721	n_ω 1.721 n_ε 1.719	Satterlyite $(Fe,Mg)_2PO_4(OH)$	0.002	Trigonal	Rad agg nodules	None	
1.723	n_ω 1.723 n_ε 1.681	Pyrochroite $Mn(OH)_2$	0.042	Trigonal	Tab rhomb, prism crystals, massive	{0001} perfect	
1.724	n_ω 1.724 n 1.720	Tetrawickmanite $MnSn(OH)_6$	0.004	Tetragonal	Minute tabular crystals {001}, globular	None	
1.726	n_ω 1.726 n_ε 1.723	Iron Gehlenite $Ca_2Fe^{3+}(SiAl)O_7$	0.003	Tetragonal	Prism. xls.	{001} distinct, {110} poor	None
1.725 1.728	n_ω 1.725–1.728 n_ε >1.81	Chlorotile $Cu_3(AsO_4)_2\cdot 6H_2O$	>0.085	Hexagonal	Prism, fibrous massive		
1.727	n_ω 1.727 n_ε 1.535	Kutnohorite $Ca(Mn,Mg,Fe)(CO_3)_2$	0.192	Trigonal	Cleavable aggregate	{10$\bar{1}$1} perfect	{0001}, {10$\bar{1}$1}
1.702 1.752	n_ω 1.702–1.752 n_ε 1.698–1.746	Idocrase (Vesuvianite) $Ca_{10}(Mg,Fe^{2+})_2Al_4(Si_2O_7)_2(SiO_4)_5(OH,F)_4$	0.001 to 0.012	Tetragonal	Pyram. xls., granular, fib. agg.	{110} or {001} poor	Sector (rare)
1.73	n_ω ~1.73 n_ε 1.69	Birnessite $(Na,Ca)Mn_7O_{14}\cdot 3H_2O$	0.04	Hexagonal	Microscopic granular		
1.730	n_ω 1.730 n_ε 1.715	Hisingerite (Canbyite) $Fe^{3+}Si_2O_5(OH)_4\cdot 2H_2O$	0.015	Hexagonal(?)	Massive, compact	None	
*1.722 1.739	n_ω 1.722–1.739 n_ε 1.717–1.735	Taaffeite $BeMgAl_4O_8$	0.005 to 0.004	Hexagonal or Trigonal	Hexagonal crystals, fine-grained agg		
1.732	n_ω 1.732 n_ε 1.712	Chalcomenite $CuSeO_3\cdot 2H_2O$	0.020	Orthorhombic	Prism acicular	None	

* n_ω variation exceeds 0.01

Color In Thin Section	Hardness Sp Gravity	Color Hand Sample	Alteration	Occurrence	Remarks	Reference
Colorless	3½ 3.75	Colorless, gray, lt blue, yel-grn		Oxide zone ore deposits with anglesite, pyromorphite	Colored var fluor U.V.-blue-green	Am Min, 61, 176
Colorless	~3.5 3.84	Colorless, white		Zn-Pb-Cu deposit. Dolomitic oil shales with shortite, searlsite, etc.		Am Min, 46, 420-429
O = Dk. brown E ~ colorless, yellow	2½-3 3.35	Dk. brown, black	Chlorite	Pegmatites. Alkali igneous rocks	Mica group. Biotite series	pp. 270, 274
Colorless	3½-4½ 3.00-3.48	White, gray, pink, yel, brown		Veins in altered serpentenite. Metasomatic schists. Bedded evaporites	Sol warm acids with effervescence	p. 56
Colorless	4-5 3.5-3.8	Colorless, gray, greenish, yellow		Disseminated in Zn-Mn ores	Apatite group. Sol weak acids	Dana II, 899-900
O = golden yellow E ~ colorless	~2½ ~4.1	Lt yellow to dk yellow		Secondary U-mineral in pegmatites, sandstone	Not fluor. U.V.	USGS Bull, 1064, 222-227
Colorless	4.33-4.35	Colorless, white		Volcanic fumeroles		Dana II, 403-404
O = brownish-red E = colorless, straw yellow	~5 3.71-3.75	Brownish-red		Quartz-sanbornite contact rocks		Am Min, 50, 1500-1503
Colorless	3½-4 2.90-3.10	White, yellowish, brownish		Veins in Fe-sediments with siderite, Fe-oxides. Schists with amphibole	Sol warm acids with efforvescence	p. 63
O = pale yellow, E = brown-yellow, E > O	4½-5 3.60-3.68	Lt yellow - lt brown		Nodules in shale with maricite, wolfeite, pyrite, quartz	Polymorph wolfeite	Can Min 16, 411-413
Colorless or O = brown E = pale brown (altered)	2½ 3.25-3.27	Colorless, white, green, blue, brn	Manganite	Low-T hydrothermal veinlets with hausmannite, rhodochrosite	Easily sol dilute HCl	Dana I, 639-641
Colorless, pale brownish	3.65-3.79	Pale yellow, brownish-orange		Pegmatites with spodumene, bavenite, eakerite, rhodochrosite	Dimorph wickmanite	Min Rec, 4, 24-29
Yellow-brown O > E	~5			High grade contact metamorphism	Melilite group	p. 159
~Colorless	Soft 3.73-4.05	Pale green, bluish-green		Oxide zones Cu-deposits		Dana (1892) 814
Colorless	3½-4 3.12	White, pink		Mn-deposits	Dolomite group	Am Min, 52, 1751-1761
Colorless, lt. green, lt. brown. O > E	6-7 3.33-3.45	Yellow-green, yellow, brown	Chlorite, epidote, calcite	Limestone contact zones with Ca-Mg silicates		p. 161
Dk brown, nearly opaque, weak pleochroism	1½ 3.0	Black, dk brown		Evaporite sediments		Min Mag, 31, 283-288
Yellow-brown	2½-3 2.50-3.0	Brownish-black		Alteration of Fe-rich igneous rocks, sulfide veins	Possible small 2V	Am Min, 46, 1412-1423
Colorless	8-8½ 3.60-3.61	Colorless, greenish, pink-lilac		Dolomite skarn with chrysoberyl, green spinel, cancrinite, fluorite	Fluor x-rays-green	Min Mag, 29, 765-772
Pale blue E > O	2-2½ 3.31-3.32	Bright blue		Oxidation of Cu or Pb-selenides	Sol acids. Small 2V	Am Min, 49, 1481-1485

n_ω Var.	Refractive Index	Mineral Name and Composition	$(n_\omega - n_\varepsilon)$	Crystal System	Habit	Cleavage	Twinning
1.732	n_ω 1.732 n_ε 1.728	Britholite-Y (Abukumalite) (Yttrobritholite) $(Ca,Y)_5(SiO_4,PO_4)_3(OH,F)$	0.004	Hexagonal	Prism flat crystals, oval forms	None	
1.733	n_ω 1.733 n_ε 1.714	Hematolite $Mn_4Al(OH)_2(AsO_4)(AsO_3)_2$	0.019	Trigonal	Thick tabular, rhombohedral	{0001} perfect	
1.735	n_ω 1.735 n_ε 1.65	Reevesite $Ni_6Fe_2(OH)_{16}CO_3 \cdot 4H_2O$	0.085	Hexagonal	Fine-grained agg, hexagonal platelets		
1.735	n_ω 1.735 n_ε 1.655	Buergerite $NaFe^{3+}_3Al_6Si_6B_3O_{30}F$	0.080	Trigonal	Prismatic	Prismatic, dist	
1.738	n_ω 1.738 n_ε 1.728	Stottite $Fe^{2+}Ge(OH)_6$	0.010	Tetragonal	Pseudo-octahedral crystals	{100} & {010} distinct, {001} fair	
*1.73 1.75	n_ω 1.73–1.75 n_ε	Coffinite $U(OH)_{4x}(SiO_4)_{1-x}$	Small	Tetragonal	Earthy aggregate	None ?	
1.747	n_ω 1.747 n_ε 1.741	Akdalaite $4Al_2O_3\ H_2O$	0.006	Hexagonal	Tabular on {0001}	None	
1.748	n_ω 1.748 n_ε 1.645	Freirinite (Lavendulan) $(Ca,Na)_2Cu_5(AsO_4)_4Cl \cdot 4\text{-}5H_2O$	0.103	Tetragonal or Orthorhombic	Agg fine flakes, botryoidal	{001} distinct {110} imperfect	
*1.733 1.76	n_ω 1.733–1.76 n_ε	Melanocerite $Ce_4CaBSi_2O_{12}(OH)$	Small ~Isotropic	Trigonal	Rhomb crystals, massive	None	
1.741 1.763	n_ω 1.741–1.763 n_ε 1.723–1.756	Cerium-vesuvianite $(Ca,Ce,RE)_{19-x}(Mg,Fe^{3+},Mn,Ti,Al)_{13-y}Si_{18}(O,OH)_{76}$	0.022 to 0.033	Tetragonal	Prismatic crystals	None	
1.755	n_ω 1.755 n_ε 1.731	Congolite $(Fe,Mg,Mn)_3ClB_7O_{13}$	0.024	Trigonal	Massive, fine agg		
1.756	n_ω 1.756 n_ε 1.680	Natisite $Na_2TiOSiO_4$	0.076	Tetragonal	Grains, rosettes	{001} very perf, {100} perfect	
1.753 1.76	n_ω 1.753–1.76 n_ε 1.700–1.717	Wiserite $Mn_4B_2O_5(OH,Cl)_4$	0.053 to 0.043	Tetragonal	Fibrous masses	{001} perfect	
1.76	n_ω ~1.76 n_ε	Cappelenite $(Ba,Ca,Ce,Na)_3(Y,Ce,La)_6(BO_3)_6Si_3O_9$	Large	Hexagonal	Hexagonal prism	None	
1.764	n_ω 1.764 n_ε 1.577	Cordylite $(Ce,La)_2Ba(CO_3)_3F_2$	0.187	Hexagonal	Short hexagonal prism	{0001} distinct (parting?)	
*1.72 1.80	n_ω 1.72–1.80 n_ε	Cronstedtite $Fe^{2+}_2Fe^{3+}_2SiO_5(OH)_4$	Large	Hexagonal-Trigonal	Micaceous	{0001} perfect	
1.765	n_ω 1.765 n_ε 1.603	Huanghoite $BaCe(CO_3)_2F$	0.162	Hexagonal	Platy masses	{0001} distinct	
1.765	n_ω 1.765 n_ε 1.735	Sherwoodite $Ca_3(V^{4+}O)_2V^{5+}_6O_{20} \cdot 15H_2O$	0.030	Tetragonal	~equant crystals, polyxline aggregate	None	
1.767 1.772	n_ω 1.767–1.772 n_ε 1.759–1.762	Corundum Al_2O_3	0.007 to 0.010	Trigonal	Crude hexagonal prisms, tab, rhomb crystals	{0001} & {10$\bar{1}$1} parting	{10$\bar{1}$1} lamellar, {0001} contact

* n_ω variation exceeds 0.01

Color In Thin Section	Hardness Sp Gravity	Color Hand Sample	Alteration	Occurrence	Remarks	Reference
Brown	5 4.25	Black		Granite pegmatites	Apatite group	Vlasov II, 297-300
~opaque, red-orange, yellow-brown. Not pleochroic	3½ 3.48-3.49	Brownish-red to black		Crystalline limestone with jacobsite, barite, fluorite	Easily sol acids	Dana II, 777-778
Golden-yellow	 2.78-2.88	Bright yellow, green-yellow		Oxide zones Ni-deposits. Weathered meteorites		Am Min, 52, 1190-1197
O = yellow-brown E = pale yellow	7 3.31	Dk brown, black		Rhyolite	Tourmaline group	Am Min, 51, 198-199
Colorless to dark brown (Zoned)	4½ 3.57-3.60	Brown		Alteration of renierite or germanite by ground waters	May show small 2V with r > v	Am Min, 43, 1006
Pale to dark brown	5-6 5.1	Black		Oxide zones U-deposits	Not fluorescent	Am Min, 47, 26-33
Colorless	? 3.66-3.70	White		Veinlets in skarn rocks		Am Min, 56, 635
O = dk greenish-blu, E = light greenish-blue	2½ 3.54	Blue, greenish-blue		Oxide zones Co-Cu-deposits with erythrite, malachite	Easily sol HCl	Am Min, 42, 123-124
Brown	6 4.13-4.29	Brown, black		Alkali pegmatites with lepidomelane, gadolinite, rare-earth minerals	Var caryocerite (Th-rich). Decomp HCl	Vlasov II, 301-302
Not pleochroic	5½-6½ 3.56-4.07	Gold, dk red, brown-black		Veinlets in altered greenstone with melanite, perovskite, clinozoisite		Am Min 64, 367
Colorless, pink	 ~3.5	Pale red			Dimorph ericaite	Am Min, 57, 1315
Colorless, pale green	3-4	Yellow-green, greenish-gray		Natrolite-ussingite veins in alkalic rocks with aegirine, chkalovite		Am Min, 61, 339
Colorless	~2½ 3.42	White to brownish or reddish		Intergrown with pyrochroite and sussexite	Possible small 2V	Am Min, 45, 258
Greenish-brown	6 4.41	Greenish-brown		Nepheline pegmatites	~Apatite group, Sol acids	Vlasov II, 248
O = pale greenish yel, E - pale brownish yellow	4½ 5.61	Colorless, yellowish		Alkali pegmatites	Sol acids	Dana II, 285-287
O = olive-green (~ opaque), E = Dr red-brown, grn	3½ 3.34-3.59	Black, greenish, brownish		Oxide zones Fe-deposits, with limonite	Septechlorite group	Nature, 180, 1066
Greenish-yellow, weak pleochroism	4½-5 4.49-4.67	Yellow, yellow-green		Hydrothermal veins in alkali plutons		Am Min, 48, 1179
O = green, E = blue	~2 2.86	Blue-black, blue-green, yellowish		Secondary U-V-ores		Am Min, 43, 749-755
Colorless, pale colors: red, blue, etc. O > E weak (often zoned)	9 3.98-4.10	Gray-blue, red, brown, grn, yel	Margarite, muscovite, diaspore, gibbsite, spinel, andalusite, etc.	Al-rich, Si-poor rocks with nepheline, micas, calcite, Al-minerals		p. 24

n_ω Var.	Refractive Index	Mineral Name and Composition	$(n_\omega - n_\varepsilon)$	Crystal System	Habit	Cleavage	Twinning
1.772	n_ω 1.772 n_ε 1.770	Swedenborgite $NaBe_4SbO_7$	0.002	Hexagonal	Short prism xls	{0001} distinct	
1.775	n_ω 1.775 n_ε 1.765	Fresnoite $Ba_2TiSi_2O_8$	0.010	Tetragonal	Elong (c), granular	{001} fair	
1.778	n_ω 1.778 n_ε 1.660	Nordenskiöldine $CaSnB_2O_6$	0.118	Trigonal	Tabular {0001}, parallel growths	{0001} perfect {10$\bar{1}$1} indistinct	
1.777 1.781	n_ω 1.777-1.781 n_ε 1.772-1.773	Britholite (Beckelite) (Lessingite) $(Ca,Ce)_5(SiO_4,PO_4)_3(OH,F)$	0.005 to 0.008	Hexagonal	Prism crystals, massive	None	
1.782	n_ω 1.782 n_ε 1.780	Claringbullite $Cu_4Cl(OH)_7 \cdot nH_2O$	0.002	Hexagonal	Platy		
1.800	n_ω 1.800 n_ε 1.743	Ferridravite $NaMg_3Fe_6^{3+}B_3Si_6(O,OH)_{30}(OH,F)$	0.057	Trigonal	Stubby trigonal xls.	None	
1.800	n_ω 1.800 n_ε 1.750	Ammoniojarosite $NH_4Fe_3^{+3}(SO_4)_2(OH)_6$	0.050	Hexagonal	Scaley plates in nodules	{0001} distinct	
1.802	n_ω 1.802 n_ε 1.740	Quetzalcoatlite $Cu_4Zn_8(TeO_3)_3$	0.062	Hexagonal	Rad acic as crusts or sprays	{10$\bar{1}$0} fair	
1.803	n_ω 1.803 n_ε 1.769	Cyrilovite (Avelinoite) $NaFe_3^{+3}(PO_4)_2(OH)_4 \cdot 2H_2O$	0.034	Tetragonal	Intergrown crystal aggregate		
1.807	n_ω 1.807 n_ε 1.79	Hibonite $(Ca,Ce)(Al,Ti,Mg)_{12}O_{18}$	0.017	Hexagonal	Tabular hexagonal prisms {0001}	{0001} distinct {10$\bar{1}$0} parting	Sector
1.815	n_ω 1.815 n_ε 1.761	Molybdophyllite $Pb_2Mg_2Si_2O_7(OH)_2$	0.054	Trigonal	Platy crystals, foliated aggregate	{0001} perfect	
1.816	n_ω 1.816 n_ε 1.785	Carphosiderite (Hydronium Jarosite) $HFe_3(SO_4)_2(OH)_6$	0.031	Trigonal	Tab., scaly, earthy, reniform	{0001} distinct	
1.816	n_ω 1.816 n_ε 1.597	Rhodochrosite $MnCO_3$	0.219	Trigonal	Cleavage aggregate, botryoidal	{10$\bar{1}$1} perfect	{01$\bar{1}$2} lamellar, rare
1.816	n_ω 1.8159 n_ε 1.7875	Painite $Al_{20}Ca_4BSiO_{38}$	0.028	Hexagonal	Pseudo-orthorhombic crystals		
1.82	n_ω 1.82 n_ε 1.78	Nanlingite $CaMg_3F_3(AsO_3)_2$	0.04	Trigonal	Tabular-rhombohedral crystals, dendritic		
1.815 1.820	n_ω 1.815-1.820 n_ε 1.713-1.715	Jarosite $KFe_3(SO_4)_2(OH)_6$	0.101 to 0.105	Trigonal	Scaly {0001}, rhomb, fibrous, coatings	{0001} distinct	Six sector
*1.816 1.827	n_ω 1.816-1.827 n_ε 1.856-1.863	Calcurmolite $Ca(UO_2)_3(MoO_4)_3(OH)_2 \cdot 11H_2O$	0.040 to 0.036	Unknown	Prism, platy, massive		
1.83	n_ω 1.83 n_ε 1.61	Gaspeite $(Ni,Mg,Fe)CO_3$	0.22	Trigonal	Rhombohedral	{10$\bar{1}$1} perfect	
1.832	n_ω 1.832 n_ε 1.750	Natrojarosite (Cyprusite) (Utahite)	0.082	Trigonal	Tiny hex scales earthy coatings	{0001} perfect	
1.83	n_ω 1.83 n_ε 1.82	Zairite $Bi(Fe,Al)_3(PO_4)_2(OH)_6$	Weak-mod	Trigonal	Small masses		

* n_ω variation exceeds 0.01

Color In Thin Section	Hardness Sp Gravity	Color Hand Sample	Alteration	Occurrence	Remarks	Reference
Colorless	~8 4.28-4.32	Colorless, wine-yellow		Skarns with bromelite, manganophyllite, richterite, calcite	Insol acids	Dana II, 1027-1029
O = colorless, E = yellow	3-4 4.43-4.45	Lemon yellow		Gneiss, quartzite with sanbornite	Fluor short λ U.V.-yel. Anomalous blue inter colors	Am Min, 50, 314-340
Colorless	5½-6 4.19-4.20	Colorless, yellow		Alkali pegmatites with meliphane, zircon, cassiterite	Decomp HCl	Dana II, 332-333
O = colorless, pale brown, E = colorless	5 3.86-4.69	Yellow, green, brown, black		Alkali igneous rocks, nepheline pegmatites, contact metamorphism	Apatite group. Var fynchenite or Th-britholite, alumobritholite, hydrobritholite	Vlasov II, 297-300
Colorless-pale blue	Soft 3.92	Blue		Oxide zone Cu-deposits with cuprite, malachite, etc.		Am Min, 63, 793
O = dk. brown, E = lt. brown	~7 3.26-3.33	Black		Fissures and cavities in schist		Am Min 64, 945-948
Pale yellow	Soft 3.028	Ocherous yellow		Veins in bituminous shales	Alunite group	Dana II, 562-563
O = blue-green, E ~ colorless	3 6.05	Capri blue		Ag-ores with hessite, galena, bornite	Sol dilute acids	Am Min, 59, 874
Dk yellow O ≥ E	? 3.08-3.09	Orange, yellow-brown		Pegmatites		Am Min, 42, 204-213
Deep reddish-brown in thin fragments	7½-8 3.84	Brownish-black		Impure marble, with corundum, spinel, thorianite		Am Min, 42, 119
Colorless	3-4 4.72	Colorless, pale green		Mn-deposits with hausmannite		Win & Win II, 479
Deep yellow	4-4½ 2.50-2.91	Deep yellow, brown, red-brown		Oxide zones of ore deposits	Jarosite group	p. 96
O = pale pink, colorless, E = colorless	3½-4 3.70	Pink-rose, brownish	Mn-Fe oxides	Hydrothermal veins with fluorite, barite, sulfides. Limestone replacement		p. 60
O = pale brownish-orange, E = ruby red	~8 3.98-4.01	Garnet red		Gem gravels Mogok, Burma	Only one crystal known	Min Mag, 31, 420-425
Deep red	~2½ 3.93-3.99	Brownish-red		Greisen with fluorite, fluorborite, sulfides	Abnormal red interference color	Am Min, 62, 1058-1059
O = deep yellow-brn, red-brown, E = pale yellow, colorless	3 3.25	Golden yellow, dark brown	Limonite, goethite	Oxide zones sulfide ores with Fe-oxides	Sol. HCl	p. 95
O = yellow, E = colorless		Honey yellow		Secondary U-mineral	Fluor U.V.-strong yellow-green	Am Min, 49, 1152-1153
Colorless	4½-5 3.71-3.75	Light green		Veins in dolomite	Calcite group	Am Min, 51, 677-684
O = pale yellow, E ~ colorless	3 2.88-3.18	Ochre yellow to brown	Limonite	Oxidation of pyrite with gypsum, jarosite, alunite	Alunite group. Slowly sol HCl	Dana II, 563-565
Colorless	4½ 4.37-4.42	Greenish		Weathered quartz veins containing mica, wolframite	Crandallite group, Fe-waylandite	Am Min, 62, 174-175

Uniaxial –

n_ω Var.	Refractive Index	Mineral Name and Composition	$(n_\omega - n_\varepsilon)$	Crystal System	Habit	Cleavage	Twinning
1.82 1.85	n_ω 1.82-1.85 n_ε 1.80-1.82	Hoegbomite (Högbomite) $Mg(Al,Fe,Ti)_4O_7$	0.02 to 0.05	Hexagonal or Tetragonal	Tab. xls., minute grains	{0001} imperfect	
1.850	n_ω 1.850 n_ε 1.625	Smithsonite $ZnCO_3$	0.225	Trigonal	Rhom crystals, botryoidal, crusts	{10$\bar{1}$1} distinct	None
1.855	n_ω 1.855 n_ε 1.60	Cobaltocalcite (Spherocobaltite) $CoCO_3$	0.255	Trigonal	Radiated spherical mass	{10$\bar{1}$1} perfect	
1.85	n_ω 1.85 n_ε 1.82	Burckhardtite $Pb_2(Fe,Mn)^{3+}AlTeSi_3O_{10}(OH_2)O_2 \cdot H_2O$	0.03	Monoclinic ? (pseudohex)	Rosettes	{001} perfect	
1.85	n_ω ~1.85 n_ε ~1.81	Beaverite $Pb(Cu,Fe,Al)_3(SO_4)_2(OH)_6$	Strong ~0.04	Trigonal	Microscopic hexagonal plates		
1.86	n_ω 1.86 n_ε 1.83	Asbecasite $Ca_3(Ti,Sn)(As_6Si_2Be_2O_{20})$	0.03	Trigonal	Rhombohedrons	{10$\bar{1}$1}	
1.875	n_ω 1.875 n_ε 1.633	Siderite $FeCO_3$	0.242	Trigonal	Cleavage aggregate, nodular, earthy	{10$\bar{1}$1} perfect	{01$\bar{1}$2} lamellar {0001} simple
1.875	n_ω 1.875 n_ε 1.786	Plumbojarosite $PbFe_6(SO_4)_4(OH)_6$	0.089	Trigonal	Microscopic hex plates	{10$\bar{1}$4} fair	
1.87	n_ω 1.87 n_ε 1.85	Dussertite $BaFe_3(AsO_4)_2(OH)_5$	0.02	Trigonal	Rosettes, micro flat crystals	None ?	
1.882	n_ω 1.882 n_ε 1.785	Argentojarosite $AgFe_3(SO_4)_2(OH)_6$	0.097	Trigonal	Micaceous scales as coatings	{0001} perfect	
1.898	n_ω 1.898 n_ε 1.815	Arseniosiderite (Mazapilite) $Ca_3Fe_4^{3+}(OH)_6(H_2O)_3(AsO_4)_4$	0.083	Monoclinic (Pseudo-uniaxial)	Fibrous, fine granular	{100} or {001} distinct	
1.91	n_ω 1.91 n_ε ~1.90	Daubréeite $BiO(OH,Cl)$	~0.01	Tetragonal	Massive, earthy	{001} perfect	
**1.80 2.15	n_ω 1.80-2.15 n_ε 1.72-2.09	Ferritungstite $Ca_2Fe_2^{2+}Fe_2^{3+}(WO_4)_7 \cdot 9H_2O$	0.08	Tetragonal	Microscopic hexagonal plates, earthy, fib		
*1.93 1.96	n_ω 1.93-1.96 n_ε	Corkite $PbFe_3PO_4SO_4(OH)_6$	Small	Trigonal	Rhombohedral, pseudocubic		
1.960 2.115	n_ω 1.960-2.115 n_ε 1.815-1.915	Tlapallite $H_6(Ca,Pb)_2(Cu,Zn)_3(SO_4)(TeO_3)_4(TeO_6)$	0.145	Monoclinic β = 91°	Films on rock fractures		
1.96	n_ω 1.96 n_ε < 1.96	Beudantite $PbFe_3AsO_4SO_4(OH)_6$	Small-mod	Trigonal	Rhombohedral	{0001} dist	Lamellar
1.977	n_ω 1.977 n_ε 1.967	Dugganite $Pb_3Zn_3(TeO_6)_x(AsO_4)_{2-x}(OH)_{6-3x}$, x = 0.94-1.33	0.010	Hexagonal	Stubby prism crystals	{11$\bar{2}$0} poor	
1.98	n_ω 1.98 n_ε 1.85	Diaboleite $Pb_2CuCl_2(OH)_4$	0.13	Tetragonal	Square tabular {001} platy agg	{001} perfect	

* n_ω variation exceeds 0.01

** n_ω variation exceeds 0.10

Color In Thin Section	Hardness Sp Gravity	Color Hand Sample	Alteration	Occurrence	Remarks	Reference
O = brown, yellow, pink. E = lt. brown, yellow. O > E	6½ 3.81	Black		Emery deposits with corundum, spinel, magnetite		Am Min 37, 600-608
Colorless	4-4½ 4.43	White, gray, lt blue, green, brn	Hemimorphite, limonite	Oxide xones Zn-deposits with secondary Zn-minerals	Sol acids with effer-vescence	p. 62
O = pale violet-red, E = pale rose red	4 4.11-4.13	Rose red		Co-Ni veins, oxide zones Co-deposits	Sol. acids	Dana II, 175-176
O = carmine red, E = pale magneta	~2 4.96	Violet red to pink		Oxide zones ore deposits with clays		Am Min 64, 355
Colorless	Soft 4.31-4.36	Canary-yellow		Oxide zones in Cu-Pb deposits in arid regions		Dana II, 568
Colorless	6½-7 3.70-3.71	Lemon-yellow		Fractures in orthogneiss	Small 2V = 0°-17°	Am Min, 52, 1583-1584
Colorless, pale yellow-brown	4-4½ 3.96	Yellow-brown, red-brown, gray	Fe-oxides	Bedded sediments with clays, chert, chamosite. Hydro thermal veins	Sol warm acids with effervescence	p. 58
O = yellow-brown, E ~ colorless	Soft 3.60-3.64	Yellow-brown to dk brown		Oxide zone Pb-deposits	Alunite group. Slowly sol acids	Dana II, 568-570
Pale yellowish-green	3½ 3.75-4.12	Green, yellowish green		Oxide zones ore deposits	Small 2V ~ 15°-20°	Dana II, 839-840
Yellow O > E	Soft 3.66-3.81	Yellow to brown		Oxide zone of ore deposits	Alunite group	Dana II, 565
E = colorless, O = dark red-brown	1½-4½ 3.58-3.60	Deep yellow-dark brown		Low T oxidation of löllingite, arsenopyrite and other ores		Am Min, 59, 48-59
Colorless	2-2½ 7.56	White, yellowish		Alteration of bismuth or bismuthinite	Sol acids	Dana II, 60-62
Pale yellow	5.2-5.8	Yellow, brownish-yellow		Oxide zones of W-deposits		Am Min, 42, 83-90
{0001} perfect	3½-4½ 4.30-4.42	Dk green, yellow-green, yellow		Oxide zones ore deposits, with limonite	Sol warm HCl	Dana II, 1002-1003
Colorless, lt green	3 5.38-5.47	Viridian green		Thin veins in altered rhyolites		Min Mag 42, 183
O = colorless, E = yellow	3½-4½ 4.1-4.3	Dk green, brown, black		Oxide zone ore deposits	Sol HCl, small 2V, anom inter colors	Am Min 35, 1055-1056
Colorless, pale grn	3 6.33	Colorless, green		Oxide zones Cu-Zn-Pb deposits with khinite and parakhinite		Am Min, 63, 1016-1019
Pale blue, O > E	2½ 5.41-5.42	Deep blue		Oxide zones Pb deposits	Sol HNO_3	Dana II, 82-83

Uniaxial –

n_ω Var.	Refractive Index	Mineral Name and Composition	$(n_\omega - n_\varepsilon)$	Crystal System	Habit	Cleavage	Twinning
*1.958 2.03	n_ω 1.958-2.03 n_ε 1.948-2.01	Hedyphane $(Ca,Pb)_5(AsO_4)_3Cl$	0.010 to 0.020	Hexagonal	Prismatic (c), pyramidal	{10$\bar{1}$1}	
~2.0	n_ω ~2.0 n_ε	Jagoite $(Pb,Ca)_3Fe^{3+}Si_3O_{10}(Cl,OH)$	0.025	Trigonal	Platy aggregate micaceous	{0001}	
2.01	n_ω 2.01 n_ε 1.99	Armangite $Mn_3(AsO_3)_2$	0.02	Trigonal	Short prismatic	{0001} fair	{02$\bar{2}$1} lamellar
2.03	n_ω 2.03 n_ε 2.00	Pseudoboléite $Pb_5Cu_4Cl_{10}(OH)_8 \cdot 2H_2O$	0.03	Tetragonal	Parallel growth on boléite	{001} perfect {101} ~perfect	
2.034	n_ω 2.034 n_ε 1.976	Simpsonite (Calogerasite) $Al_4Ta_3O_{13}OH$	0.058	Hexagonal	Tabular or short prism (c) crystals	None ?	
2.041	n_ω 2.041 n_ε 1.926	Cumengéite (Cumengite) $Pb_4Cu_4Cl_8(OH)_8 \cdot H_2O$	0.115	Tetragonal	"Octahedral"	{101} & {110} distinct {001} poor	
2.05	n_ω 2.05 n_ε 2.03	Boléite $Pb_9Cu_8Ag_3Cl_{21}(OH)_{16} \cdot H_2O$	0.02	Tetragonal	Pseudo-isometric	{001} perfect {101} distinct {100} poor	
*2.03 2.07	n_ω 2.03-2.07 n_ε 2.015-2.05	Barysilite $MnPb_8(Si_2O_7)_3$	0.02	Trigonal	Massive	{0001} distinct	
2.058 (Na)	n_ω 2.058 n_ε 2.048 (Na)	Pyromorphite $Pb_5(PO_4)_3Cl$	0.010	Hexagonal	Short hexagonal prisms. Globular, fibrous, earthy	{10$\bar{1}$1} trace	{11$\bar{2}$2} very rare
2.075	n_ω 2.075 n_ε 1.786	Salesite $CuIO_3OH$	0.289	Orthorhombic	Tiny prism crystals	{110} perfect	
2.09	n_ω 2.09 n_ε 1.94	Hydrocerussite $Pb_3(CO_3)_2(OH)_2$	0.15	Hexagonal	Hexagonal tabular {0001}	{0001} perfect	
2.125	n_ω 2.125 n_ε 2.059	Yedlinite $Pb_6Cl_6CrO_6 \cdot 2H_2O$?	0.066	Trigonal	Prism crystals	{11$\bar{2}$0} distinct	
2.13	n_ω 2.13 n_ε 1.97	Beyerite $Ca(BiO)_2(CO_3)_2$	0.16	Tetragonal	Platy, earthy	None ?	
2.145	n_ω 2.145 n_ε 2.006	Matlockite $PbFCl$	0.139	Tetragonal	Tabular crystals {001}, platy, massive	{001} perfect	
2.15	n_ω 2.15 n_ε ?	Bismoclite $BiOCl$		Tetragonal	Massive, microscopic fibrous, scaly	{001} perfect	
2.155	n_ω 2.155 n_ε 2.120	Parakhinite $Cu_3PbTeO_4(OH)_2$	0.035	Hexagonal	Hexagonal tabular	{0001} poor	
**2.124 2.263	n_ω 2.124-2.263 n_ε 2.106-2.239	Mimetite $Pb_5(AsO_4)_3Cl$	0.018 to 0.024	Hexagonal or Monoclinic	Prism hexagonal crystals, acicular, botryoidal	{10$\bar{1}$1} traces	{11$\bar{2}$2} rare

* n_ω variation exceeds 0.01
** n_ω variation exceeds 0.10

Color In Thin Section	Hardness Sp Gravity	Color Hand Sample	Alteration	Occurrence	Remarks	Reference
Colorless	4½ 5.82	White, yellowish, bluish		Hydrothermal ore veins with willemite, native metals	Svabite series - Apatite group	Dana II, 900-902
Pale yellow-green, non-pleochroic	3 5.43	Yellow-green		Hematite ores		Am Min, 43, 387
Yellow to brown, nearly opaque. Not pleochroic	~4 4.43-4.47	Black		In veins with calcite and Fe-Mn minerals	Soluble HCl	Dana II, 1031-1032
Indigo blue. Not pleochroic	2½ 4.85-4.89	Indigo blue		Oxide zones ore deposits with boléite and cumengite	Soluble HNO_3	Dana II, 80-81
Colorless	7-7½ 5.92-6.84	Colorless, lt yellow, lt brown		Granite pegmatites with tantalite, microlite, tapiolite, beryl	Fluor U.V.-pale yellow, bluish-white, yellowish-blue	Vlasov II, 446-449
O = Dk blue (greenish) E = blue	2½ 4.60-4.67	Indigo blue		Oxide zones Cu-Pb-deposits	Soluble HNO_3	Dana II, 79-80
Bluish-green, not pleochroic	3-3½ 5.05-5.10	Indigo blue		Oxide zone Pb-deposits	Soluble HNO_3	Dana II, 78-79
Colorless	3 6.72	White, pinkish		Veins in Zn ores		Am Min, 49, 1485-1488
Colorless or pale tints (E > O)	3½-4 7.04-7.10	Green, yellow, orange, brown, gray, white		Oxide zones Pb-deposits with cerussite and limonite	Pyromorphite-mimetite series. Sol HNO_3 and KOH	Dana II, 889-895
O = bluish-green, E = colorless	3 4.77-4.89	Bluish-green		Oxide zones Cu-deposits in arid regions	Easily soluble HNO_3	Dana II, 315-316
Colorless	3½ 6.80-6.94	Colorless, white, gray		Oxide zones Pb-deposits	Soluble acids, effervescence	Dana II, 270-271
O = pale blue, E = lavender	2½ 5.81-5.85	Red-violet		Oxide zones Pb-deposits with diaboleite, phosgenite, wulfenite		Am Min, 59, 1157-1159
Colorless, pale yel	2-3 6.56-6.58	White, yellow, gray		Secondary min in pegmatites	Possible small 2V	Dana II, 281-282
Colorless	2½-3 7.12-7.13	Colorless, yellow yellow-brown, green		Oxide zones Pb-deposits	Possible small 2V. Soluble HNO_3	Dana II, 59-60
Colorless	2-2½ 7.72-7.76	White, yellow-brown		Alter Bi-minerals	Soluble acids	Dana II, 60-62
O = yellowish green, E = emerald green O > E	3½ 6.69	Dk green		Oxide zones Cu-Zn-Pb deposits with khinite and dugganite	Polymorph khinite. Easily sol dilute acids	Am Min, 63, 1016-1019
Colorless, pale yel, other pale tints E > O	3½-4 7.28	Yellow, yel-brn, orange, white		Oxide Pb-deposits	Pyromorphite-mimetite series. Apatite group	Am Min, 54, 993

n_ω Var.	Refractive Index	Mineral Name and Composition	$(n_\omega - n_\varepsilon)$	Crystal System	Habit	Cleavage	Twinning
>>2.10	n_ω >>2.10 n_ε	Schuetteite $Hg_3(SO_4)O_2$	Mod to large	Trigonal	Minute hexagonal plates, crusts		
2.22	n_ω 2.22 n_ε 2.11	Vauquelinite $Pb_2Cu(CrO_4)(PO_4)(OH)$	0.11	Monoclinic β =94°	Fibrous-colliform, granular, wedge crystals	None	{120}
2.26	n_ω 2.26 n_ε 2.10	Hydrohetaerolite $Zn_2Mn_4O_8 \cdot H_2O$	0.16	Tetragonal	Massive, fibrous, botryoidal	Parallel elongation	
2.27	n_ω 2.27 n_ε 2.19	Stolzite $PbWO_4$	0.08	Tetragonal	Dipyramidal, tab, prism crystals	{001} imperfect {011} indistinct	
2.295	n_ω 2.295 n_ε 2.285	Finnemanite $Pb_5(AsO_3)_3Cl$	0.010	Hexagonal	Prism, crusts	{10$\bar{1}$1} distinct	
2.31	n_ω 2.31 n_ε 1.95	Geikielite $MgTiO_3$	0.36	Trigonal	Massive, short prism	{10$\bar{1}$1} distinct	
2.32	n_ω 2.32 n_ε 2.12	Seeligerite $Pb_3(IO_3)Cl_3O$	0.20	Ortho (Pseudo-tetragonal)	Thin plates	{001} perfect {110} distinct {100} & {010} rare	
2.32	n_ω 2.32 n_ε 2.25	Ecdemite (Ekdemite) $Pb_3Cl_2(AsO_4)$	0.07	Tetragonal	Tabular crystals, foliated masses	{001} perfect	
2.34	n_ω 2.34 n_ε 2.14	Hetaerolite $ZnMn_2O_4$	0.20	Tetragonal	Pseudo-octahedral, massive	{001} indistinct	{112} fivelings
2.35	n_ω 2.35 n_ε 2.18	Valentinite Sb_2O_3	0.17	Orthorhombic	Prism or tabular crystals, granular	{110} perfect {010} imperfect	
2.36	n_ω 2.36 n_ε 2.20	Embreyite $Pb_5(CrO_4)_2(PO_4)_2 \cdot H_2O$	0.16	Monoclinic β = 103°	Drusy crusts, tabular crystals	None	Multiple
2.36	n_ω 2.36 n_ε 2.31	Långbanite $Mn_4^{2+}Mn_9^{3+}Sb^{5+}Si_2O_{24}$	0.05	Hexagonal	Prismatic	None	
2.35 2.40 (Li)	n_ω 2.35-2.40 n_ε 2.33-2.37 (Li)	Lorettoite (Chubutite) $Pb_7Cl_2O_6$	0.02 to 0.03	Orthorhombic	Coarse fibrous or blades	{001} perfect	
2.4053	n_ω 2.4053 n_ε 2.2826	Wulfenite $PbMoO_4$	0.123	Tetragonal	Square tabular crystals, massive, granular	{011} distinct {001} & {013} indistinct	{001}
2.416	n_ω 2.416 n_ε 2.350	Vanadinite $Pb_5(VO_4)_3Cl$	0.066	Hexagonal	Prism acicular crystals, skeletal, globular	None	
**2.398 2.515	n_ω 2.398-2.515 n_ε 2.260-2.275	Goethite (Xanthosiderite) α-FeO(OH)	0.138 to 0.140	Orthorhombic	Fib-acic (c), rad, botryoidal, scaly	{010} perfect {100} imperfect	Cruciform, rare
2.46 (Li)	n_ω 2.46 n_ε 2.15 (Li)	Hausmannite $Mn_2^{2+}Mn^{4+}O_4$	0.31	Tetragonal	Pseudo-octahedral, granular	{001} perfect {112} & {011} indistinct	{112} lamellar, cyclic (fivelings)
2.481 (Na)	n_ω 2.481 n_ε 2.21 (Na)	Pyrophanite $MnTiO_3$	0.27	Trigonal	Tabular crystals, massive, skeletal, granular	{02$\bar{2}$1} perfect, {10$\bar{1}$2} good	

** n_ω variation exceeds 0.10

Color In Thin Section	Hardness Sp Gravity	Color Hand Sample	Alteration	Occurrence	Remarks	Reference
O = greenish yellow, E = orange-yellow	~3 8.18-8.36	Yellow, greenish to orange-yel		Oxide zones Hg-deposits, ore dumps and Hg-furnaces		Am Min, 44, 1026-1038
O = pale brown, E = pale green	2½-3 5.98-6.06	Green to brown, ~black		Oxide zones Pb-Cu-deposits with crocoite, pyromorphite, mimetite	Partly soluble HNO_3	Dana II, 650-652
Lt to dk brown	5-6 4.64	Dk brown, black		Oxide zones Zn-deposits	Easily soluble HCl	Dana I, 717
Pale yellow, brownish	2½-3 7.9-8.40	Yellow, brown, red-brown, green		Oxide zone Pb-W-deposits with wulfenite, cerussite	Dimorph raspite. Decomp HCl	Dana II, 1087-1089
Olive-green	2½-7 7.27-7.55	Gray to black		Veins in hematite		Dana II, 1038-1039
~opaque, red-brown, purple O > E	5-6 4.03-4.05	Iron-black	Leucoxene (Ti-oxides)	Dolomite marbles, serpentenites	Ilmenite group	Dana I, 535-541
Pale yellow	6.83-7.05	Bright yellow		Small amounts with boleite, paralaurionite, schwartzembergite		Am Min, 57, 327-328
Colorless, pale yellow	2½-3 7.14	Greenish-yellow, yellow, orange		Oxide zones Pb-deposits	Soluble HNO_3, dimorph heliophyllite	Dana II, 1036-1037
Dark brown in thin splinters. ~opaque	6 5.18-5.23	Black		Hydrothermal Zn-oxide deposits		Dana I, 715-717
Colorless	2½-3 5.76-5.83	Colorless, white, yellowish		Alteration of stibnite or other Sb-minerals with kermesite, etc.	Dimorph senarmontite. Soluble HCl	Dana I, 547-550
O = amber, E = honey yellow	3½ 6.41-6.45	Orange		Oxide zones Pb-Cr-deposits	Small 2V = 0°-11°	Min Mag, 38, 790-793
Dark red-brown O > E	6½ 4.6-4.8	Black		Fe-Mn ores	Slow soluble HCl	Win & Win II, 529
Pale yellow	2½-3 7.39-7.95	Yellow, reddish-yellow		Oxide zones Pb-deposits	Soluble hot acids	Dana II, 56
Yellow to pale orange. O > E	2½-3 6.5-7.0	Yellow, orange-brown, brown		Oxide zones Pb-deposits with pyromorphite, mimetite, cerussite	Decomp HCl	Dana II, 1081-1086
O = pale tints, E = colorless	2½-3 6.88-6.93	Orange-red, brown, yellow		Oxide zones Pb-deposits with pyromorphite, mimetite, etc.	Pyromorphite series. Apatite group	Dana II, 895-898
O = yellow-brown, dk red-orange, E = yellow, colorless	5-5½ ~4.3	Yellow-brown, red-brown, black		Weathering of Fe-sulfides and Fe-oxides with lepidocrocite		p. 46
~Opaque, deep red-brn in thin splinters, not pleochroic	5½ 4.84	Brownish-black		High-T hydrothermal veins, contact zones with Mn-minerals	Soluble hot HCl	Dana I, 712-715
~Opaque. Reddish-orange on thin edges. E > O	5-6 4.54-4.58	Iron-black. Dk blood-red	Leucoxene (Ti-oxides)	Accessory in many igneous and metamorphic rocks. Mn-ore veins	Ilmenite group	p. 28

n_ω Var.	Refractive Index	Mineral Name and Composition	$(n_\omega - n_\varepsilon)$	Crystal System	Habit	Cleavage	Twinning
2.561	n_ω 2.561 n_ε 2.488	Anatase (Octahedrite) TiO_2	0.073	Tetragonal	Stubby dipyram.	{001} & {111} perfect	{112} rare
2.665 (Li)	n_ω 2.665 n_ε 2.535 (Li)	Litharge PbO	0.130	Tetragonal	Tabular crystals, crusts	{110} perfect	
~2.7	n_ω ~2.7 n_ε	Ilmenite $FeTiO_3$	Very large	Trigonal	Tabular anhedral crystals, massive, skeletal	None. {0001} or {10$\bar{1}$1} parting	{0001} simple, {10$\bar{1}$1} lamellar
>>2.72	n_ω >>2.72 n_ε 2.72	Chalcophanite $ZnMn_3O_7 \cdot 3H_2O$	Large	Trigonal	Tabular, massive, drusy	{0001} perfect	
3.084 (Li)	n_ω 3.084 n_ε 2.881 (Li)	Pyrargyrite Ag_3SbS_3	0.203	Trigonal	Massive, crusts, prism crystals	{10$\bar{1}$1} distinct {01$\bar{1}$2} indist	{10$\bar{1}$4} common, {10$\bar{1}$1}, {01$\bar{1}$2}, {11$\bar{2}$0}
3.088 (Na)	n_ω 3.0877 n_ε 2.7924 (Na)	Proustite Ag_3AsS_3	0.295	Trigonal	Massive, crusts, prism, rhomb xls	{10$\bar{1}$1}	{10$\bar{1}$4} & {10$\bar{1}$1} common
*3.15 3.22	n_ω 3.15-3.22 n_ε 2.87-2.94	Hematite Fe_2O_3	~0.28	Trigonal	Botryoidal, scaly, earthy	{0001} or {10$\bar{1}$1} parting	{0001} or {10$\bar{1}$1} lamellar

* n_ω variation exceeds 0.01

Color In Thin Section	Hardness Sp Gravity	Color Hand Sample	Alteration	Occurrence	Remarks	Reference
Lt yellow-brown or red-brown to dk brown. E > O	5½ 3.90	Yellow-brown or red-brown to black	Rutile	Cavities in pegmatites, granites, volcanics, schists, gneisses	Optical properties vary greatly with λ. Trimorph TiO_2	p. 21
Red, orange-red	2 9.355	Red		Altered coatings on massicot	Dimorph massicot. Soluble HCl, HNO_3	Dana I, 514-515
Opaque. Deep red on thin edges	5-6 4.79	Iron-black	Leucoxene (Ti-oxides)	Accessory in many igneous and metamorphic rocks. Ore veins Pegmatites		p. 28
O ~ opaque, E = Deep red	2½ 3.83-3.98	Iron black		Oxide zone Zn-deposits		Dana I, 739-740
Deep red	2½ 5.82-5.85	Deep red	Argentite, silver, cerargyrite, stibnite	Low-T hydrothermal veins with sulfosalts, silver, galena	Decomp HNO_3	Dana I, 362-366
O = blood-red, E = cochineal red	2-2½ 5.55-5.64	Scarlet to vermilion red	Argentite, silver, cerargyrite, orpiment	Low-T hydrothermal veins with sulfo-salts	Dimorph xanthoconite. Decomp HNO_3	Dana I, 366-369
O = deep red-brown, E = yellow-brown	5-6 5.26	Steel-gray, red, red-brown	Magnetite, limonite, siderite, pyrite	Alter Fe-minerals. Silicic plutons with ilmenite. Contact deposits		p. 27

n_β Var.	Refractive Index	Mineral Name and Composition	$(n_\gamma - n_\alpha)$	2V angle Dispersion	Crystal System	Habit	Cleavage
1.3012	n_β 1.3012 n_α 1.301 n_γ 1.3068	Ferruccite $NaBF_4$	0.006	11°	Orthorhombic	Minute tabular crystals	{100}, {010}, {001}
1.338	n_β 1.338 n_α 1.338 n_γ 1.339	Cryolite Na_3AlF_6	0.001	~43° v>r weak, horizontal	Monoclinic β = 90.2°	Coarse granular, pseudo-cubic crystals	{001} & {110} parting, also {$\bar{1}$01}
1.348	n_β 1.348 n_α 1.346 n_γ 1.350	Weberite Na_2MgAlF_7	0.004	~83°	Orthorhombic	Fine-grained masses	{110} poor
1.413	n_β 1.413 n_α 1.411 n_γ 1.420	Pachnolite $NaCaAlF_6 \cdot H_2O$	0.009	76° v>r weak, horizontal strong	Monoclinic	Prism (c), massive, granular	{001} indist.
1.416	n_β 1.416 n_α 1.411 n_γ 1.422	Carlhintzeite $Ca_2AlF_7 \cdot H_2O$	0.011	77°	Triclinic <α = 91°8' <β = 104°51' <γ = 90°	Elong. on [101]	
1.428	n_β 1.428 n_α 1.425 n_γ 1.432	Calcjarlite $NaSr_3Al_3(F,OH)_{16}$	0.007	72°	Monoclinic	Tabular prism crystals	None
1.433	n_β 1.433 n_α 1.429 n_γ 1.436	Jarlite (Metajarlite) $NaSr_3Al_3F_{16}$	0.007	~90°	Monoclinic β = 101.8°	Tabular elong (c), radiated mass	
1.435	n_β 1.435 n_α 1.422 n_γ 1.480	Santite $KB_5O_8 \cdot 4H_2O$	0.058	~58°	Orthorhombic	Irregular grains	
1.438	n_β 1.438 n_α 1.431 n_γ 1.507	Sborgite $NaB_5O_8 \cdot 5H_2O$	0.076	35°	Triclinic β = 113.2°	Sugary aggregate, stalactitic, crusts	
1.442	n_β 1.442 n_α 1.439 n_γ 1.469	Stercorite $Na(NH_4)H(PO_4) \cdot 4H_2O$	0.030	36° r>v strong	Triclinic (pseudomono) β = 98.5°	Xline masses and nodules, prism xls	None
1.442	n_β 1.442 n_α 1.441 n_γ 1.444	Usovite $Ba_2MgAl_2F_{12}$	0.003	70°	Orthorhombic ?	Irregular grains, bladed	One perfect
1.460	n_β 1.460 n_α 1.445 n_γ 1.491	Mercallite $KHSO_4$	0.046	56° v>r weak	Orthorhombic	Stalactitic, minute tabular crystals	None
1.463	n_β 1.4629 n_α 1.4607 n_γ 1.4755	Picromerite $K_2Mg(SO_4)_2 \cdot 6H_2O$	0.015	48° r>v weak	Monoclinic β = 104.8°	Massive, crusts, prism crystals	{$\bar{2}$01} perfect
1.464	n_β 1.464 n_α 1.459 n_γ 1.470	Aluminite $Al_2SO_4(OH)_4 \cdot 7H_2O$	0.011	~90°	Monoclinic or Orthorhombic	Friable nodular masses of fibers	
1.466	n_β 1.466 n_α 1.462 n_γ 1.469	Bøggildite $Na_2Sr_2Al_2PO_4F_9$	0.007	78°-80°	Monoclinic β = 107.5°	Massive	

Twinning	Orientation	Color in thin section	Hardness Sp Gravity	Color hand sample	Alteration	Occurrence	Remarks	Reference
	a = Z, b = Y, c = X	Colorless	~3 2.50-2.51	Colorless, white		Volcanic sublimate	Sol H_2O, bitter-acid taste	Dana II, 98-99
Several poly-synthetic laws	b=X, c^Z = +44° a^Y = + 44°	Colorless	2½ 2.97	Colorless, white, brown	Auminum-fluorides	Cryolite pegmatites with fluorite, quartz, zircon, wolframite	Slight sol H_2O. Sol H_2SO_4	p. 16
	a = X, b = Y, c = Z	Colorless	3½ 2.96-2.97	Grayish-white		Cryolite pegmatites with fluorite, topaz, thomsenolite, pachnolite	Slight sol H_2O. Easily sol $AlCl_3$ solution	Dana II, 127-128
{100} common	b = X, c ∧ Z = 69°	Colorless	3 2.97-2.98	Colorless, white		Alteration of cryolite with elpasolite in cryolite pegmatites	Easily sol H_2SO_4	Dana II, 114-116
Tw. axis [101]	b ≈ X, c ∧ Z = 10°	Colorless	2.86	Colorless		Pegmatite		Am Min 65, 205
	b = Y, c ∧ Z = 10° - 15°	Colorless	4 3.51	White		Cryolite pegmatites with fluorite, usovite		Am Min 59, 873-874
	b=Y, c^X = +6°, c^Z = -84°	Colorless	4-4½ 3.61-3.87	Colorless, white, gray		Cryolite pegmatites with thomsenolite, chiolite, topaz, fluorite	Sol $AlCl_3$ solution	Dana II, 118-119
		Colorless	1.735-1.740	Colorless		Thermal springs and fumeroles with larderellite and sassolite		Am Min 56, 636
		Colorless	1.713	Colorless		Borate evaporites with ulexite, thenardite, colemanite, halite		Am Min 43, 378
{010} poly-synthetic	b~X, Z~⊥ {001}	Colorless	2 1.554	Colorless, white, yellowish		Guano deposits	Soluble H_2O	Dana II, 698-699
	Extinction ∥ cleavage, Optic plane ⊥ cleavage	X=Y=brownish yel, Z = pale yellow	3½	Brown to dk brown		Fluorite veins with muscovite, zeolites, halloysite		Am Min 52, 1582
	a = Z, b = X, c = Y	Colorless	2.32	Colorless, sky blue		Fumeroles with other sulfates, fluorides	Sol H_2O. Acid taste	Dana II, 395
	b = Y, a^X ~ +1°	Colorless	2½ 2.03	Colorless, white, yel-lowish		Marine salt beds with kainite, anhydrite. Fumerole deposits	Easily sol H_2O. bitter taste	Dana II, 453-454
	Fibers are length-fast (X)	Colorless (~opaque)	1-2 1.66-1.82	White, earthy		Coatings on carbonates and evaporites	Insoluble in H_2O	Dana II, 600-601
			4-5 3.66	Flesh red		Hydrothermal, greisen		Am Min 41, 959

n_β Var.	Refractive Index	Mineral Name and Composition	$(n_\gamma - n_\alpha)$	2V angle Dispersion	Crystal System	Habit	Cleavage
1.468	n_β 1.468 n_α 1.465 n_γ 1.507	Lansfordite $MgCO_3 \cdot 5H_2O$	0.042	60°	Monoclinic β = 101.75°	Short prism crystals	{001} perfect {100} dist
1.470	n_β 1.470 n_α 1.460 n_γ 1.484	Lapparentite $Al_2(SO_4)_2(OH)_2 \cdot 9H_2O$	0.024	80°	Orthorhombic ?	Rhomboidal tablets, earthy	
*1.463 1.478	n_β 1.463-1.478 n_α 1.459-1.475 n_γ 1.473-1.485	Alunogen $Al_2(SO_4)_3 \cdot 16H_2O$	0.010 to 0.014	31° - 69°	Triclinic α=90° β=97.5° γ=92°	Tabular {010}, fibrous on c	{010} perfect
1.471	n_β~1.471 n_α 1.467 n_γ 1.476	Kirovite $(Fe,Mg)SO_4 \cdot 7H_2O$	0.009	Large r>v weak, inclined	Monoclinic β = 104.3°	Prism, tabular, fibrous crusts	{001} perfect {110} dist
1.472 1.475	n_β 1.472-1.475 n_α 1.465-1.467 n_γ 1.494-1.497	Carnallite $KMgCl_3 \cdot 6H_2O$	~0.030	~70° v>r weak	Orthorhombic	Coarse-fine granular. Pseudo-hexagonal crystals	None
1.472 1.473	n_β 1.472-1.473 n_α 1.470-1.472 n_γ 1.479	Boussingaultite $(NH_4)_2Mg(SO_4)_2 \cdot 6H_2O$	0.007	51° r>v perceptible	Monoclinic β = 107°	Crusts, prismatic	{$\bar{2}$01} perfect
1.473	n_β 1.473 n_α 1.458 n_γ 1.501	Barringtonite $MgCO_3 \cdot 2H_2O$	0.043	74°	Trigonal α=94° β=96° γ=109°	Fibrous, radiated acicular	{001}, {100} & {010} dist
1.473	n_β 1.473 n_α 1.469 n_γ 1.491	Meta-alunogen $Al_4(SO_4)_6 \cdot 27H_2O$	0.022	Mod ~50°	Monoclinic	Massive	{010} perfect
1.473 1.477	n_β 1.473-1.477 n_α 1.464-1.471 n_γ 1.481-1.485	Thenardite Na_2SO_4	0.017 to 0.014	~83°	Orthorhombic	Tabular {010}, pyram, prism crystals, crusts	{010} perfect {101} fair {100} imperf
1.469 1.480	n_β 1.469-1.480 n_α 1.469-1.479 n_γ 1.473-1.483	α-Tridymite SiO_2	0.002 to 0.004	35° - 90°	Orthorhombic (pseudohex)		None
1.478	n_β 1.478 n_α 1.471 n_γ 1.486	Melanterite $FeSO_4 \cdot 7H_2O$	0.015	85° r>v weak, inclined	Monoclinic β = 105.6°	Prism, tabular, fibrous, stalactitic	{001} perfect {110} dist
1.479	n_β 1.479 n_α 1.472 n_γ 1.487	Pisanite $(Fe,Cu)SO_4 \cdot 7H_2O$	0.015	~86° r>v weak, inclined	Monoclinic β = 104.3°	Pseudorhombic xls, long prism	{001} perfect {110} dist
1.479	n_β 1.479 n_α 1.478 n_γ 1.482	Ferrierite $(Na,K)_2MgAl_3Si_{15}O_{36}(OH) \cdot 9H_2O$	0.004	~50°	Orthorhombic	Radiated tabular {010}	{010} perfect

* n_β variation exceeds 0.01

Twinning	Orientation	Color in thin section	Hardness Sp Gravity	Color hand sample	Alteration	Occurrence	Remarks	Reference
	b = X, c ~ Z, a ∧ X ~ +12°	Colorless	2½ 1.70	Colorless, white		Altered serpentenite with hydromagnesite. Mine efflorescence	Effloresces readily. Soluble dilute acids	Dana II, 228-230
	Z ⊥ tablets, Y bisects acute rhomboid angle	Colorless	~3 1.892	White		Sublimate on burning coal with alum and copiapite		Dana II, 601
{010}	b = X, c∧Z = 42°	Colorless	1½-2 1.64-1.78	Colorless, white, yellowish		Efflorescence or cavity filling in clay rocks, coals, fumeroles	Soluble in H_2O; acid taste	Dana II, 537-540
	b = Y, c ∧ Z = 12°	Colorless, pale green	2 ~1.90	Green, blue, white		Alteration of pyrite, marcasite. Oxide zones pyrite deposits	Var Melanterite. Sol H_2O. Taste astringent	Dana II, 499-504
{hk0} lamellar	a = Z, b = Y, c = X	Colorless	2½ 1.60	Colorless, white, reddish		Bedded evaporite deposits with halite, K-salts	Highly sol H_2O. Bitter taste.	p. 15
	b = Y, c ∧ Z ~ 12°	Colorless	2 1.72	Colorless, pink		Acid fumeroles, efflorescences	Soluble H_2O	Dana II, 455-456
	+ elongation, ext 17° - 34°	Colorless	2.83	Colorless		Evaporative crusts		Am Min 50, 2103
	X ⊥ {010}	Colorless		White		Dehydration of alunogen		Am Min 28, 61-62
{110} common, {011} contact	a = Z, b = Y, c = X	Colorless	2½-3 2.66-2.67	Colorless, white, yellowish		Playa-lake evaporites. Soil efflorescence. Fumeroles	Sol H_2O. Faintly saline taste	Dana II, 405-407
{10$\bar{1}$6} as wedges	a = Y, b = X, c = Z	Colorless	7 2.27	Colorless, white	Fine-grained α-quartz	Vugs & lithophysae in siliceous tuffs, glasses & flows with cristobalite	Metastable below 870°. Seldom older then Tertiary	p. 319
	b = Y, c∧Z = 61°	Colorless, pale green	2 1.90	Green, blue, white		Efflorescence in pyrite ore bodies	Sol H_2O. Taste-sweet, metallic, astringent	Dana II, 499-504
	b = Y	Pale green	2 1.90	Green, green-blue, blue		Oxide zones Cu-deposits. Alter pyrite ores	Var Melanterite. Sol H_2O. Taste metallic	
	a = X, b = Y, c = Z	Colorless	3-3½ 2.11-2.14	Colorless, white		Cavities in mafic volcanics	Zeolite group	Am Min 54, 887-895

n_β Var.	Refractive Index	Mineral Name and Composition	$(n_\gamma - n_\alpha)$	2V angle Dispersion	Crystal System	Habit	Cleavage
1.480	n_β 1.480 n_α 1.475 n_γ 1.487	Misenite $KHSO_4$	0.012	Large ~80°	Monoclinic β = 102.1°	Acicular, fibrous, lath-like	{010} dist
1.480	n_β 1.480 n_α 1.475 n_γ 1.488	Dietrichite $(Zn,Fe,Mn)Al_2(SO_4)_4 \cdot 22H_2O$	0.013	Large	Monoclinic	Fibrous crusts	
1.480	n_β 1.480 n_α 1.478 n_γ 1.481	Clinoptilolite $(Na,K,Ca)_{2-3}Al_3(Al,Si)_2Si_{13}O_{36} \cdot 12H_2O$	0.003	32° - 48° r > v strong	Monoclinic β = 91.5°	Platy clusters	{010} perfect
1.475 1.485	n_β 1.475-1.485 n_α 1.472-1.483 n_γ 1.477-1.487	Mordenite (Ptilolite) $(Na_2,K_2,Ca)(Al_2Si_{10})O_{24} \cdot 7H_2O$	0.002 to 0.005	90° - 76°	Orthorhombic	Radiated acicular-fibrous (c), cottony	{010} and {100} perfect
1.476 1.486	n_β 1.476-1.486 n_α 1.473-1.483 n_γ 1.485-1.496	Natrolite $Na_2(Al_2Si_3)O_{10} \cdot 2H_2O$	0.012 to 0.013	58° - 64° r > v weak	Orthorhombic (pseudo-tetragonal)	Radiated acicular or bladed. Fibrous	{110} perfect and {010} parting
1.481	n_β 1.481 n_α 1.470 n_γ 1.497	Rivadavite $Na_6MgB_{24}O_{40} \cdot 22H_2O$	0.027	80° r > v	Monoclinic β = 106°	Prism (b), splintery, nodular	{100} & {$\bar{1}$01} perfect {010} poor
1.482	n_β 1.482 n_α 1.475 n_γ 1.489	Bieberite $CoSO_4 \cdot 7H_2O$	0.014	88°	Monoclinic β = 105°	Crusts	{001} perfect {110} dist
1.482	n_β 1.482 n_α 1.479 n_γ 1.487	Leonite $K_2Mg(SO_4)_2 \cdot 4H_2O$	0.008	~90°	Monoclinic β = 95.3°	Tabular, elong (c)	None
1.482	n_β 1.482 n_α 1.481 n_γ 1.483	Svetlozarite $(Ca,K_2,Na_2)Al_2(Si,Al)_{12}O_{28} \cdot 6H_2O$	0.002	23° r > v distinct	Orthorhombic	Spherulitic	3 perfect at 90°
1.483	n_β 1.483 n_α 1.479 n_γ 1.488	Zinc-Melanterite $(Zn,Cu,Fe)SO_4 \cdot 7H_2O$	0.009	~90° Slight	Monoclinic β ~ 90°	Massive, columnar	
1.48	n_β 1.48 n_α 1.476 n_γ 1.500	Uklonskovite $NaMg(SO_4)(OH) \cdot 2H_2O$	0.024		Monoclinic β = 90.6°	Prism (b) flattened crystals	One dist. One indist ∥ b
1.486	n_β 1.486 n_α 1.484 n_γ 1.497	Tamarugite $NaAl(SO_4)_2 \cdot 6H_2O$	0.013	~60°	Monoclinic β = 95.2°	Fine grained, fibrous, tabular or prism xls.	{010} perfect
1.486	n_β 1.486 n_α 1.484 n_γ 1.502	Cyanochroite $K_2Cu(SO_4)_2 \cdot 6H_2O$	0.018	47° v > r strong	Monoclinic β = 104.5°	Tabular, crusts	{$\bar{2}$01} perfect
1.490 1.482	n_β 1.490-1.482 n_α 1.486-1.480 n_γ 1.497-1.486	Mohrite $(NH_4)_2Fe(SO_4)_2 \cdot 6H_2O$	0.011 to 0.006	75°	Monoclinic β = 106.9°	Irregular laminae, minute crystals	{$\bar{1}$02} perfect {010} dist

* n_β variation exceeds 0.01

Twinning	Orientation	Color in thin section	Hardness Sp Gravity	Color hand sample	Alteration	Occurrence	Remarks	Reference
	b = Z, c ∧ X = -29°	Colorless	2.32	Grayish-white		Efflorescence in fumeroles with K-alum, aluogen	Soluble H_2O	Dana II, 396-397
	b = X, c ∧ Z ~ 29°	Colorless	2	White, brownish-yel		Recent efflorescence in mines	Halotrichite group, Sol H_2O	Dana II, 528-529
	b = Y, a ∧ Z = 30°-45°	Colorless	3½-4 2.1-2.2	Colorless, white		Amygdules in basalt, altered pyroclastics	Zeolite group	Am Min 54, 887-895
None	a = Y, b = Z, c = X, ‖ extinction, - elongation	Colorless	3-4 2.12-2.22	White, yellowish, pinkish	None	Amygdules in basalts. Altered glasses and tuffs. Marine sediments	Zeolite group	p. 385
{110},{011} & {031} all rare	a = X, b = Y, c = Z	Colorless	5 2.20-2.26	White, gray, yellow, reddish	Beidellite	Amygdules in basalt with other zeolites. Alter nepheline, Na-feldspars	Zeolite group	p. 379
	b = Y, c ∧ Z = -32°	Colorless	3½ 1.90-1.91	Colorless, white		Embedded in borax		Am Min 52, 326-335
	b = Y, c ∧ Z = -29°	Colorless, pink	~2 1.94-1.95	Rose red, flesh		Oxide zone Co-deposits, efflorescence	Sol H_2O, dehydrates	Dana II, 505-507
{100} lamellar	b = Y, a ∧ Z = small	Colorless	2½-3 2.20	Colorless, yellowish		Evaporite marine sediments with halite, K, Mg-salts	Sol H_2O. Bitter taste	Dana II, 450-451
	- elongation	Colorless	4 2.166-2.174	Colorless, white		Chalcedony veinlets with ferrierite, clinoptilolite, mordenite	Zeolite group	Am Min 62, 1060
	Y ∧ elong ~ 34°	Very pale blue-green	~2 2.02	Pale greenish blue		Oxide zones of pyrite-chalcopyrites-sphalerite ores	Dehydrates readily. Readily soluble H_2O	Dana II, 508
		Colorless	2.40-2.50	Colorless		Saline, lacustrine clays with glauberite, polyhalite		Am Min 50, 520-521
		Colorless	~3 2.06	Colorless	Alteration of mendozite	Oxide zones sulfide deposits in aluminous-alkali environments	Sol H_2O. Sweet-astringent taste	Am Min 54, 19-30
	b = Y, c ∧ X = -19°	Colorless	2.22	Greenish-blue		Fumerole sublimate	Soluble H_2O	Dana II, 454-455
		Pale green	1.80-1.87	Pale green		Vents, springs, ponds in volcanic areas		Am Min 50, 805

n_β Var.	Refractive Index	Mineral Name and Composition	$(n_\gamma - n_\alpha)$	2V angle Dispersion	Crystal System	Habit	Cleavage
1.491	n_β 1.491 n_α 1.488 n_γ 1.505	Inderite (Lesserite) $Mg_2B_6O_{11} \cdot 15H_2O$	0.017	37° r > v weak	Monoclinic β = 104.5°	Prism crystals, reniform-nodular	{010} perfect {110} distinct {001} indist
1.491	n_β 1.491 n_α 1.490 n_γ 1.497	Starkeyite $MgSO_4 \cdot 4H_2O$	0.007	50°	Monoclinic β = 90.8°	Massive, powdery efflorescence	
1.490 1.496	n_β 1.490-1.496 n_α 1.473-1.490 n_γ 1.506-1.511	Fluellite (Kreuzbergite) $Al_2PO_4F_2OH \cdot 7H_2O$	0.011 to 0.021	Large 65° - 90° r > v moderate	Orthorhombic	Dipyramidal crystals	{010} & {111} indistinct
1.493	n_β 1.493 n_α 1.509 n_γ 1.561	Larderellite $NH_4B_5O_6(OH)_4$	0.052	64° - 58° v > r	Monoclinic β = 96.75°	Microscopic rhomboid tabular {100}	{100} perfect
1.494	n_β 1.494 n_α 1.485	Canavesite $MgCO_3(HBO_3) \cdot 5H_2O$	0.020	Very large Very weak disp	Monoclinic β = 114.9°	Fibrous rosettes, pseudohexagonal xls	{h0ℓ} cleavages or partings
1.495	n_β 1.4947 n_α 1.4935 n_γ 1.4973	Arcanite K_2SO_4	0.004	67° r > v moderate	Orthorhombic	Thin pseudohexagonal plates	{010}, {001} distinct
1.496	n_β 1.496 n_α 1.495 n_γ 1.504	Struvite $NH_4MgPO_4 \cdot 6H_2O$	0.009	37° v > r strong	Orthorhombic	Prism, tab, wedge-shaped crystals	{001} dist {100} poor
*1.487 1.505	n_β 1.487-1.505 n_α 1.487-1.505 n_γ 1.488-1.512	Heulandite $(Ca,Na_2)(Al_2Si_7)O_{18} \cdot 6H_2O$	0.001 to 0.007	30° r > v, crossed	Monoclinic β = 91°	Tabular plates {010}, granular	{010} perfect
*1.484 1.509	n_β 1.484-1.509 n_α 1.483-1.504 n_γ 1.486-1.514	Phillipsite $(Ca,K_2,Na_2)_6[(Al_3Si_5)O_{16}]_2 \cdot 12H_2O$	0.003 to 0.010	60° - 80° v > r weak, horizontal	Monoclinic β = 126°	Pseudo-ortho or pseudotetragonal crystals	{010} & {100} distinct
1.496	n_β 1.496 n_α 1.494 n_γ 1.499	Dachiardite $(K,Na,Ca)_5Al_5Si_{19}O_{48} \cdot 18H_2O$	0.005	70°	Monoclinic β = 105°	Prismatic	{001} & {100} perfect
1.497	n_β 1.497 n_α 1.496 n_γ 1.504	Yugawaralite $Ca(Al_2Si_6)O_{16} \cdot 4H_2O$	0.008	78° r > v weak	Monoclinic β = 111.5°	Tabular {010} xls	{010} imperf
*1.491 1.510	n_β 1.491-1.510 n_α 1.488-1.489 n_γ 1.505-1.525	Kurnakovite $Mg_2B_6O_{11} \cdot 15H_2O$	0.017 to 0.036	37° - 90°	~Monoclinic β = 104.8°	Blocky crystals, granular aggregate	{010} indist
1.500	n_β ~1.500 n_α 1.498 n_γ 1.502	Wairakite $Ca(AlSi_2O_6)_2 \cdot 2H_2O$	0.004	70° - 90° r > v	Monoclinic (Pseudocubic) β = 90.5°	Subhedral crystals, ~octahedral	{100} distinct
1.501	n_β 1.501 n_α 1.484 n_γ 1.550	Gowerite $CaB_6O_{10} \cdot 5H_2O$	0.066	63° r > v weak	Monoclinic β = 91°	Radiated globular, acicular (c)	{001?} dist {100?} imperf
1.502	n_β 1.502 n_α 1.497 n_γ 1.539	Liebigite (Flutherite) (Uranothallite) $Ca_2U(CO_3)_4 \cdot 10H_2O$	0.042	40° r > v moderate	Orthorhombic	Short prism (c), scaly crusts	{100} dist

* n_β variation exceeds 0.01

Twinning	Orientation	Color in thin section	Hardness Sp Gravity	Color hand sample	Alteration	Occurrence	Remarks	Reference
	c ∧ Z = -9°	Colorless	2½ 1.76-1.79	Colorless, white, pink		Evaporite borate deposits	Dimorph kurnakovite	Am Min 41, 927
		Colorless	2.007	White		Efflorescence in mines with sanderite, pentahydrite, hexahydrite		Am Min 41, 662
	a = X, b = Z, c = Y	Colorless	3 2.16-2.18	Colorless, white, pale yellow		Secondary min pegmatites. Alter triplite		Dana II, 124-126
	b = X, c ∧ Z ~ 15°	Colorless	1.89-1.91	White, yellowish		Calderas and thermal waters with ammonioborite and sassolite	Decomposed hot H_2O. Tasteless	Am Min 45, 1087-1093
	b = Z	Colorless	1.79	Milky white		Fe-skarns with ludwigite		Can Min 16, 69-73
Cyclic {110}?	a = Y, b = X, c = Z	Colorless	2 2.66-2.70	Colorless, white		Efflorescence	Soluble in H_2O	Dana II, 399-400
{001} common	a = Z, b = X, c = Y	Colorless	2 1.71	Colorless, white		Bat guano with newberyite, hannayite, brushite	Slightly sol H_2O. Easily sol acids	Dana II, 715-717
Uncommon	b = Z, c ∧ Y = +8° to + 32°, a ∧ X = +9° to + 33°	Colorless	3½-4 2.2	White, yel-reddish, brn	Beidellite	Cavities & veins in basalt. Altered glass & tuffs. Schists and gneisses	Zeolite group	p. 388
Cyclic on {001}{110} & {021}	b = X, c ∧ Z = +10° to + 30°, a ∧ Y = +46° to + 65°	Colorless	4-4½ 2.2	White	Montmorillonite clays	Amygdules in Si-poor volcanics with chabazite. Deep sea redclays	Zeolite group	p. 391
{110} ubiquitous, cyclic		Colorless	4-4½ 2.14-2.21	Colorless		With other zeolites in granite pegmatite	Zeolite group	Am Min 47, 190-191
None	b = Z, c ∧ Y = +6°, a ∧ X ~ +26°	Colorless	4½ 2.23-2.26	Colorless, white		Cavities in andesite tuffs altered by thermal springs	Zeolite group. Irridescent	Am Min 54, 306-309
Rare	b ~ Y, optic axis ~ ⊥{001}	Colorless	3 1.76-1.85	Colorless, white (alter)		Evaporite lacustrine borate deposits	Dimorph inderite Sol warm acids	Dana II, 360
{110} lamellar (two sets at 90°)		Colorless	5½-6 2.26	Colorless, white		Tuffaceous rocks and breccias. Graywacke. Hydrothermal alter	Zeolite group (Ca-analcime)	Min Mag 30, 691-708
	b = Y, c ∧ Z = 27°	Colorless	3 1.98-2.00	Colorless		Evaporite borate deposits		Am Min 45, 230-234
	?	X ~ colorless, Y = Z = pale greenish-yellow	2½-3 2.41-2.43	Green, yellow-green		Secondary U-mineral Efflorescence. Alter. uraninite	U.V. fluor-green or blue-green	USGS Bull 1064, 108-112

Biaxial +

n_β Var.	Refractive Index	Mineral Name and Composition	$(n_\gamma - n_\alpha)$	2V angle Dispersion	Crystal System	Habit	Cleavage
	n_γ 1.518-1.544						
1.50	n_β 1.50 n_α 1.498 n_γ 1.503	Wellsite $(Ba,Ca,K_2)(Al_2Si_3O_{10})\cdot 3H_2O$	0.005	39°	Monoclinic β = 126.6°	Penetration twins	{010} & {100} good
1.503	n_β 1.503 n_α 1.501 n_γ 1.510	Prosopite $CaAl_2(F,OH)_8$	0.009	63° r > v strong	Monoclinic β = 95°	Tiny tabular crystals, bladed, massive, earthy	{111} perfect
1.503	n_β 1.503 n_α 1.498 n_γ 1.510	Hexahydroborite $Ca[B(OH)_4]_2\cdot 2H_2O$	0.012	83° r > v very strong	Monoclinic β = 104°	Agg of flat prisms	1 perfect, & 1 imperf
1.504	n_β 1.5037 n_α 1.5008 n_γ 1.5062	Taylorite (NH_4-arcanite) $K_{2-x}(NH_4)_xSO_4$ (x ~ 0.35)	0.005	Large ~ 90°	Orthorhombic	Massive, concretionary	
1.504	n_β 1.504 n_α 1.501 n_γ 1.520	Hydrochlorborite $Ca_4B_8O_{15}Cl_2\cdot 22H_2O$	0.019	~45°	Unknown	Massive	
1.504 1.506	n_β 1.504-1.506 n_α 1.491-1.496 n_γ 1.519-1.520	Ulexite $NaCaB_5O_9\cdot 8H_2O$	0.023	73° - 78°	Triclinic α=90.3° β=109.2° γ=105.1°	Acic (c), nodular, fibrous veins	{010} perfect {1$\bar{1}$0} good {110} poor
1.505	n_β 1.505 n_α 1.500 n_γ 1.525	Loughlinite $Na_2Mg_3Si_6O_{16}\cdot 8H_2O$	0.025	60°	Unknown	Fibrous, massive	
1.505	n_β 1.505 n_α 1.501-1.502 n_γ 1.513-1.515	Rhodesite $(Ca,Na_2,K_2)_8Si_{16}O_{40}\cdot 11H_2O$	0.012	Small	Orthorhombic	Matted silky fibers (c) in rosettes	{100} dist
1.504 1.508	n_β 1.504-1.508 n_α 1.504-1.508 n_γ 1.505-1.509	Mesolite $Na_2Ca_2[(Al_2Si_3)O_{10}]_3\cdot 8H_2O$	~0.001	~80° v > r strong, inclined	Mono (pseudo-ortho) β~ 90°	Fibrous acicular (b), hair-like	?
1.507	n_β 1.507 n_α 1.495 n_γ 1.528	Bischofite $MgCl_2\cdot 6H_2O$	0.033	79° r > v weak	Monoclinic β = 93.5°	Foliated, granular, fibrous	None
1.505 1.509	n_β 1.505-1.509 n_α 1.503-1.508 n_γ 1.508-1.514	Harmotome $Ba(Al_2Si_6)O_{16}\cdot 6H_2O$	0.005 to 0.008	~80° Crossed weak	Monoclinic β = 125°	Cruciform twins	{010} dist, {001} poor
1.508	n_β 1.508 n_α 1.502 n_γ 1.525	Rabbittite $Ca_3Mg_3(UO_2)_2(CO_3)_6(OH)_4\cdot 18H_2O$	0.023	62°	Monoclinic	Acicular-fibrous (c)	{hk0} perfect {001} good
1.509	n_β 1.509 n_α 1.504 n_γ 1.545	Ussingite $Na_4Al_2Si_6O_{17}\cdot H_2O$	0.041	36°	Triclinic α=102.6° β=115.7° γ= 79.2°	Massive	{110} & {1$\bar{1}$0} perfect ~ 90°
1.510	n_β 1.510 n_α 1.470 n_γ 1.579	Strontioborite $SrB_8O_{13}\cdot 2H_2O$?	0.109	~80°	Monoclinic β = 107.8°	Small plates	

* n_β variation exceeds 0.01

Twinning	Orientation	Color in thin section	Hardness Sp Gravity	Color hand sample	Alteration	Occurrence	Remarks	Reference
						spars. Ls contacts		
{001}, {021}, {110} pene-tration	b = Z, c ∧ X = +52°	Colorless	4-4½ 2.248	Colorless, white		Fissures & cavities with corundum, anal-cime, leonhardite, calcite	Zeolite group	DHZ 4, 387-389
	b = Y, c ∧ Z = +35°	Colorless	4½ 2.89-2.90	Colorless, white, gray		Greisens, pegmatites with fluorite, apatite, cassiterite, topaz	Decomposed H_2SO_4	Dana II, 121-123
	- elongation c ∧ X = 14°	Colorless	2½ 1.87	Colorless		Veinlets with pentahydroborite	Soluble alcohol and weak acids	Am Min 63, 1283
	a = Y, b = X, c = Z	Colorless	2	White, creamy		Guano deposits	Sol H_2O. Bitter-pungent taste	Am Min 36, 590-602
		Colorless	2½ 1.83	Colorless		Evaporite borate deposits with ulexite, gypsum, halite	Sol hot H_2O or cold acids	Am Min 50, 2099-2100
Polysynthetic usually on {010} or {100}	c ∧ Y ~ 20°, Z ~ b, optic plane ~ ⊥ {010}	Colorless	2½ 1.955-1.959	Colorless, white		Arid, saline lake deposits with anhy-drite, glauberite, borates	Sol acid. Slightly sol H_2O. Tasteless	Am Min 44, 712-719
None?	Parallel ext., + elongation	Colorless	2.165	Pearly white		Veinlets in dolomitic oil shale	Strongly resem-bles sepiolite	Am Min 45, 270-281
	c = Z	Colorless	2.36	White		Diamond pipes with mountainite. Altered silicic lavas		Min Mag 31, 607-610
{100} ubiquitous	b = Y, c ∧ X = -8°, a ∧ Z = -8°	Colorless	5 2.26	White, yellowish	Beidellite	Amygdules & veins in volcanic rocks with calcite & other zeolites	Zeolite group	p. 381
Polysynthetic	b = X, c ∧ Y = 9.5°	Colorless	1-2 1.56	Colorless, white		Salt deposits	Soluble H_2O	Dana II, 46-47
Cyclic on {001}, {110} & {021}	b = Z, c∧Y = +28° to +32°, a ∧ X = +63° to +67°	Colorless	4½ 2.4-2.5	White, brownish, reddish	Montmoril-lonite clays	Hydrothermal veins & voids in basalt and Si-poor volcanics. Laterites	Zeolite group	p. 392
	b = Y, c ∧ Z ~ 15°	~Colorless	~2½ 2.57	Pale green		Efflorescence in U-mines	Fluor short λ U.V.-cream-yellow	Am Min 40, 201-212
{010} common	c ∧ Z' ~ 33° in (010)	Pale violet-red, Weak pleochroism	6½ 2.46-2.53	Pale to dk violet-red		Pegmatites	Gelat HCl	Win & Win II, 443
	Inclined extinction, ± elongation	Colorless	2.806	Colorless		Fine-lamellar rock salt		Am Min 50, 1508

Biaxial +

n_β Var.	Refractive Index	Mineral Name and Composition	$(n_\gamma - n_\alpha)$	2V angle Dispersion	Crystal System	Habit	Cleavage
1.510	n_β 1.510 n_α 1.504 n_γ 1.519	Mountainite $(Ca,Na_2,K_2)_2Si_4O_{10}\cdot 3H_2O$	0.015	~80°	Monoclinic β = 104°	Matted silky fibers (b), rosettes	One prismatic?
1.510	n_β 1.5095 n_α 1.5043 n_γ 1.5751	Pirssonite $Na_2Ca(CO_3)_2\cdot 2H_2O$	0.071	~31° v > r slight	Orthorhombic	Short prism (c), tabular {010} crystals	None
1.510	n_β 1.510 n_α 1.508 n_γ 1.516	Dypingite $Mg_5(CO_3)_4(OH)_2\cdot 5H_2O$	0.008	~60°	Unknown	Globular, radiated fibrous	
1.510	n_β 1.510 n_α 1.510 n_γ 1.545	Solongoite $Ca_4B_6O_8Cl(OH)_9$	0.035	Small r > v weak	Monoclinic β = 94°	Prism crystals	
1.510 1.513	n_β 1.510-1.513 n_α 1.504-1.507 n_γ 1.516-1.523	Petalite $Li(AlSi_4)O_{10}$	0.011 to 0.017	83° r > v weak, crossed	Monoclinic β = 113°	Blocky crystals, fibrous	{001} perfect {201} dist
1.512	n_β 1.512 n_α 1.510 n_γ 1.523	Brewsterite $Na(Sr,Ba,Ca)_5(Al_{11}Si_{29})O_{80}\cdot 25H_2O$	0.011	47°-65° r > v weak, crossed	Monoclinic β = 93°	Platy {010}	{010} perfect
1.513	n_β 1.513 n_α 1.510	Monteregianite $(Na,K)_6Y_2Si_{16}O_{38}\cdot 10H_2O$	0.007	87°	Orthorhombic	Acic. to tabular crystals	{010} perfect, {001} v. good, {100}
1.515	n_β 1.515 n_α 1.508 n_γ 1.523	Schertelite $(NH_4)_2MgH_2(PO_4)_2\cdot 4H_2O$	0.015	~90°	Orthorhombic	Tabular crystals {100}, acicular (c)	None
1.516	n_β 1.516 n_α 1.513 n_γ 1.518	Lovdarite $(Na,K,Ca)_4(Be,Al)_2Si_6O_{16}\cdot 4H_2O$	0.005	~90°	Orthorhombic	Massive, prism crystals	{100}, {010}, {001} dist. {110} poor
1.517	n_β 1.517 n_α 1.514 n_γ 1.533	Newberyite $MgHPO_4\cdot 3H_2O$	0.019	~45° v > r perceptible	Orthorhombic	Prism, tabular, equant crystals	{010} perfect {001} imperf
1.520	n_β 1.520 n_α 1.500 n_γ 1.554	Nobleite $CaB_6O_{10}\cdot 4H_2O$	0.054	76° r > v weak	Monoclinic β = 111.8°	Platy crystals, mammillary coatings	{100} perfect {001} indist
*1.513 1.533	n_β 1.513-1.533 n_α 1.497-1.530	Thomsonite $NaCa_2[(Al,Si)_5O_{10}]_2\cdot 6H_2O$	0.006 to 0.021	42° - 75° r > v moderate-strong	Ortho(pseudo-tetragonal)	Columnar, acicular, bladed (c)	{010} perfect, {100} dist
1.523	n_β 1.523 n_α 1.520 n_γ 1.533	Mascagnite $(NH_4)_2SO_4$	0.013	52° r > v weak	Orthorhombic	Crusts, stalactitic, prism crystals	{001} dist
1.524	n_β 1.524 n_α 1.512 n_γ 1.577	Strontioginorite (Volkolvite) $(Sr,Ca)_2B_{14}O_{23}\cdot 8H_2O$	0.065	53°	Monoclinic β = 101.6°	Tabular crystals (c)	{010} perfect {001} less perfect

* n_β variation exceeds 0.01

Twinning	Orientation	Color in thin section	Hardness Sp Gravity	Color hand sample	Alteration	Occurrence	Remarks	Reference
	b = Y = elongation. Parallel ext ± elongation	Colorless	2.36	White		Diamond pipes with rhodesite		Min Mag 31, 611-623
	a = X, b = Z, c = Y	Colorless	3-3½ 2.35-2.37	Colorless, white		Clay beds in evap lakes with northupite, trona, gaylussite	Sol dilute acids with efferves-cence	Dana II, 232-233
	Elongation Y, ± elongation	Colorless	~2.15	White		Serpentine-magnesite rocks	Fluor U.V.-pale blue, Phosphor-escence yellow-green. Sol with efforvesc. HCl	Am Min 55, 1457-1465
	c ∧ Z = 25°, elongation +	Colorless	3.5 2.51-2.53	Colorless		Contact limestone with szaibelyite, grossular, kurchatovite	Readily sol cold HCl or H_2SO_4	Am Min 60, 162-163
{001} lamellar	b = Z, c ∧ Y = -24° to -30°, a ∧ X = -1° to -7°	Colorless	6½ 2.41-2.42	White, lt gray pink, green	Montmoril-lonite clays (hectorite), zeolites (heulan-dite, stilbite)	Granite pegmatites with spodumene and other Li-minerals		p. 397
	b = Z, c ∧ X ~ -22°, a ∧ Y ~ -19°	Colorless	5 ~2.4	White, yellowish, greenish	Montmoril-lonite clays	Cavities and fractures in basalts	Zeolite group	p. 377
	a = Y, b = Z, c = X	Colorless	3½ 2.42	Colorless, white, gray		Cavities and inclusions in nepheline syenite		Am Min 65, 207
	a = Z, b = X, c = Y	Colorless	1.83	Colorless		Bat guano with stru-vite and newberyite	Soluble H_2O	Dana II, 699
	|| extinction	Colorless	2.32-2.33	White to yellowish		Late hydrothermal mineral in alkali pegmatites. Replace chkalovite		Am Min 59, 874
	a = X, b = Y, c = Z	Colorless	3-3½ 2.10	Colorless, grayish		Bat guano with hannayite, struvite, magnesite, collophane	Slight sol H_2O. Easily sol dilute HCl	Dana II, 709-711
{100} contact	b = Y, c ∧ Z = +7 , a ∧ X = +29	Colorless	3 2.09	Colorless		Weathering of colemanite and priceite in borate veins	Sectile crystals	Am Min 46, 560-571
{110} rare	a = X, b = Z, c = Y ± elongation	Colorless	5-5½ 2.10-2.39	White to brn, reddish	Beidellite	Cavities & veins in basalts. Alter nepheline, Ca-feld-	Zeolite group	p. 383
{110} cyclic, polysynthetic	a = Z, b = Y, c = X	Colorless	2-2½ 1.77	Colorless, gray, yellowish		Volcanic sublimate, guano deposits, burning coal	Sol H_2O. Bitter taste	Dana II, 398-399
	b = Y	Colorless	2-3 2.25	Colorless		Evaporite deposits with halite, K-salts	Not soluble H_2O	Am Min 45, 478

n_β Var.	Refractive Index	Mineral Name and Composition	$(n_\gamma - n_\alpha)$	2V angle Dispersion	Crystal System	Habit	Cleavage
1.524	n_β 1.524 n_α 1.514 n_γ 1.543	Probertite (Kramerite) $NaCaB_5O_9 \cdot 5H_2O$	0.029	73° r > v	Monoclinic β = 100.3°	Radiated acicular (c) aggregate, spherulitic	{110} perfect
1.524	n_β 1.524 n_α 1.518 n_γ 1.530	Macdonaldite $BaCa_4Si_{15}O_{35} \cdot 11H_2O$	0.012	~90°	Orthorhombic	Acicular crystals (a)	{010} perfect {001} dist {100} poor
1.524	n_β 1.524 n_α 1.517–1.520 n_γ 1.577–1.580	Ginorite (Cryptomorphite) $Ca_2B_{14}O_{23} \cdot 8H_2O$	0.060	42°	Orthorhombic	Rhombohedral-shaped plates (~79°)	{010}
1.522 1.526	n_β 1.522–1.526 n_α 1.519–1.521 n_γ 1.529–1.531	Gypsum $CaSO_4 \cdot 2H_2O$	0.010	~58° r > v strong, inclined	Monoclinic β = 127.4°	Prism-acic-fib xls, platy, granular	{010} perfect {100} & {$\bar{1}$11} distinct
1.525	n_β 1.525 n_α 1.508 n_γ 1.586	Sideronatrite $Na_2Fe^{3+}(SO_4)_2OH \cdot 3H_2O$	0.078	~58° r > v strong	Orthorhombic	Minute acicular fibrous crusts, nodular	{100} perfect
1.525	n_β 1.525 n_α 1.521 n_γ 1.545	Spadaite $MgSiO_2(OH)_2 \cdot H_2O$?	0.024	~50°	Unknown	Minute platy, columnar crystals, massive	
1.526	n_β 1.526 n_α 1.508 n_γ 1.550	Kaliborite (Paternoite) $HKMg_2B_{12}O_{21} \cdot 9H_2O$	0.042	~80°	Monoclinic β = 100°	Minute crystal aggregate, nodular	{001} & {$\bar{1}$01} perfect {100} dist
1.527	n_β 1.527 n_α 1.523 n_γ 1.545	Hydromagnesite $Mg_5(CO_3)_4(OH)_2 \cdot 4H_2O$	0.022	Moderate	Monoclinic β = 114.5°	Earthy, acicular or bladed (c), flat {100}	{010} perf
1.528	n_β 1.528 n_α 1.525 n_γ 1.544	Sazhinite $Na_3CeSi_6O_{15} \cdot 6H_2O$	0.019	47°	Orthorhombic (Pseudo-tetragonal)	Tabular crystals, fine-grained agg	{100}, {010}, {001} perf
1.528	n_β 1.528 n_α 1.526 n_γ 1.551	Teruggite $Ca_4MgB_{12}As_2O_{28} \cdot 18H_2O$	0.025	33° r > v weak	Monoclinic β = 100.1°	Acicular crystals (c), rhombohedral sections	{001} dist {110} fair
1.529 1.535	n_β 1.529–1.535 n_α 1.507–1.510 n_γ 1.576–1.575	Magnesiocopiapite (Knoxvillite) $MgFe_4(SO_4)_6(OH)_2 \cdot 20H_2O$	0.069 to 0.065	67° – 78° r > v strong	Triclinic α = 93.9° β=101.5° γ = 99.4°	Tabular crystals, scaly, granular	{010} perf {$\bar{1}$01} imperf
1.530	n_β 1.530 n_α 1.483 n_γ 1.576	Chalconatronite $Na_2Cu(CO_3)_2 \cdot 3H_2O$	0.093	Large ~ 87°	Monoclinic?	Fine crusts, acicular or hexagonal plates	
1.530	n_β 1.530 n_α 1.515 n_γ 1.580	Earlandite $Ca_3(C_6H_5O_7)_2 \cdot 4H_2O$	0.065	60°	Unknown	Fine-grained nodules	
1.53	n_β 1.53 n_α 1.525 n_γ 1.550	Roesslerite $MgH(AsO_4) \cdot 7H_2O$	0.025	Small	Monoclinic β = 95.4°	Fine-granular, fibrous crusts, prism crystals	{111} imperf

Twinning	Orientation	Color in thin section	Hardness Sp Gravity	Color hand sample	Alteration	Occurrence	Remarks	Reference
	b = Y, c ∧ Z = 12°	Colorless	3½ 2.13-2.14	Colorless		Evaporite deposits with colemanite, ulexite, kernite	Easily sol dilute acids. Partly sol H_2O	Am Min 44, 712-719
	a = Z, b = Y, c = X	Colorless	3½-4 2.27	White, colorless		Contact metamorphic rocks with sanbornite		Am Min 50, 314-340
	X = short rhombohedral diagonal, Z = long rhombohedral diagonal	Colorless	3½ 2.07-2.09	White		Small pellets in sassolite and clay. Efflorescence in borate deposits		Am Min 42, 56-61
{100} simple, common, {101̄} rare	b = Y, c ∧ Z ~ -52°, a ∧ X ~ -15°	Colorless	2 2.31	Colorless, white, yellowish, pink	Anhydrite, aragonite	Evaporite beds with anhydrite, halite, etc. Fumeroles. Soil efflorescence	Soluble HCl	p. 88
	a = X, b = Y, c = Z	X ~ colorless, Y ~ very pale amber, Z = pale amber	1½-2½ 2.28	Yellow, yel-brn, lt orange	Ferrinatrite	Oxide zones ore-deposits in arid regions with voltaite, ferrinatrite	Soluble boiling H_2O	Dana II, 604-605
	Parallel to slightly inclined extinction, + elong	Colorless	2½ ~2.2	Cream, pinkish		Contact limestone deposits with wollastonite, garnet, Cu-sulfides		Am Min 16, 231-236
	b = Y, c ∧ Z = 64°	Colorless	4-4½ 2.13	Colorless, white, red-brown		Evaporite marine salts with kainite, boracite, pinnoite	Slightly sol H_2O. Soluble acids	Am Min 50, 1079-1083
{100} polysynthetic, very common	b = Z, c ∧ X = 47°	Colorless	3½ 2.25	Colorless, white		Low T, hydrothermal alteration serpentinite and ultramafic rocks	Soluble acids, effervescence	Min Rec 4, 18-20
	a = Z, b = Y, c = X	Colorless	~2½	Light gray, white, cream		Alkali pegmatites with natrolite, neptunite, steenstrupine		Am Min 60, 162
	b = Z, c ∧ X = +26°	Colorless	2½ 2.14-2.15	Colorless, white		Cauliflower nodules with inyoite, calcite, ulexite, aragonite, realgar		Am Min 53, 1815-1827
{010} contact rare	On {010}, Y & Z ~ ∥ rectangular diag of plates	X = yellow, Y = colorless, Z = sulfur-yellow	2½-3 2.08-2.17	Yellow, greenish-yel		Oxide zones sulfide ore deposits	Copiapite group	Dana II, 623-627
	Y ∥ elong c ∧ Z = small	X = colorless, Y = pale blue, Z = blue	Soft 2.27	Greenish-blue		Weathering of Cu-deposits		Am Min 40, 943
		Colorless	1.95	White, pale yellow		Arctic marine sediments		Dana II, 1105-1106
	b = Z, c ∧ X = 14°	Colorless	2-3 1.913-1.930	Colorless, white		Oxide zones As-bearing ore deposits	Slightly sol H_2O, readily soluble HCl	Dana II, 712-713

n_β Var.	Refractive Index	Mineral Name and Composition	$(n_\gamma - n_\alpha)$	2V angle Dispersion	Crystal System	Habit	Cleavage
1.531	n_β 1.531 n_α 1.530 n_γ 1.534	Buddingtonite $NH_4AlSi_3O_8 \cdot nH_2O$	0.004	~60°	Monoclinic β = 112.75°	Tiny crystals ~ feldspar, compact	{001} & {010} dist
1.532	n_β~1.532 n_α 1.528 n_γ 1.536	Aplowite $CoSO_4 \cdot 4H_2O$	0.008		Monoclinic β = 90.5°	Very fine-grained	
1.530 1.534	n_β 1.534-1.530 n_α 1.531-1.525 n_γ 1.538	Minyulite $KAl_2(PO_4)_2(OH,F) \cdot 4H_2O$	0.007	Large ~ 80°	Orthorhombic	Radiated acicular aggregates	{001} perfect
1.533	n_β 1.533 n_α 1.520 n_γ 1.584	Kieserite (Wathlingenite) $MgSO_4 \cdot H_2O$	0.064	55° r > v moderate	Monoclinic β = 116.1°	Granular coarse to fine	{110} & {111} perfect, {$\bar{1}$11}, {101}, {011} imperfect
1.534	n_β 1.534 n_α 1.519 n_γ 1.569	Tanellite $SrB_6O_{10} \cdot 4H_2O$	0.050	68° r > v weak	Monoclinic β = 114°	Prism (c), tabular {100} crystals	{100} perfect {001} distinct
1.534 1.535	n_β 1.534-1.535 n_α 1.520-1.523 n_γ 1.569-1.571	Hydroboracite $CaMgB_6O_{11} \cdot 6H_2O$	0.049	60° - 66° v > r perceptible	Monoclinic β = 102.8°	Acicular (c), radiated fibrous, prism, lamellar	{010} perfect {100} distinct
*1.526 1.543	n_β 1.526-1.543 n_α 1.520-1.535 n_γ 1.545-1.561	Wavellite $Al_3(OH)_3(PO_4)_2 \cdot 5H_2O$	0.025 to 0.027	~72° r > v weak	Orthorhombic	Radiated acicular crystals (c), spherical	{110} perfect {101} less perfect, {010} dist
1.531 1.537	n_β 1.531-1.537 n_α 1.527-1.533 n_γ 1.538-1.543	Albite $Na(AlSi_3)O_8$	0.010	77° - 84° v > r weak	Triclinic α = 94° β = 117° γ = 88°	Anhedral to euhedral laths	{001} perfect {010} dist, {110} poor
1.535	n_β 1.535 n_α n_γ	Nekoite $CaSi_2O_5 \cdot 2H_2O$	Very small	~70°	Triclinic α = 111.7° β = 86.2° γ = 103.9°	Acicular, fine fibrous	{100} distinct
1.535	n_β 1.535 n_α n_γ 1.585	Aluminocopiapite $(Mg,Fe^{2+})(Fe^{3+},Al)_4(SO_4)_6(OH)_2 \cdot 2H_2O$		Moderate	Triclinic α = 94° β = 102° γ = 98°	Minute scales as surface coatings	
1.536 1.537	n_β 1.536-1.537 n_α 1.526-1.528 n_γ 1.541-1.545	Rozenite $FeSO_4 \cdot 4H_2O$	0.015 to 0.008	~90°	Monoclinic β = 90.5°	Minute tabular crystals crusts	
1.536	n_β 1.536 n_α 1.531 n_γ 1.544	Pentahydroborite $CaB_2O_4 \cdot 5H_2O$	0.013	73°	Unknown	Small grains	None
*1.520 1.553	n_β 1.520-1.553 n_α 1.510-1.547 n_γ 1.543-1.582	Bobierrite (Haueteuillite) $Mg_3(PO_4)_2 \cdot 8H_2O$	0.033 to 0.035	71° - 53° v > r perceptible	Monoclinic β = 104°	Micro acicular, fibrous aggregate	{010} perfect
*1.525 1.547	n_β 1.547-1.525 n_α 1.541-1.513 n_γ 1.564-1.542	Voglite $Ca_2Cu(UO_2)(CO_3)_4 \cdot 6H_2O$	0.023 to 0.029	60° - 90° v > r	Triclinic or Monoclinic	Tiny rhombohedral scales, coatings	{010} perfect

* n_β variation exceeds 0.01

Twinning	Orientation	Color in thin section	Hardness Sp Gravity	Color hand sample	Alteration	Occurrence	Remarks	Reference
	b = Z, c ∧ Y = 19°, a ∧ X = 4°	Colorless	5½ 2.32-2.39	White, gray		Cavity lining in volcanics, hydro-thermal alter	Feldspar group	Am Min 49, 831-850
		Colorless, pink	3 2.33-2.36	Pink		Efflorescence with moorhouseite	Soluble in water	Can Min 8, 166-171
	a = Y, b = Z, c = X	Colorless	3½ 2.45-2.46	Colorless, white		Altered glauconitic phosphate deposits	Soluble warm HCl or HNO_3	Min Mag 26, 313
{001} rare	b = Y, c ∧ Z = +76°	Colorless	3½ 2.57	Colorless, white, gray		Marine evaporite beds with halite, other salts. Fumeroles	Slowly soluble H_2O	Dana II, 477-479
	b = Y, c ∧ Z = +5°, a ∧ X = +29°	Colorless	2½ 2.38-2.40	Colorless		Arid lake beds with other borates with realgar, adularia, analcime	Soluble HCl	Am Min 49, 1549-1568
	b = Y, c ∧ Z = 57°, c ∧ X = 33°	Colorless	2-3 2.17	Colorless, white		Evaporite borate deposits	Readily soluble acids	Dana II, 353-355
None	a = Y, b = X, c = Z	Colorless, pale tints X > Y > Z	3½-4 2.32-2.37	White, gray, yellow, green, brown		Open cavities or veins in Fe-ores, phosphate rocks. Pegmatites, hydro veins	Easily soluble acids	p. 85
Multiple on {010} very common. Numerous other laws	See figure page 352	Colorless	6 2.61-2.63	White	Sericite, kaolin, zeolite	Pegmatites, feldspath-oidal rocks, spilite, keratophyre, green-schists	Feldspar group	p. 352
Polysynthetic ∥ fibers	Extinction on {100} = 26°, - elong. X ∧ fib axis~26°, Z ~ [100]	Colorless	2.20-2.28	White		Limestone contact zone with wilkeite		Min Mag 31, 5-20
		Colorless, to yellowish Z > Y	2.16	Yellowish		Efflorescence, oxida-tion of pyrite		Am Min 52, 1220-1223
		Colorless	2.19-2.29	Colorless, white		Alteration of pyrite, efflorescence		Can Min 7, 751-763
		Colorless	2.5 2.00	Colorless		Limestone contact rocks with garnet, szaibelyite, magnetite	Fluor short λ U.V. violet	Am Min 47, 1482
	b = Y, c ∧ Z = 29°	Colorless	2-2½ 2.17-2.20	Colorless, white		Organic PO_4 (guano etc.)		Dana II, 753-754
Lamellar {100}	a = Z, b = X, c = Y	X = Y = deep bluish green, Z = yellow		Emerald green-grass green		Secondary U-mineral. Alter uraninite with liebigite	Soluble acids with efferves-cence	USGS Bull. 1064, 126-128

n_β Var.	Refractive Index	Mineral Name and Composition	$(n_\gamma - n_\alpha)$	2V angle Dispersion	Crystal System	Habit	Cleavage
1.537 1.541	n_β 1.537-1.541 n_α 1.533-1.537 n_γ 1.543-1.547	Oligoclase $(NaSi,CaAl)Si_2AlO_8$	0.010 to 0.009	84° - 90° v > r weak	Triclinic α = 94° β = 116° γ = 88°	Anhedral to euhedral laths	{001} perfect {010} dist, {110} poor
1.539	n_β 1.539 n_α 1.536 n_γ 1.603	Volkovskite $(Ca,Sr)B_6O_{10}\cdot 3H_2O$	0.067	24°	Monoclinic β = 119.1°	Platy (elong or rhomb)	{010} perfect {001} dist
*1.530 1.548	n_β 1.530-1.548 n_α 1.523-1.544 n_γ 1.582-1.572	Botryogen (Quetenite) (Rubrite) $MgFe^{3+}(SO_4)_2(OH)\cdot 7H_2O$	0.059 to 0.028	42° - 41° r > v strong	Monoclinic β = 100°	Botryoidal, radiated prism (c)	{010} perfect {110} dist
1.54	n_β 1.54 n_α 1.529 n_γ 1.556	Paucilithionite (Trilithionite) $K_2(Li_3Al_3)(Si_6Al_2)O_{20}(OH,F)_4$	0.027	~25°	Monoclinic β ~ 100°	Micaceous, scaly	{001} perf
*1.532 1.550	n_β 1.532-1.550 n_α 1.509-1.546 n_γ 1.577-1.597	Copiapite $(Fe,Mg)Fe_4^{3+}(SO_4)_6(OH)_2\cdot 20H_2O$	0.068 to 0.051	73° - 52° r > v	Triclinic α = 94° β = 101.5° γ = 99°	Tabular {010}, scaly aggregate	{010} perfect {$\bar{1}$01} imperf
1.541 1.544	n_β 1.541-1.544 n_α 1.536-1.542 n_γ 1.542-1.548	Epididymite $NaBeSi_3O_7(OH)$	0.006	30° - 32°	Orthorhombic	Micaceous, tabular crystals, radiated fibrous	{001} perfect {100} dist
1.541 1.544	n_β 1.541-1.544 n_α 1.536-1.542 n_γ 1.542-1.548	Eudidymite $NaBeSi_3O_7(OH)$	0.006	30° - 32°	Monoclinic β = 103.75°	Tabular crystals, radiated fibrous	{001} perfect {100} dist
1.543	n_β 1.543 n_α 1.534 n_γ 1.558	Gordonite $MgAl_2(PO_4)_2(OH)_2\cdot 8H_2O$	0.024	73° r > v moderate	Triclinic α = 107.5° β = 111° γ = 72.4°	Minute prism-platy sheaf-like aggregate	{010} perfect {100} fair, {001} poor
1.544	n_β 1.544 n_α 1.540 n_γ 1.559	Farringtonite $Mg_3(PO_4)_2$	0.010 to 0.014	~55°	Monoclinic β = 94°	Massive	{100} & {010} distinct
1.544	n_β 1.544 n_α 1.542 n_γ 1.551	Cavansite $Ca(VO)(Si_4O_{10})\cdot 4H_2O$	0.009	52° v > r extreme	Orthorhombic	Prism (c), rosettes	{010} dist
1.545	n_β 1.545 n_α 1.532 n_γ 1.572	Halurgite $Mg_2B_8O_{14}\cdot 5H_2O$	0.040	70°	Orthorhombic ?	Fine-grained masses, platy	
1.546	n_β 1.546 n_α 1.539 n_γ 1.551	Brushite (Epiglaubite) (Stofferite) $CaHPO_4\cdot 2H_2O$	0.012	86° r > v, crossed perceptible	Monoclinic β = 117.5°	Acicular, tabular {010}, earthy	{010} & {001} perfect
1.546	n_β 1.546 n_α 1.544 n_γ 1.549	Chkalovite $Na_2BeSi_2O_6$	0.005	78°	Orthorhombic	Massive	{100} imperf
*1.524 1.574	n_β 1.524-1.574 n_α 1.522-1.560 n_γ 1.527-1.578	Cordierite $Mg_2Al_3(Si_5Al)O_{18}$	0.005 to 0.018	90°-75° r > v weak	Orthorhombic (Pseudohex.)	Subhedral xls	{010} dist., {100} & {001} poor

* n_β variation exceeds 0.01

Twinning	Orientation	Color in thin section	Hardness Sp Gravity	Color hand sample	Alteration	Occurrence	Remarks	Reference
Multiple on {010} very common. Numerous other laws	See page 352	Colorless	6–6½ 2.63–2.66	White, pink, etc.	Sericite, kaolin, saussurite, zeolite	Granite, rhyolite, amphibolite, schist, gneiss	Plagioclase feldspar group	p. 352
	b = Y, a ∧ Z = 31°	Colorless	2.29–2.39	Colorless		Borate evaporites with anhydrite, sylvite, hilgardite, boracite		Am Min 51, 1550
	b = X, c ∧ Z = −12°	X = colorless, Y = brown, Z = yellow	2–2½ 2.11–2.14	Lt to dk red-orange		Oxide zones pyrite deposits, arid regions	Soluble HCl	Dana II, 617–618
	b = Y, c ∧ X ~ 0°	Colorless, pale pink or violet Z = Y > X	~3 2.8–3.3	Colorless, pink, violet		Pegmatites with topaz, tourmaline etc.	Mica (Lepidolite) group	W&W II, 370–372
{010} rare	b ~ X	X = yellow, Y = colorless, Z = yellow-green	2½–3 2.08–2.22	Golden yellow, olive green		Oxidation of pyrite and other sulfides		Dana II 623–627
{001} common, trillings	Parallel ext	Colorless	6–7 2.55–2.58	White, colorless, yellow, blue		Nepheline-syenite pegmatites	Dimorph eudidymite	Am Min 55, 1541–1549
{001} lamellar	Inclined extinction	Colorless	6–7 2.55–2.58	White, yellow, blue, violet		Nepheline syenite, alkali pegmatites	Dimorph epididymite	Vlasov II, 135–140
	X ~ ⊥{010}	Colorless	3½ 2.22–2.23	Colorless, white, pink, lt green		Phosphate nodules (Variscite) in clays	Soluble acids	Dana II, 975–976
	c ∧ Z ~ 16°–17°	Colorless	2.76	White, yellow		Meteorites		Am Min 46, 1513
	a = Y, b = X, c = Z	X = Z = colorless, Y = blue	3–4 2.21–2.33	Greenish-blue		Cavities and amygdules in basalt, breccia, with zeolites	Dimorph pentagonite	Am Min 58, 405–411
	Z ∥ long diagonal rhomb plates. X ∥ short diagonal	Colorless	2.5–3 2.19	White		Evaporite salts		Am Min 47, 1217–1218
	b = Z, c ∧ X = +30°	Colorless	2½ 2.26–2.33	White, yellowish		Efflorescence on guano, phosphorite		Dana II, 704–706
		Colorless	~6 2.66–2.70	White		Syenite pegmatites with sodalite	Soluble acids	Vlasov II, 116–117
{110} or {130} polysynthetic	a = Y, b = Z, c = X	Colorless, pale blue Z > Y > X	7–7½ 2.53–2.78	Gray, blue, indigo	Pinite (chlorite, muscovite, biotite)	Pelitic hornfels. High grade regional met. with andalusite, garnet		p. 167

n_β Var.	Refractive Index	Mineral Name and Composition	$(n_\gamma - n_\alpha)$	2V angle Dispersion	Crystal System	Habit	Cleavage
1.550	n_β 1.550 n_α 1.525 n_γ 1.576	Albrittonite $CoCl_2 \cdot 6H_2O$	0.051	53°	Monoclinic β = 97°	Tabular {100}, "bipyramidal" crystals	{010} dist {110} indist
1.551	n_β 1.551 n_α 1.542 n_γ 1.587	Zincobotryogen $(Zn,Mg,Mn^{2+})Fe^{3+}(SO_4)_2(OH) \cdot 7H_2O$	0.045	54° r ~ v	Monoclinic β = 100.8°	Prism crystals, rad crystalline aggregate	{010} perfect {110} dist
1.551	n_β 1.551 n_α 1.549 n_γ 1.621	Veatchite-A $Sr_2B_{11}O_{16}(OH)_5 \cdot H_2O$	0.072	25° v > r strong	Triclinic α = 90° β = 91° γ = 92°	Colliform nodules. Platy crystals {100}	{100} perf, {011} & {01$\bar{1}$} good
1.552	n_β 1.552 n_α 1.540 n_γ 1.570	Rauenthalite $Ca_3(AsO_4)_2 \cdot 10H_2O$	0.030	~85°	Monoclinic or Triclinic	Spherulites, tiny crystals	None ?
1.553	n_β 1.553 n_α 1.489 n_γ 1.649	Whewellite (Thierschite) $CaC_2O_4 \cdot H_2O$	0.160	~80°	Monoclinic β = 107.1°	Coarse xline, short prism crystals (c)	{$\bar{1}$01} perfect {001} & {010} dist {110} indistinct
1.553	n_β 1.553 n_α 1.540 n_γ 1.570	Lemoynite $(Na,Ca)_3Zr_2Si_8O_{22} \cdot 8H_2O$	0.030	80° - 83°	Monoclinic β = 105.3°	Prismatic crystals	{001} & {010} distinct
1.553	n_β 1.553 n_α 1.550 n_γ 1.620	P-Veatchite $Sr_2B_{11}O_{16}(OH)_5 \cdot H_2O$	~0.070	25°	Monoclinic β = 119.1°	Platy {100}, prism-fibrous (c), divergent	{010} perfect {001} imperf
1.553	n_β 1.553 n_α 1.550 n_γ 1.559	Charoite $(K,Ba,Sr)(Ca,Na)_2Si_4O_{10}(OH,F) \cdot H_2O$	0.009	28° - 30°	Monoclinic		{110} fair at 124°, {001} fair
1.553	n_β 1.553 n_α 1.551 n_γ 1.62	Veatchite $Sr_2B_{11}O_{16}(OH)_5 \cdot H_2O$	~0.07	37°	Monoclinic β = 92°	Platy {100}, prism-fibrous (c), divergent	{100} perfect
1.548 1.558	n_β 1.548-1.558 n_α 1.544-1.555 n_γ 1.551-1.563	Andesine $(NaSi,CaAl)Si_2AlO_8$	0.008	90° - 77° r > v weak	Triclinic α = 94° β = 116° γ = 88°	Anhedral to euhedral laths	{001} perfect {010} dist, {110} poor
1.553 1.555	n_β 1.553-1.555 n_α 1.534-1.533 n_γ 1.638-1.635	Rhomboclase $HFe^{3+}(SO_4)_2 \cdot 4H_2O$	0.104 to 0.102	27°	Orthorhombic	Thin plates, radiated bladed, stalactitic	{001} perfect {110} dist
1.554	n_β 1.554 n_α 1.534 n_γ 1.586	Zincocopiapite $ZnFe_4(SO_4)_6(OH)_2 \cdot 18H_2O$	0.052	78° r > v	Triclinic α = 93.8° β=101.5° γ = 99.4°	Massive crystal aggregates	{010} perfect {$\bar{1}$01} imperf
1.554	n_β 1.554 n_α 1.545 n_γ 1.565	Lacroixite $NaAl(PO_4)(OH,F)$	0.020	~90° Slight	Monoclinic β = 115.5°	Fragmentary crystals	{111} & {1$\bar{1}$1} indistinct
1.555	n_β 1.555 n_α 1.532 n_γ 1.591	Amarillite $NaFe^{3+}(SO_4)_2 \cdot 6H_2O$	0.059	Large v > r strong	Monoclinic β = 95.5°	Equant crystals	{110} good

Twinning	Orientation	Color in thin section	Hardness Sp Gravity	Color hand sample	Alteration	Occurrence	Remarks	Reference
	b = Y, a ∧ Z = -3°	X = pink, Y = pale red Z = reddish-violet	1½ 1.90-1.91	Magenta, reddish-violet		Oxide zone Co-deposits with erythrite, annabergite, zaratite		Am Min 63, 410-412
	b = X, elongation	Orange-red. Strong pleochroism	~2.5 2.201	Bright orange -red		Oxide zones Zn-deposits with pickeringite	Botryogen group	Am Min 49, 1776-1777
Polysynthetic {100}	b = Z, c = X	Colorless	2.73-2.77	Colorless		Borate evaporites with colemanite, hydroboracite, ulexite		Am Min 64, 362
	X and Z lie in plane of flattening. Y ∧ elong = 5°-10°	Colorless	2.36	Colorless, white		Oxide zones ore deposits		Am Min 50, 805-806
Very common, pseudo-ortho	b = X	Colorless	2½-3 2.21-2.23	Colorless, white, yellowish		Lignite coals, calcite concretions		Am Min 53, 455-463
	b = Y, a ∧ Z = +5°, c ∧ X = -10°, + elongation	Colorless	~4 2.26-2.29	White, yellowish		Pegmatites in alkaline rocks		Can Min 9, (Pt 5), 585
None	b = Y, a ∧ Z = 30°	Colorless	2 2.60-2.65	White		Borate deposits	Dimorph veatchite	Am Min 45, 1221
	b = X, c ∧ Z = 5°	Colorless	2.54	Lilac to violet		Limestone contact zones with canasite, tinaksite		Am Min 63, 1282
None	b = Z, a ∧ Y = 2°, c ~ X	Colorless	2 2.62	Colorless, white		Cross-fiber veinlets in borate deposits with howlite, colemanite	Dimorph paraveatchite	Am Min 45, 1221-1229
Multiple on {010} very common. Numerous other laws	See figure page 352	Colorless	6-6½ 2.66-2.68	White, gray, etc	Sericite, kaolin, saussurite, calcite, zeolite	Granodiorite, andesite, amphibolite	Plagioclase feldspar group	p. 352
	a = Y, b = Z, c = X	Colorless	2 2.23	Colorless, white, pale tints		Incrustations in mines with roemerite and other sulfates	Easily soluble acids	Dana II, 436-437
		Yellowish-green. Strong pleochroism	~2 2.181	Yellowish green		Oxide zones Pb-Zn-deposits with copiapite, melanterite, coquimbite		Am Min 49, 1777
	Large extinct angles	Colorless	4½ 3.126	Pale yellow, pale green		Druses in granite with apatite, childrenite, tourmaline	Easily soluble HCl, H_2SO_4	Dana II, 783
	b = Y c ∧ X = +39°	Colorless	2½-3 2.19	Pale yellow		Veinlets of soluble minerals, with coquimbite	Soluble in H_2O. Astringent taste	Dana II, 468-469

n_β Var.	Refractive Index	Mineral Name and Composition	$(n_\gamma - n_\alpha)$	2V angle Dispersion	Crystal System	Habit	Cleavage
1.555	n_β 1.555 n_α 1.545-1.537 n_γ 1.680	Studtite $UO_4 \cdot 4H_2O$	0.135 to 0.153	Moderate-Large	Monoclinic β = 93.9°	Flexible fibrous (c), radiated aggregate	
1.555	n_β 1.555 n_α 1.551 n_γ 1.562	Vauxite $Fe^{2+}Al_2(PO_4)_2(OH)_2 \cdot 6H_2O$	0.011	32° r > v distinct	Triclinic α = 98° β = 92° γ = 108.1°	Minute tabular {010} & elongated (c), radiated	None
1.555	n_β 1.555 n_α 1.552-1.571 n_γ 1.565-1.576	Woodwardite $Cu_4Al_2SO_4(OH)_{12} \cdot 4\text{-}6H_2O$(?)	0.013 to 0.005	~58°		Fine-fibrous crystals, spherulitic	
1.55	n_β 1.55 n_α 1.55 n_γ 1.57	Bassanite $2CaSO_4 \cdot H_2O$	0.02	Small	Monoclinic β = 90.6°	Parallel microscopic needles on c	
1.557 1.558	n_β 1.557-1.558 n_α 1.550-1.553 n_γ 1.566-1.567	Lithiophosphate Li_3PO_4	0.016 to 0.014	68° - 80°	Orthorhombic	Prism (b)	{010} perfect {110} dist
1.559	n_β 1.559 n_α 1.536 n_γ 1.697	Zellerite $Ca(UO_2)(CO_3)_2 \cdot 5H_2O$	0.161	30° - 40° v > r weak	Orthorhombic	Fibrous-acicular, clumps	None
1.559	n_β 1.559 n_α 1.552 n_γ 1.572	Paravauxite $Fe^{2+}Al(PO_4)_2(OH)_2 \cdot 10H_2O$	0.020	72°	Triclinic α = 107.3° β = 111.4° γ = 72.5°	Tabular or prism, parallel or radiated	{010} perfect
1.558 1.561	n_β 1.558-1.561 n_α 1.551 n_γ 1.582-1.585	Metavariscite $AlPO_4 \cdot 2H_2O$	0.031	55° v > r percepti-ble	Monoclinic β ~90°	Minute tabular or prism crystals, granular	{010}
1.560	n_β 1.560 n_α 1.541 n_γ 1.567	Fenaksite $(K,Na,Ca)_4(Fe^{2+},Fe^{3+},Mg,Mn)_2(Si_4O_{10})_2(OH,F)$	0.026	84°	Triclinic α = 99° β = 114.8° γ = 105°	Granular	{100} & {010} distinct (∧122°)
1.561	n_β 1.561 n_α 1.494 n_γ 1.692	Humboldtine $FeC_2O_4 \cdot 2H_2O$	0.198	Large ~ 75°	Orthorhombic	Fibrous, fine-granular, earthy	{110} perfect {100} & {010} imperfect
1.561	n_β 1.561 n_α 1.550 n_γ 1.577	Metavauxite $Fe^{2+}Al_2(PO_4)_2(OH)_2 \cdot 8H_2O$	0.027	Large ~ 80°	Monoclinic β = 98°	Long prism, acicular (c), radiated aggregate	None
*1.558 1.569	n_β 1.558-1.569 n_α 1.555-1.565 n_γ 1.563-1.574	Labradorite $(CaAl,NaSi)Si_2AlO_8$	0.008	77° - 87° r > v weak	Triclinic α = 94° β = 116° γ = 89°	Anhedral to euhedral laths	{001} perfect {010} dist, {110} poor
1.560 1.566	n_β 1.560-1.566 n_α 1.558-1.564 n_γ 1.563-1.571	Dickite $Al_2Si_2O_5(OH)_4$	0.004 to 0.008	50° - 80° v > r weak	Monoclinic β = 90° - 104°	Very fine-grain, earthy scaly	{001} perfect
1.564	n_β 1.564 n_α 1.554 n_γ 1.595	Baricite $(Mg,Fe)_3(PO_4)_2 \cdot 8H_2O$	0.031	59° v > r weak	Monoclinic β = 104.9°	Platy	{010} perfect

* n_β variation exceeds 0.01

Twinning	Orientation	Color in thin section	Hardness Sp Gravity	Color hand sample	Alteration	Occurrence	Remarks	Reference
	c = Z, parallel ext, + elong.	Colorless	3.58-3.64	Pale yellow to yellow		Secondary U-mineral with uranophane, rutherfordite	Does not fluor U.V.	Am Min 59, 166-171
{010} normal	Z ~ ⊥{010}	X = Z = colorless, Y = blue	3½ 2.40-2.41	Sky-blue to dark blue		Tin deposits with paravauxite and wavellite		Am Min 53, 1026-1028
	∥ extinction	Colorless, pale greenish-blue	2.38	Greenish blue		High-T veins?		Dana II, 580
	c ~ Z	Colorless	Soft 2.55-2.76	White	Alteration product of gypsum	Gas deposition in cavities, fumeroles		Dana II, 476
	a = Y, b = Z, c = X	Colorless	4 2.48	Colorless, white, pink		Granite pegmatites. Alteration of montebrasite	Soluble strong acids	Am Min 54, 1467-1469
	c = Z	X = Y = colorless, Z = pale yellow	~2 3.24-3.25	Pale yellow		Oxidation of uraninite-coffinite with gypsum and limonite	Fluor U.V.-weak mottled green	Am Min 51, 1567-1578
		Colorless	3 2.36-2.38	Colorless, pale green		Tin veins with vauxite, metavauxite, wavellite		Am Min 47, 1-8
{102} contact	b = Y	Colorless	~3½ 2.53-2.54	Pale green		Microcrystals in cavities in massive variscite	Soluble alkalies	Dana II, 767-769
	b = Z, a ~ X	Colorless	5-5½ 2.74	Light rose		Pegmatites		Am Min 45, 252-253
	a = X, b = Y, c = Z	X = very pale yellow-green, Y = pale green-yellow, Z = deep yellow	1½-2 2.32-2.33	Yellow, brownish-yellow		Coal deposits	Soluble acids	Dana II, 1102-1103
	b = X, c ∧ Z ~ +17	Colorless	3 2.35	Colorless, white, lt green		Tin-bearing veins with vauxite, paravauxite, wavellite		Dana II, 971-972
Multiple on {010} very common. Numerous other laws	See figure page 352	Colorless	6-6½ 2.69-2.71	Gray, black, etc.	Sericite, kaolin, saussurite, calcite, zeolite	Gabbro, anorthosite, basalt	Plagioclase feldspar group	p. 352
{001} rare?	b = Z, c ∧ X = +7° to + 13°, a ∧ Y = +14° to + 20°	Colorless, ~opaque	2-3 2.5-2.7	White, Fe-stained	Diaspore, gibbsite	Hydrothermal alter. Al-silicates with quartz, chalcedony, sulfides	Hexagonal plates (electron microscope)	p. 254
	b = X, c ∧ Z = -32°	Colorless	1½-2 2.42-2.44	Pale blue, colorless		Fissures in siderite Fe-formations	Mg-vivianite	Am Min 16, 1053

n_β Var.	Refractive Index	Mineral Name and Composition	$(n_\gamma - n_\alpha)$	2V angle Dispersion	Crystal System	Habit	Cleavage
1.566	n_β 1.566 n_α 1.547 n_γ 1.594	Quenstedtite $Fe_2(SO_4)_3 \cdot 10H_2O$	0.047	70° v > r strong, horizontal	Triclinic α = 94.2° β = 101.8° γ = 96.3°	Minute tabular	{010} perfect {100} distinct
1.566	n_β 1.566 n_α 1.562 n_γ 1.587	Senegalite $Al_2(PO_4)(OH)_3 \cdot H_2O$	0.025	48° - 53° r > v weak	Orthorhombic	Blocky crystals	{100} imperf
1.567	n_β 1.567 n_α 1.562 n_γ 1.614	Cuprohydromagnesite $(Cu,Mg)_5(CO_3)_4(OH)_2 \cdot 4H_2O$	0.052	36°-38°	Monoclinic β = 115.4°	Bladed xls (c)	{010} perf
1.565 1.569	n_β 1.565-1.569 n_α 1.550-1.560 n_γ 1.583-1.584	Görgeyite (Mikheevite) $K_2Ca_5(SO_4)_6 \cdot H_2O$	0.033 to 0.024	85° - 75°	Monoclinic β = 113.3°	Thin tabular	{100} dist
1.569	n_β 1.569 n_α 1.563 n_γ 1.577	Elpidite $Na_2ZrSi_6O_{15} \cdot 3H_2O$	0.014	76° - 89° v > r	Orthorhombic	Thin acicular, splintery, radiated fibrous	{hk0} perfect
1.569 1.572	n_β 1.569-1.572 n_α 1.565-1.567 n_γ 1.574-1.577	Bytownite $(CaAl,NaSi)Si_2AlO_8$	0.009	87° - 90° r > v weak	Triclinic α = 94° β = 116° γ = 89°	Anhedral to euhedral laths	{001} perfect {010} dist, {110} poor
1.571	n_β 1.571 n_α 1.563 n_γ 1.596	Hoernesite $Mg_3(AsO_4)_2 \cdot 8H_2O$	0.033	60°	Monoclinic β = 104.9°	Bladed elongated (c), flat {010}	{010} perfect {100} poor
1.571	n_β 1.571 n_α 1.566 n_γ 1.578	Picropharmacolite $H_2Ca_4Mg(AsO_4)_4 \cdot 12H_2O$	0.012	~50°	Monoclinic β = 140°	Radiated acicular (c) nodules	{100} & {010} perfect
1.571	n_β 1.571 n_α 1.5695 n_γ 1.5775	Banalsite $Na_2BaAl_4Si_4O_{16}$	0.008	41°	Orthorhombic	Massive	{110} & {001} distinct
1.571	n_β 1.571 n_α 1.570 n_γ 1.575	Tobermorite $Ca_5Si_6O_{17} \cdot 5H_2O$	0.005	Small ~50°	Orthorhombic	Fibrous, platy {001}, granular, massive	{001} & {100} distinct
*1.566 1.579	n_β 1.566-1.579 n_α 1.561-1.567 n_γ 1.587-1.593	Norbergite $Mg_2SiO_4 \cdot Mg(OH,F)_2$	0.026 to 0.027	44° - 50° r > v weak	Orthorhombic	Massive, granular, tabular crystals	{001} poor
1.572	n_β 1.5719 n_α 1.5678 n_γ 1.5824	Wagnerite $(Mg,Fe,Mn,Ca)_2PO_4F$	0.015	28° r > v weak, inclined	Monoclinic β = 108.1°	Prism crystals (c), tabular {100}, massive	{100} & {120} imperfect {001} traces
*1.568 1.580	n_β 1.568-1.580 n_α 1.568-1.580 n_γ 1.587-1.600	Gibbsite (Hydrargillite) $Al(OH)_3$	~0.019	0° - 40° r > v or v > r strong	Monoclinic β = 94.6°	Tiny tabular crystals {001}, compact aggregates	{001} perfect
1.573 1.576	n_β 1.573-1.576 n_α 1.573-1.576 n_γ 1.594-1.596	Preobrazhenskite $Mg_3B_{11}O_{15}(OH)_9$	0.020	Small	Orthorhombic (Pseudohex)	Massive, fine-grained, nodular	

* n_β variation exceeds 0.01

Twinning	Orientation	Color in thin section	Hardness Sp Gravity	Color hand sample	Alteration	Occurrence	Remarks	Reference
{010} common	Ext ∧ c = 30° (on 010)	Colorless, pale rose	2½ 2.14-2.15	Pale violet, red-violet		Oxide zones ore deposits with coquimbite & copiapite	Soluble H_2O	Dana II, 535-536
	a = Z, b = X, c = Y	Colorless	5½ 2.551	Colorless, pale yellow		Oxidation zone Fe-deposit with turquoise, crandallite, augelite		Am Min 62, 595-596
Poor	b = Z, c ∧ X = 48°	X = colorless, Y = Z = pale blue	~3½ 2.54	Blue, pale blue		Supergene alter Cu-sulfides in magnesite-brucite-serpentine rock	Fluor short-λ UV = lt blue	Am Min 64, 886-889
	b = Y, c ∧ Z = 11° - 17°	Colorless	3½ 2.90-2.95	Colorless, yellowish		Evaporite sediments with common salts		Am Min 39, 403
	c = X	Colorless	5 2.52-2.62	Colorless, yellow-white, red		Albitized nepheline syenites		Vlasov II, 365-367
Multiple on {010} very common. Numerous other laws	See page 352	Colorless	6-6½ 2.71-2.74	Gray, white, etc.	Sericite, kaolin, saussurite, calcite, zeolite	Gabbro, anorthosite, meteorites, basalt	Plagioclase feldspar group	p. 352
	b = X, c ∧ Z = 31°	Colorless	1 2.57	White		Metamorphosed limestone	Soluble acids	Dana II, 755-756
	c ∧ Z' = 8° (on cleavage)	Colorless	2.60-2.62	Colorless, white		Oxide zones sulfide deposits		Am Min 47, 1222
	a = Y, b = Z, c = X	Colorless	6 3.065	White		Veins in Mn ores	Decomposed by HCl Feldspathoid group	Win & Win II, 260
	a = Z, b = Y, c = X	Colorless	2½ 2.42-2.44	White, pinkish		Limestone contact zones with wilkeite, monticellite, idocrase	Soluble HCl	Min Mag 31, 361-370
	a = X, b = Z, c = Y	Colorless to yellow or brown X > Y ~ Z	6-6½ 3.18-3.19	Tan, yellow, orange, brown	Serpentine, Mg-chlorite	Limestone or dolomite contact zones with Ca-Mg-silicates. Ore veins	Humite group	p. 142
	b = Y, c ∧ Z = +21°	Colorless	5-5½ 3.15	Yellow, greenish, brick red		Quartz veins and pegmatites with magnesite, lazulite, chlorite. K-lavas	Fe-Wagnerite = talktriplite. Soluble acids	Dana II, 845-848
{001} common, {100}, {110}	b = X, c ∧ Z = +21°, a ∧ Y = +25½°	Colorless	2½-3½ 2.38-2.42	White, brownish, green, pink	Boehmite	Intense chemical weathering of Al-rocks with diaspore, boehmite, Fe-oxides	Trimorph. bayerite, nordstrandite	p. 42
		Colorless	4½-5	Colorless, yellow, dk gray	Inyoite	Evaporite beds with halite, polyhalite, kaliborite, boracite, inyoite		Am Min 55, 1071

Biaxial +

n_β Var.	Refractive Index	Mineral Name and Composition	$(n_\gamma - n_\alpha)$	2V angle Dispersion	Crystal System	Habit	Cleavage
1.575	n_β 1.575 n_α 1.543 n_γ 1.634	Metasideronatrite $Na_4Fe_2(OH)_2(SO_4)_4 \cdot 3H_2O$	0.091	60° r > v strong	Orthorhombic	Fibrous aggregate, prism crystals	{100} & {010} perfect, {001} distinct
1.575	n_β 1.575 n_α 1.558 n_γ 1.620	Cuprocopiapite $CuFe_4(SO_4)_6(OH)_2 \cdot 20H_2O$	0.062	63° r > v strong	Triclinic α = 94° β = 101.5° γ = 99.5°	Tabular, scaly aggregates, granular	{010} perf {$\bar{1}$01} imperf
1.576	n_β 1.576 n_α 1.570 n_γ 1.614	Anhydrite $CaSO_4$	0.044	~43° v > r strong	Orthorhombic	Fine-course granular, fibrous-blocky crystals	Pseudocubic {010}, {100}, {001}
1.576	n_β 1.576 n_α 1.558 n_γ 1.593	Xiangjiangite $(Fe,Al)(UO_2)_4(PO_4)_2(SO_4)_2(OH) \cdot 22H_2O$	0.035	~90°	Orthorhombic ?	Earthy	Perfect basal ?
1.576	n_β~1.576 n_α~1.570 n_γ~1.630	Gunningite $ZnSO_4 \cdot H_2O$	0.060	~40°	Monoclinic β = 116°	Efflorescence, micro-crystalline	
1.576	n_β 1.576 n_α 1.573 n_γ 1.579	Ilmajokite $(Na,Ba,RE)_{10}Ti_5(Si,Al)_{14}O_{22}(OH)_{44} \cdot nH_2O$	0.006	~90°	Monoclinic?	Granular, crusts, radiated fibrous	{110} and pinacoidal at 72°
1.576	n_β 1.5759 n_α 1.5736 n_γ 1.5877	Augelite (Amphitalite) $Al_2PO_4(OH)_3$		48° - 51° No disp.	Monoclinic β = 112°	Tabular to acicular, massive	{110} perfect, {$\bar{2}$01}, {001}, {$\bar{1}$01} dist
1.577	n_β 1.577 n_α 1.554-1.559 n_γ 1.618-1.601	Bonattite $CuSO_4 \cdot 3H_2O$	0.064 to 0.056	75°	Monoclinic β = 97°	Concretions, micro crystals	{010} dist
1.578	n_β 1.578 n_α 1.575 n_γ 1.584	Nifontovite $CaB_2O_4 \cdot 3H_2O$	0.009	76° r > v strong	Monoclinic or Triclinic	Massive	‖ elong poor
1.579	n_β 1.579 n_α 1.567 n_γ 1.581	Agrellite $NaCa_2Si_4O_{10}F$	0.014	47°	Triclinic α = 90.1° β = 116.8° γ = 94.1°	Prism crystals (c), flattened {010} or {110}	{110} & {1$\bar{1}$0} perfect, {010} poor
1.580 1.581	n_β 1.580-1.581 n_α 1.580 n_γ 1.596-1.631	Nordstrandite $Al(OH)_3$	0.016 to 0.051	Small ~18°	Triclinic α = 93° β = 110.4° γ = 90.5°	Tabular {001} xls Rhombic outline	{001} perfect
1.581 1.586	n_β 1.581-1.586 n_α 1.567-1.572 n_γ 1.638-1.640	Kornelite $Fe_2(SO_4)_3 \cdot 7H_2O$	0.071 to 0.068	49° - 62° r > v perceptible	Monoclinic β = 97°	Radiated fibrous acicular (c), crusts, globular	{010} dist
1.579 1.584	n_β 1.579-1.584 n_α 1.572-1.576 n_γ 1.589-1.600	Cookeite $LiAl_4Si_3AlO_{10}(OH)_8$	0.017 to 0.024	0° - 80°	Monoclinic β = 99°	Micaceous, scales in spherulites	{001} perfect
*1.56 1.60	n_β 1.56-1.60 n_α 1.56-1.59 n_γ 1.57-1.60	Penninite (Pennine) $(Mg,Al)_6(Si,Al)_4O_{10}(OH)_8$	0.000 to 0.005	Small v > r	Monoclinic β = 97.1°	Micaceous, hexagonal-tabular crystals, scaly	{001} perfect

* n_β variation exceeds 0.01

Twinning	Orientation	Color in thin section	Hardness Sp Gravity	Color hand sample	Alteration	Occurrence	Remarks	Reference
	a = X, b = Y, c = Z	X = colorless, Y = lt yellow Z = brownish-yellow	2½ 2.46	Golden yellow, straw yellow		Oxide zones porphyry copper deposits with other sulfates	Soluble boiling H_2O, dilute acids	Dana II, 603-604
{010} rare	X ~ ⊥{010}	X = Z = green, Y = yellow	2½-3 2.08-2.23	Yellow, greenish-yellow		Oxide zones Cu-deposits	Copiapite group	Dana II, 623-627
{011} simple or polysynthetic	a = Y, b = X, c = Z	Colorless	3½ 3.0	White, gray, blue, violet	Gypsum	Evaporite beds with gypsum, dolomite. Altered sulfides. Fumeroles	Soluble acids	p. 86
	Extinction is ‖ or symmetrical	Pale yellow, weak pleochroism	1-2 2.87-3.1	Yellow		Oxide zone U-deposits with sabugalite, variscite, pyrite	Sol dilute HCl or H_2SO_4. Not fluor U.V.	Am Min 64, 466
		Colorless	2½ 3.20-3.32	White		Oxide zones Zn-deposits. Efflorescence mines and dumps	Easily soluble H_2O	Am Min 47, 1218-1219
		Yellow	1 2.20	Bright yellow		Cavities in pegmatites		Am Min 58, 139-140
	b = X c ∧ Z = +34°, c ∧ Y = -56°	Colorless	4½-5 2.70	White, pale tints		Granite pegmatites, high T ore veins		Am Min 53, 1096
{100} common		Colorless	3 2.68	Blue, salmon		Oxide zones Cu-deposits	Soluble H_2O	Can Min 7, 245-252
	Z ∧ cleavage = 32°, + elong, inclined extinction	Colorless	3½ 2.36	Colorless		Scarn rocks with andradite-grossular and szaibelyite	Fluor U.V. long λ-violet. Anomalous inter colors	Am Min 47, 172
	Optic plane ~ ‖{010}	Colorless	5½ 2.89-2.90	White, lt gray lt green		Mafic-alkalic gneisses with albite, nepheline aegirine-augite		Can Min 14, 120-126
	c ∧ Z ~ 32°, inclined extinction, elongation	Colorless	3 2.43-2.51	Colorless, pink		Solution cavities in limestone	Trimorph gibbsite and bayerite	Nature 196, 264-266
{100} polysynthetic	b = Z, c ∧ X = 20°	Colorless, pink	2.31	Pale rose, violet		Oxide zones ore-deposits with voltaite, coquimbite	Slowly soluble H_2O, soluble acids	Dana II, 530-532
	X ~ ⊥{001}	Colorless	2½-3½ 2.58-2.69	White, pink, green, yellow, brown		Li-rich granite pegmatites	Chlorite group	Vlasov II, 32-35
{001}	b = Y, c ∧ Z = Small	Colorless, X~Y = pale green, Z = pale yellow	2-2½ 2.6-2.85	Emerald, olive, white		Mg-rich schists, serpentenite	Chlorite group. Anomalous interference colors	p. 289

Biaxial +

n_β Var.	Refractive Index	Mineral Name and Composition	$(n_\gamma - n_\alpha)$	2V angle Dispersion	Crystal System	Habit	Cleavage
1.583	n_β 1.583 n_α 1.583 n_γ 1.593	Xonotlite (Jurupaite)(Eakleite) $Ca_3Si_3O_8(OH)_2$	0.010	Very small	Monoclinic $\beta \sim 90°$	Acicular fibrous crystals, matted fibers, massive	One good
1.579 1.588	n_β 1.579–1.588 n_α 1.578–1.586 n_γ 1.583–1.593	Bavenite (Duplexite)(Pilinite) $Ca_4(Be,Al)_4Si_9(O,OH)_{28}$	0.005 to 0.007	22° – 60°	Orthorhombic	Radiated fibrous aggregates	{100} dist {001} indist
*1.57 1.60	n_β 1.57–1.60 n_α 1.57–1.59 n_γ 1.58–1.61	Clinochlore $(Mg,Al)_6(Si,Al)_4O_{10}(OH)_8$	0.002 to 0.009	15° – 45° v > r	Monoclinic $\beta = 97°$	Micaceous, scaly aggregate, massive	{001} perfect
1.585	n_β 1.585 n_α 1.580 n_γ 1.590	Whiteite $Ca(Fe^{2+},Mn^{2+})Mg_2Al_2(OH)_2(PO_4)_4 \cdot 8H_2O$	0.010	40° – 50°	Monoclinic $\beta = 113°$		
1.586	n_β 1.586 n_α 1.563 n_γ 1.619	Sigloite $(Fe^{2+},Fe^{3+})Al_2(PO_4)_2(O,OH) \cdot 8H_2O$	0.056	76° v > r strong	Triclinic $\alpha = 107°$ $\beta = 111.5°$ $\gamma = 69.5°$	Short prism (c) to thick tabular crystals	{010} perfect {001} dist
1.586	n_β 1.586 n_α 1.581 n_γ 1.596	Scholzite $CaZn_2(PO_4)_2 \cdot 2H_2O$	0.015	70°	Orthorhombic	Prism, platy crystals	{100} dist
1.586	n_β 1.586 n_α 1.584 n_γ 1.600	Eakerite $Ca_2SnAl_2Si_6O_{16}(OH)_6$	0.016	~35°	Monoclinic $\beta = 101.3°$	Prism (c)	None
1.586	n_β 1.586 n_α 1.585 n_γ 1.590	Kotschubeite (Kochubeite) $Mg_5(Cr,Al)(Si_3Al)O_{10}(OH)_8$	0.005	45° – 55° v > r	Monoclinic	Micaceous, scaly, pseudohexagonal crystals	{001} perfect
1.587	n_β 1.587 n_α 1.574 n_γ 1.599	Krauskopfite $BaSi_2O_5 \cdot 3H_2O$	0.025	~90° v > r distinct	Monoclinic $\beta = 94.5°$	Short prism, grains	{010} & {001} perfect
1.587	n_β 1.587 n_α 1.586 n_γ 1.593	Rhodophylite $Mg_{4.5}(Cr,Al)_{1.5}(Si_{2.5}Al_{1.5})O_{10}(OH)_8$	0.007	20° – 30° v > r	Monoclinic	Micaceous, pseudo-hexagonal crystals	{001} perfect
1.587 1.589	n_β 1.589–1.587 n_α 1.579–1.576 n_γ 1.609–1.606	Manganese-hoernesite $(Mn,Mg)_3(AsO_4)_2 \cdot 8H_2O$	0.030	65° – 70°	Monoclinic $\beta = 105.7°$	Massive	{010} perfect {100} poor
1.588	n_β 1.588 n_α 1.579 n_γ 1.604	Heidornite $Na_2Ca_3B_5O_8(SO_4)_2Cl(OH)_2$	0.025	63° – 77° v > r	Monoclinic $\beta = 93.5°$	Acicular	{001} perfect
1.593 1.583	n_β 1.593–1.583 n_α 1.587–1.579 n_γ 1.600–1.588	Celsian $Ba(Al_2Si_2)O_8$	0.013 to 0.009	83° – 90° Imperceptible	Monoclinic $\beta = 115°$		{001} perfect {010} dist, {110} poor
1.587 1.591	n_β 1.587–1.591 n_α 1.554–1.560 n_γ 1.628–1.631	Hambergite $Be_2(OH,F)BO_3$	0.08	87° r > v weak	Orthorhombic	Prism (c)	{010} perfect {100} dist

* n_β variation exceeds 0.01

Twinning	Orientation	Color in thin section	Hardness Sp Gravity	Color hand sample	Alteration	Occurrence	Remarks	Reference
	b = Z, c ~ X	Colorless	6-6½ 2.71	White, colorless, gray, pink		Veinlets in serpentine, limestone contact zones	Soluble HCl	Win & Win II, 455
{100}	a = Z, b = Y, c = X	Colorless	5½ 2.71-2.81	White, pinkish greenish		Granite pegmatites, skarns		Vlasov II, 143-147
{001}	b = Y, Z ∧ {001} ~ 0° -3°	Colorless, X~Y = pale green, Z = pale yellow	2-3 4.33-4.35	White, lt to dk green, olive		Schists etc. Hydrothermal alter biotite, amphibole in igneous rocks	Chlorite group. Anomalous interference colors	p. 289
	b = X, a = Y	Colorless	2.58	Tan		Veins in quartz and albite with other phosphates. In siderite	Al-jahnsite	Min Mag 42, 309
		Colorless	3 2.35-2.36	Pale yellow to lt brown		Vein cavities with wavellite, pseudomorphs after paravauxite		Am Min 47, 1-8
	a = X, b = Y, c = Z	Colorless	3-3½ 3.11-3.14	Colorless, white		Pegmatite with triplite, sphalerite. Crusts on gossan		Am Min 46, 1519
		Colorless	5½ 2.93	Colorless, white		Granite pegmatites		Min Rec 1, 92-96
	b = Y, Z ∧ ⊥{001} ~ 6°	Colorless, pale violet	2-3 2.76	Pale lavender (zoned)		With chromite	Chlorite group. Anomalous interference colors	p. 290
	b = X, c ∧ Z = 10°, a ∧ Y = 6°, ± elongation	Colorless	~4 3.10-3.14	Colorless, white		Veins in metamorphic rocks, with macdonaldite, opal, witherite		Am Min 50, 314-340
	b = Y, Z ∧ ⊥{001} ~ 2°	Colorless	2-3 2.60	Pale lavender (zoned)		Ultramafic igneous rocks with black spinel	Chlorite group	p. 290
	b = X, c ∧ Z = 31°	Colorless	1 2.64-2.76	White		Crusts on Mn ores		Am Min 39, 159
	b = Y, a ∧ Z = 23°	Colorless	4-5 2.70-2.75	Colorless		Cavities in anhydrite with soluble salts		Am Min 42, 120-121
Simple on {010} (Carlsbad) {001} (Manebach)	b = Y, c ∧ X = +3° to +5°, a ∧ Z = +28° to +30°	Colorless	6-6½ 3.96-4.10	White, yellowish		Hydrothermal veins with Mn-minerals	Feldspar group	p. 360
{110}	a = X, b = Y, c = Z	Colorless	7½ 2.37	Colorless, white, yellowish		Alkali pegmatites		Am Min 50, 85-95

n_β Var.	Refractive Index	Mineral Name and Composition	$(n_\gamma - n_\alpha)$	2V angle Dispersion	Crystal System	Habit	Cleavage
1.589	n_β 1.589 n_α 1.587 n_γ 1.594	Miserite $KCa_4Si_5O_{13}(OH)_3$	0.007	65°	Triclinic α = 96.4° β = 111.1° γ = 76.7°	Massive, fine fibrous, scaly	{100} perfect {010} imperf
1.590	n_β 1.590 n_α 1.565 n_γ 1.650	Ajoite $Cu_6Al_2Si_{10}O_{29}\cdot 5H_2O$	0.085	68°	Monoclinic	Massive, {010} plates, elongated on c	
1.590	n_β 1.590 n_α 1.587 n_γ 1.597	Bultfonteinite $Ca_2SiO_2(OH,F)_4$	0.010	~67° - 70°	Triclinic α = 92° β = 94.25° γ = 90.75°	Radiated acicular, granular	{010} & {100} distinct
1.592	n_β 1.592 n_α 1.569 n_γ 1.620	Lokkaite $(Y,Ca)_2(CO_3)_3\cdot 2H_2O$	0.051	~85°	Orthorhombic	Radiated fibrous aggregates (c)	
1.592	n_β 1.592 n_α 1.586 n_γ 1.614	Colemanite $Ca_2B_6O_{11}\cdot 5H_2O$	0.028	~55° v > r weak	Monoclinic β = 110.1°	Stubby prism crystals, massive, granular	{010} perfect {001} dist
1.592	n_β 1.592 n_α 1.588 n_γ 1.598	Althausite $Mg_2(PO_4)(OH,F,O)$	0.010	~70°	Orthorhombic	Cleavable masses	{001} perfect {101} dist
1.592	n_β 1.592 n_α 1.591-1.596 n_γ 1.626-1.625	Catapleiite $(Na_2,Ca)ZrSi_3O_9\cdot 2H_2O$	0.036 to 0.029	~25°	Monoclinic	Thin "hexagonal" plates	{110} perfect {111} imperf {001} parting
*1.58 1.61	n_β 1.58-1.61 n_α 1.58-1.60 n_γ 1.59-1.62	Sheridanite $(Mg,Al)_6(Si,Al)_4O_{10}(OH)_8$	0.005 to 0.012	15° - 45° v > r	Monoclinic β = 97.2°	Micaceous, scaly aggregate	{001} perfect
1.592 1.597	n_β 1.592-1.597 n_α 1.572-1.577 n_γ 1.612-1.616	Johannite (Gilpinite) (Peligotite) $Cu(UO_2)_2(SO_4)_2(OH)_2\cdot 6H_2O$	0.040	~90° v > r strong	Triclinic α = 110° β = 112° γ = 100.3°	Tabular {100}, prism (c), scaly, fibrous	{100} dist
1.595	n_β 1.595 n_α 1.562 n_γ 1.632	Szmikite $MnSO_4\cdot H_2O$	0.070	~90°	Monoclinic β = 115.7°	Stalactitic	Splintery
1.596	n_β 1.596 n_α 1.586-1.592 n_γ 1.598-1.606	Cuspidine (Custerite) $Ca_4Si_2O_7(F,OH)_2$	0.011 to 0.014	~76° r > v moderate	Monoclinic β = 110°	Fine granular, acicular	{001} & {110} distinct
1.594 1.600	n_β 1.594-1.600 n_α 1.593-1.599 n_γ 1.597-1.604	Coesite SiO_2	0.004 to 0.005	54° - 64°	Monoclinic β = 120°	Fine granular	None
1.598	n_β 1.598 n_α 1.584 n_γ 1.602	Millisite $(Na,K)CaAl_6(PO_4)_4(OH)_9\cdot 3H_2O$	0.018	moderate ~55°	Tetragonal?	Sperulites, crust, fine fibrous	{001} ?
1.597 1.60	n_β 1.60-1.597 n_α 1.595-1.592 n_γ 1.628-1.630	Cebollite $Ca_4Al_2Si_3O_{14}(OH)_2$	0.033 to 0.038	58°	Orthorhombic	Fibrous	

* n_β variation exceeds 0.01

Twinning	Orientation	Color in thin section	Hardness Sp Gravity	Color hand sample	Alteration	Occurrence	Remarks	Reference
Lamellar	a ~ Z, b ~ X, c ~ Y	Pink, brown	5½-6 2.84-2.93	Pink to lavender, red-brown		Contact carbonate zones, carbonatites with eudialyte, scapolite		Can Min 11, 569
	b = X, c ∧ Z = 15°		2.96	Bluish-green		Secondary copper mineral		Am Min 43, 1107-1111
{010}, {100} polysynthetic		Colorless	4½ 2.73-2.75	Colorless, pink		Low T veins		Am Min 40, 900-904
	c = Z, + elong	Colorless		White		Alteration of rare-earth minerals in pegmatites		Am Min 56, 1838
None	b = X, c ∧ Y ~ +6°, a ∧ Z ~ +26°	Colorless	4-4½ 2.46	Colorless, white, yellow	Calcite	Borate playas. Alter borax or ulexite	Soluble hot HCl	p. 98
	a = Z, b = X, c = Y	Colorless	3½-4	Colorless, gray	Apatite	Serpentine-magnesite masses		Am Min 61, 502
{110} or {001} polysynthetic	b = Y, c ~ Z	Colorless	5-6 2.65-2.79	Yellow, brown, orange, sky blue		Alkali igneous rocks and pegmatites		Vlasov II, 367-370
{001}	b = Y, Z ∧ {001} = 2°-6°	Colorless, X ~ Y = green, Z = pale yellow-green	2-3 2.68-2.80	White, pale green, olive-green	Schists and other metamorphic rock		Chlorite group. Amesite-antigorite series	p. 289
{010} axis c, simple or multiple	Y ∧ elong ~ 5°-8°, X ~ ⊥ cleavage	X = colorless, Y = pale yellow X = green	2-2½ 3.27-3.32	Emerald green, yellow-green		Secondary U-mineral	Soluble HCl, Taste bitter, Not fluor U.V.	USGS Bull 1064, 130-135
	b = Z	Colorless	1½ 2.84-3.21	White, grayish brownish, rose		Efflorescence		Dana II, 481
{001} lamellar	b = Y, c ∧ Z ~ 6°	Colorless	5-6 2.8-2.99	Colorless, white, green, pink		Limestone contact zones		Win & Win II, 480
{100} or {021}		Colorless	7½ 2.93	Colorless		Meteorite craters	High P polymorph of SiO_2	Dana III, 310-316
	- elongation	Colorless	5½ 2.83-2.87	White, gray, greenish		Phosphate nodules with variscite, wardite, crandallite		Dana II, 941-942
	Z = elongation	Colorless	5 2.96	Colorless		Alteration of melilite		Wash Acad Sci J. 4, 480-482

Biaxial +

n_β Var.	Refractive Index	Mineral Name and Composition	$(n_\gamma - n_\alpha)$	2V angle Dispersion	Crystal System	Habit	Cleavage
~1.60	n_β n_α 1.598 n_γ 1.614	Kleemanite $ZnAl_2(PO_4)_2(OH)_2$	0.016		Monoclinic β = 110.2°	Tiny fibers	
1.60	n_β 1.60 n_α 1.59 n_γ 1.62	Fukalite $Ca_4Si_2O_6(OH,F)_2(CO_3)$	0.03		Orthorhombic	Flaky	
1.602	n_β 1.602 n_α 1.582 n_γ 1.629	Goldichite $KFe(SO_4)_2 \cdot 4H_2O$	0.047	82°	Monoclinic β = 101.8°	Radiating prism laths {100}	{100} perfect
*1.595 1.610	n_β 1.595-1.610 n_α 1.585-1.600 n_γ 1.595-1.613	Meta-autunite $Ca(UO_2)_2(PO_4)_2 \cdot 2\text{-}6H_2O$	0.010 to 0.013	Small r > v strong	Tetragonal	Pseudomorphic after autunite	{001} perfect {100} indist
1.602	n_β 1.602 n_α 1.590 n_γ 1.638	Haidingerite $CaHAsO_4 \cdot H_2O$	0.048	~ 58° r > v weak	Orthorhombic	Fibrous, botryoidal coatings	{010} perfect
1.603	n_β 1.603 n_α 1.579 n_γ 1.629	Metavivianite $Fe_3(PO_4)_3 \cdot 8H_2O$	0.050	85°	Triclinic α = 94.8° β = 97.2° γ = 107.4°	Flat prism crystals (c)	{110}
1.603	n_β 1.603 n_α 1.594 n_γ 1.615	Natromontebrasite (Fremontite) $(Na,Li)Al(PO_4)(OH,F)$	0.021	Large ~ 82°	Triclinic α = 112° β = 97.8° γ = 68.1°	Short prism crystals, massive	{100} perfect {110} & {0$\bar{1}$1} distinct, {001} imperf
1.604	n_β 1.604 n_α 1.598 n_γ 1.626	Metaschoderite $Al_2(PO_4)(VO_4) \cdot 6H_2O$	0.028	59°	Monoclinic β = 79°	Microscopic bladed crystals {001}	
1.605	n_β 1.605 n_α 1.580 n_γ 1.644	Euchlorin $(K,Na)_8Cu_9(SO_4)_{10}(OH)_6$?	0.064	Large ~ 80°	Orthorhombic	Rectangular tabular	Two directions
*1.59 1.62	n_β 1.59-1.62 n_α 1.59-1.60 n_γ 1.61-1.62	Amesite $Mg_4Al_2(Si_2Al_2)O_{10}(OH)_8$	0.012 to 0.018	10° - 20° v > r	Monoclinic-Hexagonal	Micaceous, hexagonal plates	{001} perfect
1.605	n_β 1.605 n_α 1.600 n_γ 1.613	Hydrophilite $CaCl_2$	0.013	Moderate ~ 75°	Orthorhombic	Massive, mealy crusts	{110} perfect
1.6	n_β 1.6 n_α n_γ	Manandonite $LiAl_4(AlBSi_2)O_{10}(OH)_8$	0.014	~25° - 30°	Monoclinic β = 97.75°	Aggregates hexagonal plates	{001} perfect
1.605 1.609	n_β 1.605-1.609 n_α 1.595-1.603 n_γ 1.622-1.618	Scawtite $Ca_7Si_6O_{18}(CO_3) \cdot 2H_2O$	0.027 to 0.015	74° - 78°	Monoclinic β = 100.7°	Bundles of thin tabular crystals	{001} perfect ?, {010} dist
1.605 1.610	n_β 1.605-1.610 n_α 1.596-1.604 n_γ 1.606-1.621	Gastunite (Weeksite) $(K,Na)_2(UO_2)_3(Si_2O_5)_4 \cdot 8H_2O$	0.010 to 0.024	Moderate ~ 75°	Orthorhombic	Radiated acicular, fibrous	{010} perfect

* n_β variation exceeds 0.01

Twinning	Orientation	Color in thin section	Hardness Sp Gravity	Color hand sample	Alteration	Occurrence	Remarks	Reference
	Inclined ext. to 40°	Colorless	 2.76	Ochre		Earthy coatings on Mn-Fe ores		Am Min 64, 1331
	− elongation	Colorless	~4 2.77	White to pale brown		Skarn rock with scawtite and hillebrandite. Alter of sparrite	Decomp by acids with efferves- cence	Am Min 63, 793
	b = X, c ∧ Z = 11°		2½ 2.42-2.43	Pale yellow- green		Oxide zones secondary U-deposits		Am Min 40, 469-480
Sector	c = X	X = colorless to pale yellow, Y = Z = dk yellow	2-2½ 3.45-3.55	Yellow, greenish-yellow		Dehydration of autunite at weathered exposures	Fluor U.V.-pale yellow-green	Am Min 48, 1389-1398
{110} rare	a = Y, b = X, c = Z	Colorless	2-2½ 2.95	Colorless, white		Oxide zones As-bearing ore deposits	Sectile. Easily soluble acids	Dana II, 708-709
	X ⊥ {110}, YZ ∥ {110}	X = pale blue-grn Y = Z = yellow to pale green	 2.69	Leek-green		Pegmatites in solu- tion cavities in triphylite with kryzhanovskite		Am Min 59, 896-899
		Colorless	5½-6 3.04-3.10	Grayish-white, white		Granite pegmatites with lepidolite, pink tourmaline	Amblygonite group	Dana II, 823-827
	b = Z c ∧ Y = +20°	Yellow	~2 1.610	Yellow-orange		Fractures in phospha- tic chert with wave- llite and vashegyite		Am Min 47, 637-648
		X = pale green, Y = grass-green, Z = yellow-green		Emerald-green		Volcanic sublimate	Somewhat soluble H_2O	Dana II, 571
	b = Y, c ∧ Z ~ 1°	Colorless	2½-3 2.77	Pale green, bluish green		Alter Mg-Al minerals with diaspore, magne- tite, chromite	Chlorite group. Amesite-antigorite series	p. 289
{110} polysynthetic		Colorless	 2.17-2.22	White		With evaporite minerals (anhydrite, halite) soil efflorescence	Soluble H_2O. Deliquescent	Dana II, 41-42
	Z ~ ⊥{001}	Colorless	~2½ 2.89	Colorless		Pegmatites with rubellite tourmaline, lepidolite	Chlorite group	Vlasov II, 35
	b = Y, a ∧ Z = 29°	Colorless	4½-5 2.74-2.77	Colorless		Limestone contact zones with diopside, wollastonite, spurrite	Soluble HCl	Am Min 40, 505-514
	Parallel ext, + or − elonga- tion	X = colorless, Y = very pale yel, Z = pale yellow	2 3.96-3.97	Yellow, greenish-yel		Secondary U-mineral	Fluor - U.V. yellow-green. Anomalous blue interfer color	Am Min 44, 1047-1056

n_β Var.	Refractive Index	Mineral Name and Composition	$(n_\gamma - n_\alpha)$	2V angle Dispersion	Crystal System	Habit	Cleavage
1.606 1.609	n_β 1.606–1.609 n_α 1.600–1.603 n_γ 1.614–1.615	Latiumite $K(Ca,Na)_3(Al,Si)_5O_{11}(SO_4,CO_3)$	0.014 to 0.012	Large 83° – 90° r > v distinct	Monoclinic β = 106°	Tabular, elongated, massive	{100} perfect
1.607	n_β 1.607 n_α 1.602 n_γ 1.615	Crandallite (Pseudowavellite) $CaAl_3(PO_4)_2(OH)\cdot H_2O$	0.013	70° – 75°	Triclinic α = 103° β = 91° γ = 90.5°		Poor ⊥ Z
1.608	n_β 1.608 n_α 1.583 n_γ 1.633	Chudobaite $(Na,K,Ca)(Mg,Zn,Mn)_2H(AsO_4)_2\cdot 4H_2O$	0.050	~90°	Triclinic α = 115° β = 96° γ = 94°	Crystals	{010} dist {100} imperf
1.609	n_β 1.609 n_α 1.602 n_γ 1.621–1.623	Brazilianite $NaAl_3(PO_4)_2(OH)_4$	0.019 to 0.021	71° – 75° v > r weak	Monoclinic β 97.3°	Radiated fibrous, short prism	{010} dist
*1.604 1.615	n_β 1.604–1.615 n_α 1.595–1.610 n_γ 1.632–1.645	Pectolite $Ca_2NaH(SiO_3)_3$	0.032 to 0.038	50° – 63° r > v weak	Triclinic α = 90° β = 96° γ = 103°	Radiated acicular crystals (b)	{100} & {001} perfect
1.609	n_β 1.609 n_α 1.604 n_γ 1.615	Uralborite $CaB_2O_4\cdot 2H_2O$	0.011	85° r > v strong	Monoclinic	Radiated fibrous aggregate	Distinct ‖ elongation
1.608 1.612	n_β 1.608–1.612 n_α 1.600–1.605 n_γ 1.645	Weinschenkite (Churchite) (Rogersite) $(Y,Er)PO_4\cdot 2H_2O$	0.045	Medium-small	Monoclinic β = 115.3°	Lath crystals (c), fibrous-radiated rosettes	{$\bar{1}$01}
1.610	n_β 1.610 n_α 1.607 n_γ 1.616	Buchwaldite $NaCaPO_4$	0.009	~65° – 71°	Orthorhombic	Fine acicular	One good cleavage or parting
1.610	n_β 1.610 n_α 1.610 n_γ 1.611	Foggite $Ca(H_2O)Al(OH)_2(PO_4)$	0.001	40° – 45°	Orthorhombic	Balls and radiated aggregate, foliated plates {010}	{010} perfect {100} good
1.611	n_β 1.611 n_α 1.610 n_γ 1.654	Pseudowollastonite β-$CaSiO_3$	0.044	~18°	Triclinic α = 90° β = 90.8° γ = 119.3°	Equant grains	{001}
1.612	n_β 1.612 n_α 1.608 n_γ 1.621	Tuhualite $(Na,K)_2(Fe^{2+},Fe^{3+},Al)_3Si_7O_{18}(OH)_2$	0.013	61° – 79° v > r strong	Orthorhombic	Prism crystals (c)	{100}, {010} and {001} good
1.612	n_β 1.6125 n_α 1.609 n_γ 1.619	Stokesite $CaSn(SiO_3)_3\cdot 2H_2O$	0.010	~70° v > r	Orthorhombic	Prism crystals	{110} perfect {010} poor
1.613	n_β 1.613 n_α 1.602 n_γ 1.649	Anapaite $Ca_2Fe^{2+}(PO_4)_2\cdot 2H_2O$	0.047	54° r > v perc	Triclinic α = 101.5° β = 104° γ = 71°	Tabular aggregates	{001} perfect {010} dist
1.614 1.615	n_β 1.614–1.615 n_α 1.600–1.587 n_γ 1.631–1.640	Monetite (Glaubapatite) $CaHPO_4$	0.031 to 0.053	Mod to large ~85° r > v weak	Triclinic α = 96.1° β = 103.9° γ = 89.2°	Tiny plates {010}, crusts, stalactites	3 dir indistinct

* n_β variation exceeds 0.01

Twinning	Orientation	Color in thin section	Hardness Sp Gravity	Color hand sample	Alteration	Occurrence	Remarks	Reference
{100} common	b = Z, c ∧ X = 16°-28°	Colorless	5½-6 2.92-2.93	White		Alkali volcanics with leucite, melilite, haüynite, garnet	Mottled extinction. Zoning	Am Min 58, 466-470
Multiple	Z ⊥ cleavage	Colorless	5 2.78-2.92	Buff yellow, white		Argillaceous sediments, with gypsum	Hexagonal dimorph. U.V. fluor-weak white	Am Min 48, 1144-1153
	c ∧ Z' on {010} = +24°, on {100} = 110°	Colorless, pink	2½-3 2.94-3.0	Pink		Oxide zones Cu-deposits		Am Min 44, 1323
	b = Y, c ∧ X = +20°	Colorless	5½ 2.98-3.03	Colorless, yellowish-grn		Granite pegmatites		Dana II 841-843
{100} parallel twin (b)	b ∧ Z ~ 2°, a ∧ Y ~ 10°, c ∧ X ~ 5°-11°. ∥ Extinction, + elongation	Colorless	4½-5 2.75-2.90	White	Stevensite	Hydrothermal veins & amygdules in basalt with zeolites, prehnite	Pyroxenoid group	p. 214
	ZY plane ~ cleavage	Colorless	~4 2.60	Colorless		Skarn rocks with garnet, magnetite, szaibelyite, frolovite	Fluor long λ U.V. violet. Anomalous interference colors	Am Min 47, 1482
	Z ⊥ plates	Colorless	3 3.26	Colorless, white, gray		Limonite deposits and oxide zones Cu-deposits with other phosphates		Dana II, 771-773
	∥ extinction, + elongation	Colorless	<3 3.21	White		Meteorites in troilite nodules with chromite		Am Min 62, 362-364
	a = Z, b = Y, c = X	Colorless	4 2.78	Colorless, white		Cavities in pegmatites with quartz, apatite, childrenite		Am Min 60, 957-964
{001} lamellar	c ∧ X ~ 9°	Colorless	5 2.90-2.91	Colorless		Contact carbonate rocks		DHZ 2, 167-175
	a = X, b = Y, c = Z	X = colorless, pale pink, Y = violet-lavender, Z = dk indigo	3-4 2.86-2.89	Dk blue to black		Alkaline volcanic rocks with aegirine, riebeckite, quartz		Min Mag 31, 96-106
	a = X, b = Y, c = Z	Colorless	6 3.18-3.21	Colorless		With axinite		Min Mag 32, 433
		Colorless to pale green	3½ 2.80-2.81	Pale green		Veins and cavities in iron sediments and bituminous clays	Easily soluble in HCl or HNO_3	Dana II, 731-732
	X ~ ⊥{11$\bar{1}$}	Colorless	3½ 2.92-2.93	Colorless, white, yellowish		Cave deposits, guano leaching with newberyite, whitlockite	Soluble acids	Dana II, 660-661

n_β Var.	Refractive Index	Mineral Name and Composition	(n_γ-n_α)	2V angle Dispersion	Crystal System	Habit	Cleavage
*1.610 1.621	n_β 1.610-1.621 n_α 1.595-1.605 n_γ 1.622-1.635	Montebrasite $LiAl(PO_4)OH$	0.027 to 0.030	90° - 70° v > r weak	Triclinic α = 112° β = 97.8° γ = 68.1°	Cleavage masses, short prism crystals	{100}, {110} distinct, {0$\bar{1}$1} dist, {001} poor
1.63 1.60	n_β 1.63-1.60 n_α 1.63-1.60 n_γ 1.64-1.61	Klementite $(Mg,Fe^{3+})_6(Si,Al)_4O_{10}(OH)_8$	0.000 to 0.001	Small r > v	Monoclinic β ~ 97°	Micaceous, scaly	{001} perf
1.615	n_β 1.615 n_α 1.600 n_γ 1.629	Lehiite $(K,Na)_2Ca_5Al_8(PO_4)_8(OH)_{12}\cdot 6H_2O$(?)	0.029	Very large ~90°	Unknown	Crusts of fibrous aggregate	
*1.60 1.63	n_β 1.60-1.63 n_α 1.60-1.63 n_γ 1.61-1.64	Pycnochlorite $(Mg,Fe^{2+},Al)_6(Si,Al)_4O_{10}(OH)_8$	0.000 to 0.005	Small-Moderate v > r	Monoclinic β ~104.5°	Massive, scaly, oolitic	{001} perfect
1.617	n_β 1.617 n_α 1.589 n_γ 1.644	Nickelbischofite $NiCl_2\cdot 6H_2O$	0.055	87°	Monoclinic <β = 122°22'	Coatings and poor crystals	
1.614 1.620	n_β 1.614-1.620 n_α 1.611-1.617 n_γ 1.632-1.639	Hemimorphite (Calamine) $Zn_4Si_2O_7(OH)_2\cdot H_2O$	0.022	44°-47° r > v strong	Orthorhombic	Sheeflike, tab., fib., colliform	{110} perf., {101} poor, {001} traces
1.617	n_β 1.617 n_α 1.616 n_γ 1.622	Kurumsakite $(Zn,Ni,Ca)_8Al_8V_2Si_5O_{35}\cdot 27H_2O$	0.006	~35°	Orthorhombic (?)	Radiated felted fibrous	
1.617 1.620	n_β 1.617-1.620 n_α 1.588-1.591 n_γ 1.654-1.655	Cyanotrichite $Cu_4Al_2SO_4(OH)_{12}\cdot 2H_2O$	0.066 to 0.064	82° v > r strong	Orthorhombic (?)	Microscopic acicular, radiated fibrous aggregate	
1.620	n_β 1.620 n_α 1.611 n_γ 1.645	Metahaiweeite $Ca(UO_2)_2Si_6O_{15}\cdot nH_2O$ (n<5)	0.034	~65°	Monoclinic (?)	Tabular crystals, spherical-radiated aggregate	{100} dist
1.620	n_β 1.6204 n_α 1.6169 n_γ 1.6336	Afwillite $CaSi_2O_4(OH)_6$	0.017	54° v > r perc	Monoclinic β = 98.4°	Elongated on b, or tabular	{001} perfect {100} good
*1.609 1.631	n_β 1.609-1.631 n_α 1.606-1.630 n_γ 1.616-1.638	Topaz $Al_2SiO_4(F,OH)_2$	0.008 to 0.011	48° - 68° r > v moderate	Orthorhombic	Stubby prism crystals (c). Granular	{001} very perfect
1.62	n_β 1.62 n_α 1.60 n_γ 1.65	Chavesite Hydrated Ca - Mn phosphate	0.05	Large	Triclinic α = 99.75° β = 108° γ = 91.3°	Platy, coatings	2 good ~90°
1.62	n_β 1.62 n_α 1.61 n_γ 1.65	Turquois (Turquoise) (Henwoodite) $CuAl_6(PO_4)_4(OH)_8\cdot 4\text{-}5H_2O$	0.04	~40° v > r strong	Triclinic α = 68.6° β = 69.7° γ = 65.1°	Fine granular, massive, veins, crusts	{001} perf {010} good
*1.59 1.65	n_β 1.59-1.65 n_α 1.58-1.63 n_γ 1.60-1.66	Corundophyllite $(Mg,Al)_6(Si,Al)_4O_{10}(OH)_8$	0.002 to 0.015	25° - 40° v > r	Monoclinic-Triclinic β = 120°	Micaceous, massive-granular	{001} perf

* n_β variation exceeds 0.01

Twinning	Orientation	Color in thin section	Hardness Sp Gravity	Color hand sample	Alteration	Occurrence	Remarks	Reference
{1̄1̄1} common		Colorless	5½–6 2.98–3.03	White, gray, pink, yellowish	Kaolin & mica	Granite pegmatites	Amblygonite-montebrasite series. Slightly soluble acids	p. 78
{001} multiple	b = Y, c ∧ Z = small	Colorless, pale green, pale brown X = Y > Z	2–3 2.6–3.0	Green-brown		Low-grade metamorphism. Oxidation of ferrous chlorites	Chlorite group	p. 294
	Extinction ~ ‖ elong − = X	Colorless	5½ 2.89	White, gray		Altered phosphate (variscite) nodules with wardite		Dana II, 942–943
{001}	b = Y, c ∧ Z = small	X = Y = pale brownish green, Z = dk green	2–3 2.8–3.1	Dk green, olive-green		Schists, altered mafic rocks	Chlorite group	p. 288
	b = Y, c ∧ X = +8°	X = pale green, Y = pale green, Z = green	1.929	Emerald green		Volcanic sublimates, ultramafic rocks		Am Min 65, 207
{001} uncommon	a = Y, b = X, c = Z	Colorless	4½–5 3.40–3.50	White, yellow, lt. green, blue	Willemite	Oxide zones Zn deposits with secondary Cu, Pb, Zn, Fe minerals		p. 162
	‖ extinction, + elongation	Colorless, yellow	4.03	Greenish-yellow, yellow		Cavities in bituminous schists		Am Min 42, 583–584
	Z ‖ elong (+ elong), X ⊥ laths	X = colorless, Y = pale blue, Z = blue	2.74–2.95	Sky blue-azure blue		Oxide zones Cu-deposits	Soluble acids	Dana II, 578–579
		Colorless, pale yellow	3½ 3.35	Pale yellow, greenish-yellow		With haiweeite in secondary U-deposits	Fluor U.V. - dull green	Am Min 44, 839–843
	b = Y, c ∧ X = 31°	Colorless	3–4 2.62	Colorless or white		Small crystals in fractures		Am Min 38, 629
None	a = X, b = Y, c = Z	Colorless	8 3.49–3.57	Colorless, yellow, wine, blue, etc.	Sericite, illite, kaolinite	Pegmatites, granites, hydrothermal veins, pneumatolitic vugs		p. 125
Multiple ‖ elongation	Ext 30° to twin plane	Colorless	~3	Colorless		Secondary mineral in pegmatites		Am Min 43, 1148–1156
		X ~ colorless, Z = pale blue to pale green	5–6 2.6–2.91	Sky blue to apple green		Near surface alteration of Al-rich rocks with clays, chalcedony	Slowly solution HCl	Am Min 50, 283
	b = Y, Z ∧ {001} ~ 2°	X ~ Y = dk bluish-green, Z = pale yellow-green	2–3 2.85	Lt to dk green		Emery deposits with margarite, magnetite, amesite	Chlorite group. Amesite-antigorite series	p. 289

n_β Var.	Refractive Index	Mineral Name and Composition	$(n_\gamma - n_\alpha)$	2V angle Dispersion	Crystal System	Habit	Cleavage
1.621	n_β 1.621 n_α 1.593 n_γ 1.666	Tinaksite $NaK_2Ca_2TiSi_7O_{19}(OH)$	0.073	74° - 78° Strong	Triclinic α = 91° β = 99.3° γ = 92.3°	Prism crystals, radiated fibrous, rosettes	{010} perf {110} imperf
1.618 1.625	n_β 1.618-1.625 n_α 1.615-1.618 n_γ 1.665-1.670	Diadochite (Destinezite) $Fe_2PO_4(SO_4)(OH)\cdot 5H_2O$	0.052 to 0.050	Small r > v moderate	Triclinic α = 99° β = 108° γ = 64°	Micro-crystalline, botryoidal, crusts	None ?
1.622	n_β 1.622 n_α 1.619-1.594 n_γ 1.631-1.604	Stanfieldite $Ca_4(Mg,Mn,Fe)_5(PO_4)_6$	0.012 to 0.010	50° - 60°	Monoclinic β = 100.5°	Massive, grains, veinlets	None
1.623	n_β 1.623 n_α 1.591 n_γ 1.663	Szomolnokite $FeSO_4\cdot H_2O$	0.072	80° r > v strong	Monoclinic β = 115.9°	Bipyramidal crystals, globular, stalactitic	None
1.623 1.624	n_β 1.623-1.624 n_α 1.621-1.622 n_γ 1.630-1.633	Celestite (Celestine) $SrSO_4$	0.008 to 0.009	~50° v > r moderate	Orthorhombic	Tabular {001}, prism (a) crystals, fibrous, earthy	{001} perf, {210} dist, {010} poor
1.625	n_β 1.625 n_α 1.614 n_γ 1.637	Parahopeite $Zn_3(PO_4)_2\cdot 4H_2O$	0.023	~90° v > r perceptible	Triclinic α = 93.3° β = 92° γ = 91.3°	Tabular, elongation (c), radiated	{010} perf
*1.602 1.655	n_β 1.602-1.655 n_α 1.592-1.643 n_γ 1.619-1.675	Chondrodite $2Mg_2SiO_4\cdot Mg(OH,F)_2$	0.025 to 0.037	64° to 90° r > v weak	Monoclinic β = 109°	Massive, granular, tabular crystals	{001} poor
1.632 1.625	n_β 1.623-1.625 n_α 1.621-1.623 n_γ 1.632-1.634	Uranopilite $(UO_2)_6SO_4(OH)_{10}\cdot 12H_2O$	0.011	~50° r > v strong	Monoclinic β = 145°	Rosettes of tiny laths (c), felty crusts	{010} perf
*1.602 1.656	n_β 1.6024-1.656 n_α 1.5788-1.616 n_γ 1.6294-1.675	Vivianite $Fe_3(PO_4)_2\cdot 8H_2O$	0.051 to 0.059	83° - 63° v > r weak, horizontal	Monoclinic β = 104.3°	Prism (c)-bladed crystals, radiated, earthy	{010} perfect {$\bar{1}$06} & {100} trace
1.628 1.630	n_β 1.628-1.630 n_α 1.622-1.623 n_γ 1.681-1.684	Guildite $CuFe(SO_4)_2(OH)\cdot 4H_2O$	0.059 to 0.061	~62°	Monoclinic β = 105.3°	Pseudo-cubic crystals	{001} & {100} perfect
1.630	n_β 1.630 n_α 1.616 n_γ 1.677	Carbonate-cyanotrichite $Cu_4Al_2(CO_3,SO_4)(OH)_{13}\cdot 2H_2O$	0.061	55° - 60° r > v strong	Orthorhombic	Elongation platelets, fibrous aggregate	
1.63	n_β 1.63 n_α 1.62 n_γ 1.64	Carpholite (Karpholite) $MnAl_2Si_2O_6(OH)_4$	0.02	67° - 87°	Orthorhombic	Radiated fibrous aggregate	{010} perfect
*1.615 1.647	n_β 1.615-1.647 n_α 1.610-1.637 n_γ 1.632-1.673	Prehnite $Ca_2Al(AlSi_3)O_{10}(OH)_2$	0.020 to 0.036	60° - 70° r > v or v > r weak to strong	Orthorhombic	Radiating platy crystals	{001} dist {110} poor
1.632 1.635	n_β 1.635-1.632 n_α 1.617-1.612 n_γ 1.652-1.653	Tilleyite $Ca_5Si_2O_7(CO_3)_2$	0.035 to 0.041	85° - 89° v > r	Monoclinic β = 105.8°	Massive, tabular, grains	{100} perfect {101} ? good {$\bar{1}$01} ? poor

* n_β variation exceeds 0.01

Twinning	Orientation	Color in thin section	Hardness Sp Gravity	Color hand sample	Alteration	Occurrence	Remarks	Reference
	X' ∧ (010) = 1°-4°, Y' ∧ (010) = 16°-18°	X = Y = colorless, Z = pale orange-yellow	6 2.82-2.85	Pale yellow		Linestone contact zones, K-feldspar metasomatites		Am Min 50, 2098-2099
	X ~ ⊥ flattening Z' ∧ elong ~ 14°	Pale yellow, pale yellow-brown	3-4 2.0-2.4	Yellow-brown, red-brown, green		Gossan zones pyrite deposits		Dana II, 1011-1013
		Colorless	4½-5 3.15	Reddish to amber		Veinlets in olivine, meteorites		Science 158, 910-911
Common	b = Y, c ∧ X = +26°	Colorless	2½ 3.03-3.09	Colorless, yellow, red-brown, bluish	Romerite, rhomboclase	Oxide zones ore deposits with copiapite and other sulfides	Soluble H_2O	Dana II, 479-480
2ndary glide twinning	a = Z, b = Y, c = X	Colorless, pale blue. Z > Y > X	3-3½ 3.95	Lt blue, white, red-brown, green	Strontianite	Veins & cavities in limestone or dolomite, evaporite beds, ore veins	Slowly soluble hot concentrated acids. May be fluor U.V.	p. 92
{100} common, polysynthetic	a ~ X, c ∧ Y on {100} ~ 30°	Colorless	3½-4 3.31	Colorless, white		Oxide zones Zn-deposits	Dimorph. hopeite	Dana II, 733-734
{001} simple or lamellar {105} and {305}	b = Z, c ∧ Y = -22° to -31°, a ∧ X = -3° to -12°	Colorless to yellow or brown X > Y ~ Z	6-6½ 3.16-3.26	Yellow, brown, red	Serpentine, Mg-chlorite	Limestone-dolomite contact zones with Ca-Mg-silicates. Carbonatites	Humite group	p. 142
	b = X, c ∧ Y = +17° to +23°	X ~ colorless, Y = Z = yellow	Soft 3.96	Bright lemon yellow		Secondary U-mineral with johannite, zippeite	Fluor U.V.-yel-green. Anomalous blue interfer colors	USGS Bull 1064, 135-140
	b = X, c ∧ Z = -28°	X = blue, Y = pale yellow-green or blue-green, Z = pale yellow-green	1½-2 2.68-2.71	Colorless to dk blue, green, violet		Oxide zones ore deposite. Alter phosphates in pegmatites	Easily sol acids. Color darkens in air or H_2O_2	Dana II, 742-746
		X = Y = pale yellow, Z = greenish-yellow	2½ 2.70	Yellow, red-brown		Oxide zones Cu-deposits with other sulfates		Am Min 55, 502-505
	∥ Extinction - elongation	X = colorless, Z = blue	~2 2.65-2.67	Lt blue-azure blue		Oxide zone Cu-deposits		Am Min 49, 441-442
	a = Y, b = X, c = Z	Colorless	5-5½ 2.9-3.04	Straw yellow		Sn-deposits		Dana (1892) 549
Two sets of fine lamellar twins	a = X, b = Y, c = Z	Colorless	6-6½ 2.80-2.95	Pale green, yellow, white	Zeolites (scolecite), chlorite, epidote, grossularite	Amygdales in basalt etc. Veins in granite. Contact metamorphic		p. 307
{101} common	b = Y, c ∧ X = 12°-18°	Colorless	2.84-2.88	Colorless, white		Limestone contact zones		Win & Win II, 480

Biaxial +

n_β Var.	Refractive Index	Mineral Name and Composition	$(n_\gamma - n_\alpha)$	2V angle Dispersion	Crystal System	Habit	Cleavage
*1.61 1.65	n_β 1.61-1.65 n_α 1.60-1.65 n_γ 1.61-1.66	Prochlorite (Ripidolite) $(Mg,Fe^{2+},Al)_6(Si,Al)_4O_{10}(OH)_8$	0.000 to 0.008	20° - 50° v > r	Monoclinic β = 97.2°	Micaceous-scaly. Hexa-gonal-tabular crystals	{001} perfect
1.633	n_β 1.633 n_α 1.630 n_γ 1.636	Danburite $CaB_2Si_2O_8$	0.006	~90° r > v strong	Orthorhombic	Prismatic (c), massive	{001} poor
1.635	n_β 1.635 n_α 1.613 n_γ 1.657	Sklodowskite (Chinkolobwite) $Mg(UO_2)_2Si_2O_7 \cdot 6H_2O$	0.043	~90°	Monoclinic β = 96°	Radiated acicular-fibrous crystals, crusts earthy	{100} perfect
1.635	n_β 1.635 n_α 1.628 n_γ 1.698	Sarmientite $Fe_2^{3+}AsO_4SO_4(OH) \cdot 5H_2O$	0.070	38°	Monoclinic β = 97.6°	Microscopic prism (c), nodular	
*1.624 1.649	n_β 1.624-1.649 n_α 1.622-1.631 n_γ 1.642-1.663	Mansfieldite $AlAsO_4 \cdot 2H_2O$	0.020 to 0.032	30° - 68° r > v strong	Orthorhombic	Fibrous crusts, cellular	{201} imperf {001} & {100} traces
1.636	n_β 1.636 n_α 1.630 n_γ 1.664	Hilgardite $Ca_2B_5O_8(OH)_2Cl$	0.034	35° r > v	Monoclinic β = 90°	Tabular crystals hemimorphic	{010} & {100} perfect
1.636	n_β 1.636 n_α 1.630 n_γ 1.664	Parahilgardite $Ca_2B_5O_8(OH)_2Cl$	0.034	35° r > v	Triclinic α = 84° β = 79.6° γ = 60.9°		{010} & {100} perfect
*1.602 1.672	n_β 1.602-1.672 n_α 1.588-1.663 n_γ 1.613-1.683	Anthophyllite $(Mg,Fe)_7Si_8O_{22}(OH)_2$	0.025 to 0.020	79° - 90° r > v or v > r weak to moderate	Orthorhombic	Fibrous, bladed, acicular	{110} perfect
*1.619 1.653	n_β 1.619-1.653 n_α 1.607-1.643 n_γ 1.639-1.675	Humite $3Mg_2SiO_4 \cdot Mg(OH,F)_2$	0.028 to 0.036	65° - 84° r > v weak	Orthorhombic	Massive, granular, tabular crystals	{001} poor
1.636 1.639	n_β 1.636-1.639 n_α 1.634-1.637 n_γ 1.646-1.649	Barite $BaSO_4$	0.010 to 0.013	36° - 40° v > r weak	Orthorhombic	Platy {001}, prism, nodular, fibrous	{001} perfect {210} dist, {010} imperf
1.638	n_β 1.638 n_α 1.622 n_γ 1.671	Koettigite (Köttigite) $Zn_3(AsO_4)_2 \cdot 8H_2O$	0.049	74°	Monoclinic β = 103.8°	Bladed {010}, fibrous (c), crusts	{010} perfect
1.639	n_β 1.639 n_α 1.638 n_γ 1.670	Strontiohilgardite $(Sr,Ca)_2Ba_5O_8(OH)_2Cl$	0.032	~19°	Triclinic α = 75.4° β = 61.2° γ = 60.5°	Small tabular xls {010}	{001} & {211} distinct
1.639	n_β 1.639 n_α 1.620 n_γ 1.686	Krautite $MnHAsO_4 \cdot H_2O$	0.066	60° - 70°	Monoclinic β = 96.5°	Lamellar	{010} perfect {̄101} good, {101} dist
*1.622 1.658	n_β 1.622, 1.658 n_α 1.618, 1.640 n_γ 1.658, 1.695	Veszelyite $(Cu,Zn)_3(PO_4)(OH)_3 \cdot 2H_2O$	0.040 to 0.055	39° - 70° v > r weak to strong	Monoclinic β = 103.4°	~ equant or octahedral crystals, granular	{001} & {110}

* n_β variation exceeds 0.01

Twinning	Orientation	Color in thin section	Hardness Sp Gravity	Color hand sample	Alteration	Occurrence	Remarks	Reference
{001}	b = Y, Z ∧ {001} ~ 4°	X ~ Y = pale yellow-green, Z = pale to dk green	2-3 2.88-3.08	Green to dk green, olive		Schists and similar metamorphic rocks. Ore veins	Chlorite group. Anomalous interference colors	p. 288
	a = X, b = Y, c = Z	Colorless	7 2.97-3.02	Colorless, white, pink, yellow, brown		Carbonate contact zones		Win & Win II, 258-259
	b = Y	Pale yellow, Not pleochroic	2-3 3.64-3.77	Pale yellow to greenish-yellow		Secondary U-mineral	Not fluor U.V. Soluble acids	USGS Bull 1064, 300-304
	b = Y, c ∧ Z = 12°	Colorless, pale yellow-orange	2.58	Pale yellowish orange		Gossans of sulfide veins as nodules with fibroferrite	Easily soluble acids	Am Min 53, 2077-2082
	a = X, b = Z, c = Y	Colorless	3½-4 3.03-3.15	White, gray		Oxide zones As-deposits with realgar, scorodite	Scorodite-Mansfieldite series	Dana II, 763-767
	b = Y, c ∧ X = -88°, c ∧ Z = -2°. ∥ ext in (001) sections	Colorless	5 2.69-2.71	Colorless		Rock salt deposits	Dimorph para-hilgardite	Dana II, 382-383
	Ext ~ 20° in {001} sections c ~ Z	Colorless	5 2.71-2.72	Colorless		Intergrowths with hilgardite. Salt deposits	Dimorph hilgardite.	Dana II, 383
None	a = X, b = Y, c = Z	Colorless to pale brown or green, weak to moderate pleo, Z > Y = X	5½-6 2.85-3.39	White, gray, green to dk brown	Talc or serpentine	Metamorphic rocks	Amphibole group	p. 223
	a = X, b = Z, c = Y	Colorless to yellow or brown X > Y ~ Z	6 3.20-3.32	Yellow to dk orange	Serpentine Mg-chlorite	Limestone or dolomite contact zones with Ca-Mg-silicates	Humite group	p. 142
2ndary glide twinning	a = Z, b = Y, c = X	Colorless, pale tints. Z > Y > X	3-3½ 4.50	White, yellow blue, reddish	Witherite	Hydrothermal veins with fluorite, galena. Vesicles. Carbonate sediments	Insoluble acids	p. 90
	b = X, c ∧ Z = 37°	Pale rose	2½-3 3.32-3.33	Carmine red, brownish		Oxide zone Zn-deposits	Soluble acids	Dana II, 751-752
	Optic axis ~ a	Colorless	5-7 2.993	Colorless, pale yellow		Insoluble residue in salt deposits with sylvite, halite, anhydrite		Am Min 53, 2084-2087
	b = X, Z ∧ {$\bar{1}$01} = 16°, + elong	Colorless	<4 3.29-3.31	Pale rose		Oxide zones ore deposits		Am Min 61, 503
	b = Y, c ∧ Z = +35° to +43°	X = greenish blue, Z = blue	3½-4 3.42	Green, green-blue, dk blue		Oxide zones Cu-deposits with secondary Cu-minerals	Soluble acids	Dana II, 916-918

n_β Var.	Refractive Index	Mineral Name and Composition	$(n_\gamma - n_\alpha)$	2V angle Dispersion	Crystal System	Habit	Cleavage
1.640	n_β 1.640 n_α 1.637 n_γ 1.662	Penkvilksite $Na_4Ti_2(Si_4O_{11})_2 \cdot 5H_2O$	0.025	42°	Monoclinic or Orthorhombic	Clotted masses, micro. platy-fibrous	{001} perfect ?
1.62 1.66	n_β 1.62-1.66 n_α 1.61-1.66 n_γ 1.63-1.67	Pargasite $NaCaMg_4Al(Si,Al)_8O_{22}(OH)_2$	0.020	70°-90° r > v weak	Monoclinic β ~ 105.5°	Prism. xls., massive	{110} perf., {001} & {100} parting
1.641	n_β 1.641 n_α 1.636 n_γ 1.651	Roscherite $(Ca,Mn,Fe)_3Be_3(PO_4)_3(OH)_3 \cdot 2H_2O$	0.015	71° to large r > v strong, horizontal	Monoclinic β = 94.8°	Tabular {100} or prism (c), radiated fibrous	{001} good, {010} dist
1.642	n_β 1.642 n_α 1.622 n_γ 1.66	Tengerite $CaY_3(CO_3)_4(OH)_3 \cdot 3H_2O$	0.04	Large	Unknown	Powdery-xln coatings, fibrous mammillary	
1.642	n_β 1.642 n_α 1.632 n_γ 1.657	Collinsite $Ca_2(Mg,Fe)(PO_4)_2 \cdot 2H_2O$	0.025	80°	Triclinic α = 97° β = 107° γ = 104.5°	Radiated tabular aggregate, prism-tabular	{001} & {010} distinct
1.642	n_β 1.642 n_α 1.637 n_γ 1.670	Tyretskite $(Ca,Sr)_2B_5O_8(OH,Cl)_3$	0.033	~46°	Triclinic α = 61.8° β = 60.3° γ = 73.5°	Massive, radiated fibrous	
1.643	n_β 1.643 n_α 1.631 n_γ 1.695	Ransomite $CuFe_2^{3+}(SO_4)_4 \cdot 6H_2O$	0.064	Small ~ 53°	Monoclinic β = 93°	Radiated acicular crystals (c), crusts	{010} perfect
1.644	n_β 1.644 n_α 1.620 n_γ 1.674	Zircosulfate $Zr(SO_4)_2 \cdot 4H_2O$	0.054	70° - 75°	Orthorhombic	Earthy, rounded forms, rhombic	None ?
1.644	n_β 1.644 n_α 1.641 n_γ 1.650	Rankinite $Ca_3Si_2O_7$	0.009	64°	Monoclinic β = 120.1°	Massive, rounded grains	None
1.65 1.64	n_β 1.65-1.64 n_α 1.641-1.637 n_γ 1.682-1.670	Kurgantaite $(Sr,Ca)_2B_4O_8 \cdot H_2O$	0.041 to 0.033	Small	Unknown	Fine-grained nodules	
1.645	n_β 1.645 n_α 1.643 n_γ 1.649	Pellyite $Ba_2Ca(Fe,Mg)_2Si_6O_{17}$	0.006	47° r > v very strong	Orthorhombic	Massive crystalline	Prismatic poor
1.644 1.651	n_β 1.644-1.651 n_α 1.630-1.638 n_γ 1.652-1.665	Tirodite $(Mg,Mn,Fe)_7(Si_4O_{11})_2(OH)_2$	0.022 to 0.027	~90°	Monoclinic β = 103.9°	Prism-acicular crystals (c)	{110} perfect {100} imperf {001} parting
1.644 1.650	n_β 1.644-1.650 n_α 1.636-1.640 n_γ 1.654-1.660	Fairfieldite $Ca_2(Mn^{2+},Fe^{2+})(PO_4)_2 \cdot 2H_2O$	0.018 to 0.020	86° r > v moderate	Triclinic α = 102° β = 108.7° γ = 90°	Lamellar, radiated fibrous, prism	{001} perfect {010} & {1$\bar{1}$0} distinct
1.647	n_β 1.647 n_α 1.642 n_γ 1.672	Korzhinskite $CaB_2O_4 \cdot H_2O$	0.030	44°	Unknown	Prism aggregate	One ∥ elong

Twinning	Orientation	Color in thin section	Hardness Sp Gravity	Color hand sample	Alteration	Occurrence	Remarks	Reference
	∥ extinction + elongation	Colorless	5 2.58	White		Alkali pegmatites with mountainite aegirine	Hydrated narsarsukite. Gelat cold 5% HCl or HNO_3	Am Min 60, 340-341
{100} simple	b = Y, c ∧ Z ≃ -26°, a ∧ X ≃ -11°	X = colorless, Y = blue-green, Z = deep blue-grn	5-6 3.07-3.18	Brown, bluish-green, black	Uralite	Many varieties of igneous and metamorphic rocks	Amphibole group. Hornblende series	DHZ II, 263-314
	b = X, c ∧ Y = +15°	X = yellow, olive-green Y = yellow-brown, green, Z = red-brown	4½ 2.93	Olive, lt to dk brown		Cavities in granite & granite pegmatites	Soluble acids. Anomalous interference colors	Am Min 43, 824-838
	- elongation	Colorless	3.12	White		Alteration coatings on gadolinite or yttrialite		Dana II, 275-276
		Colorless, pale yellow-brown	3½ 2.99-3.04	Colorless, white, lt brn		Cavities in pegmatites, veins with asphaltum	Soluble acids	Dana II, 722-723
		Colorless	2.189	White, brownish		Evaporite sediments with dolomite, sylvite, carnallite, halite, anhydrite		Am Min 53, 2084-2087
		Pale blue	2½ 2.63	Bright sky blue		Efflorescence in fire zone in mine		Am Min 55, 729-734
		Colorless	2½-3 2.83-2.85	Colorless, white		Alkali pegmatites with hisingerite, smithsonite, altered eudialyte		Am Min 51, 529
	b = Y, a ∧ X = 15°	Colorless	5½ 2.96-3.00	Colorless, white	Afwillite	Limestone contact zones with larnite, wollastonite, melilite	Gelat HCl	Win & Win II, 477
		Colorless	>6 ~3	White		Gypsum-anhydrite rocks with sylvite, carnallite, halite		Vlasov II, 202
		Colorless	6 3.48-3.51	Colorless, pale yellow		Limestone skarns with gillespite, sanbornite, barite		Can Min 11, 444-447
{100} common, may be multiple	b = Y, a ~ X, c ∧ Z = 16°-22°	X = Z = colorless, Y = yellowish	6-6½ 3.07-3.13	Pale green to rose red, tan		Mn ore deposits with anthophyllite	Amphibole group Mn-cummingtonite	Am Min 49, 963-982
		Colorless	3½ 3.08-3.09	White, yellowish, greenish		Granite pegmatite	Mn-messelite. Soluble acids	Dana II, 720-722
Twin axis 6° to Z	∥ extinction X ⊥ cleavage	Colorless		Colorless		Contact carbonate deposits with calciborite, sibirskite		Am Min 49, 441

n_β Var.	Refractive Index	Mineral Name and Composition	$(n_\gamma - n_\alpha)$	2V angle Dispersion	Crystal System	Habit	Cleavage
1.648	n_β 1.648 n_α 1.637 n_γ 1.676	Loseyite $(Zn,Mn)_7(CO_3)_2(OH)_{10}$	0.039	64° r > v weak	Monoclinic β = 95.4°	Radiated lath-like crystals (b)	None
1.648	n_β 1.648 n_α 1.647 n_γ 1.650	Kilchoanite $Ca_3Si_2O_7$	0.003	60° r > v dist	Orthorhombic	Massive	None
1.64	n_β 1.64 n_α 1.64 n_γ 1.66	Roeblingite $Pb_2Ca_7Si_6O_{14}(OH)_{10}(SO_4)_2$	0.02	Small	Monoclinic β = 103.9°	Minute prism crystals, fibrous, platy crystals	{001} perfect {100} imperf
*1.622 1.676	n_β 1.622-1.676 n_α 1.610-1.667 n_γ 1.632-1.684	Gedrite $(Mg,Fe^{2+})_5Al_2(Si_3AlO_{11})_2(OH)_2$	0.022 to 0.016	70° - 90° r > v or v > r weak to moderate	Orthorhombic	Fibrous, bladed, acicular	{110} perfect
1.650	n_β 1.650 n_α 1.588 n_γ 1.722	Krausite $KFe^{3+}(SO_4)_2 \cdot H_2O$	0.134	Large ~ 90°	Monoclinic β = 102.75°	Short prism, tabular	{001} perfect {100} dist
1.650	n_β 1.650 n_α 1.626 n_γ 1.686	Pseudolaueite $MnFe_2(PO_4)_2(OH)_2 \cdot 8H_2O$	0.060	80°	Monoclinic β = 105°	Prism, thick tabular	
1.650	n_β 1.650 n_α 1.645 n_γ 1.655	Samuelsonite $(Ca,Ba)Fe_2^{2+}Mn_2^{2+}Ca_8Al_2(OH)_2(PO_4)_{10}$	0.010	70° - 80°	Monoclinic β = 112.8°	Flattened prism crystals (b)	{001} fair
1.651	n_β 1.651 n_α 1.639 n_γ 1.681	Vuonnemite $Na_4TiNb_2Si_4O_{17} \cdot 2Na_3PO_4$	0.042	53°	Triclinic α = 93.7° β = 89.5° γ = 87.5°	Platy	{001} very perfect. Two other perfect
*1.644 1.658	n_β 1.644-1.658 n_α 1.644-1.658 n_γ 1.697-1.709	Dioptase $CuSiO_2(OH)_2$	~0.053	Small to 45°	Trigonal	Prism, rhombohedral, massive	{10$\bar{1}$1} perfect
*1.635 1.670	n_β 1.635-1.670 n_α 1.634-1.666 n_γ 1.644-1.690	Mullite $3Al_2O_3 \cdot 2SiO_2$-$2Al_2O_3 \cdot SiO_2$	0.010 to 0.024	20° - 50° r > v weak	Orthorhombic	Prism-acicular xls (c). Fibrous aggregate	{010} dist
1.653	n_β 1.653 n_α 1.652 n_γ 1.673	Kotoite $Mg_3B_2O_6$	0.021	21° - 24° r > v	Orthorhombic	Massive granular	{110} perfect {101} parting
1.649 1.659	n_β 1.649-1.659 n_α 1.644-1.653 n_γ 1.663-1.676	Messelite (Neomesselite) (Parbigite) $Ca_2(Fe^{2+},Mn^{2+})(PO_4)_2 \cdot 2H_2O$	0.019 to 0.023	20° - 35°	Triclinic α = 102° β = 109° γ = 90°	Lamellar foliated, radiated fibrous, prism	{001} perfect {010} good {1$\bar{1}$0} dist
1.65 1.66	n_β 1.65-1.66 n_α 1.64-1.65 n_γ 1.65-1.67	Boehmite γ-AlO(OH)	~0.015	~80°	Orthorhombic	Aggregate tiny scales {010}. Submicroscopic	{010} perfect
1.65	n_β 1.65 n_α 1.64 n_γ 1.66	Hellandite $Ca_3(Y,Yb...)_4B_4Si_6O_{27} \cdot 3H_2O$	~0.01	~80°	Monoclinic β = 109.75°	Prism and tabular {010}	None

* n_β variation exceeds 0.01

Twinning	Orientation	Color in thin section	Hardness Sp Gravity	Color hand sample	Alteration	Occurrence	Remarks	Reference
	b = Y	Colorless	~3 3.27-3.37	Bluish white, brownish		Veinlets in massive Zn-oxide ores with pyrochroite, etc.		Dana II, 244-245
		Colorless	2.99	Colorless		Limestone contact zones. Replaces rankinite	Dimorph rankinite Abnormal interference colors	Nature 189, (4766), 743
	- elongation	Colorless	3 3.43	White, pink		Fracture filling with garnet, axinite, esperite, clinohedrite	Gelat. acids	Am Min 51, 504-508
None	a = X, b = Y, c = Z	Colorless, pale brown to green, weak to moderate pleo, Z > Y = X	5½-6 2.85-3.30	White, gray, yellow-brown, dark brown	Talc or serpentine	Metamorphic rocks	Amphibole group	p. 223
	b = Z, c ∧ Y = +35°	X = colorless, Y = Z = pale yel	2½ 2.84	Pale lemon yellow		Oxide zones ore-deposits with voltaite, coquimbite, halotrichite	Slowly decomp H_2O Slowly soluble HCl	Am Min 50, 1929-1936
	b = Z, c ∧ X ~ 2°, a ∧ Y ~ 12°	X = Y = pale yel, Z = yellow	3 2.46-2.51	Orange-yellow		Pegmatites with Mn-Fe oxides. Cores of stewartite crystals		Am Min 54, 1312-1323
	O.A. ⊥ {001}	Colorless	5+ 3.27-3.35	Colorless to pale pink	Laueite	Cavities in pegmatites with whitlockite, apatite, childrenite		Am Min 60, 957-964
	Optic plane = {011}, c ∧ Y = 4°	Colorless	2-3 3.13	Light yellow		Albitized alkali rocks with nepheline, aegirine, lorenzenite	Decomp. H_2O	Am Min 59, 875
	c = Z	Pale green X ~ Y > Z	5 2.28-3.35	Green, dk blue-green		Oxide zones Cu-deposits	Gelat. HCl	Win & Win II, 453
	a = X, b = Y, c = Z	Colorless	6-7 3.12-3.26	White, lt gray or brown, blue, green	Sillimanite andalusite, kyanite	High T-low P buchites and hornfels with fused glass. Slags		p. 127
{101} polysynthetic	a = X, b = Y, c = Z	Colorless	6½ 3.04-3.10	Colorless		Carbonate contact zones	Soluble HCl or H_2SO_4	Dana II, 328-329
	Ext angle = 20° - 23°	Colorless	3½ 3.16	Colorless, white, greenish		Late hydrothermal mineral in granite pegmatites	Fe-fairfieldite	Am Min 40, 828-833
None	Probably a = X, b = Y, c = Z	Colorless	3½-4 3.01-3.06	White, yellowish, brown		Bauxite ores, lateritic soils, Al-clays with diaspore, gibbsite	Dimorph. Diaspore	p. 44
{001} and {100}	b = X, c ∧ Z = 43°	Red-brown	5½ 3.35-3.60	Red, brown, blackish		Granite pegmatites	Soluble HCl. Maybe metimict	Am Min 51, 534

n_β Var.	Refractive Index	Mineral Name and Composition	$(n_\gamma - n_\alpha)$	2V angle Dispersion	Crystal System	Habit	Cleavage
1.652	n_β 1.652 n_α 1.632 n_γ 1.693	Koritnigite $Zn[H_2O \cdot HOAsO_3]$	0.061	70°	Triclinic $\angle\alpha$ = 90°52' $\angle\beta$ = 96°34' $\angle\gamma$ = 90° 3'		{010} perfect
1.655	n_β 1.655 n_α 1.649 n_γ 1.714	Natrochalcite $NaCu_2(SO_4)_2(OH) \cdot H_2O$	0.065	~37° v > r strong, inclined weak	Monoclinic β = 112.9°	Pyramidal crystals, cross-fiber veinlets, crusts	{001} perfect
1.655	n_β 1.655 n_α 1.651 n_γ 1.671	Euclase $BeAlSiO_4OH$	0.020	50° r > v	Monoclinic β = 100.25°	Prismatic	{010} perfect {110} & {001} imperfect
*1.648 1.664	n_β 1.648-1.664 n_α 1.643-1.658 n_γ 1.674-1.685	Reddingite $(Mn,Fe)_3(PO_4)_2 \cdot 3H_2O$	0.031 to 0.027	41° - 65° r > v distinct	Orthorhombic	Bipyramidal, tabular {010} crystals, massive, fibrous	{010} poor
*1.645 1.667	n_β 1.667-1.645 n_α 1.662-1.643 n_γ 1.681-1.651	Mosandrite (Rinkolite)(Rinkite)(Johnstrupite)(Lovchorrite) $(Na,Ca,Ce)_3Ti(SiO_4)_2F$	0.019 to 0.008	43° - 87° v > r strong	Monoclinic β = 93°	Long prism or tabular crystals, massive	{100} dist
*1.654 1.670	n_β 1.654-1.670 n_α 1.653-1.661 n_γ 1.669-1.684	Sillimanite Al_2SiO_5	0.020 to 0.023	20° - 30° r > v strong	Orthorhombic	Acicular crystals (c). Matted fibrous	{010} perfect
1.658	n_β 1.658 n_α 1.622 n_γ 1.687	Annabergite (Cabrerite) $Ni_3(AsO_4)_2 \cdot 8H_2O$	0.065	84° - 90° r > v	Monoclinic β = 104.75°	Bladed {010}, earthy crusts	{010} perfect {100}, {$\bar{1}$02} poor
*1.640 1.676	n_β 1.640-1.676 n_α 1.628-1.658 n_γ 1.652-1.692	Cummingtonite $(Mg,Fe)_7Si_8O_{22}(OH)_2$	0.024 to 0.034	70° - 90° v > r weak	Monoclinic β = 102°	Bladed column, fibrous	Distinct {110}
1.655 1.662	n_β 1.655-1.662 n_α 1.648-1.658 n_γ 1.662-1.671	Dickinsonite $H_2Na_6(Mn,Fe,Ca,Mg)_{14}(PO_4)_{12} \cdot H_2O$	0.014 to 0.013	Moderate-large r > v strong	Monoclinic β = 105°	Tabular {001}, scaly, radiated	{001} perfect
1.658	n_β 1.658 n_α 1.652 n_γ 1.665	Hiortdahlite $(Ca,Na)_{13}Zr_3Si_9(O,OH,F)_{33}$	0.013	~90° v > r strong	Triclinic α = 90.5° β = 109° γ = 90°	Tabular {100}, pseudo-tetragonal	{110} & {1$\bar{1}$0} distinct (~90°)
*1.642 1.675	n_β 1.642-1.675 n_α 1.642-1.680 n_γ 1.661-1.690	Nagelschmidtite $Ca_{3-4}[(Si,P)O_4]_2$	~0.010	0° - 20°		Anhedral grains	{001} good, {110} fair
1.660	n_β 1.660 n_α 1.628 n_γ 1.705	Parasymplesite $Fe_3(AsO_4)_2 \cdot 8H_2O$	0.077	Large ~ 85°	Monoclinic β = 103.8°	Prism crystals	{010} very perfect
1.660	n_β 1.660 n_α 1.656 n_γ 1.668	Magnesioaxinite $(Ca,Mn,Fe,Mg)_3Al_2BSi_4O_{15}(OH)$	0.012	~70°	Triclinic α = 103° β = 98.5° γ = 88.1°	Tabular crystals	{100} good, {001}, {110} & {011} poor
1.658 1.663	n_β 1.658-1.663 n_α 1.622-1.629 n_γ 1.681=1.701	Erythrite $Co_3(AsO_4)_2 \cdot 8H_2O$	0.059 to 0.072	~90° r > v	Monoclinic β = 105°	Prism-acicular, bladed, fibrous, earthy	{010} perfect {100} & {$\bar{1}$02} indistinct

* n_β variation exceeds 0.01

Twinning	Orientation	Color in thin section	Hardness Sp Gravity	Color hand sample	Alteration	Occurrence	Remarks	Reference
	b = X, c ∧ Z = 22°	Colorless	3.54	Colorless		Cavities in tennantite, Tsumeb, SW Africa		Am Min 65, 206
	b = Y, c ∧ X = +12°	Green	4½ 3.49-3.54	Emerald green		Oxide zones Cu-deposits with antlërite, atacamite, bloedite	Easily sol acids Slowly sol H_2O	Dana II, 602-603
	b = Y, c ∧ Z = +41°	Colorless, very pale blue or grn pleochroic	7½ 2.99-3.12	Colorless, lt green, lt blue		Granite pegmatites, greisen		Vlasov II, 85-87
	a = X, b = Y, c = Z	X = colorless, Y = pinkish-brown, Z = pale yellow	3-3½ 3.23-3.24	Pink, yellow, colorless, brown		Alteration of triphylite & other phosphates in granite pegmatites	Reddingite-phosphoferrite series. Sol acids	Dana II, 727-729
{100} polysynthetic	b = Y, c ∧ X = 3°	Yellow, slightly pleochroic Z > Y > X	5 2.93-3.50	Yellow, yellow-green, red-brown		Alkali massifs and alkali pegmatites with nepheline		Vlasov II, 312-316
None	a = X, b = Y, c = Z	Colorless, lt brn or yellow. Z > Y > Z	6½-7½ 3.25	Colorless, white to dk brown	Sericite, pyrophyllite, kaolin or montmorillonite clays	High T metamorphism of pelitic rocks. Pegmatites (rare).		p. 123
	b = X, c ∧ Z = 36°	Colorless, pale green	1½-2½ 3.07-3.25	White, gray, yellow-green		Oxide zones of Ni-deposits	Erythrite-Annabergite series. Sectile	Dana II, 746-750
Simple or multiple on {100}	b = Y, c ∧ Z = -21° to -16°, a ∧ X = -9° to -3°	Colorless to pale green or brown. Weak pleo in yel, brn, grn, Z > Y ≥ X	5-6 3.10-3.40	Dk green to brown	Hornblende, talc, serpentine	Metamorphic rocks	Amphibole group	p. 225
	b = X, c ∧ Y ~ -15°	Light green	3½-4 3.38-3.42	Green, yel-green, brown-green		Granite pegmatite	Soluble acids	Am Min 50, 1647-1669
{100} polysynthetic	Inclined extinctions	X = colorless, Y = yellowish, Z = wine-yellow	5½ 3.25-3.27	Pale yellow, yellow-brown		Alkali igneous rocks and pegmatites with nepheline, aegirine	Gelat acids	Vlasov II, 379-381
Complex lamellae at 60°		Colorless	3.07	Colorless, yellow, brownish		High-T limestone contact zone with gehlenite, rankinite, perovskite	Mixture of two polymorphs	Am Min 63, 425-426
	c ∧ Z = 31°	X = bluish green, Y = yellowish, Z = brownish-yel	2 3.07-3.10	Lt greenish-blue			Dimorph. symplesite	Am Min 40, 368
		Pleochroic lt blue-lt violet-gray	6½ 3.178	Pale blue-pale violet		Contact metamorphic deposits	Fluor U.V.-red-orange	Am Min 61, 503-504
	b = X, c ∧ Z = -30° to -36°	X = pale pink, Y = pale violet, Z = red	1½-2½ 3.18	Violet-red, pink		Oxide zones Co-deposits	Erythrite-Annabergite series. Sectile	Dana II, 746-750

n_β Var.	Refractive Index	Mineral Name and Composition	$(n_\gamma-n_\alpha)$	2V angle Dispersion	Crystal System	Habit	Cleavage
1.660 1.663	n_β 1.660-1.663 n_α 1.589-1.598 n_γ 1.750-1.737	Parabutlerite $Fe^{3+}SO_4OH \cdot 2H_2O$	0.139 to 0.161	44° - 87° r > v moderate	Orthorhombic	Prism crystals	{110} poor
1.662	n_β 1.662 n_α 1.629 n_γ 1.727	Lindackerite $H_2Cu(AsO_4)_4 \cdot 8\text{-}9H_2O$	0.098	~73° v > r strong	Monoclinic	Lath crystals, rosettes	{010} perfect
*1.651 1.673	n_β 1.651-1.673 n_α 1.635-1.653 n_γ 1.670-1.690	Forsterite Mg_2SiO_4	~0.035	82° - 90° v > r weak	Orthorhombic	Granular, stubby crystals	{010} & {110} poor
*1.655 1.669	n_β 1.655-1.669 n_α 1.654-1.664 n_γ 1.665-1.675	Enstatite $Mg_2(SiO_3)_2$	0.011 to 0.008	35° - 90° r > v weak	Orthorhombic	Short prism (c) crystals, bladed	{110} dist, {010} or {100} parting
1.662	n_β 1.662 n_α 1.660 n_γ 1.670	Götzenite (Ca-rinkite) $(Ca,Na)_7(Ti,Al)_2Si_4O_{15}(F,OH)_3$	0.010	52° r > v strong	Triclinic α = 90° β = 101.3° γ = 101.1°	Prismatic crystals	{100} perfect {001} dist
*1.655 1.670	n_β 1.655-1.670 n_α 1.648-1.663 n_γ 1.662-1.679	Spodumene $LiAl(SiO_3)_2$	0.014 to 0.027	54° - 69° v > r weak	Monoclinic β = 110°	Acicular, prism (c) to huge crystals	{110} dist, {010} & {100} parting
1.663	n_β 1.663 n_α 1.640 n_γ 1.665	Seamanite $Mn_3^{2+}(OH)_2B(OH)_4(PO_4)$	0.025	40° v > r	Orthorhombic	Acicular crystals (c)	{001} dist
1.664	n_β 1.664 n_α 1.655 n_γ 1.675	Attakolite (Attacolite) $(Ca,Mn,Sr)_3Al_6(PO_4,SiO_4)_7 \cdot 3H_2O$	0.020	84° v > r moderate	Orthorhombic	Indistinctly crystalline	
1.664	n_β 1.664 n_α 1.656 n_γ 1.672	Junitoite $CaZn_2Si_2O_7 \cdot H_2O$	0.016	86° very weak	Orthorhombic	Crystal sprays, tabular hemimorphic crystals	{010} good, {100} & {101} poor
1.664	n_β 1.664 n_α 1.660 n_γ 1.688	Serandite $Mn_2NaSi_3O_8(OH)$	0.028	~35°	Triclinic β = 94.5°	Thick tabular or prism crystal aggregate	{001} & {100} perfect
1.664	n_β 1.664 n_α 1.663 n_γ 1.675	Giannettite $\sim Na_3Ca_3Mn(Zr,Fe)TiSi_6O_{21}Cl$	0.012	30°	Triclinic	Small prismatic crystals	{100} perfect {010} & {001} imperfect
1.66	n_β 1.66 n_α 1.655 n_γ 1.670	Salmonsite $Mn_9^{2+}Fe_2^{3+}(PO_4)_8 \cdot 14H_2O$	0.015	Large ~ 70° r > v strong	Orthothrombic (?)	Fibrous masses	Two pinacoidal at 90°
1.665	n_β 1.665 n_α n_γ	Dimorphite II As_4S_3	Moderate	r > v	Orthothrombic	Tiny dipyramidal crystals	None
1.662 1.667	n_β 1.662-1.667 n_α 1.658-1.662 n_γ 1.668-1.673	β-Boracite $Mg_3B_7O_{13}Cl$	0.010 to 0.011	82½°	Orthothrombic	Pseudo-isometric crystals	None

* n_β variation exceeds 0.01

Twinning	Orientation	Color in thin section	Hardness Sp Gravity	Color hand sample	Alteration	Occurrence	Remarks	Reference
{142} rare	a = Z, b = X, c = Y	X = pale yellow, Y = greenish-yel, Z = brownish-yel	2½ 2.54-2.55	Lt orange to brown-orange		Oxide zones ore-deposits with copiapite, jarosite, butlerite	Dimorph. butlerite. Soluble dilute acids	Dana II, 610-611
	b = Y, X ∧ elong = 26°	Colorless, pale green	2-2½ 3.20	Apple green		Oxide zones Cu-deposits with annabergite, erythrite		Dana II, 1007-1008
{100}, {011} & {012} simple, {031} cyclic	a = Z, b = X, c = Y	Colorless to pale green. X = Z = pale yellow-green, Y = yellow-orange	7 3.22-3.30	Pale olive green	Serpentine, iddingsite, chlorophaeite	Mafic and ultramafic igneous rocks. Dolomitic marble	Olivine group	p. 105
{101} normal	a = X, b = Y, c = Z	Colorless	5-6 3.20-3.30	Lt gray, yel, greenish, brn	Bastite (serpentine), uralite, talc, magnesite	Peridotites & dunites with olivine, phlogopite. Granulites with cordierite	Orthopyroxene group	p. 187
⊥{001}, b-axis, lamellar		Colorless	3.14	Colorless		With feldspathoids in melilite-nephelinite lavas	Soluble hot HCl	Min Mag 31, 503-510
{100} rare	b = Y, c ∧ Z = -22° to -27°, a ∧ X = -2° to -7°	Colorless	6½-7 3.0-3.2	White, yellow, lt green, lilac	Li-micas, albite, eucryptite, kaolinite, cookite	Granite pegmatites with lepidolite, beryl, tourmaline. Gneisses		p. 206
	a = X, b = Y, c = Z	Colorless	4 3.13	Pink, yellow, yellow-brown		Micro-crystals with sussexite, calcite. Mn-oxyhydroxides	Soluble dilute acid	Am Min 56, 1527-1538
			3.23	Pale red		With berlinite and lazulite in pegmatites		Am Min 51, 534
	a = X, b = Y, c = Z	Colorless	4½ 3.5	Colorless		Skarns with kinoite, apophyllite, smectite, sphalerite	Easily soluble weak HCl or HNO_3	Am Min 61, 1255-1258
	b ~ Z, c ∧ X = +57°	Colorless, pinkish	4½-5 3.32	Rose red, pink		Carbonatite, nepheline syenite with analcite aegirine		Win & Win II, 443
{100} polysynthetic	c ∧ Z ~ 22°	Colorless		Colorless, pale yellow		Alkali plutons		Vlasov II, 385-386
	+ elongation	X ~ colorless, Y = pale yellow, Z = orange-yellow	4 2.88	Pale tan		Granite pegmatites, alter. hureaulite with strengite		Dana II, 730-731
	a = Z, b = Y, c = X	Orange-yellow, pleochroic	1½ 3.58-3.60	Orange-yellow		Solfatara fumeroles		Dana I, 197
{111} penetration, rare		Colorless	7-7½ 2.95-2.97	Colorless, yellow, lt green, blue		Evaporite salt deposits	Isometric above 265°C (α-boracite)	Dana II, 378-381

n_β Var.	Refractive Index	Mineral Name and Composition	$(n_\gamma - n_\alpha)$	2V angle Dispersion	Crystal System	Habit	Cleavage
*1.65 1.68	n_β 1.65-1.68 n_α 1.65-1.68 n_γ 1.66-1.69	Pseudothuringite $(Fe^{2+},Mg,Al)_6(Si,Al)_4O_{10}(OH)_8$	0.000 to 0.005	Small-moderate v > r	Monoclinic	Massive, foliated	{001} perfect
**1.618 1.714	n_β 1.618-1.714 n_α 1.615-1.705 n_γ 1.632-1.730	Edenite	0.016 to 0.023	90°-50° r > v weak	Monoclinic β ~ 105°	Prism. xls	{110} perf., {001} or {100} parting
*1.659 1.674	n_β 1.659-1.674 n_α 1.654-1.665 n_γ 1.667-1.688	Jadeite $NaAl(SiO_3)_2$	0.012 to 0.023	70° - 75° v > r moderate	Monoclinic β = 107.4°	Fibrous, granular, stubby	Distinct {110}
*1.618 1.714	n_β 1.618-1.714 n_α 1.610-1.700 n_γ 1.630-1.730	Hornblende $(Ca,Na,K)_{2-3}(Mg,Fe^{2+},Fe^{3+},Al)_5$ $[Si_6(Si,Al)_2O_{22}](OH)_2$	0.015 to 0.034	85° - 90° v > r or r > v moderate	Monoclinic β = 105°	Prismatic	Good {110}
*1.636 1.709	n_β 1.636-1.709 n_α 1.623-1.702 n_γ 1.651-1.728	Clinohumite $4Mg_2SiO_4 \cdot Mg(OH,F)_2$	0.028 to 0.045	52° - 90° r > v weak	Monoclinic β = 101°	Massive, granular, tabular crystals	{001} poor
1.665 1.674	n_β 1.665-1.674 n_α 1.593-1.604 n_γ 1.741-1.731	Butlerite $Fe^{3+}SO_4(OH) \cdot 2H_2O$	0.148 to 0.127	Large	Monoclinic β = 108.5°	Thick tabular, octahedral	{100} perfect
1.67	n_β~1.67 n_α 1.633 n_γ 1.72	Sharpite $(UO_2)(CO_3) \cdot H_2O$?	0.09		Orthorhombic ?	Crusts of radiated fibrous rosettes	
1.667 1.675	n_β 1.667-1.675 n_α 1.650-1.653 n_γ 1.688-1.697	Ludlamite (Lehnerite) $Fe_3(PO_4)_2 \cdot 4H_2O$	0.038 to 0.044	82° r > v perc	Monoclinic β = 100.6°	Tabular wedge xls, massive, granular	{001} perfect {100} indist
1.672	n_β 1.672 n_α 1.669 n_γ 1.677	Magnesium-chlorophoenicite $(Mg,Mn)_5(AsO_4)(OH)_7$	0.008	76° v > r strong	Monoclinic	Radiated fibrous aggregates, rosettes	One perfect ‖ elongation
1.672	n_β 1.672 n_α 1.671 n_γ 1.673	Baratovite $KCa_8Li_2Si_{12}O_{37}F$	0.002	60° r > v strong	Monoclinic β = 112.5°	Platy	{001} perfect
1.673	n_β 1.673 n_α 1.662 n_γ 1.692	Goedkenite $(Sr,Ca)_2Al(OH)(PO_4)_2$	0.030	45° - 50°	Monoclinic β = 113.7°	Lozenge-shaped tabular crystals {001}	{100} fair
1.674	n_β 1.674 n_α 1.671 n_γ 1.684	Natrophilite $NaMnPO_4$	0.013	~75° v > r strong	Orthorhombic	Granular or cleavable masses, prismatic	{100} dist {010} indist {021} traces
1.672 1.676	n_β 1.672-1.676 n_α 1.665 n_γ 1.684-1.686	Lawsonite $CaAl_2Si_2O_7(OH)_2 \cdot H_2O$	0.019 to 0.021	76° - 87° r > v strong	Orthorhombic	Tabular, prismatic	{001} & {010} perfect {110} imperfect
1.675	n_β 1.675 n_α 1.661 n_γ 1.689	Liskeardite $(Al,Fe^{3+})_3AsO_4(OH)_6 \cdot 5H_2O$	0.028	~90°	Monoclinic or Orthorhombic	Radiated fibrous aggregates, crusts	One ‖ elong

* n_β variation exceeds 0.01   ** n_β variation exceeds 0.10

Twinning	Orientation	Color in thin section	Hardness Sp Gravity	Color hand sample	Alteration	Occurrence	Remarks	Reference
	b = Y, Z ∧ {001} = small	X ~ Y = dk brownish-green, Z = pale brown-grn	2-3 ~3.2	Dk green		Hydrothermal veins	Chlorite group	p. 292
{100} common	b = Y, c ∧ Z = -17° to -27°, a ∧ X = -2° to -12°	X = pale yellow, Y = green, Z = blue green	5-6 3.0-3.1	Lt. to dk. green	Uralite	Many varieties of igneous and metamorphic rocks	Amphibole group. Hornblende series	
Fine multiple on {100}	b = Y, c ∧ Z = -30° to -40°, a ∧ X = -13° to -23°	Colorless	6 3.25-3.40	White, pale green	Analcite. High T recrystallization to albite and nepheline	Low T-high P metamorphic rocks, low grade met with albite	Pyroxene group	p. 203
Simple on {100}	b = Y, c ∧ Z = -12° to -34°, a ∧ X = +3° to -19°	Green, blue-green, brown, strong pleochroic	5-6 3.02-3.45	Dk green to black	Chlorite with calcite and epidote	Ubiquitous	Amphibote group	p. 232
{001} simple or lamellar, {105} & {305}	b = Z, c ∧ Y = -7° to -15°, a ∧ X = +4° to -4°	Colorless to yellow or brown. X > Y ~ Z	6 3.17-3.35	Yellow, brn, white	Serpentine, Mg-chlorite	Dolomite contact zones. Serpentenite and talc schist	Humite group	p. 142
{$\bar{1}$05} common	b = Z, c ∧ X = +18°	X = colorless, Y ~ Z = Pale yel	2½ 2.55-2.60	Orange		Oxide zones ore deposits. Fumeroles	Dimorph. parabutlerite	Dana II, 608-609
	Parallel ext, + elongation	X = brownish, Z = pale greenish yellow	~2½ >3.33	Greenish yel to olive grn		Secondary U-mineral		USGS Bull 1064, 106-108
	b = Y, c ∧ Z = +67	Colorless	3½ 3.19-3.21	Green, white, colorless		Oxide zones ore-deposits. Alter. Fe-phosphates in pegmatites	Soluble acids	Dana II, 952-953
	Y = elongation	Colorless	3.37	Colorless, white		Veins in hydrothermal Zn-O deposits with zincite, calcite		Dana II, 780
	X ∧ ⊥(001) ~ 50°	Colorless	3½ 2.92	White		Alkalic veinlets with quartz, albite, aegirine, sphene, miserite		Am Min 61, 1053
	b = X	Colorless	5 3.83	Colorless to pale yellow		Pegmatites with palermoite, goyazite		Am Min 60, 957-964
	a = Y, b = Z, c = X	Colorless, wine-yellow	4½-5 3.41-3.47	Deep wine yellow	Fine fibrous, yellow mineral	Granite pegmatites with lithiophilite, secondary phosphates	Soluble acids	Am Min 50, 1096-1097
Simple or multiple on {110}	a = X, b = Y, c = Z	Colorless, rarely pale blue-green X > Y > Z	7-8 3.1	Colorless, blue-green, gray	Pumpellyite plagioclase	High P-low T metamorphism (blue-schist)		p. 157
	Z = elongation	Colorless	Soft 3.01	White, greenish, brownish		Secondary coatings on arsenopyrite, pyrite, chalcopyrite, quartz		Dana II, 924

Biaxial +

n_β Var.	Refractive Index	Mineral Name and Composition	$(n_\gamma - n_\alpha)$	2V angle Dispersion	Crystal System	Habit	Cleavage
1.675	n_β 1.675 n_α 1.665 n_γ 1.678	Gerstmannite $(Mn,Mg)Mg(OH)_2 \cdot ZnSiO_4$	0.013	50° – 60°	Orthorhombic	Radiated prismatic crystals	{010} good
*1.647 1.704	n_β 1.647–1.704 n_α 1.643–1.696 n_γ 1.665–1.713	Triplite $(Mn,Fe,Mg,Ca)_2(PO_4)(F,OH)$	0.022 to 0.017	25° – 76° r > v moderate to strong	Monoclinic β = 105.6°	Poor crystals, massive	{001} good {010} fair {100} poor
1.676	n_β 1.676 n_α 1.672 n_γ 1.683	Akrochordite $Mn_4Mg(AsO_4)_2(OH)_4 \cdot 4H_2O$	0.011	Medium v > r moderate	Monoclinic β 99.8°	Spherical aggregates of tiny crystals	Two directions at 90°
1.672 1.681	n_β 1.672–1.681 n_α 1.664–1.672 n_γ 1.694–1.702	Diopside $CaMg(SiO_3)_2$	0.030	58° – 57° r > v weak	Monoclinic β = 106°	Stubby prismatic (c) crystals. Granular	{110} dist, {001} & {100} parting
*1.654 1.706	n_β 1.654–1.706 n_α 1.651–1.705 n_γ 1.660–1.726	Clinoenstatite $(Mg,Fe)_2(SiO_3)_2$	0.009 to 0.021	53° – 25°	Monoclinic β 108°	Stubby prism xls, massive	{110} dist
1.680	n_β 1.680 n_α 1.676 n_γ 1.693	Curetonite $Ba_4Al_3Ti(PO_4)_4(O,OH)_6$	0.017	60° r > v strong	Monoclinic ∠β = 102°		{011} good, {010} parting
1.680	n_β 1.680 n_α 1.672 n_γ 1.700	Phosphoferrite $(Fe,Mn)_3(PO_4)_2 \cdot 3H_2O$	0.028	68° r > v distinct	Orthorhombic	Bipyram, tabular {010} crystals, massive, fibrous	{010} poor
1.680	n_β 1.680 n_α 1.678 n_γ 1.683	Harstigite $MnCa_6(Be_2OOH)_2(Si_3O_{10})_2$	0.005	52° v > r weak	Orthorhombic	Small prismatic crystals	None
1.681	n_β 1.681 n_α 1.676 n_γ 1.690	Satpaevite $Al_{12}(V^{4+}O)_2V_6O_{35} \cdot 30H_2O$	0.014	~70°	Orthorhombic ?	Earthy veinlets and crusts, foliated	Pinacoidal perfect
*1.677 1.689	n_β 1.677–1.689 n_α 1.670–1.689 n_γ 1.684–1.696	Lithiophilite $Li(Mn,Fe)PO_4$	0.014 to 0.007	63°– 0° v > r strong	Orthorhombic	Cleavage masses	{001} perfect {010} dist, {110} imperf
1.685	n_β 1.685 n_α 1.646 n_γ 1.745	Eriochalcite (Erythrocalcite) (Antofagastite) $CuCl_2 \cdot 2H_2O$	0.060	75° v > r strong	Orthorhombic	Wool-like, elongation (c)	{110} perf {001} dist
*1.670 1.700	n_β 1.670–1.700 n_α 1.662–1.691 n_γ 1.688–1.718	Omphacite $(Ca,Na)(Mg,Fe^{+2},Fe^{+3},Al)(SiO_3)_2$	0.018 to 0.027	58° – 83° r > v moderate	Monoclinic β = 106°	Anhedral granules	{110} dist
1.685	n_β 1.685 n_α 1.672–1.695 n_γ 1.698–1.73	Talmessite (Arsenate-belovite) $Ca_2Mg(AsO_4)_2 \cdot 2H_2O$	0.026 to 0.035	~90°	Triclinic α = 112.6° β = 70.8° γ = 119.4°	Fine crystalline aggregates, drusy	
1.686	n_β 1.686 n_α 1.659 n_γ 1.785	Nenadkevichite $(Na,Ca)(Nb,Ti)Si_2O_7 \cdot 2H_2O$	0.126	~50°	Orthorhombic ?	Foliated masses	{001} poor

* n_β variation exceeds 0.01

Twinning	Orientation	Color in thin section	Hardness Sp Gravity	Color hand sample	Alteration	Occurrence	Remarks	Reference
	a = Z, b = X, c = Y	Colorless	4½ 3.66-3.68	White, pale pink		Hydrothermal veins with calcite, mangan-pyrosmalite, sphalerite		Am Min 62, 51-59
	b = Y, c ∧ Z = +41°	Yellow-brown to red-brown. X > Z = Y	5-5½ 3.55-3.87	Red-brown to dk brown, black		Granite pegmatites, high-T veins	Triplite- wieselite series. Soluble acids	Dana 849-852
	b = X, c ∧ Y ~ 45°	Brownish	3½ 3.20-3.29	Reddish-brown		Manganese ores		Am Min 53, 1779
{100} or {001} lamellar	b = Y, c ∧ Z = -38° to -39°	Colorless, pale green. Weak pleochroism	6½ 3.2	White, lt grn, gray-green	Uralite, talc, serpentine, chlorite	Calc-silicate skarns, pyroxene granulites, olivine & alkali basalts	Pyroxene group. Diopside-hedenbergite series	p. 191
{100} common	b = X, c ∧ Z ~ 22°	Colorless	5-6 3.19	White, yellow-brown, green	Inverts to Mg-ortho-pyroxene	Rare mineral in volcanic rocks and meteorites	Pyroxene group. Clinoenstatite-clinoferrosalite series	Am J. Sci. 229, 151-217
Polysynthetic {100}	b = X, c ∧ Z = +30°	Colorless to yellow, X > Y = Z	3½ 4.42	Yellow-green		Veins with barite and adularia		Am Min 65, 206
	a = X, b = Y, c = Z	Colorless	3-3½ 3.29-3.34	Colorless, lt green, red-brn		Alteration of triphylite or graftonite in granite pegmatites	Reddingite-Phosphoferrite series. Sol acids	Dana II, 727-729
	a = Z, b = Y, c = X	Colorless	5½ 3.16	Colorless		Contact Mn-deposits		Am Min 53, 309-315; 1418-1420
	Parallel extinction	Pale greenish-yellow to pale olive, Z > X	1½ 2.4	Yellow		Carbonaceous shales with gypsum, steigerite, hewettite	Readily soluble dilute acids. Partly isotropic	Am Min 44, 1325-1326
None	a = X, b = Z, c = Y	Colorless, X = deep pink, Y = pale greenish yellow, Z = pale buff	4½-5 3.34	Salmon-pink to clove-brown	Rare phosphates	Granite pegmatites with Li-minerals and phosphates	Lithiophilite-Triphylite series	p. 79
	a = Z, b = X, c = Y	X = pale green, Y = pale olive-grn Z = pale blue	2½ 2.47-2.55	Greenish-blue		Volcanic sublimate, oxide zones Cu-deposits arid regions	Easily soluble H_2O or NH_4OH	Dana II, 44-46
	b = Y, c ∧ Z = -36° to -48°, a ∧ X = -20° to -32°	X = colorless, Y = Z = very pale green	5-6 3.29-3.37	Green, dark green	Glaucophane, green hornblende (smaragdite)	Eclogites, rarely in granulites, migmatites, gneisses	Pyroxene group	p. 204
		Colorless	5 3.2-3.5	Colorless, white, pale green		Oxide zones ore-deposits	Isomorph beta-roselite	Am Min 50, 813
	a = X, b = Z, c = Y	X = colorless, Y = pale yellow, Z = pale rose	~5 2.84-2.89	Dk brown to rose		Alkali pegmatites with aegirine, eudialyte, microcline	Easily soluble H_2SO_4	Am Min 40, 1154

Biaxial +

n_β Var.	Refractive Index	Mineral Name and Composition	$(n_\gamma - n_\alpha)$	2V angle Dispersion	Crystal System	Habit	Cleavage
1.687	n_β 1.687 n_α 1.678 n_γ 1.705	Rosenbuschite $(Ca,Na)_3(Zr,Ti)Si_2O_8F$	0.027	78°	Triclinic α = 91.4° β = 99.6° γ = 111.9°	Prism-acicular, radiated fibrous (c)	{100} perfect
1.688	n_β 1.688 n_α 1.684 n_γ 1.705	Penikisite $Ba(Mg,Fe)_2Al_2(PO_4)_3(OH)_3$	0.021	52° - 56° r > v extreme	Triclinic α ~ 90° β = 101.5° γ ~ 90°	Crystals	{010} & {100} good
*1.672 1.710	n_β 1.672-1.710 n_α 1.663-1.694 n_γ 1.683-1.722	Ferroanthophyllite $(Fe,Mg)_7Si_8O_{22}(OH)_2$	0.028 to 0.020	70° - 79° r > v or v > r, weak to moderate	Orthorhombic	Fibrous, bladed, acicular	{110} perfect
1.695 1.687	n_β 1.695-1.687 n_α 1.682-1.683 n_γ 1.720-1.718	Bornemanite $BaNa_4Ti_2NbSi_4O_{17}(F,OH)\cdot Na_3PO_4$	~0.036	40° - 66°	Orthorhombic	Platy aggregates, fibrous	{001} very perfect
1.691	n_β 1.691 n_α 1.688 n_γ 1.696	Wyllieite $Na_2Fe_2^{2+}Al(PO_4)_3$	0.008	~50°	Monoclinic β = 114.9°	Large, crude crystals	{010} perf {$\bar{1}$01} dist
1.691	n_β 1.691 n_α 1.689 n_γ 1.715	Yoderite $(Al,Mg,Fe)_2Si(O,OH)_5$	0.026	25°	Monoclinic β = 106°	Anhedral grains	{100} poor, {001} parting
1.691	n_β 1.691 n_α 1.686 n_γ 1.708	Marsturite $Mn_3CaNaHSi_5O_{15}$	0.022	60° r > v weak	Triclinic α = 85° β = 94° γ = 111°	Prismatic crystals	{100} & {001} imperfect
1.689 1.695	n_β 1.689-1.695 n_α 1.689-1.698 n_γ 1.696-1.671	Triphylite $Li(Fe,Mn)PO_4$	0.007	0° - 90° r > v	Orthorhombic	Cleavage masses	{001} perf, {010} dist, {110} imperf
1.693	n_β 1.693 n_α 1.672 n_γ 1.710	Jagowerite $BaAl_2(PO_4)_2\cdot(OH)_2$	0.038	~80°	Triclinic α = 116.8° β = 86.1° γ = 113°	Crystalline masses	{100} & {0$\bar{1}$1} good, {0$\bar{2}$1} fair
*1.676 1.710	n_β 1.676-1.710 n_α 1.667-1.694 n_γ 1.684-1.722	Ferrogedrite $(Fe^{2+},Mg)_5Al_2(Si_3AlO_{11})_2(OH)_2$	0.028 to 0.018	78° - 90° r > v or v > r, weak to moderate	Orthorhombic	Fibrous, bladed, acicular	{110} perf
*1.681 1.706	n_β 1.681-1.706 n_α 1.672-1.697 n_γ 1.702-1.726	Salite $Ca(Mg,Fe)(SiO_3)_2$	0.030 to 0.029	57° - 58° r > v moderate	Monoclinic β = 105°	Prism (c) crystals. Granular	{110} dist, {001} & {100} parting
1.695	n_β 1.695 n_α 1.692 n_γ 1.710	Bjarebyite $(Ba,Sr)(Mn,Fe,Mg)_2Al_2(OH)_3(PO_4)_3$	0.018	~35°	Monoclinic β = 100°	Complex crystals	{010} & {100} perfect
1.693 1.700	n_β 1.693-1.700 n_α 1.690-1.691 n_γ 1.719-1.736	Neptunite $(Na,K)_2(Fe^{2+},Mn)TiSi_4O_{12}$	0.029 to 0.045	49° - 36° v > r extreme	Monoclinic β = 115.4°	Prismatic square crystals	{110} perfect at 80°
1.698	n_β 1.698 n_α 1.695 n_γ 1.733	Euchroite $Cu_2AsO_4(OH)\cdot 3H_2O$	0.038	29° r > v moderate	Orthorhombic	Prismatic (b), tabular	{101} & {110} indistinct

* n_β variation exceeds 0.01

Twinning	Orientation	Color in thin section	Hardness Sp Gravity	Color hand sample	Alteration	Occurrence	Remarks	Reference
	c ~ X, Z ∧ ⊥(001)~28°	Pale yellow, Z > Y > X	5-6 3.30-3.32	Gray to orange to brownish		Nepheline-syenite with aegirine, woehlerite, mosandrite	Fluor. U.V.-yel to greenish-white. Soluble HCl	Vlasov II 388-390
	b ~ Y, c ∧ Z = +6°, b ∧ Y = 0° to +19°	X = grass green, Y = blue-green, Z = pale pink X ~ Y > Z	~4 3.79-3.82	Blue to green		Kulanite-penikisite crystals	Kulanite-penikisite series	Can Min 15, 393
None	a = X, b = Y, c = Z	Colorless, pale brown or green. Moderate pleo Z > Y = X	5½-6 3.39-3.57	Green to dk brown	Talc or serpentine	Metamorphic rocks	Amphibole group	p. 223
	a = Z, b = Y, c = X	X = Y = colorless, Z = brownish	3½-4 ?	Pale yellow		Alkali pegmatites with lomonosovite, natrolite		Am Min 61, 338
			>4 3.60	Dk bluish-grn to greenish-black		Granite pegmatites with perthite, schorl, arrojadite, scorzalite		Min Rec. 4, 131-136
	b = Y, c ∧ Z = -7°, a ∧ X = +9°	X = Prussian blue, Y = indigo, Z = pale olive grn Y > X > Z	6 3.39	Purple		Schist with quartz, kyanite, talc		Min Mag 32, 283-307
	b ∧ Z' = 4° (on {001}) and b ∧ Z' = 15° (on {100})	Colorless	~6 3.46	White, lt pink		Hydrothermal ore veins with rhodonite, willemite, manganaxinite	Nambulite group	Am Min 63, 1187-1189
None	a = Y, b = Z, c = X	Colorless	4½-5 3.48-3.53	Blue-gray	Rare phosphates	Granite pegmatites with Li-minerals and other phosphates	Lithiophilite-triphylite <75% $LiFePO_4$	p. 79
		Colorless	4½ 4.01-4.05	Pale green		Quartz veins with pyrite, hinsdalite	Fluor U.V.-greenish white	Am Min 61, 175
None	a = X, b = Y, c = Z	Pale brn to grn, colorless, weak to moderate pleo, Z > Y = X	5½-6 3.30-3.57	Yellow-brown, dk brown	Talc or serpentine	Metamorphic rocks	Amphibole group	p. 223
{100} or {001} lamellar	b = Y, c ∧ Z = -39° to -43°	X = pale blue-grn, Y = pale brown-grn, blue-green, Z - yellow-green	~6 3.3	Gray-green, lt brown	Uralite, talc, serpentine, chlorite	Calc-silicate skarns, alkali basalts and intrusives with olivine, nepheline	Pyroxene group. Diopside-hedenbergite series	p. 191
		Colorless, pale green	4+ 4.02	Emerald green		Cavities in pegmatite		Min Record 4, 282-285
	b = Y, c ∧ Z = +20°	X = pale yellow, Y = orange-red, Z = brownish-red, ochre	5-6 3.19-3.24	Dk red-brown, black		Nepheline-syenite, carbonatite, with benitoite, eudialyte		Win & Win II, 463
	a = Y, b = Z, c = X	Pale blue-green, weak pleochroism	3½-4 3.44-3.45	Emerald-green		Oxide zones Cu-deposits	Soluble acids	Dana II, 394-395

Biaxial +

n_β Var.	Refractive Index	Mineral Name and Composition	$(n_\gamma - n_\alpha)$	2V angle Dispersion	Crystal System	Habit	Cleavage
*1.690 1.709	n_β 1.690–1.709 n_α 1.675–1.702 n_γ 1.735–1.740	Legrandite $Zn_2(OH)AsO_4 \cdot H_2O$	0.060 to 0.038	35° – 50° v > r distinct	Monoclinic β = 104.2°	Long prismatic (c), radiated acicular	{100} fair-poor
1.7	n_β~1.7	Hainite Silicate of Na,Ca,Ti and Zr	0.012	Large r > v very strong	Unknown	Prism-acicular (c)	{010} dist
*1.688 1.711	n_β 1.688–1.711 n_α 1.685–1.707 n_γ 1.697–1.725	Zoisite $Ca_2Al_3O \cdot SiO_4 \cdot Si_2O_7 \cdot (OH)$	0.005 to 0.020	0° – 60° v > r (α) or r > v (β) strong	Orthorhombic	Columnar, fibrous, granular	{010} perfect
1.698 1.702	n_β 1.698–1.702 n_α 1.689–1.695 n_γ 1.795–1.825	Labuntsovite $(K,Ba,Na)(Ti,Nb)(Si,Al)_2(O,OH)_7 \cdot H_2O$	0.106 to 0.130	20° – 44°	Monoclinic β = 116.9°	Primatic crystals	{$\bar{1}$02}
*1.683 1.719	n_β 1.683–1.719 n_α 1.676–1.716 n_γ 1.702–1.736	Fassaite $Ca(Mg,Fe^{3+},Al)(Si,Al)_2O_6$	0.018 to 0.028	51° – 62° r > v moderate	Monoclinic β = 106°	Short prism, massive	{110} dist {100} parting
*1.690 1.712	n_β 1.690–1.712 n_α 1.690–1.702 n_γ 1.702–1.719	Riebeckite $Na_2Fe_3^{2+}Fe_2^{3+}(Si_4O_{11})_2(OH)_2$	0.012 to 0.017	0° to 90° v > r strong	Monoclinic β =104°	Acicular, columnar (c)	{110} dist
1.703	n_β 1.703 n_α 1.701 n_γ 1.706	Serendibite $Ca_4(Mg,Fe,Al)_6(Al,Fe)_9(Si,Al)_6B_3O_{40}$	0.005	Large ~83°	Triclinic	Massive, irregular grains	None
1.704	n_β 1.704 n_α 1.696 n_γ 1.713	Zwieselite $(Fe,Mn,Mg,Ca)_2(PO_4)(F,OH)$	0.017	87° r > v strong	Monoclinic β = 105.7°	Rough crystals, massive	{001} good, {010} fair {100} poor
1.704	n_β 1.704 n_α 1.698 n_γ 1.720	Akatoreite $Mn_9(Si,Al)_{10}O_{23}(OH)_9$	0.022	65.5°	Triclinic α = 104.3° β = 93.6° γ = 104°	Prismatic, striated crystals ∥ (a)	{010} perfect {0$\bar{1}$2} imperf
*1.700 1.711	n_β 1.700–1.711 n_α 1.700–1.708 n_γ 1.715–1.723	Beusite $(Mn,Fe,Ca,Mg)_3(PO_4)_2$	0.020 to 0.015	25° – 45° r > v strong	Monoclinic β = 99°	Prismatic	{010} dist {100} fair
1.705	n_β 1.705 n_α 1.703 n_γ 1.723	Kulanite $Ba(Fe,Mn,Mg,Ca)_2(Al,Fe)_2(PO_4)_3(OH)_3$	0.020	32° r > v very strong, triclinic	Triclinic α ~90° β = 100.4° γ ~90°	Platy {$\bar{1}$01}, rosettes	{010} & {100} good
*1.672 1.741	n_β 1.672–1.741 n_α 1.671–1.735 n_γ 1.703–1.761	Augite $(Ca,Mg,Fe^{2+},Na)(Mg,Fe^{2+},Fe^{3+},Al,Ti)(Si,Al)_2O_6$	0.018 to 0.030	25° – 60° r > v weak, inclined	Monoclinic β = 105°	Prism crystals (c), octagonal, granular	{110}, dist, {100} & {001} parting
*1.699 1.714	n_β 1.699–1.714 n_α 1.695–1.709 n_γ 1.719–1.736	Graftonite (Repossite) $(Fe^{2+},Mn^{2+},Ca)_3(PO_4)_2$	0.024 to 0.027	43° – 60° r ≷ v strong	Monoclinic β = 99°	Massive	{010} dist {100} indist

* n_β variation exceeds 0.01

Twinning	Orientation	Color in thin section	Hardness Sp Gravity	Color hand sample	Alteration	Occurrence	Remarks	Reference
	b = X, c ∧ Z = -36° to -40°	X = Y = colorless, pale yellow, Z = yellow	4½-5 3.98-4.02	Colorless, yellow		Oxide zones Zn-deposits with adamite, limonite, mimetite		Am Min 48, 1258-1265
{100}	b ~ Z. Ext ~ ∥ cleavage on {100}	X ~ colorless, Y = yellowish, Z = pale yellow	5 3.18	Yellow-brown		Alkali volcanics		Vlasov II, 385
None	a = Z, b = Y, c = X (α-zoisite); a = Z, b = X, c = Y (β-zoisite)	Colorless	6 3.15-3.36	Gray, green, brown	High T recrystallization to Ca-feldspar	Schists, granulites; hydrothermal alteration; mafic igneous		p. 150
	b = X, a ~ Z, c ~ Y	X = yellowish, Z = brownish-yel	~6 2.90-2.96	Rose, brownish yellow		Voids in alkali plutons with albite, natrolite	Epidote group	Am Min 41, 163
{100} simple, lamellar	b = Y, c ∧ Z = -41° to -47°	X = Z = pale grn, Y = pale yellow-green	6 2.96-3.34	Lt-dk green, black		Carbonate contact metamorphic zones	Pyroxene group	DHZ 2, 161-166
{100} simple or multiple	b = Y, c ∧ X = -8° to -7°, a ∧ Z = +6° to +7°	X = dk blue, Y = dk gray-blue, Z = yellow-brown	6 3.1-3.4	Dk blue-green to black	Limonite, siderite, quartz	Alkali granites and syenites with aegirine. Pegmatites. Gneiss-quartzite	Amphibole group	p. 237
Polysynthetic		X = very pale yellowish green, Y ~ colorless, Z = Prussian blue	6½-7 3.42	Blue-green, dk blue, grayish		Contact limestone deposits with idocrase, diopside, scapolite, spinel		Am Min 17, 457-465
	b = Y, c ∧ Z ~ +41°	Yellow-brown to red-brown X > Z-Y	5-5½ 3.93-3.97	Clove brown to black		Granite pegmatites	Triplite-Zwieselite series Soluble acids	Dana II, 849-852
{0$\bar{2}$1}	X ∧ {010} = 58°, Y ∧ {010} = 30°, Z ∧ {010} = 13°	X = colorless, Y = pale yellow, Z = light canary yellow	6 3.47-3.49	Yellow-orange to brown		With other Mn-minerals in metachert		Am Min 56, 416-426
	b = X, c ∧ Z = 37°	Reddish-brown	5 3.70-3.72	Red-brown		Granite pegmatites	Beusite-graftonite series	Am Min 53, 1799-1814
	b ~ Y, c ∧ Z = -8°	X = brownish green, Y = green, Z = pale brown	4 3.91-3.92	Blue to green		Siderite Fe formations	Fe-bjarebyite	Can Min 14, 127-131
	b = Y, c ∧ Z = -35° to -50°, a ∧ X = -20° to -35°	Colorless, neutral gray, pale green, brown-violet	5-6 3.2-3.6	Gray-green, black, brown	Hornblende, uralite, chlorite, epidote, carbonates, biotite	Gabbro, granodiorite, basalt, ultramatics, mafic gneisses	Pyroxene group	p. 196
	b = X, c ∧ Z = +36°	X = Y = colorless, Z = pale pink	5 3.67-3.79	Salmon pink, red-brown		Granite pegmatites	Easily soluble acids	Am Min 53, 742-750

n_β Var.	Refractive Index	Mineral Name and Composition	$(n_\gamma - n_\alpha)$	2V angle Dispersion	Crystal System	Habit	Cleavage
1.711 1.712	n_β 1.711-1.712 n_α 1.706-1.708 n_γ 1.724-1.718	Merwinite $Ca_3MgSi_2O_8$	0.018 to 0.010	67° - 71° r > v weak	Monoclinic β = 91.9°	Massive, granular, rounded crystals	{100} poor, {010} very poor
1.711	n_β 1.711 n_α 1.709-1.707 n_γ 1.724-1.729	Brandtite $Ca_2Mn(AsO_4)_2 \cdot 2H_2O$	0.015 to 0.022	23° v > r strong	Monoclinic β = 99.5°	Prism (c), radiated fibrous	{010} perf {001} dist
1.712	n_β 1.712 n_α 1.707 n_γ 1.723	Manganhumite $(Mn_{0.68}Mg_{0.30},F_{0.01}Ca_{0.01})(OH)_2(SiO_4)_3$	0.016	37° r > v perc	Orthorhombic		{010} perfect
*1.704 1.728	n_β 1.704-1.728 n_α 1.694-1.725 n_γ 1.719-1.735	Roselite $Ca_2(Co,Mg)(AsO_4)_2 \cdot 2H_2O$	0.025 to 0.010	75° - 60° v > r weak	Monoclinic β = 100.9°	Prism (c) to tabular {001}, spherical aggregates	{010} perf
**1.795 1.649	n_β 1.795-1.649 n_α 1.784-1.631 n_γ 1.814-1.663	Scorodite $Fe^{3+}AsO_4 \cdot 2H_2O$	~0.030	~75° r > v strong	Orthorhombic	Bipyramids, tabular, prism (b), massive	{201} imperf {001} & {100} traces
1.713	n_β 1.713 n_α 1.703 n_γ 1.722	Gerhardtite $Cu_2NO_3(OH)_3$	0.019	Large ~ 90° v > r very strong	Orthorhombic	Thick tabular	{001} perf, {100} good
1.711 1.716	n_β 1.711-1.716 n_α 1.703-1.716 n_γ 1.732-1.745	Johannsenite $Ca(Mn,Fe^{2+})(SiO_3)_2$	0.029	68° - 73° r > v	Monoclinic β = 105°	Stubby prism crystals. Radiated acicular crystals	{110} dist {001}, {010} & {100} parting
*1.705 1.725	n_β 1.705-1.725 n_α 1.682-1.706 n_γ 1.730-1.752	Diaspore α-AlO(OH)	0.052 to 0.046	84° - 86° v > r weak	Orthorhombic	Aggregate tiny scales {010}, fibrous	{010} perf {110} imperf {100} poor
1.715	n_β 1.715 n_α 1.700 n_γ 1.732	Eveite $Mn(OH)(AsO_4)$	0.032	65° r > v moderate	Orthorhombic	Tabular sheaf-like aggregate	{101} dist
*1.675 1.754	n_β 1.675-1.754 n_α 1.674-1.748 n_γ 1.688-1.764	Pumpellyite $Ca_2Al_2(Mg,Fe^{2+},Fe^{3+},Al)(SiO_4)(Si_2O_7)(OH)_2(H_2O,OH)$	0.002 to 0.022	10° - 85° (rarely -) v > r, rarely r > v	Monoclinic β = 97.6°	Acicular, bladed, radiated	{001} dist, less so {100}
*1.707 1.725	n_β 1.707-1.725 n_α 1.703-1.715 n_γ 1.709-1.734	Clinozoisite $Ca_2Al_3O \cdot SiO_4 \cdot Si_2O_7 \cdot OH$	0.004 to 0.012	14° - 90° v > r strong	Monoclinic β = 115.4°	Columnar, acicular, granular	{001} perfect
1.72 1.715	n_β 1.72-1.715 n_α 1.64-1.68 n_γ 1.75-1.77	Zippeite $(UO_2)_2(SO_4)(OH)_2 \cdot 4H_2O$	0.11 to 0.09	80° - 90°	Orthorhombic	Microscopic crystals, coatings	{010} perfect (?)
1.717	n_β 1.717 n_α 1.715 n_γ 1.728	Kanoite $(Mn^{2+},Mg)_2(Si_2O_6)$	0.011	40° - 42°	Monoclinic	Grains	{110} perfect

* n_β variation exceeds 0.01

** n_β variation exceeds 0.10

Twinning	Orientation	Color in thin section	Hardness Sp Gravity	Color hand sample	Alteration	Occurrence	Remarks	Reference
{100}, {611}, {6$\bar{1}$1} poly-synthetic	b = Y, c ∧ X = 13°	Colorless	6 3.15-3.33	Colorless, pale greenish		Limestone contact zones with spurrite, gehlenite, larnite		Am Min 59, 1117-1120
{100} common	b = X, c ∧ Y = 6°, c ∧ Z = 84°	Colorless	3½ 3.67-3.70	Colorless, white		Hydrothermal ore veins	Soluble dilute acids	Dana II, 725-727
			4 3.83			Mn skarns with katoptrite, mangano-stibite, galaxite, etc.	Mn-humite	Min Mag 42, 133
{100} common, fourlings	b = Y or X, c ∧ Z = 90° to +70	Pale rose, X > Y > Z	3½ 3.50-3.74	Pink to dark rose		Drusy cavities with quartz, chalcedony, erythrite	Dimorph. β-roselite	Dana II, 723-725
	a = X, b = Z, c = Y	Colorless, lt green, green-brown Z > X = Y	3½-4 3.31-3.15	Green, brown, violet, yellow	Limonite	Oxide zones (gossans) As-deposits. Also primary hydrothermal	Scorodite-Mansfieldite series	Dana II, 763-767
	a = X, b = Y, c = Z	X ~ Y = green, Z = blue	2 3.40-3.43	Emerald green, dark green		Oxide zones Cu-deposits	Soluble dilute acids	Dana II, 308-309
{100} simple or lamellar	b = Y, c ∧ Z = -46° to -49°, a ∧ X = -31° to -34°	Colorless	6 3.44-3.55	Clove-brown, gray, green, black	Rhodonite, xonotlite	Limestone contact zones with rhodonite. Veins in rhyolite	Pyroxene group. Johannsenite-hedenbergite series	DHZ 2, 75-78
{061} or {021} common	a = Z, b = Y, c = X	Colorless, rarely pale tints	6½-7 3.3-3.5	White		Bauxite, laterites, Al-clays. Meta-morphosed Al-rocks. Hydro alteration	Dimorph boehmite	p. 44
		Colorless	4 3.76	Apple-green		Fissures in Fe-Mn ores	Mn-adamite	Am Min 55, 319-320
Sectored on {001} and {100}	b = Y, a ∧ X = +4° to +32°, c ∧ Z = -4° to +34° (p. 155)	Blue-green to yellow, red-brown Y > Z > X	6 3.2-3.3	Green, brownish	Epidote, actinolite	Low-grade regional metamorphic rocks, blue or green-schists	Anomalous interference colors due to dispersion	p. 155
Multiple on {100}	b = Y, c ∧ X = -85° to 0°, a ∧ Z = -60° to +25°	Colorless	7-6 3.21-3.49	Yellow, gray, green	High T re-crystallization to Ca-feldspar, pyroxene	Low grade matamorphic (regional and contact), plutonic igneous	Anomalous interference colors	p. 148
Common {001} (?)	a = Z, b = X, c = Y	X ~ colorless, Y = very pale yellow or yellow-orange, Z = pale yellow or yellow-orange	3.66-3.68	Yellow, orange, red, red-brown		Efflorescence in mine tunnels and dump heaps	Fluor U.V.-variable and undependable	USGS Bull 1064, 141-147
{100} polysynthetic	b = Y, c ∧ Z = 42°	Colorless, pinkish-brown	6 3.60-3.66	Lt pinkish-brown		Seams in metamorphic rock with spessartite, cummingtonite, pyroxmangite	Pyroxene group	Am Min 63, 598

Biaxial +

n_β Var.	Refractive Index	Mineral Name and Composition	$(n_\gamma - n_\alpha)$	2V angle Dispersion	Crystal System	Habit	Cleavage
1.717	n_β 1.717 n_α 1.709 n_γ 1.729	Stringhamite $CuCaSiO_4 \cdot 2H_2O$	0.020	80°	Monoclinic	Radiated aggregate tiny crystals, botryoidal	
1.717	n_β 1.717 n_α 1.712 n_γ 1.726	Johachidolite $CaAl(B_3O_7)$	0.014	70° r > v strong	Orthorhombic	Equant anhedral grains	None, {201} parting
*1.706 1.730	n_β 1.706-1.730 n_α 1.697-1.722 n_γ 1.726-1.750	Ferrosalite $Ca(Fe,Mg)(SiO_3)_2$	0.029 to 0.035	58° - 62° r > v moderate	Monoclinic β = 105°	Prism-acicular xls (c)	{110} dist, {001} & {100} parting
*1.703 1.732	n_β 1.703-1.732 n_α 1.701-1.729 n_γ 1.705-1.734	Sapphirine $(Mg,Fe)_2Al_4O_6(SiO_4)$	0.004 to 0.007	66° - 90° v > r strong, inclined dist	Monoclinic β = 125.3°	Granular, prism crystals (c)	{010}, {001} and {100} poor
1.718	n_β 1.718 n_α 1.709 n_γ 1.734	Metahohmannite $Fe_2(SO_4)_2(OH)_2 \cdot 3H_2O$	0.025	~70°	Unknown	Massive	
1.719	n_β 1.719 n_α 1.707 n_γ 1.741	Strengite $FePO_4 \cdot 2H_2O$	0.034	Mod (75°) to small v > r strong	Orthorhombic	Radiated-fibrous, crusts, tab-prism crystals	{010} dist {001} poor
1.715 1.722	n_β 1.715-1.722 n_α 1.707-1.715 n_γ 1.730-1.740	Larnite β-Ca_2SiO_4	0.023 to 0.025	63° - 73°	Monoclinic β = 94.5°	Tabular, massive, granular	{100} dist
1.72	n_β 1.72 n_α 1.71 n_γ 1.73	Haradaite $SrVSi_2O_7$	0.02	~90°	Orthorhombic	Massive	{010} perfect
*1.71 1.73	n_β 1.71-1.73 n_α 1.708-1.720 n_γ 1.722-1.732	Varulite $(Na_2,Ca)(Mn^{2+},Fe^{2+})_2(PO_4)_2$	0.014 to 0.012	Large ~ 70° r > v	Monoclinic	Massive, granular	{001} & {010} distinct
1.721	n_β 1.721 n_α 1.707 n_γ 1.739	Leucophosphite $KFe_2^{3+}(PO_4)_2(OH) \cdot 4\frac{1}{2}H_2O$	0.032	~84°	Monoclinic (Pseudo-ortho) β = 102.4°	Short prism, lamellar, earthy	{100} perfect
1.721	n_β 1.721 n_α 1.712 n_γ 1.731	Adelite $CaMgAsO_4(OH)$	0.019	69° - 90° v > r perc	Orthorhombic	Massive, crystals elongated on a	None
1.721	n_β 1.721 n_α 1.713 n_γ 1.734	Barium-haradaite $(Ba,Sr)VSi_2O_7$	0.021	77° v > r very strong	Orthorhombic	Veinlets	{010} perf, {100} & {001} distinct
*1.716 1.728	n_β 1.716-1.728 n_α 1.712-1.725 n_γ 1.725-1.740	Bredigite α-Ca_2SiO_4	0.013 to 0.019	30° - 34°	Orthorhombic	Rounded grains	{130} dist
*1.716 1.730	n_β 1.716-1.730 n_α 1.700-1.723 n_γ 1.755-1.795	Rutherfordine (Diderichite) (Rutherfordite) UO_2CO_3	0.055 to 0.072	37° - 67°	Orthorhombic	Lath-like crystals (c), radiated fibrous aggregates	{010} perfect {001} dist

* n_β variation exceeds 0.01

Twinning	Orientation	Color in thin section	Hardness Sp Gravity	Color hand sample	Alteration	Occurrence	Remarks	Reference
	b = X, c ∧ Y = 2.5°	X = lt gray blue, Y = lt blue, Z = dk blue	3.16-3.18	Azure blue		Diopside-magnetite skarn with thaumasite, bornite		Am Min 61, 189-192
	a = X, b = Z, c = Y	Colorless	7½ 3.37-3.43	Colorless		Nepheline dikes with albite, diopside, scapolite, phlogopite	Fluor U.V.-pale blue	Am Min 62, 327-329
{100} or {001} lamellar	b = Y, c ∧ Z = -43° to -47°	X = blue-green, green, Y = brown-green, blue-green, Z = yellow-green	~5½ 3.4	Green to brown	Uralite	Calc-silicate skarns. Alkali granites with forsterite	Pyroxene group. Diopside-hedenbergite series	p. 191
{010} & {100} lamellar	b = Y, c ∧ Z = +6° to +15°, a ∧ X = +41° to +50°	Colorless to pale blue Z > Y > X	7½ 3.40-3.58	Pale blue, green or gray	Corundum, biotite & talc	High T hornfels or gneiss-granulite with cordierite, spinel		p. 138
		X = pale yellow, Y = orange, Z = reddish-brown		Orange		Alteration of hohmannite in oxide zones of ore-deposits		Dana II, 608
{201} rare	a = X, b = Z, c = Y	Colorless to pale pink	3½-4½ 2.85-2.87	Colorless, dk to lt violet, red		Pegmatites as alteration of Fe-phosphates with metastrengite, cacoxenite	Variscite-strengite series	Dana II, 756-761
{100} polysynthetic, common	b = Z, c ∧ X = 13°-14°	Colorless	~6 3.28-3.31	Colorless, gray		High T limestone contact zones with spurrite, bredigite, shannonite	Dimorph bredigite Gelat weak acids	Am Min 51, 1766-1774
		Colorless, pale blue	3.75-3.80	Light blue				Am Min 56, 1123
	b = Y	X = yellow-green, Z = grass-green	5 3.58	Yellow-green	Mangan-alluaudite	Granite pegmatites	Hühnerkobelite-varulite series	Dana II, 669-670
			3½ 2.30-2.95	Buff, brown, green, red, white		Pegmatites with rockbridgeite, triphylite. Earthy in sed rocks	Soluble HCl	Am Min 57, 397-410
	a = X, b = Z, c = Y	Colorless	5 3.70-3.79	Colorless, gray, yellowish		Mn-deposits		Dana II, 804-806
	a = X, b = Y, c = Z	X ~ colorless, Y = colorless, yellowish-green, Z = bluish green	4½ 3.80-3.83	Bright green		Veinlets in Mn-ores with rhodonite, Mn-goldmanite, quartz		Am Min 60, 340
{110} pseudo-hexagonal	a = Y, b = X, c = Z	Colorless	3.38-3.40	Colorless, gray		Contact metamorphic zones in limestone. Slags.	Inverts to larnite at atm. T&P	Am Min 51, 1766-1774
	a = Z, b = X, c = Y	Colorless, pale yellow. Weak pleochroism	Soft 5.72	White, yellow, orange, brn		Pegmatites as alteration of uraninite	Soluble acid	USGS Bull 1064, 104-106

n_β Var.	Refractive Index	Mineral Name and Composition	$(n_\gamma - n_\alpha)$	2V angle Dispersion	Crystal System	Habit	Cleavage
*1.703 1.746	n_β 1.703-1.746 n_α 1.678-1.740 n_γ 1.733-1.765	Astrophyllite $(K,Na)_3(Fe,Mn)_7Ti_2Si_8O_{24}(O,OH)_7$	0.025 to 0.055	70° - 90° r > v	Triclinic α = 90° β = 94° γ = 103°	Bladed, stellate groups	{010} perfect {100} poor
1.725	n_β 1.725 n_α 1.712 n_γ 1.760	Krinovite $NaMg_2CrSi_3O_{10}$	0.048	61° - 64°	Monoclinic β = 103°	Minute grains	None
1.725	n_β 1.725 n_α 1.715 n_γ 1.738	Homilite $(Ca,Fe)_3B_2Si_2O_{10}$	0.023	80° r > v distinct, horizontal strong	Monoclinic β = 90.4°	Tabular {001}	None
1.72 1.73	n_β 1.72-1.73 n_α 1.718-1.727 n_γ 1.731-1.738	Hühnerkobelite $(Na_2,Ca)(Fe^{2+},Mn^{2+})_2(PO_4)_2$	0.013 to 0.011		Monoclinic β = 97.5°	Prism (b) or (c), granular	{001} & {010} distinct
1.726	n_β 1.726 n_α 1.725 n_γ 1.730	Triploidite $(Mn^{2+},Fe^{2+})_2PO_4(OH)$	0.005	Moderate ~ 55° r > v extreme, horizontal mod	Monoclinic β = 108.1°	Prism-fibrous (c), parallel-radiated aggregate	{010} dist {120} fair
1.730	n_β 1.730 n_α 1.702 n_γ 1.823	Kanonaite $(Mn^{3+}_{0.76}Al_{0.23}Fe^{3+}_{0.02})AlOSiO_4$	0.121	60°	Orthorhombic		{110} poor
*1.742 1.710	n_β 1.742-1.710 n_α 1.722-1.700 n_γ 1.758-1.730	Aegirine-augite $(Na,Ca)(Fe^{3+},Fe^{2+}Mg,Al)(SiO_3)_2$	0.036 to 0.030	90° - 70° r > v moderate-strong	Monoclinic β = 106°	Stubby, eight-sided crystals	{110} dist, {001} & {100} parting
*1.715 1.741	n_β 1.715-1.741 n_α 1.711-1.738 n_γ 1.724-1.751	Rhodonite $MnSiO_3$	0.011 to 0.014	61° - 76° v > r weak, inclined	Triclinic α = 85° β = 94° γ = 112°	Tabular (001) xls, massive	{100} & {010} perfect, {001} dist
1.73	n_β 1.73 n_α 1.73 n_γ 1.74	Ottrelite $(Mn,Fe^{2+})_2(Al,Fe^{3+})Al_3O_2(SiO_4)_2(OH)_4$	~0.01	Moderate r > v strong	Triclinic?	Foliated on {001}, platy agg	{001} perfect, {110} imperf
*1.706 1.761	n_β 1.706-1.761 n_α 1.705-1.760 n_γ 1.726-1.792	Clinoferrosilite $(Fe,Mg)_2(SiO_3)_2$	0.021 to 0.032	25° - 20°	Monoclinic β = 108°	Tiny acicular xls (c)	{110} dist
1.731	n_β 1.731 n_α 1.720 n_γ 1.752	Babingtonite $Ca_2Fe^{2+}Fe^{3+}Si_5O_{14}(OH)$	0.032	76° r > v strong	Triclinic α = 91.5° β = 93.9° γ = 104.1°	Short prismatic	{001} perfect {1$\bar{1}$0} imperf
1.732	n_β 1.732 n_α 1.727 n_γ 1.789	Batisite $Na_2BaTi_2(Si_2O_7)_2$	0.062	7° - 40° v > r strong	Orthorhombic	Elongated prism (c)	{100}dist
1.733	n_β 1.733 n_α 1.723 n_γ 1.755	Mukhinite $Ca_2(Al_2V)(SiO_4)_3(OH)$	0.032	88°	Monoclinic	Short prism crystals	{001} very perfect {100} perfect
1.730 1.736	n_β 1.730-1.736 n_α 1.722-1.728 n_γ 1.750-1.757	Hedenbergite $CaFe(SiO_3)_2$	0.028 to 0.029	62° - 63° r > v strong	Monoclinic β = 104°	Radiated acicular-column crystals (c)	{110} dist {001} & {100} parting

* n_β variation exceeds 0.01

Twinning	Orientation	Color in thin section	Hardness Sp Gravity	Color hand sample	Alteration	Occurrence	Remarks	Reference
	a ~ Y, b ~ X, c ~ Z	X = deep yellow, red-brown, Y = orange-yellow, Z = pale yellow	3 3.3-3.4	Golden yellow		Alkali-rich plutons		Win & Win II, 480-481
Multiple common	b = X	X = yellow-green, Y = blue-green, Z = greenish-blk	5½-7 3.38-3.44	Deep emerald green		Graphite nodules in octahedrite meteorites		Science 161 786-787
{001}	b = Z, c ∧ Y = +1°	†X = bluish-green, Y = red-brown, Z = yellow-brown	5 3.36-3.58	Dk brown, black	Metamict	Pegmatites	Gelat HCl. No pleochroism in thin section	Win & Win II, 356-357
	b = Y	X = lt brown, Z = blue-green	5 3.5-3.6	Greenish-black olive green	Alluaudite	Pegmatites with other phosphates. Alter triphylite	Hühnerkobelite-varulite series	Am Min 50, 713-717
	b = X, c ∧ Z = 4°	Pale pink to pale brown. Z > X = Y	4½-5 3.697	Pink, yellow to brown		Granite pegmatites with alluaudite, loellingite, lithiophilite, triplite.	Triploidite-wolfeite. Sol acids	Dana II, 853-855
	a = X, b = Y, c = Z	X = yellow-green, Y = bluish green, Z = golden yellow	6½ 3.395	Greenish black		Schist with gahnite, chlorite, coronadite, quartz		Con Min Pet 66, 325-332
Common on {100}	b = Y, c ∧ X = +12° to +20°, a ∧ Z = +6° to -4°	X = bright green, Y = yellow-green, Z = pale yellow	6 3.4-3.6	Dark green to black	Chlorite, epidote, Na-uralite, Fe-oxides	Feldspathoidal intrusives, Na-Fe-rich schists & gneisses	Pyroxene group	p. 199
{010} lamellar, rare	a ∧ X ~ 5°, b ∧ Y ~ 20°, c ∧ Z ~ 25°. Extinction is inclined	Colorless to pale pink. X = orange, Y = rose pink, Z = lt yellow-orange	5½-6½ 3.57-3.76	Rose-red, pink, red-brown	Rhodochrosite, Mn-oxides	Hydrothermal ore veins with metal sulfides. Mn-skarns	Pyroxenoid group	p. 211
{001} multiple	Optic plane ~ ∥ {010}, c ∧ Z ~ 12°	X = olive green, Y = blue, Z = yellow-green	6-7 3.3	Gray-dk. green	Chlorite, sericite	Porphyroblasts in schists	Mn chloritoid	p. 140
{100}	b = X, c ∧ Z ~ 34°	Colorless	5-6 4.068	Colorless, amber	Inverts to Fe-orthopyroxene	Lithophysae in obsidian	Pyroxene group. Clinoenstatite-clinoferrosalite series	p. 189
	Optic plane ~ ∥ {1$\bar{1}$0} and ~ ⊥{001}	X = dk green, Y = pale lilac, Z = brown, green	5½-6 3.26-3.36	Green-black, brown-black		Cavities and fractures in granite, gneiss, trap rock		Win & Win II, 462
	a = X, b = Y, c = Z	X = colorless, Y = yellow-brown, Z = red-brown	5.9 3.43	Dk brown		Nepheline pegmalites		Am Min 45, 908-909
{100}			8	Brownish-black		Contact marbles with goldmanite, sulfide minerals	V-clinozoisite (V:Al=1:2). Epidote group	Am Min 55, 321-322
{100} or {001} lamellar	b = Y, c ∧ Z = -47° to -48°	X = dk green, blue-green, Y = brown-green, blue-green, Z = yellow-green	~5½ 3.5	Dk green or brown to blk	Uralite	Skarns and siliceous hornfels with grunerite. Alkali granites	Pyroxene group. Diopside-hedenbergite series	p. 191

n_β Var.	Refractive Index	Mineral Name and Composition	$(n_\gamma - n_\alpha)$	2V angle Dispersion	Crystal System	Habit	Cleavage
1.737	n_β 1.737 n_α 1.726 n_γ 1.766	Innelite $Na_2(Ba,K)_4(Ca,Mg,Fe)Ti_3SiO_4O_{18}(OH,F)_{1.5}(SO_4)$	0.040	~82° r > v strong	Triclinic α ~99° β = 95° γ ~90°	Platy	{010}, {110} & {101} perf {001} dist
**1.786 1.687	n_β 1.786-1.687 n_α 1.775-1.72 n_γ 1.815-1.78	Beraunite $Fe^{2+}Fe^{3+}_5(OH)_5(H_2O)_4(PO_4)_4 \cdot 2H_2O$	0.040 to 0.083	medium large r > v moderate	Monoclinic β = 93.5°	Radiated fibrous aggregate, crusts, tabular {100}	{100} dist
1.738	n_β 1.738 n_α 1.726 n_γ 1.789	Antlerite $Cu_3SO_4(OH)_4$	0.063	53° v > r very strong	Orthorhombic	Nearly equant crystals	{010} perfect {100} poor
*1.728 1.750	n_β 1.728-1.750 n_α 1.726-1.748 n_γ 1.744-1.764	Pyroxmangite $(Mn,Fe)SiO_3$	0.018 to 0.016	35° - 46° r > v moderate	Triclinic α = 84° β = 94.3° γ = 113.7°	Tabular crystals, massive	{110} & {1$\bar{1}$0} perf {010} & {001} poor
*1.718 1.764	n_β 1.718-1.764 n_α 1.697-1.743 n_γ 1.741-1.795	Planchéite $Cu_8(Si_4O_{11})_2(OH)_4 \cdot H_2O$	0.044 to 0.052	90° - 80°	Orthorhombic	Radiated fibrous aggregates	
1.740	n_β 1.740 n_α 1.740 n_γ 1.760	Andremeyerite $BaFe_2Si_2O_7$	0.020	0° - 40° very strong	Monoclinic β = 118.3°	Prismatic crystals	{100} & {010} perfect
*1.734 1.749	n_β 1.734-1.749 n_α 1.732-1.747 n_γ 1.741-1.758	Chrysoberyl $BeAl_2O_4$	0.008 to 0.011	70° - 10° r > v weak	Orthorhombic	Prism (c) crystals, tabular {100}, granular	{011} dist, {010} indist {100} poor
1.742	n_β 1.742 n_α 1.741 n_γ 1.746	Wolfeite $(Fe^{2+},Mn^{2+})_2PO_4(OH)$	0.005	Moderate ~ 55° r > v strong, horiz. moderate	Monoclinic β = 108°	Prism-fibrous (c), parallel-radiated aggregate	{010} dist {120} fair
1.743	n_β 1.743 n_α 1.701 n_γ 1.787	Libethenite Cu_2PO_4OH	0.086	~90° v > r strong	Orthorhombic	Short prism (c), crusts	{100} & {010} indistinct
1.742 1.744	n_β 1.742-1.744 n_α 1.708-1.722 n_γ 1.763-1.773	Adamite $Zn_2AsO_4(OH)$	0.041 to 0.056	86° - 90° r > v strong	Orthorhombic	Drusy crystal groups	{101} dist {010} poor
1.74	n_β>1.74	Rooseveltite $BiAsO_4$			Monoclinic	Thin botryoidal crusts	
1.747	n_β 1.747 n_α 1.737 n_γ 1.768	Creaseyite $Cu_2Pb_2(Fe,Al_2)Si_5O_{17} \cdot 6H_2O$	0.031	~70° v > r weak	Orthorhombic	Radiated fibrous (c), matted	
*1.740 1.754	n_β 1.740-1.754 n_α 1.736-1.747 n_γ 1.745-1.762	Staurolite $Fe^{2+}_2Al_9O_6(SiO_4)_4(O,OH)_2$	0.009 to 0.015	80° - 90° r > v weak-mod	Monoclinic (pseudo-ortho) β ~ 90°	Prism crystals (c), penetration twins	{010} poor
1.748	n_β 1.748 n_α 1.746 n_γ 1.768	Prosperite $HCaZn_2(AsO_4)_2(OH)$	0.020	34° r >> v	Monoclinic <β = 104°32'	Rad. prism. crystals	None

* n_β variation exceeds 0.01

** n_β variation exceeds 0.10

Twinning	Orientation	Color in thin section	Hardness Sp Gravity	Color hand sample	Alteration	Occurrence	Remarks	Reference
Polysynthetic Manebach	a ~ Z, optic plane ~ 12° to {001}	X = Y = pale yel, Z = brownish-yel	4½-5 3.96	Pale yellow, brown		Alkali plutons and pegmatites		Am Min 47, 805-806
{100} inter-penetration	b = Z, c ∧ Y = 2° to 5°	X ~ Y = colorless, Z = red	2-4 2.87-3.01	Red-brown, greenish-brn		Oxide zones of ore deposits, altered pegmatites	Soluble HCl, var eleonorite	Am Min 55, 135-169
	a = Y, b = X, c = Z	X = yellow-green, Y = blue-green, Z = green	3½ 3.88-3.94	Light to dark green		Oxide zones of Cu deposits in arid regions	Soluble in weak H_2SO_4	Dana II, 544-546
{110} lamellar, {001} simple	Optic plane ~ ⊥{1̄10}	Colorless to pale lilac	5½-6 3.61-3.91	Pink, purplish, red-brown, brown		Metamorphic or metasomatic rocks with Mn-minerals	Pyroxene group	DHZ 2, 196-202
	a = Y, b = X, c = Z	X = very pale blue Y = pale blue, Z = deep blue	5½ 3.65-3.85	Pale blue to dk blue		Oxide zone Cu-deposits with malachite		Science 154, 506-507
{100} multiple	b = Z, c ∧ X ~ 30°	X = pale bluish green, Y ~ Z = colorless	5½ 4.14-4.15	Pale emerald-green		Vesicles in alkali lavas with nepheline, leucite, melilite		Am Min 59, 381
{031} simple or cyclic	a = X, b = Y, c = Z	Colorless. X = pale rose, Y = greenish-yel, Z = pale green	8½ 3.68-3.75	Yellow to yellow-green	Dolomite, mica, kaolin	Granite pegmatites. Contact dolomite marbles. Gem placers	Var alexandrite red in artificial light	p. 39
	b = X	Pale brown. Z ~ X=Y	4½-5 3.79	Clove to red-brown, green		Granite pegmatites. Alter triphylite	Triploidite-wolfeite series. Soluble acids	Dana II, 853-855
None observed	a = Z, b = X, c = Y	Pale bluish-green to pale green. Weak pleochroism	4 3.93-3.97	Dk - lt green, dk olive		Oxide zones Cu-deposits	Easily soluble acids or NH_4OH	Dana II, 862-864
	a = X, b = Z, c = Y	Colorless	3½ 4.32-4.48	Yellowish-green		Oxide zones of ore deposits		Dana II, 864-866
		Colorless	4-4½ 6.86	Gray			Soluble acids. Isostructural with monazite	Dana II, 697
	a = X, b = Y, c = Z	X = yellow, Y = lt green, Z = yellow-green X = Z > Y	2½ 4.0-4.1	Green		Andesite breccia with wulfenite, chrysocolla, descloizite		Min Mag 49 227-231
{023} or {232} penetration, {031}	a = Y, b = X, c = Z	X = colorless, lt yellow, Y = yellow, yellow-brown, Z = golden yellow, red-brown	7-7½ 3.74-3.83	Dk brown, red-brown, yellow-brown	Sericite, chlorite, limonite	Pelitic schists with almandine, kyanite, micas, tourmaline		p. 136
	b = Y, c ∧ Z = 27°	Colorless	4½ 4.31	White to colorless		Vugs in partly altered sulfide ore		Am Min 65, 208

n_β Var.	Refractive Index	Mineral Name and Composition	$(n_\gamma - n_\alpha)$	2V angle Dispersion	Crystal System	Habit	Cleavage
1.75	n_β 1.75 n_α 1.715 n_γ 1.80	Erythrosiderite $K_2FeCl_5 \cdot H_2O$	0.085	62° r > v strong	Orthorhombic	Tabular {100}	{210} & {011} perfect
1.75	n_β 1.75 n_α 1.715 n_γ 1.80	Kremersite $(NH_4,K)_2FeCl_5 \cdot H_2O$	0.085	62° r > v strong	Orthorhombic	Pseudo-octahedral	{210} & {011} perfect
*1.737 1.766	n_β n_α 1.737 n_γ 1.766	Arsenuranylite $Ca(UO_2)_4(AsO_4)_2(OH)_4 \cdot 6H_2O$	0.029		Orthorhombic	Tiny scales	{001} perfect
1.751	n_β 1.751 n_α 1.750 n_γ 1.761	Synadelphite $Mn_9^{2+}(OH)_9(H_2O)_2(AsO_3)(AsO_4)_2$	0.011	37° r > v	Orthorhombic	Short prism crystals, crusts, massive	{010} imperf
1.754	n_β 1.754 n_α 1.742-1.743 n_γ 1.776-1.778	Barytolamprophyllite $Na_2(Ba,Sr)_2Ti_3(SiO_4)_4(OH,F)_2$	0.034	29° - 30° r > v distinct	Monoclinic β = 96.5°	Foliated aggregate	{100} perfect {011} distinct {010} imperf
1.754	n_β 1.754 n_α 1.743 n_γ 1.764	Caracolite $\sim Na_2PbSO_4Cl(OH)$	0.021	Large r > v strong	Orthorhombic ?	Hexagonal prism	
1.754	n_β 1.754 n_α 1.746 n_γ 1.779	Lamprophyllite (Molengraafite) $Na_2(Sr,Ba)_2Ti_3(SiO_4)_4(OH,F)_2$	0.033	21° - 43° r > v strong	Monoclinic-Orthorhombic β = 96.5°	Micaceous, tabular, stellate acicular	{100} perfect
1.755	n_β 1.755 n_α 1.753 n_γ 1.766	Pyroxferroite $(Fe,Ca,Mg,Mn)SiO_3$	0.013	35°	Triclinic α = 114.4° β = 82.7° γ = 94.5°	Embedded grains	
**1.700 1.815	n_β 1.700-1.815 n_α 1.690-1.791 n_γ 1.706-1.828	Allanite (Orthite) $(Ca,Ce,La)_2(Al,Fe^{3+}Fe^{2+})_3O \cdot SiO_4 \cdot Si_2O_7(OH)$	0.013 to 0.036	90°-57° r > v strong (also v > r)	Monoclinic β = 115°	Tab. {100}, acic., gran	{001} imperf., {100} & {110} poor
1.758	n_β 1.758 n_α 1.725 n_γ 1.830	Seidozerite (Seidoserite) $Na_2(Zr,Ti,Mn)_2(SiO_4)_2F_2$	0.105	68° r > v strong	Monoclinic β = 102.7°	Radiated prism-acicular crystals (b), fibrous	{001} perfect
1.755 1.765	n_β 1.755-1.765 n_α 1.725-1.775 n_γ 1.815-1.835	Olmsteadite $K_2Fe_2^{2+}[Fe_2^{2+}(Nb,Ta)_2^{5+}O_4(H_2O)_4(PO_4)_4]$	0.090	~60°	Orthorhombic	Thick prism crystals	{100} & {001} good
1.766 1.756	n_β 1.766-1.756 n_α 1.761-1.750 n_γ 1.772-1.763	Tadzhikite $Ca_3(Ce,Y)_2(Ti,Al,Fe)Be_4Si_4O_{22}$	0.011 to 0.013	80° - 90°	Triclinic α = 90° β = 100.7° γ = 90°	Prism crystals flattened {010}, plates	{010} dist
1.760 1.767	n_β 1.760-1.767 n_α 1.748-1.754 n_γ 1.762-1.823	Joaquinite $Ba_2NaCe_2Fe(Ti,Nb)_2Si_8O_{26}(OH,F)_2$	0.014 to 0.069	40°	Orthorhombic	Minute tabular equant crystals	None ?
1.763	n_β 1.763 n_α 1.759 n_γ 1.783	Austinite $CaZnAsO_4(OH)$	0.024	47°	Orthorhombic	Acicular on a, fibrous masses	{011} dist

* n_β variation exceeds 0.01 ** n_β variation exceeds 0.10

Twinning	Orientation	Color in thin section	Hardness Sp Gravity	Color hand sample	Alteration	Occurrence	Remarks	Reference
	a = X, b = Z, c = Y	Ruby red	2.37	Red, reddish-brown		Efflorescence, volcanic sublimate	Erythrosiderite-kremersite series Easily soluble H_2O	Dana II, 101-103
	a = X, b = Z, c = Y	Yellowish, brownish-red	2.00	Red, brownish red		Volcanic fumeroles	Erythrosiderite-kremersite series Easily soluble H_2O, Deliquescent	Dana II, 101-103
		Yellow, no pleochroism	2½ 4.25	Orange		Secondary U-mineral		Am Min 44, 208
Cyclic {100} & {010} (fourlings)		X ~ colorless, Y = colorless, brn Z = pale brown, dk red-brown	4½ 3.57-3.59	Colorless, red, brown, black (zoned)	Oxidizes readily in air	Low-temp hydrothermal veins cutting Mn ore bodies	Easily soluble acids	Am Min 55, 2023-2037
	b = X, $c \wedge Z = 6^\circ - 7^\circ$	Brown to lt yel Z > Y > X	2-3 3.61-3.66	Dk brown		Alkali plutons	Ba analogue of lamprophyllite	Am Min 51, 1549
Pseudohexagonal trillings		Colorless	4½ ~5.1	Colorless, gray, greenish		Incrustations oxide zones Pb-deposits		Dana II, 546-547
{100} sometimes polysynthetic	b = X, $c \wedge Z = 0^\circ - 8^\circ$	X = pale yellow, Y = straw yellow, Z = orange-yellow	2-3 3.44-3.53	Gold-brown, dk brown		Nepheline syenites and pegmatites with aegirine, eudialyte		Vlasov II, 207-209
		Colorless, pale yellow	3.68-3.83	Yellow		Lunar rocks (micro-gabbro, diabase) with clinopyroxene, ilmenite	Pyroxene group	Geochim Cosmochim Acta Suppl Proc Apollo XI Lunar Sci Conf. 1: 65-79
{100} rare	b = Y, $c \wedge X = +1^\circ$ to $+47^\circ$, $a \wedge Z = +26^\circ$ to $+72^\circ$	Lt. to dk. brown, yellow, green Z > Y > X	5-6½ 3.4-4.2	Dk. brown, black	Limonite, silica, alumina	Pegmatites, granitic plutons with epidote & Fe-silicates	Epidote group. Often metamict glass	p. 153
	b = X, a ~ Z, $c \wedge Y = -13^\circ$	X = dk red, Y = red, Z = pale yellow	4-5 3.47-3.59	Brownish-yellow, orange		Syenite pegmatites with aegirine, nepheline, prochlore, eudialite	Slightly soluble HCl	Vlasov II, 386-388
	a = Y, b = Z, c = X	X = dk blue-green, Y = yellow, pale brown, Z = brown	4 3.31-3.36	Olive-green, dk brown to black		Granite pegmatites, hydrothermal alter triphylite		Am Min 61, 5-11
Polysynthetic	$a \wedge Y = 7^\circ$, $b \wedge Z = 4^\circ$, $c \wedge Z = 23^\circ$	Colorless, brown. Not pleochroic	6 3.73-3.86	Pale gray-brown to dk brown		Zoned alkali pegmatites with albite, aegirine, riebeckite	Hellandite group	Am Min 56, 1838-1839
	c = Z	Yellow-green Z > X	5-5½ 3.89-3.93	Amber to brown		Alkali plutons, schists, pegmatites with Na-amphiboles, benitoite		Am Min 52, 1762-1769
	a = X, b = Z, c = Y	Colorless	4-4½ 4.13-4.37	Colorless, white, yellowish	Carbonates	Fibrous veins in ozidized ore deposits		Am Min 56, 1359-1365

n_β Var.	Refractive Index	Mineral Name and Composition	$(n_\gamma - n_\alpha)$	2V angle Dispersion	Crystal System	Habit	Cleavage
1.765	n_β 1.765 n_α 1.747 n_γ 1.78	Joesmithite $(Ca,Pb)Ca_2(Mg,Fe^{2+},Fe^{3+})_5(Si_6Be_2O_{22})(OH)_2$	0.033	large ~ 85°	Monoclinic β = 105.6°	Prismatic crystals	{110} perfect
*1.765 1.802	n_β 1.765–1.802 n_α 1.760–1.782 n_γ 1.775–1.835	Alluaudite $(Na,Fe^{3+},Mn^{2+})PO_4$	0.015 to 0.060	~78° r > v moderate, crossed	Monoclinic β = 114.4°	Granular, radiating fibrous	{100}, {010} distinct
1.766	n_β 1.766 n_α 1.750 n_γ 1.85	Komarovite $(Ca,Mn)Nb_2Si_2O_9(O,F)\cdot 3.5H_2O$	0.01	48°	Orthorhombic	Foliated, platy, massive	{001} dist
1.765 1.771	n_β 1.765–1.771 n_α 1.755–1.768 n_γ 1.775–1.788	Orthoferrosilite $Fe_2(SiO_3)_2$	0.020	90° – 35°	Orthorhombic	Short prism (c) xls, massive	{110} dist, {010} or {100} parting
*1.730 1.807	n_β 1.730–1.807 n_α 1.725–1.794 n_γ 1.750–1.832	Piemontite $Ca_2(Al,Fe^{3+},Mn^{3+})_3O\cdot SiO_4\cdot Si_2O_7(OH)$	0.025 to 0.082	50°–86° r > v strong (v > r rare)	Monoclinic β = 115.7°	Columnar, acic. xls. Rad.	{001} perfect
1.770	n_β 1.770 n_α 1.769 n_γ 1.785	Holdenite $(Mn,Zn)_6(AsO_4)(OH)_5O_2$	0.016	~30° r > v perc	Orthorhombic	Thick tabular crystals	{010} poor
1.770	n_β 1.770 n_α 1.710 n_γ 1.840	Rossite $CaV_2O_6\cdot 4H_2O$	0.130	~90° Very strong	Triclinic α = 101.5° β = 115° γ = 103.4°	Bladed (c), tabular {001}, acicular (c)	{010} dist
1.774	n_β 1.774 n_α 1.77 n_γ 1.83	Taramellite $Ba_4(Fe,Mg)Fe_2^{3+}Ti(SiO_3)_8(OH)_2$	0.06	Small r > v distinct	Orthorhombic	Massive, fibrous	{100} perfect ⊥ {100} parting
1.775	n_β 1.775 n_α 1.752 n_γ 1.796	Luetheite $Cu_2Al_2(AsO_4)_2(OH)_4\cdot H_2O$	0.044	~88° v > r moderate	Monoclinic β = 101.8°	Minute tabular crystals	{100} dist
1.777	n_β 1.777 n_α 1.763 n_γ 1.785	Yoshimuraite $(Ba,Sr)_2TiMn_2(SiO_4)_2(PO_4,SO_4)(OH,Cl)$	0.022	85° – 90°	Triclinic α = 93.5° β = 90.2° γ = 95.3°	Bladed-micaceous crystals	{010} perfect {10$\bar{1}$} & {101} distinct
1.78	n_β 1.78 n_α 1.76 n_γ >1.85	Carpathite (Karpatite) (Pendletonite) (Coronene) $C_{24}H_{12}$	0.1	65° – 84° v > r extreme	Monoclinic	Thin tabular {001}, fibrous (b)	{100}, {001} & {$\bar{2}$01} very perfect
1.780	n_β 1.780 n_α 1.776 n_γ 1.805	Caryinite $(Ca,Na,Pb,Mn)_3(Mn,Mg)_2(AsO_4)_{3-y}(OH)_x$	0.029	41° r > v weak	Monoclinic β = 99°	Massive, granular	{110} & {010} distinct
1.780	n_β 1.780 n_α 1.779 n_γ 1.816	Cheralite $(Th,Ca,Ce,La,U,Pb)(PO_4,SiO_4)$	0.037	17° – 19° No disp.	Monoclinic β = 104°	Massive	{010} dist {100} diff, {001} parting
1.782	n_β 1.782 n_α 1.752 n_γ 1.815	Shattuckite $Cu_5(SiO_3)_4(OH)_2$	0.063	~90°	Orthorhombic	Radiated acicular-fibrous (c), granular, compact	{010} & {100} perfect
*1.77 1.80	n_β 1.80–1.77 n_α 1.78–1.77 n_γ 1.82–1.78	Gadolinite $Be_2FeY_2Si_2O_{10}$	0.04	85° v > r strong	Monoclinic (metamict) β = 90.5°	Prismatic, massive	None

* n_β variation exceeds 0.01

Twinning	Orientation	Color in thin section	Hardness Sp Gravity	Color hand sample	Alteration	Occurrence	Remarks	Reference
		Y = olive-brown, X = Z = brown	5½ 3.83	Black		Hematite-magnetite-schefferite skarn		Am Min 54, 577-578
Polysynthetic {$\bar{1}$01}	b = Z, c ∧ Y = -18°	Yellow, olive, brown. Z ~ Y > X	5-5½ 3.52-3.58	Brownish-yellow to black		Pegmatite deposits		Am Min 56, 1955-1975
	a = X, b = Z, c = Y	Colorless, pale rose	1½-2 2.96-3.0	Pale rose		Alkali rocks with albite, natrolite		Am Min 57, 1315-1316
{101} normal	a = X, b = Y, c = Z	Deep green to brown	5-6	Dk green, dk brown, black	Uralite, Fe-oxides	Thermal metamorphism of Fe-rich rocks	Orthopyroxene group	p. 187
{100} lamellar (rare)	b = Y, c ∧ X = +2° to +9°, a ∧ Z = +27° to +35°	X = pale yellow, pink, Y = pale violet, red-violet, Z = dk red-brown	6-6½ 3.40-3.52	Maroon, red-brown, black		Low grade schists with chlorite, quartz, glaucophane	Epidote group. Negative varieties are Manganepidote	p. 152
	a = Z, b = Y, c = X	Pink	4 4.11-4.12	Pink, orange, deep red		Hydrothermal Mn-Zn oxide veins		Dana II, 775-777
{100}	c ~ Z, b ∧ Y ~ 45°	Yellow	2-3 2.45	Yellow		Glassy grains in metarossite in secondary U-veins	Slowly soluble H_2O	Dana II, 1053-1054
	∥ extinction, + elongation. Optic plane ⊥ {100}	X = Y = flesh red, Z = dk brown	5½ 3.9	Reddish brown		Assoc with sanbornite, gillespite, diopside, witherite, pyrrhotite		Win & Win II, 401
	b = X, c ∧ Z = -10°	Pale blue, weak pleochroism	3 4.28-4.40	Indian blue		Tiny crystals in rhyolite porphyry with chenevixite and hematite	Al-chenevixite	Min Mag 41, 27-32
(010) poly- synthetic			4½ 4.13-4.21	Orange-brown		Pegmatites in massive Mn-ores		Min J (Japan) 3, 156-167
	b = X, c ∧ Z = 21°	Colorless	<1 1.35	Pale yellow		Silica veins in Hg-deposits		Am Min 52, 611-616
	a = Y, b = Z, c = X	Pale yellow-brown. Not pleochroic	4 4.29	Yellow-brown, brown	Berzeliite	Veins in limestone skarn	Soluble HNO_3	Dana II 683-684
	b = X, c ∧ Z = 7°	X = Y = pale green, Z = pale yellow-green	5 5.3-5.4	Lt-dk green		Pegmatite, altered granite-gneiss	Monazite group	Am Min 52, 13-19
	a = Y, b = X, c = Z	X = very pale blue, Y = pale blue, Z = deep blue	~3½ 4.11-4.14	Blue to dark blue		Alteration of other secondary Cu-minerals		Am Min 51, 266-267
	b = Y, c ∧ Z = +10°	X = olive green, Y = Z = grass green	6½-7 4.0-4.65	Black, green-black, brown		Granite, granite-pegmatite with fluorite, allanite	Isotropic (metamict) in areas	Vlasov II, 235-239

n_β Var.	Refractive Index	Mineral Name and Composition	$(n_\gamma - n_\alpha)$	2V angle Dispersion	Crystal System	Habit	Cleavage
1.78 1.79	n_β 1.78–1.79 n_α 1.76–1.77 n_γ 1.80	Vandenbrandeite (Uranolepidite) $CuUO_4 \cdot 2H_2O$	0.04	~90° Strong	Triclinic α = 91.9° β = 102° γ = 89.6°	Radiated fibrous rosettes, lamellar, tabular {001}	{110} perfect
1.788	n_β 1.788 n_α 1.777 n_γ 1.800	Retzian $Mn_2Y(AsO_4)(OH)_4$	0.023	Large ~ 88° v > r weak	Orthorhombic	Prism (c) or tabular {010}	None
*1.777 1.801	n_β 1.777–1.801 n_α 1.774–1.800 n_γ 1.828–1.851	Monazite $(Ce,La,Th)PO_4$	0.045 to 0.075	6° to 19° v > r or r > v weak, horizontal	Monoclinic β = 104°	Blocky crystals, flattened {100}, elongation (b)	{100} dist, {010} indist, {001} parting
1.79	n_β 1.79 n_α 1.77 n_γ 1.835	Barbosalite $Fe^{2+}Fe^{3+}_2(PO_4)_2(OH)_2$	0.065	70° – 80°	Monoclinic β = 120.25°	Prismatic, earthy	None ?
1.794	n_β 1.794 n_α 1.792 n_γ 1.821	Jimboite $Mn_3B_2O_6$	0.029	30° r > v	Orthorhombic	Massive	{110} perfect {101} parting
*1.684 1.722	n_β 1.684–1.722 n_α 1.682–1.722 n_γ 1.705–1.751	Pigeonite $(Mg,Fe^{2+},Ca)(Mg,Fe^{2+})(SiO_3)_2$	0.023 to 0.029	0° – 32° r > v or v > r weak to moderate	Monoclinic β = 108°	Prism crystals (c), granular	{110} dist, {001} parting
1.80	n_β 1.80 n_α 1.743 n_γ 1.88	Tundrite $Na_2Ce_2(Ti,Nb)SiO_8 \cdot 4H_2O$	0.137	Large ~ 85°	Triclinic α = 105.3° β = 102.8° γ = 70.8°	Acicular crystals, spherulitic	None ?
1.800	n_β 1.800 n_α 1.775 n_γ 1.846	Botallackite $Cu_2Cl(OH)_2$	0.071	Moderate-large r > v strong	Monoclinic β = 92.75°	Microscopic columnar crystals	1 direction
1.801	n_β 1.801 n_α 1.783 n_γ 1.834	Flinkite $Mn^{2+}_2Mn^{3+}AsO_4(OH)_4$	0.051	Large ~ 75° r > v weak	Orthorhombic	Tabular aggregates {001}	None ?
1.80	n_β 1.80 n_α 1.79 n_γ 1.82	Kryzhanovskite $MnFe_2(PO_4)_2(OH)_2 \cdot H_2O$	0.03	40° – 45° v > r strong	Monoclinic	Prism crystals	{001} perfect
1.80	n_β ~1.80 n_α 1.791 n_γ 1.886	Aluminous-ludwigite $(Mg,Fe^{2+})_2(Fe^{3+},Al)BO_3 \cdot O_2$	0.095	Small r > v extreme	Orthorhombic	Prism-acicular xls (c)	{001} perfect
1.801	n_β 1.801 n_α 1.793 n_γ 1.816	Volborthite (Uzbekite)(Vesbine) $Cu_3(VO_4)_2 \cdot 3H_2O$	0.023	Large r > v inclined	Monoclinic	Scaly, radiated fibrous, spherulitic, spongy	{001} perfect
1.809	n_β 1.809 n_α 1.806 n_γ 1.830	Warwickite $(Mg,Fe)_3TiB_2O_8$	0.024	Small, variable	Orthorhombic	Prism-acicular crystals	{100} perfect

* n_β variation exceeds 0.01

Twinning	Orientation	Color in thin section	Hardness Sp Gravity	Color hand sample	Alteration	Occurrence	Remarks	Reference
	Z ∧ elong ~ 40°, optic axis ~ ⊥{001}	Absorption green to colorless	4 5.08-5.26	Dk green to greenish-black		Oxidation of uraninite with curite, uranophane cuprosklodowskite	Soluble warm, dilute acids	USGS Bull. 1064, 100-103
{150} doubtful	a = Z, b = Y, c = X	X ~ colorless, orange-yellow, Y = yellow-brown, dk brown, Z = red-brown	4 4.14-4.15	Dk brown		Mn-ores in cavities in dolomite with jacobsite	Soluble acids	Dana 53, 1779
{100} simple, {100} lamellar, rare	b = X, c ∧ Z = -2° to -7°, a ∧ Y = +12° to +7°	Colorless, yellow, yellow-brown. Y > X ~ Z	5-5½ 4.6-5.4	Yellow-brown to red-brown	Yellowish brown, earthy	Granites, syenites, granite pegmatites with rare earth minerals	Slowly decomp acids	p. 75
		(~opaque) X = Y = dk blue-green, Z = dk olive green	5½-6 3.60	Green-blue to black		Pegmatites	Dimorph stichite	Am Min 40, 952-966
{101}	a = X, b = Y, c = Z	Colorless	5½ 3.98-4.09	Pale violet-brown		Contact ores with Mn-carbonates and Mn-silicates	Mn-kotoite	Am Min 48, 1416-1417
{100} simple or multiple	b = X, c ∧ Z = -37° to -44° or b = Y, c ∧ Z = -40° to -44°	Colorless to pale green or brown. Weak pleochroism Y > X ~ Z	6 3.30-3.46	Green-brown to black	Uralite, serpentine, chlorite	Andesite flows and shallow intrusions with augite, hypersthene, olivine	Pyroxene group	p. 194
	c ∧ Z ~ 14°	X = pale yellow, Z = greenish yellow	~3 3.70	Brownish-yel, brown-green	Rhabdophane	Nepheline syenite pegmatites with aegirine, lamprophyllite, ramsayite		Am Min 53, 1780
	X ⊥ cleavage	Pale bluish-green, weak pleochroism	3.6	Pale green, bluish		Oxide zones Cu-deposits		Am Min 36, 384
	a = Z, b = X, c = Y	X = pale brown, Y = yellow-green, Z = orange-brown	4½ 3.73-3.78	Dk greenish brown		Rare secondary mineral Mn-deposits	Soluble HCl	Am Min 52, 1603-1613
	Optic axial plane ⊥ {001}, b = Y (?)	X = wine-yellow, Y = orange-brown, Z = reddish-brown X > Y > Z	3½-4 3.31	Brown, greenish-brown		Alter triphylite or sicklerite in pegmatites		Am Min 56, 1-17
	a = X, b = Y, c = Z	X = pale green, Y = slightly darker green, Z = dk brown	5 ~3.6	Dark green	Limonite	High-T contact zones in limestone	Ludwigite group	Cal Jour Mines & Geol 39, 333-365
Lamellar common	b = Y, elongation-	Pale green to yellowish green. Faint pleochroism	3½ 3.5-3.8	Yellow to olive green, brown		Secondary mineral in detrital sediments	Soluble acids	Am Min 45, 1307-1309
	a = Z, b = Y, c = X	X = yellow-brown, Y = reddish brown, Z = cinnamon brown	3½-4 3.35-3.43	Dk brown to black		Crystalline limestone with spinel, ilmenite, chondrodite, diopside	Decomp H_2SO_4	Dana II, 326-327

n_β Var.	Refractive Index	Mineral Name and Composition	$(n_\gamma - n_\alpha)$	2V angle Dispersion	Crystal System	Habit	Cleavage
*1.795 1.831	n_β 1.795–1.831 n_α 1.730–1.800 n_γ 1.771–1.846	Conichalcite $CuCaAsO_4(OH)$	~0.045	90° – small r > v or v > r strong	Orthorhombic	Coliform masses, radiated fibrous	None
1.810 1.820	n_β 1.810–1.820 n_α 1.772–1.780 n_γ 1.863–1.865	Olivenite (Leucochalcite) $Cu_2AsO_4(OH)$	0.091 to 0.085	~90° v > r strong	Orthorhombic	Prism acicular-fibrous (c or a), reniform	{001} & {110} indist
*1.808 1.827	n_β 1.808–1.827 n_α 1.791–1.806 n_γ 1.997–2.005	Ferrimolybdite $Fe_2(MoO_4)_3 \cdot 8H_2O$	0.206 to 0.199	Small ~ 28° v > r moderate	Orthorhombic ?	Microscopic radiated fibrous, earthy	
1.815	n_β 1.815 n_α 1.81 n_γ 1.85	Cornwallite $Cu_5(AsO_4)_2(OH)_4 \cdot H_2O$	0.04	Small	Monoclinic β = 92°	Radiated botryoidal	None ?
1.816	n_β 1.816 n_α 1.770 n_γ 1.856–1.863	Calcurmolite $Ca(UO_2)_3(MoO_4)_3(OH)_2 \cdot 11H_2O$	0.080 to 0.093	Large	Unknown	Prismatic, platy, massive	
1.82	n_β 1.82 n_α 1.81 n_γ 1.88	Aenigmatite $Na_2Fe_5TiSi_6O_{20}$	0.07	32°	Triclinic	Long prismatic crystals	{010} and {100} perfect
1.820	n_β 1.820 n_α 1.715 n_γ 1.880	Dolerophane (Dolerophanite) Cu_2SO_5	0.065	85° r > v extreme, crossed	Monoclinic β = 122°	Minute elongated crystals	{$\bar{1}$01} perfect
1.820	n_β 1.820 n_α 1.813 n_γ 1.830	Andrewsite $(Cu,Fe^{2+})Fe^{3+}_3(PO_4)_3(OH)_2$	0.017	Moderate-large v > r extreme, crossed	Orthorhombic	Radiating fibrous aggregates	Two directions ∥ fibers
1.821	n_β 1.821 n_α 1.809 n_γ 1.857	Betpakdalite $CaFe_2H_8(AsO_4)_2(MoO_4)_5 \cdot 10H_2O$	0.048	60°	Monoclinic ?	Short prism, aggregates	
1.822	n_β 1.822 n_α 1.799 n_γ 1.855	Azoproite $(Mg,Fe^{2+})_2(Fe^{3+},Ti,Mg)BO_5$	0.056	70° – 80°	Orthorhombic	Prismatic	{010} dist, {001} poor
1.83	n_β 1.83 n_α 1.81 n_γ	Welshite $Ca_2Mg_4Fe^{3+}O_2Si_4Be_2O_{18}$	>0.02	~24°	Triclinic α = 106° β = 96° γ = 124°	Prismatic crystals	None
1.833	n_β 1.833 n_α 1.807 n_γ 1.89	Orthoericssonite $BaMn_2Fe(Si_2O_7)(O,OH)$	~0.08	43° r > v perc	Orthorhombic	Massive as embedded plates {100}	{100} perf {011} dist
1.83 1.84	n_β 1.83–1.84 n_α 1.83–1.84 n_γ 1.97–2.00	Ludwigite (Collbranite) $(Mg,Fe^{2+})_2Fe^{3+}BO_3 \cdot O_2$	0.14 to 0.16	Small 20° – 45° r > v extreme	Orthorhombic	Acicular-fibrous aggregates (c)	{001} perf
1.840	n_β 1.840 n_α 1.792 n_γ 1.888	Lautarite $Ca(IO_3)_2$	0.096	~90°	Monoclinic β = 106.4°	Short prism, radiated aggregates	{011} dist {100} & {110} trace
*1.830 1.850	n_β 1.830–1.850 n_α 1.820–1.842 n_γ 1.875–1.925	Dufrenite $Fe^{2+}Fe^{3+}_4(PO_4)_3(OH)_5 \cdot 2H_2O$	0.055 to 0.083	Small ~40°–50° v > r or r > v extreme	Monoclinic β = 111°	Botryoidal, radiated fibrous	{100} perf

* n_β variation exceeds 0.01

Twinning	Orientation	Color in thin section	Hardness Sp Gravity	Color hand sample	Alteration	Occurrence	Remarks	Reference
	a = Z, b = Y, c = X	Pale green, blue-green	4½ 4.1-4.3	Green, yellow-green		Oxide zone Cu-deposits		Dana II, 806-808
	c = Y	Pale green to yellow. Pleochroism weak Y > X = Z	3 4.38-4.45	Olive, brown, yellow, white		Oxide zones Cu-deposits with secondary Cu-minerals	Soluble acids, NH_4OH	Dana II, 859-861
	Z = elong	X = Y = colorless, Z = pale yellow, gray	1-2 3.06-4.46	Yellow		Oxidation of molybdenite	Soluble HCl	Am Min 48, 14-32
	Elongation - inpart (+)	Emerald green	4½ 4.52-4.65	Lt green-dk green		Oxide zones Cu-deposits		Am Min 36, 484-503
		X = colorless, Y = pale yellow, Z = yellow		Honey yellow		Secondary U-mineral	U.V. fluor strong yellow-green	Am Min 49, 1152-1153
{1$\bar{1}$0}	b = Y, c ∧ Z = 45°	Nearly opaque, dark reddish brown Z > Y > X	5½ 3.74-3.85	Black		Small phenocrysts in Si-poor lavas		Am Min 56, 427-446
	b = Y, c ∧ Z = +10°	X = Dk brown, Y = brownish yellow Z = lemon yellow	3 4.17	Brown, ~ black		Volcanic sublimate	Soluble H_2O	Dana II, 551-553
		X = pale yellow, Y = emerald green, Z = olive green	4 3.475	Dark green, bluish-green		Oxide zones of ore deposits	Anomalous interference colors due to dispersion	Am Min 34, 534
	b = Y, c ∧ X = 12°	X = pale yellow, Y = greenish-yellow Z = greenish, Z > Y > X	~3 2.98-3.05	Lemon yellow		Oxide zones W-Mo-deposits		Am Min 47, 172-173
		X = pale green, Y = dark green (nearly opaque), Z = red-brown	5½ 3.63	Black		Magnesian Skarns	Ludwigite group. Soluble in dilute HCl	Am Min 56, 360
		No pleochroism	6 3.71-3.77	Reddish-black		Dolomite skarn with berzeliite	Aenigmatite group. May be biaxial -	Min Mag 42, 129
	a = Z, b = X, c = Y	X = pale green-brn Y = red-brown, Z = deep brown	4½ 4.21	Reddish-black		Intergrown with ericssonite in skarn with schefferite, rhodonite, tephroite	Dimorph ericssonite. Lamprophyllite group. Weakly magnetic	Lithos 4: 137-145
	a = X, b = Y, c = Z	X ~ Y = dk green, Z = dk brown	5 3.6-4.2	Dark green, black	Limonite	High-T skarn rocks with magnetite, forsterite	Ludwigite-vonsenite series. Slowly soluble acids	p. 99
		Colorless	3½-4 4.48-4.52	Colorless, yellowish		Evaporite nitrate deposits with gypsum		Dana II, 312-313
		X = Blue, pale brn, Y = Buff, pale brn, Z = Dk brown	3½-4½ 3.1-3.3	Dk green-black, brownish		Oxide zones Fe-deposits	Oxidation of Fe^{2+} results in high IR & color change	Am Min 55, 135-169

n_β Var.	Refractive Index	Mineral Name and Composition	$(n_\gamma - n_\alpha)$	2V angle Dispersion	Crystal System	Habit	Cleavage
1.842	n_β 1.842 n_α 1.842 n_γ 1.848	Paratacamite $Cu_2(OH)_3Cl$	0.006	0° - 50° r > v	Trigonal	Rhombohedral crystals, massive granular	{10$\bar{1}$1} dist
1.847	n_β 1.847 n_α 1.840 n_γ 1.892	Laubmannite $Fe_3^{2+}Fe_6^{3+}(PO_4)_4(OH)_{12}$	0.052	Moderate, variable ~ 45° v > r extreme	Orthorhombic	Botryoidal, radiated fibrous, ‖ fibers	Two probable ‖ filer elong
1.84 1.85	n_β 1.84-1.85 n_α 1.84-1.85 n_γ 1.99-2.02	Vonsenite (Paigeite) $(Fe^{2+},Mg)_2Fe^{3+}BO_3 \cdot O_2$	0.15 to 0.17	Small 20° - 45° r > v extreme	Orthorhombic	Acicular fibrous aggregate (c), granular	{001} perfect
1.85	n_β>1.85 n_α 1.840 n_γ>1.85	Metarossite $CaV_2O_6 \cdot 2H_2O$	>0.01	Large Strong	Triclinic α = 96.7° β = 105.8° γ = 93°	Platy to flaky masses	
1.852	n_β 1.852 n_α 1.845 n_γ 1.878	Törnebohmite $(Ce,La)_3(SiO_4)_2(OH)$	0.033	Small ~ 55° r > v very strong	Hexagonal or pseudohex	Fine-grained massive	
1.85	n_β 1.85 n_α 1.85 n_γ 1.88	Parwelite $(Mn,Mg)_5Sb(Si,As)_2O_{10-11}$	0.03	~27° r > v strong	Monoclinic β = 95.9°	Stubby prism crystals	{010} poor to fair
1.85	n_β 1.85 n_α 1.85 n_γ 1.88	Anandite $(Ba,K)(Fe,Mg)_3(Si,Al,Fe)_4O_{10}(O,OH)_2$	0.03	Small strong	Monoclinic β = 95°	Micaceous	{001} perfect
1.86	n_β 1.86 n_α 1.85 n_γ 1.92	Purpurite $(Mn^{3+},Fe^{3+})PO_4$	0.07	Moderate ~ 45° very strong	Orthorhombic	Massive	{100} dist {010} imperf
*1.835 1.910	n_β n_α~1.835 n_γ~1.910	Demesmaekerite $Pb_2Cu_5(UO_2)_2(SeO_3)_6(OH)_6 \cdot 2H_2O$	0.075		Triclinic α = 90 β = 100 γ = 92	Elongate crystals (a)	None
1.87	n_β 1.87 n_α 1.85 n_γ~1.89	Dumontite $Pb_2(UO_2)_3(PO_4)_2(OH)_4 \cdot 3H_2O$	~0.04 to 0.010	Large v > r	Monoclinic or Orthorhombic	Minute crystals, elong (c), flat {010}	
**1.74 2.0	n_β 1.74-2.0 n_α 1.74-2.0 n_γ 1.76-2.0	Ardennite $Mn_5Al_5(As,V)O_4Si_5O_{20}(OH)_2 \cdot 2H_2O$	0.015 to 0.020	0° - 70° r > v strong	Orthorhombic	Acicular on c, radiating aggregates	{010} perfect {110} dist, {001} parting
1.879	n_β 1.879 n_α 1.817 n_γ 2.057	Uvanite $(UO_2)_2V_6O_{17} \cdot 15H_2O$	0.240	52°	Orthorhombic	Microcrystalline coatings	Two pinacoidal
1.880	n_β 1.880 n_α 1.87 n_γ 1.98	Leiteite $ZnAs_2O_4$	0.11	~26° v > r very strong	Monoclinic β = 91°	Cleavable masses	{100} perfect
1.880	n_β 1.880 n_α 1.875 n_γ 1.897	Rockbridgeite $(Fe^{2+},Mn^{2+})Fe_4^{3+}(PO_4)_3(OH)_5$	0.022	Moderate ~ 60° v > r	Orthorhombic	Radiated fibrous, botryoidal	{100} perfect {010} & {001} distinct

* n_β variation exceeds 0.01

** n_β variation exceeds 0.1

Twinning	Orientation	Color in thin section	Hardness Sp Gravity	Color hand sample	Alteration	Occurrence	Remarks	Reference
{10$\bar{1}$1} common	c = X	Deep green. Not pleochroic	3 3.74-3.75	Green, greenish-black		Oxide zones Cu-deposits as xl coatings	Dimorph atacamite Patchy extinction	Dana II, 74-76
		X = pale buff, Y = olive-brown, Z = reddish-brown, Z > Y > X	3½-4 3.33	Yel-green, gray-green, brown		Crevices in novaculite, gossan zones Fe-deposits		Am Min 55, 135-169
	a = X, b = Y, c = Z	X ~ Y = dk-green, Z = dk red-brown, ~ opaque	5 4.2-4.77	Coal-black	Limonite	High T skarn rocks with magnetite, forsterite	Ludwigite-vonsenite series. Slowly soluble acids	Am Min 50, 249-254
{0$\bar{1}$0}		Colorless to pale yellow	Soft 2.45	Light yellow		Dehydration of rossite in secondary uranium ores	Soluble H_2O	Dana II, 1054-1055
		X = pink, greenish yellow, Y = bluish-green, Z = pink	4½ 4.9	Olive-green to green		Inclusions in cerite. Alkali granite pegmatites		Vlasov II, 319-320
		Pale yellow-brown	5½ 4.62	Yellowish-brn		Mn-carbonatite with langbanite, spessartine, hausmannite		Am Min 55, 323
	b = Y, a ∧ X = 12°	~opaque. Y = green, Z = brown	3-4 3.91-3.94	Black		Lenses and veins in iron deposits	Mica group	Am Min 52, 1586
	a = Y, b = Z, c = X	X = greenish-gray, rose-red, Y ~ Z = deep blood-red, purple-red	4-4½ 3.69	Dk rose, red-purple, dk brown		Weathering of lithiophilite in granite pegmatites	Heterosite-purpurite. Easily soluble HCl. Anomalous colors	Dana II, 675-677
		Pale green	3-4 5.28-5.45	Bottle green, brownish		Oxide zones Cu-deposits		Am Min 51, 1815-1816
	a = X, b = Z, c = Y	Pale yellow Y > X	5.65	Pale golden yellow		Secondary U-mineral	U.V. fluor weak green	USGS Bull 1064, 236-238
	a = Y, b = Z, c = X or a = X, b = Y, c = Z	X = dark brownish yellow, Y = golden yellow, Z = pale yellow	6-7 3.60-3.65	Yellow, brown, black		Veins and pegmatites		Dana II, 1010-1011
		X = Lt brown, Y = dk brown, Z = greenish yellow		Brownish yellow		Secondary U-mineral in asphaltic sandstone with gypsum, halite	Not fluor U.V. Soluble $(NH_4)_2CO_3$	USGS Bull 1064, 261-263
	b = Y, a ∧ X = +11°, c ∧ Z = +10°	Colorless	1½-2 4.31-4.61	Colorless to brown		Oxide zones Zn-Cu deposits with tennantite, smithsonite, chalcocite		Am Min 62, 1259-1260
	c = X	X = pale yel-brown, Y = bluish-green, Z = dk bluish-green	4½ 3.3-3.49	Lt-dk green, ~ black		Alteration Fe-Mn phosphates in pegmatites. "Limonite" beds	Frondelite-rockbridgeite series Soluble HCl	Am Min 55, 135-169

Biaxial +

n_β Var.	Refractive Index	Mineral Name and Composition	$(n_\gamma - n_\alpha)$	2V angle Dispersion	Crystal System	Habit	Cleavage
1.883	n_β 1.883 n_α 1.878 n_γ 1.895	Anglesite $PbSO_4$	0.017	60° - 75° v > r strong	Orthorhombic	Prism-tabular xls, granular, nodular	{001} & {210} dist, {010} poor
1.89	n_β 1.89 n_α 1.86 n_γ 1.91	Heterosite $(Fe^{3+},Mn^{3+})PO_4$	0.05	~90° very strong	Orthorhombic	Massive	{100} dist {010} imperf
1.895	n_β 1.895 n_α 1.882 n_γ 1.915	Hallimondite $Pb_2(UO_2)(AsO_4)_2$	0.033	~80° r > v	Triclinic α = 100.5° β = 94.8° γ = 91.2°	Tabular elongation (c)	None
>1.9	n_β>1.9 n_α n_γ	Paulmooreite $Pb_2As_2O_5$	0.080 to 0.110	65° r > v very strong	Monoclinic β = 109°	Tab crystals {001} or {100}	{001} perfect
1.900	n_β 1.900 n_α 1.810 n_γ 2.01	Duttonite $VO(OH)_2$	0.19	~60° v > r moderate	Monoclinic β = 90.6°	Platy {001}, crusts	{100} dist
1.900	n_β 1.900 n_α 1.898 n_γ 1.922	Huttonite $ThSiO_4$	0.024	25° v > r moderate	Monoclinic β = 105°	Anhedral grains	{001} dist {100} indist
1.90	n_β 1.90 n_α 1.89 n_γ 1.99	Bellingerite $3Cu(IO_3)_2 \cdot 2H_2O$	0.10	Medium r > v strong	Triclinic α = 105.1° β = 97° γ = 92.9°	Prismatic on c, tabular {100}	None
1.90 1.91	n_β 1.90-1.91 n_α 1.89-1.90 n_γ 1.95-1.97	Kasolite $PbUO_2SiO_4 \cdot H_2O$	0.06	43°	Monoclinic β = 104.3°	Bladed (b), flat {001}, radiated fibrous aggregates	{001} perf {100} & {010} indistinct
1.90	n_β~1.90 n_α 1.898 n_γ 1.915	Hugelite $Pb_2(UO_2)_3(AsO_4)_2(OH)_4 \cdot 3H_2O$	0.017	Small strong	Monoclinic	Tabular {100}, elong (c)	{100} good
1.901	n_β 1.901 n_α 1.899 n_γ 1.903	Queitite $Pb_4Zn_2(SO_4)(SiO_4)(Si_2O_7)$	0.004	~90°	Monoclinic β = 108.2°	Tab. xls. {001}	{010} & {001} poor
1.91	n_β 1.91 n_α 1.85 n_γ 2.20	Spiroffite $(Mn,Zn)_2Te_3O_8$	0.35	55°	Monoclinic β = 98°	Cleavable masses	Two directions
1.910	n_β 1.9097 n_α 1.8903 n_γ 1.9765	Schultenite $PbHAsO_4$	0.086	58° strong	Monoclinic β = 95.5°	Tabular crystals 010 , rhomb outline	{010} dist
1.92	n_β 1.92 n_α 1.871 n_γ 2.01	Claudetite As_2O_3	0.14	58° v > r	Monoclinic β = 94°	Micaceous {010}, fibrous	{010} perfect

Twinning	Orientation	Color in thin section	Hardness Sp Gravity	Color hand sample	Alteration	Occurrence	Remarks	Reference
Rare	a = Z, b = Y, c = X	Colorless. Transparent to opaque	2½–3 6.38	Colorless, white to dk gray	Cerussite	Oxide zones Pb-deposits with cerussite, pyromorphite	Slightly soluble HNO_3	p. 93
	a = X	X = greenish-gray, Y ~ Z = red-violet	4–4½ 3.41–3.70	Red-violet, dk brown		Alteration of triphylite in granite pegmatites	Heterosite-purpurite series Easily sol HCl	Dana II, 675–677
	on (100), c ∧ Z' = 11°, on (010), c ∧ X' = 11°, on (001), b ∧ X' = 9°	X = pale yellow, Z = colorless	2 1/3–3 6.39–6.40	Yellow		Coatings in cavities, oxide zones ore deposits	Anomalous blue & brown interference colors	Am Min 50, 1143–1157
	b = Y, c ∧ Z = +10°	Colorless	~3 6.86–7.01	Colorless, lt orange		Vugs with hematite, calcite, andradite	Sol 1:1 HCl	Am Min 64, 352
	a = X, b = Z, c = Y	X = pale pinkish-brown, Y = pale yellow-brown, Z = pale brown	~2½ 3.24	Pale brown		Secondary U-V-deposits		Am Min 42, 455–460
	b = Y, c ∧ Z = small	Colorless, very pale yellow	7.18	Colorless, white		Heavy sands with scheelite, cassiterite, zircon, ilmenite	Dimorph thorite U.V. fluor-white, pinkish	Am Min 36, 60–69
{$\bar{1}$01}		Pale bluish-green Z > X = Y	4 4.89–4.93	Pale green		Secondary mineral Cu deposits in arid regions		Dana II, 313–315
	b = X, c ∧ Z = -1/2°	X = Y = very pale yellow. Z ~ colorless	~4½ 5.83–6.5	Ocher, red-orange, greenish		Secondary U-mineral, alter uraninite	Not fluor U.V.	USGS Bull 1064, 315–319
		X = yellow, Y = orange-yellow, Z = colorless	~5.1	Brown, orange-yellow		Secondary U-mineral	Anomalous interference colors	Am Min 47, 418–419
	b = X, a ∧ Z ~ 20°	Colorless	4 6.07	Colorless, pale yellow		Oxide zones Pb-Zn ores with larsenite, alamosite, leadhillite		Am Min 64, 1331
		Red to purple	~3½ 4.97–5.01	Red to purple		Associate with tellurite, paratellurite, native Te		MSA Spec. pap 1 305–309
	b = X, c ∧ Y = +24°, c ∧ Z = -66°	Colorless	2½ 5.94–6.06	Colorless		Oxide zones Pb-deposits with anglesite, bayldonite		Dana II, 661–663
{100} contact or penetration	b = Y, c ∧ X = 85°, c ∧ Z = 6°	Colorless	2½ 4.15–4.26	Colorless, white		Oxide zones As-deposits	Dimorph arsenolite	Dana I, 545–547

n_β Var.	Refractive Index	Mineral Name and Composition	$(n_\gamma - n_\alpha)$	2V angle Dispersion	Crystal System	Habit	Cleavage
1.920	n_β 1.920 n_α 1.885 n_γ 1.956	Tsumebite $Pb_2Cu(PO_4)(SO_4)(OH)$	0.071	~90° v > r strong	Monoclinic β = 111.5°	Thick tabular crystals as twinned groups	None
1.928	n_β 1.928 n_α 1.880 n_γ 2.029	Cesbronite $Cu_5(TeO_3)_2(OH)_6 \cdot 2H_2O$	0.149	72° r > v moderate	Orthorhombic	Prismatic crystals (a)	{021} good, {101} poor
**1.870 2.034	n_β 1.870–2.034 n_α 1.840–1.950 n_γ 1.943–2.110	Sphene (Titanite) $CaTiSiO_5$	0.100 to 0.192	17° – 56° r > v strong, inclined	Monoclinic β = 120°	Small euhedral diamond sections	{110} distinct
1.95	n_β 1.95 n_α 1.89 n_γ 2.02	Cerotungstite $(Ce,Nd)W_2O_6(OH)_3$	0.13	Large ~ 85°	Monoclinic β = 105.5°	Radiated bladed	{100} perf
1.960	n_β 1.960 n_α 1.920 n_γ 2.20	Graemite $CuTeO_2 \cdot H_2O$	0.28	48° No disp	Orthorhombic	Prism crystals	{010} good, {100} parting
1.963	n_β 1.963 n_α 1.963 n_γ 1.966	Hyalotekite $(Pb,Ca,Ba)_4BSi_6O_{17}(OH,F)$	0.003	Small v > r strong	Orthorhombic ?	Massive coarse crystalline	Two at 90°
1.97	n_β 1.97 n_α 1.93 n_γ 2.01	Bismutoferrite $BiFe_2(SiO_4)_2(OH)$	0.08	Large	Orthorhombic	Massive, earthy	
1.97	n_β 1.97 n_α 1.95 n_γ 1.99	Bayldonite (Cuproplumbite) $(Pb,Cu)_3(AsO_4)_2(OH)_2$	0.04	Large v > r strong	Monoclinic	Tiny fibrous concretions	
1.981 1.985	n_β 1.985–1.981 n_α 1.955–1.959 n_γ 2.05–2.060	Uranospherite (Uranosphaerite) $Bi_2U_2O_9 \cdot 3H_2O$	0.095 to 0.101	Large 70° ~ 55° v > r strong	Orthorhombic or Monoclinic	Tiny prism (c) xls, globular	{100} dist
>2.0	n_β>2.0 n_α>2.0 n_γ>2.0	Molybdite MoO_3	~0.2	Large v > r distinct	Orthorhombic	Flat needles (c) or platy {010}	{100} & {010} perfect, {001} distinct
2.01 2.05	n_β 2.05–2.01 n_α 2.01–2.00 n_γ 2.09–2.02	Calciovolborthite (Tangeite) $CaCuVO_4(OH)$	0.08 to 0.02	83° r > v strong	Orthorhombic	Scaly, fibrous, aggregates	One perfect
2.038	n_β 2.038 n_α 1.958 n_γ 2.245	Sulfur S	0.287	~69° v > r weak	Orthorhombic	Dipyram crystals, colliform, massive	{001}, {110} {111} poor, {111} parting
2.040	n_β 2.040 n_α 2.016 n_γ 2.130	Gamagarite $Ba_4(Fe,Mn)_2V_4O_{15}(OH)_2$	0.114	46° – 62° v > r distinct	Monoclinic β = 117.3°	Prism-acicular (b)	{001} & {100} distinct, {$\bar{1}$01} indist
2.055	n_β 2.055 n_α 1.96 n_γ>2.11	Ludlockite $(Fe,Pb)As_2O_6$	~0.1		Triclinic α = 113.9° β = 99.7° γ = 82.7°	Crystals elongation (a), flattened {0$\bar{1}$1}	{0$\bar{1}$1} perfect {0 ℓ} distinct

** n_β variation exceeds 0.10

Twinning	Orientation	Color in thin section	Hardness Sp Gravity	Color hand sample	Alteration	Occurrence	Remarks	Reference
{$\bar{1}$22} cyclic (trillings)		Pale green Z > X	3½ 6.133	Emerald green		Oxide zones Pb-Cu-deposits with smithsonite, cerussite, azurite	Easily soluble HCl	Dana II, 918-919
	a = X	X = lt blue-green, Y = yellow-green, Z = dk green, Z ~ Y >> X	3	Green		Ore veins with electrum, teineite, carlfriesite	Soluble 1:1 HCl or HNO_3	Min Mag 39, 744-746
Common, simple on {100}	b = Y, c ∧ Z = -36° to -51°, a ∧ X = -6° to -21°	Pale gray-brown, yellow-brown. Z > Y > X	5-5½ 3.45-3.56	Dark brown, yellow	Lucoxene (fine aggregates of rutile or anatase)	Most plutonic igneous rocks, schists, gneisses, skarn		p. 133
{001}	b = Z, Y ~ ⊥{001}, a ~ X	Yellow	~1	Orange-yellow		Hydrothermal W-veins		Am Min 57, 1558-1559
		X = yellowish-grn, Y = Z = blue-green	3-3½ 4.13-4.24	Blue-green, jade green		Oxide zones Cu-deposits with teineite, malachite, cuprite	Easily soluble cold dilute acids. Not fluor U.V.	Am Min 60, 486
	Optic plane is parallel to cleavage intersection	Colorless	5-5½ 3.81	White, gray		Iron ores with barylite		Win & Win II, 401
		Colorless, yellow	4.47	Yellow-green		Primary ore veins		Am Min 43, 656-670
	b = X, Y ∧ elong ~ 45°	Greenish	4½ 4.35-5.5	Siskin-green to yellow-grn		Oxide zones of Cu deposits	Soluble HNO_3	Dana II, 929-930
	a = X, b = Y, c = Z	Colorless, pale orange. No pleochroism	2-3 6.36	Red-orange, yellow-orange		Secondary U-mineral in oxide zones ore veins		USGS Bull 1064, 98-99
	a = Y, b = X, c = Z	Colorless	1-2 4.72	Yellow, greenish, white		Alteration of molybdenite		Am Min 49, 1497-1498
		Pale green, X = Y > Z	3½ 3.75-3.82	Yellow-green, dk green		Oxide zones V-deposits	Soluble acids	Dana II, 817-818
{101}, {011}, {110} simple, uncommon	a = X, b = Y, c = Z	Pale yellow. Weakly pleochroic	1½-2½ 2.07	Yellow, brownish, reddish, greenish	Oxidizes to sulfurous & sulfuric acids	Fumeroles, hot springs, bedded evaporite deposits with gypsum, aragonite	Easily fused giving SO_2 gas.	p. 5
	b = Y, c ∧ X ~ -41°	X = red-brown, Y = dk red-brown, Z = lt salmon buff	4½-5 4.62	Dk brown, black		Mn-ores with diaspore, bixbyite, ephesite		Am Min 28, 329-335
{0$\bar{1}$1} lamellar	Optic plane ~ ⊥{0$\bar{1}$1}, a ~ Z	X = yellow, Y = deep yellow, Z = orange-yellow	1½-2 4.35-4.40	Red		Cavities in Pb-Zn ore-veins	Sectite, flexible fibers	Am Min 57, 1003-1004

n_β Var.	Refractive Index	Mineral Name and Composition	$(n_\gamma - n_\alpha)$	2V angle Dispersion	Crystal System	Habit	Cleavage
2.05 2.07	n_β 2.05–2.07 n_α 2.05–2.07 n_γ 2.06–2.08	Carminite $PbFe_2(AsO_4)_2(OH)_2$	0.01	Medium v > r strong	Orthorhombic	Acicular (c), radiated fibrous aggregates	{110} dist
>2.05	n_β>2.05 n_α>2.05 n_γ>2.19	Rynersonite $Ca(Ta,Nb)_2O_6$	~0.14	Moderate	Orthorhombic	Fibrous, felted masses, bladed	None
2.08	n_β 2.08 (Li) n_α 2.07 n_γ 2.19	Fersmite $(Ca,Ce,Na)(Nb,Ti,Fe,Al)_2(O,OH,F)_6$	0.11	20° – 25°	Orthorhombic	Prismatic, tabular	None
2.11	n_β~2.11 n_α~2.11 n_γ~2.13	Schneiderhöhnite $8FeO\cdot 5As_2O_3$	Weak		Triclinic α = 63° β = 116.2° γ = 81.8°		{100} perfect Two others distinct
2.112	n_β 2.112 n_α 2.110 n_γ 2.165	Khinite $Cu_3PbTeO_4(OH)_6$	0.055	20°	Orthorhombic	Bipyram crystals	{001} fair
2.116	n_β 2.116 n_α 2.077 n_γ 2.158	Laurionite PbCl(OH)	0.081	Large ~ 90°	Orthorhombic	Tabular {100}, elongation (c)	{101} dist
2.12	n_β~2.12 n_α~2.05 n_γ~2.20	Blixite $Pb_2Cl(O,OH)_{2-x}$	0.15	80°	Orthorhombic	Coating	One dist
2.135	n_β 2.135 n_α 2.115 n_γ 2.26	Rajite $CuTe_2O_5$	0.15	40°	Monoclinic β = 109.1°	Small xls	
2.14 2.24	n_β n_α 2.14 n_γ 2.24	Fornacite $(Pb,Cu)_3[(Cr,As)O_4]_2(OH)$	0.10		Monoclinic	Small prism crystals	None
2.15	n_β 2.15 n_α 2.14 n_γ 2.18	Atelestite (Rhagite) $Bi_8(AsO_4)_3O_5(OH)_5$	0.04	44° v > r rather strong	Monoclinic β 107.2°	Minute tabular {100} crystals	{001} poor
2.17	n_β 2.17 n_α 2.12 n_γ 2.31	Melanotekite $Pb_2Fe^{3+}_2Si_2O_9$	0.019	67° v > r rather strong	Orthorhombic	Prism crystals, massive	Two-unequal
2.17	n_β 2.17 n_α 2.17 n_γ 2.18	Georgiadesite $Pb_3AsO_4Cl_3$	0.01	Large v > r strong	Monoclinic β = 102.5°	Pseudo-hexagonal tabular	None
2.18	n_β 2.18 n_α 2.00 (Li) n_γ 2.35	Tellurite TeO_2	0.35	~90° v > r moderate	Orthorhombic	Radiated acicular crystals (c), plates, coatings	{010} perfect
2.20	n_β 2.20 n_α 2.10 n_γ 2.31	Kentrolite $Pb_2Mn^{3+}_2Si_2O_9$	0.21	88° v > r strong	Orthorhombic	Massive, minute prism crystals	{110} dist
2.19 2.21	n_β 2.19–2.21 n_α 2.10–2.13 n_γ 2.21–2.24	Natroniobite $NaNbO_3$	0.11	10° – 30° v > r	Monoclinic ?	Fine-grained aggregate	

Twinning	Orientation	Color in thin section	Hardness Sp Gravity	Color hand sample	Alteration	Occurrence	Remarks	Reference
	a = Y, b = Z, c = X	X = pale orange, Y = Z = dark carmine red	3½ 5.22-5.46	Red, reddish-brown		Oxide zones Pb-deposits	Soluble HNO_3	Am Min 48, 1-13
None ?	a = Z, b = Y, c = X. + elongation	X = Y = straw-yellow, Z = pale straw-yel	~4½	White, pink		Pegmatite-aplite dikes		Am Min 63, 709-714
		Greenish-yellow, X > Y = Z	4-4½ 4.69-4.79	Dk brown, black		Rare earth pegmatites		Am Min 44, 1-8
		Red-brown to bright yellow X > Y < Z	3	Brown-black		Oxide zones Cu-Zn-deposits with chalcocite, zincian stottite		Am Min 59, 1139
		X = emerald green, Y = Z = yellowish-green, Z = Y > X	3½ 6.69	Dk green		Oxide zones Cu-Zn-Pb deposits with para-khinite and dugganite	Polymorph para-khinite. Easily soluble dilute acids	Am Min 63, 1016-1019
	a = Y, b = Z, c = X	Colorless	3-3½ 6.14-6.25	Colorless, white		Oxide zones Pb-deposits with para-laurionite, phosgenite	Dimorph Para-laurionite. Soluble HNO_3	Dana II, 62-64
	Ext parallel to cleavage. Optic plane is normal to cleavage	Colorless	~3 7.35	Pale yellow		Fissure in Mn-skarn	Soluble dilute acids	Am Min 45, 908
	b = Y, c ∧ X = -22°	Green Z > Y > X	4 5.75-5.77	Green		Rhyolite with mackayite	Sol dilute acids	Min Mag 43, 91-92
			6.27-6.40	Olive-green		Oxide zones Cu-Pb-deposits		Am Min 49, 447
		Colorless, pale yellow	4½-5 6.82-6.95	Sulfur-yellow, greenish		Oxide zones of Bi deposits	Soluble in HCl	Dana II, 792-793
		X ~ colorless, Y = pale red-brown, Z = deep red-brown	6½ 5.7-2.23	Black		Pb-ores with native Pb	Decomp HNO_3	Am Min 52, 1085-1093
{100} & $\{\bar{1}04\}$ lamellar	b = Y, c ~ Z	Colorless	3½ 7.1	White, brownish-yellow		Altered Pb slag.	Soluble HNO_3	Dana II, 791-792
	a = Y, b = X, c = Z	Colorless	2 5.83-5.90	White, yellow, yellow-orange		Oxidation of tellurium minerals with native Te	Dimorph paratel-lurite. Easily soluble HCl or HNO_3	Dana I, 593-595
	a = X, b = Y, c = Z	Reddish-brown, Z > Y > X strong pleochroism	5 6.19-6.24	Dk reddish-brn		Mn-deposits with calcite, willemite		Am Min 52, 1085-1093
	c ∧ X = 10° - 15°	Yellow to brown	5½-6 4.40	Yellow, brownish, black		Dolomite carbonatites with apatite, pyro-chlore, dysanalyte	Dimorph lueshite	Am Min 47, 1483

Biaxial +

n_β Var.	Refractive Index	Mineral Name and Composition	$(n_\gamma - n_\alpha)$	2V angle Dispersion	Crystal System	Habit	Cleavage
2.20	n_β 2.20 n_α 2.13 n_γ 2.40	Angelellite $Fe_4As_2O_{11}$	0.27	Medium large	Triclinic α = 114° β = 116° γ = 81°	Tabular {001}	{001}
2.20	n_β 2.20 n_α 2.19 n_γ 2.33	Tripuhyite (Flajolotite) (Jujuyite) $FeSb_2O_6$	0.14	Small ~ 30° v > r very strong	Tetragonal	Microcrystalline aggregates, earthy, nodular	
2.212	n_β 2.212 n_α 2.194 n_γ 2.248	Taiyite (Aeschynite-Y) Rare earth-Ti silicate	0.054	~70°	Orthorhombic	Tabular prismatic, granular	{100}, {010} & {001} perf
2.217	n_β 2.217 n_α 2.199 n_γ 2.260	Cotunnite $PbCl_2$	0.061	67°	Orthorhombic	Massive granular, elongation (c)	{010} perf
2.219	n_β 2.219 n_α 2.185 n_γ 2.266	Heyite $Pb_5Fe_2(VO_4)_2O_4$	0.081	82° - 90° r > v weak	Monoclinic β = 112°	Elongation (b), tabular {100}	None
2.22	n_β 2.22 n_α 2.17-2.20 n_γ 2.30-2.32	Huebnerite $MnWO_4$	0.013	~73°	Monoclinic β = 91°	Prism, tabular, radiated bladed	{010} perf
*2.17 2.30	n_β 2.17-2.30 n_α 2.15-2.20 n_γ 2.25-2.35	Tantalite $(Fe,Mn)(Ta,Nb)_2O_6$	0.05 to 0.2	90° - moderate v > r moderate	Orthorhombic	Tabular, massive	{010} dist {100} poor
2.25	n_β 2.25 n_α 2.19 n_γ 2.34	Manganotantalite $(Mn,Fe)(Ta,Nb)_2O_6$	0.15	Large ~ 80° v > r moderate	Orthorhombic	Short prism, thick tabular, massive	{010} dist {100} indist
2.25	n_β 2.25 (Li) n_α 2.25 n_γ 2.35	Manganite MnO(OH)	0.10	Small r > v very strong	Monoclinic β ~ 90°	Prism, fibrous, massive, granular	{010} perf {110} & {001} imperfect
2.265	n_β 2.265 n_α 2.185 n_γ 2.35	Descloizite $Pb(Zn,Cu)VO_4(OH)$	0.165	~90° v > r strong	Orthorhombic	Prism, tabular, granular, botryoidal	None
2.27	n_β 2.27 n_α 2.24 n_γ 2.31	Mendipite $Pb_3Cl_2O_2$	0.07	~90° v > r very strong	Orthorhombic	Columnar, fibrous, radiated, nodular	{110} perf {100} & {010} distinct
2.27	n_β 2.27 n_α 2.27 n_γ 2.30	Raspite $PbWO_4$	0.03	~0°	Monoclinic β = 96.3°	Tabular elongation (b)	{100} perf
2.30	n_β~2.30	Quenselite $PbMnO_2(OH)$			Monoclinic β = 93.3°	Tabular {010}, prism (c or a) crystals	{001} perf
2.305	n_β 2.305 n_α 2.255 n_γ 2.414	Ferberite $FeWO_4$	0.159	68°	Monoclinic β ~ 90°	Bladed	{010} perf
*2.22 2.40	n_β 2.22-2.40 (Li) n_α 2.17-2.31 n_γ 2.30-2.46	Wolframite $(Fe,Mn)WO_4$	0.13 to 0.15	73° - 79°	Monoclinic β = 89.6°	Long prism (c) to tabular 100 , radiated acicular	{010} perf

* n_β variation exceeds 0.01

Twinning	Orientation	Color in thin section	Hardness Sp Gravity	Color hand sample	Alteration	Occurrence	Remarks	Reference
		Deep reddish-brown Z > X	5½ 4.86-4.95	Dark brown		Fumerole incrustations in volcanics		Am Min 44, 1322-1323
	c = Z	Bright yellow. Not pleochroic	~7 5.82-6.14	Yellow to brownish-black		Opaline silica veins, placer deposits with cinnabar, lewisite	Insoluble acids	Dana II, 1024
	- elongation	Pale yellow, not pleochroic	4½-5 4.4	Yellow, orange, green-yellow		Muscovite granite with other rare earth minerals	Aeschynite-priorite group. Soluble H_2SO_4	Am Min 61, 178
{120}	a = Y, b = X, c = Z	Colorless	2½ 5.80-5.88	Colorless, white, yellow, green		Alteration of galena in saline environ. Volcanic sublimate	Soluble H_2O	Dana II, 42-44
{110} simple	b = Y, c ∧ X = -36°	Orange-yellow, no pleochroism	4 6.28	Yellow-orange		Oxide zones Pb-deposits		Min Mag 39, 65-68
{100} common	b = X, c ∧ Z = 17° to 21°	Yellow, orange, red-brown, green, red. Z > Y > X	4-4½ 7.18-7.23	Yellow-brown, red-brown		High T hydrothermal veins in granitic rocks	Wolframite series	Dana II, 1064-1072
{201} contact, {010} trillings {203} or {501} rare	a = X, b = Y, c = Z	Red-brown, Z > Y = X, ~opaque	6-6½ 8.0-6.5	Dk brown, black		Granite pegmatites, granite	Columbite-tantalite series	Dana I, 780-787
{201} common	a = X, b = Y, c = Z	Red-brown ~ opaque. Z > X	6-6½ 8.00	Brownish-black		Granite pegmatites with lepidolite, albite microcline	Columbite-tantalite series	Dana I, 780-787
{011} contact, penetration, {100} lamellar	b = Y, a ~ X, c ~ Z	Red-brown. ~opaque Z > X = Y	4 4.30-4.33	Black		Low T hydrothermal vein mineral	Trimorph feit-knechtite, groutite	Dana I, 646-650
	a = Z, b = Y, c = X	X = Y = yellow, Z = brownish-yel	3-3½ 6.14-6.26	Orange-red, dk brown, green		Oxide zones ore-deposits	Descloizite-mottramite series Soluble acids	Dana II, 811-815
	a = X, b = Y, c = Z	~Colorless	2½ 7.22-7.24	Colorless, white, gray		Oxide zones Pb-deposits	Soluble dilute HNO_3	Dana II, 56-58
{100} common, {$\bar{1}$02} rare	b = Y, c ∧ Z ~ 30°	Pale yellow	2½-3 8.46-8.52	Yellow, yellow-brown, gray		Oxide zones Pb-deposits with stolzite	Dimorph stolzite	Dana II, 1089-1090
	b = X, c ∧ Y = small	Deep brown, Z > X	2½ 6.84-7.09	Pitch-black		Mn-ores with barite & calcite	Soluble dilute acids	Dana I, 729-730
{100} and {023} common	b = X, c ∧ Z = 17° to 21°	~opaque, dk brown in thin splinters Z > Y > X	4-4½ 7.51-7.52	Brownish-black, black		High T hydrothermal veins in granitic rocks	Wolframite series	Dana II, 1064-1072
{100} or {023} usually simple contact	b = X, c ∧ Z = +17° to +21°	X = Yellow, orange, Y = greenish yel, red-brown, Z = olive, dk red-brown	4-4½ 7.12-7.60	Yellow-or red-brown to black		High T hydrothermal veins with cassiterite, molybdenite, tourmaline	Wolframite Series	Dana II, 1063-1072

Biaxial +

n_β Var.	Refractive Index	Mineral Name and Composition	$(n_\gamma - n_\alpha)$	2V angle Dispersion	Crystal System	Habit	Cleavage
2.32	n_β 2.32 n_α 2.11 n_γ 2.65	Hemihedrite $ZnF_2[Pb_5(CrO_4)_3SiO_4]_2$	0.54	88° - 90° Perc horizontal	Triclinic α = 103.5° β = 92° γ = 56°	Elongated hemihedral	{110} poor
*2.32 2.40	n_β 2.32-2.40 n_α 2.25-2.30 n_γ 2.40-2.50	Iranite $PbCrO_4 \cdot H_2O$	0.15 to 0.20	Large	Triclinic α = 104.5° β = 66° γ = 108.5°	Elongated crystals	
2.35	n_β 2.35 (Li) n_α 2.30 n_γ 2.40	Nadorite (Ochrolite) $PbSbO_2Cl$	0.10	Large ~ 90° r > v strong	Orthorhombic	Prism (a) or tabular {010}, divergent	{010} perfect
2.36	n_β 2.36 (Li) n_α 2.29 n_γ 2.66	Crocoite $PbCrO_4$	0.37	57° r > v strong, inclined	Monoclinic β = 102.5°	Prism (c), crystal aggregates	{110} dist {001} & {100} indistinct
2.36	n_β 2.36 n_α 2.33 n_γ 2.40	Magnocolumbite $(Mg,Fe,Mn)(Nb,Ta)_2O_6$	0.007	~80°	Orthorhombic	Acicular, tabular	{010} & {100}
2.38	n_β 2.38 n_α 2.28 n_γ 2.48	Brackenbuschite $Pb_4(Mn,Fe)(VO_4)_4 \cdot 2H_2O$	0.20	Large r > v strong	Monoclinic β = 111.75°	Acicular, botryoidal, tufts	
2.39	n_β 2.39 (Li) n_α 2.38 n_γ 2.42	Pseudobrookite Fe_2TiO_5	0.04	50° v > r	Orthorhombic	Tabular, prism, acicular crystals	{010} dist
2.403	n_β 2.403 n_α 2.388 n_γ 2.428	Bismutotantalite $(Bi,Sb)(Ta,Nb)O_4$	0.040	80° v > r	Orthorhombic	Prism, massive	{010} perf, {101} dist (∧83°)
2.404	n_β 2.4039 (Na) n_α 2.3742 n_γ 2.4568	Stibiotantalite $Sb(Ta,Nb)O_4$	0.83	75° v > r strong	Orthorhombic	Prism (c) to thin tabular crystals	{010} dist {100} indist
2.419	n_β 2.4190 (Na) n_α 2.3977 n_γ 2.4588	Stibiocolumbite $SbNbO_4$	0.061	73° v > r strong	Orthorhombic	Prism-bladed crystals	{010} dist {100} indist
2.42	n_β 2.42 n_α 2.39 n_γ 2.52	Thoreaulite $SnTa_2O_7$	0.13	25° - 35°	Monoclinic β = 91°	Short prism crystals (c)	{100} perf {011} imperf
2.44	n_β 2.44 n_α 2.38 n_γ 2.65	Phoenicochroite $Pb_2O(CrO_4)$	0.27	58° - 60° r > v moderate	Monoclinic β = 115.2°	Polycrystalline masses	{$\bar{2}$01} good, {001}, {010} & {011} fair
2.45	n_β 2.45 (Li) n_α 2.45 n_γ 2.51	Derbylite $Fe_6Ti_6Sb_2O_{23}$	0.06	~0°	Monoclinic β = 104.6°	Prismatic (c)	None
2.5	n_β 2.5 (Li) n_α 2.37 n_γ 2.65	Montroydite HgO	0.28	Large	Orthorhombic	Long prism, vermicular, massive	{010} perfect

* n_β variation exceeds 0.01

Twinning	Orientation	Color in thin section	Hardness Sp Gravity	Color hand sample	Alteration	Occurrence	Remarks	Reference
{$\bar{2}\bar{2}3$} common, {$0\bar{1}0$} or {$0\bar{1}2$} rare		X = Y = yellow, Z = orange	3 6.42-6.50	Orange, brown, black		Oxide zones Pb-deposits		Am Min 55, 1088-1102
	Extinction 5° to elongation	Brownish-orange ‖ elongation, yellow-orange ⊥ elongation	5.8	Saffron yellow		Oxide zones Pb-deposits		Am Min 48, 1417
{101} common	a = Z, b = X, c = Y	Pale brown, yellow	3½-4 7.02-7.07	Brown, brown-yellow, yellow		Oxide zones Pb-Zn deposits with bindheimite, jamesonite		Dana II, 1039-1040
	b = Y, c ∧ Z = -5½°	X = Y = orange-red, Z = blood-red	2½-3 5.99-6.11	Orange-red		Oxide xones Pb-Cr deposits		Dana II, 646-649
Common	a = Z, b = Y, c = X	~Opaque, X = brownish-yellow Z = brownish-red	~6 5.17-5.23	Black, brownish black		Pegmatites often intergrown with ilmenorutile		Am Min 48, 1182-1183
		X ~ colorless, Y = dk red-brown, Z = red-brown	6.05-6.11	Dk brown-black		Oxide zones Pb-Zn-deposits		Dana II, 1052-1053
{hk0}	a = Z, b = X, c = Y	Brown Y > X = Z	6 4.39	Black, brown-black, red-brown		Cavities in basalt or rhyolite with topaz, bixbyite		Dana I, 736-738
	a = X, b = Y, c = Z	Yellow-brown, dark brown ~opaque	5 8.51-8.98	Lt brown to black		Pegmatites		Am Min 42, 178-183
{010} polysynthetic	a = X, b = Y, c = Z	Pale yellow-brown to brown. May be zoned	5-5½ 7.34-7.53	Brown to yellow-brown, reddish, greenish		Pegmatites with beryl, pink tourmaline, lepidolite, cassiterite	Stibiotantalite-stibiocolumbite series	Dana I, 767-769
{010} polysynthetic	a = X, b = Y, c = Z	Pale yellow-brown to brown. May be zoned	5½ 5.68-5.98	Red-brown to yellow-brown		Pegmatites with beryl, pink tourmaline, lepidolite	Stibiotantalite-stibiocolumbite series	Dana I, 767-769
{010} lamellar	b = Y, c ∧ Z = 58°-63°	Yellow, Z > X	5½-6 7.5-7.9	Brown to yellowish		Granite pegmatites with cassiterite, microlite, tantalite		Vlasov II, 539-541
	b = X, c ∧ Y = +2°, c ∧ Z = -88°	Orange to yellow-orange, weak pleochroism, Z > Y > X	2½ 7.01-7.08	Dark red		Oxide zones Pb-deposits with crocoite, vauquelinite		Am Min 55, 784-792
{011} common (cruciform)		~Opaque, dk brown, not pleochroic	5 4.53	Black		Heavy fraction in gravels		Dana II, 1025-1026
	a = X(?), b = Y(?), c = Z	Orange-red, pale yellow	2½ 11.21	Dk red, brownish-red, brown		Hg-deposits with native Hg, eglestonite, calomel, cinnabar	Sectile, flexible inelastic. Readily sol HCl, HNO_3	Dana I, 511-514

Biaxial +

n_β Var.	Refractive Index	Mineral Name and Composition	$(n_\gamma - n_\alpha)$	2V angle Dispersion	Crystal System	Habit	Cleavage
2.584	n_β 2.584 n_α 2.583 n_γ 2.700	Brookite (Pyromelane) TiO_2	0.117	$0° - 30°$ Crossed very strong	Orthorhombic	Tabular {010}, prism (c) crystals	{120} indist {001} poor
2.61	n_β 2.61 (Li) n_α 2.51 n_γ 2.71	Massicot PbO	0.20	~90° Strong	Orthorhombic	Massive, earthy, scaly, tabular	{100} & {010} dist {110} traces
2.684	n_β 2.684 (Na) n_α 2.538 n_γ 2.704	Realgar AsS	0.116	39° r > v strong, inclined	Monoclinic $\beta = 106.6°$	Prism crystals, granular, massive	{010} dist {$\bar{1}$01}, {100} {120} indist
2.72	n_β>2.72 (Li) n_α>2.11 (white light)	Getchellite $AsSbS_3$	Extreme	<46° r> v strong, crossed	Monoclinic $\beta = 116.2°$	Grains, micaceous	{001} perfect
2.74	n_β 2.74 n_α>2.72	Kermesite Sb_2S_2O	Very large	Small	Monoclinic $\beta = 101.75°$	Lath crystals, radiated aggregate, hairlike (b)	{001} perfect {100} dist

Twinning	Orientation	Color in thin section	Hardness Sp Gravity	Color hand sample	Alteration	Occurrence	Remarks	Reference
{120} rare	a = X, b = Z, c = Y (a = Y, c = X for blue light)	Yellow-brown or red-brown to deep brown. Z > Y > X	5½-6 4.12	Yellow-brown to red-brown, black	Rutile	Gneisses, hydrothermal veins, pegmatites with rutile and anatase	Trimorph rutile, anatase. Properties vary greatly with λ.	p. 22
Rare	a = Y ?	X ~ colorless, Y = pale yellow, Z = deep yellow	2 9.642	Yellow, orange		Oxidation of Pb-minerals	Dimorph litharge Soluble HCl, HNO_3	Dana I, 516-517
{100} contact	b = Y, c ∧ X = +11°	X ~ Colorless Y = pale yellow, Z = pale yellow	1½-2 3.56-3.59	Red, orange-red, yellow-orange	Orpiment and arsenolite	Low T hydrothermal veins with orpiment, stibnite. Hot springs	Decomp HNO_3	Dana I, 255-258
	b = Z, a ∧ Y = 15°, c ∧ Y = 101°	Blood-red, little pleochroism	1½-2 3.92-4.01	Dark blood red		Epithermal veins with orpiment, realgar, stibnite, cinnabar		Am Min 50, 1817-1826
	b = Z	Red in thin splinters	1-1½ 4.68-4.69	Cherry red		Alteration of stibnite or native antimony	Sectile, flexible laminae	Dana I, 279-280

n_β Var.	Refractive Index	Mineral Name and Composition	$(n_\gamma - n_\alpha)$	2V angle Dispersion	Crystal System	Habit	Cleavage
1.325	n_β 1.325 n_α 1.324 n_γ 1.325	Avogadrite $(K,Cs)BF_4$	0.001	~90°	Orthorhombic	Tiny platy {001} aggregates	
1.410	n_β 1.410 n_α 1.396 n_γ 1.419	Mirabilite $Na_2SO_4 \cdot 10H_2O$	0.023	76° v > r strong, crossed	Monoclinic β = 107.7°	Prism-acicular, granular, crusts	{100} perf. {001}, {010} & {011} traces
1.414	n_β 1.4136 n_α 1.4072 n_γ 1.4150	Thomsenolite $NaCaAlF_6 \cdot H_2O$	0.008	50° v > r weak	Monoclinic β = 96.4°	Pseudocubic, tabular {001}, prism crystals	{001} perfect {110} dist
1.417	n_β 1.417 n_α 1.413 n_γ 1.423	Yaroslavite $Ca_3Al_2F_{10}(OH)_2 \cdot H_2O$	0.010	74°	Orthorhombic	Radiated fibrous, spherulitic	Pinacoidal
1.425	n_β 1.425 n_α 1.405 n_γ 1.440	Natron $Na_2CO_3 \cdot 10H_2O$	0.035	Large ~ 80° r > v perc crossed	Monoclinic β = 121.4°	Efflorescent coatings, tabular crystals {010}	{001} dist {010} imperf {110} traces
1.433	n_β 1.433 n_α 1.429 n_γ 1.436	Jarlite (Metajarlite) $NaSr_3Al_3F_{16}$	0.007	82° - 90°	Monoclinic β = 101.8°	Tabular, elongation(c), radiated, massive	
1.452	n_β 1.452 n_α 1.429-1.430 n_γ 1.456-1.458	Kalinite $KAl(SO_4)_2 \cdot 11H_2O$	0.027	Small - 52° Disp weak	Monoclinic	Fibrous (c)	
1.453	n_β 1.453 n_α 1.426 n_γ 1.456	Hexahydrite $MgSO_4 \cdot 6H_2O$	0.030	38°	Monoclinic β = 98.5°	Columnar-fibrous, tabular crystals	{100} perfect
1.454	n_β 1.454 n_α 1.440 n_γ 1.455	Lecontite $Na(NH_4,K)SO_4 \cdot 2H_2O$	0.015	~30°	Orthorhombic	Long prism, fine granular	{001} dist
1.454	n_β 1.454 n_α 1.448 n_γ 1.456	Gearksutite (Gearksite) $CaAlF_4(OH) \cdot H_2O$	0.008	Moderate	Monoclinic ?	Massive, earthy	None
1.455	n_β 1.455 n_α 1.435 n_γ 1.459	Wattevilleite $Na_2Ca(SO_4)_2 \cdot 4H_2O(?)$	0.024	~48°	Orthorhombic or Monoclinic	Minute acicular fibrous crystals	
1.456	n_β 1.456 n_α 1.452 n_γ 1.458	Tikhonenkovite $SrAlF_4(OH) \cdot H_2O$	0.006	70°	Monoclinic β = 102.7°	Small prism crystals	{001} perfect
1.452 1.462	n_β 1.452-1.462 n_α 1.430-1.440 n_γ 1.457-1.469	Epsomite (Seelandite) $MgSO_4 \cdot 7H_2O$	~0.028	~50° v > r weak	Orthorhombic	Fibrous-acicular crusts	{010} perfect {101} dist
1.457 1.461	n_β 1.457-1.461 n_α 1.337-1.340 n_γ 1.459-1.462	Sassolite $B(OH)_3$	0.119 to 0.122	10° - 17°	Triclinic α = 92.5° β = 101.2° γ = 120°	Pseudo-hexagonal plates {001}, acicular	{001} perfect

Twinning	Orientation	Color in thin section	Hardness Sp Gravity	Color hand sample	Alteration	Occurrence	Remarks	Reference
	a = Z, b = Y, c = X	Colorless	2.51-3.30	Colorless, yellowish, reddish		Sublimation around fumeroles	Slightly soluble in H_2O	Dana II, 97-98
{001} penetration, also {100}	b = X, c ∧ Z = −31°	Colorless	1½-2 1.465	Colorless, white		Saline lake deposits, hot springs, soil efflorescence	Easily soluble H_2O. Bitter taste	Dana II, 439-442
	b = Z, c ∧ X = +52°	Colorless	2 2.98-2.99	Colorless, white, reddish		Cryolite pegmatites with ralstonite, pachnolite. Alter cryolite	Easily soluble H_2SO_4	Dana II, 116-118
		Colorless	3.09	White		Hydrothermal veins with fluorite, sellaite, tourmaline	Fluor U.V. pale violet	Am Min 51, 1546-1547
{001}	b = X, c = Z ~ 41°	Colorless	1-1½ 1.44-1.46	Colorless, white, yellowish	Thermonatrite	Low T deposits in soda lakes with trona, gaylussite, thermona trite	Easily soluble H_2O, alkaline taste. Effer acids	Dana II, 230-231
	b = Y, c ∧ X = +6°, c ∧ Z = −84°	Colorless	4-4½ 3.61-3.87	Colorless, white, gray		Cryolite pegmatites with thomsenolite, chiolite, topaz, fluorite	Soluble in $AlCl_3$ solution	Dana II, 118-119
	b = Z, c ∧ Y ~ 13°	Colorless	2-2½	Colorless, white		Efflorescence on clays and aluminous shales	Solution H_2O. Taste sweetish-astringent	Dana II, 471
{001} and {110}	b = Y, c ∧ X = +25°	Colorless	1.745-1.757	Colorless, white, greenish	Epsomite	Evaporite salt deposits Dehydration of epsomite	Soluble H_2O. Bitter salty taste	Dana II, 494-495
		Colorless	2-2½ 1.74	Colorless		Bat guano	Soluble H_2O. Bitter-saline taste	Am Min 48, 180-188
	b = X, Z ∧ elong = large	Colorless	2 2.77	White		Hydrothermal veins, pegmatites	Easily soluble dilute acids	Dana II, 119-120
		Colorless	1.81	White		Pyrite-bearing lignite	Soluble H_2O. Taste sweet-astringent	Dana II, 452
		Colorless	3½ 3.26	Colorless, pinkish		Fissures or cavities in Fe-ores with gearksutite, celestite, barite		Am Min 49, 1774-1775
{110} rare	a = X, b = Z, c = Y	Colorless	2-2½ 1.677-1.678	Colorless, white, pinkish	Hexahydrite	Efflorescences in mines and caves, saline evaporites, fumeroles	Very soluble H_2O	Dana II, 509-513
	X ∧ {001}	Colorless	1 1.46-1.52	White, gray				Am Min 42, 56-61

Biaxial –

n_β Var.	Refractive Index	Mineral Name and Composition	$(n_\gamma - n_\alpha)$	2V angle Dispersion	Crystal System	Habit	Cleavage
1.460	n_β 1.460 n_α 1.44 n_γ 1.487	Khademite $Al(SO_4)(OH)\cdot 5H_2O$	0.047	68°	Orthorhombic	Elongation crystals (a)	None
1.461	n_β 1.461 n_α 1.449 n_γ 1.463	Mendozite $NaAl(SO_4)_2\cdot 11H_2O$	0.014	56° Crossed distinct	Monoclinic β = 109°	Fibrous crusts, prism crystals (c)	{100} dist {001} & {010} indist
1.465	n_β 1.465	Zinc-fauserite $(Mn,Mg,Zn)SO_4\cdot 7H_2O$		Large r > v	Orthorhombic	Efflorescent crusts	{010} good
1.469	n_β 1.469 n_α 1.447 n_γ 1.472	Borax $Na_2B_4O_7\cdot 10H_2O$	0.025	~40° r > v strong, crossed	Monoclinic β = 106.6°	Stubby prism crystals, granular-earthy	{100} perf {110} dist
1.472	n_β 1.472 n_α 1.454 n_γ 1.488	Kernite $Na_2B_4O_7\cdot 4H_2O$	0.034	80° r > v distinct	Monoclinic β = 97.5°	Fibrous, cleavage masses	{100} & {001} perfect {$\bar{2}$01} distinct
*1.465 1.480	n_β 1.465–1.480 n_α 1.447–1.463 n_γ 1.470–1.485	Goslarite $ZnSO_4\cdot 7H_2O$	0.023 to 0.022	Mod-small 46° – 65° r > v weak	Orthorhombic	Fibrous-acicular, massive, crusts	{010} perfect
1.473	n_β 1.473 n_α 1.459 n_γ 1.483	Jurbanite $AlSO_4(OH)\cdot 5H_2O$	0.024	80°	Monoclinic β = 102.2°	Prism crystals	None
1.473	n_β 1.473 n_α 1.465 n_γ 1.477	Sasaite $(Al,Fe^{3+})_{14}(PO_4)_{11}(OH)_7SO_4$ $\cdot 83H_2O$	0.012	~70°	Orthorhombic	Earthy nodules, rhombic plates	
1.478	n_β 1.478 n_α 1.461 n_γ 1.485	Creedite $Ca_3Al_2SO_4(F,OH)_{10}\cdot 2H_2O$	0.024 to 0.021	64° r > v strong	Monoclinic β = 94.5°	Prism-acicular, radiated aggregates	{100} perfect
1.475 1.481	n_β 1.475–1.481 n_α 1.471–1.480 n_γ 1.477–1.482	Sodium-dachiardite $(Na_2,Ca,K_2)_{4-5}(Si,Al)_{48}O_{96}$ $\cdot 25\text{–}27H_2O$	0.002 to 0.006	76° – 80°	Monoclinic β = 109°		
1.479	n_β 1.479 n_α 1.388 n_γ 1.486	Darapskite $Na_3NO_3SO_4\cdot H_2O$	0.098	27° r > v strong	Monoclinic	Prism (c), tabular {100}, granular	{100} perfect {010} dist
1.479	n_β 1.479 n_α 1.476 n_γ 1.481	Tveitite $Ca_{1-x}(Y,RE)_xF_{2+x}(x \sim 0.3)$	0.005	34°	Monoclinic (Pseudocubic) β = 90.3°		
1.48	n_β 1.48 n_α 1.47 n_γ 1.49	Boothite $CuSO_4\cdot 7H_2O$	0.02	~90°	Monoclinic β = 105.5°	Massive, fibrous	{001} imperf
1.480	n_β 1.480 n_α 1.475 n_γ 1.483	Pickeringite $MgAl_2(SO_4)_4\cdot 22H_2O$	0.008	60°	Monoclinic β = 95°	Acicular, fibrous aggregates, efflorescence	{010} poor

* n_β variation exceeds 0.01

Twinning	Orientation	Color in thin section	Hardness Sp Gravity	Color hand sample	Alteration	Occurrence	Remarks	Reference
	a = Z, b = Y, c = X	Colorless	1.925	Colorless		With other secondary Fe-Al sulfates (copiapite, butlerite, jarosite)		Am Min 60, 486
	b = X, c ∧ Y = 30°	Colorless	~3 1.73-1.77	Colorless, white		Efflorescence in tunnels, fumeroles	Soluble H_2O. Monoclinic alum	Dana II, 469-471
		Colorless	2½ 1.997	Pale rose	Chalky crust	Efflorescence in Zn-mines	Readily soluble H_2O	Am Min 35, 333
{100} uncommon	b = X, c ∧ Y = -33° to -36°, a ∧ Z = -16° to -19°	Colorless	2-2½ 1.71	White, gray, greenish, bluish	Tincalconite	Saline playas. Efflorescence-desert soils. Thermal springs	Highly soluble H_2O. Weak alkaline taste	p. 97
{110}	b = Z c ∧ X = -38°, c ∧ Y = +52°	Colorless	2½-3 1.91-1.93	Colorless, white	Tincalconite	Playa lake evaporites with borax and other borates	Soluble hot H_2O or acids	Dana II, 335-337
	a = Z, b = X, c = Y	Colorless	2-2½ 1.94-1.98	Colorless, brownish, blue, green		Efflorescence in mines and oxide zones Zn-deposits	Soluble H_2O. Astringent-metallic-nauseous taste	Dana II, 513-516
	b = Y, a ∧ Z = -5°	Colorless	2½ 1.78-1.83	Colorless		Mine stalactites with epsomite, hexahydrite, pickeringite, starkeyite	Easily soluble H_2O	Am Min 61, 1-4
		Colorless	Soft 1.75	White		Efflorescence in dolomite cave	Dehydrates rapidly	Am Min 64, 464
	b = Y, c ∧ Z = 42°	Colorless	4 2.71	Colorless, white, violet		Hydrothermal veins with fluorite, barite, sulfides	Soluble acids	Dana II, 129-130
	- elongation	Colorless	2.16	Colorless, white		With high-silica zeolites		Am Min 64, 244
{100} polysynthetic	b = X, c ∧ Z = 12°	Colorless	2½ 2.20	Colorless		Nitrate evaporite deposits	Soluble H_2O	Am Min 55, 1500-1517
Polysynthetic		Colorless		White, pale yellow		Pegmatites with beryl, monazite	Fluor UV short λ-yellow-orange	Am Min 62, 1060
	b = Y, c ∧ X = small	Pale blue	2-2½ ~2.1	Blue		Oxide zones Cu-deposits	Soluble H_2O. Metallic taste	Dana II, 504-505
	b = Y, c ∧ Z = 36°	Colorless	1½ 1.79-1.84	Colorless, white, yellowish		Weathering of prite ores, coal veins and in caverns	Pickeringite-halotrichite series. Soluble H_2O-astringent	Dana II, 523-527

Biaxial −

n_β Var.	Refractive Index	Mineral Name and Composition	$(n_\gamma - n_\alpha)$	2V angle Dispersion	Crystal System	Habit	Cleavage
1.475 1.485	n_β 1.475–1.485 n_α 1.472–1.483 n_γ 1.477–1.487	Mordenite (Ptilolite) $(Na_2,K_2,Ca)(Al_2Si_{10})O_{24}\cdot 7H_2O$	0.002 to 0.005	76° − 90°	Orthorhombic	Radiated acicular-fibrous (c), cottony	{010} & {100} perf
1.482	n_β 1.482 n_α 1.462 n_γ 1.495	Aubertite $CuAl(SO_4)_2Cl\cdot 14H_2O$	0.033	71° r > v mod	Triclinic <β 91°52' <β 94°40' <β 82°27'	Crusts	{010} perfect
1.482	n_β 1.482 n_α 1.380 n_γ 1.578	Kalicine (Kalicinite) $KHCO_3$	0.198	~ 81°	Monoclinic β = 104.5°	Prism (c), fine crystalline aggregate	{100}, {001}, {101}
1.482	n_β 1.482 n_α 1.462 n_γ 1.490	Moraesite $Be_2PO_4(OH)\cdot 4H_2O$	0.028	65°	Monoclinic β = 97.7°	Fibrous, spherulitic, crusts	2 perfect ∥ to b & c axes
1.482	n_β 1.482 n_α 1.478 n_γ 1.482	Apjohnite $Mn^{2+}Al_2(SO_4)_4\cdot 22H_2O$	0.004	Small	Monoclinic	Fibrous on c	Probably {010}, {001}
1.482	n_β 1.482 n_α 1.479 n_γ 1.487	Leonite $K_2Mg(SO_4)_2\cdot 4H_2O$	0.008	~ 90°	Monoclinic β = 95.3°	Tabular, elongation (c)	None
1.483	n_β 1.483 n_α 1.479 n_γ 1.488	Zinc-melanterite $(Zn,Cu,Fe)SO_4\cdot 7H_2O$	0.009	~ 90° Slight	Monoclinic β ~ 90°	Massive, columnar	
1.485	n_β 1.485 n_α 1.442 n_γ 1.490	Hungchaoite $MgB_4O_7\cdot 9H_2O$	0.048	15° − 36°	Triclinic α = 103° β = 109° γ = 97°	Tab. xls {100}	
1.485	n_β 1.485 n_α 1.477 n_γ 1.486	Phosphorroesslerite $MgH(PO_4)\cdot 7H_2O$	0.009	38° r > v	Monoclinic β = 95°	Short prism, equant crystals	None ?
1.486	n_β 1.486 n_α 1.480 n_γ 1.490	Halotrichite $Fe^{2+}Al_2(SO_4)_4\cdot 22H_2O$	0.010	35°	Monoclinic β = 96.8°	Fibrous, acicular, aggregate	{010} poor
1.486	n_β 1.486 n_α 1.483 n_γ 1.487	Bloedite (Blödite) $Na_2Mg(SO_4)_2\cdot 4H_2O$	0.004	71°	Monoclinic β = 100.7°	Prism, massive	None
1.488	n_β 1.488 n_α 1.486 n_γ 1.489	Vanthoffite $Na_6Mg(SO_4)_4$	0.003	83° v > r weak	Monoclinic β = 113.5°	Massive, grains	
1.489	n_β 1.489 n_α 1.448 n_γ 1.493	Burkeite $Na_6CO_3(SO_4)_2$	0.045	34° r > v distinct	Orthorhombic	Tabular, massive, platy	None ?
1.490 1.492	n_β 1.490–1.492 n_α 1.453–1.455 n_γ 1.498–1.502	Bayleyite $Mg_2UO_2(CO_3)_3\cdot 18H_2O$	0.045 to 0.047	30°	Monoclinic β = 93°	Prism, acicular (c)	
1.492	n_β 1.492 n_α 1.412–1.418 n_γ 1.540–1.543	Trona $Na_3H(CO_3)_2\cdot 2H_2O$	0.128 to 0.125	76° v > r strong	Monoclinic β = 103.1°	Fibrous-columnar (b)-bladed {001}, massive	{100} perfect {$\bar{1}$11} & {001} indistinct

Twinning	Orientation	Color in thin section	Hardness Sp Gravity	Color hand sample	Alteration	Occurrence	Remarks	Reference
None	a = Y, b = Z, c = X ‖ extinction, -elongation	Colorless	3-4 2.12-2.22	White, yellowish, pinkish	None	Amygdules in basalts. Altered glasses and tufts. Marine sediments	Zeolite group	p. 385
			1.815	Azure blue		Oxidized zone of copper deposit		Am Min 65, 205
	b = Y, c ∧ X = -30°	Colorless	Soft 2.17	Colorless, white, yellowish		Evaporite with trona	Soluble H_2O	Dana II, 136
	b = Z, c ∧ Y = 11°	Colorless	1.805	White		Vugs in pegmatites with beryl, albite, quartz, muscovite		Am Min 38, 1126-1133
	b = Y, c ∧ Z = 29°	Colorless	1½ 1.78-2.30	White, pale pink, green, yellow		Efflorescence	Soluble H_2O	Dana II, 527-528
{100} lamellar	b = Y, a ∧ Z = small	Colorless	2½-3 2:20	Colorless, yellowish		Evaporite marine sediments with halite, K, Mg-salts	Soluble H_2O. Bitter taste	Dana II, 450-451
	Y ∧ elong ~ 34°	Very pale blue-grn	~2 2.02	Pale greenish-blue		Oxide zones of pyrite-chalcopyrite-sphalerite ores	Dehydrates readily. Readily soluble H_2O	Dana II, 508
None?	X ∧ a = +28° Z ∧ b = +37° Y ∧ c = -66°	Colorless	2½ 1.70	White		Borate deposits in ulexite nodules		Am Min 64, 369
	b = X, c ∧ Z ~ -6°	Colorless	2½ 1.72	Colorless, yellowish		Mine efflorescence		Dana II, 713-714
	b = Y, c ∧ Z = -38°	Colorless	1½ 1.90-1.95	Colorless, white, yellow, green		Weathering of pyrite-bearing clays. Volcanic sublimate	Soluble H_2O. Astringent taste	Dana II, 523-527
	b = Y, c ∧ X = 37°	Colorless	2½-3 2.25-2.27	Colorless, reddish, etc.		Evaporite salt deposits	Soluble H_2O, saline-bitter	Dana II, 447-450
		Colorless	3½ 2.694	Colorless		Marine salt deposits with bloedite, loeweite, langbeinite		Dana II, 430
{110} cross-shaped	a = Y, b = Z, c = X	Colorless	3½ 2.56-2.57	White, tan, grayish		Evaporites, efflorescences	Soluble H_2O	Dana II, 633-634
	c ∧ X = 8°-15°	X = Pink, Y = Z = pale yellow	2.05-2.06	Yellow		Efflorescence in oxide zones U-deposits	Soluble H_2O, Decomposes in air	USGS Bull, 1064, 112-115
	b = X, c ∧ Z = 83°	Colorless	2½-3 2.11-2.13	Colorless, gray, yellow, lt brn		Soil efflorescence, arid lake deposits with borax, thenardite	Soluble H_2O. Alkaline taste. Effervescent acids	Am Min 44, 274-281

n_β Var.	Refractive Index	Mineral Name and Composition	$(n_\gamma - n_\alpha)$	2V angle Dispersion	Crystal System	Habit	Cleavage
1.49	n_β 1.49 n_α 1.47 n_γ 1.49	Morenosite $NiSO_4 \cdot 7H_2O$	0.02	42° r > v weak	Orthorhombic	Stalactitic, efflorescent crusts	{010} dist
1.492 1.495	n_β 1.492-1.495 n_α 1.482-1.512 n_γ 1.493-1.518	Pentahydrite $MgSO_4 \cdot 5H_2O$	0.011 to 0.006	45° - 55° v > r	Triclinic α = 98.5° β = 109° γ = 75.1°	Prism (c), massive, granular	None
1.494	n_β 1.494 n_α 1.465 n_γ 1.495	Bianchite $(Zn,Fe)SO_4 \cdot 6H_2O$	0.030	10°	Monoclinic β = 98.5°	Crusts	
1.494	n_β 1.494 n_α 1.485 n_γ 1.505	Canavesite $MgCO_3(HBO_3) \cdot 5H_2O$	0.020	Very large Very weak disp	Monoclinic β = 114.9°	Fibrous rosettes, pseudohex crystals	{h0ℓ} cleavages or
1.497	n_β 1.497 n_α 1.470 n_γ 1.497	Moorhouseite $CoSO_4 \cdot 6H_2O$	0.027	Very small	Monoclinic β = 98.4°	Granular, fine columnar	None
1.498	n_β 1.498 n_α 1.465 n_γ 1.504	Nitrocalcite $Ca(NO_3)_2 \cdot 4H_2O$	0.039	~ 50° Not perceptible	Monoclinic β = 98°	Efflorescence, prism, crystals (c)	One perfect
*1.491 1.507	n_β 1.491-1.507 n_α 1.482-1.500 n_γ 1.493-1.513	Stilbite (Desmine) $(Ca,Na_2,K_2)(Al_2Si_7)O_{18} \cdot 7H_2O$	0.006 to 0.013	30° - 49° v > r, inclined	Monoclinic β = 129°	Platy in sheaf-like masses	{010} dist {100} poor
1.50	n_β ~1.50 n_α ~1.488 n_γ ~1.500	Hydroglauberite $Na_4Ca(SO_4)_3 \cdot 2H_2O$	0.012	Small	Orthorhombic ?	Dense fibrous	One perfect, one distinct, one poor
1.491 1.510	n_β 1.491-1.510 n_α 1.488-1.489 n_γ 1.505-1.525	Kurnakovite $Mg_2B_6O_{11} \cdot 15H_2O$	0.017 to 0.036	90° - 80°	~Monoclinic β=104.8°	Blocky crystals, granular aggregates	{010} indist
1.500	n_β 1.500 n_α 1.442 n_γ 1.504	Admontite $Mg_2B_2O_5 \cdot 15H_2O$	0.062	~30°	Monoclinic β ≈ 110°	Elong. on c	
1.500	n_β ~1.500 n_α 1.498 n_γ 1.502	Wairakite $Ca(AlSi_2O_6)_2 \cdot 2H_2O$	0.004	70° - 90° v > r	Monoclinic (Pseudocubic) β = 90.5°	Subhedral crystals ~ octahedral	{100} dist
1.502	n_β 1.502 n_α 1.490 n_γ 1.502	Bearsite $Be_2(AsO_4)(OH) \cdot 4H_2O$	0.012	Small	Monoclinic β = 98°	Fibrous aggregates	
1.498 1.507	n_β 1.498-1.507 n_α 1.497-1.506 n_γ 1.499-1.508	Gonnardite $Na_2Ca[(Al,Si)_5O_{10}]_2 \cdot 6H_2O$	~0.002	~50°	Orthorhombic (pseudo-tetragonal)	Radiated acicular spherulites	Good ∥ c

* n_β variation exceeds 0.01

Twinning	Orientation	Color in thin section	Hardness Sp Gravity	Color hand sample	Alteration	Occurrence	Remarks	Reference
	a = X, b = Z, c = Y	Pale green	2-2½ 1.91-1.98	Light green, green		Oxide zones Ni-deposits with secondary minerals of Co, Hg, Cu	Soluble H_2O. Taste astringent-metallic	Dana II, 516-519
	X ~ ⊥{010}	Colorless	1.718	White, lt blue, lt green-blue		Efflorescence in mines thermal springs with pickeringite	Unstable in air	Dana II, 492-493
{001}	b = Y, c ∧ X = +26°	Colorless	2½ 2.03-2.07	White, yellow		Efflorescence	Soluble H_2O	Dana II, 495-496
	b = Z	Colorless	1.79	Milky white		Fe-skarns with ludwigite		Can Min 16, 69-73
	b = Z	Pale pink X >> Z	2½ 1.97-2.04	Pink		Soluble efflorescences on Co-Ni ores	Soluble H_2O.	Can Min 8, 166-171
	X ⊥ cleavage	Colorless	Soft 1.90	White, gray		Efflorescence in caves and on limestone and limy soils	Easily soluble H_2O. Taste bitter and sharp	Dana II, 306-307
{001} cruciform penetration	b = Y, c ∧ X = 0° to -7°, a ∧ Z = +39° to +32°	Colorless	3½-4 2.1-2.2	White, yellowish, brownish	Montmorillonite clays	Hydrothermal veins & cavities in many igneous rocks. Hot springs	Zeolite group	p. 389
	∥ extinction, + elongation	Colorless		White		Alteration of glauberite in evaporite deposits	Soluble H_2O salty-bitter taste, decomp. HCl	Am Min 55, 321
Rare	b ~ Y, optic axis ~ ⊥{001}	Colorless	3 1.76-1.85	Colorless, white (alter)		Evaporite, lacustrine borate deposits	Dimorph inderite. Soluble warm acids	Dana II, 360
	b = Y	Colorless	2-3 1.82	Colorless		Evaporite deposits		Am Min 65, 205
{110} lamellar (two sets at 90°)		Colorless	5½-6 2.26	Colorless, white		Tuffaceous rocks and breccias. Graywacke. Hydrothermal alter	Zeolite group (Ca-analcime)	Min Mag 30, 691-708
	c ∧ Z = 8°-10°	Colorless	1.8-2.2	White		Oxide zones of ore deposits	As-moraesite	Am Min 48, 210-211
Uncommon	c = Z, ∥ extinction, + elongation	Colorless	4½-5 ~2.3	White	Thomsonite	Cavities in Si-poor volcanics with other zeolites. Alter nepheline	Zeolite group	p. 384

n_β Var.	Refractive Index	Mineral Name and Composition	$(n_\gamma - n_\alpha)$	2V angle Dispersion	Crystal System	Habit	Cleavage
1.503	n_β 1.503 n_α 1.377 n_γ 1.583	Nahcolite $NaHCO_3$	0.206	~ 75° v > r weak	Monoclinic β = 93.3°	Prism (c), fibrous aggregate-veinlets	{101} perfect {111} & {100} distinct
1.503	n_β 1.503 n_α 1.417 n_γ 1.527	Nesquehonite $MgCO_3 \cdot 3H_2O$	0.110	53° v > r weak	Monoclinic β = 90.8°	Prism (c), radiated acicular, felted	{110} perfect {001} distinct
1.503	n_β 1.503 n_α 1.455 n_γ 1.549	Hellyerite $NiCO_3 \cdot 6H_2O$	0.094	85°	Monoclinic or Triclinic	Fine-grained coatings	One perfect, two distinct at 112°
1.503	n_β 1.503 n_α 1.472 n_γ 1.526	Aksaite $MgB_6O_{10} \cdot 5H_2O$	0.054	73° - 88°	Orthorhombic	Crystals elongated on c, flattened on (100)	{100}, {010} probable
1.504	n_β 1.5037 n_α 1.5008 n_γ 1.5062	Taylorite (NH_4-arcanite) $K_{2-x}(NH_4)_xSO_4$ (x~0.35)	0.005	Large ~ 90°	Orthorhombic	Massive, concretionary	
1.504 1.506	n_β 1.504-1.506 n_α 1.332-1.335 n_γ 1.504-1.506	Niter (Nitre) KNO_3	0.172	~7° v > r weak	Orthorhombic	Thin crusts, tufts, massive, earthy	{011} perfect {010} dist {110} imperf
1.505	n_β 1.505 n_α 1.494 n_γ 1.516	Kainite $KMgSO_4Cl \cdot 3H_2O$	0.022	~90° r > v weak	Monoclinic β = 95°	Tabular-equant crystals, granular	{001} perfect
*1.497 1.515	n_β 1.497-1.515 n_α 1.485-1.505 n_γ 1.497-1.519	Epistilbite $CaAl_2Si_6O_{16} \cdot 5H_2O$	0.012 to 0.014	~44° v > r	Monoclinic β = 124.3°	Radiated prism aggregates, granular	{010} perfect
1.507	n_β 1.507 n_α 1.468 n_γ 1.529	Ezcurrite $Na_4B_{10}O_{17} \cdot 7H_2O$	0.061	74° r > v	Triclinic α = 103° β = 107.5° γ = 71.5°	Bladed-fibrous (c)	{110} perfect {010} dist {100} & {$\bar{1}\bar{2}6$} fair
1.506 1.509	n_β 1.506-1.509 n_α 1.420 n_γ 1.524-1.525	Thermonatrite $Na_2CO_3 \cdot H_2O$	~0.105	48° v > r weak	Orthorhombic	Crusts, platy needles (c)	{100} difficult
1.505 1.510	n_β 1.505-1.510 n_α 1.492-1.495 n_γ 1.516-1.520	Inyoite $Ca_2B_6O_{11} \cdot 13H_2O$	0.025	70° - 84° v > r slight	Monoclinic β = 114°	Short prism (c), tabular {001}, granular	{001} dist {010} indist
1.510	n_β 1.510 n_α 1.470 n_γ 1.579	Strontioborite $SrB_8O_{13} \cdot 2H_2O$(?)	0.109	~90°	Monoclinic β = 107.8°	Small plates	
1.510	n_β 1.510 n_α 1.498 n_γ 1.517	Jokokuite $MnSO_4 \cdot 5H_2O$	0.019	70° - 80 Very weak disp	Triclinic α = 98.8° β = 110° γ = 77.8°	Stalactitic	None
1.510	n_β 1.510 n_α 1.49 n_γ 1.521	Uranospathite $Cu(UO_2)_2[(As,P)O_4]_2 \cdot 12(?)H_2O$	0.03	69°	Tetragonal (?)	Platy {001}, elongation (b)	{001} perfect {100} good

* n_β variation exceeds 0.01

Twinning	Orientation	Color in thin section	Hardness Sp Gravity	Color hand sample	Alteration	Occurrence	Remarks	Reference
	b = Y, c ∧ X = 27° optic axis ~ ⊥{101}	Colorless	2½ 2.16-2.21	White, grayish		Volcanic-hot spring efflorescence, playa evaporites with salts	Easily soluble H_2O	Dana II, 134-135
	a = X, b = Z, c = Y	Colorless	2½ 1.852-1.856	Colorless, white		Coal mine efflorescence, cavities in serpentine, mineral springs	Easily soluble acids with effervescence	Dana II, 225-227
Lamellar ‖ perfect cleavage	Optic axes ~ ⊥ distinct cleavages	X = Y = very pale greenish-blue, Z = pale greenish-blue	2½ 1.97	Pale blue		Shear planes in serpentenite with zaratite		Am Min 44, 533-538
	a = X, b = Z, c = Y	Colorless	2½ 1.97-1.99	Light gray		Salt deposits		Am Min 48, 930-935
	a = Z, b = Y, c = X	Colorless	2	White, creamy		Guano deposits	Soluble H_2O. Bitter-pungent taste	Am Min 36, 590-602
{110} cyclic	a = Y, b = Z, c = X	Colorless	2 2.08-2.11	Colorless, white		Efflorescences on arid surfaces, caves. With soda-niter, gypsum	Easily soluble H_2O. Saline, cooling taste	Dana II, 303-305
	b = Y, c ∧ Z = 13°	Colorless	2½-3 2.15-2.24	Colorless, blue, violet, red, yellow	Epsomite, sylvite	Bedded marine salt deposits, often metamorphosed	Soluble H_2O, taste saline, bitter	Dana II, 594-596
{100} or {010} ubiquitous	b = Y, c ∧ Z ~ +10°	Colorless	4 2.25-2.27	Colorless, white, pinkish		Cavities in basalt, dike rocks, pegmatites	Zeolite group	DHZ 4, 377-385
		Colorless	3-3½ 2.05	Colorless		Evaporites with borax, kernite		Am Min 52, 1048-1059
	a = Z, b = X, c = Y	Colorless	1-1½ 2.25-2.26	Colorless, grayish, yellowish		Arid lake evaporites. Soil efflorescence. Fumeroles	Soluble H_2O. Alkaline taste	Dana II, 224-225
	b = Y, c ∧ X = -37°, c ∧ Z = +53°	Colorless	2 1.875	Colorless, white		Evaporite borate deposits with, colemanite, priceite	Soluble hot H_2O, dilute acids	Dana II, 358-360
	Inclined extinction, ± elongation	Colorless	2.806	Colorless		Fine-lamellar rock salt		Am Min 50, 1508
		Colorless	~2½ 2.03-2.09	Pale pink		Stalactites in mines with gypsum, szmikite, ilesite, rozenite	Dehydrates to ilesite	Am Min 64, 655
{110}	a = Y, b = Z, c = X	X = pale yellow, Y = Z = deep yellow	2.50	Yellow, lt green, blue-green		Secondary U-mineral with bassetite		USGS Bull 1064, 194-195

Biaxial —

n_β Var.	Refractive Index	Mineral Name and Composition	$(n_\gamma - n_\alpha)$	2V angle Dispersion	Crystal System	Habit	Cleavage
1.51	n_β 1.51 n_α 1.465 n_γ 1.540	Swartzite $CaMgUO_2(CO_3)_3 \cdot 12H_2O$	0.075	40°	Monoclinic β = 99.4°	Minute prism crystals (c), crusts	
1.512	n_β 1.512 n_α 1.482 n_γ 1.530	Inderborite $CaMgB_6O_{11} \cdot 11H_2O$	0.048	77°	Monoclinic β = 90.8°	Prism crystals, crystal aggregates	{100} dist
1.512	n_β 1.512 n_α 1.497 n_γ 1.513	Meta-aluminite $Al_2(SO_4)(OH)_4 \cdot 5H_2O$	0.016	Small ~ 30°	Monoclinic	Minute laths, crypto-crystalline	
1.512	n_β 1.512 n_α 1.494 n_γ 1.524	Nasinite $Na_4B_{10}O_{17} \cdot 7H_2O$	0.030	~67°	Monoclinic	Microcrystalline aggregates	
1.514	n_β 1.514 n_α 1.509 n_γ 1.514	Carnegieite $Na_3KAl_4Si_4O_{16}$	0.005	12°-15°	Pseudoisometric		Poor
1.515	n_β 1.515 n_α 1.508 n_γ 1.523	Schertelite $(NH_4)_2MgH_2(PO_4)_2 \cdot 4H_2O$	0.015	~90°	Orthorhombic	Tabular crystals {100}, acicular (c)	None
1.515	n_β 1.515 n_α 1.512 n_γ 1.517	Cowlesite $Ca_{0.96}Na_{0.09}Al_2Si_3O_8 \cdot 5\text{-}6H_2O$	0.005	44° - 53°	Orthorhombic	Thin pointed blades {010}	{010} perfect
1.516	n_β 1.516 n_α 1.445 n_γ 1.522	Gaylussite (Gay-lussite) $Na_2Ca(CO_3)_2 \cdot 5H_2O$	0.077	34° v > r strong, crossed	Monoclinic β = 102°	Wedge-shaped crystals	{110} perfect {001} indist
1.516	n_β 1.516 n_α 1.501-1.499 n_γ 1.525-1.526	Letovicite $(NH_4)_3H(SO_4)_2$	0.024	75°	Monoclinic β = 102.1°	Tiny pseudohexagonal plates {001}, granular	{001} dist
1.516	n_β 1.516 n_α 1.513 n_γ 1.518	Lovdarite $(Na,K,Ca)_4(Be,Al)_2Si_6O_{16} \cdot 4H_2O$	0.005	~90°	Orthorhombic	Massive, prismatic crystals	{100}, {010}, {001} dist {110} poor
1.517	n_β 1.5166 n_α 1.5010 n_γ 1.5176	Syngenite $K_2Ca(SO_4)_2 \cdot H_2O$	0.017	28° v > r very strong	Monoclinic β = 105°	Tabular or prism crystals	{110} & {100} perfect {010} distinct
*1.505 1.530	n_β n_α 1.490-1.522 n_γ 1.505-1.530	Sepiolite (Meerschaum) $(Mg,Al,Fe^{3+})_8Si_{12}O_{30}(OH)_4(OH_2)_4 \cdot 8H_2O$	0.009 to 0.020	20° - 70°	Orthorhombic	Matted fibers (c), felted	{110} dist
1.517	n_β 1.517 n_α 1.507 n_γ 1.521	Natrosilite $Na_2Si_2O_5$	0.014	~50° v > r	Monoclinic β = 104.2°	Pseudohexagonal tabular	{100} perfect {001} dist, {011} poor
1.512 1.522	n_β 1.512-1.522 n_α 1.502-1.514 n_γ 1.514-1.525	Laumontite $Ca(Al_2Si_4)O_{12} \cdot 4H_2O$	0.008 to 0.016	25° - 47° v > r strong, inclined	Monoclinic β = 111°	Prism crystals (c), radiated fibrous	{010} & {110} perfect

* n_β variation exceeds 0.01

Twinning	Orientation	Color in thin section	Hardness Sp Gravity	Color hand sample	Alteration	Occurrence	Remarks	Reference
		X = colorless, Y = Z = yellow	2.3	Bright green		Efflorescence in U-mines with bayleyite, andersonite	Fluor U.V.-bright yellowish-green. Soluble H_2O	USGS Bull 1064, 117-119
	b = Z, c ∧ X ~ 2°	Colorless	3½ 2.004	Colorless, white		Evaporite borate deposits with colemanite, ulexite	Slowly soluble H_2O	Dana II, 355-356
	b = Z(?), ext to elong = 43°	Colorless	1.85	White		Veinlets in sandstone with basaluminite and gypsum		Am Min 53, 717-721
	b = Y, a ∧ Z ~ 7°	Colorless		Colorless, white		Volcanic vents inseparably mixed with biringuccite		Am Min 48, 709-711
Multiple		Colorless	2.34	White	Inverts to nepheline at low temperature	No natural occurrences	Synthetic polymorph of nephaline	
	a = Z, b = X, c = Y	Colorless	1.83	Colorless		Bat guano with struvite & newberyite	Soluble H_2O	Dana II, 699
	a = Y, b = X, c = Z	Colorless	~2 2.14	Colorless, white, gray		Amygdules in basalt with other zeolites	Zeolite group	Am Min 60, 951-956
	b = X, c ∧ Z = +15°	Colorless	2½-3 1.99-2.00	Colorless, yellowish		Evaporite sediments, muds	Slightly soluble H_2O	Am Min 52, 1570-1572
Lamellar, common	b = Z, c ∧ X = -78°, c ∧ Y = +12°	Colorless	1.83	Colorless, white		Formed by burning coal wastes	Volatile, easily soluble H_2O	Dana II, 397
	∥ extinction	Colorless	2.32-2.33	White to yellowish		Late hydrothermal mineral in alkali pegmatites. Replace chkalovite		Am Min 59, 874
{100} common	b = Z, c ∧ X = +2°	Colorless	2½ 2.60	Colorless, white		Marine salt beds. Volcanic eminations	Decomp H_2O	Dana II, 442-444
	c = Z, ∥ extinction, + elongation	Colorless	2-2½ 1.0-2.6	White, gray, yellow, gray-green	Montmorillonite, kaolin	Desert (playa) soils, marls. Veins or cavities in altered mafic rocks		p. 265
{100}	c ~ Y	Colorless		Colorless		Alkali pegmatites intergrown with microcline, natrolite, analcime	Decomp. H_2O. Anom interference colors	Am Min 61, 340
{100} simple and common	b = Y, c ∧ Z = +8° to +40°, a ∧ X = +30° to +62°	Colorless	3-4 2.23-2.41	Colorless, white, yellow, reddish	Leonhardite	Cavities and veins in many types of igneous rocks. High P zeolite	Zeolite group	p. 386

n_β Var.	Refractive Index	Mineral Name and Composition	$(n_\gamma - n_\alpha)$	2V angle Dispersion	Crystal System	Habit	Cleavage
1.516 1.520	n_β 1.516-1.520 n_α 1.507-1.513 n_γ 1.517-1.521	Scolecite $Ca(Al_2Si_3)O_{10}\cdot 3H_2O$	0.010 to 0.007	36° - 56° v > r strong, horizontal	Monoclinic (pseudo-tetra- gonal) β = 91°	Fibrous, acicular (c), hair-like	{110} dist
1.518	n_β 1.518 n_α 1.513 n_γ 1.520	Nickelblödite $Na_2Ni(SO_4)_2\cdot 4H_2O$	0.007	60° - 70°	Monoclinic β = 100.7°	Tabular crystallites	
1.518 1.519	n_β 1.518-1.519 n_α 1.514-1.516 n_γ 1.521-1.522	Microcline $KAlSi_3O_8$	0.007	66° - 68° r > v perc	Triclinic α = 90° β = 116° γ = 87°	Anhedral-lattice intergrowths	{001} perfect, {010} dist, {110} poor
1.519	n_β 1.519 n_α 1.433 n_γ 1.528	Wegscheiderite $Na_5H_3(CO_3)_4$	0.095	~32° v > r	Triclinic α = 91.9° β = 95.8° γ = 108.7°	Acicular-bladed crystals	Prismatic distinct
1.519	n_β 1.519 n_α 1.511 n_γ 1.521	Ilesite $(Mn,Zn,Fe)SO_4\cdot 4H_2O$	0.010	Moderate ~ 53°	Monoclinic β = 90.8°	Prism crystalline aggregates	
1.52	n_β~1.52 n_α~1.49 n_γ 1.52	Hectorite $Na_{0.33}(Mg,Li)_2Si_4O_{10}(F,OH)_2$	0.03	Small	Monoclinic β ~99°	Massive, earthy micro flakes	{001} perfect
1.521	n_β 1.521 n_α 1.510 n_γ 1.523	Bikitaite $LiAlSi_2O_6\cdot H_2O$	0.013	45° v > r	Monoclinic β = 114.5°	Prismatic, massive	{100} perf {001} dist
1.522 1.524	n_β 1.522-1.524 n_α 1.518-1.520 n_γ 1.522-1.525	Orthoclase $K(AlSi_3)O_8$	0.005	35° - 50° r > v distinct, horizontal	Monoclinic β = 116°	Anhedral-lattice intergrowths	{001} perf, {010} dist, {110} poor
1.523 1.525	n_β 1.523-1.525 n_α 1.518-1.521 n_γ 1.524-1.526	High Sanidine $(K,Na)AlSi_3O_8$	0.006	63° - 15° v > r weak, inclined	Monoclinic β = 116°	Phenocrysts. Radiated acicular	{001} perf {010} dist
1.522 1.529	n_β 1.522-1.529 n_α 1.518-1.524 n_γ 1.522-1.530	Sanidine $(K,Na)AlSi_3O_8$	0.005 to 0.006	18° - 42° r > v weak, horizontal	Monoclinic β = 116°	Phenocrysts. Radiated acicular	{001} perf {010} dist
1.524	n_β 1.524 n_α 1.518 n_γ 1.530	Macdonaldite $BaCa_4Si_{15}O_{35}\cdot 11H_2O$	0.012	~90°	Orthorhombic	Acicular crystals (a)	{010} perf {001} dist {100} poor
1.525	n_β 1.525 n_α 1.510 n_γ 1.536	Uralolite $CaBe_3(PO_4)_3(OH)_2\cdot 4H_2O$	0.026	80°	Monoclinic β = 95°	Radiated fibrous concretions	∥ and ⊥ elongation
1.525 1.526	n_β 1.525-1.526 n_α 1.513-1.515 n_γ 1.534-1.536	Siderotil $FeSO_4\cdot 5H_2O$	0.021	50° - 60° r > v	Triclinic α = 97.3° β = 109.7° γ = 75°	Diverg acicular-fibrous crystals, crusts	
1.527	n_β 1.527 n_α 1.452 n_γ 1.538	Stenonite $Sr_2AlCO_3F_5$	0.086	43°	Monoclinic β = 98.3°	Massive	{001}, {120}

Twinning	Orientation	Color in thin section	Hardness Sp Gravity	Color hand sample	Alteration	Occurrence	Remarks	Reference
{100} parallel (c), {010} multiple	b = Z, c ∧ X ~ -18°, a ∧ Y ~ -17°	Colorless	5 2.25-2.29	White, yellowish	Beidellite	Amygdules and veins in basaltic rocks. Meta-carbonates with albite	Zeolite group	p. 382
		Colorless, pale green	2.43	Lt green		Surface efflorescence on Ni-sulfide ores		Min Mag 41, 37-41
Grid pattern (double multiple)	See page 348	Colorless	6-6½ 2.56-2.63	White, salmon, green, etc.	Kaolin, illite, sericite	Granite, granodiorite, pegmatites, schists, arkose	Alkali feldspar group	p. 348
		Colorless	2½-3 2.34	Colorless		Lake sediments with trona, nahcolite, halite	Soluble weak acids with effervescence	Am Min 48, 400-406
		Pale green, colorless	2.26	Green, white		Oxide zones sulfide ore deposits	Soluble H_2O	Dana II, 486-487
None?	b = Y, c ∧ X = small, a ∧ Z = small	Colorless, pale pink, brown, Z = Y > X	1-2 2.0-2.7	White, Fe-stained		Weathering of alkali zeolites, bentonitic clays	Montmorillonite Group	p. 257
	b = Z, c ∧ X = 28°	Colorless	6 2.29	Colorless		Li-pegmatites		Am Min 43, 768-770
Simple {010} (Carlsbad) Other twin laws	b = Z, c ∧ Y = -13 to -21°, a ∧ X = +14° to +6°	Colorless	6 2.55	White, Salmon, etc.	Sericite, illite, kaolin, alunite	Granites (perthite) feldspathoidal rocks, gneisses, granulites	Alkali feldspar group	p. 345
{010} parallel twin (c)	b = Y, c ∧ Z ~ -20°, optic plane ∥ {010}	Colorless	6 ~2.56	Transparent, white, yellowish	Kaolin clays, sericite, cryptocrystalline silica	High T, siliceous volcanic rocks	Alkali feldspar group	p. 342
{010} parallel twins (c)	b = Z, c ∧ Y ~ -21°, a ∧ X ~ +5°, optic plane ⊥ {010}	Colorless	6 ~2.56	Colorless, white, yellowish	Kaolin clays, sericite, cryptocrystalline silica	Siliceous volcanic rocks. Spherulites in volcanic glasses	Alkali feldspar group	p. 342
	a = Z, b = Y, c = X	Colorless	3½-4 2.27	White, colorless		Contact metamorphic rocks with sanbornite		Am Min 50, 314-340
	Maximum extinction 20 , + elongation	Colorless	2½ 2.04-2.14	Colorless, white, brownish		Kaolin-hydromuscovite rock with fluorite, beryl, apatite, crandallite		Am Min 49, 1776
		Colorless	2.1-2.21	White, yellowish, greenish		Dehydration of melanterite in altered ore-deposits	Cu may partially replace Fe. Soluble H_2O	Can Min 7, 751-763
	b = X, c ∧ Z = +32°	Colorless	~3.5 3.86	Colorless, white	Jarlite	Cryolite pegmatite contact zone with jarlite, weberite, fluorite		Am Min 48, 1178

n_β Var.	Refractive Index	Mineral Name and Composition	$(n_\gamma - n_\alpha)$	2V angle Dispersion	Crystal System	Habit	Cleavage
*1.500 1.555	n_β n_α 1.500-1.520 n_γ 1.540-1.555	Palygorskite (Attapulgite) (Pilolite) $(Mg,Al)_4Si_8O_{20}(OH)_2(OH_2)_4 \cdot 4H_2O$	0.020 to 0.035	0° - 60°	Monoclinic β = 96°	Matted fibers (c), felted	{110} dist
1.528	n_β 1.528 n_α 1.429 n_γ 1.538	Ameghinite $NaB_3O_5 \cdot 2H_2O$	0.109	33° v > r	Monoclinic β = 104°	Elongate on b, flattened on {001}	{100} good, {010} & {001} poor
*1.49 1.57	n_β 1.49-1.57 n_α 1.48-1.53 n_γ 1.50-1.59	Saponite $(\frac{1}{2}Ca,Na)_{0.67}Mg_6(Si_{7.33}Al_{0.67})O_{20}(OH)_4 \cdot nH_2O$	0.01 to 0.04	Moderate	Monoclinic β ~90°	Micro flakes or granules	{001} perfect
1.530	n_β 1.530 n_α 1.522 n_γ 1.531	Fedorite $(Na,K)Ca(Si,Al)_4(O,OH)_{10} \cdot 1.5H_2O$	0.009	32°	Monoclinic?	Pseudohexagonal micaceous sheets	{001} perfect
1.529 1.532	n_β 1.529-1.532 n_α 1.524-1.526 n_γ 1.530-1.534	Anorthoclase $(Na,K)AlSi_3O_8$	0.006 to 0.007	42° - 52° r > v weak	Triclinic α~ 93° β = 116° γ = 90°	Phenocrysts	{001} perfect {010} dist
1.531	n_β 1.531 n_α 1.522 n_γ 1.536	Boyleite $(Zn_{0.84}Mg_{0.16})SO_4 \cdot 4H_2O$	0.014	~70°	Monoclinic β ~ 90°	Earthy, reniform	None?
1.531	n_β 1.531 n_α 1.516 n_γ 1.535	Searlesite $NaBSi_2O_6 \cdot H_2O$	0.019	55°	Monoclinic β = 94°	Prism crystals, spherulites (radiated fibrous)	{100} perfect {$\bar{1}$02} & {010} imperfect
1.532	n_β 1.532 n_α 1.524 n_γ 1.536	Koktaite $(NH_4)_2Ca(SO_4)_2 \cdot H_2O$	0.012	72°	Monoclinic β = 102.75°	Acicular-fibrous crystals	None
1.532	n_β~1.532 n_α 1.528 n_γ 1.536	Aplowite $CoSO_4 \cdot 4H_2O$	0.008		Monoclinic β = 90.5°	Very fine-grained	
1.532	n_β~1.532 n_α~1.532 n_γ~1.532	Delhayelite $(Na,K)_{10}Ca_5Al_6Si_{32}O_{80}(Cl_2,F_2,SO_4)_3 \cdot 18H_2O$	0.002 to 0.003	83°	Orthorhombic	Platy crystals	{010} dist
1.532 1.533	n_β 1.532-1.533 n_α 1.526-1.527 n_γ 1.534-1.541	High Albite $Na(AlSi_3)O_8$	0.007 to 0.008	52° - 54° r > v weak	Triclinic α = 93.4° β = 116.3° γ = 90.3°	Elongated laths	{001} perfect, {010} dist
1.533	n_β 1.5333 n_α 1.5112 n_γ 1.5345	Nyerereite $Ca_4(Na,K)_8(CO_3,SO_4,F)_8$	0.023	29°	Orthorhombic	Tabular	
1.530 1.536	n_β 1.536-1.530 n_α 1.513-1.518 n_γ 1.545-1.542	Minasragrite $VOSO_4 \cdot 5H_2O$	0.032 to 0.024	Med-large ~ 65°	Monoclinic β = 110.9°	Minute crystals, spherulitic aggregates	None
1.534	n_β 1.534 n_α 1.488 n_γ 1.556	Artinite $Mg_2CO_3(OH)_2 \cdot 3H_2O$	0.068	70°	Monoclinic β = 99.1°	Silky fibrous aggregates	{100} perfect, {001} good

* n_β variation exceeds 0.01

Twinning	Orientation	Color in thin section	Hardness Sp Gravity	Color hand sample	Alteration	Occurrence	Remarks	Reference
None	c ~ Z, ‖ extinction, + elongation	Colorless	2-2½ 1.0-2.6	White, gray, yellow, gray-green	Montmorillonite, kaolin	Desert (playa) soils, marls. Veins & cavities in altered mafic rocks	Often called mineral cork, leather, cardboard or skin	p. 265
	b = Z, c ∧ X = 9°	Colorless	2½ 2.03-2.04	Colorless		Borax deposits		Am Min 52, 935-945
Not visible	b = Y, c ∧ X = small	Colorless, pale tints, Z = Y > X	1-2 2.0-2.7	White, yellow, red, brown, green, black		Hydrothermal veins and cavities	Montmorillonite group	p. 257
		Colorless	2.58	Colorless, pale red				Am Min 52, 561-562
Polysynthetic on {010} & ~{001}	b ∧ Z ~ 5°, a ∧ X ~ 10°, c ∧ Y ~ 20°, optic plane ~ ⊥{010}	Colorless	6 ~2.6	White, yellowish, etc	Kaolin clays, sericite, cryptocrystalline silica	High T. Na-rich volcanic rocks	Alkali feldspar group	p. 342
		Colorless	~2 2.41	White		Alteration of sphalerite	Sol H_2O. Dehydrates to gunningite	Am Min 64, 241
	b = Z, c ∧ X = +32°, + elongation	Colorless	Soft 2.46	White, pale brown		Spherulites in borate evaporites	Soluble HCl	Am Min 35, 1014-1020
{100} common	b = Y, c ∧ Z' on {110} = 2°	Colorless	2.09	Colorless, white		Pseudomorphs after gypsum, with mascagnite. NH_4-alum	Soluble H_2O	Dana II, 444
		Colorless, pink	3 2.33-2.36	Pink		Efflorescence with moorhouseite	Soluble in water	Can Min 8, 166-171
	a = X, b = Z, c = Y	Colorless	2.60	Colorless		Kalsilite-melilite-nephelinite lava	Bluish gray interference colors	Min Mag 32, 6-9
Multiple on {010} very common. Numerous other laws	See figure p. 342	Colorless	6 2.62	White(glassy)	Kaolin, sericite	High temperature lavas	Feldspar group	p. 342
Ubiquitous	a = Y, b = Z, c = X	Colorless	Soft 2.541	Colorless		Carbonate lavas		Am Min 63, 600
	b = X, c ∧ Z = small	X = deep blue, Y = blue, Z = colorless	Soft 2.03-2.04	Blue		Oxide zones V-deposits as delicate efflorescence on patronite	Easily soluble H_2O	Am Min 58, 531-534
	b = Y, c ∧ Z = 30°	Colorless	2½ 2.02-2.05	White		Low T hydrothermal veins in serpentinite	Effervesces in cold acid	Dana II, 263-264

Biaxial –

n_β Var.	Refractive Index	Mineral Name and Composition	$(n_\gamma - n_\alpha)$	2V angle Dispersion	Crystal System	Habit	Cleavage
*1.524 1.545	n_β 1.545-1.524 n_α 1.542-1.520 n_γ 1.547-1.526	Hyalophane $(BaAl,KSi)Si_2AlO_8$	0.010 to 0.005	48° - 79° r > v perc	Monoclinic β = 116°		{001} perfect, {010} dist, {110} poor
1.535	n_β 1.535 n_α 1.460 n_γ 1.545	Pentahydrocalcite $CaCO_3 \cdot 5H_2O$(?)	0.085	38°	Monoclinic	Acute rhombohedral crystals	
1.535	n_β 1.535 n_α 1.500 n_γ 1.560	Meyerhofferite $Ca_2B_6O_{11} \cdot 7H_2O$	0.060	78° Weak	Triclinic α = 91° β = 101.5° γ = 86.9°	Prism crystals (c), fibrous	{010} perfect {100} & {1$\bar{1}$0} imperfect
1.535	n_β 1.535 n_α 1.515 n_γ 1.536	Glauberite $Na_2Ca(SO_4)_2$	0.021	~7° r > v strong, horizontal	Monoclinic β = 112.2°	Tabular, prism, dipyramidal	{001} perfect {110} indist
1.535	n_β 1.535 n_α 1.531 n_γ 1.541	Vertumnite $Ca_4Al_4Si_4O_6(OH)_{24} \cdot 3H_2O$	0.010	62° None	Monoclinic (Pseudohexa) β = 119.7°	Tabular hexagonal crystals	None
1.536	n_β 1.5358 n_α 1.4227 n_γ 1.5545	Teschemacherite $(NH_4)HCO_3$	0.132	~42° v > r weak	Orthorhombic	Massive, compact crystalline	{110} perfect
1.536	n_β 1.536 n_α 1.527 n_γ 1.551	Sulfoborite $Mg_3SO_4B_2O_4(OH)_2 \cdot 4H_2O$	0.024	70° - 80°	Orthorhombic	Short to long prism crystals	{110} dist, {001} fair
1.536	n_β 1.536 n_α 1.530 n_γ 1.541	Okenite $CaSi_2O_4(OH)_2 \cdot H_2O$	0.011	Large	Triclinic α = 90° β = 103.9° γ = 111.5°	Bladed {010}, fibrous	{010} perfect
1.536 1.537	n_β 1.536-1.537 n_α 1.526-1.528 n_γ 1.541-1.545	Rozenite $FeSO_4 \cdot 4H_2O$	0.015 to 0.008	~90°	Monoclinic β = 90.5°	Minute tabular crystals, crusts	
1.538	n_β 1.538 n_α 1.455 n_γ 1.545	Ikaite $CaCO_3 \cdot 6H_2O$	0.090	38° - 45° Inclined perc	Monoclinic β = 110.2°	Columnar, earthy	
1.538	n_β 1.538 n_α 1.522 n_γ 1.542	Arsenuranospathite I $(HAl)_{0.5}(UO_2)_2(AsO_4)_2 \cdot 20H_2O$	0.020	52°	Orthorhombic	Lath crystals	{001} perfect, {100} & {010} good
1.538	n_β 1.538 n_α 1.534 n_γ 1.543	Canasite $(Na,K)_5(Ca,Mn,Mg)_4(Si_2O_5)_5(OH,F)_3$	0.009	~58°	Monoclinic β = 112°	Granular	1 very perf, 1 perf at 118°
1.539	n_β 1.539 n_α 1.496 n_γ 1.557	Biringuccite (Hoeferite) $Na_4B_{10}O_{17} \cdot 4H_2O$	0.061	63° - 65°	Monoclinic ?	Micro laminae or acicular	{001} & {100} distinct
1.539	n_β 1.539 n_α 1.516 n_γ 1.546	Chalcanthite $CuSO_4 \cdot 5H_2O$	0.030	56° v > r	Triclinic α = 97.5° β = 107.1° γ = 77.5°	Prism, tabular, fibrous, massive	{1$\bar{1}$0} imperf {110} traces

* n_β variation exceeds 0.01

Twinning	Orientation	Color in thin section	Hardness Sp Gravity	Color hand sample	Alteration	Occurrence	Remarks	Reference
Simple Carlsbad, Manebach, Baveno	b = Z, c ∧ Y = -25° to -45°, a ∧ X = +1° to +19°	Colorless	6-6½ 2.82-2.55	White, yellowish		Hydrothermal veins with Mn-minerals, granite gneiss	Feldspar group	p. 360
		Colorless	1.75	Colorless		Recent deposits in water pipes	Decomposes rapidly to calcite and H_2O	Dana II, 228
	On {010} c ∧ X' ~ 30°; on {100} c ∧ Z' ~ 25°	Colorless	2 2.12-2.13	Colorless, white		Alter of inyoite or colemanite in borate sediments	Easily soluble acids	Dana II, 356-358
	b = Z, c ∧ Y = 12°	Colorless	2½-3 2.78-2.80	Colorless, white, yellowish		Evaporite sediments, cavities in basalts, volcanic sublimates	Soluble HCl, slightly saline taste	Am Min 52, 1272-1277
	c = X, b ∧ Z = -16°	Colorless	5 2.15	Colorless		Voids in phonolite lavas with tobermorite	Soluble cold HCl	Am Min 62, 1061
	a = X, b = Y, c = Z	Colorless	1½ 1.58	Colorless, white, yellowish		Guano deposits	Soluble H_2O Effervescence in weak acids	Dana II, 137-138
	a = Z, b = Y, c = X	Colorless	4-4½ 2.440	Colorless		Evaporite salt deposits	Decomp. H_2O. Soluble acids	Dana II, 387-388
Parallel twins {010} & (c)	Parallel ext. on {010}, inclined on fibers, + elong	Colorless	4½-5 2.28-2.33	White, yellowish, bluish		Amygdules in basalt with apophyllite, zeolites	Gelat. HCl	Win & Win II, 358
		Colorless	2.19-2.29	Colorless, white		Alteration of pyrite, efflorescence		Can Min 7, 751-763
	b = Y, c ∧ Z = -17°	Colorless	1.80	Chalky white		Cold marine waters	Inverts to calcite and H_2O in air	Am Min 49, 439
		Colorless	2.54	White, pale yellow	Dehydrates rapidly to arsenuranospathite III	Secondary U-mineral		Min Mag 42, 117
Polysynthetic	b = Y, Z ∧ less perf cleavage = 2°	Pale greenish-yellow	2.71	Greenish-yel		Pegmatite		Am Min 45, 253-254
	b = Y, a ∧ Z = 5°-7°, + elongation	Colorless		White		Incrustations, low T pneumatolitic		Am Min 48, 709-711
Cruciform-rare		Pale blue	2½ 2.29	Deep blue		Oxide zones Cu-deposits efflorescence	Soluble H_2O, metallic taste	Dana II, 488-491

n_β Var.	Refractive Index	Mineral Name and Composition	$(n_\gamma - n_\alpha)$	2V angle Dispersion	Crystal System	Habit	Cleavage
1.537 1.542	n_β 1.537-1.542 n_α 1.462-1.466 n_γ 1.589-1.596	Dawsonite $NaAlCO_3(OH)_2$	0.127 to 0.130	77° v > r perc	Orthorhombic	Radiated acicular	{110} perf
1.543 1.537	n_β 1.543-1.537 n_α 1.495-1.490 n_γ 1.544-1.538	Schroeckingerite (Dakeite) $NaCa_3UO_2SO_4(CO_3)_3F \cdot 10H_2O$	0.049 to 0.047	16° - 22°	Orthorhombic	Pseudohexagonal flakes, globular, rosettes	{001} perf
1.540 1.541	n_β 1.540-1.541 n_α 1.520-1.522 n_γ 1.545-1.548	Lüneburgite (Lueneburgite) $Mg_3(PO_4)_2B_2O_3 \cdot 8H_2O$	0.025 to 0.026	53° - 62°	Monoclinic β = 97.4°	Minute pseudohexagonal tabular, nodules	{110} fair (73°)
1.540 1.543	n_β 1.540-1.543 n_α 1.531-1.538 n_γ 1.548	Gismondine (Gismondite) $Ca(Al_2Si_2)O_8 \cdot 4H_2O$	0.017	82° - 86°	Monoclinic β = 92.5°	Pseudotetragonal bipyramids	{101} dist
1.542	n_β 1.542 n_α 1.535 n_γ 1.549	Foshallasite $Ca_3Si_2O_7 \cdot 3H_2O$?	0.014 to 0.018	12° - 18°	Orthorhombic	Tabular, radiated aggregates	{100} perf
1.544	n_β 1.544 n_α 1.533 n_γ 1.547	Pentagonite $CaVO(Si_4O_{10}) \cdot 4H_2O$	0.014	50° r > v strong	Orthorhombic	Prismatic (c)	{010} dist
1.544	n_β 1.544 n_α 1.533 n_γ 1.548	Behoite β-$Be(OH)_2$	0.015	82° v > r strong	Orthorhombic	Pseudo-octahedral	None
*1.52 1.568	n_β 1.52-1.568	Chlorocalcite $KCaCl_3$	Small		Orthorhombic	"Cubic"	"Cubic" 3-dir ~ 90°
1.541 1.548	n_β 1.541-1.548 n_α 1.537-1.544 n_γ 1.547-1.551	Oligoclase $(NaSi,CaAl)Si_2AlO_8$	0.009 to 0.008	90° - 86° r > v weak	Triclinic α = 94° β = 116° γ = 88°	Anhedral to euhedral laths	{001} perfect, {010} dist, {110} poor
1.545	n_β 1.545 n_α 1.533 n_γ 1.547	Mooreite $(Mg,Zn,Mn)_8SO_4(OH)_{14} \cdot 3\text{-}4H_2O$	0.014	~50° r > v perc	Monoclinic β = 92.9°	Tabular-platy crystals	{010} perfect
*1.50 1.59	n_β 1.50-1.59 n_α 1.48-1.57 n_γ 1.50-1.60	Montmorillonite $(\frac{1}{2}Ca,Na)_{0.67}(Al_{3.33}Mg_{0.67})Si_8O_{20}(OH)_4 \cdot nH_2O$	0.02 to 0.03	0° - 30°	Monoclinic β ~90°	Microscopic flakes or granules	{001} perf
1.546	n_β 1.546 n_α 1.507 n_γ 1.569	Carboborite $Ca_2Mg(CO_3)(B_2O_5) \cdot 10H_2O$	0.062	75°	Monoclinic β = 91.7°	~long rhombohedrons	{100} perfect {$\bar{1}$11} dist {001} imperf
*1.530 1.564	n_β 1.530-1.564 n_α 1.529-1.559 n_γ 1.537-1.567	Chrysotile $Mg_3Si_2O_5(OH)_4$	~0.008	30° - 50°	Monoclinic β ~90°	Flexible fibrous elongation a, massive	Fibers (elong a)
1.547	n_β 1.547 n_α 1.438 n_γ 1.595	Oxammite $(NH_4)_2C_2O_4 \cdot H_2O$	0.157	62° v > r distinct	Orthorhombic	Lamellar masses, pulverulent	{001} dist

* n_β variation exceeds 0.01

Twinning	Orientation	Color in thin section	Hardness Sp Gravity	Color hand sample	Alteration	Occurrence	Remarks	Reference
	a = X, b = Z, c = Y	Colorless	3 2.44-2.46	Colorless, white		Arid, saline soils & clays. Alter nepheline	Soluble acids with effer- vescence	Dana II, 276-278
	a = Z, b = Y, c = X	X = very pale yellow, Y = Z = yellow-grn	2½ 2.55	Greenish-yellow		Secondary U-mineral, mine efflorescence	Fluor U.V.- yellowish-green	USGS Bull 1064, 121-126
	b = Y	Colorless	~2 2.05	White, lt brown, green		Clays and marls with boracite, gypsum, halite, sylvite. Guano		Dana II, 385
{110} & {001} common	b = X, c ~ Y, a ~ Z	Colorless	4½ 2.27-2.28	Colorless, white, reddish		Cavities in basalt, leucite volcanics, altered granite		Am Min 48, 664-672
	X ~ ⊥{100}	Colorless	2½-3 2.5	White		Calcite-zeolite veins		Win & Win II, 478
Cyclic	a = Y, b = X, c = Z	X = Z = colorless, Y = blue	3-4 2.33	Bluish-green		Cavities and amygdules in basalt, breccia, with zeolites	Dimorph. Cavan- site	Am Min 58, 405-424
{010}	a = Y, b = X, c = Z	Colorless	~4 1.92	Colorless, brown		Alteration in pegmatites, tuffs		Am Min 55, 1-9
Lamellar ∥ "cube" face	Optic plane ∥ "cubic" face	Colorless	2½-3 2.16	White		Volcanic sublimate	Pseudoisometric, deliquescent, soluble H_2O, bitter	Dana II, 91-92
Multiple on {010} very common. Numer- ous other laws	See page 352,	Colorless	6-6½ 2.63-2.66	White, pink, etc.	Sericite, kaolin, saussurite, zeolite	Granite, rhyolite, amphilbolite, schist, gneiss	Plagioclase feldspar group	p. 352
	b = X, c ∧ Z = 44°	Colorless	3 2.47-2.52	Colorless		Veinlets in franklinite willemite ore with rhodochrosite, zincite		Am Min 54, 973-975
Not visible	b = Y, c ∧ X = small	Colorless, tints of yellow, brown or pink, Z = Y > X	1-2 2.0-2.7	White, yellow, red, brown, green, black		Weathering of mafic igneous rocks. Bentonite clays. Hydrothermal alter		p. 257
	b = Y, c ∧ Z = 12°, a ∧ X = 10°, + elongation	Colorless	2 2.11-2.12	Colorless		Evaporites	Fluor U.V. white	Am Min 50, 262-263
	b = Y, a ~ Z, c ~ X	X ~ Y = colorless, pale yellow-green, Z = yellow, pale green	2½ ~2.55	White, yellow, green, brown	Replaced by silica	Alter ultramafic rocks with chromite, anti- gorite. Contact met.	Serpentine group (asbestos)	p. 298
	a = Y, b = Z, c = X	Colorless	2½ 1.54	Colorless, white, yellowish		Guano deposits with mascagnite		Dana II, 1103-1104

n_β Var.	Refractive Index	Mineral Name and Composition	$(n_\gamma - n_\alpha)$	2V angle Dispersion	Crystal System	Habit	Cleavage
*1.522 1.572	n_β 1.522–1.572 n_α 1.504–1.555 n_γ 1.539–1.579	Hannayite $(NH_4)_2Mg_3H_4(PO_4)_4 \cdot 8H_2O$	0.035 to 0.024	45° – 90° Not perceptible	Triclinic α = 76° β = 100° γ = 116°	Elong (c), tabular {100}	{001} perf {110}, {1$\bar{1}$0} & {130} poor
*1.524 1.574	n_β 1.524–1.574 n_α 1.522–1.560 n_γ 1.527–1.578	Cordierite $Mg_2Al_3(Si_5Al)O_{18}$	0.005 to 0.018	40°–90° v > r weak	Orthorhombic (Pseudohex.)	Subhedral xls	{010} dist., {100} & {001} poor
*1.540 1.558	n_β n_α 1.538–1.554 n_γ 1.546–1.560	Lizardite $Mg_3Si_2O_5(OH)_4$	0.008 to 0.006		Monoclinic β ~90°	Massive, compact, granular	{001} perfect
*1.543 1.556	n_β n_α 1.543 (Mg)– 1.548 (Ca) n_γ 1.550 (Mg)– 1.556 (Ca)	Ursilite $(Ca,Mg)_2(UO_2)_2Si_5O_{14} \cdot 9H_2O$?	0.007 to 0.008		Unknown	Radiated spherulites, earthy, nodular	None ?
1.55	n_β ~1.55 n_α 1.49 n_γ 1.56	Bradleyite $Na_3MgPO_4CO_3$	0.07	~45°	Monoclinic β = 90.5°	Fine-grained masses, fibrous (b)	
1.550	n_β 1.550 n_α 1.537 n_γ 1.556	Naujakasite $(Na,K)_6(Fe,Mn,Ca)(Al,Fe)_4Si_8O_{26} \cdot H_2O$	0.019	52° – 71°	Monoclinic β = 113.7°	Platy crystals, micaceous	{001} perf {$\bar{4}$01} & {010} dist
1.552	n_β 1.552 n_α 1.529 n_γ 1.555	Polylithionite $KLi_2Al(Si_4O_{10})(F,OH)_2$	0.026	0° – 30° r > v weak	Monoclinic β = 100.4°	Micaceous, scaly aggregates	{001} perf
1.552	n_β 1.552 n_α 1.535 n_γ 1.552	Satimolite $KNa_2Al_4B_6O_{15}Cl_3 \cdot 13H_2O$	0.017	Very small	Orthorhombic	Earthy aggregates, tabular-rhombic	None ?
1.553	n_β 1.553 n_α 1.535 n_γ 1.557	Hydrocalumite $Ca_2Al(OH)_7 \cdot 3H_2O$	0.022	24°	Monoclinic β = 111°	Massive	{001} perf
1.553	n_β 1.553 n_α 1.541 n_γ 1.557	Edingtonite $Ba(Al_2Si_3)O_{10} \cdot 4H_2O$	0.016	~54° v > r strong	Orthorhombic (pseudo-tetragonal)	Tiny pyramidal crystals, fibrous (c)	{110} perf
1.554	n_β 1.554 n_α 1.545 n_γ 1.565	Lacroixite $NaAl(PO_4)(OH,F)$	0.020	~90° Slight	Monoclinic β = 115.5°	Fragmentary crystals	{111} & {1$\bar{1}$1} indistinct
1.554 1.555	n_β 1.555–1.554 n_α 1.501–1.498 n_γ 1.597–1.594	Minguzzite $K_3Fe(C_2O_4)_3 \cdot 3H_2O$	0.096	~80°	Monoclinic β = 94.2°	Tabular crystals (b)	{010} perfect
1.555	n_β 1.555 n_α 1.531 n_γ 1.570	Shortite $Na_2Ca_2(CO_3)_3$	0.039	75° v > r moderate	Orthorhombic	Wedge-shaped, tabular, prism crystals	{010} dist
1.553 1.560	n_β 1.553–1.560 n_α 1.485–1.500 n_γ 1.570–1.584	Alumohydrocalcite $CaAl_2(CO_3)_2(OH)_4 \cdot 3H_2O$	0.084	50° – 64°	Monoclinic	Chalky masses of fibrous crystals	{100} perf, {010} imperf

* n_β variation exceeds 0.01

Twinning	Orientation	Color in thin section	Hardness Sp Gravity	Color hand sample	Alteration	Occurrence	Remarks	Reference
	X ⊥{001} (?), c ∧ Y ~ 33°(?)	Colorless	Soft 1.89-2.03	White, yellowish		Bat guano with struvite, newberyite, brushite		Am Min 48, 635-641
{110} or {130} polysynthetic	a = Y, b = Z, c = X	Colorless, pale blue, Z > Y > X	7-7½ 2.53-2.78	Gray, blue, indigo	Pinite (chlorite, muscovite, biotite)	Pelitic hornfels. High grade regional met. with andalusite, garnet		p. 167
	b = Y, a ~ Z, c ~ X	Colorless, pale green	2½ 2.55-2.58	Green, white		Massive serpentine with chrysotile	Serpentine group	DHZ 3, 107-190
	∥ extinction, + elongation	X = pale green, Z = green	3 3.034(Ca) 3.254(Mg)	Lemon yellow		Secondary U-mineral in veins with uranophane, sklodowskite	Fluor U.V.-greenish yellow	Am Min 44, 464-465
	b = X, c ∧ Y = 7°	Colorless	3½ 2.72	Colorless, white		Clay rocks, evaporites		USGS PP 405, 34-35
		Colorless	2-3 2.62-2.66	Gray, silver-white		Alkali plutonic rocks with arfvedsonite		Am Min 53, 1780
{001} complex	b = Y, a ∧ Z ~ 0°-7°	Colorless	2-3 2.58-2.82	Colorless, pink, pale yellow		Carbonatites with natrolite, microlite, tainolite	Mica (Lepidolite) group	p. 278
		Colorless		White		Clays and burate evaporites with polyhalite, boracite, kieserite		Am Min 55, 1069
	b = Y, c ∧ X ~ 3°	Colorless	3 2.15	Colorless, pale green		Limestone contact zones with larnite	Soluble weak HCl	Dana I, 667-668
None	a = Z, b = Y, c = X	Colorless	4-4½ 2.69-2.8	White, gray, pink	Beidellite	Cavities in mafic volcanics with Mn-minerals, harmotome	Zeolite group	p. 387
	Large ext. angles	Colorless	4½ 3.126	Pale yellow, pale green		Druses in granite with apatite, childrenite, tourmaline	Easily soluble HCl, H_2SO_4	Dana II, 783
		X = yellow-green, Z = emerald green	2.08-2.09	Green to yellow-green		Associated with humboltine		Am Min 41, 370
	a = Y, b = Z, c = X	Colorless	3 2.60	Colorless, pale yellow		Crystals or veins in limestone or clays with calcite, pyrite	Fluor U.V.-amber Decomp H_2O	Dana II, 222-223
	b = X, extinction <7° - 10°	Colorless	2½ 2.23-2.24	White, pale blue, yellowish		Alteration of allophane	Effervesses with acid. Dawsonite group	Dana II, 280-281

Biaxial —

n_β Var.	Refractive Index	Mineral Name and Composition	$(n_\gamma - n_\alpha)$	2V angle Dispersion	Crystal System	Habit	Cleavage
1.558	n_β 1.558 n_α 1.552 n_γ 1.561	Beryllonite $NaBePO_4$	0.009	68° v > r weak	Monoclinic β = 90°	Tabular {010}, prismatic on b	{010} perf, {100} dist, {101} poor
1.559	n_β 1.559 n_α 1.514 n_γ 1.562	Zincsilite $\sim Zn_3Si_4O_{10}(OH)_2 \cdot nH_2O$	0.048	0° - 22°	Monoclinic ?	Fine foliae or lamellae	{001} perfect
1.559	n_β 1.559 n_α 1.551 n_γ 1.562	Armenite $BaCa_2Al_6Si_9O_{30} \cdot 2H_2O$	0.011	60° ?	Orthorhombic	Prismatic	{010} perfect {110} dist
1.56	n_β ~1.56 n_α 1.541 n_γ 1.560	Beryllite (Berillite) $Be_3SiO_4(OH)_2 \cdot H_2O$	0.019	Small	Orthorhombic or Monoclinic	Radiated fibrous	
1.560	n_β 1.560 n_α 1.34 n_γ 1.560	Nitromagnesite $Mg(NO_3)_2 \cdot 6H_2O$	0.22	~5° v > r perc	Monoclinic β = 93°	Efflorescence, long prism crystals (c)	{110} perf
1.560	n_β 1.560 n_α 1.547 n_γ 1.567	Polyhalite $K_2MgCa_2(SO_4)_4 \cdot 2H_2O$	0.020	62° - 70°	Triclinic α = 104.1° β = 113.9° γ = 101.2°	Massive, fibrous, foliated {010}	{10$\bar{1}$} perf. {010} parting
1.562 1.563	n_β 1.562-1.563 n_α 1.557-1.560 n_γ 1.563-1.566	Nacrite $Al_2Si_2O_5(OH)_4$	0.006	40° - 90°	Monoclinic β = 90°-104°	Very fine-grain, earthy, scaly	
1.563	n_β 1.563 n_α 1.521 n_γ 1.585	Sidorenkite $Na_3Mn(PO_4)(CO_3)$	0.064	68°-70°	Monoclinic β = 90.06°	Large blocky xls., gran	{100} & {010} perfect, {001} imperf
1.563	n_β 1.563 n_α 1.551 n_γ 1.565	Morinite (Jezekite) $Ca_4Na_2Al_4(PO_4)_4O_2F_6 \cdot 5H_2O$	0.014	43° v > r weak	Monoclinic β = 105.5°	Long prism or tabular crystals, radiated-fibrous	{100} perf. {001} imperf
1.563	n_β 1.563 n_α 1.560 n_γ 1.565	Schoderite $Al_2(PO_4)(VO_4) \cdot 8H_2O$	0.005	42° No dispersion	Monoclinic β = 91.8°	Microxln. platelets, rosettes	{010} dist
1.564	n_β 1.564 n_α 1.552 n_γ 1.571	Jennite $Na_2Ca_8(SiO_3)_3(Si_2O_7)(OH)_6 \cdot 8H_2O$	0.019	74°	Triclinic α = 99.7° β = 97.7° γ = 110°	Bladed, fibrous aggregates	{001} dist
*1.545 1.583	n_β 1.545-1.583 n_α 1.525-1.564 n_γ 1.545-1.583	Vermiculite $(Mg,Ca)[(Mg,Fe^{2+})_5(Fe^{3+},Al)]$ $(Si_5Al_3)O_{20}(OH)_4 \cdot 8H_2O$	0.02 to 0.03	0° - 8° v > r weak	Monoclinic β ~97°	Micaceous, fine to large books	{001} perf
1.559 1.569	n_β 1.559-1.569 n_α 1.553-1.565 n_γ 1.560-1.570	Kaolinite (Anauxite) $Al_2Si_2O_5(OH)_4$	0.008 to 0.005	23° - 60° r > v weak	Monoclinic β = 90° - 104°	Earthy, very fine-grain, scaly	{001} perfect

* n_β variation exceeds 0.01

Twinning	Orientation	Color in thin section	Hardness Sp Gravity	Color hand sample	Alteration	Occurrence	Remarks	Reference
{101} simple or poly-synthetic	a = Y, b = X, c = Z	Colorless	5½-6 2.79-2.84	Colorless, white, yellowish	Herderite	Granite pegmatites	Pseudo-orthorhombic	Dana II, 677-679
	b = Y, c ∧ Z = 3°	Colorless	1½-2 2.61-2.71	White to bluish		Oxide zones galena-sphalerite-chalcopyrite skarn deposits	Montmorillonite group	Am Min 46, 241
Pseudohexagonal	a = Y, b = Z, c = X	Colorless, greenish	7.5 2.76-2.79	Grayish-green		High T ore veins		Am Min 26, 235
	∥ extinction, + elongation	Colorless	1 2.20	White		Cavities in pegmatites		Vlasov II, 96-98
		Colorless	1.58-1.64	Colorless		Efflorescence in limestone caves	Easily soluble H_2O. Taste bitter	Dana II, 307
{010} & {100} polysynthetic		Colorless	3½ 2.68-2.78	Colorless, white, salmon to red		Marine salt beds with halite, anhydrite, sylvite. Fumeroles	Decomp H_2O	Dana II, 458-460
{001} rare?	b = Z, c ∧ X = +7° to +10°, a ∧ Y = +7° to +10°	Colorless, ~ opaque	2-3 2.5-2.7	White, Fe-stained	Diaspore, gibbsite	Hydrothermal alter Al-silicates, with quartz, chalcedony, sulfides	Hexagonal plates (electron microscope)	p. 254
		Colorless	~2 2.90-2.98	Pale rose		Alkali pegmatites with nepheline, sodalite, concrinite	Mn-bradleyite	Am Min 64, 1332
	b = Z, c ∧ Y = -60°	Colorless	4-4½ 2.91-2.96	Colorless, white, pink		Granite pegmatites with apatite, augelite, wardite, montebrasite		Am Min 43, 585-594
	b = X, c ∧ Y = -26°	X = pale yellow, Y = dk. yellow, Z = yellow	1.92-1.93	Yellow-orange		Phosphatic chert, V-ores with montroseite, fervanite, hewettite	Anomalous interference colors	Am Min 64, 713-720
	X ⊥ {001}, Y ∧ elong = 35° - 40°	Colorless	2.31-2.32	White		Carbonate contact zones with monticellite, vesuvianite		Am Min 51, 56-74
Mica laws?	b = Y, c ∧ X = -3° to -6°, a ∧ Z = +4° to +1°	X = colorless, pale green, Y = Z = yellow brown, green brown	1.5 ~2.4	Dk brown, green		Alter biotite in ultramafic rocks with serpentine, chlorite, talc	Volume increase when heated. Interlayered with biotite	p. 263
{001} rare	b = Z, c ∧ X = -13° to -10°, a ∧ Y = +1° to +4°	Colorless, ~ opaque	2-3 2.5-2.7	White, Fe-stained	Diaspore, gibbsite	Alter Al-minerals (e.g., feldspar); weathering, hydro alter. Shales, soils	Soapy feel. Hex plates (electron microscope)	p. 254

Biaxial −

n_β Var.	Refractive Index	Mineral Name and Composition	$(n_\gamma - n_\alpha)$	2V angle Dispersion	Crystal System	Habit	Cleavage
1.564 1.567	n_β 1.564–1.567 n_α 1.543–1.549 n_γ 1.570–1.575	Furongite $Al_2(UO_2)(PO_4)_2(OH)_2 \cdot 8H_2O$	0.027	65°	Triclinic α = 68° β = 78° γ = 80°	Tabular crystals	One perfect, two distinct
1.565 1.570	n_β~1.565–1.570 n_α~1.565 n_γ~1.560	Zeophyllite (Radiophyllite) $Ca_4(Si_3O_7)(OH)_4F_2$	~0.005	0° – 10° v > r	Triclinic α = 90° β = 110° γ = 120°	Platy crystals, radiated, foliated, spherical	{001} perf
1.568	n_β 1.568 n_α 1.541 n_γ 1.570	Pseudoautunite $(H_3O)_4Ca_2(UO_2)_2(PO_4)_4 \cdot 5H_2O$	0.029	32° r > v distinct	Orthorhombic (?)	Micaceous-platy crystals, hexagonal sections	{001} perfect
*1.551 1.585	n_β 1.551–1.585 n_α 1.525–1.548 n_γ 1.554–1.587	Lepidolite $K_2(Li_{4-3}Al_{2-3})_{5-6}(Si_{8-6}Al_{0-2})O_{20}(OH,F)_4$	0.018 to 0.038	0° – 58° r > v weak	Monoclinic β = 100°, Hexagonal	Micaceous, scaly	{001} perfect
1.568	n_β 1.568 n_α 1.557 n_γ 1.570	Olshanskyite $Ca_3[B(OH)_4]_4(OH)_2$	0.013	~54° r > v weak	Monoclinic or Triclinic	Transverse fibers	None ?
1.569	n_β 1.569 n_α 1.534 n_γ 1.570	Masutomilite $K_2(Mn^{2+},Fe^{2+})_{1-2}Li_{2-3}Al_2(Si_{6-7}Al)_{2-1}O_{20}(OH,F)_4$		29° – 31° r > v weak	Monoclinic β = 100°		{001} perfect
1.569	n_β 1.569 n_α 1.549 n_γ 1.571	Dittmarite $(NH_4)MgPO_4 \cdot H_2O$	0.022	40°	Orthorhombic	Minute crystals	
1.569	n_β 1.569 n_α 1.567 n_γ 1.570	Phosinaite $H_2Na_3(Ca,Ce)(SiO_4)(PO_4)$	0.003	68° – 70°	Orthorhombic	Columnar crystals (c)	{100} perfect {010} & {110} imperfect
*1.56 1.58	n_β 1.56–1.58	Zinalsite $Zn_7Al_4(SiO_4)_6(OH)_2 \cdot 9H_2O$	?	?	Unknown	Cryptocrystalline aggregates	
1.570 1.571	n_β 1.570–1.571 n_α 1.519–1.524 n_γ 1.578–1.583	Röemerite (Weslienite) $Fe^{2+}Fe^{3+}_2(SO_4)_4 \cdot 14H_2O$	0.059	45° – 51° r > v strong, crossed	Triclinic α = 90.5° β = 101.1° γ = 85.7°	Thick tabular, cuboid crystals, stalactitic	{010} perfect {001} dist
*1.55 1.59	n_β 1.55–1.59 n_α 1.54–1.58 n_γ 1.55–1.59	Antigorite $Mg_6Si_4O_{10}(OH)_8$	0.000 to 0.005	~20° r > v	Monoclinic β = 91.1°	Massive, fine scaly, micaceous	{001} perfect {010} & {100} distinct
1.572	n_β 1.572 n_α 1.561 n_γ 1.576	Sekaninaite $Al_3(Fe,Mg)_2(Si_5Al)O_{18} \cdot H_2O$	0.015	66° v > r weak	Orthorhombic	Pseudohexagonal crystals	{100} imperf {001} parting
1.572	n_β 1.572 n_α 1.570 n_γ 1.572	Englishite $K_2Ca_4Al_8(PO_4)_8(OH)_{10} \cdot 9H_2O$	0.002	Small	Monoclinic ?	Micaceous, scaly aggregates	{001} perfect
*1.553 1.594	n_β n_α 1.570–1.553 n_γ 1.594–1.569	Karpinskite $(Mg,Ni)_2Si_2O_5(OH)_2$	0.024 to 0.016		Monoclinic ?	Cryptocrystalline, platy, prismatic	

* n_β variation exceeds 0.01

Twinning	Orientation	Color in thin section	Hardness Sp Gravity	Color hand sample	Alteration	Occurrence	Remarks	Reference
	Small extinction angle	Pale yellow. Pleochroic	2.82-2.90	Bright yellow		Oxide zone U-deposits with variscite, evansite, opal, autunite	Fluor U.V. = yellowish-green	Am Min 63, 424-427
	b ~ Z, a ~ Y	Colorless	3 2.61-2.76	White		Cavities in basalt with apophyllite and zeolites		Min Mag 31, 726-735
	c = Y	Colorless	3.28-3.29	Pale yellow, white		Cavities oxidized sulfide veins with albite-acmite, pyrochlore	Fluor short λ U.V. intense yellow-green	Am Min 50, 1505-1506
{001} complex	b = Y, a ∧ Z = 0° to +7°, c ∧ X = -10° to -3°	Colorless	2½-4 2.18-3.3	Colorless, gray, violet, pink, yellow	Hydromica	Granite pegmatites with spodumene, amblygonite, topaz	Mica group	p. 278
Polysynthetic	Inclined extinction, -elongation	Colorless	4 2.23	Colorless		Veinlets in sakhaite in "magnesian" skarn		Am Min 54, 1737-1738
	b = Y, a ∧ Z = +3°	X = Y = colorless, pale pink, Z = purple	2.5 2.94	Purplish-pink		Granite pegmatites with cassiterite, topaz, tourmaline, albite	Mn-zinnwaldite. Mica group	Am Min 62, 594
		Colorless	Soft 2.19	Colorless		Bat guano		USGS PP 750-A, p A115
	a = Z, b = X, c = Y	Colorless	3½ 2.62	Colorless, pale rose, brownish		Alkali pegmatites with nepheline, aegirine, ussingite, neptunite		Am Min 60, 488
		Colorless	2½-3 3.007	White, rose, red-brown		Oxide zones Zn-deposits with hydromicas and Zn-clays	Too fine grained for good optical data	Am Min 44, 208-209
	X or Z ∧ c ~ 33°	X = orange, Y = pale yellow, Z = yellow-brown	3-3½ 2.17-2.18	Brown to yellow, violet-brown		With coquimbite and other soluble sulfates Alter pyrrhotite	Soluble H_2O. Taste saline, astringent. Anomalous colors	Am Min 55, 78-89
Rare	b = Y, c ∧ X ~ 1°	Colorless, very pale green	2-3 2.58-2.61	White, yellowish, shades of green		Alter ultramafic rocks, as serpentenite often with chrysotile	Chlorite and serpentine groups. Smooth-soapy feel	p. 298
{110} & {310}	a = Z, b = Y, c = X	X = colorless, Y = blue, Z = pale blue	7-7½ 2.77	Blue to violet-blue	Chlorite Fe-anthophyllite	Pegmatites, granulites, gneisses	Fe-cordierite	Am Min 62, 395
	a ~ Y, b ~ Z, c = X	Colorless	~3 ~2.65	Colorless		Phosphate nodules with variscite		Dana II, 957-958
	Inclined ext up to 12°, + elongation	Colorless, pale green-blue. Not pleochroic	2½-3 2.53-2.63	Colorless, lt-blue, dk green-blue		Veintets in "kerolitized" serpentinites	Deep colored varieties have large I.R. and Sp. Gr.	Am Min 42, 584

Biaxial —

n_β Var.	Refractive Index	Mineral Name and Composition	$(n_\gamma - n_\alpha)$	2V angle Dispersion	Crystal System	Habit	Cleavage
1.574	n_β 1.574 n_α 1.568 n_γ 1.580	Overite $CaMgAl(PO_4)_2(OH)\cdot 4H_2O$	0.012	75° r > v weak	Orthorhombic	Tiny plates-laths, massive	{010} perfect {100} poor
1.575	n_β 1.575 n_α 1.571 n_γ 1.578	Haiweeite $CaU_2^{4+}Si_6O_{19}\cdot 5H_2O$	0.007	Small ~ 15° r > v strong	Monoclinic β= 108°	Spher acicular aggregates (c), flaky	{100} dist
*1.570 1.582	n_β 1.570–1.582 n_α 1.544–1.565 n_γ 1.571–1.585	Saléeite $Mg(UO_2)_2(PO_4)_2\cdot 8\text{-}10H_2O$	0.027 to 0.020	0° – 65° r > v	Tetragonal	Platy rect crystals {001}, scaly, radiated	{001} perf {010} & {110} indistinct
1.576	n_β 1.576 n_α 1.567 n_γ 1.579	Panethite $(Na,Ca,K)_2(Mg,Fe,Mn)_2(PO_4)_2$	0.012	51°	Monoclinic β= 91.5°	Massive	
1.576	n_β 1.576 n_α 1.558 n_γ 1.593	Xiangjiangite $(Fe,Al)(UO_2)_4(PO_4)_2(SO_4)_2(OH)\cdot 22\ H_2O$	0.035	87°	Orthorhombic ?	Earthy	Perfect basal ?
1.572 1.580	n_β 1.572–1.580 n_α 1.567–1.573 n_γ 1.577–1.585	Bytownite $(CaAl,NaSi)Si_2AlO_8$	0.010 to 0.012	90° – 76° v > r weak	Triclinic α= 93° β= 116° γ= 90°	Anhedral to euhedral laths	{001} perfect, {010} dist, {110} poor
1.576	n_β 1.576 n_α 1.573 n_γ 1.579	Ilmajokite $(Na,Ba,RE)_{10}Ti_5(Si,Al)_{14}O_{22}(OH)_{44}\cdot nH_2O$	0.006	~90°	Monoclinic ?	Granular, crusts, radiated fibrous	{110} & pinacoidal at 72°
1.577	n_β 1.577 n_α 1.542 n_γ 1.578	Eastonite $K_2(Mg_5Al)(Si_5Al_3)O_{20}(OH,F)_4$	0.036	Small r > v	Monoclinic β = 100°	Micaceous	{001} perf.
1.578	n_β 1.578 n_α 1.544 n_γ 1.601	Kroehnkite (Kröhnkite) (Salvadorite) $Na_2Cu(SO_4)_2\cdot 2H_2O$	0.057	79° v > r weak, inclined	Monoclinic β= 108.5°	Prism, fibrous aggregates, octahedral	{010} perf {$\bar{1}$01} imperf
1.578	n_β 1.578 n_α 1.572 n_γ 1.582	Montgomeryite $Ca_2Al_2(PO_4)_3(OH)\cdot 7H_2O$	0.010	75°	Monoclinic β= 91.5°	Minute laths or plates, massive	{010} perf {100} poor
*1.570 1.589	n_β 1.570–1.589 n_α 1.535–1.558 n_γ 1.572–1.590	Zinnwaldite $K_2(Fe^{2+}_{1-2}Li_{2-3}Al_2)(Si_{6-7}Al_{2-1})O_{20}(OH,F)_4$	~ 0.035	0° – 40°	Monoclinic β= 100°	Micaceous	{001} perf
*1.56 1.60	n_β 1.56–1.60 n_α 1.53–1.57 n_γ 1.56–1.61	Hydromuscovite $(K,H_3O)_2Al_4(Si_6Al_2)O_{20}(OH)_4$	~ 0.03	Small (0° – 5°)	Monoclinic β ~90°	Fine scales	{001} perf
1.58	n_β~1.58 n_α 1.565 n_γ 1.583	Calciotalc $CaMg_2Si_4O_{10}(OH)_2$	0.018	Small	Orthorhombic	Acicular, columnar	
1.581 1.583	n_β 1.581–1.583 n_α 1.564–1.565 n_γ 1.582–1.584	Sabugalite $HAl(UO_2)_4(PO_4)_4\cdot 16H_2O$	0.018 to 0.019	0 to moderate	Tetragonal	Thin plates or laths (square-rect.)	{001} perf

* n_β variation exceeds 0.01

Twinning	Orientation	Color in thin section	Hardness Sp Gravity	Color hand sample	Alteration	Occurrence	Remarks	Reference
	a = Y, b = Z, c = X	Colorless	3½-4 2.48-2.53	Lt green, colorless		Phosphate nodules with variscite, crandallite	Al-segelerite. Easily soluble hot HNO_3	Am Min 59, 48-59
	X ~ ⊥{100}	Colorless, pale yellow. Weak pleochroism	3½ 3.35	Pale yellow, green-yellow		Secondary U-mineral. Small veins in granitic rocks	Fluor-U.V.-dull green	Am Min 44, 839-843
	a ~ b = Z or Y, c = Z	Colorless to pale yellow	2½ 3.27	Yellow, yellow-green		Secondary U-mineral with carnotite, autu-ite	Alunite group. Fluor U.V.-bright yellow-green	USGS Bull. 1064, 177-183
Simple		Colorless	2.9-3.0	Pale amber		Octahedrite meteorites with whitlockite, brianite, albite		Am Min 53, 509
	Extinction is ∥ or sym-metrical	Pale yellow, weak pleochroism	1-2 2.87-3.1	Yellow		Oxide zone U-deposits with sabugalite, variscite, pyrite	Sol dilute HCl or H_2SO_4. Not fluor U.V.	Am Min 64, 466
Multiple on {010} very common. Numerous other laws	See page 352	Colorless	6-6½ 2.71-2.74	Gray, white, etc.	Sericite, kaolin, saussurite, calcite, zeolite	Gabbro, anorthosite, meteorites, basalt	Plagioclase feldspar group	p. 352
		Yellow	1 2.20	Bright yellow		Cavities in pegmatites		Am Min 58, 139-140
{110} normal, {001} complex	b = Y, c ∧ X = small	Colorless, yellow, brown Z ~ Y > X	2½-3 2.86	Colorless, brown	Chlorite	Ultramafic igneous rocks	Mica group. Biotite series	p. 270-274
{101} common	b = Y, c ∧ Z = +48°	Colorless, pale blue	2½-3 2.90-2.95	Lt blue- dk blue, greenish		Oxide zones Cu-deposits	Easily soluble H_2O	Dana II, 444-446
			4 2.46-2.53	Dk to lt green white, colorless		Variscite nodules with mitridatite, crandal-ite, etc. Altered triphylite		USGS PP 6000 p. D204-D206
	b = Y, a ∧ Z = 0° to +2°, c ∧ X = -10° to -8°	X = colorless, pale yellow, Y = gray-brown, Z = gray	~3 ~3.0	Yellow to dk brown, gray		Granite pegmatites, high T hydrothermal veins with cassiterite	Mica group	p. 281
Not visible	b = Z, c ∧ X = small	Colorless	1-2 2.6-2.9	White, Fe-stained		Alter K-feldspar or muscovite. Shale, soils.		p. 259
	∥ extinction, + elongation	Colorless	2-3	Greenish gray		Hydrothermal alter diopside-actinolite		Am Min 45, 476-477
		X = colorless, Y ~ Z = pale yellow	2½ 3.15-3.20	Bright yellow		Secondary U-mineral in sandstone-type deposits	Fluor U.V.-lemon yellow	USGS Bull 1064, 196-200

Biaxial −

n_β Var.	Refractive Index	Mineral Name and Composition	$(n_\gamma - n_\alpha)$	2V angle Dispersion	Crystal System	Habit	Cleavage
1.582	n_β 1.5824 n_α 1.5702 n_γ 1.5869	Paracelsian $Ba(Al_2Si_2)O_8$	0.017	53°	Monoclinic (pseudo-ortho) β ~90°	Large prism crystals, pseudo-orthorhombic	{110} poor
*1.582 1.598	n_β 1.582-1.598 n_α 1.574-1.589 n_γ 1.582-1.599	Hopeite (Hibbenite) $Zn_3(PO_4)_2 \cdot 4H_2O$	0.008 to 0.010	Small - 37° v > r perceptible	Orthorhombic	Prism, tabular, massive	{010} perf {100} dist {001} poor
1.581	n_β 1.582 n_α 1.576 n_γ 1.584	Guerinite $Ca_5H_2(AsO_4)_4 \cdot 9H_2O$	0.008	Small ~ 10° r $\gtrless$ v strong	Monoclinic or Triclinic	Acicular rosettes	One perfect, one distinct, one imperfect
1.580 1.585	n_β 1.580-1.585 n_α 1.573-1.577 n_γ 1.585-1.590	Anorthite $Ca(Al_2Si_2)O_8$	0.012 to 0.013	77° - 78° v > r weak	Triclinic α = 93° β = 116° γ = 91°	Anhedral to euhedral laths	{001} perfect, {010} dist, {100} poor
1.583	n_β 1.583 n_α 1.574 n_γ 1.588	Uranocircite II $Ba(UO_2)_2(PO_4)_2 \cdot 10H_2O$	0.014	Highly variable 0° - 70°	Tetragonal ?	Thin rectangular plates fan-like	{001} perf {100} dist
1.581 1.585	n_β 1.581-1.585 n_α 1.576-1.579 n_γ 1.584-1.586	Sorensenite $Na_4SnBe_2Si_6O_{16}(OH)_4$	0.008 to 0.007	Small to 75° Inclined strong	Monoclinic β = 98.2°	Acicular or elongated tabular crystals	Two dir at 63° distinct
1.583	n_β 1.583 n_α 1.573 n_γ 1.583	Threadgoldite $Al(UO_2)_2(PO_4)_2(OH) \cdot 8H_2O$		70°	Monoclinic <β = 111°24'	Tabular, micaceous	{100}, {010}, {001}, {012}
1.584	n_β 1.584 n_α 1.568 n_γ 1.585	Laplandite $Na_4CeTiPSi_7O_{22} \cdot 5H_2O$	0.017	28°	Orthorhombic	Radiated fibrous aggregates, platy (001)	
1.584	n_β 1.584 n_α 1.570 n_γ 1.585	Torreyite (Delta-Mooreite) $(Mg,Zn,Mn)_7SO_4(OH)_{12} \cdot 4H_2O$	0.015	~40°	Monoclinic	Massive granular	{010} dist
1.584	n_β 1.584 n_α 1.582 n_γ 1.584	Zektzerite $LiNa(Zr,Ti,Hf)Si_6O_{15}$	0.002	Small ~ 0° r > v very weak	Orthorhombic (Pseudohex)	Tabular crystals {100}	{100} & {010} perfect
*1.575 1.594	n_β 1.575-1.594 n_α 1.538-1.550 n_γ 1.565-1.600	Talc (Steatite) $Mg_3Si_4O_{10}(OH)_2$	0.04 to 0.05	0° - 30° r > v perc	Triclinic α ~ 90.5° β = 98.9° γ ~ 90°	Foliated, fibrous, radiated, massive	{001} perfect
1.587	n_β 1.587 n_α 1.52 n_γ 1.613	Lanthanite $(La,Ce)_2(CO_3)_3 \cdot 8H_2O$	0.093	~63° v > r weak	Orthorhombic	Tabular, platy, scaly, granular, earthy	{010} perfect
1.587	n_β 1.587 n_α 1.574 n_γ 1.599	Krauskopfite $BaSi_2O_5 \cdot 3H_2O$	0.025	88° r > v distinct	Monoclinic β = 94.5°	Short prism grains	{010} & {001} perfect
1.587	n_β 1.587 n_α 1.578 n_γ 1.595	Leightonite $K_2Ca_2Cu(SO_4)_4 \cdot 4H_2O$	0.017	~86° r > v distinct	Orthorhombic	Lath-like crystals (c), fibrous veinlets	None

* n_β variation exceeds 0.01

Twinning	Orientation	Color in thin section	Hardness Sp Gravity	Color hand sample	Alteration	Occurrence	Remarks	Reference
Multiple and complex		Colorless	~6 3.31-3.34	Colorless		Low-T veins in sediments	Feldspar group (dimorph. celsian)	DHZ 4, 166-178
None?	a = X, b = Z, c = Y	Colorless	3-3½ 3.05-3.08	Colorless, white, pale yellow		Oxide zones Zn-deposits	Dimorph. parahopeite. Easily soluble HCl	Dana's II, 734-737
	Z = elongation (+ elongation)	Colorless	1½ 2.68	Colorless		Oxide zones ore-deposits		Am Min 50, 812
Multiple on {010} very common. Numerous other laws	See page 352	Colorless	6-6½ 2.75-2.76	White, gray	Calcite, saussurite, kaolin, zeolite	Marble and calc-silicates, norites, basalt	Plagioclase feldspar group	p. 352
Polysynthetic (~microcline)	a = Y or Z, c = X	Y = Z = pale yel X = colorless	2-2½ 3.46-4.10	Yellow		Secondary U-mineral	Autunite group. Fluor U.V.-green	USGS Bull. 1064, 211-215 & 177
	b = Y	Colorless	5½ 2.90	Colorless, pinkish		Alkali veins and plutons with nepheline, analcime, neptunite, aegirine	Anomalous yellow-brown & blue interference colors	Am Min 51, 1547-1548
	b = Y, c ∧ Z = 4°	Colorless to v. pale yellow	3.4	Greenish-yellow		Pegmatite		Am Min 65, 209
		Colorless	2-3 2.83	Lt gray, yellowish, bluish		Alkali pegmatites with natrolite, belovite, nordite, serandite		Am Min 60, 487
Polysynthetic in zone {010}	b = X	Colorless	3 2.665	Bluish white		Cavities in pyrochroite in Mn-Zn ore veins	Soluble acids	Dana II, 575-576
	a = X, b = Y, c = Z	Colorless	~6 2.79	Colorless, pink		Riebeckite granite	Osumilite group. Fluor U.V. short λ-light yellow	Am Min 62, 416
Rare	b = Y, a ~ Z, c ∧ X ~ -10°	Colorless	1 2.6-2.8	White, lt to dk green	Magnesite, quartz	Talc-tremolite schists		p. 301
{101} normal	a = Z, b = X, c = Y	Colorless	2½-3 2.69-2.75	Colorless, white, yellowish		Oxide zones Zn-deposits. Alter allanite, cerite in pegmatites	Soluble acids	Dana II, 241-243
	b = X, c ∧ Z = 10°, a ∧ Y = 6°, ± elongation	Colorless	~4 3.10-3.14	Colorless, white		Veins in metamorphic rocks, with macdonaldite, opal, witherite		Am Min 50, 314-340
{100} & {010} lamellar	a = Z, b = X, c = Y	Pale blue	3 2.95	Lt blue, blue-green		Oxide zones Cu-deposits with kroehnkite, atacamite		Dana II, 461-462

Biaxial –

n_β Var.	Refractive Index	Mineral Name and Composition	$(n_\gamma - n_\alpha)$	2V angle Dispersion	Crystal System	Habit	Cleavage
1.586 1.589	n_β 1.586-1.589 n_α 1.534-1.556 n_γ 1.596-1.601	Pyrophyllite $Al_2Si_4O_{10}(OH)_2$	~0.050	53°-62° r > v weak	Monoclinic β = 100°	Micaceous, rosettes of flakes	{001} perfect
1.588	n_β 1.588 n_α 1.550 n_γ 1.590	Coconinoite $Fe_2Al_2(UO_2)_2(PO_4)_4(SO_4)(OH)_2 \cdot 20H_2O$	0.040	28° - 43°	Monoclinic	Micro-crystalline aggregates, laths	
1.588	n_β 1.588 n_α 1.563 n_γ 1.594	Variscite (Elroquite)(Lucinite)(Phosphochromite)(Peganite) $AlPO_4 \cdot 2H_2O$	0.031	Moderate ~ 52° v > r perc	Orthorhombic	Massive veinlets-nodules-crusts	{010} good {001} poor
1.589	n_β 1.589 n_α 1.583 n_γ 1.594	Pharmacolite $CaHAsO_4 \cdot 2H_2O$	0.011	79° r > v	Monoclinic β = 114.8°	Acicular (c), silky fibrous, botryoidal	{010} perf
*1.57 1.61	n_β 1.57-1.61 n_α 1.54-1.57 n_γ 1.57-1.61	Illite $K_{1-1.5}Al_4(Si_{7-6.5}Al_{1-1.5})O_{20}(OH)_4$	0.03	0° - 30°	Monoclinic β ~90°	Fine scales	{001} perf
1.590	n_β 1.590 n_α 1.572 n_γ 1.601	Kolbeckite (Sterrettite)(Eggonite) $ScPO_4 \cdot 2H_2O$	0.029	60°	Monoclinic β = 90.75°	Short prism (a)	{110} dist {100} & {001} poor
1.590	n_β 1.590 n_α 1.525 n_γ 1.593	Brenkite $CaF_2 \cdot CaCO_3$	0.068	26° - 28°	Orthorhombic	Lath-like crystals, radiated	None
1.590	n_β 1.590 n_α 1.575 n_γ 1.601	Dalyite $K_2ZrSi_6O_{15}$	0.026	~72° r > v weak	Triclinic α = 106° β = 113.5° γ = 99.5°	Short prismatic	{101} & {010} distinct, {100} imperf
1.590	n_β 1.590 n_α 1.581 n_γ 1.591	Tuscanite $[K_{0.8}(H_2O)_{1.1}](Ca_{5.3}Na_{0.5}Fe^{3+}_{0.1}Mn_{0.1})(Si_{6.3}Al_{3.7})O_{22}[(SO_4)_{1.4}(CO_3OH)_{0.6}(O_4H_4)0.1]$	0.010	40°	Monoclinic β = 107°	Tabular {100}	{100} dist
*1.56 1.63	n_β 1.56-1.63 n_α 1.53-1.61 n_γ 1.56-1.64	Nontronite $(\frac{1}{2}Ca,Na)_{0.67}Fe^{3+}_4(Si_{7.33}Al_{0.67})O_{20}(OH)_4 \cdot nH_2O$	0.03 to 0.04	25° - 70°	Monoclinic β ~ 90°	Fine-grain scaly, fibrous, granular	{001} perf
1.590 1.591	n_β 1.590-1.591 n_α 1.571-1.573 n_γ 1.593-1.594	Priceite (Pandermite) $Ca_4B_{10}O_{19} \cdot 7H_2O$	0.022	~32° v > r strong	Triclinic β = 110°	Nodules, massive, rhombic crystals	{001} perf
*1.585 1.598	n_β 1.598-1.585 n_α 1.597-1.580 n_γ 1.600-1.586	Kämmererite $Mg_5Cr(Si_3Al)O_{10}(OH)_8$	0.003 to 0.006	30° - 40°	Monoclinic β = 97.5°	Micaceous, hexagonal plates	{001} perf
*1.574 1.610	n_β 1.610-1.574 n_α 1.603-1.56 n_γ 1.617-1.580	Bassetite $Fe^{2+}(UO_2)_2(PO_4)_2 \cdot 8H_2O$	0.014	~90° - 62° r > v strong	Monoclinic β = 90.5°	Platy {010}	{010} perfect

* n_β variation exceeds 0.01

Twinning	Orientation	Color in thin section	Hardness Sp Gravity	Color hand sample	Alteration	Occurrence	Remarks	Reference
	b = Z, a ~ Y, c ∧ X = -10°	Colorless	1-2 2.8-2.9	White, pale green, brownish		Alter. of feldspars. Schists with kyanite, muscovite	Talc group	p. 303
	Y ∥ laths with + elong. Y ∧ laths with -elong = 8° - 25°	X = colorless, Y = Z = pale	Soft 2.68-2.70	Pale yellow		Oxide zones U-deposits	Nonfluorescent	Am Min 51, 651
{201} rare	a = X, b = Z, c = Y	Colorless, pale green	3½-4½ 2.57-2.61	Pale green to green, white		Nodules in clay rocks with crandallite and other phosphates	Variscite-strengite series. Soluble alkalies	Dana II, 756-761
	b = Z, c ∧ X = +29°	Colorless	2-2½ 2.68	Colorless, white, gray		Oxide zones As-deposits with other arsenates (haidingerite)	Readily soluble acids	Dana II, 706-708
Not visible	b = Z, c ∧ X = small	Colorless	1-2 2.6-2.9	White, Fe-stained		Alter K-feldspar or muscovite. Shale, soil with other clays. Ore veins		p. 259
{001} and {031} (?)	b = Y, a ~ X, c ~ Z	Strong pleochroism Bright blue, colorless	4-5 2.35-2.47	Colorless, yellow, blue, gray-blue		Phosphate nodules with variscite. Ore veins with Ag W minerals	Decomp acids	GSA Bull 70, 1648-1649
	a = Y, b = Z, c = X	Colorless	5 3.10-3.12	White, colorless			Sol with effor-vescence-dilute HCl	Am Min 64, 241
{100}	c ∧ X = 7°, optic plane ∧ {100} = 18°	Colorless	7½ 2.84	Colorless		Accessory in alkali granites		Min Mag 29, 850-855
{100} common	b = Z, c ∧ X = 40°	Colorless	5½-6 2.77-2.83	Colorless		Ejecta in pumice ash with latiumite, ido-crase, garnet, pyroxene		Am Min 62, 1110-1113
Not visible	b = Y, c ∧ X = small	X = yellow green, Y = Z = bright green, brownish green	1-2 2.0-2.7	Green		Hydrothermal veins or vesicles		p. 257
	Z' ∧ elong ~ 25°	Colorless	3-3½ 2.42	White		Nodules in clay with gypsum, colemanite, aragonite. Hot springs	Easily soluble acids	Dana II, 341-343
	b = Y, c ∧ Z ~ 3°	Pale pinkish, non-pleochroic	2-2½ 2.64	Reddish-violet		Serpentenite-chromite ores with uvarovite	Chlorite group	DHZ 3, 145-146
	b = X, c ∧ Y = -18½°	X = Y = dk yel Z = dk brown	2½ 3.4-3.6	Green, yellow, brown		Oxide zone U-deposits		USGS Bull 1064 p 200-204

n_β Var.	Refractive Index	Mineral Name and Composition	$(n_\gamma - n_\alpha)$	2V angle Dispersion	Crystal System	Habit	Cleavage
1.590 1.595	n_β 1.590-1.595 n_α 1.518 n_γ 1.601	Dresserite $Ba_2Al_4(CO_3)_4(OH)_8 \cdot 3H_2O$	0.083	30° - 40°	Orthorhombic	Spherical aggregate fibers (c)	
*1.567 1.619	n_β 1.567-1.619 n_α 1.55 -1.596 n_γ 1.572-1.621	Uranospinite $Ca(UO_2)_2(AsO_4)_4 \cdot 10H_2O$	0.022 to 0.025	0° - 62° r > v moderate	Tetragonal	Thin rectangular plates {001}	{001} perf {100} dist
1.594	n_β 1.594 n_α 1.564 n_γ 1.596	Arsenuranospathite III $(HAl)_{0.5}(UO_2)_2(AsO_4)_2 \cdot 10H_2O$	0.032	~28°	Orthorhombic		
1.594	n_β 1.594 n_α 1.55 n_γ 1.594	Alurgite $K_2(Mg,Mn)Al_3(AlSi_7)O_{22}(OH)_2$	0.04	Small r > v weak	Monoclinic β ~ 100°	Micaceous	{001} perf.
1.594 1.595	n_β 1.595-1.594 n_α 1.570-1.571 n_γ 1.596-1.598	Leucophane (Leucophanite) $(Ca,Na)_2BeSi_2(O,F,OH)_7$	0.026 to 0.027	39° - 40° r > v	Orthorhombic (Pseudotet.)	Tabular pseudo-tetragonal, spherulitic	{001} perf {100} & {010} distinct
1.594	n_β 1.594 n_α 1.501 n_γ 1.595	Hydrodresserite $BaAl_2(CO_3)_2(OH)_4 \cdot 3H_2O$	0.094	17°	Triclinic α = 95.7° β = 92.4° γ = 115.8°	Radiating-fibrous crystals (c)	{010} & $\{2\bar{1}0\}$ perfect
1.592 1.597	n_β 1.592-1.597 n_α 1.572-1.577 n_γ 1.612-1.616	Johannite (Gilpinite) (Peligotite) $Cu(UO_2)_2(SO_4)_2(OH)_2 \cdot 6H_2O$	0.040	~90° r > v strong	Triclinic α = 110° β = 112° γ = 100.3°	Tabular {100}, prism (c), scaly, fibrous	{100} dist
1.595	n_β 1.595 n_α 1.562 n_γ 1.632	Szmikite $MnSO_4 \cdot H_2O$	0.070	~90°	Monoclinic β = 115.7°	Stalactitic	Splintery
*1.582 1.610	n_β 1.582-1.610 n_α 1.552-1.574 n_γ 1.587-1.610	Muscovite $KAl_2(Si_3Al)O_{10}(OH)_2$	0.036 to 0.049	30° - 47° r > v distinct, horizontal	Monoclinic β = 95.5°	Micaceous, scaly	{001} perf
1.596	n_β 1.596 n_α 1.581 n_γ 1.601	Balipholite $BaMg_2LiAl_3(SiO_3)_4(OH)_8$	0.020	68° - 72°	Orthorhombic	Acicular-fibrous (c), radiated to ∥ aggregates	{010}
1.596	n_β 1.596 n_α 1.544 n_γ 1.627	Cuproartinite $(Cu,Mg)_2(CO_3)(OH)_2 \cdot 3H_2O$	0.083	70°-73°	Monoclinic β = 100.7°	Acicular, rad fibers (b)	{100} perfect
*1.557 1.637	n_β 1.557-1.637 n_α 1.530-1.590 n_γ 1.558-1.637	Phlogopite $KMg_3(Si_3Al)O_{10}(OH)_2$	0.028 to 0.047	0° - 15° v > r	Monoclinic β = 100°	Micaceous, pseudohex crystals	{001} perf
1.597	n_β 1.597 n_α 1.568 n_γ 1.600	Thadeuite $Mg(Ca,Mn)(Mg,Fe,Mn)_2(PO_4)_2(OH,F)_2$	0.032	33°	Orthorhombic	Massive	{010} dist
1.597	n_β 1.597 n_α 1.575-1.585 n_γ 1.598-1.635	Chrysocolla $Cu_2H_2Si_2O_5(OH)_4$	0.023 to 0.050	~24°	Orthorhombic ?	Cryptocrystalline microscopic acicular	None

* n_β variation exceeds 0.01

Twinning	Orientation	Color in thin section	Hardness Sp Gravity	Color hand sample	Alteration	Occurrence	Remarks	Reference
	a = X, b = Y, c = Z	Colorless	2½-3 2.96-3.06	White		Cavities in alkalic sill	Ba analogue dundasite, soluble effervesc HCl	Can Min 10 84-89
	c = X	X ~ colorless, Y = Z = pale yel	2-3 3.45-3.65	Lemon yellow, siskin green		Secondary U-mineral, alter uraninite and arsenides	Autunite group. Fluor U.V.- bright yellow	USGS Bull 1064, 183-187
		Colorless	3.20	White, pale yellow		Secondary U-mineral	Dehydration of arsenurano-spathite I	Min Mag 42, 117
	b = Y, c ∧ X = small	Pink Faint pleochroism	2½-3 2.84	Copper-red		Mn ore deposits	Mica group	p. 270,274
{001} or {110} common, penetration	a = Y, b = Z, c = X	Colorless	4 2.96-2.98	White, yellow-green, yellow		Alkali pegmatites with apatite, fluorite, polylithionite	Fluor-U.V. pinkish-red	Vlasov II, 129-134
		Colorless	3-4 2.79-2.81	Colorless		Alkalic plutons	Dehydrates to dresserite	Can Min 15, 399-404
{010} axis c, simple or multiple	Y ∧ elong ~ 5° - 8°, X ~ ⊥ cleavage	X = colorless, Y = pale yellow, Z = green	2-2½ 3.27-3.32	Emerald green, yellow-green		Secondary U-mineral with gypsum, zippeite	Soluble HCl. Taste bitter. Not fluor. U.V.	USGS Bull 1064, 130-135
	b = Z	Colorless	1½ 2.84-3.21	White, grayish brownish, rose		Efflorescence		Dana II, 481
{001} complex, {110} normal	b = Z, c ∧ X = 0° to -5°	Colorless	2½-4 2.77-2.88	Colorless, lt green, brown, red		Schists, phillites, gneisses. Granites, pegmatites, greisens. Shales	Mica group	p. 272
	c = Y, optic plane = 001	Colorless	3.33-3.35	Pale yellow		Cavities in quartz with Li-mica		Min Abs 26, 325
	b = Y, c ∧ Z = +31°	X = colorless, Y = colorless, pale blue, Z = pale blue	~2½ 2.19	Blue, lt. blue		Supergene Cu-deposits with cuprohydromagnesite	Fluor short λ UV = pale blue	Am Min 64, 886-889
{110} normal, {001} complex	b = Y, c ∧ X = -10° to -5°, a ∧ Z = 0° to +5°	X = colorless, Y ~ Z = pale yel, pale pink-brown, pale green	2-2½ 2.76-2.90	Bronze brown, colorless, greenish		Metamorphosed dolomites, peridotites, lamprophyres, leucite volcanics	Mica (biotite) group	p. 274
	a = Z, b = Y, c = X	Colorless, yellow	~4 3.21-3.25	Yellow-orange		Hydrothermal Sn-W veins		Am Min 64, 359
		Colorless, pale bluish-green X > Z	2-4 2.0-2.4	Sky blue, blue-green, brown, black		Oxide zones Cu-deposits	Properties vary with H_2O content	Win & Win II, 420

Biaxial –

n_β Var.	Refractive Index	Mineral Name and Composition	$(n_\gamma - n_\alpha)$	2V angle Dispersion	Crystal System	Habit	Cleavage
1.597	n_β 1.597 n_α 1.596 n_γ 1.597	Emeleusite $Li_2Na_4Fe_2^{3+}Si_{12}O_{30}$	0.001	Small	Orthorhombic	Pseudohexagonal crystals	Triplets {110}
*1.586 1.611	n_β 1.586-1.611 n_α 1.544-1.574 n_γ 1.588-1.612	Bannisterite $(K,Na,Ca)Mn_{10}Al_2Si_{15}O_{44}\cdot 10H_2O$	0.044	Small ~ 25° v > r weak	Monoclinic β = 94.3°	Platy aggregates	{001} perf
*1.587 1.610	n_β 1.587-1.610 n_α 1.575-1.595 n_γ 1.590-1.622	Amblygonite $LiAl(PO_4)F$	0.014 to 0.027	50° - 90° r > v weak, crossed-inclined	Triclinic α = 112° β = 97.8° γ = 68.3°	Cleavage aggregates, rough prism crystals	{100} & {110} perfect, {0$\bar{1}$1} dist, {001} poor
1.596 1.598	n_β 1.596-1.598 n_α 1.583-1.586 n_γ 1.605	Howlite $Ca_2B_5SiO_9(OH)_5$	0.019 to 0.022	Large ~ 80°	Monoclinic β = 104.8°	Compact nodular, earthy, tabular {100}	None
1.594 1.603	n_β 1.594-1.603 n_α 1.597 n_γ 1.598-1.605	Foshagite $Ca_4Si_3O_9(OH)_2$	0.008	~60°	Monoclinic β = 106.4°	Fibrous (c)	{001} dist
1.595 1.604	n_β 1.595-1.604 n_α 1.559-1.572 n_γ 1.595-1.612	Fuchsite $K(Al,Cr)_2AlSi_3O_{10}(OH)_2$	0.036 to 0.042	~40° r > v strong	Monoclinic β ~ 100°	Micaceous, scaly	{001} perfect
*1.586 1.612	n_β 1.586-1.612 n_α 1.545-1.573 n_γ 1.589-1.612	Ganophyllite $NaMn_3(OH)_4(Si,Al)_4(OH)_{10}$	0.030 to 0.044	Small-moderate ~30°	Monoclinic β = 94.2°	Micaceous, rosettes, tabular, prismatic	{001} perfect easy {010} indistinct
~1.60	n_β n_α 1.598 n_γ 1.614	Kleemanite $ZnAl_2(PO_4)_2(OH)_2$	0.016		Monoclinic β = 110.2°	Tiny fibers	
*1.594 1.609	n_β 1.594-1.609 n_α 1.564-1.580 n_γ 1.600-1.609	Paragonite $NaAl_2(Si_3Al)O_{10}(OH)_2$	0.036 to 0.029	0° - 40° r > v moderate	Monoclinic β = 95°	Micaceous, fine scaly, massive	{001} perfect
1.598 1.605	n_β 1.598-1.605 n_α 1.510-1.516 n_γ 1.611-1.621	Amarantite $Fe^{3+}SO_4(OH)\cdot 3H_2O$	0.100 to 0.105	28° - 30° v > r strong, horizontal	Triclinic α = 95.5° β = 90.5° γ = 97°	Prismatic or acicular (c)	{010} & {100} perfect
1.601	n_β 1.601 n_α 1.595 n_γ 1.604	Hurlbutite $CaBe_2(PO_4)_2$	0.009	70° r > v weak	Orthorhombic	Stout prismatic (c)	None
1.602	n_β 1.602 n_α 1.526 n_γ 1.602	Anthonyite $Cu(OH,Cl)_2\cdot 3H_2O$	0.076	very small ~ 3°	Monoclinic β = 112.5°	Tiny bent prisms on c	{100} dist
1.602	n_β 1.602 n_α 1.586 n_γ 1.606	Spencerite $Zn_4(PO_4)_2(OH)_2\cdot 3H_2O$	0.020	49° r > v moderate	Monoclinic β = 116.8°	Platy to column crystals, massive, stalactitic	{100} perfect {010} good, {001} dist
1.603	n_β ~1.603 n_α 1.564 n_γ 1.605	Sauconite $Zn_3(Al,Si)_4O_{10}(OH)_2\cdot 4H_2O$	0.041	Small	Monoclinic	Clay-like, minute micaceous plates	{001} perfect

* n_β variation exceeds 0.01

Twinning	Orientation	Color in thin section	Hardness Sp Gravity	Color hand sample	Alteration	Occurrence	Remarks	Reference
	a = Y, b = X, c = Z	Colorless	5-6 2.76-2.78	Colorless		Peralkaline trachyte dike		Min Mag 42, 31
None observed	b = Y, $c \wedge X$ = small	X = colorless, Y = Z = pale yellow, brown	4 2.92-2.94	Dk brown		Mn ore veins	Dimorph of ganophyllite	Min Mag 36, 893-913
$\{\bar{1}\bar{1}1\}$, {110} lamellar, {111} rare	$a \wedge Y \sim 15^\circ$-20° $b \wedge X \sim 40^\circ$ - 30°, $c \wedge Z \sim$ 30° - 20°	Colorless	5½-6 3.0-3.1	White, gray, pale violet, pink	Kaolin and mica	Granite pegmatites with Li-minerals. Hydrothermal veins. Greisen	Amblygonite-montebrasite series. Slightly soluble acids	p. 77
	b = X, $c \wedge Z \sim 44^\circ$	Colorless	3½ 2.43-2.59	White		Evaporite borate deposits with borax, ulexite, bakerite	Easily soluble diluted acids	Am Min 55, 716-728
	$c \sim Z$, + elongation	Colorless	3 2.36-2.67	White		Contact carbonate zones	Gelatinizes HCl	Am Min 43, 1-15
	b = Y, $c \wedge X$ = small	Colorless, X = pale blue-grn Y = yellow-green, Z = dk. blue-grn	2½-3 2.86	Emerald green		Quartzite, mica schist	Mica group	pp. 270,272
	b = Y or Z, $c \sim X$	X = yellow-brown, Y = Z = colorless	4-4½ 2.84-2.92	Brown, yellow-brown		Hydrothermal Mn-veins	Dimorph bannisterite. Gelatinizes HCl	USGS PP 180, 114
	Inclined ext. to 40°	Colorless	2.76	Ochre		Earthy coatings on Mn-Fe ores		Am Min 64, 1331
{001} complex	b = Z, $a \sim Y$, $c \wedge X \sim -5^\circ$	Colorless	2½ 2.78-2.90	Colorless, pale yellow		Phyllite, schist, gneiss. Quartz veins. Fine sediments	Mica group	p. 271
	X nearly ⊥{100}	X = colorless, Y = pale orange-yellow, Z = reddish brown	2½ 2.11-2.29	Red to brownish-red		Groundwater deposit in mines	Soluble in HCl	Dana II, 611-612
	a = Z, b = X, c = Y	Colorless	6 2.88	Colorless, greenish-white		Pegmatites	Slowly soluble acids	Am Min 37, 931-940
	b = Y, $c \wedge Z = +13^\circ$	X = lavender, Y = Z = deep smoky blue	2	Lavender		In cavities and incrustations in basalt	Easily soluble in cold dilute acids	Am Min 48, 614-619
{100} polysynthetic	b = Z, $a \sim X$	Colorless	3 3.14-3.25	White		Oxide zones Zn-deposits with hemimorphite, hopeite	Easily soluble acids	Dana II, 931-933
		Pale brown	1-2	Red-brown, brownish-yellow		Altered Pb-Zn deposits	Montmorillonite (Smectite) Group	Am Min 31, 411

n_β Var.	Refractive Index	Mineral Name and Composition	$(n_\gamma - n_\alpha)$	2V angle Dispersion	Crystal System	Habit	Cleavage
1.603	n_β 1.603 n_α 1.585 n_γ 1.604	Wightmanite $Mg_9B_2O_{12} \cdot 8H_2O$	0.019	33° v > r strong	Triclinic α = 96.2° β = 97.8° γ = 105.9°	Prism pseudohexagonal crystals, radiated crystals	{010} perfect {100} fair
1.603	n_β 1.603 n_α 1.596 n_γ 1.606	Weeksite (Gastunite) $K_2(UO_2)_2(Si_2O_5)_3 \cdot 4H_2O$	0.010	50° - 66° r > v strong	Orthorhombic	Radiated crystals in small spherulites	Two prism good
1.602 1.605	n_β 1.602-1.605 n_α 1.584-1.591 n_γ 1.611-1.614	Bertrandite (Hessenbergite) $Be_4Si_2O_7(OH)_2$	0.023 to 0.027	73° - 81° v > r weak	Orthorhombic	Tabular or prismatic	{001} perfect {110}, {010} & {100} dist
1.604	n_β 1.604 n_α 1.585 n_γ 1.612	Nakauriite $Cu_8(SO_4)_4(CO_3)(OH)_6 \cdot 48H_2O$	0.027	65°	Orthorhombic	Acicular-fibrous aggregates	
1.605	n_β 1.605 n_α 1.580 n_γ 1.605	Montdorite $K(Fe,Mn,Mg,Ti)_5(Si,Al)_8O_{20}(F,OH)_4$	0.025	Very small	Monoclinic β = 99.9°		
1.605	n_β 1.605 n_α 1.598 n_γ 1.608	Brianite $Na_2CaMg(PO_4)_2$	0.010	63° - 66°	Orthorhombic	Grains	
1.600 1.610	n_β 1.600-1.610 n_α 1.580-1.589 n_γ 1.610-1.619	Herderite $CaBePO_4(F,OH)$	0.030	67° - 70° r > v, inclined	Monoclinic β ~90°	Thick tabular crystals	{110} inter-rupted
1.606 1.609	n_β 1.606-1.609 n_α 1.600-1.603 n_γ 1.614-1.615	Latiumite $K(Ca,Na)_3(Al,Si)_5O_{11}(SO_4,CO_3)$	0.014 to 0.012	72° - 90° v > r distinct	Monoclinic β = 106°	Tabular, elongation	{100} perfect
1.608	n_β 1.608 n_α 1.583 n_γ 1.633	Chudobaite $(Na,K,Ca)(Mg,Zn,Mn)_2H(AsO_4)_2 \cdot 4H_2O$	0.050	89°	Triclinic α = 115° β = 96° γ = 94°	Crystals	{010} dist {100} imperf
1.608	n_β 1.608 n_α 1.590 n_γ 1.611	Lawsonbauerite $(Mn,Mg)_5Zn_2(SO_4)(OH)_{12} \cdot 4H_2O$	0.021	42°-45° r > v strong	Monoclinic β = 95.2°	Tiny bladed-acicular xls. (b)	None
1.609	n_β 1.609 n_α 1.597 n_γ 1.615	Magbasite $KBa(Al,Sc)(Mg,Fe^{2+})_6Si_6O_{20}F_2$	0.018	70°	Unknown	Radiated acicular crystals	
*1.58 1.64	n_β 1.58-1.64 n_α 1.57-1.62 n_γ 1.58-1.65	Delessite (Melanolite) $(Mg,Al,Fe^{3+})_6(Si,Al)_4O_{10}(OH)_8$	0.001 to 0.020	0° - 25° r > v	Monoclinic β = 97°	Massive, granular, foliated	{001} perfect
1.610	n_β 1.610 n_α 1.600 n_γ 1.616	Sainfeldite $H_2Ca_5(AsO_4)_4 \cdot 4H_2O$	0.016	~80°	Monoclinic β ~97°	Radiated lath crystals (c), rosettes	
1.611	n_β 1.611 n_α 1.413 n_γ 1.637	Ammonia-niter NH_4NO_3	0.224	35° v > r weak	Orthorhombic	Delicate acicular crystals on c	{010} dist

* n_β variation exceeds 0.01

Twinning	Orientation	Color in thin section	Hardness Sp Gravity	Color hand sample	Alteration	Occurrence	Remarks	Reference
	c ∧ Z = 5°	Colorless	5.5 2.58-2.60	Colorless, pale greenish		Contact crystalline dolomite-limestone with fluoborite & ludwigite	Soluble acids	Am Min 47, 718-722
	a = Z, b = X, c = Y	X = colorless, Y = pale yellow-green, Z = yellow-green	Soft 4.1	Yellow		Opal veinlets in rhyolites and tuffs	Not fluor U.V.	Am Min 45, 39-52
{011} or {021} common	a = X, b = Y, c = Z	Colorless	6-7 2.61	Colorless, yellow		Granite pegmatites, greisens, aplites, hydrothermal veins		Vlasov II, 89-96
	+ elongation	X = colorless, Y = lt greenish-blue, Z = lt blue	2.35-2.39	Sky blue		Veins in serpentenite with chrysotile, magnetite, brochantite		Am Min 62, 594
			3.15-3.16	Green, brownish green		Peralkaline rhyolite		Am Min 64, 1331
Lamellar fine		Colorless	4-5 3.10-3.17	Colorless		Meteorites		Am Min 53, 508-509
{001} or {100} common	b = Y, c ∧ X ~ -87°, c ∧ Z ~ +3°	Colorless	5-5½ 2.94-3.01	Colorless, yellow, greenish		Late-stage hydrothermal mineral in granite pegmatites		Am Min 63, 913-917
{100} common	b = Z, c ∧ X = 16° - 28°	Colorless	5½-6 2.92-2.93	White		Alkali volcanics with leucite, melilite, haüynite, garnet	Mottled extinction. Zoning	Am Min 58, 466-470
	c ∧ Z' on {010} = +24°, on {100}=110°	Colorless, pink	2½-3 2.94-3.0	Pink		Oxide zones Cu-deposits		Am Min 44, 1323
None	b = Y, c ∧ Z = +7°	Colorless	4½ 2.87-2.92	Colorless, white		Secondary Zn ores with sussexite and	Black coating on xls	Am Min 68, 949-952
	c ∧ Z = 10°	X = Y = colorless, Z = lilac	~5 3.41	Colorless, rose-violet		Alkali plutons with barkevikite, fluorite, barite, parisite		Am Min 51, 530-531
{001}	b = Y, Z ∧ {001} ~ 2°	Colorless, X = lt yellow-grn, Y = Z = dk green	2-3 2.73	Lt Yellow, dk green, black		Altered mafic rocks	Chlorite group	p. 294
	b = X, c ∧ Y ~ 20°	Colorless	3.00-3.04	Colorless, pink		Oxide zones ore deposits		Am Min 50, 806
	a = Y, b = X, c = Z	Colorless	2 1.725	Colorless		Efflorescence in caves. May not occur in nature	Highly soluble H_2O	Dana II, 305

n_β Var.	Refractive Index	Mineral Name and Composition	$(n_\gamma - n_\alpha)$	2V angle Dispersion	Crystal System	Habit	Cleavage
1.613	n_β 1.613 n_α 1.511 n_γ 1.623	Ktenasite $(Cu,Zn)_3(SO_4)(OH)_4 \cdot 2H_2O$	0.112	51°	Monoclinic β = 95.4°	Groups tabular crystals	
1.614	n_β 1.614 n_α 1.585 n_γ 1.614	Vimsite $CaB_2O_2(OH)_4$	0.029	~28°	Monoclinic β = 92°	Crystals	Perfect ‖ elongation
1.614 1.616	n_β 1.614-1.616 n_α 1.586-1.589 n_γ 1.621-1.626	Caysichite $(Y,Ca,RE)_4(Si,Al)_4O_{10}(CO_3)_3 \cdot 4H_2O$	0.035 to 0.037	53° - 72°	Orthorhombic	Thin incrustations, radiated, columns (c)	{010} (?)
1.614	n_β 1.614 n_α 1.595 n_γ 1.616	Phosphophyllite $Zn_2(Fe,Mn)(PO_4)_2 \cdot 4H_2O$	0.021	~45° r > v perc	Monoclinic β = 120.2°	Prismatic or thick tabular crystals	{100} perfect {010} & {$\bar{1}$02} distinct
1.61	n_β 1.61 n_α 1.57-1.60 n_γ 1.62	Montebrasite $(Li,Na)Al(PO_4)(OH,F)$	0.005	~80°	Triclinic α = 112° β = 97.8° γ = 68.1°	Cleavable masses, ~ equant crystals	{100} perf, {110} & {0$\bar{1}$1} distinct {001} poor
1.614 1.615	n_β 1.614-1.615 n_α 1.600-1.587 n_γ 1.631-1.640	Monetite (Glaubapatite) $CaHPO_4$	0.031 to 0.053	Moderate to large ~ 85° v > r weak	Triclinic α = 96.1° β = 103.9° γ = 89.2°	Tiny plates {010}, crusts, stalactites	3 dir indist
1.63 1.60	n_β 1.63-1.60 n_α 1.64-1.61 n_γ 1.64-1.61	Klementite $(Mg,Fe^{3+})_6(Si,Al)_4O_{10}(OH)_8$	0.000 to 0.001	Small v > r	Monoclinic β ~ 97°	Micaceous, scaly	{001} perf.
1.615	n_β 1.615 n_α 1.600 n_γ 1.629	Lehiite $(K,Na)_2Ca_5Al_8(PO_4)_8(OH)_{12} \cdot 6H_2O$ (?)	0.029	Very large ~ 87°	Unknown	Crusts of fibrous aggregates	
*1.59 1.64	n_β 1.59-1.64 n_α 1.58-1.65 n_γ 1.60-1.66	Diabantite $(Mg,Fe^{2+},Al)_6(Si,Al)_4O_{10}(OH)_8$	0.000 to 0.015	15° - 35° r > v	Monoclinic β ~97°	Massive, foliated, scaly, radiated	{001} perfect
1.616	n_β 1.616 n_α 1.559 n_γ 1.624	Phuralumite $Al_2(UO_2)_3(PO_4)_2(OH)_6 \cdot 10H_2O$		40°	Monoclinic <β = 112°	Prism. crystals	
1.617	n_β 1.617 n_α 1.598 n_γ 1.625	Sanbornite $BaSi_2O_5$	0.027	67°	Orthorhombic	Anhedral plates	{001} perf {100} dist {010} indist
1.61	n_β ~1.61 n_α 1.605 n_γ 1.612	Hillebrandite $Ca_2SiO_4 \cdot H_2O$	0.007	~60° v > r strong	Monoclinic β = 90°	Radiated fibrous	{hk0}
1.620	n_β 1.620 n_α 1.584 n_γ 1.621	Nissonite $Cu_2Mg_2(PO_4)_2(OH)_2 \cdot 5H_2O$	0.037	19° r > v very strong	Monoclinic β = 99.3°	Tabular {100}, elong (c), diamond crystals	{100} fair

* n_β variation exceeds 0.01

Twinning	Orientation	Color in thin section	Hardness Sp Gravity	Color hand sample	Alteration	Occurrence	Remarks	Reference
	b = Z	Blue-green	2-2½ 2.97	Blue-green		Oxide zones Cu-Zn deposits	Soluble dilute acids or NH_4OH	Am Min 36, 381
		Colorless	4 2.54-2.56	Colorless		Skarns with uralborite		Am Min 54, 1219-1220
	a = Y, b = X, c = Z	Colorless	4½ 3.03	Colorless, white, pale yellow		Granite pegmatites	Effervesces in cold, dilute HCl	Am Min 61, 174-175
{100} common, {$\bar{1}$02} rare	b = Z, c ∧ Y = -50°	Colorless	3-3½ 3.13-3.15	Colorless, blue-green		Alteration sulfides in sulfide ores or phosphates in pegmatites	Soluble acids	Acta Cryst. 14, 794
{$\bar{1}\bar{1}$1} common		Colorless	5½-6 2.98-3.03	White, pale colors		Granite pegmatites with Li-minerals and phosphates	Amblygonite	Dana II, 823-827
	X ~ ⊥{11$\bar{1}$}	Colorless	3½ 2.92-2.93	Colorless, white, yellowish		Cave deposits, guano leaching with newberyite, whitlockite	Soluble acids	Dana II, 660-661
{001} multiple	b = Y, c ∧ X = small	Colorless, pale green, pale brown Z = Y > X	2-3 2.6-3.0	Green-brown		Low grade schists. Oxidation of ferrous chlorites	Chlorite group	p. 294
	Extinction ~ ∥ elongation (-) = X	Colorless	5½ 2.89	White, gray		Altered phosphate (variscite) nodules with wardite		Dana II, 942-943
{001}	b = Y, Z ∧ {001} ~ 2°	X = green-brown, Y = Z = dk green	2-2½ 2.77-3.03	Dk green, blackish green		Altered diabase, mafic rocks	Chlorite group	p. 288
	b = X	Y = colorless, Z = yellow	~3 3.5	Lemon yellow		Pegmatite		Am Min 65, 208
Polysynthetic {010}	a = Y, b = X, c = Z	Colorless	~5 3.71-3.77	Colorless, white		With gillespite, quartz, witherite		Am Min 43, 517-536
	a ~ X, b = Y, c ~ Z	Colorless	5½ 2.63-2.66	Colorless, white, greenish		Contact carbonate zones with wollastonite, garnet	Soluble HCl	Win & Win II, 506
	b = Z, a ∧ X = 15°, c ∧ Y = 6°	X ~ colorless, Y = Z = turquoise blue	2½ 2.73-2.74	Bluish-green		Metamorphic Cu-deposits with secondary Cu-minerals	Easily soluble dilute acids	GSA Progmam Ann Mtg (Abst.) Nov 14-16-1966 p145-146

n_β Var.	Refractive Index	Mineral Name and Composition	$(n_\gamma - n_\alpha)$	2V angle Dispersion	Crystal System	Habit	Cleavage
1.620	n_β 1.620 n_α 1.610 n_γ 1.623	Suolunite $Ca_2H_2Si_2O_7 \cdot H_2O$	0.013	30° - 50°	Orthorhombic	Fine granular	None
*1.612 1.629	n_β 1.612-1.629 n_α 1.600-1.616 n_γ 1.626-1.640	Tremolite $Ca_2Mg_5Si_8O_{22}(OH)_2$	0.026 to 0.024	88° - 84° v > r weak	Monoclinic β = 105°	Fibrous, bladed	{110} dist
*1.61 1.63	n_β 1.61-1.63 n_α 1.60-1.62 n_γ 1.61-1.63	Pycnochlorite $(Mg,Fe^{2+},Al)_6(Si,Al)_4O_{10}(OH)_8$	0.000 to 0.005	Small-moderate r > v	Monoclinic β ~104.5°	Massive, scaly, oolitic	{001}
*1.610 1.634	n_β 1.634-1.610 n_α 1.615-1.589 n_γ 1.644-1.619	Hydroxyl-herderite $CaBePO_4(OH,F)$	~0.030	77° - 70° r > v, inclined	Monoclinic β ~90°	Radiated fibrous, tabular crystals in cavities	{110} inter-rupted
1.622	n_β 1.622 n_α 1.592 n_γ 1.623	Minnesotaite $(Fe,Mg)_3Si_4O_{10}(OH)_2$	0.040	Small	Monoclinic β = 99.5°	Micro plates or needles	{001} perfect
1.623	n_β 1.623 n_α 1.606 n_γ 1.628	Vlasovite $Na_2ZrSi_4O_{11}$	0.021	50° - 56° r > v distinct	Monoclinic β = 100.4°	Monocrystalline masses	{010} dist, another at 88°
1.623	n_β 1.623 n_α 1.610 n_γ 1.623	Meta-uranocircite $Ba(UO_2)_2(PO_4)_2 \cdot 6\text{-}8H_2O$	0.013	Small	Tetragonal	Thin flexible plates {001}, radiated	{001} perf {100} dist
1.624	n_β 1.624 n_α 1.592 n_γ 1.625	Ephesite $Na(LiAl_2)(Al_2Si_2)O_{10}(OH)_2$	0.033	~20° v > r	Monoclinic β = 95°	Indistinct crystals, scaly	{001} perf
*1.610 1.641	n_β 1.641-1.610 n_α 1.610-1.606 n_γ 1.641-1.610	Celadonite $K(Mg,Fe^{2+})(Fe^{3+},Al)Si_4O_{10}(OH)_2$	0.014 to 0.032	Small 0° - 20°	Monoclinic β = 100.1°	Micaceous scales	{001} perf
1.625	n_β 1.625 n_α 1.614 n_γ 1.637	Parahopeite $Zn_3(PO_4)_2 \cdot 4H_2O$	0.023	~90° r > v perc	Triclinic α = 93.3° β = 92° γ = 91.3°	Tabular, elongagated (c), radiated	{010} perf
1.626	n_β 1.626 n_α 1.605 n_γ 1.633	Jimthompsonite $(Mg,Fe)_5Si_6O_{16}(OH)_4$	0.028	62° r > v weak	Orthorhombic	Radiated acicular, prism-fibrous	{210}perf (38°), {100} & {010} parting
1.618 1.636	n_β 1.618-1.636 n_α 1.590-1.602 n_γ 1.623-1.639	Grandidierite $(Mg,Fe)Al_3BSiO_9$	0.033 to 0.037	30° v > r very strong	Orthorhombic	Elongation prismatic crystals	{100} perf {010} dist
1.628	n_β 1.628 n_α 1.598 n_γ 1.654	Lausenite $Fe_2(SO_4)_3 \cdot 6H_2O$	0.056	Large ~ 85°	Monoclinic	Fibrous aggregates (c)	

* n_β variation exceeds 0.01

Twinning	Orientation	Color in thin section	Hardness Sp Gravity	Color hand sample	Alteration	Occurrence	Remarks	Reference
	a = X, b = Z, c = Y	Colorless	2.683	White		Veins in ultramafic rocks		Am Min 53, 349
Simple or multiple on {100}	b = Y, c ∧ Z = -21° to -17°, a ∧ X = -6° to -2°	Colorless	5-6 3.0-3.1	White, gray, lilac	Talc, chlorite, calcite-dolomite	Metamorphic rocks	Amphibole group	p. 229
{001}	b = Y, c ∧ X = small	X = pale green-brown, Y = Z = dk green	2-3 2.8-3.1	Dk green, dk olive-green		Schists, altered mafic rocks	Chlorite group	p. 288
{001} or {100} common	b = Y, c ∧ X ~ -87°, c ∧ Z ~ +3°	Colorless	5-5½ 2.94-3.01	Colorless, yellow, greenish		Alter of beryl or beryllonite in granite pegmatites		Am Min 63, 913-917
	b = Y	X = pale green, Y = Z = colorless to pale greenish-yellow	<3 3.01-3.21	Greenish-gray		Cherty Fe-ores with siderite, stilpnomelane, greenalite	Fe-talc	Am Min 50, 148-169
	b = Y	Colorless	6 2.97	Colorless, lt brown (zoned)		Contact zones in alkali rocks with aegirine, arfvedsonite, eudialyte	Fluor U.V.-strong orange-yellow	Vlasov II, 373-375
{100} & {010} polysynthetic	a = Y, b = Z, c = X	X ~ colorless, Y = Z = pale yel	2-2½ 3.95-4.00	Yellow-green		Secondary U-mineral also primary in low T veins	Fluor U.V.-green. Uniaxial	USGS Bull 1064, 211-215
Rotation about [310] or [3̄10]	b = Z, c ~ X	Colorless	3½-4½ 2.96-2.98	Pink		Emery deposits, Mn-ores	Margarite group	Am Min 52, 1689-1696
	b = Y, a ~ Z, c ∧ X = +10°	X = pale yellow-green, Y = Z = dk green, blue-green	2 2.95-3.05	Green, blue-green		Altered volcanic rocks basalt to intermediate	Mica group, essentially glauconite	Am Min 49, 1031-1083
{100} common, polysynthetic	a ~ X, c ∧ Y on {100} ~ 30°	Colorless	3½-4 3.31	Colorless, white		Oxide zones Zn-deposits	Dimorph. hopeite	Dana II, 733-734
	a = X, b = Y, c = Z	Colorless		Colorless, lt pinkish-brown		Ultramafic metamorphic rocks with chesterite, talc, anthophyllite	Biopyribole. Clinojimthompsonite is monoclinic polymorph	Am Min 63, 1000-1009
	a = X, b = Z, c = Y	X = dk blue-green, Y = colorless, Z = green, lt-blue-green	7½ 2.98-3.10	Greenish-blue	~Kryptotile	Pegmatites		Win & Win II, 497
	c ∧ X = 27°	Colorless		White		Formed by mine fires. With copiapite		Dana II, 530

n_β Var.	Refractive Index	Mineral Name and Composition	$(n_\gamma - n_\alpha)$	2V angle Dispersion	Crystal System	Habit	Cleavage
*1.61 1.65	n_β 1.61-1.65 n_α 1.56-1.61 n_γ 1.61-1.65	Glauconite $(K,H_3O)_2(Fe^{3+},Al,Fe^{2+},Mg)_4(Si_{7-7.5}Al_{1-0.5})O_{20}(OH)_4$	0.014 to 0.032	0° - 20° r > v, inclined	Monoclinic β ~100°	Fine flakes in pellets or granules	{001} perfect
1.630	n_β 1.630 n_α 1.618 n_γ 1.632	Parawollastonite α-$CaSiO_3$	0.014	38° - 60°	Monoclinic β = 95.4°	Tabular, fibrous crystals, massive, granular	{100} perfect {001} & {$\bar{1}$02} distinct
*1.620 1.641	n_β 1.620-1.641 n_α 1.620-1.625 n_γ 1.620-1.641	Novacekite $Mg(UO_2)_2(AsO_4)_2 \cdot 12H_2O$	Very small to 0.016	0° - 20° r > v	Tetragonal	Rectangular plates {001}, crusts	{001} perfect {010} & {110} indistinct
1.63	n_β 1.63 n_α 1.610 n_γ 1.639	Gatumbaite $CaAl_2(PO_4)_2(OH)_2 \cdot H_2O$	0.029	65°	Monoclinic β = 91°	Fibrous, radiated, rosettes	
1.63	n_β 1.63 n_α 1.60 n_γ 1.630	Mariposite $K_2(Al,Cr)_3(Mg,Fe^{2+})Si_7AlO_{20}(OH,F)_4$	0.03	Small r > v	Monoclinic β ~ 100°	Micaceous	{001} perf.
1.630 1.633	n_β 1.633~1.630 n_α 1.619-1.615 n_γ 1.635-1.630	Kinoshitalite $(Ba,K,Na)(Mg,Mn,Al,Fe)_3(Si,Al)_4O_{10}(OH,F,O)_2$	0.016	Small, 23°	Monoclinic β = 100°	Small scales	{001} perfect
1.632	n_β 1.632 n_α 1.617 n_γ 1.640	Chesterite $(Mg,Fe)_{17}Si_{20}O_{54}(OH)_6$	0.023	71° r > v weak	Orthorhombic	Radiated acicular, prism-fibrous	{110} perf (45°), {100} & {010} parting
1.632 1.635	n_β 1.635-1.632 n_α 1.617-1.612 n_γ 1.652-1.653	Tilleyite $Ca_5Si_2O_7(CO_3)_2$	0.035 to 0.041	~90° r > v	Monoclinic β = 105.8°	Massive, tabular, grains	{100} perf {101} ? good {$\bar{1}$01} ? poor
1.633	n_β 1.633 n_α 1.620 n_γ 1.640	Garrelsite $Ba_3NaSi_2B_6O_{13}(OH)_7$	0.020	72°	Monoclinic β = 114.3°	Bipyramidal crystals	
1.633	n_β 1.633 n_α 1.622 n_γ 1.638	Liberite Li_2BeSiO_4	0.016	66°	Monoclinic	Minute crystal aggregate	{010} perfect {100} & {001} distinct
1.633	n_β 1.633 n_α 1.630 n_γ 1.636	Danburite $CaB_2Si_2O_8$	0.006	~88° v > r strong	Orthorhombic	Prismatic (c), massive	{001} poor
1.635	n_β 1.635 n_α 1.611 n_γ 1.643	Burangaite $2(Na,Ca)_2(Fe,Mg)_2Al_{10}(PO_4)_8(O,OH)_{12} \cdot 4H_2O$	0.032	58° - 60° r > v	Monoclinic β = 110.9°	Prismatic crystals	{100} perfect
1.635	n_β 1.635 n_α 1.613 n_γ 1.657	Sklodowskite (Chinkolobwite) $Mg(UO_2)_2Si_2O_7 \cdot 6H_2O$	0.043	~90°	Monoclinic β = 96°	Radiated acicular fibrous crystals, crusts, earthy	{100} perf
1.635	n_β 1.635 n_α 1.618 n_γ 1.650	Segelerite $CaMg(H_2O)_4Fe^{3+}(OH)(PO_4)_3$	0.032	~85°	Orthorhombic	Long prism crystals (c)	{010} perfect

* n_β variation exceeds 0.01

Twinning	Orientation	Color in thin section	Hardness Sp Gravity	Color hand sample	Alteration	Occurrence	Remarks	Reference
Not visible	b = Y, a ~ Z, c ∧ X ~ -10°	X = pale yellow, green, Y = Z = yellow-green, olive green, blue-green	2 2.4-3.0	Olive-green, blue-green	Limonite, goethite	With clays in diagenetic pellets in marine sediments	Mica Group	p. 261
{100} common	b = Y, c ∧ X = 38°	Colorless	4½-5 2.91-2.93	White, gray, yellowish		Limestone contact zones		DHZ 2, 167-175
	X = c	Pale yellow	2½ ~3.7	Yellow		Secondary U-mineral	Novacekite-saléeite series, Autunite group. Fluor. U.V.-dull green	USGS Bull 1064, 177-183
	b = Z	Colorless	2.92-2.95	White		Pegmatites with trolleite, scorzalite, apatite, bjarebyite		Am Min 63, 793-794
	b = Y, c ∧ X = small	Colorless, pale green Y ~ X > Z	2½-3 2.79	Apple-green			Mica group. Cr-phengite	p. 270-272
		X = very pale yel, Y = Z = pale yel	2½-3 3.30-3.33	Yellow-brown		Mn-ores with hausmannite, tephroite, celsian, rhodonite	Mica group (Mg-anandite)	Am Min 60, 486-487
	a = X, b = Y, c = Z	Colorless		Colorless, lt pinkish-brown		Ultramafic metamorphic rocks with jimthompsonite, talc anthophyllite	Biopyribole. Monoclinic polymorph exists	Am Min 63, 1000-1009
{101} common	b = Y, c ∧ X = 12°-18°	Colorless	2.84-2.88	Colorless, white		Limestone contact zones		Win & Win II, 480
	b = Z	Colorless	3.68-3.73	Colorless		Shales with borate salts		Am Min 41, 672
	c ∧ Z = 41°, + elongation	Colorless, yellow, brown	7 2.69	Pale yellow, brown		Hydrothermal veins with lepidolite, cassiterite, scheelite		Am Min 50, 519
	a = X, b = Y, c = Z	Colorless	7 2.97-3.02	Colorless, white, pink, yellow, brown		Carbonate contact zones		Win & Win II, 258-259
	b = Z, c ∧ X = +11°	X = pale blue, Y = dk blue, Z = colorless	5 3.05	Pale blue, bluish-green		Pegmatites with bertossaite, trolleite, scorzalite, apatite		Am Min 63, 793
	b = Y	Pale yellow. Not pleochroic	2-3 3.64-3.77	Pale yellow to greenish-yellow		Secondary U-mineral	Not fluor. U.V. Soluble acids	USGS Bull 1064, 300-304
	a = Y, b = X, c = Z	X = Y = colorless, Z = yellow	4 2.61-2.67	Pale yellow-green to colorless		Pegmatites with collinsite, apatite, robertsite	Fe-overite. Soluble dilute HCl	Am Min 59, 48-59

Biaxial –

n_β Var.	Refractive Index	Mineral Name and Composition	$(n_\gamma - n_\alpha)$	2V angle Dispersion	Crystal System	Habit	Cleavage
1.63 1.64	n_β 1.63-1.64 n_α 1.619-1.621 n_γ 1.642-1.655	Nordite $Na_3Ce(Sr,Ca)(Mn,Mg,Fe,Zn)_2Si_6O_{18}$	0.023 to 0.034	30° - 32°	Orthorhombic	Tabular crystals {100}, radiated fibrous	{100} dist
1.635	n_β 1.635 n_α 1.624 n_γ 1.654	Bakerite $Ca_4B_4(BO_4)(SiO_4)_3(OH)_3 \cdot H_2O$	0.030	85°	Monoclinic β = 90.2°	Tabular, prism	None ?
*1.625 1.648	n_β 1.625-1.648 n_α 1.595-1.638 n_γ 1.627-1.650	Margarite $CaAl_2(Si_2Al_2)O_{10}(OH)_2$	0.032 to 0.012	26° - 67° v > r distinct, horizontal	Monoclinic β = 95°	Micaceous, platy or scaly aggregates	{001} perf
1.636	n_β 1.636 n_α 1.622 n_γ 1.636	Manganophyllite $K(Mg,Mn)_3Fe^{3+}Si_3O_{10}(OH)_2$	0.014	14° r > v	Monoclinic β ~ 100°	Micaceous, scaly	{0001} perf
1.636	n_β 1.636 n_α 1.624 n_γ 1.642	Bertossaite $(Li,Na)_2(Ca,Fe,Mn)Al_4(PO_4)_4(OH,F)_4$	0.018	Medium-large v > r moderate	Orthorhombic	Massive	{100} dist
1.636 1.637	n_β 1.637-1.636 n_α 1.608-1.612 n_γ 1.650-1.653	Fabianite $CaB_3O_5(OH)$	0.042	65° v > r perc	Monoclinic β = 113.4°	Prismatic crystals	{011}
*1.602 1.672	n_β 1.602-1.672 n_α 1.588-1.663 n_γ 1.613-1.683	Anthophyllite $(Mg,Fe)_7Si_8O_{22}(OH)_2$	0.025 to 0.020	78° - 90° r > v or v > r weak to moderate	Orthorhombic	Fibrous, bladed, acicular	{110} perf
*1.612 1.663	n_β 1.612-1.663 n_α 1.594-1.647 n_γ 1.618-1.663	Glaucophane $Na_2Mg_3Al_2(Si_4O_{11})_2(OH)_2$	0.024 to 0.016	50° to 0° v > r weak	Monoclinic β = 104°	Acicular, columnar (c)	{110} dist
*1.618 1.656	n_β 1.618-1.656 n_α 1.604-1.648 n_γ 1.624-1.664	Richterite $Na_2CaMg_5Si_8O_{22}(OH)_2$	0.020 to 0.016	68° - 50° v > r	Monoclinic β = 104°	Prismatic	{110} perf
*1.625 1.652	n_β 1.625-1.652 n_α 1.610-1.638 n_γ 1.630-1.654	Eckermannite $Na_3Mg_4AlSi_8O_{22}(OH)_2$	0.016 to 0.020	80° - 15° r > v strong	Monoclinic β = 105°	Prismatic	{110} perf
*1.633 1.644	n_β 1.633-1.644 n_α 1.629-1.640 n_γ 1.638-1.650	Andalusite Al_2SiO_5	0.009 to 0.011	71° - 86° v > r weak, rarely, r > v	Orthorhombic	Prismatic (c) crystals. Fibrous, radiated	{110} good at ~90°
1.639	n_β 1.639 n_α 1.596 n_γ 1.670	Suanite $Mg_2B_2O_5$	0.074	70° r > v weak	Monoclinic β = 104.3°	Fibrous aggregates	{hk0} perf
1.639	n_β 1.639 n_α 1.618 n_γ 1.652	Inesite $Ca_2Mn_7Si_{10}O_{28}(OH)_2 \cdot 5H_2O$	0.034	~75° r > v weak	Triclinic α = 91.8° β = 132.5° γ = 94.4°	Prism (c) or tabular, acicular radiated aggregates	{010} perfect {100} dist

* n_β variance exceeds 0.01

Twinning	Orientation	Color in thin section	Hardness Sp Gravity	Color hand sample	Alteration	Occurrence	Remarks	Reference
	a = X, b = Y, c = Z	Pale brown	5 3.43-3.49	Dk - lt brown		Sodalite-syenites, ussingite pegmatites with steenstrupine		Am Min 55, 1167-1181
	b = Y, c ∧ Z = 44°	Colorless	4½ 2.88	Colorless, white		Evaporite veins		Am Min 47, 919-923
{001} complex [310] twin axis	b = Z, a ∧ Y = -6° to -8°, c ∧ X = -11° to -13°	Colorless	3½-4½ 3.0-3.1	Gray-pink, lt yellow, lt green	Vermiculite	Contact limestones with corundum, diaspore. Mica or chlorite schists	Brittle mica group	p. 282
	b = Y, c ∧ X = small	X = dk. red, Y = lt. red	2½-3 2.95	Bronze, copper red		Mn ore deposits	Mica group	DHZ 3, 44
	a = X, b = Z, c = Y	Colorless	6 3.10	Pink		Pegmatites		Can Min 8, 668
	b = Y, a ∧ X = -22°, c ∧ Z = +45°	Colorless	6 2.77-2.79	Colorless		Evaporite salts	Fluor-U.V. brownish-yellow	Can Min 10, 108-112
None	a = X, b = Y, c = Z	Colorless to pale brown or green. Weak to moderate pleo, Z = Y > X	5½-6 2.85-3.39	White, gray, green to dk brown	Talc or serpentine	Metamorphic rocks	Amphilbole group	p. 223
{100} simple or multiple	b = Y, c ∧ Z = -6° to -9°, a ∧ X = +8° to +5°	X = yellow, Y = violet, Z = blue	6 3.0-3.3	Inky blue to gray	Actinolite	Low T-high P metamorphic rocks (blueschists) with ophiolite rocks	Amphibole group	p. 237
Simple or multiple on {100}	b = Y, c ∧ Z = -15° to -40°	Colorless, pale yellow. Pleochroic yellow to red, Y > Z > X	5-6 2.97-3.23	Yellow, brown to red, green		Skarns, hydrothermal veins in alkaline igneous rocks	Amphibole group	p. 220
Simple or multiple on {100}	b = Y, c ∧ X = +30° to +55°, a ∧ Z = +45° to +70°	Pleo yellow, green to indigo, X > Y > Z	5-6 3.03-3.16	Black to dk green or blue-green	Uralite, limonite, siderite	Alkaline igneous rocks	Amphibole group	p. 243
{101} rare	a = Z, b = Y, c = X	Colorless. X = pink, yellow, Y = colorless, lt green, Z = colorless, olive	6½-7½ 3.13-3.16	Pink, white, tan, bluish, yellow	Sericite. Fine aggregate spinel, corundum & feldspar	Pelitic hornfels with cordierite. Pegmatites, granites		p. 119
	b = X, Y ∧ cleavage = 23°	Colorless	5½ 2.91	White		Marble with kotoite, szaibelyite, spinel, clinohumite, calcite		Am Min 48, 915
	c ∧ X = 74°, c ∧ Y = 32°, c ∧ Z = 62°	Colorless, pinkish, no pleochroism	5½ 3.03-3.04	Pink, pinkish-orange, brown		Post ore veins in Pb-Zn-Mn ore deposits		Am Min 53, 1614-1634

n_β Var.	Refractive Index	Mineral Name and Composition	$(n_\gamma - n_\alpha)$	2V angle Dispersion	Crystal System	Habit	Cleavage
1.639	n_β 1.639 n_α 1.619 n_γ 1.643	Trolleite $Al_5(PO_4)_3(OH)_4$	0.024	49° r > v weak	Monoclinic or Triclinic	Massive, lamellar	Two dir dist at 111°
1.639	n_β 1.639 n_α 1.624 n_γ 1.643	Roscherite $(Ca,Mn,Fe)_3Be_3(PO_4)_3(OH)_3 \cdot 2H_2O$	0.019	55° to large v > r strong, crossed	Monoclinic β = 94.8°	Tabular {100} or prismatic (c), radiated fibrous	{001} good, {010} dist
1.640	n_β 1.640 n_α 1.625 n_γ 1.646	Rosenhahnite $(CaSiO_3)_3 \cdot H_2O$	0.021	~64°	Triclinic α = 108.6° β = 94.8° γ = 95.7°	Tabular to lath-like {010} crystals	{001} perf {100} & {010} distinct
1.641	n_β 1.641 n_α 1.583 n_γ 1.648	Serpierite $Ca(Cu,Zn)_4(SO_4)_2(OH)_6 \cdot 3H_2O$	0.065	37°	Monoclinic β = 113.4°	Minute lath-like crystals (c), tufts	{001} perfect
*1.632 1.650	n_β 1.632-1.650 n_α 1.602-1.624 n_γ 1.632-1.650	Bementite (Caryopilite) (Ectropite) $Mn_8Si_6O_{15}(OH)_{10}$	0.030 to 0.026	Small	Monoclinic	Radiated fibrous (c), platy {001}, massive	{001} & {010} & {100} perf
1.641	n_β 1.641 n_α 1.607 n_γ 1.672	Papagoite $CaCuAlSi_2O_6(OH)_3$	0.065	78° r > v weak	Monoclinic β = 100.6°	Equant crystals, microcrystalline coatings	{100} dist
1.641	n_β 1.641 n_α 1.608 n_γ 1.642	Metakahlerite $Fe(UO_2)_2(AsO_4)_2 \cdot 8H_2O$	0.034	0° - 22°	Tetragonal	Scaly aggregates	{001} perf {100} dist
1.641	n_β 1.641 n_α 1.635 n_γ 1.642	Killalaite $Ca_6(Si_2O_7) \cdot H_2O$	0.007	26°	Monoclinic β = 105°	Prismatic crystals	{100} perf {010} & {001} good
1.642	n_β 1.642 n_α 1.618 n_γ 1.652	Souzalite $(Mg,Fe^{2+})_3(Al,Fe^{3+})_4(PO_4)_4(OH)_6 \cdot 2H_2O$	0.034	68° r > v extreme	Monoclinic ?	Coarse fibrous masses	One good, one fair at ~90°
1.642	n_β 1.642 n_α 1.627 n_γ 1.644	Palermoite $(Li,Na)_2(Sr,Ca)Al_4(PO_4)_4(OH)_4$	0.017	~20°	Orthorhombic	Long prism, fibrous	{100} perfect {001} fair
1.643	n_β 1.643 n_α 1.555 n_γ 1.658	Sibirskite $CaHBO_3$	0.103	43°	Orthorhombic?	Irregular grains, diamond shapes	
1.643	n_β 1.643 n_α 1.559 n_γ 1.655	Hohmannite (Castanite) $Fe_2(OH)_2(SO_4)_2 \cdot 7H_2O$	0.096	40° r > v extreme	Triclinic α = 90.1° β = 90.5° γ = 107°	Granular aggregates, prismatic	{010} perfect {110} & {1$\bar{1}$0} distinct
1.643	n_β 1.643 n_α 1.617 n_γ 1.644	Metakirchheimerite $Co(UO_2)_2(AsO_4)_2 \cdot 8H_2O$	0.027	0° - 20°	Tetragonal	Tabular crystals, crusts	{001} perfect

* n_β variation exceeds 0.01

Twinning	Orientation	Color in thin section	Hardness Sp Gravity	Color hand sample	Alteration	Occurrence	Remarks	Reference
	Optic plane ∥ best cleavage	Colorless-pale green	5½-6 3.09-3.10	Pale green		Iron ores with berlinite and other phosphates	~ insoluble acids	Dana II, 911
	b = X, c ∧ Y = +15°	X = yellow, olive-green, Y = yellow-brown, green, Z = red-brown	4½ 2.93	Olive, lt to dk brown		Cavities in granite & granite pegmatites	Soluble acids. Anomalous interference colors	Am Min 43, 824-838
		Colorless	4½-5 2.89-2.91	Colorless to buff		Veins in diopside-garnet metasedimentary rock		Am Min 52, 336-351
	b = Y, c ~ Z, a ∧ X = +24°	X = pale green, Y = bluish green, Z = blue green	3.07-3.08	Sky blue		Oxide zones ore deposits with smithsonite, cyanotrichite, linarite		Am Min 54, 328-329
	c ~ X	Lt brown, yellow, X < Y = Z	6 2.9-3.1	Brown, yellow, gray		Primary Mn-deposits		Am Min 49, 446-447
	b = Z, c ∧ X = +44°, a ∧ Y = +55°	X ~ colorless, Y = cendre blue, Z = Venice green, Z > Y > X	5-5½ 3.25	Cerulean blue		Veinlets in metasomatic rocks with ajoite		Am Min 45, 599-611
	c = X	O = pale yellow, E ~ colorless	3.84	Sulfur yellow		Secondary U-mineral		Am Min 45, 254
{h0ℓ} complex penetration	b = Y, c ∧ Z = 16°	Colorless		White		Contact limestone with afwillite, tilleyite, spurrite, scawtite, cuspidine		Min Mag 39, 544-548
Polysynthetic ∥ good cleavage	X ⊥ cleavage plates, Z = elongation	X = green, Y = blue, Z = yellow	5½-6 3.087	Dk green		Pegmatites as alteration of scorzalite		Am Min 34, 83
		Colorless	5½ 3.22-3.24	Colorless, white		Cavities in pegmatites with childrenite, beraunite, brazilianite		Am Min 50, 777-779
	Optic plane ~ (100)	Colorless		Colorless to dk gray		Contact limestone deposits with korzhinskite, calciborite, chlorite, pyrite		Am Min 48, 433
	In (010) sections, c ∧ Y' ~ 22°	X = Pale yellow, Y = Pale green-yel, Z = Dk green-yellow	3 2.28	Brown, red-brown	Dehydrates & crumbles to metahohmannite	Oxide zones ore deposits	Easily soluble HCl	Dana II, 613-614
	c = X	Colorless, greenish yellow, not pleochroic	2-2½ <3.33	Pale rose		Secondary U-mineral with pitchblende, erythrite	Meta-torbenite group	Am Min 44, 466

n_β Var.	Refractive Index	Mineral Name and Composition	$(n_\gamma - n_\alpha)$	2V angle Dispersion	Crystal System	Habit	Cleavage
*1.627 1.659	n_β 1.627-1.659 n_α 1.615-1.646 n_γ 1.629-1.662	Wollastonite $CaSiO_3$	0.013 to 0.017	35° to 63° r > v weak	Triclinic α = 90° β = 96° γ = 104°	Columnar, bladed, fibrous, crystals (b)	{100} perfect {001} & {$\bar{1}$02} good
1.643	n_β 1.643 n_α 1.637 n_γ 1.648	Koashvite $Na_6(Ca,Mn)(Ti,Fe)(SiO_3)_6$?	0.011	83° r > v weak	Orthorhombic		None
*1.633 1.654	n_β 1.633-1.654 n_α 1.604-1.626 n_γ 1.642-1.663	Lazulite (Gersbyite) $(Mg,Fe^{2+})Al_2(PO_4)_2(OH)_2$	0.009 to 0.037	70° - 64° v > r perc	Monoclinic β = 120.6°	Pyramidal, tabular {101} or {$\bar{1}$11}, granular	{110} & {101} indistinct
1.644	n_β 1.644 n_α 1.628 n_γ 1.647	Ferrocarpholite $(Fe,Mg)Al_2Si_2O_6(OH)_4$	0.019	49° r > v weak	Orthorhombic	Prismatic (c), fibrous	{010} perfect {110} indist
*1.63 1.66	n_β 1.63-1.66 n_α 1.613-1.645 n_γ 1.645-1.672	Sodium Boltwoodite $H_3O(Na,K)UO_2SiO_4 \cdot H_2O$	~0.030	Large	Orthorhombic	Earthy crusts, platy, radiated fibrous	{010} perfect {001} imperf
1.645	n_β 1.645 n_α 1.640-1.642 n_γ 1.654	Shannonite γ-Ca_2SiO_4	0.015	52° - 60°	Orthorhombic	Fibrous	
1.646	n_β 1.646 n_α 1.551-1.561 n_γ 1.652	Donnayite $NaCaSr_3Y(CO_3)_6 \cdot 3H_2O$	0.101 to 0.111	0° to 20°	Triclinic (pseudo-rhomb)	Pseudo-hex crystals, tab, prism, granular	{001} indist
1.646	n_β 1.646 n_α 1.558-1.553 n_γ 1.648-1.654	Weloganite $Sr_5Zr_2(CO_3)_9 \cdot 4H_2O$	0.090	~15°	Trigonal	Hexagonal prism crystals (c), massive	{0001} perf
1.646	n_β 1.646 n_α 1.575 n_γ 1.650	Szaibelyite (Camsellite) $(Mg,Mn)BO_3H$	0.075	~25° r > v	Orthorhombic	Fibrous veinlets, nodules, earthy	
1.646	n_β 1.646 n_α 1.639 n_γ 1.646	Kellyite $Mn_4Al_2(Si_2Al_2)O_{10}(OH)_8$	0.007	16° - 30° r > v moderate	Hexagonal-trigonal	Irregular grains, micaceous	{001} perf
*1.628 1.666	n_β 1.628-1.666 n_α 1.602 n_γ 1.632-1.670	Switzerite $(Mn,Fe)_3(PO_4)_2 \cdot 4H_2O$	0.030	42° v > r slight	Monoclinic β = 95.9°	Flaky-bladed crystals {100}, elongation (c)	{100} perf {010} fair
1.644 1.651	n_β 1.644-1.651 n_α 1.630-1.638 n_γ 1.652-1.665	Tirodite $(Mg,Mn,Fe)_7(Si_4O_{11})_2(OH)_2$	0.022 to 0.027	74° - 90°	Monoclinic β = 103.9°	Prism-acicular crystals (c)	{110} perf {100} imperf {001} parting
1.647	n_β 1.647 n_α 1.637 n_γ 1.648	Nimite (Schuchardite) $(Ni,Mg,Fe,Al)_6(AlSi_3)O_{10}(OH)_8$	~0.010	~15°	Monoclinic β = 97.2°	Massive, minute veins	{001} dist
*1.622 1.676	n_β 1.622-1.676 n_α 1.610-1.667 n_γ 1.632-1.684	Gedrite $(Mg,Fe^{2+})_5Al_2(Si_3AlO_{11})_2(OH)_2$	0.022 to 0.016	75° - 90° r > v or v > r weak to moderate	Orthorhombic	Fibrous, bladed, acicular	{110} perf

* n_β variation exceeds 0.01

Twinning	Orientation	Color in thin section	Hardness Sp Gravity	Color hand sample	Alteration	Occurrence	Remarks	Reference
{100} common	b ∧ Y ~ 5° Optic plane ~ (010). Elong ± (usually +)	Colorless	4½-5 2.9-3.1	White, gray, lt green	Calcite, pectolite	Limestone contact zones with Ca-Mg-silicates. Alkaline igneous rocks	Pyroxenoid group	p. 207
		Colorless, pale yellow	6	Pale yellow		Veinlets replacing lomonosovite in alkali pegmatite		Am Min 60, 487
{100} contact, polysynthetic	b = Y, c ∧ X = 9½°	X = colorless, Y = blue, Z = dark blue	5½-6 3.08-3.14	Dk to lt azure blue, blue-grn		Quartz veins, granite pegmatites, high grade quartzites	Lazulite-scorzalite series	Dana II, 908-911
	a = Y, b = X, c = Z	X = Y = yellow-grn, Z = pale blue-grn	5½ 3.04	Dk green		Quartz veins with rutile, zircon, tourmaline		Am Min 36, 736-745
		X = colorless, Z = pale yellow	~4 ~3.6	Pale yellow		Secondary U-mineral, near surface in arid regions	Anomalous interference colors	Am Min 61, 1054-1055
Fine polysynthetic	c = Y	Colorless	5-6 2.97-3.0	Colorless, white		Limestone contact zones	Dimorph larnite	Am Min 51, 1766-1774
{010}, {30$\bar{1}$}, {$\bar{3}\bar{3}$1}	c ~ X	Colorless, ~ opaque	3 3.27-3.30	Colorless, yellow, gray		Cavities in pegmatites, nepheline syenite with ewaldite	Soluble with effervescence in 1:1 HCl	Can Min 16, 335-340
	c = X	Colorless	3½ 3.16-3.22	White, yellow, amber		Alkali sill with quartz, dawsonite, dresserite, siderite, barite, fluorite	May appear uniaxial	Can Am. 9, 468-477
	Parallel ext., - elongation	Colorless	3-3½ 2.60	White, straw yellow		Veinlets or coatings in Fe-ore or serpentine	Slowly soluble acids	Dana II, 375-377
Sector (?)	X ⊥ (001)	X = colorless to greenish-yellow, Y = Z = pale yellow to red-brown	~3 3.07-3.11	Golden yellow		Carbonate Mn-deposits with alleghanyite, galaxite, kutnahorite	Serpentine group	Am Min 59, 1153-1156
	b = Y, c ∧ Z = 10°	Pale yellow-brown to dk red-brown	~2½ 2.95-3.18	Pink, golden-brown, chocolate		Pegmatites with spodumene, vivianite	I.R., pleochroism and dispersion increase with oxidation	Am Min 52, 1595-1602
{100} common, may be multiple	b = Y, a ~ X, c ∧ Z = 16°-22°	X = Z = colorless, Y = yellowish	6-6½ 3.07-3.13	Pale green to rose red, tan		Mn ore deposits with anthophyllite	Amphibole group. Mn-cummingtonite	Am Min 49, 963-982
	b = Y, X ~ ⊥{001}	X = yellow-green, Y = Z = apple green	3 3.19-3.20	Yellowish-green		Veinlets in talc in Ni-rocks with ultramafic rocks	Chlorite group	Am Min 54, 1739-1740
None	a = X, b = Y, c = Z	Colorless, pale brn to green, weak to moderate pleochroic, Z = Y > X	5½-6 2.85-3.30	White, gray, yellow, brn, dk brown	Talc or serpentine	Metamorphic rocks	Amphibole group	p. 223

Biaxial –

n_β Var.	Refractive Index	Mineral Name and Composition	$(n_\gamma - n_\alpha)$	2V angle Dispersion	Crystal System	Habit	Cleavage
*1.605 1.696	n_β 1.605–1.696 n_α 1.565–1.625 n_γ 1.605–1.696	Biotite $K(Mg,Fe^{2+})_{6-5}Al_{0-1}(Si_{6-5}Al_{2-3})O_{20}(OH,F)_4$	0.04 to 0.07	0° – 25° v > r	Monoclinic β = 90°–100°, Trigonal	Micaceous, pseudohexagonal crystals	{001} perf
1.650	n_β 1.650 n_α 1.588 n_γ 1.722	Krausite $KFe^{3+}(SO_4)_2 \cdot H_2O$	0.134	Large ~ 90°	Monoclinic β = 102.75°	Short prism, tabular	{001} perf {100} dist
1.649 1.652	n_β 1.652–1.649 n_α 1.62 –1.585 n_γ 1.656–1.660	Devillite (Devilline) $CaCu_4(SO_4)_2(OH)_6 \cdot 3H_2O$	0.036 to 0.075	37° – 39° v > r moderate	Monoclinic β = 105.5°	"Hex" plates, rosettes	{001} perf {110}, {101} &{10$\bar{1}$} dist
1.651 1.653	n_β 1.651–1.653 n_α 1.590–1.594 n_γ 1.657–1.660	Borcarite $Ca_4MgH_6(BO_3)_4(CO_3)_2$	0.067	~30° v > r perc	Monoclinic β = 92.5°	Massive	{100} & {110} perf, {hkℓ} & {h0ℓ}
1.649 1.654	n_β 1.649–1.654 n_α 1.622–1.626 n_γ 1.666–1.670	Datolite $CaB(SiO_4)(OH)$	0.044 to 0.046	72° – 75° r > v weak	Monoclinic β ~90°	Blocky crystals	None
*1.642 1.660	n_β 1.642–1.660 n_α 1.622–1.642 n_γ 1.646–1.666	Holmquistite $Li_2(Mg,Fe^{2+})_3(Al,Fe^{3+})_2(Si_4O_{11})_2(OH)_2$	0.024	49° r > v slight	Orthorhombic	Prism, acicular, radiated fibrous aggregates	{110} perf {001}, {112} {113} parting
1.651	n_β 1.651 n_α 1.640 n_γ 1.653	Jeremejevite (Eremeyevite) $Al_6B_5O_{15}(OH)_3$	0.013	Small to ~ 50°	Hexagonal- Monoclinic ?	Elongated hexagonal prisms (c)	None
*1.645 1.658	n_β 1.645–1.658 n_α 1.637–1.652 n_γ 1.649–1.663	Hureaulite $H_2(Mn,Fe)_5(PO_4)_4 \cdot 4H_2O$	0.012 to 0.011	75° v > r strong, crossed strong	Monoclinic β = 96.6°	Prism, tabular, massive, scaly	{100} dist
1.652	n_β 1.652 n_α 1.600 n_γ 1.655	Willemseite $(Ni,Mg)_3Si_4O_{10}(OH)_2$	0.055	~27°	Monoclinic β = 100°	Massive fine-grained	{001} perfect
1.652	n_β 1.652 n_α 1.612 n_γ 1.675	Liroconite $Cu_2Al(As,P)O_4(OH)_4 \cdot 4H_2O$	0.063	~72° v > r moderate	Monoclinic β = 91.4°	Flattened, wedge-shaped crystals	{110} & {011} indistinct
1.652	n_β 1.652 n_α 1.618 n_γ 1.683	Schoonerite $ZnMnFe_2^{2+}Fe^{3+}(OH)_2(H_2O)_7(PO_4)_3 \cdot 2H_2O$	0.065	70° – 80°	Orthorhombic	Rosettes, scales, laths, tabular {010}	{010} perf, {001} good
1.653	n_β 1.653 n_α 1.559 n_γ 1.680	Callaghanite $Cu_2Mg_2(CO_3)(OH)_6 \cdot 2H_2O$	0.121	~55° r > v strong	Monoclinic β = 107.3°	Tiny pyramidal crystals	{111} & {$\bar{1}$11} perfect
*1.637 1.670	n_β 1.637–1.670 n_α 1.581–1.582 n_γ 1.637–1.670	Ekmanite $(Fe,Mn,Mg)_6(Si,Al)_8O_{20}(OH)_8 \cdot 2H_2O$	0.088 to 0.056	~0°	Orthorhombic	Foliated, columnar, scaly, massive	{001} perfect

* n_β variation exceeds 0.01

Twinning	Orientation	Color in thin section	Hardness Sp Gravity	Color hand sample	Alteration	Occurrence	Remarks	Reference
{110} normal, {001} complex	b = Y, c ∧ X = -10° to +9°, a ∧ Z = 0° to +9°	X = pale yellow, pale green, pale brown, Y ~ Z = dk brown, dk green, dk red-brown	2½-3 2.7-3.3	Black to dk brown or grn	Chlorite, vermiculite, hydrobiotite	Granitic plutons, silicic volcanics, schists and gneisses	Mica group	p. 274
	b = Z, c ∧ Y = +35°	X = colorless, Y = Z = pale yel	2½ 2.84	Pale lemon yellow		Oxide zones ore-deposits with voltaite, coquimbite, halo-trichite	Slowly decomp. H_2O. Slowly soluble HCl	Am Min 50, 1929-1936
{010}	b = Z, c ~ X	X ~ colorless, Y = Venice-green, Z = blue-green	2½ 3.31	Dk green, bluish-green		Oxide zones Cu-deposits	Soluble HNO_3	Dana II, 590-592
	b = Z, c ∧ Y = 28°	Colorless, greenish blue	4 2.77	Green-blue, colorless		Contact metasomatic rocks		Am Min 50, 2097
None	b = Y, c ∧ Z = +1° to +4°	Colorless	5-5½ 2.9-3.0	Colorless, white, pale yellow	Very stable	Amygdules and veins in basalt with zeolites. Skarns, granite, schists		p. 135
	b = Y, c ∧ Z = 0°-4°	Colorless, bluish, violet tints Z > Y > X	5-6 3.09-3.13	Dk violet to pale blue		Contact Li-pegmatites and basic country rock	Amphibole group	Vlasov II, 13-17
Sector	c ~ X	Colorless	6½ 3.27-3.28	Colorless, pale yellow-brown		Granite plutons with quartz and orthoclase	Hexagonal crystals biaxial cores	Dana II, 330-332
	b = X, c ∧ Z = 75°	X = colorless, Y = yellow, pale rose, Z = orange, red-brn	3½ 3.19-3.23	Rose, red, orange, yel, brown		Secondary phosphate in granite pegmatites	Easily soluble acids	Dana II, 700-702
		Colorless	2 3.31-3.35	Lt green		Ultramafic contact zone with trevorite, nimite, violarite, millerite	Ni-talc	Am Min 55, 31-42
	b = Y, a ∧ Z = -25°	Pale blue, lt bluish-green, not pleochroic	2-2½ 2.92-3.01	Sky blue, verdigris grn		Oxide zones Cu-deposits	Soluble acids	Dana II, 921-922
	a = Z, b = X, c = Y	X = pale yellow, Y = pale brown, Z = brown	~4 2.87-2.92	Brown, reddish-brown		Pegmatites with siderite, mitridatite, jahnsite, witmoreite	Rapidly soluble cold 1:1 HCl	Am Min 62, 246-249
	c ∧ Z = 18°	Blue Z > Y > X	3-3½ 2.71-2.78	Azure blue		Encrustations and veins in magnesite and dolomite beds		Am Min 39, 630-635
	c = X	X = colorless, Y = Z = pale green, dk greenish brown	2-2½ 2.79	Gray to black, greenish		Magnetite ores, with pyrochroite		Am Min 39, 946-956

Biaxial –

n_β Var.	Refractive Index	Mineral Name and Composition	$(n_\gamma - n_\alpha)$	2V angle Dispersion	Crystal System	Habit	Cleavage
*1.627 1.653	n_β 1.627–1.653 n_α 1.619–1.643 n_γ 1.629–1.653	Fedorovskite $Ca_2(Mg,Mn)_2(OH)_4[B_4O_7(OH)_2]$	~0.010	40° v > r strong	Orthorhombic	Fibrous	{100} perfect
1.654	n_β 1.654 n_α 1.567 n_γ 1.722	Tatarskite $Ca_6Mg_2(SO_4)_2(CO_3)_2Cl_4(OH)_4 \cdot 7H_2O$	0.155	83°	Orthorhombic ?	Massive, coarse crystalline	Two pina-coidal
*1.629 1.680	n_β 1.629–1.680 n_α 1.616–1.669 n_γ 1.640–1.688	Actinolite $Ca_2(Mg,Fe^{2+})_5Si_8O_{22}(OH)_2$	0.024 to 0.019	84° – 75° v > r weak	Monoclinic β = 105°	Fibrous, bladed	{110} dist
1.65	n_β 1.65 n_α 1.645 n_γ 1.656	Koninckite $FePO_4 \cdot 3H_2O$	0.011	Small	Orthorhombic (?)	Radiated acicular spherical aggregate	One trans-verse to elongation
*1.64 1.67	n_β 1.64–1.67 n_α 1.62–1.65 n_γ 1.65–1.67	Brunsvigite $(Fe^{2+},Mg,Al)_6(Si,Al)_4O_{10}(OH)_8$	0.006 to 0.015	0° – 25° r > v	Monoclinic β ~97°	Massive, scaly, foli-ated, hexagonal tabulated crystals	{001} perfect
*1.64 1.67	n_β 1.64–1.67 n_α 1.64–1.66 n_γ 1.65–1.67	Chamosite $(Fe^{2+},Mg,Al,Fe^{3+})_6(Si,Al)_4O_{10} \cdot (OH)_8$	0.002 to 0.010	Small r > v	Monoclinic-Hexagonal β = 104.5°	Massive, scaly, oolitic	{001} perfect
1.654 1.656	n_β 1.654–1.656 n_α 1.650 n_γ 1.661	Vladimirite $Ca_5H_2(AsO_4)_4 \cdot 5H_2O$	0.011	70° r > v strong	Monoclinic β = 97.3°	Acicular radiated aggregates	One good
1.653 1.658	n_β 1.653–1.658 n_α 1.612–1.63 n_γ 1.660–1.681	Stewartite $MnFe_2(OH)_2(PO_4)_2 \cdot 8H_2O$	0.048 to 0.051	Large-moderate v > r strong	Triclinic α = 90.6° β = 110° γ = 71.4°	Minute crystals, fibrous tufts	One distinct
1.657	n_β 1.657 n_α 1.569 n_γ 1.686	Calkinsite $(La,Ce)_2(CO_3)_3 \cdot 4H_2O$	0.117	54° – 57° v > r perc	Orthorhombic	Platy {010}	{010} perf, {101} dist {001} parting
* 1.648 1.666	n_β 1.648–1.666 n_α 1.628–1.639 n_γ 1.657–1.674	Eosphorite $(Mn,Fe)AlPO_4(OH)_2 \cdot H_2O$	0.029 to 0.035	57° – 67° v > r strong	Orthorhombic	Pyramidal,tabular, platy	{100} poor
1.658	n_β 1.658 n_α 1.640 n_γ 1.670	Jahnsite $CaMn^{2+}Mg_2(H_2O)_8Fe_2^{3+}(OH)_2(PO_4)_4$	0.030	Large ~ 78°	Monoclinic β = 110.2°	Prism (b), tabular {100}	{001} dist
1.655 1.662	n_β 1.655–1.662 n_α 1.643–1.648 n_γ 1.655–1.663	Clintonite (Seybertite) (Brandisite) $Ca_2(Mg_{4.6}Al_{1.4})(Si_{2.5}Al_{5.5})O_{20}(OH)_4$	0.012	~32° v > r weak	Monoclinic β = 100.3°	Micaceous, radiated, massive	{001} perfect
1.655 1.662	n_β 1.655–1.662 n_α 1.643–1.648 n_γ 1.655–1.663	Xanthophyllite $Ca_2(Mg_{4.6}Al_{1.4})(Si_{2.5}Al_{5.5})O_{20}(OH)_4$	0.012	0° – 23° v > r weak, inclined	Monoclinic β = 100.3°	Micaceous, radiated, massive	{001} perfect
1.658	n_β 1.658 n_α 1.652 n_γ 1.665	Hiortdahlite $(Ca,Na)_{13}Zr_3Si_9(O,OH,F)_{33}$	0.013	~90° v > r strong	Triclinic α = 90.5° β = 109° γ = 90°	Tabular {100}, pseudotetragonal	{110} & {1$\bar{1}$0} distinct (~90°)

* n_β variation exceeds 0.01

Twinning	Orientation	Color in thin section	Hardness Sp Gravity	Color hand sample	Alteration	Occurrence	Remarks	Reference
{100} poly-synthetic	a = Z, b = Y, c = X	X = colorless, pale yellow, Z = yellow	4½ 2.65-2.90	Brown		Veins in garnet-idocrase-svabite skarns with datolite	Soluble dilute HCl	Am Min 62, 173
	∥ extinction, + elongation	Colorless	2½ 2.341	Colorless, yellowish		Anhydrite evaporite deposits with bischofite, hilgardite, halite		Am Min 49, 1151-1152
Simple or multiple on {100}	b = Y, c ∧ Z = -17° to -13°, a ∧ X = -2° to +2°	Pale yellow green, pleo in yellow and green, Z > Y ≥ X	5-6 3.1-3.4	Bright green to dk green	Talc, chlorite, calcite-dolomite	Metamorphic rocks	Amphibole group	p. 229
	- elongation	Pale yellow, colorless	3½ 2.3	Yellow		With richellite, halloysite, allophane	Easily soluble hot HCl, HNO_3	Dana II, 763
{001}	b = Y, Z ∧ {001} = small	X = pale green-brn Y ~ Z = very dk green	~2 2.99-3.08	Olive green, dk green		Hydrothermal veins. Veins and voids in granite, gabbro, spilite	Chlorite group	p. 292
	b = Y, Z ∧ {001} small	X = dk green-brown Y ~ Z = brownish-green	2-3 3.03-3.4	Dk olive green, dk green		Sedimentary iron-stones or laterites with siderite, clays, Fe-oxides	Chlorite group	p. 292
	c ∧ Z = 37°	Colorless, pale rose	3½ 3.14-3.17	Pale rose		Oxide zone ore deposits		Am Min 50, 813
	Inclined extinction	X = colorless, Y = pale yellow, Z = yellow	2.94	Yellow, brownish-yel		Granite pegmatites with rockbridgeite, metastrengite, triphylite		Am Min 48, 913-914
{010}	a = Z, b = X, c = Y	X = pale yellow, Y = Z = colorless	~2½ 3.28	Pale yellow		Hydrothermal veins		Am Min 38, 1169-1183
{100} and {001} contact	a = Y, b = X, c = Z	X = Yellow, Y = pink, Z = colorless	5 3.06	Pink, rose-red		Granite pegmatites, hydrothermal veins	Childrenite-eosphorite series Soluble acids	Dana II, 936
{001} simple	b = Z, c ∧ X ~ 18°	X = pale violet, Y = dk violet-brn, Z = yellow	4 2.71	Brown, yellow, yel-orange, yellow-green		Granite pegmatites with secondary phosphates	Soluble dilute HCl	Am Min 59, 48-59
{001} complex [310] twin axis	b = Z, a ~ Y, c ∧ X ~ -5°	X = colorless, pale orange, red-brown, Y = Z = pale green, brown	3½-6 3.1	Colorless, yellow, copper-red	Vermiculite	Contact limestones with idocrase, phlogopite, grossular. Schists	Brittle mica group	p. 282
{100} complex [310] twin axis	b = Y, a ~ Z, c ∧ X ~ -10°	X = colorless, pale orange, red-brown, Y = Z = pale green, brown	3½-6 3.1	Colorless, yellow, copper-red	Vermiculite	Contact limestones with idocrase, phlogopite, grossular. Chlorite schist	Brittle mica group	p. 282
{100} polysynthetic	Inclined extinctions	X = colorless, Y = yellowish, Z = wine-yellow	5½ 3.25-3.27	Pale yellow, yellow-brown		Alkali igneous rocks and pegmatites with nepheline, aegirine	Gelat acids	Vlasov II, 379-381

Biaxial —

n_β Var.	Refractive Index	Mineral Name and Composition	$(n_\gamma - n_\alpha)$	2V angle Dispersion	Crystal System	Habit	Cleavage
1.658	n_β 1.658 n_α <1.612 n_γ 1.682	Laueite $MnFe_2(PO_4)_2(OH)_2 \cdot 8H_2O$	> 0.070	50°	Triclinic α = 107.9° β = 111° γ = 71.1°	Wedge-shaped crystals	{010} perf
1.659	n_β 1.659 n_α 1.651 n_γ 1.661	Bityite (Bowleyite) $Ca(Al,Li)_2[(Al,Be)_2Si_2(O,OH)_{10}] \cdot H_2O$	0.010	35° - 52°	Monoclinic β ~90°	Radiated, tabular, micaceous aggregates	{001} perf
1.660	n_β 1.660 n_α 1.628 n_γ 1.705	Parasymplesite $Fe_3(AsO_4)_2 \cdot 8H_2O$	0.077	Large	Monoclinic β = 103.8°	Prism crystals	{010} very perfect
**1.576 1.745	n_β 1.576-1.745 n_α 1.543-1.634 n_γ 1.576-1.745	Stilpnomelane $(Fe^{3+},Fe^{2+},Mg,Mn,Al)_{5-6}(Si_4O_{10})_2(OH)_4$	0.030 to 0.110	~0°	Triclinic α = 124° β = 96° γ = 120°	Micaceous, radiated	{001} perfect
1.658 1.662	n_β 1.658-1.662 n_α 1.638-1.640 n_γ 1.664-1.667	Väyrynenite $BeMnPO_4(OH,F)$	0.026 to 0.027	54°	Monoclinic β = 102.8°	Prism crystals, drusy, massive, granular	{001} dist
1.660	n_β 1.660 n_α 1.640 n_γ 1.675	Tilasite $CaMgAsO_4F$	0.035	83° v > r slight	Monoclinic β = 121°	Elongated-bladed {010} crystals, massive	{10$\bar{1}$} dist {1$\bar{3}$3}, {10$\bar{2}$} &{0$\bar{1}$1} parting
1.659 1.661	n_β 1.659-1.661 n_α 1.644-1.645 n_γ 1.687-1.688	Leucosphenite $BaNa_4Ti_2B_2Si_{10}O_{30}$	0.043	75° - 77° r > v distinct	Monoclinic β = 93.4°	Tabular or wedge-shaped crystals	{010} dist {001} fair
1.660	n_β 1.660 n_α 1.648 n_γ 1.663	Roweite $Ca_2Mn_2^{2+}(OH)_4[B_4O_7(OH)_2]$	0.015	15° v > r strong	Orthorhombic	Lath-like crystals (c) flattened {010}	{101} poor
*1.646 1.674	n_β 1.646-1.674 n_α 1.639-1.663 n_γ 1.653-1.680	Monticellite $CaMgSiO_4$	0.014 to 0.020	90° - 70° r > v distinct	Orthorhombic	Granular, anhedral crystals	{010} poor
1.658 1.663	n_β 1.658-1.663 n_α 1.622-1.629 n_γ 1.681-1.701	Erythrite $Co_3(AsO_4)_2 \cdot 8H_2O$	0.059 to 0.072	~90° r > v	Monoclinic β = 105°	Prism-acicular, bladed, fibrous, earthy	{010} perfect {100} & {$\bar{1}$02} indistinct
*1.654 1.670	n_β 1.654-1.670 n_α 1.626-1.639 n_γ 1.663-1.680	Scorzalite $(Fe^{2+},Mg)Al_2(PO_4)_2(OH)_2$	0.037 to 0.041	64° - 58° v > r perc	Monoclinic β = 119°	Massive, granular	{110} & {101} distinct
1.664	n_β 1.664 n_α 1.646 n_γ 1.664	Gonyerite $(Mn,Mg,Fe)_6Si_4O_{10}(OH)_8$	0.018	~0	Orthorhombic	Radiated aggregates pseudo-hexagonal scales	{001} perfect
1.661 1.667	n_β 1.661-1.667 n_α 1.642-1.648 n_γ 1.667-1.675	Uranophane (Lambertite) $Ca(UO_2)_2Si_2O_7 \cdot 6H_2O$	0.025 to 0.027	32° - 45° v > r strong	Monoclinic β = 97.3°	Prism-acicular-fibrous, radiated, earthy coatings	{100} perfect

* n_β variation exceeds 0.01

** n_β variation exceeds 0.10

Twinning	Orientation	Color in thin section	Hardness Sp Gravity	Color hand sample	Alteration	Occurrence	Remarks	Reference
		Yellow	3 2.44-2.56	Honey-brown, yellow-orange		Pegmatites with rock-bridgeite, strunzite, stewartite, siderite	Dimorph strunzite	Am Min 54, 1312-1323
Sector, lamellar	Ext 30°	Colorless	5½ 3.02-3.14	White, yellow brownish		Li-pegmatites		Am Min 35, 1091
	$c \wedge Z = 31°$	X = bluish green, Y = yellowish, Z = brownish yellow	2 3.07-3.10	Lt greenish-blue			Dimorph. symplesite	Am Min 40, 368
	b = Y, a ~ Z, $c \wedge X \sim -7°$	X = colorless, yel, Y = Z = black, dk brown, dk green	3-4 2.59-2.96	Black, red-brown, deep green	Iron oxides and clays	Low grade or low T-high P metamorphic rocks		p. 305
		Colorless, pale pink	5 3.18-3.23	Pale pink to rose red		Li-pegmatites with herderite, hurlbutite, beryllonite, morinite		Am Min 41, 371
{001} common, sym contact	b = Z, $c \wedge X \sim 30°$	Colorless, pale green	5 3.77-3.78	Violet-gray, green, olive green		Veinlets in crystal-line limestone with braunite, barite, conichalcite	Readily soluble HCl, HNO_3	Dana II, 827-829
{001} common	b = Z, a ~ X, $c \wedge Y = +3°$ to $-9°$	Colorless	6-6½ 3.05-3.10	Colorless, white, pale blue		Clay deposits with shortite, analcite		Am Min 57, 1801-1822
	a = X, b = Z, c = Y	Colorless	~5 2.93	Light brown		Oxide ores Zn with willemite, thomsonite	Easily soluble dilute HCl	Dana II, 377-378
{031} cyclic	a = Z, b = X, c = Y	Colorless	5½ 3.1-3.3	Colorless to gray	Serpentine and augite	High T dolomite mar-bles. Peridotites, carbonatites	Olivine group	p. 110
	b = X, $c \wedge Z$ = $-30°$ to $-36°$	X = pale pink, Y = pale violet, Z = red	1½-2½ 3.18	Violet-red, pink		Oxide zones Co-deposits	Erythrite-anna-bergite series. Sectile	Dana II, 746-750
{100} contact, polysynthetic	b = Y, $c \wedge X \sim 10°$	X = colorless, Y = blue, Z = dark blue	5½-6 3.38-3.39	Dk azure blue, blue-green		Granite pegmatites with apatite, brazilianite, tourmaline	Lazulite-scorzalite series	Dana II, 908-911
	c = X	X = dk brown Z = Y = lt brown	2½ 3.01-3.03	Deep brown		Hydrothermal veins with Mn-minerals	Chlorite group	Am Min 40, 1090-1094
	b = Z	X = colorless, Y = pale yellow, Z = yellow	2½ 3.83-3.85	Yellow, yel-green, yel-orange		Alter uraninite in pegmatites, oxide zones of veins	Fluor U.V.-weak yellowish-green	USGS Bull 1064, 294-300

n_β Var.	Refractive Index	Mineral Name and Composition	$(n_\gamma - n_\alpha)$	2V angle Dispersion	Crystal System	Habit	Cleavage
1.664	n_β 1.664 n_α 1.516 n_γ 1.666	Strontianite $SrCO_3$	0.150	~7° v > r weak	Orthorhombic	Pseudohexagonal prism-acicular-fibrous crystals	{110} dist {021} & {010} poor
*1.652 1.677	n_β 1.652-1.677 n_α 1.638-1.672 n_γ 1.654-1.684	Magnesioarfvedsonite $Na_3(Mg,Fe^{2+})_4(Fe^{3+},Al)Si_8O_{22}(OH)_2$	0.014 to 0.018	0° - 80° r > v or v > r	Monoclinic β 105°	Prismatic	{110} perf
1.665	n_β 1.665 n_α 1.638 n_γ 1.676	Kinoite $Cu_2Ca_2Si_3O_{10}\cdot 2H_2O$	0.038	68° v > r distinct	Monoclinic β = 96.1°	Tabular {100}, elongation (c)	{010} perfect {100} & {001} distinct
1.664 1.667	n_β 1.664-1.667 n_α 1.654-1.655 n_γ 1.664-1.667	Cuprosklodowskite (Jachimovite) $Cu(UO_2)_2(SiO_3)_2(OH)_2\cdot 5H_2O$	0.010 to 0.012	Very small r > v	Triclinic α = 90° β = 110° γ = 108.5°	Acicular (c), radiated fibrous	{100} dist
*1.64 1.68	n_β 1.64-1.68 n_α 1.63-1.67 n_γ 1.64-1.68	Daphnite $(Fe^{2+},Mg,Al)_6(Si,Al)_4O_{10}(OH)_8$	0.000 to 0.006	15° - 25° r > v	Monoclinic β = 94.1°	Massive, foliated, botryoidal	{001} perf
1.666	n_β 1.666 n_α 1.649 n_γ 1.676	Upalite $Al(UO_2)_3(PO_4)_2(OH)_3$	0.027	74°	Orthorhombic	Acicular	
*1.618 1.714	n_β 1.618-1.714 n_α 1.610-1.700 n_γ 1.630-1.730	Hornblende $(Ca,Na,K)_{2-3}(Mg,Fe^{2+},Fe^{3+},Al)_5[Si_6(Si,Al)_2O_{22}](OH)_2$	0.015 to 0.034	35° - 90° v > r or r > v moderate	Monoclinic β = 105°	Prismatic	Good {110}
**1.618 1.714	n_β 1.618-1.714 n_α 1.615-1.705 n_γ 1.632-1.730	Edenite-Ferroedenite $NaCa_2(Mg,Fe^{2+})_5Si_7AlO_{22}(OH,F)_2$	0.016 to 0.023	90°-27° r $\gtrless$ v weak	Monoclinic β ~ 105°	Short prism. xls	{110} perf., {001} or {100} parting
1.667	n_β 1.667 n_α 1.658 n_γ 1.667	Grovesite $(Mn,Fe,Al)_{13}(Al,Si)_8O_{22}(OH)_{14}$	0.009	Small	Orthorhombic	Small rosettes, micaceous	{001} perfect
1.667	n_β 1.667 n_α 1.662 n_γ 1.669	Clinohedrite $CaZnSiO_3(OH)_2$	0.007	Large r > v perc	Monoclinic β = 104°	Prismatic, tabular, massive	{010} perfect
1.668	n_β 1.668 n_α 1.635 n_γ 1.702	Symplesite $Fe_3(AsO_4)_2\cdot 8H_2O$	0.067	~86° r > v strong	Triclinic α = 99.9° β = 97.6° γ = 106°	Sperulitic radiated-fibrous, earthy	{1$\bar{1}$0} perfect
1.667 1.669	n_β 1.667-1.669 n_α 1.662-1.664 n_γ 1.672-1.675	Arrojadite $(Na,Ca)_2(Fe^{2+},Mn^{2+})_5(PO_4)_4$	0.010	71° - 86° v > r	Monoclinic β = 93.6°	Cleavable masses	{001} dist, {201} poor
1.665 1.674	n_β 1.665-1.674 n_α 1.593-1.604 n_γ 1.741-1.731	Butlerite $Fe^{3+}SO_4(OH)\cdot 2H_2O$	0.148 to 0.127	Large	Monoclinic β = 108.5°	Thick tabular, octahedral	{100} perfect

* n_β variation exceeds 0.01

** n_β variation exceeds 0.10

Twinning	Orientation	Color in thin section	Hardness Sp Gravity	Color hand sample	Alteration	Occurrence	Remarks	Reference
{110} simple, cyclic, lamellar	a = Z, b = Y, c = X	Colorless	3½ 3.75	White, gray, yellow, brown, green	Celestite	Low T hydrothermal veins in carbonates & clays with celestite, barite	Easily soluble cold acids with effervescence	p. 68
Simple or multiple on {100}	b = Z, c ∧ X = +5° to +30°, a ∧ Y = +20° to +45°	Pleochroic yellow, green to indigo, X > Y > Z	5-6 3.16-3.30	Black to dk green or blue-green	Uralite, limonite, siderite	Alkaline igneous rocks	Amphibole group	p. 243
	b = X, c ∧ Z ~ 0°	X = pale greenish-blue, Y = blue, Z = deep blue	~5 3.16-3.19	Deep azure blue		Contact Cu-deposits with apophyllite, Cu-sulfides		Am Min 55, 709-715
	a = X, b = Z, c = Y	X = colorless, Y = Z = pale yellow-green	4 3.83-3.85	Green, yellow-green		Oxide zones U-deposits	Not fluorescent. Abnormal interference colors	USGS Bull 1064, 304-307
{001}	b = Y, Z ∧ {001} ~ 2°	X = pale green-brn, Y = Z = dk green	2-3 3.20	Dk green		Ore veins with quartz, arsenopyrite	Chlorite group	p. 292
	a = Y, b = X, c = Z	X = colorless, Y = Z = canary yellow		Amber-yellow		Pegmatite		Am Min 65, 208
Simple on {100}	b = Y, c ∧ Z = -12° to 34°, a ∧ X = +3° to -19°	Green, blue-green, brown, strong pleochroism	5-6 3.02-3.45	Dk green to black	Chlorite with calcite and epidote	Ubiquitous	Amphibole group	p. 232
{100} common	b = Y, c ∧ Z ≈ -19°, a ∧ X ≈ -4°	X = yellow, Y = green, Z = deep green	5-6 3.0-3.1	White, lt. to dk. green	Uralite	Many varieties of igneous and metamorphic rocks	Amphibole group. Hornblende series	DHZ II, 263-314
	c = X, Y & Z ∥ cleavage	X = red-brown, Y = Z = dk brown	~2-3 3.15	Blackish-brn		Fe-ore deposits	Septechlorite group	Am Min 41, 164
	b = Z, c ∧ Y = -28°	Colorless	5½ 3.25-3.34	Colorless, white, violet		Contact metamorphic zones		USGS P.P. 180, 106-108
	X ⊥ (1$\bar{1}$0), c ∧ Z = 3½°	X = deep blue, Y ~ colorless, Z = yellowish	~2½ 3.01-3.02	Lt to dk grn or blue (indigo)		Secondary mineral in pegmatite cavities or ore veins	Dimorph parasymplesite. Soluble acids	Dana II, 752-753
	b = X, c ∧ Y = -21°	X = colorless, Y = pale green, Z = pale yellow-grn	5 3.55	Dark green		Granite pegmatites	Soluble in dilute acids. Fe analogue of dickinsonite	Dana II, 679-681
{$\bar{1}$05} common	b = Z, c ∧ X = +18°	X = colorless, Y ~ Z = Pale yel	2½ 2.55-2.60	Orange		Oxide zones ore deposits. Fumeroles	Dimorph parabutlerite	Dana II, 608-609

n_β Var.	Refractive Index	Mineral Name and Composition	$(n_\gamma - n_\alpha)$	2V angle Dispersion	Crystal System	Habit	Cleavage
*1.658 1.680	n_β 1.658-1.680 n_α 1.639-1.670 n_γ 1.660-1.685	Magnesiokatophorite $Na_2CaMg_4Fe^{3+}Si_7AlO_{22}(OH)_2$	0.021 to 0.015	55° - 40° v > r strong	Monoclinic β = 105°	Prismatic	{110} perfect
*1.64 1.70	n_β 1.64-1.70 n_α n_γ 1.64-1.70	Hendricksite $K(Zn,Mn)_3(Si_3Al)O_{10}(OH)_2$		2° - 5°	Monoclinic β = 99°	Micaceous, tabular crystals	{001} perf
1.670	n_β 1.670 n_α 1.616 n_γ 1.670	Siderophyllite $K_2(Fe^{2+}_5Al)_2(Si_5Al_3)O_{20}(OH,F)_4$	0.064	Small r > v	Monoclinic β = 100°	Micaceous	{001} perf.
1.670	n_β 1.670 n_α 1.619 n_γ 1.720	Strunzite $MnFe_2(PO_4)_2(OH)_2 \cdot 8H_2O$	0.101	Moderate - 85°	Monoclinic β = 100.2°	Tufts hair-like crystals, laths {010}	
*1.63 1.71	n_β 1.63-1.71 n_α 1.62-1.70 n_γ 1.64-1.72	Tschermakite-Ferrotschermakite $Ca_2Mg_3(Al,Fe)_2(Al_2Si_6)O_{22}(OH,F)_2$	0.018 to 0.020	90°-50° r $\gtrless$ v med	Monoclinic β ~ 105°	Short prism. xls., massive	{110} perf., {001} or {100} parting
1.671	n_β 1.6710 n_α 1.5261 n_γ 1.6717	Alstonite $CaBa(CO_3)_2$	0.146	6°	Orthorhombic	Pseudohexagonal dipyramids	{110} imperf
1.671	n_β 1.671 n_α 1.626 n_γ 1.699	Poitevinite $(Cu,Fe,Zn)SO_4 \cdot H_2O$	0.073	75°	Monoclinic β = 114.7°	Massive, earthy	None
1.673	n_β 1.673 n_α 1.634 n_γ 1.685	Durangite $NaAlAsO_4F$	0.051	45° - 55° r > v perc horizontal dist	Monoclinic β = 115°	Prismatic, tabular	{110} dist
1.674	n_β 1.674 n_α 1.640 n_γ 1.679	Spurrite $Ca_5(SiO_4)_2CO_3$	0.039	~40° r > v weak, crossed distinct	Monoclinic β = 101.3°	Anhedral crystals, massive, granular	{001} dist, {100} poor at 79°
*1.666 1.683	n_β 1.683-1.666 n_α 1.649-1.639 n_γ 1.691-1.674	Childrenite $(Fe,Mn)AlPO_4(OH)_2 \cdot H_2O$	0.035 to 0.042	40° - 55° v > r strong	Orthorhombic	Pyramidal, tabular, platy	{100} poor
1.675	n_β 1.675 n_α 1.661 n_γ 1.689	Liskeardite $(Al,Fe^{3+})_3AsO_4(OH)_6 \cdot 5H_2O$	0.028	~90°	Monoclinic or Orthorhombic	Radiated fibrous aggregates, crusts	One ∥ elong
*1.66 1.69	n_β 1.66-1.69 n_α 1.65-1.67 n_γ 1.67-1.69	Thuringite $(Fe^{2+},Fe^{3+},Mg)_6(Si,Al)_4O_{10}(OH)_8$	0.015	10° - 20° r > v	Monoclinic	Massive, scaly aggregates	{001} perf
1.676	n_β 1.676 n_α 1.529 n_γ 1.677	Witherite $BaCO_3$	0.148	16° v > r very weak	Orthorhombic	Pseudo-hexagonal bipyramidal crystals, globular	{010} dist {110} & {012} poor

* n_β variation exceeds 0.01

Twinning	Orientation	Color in thin section	Hardness Sp Gravity	Color hand sample	Alteration	Occurrence	Remarks	Reference
Simple or multiple on {100}	b = Y	Yellow to red-brn, Y > Z ~ X	5 3.2-3.4	Black or dk blue-green	Uralite, limonite, siderite	Alkali igneous rocks	Amphibole group	p. 242
	b = Y, a ∧ Z ~ small, optic plane ⊥ (010) rare	Pink Y ~ Z > X	2.5-3 3.4	Copper-red, red-black		Skarn zones near Zn-oxide ore bodies	Mica group	Am Min 51, 1107-1123
{110} normal, {001} complex	b = Y, c ∧ X = small	Colorless, yellow, brown Z ~ Y > X	2½-3 3.19	Brown, black	Chlorite	Pegmatites. Alkali igneous rocks	Mica group. Biotite series	p. 270-274
{100}	c ∧ Z = 10°	Yellow, weak pleochroism, Z > X = Y	2.47-2.56	Straw, brownish-yellow		Weathering of phosphate minerals in pegmatites. Phosphate rock	Dimorph laueite	Am Min 43, 793-794
{100} simple	b = Y, c ∧ Z ≃ -19°, a ∧ X ≃ -4°	X = colorless to yellow, Y = yellow-green, Z = green to blue green	5-6 3.14-3.35	Green, black	Uralite	Many varieties of igneous and metamorphic rocks	Amphibolite group. Hornblende series	DHZ II, 263-314
{110} and {130} ubiquitous	a = Y, b = Z, c = X	Colorless	4-4½ 3.67-3.71	Colorless, white, pinkish		Low temperature hydrothermal veins		Dana II, 218-219
		Pale blue	3-3½ 3.30	Salmon pink, pale blue		Oxide zone ore deposits with bonattite	Soluble H_2O	Can Min 8, 109-110
{001} penetration	b = Z, c ∧ X = +25°	X = orange-yellow, Y = pale orange-yellow, Z ~ colorless	5 3.61-4.07	Lt-dk red-orange		Pegmatites, high T hydrothermal veins	Soluble H_2SO_4	Dana II, 829-831
{20$\bar{5}$} simple, {001} polysynthetic	b = X, a ~ Z	Colorless	5 3.0	Gray, lavender-gray	Afwillite	Contact limestone deposits with gehlenite and merwinite	Effervesces and gelat. HCl	Win & Win II, 516
{100} and {001} contact	a = Y, b = X, c = Z	X = yellow, Y = pink, Z = colorless	5 3.25	Brown, yellow-brown		Granite pegmatites, hydrothermal veins	Childrenite-eosphorite series Soluble acids	Am Min 35, 793-805
	Z = elong	Colorless	Soft 3.01	White, greenish, brownish		Secondary coatings on arsenopyrite, pyrite, chalcopyrite, quartz		Dana II, 924
{001}	b = Y, Z ∧ {001} ~ 3°	X = pale yellow-brn Y = Z = dk green	2-3 2.96-3.31	Dk olive-grn, dk brown		Hydrothermal veins, Fe-rich metamorphic rocks	Chlorite group	p. 296
{110} cyclic (Pseudo-hex) ubiquitous	a = Z, b = Y, c = X	Colorless	3½ 4.29	White, gray, yellow, brownish	Barite	Low T hydrothermal veins in limestone with galena, barite	Easily soluble weak acids with efforvescence	p. 69

Biaxial –

n_β Var.	Refractive Index	Mineral Name and Composition	$(n_\gamma - n_\alpha)$	2V angle Dispersion	Crystal System	Habit	Cleavage
1.677	n_β 1.677 n_α 1.629 n_γ 1.679	Sampleite $NaCaCu_5(PO_4)_4Cl \cdot 5H_2O$	0.050	~23° r > v	Orthorhombic	Lath-like crystals (c) flattened {010}	{010} perf {100} & {001} distinct
1.677	n_β 1.677 n_α 1.664 n_γ 1.688	Weilite $CaHAsO_4$	0.024	~82°	Triclinic α = 94.3° β = 101.6° γ = 87.4°	Powdery crusts or pseudomorphs	
*1.656 1.700	n_β 1.656-1.700 n_α 1.648-1.685 n_γ 1.664-1.712	Ferrorichterite $Na_2CaFe^{2+}_5Si_8O_{22}(OH)_2$	0.016 to 0.028	50° - 35° v > r	Monoclinic β = 104°	Prismatic	{110} perf
1.679	n_β 1.679 n_α 1.569 n_γ 1.708	Carbocernaite $(Na,Ca,Sr,Ce)CO_3$	0.139	52° r > v distinct	Orthorhombic	Tiny grains or crystals	{100}, {021} & {010} poor
1.680	n_β 1.680 n_α 1.625 n_γ 1.706	Posnjakite $Cu_4SO_4(OH)_6 \cdot H_2O$	0.081	57°	Monoclinic β = 102.9°	Tabular crystals, grains, thin films	
1.680	n_β 1.680 n_α 1.530 n_γ 1.685	Aragonite $CaCO_3$	0.155	18° v > r weak	Orthorhombic	Columnar-acicular-fibrous crystals, coral-like	{010} imperf {110} & {011} very poor
1.680	n_β 1.680 n_α 1.668 n_γ 1.686	Davreuxite $Mn_2Al_{12}Si_7O_{31}(OH)_6$	0.018	70°	Monoclinic β = 116°	Long fibers	
1.676 1.684	n_β 1.676-1.684 n_α 1.658-1.664 n_γ 1.692-1.700	Cummingtonite $(Mg,Fe)_7Si_8O_{22}(OH)_2$	0.034 to 0.036	90° - 85° v > r weak	Monoclinic β = 102°	Bladed, columnar, acicular	{110} dist
1.681	n_β 1.681 n_α 1.635 n_γ 1.698	Kurchatovite $Ca(Mg,Mn)B_2O_5$	0.063	66° r > v weak	Orthorhombic	Massive	{010} perf Two imperfect
*1.663 1.690	n_β 1.663-1.690 n_α 1.647-1.690 n_γ 1.663-1.702	Crossite $Na_2Mg_3(Fe^{3+},Al)_2(Si_4O_{11})_2(OH)_2$	0.016 to 0.012	0° to 90° r > v extreme	Monoclinic β = 104°	Acicular, columnar (c)	{110} dist
*1.669 1.695	n_β 1.669-1.695 n_α 1.664-1.686 n_γ 1.675-1.699	Bronzite $(Mg,Fe)_2(SiO_3)_2$	0.011 to 0.013	90° - 64°	Orthorhombic	Short prism (c) crystals. Bladed	{110} dist, {010} or {100} parting
1.684	n_β 1.684 n_α 1.525 n_γ 1.686	Barytocalcite $BaCa(CO_3)_2$	0.161	15° r > v perc	Monoclinic β = 106°	Prismatic	{210} perf {001} imperf
1.684	n_β 1.684 n_α 1.593 n_γ 1.698	Yavapaiite $KFe(SO_4)_2$	0.105	31° r > v strong	Monoclinic β = 94.4°	Short prismatic crystals (b)	{001} & {100} perfect, {110} distinct

* n_β variation exceeds 0.01

Twinning	Orientation	Color in thin section	Hardness Sp Gravity	Color hand sample	Alteration	Occurrence	Remarks	Reference
	a = Y, b = X, c = Z	X = dk blue, Y = lt blue, Z = colorless	~4 3.20-3.27	Lt blue to bluish-green		Oxide zones Cu-deposits with gypsum, limonite, jarosite	Easily soluble acids	Dana II, 945-946
	$X \wedge (001) = 20°$, $Y \wedge (001) = 27°$, $Z \wedge \perp (001) = 34°$	Colorless	3.45-3.48	White		Oxide zones As-rich veins		Am Min 49, 816
Simple or multiple on {100}	b = Y, $c \wedge Z = -15°$ to $-40°$	Pale yellow, violet Pleochroic yellow to red, Y > Z > X	5-6 3.23-3.45	Yellow, brown to red, green		Skarns, hydrothermal veins in alkaline igneous rocks	Amphibole group	Econ Geol 51, 77-87
	a = Y, b = X, c = Z	Colorless	3 3.53	Colorless, white, yellow, pink, brown		Carbonatite veins		Am Min 46, 1202
		X ~ colorless, Y = blue, Z = greenish-blue Y > Z >> X	2-3 3.35-3.36	Blue to dk blue		Oxide zones Cu-deposits with aurichalcite		Am Min 52, 1582-1583
{110} cyclic (pseudo-hex) common	a = Y, b = Z, c = X	Colorless	3½-4 2.95	Colorless, white, yel, violet, etc.	Inverts to calcite	Sinter in hot springs and caves. Amygdules. Marine sediments. Blueschists		p. 66
	+ Elongation	Colorless		White to pale rose		With quartz and pyrophyllite		Am Min 63, 795
Simple or multiple on {100}	b = Y, $c \wedge Z = -17°$ to $-16°$, $a \wedge X = -9°$ to $-3°$	Colorless to pale green or brown. Weak pleo in yellow brown, green, Z > Y ≥ X	5-6 3.10-3.40	Dk green to brown	Hornblende, talc, serpentine	Metamorphic rocks	Amphibole group	p. 225
	a = Y, b = X, c = Z	Colorless	4½ 3.02-3.03	Pale gray		Skarns with idocrase, garnet, magnetite, svabite	U.V. fluor long λ-bright violet	Am Min 51, 1817-1818
{100} simple or multiple	b = Z, $c \wedge Y = -10°$ to $-8°$, $a \wedge X = +4°$ to $+6°$	X = yellow, Y = blue, Z = violet	6 3.1-3.3	Inky blue to black	Actinolite	Low T-high P metamorphic rocks (blueschists) with ophiolite rocks	Amphibole group	p. 237
{101} normal	a = X, b = Y, c = Z	X = pale orange, Y = pale yellow, Z = pale green	5-6 3.30-3.44	Bronze yellow greenish, brn	Bastite (serpentine) uralite, talc, magnesite	Peridotites & dunites with olivine, phlogopite. Charnokites & granulites	Orthopyroxene group. Enstatite-orthoferrosilite series	p. 187
	b = Z $c \wedge X = -64°$, $c \wedge Y = +26°$	Colorless	4 3.65-3.71	White, yellow, green		Veins in limestone	Dimorph alstonite	Dana II, 220-221
	b = Z, $c \wedge X = -6°$, a ~ Y	Colorless	2½-3 2.88-2.92	Pale pink		Oxide zones ore deposits with sulfur, voltaite, other sulfates	Readily soluble HCl	Am Min 44, 1105-1114

n_β Var.	Refractive Index	Mineral Name and Composition	$(n_\gamma - n_\alpha)$	2V angle Dispersion	Crystal System	Habit	Cleavage
1.684	n_β 1.684 n_α 1.668 n_γ 1.685	Walstromite $BaCa_2Si_3O_9$	0.017	30° Weak	Triclinic α = 69.8° β = 102.2° γ = 97.1°	Short prismatic crystals	{011}, {010} & {100} ~ perfect
1.685	n_β 1.685 n_α 1.610 n_γ 1.704	Roscoelite $K(V,Al,Mg)_3(AlSi_3)O_{10}(OH)_2$	0.094	24° - 40°	Monoclinic β = 101°	Minute scales, stellate	{001} perf
*1.650 1.720	n_β 1.650-1.720 n_α 1.610 n_γ 1.682-1.770	Epistolite $Na_2(Nb,Ti)_2Si_2O_9 \cdot nH_2O$	0.072	~60° - 80°	Monoclinic β = 74.75°	Rectangular plates, lamellar aggregates	{001} perfect {110} dist
1.685	n_β 1.685 n_α 1.650-1.654 n_γ 1.699-1.715	Soddyite (Soddite) $(UO_2)_5(SiO_4)_2(OH)_2 \cdot 5H_2O$	0.049 to 0.061	Large ~ 65° r > v strong	Orthorhombic	Bipyramidal, tabular crystals, radiated fibrous, earthy	{001} perf {111} dist
1.681 1.689	n_β 1.681-1.689 n_α 1.658-1.667 n_γ 1.683-1.692	Cenosite (Kainosite) $Ca_2(Ce,Y)_2(SiO_4)_3(CO_3) \cdot H_2O$	0.016 to 0.030	40° v > r distinct	Orthorhombic	Pseudotetragonal prismatic	2 cleavages at 90°
1.685	n_β 1.685 n_α 1.672 n_γ 1.698	Talmessite (Arsenate-belovite) $Ca_2Mg(AsO_4)_2 \cdot 2H_2O$	0.026	~90°	Triclinic α = 112.6° β = 70.8° γ = 119.4°	Fine crystalline aggregates, drusy	
*1.680 1.695	n_β 1.680-1.695 n_α 1.670-1.690 n_γ 1.685-1.700	Katophorite $Na_2CaFe_4^{2+}Fe^{3+}Si_7AlO_{22}(OH)_2$	0.015 to 0.010	40° - 0° r > v strong	Monoclinic β = 105°	Prismatic	Perfect {110}
*1.677 1.696	n_β 1.677-1.696 n_α 1.665-1.682 n_γ 1.677-1.699	Kornerupine $Mg_3Al_6(Si,Al,B)_5O_{21}(OH)$	0.012 to 0.017	3° - 48° r $\gtrless$ v weak	Orthorhombic	Prismatic (c), fibrous aggregates	{110} dist to imperfect
*1.680 1.696	n_β 1.680-1.696 n_α 1.669-1.686 n_γ 1.688-1.704	Ferroactinolite $Ca_2Fe_5^{2+}Si_8O_{22}(OH)_2$	0.019 to 0.018	75° - 72° v > r weak	Monoclinic β = 105°	Fibrous, bladed	Distinct {110}
1.677 1.701	n_β 1.677-1.701 n_α 1.672-1.693 n_γ 1.681-1.704	Axinite $(Ca,Fe^{2+},Mn)_3Al_2BO_3(SiO_3)_4($(	0.009 to 0.013	63°-90° v > r strong	Triclinic α = 91.8° β = 98.2° γ = 77.3°	Wedge shaped xls	{100} dist., {001}, {110} & {011} poor
1.689	n_β 1.689 n_α 1.658 n_γ 1.714	Alvanite $Al_6VO_8(OH)_{12} \cdot 5H_2O$	0.056	80° - 85° v > r strong	Monoclinic β = 115°	Hexagonal platelets {010}	{010} perfect
*1.674 1.705	n_β 1.674-1.705 n_α 1.662-1.692 n_γ 1.676-1.707	Bustamite $(Ca,Mn)Si_2O_6$	0.014 to 0.015	30°-44° v > r weak, crossed	Triclinic α = 89.6° β = 94.9° γ = 102.8°	Tab {001}, fib., massive	{001} perf., {110} & {1$\bar{1}$0} good, {010} poor
1.690	n_β 1.690 n_α 1.660 n_γ 1.695	Bergenite $Ba(UO_2)_4(PO_4)_2(OH)_4 \cdot 8H_2O$	0.035	45°	Orthorhombic	Platy	

* n_β variation exceeds 0.01

Twinning	Orientation	Color in thin section	Hardness Sp Gravity	Color hand sample	Alteration	Occurrence	Remarks	Reference
	b near Z	Colorless	~3½ 3.67	White, colorless		Quartz-sanbornite rock in contact zones	U.V. fluor-pink	Am Min 50, 314-340
		X = green-brown, Y = Z = olive-green	2½ 2.97	Clove-brown to dk green		Secondary U-V-deposits. Au ores with tellurides	Mica group	DHZ 3, 11-30
	b = Y, c ∧ Z = 7°	Colorless	1-1½ 2.65-2.89	White, yellowish, gray, lt brn		Alkali plutons and alkali pegmatites		Vlasov II, 562-564
	a = Z, b = Y, c = X	X = colorless, Y = very pale yel, Z = pale grn-yellow	3½ 4.70-4.75	Amber yellow, yellow-green		Pegmatites with urano-phane, kasolite, curite, sklodowskite, kasolite	Not fluor. U.V.	USGS Bull 1064, 312-315
	Optic plane = {010}	Pale brown or rose Not pleochroic	5-6 3.34-3.61	Lt-dk brown, yellow, red		Granite pegmatite, skarns	Effervesces slowly in cold, dilute HCl	Vlasov II, 246-247
		Colorless	5 3.2-3.5	Colorless, white, pale green		Oxide zones ore-deposits	Isomorph beta-roselite	Am Min 50, 813
Simple or multiple on {100}	b = Z, c ∧ Y = -54° to -20°, a ∧ X = -39° to -5°	Yellow to green, Z > Y > X	5 3.4-3.5	Black or dk blue-green	Uralite, limonite, siderite	Alkali igneous rocks	Amphibole group	p. 241
	a = Y, b = Z, c = X	Colorless or X = pale brown, green, Y = yellow, green, Z = dk grn, brown	6-7 3.27-3.45	Colorless, pink, yellow, green, brown, black		Granulites, schists, pegmatites with tour-maline, rutile, garnet		Am Min 37, 531-541
Simple or multiple on {100}	b = Y, c ∧ Z = -13°, a ∧ X = +2°	Yellow-green or blue green. Pleo-chroic in yellow, green, blue-green, Z > Y ≥ X	5-6 3.4-3.5	Dk green	Talc chlorite	Metamorphic rocks	Amphibole group	p. 230
{110} normal repeated	X ~ ⊥ $\{\bar{1}11\}$	Pale purple, brn, yellow. Weak pleochroism Y > X > Z	6½-7 3.26-3.36	Lt. to dk. brown-purple	Chlorite, calcite	Contact aureoles in carbonates with Ca-Mg silicates		p. 174
Polysynthetic {010}	Positive or negative elongation b = Y	Blue-green	3-3½ 2.41	Bluish-green to bluish-black		Oxide zone in vanadium deposits		Am Min 44, 1325-1326
	Optic plane ~ ⊥ {100}	Colorless	5½-6½ 3.32-3.43	Pink, brownish red, white		Metasomatic Mn ore deposits	Pyroxene group	DHZ II, 191-195
		Colorless	2.72-4.1	Yellow		Secondary U-mineral	Weak fluor U.V. orange-brown	Am Min 45, 909

n_β Var.	Refractive Index	Mineral Name and Composition	$(n_\gamma - n_\alpha)$	2V angle Dispersion	Crystal System	Habit	Cleavage
1.690	n_β 1.690 n_α 1.666 n_γ 1.690	Calumetite $Cu(OH,Cl)_2 \cdot 2H_2O$	0.024	~2°	Orthorhombic ?	Scaly, spherules	{001} dist
*1.68 1.70	n_β 1.68-1.70 n_α 1.68-1.70 n_γ 1.70-1.72	Strigovite $(Fe^{2+},Fe^{3+})_6(Si,Al)_4O_{10}(OH)_8$	0.017 to 0.020	Small r > v strong	Monoclinic β ~ 97°	Scaly agg	{001} perfect
1.690	n_β 1.690 n_α 1.675 n_γ 1.693	Sincosite $Ca(VO)_2(PO_4)_2 \cdot 5H_2O$	0.018	Small ~ 45° r > v perc	Tetragonal	Tabular (square) crystals, scaly, rosettes	{001} perf {100} & {110} distinct
1.690	n_β 1.690 n_α 1.625-1.631 n_γ 1.691-1.697	Annite $K_2Fe_6^{2+}(Si_6Al_2)O_{20}(OH,F)_4$	0.066	Small r > v	Monoclinic β ~ 100°	Micaceous	{001} perf.
1.690	n_β 1.690 n_α 1.682 n_γ 1.697	Chlorophoenicite $(Mn,Zn)_5AsO_4(OH)_7$	0.015	~83° r > v strong	Monoclinic β = 106°	Prismatic, acicular (b)	{100} dist
1.692	n_β 1.692 n_α 1.670 n_γ 1.713	Parakeldyshite $NaZrSi_2O_7$	0.043	84° - 88°	Triclinic α = 71.5° β = 87.2° γ = 85.6°	Cleavage masses	{001} perf, {110} & {1$\bar{1}$0} dist, {011} good
1.692	n_β 1.692 n_α n_γ 1.699	Keckite $(Ca,Mg)(Mn,Zn)_2(Fe^{3+})(OH)_3(PO_4)_4 \cdot 2H_2O$			Monoclinic β = 110.5°	Xl. aggregates	{001}, {100}
1.693	n_β 1.693 n_α 1.672 n_γ 1.710	Jagowerite $BaAl_2$	0.038	~83°	Triclinic α = 116.8° β = 86.1° γ = 113°	Crystalline masses	{100} & {0$\bar{1}$1} good, {0$\bar{2}$1} fair
*1.676 1.710	n_β 1.676-1.710 n_α 1.667-1.694 n_γ 1.684-1.722	Ferrogedrite $(Fe^{2+},Mg)_5Al_2(Si_3AlO_{11})_2(OH)_2$	0.028 to 0.018	82° - 90° r > v or v > r weak to moderate	Orthorhombic	Fibrous, bladed, acicular	{110} perf
*1.677 1.710	n_β 1.677-1.710 n_α 1.672-1.700 n_γ 1.684-1.715	Arfvedsonite $Na_3Fe_4^{2+}Fe^{3+}Si_8O_{22}(OH)_2$	0.012 to 0.015	0° - 70° v > r very strong	Monoclinic β = 105°	Prismatic	{110} perf
1.694	n_β 1.694 n_α 1.670 n_γ 1.710	Keldyshite $(Na,H)_2ZrSi_2O_7$	0.040	78° v > r extreme	Triclinic α = 92.75° β = 94.2° γ = 72.3°		Two poor at 90°
*1.66 1.73	n_β 1.66-1.73 n_α 1.65-1.70 n_γ 1.67-1.73	Hastingsite-Ferrohastingsite $NaCa(Mg,Fe^{2+})_4(Al,Fe^{3+})[(Si_3Al)O_{11}]_2(OH)_2$	0.020 to 0.024	90°-10° v > r moderate	Monoclinic β ~ 105°	Prismatic xls., massive	{100} perf., {100} or {001} parting
1.695 1.696	n_β 1.695-1.696 n_α 1.668-1.670 n_γ 1.698-1.703	Boltwoodite $K_2(UO_2)_2(SiO_3)_2(OH)_2 \cdot 5H_2O$	0.030 to 0.033	Large	Orthorhombic or Monoclinic	Radiated acicular, fibrous	{010} perf {001} imperf
*1.673 1.717	n_β 1.673-1.717 n_α 1.653-1.691 n_γ 1.690-1.731	Crysolite $(Mg,Fe)_2SiO_4$	~0.040	90° - 82° r > v weak	Orthorhombic	Granular, anhedral crystals	{010} & {110} poor

* n_β variation exceeds 0.01

Twinning	Orientation	Color in thin section	Hardness Sp Gravity	Color hand sample	Alteration	Occurrence	Remarks	Reference
	a = Y, b = Z, c = X	Blue Z ≥ Y > X	2	Azure blue		Encrustations oxide zones Cu-deposits	Soluble dilute acids	Am Min 48, 614-619
	b = Y, c ∧ X = small, a ∧ Z = small	Yellow-brown, deep green Z = Y > X	1.5-2.5 2.9-3.3	Dk. green, dk. brown, black		Fe-rich sediments, hydrothermal veins	Chlorite group	p. 296
{110} rare	c = X	X = colorless to pale yellow, Y = Z = gray-green	Soft ~2.84	Yellow-green to brown-green		Vugs, cavities, veins	Soluble dilute acids	Dana II, 1057-1058
{110} normal, {001} complex	b = Y, c ∧ X = small	X ~ colorless, yellow, X ~ Z = Dk brown Z ~ Y >> X	2½-3 3.35	Dk. brown, black	Chlorite	Pegmatites. Alkali igneous rocks	Mica group. Biotite series	p. 270-274
	b = Y	~colorless	3-3½ 3.46-3.47	Lt green (daylight), red-violet (artificial)		Oxide zones Zn-Mn-deposits		Am Min 53, 1110-1119
{100} polysynthetic	a ~ Z, b ~ X, c ~ Y	Colorless	5½-6 3.39-3.40	White, bluish	Easily weathered	Nepheline syenite pegmatites with aegirine, pyrophanite, loparite	Fluor U.V. short λ - cream color	Can Min 15, 102-107
	b = Z, c ∧ X = 15°-22°	X = red-brown, Y = yellow, Z = bright yellow	4½ 2.68	Brown, yellow-brown		Weathering of phosphophyllite or rockbridgeite in pegmatites		Am Min 64, 1330
		Colorless	4½ 4.01-4.05	Pale green		Quartz veins with pyrite, hinsdalite	Fluor U.V.- greenish white	Am Min 60, 945
None	a = X, b = Y, c = Z	Pale brown to grn, colorless. Weak to moderate pleo, Z = Y > X	5½-6 3.30-3.57	Yellow brown dk brown	Talc or serpentine	Metamorphic rocks	Amphibole group	p. 223
Simple or multiple on {100}	b = Z, c ∧ X = +5° to +30°, a ∧ Y = +20° to +45°	Pleo yellow, green to indigo, X > Y > Z	5-6 3.30-3.50	Black to dk green or blue-green	Uralite, limonite, siderite	Alkaline igneous rocks	Amphibole group	p. 243
Fine polysynthetic		Colorless	~4 ~3.30	White		Alkali plutons with nepheline, sodalite, eudialyte, ramsayite	Decomp Acids	Am Min 55, 1072
{100} common	b = Y, c ∧ Z ~ -19° a ∧ X ~ -4°	X = yellow, Y = green, Z = dk. green	5-6 ~3.30	Dark green	Uralite	Igneous and metamorphic rocks	Amphibole group. Hornblende series	p. 221
	∥ extinction, + elongation	X = colorless, Y = Z = pale yellow	3½-4 ~3.6	Pale yellow		Secondary U-mineral	Fluor U.V. dull green. Anomalous blue interference colors	Am Min 46, 12-25
{100}, {011} & {012} simple. {031} cyclic	a = Z, b = X, c = Y	Colorless to pale yellow-green Y > X = Z	7 3.30-3.53	Pale olive-green	Serpentine, iddingsite, chlorophaeite	Mafic and ultramafic igneous rocks (peridotite, gabbro, basalt)	Olivine group	p. 105

n_β Var.	Refractive Index	Mineral Name and Composition	$(n_\gamma - n_\alpha)$	2V angle Dispersion	Crystal System	Habit	Cleavage
1.695	n_β 1.695 n_α 1.676 n_γ 1.698	Marićite $NaFePO_4$	0.022	43° r > v weak	Orthorhombic	Radiating grains in nodules	None
1.695	n_β 1.695 n_α 1.684 n_γ 1.698	Kempite $Mn_2Cl(OH)_3$	0.014	55°	Orthorhombic	Minute prismatic (c) crystals	None
1.685 1.702	n_β 1.685–1.702 n_α 1.681–1.700 n_γ 1.695–1.708	Barylite $BaBe_2Si_2O_7$	0.012 to 0.014	40° – 70° r > v weak	Orthorhombic	Tabular, prismatic	{001} & {100} perfect, {010} poor
*1.682 1.712	n_β 1.682–1.712 n_α 1.664–1.688 n_γ 1.700–1.732	Grunerite $(Fe,Mg)_7Si_8O_{22}(OH)_2$	0.036 to 0.044	80° – 90° r > v, inclined weak	Monoclinic β = 102°	Bladed, columnar, fibrous	{110} dist
1.700 1.696	n_β 1.700–1.696 n_α 1.691–1.685 n_γ 1.707–1.701	Barkevikite $(Na,K)Ca_2(Fe^{2+},Mg,Fe^{3+},Mn)_5(Si_7Al)O_{22}(OH)_2$	0.012 to ~0.018	40°–50° r > v weak to strong	Monoclinic β ~ 105°	Prism. xls	{110} perf., {100} or {001} parting
*1.675 1.722	n_β 1.675–1.722 n_α 1.655–1.686 n_γ 1.684–1.723	Dumortierite $(Al,Fe^{3+})_7O_3BO_3(SiO_4)_3$	0.011 to 0.027	15° – 52° v > r strong Rarely r > v	Orthorhombic	Fibrous, bladed, prismatic crystals (c)	{100} dist, {110} poor
1.699	n_β 1.699 n_α 1.656 n_γ 1.731	Kupletskite $(K,Na)_2(Mn,Fe)_4(Ti,Nb)Si_4O_{14}(OH)_2$	0.075	79° r > v strong	Monoclinic	Lamellar, platy	{100} perfect
1.695 1.705	n_β 1.695–1.705 n_α 1.698–1.710 n_γ 1.671–1.720	Triphylite $Li(Fe,Mn)PO_4$	~0.015	90° – 0° v > r	Orthorhombic	Cleavage masses	{001} perf, {010} dist, {110} imperf
1.700	n_β 1.700 n_α 1.625 n_γ 1.735	Ancylite $(Ce,La)_4(Sr,Ca)_3(CO_3)_7(OH)_4 \cdot 3H_2O$	0.110	66°	Orthorhombic	Pseudo-octahedral, prismatic on c	None
1.697 1.704	n_β 1.697–1.704 n_α 1.667–1.677 n_γ 1.705–1.712	Sinhalite $MgAlBO_4$	0.038 to 0.035	~55°	Orthorhombic		None
1.701	n_β 1.701 n_α 1.693 n_γ 1.704	Tinzenite $Ca_2MnAl_2BO_3(SiO_3)_4(OH)$	0.011	63°	Triclinic α = 137°, β = 105°, γ = 87°	Platy	{100} perfect
*1.690 1.712	n_β 1.690–1.712 n_α 1.690–1.702 n_γ 1.702–1.719	Riebeckite $Na_2Fe_3^{2+}Fe_2^{3+}(Si_4O_{11})_2(OH)_2$	0.012 to 0.017	90° to 50° r > v strong	Monoclinic β = 104°	Acicular, columnar (c)	{110} dist
*1.682 1.723	n_β 1.682–1.723 n_α 1.660–1.678 n_γ 1.689–1.730	Beta-uranophane (β-uranotile) $Ca(UO_2)_2(SiO_3)_2(OH)_2 \cdot 5H_2O$	0.029 to 0.052	35° – 71° r > v, crossed strong	Monoclinic β = 91°	Radiated fibrous aggregates	{010} perf {100} poor
1.703	n_β 1.703 n_α 1.701 n_γ 1.706	Serendibite $Ca_4(Mg,Fe,Al)_6(Al,Fe)_9(Si,Al)_6B_3O_{40}$	0.005	Large	Triclinic	Massive, irregular grains	None

* n_β variation exceeds 0.01

Twinning	Orientation	Color in thin section	Hardness Sp Gravity	Color hand sample	Alteration	Occurrence	Remarks	Reference
	a = X, b = Y, c = Z	Colorless	4-4½ 3.64-3.66	Colorless gray, lt brown		Shale with quartz, ludlamite, vivianite, wolfeite, satterlyite		Can Min 15, 518-521
	a = Z, b = Y, c = X	Green	~3½ ~2.94	Emerald green		Altered Mn-ores with pyrochroite, hausmannite	Soluble dilute acids	Dana II, 73-74
	a = Y, b = X, c = Z	Colorless	7 4.04-4.05	Colorless, white		Calcite veins in alkali plutons		Am Min 41, 512
Simple or multiple on {100}	b = Y, c ∧ Z = -16° to -12°, a ∧ X = -3° to +2°	Colorless to pale brown or green. Weak pleo in yel, brown, green, Z > Y ≥ X	5-6 3.40-3.60	Dk green to brown	Hornblende, talc, serpentine	Metamorphic rocks	Amphibole group	p. 225
{100} simple	b = Y, c ∧ Z ~ -11° to -18°, a ∧ X ~ +4° to -3°	X = yellow-brn, Y = red-brown, Z = dk brown	5-6 3.35-3.44	Dk. brown, black	Uralite, Na-pyroxene, Fe-oxides	Alkaline plutons with nepheline	Amphibole group	p. 245
{110} cyclic (trillings)	a = Z, b = Y, c = X	X = dk blue, Y ~ Z = colorless	7-8½ 3.26-3.41	Deep blue, violet, green-blue	Sericite	Pegmatites, quartz veins. Granite gneiss, schists		p. 129
	X = orange-yellow, Z = brown	+ elongation	~3 3.20-3.23	Dk brown, black	Black Mn-oxides	Alkali pegmatites with nepheline, aegirine, schizolite	Mn-astrophyllite, soluble acids	Am Min 42, 118-119
None	a = Z, b = Y, c = X	Colorless	4½-5 3.53-3.58	Blue-gray	Rare phosphates	Granite pegmatites with Li-minerals and other phosphates	Lithiophilite-triphylite series >75% $LiFePO_4$	p. 79
	a = X, b = Y, c = Z	Colorless	4-4½ 3.95	Pale yellow to brown		Nepheline-syenite pegmatites, hydrothermal veins, carbonatite		Dana II, 291-293
		X = brown, Y = green, Z = pale brown	~7 3.47-3.50	Brown, yellow, yellow-green		Gem gravels	Insoluble acids	Min Mag 29, 841-849
	Optic plane near {010}	X = lt yellow-grn, Y = lt green, Z = colorless	6½-7 3.3	Yellow, orange-red, pink	Chlorite, calcite	Mn ore zones	Mn-axinite	p. 174
{100} simple or multiple	b = Y, c ∧ X = -8° to -7°, a ∧ Z = +6° to +7°	X = dk blue, Y = dk gray-blue, Z = yellow-brown	6 3.1-3.4	Dk blue-green to black	Limonite, siderite, quartz	Alkali granites and syenites with aegirine. Pegmatites. Gneiss-quartzite	Amphibole group	p. 237
	b = X c ∧ Z = -18° to +57°	X = colorless, Y = Z = lemon yel	2½-3 3.90-3.93	Yellow, yellow-green		Secondary U-mineral	Weak fluor U.V. green	USGS Bull 1064, 307-311
Polysynthetic		Y = very pale yellowish-green, Y ~ colorless, Z = Prussian blue	6½-7 3.42	Blue-green, dk blue, grayish		Contact limestone deposits with idocrase, diopside, scapolite, spinel		Am Min 17, 457-465

n_β Var.	Refractive Index	Mineral Name and Composition	$(n_\gamma - n_\alpha)$	2V angle Dispersion	Crystal System	Habit	Cleavage
1.705	n_β 1.705 n_α 1.695-1.70 n_γ 1.70-1.715	Tawmawite $Ca_2(Al,Fe^{3+},Cr)_3O\cdot SiO_4\cdot Si_2O_7(OH)$	~0.008	50° v > r dist	Monoclinic		{001} perf., {100} dist
1.705	n_β 1.705 n_α 1.660 n_γ 1.713	Tarbuttite $Zn_2PO_4(OH)$	0.053	50° Strong bisectrix	Triclinic α = 102.5° β = 87.7° γ = 102.6°	Short prismatic crys-tals, aggregate or crusts	{010} perfect
1.705	n_β 1.705 n_α 1.690 n_γ 1.710-1.712	Yftisite $(Y,TR)_4(F,OH)_6TiO(SiO_4)_2$	~0.020	Large r > v distinct	Orthorhombic	Prismatic crystals	None
1.706	n_β 1.706 n_α 1.678 n_γ 1.712	Ernstite $(Mn^{2+}_{1-x}Fe^{3+}_x)Al(PO_4)(OH)_{2-x}O_x$, (x=0-1)	0.043	74° r > v	Monoclinic β = 90.4°	Radiated fibrous	{010} & {100} distinct
*1.695 1.722	n_β 1.695-1.722 n_α 1.686-1.710 n_γ 1.699-1.725	Hypersthene $(Mg,Fe)_2(SiO_3)_2$	0.013 to 0.015	64° - 53°	Orthorhombic	Short prismatic (c) crystals. Massive	{110} dist, {010}or {100} parting
*1.69 1.75	n_β 1.69-1.75 n_α 1.67 n_γ 1.72	Låvenite (Laavenite)(Lovenite) $(Na,Ca,Mn)_3(Zr,Ti,Fe)(SiO_4)_2F$	0.05 to 0.03	40° - 70°	Monoclinic β = 108.2°	Prismatic, radiated, acicular, massive	{100} perf
1.704	n_β 1.704 n_α 1.650 n_γ 1.712	Parnauite $Cu_9(AsO_4)_2SO_4(OH)_{10}\cdot 7H_2O$	0.062	60° No disp	Orthorhombic	Bladed crystals, rosettes, crusts	{010}
*1.699 1.724	n_β 1.699-1.724 n_α 1.658-1.690 n_γ 1.699-1.724	Phosphuranylite $Ca(UO_2)_4(PO_4)_2(OH)_4\cdot 7H_2O$	0.041 to 0.034	0° - 51° r > v strong	Orthorhombic	Tiny scales, plates, laths. Earthy	{100} perf, {010} indist
1.713	n_β 1.713 n_α 1.654 n_γ 1.722	Langite $Cu_4SO_4(OH)_6\cdot 2H_2O$	0.068	70°	Orthorhombic	Small crystals elongation (a), earthy	{001} perf {010} dist
1.713	n_β 1.713 n_α 1.705 n_γ 1.715	Mboziite $Na_2CaFe^{2+}_3Fe^{3+}_2(Si_3AlO_{11})_2(OH)_2$	0.010	53°	Monoclinic β = 105.8°	Prismatic crystals	{110} perf
1.714	n_β 1.714 n_α 1.701 n_γ 1.720	Niocalite $Ca_4NbSi_2O_{10}(O,F)$	0.019	56°	Monoclinic β = 109.7°	Prismatic crystals	None
*1.69 1.74	n_β 1.69-1.74 n_α 1.67-1.69 n_γ 1.70-1.77	Kaersutite $NaCa_2(Mg,Fe^{2+})_4Ti[(Si_3Al)O_{11}]_2(OH)_2$	0.019 to 0.083	66°-82° r > v strong	Monoclinic β ~ 106°		{110} perf., {100} or {001} parting
1.715	n_β 1.715 n_α 1.66 n_γ 1.734-1.725	Lavendulan (Freirinite) $(Ca,Na)_2Cu_5(AsO_4)_4Cl\cdot 4\text{-}5H_2O$	0.074	33	Orthorhombic	Aggregate fine flakes, botryoidal, fibrous	{001} dist {110} imperf

* n_β variation exceeds 0.01

Twinning	Orientation	Color in thin section	Hardness Sp Gravity	Color hand sample	Alteration	Occurrence	Remarks	Reference
	b = Z, c = X = -24°	X = Z = emerald green, Y = yellow, brown	6-7 ~3.3	Bright green	Alters easily	Limestone contact zones	Cr-epidote. Abnormal interference colors	pp. 146, 148
		Colorless	3½-4 4.14-4.21	Colorless, white, pale tints		Oxide zones Zn-deposits with hopeite, hemimorphite, pyromorphite		Am Min 51, 1218-1220
		Colorless	3½-4 3.96	Pale yellow		Alkali granites	Anomalous blue interference colors	Am Min 62, 396
None ?	b = Z, c ∧ Y = 4°	X = yellow-brown, Y = red-brown, Z = pale yellow	3-3½ 3.07-3.09	Yellow-brown		Oxidation of eosphorite in pegmatites		Am Min 56, 637
{101} normal	a = X, b = Y, c = Z	X = salmon-pink, Y = pale brown, yellow, Z = pale bluish-green	5-6 3.44-3.58	Dk green or brown to black	Bastite, uralite, talc	Norites, gabbros, andesites. Amphibolite gneisses and granulites	Orthopyroxene group. Enstatite-orthoferrosilite series	p. 187
{100} polysynthetic	b = Y, c ∧ Z = -70°	Colorless, X = lt yellow, Y = greenish-yellow Z = brown-yellow, orange-red	6 3.47-3.55	Colorless, yellow, red-brown, dk brown		Alkali plutons, volcanics & pegmatites	May be metamict	Vlasov II, 381-384
	a = Y, b = X, c = Z	X = pale green, Y = yellow-green, Z = blue-green, Z > Y > X	~2 3.09	Pale blue, green, blue-green		Secondary Cu-mineral. Surface coatings with chrysocolla	Readily soluble HCl	Am Min 63, 704-708
	c = Z	X ~ colorless, Y = Z = golden yel	~2½ ~4.1	Lt yellow to dk yellow		Secondary U-mineral in pegmatites, sandstones	Not fluor U.V.	USGS Bull 1064, 222-227
{110} repeated cyclic	a = Z, b = Y, c = X	X = Lt yellow-green Y = blue-green, Z = sky-blue	2½-3 3.26-3.31	Sky blue, green-blue		Oxide zones Cu-deposits with gypsum	Easily soluble acids, NH_4OH	Dana II, 583-585
	b = Z, a ~ X, c ∧ Y = 9°-12°	X = pale yellow, Y = deep blue-grn, Z = deep blue, green, black	5-6	Black, dk blue-green		Alkali plutons & dikes with nepheline, sodalite, aegirine-augite	Amphibole group. Anomaous red-orange interfer colors	Min Mag 33, 1057-1065
{010} normal contact, very common	b = X, a ∧ Y = -32°, c ∧ Z = -12°	Colorless	~6 3.32	White, pale yellow		Carbonatite with pyrochlore, Nb-perovskite in calcite		Am Min 41, 785-786
	b = Y, c ∧ Z = 0° to -19°, a ∧ X = +16° to -3°	X = pale yellow, Y = red-brown, Z = dk. red-brown	5-6 3.2	Dk. brown, black		Volcanic rocks, especially basalt	Amphibole group. Ti-oxyhornblende	p. 222
	Z ∥ elong, inclined ext	Blue, slightly pleochroic	2½ 3.54	Blue, lavender-blue		Oxide zones Co-Cu deposits with erythrite, malachite		Am Min 42, 123-124

Biaxial —

n_β Var.	Refractive Index	Mineral Name and Composition	$(n_\gamma - n_\alpha)$	2V angle Dispersion	Crystal System	Habit	Cleavage
1.715	n_β 1.715 n_α 1.704 n_γ 1.724	Xanthoxenite $Ca_4Fe_2(PO_4)_4(OH)_2 \cdot 3H_2O$	0.020	Large ~ 85° v > r strong	Triclinic	Platy-lath crystals, radiated-fibrous	One perfect
1.715	n_β 1.715 n_α 1.663 n_γ 1.734	Ruizite $CaMn^{3+}(SiO_3)_2(OH) \cdot 2H_2O$	0.071	60° r > v strong, inclined	Monoclinic β = 91°	Elongation crystals(b)	
1.716	n_β ~1.716 n_α 1.602 n_γ 1.750	Dundasite $Pb_2Al_4(CO_3)_4(OH)_8 \cdot 3H_2O$	0.048	Large ~ 50°	Orthorhombic	Radiated crystals, crusts	{010} perf
1.716	n_β 1.716 n_α 1.700 n_γ 1.726	Woehlerite (Wöhlerite) $NaCa_2(Zr,Nb)(SiO_4)_2(O,OH,F)$	0.026	72° - 77° v > r weak	Monoclinic β = 109°	Prismatic, tabular {100} crystals, grains	{010} dist {100} & {110} indistinct
1.714 1.720	n_β 1.720-1.714 n_α 1.700-1.690 n_γ 1.735	Schoepite (Epi-ianthinite) $UO_3 \cdot 2H_2O$	0.035 to 0.045	~75°	Orthorhombic	Tabular {001}, prism (c), pseudohexagonal	{001} perf, {010} imperf
1.715 1.712	n_β 1.72-1.715 n_α 1.64-1.68 n_γ 1.75-1.77	Zippeite $(UO_2)_2SO_4(OH)_2 \cdot 4H_2O$	0.11 to 0.09	60° - 90°	Orthorhombic	Microscopic crystals, coatings	{010} perf (?)
1.717	n_β 1.717 n_α 1.685 n_γ 1.720	Kôzulite $(Na,K,Ca)_3(Mn,Mg,Fe,Al)_5$ $(Si_4O_{11})_2(OH,F)_2$	0.035	34° - 36° very weak	Monoclinic β = 104.6°	Short prismatic crystals	{110} perf
*1.703 1.732	n_β 1.703-1.732 n_α 1.701-1.729 n_γ 1.705-1.734	Sapphirine $(Mg,Fe)_2Al_4O_6SiO_4$	0.004 to 0.007	90° - 50° v > r strong. Inclined dist	Monoclinic β = 125.3°	Granular, prismatic crystals (c)	{010}, {001} & {100} poor
**1.670 1.770	n_β 1.670-1.770 n_α 1.650-1.700 n_γ 1.680-1.800	Oxyhornblende (Basaltic hornblende) $Ca\ Na(Mg,Fe^{2+},Fe^{3+},Al,Ti)_5Si_6$ $Al_2O_{22}(OH,O)_2$	0.018 to 0.083	56° - 88° v > r weak to r > v strong	Monoclinic β = 106°	Prismatic crystals	{110} dist
1.720	n_β 1.720 n_α 1.660 n_γ 1.728	Hydroastrophyllite $\sim(K,Na,H_3O)_3(Fe,Mn)_7Ti_2Si_8O_{24}$ $(O,OH)_7$	0.048	40°	Triclinic α = 103.4° β = 95.2° γ = 112.2°	Aggregates of blocky crystals	Two cleavages
1.720	n_β 1.720 n_α 1.701 n_γ 1.734	Howieite $Na(Fe,Mn)_{10}(Fe,Al)_2Si_{12}O_{31}(OH)_{13}$	0.033	65° v > r strong	Triclinic α = 91.3° β = 70.7° γ = 109°	Bladed crystals	{010} dist, {100} fair, {2̄10} poor
1.720	n_β 1.720 n_α 1.715 n_γ 1.725	Trimerite $CaMn_2(BeSiO_4)_3$	0.010	83° No disp	Monoclinic β = 90.2°	Prismatic crystals, pseudohexagonal	{001} dist

* n_β variation exceeds 0.01
** n_β variation exceeds 0.10

Twinning	Orientation	Color in thin section	Hardness Sp Gravity	Color hand sample	Alteration	Occurrence	Remarks	Reference
		Pale yellow. Weak pleochroism	~2½ . 2.8-2.97	Pale yellow to brown-yel		Alteration of triphylite in pegmatites with other phosphates		Dana II, 977-978
{100}	b = Y, c ∧ Z = +44°	Colorless, pale orange	5 2.9-3.0	Orange to brown		Calc-silicate skarns with kinoite, apophyllite, smecite, junitoite		Min Mag 41, 429-432
	a = X, b = Y, c = Z	Colorless	2 3.41-3.81	White		Oxide zones Pb-deposits	Soluble with effor in acids	Dana II, 279-280
{100} common, sometimes complex	b = Z, c ∧ X = 45°	X = Y ~ colorless, pale yellow, Z = dk wine yellow	5½-6 3.41-3.44	Dk to lt yellow, gray, brown		Alkali plutons & pegmatites with nepheline, aegirine, barkevikite	Soluble HCl	Vlasov II, 377-379
	a = Z, b = Y, c = X	X ~ colorless, Y = Z = yellow	~2½ 4.83	Yellow, brownish-yel		Alter of uraninite with secondary U-minerals		Am Min 45, 1026-1061
Common {001} (?)	a = Z, b = X, c = Y	X ~ colorless, Y = very pale yellow or yellow-orange, Z = pale yellow or yellow-orange	3.66-3.68	Yellow, orange, red, red-brown		Efflorescence on mine walls and dumps	Fluor U.V. variable & undependable	USGS Bull 1064, 141-147
	c ∧ X = 25°	X = yellow-brown, Y = reddish-brown, Z = dark brown	5 3.30-3.36	Reddish-black, black		Contact Mn-deposits with braunite, rhodonite, Mn-pyroxene	Amphibole group	Am Min 55, 1815-1816
{010} & {100} lamellar	b = Y, c ∧ Z = +6° to +15°, a ∧ X = +41° to +50°	Colorless to pale blue. Z > Y > X	7½ 3.40-3.58	Pale blue, green or gray	Corundum, biotite & talc	High T hornfels or gneiss-granulite with cordierite, spinel.		p. 138
Simple or multiple on {100}	b = Y, c ∧ Z = -19° to 0°, a ∧ X = -3° to +16°	Intense brown, brown-red. Strong pleochroism Z > Y > X	5-6 3.2-3.3	Brown to black	Pyroxenes, plagioclase, Fe-oxides	Volcanic rocks	Amphibole group	p. 235
		X = bright yellow, Y = orange yellow, Z = dull yellow. Z > Y > X	3.151	Dk brown		Weathering product in alkalic pegmatites		Am Min 60, 736-737
		X = pale golden yel Y = dk lilac gray, Z = dull green	3.38	Dk green, blk		Metamorphosed shale, impure limestones, ironstones		Am Min 50, 278
{100}, {101} & {10$\bar{1}$} polysynthetic	b = Z, a ~ Y, c ~ X	Colorless	6-7 3.47-3.51	Colorless, pink, orange		Mn ores with magnetite, pyroxene, garnet, calcite, hematite		Vlasov II, 117-118

n_β Var.	Refractive Index	Mineral Name and Composition	$(n_\gamma - n_\alpha)$	2V angle Dispersion	Crystal System	Habit	Cleavage
1.720	n_β 1.720 n_α 1.715 n_γ 1.726	Johachidolite $CaAlB_3O_7$	0.011	72° r > v strong	Orthorhombic	Grains and lamellar masses	None
1.72	n_β 1.72 n_α 1.71 n_γ 1.73	Haradaite $SrVSi_2O_7$	0.02	~90°	Orthorhombic	Massive	{010} perf
1.721	n_β 1.721 n_α 1.713 n_γ 1.734	Barium-Haradaite $(Ba,Sr)VSi_2O_7$	0.021	77° v > r Very strong	Orthorhombic	Veinlets	{010} perf., {100} & {001} dist
1.722	n_β 1.722 n_α 1.687 n_γ 1.731	Tarapacaite K_2CrO_4	0.044	52° r > v weak	Orthorhombic	Thick tabular {001} crystals	{001} & {010} distinct
1.720 1.725	n_β 1.720–1.725 n_α 1.712–1.718 n_γ 1.727–1.734	Kyanite Al_2SiO_5	0.012 to 0.016	82° r > v weak	Triclinic α = 90.1° β = 101° γ = 105.7°	Bladed to columnar (c)	{100} perf, {010} dist, {001} parting
1.723	n_β 1.723 n_α 1.685 n_γ 1.736	Glaucochroite $CaMnSiO_4$	0.051	61°	Orthorhombic	Long prismatic, massive	{001} dist
1.725	n_β 1.725 n_α 1.676 n_γ 1.745	Whitmoreite $Fe^{2+}Fe^{3+}_2(OH)_2(PO_4)_2(H_2O)_4$	0.069	60° – 65°	Orthorhombic ?	Long prismatic acicular (c)	{100} fair
1.725	n_β 1.725 n_α 1.692 n_γ 1.738	Metastrengite (Phosphosiderite) $FePO_4 \cdot 2H_2O$	0.046	62° r > v very strong	Monoclinic β = 90.6°	Tabular, prismatic, botryoidal, radiated-fibrous	{010} dist {001} indist
1.725	n_β 1.725 n_α 1.700 n_γ 1.730	Vuagnatite $CaAl(OH)SiO_4$	0.030	48° v > r very strong	Orthorhombic	Tiny prismatic crystals (c) granular	
*1.719 1.734	n_β 1.719–1.734 n_α 1.713–1.730 n_γ 1.723–1.740	Chloritoid $(Fe^{2+},Mg,Mn)_2(Al,Fe^{3+})Al_3O_2(SiO_4)_2(OH)_4$	0.006 to 0.022	36° – 90° r > v strong	Monoclinic or triclinic β = 102°	Foliated-tabular {001} aggregates	{001} perf, {110} imperf parting {010}
1.726	n_β 1.726 n_α 1.694 n_γ 1.730	Tyrolite (Trichalcite) $Ca_2Cu_9(AsO_4)_4(OH)_{10} \cdot 10H_2O$	0.036	~36° r > v strong	Orthorhombic	Reniform, radiated, fibrous, scaly, foliated	{001} perfect
1.727	n_β 1.727 n_α 1.711 n_γ 1.740	Picrotephroite $(Mg,Mn)_2SiO_4$	0.029	85° r > v weak	Orthorhombic	Granular, massive	{010} dist., {100} poor
1.728	n_β 1.728 n_α 1.670 n_γ 1.732	Sussexite $MnBO_3H$	0.062	~25° r > v	Orthorhombic	Fibrous veinlets, porcelaneous, chalky	

* n_β variation exceeds 0.01

Twinning	Orientation	Color in thin section	Hardness Sp Gravity	Color hand sample	Alteration	Occurrence	Remarks	Reference
		Colorless	6½-7 3.44	Colorless		Nepheline dikes	U.V. fluor- intense blue	Dana II, 384
		Colorless, pale blue	3.75-3.80	Light blue				Am Min 56, 1123
	a = X, b = Y, c = Z	X ~ colorless, Y = colorless, yellowish-green, Z = bluish green	4½ 3.80-3.83	Bright green		Veinlets in Mn-ores with rhodonite, Mn-goldmanite, quartz		Am Min 60, 340
{110} cyclic (pseudohexa- gonal)	a = Y, b = X, c = Z		2.74	Bright yellow		Nitrate deposits with dietzeite & lopezite	Easily soluble H_2O	Dana II, 644-645
{100} simple, {001} lamellar	a ~ X in {001}. Maximum extinction ~30°. + elong	Colorless to pale blue. Z > Y > X	4-7½ ~3.6	Lt blue, white, gray, green	Sericite, pyrophyl- lite, chlorite	Pelitic schists, gneisses with alman- dine, staurolite. Pegmatites		p. 121
{001} contact, penetration	a = Z, b = X, c = Y	Colorless, pale bluish-green	6 3.48-3.49	Bluish-green, white, pink		Hydrothermal Mn- deposits	Gelatinizes HCl	USGS PP 180, 79-80
{100} contact persistent. Rarely re- peated	a = X, b = Y, c = Z	X = Y = Lt. green- ish-brown, Z = Dk greenish-brown	3 2.85-2.87	Lt to dk brn, green-brown		Hydrothermal alter of triphylite in pegmatites		Am Min 59, 900-905
{101} penetration	b = Y, c ∧ X ~ 4°	X = pale rose, Y = carmine-red, Z = colorless	3½-4 2.76	Red-violet, pink, yellow, colorless		Pegmatites with strengite, rockbridge- ite, bermanite, barbosalite	Dimorph stren- gite	Dana II, 769-771
None	a = Z, b = Y, c = X	Colorless	3.20-3.42	White		Rodingitic dikes in ophiolites with preh- nite, hydrogrossular, idocrase		Am Min 61, 825-830
{001} simple or multiple	b = Y or X, c ∧ Z = -2° to -30°	X = gray-green, olive-green, Y = slate-blue, indigo, Z = color- less, lt green	6½ 3.26-3.80	Drk gray to dk green	Chlorite, sericite, kaolin clays	Greenschists with chlor- ite, albite. Hydro- thermal veins & cavities		p. 140
{101} cyclic (pseudohex)	a = Z, b = X, c = Y	X = Z = pale green, Y = pale yellow- green	~2 3.18-3.27	Pale apple green to sky blue		Oxide zones Cu-deposits with chalcophyllite, malachite, cuprite, etc	Sectile	Dana II, 925-926
	a = Z, b = X, c = Y	Pale brown	6½-7 4.0	Red-brown	Serpentine	Mn ore deposits	Olivine group	pp. 102,109
	x = elong, parallel extin- ction, Z ⊥ flattening	Colorless	3-3½ 3.30-3.43	White, pinkish, yellowish		Hydrothermal veinlets in Mn-ores with leuco- phoenicite, pyrochroite	Sussexite-szaibel yite series. Slowly soluble acids	Dana II, 375-377

Biaxial —

n_β Var.	Refractive Index	Mineral Name and Composition	$(n_\gamma - n_\alpha)$	2V angle Dispersion	Crystal System	Habit	Cleavage
1.728	n_β 1.728 n_α 1.720 n_γ 1.735	Landesite $Mn_{10}Fe^{3+}_3(PO_4)_8(OH)_5 \cdot 11H_2O(?)$	0.015	Large ~ 85°	Orthorhombic	Pseudo-octahedral	{010} dist
1.728 1.730	n_β 1.728-1.730 n_α 1.670-1.676 n_γ 1.732-1.734	Sarcopside $(Fe,Mn,Mg)_3(PO_4)_2$	0.062 to 0.058	26° - 28° r > v	Monoclinic β ~90°	Fibrous masses	{100} & {001} dist, {010}
1.730	n_β 1.730 n_α 1.690 n_γ 1.749	Phurcalite $Ca_2(UO_2)_3(PO_4)_2(OH)_4 \cdot 4H_2O$	0.059	68°	Orthorhombic	Tabular {010}, elongation c	{001} & {010} perfect, {100} dist
1.731 1.732	n_β 1.732-1.731 n_α 1.712-1.710 n_γ 1.732	Chalcomenite $CuSeO_3 \cdot 2H_2O$	0.020	~0° v > r or r > v	Orthorhombic	Prismatic acicular	None
1.732	n_β ~1.732 n_α 1.728 n_γ 1.732	Britholite-y (Abukumalite) (Yttrobritholite) $(Ca,Y)_5(SiO_4,PO_4)_3(OH,F)$	0.004	Small to 44°	Hexagonal ?	Prismatic crystals, oval forms	None
1.733	n_β 1.733 n_α 1.723 n_γ 1.755	Mukhinite $Ca_2(Al_2V)(SiO_4)_3(OH)$	0.032	~90°	Monoclinic β ~90°	Short prismatic crystals	{001} very perfect, {100} perfect
1.733	n_β 1.733 n_α 1.724 n_γ 1.739	Chalcocyanite (Chalcokyanite) $CuSO_4$	0.015	Large r > v extreme	Orthorhombic	Tabular crystals	
1.734	n_β 1.734 n_α 1.679 n_γ 1.742	Huemulite $Na_4MgV_{10}O_{28} \cdot 24H_2O$	0.063	20° - 30° r > v strong	Triclinic α = 107.2° β = 112.1° γ = 101.5°	Microscopic fibrous aggregates, tabular elongation (c)	{001} perf {010} dist
1.734	n_β 1.734 n_α 1.696 n_γ 1.743	Kirschsteinite $CaFeSiO_4$	0.047	50°	Orthorhombic	Massive	
*1.725 1.744	n_β 1.725-1.744 n_α 1.685-1.690 n_γ 1.748-1.751	Bermanite $Mn^{2+}Mn^{3+}_2(PO_4)_2(OH)_2 \cdot 4H_2O$	0.061 to 0.059	72° - 75° v > r strong	Monoclinic β = 110.5°	Tabular, rosette, massive	{001} perf {110} imperf
1.735	n_β 1.735 n_α 1.715 n_γ 1.745	Sicklerite $(Li,Mn^{2+},Fe^{3+})PO_4$	0.030	Med-large ~ 70° r > v extreme	Orthorhombic	Massive	{100} dist
*1.722 1.747	n_β 1.722-1.747 n_α 1.710-1.734 n_γ 1.725-1.752	Ferrohypersthene $(Fe,Mg)_2(SiO_3)_2$	0.015 to 0.018	53° - 62°	Orthorhombic	Short prism (c) crystals. Massive	{110} dist, {010} or {100} parting
1.732 1.738	n_β 1.732-1.738 n_α 1.714-1.720 n_γ 1.805-1.820	Lopezite $K_2Cr_2O_7$	0.091 to 0.100	50° - 51° r > v moderate	Triclinic α = 98° β = 90.8° γ = 96.2°	Prism (c), spherical aggregates	{010} perf {100} & {001} distinct
1.736 1.741	n_β 1.736-1.741 n_α 1.715-1.721 n_γ 1.739-1.745	Renardite $Pb(UO_2)_4(PO_4)_2(OH)_4 \cdot 7H_2O$	0.024	40° - 45° r > v	Orthorhombic	Platy {100}, lath-like (c), radiated fibrous	{100} perf

* n_β variation exceeds 0.01

Twinning	Orientation	Color in thin section	Hardness Sp Gravity	Color hand sample	Alteration	Occurrence	Remarks	Reference
	Z ⊥ {010}	X = dk brown, Y = lt brown, Z = yellow	3-3½ 3.026	Brown		Granite pegmatites with lithiophilite, apatite, eosphorite		Am Min 49, 1122-1125
{001} polysynthetic	"triclinic" orientation b ~ Y, c ∧ Z ~ 45°	Colorless	4 3.79-3.80	Colorless, gray to brown		Granite pegmatites intergrown with graftonite		Am Min 54, 969-972
	c = Z	X = bright yellow, Y = pale yellow	3 4.03-4.14	Yellow		With specular hematite		Am Min 64, 243
	a = X b = Z c = Y	Pale blue, Z = Y > X	2-2½ 3.31-3.32	Bright blue		Oxidation of Cu or Pb-selenides	Soluble acids	Am Min 49, 1481-1485
	c = X	Brown	5 4.25	Black		Granite pegmatites	Apatite group	Vlasov II, 297-300
{100}			8	Brownish-black		Contact marbles with goldmanite, sulfide minerals	V-clinozoisite (V:Al = 1:2) Epidote group	Am Min 55, 321-322
	a = Y, b = X, c = Z	Colorless, pale blue Z > X = Y	3½ 3.65-3.90	Lt green, lt blue, yellowish		Sublimation, fumeroles	Soluble H_2O (hygroscopic)	Am Min 46, 758-759
		X = pale yellow, Y = golden yellow, Z = yellow-orange	2½-3 2.39-2.40	Yellowish-orange		Botryoidal masses and films in "sandstone-type" U-deposits.		Am Min 51, 1-13
		Colorless	3.43	White, greenish		Alkali lavas with melilite, nepheline, kalsilite, sodalite	Fe-monticellite Olivine group	Min Mag 31, 698-699
Common {101}	b = X, c ∧ Y = 36½°	X = pink, Y = pale yellow, Z = dk red	3½ 2.84	Lt red, dk red-brown		Pegmatites		Am Min 53, 416-431
	a = X	Red-pink, X > Y > Z	~4 3.45	Yellow-brown, Dk brown		Alteration of triphylite in granite pegmatites	Sicklerite-ferrisicklerite series	Dana II, 672-673
{101} normal	a = X, b = Y, c = Z	X = salmon-pink, Y = yellow, Z = blue-green	5-6 3.58-3.73	Dk green, brown, black	Bastite, uralite, talc	Gabbros, diorites, monzonites. Metamorphosed Fe-sediments	Orthopyroxene group. Enstatite-orthoferrosilite series	p. 187
		Orange-red	2½ 2.66-2.69	Orange, red		Nitrate evaporites with dietzeite, tarapacaite, ulexite	Easily soluble H_2O	Dana II, 645-646
	a = X, b = Z, c = Y	X = colorless, Y = Z = yellow	3½ 4.34-4.35	Yellow, brownish-yellow		Secondary U-minerals, alteration of uraninite	Not fluor U.V.	USGS Bull 1064, 227-230

n_β Var.	Refractive Index	Mineral Name and Composition	$(n_\gamma - n_\alpha)$	2V angle Dispersion	Crystal System	Habit	Cleavage
1.736	n_β 1.736 n_α 1.635-1.650 n_γ 1.740-1.750	Hydrozincite $Zn_5(CO_3)_2(OH)_6$	~0.10	Moderate ~ 40° v > r strong	Monoclinic β = 95.5°	Massive, earthy, reniform, lath crystals	{100} perf
1.737	n_β 1.737 n_α 1.723 n_γ 1.756	β-roselite $Ca_2(Co,Mg)(AsO_4)_2 \cdot 2H_2O$	0.033	80° - 90° v > r strong, crossed distinct	Triclinic α = 112° β = 71° γ = 119.5°	Granular	{010} perf
1.738	n_β 1.738 n_α 1.731 n_γ 1.744	Thalenite (Yttrialite) (Rowlandite) $Y_2Si_2O_7$	0.013	68°	Monoclinic β = 97.1°	Tabular or prismatic crystals, massive	None
*1.717 1.760	n_β 1.717-1.760 n_α 1.691-1.730 n_γ 1.731-1.774	Hyalosiderite $(Mg,Fe)_2SiO_4$	~0.042	82° - 72° r > v weak	Orthorhombic	Granular, anhedral crystals	{010} & {110} poor
1.740	n_β 1.740 n_α 1.726 n_γ 1.747	Strashimirite $Cu_8(AsO_4)_4(OH)_4 \cdot 5H_2O$	0.021	70°	Monoclinic β = 97.2°	Platy to fibrous aggregates in spherulites	
*1.718 1.764	n_β 1.178-1.764 n_α 1.697-1.743 n_γ 1.741-1.795	Planchéite $Cu_8(Si_4O_{11})_2(OH)_4 \cdot H_2O$	0.044 to 0.052	~90°	Orthorhombic	Radiated fibrous aggregates	
1.741 1.742	n_β 1.741-1.742 n_α 1.720-1.724 n_γ 1.746	Hodgkinsonite $MnZn_2(OH)_2SiO_4$	0.026 to 0.022	50° - 60° r > v distinct	Monoclinic β = 95.5°	Prismatic, pyramidal, massive granular	{001} perfect
1.742	n_β 1.742 n_α 1.735 n_γ 1.745	Hagendorfite $(Na,Ca)(Fe^{2+},Mn^{2+})_2(PO_4)_2$	0.010	~65°	Monoclinic β 98°	Massive	One perfect, one fair, one poor
1.743	n_β 1.743 n_α 1.701 n_γ 1.787	Libethenite $Cu_2PO_4(OH)$	0.086	~90° r > v strong	Orthorhombic	Short prismatic (c), crusts	{100} & {010} indist
1.743	n_β 1.743 n_α 1.738 n_γ 1.746	Surinamite $(Mg,Fe,Mn)_3(Al,Fe)_{3.99}Si_{3.01}O_{15}$	0.008	67° - 68° very strong	Monoclinic β = 109°	Platy {010}	⊥ {010} distinct
1.740 1.749	n_β 1.740-1.749 n_α 1.654-1.661 n_γ 1.743-1.756	Aurichalcite (Zeyringite) $(Zn,Cu)_5(CO_3)_2(OH)_6$		Very small v > r strong	Orthorhombic	Acicular on c, tufted groups	{010} perf
1.74	n_α>1.74	Klebelsbergite Sb-sulfate		v > r	Monoclinic β = 91.8°	Acicular tufts	
1.745	n_β 1.745 n_α 1.707 n_γ 1.776	Shcherbakovite $Na(K,Ba)_2(Ti,Nb)_2(Si_2O_7)_2$	0.069	82°	Orthorhombic	Prismatic crystals	Two ?

* n_β variation exceeds 0.01

Twinning	Orientation	Color in thin section	Hardness Sp Gravity	Color hand sample	Alteration	Occurrence	Remarks	Reference
	b = X, c ∧ Z ~ small -40°	Colorless	2-2½ 3.5-3.8	White, lt yellow, pink, brown		Oxide zones Zn-deposits pegmatites	May fluor U.V.-blue	Dana II, 247-249
	Fragments on {010} show off-center Bxa(X)	Colorless-pink X > Y > Z	3½-4 3.71	Dk rose red		Oxide zone Co-deposits		Am Min 40, 828-833
		Brownish	6 4.3-4.6	Pink, brown, brownish-grn		Granite pegmatites with fergusonite, allanite, gadolinite, cyrtolite	May be metamict	Vlasov II, 243-246
{100}, {011} & {012} simple. {031} cyclic	a = Z, b = X, c = Y	Colorless to pale yellow-green Y > X = Z	6½-7 3.53-3.77	Pale olive-green	Serpentine, iddingsite, chloro-phaeite	Mafic & ultramafic igneous rocks (basalt, gabbro, norite)	Olivine group	p. 105
	Z ∧ elong ~ 5°, + elongation	Y = very pale yellowish green, Z = yellowish-green	3.81	White to pale green		Oxide zones Cu-deposits Replaces tyrolite and cornwallite		Am Min 54, 1221
	a = Y, b = X, c = Z	X = very pale blue, Y = pale blue, Z = deep blue	5½ 3.65-3.85	Pale blue to dk blue		Oxide zone Cu-deposits with malachite		Science 154, 506-507
	b = Y, c ∧ Z = 38°	X = Z = lavender, Y ~ colorless	4½-5 3.91-4.08	Pink, reddish-brown		Hydrothermal Mn-Zn oxide veins	Gelat HCl	Am Min 49, 415-420
	b = Y	X = yellow-brown, Y = green, Z = blue-green	3½ 3.71	Greenish-black		Pegmatites	Varulite series	Am Min 40, 553
None observed	a = Z, b = X, c = Y	Pale bluish-green to pale green. Weak pleochroism	4 3.93-3.97	Dk-lt green, dk olive		Oxide zones Cu-deposits	Easily soluble acids or NH_4OH	Dana II, 862-864
	b = Y, Z ∧ cleavage = 44°	In {101} ∥ cleavage = blue-green, ⊥ cleavage = colorless, lt green-brown	>3.3	Blue		Mylonitic mesoperthitic gneiss with spinel, kyanite, sillimanite		Am Min 61, 193-199
	a = X, b = Y, c = Z	X = colorless, Y = Z = pale blue-green	1-2 3.64-4.23	Pale green or blue		Oxide zones of Cu deposits	Soluble in acids or ammonia	Dana II, 249-250
	b = X, c ∧ Y ~ 2°	Colorless, yellow. Not pleochroic		Dark yellow		Interstices in stibnite aggregates	Soluble HCl	Dana II, 583
	+ elongation	X = pale yellow, Y = yellow, Z = brownish-yellow	6½ 2.968	Dk brown		Alkali pegmatites with natrolite-pectolite aggregates		Am Min 40, 788

n_β Var.	Refractive Index	Mineral Name and Composition	$(n_\gamma - n_\alpha)$	2V angle Dispersion	Crystal System	Habit	Cleavage
1.749	n_β 1.749 n_α 1.739 n_γ 1.752	Przhevalskite $Pb(UO_2)_2(PO_4)_2 \cdot 2H_2O$	0.013	~30°	Orthorhombic	Foliated aggregate tabular crystals	{001} dist
1.750	n_β 1.750 n_α 1.670 n_γ 1.778	Lomonosovite $Na_8(Mn,Fe,Ca,Mg)Ti_3Si_4P_2O_{24}$	0.108	56°	Triclinic α = 100° β = 96° γ = 90°	Tabular crystals	{100} perf
1.754	n_β 1.754 n_α 1.743 n_γ 1.764	Caracolite $\sim Na_2PbSO_4Cl(OH)$	0.021	Large r > v strong	Orthorhombic ?	Hexagonal prismatic	
1.753 1.755	n_β 1.753-1.755 n_α 1.735-1.736 n_γ 1.766-1.767	Sursassite $Mn_5Al_4Si_5O_{21} \cdot 3H_2O$	0.031	~65° r > v strong	Monoclinic β = 108.4°	Massive, fibrous (b), botryoidal	{001} dist
1.752	n_β 1.752 n_α 1.709 n_γ 1.787	Ahlfeldite $(Ni,Co)SeO_3 \cdot 2H_2O$	0.078	85° v > r strong	Monoclinic β = 99.1°	Crystalline crusts	{110} fair, {103} fair
*1.725 1.784	n_β 1.725-1.784 n_α 1.715-1.751 n_γ 1.734-1.797	Epidote (Pistacite) $Ca_2Al_2Fe^{3+}O \cdot SiO_4 \cdot Si_2O_7(OH)$	0.012 to 0.049	90° - 64° r > v strong	Monoclinic β = 115.4°	Columnar, acicular, granular	{001} perf, {100} imperf
*1.737 1.766	n_β n_α 1.737 n_γ 1.766	Arsenuranylite $Ca(UO_2)_4(AsO_4)_2(OH)_4 \cdot 6H_2O$	0.029		Orthorhombic	Tiny scales	{001} perfect
1.747 1.765	n_β 1.747-1.765 n_α 1.734-1.755 n_γ 1.752-1.775	Eulite $(Fe,Mg)_2(SiO_3)_2$	0.018 to 0.020	62° - 90°	Orthorhombic	Short prism (c) crystals. Massive	{110} dist, {010} or {100} parting
1.750 1.760	n_β 1.750-1.760 n_α 1.700-1.705 n_γ 1.770	Paraschoepite (Schoepite III) $UO_3 \cdot 2H_2O$ (?)	0.070	40°	Orthorhombic	Pseudo-hexagonal, tabular {001}, prismatic (b)	{001} perfect {010} dist
1.755	n_β 1.755 n_α 1.710 n_γ 1.775	Schuilingite $Pb_3Ca_6Cu_2(CO_3)_8(OH)_6 \cdot 6H_2O$	0.065	66°	Monoclinic	Acicular crystals, crusts	{110} perf, {100} poor
1.754 1.758	n_β 1.754-1.758 n_α 1.730 n_γ 1.835-1.838	Azurite $Cu_3(OH)_2(CO_3)_2$	0.105	~68° r > v distinct horizontal dist	Monoclinic β = 92.4°	Tabular or prismatic crystals, globular-radiated	{011} perf, {100} dist, {110} poor
1.756	n_β 1.756 n_α 1.740-1.748 n_γ 1.762-1.765	Uregite $NaCr(SiO_3)_2$	0.022 to 0.017	60° - 70°	Monoclinic β = 107.6°	Polycrystalline aggregates	{110} dist, {001} parting
1.756 1.759	n_β 1.756-1.759 n_α 1.743-1.746 n_γ 1.758-1.761	Holtite $(Al,Sb,Ta)_7(B,Si)_4O_{18}$(?)	0.015	49° - 55° v > r	Orthorhombic	Prismatic, acicular, pseudo-hexagonal	{010} dist

* n_β variation exceeds 0.01

Twinning	Orientation	Color in thin section	Hardness Sp Gravity	Color hand sample	Alteration	Occurrence	Remarks	Reference
	∥ extinction, - elongation	X = colorless, Y = pale yellow, Z = deep yellow		Bright yellow		Oxide zones pitchblende sulfide deposits		Am Min 43, 381-382
Fine polysynthetic	X ∧ cleavage = 61°-66°, Y ∧ cleavage = 59°-65°, Z ∧ cleavage = 37°-41°	Lt cinnamon-brown, cinnamon-yellow	3-4 3.15	Dk red-brown, black		Alkali pegmatites with hackmanite, eudialyte, aegirine		Am Min 35, 1092-1093
Pseudohexagonal trillings		Colorless	4½ ~5.1	Colorless, gray, greenish		Incrustations oxide zones Pb-deposits		Dana II, 546-547
	b = Y, c ∧ X = 55°	X = Z ~ colorless, Y = dk brown	3.19-3.26	Dk red-brown to copper-red		Veinlets in Mn-ores with barite, calcite, quartz		Am Min 49, 168-173
	b = Y, c ∧ Z = +16°	X = pale green, Y = pale pink, Z = pink	2-2½ 3.37-3.51	Brownish to reddish		Alteration product of penroseite		Am Min 54, 448-456
Multiple on {100}	b = Y, c ∧ X = 0° to +15°, a ∧ Z = +25° to +40°	Colorless, pale yellow-green Y > Z > X	7-6 3.21-3.49	Olive green		Low grade metamorphic rocks (regional and contact), plutonic igneous		p. 148
		Yellow, no pleochroism	2½ 4.25	Orange		Secondary U-mineral		Am Min 44, 208
{101} normal	a = X, b = Y, c = Z	X = brown, Y = yellow, Z = blue-green	5-6 3.73-3.87	Dk green, brown, black	Bastite, uralite	Metamorphosed Fe-sediments with Fe-olivine. Diorites, monzonites	Orthopyroxene group. Enstatite-orthoferrosilite series	p. 187
	a = Z, b = Y, c = X	X = colorless, Y = yellow, Z = yellow	2-3	Yellow		Dehydration of schoepite in secondary U-deposits		Am Min 45, 1026-1061
	a ~ X, b = Z, c ~ Y. Extinction = 0° to 10°	Pale blue, weak pleochroism	3-4 5.2	Turquoise to azure blue		Oxide zones Cu-Pb-deposits with cerussite, pyrite, calcite	Decomp dilute acids with effervescence	Am Min 43, 796
{̄101}, {̄102}, {001} all rare	b = X, c ∧ Z = +12°, a ∧ Y = +15°	X = clear blue, Y = azure blue, Z = deep violet blue	3½-4 3.77-3.89	Lt to dk azure blue	Malachite	Oxide zones Cu-deposits with malachite, cuprite	Soluble dilute acids	p. 73
	b = Y, c ∧ X = 8°-22°	X = dk green, Y = yellow green, Z = emerald green	~5½ 3.60	Emerald green		Meteorites with daubreelite, cliftonite, troilite	Pyroxene group	Science 149, 742-744
{110} multiple	a = Z, b = Y, c = X	X = yellow, Y = colorless, Z = colorless	8½ 3.90	White, brown, greenish		Heavy mineral in alluvial deposit with tantalite	Fluor-U.V.-orange (short λ), yellow (long λ)	Min Mag 38, 21-25

n_β Var.	Refractive Index	Mineral Name and Composition	$(n_\gamma - n_\alpha)$	2V angle Dispersion	Crystal System	Habit	Cleavage
**1.700 1.815	n_β 1.700-1.815 n_α 1.690-1.791 n_γ 1.706-1.828	Allanite (Orthite) $(Ca,Ce,La)_2(Al,Fe^{3+},Fe^{2+})_3O\cdot SiO_4\cdot Si_2O_7(OH)$	0.013 to 0.036	40°-90° r > v strong	Monoclinic β = 115°	Tab {100}, acic., gran	{001} imperf., {100} & {110} poor
1.76	n_β 1.76 n_α 1.72 n_γ 1.76	Meta-uranopilite $(UO_2)_6SO_4(OH)_{10}\cdot 5H_2O$	0.04	Small	Orthorhombic	Fibrous, acicular, laths	
1.760	n_β 1.760 n_α 1.724 n_γ 1.772	Niobophyllite $(K,Na)_3(Fe,Mn)_6(Nb,Ti)_2S_8(O,OH,F)_{31}$	0.048	60°	Triclinic α = 113.1° β = 94.5° γ = 103.1°	Aggregates small flakes	{001} perf, {0kℓ} poor
1.757 1.767	n_β 1.767-1.757 n_α 1.761-1.750 n_γ 1.772-1.763	Tadzhikite $Ca_3(Ce,Y)_2(Ti,Al,Fe)Be_4Si_4O_{22}$	0.011 to 0.013	~90°	Triclinic α = 90° β = 100.7° γ = 90°	Prismatic crystals flattened {010}, plates	{010} dist
1.767	n_β 1.767 n_α 1.705 n_γ 1.769	Simplotite $CaV_4O_9\cdot 5H_2O$	0.064	~25° r > v weak, crossed	Monoclinic β = 90.4°	Micaceous plates-flakes	{010} perf
1.765 1.770	n_β 1.765-1.770 n_α 1.735-1.682 n_γ 1.807-1.839	Murmanite $Na_2(Ti,Nb)_2Si_2O_9\cdot H_2O$	0.072 to 0.157	50° - 75°	Monoclinic β ~90°	Platy crystals, flaky, lamellar	{001} perf
1.767 1.768	n_β 1.768-1.767 n_α 1.760-1.762 n_γ 1.770-1.768	Dewindtite (Stasite) $Pb(UO_2)_2(PO_4)_2\cdot 3H_2O$	0.010 to 0.006	Moderate r > v	Orthorhombic	Microscopic tabular {100}	{100} perf
1.786	n_β 1.786 n_α 1.779 n_γ 1.790	Kolicite $Mn_7Zn_4(AsO_4)_2(SiO_4)_2(OH)_8$	0.011	78° v > r strong	Orthorhombic	Grains	None
1.770	n_β 1.770 n_α 1.670 n_γ 1.779	β-lomonosovite $Na_2Ti_2Si_2O_9\cdot NaH_2(PO_4)$?	0.109	10° - 24° r > v	Triclinic α = 102.5° β = 97° γ = 90°	Tabular-platy	One perfect
1.770	n_β 1.770 n_α 1.710 n_γ 1.840	Rossite $CaV_2O_6\cdot 4H_2O$	0.130	~90° Very strong	Triclinic α = 101.5° β = 115° γ = 103.4°	Bladed (c), tabular {001}, acicular (c)	{010} dist
1.770	n_β 1.770 n_α 1.720 n_γ 1.800	Melonjosephite $CaFe^{2+}Fe^{3+}(PO_4)_2(OH)$	0.080	80° - 85° Strong	Orthorhombic	Fibrous, splintery, massive	{110} perf, {010} imperf
1.770	n_β 1.770 n_α 1.745 n_γ 1.775	Vinogradovite $(Na,Ca,K)_4Ti_4AlSi_6O_{23}\cdot 2H_2O$	0.030	41° r > v	Monoclinic (Pseudo-ortho) β = 92°	Prismatic-fibrous crystals, spherulitic	{010} perf

** n_β variation exceeds 0.10

Twinning	Orientation	Color in thin section	Hardness Sp Gravity	Color hand sample	Alteration	Occurrence	Remarks	Reference
{100} rare	b = Y, c ∧ X = +1° to +47°, a ∧ Z = +26° to +72°	Lt. to dk. brown, yellow, green. Z > Y > X	5-6½ 3.4-4.2	Dk. brown, black	Limonite, silica, alumina	Pegmatites, granitic plutons with epidote and Fe-silicates	Epidote group. Often metamict glass	p. 153
	Elong = Y, X ⊥ laths, ∥ extinction	X = colorless, Y = Z = yellow		Gray, greenish, brownish yellow		Secondary U-mineral	Fluor U.V.-yellowish-green	Am Min 37, 950-959
{001} contact	Z ~ ⊥{001}, a = Y, b ∧ X = -13°	X = Y = brownish-yellow, Z = orange-red	3.41-3.42	Brown		Alkali plutons with aegirine-augite, barylite, pyrochlore, albite	Nb-astrophyllite	Can Min 8 (pt 1), 40-52
Polysynthetic	a ∧ Y = 7°, b ∧ Z = 4°, c ∧ Z = 23°	Colorless to brown. Not pleochroic	6 3.73-3.86	Pale gray-brown to dk brown		Zoned alkali pegmatites with albite, aegirine, riebeckite	Hellandite group	Am Min 56, 1838-1839
	b = X, c ∧ Z ~ +58°	X = yellow, Y = Z = green	~1 2.64	Greenish-black, black		Fractures in sandstone with duttonite, melanovanadite, selenium		Am Min 43, 16-24
	b = Z, a ∧ X = small	X = pink, Y = lt brown, Z = dk violet-brown	2-3 2.76-2.84	Lilac to pink Altered yel-brown		Nepheline-syenite & alkali pegmatites with aegirine, eudialyte	Soluble HCl	Vlasov II, 558-562
	a = X (?), b = Z (?), c = Y	X = colorless, Y = Z = golden yel	5.01-5.03	Canary yellow		Secondary U-mineral	Fluor U.V. green	USGS Bull 1064, 230-232
	a = Z, b = X, c = Y	X = colorless, Y = yellow-orange, Z = pale yellow, Z = Y > X	~4½ 4.17-4.20	Yellow-orange		Secondary Zn-ores with willemite, franklinite, sonolite, friedelite	Easily sol. 1:1 HCl	Am Min 64, 708-712
	Extinction inclined to cleavage	Colorless, yellow, brown	~4 2.95-2.98	Yellow-brown, brown, rose		Alkali pegmatites	Var lomonosovite	Am Min 48, 1413-1414
{100}	c ~ Z, b ∧ Y ~ 45°	Yellow	2-3 2.45	Yellow		Glassy grains in metarossite in secondary U-veins	Slowly soluble H_2O	Dana II, 1053-1054
	a = Y, b = Z, c = X	X = dk brown, ~ opaque, Y = greenish-brown, Z = yellow	>5 3.61-3.65	Dk green to black		Pegmatites with triphylite, alluaudite	Soluble HCl	Am Min 60, 946
	b = X, c ∧ Z = 7°	Colorless	~4 2.878	Colorless, white		Alkali pegmatites, contact zones with natrolite, analcite, ramsayite		Am Min 42, 308

n_β Var.	Refractive Index	Mineral Name and Composition	$(n_\gamma - n_\alpha)$	2V angle Dispersion	Crystal System	Habit	Cleavage
1.770	n_β 1.770 n_α 1.750 n_γ 1.780	Ferri-sicklerite $(Li,Fe^{3+},Mn^{2+})PO_4$	0.030	Med-large ~ 70° r > v extreme	Orthorhombic	Massive	{100} dist
1.770	n_β 1.770 n_α 1.762 n_γ 1.774	Esperite (Ca-larsenite) $(Ca,Pb)ZnSiO_4$	0.012	40°	Monoclinic β = 90°	Massive, granular	{010} & {100} distinct {101} poor
1.771	n_β 1.771 n_α 1.726 n_γ 1.780	Paradamite $Zn_2AsO_4(OH)$	0.054	50°	Triclinic α = 104.2° β = 87.9° γ = 103.2°	Sheaf-like crystal aggregates	{010} perfect
1.771	n_β 1.771 n_α 1.727 n_γ 1.798	Margarosanite $(Ca,Mn)_2Pb(SiO_3)_3$	0.071	78° v > r strong	Triclinic α = 110.6° β = 102° γ = 88.5°	Platy, rhombic outline	{010} perf {100} dist {001} fair
1.771	n_β 1.771 n_α 1.728 n_γ 1.800	Brochantite (Kamarezite) $Cu_4SO_4(OH)_6$	0.072	~77° v > r distinct	Monoclinic β = 103.3°	Acicular, tabular, crusts	{100} perf
1.771 1.778	n_β 1.771-1.778 n_α 1.751-1.760 n_γ 1.782-1.790	Leucophoenicite (Leucophenicite) $Mn_7Si_3O_{12}(OH)_2$	0.031	~74° r > v weak	Monoclinic β = 103.9°	Massive granular, prismatic (b), tabular	{001} imperf
1.775	n_β 1.775 n_α 1.772 n_γ 1.777	Britholite (Beckelite) (Lessingite) $(Ca,Ce)_5(SiO_4,PO_4)_3(OH,F)$	0.005	Small to 44°	Hexagonal	Prismatic crystals, massive	None
1.776	n_β 1.776 n_α 1.679 n_γ 1.807	Sahamalite $(Mg,Fe^{2+})(Ce,La)_2(CO_3)_4$	0.128	57° v > r perc	Monoclinic β = 106.8°	Tabular crystals {$\bar{2}$01}	{010} poor
^ 1.775 1.780	n_β 1.775-1.780 n_α 1.740 n_γ ~1.778	Belyankinite $Ca(Ti,Zr,Nb)_6O_{13}\cdot 14H_2O$	~0.04	Small	Orthorhombic or Monoclinic	Massive	1 dir perf
1.777	n_β 1.777 n_α 1.756 n_γ 1.794	Orientite $Ca_2Mn_3^{3+}Si_3O_{12}(OH)$	0.038	68° - 83° v > r very strong	Orthorhombic	Prismatic (c) to tabular crystals	{001} & {120} imperfect
1.777	n_β 1.777 n_α 1.763 n_γ 1.785	Yoshimuraite $(Ba,Sr)_2TiMn_2(SiO_4)_2(PO_4,SO_4)(OH,Cl)$	0.022	~90°	Triclinic α = 93.5° β = 90.2° γ = 95.3°	Bladed-micaceous crystals	
*1.760 1.804	n_β 1.760-1.804 n_α 1.730-1.768 n_γ 1.774-1.816	Hortonolite $(Fe,Mg)_2SiO_4$	~0.044	72° - 62° r > v weak	Orthorhombic	Granular, anhedral crystals	{010} fair, {110} poor
1.778 1.779	n_β 1.778-1.779 n_α 1.765-1.763 n_γ 1.787-1.793	Sonolite $Mn_9(SiO_4)_4(OH,F)_2$	0.022 to 0.030	75° - 82° r > v	Monoclinic β = 100.6°	Prismatic crystals, massive, fine-grained	None
1.780 1.782	n_β 1.780-1.782 n_α 1.756-1.770 n_γ 1.792-1.795	Alleghanyite $Mn_5Si_2O_8(OH)_2$	0.036	72° r > v	Monoclinic β = 109.1°	Platy crystals	None

* n_β variation exceeds 0.01

Twinning	Orientation	Color in thin section	Hardness Sp Gravity	Color hand sample	Alteration	Occurrence	Remarks	Reference
	a = X	Red-brown X > Y > Z	~4 3.2-3.4	Yel-brown, dk brown		Alteration of triphylite in granite pegmatites	Sicklerite-ferri-sicklerite series	Dana II, 672-673
		Colorless	5+ 4.25-4.28	White		Hydrothermal Zn deposits	Fluor-U.V. bright yellow	Am Min 50, 1170-1178
		Colorless	3½ 4.55-4.67	Pale yellow		Oxide zones Zn-deposits with adamite, mimetite	Dimorph adamite	Am Min 51, 1218-1220
		Colorless	2.5-3 4.30-4.33	Colorless		Mn-deposits with franklinite, roeblingite, apophyllite		Am Min 49, 781-782
{100} pseudo-ortho	b = Y, a ∧ X ~ 13°, c ~ Z	Bluish-green, weak pleochroism Z > Y = X	3½-4 3.97-4.09	Lt green-dk green		Oxide zones Cu-deposits	Soluble in acids	Dana II, 541-544
{001} contact or penetration	X ⊥ {001}	Colorless to pale rose	5½-6 3.85	Pink, red-violet, brown		Late hydrothermal with willemite, franklinite. Skarns	Gelat HCl	Am Min 55, 1146-1166
	c = X	X = colorless, Y = Z = colorless, pale brown	5 3.86-4.69	Yellow, green, brown, black		Alkali igneous rocks, nepheline pegmatite, contact met	Apatite group. Var fynchenite or Th-britholite, alumobritholite, hydrobritholite	Vlasov II, 297-300
	b = Y, c ∧ Z = -29°	Colorless	4.30	Colorless		Barite-dolomite rock with bastnaesite, parisite	~Insoluble acids	Am Min 38, 741-754
		X = brown, Y = Z = lt brown	2-3 2.32-2.40	Lt yellow, brownish yel		Nepheline pegmatite		Am Min 43, 1220
	a = Z, b = X, c = Y	X = yellow, Y = yellow-brown, Z = red-brown	4½-5 3.05-3.09	Dk red-brown to black		Veinlets and cavities in Mn-oxides		Am Min 46, 226-232
(010) polysynthetic			4½ 4.13-4.21	Orange-brown		Pegmatites in massive Mn-ores		Min J (Japan) 3, 156-167
{100}, {011} & {012} simple, {031} cyclic	a = Z, b = X, c = Y	Colorless to pale yellow-green Y > X = Z	6½-7 3.77-4.01	Pale yellow-green	Serpentine, iddingsite, chlorophaeite	Mafic and alkaline igneous rocks	Olivine group	p. 105
{011} simple or polysynthetic	a ∧ X ~ 10°	Colorless	5.5 3.82-3.97	Red-orange, red-brown, dk brown		Mn-ores with rhodochroaite, galaxite, pyrochroite, willemite, zincite	Mn-clinohumite. Decomp HCl	Am Min 54, 1392-1398
Multiple {001} common		Pinkish, non-pleochroic	5½ 4.02	Pink to brownish-pink		With other primary Mn-minerals (e.g. tephroite, willemite)	Mn analogue of chondrodite	Am Min 54, 1392-1398

n_β Var.	Refractive Index	Mineral Name and Composition	$(n_\gamma - n_\alpha)$	2V angle Dispersion	Crystal System	Habit	Cleavage
1.78	n_β 1.78	Arthurite $Cu_2Fe_4(AsO_4,PO_4,SO_4)_4(O,OH)_8 \cdot 8H_2O$	Low-medium		Monoclinic β = 92.2°	Radiated fibrous	
*1.742 1.820	n_β 1.820-1.742 n_α 1.776-1.722 n_γ 1.836-1.758	Aegirine (Acmite) $NaFe^3(SiO_3)_2$	0.060 to 0.036	58° - 90° r > v moderate-strong	Monoclinic β = 106°	Stubby to acicular	{110}, dist, {001} & {100} parting
1.782	n_β 1.782 n_α 1.752 n_γ 1.815	Shattuckite $Cu_5(SiO_3)_4(OH)_2$	0.063	~90°	Orthorhombic	Radiated acicular-fibrous (c), granular compact	{010} & {100} perfect
1.782	n_β 1.782 n_α 1.767 n_γ 1.791	Teineite $Cu(Te,S)O_3 \cdot 2H_2O$	0.024	36°	Orthorhombic	Prismatic crystals, intergrown crystals, crusts	{001} dist, {100} indist
1.780 1.785	n_β 1.780-1.785 n_γ 1.780-1.785	Marthozite $Cu(UO_2)_3(SeO_3)_3(OH)_2 \cdot 7H_2O$	Small	39°	Orthorhombic	Thick tabular {100}	{100} perf, {010} imperf
1.784	n_β 1.784 n_α 1.765 n_γ 1.799	Malayite $CaSnSiO_5$	0.034	~85°	Monoclinic or Triclinic	Massive	
1.78 1.79	n_β 1.78-1.79 n_α 1.76-1.77 n_γ 1.80	Vandenbrandeite (Uranolepidite) $CuUO_4 \cdot 2H_2O$	0.04	~90° Strong disp	Triclinic α = 91.9° β = 102° γ = 89.6°	Radiated fibrous rosettes, lamellar, tabular {001}	{110} perf
1.786	n_β 1.786 n_α 1.779 n_γ 1.790	Kolicite $Mn_7Zn_4(AsO_4)_2(SiO_4)_2(OH)_8$	0.011	78° v > r strong	Orthorhombic	Grains	None
1.79	n_β 1.79 n_α 1.74 n_γ 1.82	Reinerite $Zn_3(AsO_3)_2$	0.08	~75°	Orthorhombic	Rough prismatic (c) crystals	{110} dist, {011} & {111} indistinct
1.789 1.793	n_β 1.789-1.793 n_α 1.750-1.756 n_γ 1.800-1.809	Thortveitite $(Sc,Y)_2Si_2O_7$	0.050 to 0.053	60° - 65° v > r distinct	Monoclinic β = 102.7°	Prismatic crystals	{110} dist, {001} parting
1.796	n_β 1.796 n_α 1.758 n_γ 1.810	Tlalocite $Cu_{10}Zn_6(TeO_4)_2(TeO_3)Cl(OH)_{25} \cdot 27H_2O$	0.052	60° - 65°	Orthorhombic or Monoclinic	Spherules, lath crystals	
1.797	n_β 1.797 n_α 1.773 n_γ 1.814	Brüggenite $Ca(IO_3)_2 \cdot H_2O$	0.041	88° r > v moderate	Monoclinic β = 95.25°	Columnar, crusts	None ?
1.798	n_β 1.798 n_α 1.720 n_γ 1.805	Guilleminite $Ba(UO_2)_3(OH)_4(SeO_3)_2 \cdot 3H_2O$	0.085	~35° r > v strong	Orthorhombic	Tabular crystals, coatings	{100} perf, {010} dist
1.798	n_β 1.798 n_α <1.790 n_γ 1.802	Compreignacite $K_2U_6O_{19} \cdot 11H_2O$	<0.012	10° - 15° No disp	Orthorhombic	Microscopic plates {001}	{001} perf

* n_β variation exceeds 0.01

Twinning	Orientation	Color in thin section	Hardness Sp Gravity	Color hand sample	Alteration	Occurrence	Remarks	Reference
		Pale olive-green	3.02	Apple-green, emerald-grn		Oxide zones of ore deposits		Min Mag 37, 519-521
Common on {100}	b = Y, c ∧ X = -10° to +12°, a ∧ Z = +28° to +6°	X = bright green, Y = yellow green, Z = yellow	6 3.4-3.6	Dark green, brown, black	Chlorite, epidote, Na-uralite, Fe-oxides	Feldspathoidal intrusives, Na-Fe-rich schists and gneisses	Pyroxene group	p. 199
	a = Y, b = X, c = Z	X = very pale blue, Y = pale blue, Z = deep blue	~3½ 4.11-4.14	Blue to dk blue		Alteration of other secondary Cu-minerals		Am Min 51, 266-267
	c = Z	X = greenish-blue, Y = blue, Z = indigo blue	2½ 3.80-3.85	Sky blue, bluish gray		Hydrothermal veins with quartz, barite, sylvanite, tetrahedrite	Soluble HCl or HNO_3	Vlasov II, 759-760
	a = X, b = Y, c = Z	X = yellow-brown, Y = greenish-yel	4.4-4.7	Yellow-green, green-brown		Oxide zones Cu-U-deposits	Soluble HNO_3,HCl	Am Min 55, 533
		Colorless	3½-4 4.3	Colorless, pale yellow		Coatings on cassiterite associated with varlamoffite	Isostructural sphene. Fluor. short λ U.V.-yellow-green	Am Min 46, 768-769
	Z ∧ elong. ~ 40°, optic axis ~ ⊥{001}	Absorption green to colorless	4 5.08-5.26	Dk green to greenish black		Oxidation of uraninite with curite, uranophane, cuprosklodowskite	Soluble warm, dilute acids	USGS Bull 1064, 100-103
	a = Z, b = X, c = Y	X = colorless, Y = yellow-orange, Z = pale yellow, Z = Y > X	~4½ 4.17-4.20	Yellow-orange		Secondary Zn-ores with willemite, franklinite, sonolite, friedelite	Easily sol. 1:1 HCl	Am Min 64, 708-712
		~colorless	5-5½ 4.27-4.28	Aqua blue, yellow-green		Oxide zones Zn-deposits	Soluble HCl	Am Min 44, 207-208
{110} common	b = Z, c ∧ X = 5°	X = greenish, Y = Z = yellowish	6-7 3.39-3.58	Grayish-green		Granite pegmatites with euxenite, monazite, fergusonite		Vlasov II, 212-217
	∥ extinction, + elongation	X = yellow-green, Y = Z = bluish-grn, Z > Y > X	1 4.55	Capri blue		Oxide zone Cu-Pb ore deposits with cerussite, teineite	Gummy, sectile	Min Mag 40, 221-226
	b = Z, a ∧ X = 9°, (-) elongation	Colorless	3½ 4.24-4.27	Colorless, yellow		Evaporites with $NaNO_3$		Am Min 57, 1911
	a = Z, b = Y, c = X	X = bright yellow, Y = yellow, Z = colorless	4.88-4.92	Canary yellow		Oxide zones Cu-Co-deposits		Am Min 50, 2103
{110} common	a = Y, b = Z, c = X	X = colorless, Y = Z = yellow	5.03-5.13	Yellow		Oxide zones U-deposits		Am Min 50, 807-808

n_β Var.	Refractive Index	Mineral Name and Composition	$(n_\gamma - n_\alpha)$	2V angle Dispersion	Crystal System	Habit	Cleavage
1.800	n_β 1.800 n_α 1.787 n_γ 1.805	Kidwellite $NaFe^{3+}_9(OH)_{10}(PO_4)_6 \cdot 5H_2O$	0.018	Large	Monoclinic β = 113°	Tufts of feathery crystals	{100} perfect
1.800	n_β 1.800 n_α 1.700 n_γ 1.84	Roubaultite $Cu_2(UO_2)_3(OH)_{10} \cdot 5H_2O$	0.14	~62°	Triclinic α = 86.5° β = 134.2° γ = 93.2°	Platy crystals {100}, rosettes	{100} perf, {010} dist
1.80	n_β 1.80 n_α 1.715 n_γ 1.87	Mandarincite $Fe^{3+}Se_3O_9 \cdot 4H_2O$	0.155	85°	Monoclinic <β = 98°18'	Elong. on c	None
*1.780 1.822	n_β 1.780-1.822 n_α 1.725-1.730 n_γ 1.790-1.829	Billietite $BaU_6O_{19} \cdot 11H_2O$	0.045	35° - 47° r > v	Orthorhombic	Tabular {001}, pseudo-hexagonal	{001} perf, {110} & {010} imperfect
1.801	n_β 1.801 n_α 1.793 n_γ 1.816	Volborthite (Uzbekite)(Vesbine) $Cu_3(VO_4)_2 \cdot 3H_2O$	0.023	Large ~ 73° r > v inclined	Monoclinic	Scaly, radiated fibrous, spherulitic, spongy	{001} perf
1.803	n_β 1.803 n_α 1.793 n_γ 1.808	Perloffite $Ba(Mn,Fe^{2+})_2Fe^{3+}_2(OH)_3(PO_4)_3$	0.015	70° - 80° v > r strong	Monoclinic β = 100.4°	Spear-shaped crystals	{100} perf
1.80 1.81	n_β 1.80-1.81	Muskoxite $Mg_7Fe^{3+}_4O_{13} \cdot 10H_2O$		10° - 40°	Trigonal ?	Hexagonal plates, aggregate minute crystals	{001} perf
1.807	n_β 1.807 n_α 1.793 n_γ 1.809	Sarkinite $Mn_2AsO_4(OH)$	0.016	83° Very weak dispersion	Monoclinic β = 108.7°	Tabular {100}, prism (b), spherical	{100} dist
1.810	n_β 1.810 n_α 1.787 n_γ 1.816	Arsenoclasite $Mn_5(OH)_4(AsO_4)_2$	0.029	26° - 53°	Orthorhombic	Massive, granular	{010} perfect
1.81	n_β 1.81	Kegelite $Pb_{12}(Zn,Fe)_2Al_4(Si_{11}S_4)O_{54}$		Very small		Tiny pseudohexagonal plates	None ?
*1.795 1.831	n_β 1.795-1.831 n_α 1.730-1.800 n_γ 1.771-1.846	Conichalcite $CuCaAsO_4OH$	~0.045	90° - small r > v or v > r strong	Orthorhombic	Coliform masses, radiated fibrous	None
1.81	n_β~1.81 n_α~1.74 n_γ~1.81	Gerasimovskite $(Mn,Ca)_2(Nb,Ti)_5O_{12} \cdot 9H_2O$	0.07	~18°	Unknown	Platy masses	One perfect
1.81	n_β 1.81 n_α 1.765 n_γ 1.82	Cornetite $Cu_3PO_4(OH)_3$	0.055	~33° v > r strong	Orthorhombic	Prismatic (c), crusts	None

* n_β variation exceeds 0.01

Twinning	Orientation	Color in thin section	Hardness Sp Gravity	Color hand sample	Alteration	Occurrence	Remarks	Reference
	b = Y	Colorless	3 3.34	Yellow-green, yellow, lt green		Arkansas novaculite with rockbridgeite, dufrenite, beraunite		Min Mag 42, 137
	Optic plane = (100), extinction angle = 20° - 24°	X' = Y' = colorless Z' = greenish-yel	~3 5.02	Green		Oxide zones U-deposits with becquerelite, vandenbrandeite		Min Abs. 22, 229
Tw. plane {100}	b = X, c ∧ Z = 2°	Colorless	2½ 2.93	Lt. green		Oxidized zone of iron-selenium occurrences		Am Min 65, 206
{110}	a = Y, b = Z, c = X	X = colorless, Y = greenish yellow Z = deep yellow	 5.27-5.28	Yellow, yellow-brown		Secondary U-mineral		Am Min 45, 1026-1061
Lamellar common	b = Y, - elongation	Pale green to yellowish green. Faint pleochroism	3½ 3.5-3.8	Yellow to olive-green, brown		Secondary mineral in detrital sediments	Soluble acids	Am Min 45, 1307-1309
	b = X, c ∧ Y ~ 42°	X = Z = dk greenish brown, Y = lt greenish brown	~5 3.996	Dk brown, green-brown, black		Pegmatites with ludlamite, hureaulite and siderite	Fe^{3+}-bjarebyite. Slowly soluble cold 1:1 HCl	Am Min 62, 1059
		Amber to deep red. Not pleochroic	~3 3.10-3.20	Dk reddish-brown		Veinlets with serpentine		Am Min 54, 684-696
	b = Y, c ∧ X = +54°	Pale rose to yellow X > Z > Y	4-5 4.04-4.18	Blood-red, flesh-pink, yellow		Mn-ores with Mn-silicates and Mn-oxides	Easily soluble dilute acids	Dana II, 855-857
	a = Y, b = X, c = Z	Red	5-6 4.16-4.27	Red-brown		Fissures in carbonate rocks with Mn minerals		Am Min 17, 251
						Oxide zones ore deposits with hematite, mimetite		Am Min 61, 175-176
	a = Z, b = Y, c = X	Pale green, blue-green	4½ 4.1-4.3	Green, yellow-green		Oxide zones Cu-deposits		Dana II, 806-808
		X = brown, Y = Z = Lt brown	2 2.52-2.58	Brown, gray		Pegmatites	Nb-belyankinite	Am Min 43, 1220-1221
{h0ℓ} rare	a = Y, b = X, c = Z	Pale green-blue. Not pleochroic	4½ 4.10	Dk blue, greenish-blue		Oxide zones Cu-deposits	Soluble HCl	Dana II 789-791

n_β Var.	Refractive Index	Mineral Name and Composition	$(n_\gamma - n_\alpha)$	2V angle Dispersion	Crystal System	Habit	Cleavage
1.81	n_β 1.81 n_α 1.77 n_γ 1.82	Tephroite Mn_2SiO_4	0.04	70° - 64° r > v weak	Orthorhombic	Stubby crystals	{010} dist, {001} poor
*1.770 1.850	n_β 1.850-1.770 n_α 1.785-1.762 n_γ 1.851-1.770	Mitridatite $Ca_3Fe_4^{3+}(PO_4)_4(OH)_6 \cdot 3H_2O$	~0.065	5° - 10°	Monoclinic β = 95.9°	Tabular, pseudo-rhomb, crusts	{100} dist
1.814	n_β 1.814 n_α 1.776 n_γ 1.836	Julgoldite $Ca_2Fe^{2+}(Fe^{3+},Al)_2(SiO_4)(Si_2O_7)(OH)_2 \cdot H_2O$	0.060	50° - 70°	Monoclinic β = 97.5°	Bladed elongation (b), flat {001}	{100} dist
1.815	n_β 1.815 n_α 1.775 n_γ 1.825	Pascoite $Ca_2V_6O_{17} \cdot 11H_2O$	0.050	~50° Crossed strong	Monoclinic β = 93.1°	Granular crusts, minute laths	{010} dist
1.810 1.820	n_β 1.810-1.820 n_α 1.772-1.780 n_γ 1.863-1.865	Olivenite (Leucochalcite) $Cu_2AsO_4(OH)$	0.091 to 0.085	~90° r > v strong	Orthorhombic	Prismatic-acicular-fibrous (c or a), reniform	{011} & {110} indistinct
*1.810 1.825	n_β 1.810-1.825 n_α 1.788 n_γ 1.830	Hancockite $(Pb,Ca,Sr)_2(Al,Fe)_3Si_3O_{12}$	0.042	~38° - 50° r > v perc	Monoclinic β = 116°	Lath-shaped crystals, massive	{001} perfect
1.820	n_β 1.820 n_α 1.772 n_γ 1.821	Robertsite $Ca_3Mn_4^{3+}(PO_4)_4(OH)_6 \cdot 3H_2O$	~0.50	~8°	Monoclinic β = 96°	Thick tabular, pseudorhombic, radiated fibrous	{100} dist
1.82	n_β 1.82 n_α 1.78 n_γ 1.83	Manganknebelite $(Mn,Fe)_2SiO_4$	0.05	64° - 51° r > v weak	Orthorhombic	Stubby crystals	{010} dist, {001} poor
*1.816 1.827	n_β 1.816-1.827 n_α 1.770 n_γ 1.856-1.863	Calcurmolite $Ca(UO_2)_3(MoO_4)_3(OH)_2 \cdot 11H_2O$	0.080 to 0.093	Large	Unknown	Prismatic, platy, massive	
1.825 1.835	n_β 1.825-1.835 n_α 1.730-1.87 n_γ 1.830-1.88	Becquerelite $CaU_6O_{19} \cdot 11H_2O$	0.10 to 0.13	~30° - 35° r > v	Orthorhombic	Prismatic (b), coatings	{001} perf, {101}, {010} & {110} imperfect
1.825	n_β 1.825 n_α 1.842 n_γ 1.857	Dietzeite $Ca_2(IO_3)_2CrO_4$	0.015	86° v > r strong, inclined	Monoclinic β = 106.5°	Fibrous crusts, tab {100}, elongation (c)	{100} imperf
1.827	n_β 1.827 n_α 1.786 n_γ 1.83	Iimoriite $Y_5(SiO_4)_3(OH)_3$	~0.04	5° - 15°	Triclinic	Massive	{011} dist
*1.804 1.848	n_β 1.804-1.848 n_α 1.768-1.806 n_γ 1.816-1.858	Ferrohortonolite $(Fe,Mg)_2SiO_4$	~0.046	62° - 51° r > v weak	Orthorhombic	Stubby anhedral crystals	{010} good, {110} poor
1.83	n_β 1.83 n_α 1.81 n_γ	Welshite $Ca_2Mg_4Fe^{3+}O_2Si_4Be_2O_{18}$	>0.02	~24°	Triclinic α = 106° β = 96° γ = 124°	Prismatic crystals	None

* n_β variation exceeds 0.01

Twinning	Orientation	Color in thin section	Hardness Sp Gravity	Color hand sample	Alteration	Occurrence	Remarks	Reference
Rare	a = Z, b = X, c = Y	X = reddish-brown, Y = pink, Z = blue-green	6-6½ 3.78-3.84	Reddish to gray, green	Mn-chlorites and serpentine	Contact or regional metamorphis rocks with other Mn-minerals	Olivine group	p. 109
	X ~ ⊥{100}	X = pale greenish-yellow, Y = Z = deep greenish-brown, red	3½ 3.08-3.24	Dk red, green, yellow-brown, black		Cavities in pegmatites with triphylite, jahnsite, collinsite	Robertsite-mitridatite series	Am Min 59, 48-59
{001} often repeated	b = Y	X = pale brown, Y = pale brownish green, Z = dk emerald green	4½ 3.60	Dark green to black		Cavities in Fe-Mn ore veins with apophyllite, barite	Soluble hot HCl	Am Min 56, 2157-2158
	b = X	X = pale yellow, Y = yellow, Z = orange	~2½ 1.87-1.88	Yellow-orange to red-orange		Secondary U-V-deposits with carnotite	Soluble H_2O	Dana II, 1055-1056
	c = Y	Pale green to yel Weak pleochroism Y > X = Z	3 4.38-4.45	Olive, brown, yellow, white		Oxide zones Cu-deposits with secondary Cu-minerals	Soluble acids, NH_4OH	Dana II, 859-861
	b = Y	Y = pale brownish yellow, X or Z = pale rose or greenish-yellow	6-7 4.03	Yellow-brown, brownish red		Contact metamorphic deposits	Epidote group	USGS PP 180, 98
{100} normal	X ~ ⊥{100}	X = pale pink, Y = Z = deep reddish brown	3½ 3.05-3.17	Lt - dk red-brown, black		Alter triphylite in granite pegmatites with whitlockite	Robertsite-mitridatite series	Am Min 59, 48-59
Rare	a = Z, b = X, c = Y	Colorless, pale green or pink, X = pinkish, Z = blue-green	6-6½ ~3.9	Dk green, reddish	Mn-chlorites and serpentine	Contact or regional metamorphic rocks with other Mn-minerals	Olivine group	p. 109
		X = colorless, Y = pale yellow Z = yellow		Honey yellow		Secondary U-mineral	Fluor U.V. strong yellow-grn	Am Min 49, 1152-1153
{101}	a = Z, b = X, c = Y	X = colorless, Y = Z = yellow	2½ 5.09-5.14	Golden yellow		Secondary U-mineral from uraninite		Am Min 45, 1026-1061
	b = Y, c ∧ Z = 6°	Yellow	3½ 3.61-3.62	Deep golden yellow		Cavities in nitrate sediments	Soluble H_2O	Dana II, 318-319
		Colorless	5½-6 4.21	Purplish-gray		Pegmatites with rare earth minerals		Am Min 58, 140
{100}, {011} & {012} simple, {031} cyclic	a = Z, b = X, c = Y	Colorless to pale yellow. Y > X = Z	6½ 4.01-4.25	Pale greenish yellow	Limonite, chlorite	Mafic and alkaline igneous rocks	Olivine group	p. 105
		No pleochroism	6 3.71-3.77	Reddish-black		Dolomite skarn with berzeliite	Aenigmatite group. May be biaxial +	Min Mag 42, 129

Biaxial —

n_β Var.	Refractive Index	Mineral Name and Composition	$(n_\gamma - n_\alpha)$	2V angle Dispersion	Crystal System	Habit	Cleavage
1.83	n_β 1.83 n_α 1.66 n_γ 1.91	Umohoite $UO_2MoO_4 \cdot 4H_2O$	0.25	65°	Monoclinic β = 99.1°	Platy crystals, fine crystalline	None
1.830	n_β~1.830 n_α 1.672-1.708 n_γ 1.832	Rosasite (Zn-malachite) $(Cu,Zn)_2CO_3(OH)_2$	0.160	Very small v > r strong	Monoclinic β ~90°	Botryoidal crusts, fibrous-felty crystals	Two dir at 90°
1.830	n_β 1.830 n_α 1.808 n_γ 1.860	Bafertisite $BaFe_2TiSi_2O_9$	0.052	54°	Monoclinic β = 94°	Acicular aggregates	1-distinct, 1-poor
*1.810 1.853	n_β 1.810-1.853 n_α 1.775-1.815 n_γ 1.826-1.867	Knebelite $(Fe,Mn)_2SiO_4$	0.051	44° - 61° r > v moderate	Orthorhombic	Massive, granular, short prismatic crystals	{010} dist, {001} & {110} imperfect
1.834	n_β 1.834 n_α 1.783 n_γ 1.866	Delrioite $CaSrV_2O_6(OH)_2 \cdot 3H_2O$	0.083	Medium-large ~78°	Monoclinic β = 102.5°	Radiated, fibrous aggregates	
1.835	n_β 1.835 n_α 1.68 n_γ 1.865	Metatyuyamunite $Ca(UO_2)_2(VO_4)_2 \cdot 3\text{-}5H_2O$	0.185	45° v > r weak	Orthorhombic	Minute scales, laths, microcrystalline	{001} perf, {010} & {100} distinct
1.838	n_β 1.838 n_α 1.809 n_γ 1.850	Linarite $PbCuSO_4(OH)_2$	0.041	80° v > r strong	Monoclinic β = 102.8°	Tabular {001}, elongation (b), crusts	{100} perf, {001} imperf
1.840	n_β 1.840 n_α 1.775 n_γ 1.844	Chalcosiderite $CuFe_6(PO_4)_4(OH)_8 \cdot 4H_2O$	0.069	~22° r > v strong, crossed	Triclinic α = 112.5° β = 115.3° γ = 69°	Prismatic, sheaf-like groups	{001} perf, {010} dist
1.840	n_β 1.840 n_α 1.792 n_γ 1.888	Lautarite $Ca(IO_3)_2$	0.096	~90°	Monoclinic β = 106.4°	Short prismatic, radiated aggregates	{011} dist, {100} & {110} trace
1.842	n_β 1.842 n_α 1.795 n_γ 1.874	Stranskiite $Zn_2Cu(AsO_4)_2$	0.079	80°	Triclinic α = 111° β = 113.5° γ = 86°	Radiated aggregates	{010} perf, {100} dist, {001} & {$\bar{1}$01}
1.85	n_β 1.85 n_α 1.65 n_γ 1.90	Vanuralite $Al(UO_2)_2(VO_4)_2(OH) \cdot 11H_2O$	0.25	44° v > r weak	Monoclinic β = 103°	Platy crystals {001}	{001} perf
1.85	n_β 1.85 n_α 1.77 n_γ 1.89	Derriksite $Cu_4(UO_2)(SeO_3)_2(OH)_6 \cdot H_2O$	0.12	68°	Orthorhombic	Elongate crystals (c), flat {100}	{010} dist
1.850	n_β 1.850 n_α 1.780 n_γ 1.860	Vandendreisscheite $PbU_7O_{22} \cdot 12H_2O$	0.080	~60° r > v	Orthorhombic	Prismatic (a) crystals, tabular {001}, pseudohexagonal	{001} perfect

* n_β variation exceeds 0.01

Twinning	Orientation	Color in thin section	Hardness Sp Gravity	Color hand sample	Alteration	Occurrence	Remarks	Reference
			~2 4.55-4.93	Bluish-black to black		Veinlets in U-ore veins and alternate zones		USGS Bull 1064, 148-149
	X = elongation Y ⊥ a cleavage	X = colorless, Y ~ Z = pale blue	~4½ 4.0-4.2	Green, blue-green, sky-blue		Oxide zones Cu-Zn-Pb ores	Soluble acids	Dana II, 251-252
		X = yellow, red, Z = pale yellow	~5 3.8-4.25	Red, light brown		Hydrothermal veins		Am Min 45, 754
{011}, {012} or {031} rare	a = Z, b = X, c = Y	Colorless, pale yellow	6½ 3.96-4.25	Gray-black, brown-black		Metamorphosed Fe-Mn deposits	Olivine group	p. 109
{100} common	∥ extinction, + elongation	X = colorless, Y = pale yellow, Z = deeper yellow	~2 3.16	Pale yellow-green		Mine efflorescence		Am Min 55, 185-200
	a = Y, b = Z, c = X	X ~ colorless, Y = very pale yel, Z = pale yellow	~2 3.8-3.9	Canary yellow to greenish-yellow		Secondary U-deposits with tyuyamunite, carnotite	Not fluor U.V.	USGS Bull 1064, 254-257
{100} common	b = Z, c ∧ X = +24°	X = Lt blue, Y = clear blue, Z = Prussian blue	2½ 5.30-5.32	Dk azure blue		Oxide zones Cu-Pb-deposits		Dana II, 553-555
		X = colorless, Z = pale green	4½ 3.22-3.26	Dark green		Oxide zones Cu-deposits	Turquoise group	Dana II, 947-951
		Colorless	3½-4 4.48-4.52	Colorless, yellowish		Evaporite nitrate deposits with gypsum		Dana II, 312-313
		Pale blue	4 5.23	Sky blue		Oxide zones Cu-deposits with chalcocite		Am Min 45, 1315
	b = Z, a ~ Y, c ~ X	X = colorless, Y = Z = pale yel	~2 3.62	Citron-yellow		Supergene U-Pb-V deposits with france-villite, chervetite		Am Min 56, 639-640
	a = X, b = Y, c = Z	Pale green	4.72	Bottle green		Oxide zones Cu-Co-deposits		Am Min 57, 1912-1913
	a = Z, b = Y, c = X	X ~ colorless, Y = Z = golden yel	~3 5.45	Orange to amber		Altered pegmatites with fourmarierite, ruther-fordine, gummite		Am Min 45, 1026-1061

Biaxial –

n_β Var.	Refractive Index	Mineral Name and Composition	$(n_\gamma - n_\alpha)$	2V angle Dispersion	Crystal System	Habit	Cleavage
1.850	n_β 1.850 n_α 1.780 n_γ 1.860	Metavandendriesscheite $PbU_7O_{22}\cdot nH_2O$ (n < 12)	0.080	~60°	Orthorhombic	Prismatic crystals (a), tabular	{001} perfect
1.85	n_β 1.85 n_α 1.79 n_γ 1.89	Mroseite $CaTe(CO_3)O_2$	0.10	74° - 80° v > r strong	Orthorhombic	Massive, radiated, prismatic (c)	
1.856	n_β 1.856 n_α 1.791 n_γ 1.867	Pseudomalachite (Tagilite) $Cu_5(PO_4)_2(OH)_4\cdot H_2O$	0.076	48° v > r strong	Monoclinic β = 91°	Radiated fibrous botryoidal, prismatic (c) crystals	{100} perf difficult
*1.848 1.869	n_β 1.848-1.869 n_α 1.806-1.827 n_γ 1.858-1.879	Fayalite Fe_2SiO_4	~0.052	51° - 46° r > v weak	Orthorhombic	Stubby crystals	{010} dist, {110} poor
**1.67 2.05	n_β 1.67-2.05	Cervantite Sb_2O_4		Small	Orthorhombic	Massive, fine-grained	{001} perf, {100} dist
1.86	n_β 1.86 n_α 1.82 n_γ 1.87	Iron Knebelite $(Fe,Mn)_2SiO_4$	0.05	~45° r > v weak	Orthorhombic	Stubby crystals	{010} dist, {001} poor
1.861	n_β 1.861 n_α 1.831 n_γ 1.880	Atacamite $Cu_2Cl(OH)_3$	0.049	75° v > r strong	Orthorhombic	Slender prismatic on c	{010} perf, {101} dist
1.86	n_β 1.86 n_α 1.85 n_γ 1.86	Parsonite $Pb_2UO_2(PO_4)_2\cdot 2H_2O$	0.01	~11° - 26°	Triclinic α = 101.4° β = 98.2° γ = 86.3°	Prismatic (c), radiated acicular, earthy	{010} indist
1.866	n_β 1.866 n_α 1.818 n_γ 1.909	Caledonite $Cu_2Pb_5(SO_4)_3CO_3(OH)_6$	0.091	~85° v > r very weak	Orthorhombic	Elongation prismatic divergent, massive	{010} perf, {100} & {101} imperfect
1.87	n_β 1.87 n_α 1.81 n_γ 1.92	Bahianite $Sb_3Al_5O_{14}(OH)_2$	0.11	Large r > v	Monoclinic β = 100°	Crystals in vugs	{100} perfect
1.87	n_β 1.87 n_α 1.84 n_γ 1.87	Deerite $(Fe^{2+},Mn,Mg)_{13}(Fe^{3+},Al)_7Si_{13}O_{44}(OH)_{11}$	0.03	Small	Monoclinic β = 107°	Acicular, amphibole-like	{110} distinct
1.874	n_β 1.874 n_α 1.756 n_γ 1.896	Clinoclase $Cu_3AsO_4(OH)_3$	0.140	50° v > r very strong	Monoclinic β = 99.5°	Prismatic, tabular, rhombic	{001} perf
1.875	n_β 1.875 n_α 1.655 n_γ 1.909	Malachite $Cu_2(OH)_2CO_3$	0.244	~43° v > r distinct, inclined weak	Monoclinic β = 98.7°	Prismatic-acicular crystals, botryoidal	{$\bar{2}$01} perf, {010} dist

* n_β variation exceeds 0.01

** n_β variation exceeds 0.10

Twinning	Orientation	Color in thin section	Hardness Sp Gravity	Color hand sample	Alteration	Occurrence	Remarks	Reference
		Orange, yellow	~3 5.45	Orange to amber		Dehydration of vandendriesscheite		Am Min 45, 1026-1061
	a = X, b = Z, c = Y	Colorless	~4 4.23-4.35	Colorless, white		With quartz, spiroffite	Effervesces cold dilute HCl	Am Min 61, 339
{100}	b = Z, c ∧ X ~ 22°	X = pale green, Y = yellow-green, Z = blue-green	4½-5 4.08-4.35	Dk green, bluish-green		Oxide zones Cu-deposits with malachite, chrysocolla, tenorite	Soluble abids	Dana II, 799-801
{100}, {011} & {012} simple, {031} cyclic	a = Z, b = X, c = Y	Colorless to pale yellow, red-brown, X = Z = pale yellow, Z = yellow-orange	6½ 4.25-4.35	Pale yellow, reddish brown	Limonite	Alkaline igneous rocks with nepheline. Alkali granite, pegmatite, rhyolite	Olivine group	p. 105
		Colorless	4-5 6.5-6.6	Yellow		Alteration of stibnite		Am Min 47, 1221
Rare	a = Z, b = X, c = Y	Colorless or pale yellow. X = lt yellow, Z = yellow-green	6-6½ ~4.2-4.3	Yellow, green	Mn-chlorites and serpentine with magnetite	Contact or regional metamorphic rocks with other Mn-minerals	Olivine group	p. 109
{110} rare	a = Y, b = X, c = Z	X = pale green, Y = yellow-green, Z = green	3-3½ 3.76-3.87	Bright green to dark green	Malachite, chrysocolla	Oxide zones of Cu deposits	Dimorph of paratacamite. Soluble in acid	Dana II, 69-73
			2½-3 5.72-6.29	Lt yellow, green-brown, brown		Secondary U-deposits, altered pegmatites	Not fluor. U.V.	USGS Bull 1064, 233-236
	a = Y, b = Z c = X	Pale blue-green. Weak pleochroism	2½-3 5.6-5.8	Dk green, blu		Oxide zones Cu-Pb deposits	Soluble with effervescence HNO_3	Dana II, 630-632
		Colorless	9 4.89-5.46	Tan, lt violet		In vugs		Min Mag 42, 179
	c = Z	Transparent on thin edges only. X = dark brown, Y = Z = dark brown black	3.837	Black		Metamorphosed siliceous ironstones, shale or limestone		Am Min 50, 278
	b = Y, a ~ Z	X = Pale blue-grn, Y = Lt blue-green, Z = Benzol green	2½-3 4.33-4.35	Dk green-blue, green-black		Oxide zones Cu-deposits with olivenite		Dana II, 787-789
{100} simple very common	b = Y, c ∧ X = +23.5°, a ∧ Z ~ +32°	X = colorless, pale green, Y = yellow-green, Z = dk green	3½-4 3.9-4.1	Bright green, dk green		Oxide zones Cu-deposits with azurite, chrysocolla, cuprite	Easily soluble dilute acids	p. 71

Biaxial –

n_β Var.	Refractive Index	Mineral Name and Composition	$(n_\gamma - n_\alpha)$	2V angle Dispersion	Crystal System	Habit	Cleavage
*1.855 1.907	n_β 1.855–1.907 n_α 1.674–1.770 n_γ 1.880–1.915	Strelkinite $Na_2(UO_2)_2(VO_4)_2 \cdot 6H_2O$	0.143	Medium	Orthorhombic	Foliated	⊥{001} perf
*1.870 1.89	n_β 1.870–1.89 n_α 1.727 n_γ 1.883–1.92	Ilvaite $CaFe^{2+}_2Fe^{3+}Si_2O_8$	0.156	20° – 30° v > r strong	Orthorhombic	Thick prismatic, massive	{001} & {010} distinct
1.880	n_β 1.880 n_α 1.860 n_γ 1.893	Frondelite $(Mn^{2+},Fe^{2+})Fe^{3+}_4(PO_4)_3(OH)_5$	0.033	Moderate ~ 77° r > v	Orthorhombic	Radiated fibrous, botryoidal	{100} perf, {010} & {001} distinct
1.886	n_β 1.886 n_α 1.873 n_γ 1.914–1.939	Fersmanite $(Na,Ca)_2(Ti,Nb)Si(O,F)_6$	0.041	Small 0° – 7°	Monoclinic β = 97.3°	Pseudotetragonal crystals	None, {001} parting in traces
1.89	n_β 1.89 n_α 1.80 n_γ 1.91	Wyartite $Ca_3(UO_2)_6U(CO_3)_2(OH)_{16} \cdot 3\text{-}5H_2O$	~0.11	48°	Orthorhombic	Small tabular crystals	{001} perf, {010}
1.89	n_β 1.89 n_α 1.86 n_γ 1.91	Heterosite $(Fe^{3+},Mn^{3+})PO_4$	0.05	~90° very strong	Orthorhombic	Massive	{100} dist, {010} imperf
*1.895 1.906	n_β 1.895–1.906 n_α –1.785 n_γ 1.915–1.917	Masuyite $UO_3 \cdot 2H_2O$	0.132	~50°	Orthorhombic	Tabular pseudo hexagonal {001}	{001} perf, {010} imperf
1.898	n_β 1.898 n_α 1.815 n_γ 1.898	Arseniosiderite (Mazapilite) $Ca_3Fe^{3+}_4(OH)_6(H_2O)_3\ AsO_{4\ 4}$	0.083	Small	Monoclinic β = 96°	Fibrous, fine granular	{100} or {001} dist
*1.885 1.920	n_β 1.885–1.920 n_α 1.863–1.865 n_γ 1.890–1.940	Fourmarierite $PbU_4O_{13} \cdot 4H_2O$	0.027	50° – 55° r > v	Orthorhombic	Pseudohexagonal tabular {001}, elongation (b)	{001} perf, {100} imperf
*1.870 1.93	n_β 1.870–1.93 n_α 1.670–1.77 n_γ 1.895–1.97	Tyuyamunite $Ca(UO_2)_2(VO_4)_2 \cdot 5\text{-}8H_2O$	0.225 to 0.20	40° – 55° v > r weak	Orthorhombic	Scaly-laths (b), microcrystalline, radiated fibrous	{001} perf, {010} & {100} distinct
1.900	n_β 1.900 n_α 1.775 n_γ 1.920	Xocomecatlite $Cu_3(TeO_4)(OH)_3$	0.145	41° Weak disp	Orthorhombic	Radiated acicular, spherules	
1.901	n_β 1.901 n_α 1.899 n_γ 1.903	Queitite $Pb_4Zn_2(SO_4)(SiO_4)(Si_2O_7)$	0.004	~90°	Monoclinic β = 108.2°	Tab. xls {001}	{010} & {001} poor
1.90 1.91	n_β 1.91–1.90 n_α 1.685–1.674 n_γ 1.93–1.92	Ianthinite $UO_2(UO_3)_5 \cdot 10H_2O$	0.245	58°	Orthorhombic	Minute plates, laths (b) flat {001}	{001} perf, {100} dist

* n_β variation exceeds 0.01

Twinning	Orientation	Color in thin section	Hardness Sp Gravity	Color hand sample	Alteration	Occurrence	Remarks	Reference
	a = Z, b = Y, c = X	Y = yellow, Z = pale yellow	2-2½ 4.22	Gold yellow, canary yellow		Veins in carbonaceous shales with Fe-hydroxides, calcite	Na-carnotite. Fluor. U.V.-weak green	Am Min 60, 488-489
	a = Y, b = X, c = Z	X = brown, brown-yellow, Y = dk brown, opaque, Z = dk green, opaque	5½-6 3.8-4.1	Black, blackish-gray	Limonite	Contact metosomatic Fe-Zn-Cu deposits	Gelat acids	Win & Win II, 511
	c = X	X = pale yel-brown, Y ~ Z = orange-brn	4½ 3.47	Olive green, green-black		Alteration Fe-Mn-phosphates in pegmatites	Frondelite-rockbridgeite series. Soluble HCl	Dana II, 867-869
{001} rare	b = Y, c ∧ Z = +38°	Pale yellow	5-5½ 3.44-3.46	Lt-dk brown		Alkali pegmatites		Vlasov II, 564-566
	a = Z, b = Y, c = X	X = gray, Y = violet, Z = lavender blue	3-4 4.69	Black to violet-black		Alteration of uraninite		Am Min 45, 200-208
	a = X	X = greenish-gray, Y ~ Z = red-violet	4-4½ 3.41-3.70	Red-violet, dk brown		Alteration of triphylite in granite pegmatites	Heterosite-purpurite series. Easily soluble HCl	Dana II, 675-677
{110} and {1$\bar{3}$0} common	a = Z, b = Y, c = X	X = pale yellow, Y = Z = deep golden yellow	5.03-5.08	Red-orange, carmine		Cavities in uraninite with uranophane, sklowdowskite		Am Min 45, 1026-1061
	c ~ X	X = colorless, Y = Z = dark red-brown	1½-4½ 3.58-3.60	Deep yellow, red-brown, black		Low T oxidation of löllingite, arsenopyrite and other ores	Essentially uniaxial	Am Min 59, 48-59
	a = Y, b = Z, c = X	X = colorless, Y = pale amber, Z = amber yellow	3-4 5.74-5.77	Reddish-orange, red, brown		Secondary U-mineral. Alteration of uraninite		Am Min 45, 1026-1061
	a = Y, b = Z, c = X	X = colorless, Y = very pale yel, Z = pale yellow	~2 3.3-3.6	Greenish-yellow, yel		Secondary U-mineral in sandstone, limestone ores	Occasionally fluor U.V.-weak yellow green	USGS Bull 1064, 248-253
		Bluish-green. Z > X = Y	4 4.42-4.65	Emerald green		Oxide zone Cu-Pb ore deposits with cerussite, teineite		Min Mag 40, 221-226
	b = X, a ∧ Z ~ 20°	Colorless	4 6.07	Colorless, pale yellow		Oxide zones Pb-Zn ores with larsenite, alamosite, leadhillite		Am Min 64, 1331
	a = Z, b = Y, c = Z	X = colorless, Y = violet, Z = dk violet	2-3 5.16	Black, violet		Secondary U-mineral in pitchblende	Not fluor	USGS Bull 1064, 56-60

n_β Var.	Refractive Index	Mineral Name and Composition	$(n_\gamma - n_\alpha)$	2V angle Dispersion	Crystal System	Habit	Cleavage
1.905	n_β 1.905 n_α 1.873 n_γ 1.910	Yeatmanite $(Mn,Zn)_{13}Sb_2Si_4Zn_2O_{28}$	0.037	~49° v > r moderate	Triclinic α = 92.8° β = 101.8° γ = 76.2°	Pseudo-orthorhombic crystals	{100} perf
1.917	n_β 1.917 n_α 1.826 n_γ 1.921	Texasite $Pr_2O_2(SO_4)$	0.095	26° - 31°	Orthorhombic	Bladed crystals, radiated	{010}
1.920	n_β 1.920 n_α 1.885 n_γ 1.956	Tsumebite $Pb_2CuPO_4SO_4(OH)$	0.071	~90° r > v strong	Monoclinic β = 111.5°	Thick tabular crystals as twinned groups	None
*1.92 1.94	n_β 1.92-1.94 n_α 1.76-1.77 n_γ 1.94-1.97	Sengierite $Cu(UO_2)_2(VO_4)_2 \cdot 6H_2O$	0.18 to 0.20	37° - 39° v > r strong	Monoclinic β = 103.6°	Tabular-platy {001}, coatings	{001} perf
*1.910 1.952	n_β 1.910-1.952 n_α 1.750-1.785 n_γ 1.945-2.002	Francevilleite $(Ba,Pb)(UO_2)_2(VO_4)_2 \cdot 5H_2O$	0.185 to 0.217	46° - 53°	Orthorhombic	Thick tabular, cryptocrystalline veinlets	{001} perf
**1.901 2.06	n_β 1.901-2.06 n_α 1.750-1.78 n_γ 1.92-2.08	Carnotite $K_2(UO_2)_2(VO_4)_2 \cdot 1\text{-}3H_2O$	0.17 to 0.30	43° - 60°	Monoclinic β = 103.6°	Micaceous, earthy	{001} perf
**1.92 2.03	n_β 1.92-2.03 n_γ 1.98-2.07	Agrinierite $K_2CaU_6O_{20} \cdot 9H_2O$	0.04	55°	Orthorhombic	Tabular {001}	{001} dist
1.95	n_β 1.95 n_α 1.70 n_γ 2.04	Hydrotungstite $H_2WO_4 \cdot H_2O$	0.34	52°	Monoclinic β ~90°	Microscopic tabular crystals	{010} imperf
1.95	n_β 1.95 n_α 1.92 n_γ 1.95	Catoptrite (Katoptrite) $Mn_{14}Sb_2(Al,Fe)_4(SiO_4)_2O_{21}$	0.03	Small r > v strong, inclined	Monoclinic β = 101.5°	Tabular, granular	{100} perf
1.95	n_β 1.95 n_α 1.92 n_γ 1.96	Manganostibite $Mn_7^{2+}Sb^{5+}As^{5+}O_{12}$	0.04	Small	Orthorhombic	Embedded grains	None
1.95	n_β 1.95 n_α 1.92 n_γ 1.96	Larsenite $PbZnSiO_4$	0.04	80° r > v distinct	Orthorhombic	Prismatic, tabular {010}	{120} dist
1.96	n_β 1.96 n_α 1.73 n_γ 1.98	Melanovanadite $Ca_2V_4^{4+}V_6^{5+}O_{25} \cdot nH_2O$	0.25	~30°	Triclinic α = 90° β = 101.8° γ = 93.2°	Divergent prismatic crystals (c), velvety	{010} perf
1.96	n_β ~1.96 n_γ 1.96	Beudantite $PbFe_3AsO_4SO_4(OH)_6$	Small-moderate	Small-medium Abnormal disp	Trigonal	Rhombohedral	{0001} dist
1.961	n_β 1.961 n_α 1.947 n_γ 1.968	Alamosite $PbSiO_3$	0.021	65° v > r extreme, inclined	Monoclinic β = 113.2°	Fibrous on b, radiating	{010} perf

* n_β variation exceeds 0.01

** n_β variation exceeds 0.10

Twinning	Orientation	Color in thin section	Hardness Sp Gravity	Color hand sample	Alteration	Occurrence	Remarks	Reference
{023} cyclic, {010} multiple	X ⊥ {100}, optic plane ~ {010}	Pale brown. No pleochroism	4 5.03-5.37	Clove brown		Thin plates in willemite	Easily soluble dilute HCl	Am Min 51, 1494-1500
	a = Z, b = Y, c = X	X = colorless, pale gray-green, Y = grayish-green, Z = pale green	2½ 5.769	Apple green		Alteration of rare-earth pegmatite minerals		Am Min 62, 1006-1008
{1̄22} cyclic (trillings)		Pale green, Z > X	3½ 6.133	Emerald green		Oxide zones Pb-Cu-deposits with smithsonite, cerussite, azurite	Easily soluble HCl	Dana II, 918-919
	b = Y, c ∧ X = 14°	X = colorless to bluish green, Y = olive green, Z = colorless to yellow-green	2½ 4.05-4.41	Green to yellowish-green		Oxide zones Cu-U ores; efflorescence		USGS Bull 1064, 258-260
	a = Z, b = Y, c = X	X = colorless, Y = Z = yellow	3 4.55	Yellow, yel-green, brn-yellow		Secondary U-mineral	Ba analogue meta-tyuyamunite. No fluor U.V.	Am Min 43, 180
	a = Z, b = Y, c = X	X = colorless, Y = Z = pale yellow	Soft 4.70-4.95	Bright yellow		Secondary U-mineral	Not fluor U.V. Soluble acids	USGS Bull 1064, 243-247
Sector twinning {110}	a = Y, b = Z, c = X	Orange	5.62-5.7	Orange		Secondary U-mineral	Pseudo-hexagonal	Min Mag 38, 781-789
Polysynthetic ∥ prism, very common		X = colorless, Y = yellow-green, Z = dark green	~2 4.60-4.64	Dk green, yellow-green		Alter ferberite	Soluble NH_4OH	Am Min 48, 935
	b = Y, c ∧ Z ~ +3°, X ~ ⊥{100}	Deep red in thin sheets, ~opaque. Strong pleochroism	5½ 4.56-4.65	Black		Metamorphosed limestone		Am Min 51, 1494-1500
		Brown. ~opaque	4.95-5.00	Black		Marble with sonolite, catoptrite		Am Min 55, 1489-1499
	a = X, b = Z, c = Y	Colorless	3 5.90-6.14	White		Hydrothermal Zn-oxide deposits with willemite	Gelat. HCl	Am Min 51, 269
	b ~ Z, c ∧ Y = 15°	~opaque, X = lt red-brown, Y = deep red-brown, Z = dk red-brown	2½ 3.48	Black	Pascoite	Veins in black shale with pascoite, prite, native Cu	Easily soluble acids	Dana II, 1058-1060
Lamellar	c = X	X = colorless, Y = Z = yellow	3½-4½ 4.1-4.3	Dk green, brown, black		Oxide zones ore deposits	Soluble HCl, anomalous interference colors	Am Min 35, 1055-1056
	b = Y	Colorless	4½ 6.49	White		Oxide zones of ore deposits		Win & Win II, 455

n_β Var.	Refractive Index	Mineral Name and Composition	$(n_\gamma - n_\alpha)$	2V angle Dispersion	Crystal System	Habit	Cleavage
1.966	n_β 1.966 n_α 1.945 n_γ 1.983	Olsacherite $Pb_2SO_4SeO_4$	0.038	80° - 84° None	Orthorhombic	Prismatic crystals (b)	{101} dist, {010} fair
1.97	n_β 1.97 n_α 1.93 n_γ 2.01	Bismutoferrite $BiFe_2(SiO_4)_2(OH)$	0.08	Large	Orthorhombic	Massive, earthy	
1.97	n_β 1.97 n_α 1.96 n_γ 1.98	Kerstenite $PbSeO_4$	0.02	50°	Orthorhombic	Prismatic, acicular, botryoidal	⊥ elongation imperfect
1.977 1.982	n_β 1.977-1.982 n_α 1.780 n_γ 2.05-2.15	Carpathite (Karpatite) (Pendletonite) $C_{24}H_{12}$	0.27 to 0.37	~70° - 80°	Monoclinic β = 69°	Radiated acicular, aggregates	{100}, {001} & {$\bar{2}$01} very perfect
*1.975 2.00	n_β 1.975-2.00 n_α 1.871-1.91 n_γ 2.005-2.06	Walpurgite (Waltherite) $(BiO)_4UO_2(AsO_4)_2 \cdot 3H_2O$	0.03 to 0.15	50° - 60° v > r slight	Triclinic α = 101.7° β = 110.8° γ = 88.3°	Lath crystals (c), {010}, radiated aggregates	{010} perfect
1.985	n_β 1.985 n_α 1.72 n_γ 1.990	Poughite $Fe_2(TeO_3)_2SO_4 \cdot 3H_2O$	0.27	15° - 20°	Orthorhombic	Tabular crystals {010}. Diamond cross sections	{010} perf, {101} dist
1.98	n_β>1.98 n_α 1.96 n_γ>2.10	Karibibite $Fe_2^{3+}As_4^{3+}O_9$	>0.04	Large	Orthorhombic	Fibrous bundles	
1.993	n_β 1.993 n_α 1.990 n_γ 1.994	Elyite $Pb_4CuSO_4(OH)_8$	0.004	~60° - 65° Moderate inclined	Monoclinic β = 100.5°	Radiated fibrous	{001} dist
2.00	n_β 2.00 n_α 1.87 n_γ 2.01	Leadhillite $Pb_4SO_4(CO_3)_2(OH)_2$	0.14	~10° v > r strong, horizontal	Monoclinic β = 90.5°	Pseudohexagonal tabular {001}, prismatic	{001} perfect easy
2.00	n_β> 2.00 n_α ~1.8 n_γ< 2.07	Richetite Hydrous oxide of U & P	~0.3	Large	Monoclinic	Pseudohexagonal tabular {010}	{010} perfect another dist at 90°
2.0	n_β>2.0 n_α 1.797 n_γ>2.0	Barnesite $Na_2V_6O_{16} \cdot 3H_2O$	>0.20		Monoclinic β = 95°	Radiated microscopic fibrous (b)	None ?
2.0	n_β 2.0 n_α 1.82 n_γ 2.0	Grantsite $Na_4CaV_2^{4+}V_{10}^{5+}O_{32} \cdot 8H_2O$	>0.20		Monoclinic β = 95.2°	Micro bladed crystals (b)	
2.002	n_β 2.002 n_α 1.930 n_γ 2.020	Lindgrenite $Cu_3(MoO_4)_2(OH)_2$	0.090	71° r > v	Monoclinic β = 98.4°	Tabular, platy, acicular, massive	{010} perf, {101} & {100} very poor

* n_β variation exceeds 0.01

Twinning	Orientation	Color in thin section	Hardness Sp Gravity	Color hand sample	Alteration	Occurrence	Remarks	Reference
	a = X, b = Z, c = Y	Colorless	3-3½ 6.55	Colorless		Oxide zones Pb-deposits as alteration of penroseite		Am Min 54, 1519-1527
		Colorless, yellow	4.47	Yellow, green		Primary ore veins		Am Min 43, 656-670
Polysynthetic, rare	∥ extinction, + elongation	Colorless	3½ 7.08	Colorless, greenish yel		Oxide zones of Pb-Cu deposits containing selenides	Soluble warm 1:1 HNO_3 with difficulty	Am Min 39, 850
	b = X, c ∧ Z = 21°	Colorless	~1 1.35-1.40	Pale yellow		Contact metamorphic zones		Am Min 52, 611-616
{010} very common	X ~ ⊥(010)	X = colorless, Y = Z = very pale greenish yellow	3½ 5.95-6.69	Colorless, pale yellow		Oxide zone U-Co veins with secondary U or Co minerals	Not fluor U.V. Decomp HNO_3	USGS Bull 1064, 239-242
	a = Y, b = X, c = Z	X = colorless, Y = pale greenish-yellow, Z = pale yellow	2½ 3.70-3.80	Yellow, brown-yellow, green-yellow		Botryoidal crusts secondary ores with emmonsite	Easily soluble HCl	Am Min 53, 1075-1080
	Z = elongation	Straw yellow to pale yellow-brown	Soft 4.04-4.07	Brownish yellow		Cavities in pegmatites with loellingite, eosphorite, scorodite	Readily soluble dilute acids. Fluor U.V.-yellow	Am Min 59, 382
{001} simple common	b = Y, c ∧ X = 45°	Pleochroic in violet Z > Y > X	2 6.0-6.3	Violet		Cavities in massive sulfides	Sectile, insoluble H_2O, decomposes HNO_3	Am Min 57, 364-367
{140}, {340} contact, cyclic lamellar	b = Z, c ∧ X = +5½°	Colorless	2½-3 6.55-6.57	Colorless, white, yellow green, blue		Oxide zone Pb-deposits	Low T polymorph susannite. May fluor U.V.yellow	Can Min 10, 141
{010} common	b = X, Z ∧ a or c ~ 6°	X = colorless to pale brown, Y = Z = dull brown		Black		Secondary U-mineral with uranophane		USGS Bull 1064, 91-92
	b = Z, a ~ Y, c ∧ X ~ 5°	X = Y = orange-yel Z = red	3.15-3.21	Dk red		Secondary V ores in sandstone	Soluble HCl	Am Min 48, 1187-1195
	b = Z, X ~ ⊥ blade, Y ~ across blade, + elongation	X = green, Y = green-brown, Z = brown Z > Y > X	Soft ~ 1 2.94-2.95	Dk olive grn, green-black		Secondary, sedimentary V-U deposits	Becomes orange-red in HCl	Am Min 49, 1511-1526
	b = Z, c ∧ X = -9°	Pale green, yellow-green	4½ 4.26-4.29	Green, yellowish-grn		Secondary mineral in porphyry Cu-deposits	Easily soluble HCl, HNO_3	Dana II, 1094-1095

Biaxial —

n_β Var.	Refractive Index	Mineral Name and Composition	$(n_\gamma - n_\alpha)$	2V angle Dispersion	Crystal System	Habit	Cleavage
2.007	n_β 2.007 n_α 1.928 n_γ 2.036	Lanarkite $PbSO_4$	0.108	~ 60° r > v strong, inclined	Monoclinic β = 116.2°	Prismatic (b), massive	{$\bar{2}$01} perf, {$\bar{4}$01} & {201} imperfect
2.00	n_β>2.00	Gabrielsonite $PbFe(AsO_4)(OH)$		80° - 90°	Orthorhombic	Aggregates equant crystals	None
2.0	n_β~2.0 n_α<2.04 n_γ>2.04	Arsenbrackebuschite $Pb_2(Fe,Zn)(OH,H_2O)(AsO_4)_2$			Monoclinic β = 112.5°		{010} perfect
2.0	n_β>2.0	Curienite $Pb(UO_2)_2(VO_4)_2 \cdot 5H_2O$		~ 66°	Orthorhombic	Micro-crystalline, earthy	
2.01	n_β 2.01 n_α 1.90-1.95 n_γ 2.02-2.06	Perrierite $(Ca,Ce,Th)_4(Mg,Fe)_2(Ti,Fe)_3Si_4O_{22}$	0.12 to 0.11	60°	Monoclinic β = 113.5°	Prismatic crystals	None
2.01	n_β 2.01 n_α 1.91 n_γ 2.03	Lorenzenite (Ramsayite) $Na_2Ti_2Si_2O_9$	0.12	38° - 40° r > v distinct	Orthorhombic	Minute acicular crystals	{010} dist
2.010	n_β 2.010 n_α 1.942 n_γ 2.024	Wherryite $Pb_4Cu(SO_4)_2CO_3(Cl,OH)_2O$	0.082	50°	Monoclinic β = 91.3°	Fibrous-acicular, fine granular, massive	None ?
2.01	n_β 2.01 n_α 1.96 n_γ 2.04	Brownmillerite $Ca_2(Al,Fe)_2O_5$	0.09 to 0.08	75°	Orthorhombic	Square platelets	
2.01	n_β 2.01 n_α 2.00 n_γ 2.02	Turanite $Cu_5(VO_4)_2(OH)_4$	0.02	Medium r > v strong	Orthorhombic ?	Reniform crusts, radiated fibrous	
2.01	n_β>2.01 n_α<2.00 n_γ>2.01	Hendersonite $Ca_2V^{+5}_{1+x}V^{+4}_{8-x}(O,OH)_{24} \cdot 8H_2O$	~ 0.02	Medium	Orthorhombic	Parallel acicular (c)-bladed {100}	
2.02	n_β 2.02 n_α 1.905 n_γ>2.02	Navajoite $V_2O_5 \cdot 3H_2O$	> 0.12	Small	Monoclinic β = 97.5°	Fibrous (b), laths {001}, coatings	None ?
2.02	n_β~2.02 n_α 1.967-1.973 n_γ~2.05	Chevkinite $Ce_4Fe^{2+}_2Ti_3(Si_2O_7)_2O_8$	~0.01 to 0.02	~75° r > v	Monoclinic β = 100.75°	Prismatic, massive	None
2.023	n_β 2.023 n_α 2.018 n_γ 2.025	Sonoraite $FeTeO_3(OH) \cdot H_2O$	0.007	20° - 25°	Monoclinic β = 108.5°	Platy crystals in sheaves, rosettes	

* n_β variation exceeds 0.01

Twinning	Orientation	Color in thin section	Hardness Sp Gravity	Color hand sample	Alteration	Occurrence	Remarks	Reference
Polysynthetic, rare	b = Y, c ∧ Z = 30°	Colorless	2-2½ 6.92-7.08	Gray, greenish-white, yellow		Oxide zones Pb-deposits	Fluor U.V.-yellow	Dana II, 550-551
		Greenish-brown	3½ 6.67-6.69	Black		Oxide zones Pb-deposits	Easily soluble HCl. Descloizite-pyrobelonite group	Am Min 53, 1063-1064
		Colorless, lt yellow	4-5 6.54	Honey yellow		Oxide zone of Pb-Zn deposits with beudantite		Am Min 63, 1282
		Pale yellow	4.88-4.94	Canary yellow		Secondary U-mineral		Am Min 54, 1220
{100} common	b = Z, a ∧ X = -24°	X = yellow, Y = opaque to violet-red, Z = opaque to dk brown	5½ 4.3	Brownish-blk		Silicic tuffs and ash.	Dimorph chevkinite	Am Min 51, 1394-1405
	a = Y, b = X, c = Z	X = Y = pale orange Z = pale yellow	6 3.41-3.43	Brown, black		Nepheline syenite, alkali pegmatites		Am Min 32, 59-63
		Colorless, pale green	6.45-7.22	Lt green, yellow, yellow-green		Oxide zones Cu-Pb deposits	Slowly soluble strong HCl, HNO_3	Am Min 55, 505-508
	Y and Z lie in plane of platelets along diagonals	X = yellow-brown, Y = Z = red-brown	3.73-3.77	Reddish-brown		Contact met zones in carbonate rocks. Portland cement		Am Min 50, 2106
	+ elongation	X ~ Y = brown, Z = green	5	Olive green		Voids in limestone with secondary U-V minerals		Dana II, 818
	a = X, b = Y, c = Z. ∥ ext, + elongation	X = yellow-green, Y = green, Z = yellow-brown	~2½ 2.77-2.80	Dk green, blk		Oxide zones in secondary U-V deposits		Am Min 47, 1252-1272
	∥ extinction, + elongation	X = Y = yellowish brown, Z = dark-brown	<2 2.56	Dark brown		Secondary U-V deposits		Am Min 44, 322-341
Rare	c ∧ Z ~11° to 26°	X ~ colorless, Y = red-brown, Z ~ opaque	5-6 4.3-4.7	Dk red-brown, black		Ash flows, alkali granites, syenite pegmatites	Dimorph perrierite	Am Min 53, 1558-1567
		Yellowish green. No pleochroism	~3 3.95-4.18	Yellowish green		Gold telluride deposit with emmonsite, anglesite, limonite, quartz		Am Min 53, 1828-1832

Biaxial —

n_β Var.	Refractive Index	Mineral Name and Composition	$(n_\gamma - n_\alpha)$	2V angle Dispersion	Crystal System	Habit	Cleavage
2.04	n_β 2.04 n_α 1.98 n_γ 2.10	Fiedlerite $Pb_3Cl_4(OH)_2$	0.12	Large ~ 87° v > r perc	Monoclinic β = 102.2°	Lath-like (c)	{100} dist
2.05	n_β 2.05 n_α 1.908 n_γ 2.065	Pinakiolite $Mg_3Mn^{2+}Mn_2^{3+}B_2O_{10}$	0.157	32° v > r moderate	Monoclinic β = 95.8° or 120.6°	Thin tabular, short prismatic crystals	{010} dist
2.05	n_β 2.05 n_γ ~2.15	Schmitterite UO_2TeO_3	>0.10	~75°	Orthorhombic	Bladed crystals (a), micaceous rosettes	{100}
*2.06 2.08	n_β 2.06-2.08 n_α 2.03-2.06 n_γ 2.08-2.10	Duftite $CuPbAsO_4(OH)$	0.06 to 0.03	Large ~ 70° - 80° r > v perc	Orthorhombic	Microscopic crystals, crusts and aggregates	
2.070	n_β 2.070 n_α 1.786 n_γ 2.075	Salesite $CuIO_3(OH)$	0.289	0° - 5° r > v extreme	Orthorhombic	Tiny prismatic crystals	{110} perfect
2.074	n_β 2.074 n_α 1.803 n_γ 2.076	Cerussite $PbCO_3$	0.273	~9° r > v strong	Orthorhombic	Tabular {010} crystals, prismatic-acicular, reticulate	{110} & {021} distinct {010} & {012} poor
2.07	n_β 2.07 n_α 2.04 n_γ 2.08	Vésignieite $BaCu_3(VO_4)_2(OH)_2$	0.04	~60°	Unknown	Lamellar aggregates	{001} dist
2.08	n_β 2.08 n_α 1.95 n_γ 2.11	Cafetite $CaFe_2Ti_4O_{12}\cdot 4H_2O$	0.16	38° Strong disp	Orthorhombic or Monoclinic β = 90°	Acicular, radiated, fibrous aggregates	Prismatic 2 dir
*2.07 2.11	n_β 2.07-2.11 n_α 2.05-2.06 n_γ 2.12-2.15	Curite $Pb_2U_5O_{17}\cdot 4H_2O$	0.07 to 0.09	Large r > v strong	Orthorhombic	Prismatic-acicular (c), massive	{100}
2.09	n_β 2.09 n_α 1.962 n_γ 2.10-2.12	Emmonsite (Durdenite) $Fe_2(TeO_3)_3\cdot 2H_2O$	0.138	Small ~ 20° r > v strong	Monoclinic	Microcrystalline, fibrous-acicular	{010} perf, Two others
2.095	n_β 2.095 n_α 1.982 n_γ 2.19	Carlfriesite $H_4Ca(TeO_3)_3$	~0.20	~80°	Monoclinic β = 115.6°	Botryoidal crusts, radiated crystals	{010} fair
2.098	n_β 2.098 n_α 1.997 n_γ 2.108	Clarkeite $(Na,Ca,Pb)_2U_2(O,OH)_7$	0.111	30° - 50°	Unknown	Microcrystalline crystalline	None ?
2.10	n_β 2.10 n_α 1.70 n_γ ~2.23	Metahewettite $CaV_6O_{16}\cdot 9H_2O$	~0.53	~52°	Monoclinic β = 95.2°	Fibrous-bladed, elongation (b), flat {001}	
2.10	n_β 2.10 n_α 2.07 n_γ 2.12	Trigonite $Pb_3Mn(AsO_3)_2[AsO_2(OH)]$	0.05	~90° v > r perc	Monoclinic β = 91.8°	Thick tabular triangular outline	{010} perfect, {101} dist

* n_β variation exceeds 0.01

Twinning	Orientation	Color in thin section	Hardness Sp Gravity	Color hand sample	Alteration	Occurrence	Remarks	Reference
{100} common	b = Z, c ∧ Y = +34°	Colorless	3½ 5.64-5.88	Colorless, white		Alteration Pb	Soluble HNO_3	Dana II, 67-69
{011} common	a = Z, b = X, c = Y	X = deep red-brown, Y ~ opaque, Z = orange	6 3.88-3.96	Black		In dolomites associated with hausmannite, manganophyllite	Dimorph orthopinakiolite. Soluble concentrated HCl	Dana II, 324-325
	a = Z, b = Y, c = X	Colorless	~1 6.88-6.92	Colorless to pale yellow		Te-bearing ores with emmonsite	Soluble dilute HCl	Am Min 56, 411-415
	Crystals elongated (c)	Pale green	3 6.40-6.49	Apple green-olive green		Oxide zones Cu-Pb deposits	Soluble acids	Dana II, 810-811
	a = X, b = Z, c = Y	X = colorless, Y = Lt bluish-green Z = bluish-green	3 4.77-4.89	Bluish-green		Oxide zones Cu-deposits in arid regions	Easily soluble HNO_3	Dana II, 315-316
{110} cyclic, lamellar, {130} simple	a = Z, b = Y, c = X	Colorless	3-3½ 6.56	Colorless, white, gray, black		Oxide zones Pb-	Soluble dilute HNO_3 with effervescence	p. 70
Polysynthetic, pseudohexagonal	Optic plane = (001), + elongation	Pale yellow-green	3-4 4.05	Yellow-green to dk olive		Cavities in Mn-ores with barite, calcite		Am Min 40, 942-943
	c ∧ Z = 2°-4°	Colorless	4-5 3.19-3.28	Colorless, lt yellow		Cavities in pegmatites		Am Min 45, 476
	a = Y, b = X, c = Z	X = pale yellow, Y = lt red-orange, Z = dk red-orange	4-5 7.37-7.4	Dk orange-red, orange		Oxide zones U-deposits	Not fluor U.V.	USGS Bull 1064, 92-95
	b = Y, X ~ ⊥ cleavage trace on {010}	Pale yellow-green, = Y > X	~5 4.52-4.53	Yellowish green		Oxide zones ore deposits containing Te	Soluble acids	Dana II, 640-641
		Pale yellow. Weak pleochroism Z > X = Y	3½ 5.93-6.3	Primrose yellow		Oxide zones ore deposits with cerussite, chlorargyrite		Am Min 61, 1053
			4-4½ 6.29-6.39	Dk brown, yellow-brown		Alteration of uraninite		Am Min 41, 131
	b = Y, a ~ Z, c ~ X	X = pale orange-yellow, Y = deep red, Z = dark red	2.51	Deep red		Oxide zones V-U-deposit with gypsum, native Se	Slightly soluble H_2O	Am Min 44, 322-341
	b = Y, c ∧ Z ~ 45°	Pale yellow to brownish yellow. Not pleochroic	2-3 6.1-7.1	Yellow to dk brown		Mn ores with hausmannite, native Pb, magnussonite, dixenite	Soluble dilute acids	Dana II, 1032-1033

Biaxial –

n_β Var.	Refractive Index	Mineral Name and Composition	$(n_\gamma - n_\alpha)$	2V angle Dispersion	Crystal System	Habit	Cleavage
2.11	n_β >2.11 n_α >2.11 n_γ >2.11	Moctezumite $PbUO_2(TeO_3)_2$		5° - 10°	Monoclinic β = 93.6°	Crystals elongation (b) and flattened {201}	{100} perfect
2.116	n_β 2.116 n_α 2.077 n_γ 2.158	Laurionite $PbCl(OH)$	0.081	Large ~ 90°	Orthorhombic	Tabular {100}, elongation (c)	{101} dist
2.13	n_β 2.13 n_α 1.95 n_γ 2.21	Kassite $CaTi_2O_4(OH)_2$	0.26	58° r > v very strong	Orthorhombic	Tabular {010}	{010} perf, {101} dist
2.137	n_β 2.137 n_α 2.131 n_γ 2.142	Yttrocrasite $(Y,RE,Th,Ca)(Ti,Fe,W)_2(O,OH)_6$	0.011	60° - 70°	Orthorhombic	Prismatic crystals, tabular	
2.14	n_β 2.14 n_α 2.12 n_γ 2.14	Molybdomenite $PbSeO_3$	0.02	80°	Monoclinic β = 112.8°	Very thin scales, blades	{001} perf, {h0ℓ} dist at 90°
2.15	n_β 2.15 n_α 2.05 n_γ 2.20	Paralaurionite $PbCl(OH)$	0.15	Med-large ~ 68° v > r strong	Monoclinic β 117.2°	Laths (c), thin tabular {100}	{001} perf easy
2.18	n_β 2.18 n_α 2.00 (Li) n_γ 2.35	Tellurite TeO_2	0.35	~90° r > v moderate	Orthorhombic	Radiated acicular crystals (c), plates, coatings	{010} perfect
2.18	n_β 2.18 n_α 1.77 n_γ 2.35	Hewettite $CaV_6O_{16} \cdot 9H_2O$	~0.60	Medium ~ 60°	Orthorhombic ?	Fibrous-bladed elongated (b), flattened {001}	
2.18	n_β 2.18 n_α 2.16 n_γ 2.18	Kleinite $Hg_2N(Cl,SO_4) \cdot nH_2O$	0.02	Small-medium v > r strong	Triclinic	Pseudo-hexagonal prismatics (c)	{001} dist, prismatic imperfect
*2.19 2.24	n_β 2.19-2.24 n_α 2.13-2.14 n_γ 2.20-2.24	Baddeleyite ZrO_2	0.07 to 0.10	28° - 30° r > v strong	Monoclinic β = 99.2°	Prismatic, tabular	{001} perf, {010} & {110} imperfect
2.20	n_β 2.20 n_α 1.94 n_γ 2.51	Lepidocrocite (Pyrosiderite) γ-$FeO(OH)$	~0.57	~83° Weak disp	Orthorhombic	Scaly aggregates, acicular-bladed (a)	{010} perf, {100} imperf, {001} fair
2.2	n_β ~2.2 n_α ~2.1 n_γ ~2.2	Rodalquilarite $H_3Fe_2(TeO_3)_4Cl$		~38°	Triclinic α = 103.2° β = 107.1° γ = 77.9°	Tiny stout crystals, crusts	One distinct
2.21	n_β ~2.21 n_γ ~2.24	Onoratoite $Sb_8O_{11}Cl_2$	~0.024		Triclinic α ~90° β = 110° γ ~90°	Acicular (b) flattened {001}	
2.219	n_β 2.219 n_α 2.185 n_γ 2.266	Heyite $Pb_5Fe_2(VO_4)_2O_4$	0.081	~90° v > r weak	Monoclinic β = 112°	Tabular {100}, elongation (b)	None

* n_β variation exceeds 0.01

Twinning	Orientation	Color in thin section	Hardness Sp Gravity	Color hand sample	Alteration	Occurrence	Remarks	Reference
	Parallel extinction	Bright orange	~3 5.73	Orange, brownish-orange		Oxide zones of deposits containing tellurides	Soluble dilute HCl or NaOH	Am Min 50, 1158-1163
	a = Y, b = Z, c = X	Colorless	3-3½ 6.14-6.24	Colorless, white		Oxide zones Pb-deposits with paralaurionite, phosgenite	Dimorph paralaurionite. Soluble HNO_3	Dana II, 62-64
{101} & {181} common	a = Y, b = Z, c = X	Colorless to pale yellow. Weak pleochroism	5 3.42	Pale yellow		Alkali pegmatites in cavities. Replaces perovskite, ilmenite		Am Min 52, 559-560
		X = colorless, amber, Y = brown, Z = green-brown, ~ opaque	5½-6 5.315	Green-brown	Dull coating	Pegmatites with other rare earth minerals	Often metamict	Am Min 62, 1009-1011
	∥ extinction, + elongation	Colorless	3½ 7.07-7.12	Colorless, yellowish-white		Oxide zones of Cu-Pb-Se deposits with chalcomenite	Soluble HNO_3	Can Min 8, 149-158
{100} contact, very common	b = Y, $c \wedge Z = -25^\circ$	Colorless, pale violet Y > X = Z	Soft 6.15-6.28	Colorless, white, violet, green		Oxide zones Pb-deposits with laurionite, diaboleite, matlockite	Dimorph laurionite. Soluble HNO_3	Dana II, 64-66
	a = Y, b = X, c = Z	Colorless	2 5.83-5.90	White, yellow, yellow-orange		Oxidation of tellurium minerals with native Te.	Dimorph, paratellurite. Easily soluble HCl or HNO_3	Dana I, 593-595
	c = Z	X = Y = pale yellow-orange, Z = dk red	2.55-2.62	Deep red		Oxide zones V-U deposits	Slightly soluble H_2O	Am Min 44, 322-341
	c ~ X	Colorless, yellow	3½-4 ~8.0	Pale yellow-orange		Hg-deposits with gypsum, calcite, barite, terlinguaite	Hexagonal above 130°C. Soluble warm HCl or HNO_3	Dana II, 87-89
{100}, {110}, {201} simple or polysynthetic	b = Y, $c \wedge X = 13^\circ$	X = dk red-brown, green; Y = red-brown, green; Z = lt brown	6½ 5.74-5.83	Colorless, yellow, green, brown, black		Placer gravels, corundum-syenite, carbonatite		Am Min 40, 275-282
None	a = Z, b = X, c = Y	X = yellow to colorless, Y = yellow-brown, red-orange, Z = yellow-orange, dk red	5-5½ ~4.1	Yellow-brown, red-brown, black		Weathering of Fe-sulfides and Fe-oxides with goethite		p. 46
		Pale yellow-green. Not pleochroic	2-3 5.05-5.15	Grass to emerald green		Oxide zones quartz veins with gold, javosite, emmonsite		Am Min 53, 2104-2105
	$a \wedge Y \sim 8^\circ$, $c \wedge X \sim 12^\circ$, $b \wedge Z \sim 0 - 14^\circ$	Colorless	5.3-5.49	White		Oxidation of stibnite	Soluble HCl	Min Mag 36, 1037-1044
{110} common	b = Y, $c \wedge X = -36^\circ$	Orange-yellow. No pleochroism	4 6.28	Yellow-orange		Oxide zones Pb-deposits		Min Mag 39, 65-68

n_β Var.	Refractive Index	Mineral Name and Composition	$(n_\gamma - n_\alpha)$	2V angle Dispersion	Crystal System	Habit	Cleavage
2.22	n_β 2.22 n_α 2.11 n_γ 2.22	Vauquelinite $Pb_2Cu(CrO_4)(PO_4)(OH)$	0.11	Small	Monoclinic β = 94°	Fibrous-colliform, granular. Wedge crystals	None
2.24	n_β 2.24 n_α 2.09 n_γ 2.26	Tungstite $WO_3 \cdot H_2O$	0.17	26° v > r strong	Orthorhombic	Minute plates, massive, earthy	{001} perfect, {110} imperf
2.24	n_β 2.24 n_α 2.16 n_γ 2.25	Chloroxiphite $Pb_3CuCl_2O_2(OH)_2$	0.09	~70° r > v strong	Monoclinic β = 97°	Bladed (c)	{$\bar{1}$01} perf, {100} dist
**2.2 2.6	n_β 2.2-2.6	Chervetite $Pb_2V_2O_7$	0.279	65° - 75° Inclined weak	Monoclinic β = 107.5°	Crystals, pseudomorphs	{100} & {010} doubtful
2.26	n_β 2.26 n_α 2.17 n_γ 2.32	Mottramite (Psittacinite) $Pb(Cu,Zn)VO_4(OH)$	0.15	~73° r > v strong	Orthorhombic	Prismatic, tabular, granular, botryoidal	None
2.265	n_β 2.265 n_α 2.185 n_γ 2.35	Descloizite $Pb(Zn,Cu)VO_4(OH)$	0.165	~90° r > v strong	Orthorhombic	Prismatic, tabular, granular, botryoidal	None
2.27	n_β 2.27 n_α 2.24 n_γ 2.31	Mendipite $Pb_3Cl_2O_2$	0.07	~90° v > r very strong	Orthorhombic	Columnar, fibrous, radiated, nodular	{110} perfect, {100} & {010} distinct
2.32	n_β 2.32 n_α 2.11 n_γ 2.65	Hemihedrite $ZnF_2[Pb_5(CrO_4)_3SiO_4]_2$	0.54	88° - 90° perceptible, horizontal	Triclinic α = 103.5° β = 92° γ = 56°	Elongation hemihedral	{110} poor
2.32	n_β 2.32 n_α 2.12 n_γ 2.32	Seeligerite $Pb_3(IO_3)Cl_3O$	0.20	~4°	Orthorhombic (Pseudotetragonal)	Thin plates	{001} perf, {110} dist, {100} & {010} rare
2.35	n_β 2.35 n_α 2.18 n_γ 2.35	Valentinite Sb_2O_3	0.17	Very small v > r strong	Orthorhombic	Prismatic or tabular crystals, granular, massive	{110} perf, {010} imperf
2.35	n_β 2.35 n_α 2.30 (Li) n_γ 2.40	Nadorite (Ochrolite) $PbSbO_2Cl$	0.10	Large ~ 90° v > r strong	Orthorhombic	Prismatic (a) or tabular {010}, divergent	{010} perfect
*2.32 2.40	n_β 2.32-2.40 n_α 2.25-2.30 n_γ 2.40-2.50	Iranite $PbCrO_4 \cdot H_2O$	0.15 to 0.20	Large	Triclinic α = 104.5° β = 66° γ = 108.5°	Elongation crystals	
**2.30 2.45	n_β 2.30-2.45 n_α 2.20-2.30 n_γ 2.35-2.45	Columbite $(Fe,Mn)(Nb,Ta)_2O_6$	0.1 to 0.2	Moderate-90° v > r moderate	Orthorhombic	Tabular, massive	{010} dist, {100} poor

* n_β variation exceeds 0.01

** n_β variation exceeds 0.10

Twinning	Orientation	Color in thin section	Hardness Sp Gravity	Color hand sample	Alteration	Occurrence	Remarks	Reference
	- elongation	O = pale brown, E = pale green	2½-3 5.98-6.06	Green to brown, ~black		Oxide zones Pb-Cu deposits with crocoite, pyromorphite, mimetite	Partly soluble HNO_3	Dana II, 650-652
	c = X	Pale yellow. Z > Y > X	1-2½ 5.5	Golden yellow, yellow-green		Alter scheelite, wolframite or other W-minerals	Soluble alkalies	Dana I, 605-606
	b = Z, X ~ ⊥{$\bar{1}01$}	Brown-green	2½ 6.93-7.07	Olive green		Oxide zones Pb-deposits	Soluble HNO_3	Dana II 84-85
{100} ubiqui- tous		Colorless	<3 6.30-6.49	Colorless, gray, brown		Oxide zones U-mines with francevillite		Am Min 48, 1416
	a = Z, b = Y, c = X	X = Y = yellow-grn, Z = brownish-green	3-3½ ~5.9	Dk green, brown, black		Oxide zones ore- deposits	Descloizite- mottramite series Soluble acids	Dana II 811-815
	a = Z, b = Y, c = X	X = Y = yellow, Z = brownish-yel	3-3½ 6.14-6.26	Orange-red, dk brown, green		Oxide zones ore- deposits	Descloizite- mottramite series Soluble acids	Dana II, 811-815
	a = X, b = Y, c = Z	~Colorless	2½ 7.22-7.24	Colorless, white, gray		Oxide zones Pb-deposits	Soluble dilute HNO_3	Dana II, 56-58
{$\bar{2}\bar{2}3$} common, {$0\bar{1}0$} or {$0\bar{1}2$} rare		X = Y = yellow, Z = orange	3 6.42-6.50	Orange, brown, black		Oxide zones Pb-deposits		Am Min 55, 1088-1102
		Pale yellow	6.83-7.05	Bright yellow		Small amounts with boleite, paralaurionite, schwartzembergite		Am Min 57, 327-328
	a = X, b = Y (Z for red), c = Z (Y for red)	Colorless	2½-3 5.76-5.83	Colorless, snow white, yellowish		Alteration of stibnite or other Sb-minerals with kermesite.	Dimorph senar- montite. Soluble HCl	Dana I, 547-550
{101} common	a = Z, b = X, c = Y	Pale brown, yellow	3½-4 7.02-7.07	Brown, brown- yellow, yellow		Oxide zones Pb-Zn deposits with bind- heimite, jamesonite		Dana II, 1039-1040
	Extinction 5° to elongation	Brownish-orange ∥ elongation, yellow- orange ⊥ elong	5.8	Saffron yel		Oxide zones Pb-deposits		Am Min 48, 1417
{201} contact, {010} trillings {203} or {501} rare	a = X, b = Y, c = Z	Red-brown, Z > Y = X, ~ opaque	6 5.15-6.5	Dark brown, black		Granite pegmatites, granitic rocks	Columbite- tantalite series	Dana I 780-787

n_β Var.	Refractive Index	Mineral Name and Composition	$(n_\gamma - n_\alpha)$	2V angle Dispersion	Crystal System	Habit	Cleavage
2.36	n_β 2.36 n_α 2.20 n_γ 2.36	Embreyite $Pb_5(CrO_4)_2(PO_4)_2 \cdot H_2O$	0.16	Small 0° - 11°	Monoclinic β = 103°	Drusy crusts, tabular crystals	None
2.36	n_β 2.36 n_α 2.32 n_γ 2.37	Pyrobelonite $PbMnVO_4(OH)$	0.05	29° r > v moderate	Orthorhombic	Acicular crystals (c), tiny grains	None
2.384	n_β 2.384 n_α 2.373 n_γ 2.388	Landauite $(Mn,Zn,Fe)(Fe,Ti)_3O_7$	0.015	~60° r > v weak	Monoclinic β = 107.6°	Fine-grain aggregates	None
*2.393 2.409	n_β 2.393-2.409 n_α 2.260-2.275 n_γ 2.398-2.515	Goethite (Xanthosiderite) α-FeO(OH)	0.138 to 0.140	0° - 27° r > v extreme	Orthorhombic	Fibrous-acicular (c), radiated, botryoidal, scaly	{010} perf, {100} imperf
2.40	n_β~2.40 n_α~2.10 n_γ~2.42	Marokite $CaMn_2O_4$	0.32	20° - 25°	Orthorhombic	Large tabular {010} crystals	{100} perf, {001} dist
2.50	n_β 2.50 n_α 2.41 n_γ 2.51	Pucherite $BiVO_4$	0.10	19° v > r extreme	Orthorhombic	Tabular {001}, acicular crystals, massive	{001} perf
2.61	n_β 2.61 n_α 2.51 (Li) n_γ 2.71	Massicot PbO	0.20	~90° Strong	Orthorhombic	Massive, earthy, scaly, tabular	{100} & {010} distinct, {110} traces
2.61	n_β 2.61 n_α 2.52 n_γ 2.67	Koechlinite $(BiO)_2MoO_4$	0.15	Large v > r strong	Orthorhombic	Square-rectangular laths, earthy, massive	{010} perf, {0kℓ} imperf
2.64	n_β 2.64 n_α 2.35 (Li) n_γ 2.66	Terlinguaite Hg_2ClO	0.31	~20° v > r extreme	Monoclinic β = 106°	Prism crystals (b), massive, powdery	{$\bar{1}$01} perf
2.81	n_β 2.81 n_α 2.4 (Li) n_γ 3.02	Orpiment As_2S_3	~0.6	76° r > v strong	Monoclinic β = 90.8°	Foliated, granular, earthy	{010} perf, {100} traces
3.0	n_β~3.0 (Li) n_α>2.72	Livingstonite $HgSb_4S_8$	Extreme		Monoclinic or Triclinic β = 104°	Acicular, fibrous, columnar (b), globular	{010} & {100} perfect,{001} poor
3.00	n_β~3.00	Xanthoconite Ag_3AsS_3	Extreme	v > r distinct	Monoclinic β = 110°	Tabular pseudohexagonal crystals, reniform	{001} dist
3.176	n_γ 3.176 n_α 3.078 n_β 3.188	Hutchinsonite $(Tl,Pb)_2(Cu,Ag)As_5S_{10}$	0.110	37° v > r extreme	Orthorhombic	Radiated prismatic, acicular	{010} dist
3.27	n_β 3.27 (Na)	Smithite $AgAsS_2$	Very large	65°	Monoclinic β = 101.2°	Equant-tabular crystals {100}, pseudohexagonal	{100} perfect

* n_β variation exceeds 0.01

Twinning	Orientation	Color in thin section	Hardness Sp Gravity	Color hand sample	Alteration	Occurrence	Remarks	Reference
Multiple	b = Y	X = honey yellow, Y = Z = amber	3½ 6.41-6.45	Orange		Oxide zones Pb-Cr deposits		Min Mag 38, 790-793
	a = X, b = Z, c = Y	Red-brown, Y > X = Z	3½ 5.38-5.39	Scarlet red		Tiny crystals with hausmannite, pyro-chroite, manganite		Can Min 10, 117-123
		X = Z = bottle grn, Y = green, Z > Y > X, ~ opaque	7½ 4.42-4.70	Black		Alkali plutons and pegmatites with brookite, monazite		Am Min 51, 1546
Cruciform rare	a = Z, b = X, c = Y	X = yellow to colorless, Y = yellow-brown, red-orange, Z = yellow-orange, dk red	5-5½ ~4.3	Yellow-brown, red-brown, black		Weathering of Fe-sul-fides and Fe-oxides with lepidocrocite		p. 46
	a = Y, b = Z, c = X	~opaque, X = saf-flower red, Y = Z ~ opaque (very dk red)	4.63-4.64	Black		Mn-deposits with hausmannite, braunite, polianite, barite		Am Min 49, 817
	a = Y, b = Z, c = X	Yellow-brown	4 6.25-6.63	Yellow-brown, red-brown		Pegmatites with native Bi. Oxide zones Bi-Ag-U-Cu ores	Soluble HCl	Dana II, 1050-1052
Rare	a = Y ?	X ~ colorless, Y = pale yellow, Z = deep yellow	2 0.642	Yellow, orange		Oxidation of Pb-minerals	Dimorph ligharge. Soluble HCl, HNO_3	Dana I, 516-517
{101} contact or penetration	a = Y, b = Z, c = X	Greenish-yellow. Weak pleochroism	8.26-8.28	Green-yellow, yellow, white		Oxide zones ore veins with bismuth, smaltite	Easily soluble HCl	Dana II, 1092-1093
	Optic plane ⊥ {010} and ∧ c = -7°	Pale yellow or grn. Slightly pleo chroic	2½ 8.73	Yellow to greenish, brn		Hg-deposits with calomel, cinnabar, eglestonite, kleinite	Decomp acids	Dana II, 52-56
{100}	b = X, c ∧ Z = ±2°	Y = pale yellow, Z = greenish yellow	1½-2 3.49-3.52	Yellow, yellow-orange		Low T hydrothermal veins with realgar, stibnite. Hot springs	Sectile inelastic cleavage plates. Soluble alkalies	Dana I, 266-269
Glide twinning		Red X > Z, ~ opaque	2 4.88-5.00	Blackish gray	Cinnabar, metacinna-bar and Sb-oxide	Epithermal veins with cinnabar, stibnite, sulfur, gypsum	Soluble warm HNO_3	Dana I, 485-486
{001} common, pseudo-ortho	b = Z, c ~ X	Lemon yellow	2-3 5.53-5.54	Yellow-orange to red-brown		Low T hydrothermal veins with ruby silver minerals	Dimorph prou-stite	Dana I, 371-372
	a = X, b = Y, c = Z	Red	1½-2 4.58	Cherry red		Crystalline dolomite with realgar, orpiment, sphalerite		Dana I, 468-469
	b = Y, c ∧ Z = 6½°	~ Opaque. Red in thin splinters. Some pleochroism	1½-2 4.88-4.93	Lt red to brownish-orange		Crystalline dolomite with realgar, orpiment, sphalerite, pyrite		Dana I, 430-432

Index

Note: An *italicized* page number indicates a detailed description of the mineral variety.